4250
60C

AGRICULTURAL ENGINEERS' HANDBOOK

AGRICULTURAL ENGINEERS' HANDBOOK

C. B. RICHEY, EDITOR-IN-CHIEF
SUBEDITOR FOR CROP PRODUCTION EQUIPMENT
Chief Research Engineer
Tractor and Implement Division
Ford Motor Company

PAUL JACOBSON
SUBEDITOR FOR SOIL AND WATER CONSERVATION
State Conservation Engineer for Iowa
Soil Conservation Service
U.S. Department of Agriculture

CARL W. HALL
SUBEDITOR FOR FARMSTEAD STRUCTURES AND EQUIPMENT
Professor of Agricultural Engineering
Michigan State University

McGRAW-HILL BOOK COMPANY

New York   Toronto   London

1961

1982 Reissue

AGRICULTURAL ENGINEERS' HANDBOOK

Copyright © 1961 by the McGraw-Hill Book Company, Inc. Printed in the United States of America. All rights reserved. This book, or parts thereof, may not be reproduced in any form without permission of the publishers. *Library of Congress Catalog Card Number* 59-13942

10 11 12 13 14 15 16   VBVB   854321

52617

LIST OF CONTRIBUTORS

KEITH BEAUCHAMP, B.S., Irrigation Engineer, Soil Conservation Service, U.S.D.A. *Surface Drainage.*

FRED W. BLAISDELL, B.S., M.S., C.E., Project Supervisor, Soil & Water Research Div., A.R.S., U.S.D.A. *Erosion-control Structures.*

B. D. BLAKELY, B.S., M.S., Head Agronomist, Soil Conservation Service, U.S.D.A. *Field Practices for Erosion Control.*

J. H. BORNZIN, Divisional Chief Engineer, Engineering Test and Development, International Harvester Co. *Balers.*

F. A. BROOKS, B.S.E.E., M.S., Sc.D., M.E., Professor of Agricultural Engineering and Agricultural Engineer in the Experiment Station, University of California, Davis. *Frost Control. Farm Climate and Solar Energy.*

DONALD P. BROWN, B.S., M.S., PhD., Associate Professor of Agricultural Engineering, Michigan State University. *Electrical Equipment* (co-editor).

ROBERT H. BROWN, B.S.E.E., B.S.A.E., M.E.D., Associate Professor of Agricultural Engineering, University of Georgia. *Electric Motors.*

THOMAS CARROLL, Staff Chief Engineer, Combine Harvester, Massey-Ferguson Ltd. *Combines.*

O. T. COLEMAN, B.S. Agr., Extension Professor of Soils, University of Missouri. *Soil Plant Food (Nutrients). Use of Fertilizer.*

WAYNE D. CRIDDLE, B.S., M.S., State Engineer, State of Utah. *Irrigation.*

CHESTER P. DAVIS, JR., B.S.A.E., M.S., Project Leader, Heat Pump Investigations, A.R.S., U.S.D.A. *Heat Pumps.*

DONALD H. DEWEY, B.S., Ph.D., Professor of Horticulture, Michigan State University. *Fruit and Vegetable Handling. Storage of Fresh Fruits and Vegetables.*

JOSEPH P. DITCHMAN, B.S.E.E., Rural Lighting Extension Specialist, General Electric Co. *Lamps and Lighting.*

FRANK L. DULEY, B.S. Agronomy, A.M., Ph.D., Formerly Soil Conservationist, A.R.S., U.S.D.A. *Field Practices for Erosion Control.*

MARTIN A. ERICKSON, B.Sc., M.M.E., Supervisor, Laboratory Test Section, Tractor and Implement Div., Ford Motor Co. *Stress Measurements and Strength Analysis.*

ROBERT L. ERWIN, B.S. Agr., B.S.A.E., Manager, Engineering Administration and Services, Tractor and Implement Div., Ford Motor Co. *Tractor Force Reactions.*

MERLE L. ESMAY, B.S., M.S., Ph.D., Professor of Agricultural Engineering, Michigan State University. *Structural Requirements for Farm Buildings.*

WESLEY W. GUNKEL, B.S., M.S., Ph.D., Associate Professor of Agricultural Engineering, Cornell University. *Sprayers and Dusters.*

CARL W. HALL, B.S. Agr., B.S.A.E., M.M.E., Ph.D., Professor of Agricultural Engineering, Michigan State University. *Livestock Production Facilities. Heat, Air, and Moisture. Drying Farm Crops. Storage of Farm Crops. Equipment for Handling Milk. Service Buildings and Equipment.*

L. L. HARROLD, B.S.C.E., Project Supervisor, A.R.S., U.S.D.A. *Principles of Agricultural Hydrology.*

PAUL JACOBSON, B.S.A.E., State Conservation Engineer of Iowa, Soil Conservation Service, U.S.D.A. *Mechanics of Water Erosion. Terraces and Diversions. Waterways for Erosion Control.*

LIST OF CONTRIBUTORS

C. F. KELLY, B.S.A.E., M.S.A.E., Professor of Agricultural Engineering, University of California, Davis. *Livestock.*

DON KIRKHAM, A.B., A.M., Ph.D. Physics, Professor of Soils and Physics, Iowa State University. *Soil Physical Properties.*

ROBERT W. KLEIS, B.S., M.S., Ph.D., Professor and Head of Agricultural Engineering, University of Massachusetts. *Electric Heating.*

H. H. KRUSEKOPF, B.S., A.M., Professor Emeritus of Soils, University of Missouri. *Soil Formation and Classification.*

JORDON H. LEVIN, B.S.A.E., B.S.M.E., M.S., Project Leader of Fruit Handling and Conditioning, A.R.S., U.S.D.A. *Fruit and Vegetable Handling. Storage of Fresh Fruits and Vegetables.*

RALPH I. LIPPER, B.S.A.E., M.S.A.E., Assistant Professor of Agricultural Engineering, Kansas State College. *Heat Pumps.*

MORRIS H. LLOYD, B.S.A.E., Rural Service Supervisor, Niagara Mohawk Power Corp. *Wiring.*

HOWARD MATSON, B.A., B.S.A.E., M.S.A.E., Head, Engineering and Watershed Planning Unit, Soil Conservation Service, U.S.D.A. *Water Management, Conservation, Use, and Legal Aspects.*

ARTHUR F. MORATZ, B.S. Arch. E., Watershed Planning Specialist, Soil Conservation Service, U.S.D.A. *Erosion-control Structures.*

JOHN T. PHELAN, B.S., Irrigation Engineer, Soil Conservation Service, U.S.D.A. *Land Leveling and Grading.*

C. B. RICHEY, B.S.A.E., B.S.M.E., Chief Research Engineer, Tractor & Implement Div., Ford Motor Co. *Section 1. Crop Production Equipment* (except as otherwise noted).

W. J. RIDOUT, JR., B.S., Editorial Director, *Electricity-on-the-Farm Magazine. Electrical Equipment* (co-editor).

J. P. SCHAENZER, B.S., Electro-Agriculture Specialist, Rural Electric Administration, U.S.D.A. *Electric Farm Equipment Data.*

KARL STALEY, B.S.E.E., Lighting Education Specialist, General Electric Co. *Lamps and Lighting.*

JOHN G. SUTTON, B.S.C.E., Drainage Engineer, Soil Conservation Service, U.S.D.A. *Agricultural Drainage.*

TRUMAN J. WAKEMAN, B.S. Ag. Ed., M.S. Ag. Ed., Professor of Agricultural Engineering, Virginia Polytechnic Institute. *A Home Farm Shop.*

WALTER WEISS, B.S. Ag. Ec., Area Conservationist, Soil Conservation Service, U.S.D.A. *Field Practices for Erosion Control.*

W. S. WORLEY, B.S.M.E., Chief Product Application Engineer, Gates Rubber Co. *V-Belts.*

WAYNE W. WORTHINGTON, B.S., formerly Director of Engineering Research, John Deere Tractor Research and Engineering Center. *Fuels and Combustion.*

AUSTIN L. ZINGG (deceased), B.S.A.E., M.S.A.E., Chief, Watershed Hydrology Branch, A.R.S., U.S.D.A. *Wind Erosion and Its Control.*

FOREWORD

Agriculture is the industry which supplies the human race with food and raw materials for food, shelter, and clothing. It is recognized as a primary industry because agricultural products are a universal need. History reveals that no country has prospered long without an adequate supply of food. This universal need can be cited as the reason for the general public interest in agriculture and in legislation, education, and research related to agriculture.

A study of the importance of agriculture indicates that it is a matter of general public concern that the agricultural industry achieve four specific objectives:

1. Production of an adequate *quantity* of agricultural products, more specifically food and materials for food, shelter, and clothing

2. Constant improvement in the *quality* of agricultural products

3. Production of agricultural products at a *low cost* so that all persons in the community concerned shall have plenty

4. Securing for those engaged in agriculture a measure of well-being comparable with that received by those in other vocations and industries

Engineering has been defined as "the art and science of organizing and directing men and utilizing the forces and materials of nature for the benefit of mankind."

If simple terms are substituted for the phrases used in the definition given, i.e., *labor* for "organizing and directing men" and *power* for "the forces of nature," and assuming that the materials are those of construction and manufacture, the three central interests of the engineer become (1) labor, (2) power, and (3) materials. It is the engineer's task to utilize these resources for the production of services and commodities which contribute to the well-being of all the people of any community.

Numerous branches of engineering have developed. These special branches may represent a particular type of engineering, as in the case of military and civil engineering, which represent the earliest differentiation in engineering.

Special branches may also be established by the segregation and organization of the engineering science and practice of any nature related to an industry. Mining, ceramic, electrical, and *agricultural engineering* are so constituted.

Although agricultural engineering is one of the latest branches of engineering to be recognized, most states provide for the licensing of agricultural engineers. The first degrees in this field were granted in 1910, but at present there are 24 [43 in 1959] colleges in the United States having curricula in agricultural engineering.

With the development of agricultural engineering five distinct branches or phases have been recognized:

1. Soil water conservation, including drainage, irrigation, and soil-erosion control
2. Farm structures and housing
3. Crop storage, conditioning, and processing
4. Rural electrification
5. Mechanical farm equipment, including power and machines

The American Society of Agricultural Engineers was organized in 1907 to further the development and application of engineering related to agriculture. The society publishes a journal, establishes certain standards of procedure and construction, and performs other customary functions of national professional engineering societies. In 1956 the membership was approximately five thousand. The office of the secretary is located at Saint Joseph, Mich.

J. B. Davidson

NOTE: Professor Davidson, who was the first president of the American Society of Agricultural Engineers, died shortly after this foreword was written.

PREFACE

This Handbook is intended to include under one cover the basic theory and practice for the various areas of agricultural engineering, the application of engineering to the problems of agricultural production.

Basic reader familiarity with such subject matter has been assumed in order to allow coverage of the theory and latest practice as thoroughly as possible in the allotted space. Elementary information, such as definitions of names of machine parts, has been held to a minimum. The amount of mathematical tables and design information readily available from other books has also been minimized. The engineering approach has been used throughout, but an effort has been made to keep the language simple and clear to allow the nonengineer reader to comprehend all but the more technical portions.

Section 1, Crop Production Equipment, deals primarily with tractors and field machines. Space does not permit coverage of tractor components such as engines, transmissions, axles, etc., which are basically similar to those used in other automotive vehicles, but several chapters are devoted to specific tractor problems such as traction, force reactions, fuels, and implement controls. In sections covering a class of implements, basic components are analyzed and then combined into the various models within the class. Unfortunately, basic engineering information is still lacking for many implements.

Section 2, Soil and Water Conservation, has been written primarily by specialists in the USDA Soil Conservation Service who are personally engaged in research or extension work in various areas. Much of the material presented has not heretofore been available to the general public.

Section 3, Farmstead Structures and Equipment, covers the areas which are usually separated into the fields of farm structures, rural electrification, and crop processing. The advent of *farmstead* mechanization requires integration of crop processing, crop storage, and feed handling in adapted structures tying in with the housing requirements for livestock. Each of these factors affects the others, and all must be considered together to secure an efficient system.

Section 4, Basic Agricultural Data, is designed to present succinctly agricultural data affecting the solutions of agricultural engineering problems. Heat and moisture production by livestock is given as well as basic food and water requirements. The fundamentals of soil formation, physical characteristics of soil, and soil fertility are presented. Basic climatic and solar-energy data affecting crop production and environmental control are also included.

Chapters which do not have the author listed have been written by the subeditor for the section in which they appear.

Because of the breadth of the field of agricultural engineering, the cooperation and combined efforts of many individuals were required to prepare this Handbook. Without exception, the authors have regular full-time jobs and their assignments represented a substantial additional burden in many cases. I hereby express my gratitude for their willing efforts and ungrudging sacrifices.

Appreciation is hereby expressed to Ray Olney, American Society of Agricultural Engineering, for encouragement and review; to George R. Shier, Golden, Colorado, who initiated the work on Section 2; and to the following individuals for reviews and many constructive suggestions:

Earl D. Anderson, Stran-Steel Corp.; Henry J. Barre, Mansfield, Ohio; John R. Davis, University of California, Davis; Orval C. French, Cornell University; Harold Gray, Cornell University; Ralph C. Hay, University of Illinois; Ernest H. Kidder, Michigan State University; Drayton T. Kinard, University of Kentucky; C. S. Morrison, Deere & Co.; Charles K. Otis, University of Minnesota; George E. Pickard, University of Illinois; George D. Scarseth, Lafayette, Ind.; E. W. Schroeder, Oklahoma State University; Glenn O. Schwab, Ohio State University; H. P. Smith, Agricultural and Mechanical College of Texas; C. W. Smith, University of Nebraska and Nebraska Tractor Testing Board; and Wayne Worthington, Deere & Co.

Gratitude is hereby expressed to the Tractor and Implement Division of the Ford Motor Company for making available its resources and to Mrs. Elizabeth Headrick of the Division for much of the final typing.

I finally wish to thank my wife, Marguerite, not only for the uncounted hours of typing from handwritten manuscript, but also for her toleration of a project which took a major share of my "spare" time for several years.

C. B. Richey

CONTENTS

Foreword by J. B. Davidson vii
Preface ix

SECTION I. CROP-PRODUCTION EQUIPMENT

1. Economics of Farm Machinery 1
2. Design of Field Machinery 18
3. Power-transmission Elements 33
4. Wheels and Tires 60
5. Stress Measurements and Strength Analysis 75
6. Tractor Force Reactions 94
7. Tractor Design Objectives 103
8. Tractor Performance Tests 107
9. Fuels and Combustion 110
10. Power Controls for Implements 117
11. Tillage Objectives 125
12. Moldboard Plows 128
13. Disk Tools 138
14. Shovel and Sweep Tools 148
15. Minor Tillage Tools 155
16. Fertilizing and Liming Machines 160
17. Seeding and Planting Machines 172
18. Sprayers and Dusters 187
19. Mowers and Crushers 200
20. Rakes 207
21. Hay-handling Equipment 213
22. Balers 216
23. Forage Harvesters and Blowers 227
24. Combines 238
25. Corn-harvesting Machines 251
26. Cotton-harvesting Machines 261
27. Root-harvesting Equipment 269
28. Specialized Harvesting Equipment 276
29. Stalk and Brush Shredders 284
30. Farm Transport Equipment 287
31. Tractor Loaders 295
32. Feed Grinders 298
33. Feed Elevating and Conveying Equipment 305

SECTION II. SOIL AND WATER CONSERVATION

34. Principles of Agricultural Hydrology 313
35. Land Leveling and Grading 347

36. Agricultural Drainage. 356
37. Mechanics of Water Erosion 401
38. Terraces and Diversions 407
39. Field Practices for Erosion Control 414
40. Waterways for Erosion Control 419
41. Erosion-control Structures 426
42. Water Management, Conservation, Use, and Legal Aspects 492
43. Wind Erosion and Its Control 504
44. Irrigation 509
45. Frost Control 532

SECTION III. FARMSTEAD STRUCTURES AND EQUIPMENT

46. Structural Requirements for Farm Buildings 541
47. Livestock Production Facilities 587
48. Heat, Air, and Moisture 622
49. Drying Farm Crops 646
50. Storage of Farm Crops 672
51. Electrical Equipment 695
52. Fruit and Vegetable Handling 728
53. Storage of Fresh Fruits and Vegetables 741
54. Service Buildings and Equipment 756

SECTION IV. BASIC AGRICULTURAL DATA

55. Livestock 771
56. Soil 788
57. Farm Climates and Solar Energy 817

Index 867

SECTION I
CROP-PRODUCTION EQUIPMENT
C. B. Richey, *Editor*

Chapter 1
ECONOMICS OF FARM MACHINERY

Food production is a basic concern of mankind. The economics of food production are of vital importance to a nation as a whole as well as to the people specifically engaged in it. Adequate production of food is less of a problem in the United States today than in any other time or place in the history of the world. This is due to progress in the sciences dealing with the growth efficiency of plants and animals and also to the multiplication of man's efforts through mechanical power and machines to perform food-production operations.

Table 1-1. Indexes of Farm Output, Volume of Farm Power, Machinery, and Equipment, Farm Employment, and Farm Population, United States, 1870–1955
(1870 = 100)

Year	Indexes of volume of farm output		Indexes of volume of farm power, machinery, and equipment		Farm output per unit of farm power, machinery, and equipment	Index of farm employment	Index of farm population	Total persons supported at home and abroad by one farm worker
	Total	Per worker	Total	Per worker				
1870	100	100	100	100	100	100	100	5.14
1880	156	125	163	130	96	125	121	5.57
1890	183	125	232	159	79	146	131	5.77
1900	240	151	295	186	81	159	139	6.95
1910	261	154	385	228	68	169	143	7.07
1920	313	197	477	300	66	159	141	8.27
1930	324	208	471	302	69	156	135	9.76
1940	373	252	434	293	86	148	135	10.81
1950	448	356	868	690	52	126	111	15.49
1953	484	417	1,120	973	43	116	100	17.82
1955	503	453	1,132	1,025	44	111	98	19.74

sources: Martin R. Cooper, Glen T. Barton, and Albert P. Brodell, Progress of Farm Mechanization, *USDA Misc. Publ.* 630, 1947. 1957 Agricultural Outlook Charts, USDA AMS and ARS, November, 1956.

PROGRESS OF FARM MECHANIZATION

In 1850, the approximate beginning of farm mechanization, one United States farm worker supported 4.68 persons at home and abroad. In 1910, mechanization with animal power had approached its peak and one farm worker supported 8 persons.[1] In 1952, the replacement of animal power by mechanical power was almost complete and, because of multiplication of output and the release of acreage for supporting work animals, one farm worker supported 18 persons.[2] As shown in Table

[1] Martin R. Cooper, Glen T. Barton, and Albert P. Brodell, Progress of Farm Mechanization, *USDA Misc. Publ.* 630, 1947.
[2] Agricultural Statistics, 1953, USDA, pp. 563, 565.

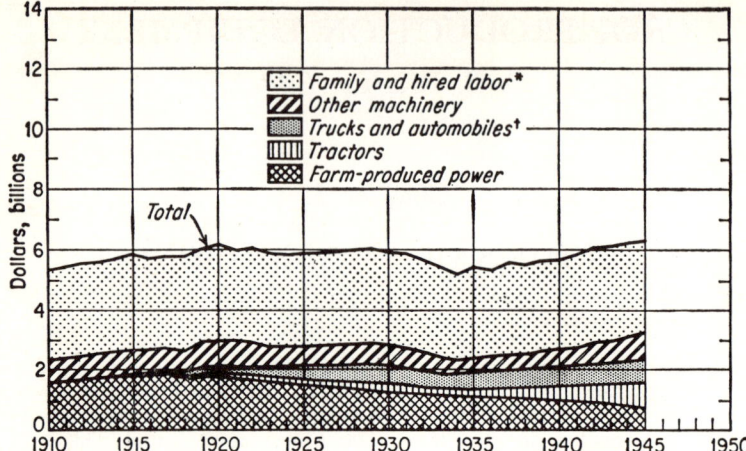

* *All farm labor costs except those for servicing and maintenance of machinery, trucks, automobiles, and tractors and for care and maintenance of and growing of feed for horses and mules, which are included in the power and machinery cost items.*

† *Includes only farm production share of automobile costs. (Cooper, USDA Misc. Publ. 630, 1947.)*

Fig. 1-1. Total production costs of farm power, machinery, and labor, United States, 1910–1945. (Costs in 1935–1939 average dollars.)

1-1, the actual output per worker had increased somewhat more than the number of persons supported, indicating that there was more food available per person.

The total *farm production costs of power, machinery, and labor* are shown in Fig. 1-1 for the period 1910 to 1945. The total cost adjusted to 1935–1939 average dollars has not varied greatly in spite of an increase in total production of approximately 70 per cent, as indicated by Table 1-1. Labor and machinery costs have re-

TABLE 1-2. MAN-HOURS AND METHODS TO PRODUCE WHEAT*

Year and location	Hr/acre	Hr/bu	Bu/acre	Methods and equipment	Source of figures
1830: U.S. average........	57.7	2.88	20	Horse-drawn walking plow, brush harrow, hand broadcast, sickle, flail, wagon, winnowing	13th Ann. Rept. of Commissioner of Labor, 1898
1896: Central Wheat Belt...	8.8	0.44	20	Horse-drawn 2-bottom plow, broadcast seeder, section harrow, binder, thresher	13th Ann. Rept. of Commissioner of Labor, 1898
1930: Great Plains.........	3.3	0.16	20	Tractor, 3-bottom plow, tandem disk, 6-section harrow, two 10-ft drills, 12-ft combine, trucks	USDA Bull. 157
1949: Great Plains.........	1.82	0.09	20	5-plow tractor, subsurface blade, disk harrow, grain drill, self-propelled combine, trucks	S. Dakota State Coll.
1950–1953: U.S. average....	4.4	0.26	17.1		USDA Statist. Bull. 144, 1954

* Bert S. Gittens, "Land of Plenty," Farm Equipment Institute, Chicago, 1950.

mained about the same. Crop-production power costs are somewhat lower since tractors have largely replaced horses, but transportation, as indicated by truck and automobile costs, has come to cost as much as tractor power. Total cost of agricultural production must include the costs of fertilizer and of the use of the land and buildings. These costs averaged 35 per cent, and labor, power, and machinery costs 65 per cent, of the total during the 1935–1939 base period.[1]

[1] Cooper, Barton, and Brodell, *op. cit.*

TABLE 1-3. MAN-HOURS AND METHODS TO PRODUCE CORN*

Year and location	Hr/acre	Hr/bu	Bu/acre	Methods and equipment	Source of figures
1855: U.S. average	33.6	0.84	40	Horse-drawn walking plow, spike-tooth harrow, hand planting, half-row shovel cultivator, hand husking, wagon	13th Ann. Rept. of Commissioner of Labor, 1898
1896: Univ. of Illinois cost studies	20.8	0.38	55	Horse-drawn 2-bottom plow, disk harrow, 2-row planter, peg-tooth harrow, single-row cultivator, hand husking, wagon	Univ. of Illinois Agr. Expt. Sta. Bull. 50
1930: Corn Belt	6.9	0.17	40	Tractor, 2-bottom plow, tandem disk, 2-row planter, 2-row tractor cultivator, 2-row mechanical picker	USDA Bull. 157
1949: Iowa	3.88	0.05	70	Tractor, 2-bottom plow, 10-ft tandem disk, tractor planter and cultivator, mechanical picker	Iowa State Univ. Agr. Expt. Sta.
1950–1953: U.S. average	13.1	0.34	38.3		USDA Statist. Bull. 144, 1954

* Bert S. Gittins, "Land of Plenty," Farm Equipment Institute, Chicago, 1950.

TABLE 1-4. MAN-HOURS AND METHODS TO PRODUCE COTTON*

Year and location	Hr/acre	Hr/bale	Yield per acre, lb	Methods and equipment	Source of figures
1841: U.S. average	148.6	...	750, seed cotton	Wooden moldboard walking plow, bull-tongue plow, hand planting, hand hoeing and chopping, hand picking	13th Ann. Rept. of Commissioner of Labor, 1898
1895: U.S. average	102.4	...	750, seed cotton	1-mule equipment, hand hoeing and chopping, hand picking	13th Ann. Rept. of Commissioner of Labor, 1898
1925–1929: U.S. average	96	268	200.6, lint	1-row mule equipment, hand chopping and hoeing, hand picking	USDA Misc. Publ. 707
1948: California	38	...	750, lint	4-row tractor equipment, mechanical picker	Univ. of Calif., Agr. Eng. Dept.
1948: Mississippi River Delta	28.3	...	500, lint	4-row tractor equipment, mechanical picker	Louisiana Agr. Expt. Sta. Circ. 84
1946: North Carolina Coastal Plains	21.4	...	423, lint	2-row tractor equipment, mechanical picker	N. Carolina State Coll. Agr. Expt. Sta. Bull. 348
1949: Texas High Plains	6.5	...	182, lint	4-row tractor equipment, 2-row stripper	Texas A&M, Agr. Econ. Dept.
1950–1953: U.S. average	70	...	286.5 lint		USDA Statist. Bull. 144, 1954

* Bert S. Gittins, "Land of Plenty," Farm Equipment Institute, Chicago, 1950.

Information on the *reduction of man-hours to produce crops* in the United States from the middle 1800s to the present is given in Table 1-2 for wheat, Table 1-3 for corn, and Table 1-4 for cotton. Table 1-5 shows progress in caring for dairy cattle. This information was assembled from various governmental sources.[1]

COST OF USE OF FARM EQUIPMENT

Cost of use of farm machines consists of charges for depreciation, repairs, interest on investment, housing, taxes, and insurance.

(*From Agr. Eng. Yearbook*, 1957, p. 96.)

TABLE 1-5. MAN-HOURS AND EQUIPMENT TO CARE FOR DAIRY CATTLE*

Year and location	Annual hours per animal	Methods and equipment	Source of figures
1909: U.S. average..............	135	Manure fork for cleaning stables, hand milking	Bur. Agr. Econ.
1932–1936: Michigan...........	115	Manure fork and spreaders, hand and machine milking	*Mich. State Univ. Agr. Expt. Sta. Bull.* 297
1948: Michigan................	70	Litter carrier, machine milking	Mich. State Coll., Dept. Agr. Econ.
1948: Michigan................	61	Pen barn, milking parlor, machine milking, tractor manure loader for cleaning lounging area	Mich. State Coll.

* Bert S. Gittins, "Land of Plenty," Farm Equipment Institute, Chicago, 1950.

NOTE: Better care of cows and higher standards of sanitation have increased dairy chores; yet according to Cornell University, since 1900, production per man has almost doubled and, since 1918, production per hour has increased 49 per cent.

Depreciation. This is the largest single item in the cost of farm machinery. It is defined as the loss in value with the passing of time, and the rate of depreciation depends on the length of the useful life of the machine.

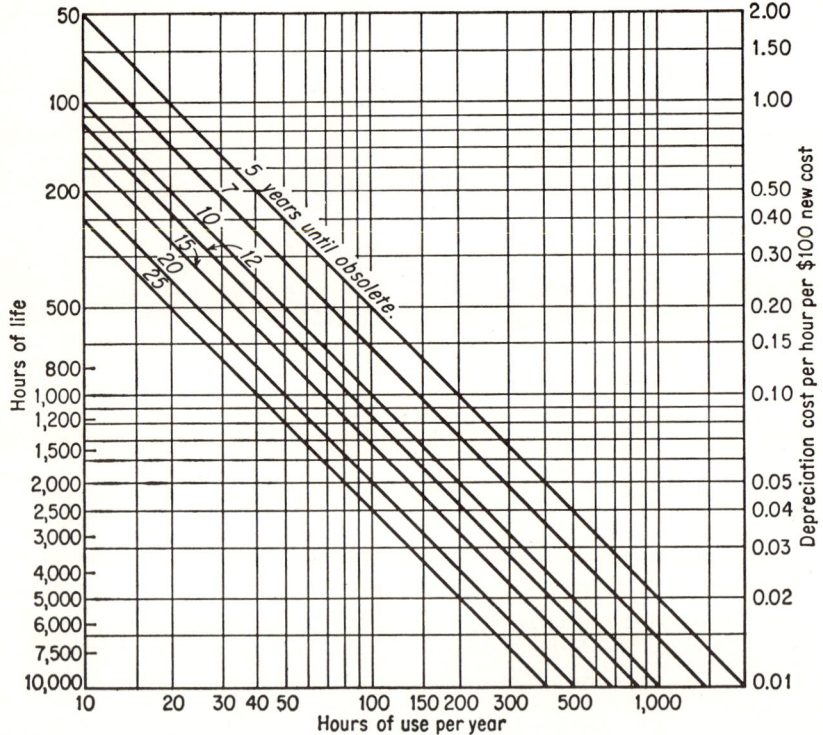

FIG. 1-2. Depreciation cost of farm machines. To find depreciation cost per hour per $100 new cost: (1) Locate intersection of hours of use per year with hours of life. (2) Locate intersection of hours of use per year with years until obsolete. (3) From upper of these intersections move to right and read. (*From Agr. Eng. Yearbook*, 1957, p. 96).

ECONOMICS OF FARM MACHINERY 5

The length of the useful life of a machine depends on:
1. Mechanical deterioration or wear, as affected by
   a. Amount of use
   b. Operating conditions
   c. Care by operator as to overloading, lubrication, adjustment, repair, and shelter
2. Obsolescence, as affected by
   a. Increased efficiency of new machines doing the same job
   b. Development of new methods eliminating the job

TABLE 1-6. COST PER HOUR OF USING FARM MACHINES (PER $100 OF NEW COST)

Machine	Years until obsolete	Hours to wear out	Total repair cost, % new cost	Cost per hour of use per $100 of new cost*						
				20 hr per yr	40 hr per yr	60 hr per yr	100 hr per yr	150 hr per yr	250 hr per yr	350 hr per yr
Tractor plow	15	2,000	80	$.600	$.319	$.226	$.152	$.120	$.108	$.103
Tractor disk harrow	15	2,000	30	.575	.295	.202	.127	.095	.083	.078
Spring-tooth harrow	20	2,000	40	.495	.253	.179	.115	.100	.088	.083
Spike-tooth harrow	20	2,500	30	.487	.250	.171	.107	.082	.070	.065
Roller	25	1,500	10	.432	.220	.149	.118	.103	.091	.086
Soil pulverizer	20	2,000	15	.483	.245	.167	.103	.088	.076	.071
Endgate seeder	20	800	30	.512	.275	.238	.208	.193	.181	.176
Grain drill	20	1,200	25	.496	.258	.179	.149	.135	.122	.117
Corn planter	20	1,200	30	.500	.263	.184	.153	.138	.126	.121
Field sprayer	10	1,500	30	.745	.383	.262	.165	.117	.105	.100
Rotary hoe	15	1,500	20	.573	.293	.200	.125	.110	.098	.093
Tractor cultivator	12	2,500	40	.662	.341	.234	.148	.106	.078	.073
Rotary cutter	12	2,000	35	.659	.338	.232	.146	.104	.086	.081
Tractor mower	12	2,000	75	.679	.358	.252	.166	.124	.106	.101
Dump rake	10	1,500	25	.742	.379	.259	.162	.114	.102	.097
Side-delivery rake	15	1,500	50	.591	.312	.219	.145	.130	.118	.113
Tractor buck rake	12	1,500	25	.657	.337	.231	.145	.114	.102	.097
Hay loader	10	1,200	25	.746	.383	.263	.166	.134	.122	.117
Forage harvester†	12	2,000	60	.671	.350	.244	.158	.116	.098	.093
Forage blower	12	2,500	25	.651	.330	.224	.138	.096	.068	.063
Pickup baler-auto. tie†	12	2,500	40	.657	.336	.230	.144	.102	.074	.069
Swather	12	1,200	25	.662	.341	.235	.149	.134	.122	.117
Combine†	10	2,000	40	.745	.383	.262	.165	.117	.088	.083
Corn binder	10	1,000	40	.765	.402	.282	.185	.170	.168	.153
Stationary silage cutter	10	1,200	30	.75	.387	.267	.170	.138	.126	.121
Husker-shredder	10	2,500	25	.735	.372	.252	.155	.107	.068	.063
Corn picker	10	1,500	30	.745	.383	.262	.165	.117	.105	.091
Spindle cotton picker	10	2,000	55	.753	.390	.270	.173	.125	.096	.091
Manure loader	10	2,000	25	.738	.375	.255	.158	.110	.086	.076
Manure spreader	15	2,500	25	.568	.289	.195	.122	.084	.068	.063
Feed grinder	15	2,000	25	.571	.292	.198	.125	.093	.081	.076
Portable elevator	15	1,500	15	.568	.289	.195	.122	.107	.095	.090
				60 hr per yr	150 hr per yr	300 hr per yr	500 hr per yr	750 hr per yr	1,000 hr per yr	1,500 hr per yr
Wagon gear and box	15	5,000	50	.196	.084	.047	.039	.036	.035	.033
Tractor	15	12,000	120	.196	.084	.047	.032	.025	.023	.021

* Based on 4½ per cent of new cost as total annual charge for interest, housing, taxes, and insurance.
† Operating costs such as fuel, oil, grease, wire, twine, etc., not included.
SOURCE: *Agr. Eng. Yearbook*, 1956, p. 97.

In this discussion, depreciation has been calculated on a straight-line basis, using a life based on whichever occurs first, wearing out or obsolescence. Depreciation cost for a particular machine can be estimated by applying the life data in Table 1-6 to the chart in Fig. 1-2.

These life data consist of estimates based on a study of the data contained in the references and those obtained by farm surveys in various states. Although there is much variation in operating conditions and annual use, as well as fallibility in farmers' estimates of life and repair costs, there is a pattern of agreement as to *hours of life* and *life repair cost*.

The estimates for *years until obsolete* are based on experience and judgment as well as on farmers' estimates of total life. Obsolescence is particularly hard to predict during periods of rapid change, such as the present, and the estimates are considered conservative. Because of the low average annual use per farm, most farm machines are not actually worn out but are discarded, primarily because of obsolescence.

Repair Cost. This has been calculated as a constant cost per hour based on the life repair cost divided by the hours of life. Actual repair costs increase as a machine

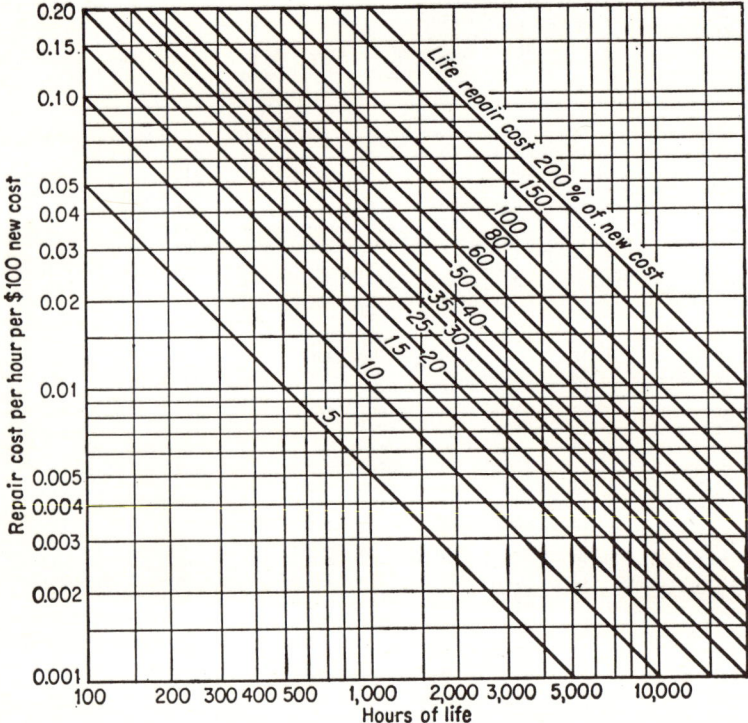

Fig. 1-3. Repair cost of farm machines. To find repair cost per hour, locate intersection of hours of life with life-repair-cost percentage, move left, and read. (*From Agr. Eng. Yearbook, 1957, p. 97.*)

grows older, but in this case, for the sake of simplicity, a uniform charge is made throughout the life. It is recognized that with more repairs a machine can be made to run longer, but the life repair cost used is estimated for the hours of life shown as based on a study of the references. Repair cost for a particular machine can be estimated by applying the life and repair data in Table 1-6 to the chart in Fig. 1-3.

These estimates, of course, assume average machine quality and operating conditions, but if desired, they can be modified according to experience in a particular situation and costs calculated accordingly.

The use of a uniform hourly charge for depreciation plus repairs appears to be a reasonable approximation for a machine which will be worn out. This infers that the actual depreciation during the early years is the straight-line depreciation charge plus the unspent balance of the repair charge and that during the late years it is the straight-line depreciation charge minus the overaverage portion of the repair cost.

Interest, Housing, Taxes, and Insurance. These charges may be grouped together since the costs of each may be expressed by a fixed annual charge. They may be best determined by considering the specific case, but if better estimates are not available, the following costs may be used as typical for farm machines in general:

	Cost per year for each $100 value
Interest................	$5.00
Housing................	1.60
Taxes.................	2.00
Insurance..............	0.40

This gives a total of $9 per year for each $100 value, or $4.50 per year for each $100 of new cost, assuming the average value of a machine to be one-half the new cost. The hourly cost for any desired total annual rate can be secured from the chart in Fig. 1-4.

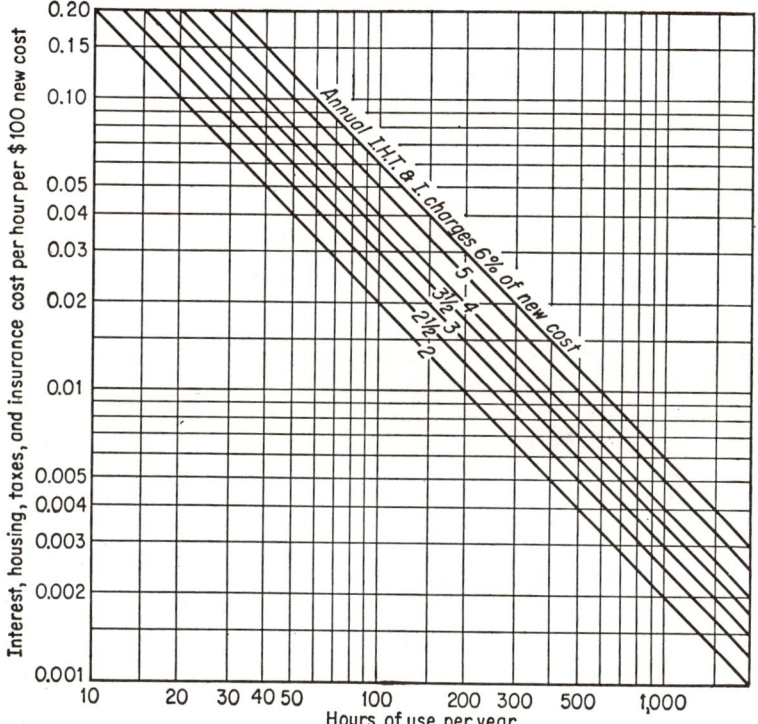

FIG. 1-4. Interest, housing, taxes, and insurance cost of farm machines. To find cost per hour of interest, housing, taxes, and insurance, locate intersection of hours of use per year with annual interest, housing, taxes, and insurance (at desired percentage of new cost), move left, and read. (*From Agr. Eng. Yearbook*, 1957, p. 97.)

Expressing all costs as a percentage of *new cost* has the advantage that data for various sizes of machines are not needed, this variable being taken care of by the new cost used, and also that changes in the general price level are taken care of by using the current new cost.

The charts in Figs. 1-2 to 1-4 may thus be used to compute the cost per hour for (1) depreciation, (2) repairs, and (3) interest, housing, taxes, and insurance. The total cost, excluding operating costs, is the sum of these three items. To get the complete cost for a particular field operation, the tractor cost, labor cost, and operating costs such as fuel, lubricant, wire, twine, etc., must also be added.

8　　　　　　　　CROP-PRODUCTION EQUIPMENT

The cost figures in Table 1-6 assume an annual charge for interest, housing, taxes, and insurance of 4½ per cent of the new cost and are intended to apply to average quality of machines and average conditions of use. Where desirable, appropriate corrections can be made.

Tractor Operating Costs. These vary greatly with the particular tractor used and the degree of loading, fuel consumption being the greatest variable. For a particular job, experience is the best guide. For an average figure, actual records of hours worked and fuel used on the particular farm should be used. Where this is not available an estimate may be based on the hourly fuel consumption of the particular tractor recorded in the Nebraska Tractor Test for Test H, rated drawbar load, 75 per cent of maximum. Many operations are lighter than this, but on the other hand,

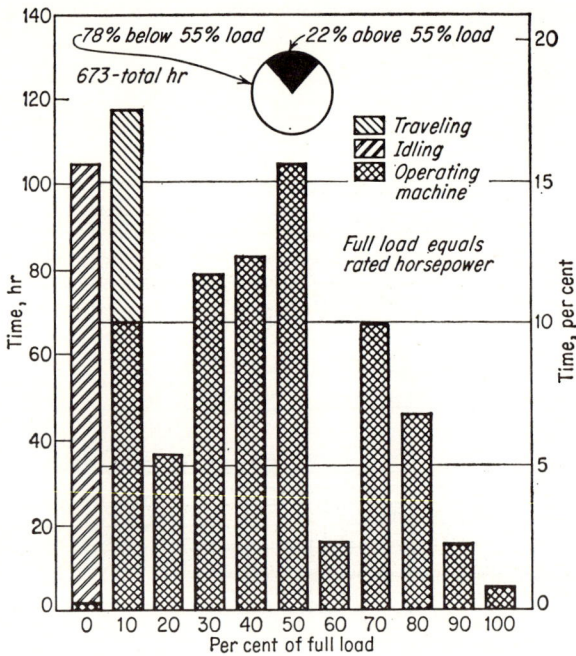

FIG. 1-5. Distribution of a season's tractor-operating time. (*Bateman, Agr. Eng., April,* 1943.)

many tractors are not kept in peak operating condition and there may be frequent warm-up periods. Records from Illinois [1], Kansas [10], and Nebraska [19] indicate that a tractor used for general farm work will have an hourly *fuel consumption* averaging from 75 to 90 per cent of that recorded in the Nebraska Tractor Test as the average for the Power Take-off Varying Power tests, formerly Test E.

Oil and grease requirements for a particular tractor can be closely estimated from experience. A nationwide survey in 1947 [3] indicated an average oil consumption of 0.038 gal per hr for all United States wheel tractors. An average grease requirement of 0.03 lb per hr was found in a survey of several states by USDA workers [20].

The total cost per drawbar horsepower-hour of field work is greatly influenced by the *degree of loading*. Nebraska Test data indicate that a tractor operating at its rated drawbar horsepower will use only about one-third more fuel than when operating at half this load. Thus the operator who has equipment of the right size to load his tractor properly will save considerable fuel and time.

The average tractor is operated at optimum load only a small proportion of the time, as is indicated by Fig. 1-5. These records are for a two-plow tractor on a typical

Illinois farm growing 72 acres of corn, 81 acres of soybeans, 5 acres of oats, 36 acres of wheat, and 38 acres of clover [1].

Where light loads are unavoidable, considerable fuel can be saved by operating in a higher gear with the engine throttled back to give the desired ground speed. This saves fuel by reducing engine friction losses and also by operating at a higher compression pressure. Of course, an engine should never be loaded to the point where it labors continually. A vacuum gage attached to the intake manifold is a valuable indicator of engine loading. The reading will range from 20 to 21 in. of mercury when idling to 0 at full load. An engine will usually operate efficiently at 5 in. of mercury without being overloaded.

REFERENCES

1. Bateman, H. P.: Effect of Full Load in Farm Machine Operating Economics, *Agr. Eng.*, 24:111–114, April, 1943.
2. Bateman, H. P.: Unpublished data, March, 1955.
3. Brodell, Albert P., and Albert R. Kendall: Farm Consumption of Liquid Petroleum Fuels and Motor Oil, *USDA BAE* FM-73, 1949.
4. Brodell, Albert P., and Albert R. Kendall: Life of Farm Tractors, *USDA BAE* FM-80, June, 1950.
5. Byers, George D., and B. T. Inman: The Use and Expense of Farm Implements, *Kentucky Agr. Expt. Sta. Bull.* 345, 1933.
6. Butz, E. L., and O. G. Lloyd: The Cost of Using Farm Machinery in Indiana, *Purdue Univ. Agr. Expt. Sta. Bull.* 437, 1939.
7. Davidson, J. B., and S. M. Henderson: Life, Service and Cost of Service on 400 Iowa Farms, *Iowa State Univ. Agr. Expt. Sta. Bull.*, 1942, p. 37.
8. Day, C. L.: Hay and Ensilage Harvesting Costs, *Missouri Agr. Expt. Sta. Bull.* 561, 1951.
9. Department of Agricultural Economics, Cornell Univ.: Individual Factors and Annual Averages from Farm Cost Accounts, 46 Farms, 1952, *Cornell Univ. Agr. Expt. Sta.*, AE-941, 1954.
10. Fenton, F. C., and E. L. Barger: The Cost of Using Farm Machinery, *Kansas State Coll. Eng. Expt. Sta. Bull.* 45, 1945.
11. Fenton, F. C., and G. E. Fairbanks: The Cost of Using Farm Machinery, *Kansas State Coll. Eng. Expt. Sta. Bull.* 74, 1954.
12. Hertel, J. P., and Paul Williamson: Costs of Farm Power and Equipment, *Cornell Univ. Agr. Expt. Sta. Bull.* 751, 1941.
13. Hedges, T. R., and W. R. Bailey: Economics of Mechanical Cotton Harvesting, *Calif. Agr. Expt. Sta. Bull.* 743, 1954.
14. Jasny, N.: "Research Methods on Farm Use of Tractors," Columbia University Press, New York, 1938.
15. Johnston, Rupert B.: Farm Mechanization in the Upland Area of Mississippi, 1949, *Mississippi State coll. Agr. Expt. Sta. Bull.* 502, 1953.
16. Jones, M. M.: unpublished data, March, 1955.
17. Kalbfleisch, W.: Cost of Operating Farm Machinery, *Can. Dept. Agr. Bull.* 118, Ottawa, 1950.
18. Marx, R. E., and J. W. Birkenhead: Hay Harvesting Methods and Costs, *USDA Circ.* 868, 1951.
19. Miller, Frank, W. L. Ruden, and C. W. Smith: Cost of Tractor Power on Nebraska Farms, *Nebraska Univ. Agr. Expt. Sta. Bull.* 324, 1942.
20. Reed, A. D.: Machinery Costs and Related Data, *Univ. Calif. Ext. Serv.*, November, 1954.
21. Reynoldson, L. A., et al.: Utilization and Cost of Power on Corn Belt Farms, *USDA Tech. Bull.* 384, 1933.
22. Richey, C. B.: Cost per Hour of Using Farm Machinery, *Ohio Agr. Ext. Serv. Bull.* 221, 1942.
23. Richey, C. B., et al.: Rental Rates for Farm Machinery, *Agr. Eng.*, 24:15–16, January, 1943.

24. Rorholm, Niels, et al.: Farm Labor and Farm Costs, 1953, *Univ. Minn. Dept. Agr. Econ. Rept.* 217, 1954.

25. Smith, D. D., and M. M. Jones: Power, Labor, and Machine Costs in Crop Production, *Missouri Univ. Agr. Expt. Sta. Research Bull.* 197, 1933.

CAPACITY OF FARM MACHINES

The capacity of field machines in acres per hour is a function of the following factors:

1. *Operating width*, as affected by
   a. Measured width of machine
   b. Percentage of width actually used
2. *Speed of travel*, as affected by
   a. Draft of machine
   b. Drawbar horsepower available
   c. Traction of power source
   d. Variations in grade and rolling resistance
   e. Operating limitations on speed such as quality of work, rough ground, obstacles, etc.
3. *Percentage of nonoperating time*, due to
   a. Idle travel, such as traveling to field, turning at ends, etc.
   b. Adding seed, fertilizer, etc.
   c. Unloading harvested products
   d. Resting animal power
   e. Lubrication, refueling, etc.
   f. Machine adjustment, resharpening, replacing wearing parts, etc.
   g. Clogging
   h. Breakdowns

Of these factors, the percentage of nonoperating time is the most difficult to evaluate. Results of three seasons' records of a two-plow tractor (equipped with a Servis-Recorder), used on a typical Illinois grain farm, are reported as follows:[1]

Machine	Percentage of nonoperating time
2-16-in. bottom tractor plow	16–22
8-ft tandem-disk harrow	9–23
18-ft spike-tooth harrow	24–30
8-ft grain drill	22–30
4-row corn planter (checkrowing)	41
2-row tractor cultivator	
First cultivation	20
Second cultivation	15
Third cultivation	12
7-ft tractor mower	31
12-ft combine	37–43
2-row pull-type corn picker	35

These figures do not include width loss and turning time at the ends, but they give an indication of average actual capacities for typical widths and speeds.

A simple formula for capacity, which assumes 17½ per cent nonoperating time, is

$$\text{Acres per hour} = \tfrac{1}{10} (\text{width, ft})(\text{speed, mph})$$

The chart in Fig. 1-6, originally devised by E. G. McKibben, makes it possible to read the resultant capacity for any normal width, speed, and percentage of nonoperating time without making calculations.

The nonoperating time due to the routine operations listed above can be closely predicted. Clogging and breakdowns, however, vary greatly with field conditions and machine reliability and are relatively unpredictable. A loss of to 5 to 10 per cent of operating time from clogging and breakdowns is not uncommon. Where several units are used together and the stoppage of any one unit halts the group, the loss in time is much more serious. A four-row cultivator, for instance, cannot accomplish twice as

---

[1] H. P. Bateman, Effect of Full Load in Farm Machine Operating Economies, *Agr. Eng.*, 24:111–114, April, 1943.

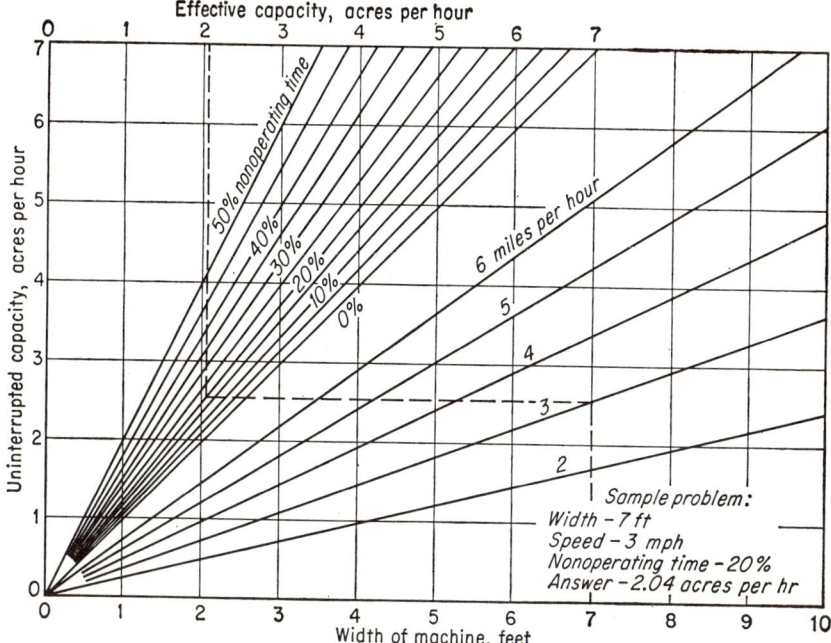

FIG. 1-6. Capacity chart for field machines. From this chart may be read the resultant capacity of field implements and machines for normal width, speed, and percentage of nonoperating time, without making calculations. (*After E. G. McKibben.*)

much as a two-row cultivator where the operator must occasionally stop to clear an individual gang of trash. This principle also applies where several different machines are used together, as in harvesting grass silage with a field chopper, two or more unloading wagons, and a blower.

The following equation has been developed for several machines in series:[1]

$$y = 100 \left(\frac{x}{100}\right)^n$$

where y = expected effective operating time for a series combination if operation of series depends on operation of each machine, per cent
x = expected effective operating time for each of individual machines of series, per cent
n = number of machines in series

If the expected effective operating time x is different for each machine, the equation takes the more general form[1]

$$y = \frac{(x_1 x_2 x_3 \cdots x_n) \, 100}{100^n}$$

A chart based on this equation is shown in Fig. 1-7. It can be seen, for example, that a series of five machines, each having a reliability of 90 per cent, will have a probable performance of 59 per cent.

This principle also applies to complex machines having a series of mechanisms performing various operations, such as combines or corn pickers. In order that the over-all reliability may be satisfactory, each element of the machine must be as re-

[1] E. G. McKibben and P. L. Dressel, Overall Performance of Series Combinations of Machines as Affected by the Reliability of Individual Units, *Agr. Eng.*, 24:121–122, April, 1943.

liable as possible. Needless to say, such reliability is often not obtained until a new model has been in production for several years. In some cases two or more simpler machines have proved more efficient than a combination machine.

The wise operator will put complex machines, and any other machines to be used in series, in the best possible condition before the season of use and will anticipate repairs in so far as possible by providing a supply of parts. A study of the routine

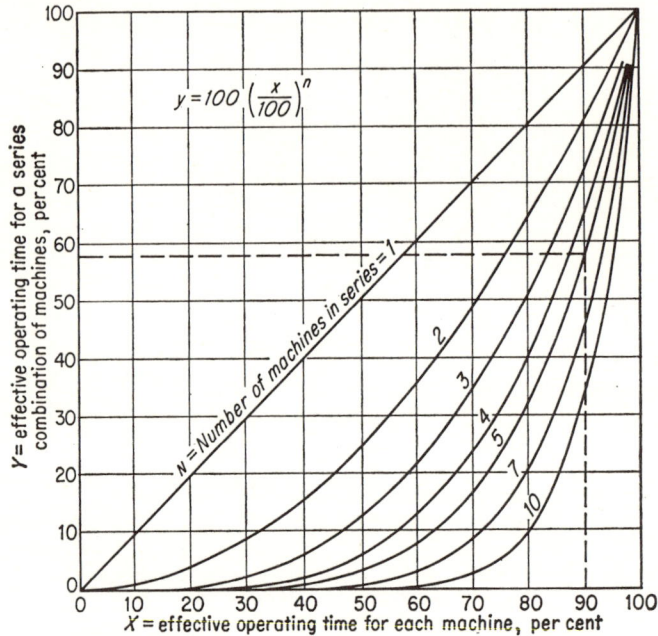

FIG. 1-7. Effect of variations in the reliability of individual machines on the performance of series combinations of different numbers of machines. (*McKibben, Agr. Eng., April,* 1943.)

operations requiring nonoperating time and the use of the most efficient method for each will also contribute greatly to increased capacity.

ESTIMATING CROP-PRODUCTION COSTS

Total crop-growing cost consists of the following items:
1. Cost of use of the land
2. Cost of materials such as seed, fertilizer, pest-control sprays or dusts, etc.
3. Cost of power and machinery
4. Cost of labor

The last two items are usually the most difficult to estimate, and this discussion is concerned with them. Since tractor and machine costs per unit of use are greatly affected by the amount of use per year, it is necessary to estimate their total use per year on the farm in question, along with the labor requirements. The steps necessary to estimate power, machinery, and labor costs for a particular farm are as follows:

1. List for each crop grown the field operations required, the acres to be covered, and the date of the optimum period for performing the operation. The last item is not necessary for a cost estimate but is needed for checking peak loads as described in the following topic, Selection of Optimum Capacity of Equipment.

2. List the power and machine to be used for each operation and estimate the total hours required. Experience on the particular farm is the best guide, but Table

ECONOMICS OF FARM MACHINERY 13

TABLE 1-7. EXAMPLE OF TABULATION OF CROP-PRODUCTION OPERATIONS

Operation	Most favorable period	Size and type machine	Est. av. speed, mph	Est. non-operating time, %	Est. acres/hr	Once-over acres	Estimated hours		
							Machine	Tractor	Man
Corn, 30 acres:									
Plow........	3/15– 4/15	2-16″ plow	3½	20	0.9	30	33.0	33.0	33.0
Disk........	4/15– 4/30	8′ tandem disk	3½	15	2.9	30	10.0	10.0	10.0
Disk and harrow.	5/1 – 5/15	8′ tandem disk, spike-tooth harrow	3½	15	2.9	30	10.0	10.0	10.0
Plant........	5/1 – 5/15	Checkrow planter, 80″	4	40	1.9	30	16.0	16.0	16.0
Cultivate......	6/1 – 6/15	Cultivator, 80″	2½	20	1.6	30	19.0	19.0	19.0
	6/16– 6/30	Cultivator, 80″	3½	20	2.4	30	12.5	12.5	12.5
	7/1 – 7/15	Cultivator, 80″	4½	12	3.2	30	9.5	9.5	9.5
Pick.........	11/1 –12/1	1-row picker	2½	35	0.65	30	46.0	46.0	46.0
Haul and store..	11/1 –12/1	2 wagons, elevator	...	...	1.5	30	20.0	20.0	20.0
Soybeans, 30 acres:									
Plow........	4/16– 4/30	2-16″ plow	3½	20	0.9	30	33.0	33.0	33.0
Disk........	5/1 – 5/15	8′ tandem disk	3½	15	2.9	30	10.0	10.0	10.0
Disk and harrow.	5/16– 5/31	8′ tandem disk, spike-tooth harrow	3½	15	2.9	30	10.0	10.0	10.0
Drill........	5/16– 5/31	12-7 fertilizer grain drill	4	30	2.4	30	12.5	12.5	12.5
Combine......	10/16–10/31	5′ combine	2½	30	1.0	30	30.0	30.0	30.0
Haul and store..	10/16–10/31	2 wagons, elevator	...	...	4.0	30	7.5	7.5	7.5
Wheat, 30 acres:									
Disk and harrow.	10/16–10/31	8′ tandem disk, harrow	3½	15	2.9	30	10.0	10.0	10.0
Drill........	10/16–10/31	12-7 fertilizer grain drill	4	30	2.4	30	12.5	12.5	12.5
Sow grass seed..	3/1 – 3/15	Hand	...	...	2.5	30	....	....	12.0
Combine......	7/1 – 7/15	5′ combine	2½	30	1.0	30	30.0	30.0	30.0
Haul and store..	7/1 – 7/15	2 wagons, elevator	...	...	4.0	30	7.5	7.5	7.5
Rake straw.....	7/16– 7/30	Side-delivery rake	4½	15	3.5	30	8.5	8.5	8.5
Bale straw.....	7/16– 7/30	Baler	...	...	3.0	30	10.0	10.0	10.0
Load straw.....	7/16– 7/30	Bale loader, wagon	...	...	5.0	30	6.0	6.0	12.0
Haul and store..	7/16– 7/30	2 wagons, elevator	...	...	5.0	30	6.0	6.0	12.0
Alfalfa, 30 acres:									
Mow........	6/1 – 6/15	7′ tractor mower	3¾	30	2.2	30	14.0	14.0	14.0
Rake........	6/1 – 6/15	Side-delivery rake	3¾	15	2.7	60	22.0	22.0	22.0
Bale........	6/1 – 6/15	Baler	...	30	2.0	30	15.0	15.0	15.0
Load bales.....	6/1 – 6/15	Bale loader, wagon	...	...	2.6	30	11.5	11.5	23.0
Haul and store..	6/1 – 6/15	2 wagons, elevator	...	...	2.6	30	11.5	11.5	23.0
Mow........	8/1 – 8/15	7′ tractor mower	3¾	30	2.2	30	14.0	14.0	14.0
Rake........	8/1 – 8/15	Side-delivery rake	3¾	15	2.7	60	22.0	22.0	22.0
Bale........	8/1 – 8/15	Baler	...	30	2.4	30	12.5	12.5	12.5
Load bales.....	8/1 – 8/15	Bale loader, wagon	...	...	4.0	30	7.5	7.5	15.0
Haul and store..	8/1 – 8/15	2 wagons, elevator	...	...	4.0	30	7.5	7.5	15.0

Total crop-production tractor use... 507 hr
Est. miscellaneous use... 93 hr
Est. total tractor use.. 600 hr

2-1, showing power requirements of various machines, and the common labor-efficiency percentages for various field operations, as listed above, should be helpful.

3. From step 2, total the hours of use for each machine and, using its new cost, determine its cost per hour and year as shown earlier under Cost of Use of Farm Equipment. In the same way, tractor overhead cost can be determined and fuel and lubricant cost added.

4. From the above figures, the total annual cost of tractor and machine use can be secured and labor hours and cost can also be totaled. By using the hourly costs calculated, the total production cost or cost per acre for any of the various crops grown can be secured.

This method of estimating crop-production costs is illustrated in the following hypothetical example:

Farmer: John Doe, Corn Belt
Crops: Corn, 30 acres; soybeans, 30 acres; wheat, 30 acres; alfalfa, 30 acres
Power: Two-plow tractor on rubber, $2,000
Equipment:

2-16" tractor plow	$ 250	8' tandem disk	$ 250
2-section drag harrow	90	12-7 fertilizer grain drill	550
2-row corn planter	250	2-row tractor cultivator	250
7' tractor mower	275	7' side-delivery rake	300
Baler	1,400	Bale loader	250
5' combine	1,400	1-row corn picker	1,050
2 wagons with 7 × 14 beds	700	Elevator	425

Labor available: One man full time, 8 hr field work per day maximum because of livestock chores. Extra help is hired only for loading, hauling, and storing bales.

Table 1-7 shows the crop-production operations and estimated hours for each.

Table 1-8 shows the cost calculations for each implement, based on the information given under Cost of Use of Farm Equipment above. The costs are based on the hours of usage totaled from Table 1-7, except for tractor, elevator, and wagons, to which some miscellaneous time has been added.

TABLE 1-8. EXAMPLE OF EQUIPMENT COST-OF-USE CALCULATIONS

Machine	New cost	Est. hr/yr	Cost per hr per $100 new cost	Overhead cost per hr	Operating cost per hr	Total cost per hr	Cost per yr
Tractor	$2,000	600	$0.029	$0.58	$0.35	$0.93	$ 558.00
2-16" tractor plow	250	66	0.22	0.55	.....	0.55	36.30
8' tandem disk	250	50	0.25	0.63	.....	0.63	31.50
2-section drag harrow	90	30	0.37	0.33	.....	0.33	9.90
12-7 fertilizer grain drill	550	25	0.44	2.42	.....	2.42	60.50
2-row corn planter	300	16	0.62	1.86	.....	1.86	29.80
2-row tractor cultivator	250	41	0.34	0.85	.....	0.85	34.80
7' tractor mower	275	28	0.50	1.37	.....	1.37	38.40
Side-delivery rake	300	52.5	0.23	0.69	.....	0.69	36.20
PTO baler	1,400	37.5	0.36	5.04	1.75	6.79	254.60
Bale loader	250	25	0.39	0.97	.....	0.97	24.30
5' PTO combine	1,400	60	0.26	3.64	.....	3.64	218.40
1-row corn picker	1,050	46	0.34	3.57	.....	3.57	164.20
Elevator	425	60	0.20	0.85	.....	0.85	51.00
Flat-bed wagon	350	150	0.084	0.29	.....	0.29	43.50
Flat-bed wagon	350	150	0.084	0.29	.....	0.29	43.50
Total	$9,440	.....	.....	.....	.....	.....	$1,634.90

Table 1-9 shows the cost calculation for each operation, using tractor and machine costs from Table 1-8 and assuming labor at $1 per hour. The figures are totaled for each crop.

These tables are included to illustrate a method of estimating the cost of machine use, power, and labor. The assumptions and results are intended to be representative, but it is understood that they vary much from farm to farm, changing with acreages, operating conditions, and managerial ability, as well as choice of equipment. This material offers a method of cost analysis which can be applied to the particular conditions on any farm.

If desired, alternative setups may be analyzed in order to compare margins of safety, total cost, etc. The final choice will also be influenced by convenience, elimination of arduous physical effort, quality and yield of crops secured by various methods, and many other factors.

TABLE 1-9. EXAMPLE OF CROP-PRODUCTION COST CALCULATIONS

Operation	Machine			Tractor @ $.93		Labor @ $1		Total cost	Cost per acre
	Hours	Cost per hr	Total cost	Hours	Cost	Hours	Cost		
Corn, 30 acres:									
Plow	33	$0.55	$ 18.15	33	$ 30.70	33	$ 33.00	$ 81.85	$ 2.73
Disk	10	0.63	6.30	10	9.30	10	10.00	25.60	0.85
Disk and harrow	10	0.96	9.60	10	9.30	10	10.00	28.90	0.96
Plant	16	1.86	29.80	16	14.90	16	16.00	60.70	2.02
Cultivate	41	0.85	34.80	41	38.10	41	41.00	113.90	3.80
Pick	46	3.86	175.55	46	42.80	46	46.00	264.35	8.81
Haul and store	20	1.43	28.60	20	18.60	20	20.00	67.20	2.24
Total	....	.....	$302.80	176	$163.70	176	$176.00	$642.50	$21.41
Soybeans, 30 acres:									
Plow	33	$0.55	$ 18.15	33	$ 30.70	33	$ 33.00	$ 81.85	$ 2.73
Disk	10	0.63	6.30	10	9.30	10	10.00	25.60	0.85
Disk and harrow	10	0.96	9.60	10	9.30	10	10.00	28.90	.96
Drill	12.5	2.42	30.25	12.5	11.60	12.5	12.50	54.35	1.81
Combine	30	3.93	117.90	30	27.90	30	30.00	175.80	5.86
Haul and store	15	1.43	21.45	7.5	6.95	7.5	7.50	35.90	1 20
Total	....	.....	$203.65	103	$ 95.75	103	$103.00	$402.40	$13.41
Wheat, 30 acres:									
Disk and harrow	10	$0.96	$ 9.60	10	$ 9.30	10	$ 10.00	$ 28.90	$ 0.96
Drill	12.5	2.42	30.25	12.5	11.60	12.5	12.50	54.35	1.81
Sow alfalfa seed	....	.....	......	....	....	12.0	12.00	12.00	0.40
Combine	30	3.93	117.90	30	27.90	30	30.00	175.80	5.86
Haul and store	15	1.43	21.45	7.5	6.95	7.5	7.50	35.90	1.20
Rake straw	8.5	0.69	5.85	8.5	7.90	8.5	8.50	22.25	0.74
Bale straw	10.0	6.79	67.90	10	9.30	10	10.00	87.20	2.91
Load straw	6	1.26	7.55	6	5.60	12	12.00	25.15	0.84
Haul and store straw	6	1.43	8.60	6	5.60	12	12.00	26.20	0.87
Total	....	.....	$269.10	90.5	$ 84.15	114.5	$114.50	$467.75	$15.59
Alfalfa, 30 acres:									
Mow	28	$1.37	$ 38.40	28	$ 26.05	28	$ 28.00	$ 92.45	$ 3.08
Rake	44	0.69	30.40	44	40.90	44	44.00	115.30	3.84
Bale	27.5	6.79	186.73	27.5	25.55	27.5	27.50	239.78	7.99
Load	19	1.26	23.95	19	17.65	38	38.00	79.60	2.65
Haul and store	19	1.43	27.15	19	17.65	38	38.00	82.80	2.76
Total 2 cuttings	....	.....	$306.63	137.5	$127.80	175.5	$175.50	$609.93	$20.32

SELECTION OF OPTIMUM CAPACITY OF EQUIPMENT

It is obvious that the economic objective in the selection of power and equipment is to secure a maximum profit, and *optimum capacity* is that capacity which gives the greatest profit over a period of years on the crop acreages to be handled. The problem is akin to that of the manufacturer who selects tooling and machine capacity for a particular product to give him the lowest over-all cost for the expected volume. The farmer's problem, however, is enormously complicated by the fact that the time available for performing field operations depends on the weather. He must get his crop planted during a relatively short period for best results. (In Iowa, records indicate that oat yields are reduced one bushel per acre for every day's delay in planting after Apr. 1.) After a crop is mature, it must be harvested quickly because of the danger of losses due to bad weather.

Factors determining optimum capacity are:
1. Days dry enough to work, as influenced by
    a. Number of rainy days
    b. Amount of rainfall
    c. Temperature
    d. Sunshine
    e. Relative humidity
    f. Topography

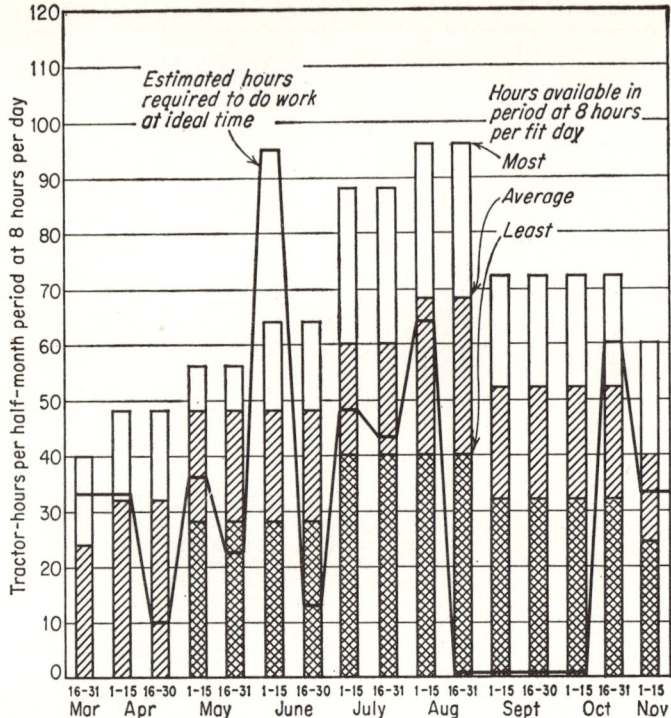

Fig. 1-8. Comparison of tractor hours required for crop-production operations with range of available hours. Hypothetical John Doe Farm in Corn Belt.

    g. Soil type
    h. Soil structure
    i. Drainage—surface and tile
2. Rate of work as influenced by
    a. Amount of power
    b. Size of machine
    c. Draft requirements of soil
    d. Crop conditions at harvest
    e. Smoothness of fields
3. Time available for actual field operation, as affected by
    a. Travel time to field
    b. Repair and maintenance time for equipment
    c. Time required for livestock and other daily chores
    d. Time for emergencies such as broken fences
    e. Possibility of securing extra labor for night operation
4. Profit margins, as affected by
    a. Ratio of labor and equipment costs
    b. Price differential for quality
    c. Price rises due to short crops resulting from adverse seasons

    The lowest costs for power and equipment will be secured with the smallest capacity that will get the job done on time. If, however, the capacity is only enough to do the job on time in an average year, as to days suitable for work, the losses in the adverse years will far outweigh the savings in the favorable years. On the other hand, it is doubtful if the gain from having enough capacity to do the job on time in the worst year compensates for the high cost during the other years. It is a recognized fact, however, that during several World War II years the extra capacity of tractor equipment resulted in bumper crops in spite of late, wet springs which would have inevitably resulted in near failures in horse-farming days.

In spite of the complexity of the problem, an analysis of time required compared with an estimate of time available is always valuable.

Referring to the hypothetical operation of John Doe above, records over several years on his farm as to days fit for field work are summarized in Table 1-10.

The chart in Fig. 1-8 shows the total tractor hours for the various periods secured from Table 1-7 and also the fewest, most, and average hours available plotted from the data in Table 1-10 on the basis of 8-hr days. A study of the chart shows that, in a normal year, John Doe will get his crops planted on time although he will not get the ground prepared as early as he would like. In a year in which each spring month is as bad as the worst he has recorded, he will not get his crops in on time because he requires a total of 134½ hr to put in his corn and soybeans whereas only

TABLE 1-10. EXAMPLE OF RECORDS OF DAYS AVAILABLE FOR CROP OPERATIONS

Month	Days available for field work		
	Fewest	Most	Average
March..................	0	5	3
April..................	0	12	8
May...................	7	14	11
June...................	7	16	12
July...................	10	22	15
August.................	10	24	17
September.............	8	18	13
October...............	8	18	13
November.............	6	15	10

7 days are available up to the end of May. If, however, by working long hours and hiring extra help he can keep the tractor going 20 hr per day, he will have available 140 hr and can get the job done.

During the first half of June, with haying and corn cultivating, he faces a very difficult situation. In a normal year, 16-hr days would be necessary to do the work in the time available. As a result much of the haying will, of necessity, be done in the next, less favorable period. That this situation is not uncommon is indicated by the popularity of the custom baler as a means of hiring haying help and also the frequent losses incident to the harvesting of first-cutting alfalfa.

Soybean harvest and wheat planting show another peak, but 16-hr days will do the job in the worst year, according to the records.

This analysis serves to indicate the extent of the capacity problem, even when many of the more unpredictable factors are omitted. In this case we have assumed that there are reliable records for the particular farm as to days fit for field work. Most farmers do not have such records, and weather records alone are not sufficient because of the variation in drying time from farm to farm.

Optimum capacity will depend on the economic situation as well as on the weather. The extra capacity necessary to produce a crop in a bad year may not be worth its cost in an era of crop surpluses and large carryover, but with no surplus and food production a matter of life and death, extra capacity must be available.

Farmers are extremely conscious of the problem of optimum capacity, and a systematic analysis of their particular farms, as is outlined here, should be of value. It is certain that their combined judgment in the matter, which in a free economy is reflected by the actual amount of equipment in use, results in a better balance and distribution of equipment than would be secured in a regimented economy.

Chapter 2

DESIGN OF FIELD MACHINERY

BASIC REQUIREMENTS

Factors which affect the value of a machine may be concerned either with efficiency of operation or with the amount of work the machine can accomplish before it becomes worn out. Some of the factors in each of these classes are outlined as follows:

A. Efficiency of operation
  1. Quality of work done
     For example, pulverization and trash coverage in a plow or getting all the grain and no foreign material in a combine
  2. Efficient use of power in doing the work
     For example, light draft in a plow due to shape of bottom and replacement of sliding friction by rolling friction, or low power requirements in a forage blower due to efficient fan design
  3. Efficient use of human labor in doing the work
     a. Minimum number of man-hours per unit of work done
     b. Minimum number of men required for operation
     c. Minimum of lost time, obtained by having convenient controls and adjustments, freedom from clogging, minimum lubrication, quick attachment, etc.
     d. Minimum operator fatigue, secured by having low muscular requirements, comfortable ride, etc.
     e. Minimum opportunity for accidents
B. Amount of work accomplished before wearing out
  1. Freedom from breakage
     a. Parts of adequate strength
     b. Safety releases
  2. Resistance to mechanical wear due to friction
     a. Wearing parts of wear-resistant materials
     b. Bearings of proper load capacity
     c. Easy and adequate lubrication
     d. Dirt excluded from bearings, gears, etc.
  3. Resistance to chemical deterioration
     a. Corrosion-resistant materials
     b. Protective coatings or treatments for vulnerable materials
  4. Low repair cost
     a. Adjustments to compensate for wear
     b. Designed to permit easy replacement of wearing parts
     c. Exchange provisions for worn assemblies
     d. Simplicity of construction

The **design of field machinery is differentiated** from many other types of equipment by the importance of the following requirements:

1. *Mobility*. In crop production the machine comes to the work rather than vice versa. It must have its own wheels or be carried by a tractor. The weight should

ordinarily be kept at a minimum to reduce transport power requirements and soil compaction.

2. *Low cost.* Because of the nature of crop production, many machines are used only a few days per year. Obsolescence may limit the number of years of use, so it is not economically sound to build a machine to the standards of durability of a machine tool, for instance. Ordinarily a machine having a life expectancy of 2,000 hr is adequate and the designer aims for the lowest-cost construction which will give this life. In normal times the sale of farm machinery is very competitive, with the farmer able to choose from several of each type, each of which functions satisfactorily.

Many manufacturers have volumes of less than 5,000 units per year for most of their models, and this makes the economic design much different from automobiles, for instance, which are often produced by the hundreds of thousands per model. The implement designer must work for the lowest cost at the probable volume for the particular machine.

3. *Ability to perform in a variety of conditions.* It is commonly acknowledged that the operation of field machinery is affected by a greater number of variables than almost any other type of machinery. For instance, corn picking is affected by endless variations of soil, crop, and climatic factors: the soil may be firm or soft, wet or dry, weedy or cleanly cultivated; the corn may be short or tall, thick or thin, standing or down, ears large or small, and easy or hard to snap or husk; the weather may range from hot and dry in Texas in August to cold and snowy in Michigan in December.

In addition to geographical differences, variations in crop growth and conditions from one season to another are often extreme, making the design and testing of a new machine a formidable problem, usually requiring several years.

Safety Considerations. These are being given more attention as it is realized that proper engineering can reduce accidents. Major precautions are as follows:

1. *Sharp corners and edges* near the operator's path of travel should be eliminated or shielded.

2. *Steps for climbing* to operating platform or to lubricate, adjust, or inspect a machine should be nonskid and supplemented by adequate handholds.

3. *Implement controls operable by children* (to drop a combine header, mounted implement, etc.) when the machine is idle should be labeled with advice to leave in the lowered position, or better, designed to be inoperative when the engine is not running.

4. *Springs* operating with dangerous amounts of potential energy should have adequate provision for safe release of energy before removal.

5. *Controls* should be easily reached and have low muscular requirements to avoid strained muscles or dangerous slips.

6. *Moving parts capable of causing injury,* such as turning shafts, chains, belts, gears, fast-moving projections, control linkages, etc., should be shielded or located out of reach as much as possible without interfering with the function of the machine. Safety warnings should be placed where needed.

7. *Unclogging with machine rotating* should be discouraged by warning signs, and the necessity eliminated as much as possible by provision of reversing controls, roll-adjustment controls, etc., operable from the seat.

8. *Manual operations involving large forces,* such as mounting heavy implements on tractors, hitching to the drawbar, erecting blower pipes, etc., should be carefully studied, and as much as possible should be done to reduce effort and risk.

DRAFT AND POWER REQUIREMENTS

Draft and power requirements of farm machines are extremely variable from one section of the country to another and on the same farm from year to year. The common ranges, however, are indicated in Table 2-1.

The average power requirement is often, however, a poor guide to maximum operational stresses for design purposes. Starting inertia, velocity fluctuations in the PTO drive line, and clogging may all cause torque peaks of several times the average

TABLE 2-1. DRAFT AND POWER REQUIREMENTS OF CROP MACHINES

Machine	Normal range	References
Tillage:*		
Plow	5–12 psi of furrow section	3, 4, 6
Lister	400–750 lb per row	1–3, 7
One-way disk	150–350 lb/ft width	1, 7
Single-disk harrow	40–130 lb/ft width	1–6
Tandem-disk harrow	80–160 lb/ft width	1–4, 6
Tandem-disk harrow, 22-in. diam, 9-in. spacing	170–225 lb/ft width, or 90% of weight	4, 11
Spike-tooth harrow	30–60 lb/ft width	1–3, 5, 6
Spring-tooth harrow	75–150 lb/ft width	1, 6, 7, 11
Duckfoot field cultivator	90–160 lb/ft width	1–3, 6, 7
Roller	30–60 lb/ft width	5, 6
Subsoiler	80–160 lb/in. depth	16, 21
Planting:		
Grain drill	30–80 lb/ft width	1
Corn planter	80–120 lb per row	2, 3
Cultivating:		
Rotary hoe	30–60 lb/ft width	2, 3, 5
Corn cultivator	22–95 lb per shovel	11
Spring-tooth weeder	25–35 lb/ft width	2, 3
Rod weeder	80–110 lb/ft width	1, 7
Harvesting:		
Mower	0.4 hp/ft width	22
Grain binder	65–150 lb/ft width	12
Thresher	0.8–1.2 hp/in. cylinder width	8
Combine, 5 and 6 ft	2–4½ (PTO) hp/ft of cutter bar	9, 10
Combine, 8–12 ft	Engine with 2–3 net hp/ft of cutter bar	Common practice
Self-propelled combine, 10–14 ft	Engine with 4–6 hp/ft of cutter bar	Common practice, 18
Corn picker, 2-row	2–9 (PTO) hp per row	2, 24
Stationary silage cutter	0.76–1.60 hp-hr/ton	13
Husker-shredder	0.25–0.35 hp-hr/bu	12
Pickup baler	Engine with 15–25 net hp	Common practice, 17
Forage harvester	1–3 (PTO) hp-hr/ton of grass silage at ½-in. cut	19, 20
Forage blower	1.1–2 hp-hr/ton of grass silage at ½-in. cut	23

* In certain southern and far-western soils, draft figures ranging up to a maximum of approximately double the figures given have been recorded [14].

requirement. A group of measurements made in the course of a study of this problem are given in Table 2-2 [15].

The perfection of electric-strain-gage techniques for measuring torque and draft makes possible field measurements with much greater ease and accuracy than heretofore, and it is anticipated that much additional data will soon become available.

REFERENCES

1. Murdock, H. E.: Mechanical Tests on Tractor Farming Equipment, *Montana State Coll. Agr. Expt. Sta. Bull.* 243, 1931.

2. Shedd, C. K., E. V. Collins, and J. B. Davidson: Labor Power and Machinery in Corn Production, *Iowa State Univ. Agr. Expt. Sta. Bull.* 365, 1937.

3. Shedd, C. K., J. B. Davidson and E. V. Collins, Machinery for Growing Corn, *USDA Circ.* 592, 1940.

4. Clyde, A. W., and R. J. McCall: Tillage Tools, *Penn. State Univ. Agr. Expt. Sta. Bull.* 465, 1944.

5. Shawl, R. I.: unpublished data, University of Illinois, Urbana, Ill., 1930.

6. Schwantes, A. J.: unpublished data, University of Minnesota, Minneapolis, 1946.

7. Promersberger, W. J.: unpublished data, Kansas State College, Manhattan, Kans., 1946.

TABLE 2-2. PTO TORSIONAL LOADS

Test no.	Approx. max. tractor, bhp	Implement make and model	Coupling in PTO drive	Max. starting torque		Max. operational torque peaks			Av. torque under normal operating conditions, lb-in.	Work being performed
				With normal clutch engagement, lb-in.	With rapid clutch engagement, lb-in.	Average conditions, lb-in.	Near plugged conditions, lb-in.			
1	35	Ensilage harvester A	Standard	4,900–6,400	10,800–15,370	4,680–6,390	5,450–7,140		2,720	Chopping heavy drilled corn
2	35	Ensilage harvester A	Special slip	....	8,660*	5,112–5,723	6,025–6,865		3,200	Chopping heavy drilled corn
3	40	Ensilage harvester A	Standard	....	11,600	4,700–4,925	6,200–8,025		3,261	Chopping heavy drilled corn
4	35	Ensilage harvester B	Standard	2,600–4,000*	....	3,520–3,820	3,960–7,630*		2,390	Chopping heavy drilled corn
5	35	Forage harvester C	Standard	....	14,600	3,730–7,200	6,370–7,200		2,870	Chopping green alfalfa
6	35	Forage harvester C	Special slip	....	9,800–7,530*	5,230–6,700	6,100–8,700		3,270	Chopping green alfalfa
7	35	Forage harvester C	Standard	....	12,500–10,900	6,000–7,460	9,500		3,600	Chopping green alfalfa
8	35	Forage harvester C	Standard	....	21,400	....	....		....	Attempting to start a plugged machine
9	35	Corn picker D	Standard	1,570–1,740	3,990	822–1,031	....		727	Picking corn
10	35	Baler E	Standard	....	18,300–20,600	5,860–7,470	12,100		1,140	Baling alfalfa
11	35	Baler E	Standard	13,100	....	6,550–8,140	11,600–15,000		1,545	Recheck of test 10
12	35	Baler E	Special slip	10,700–12,100*	10,700–12,100*	7,250–8,920	11,500–13,300*		2,250	Baling alfalfa
13	35	Baler E	Special slip	10,100*	....	8,600–11,100	10,350–12,600		1,580	Baling straw
14	40	Baler E	Standard	12,250	....	7,749–10,945	10,960–12,095		1,938	Baling alfalfa
15	35	Baler E	Standard and universal joints aligned	....	....	4,601–5,867	....		1,383	Baling alfalfa
16	35	Baler F	Standard	16,500	....	8,600	22,700		....	Baling alfalfa
17	35	Baler F	Special slip	5,000*	5,000*	5,000*	5,000*		....	Baling alfalfa
18	35	Combine G	Standard	....	10,100–16,600*	3,760	9,380		1,890	Combining windrows
19	35	Combine G	Special slip	....	....	7,150	7,760–9,130		1,700	Combining windrows
20	35	Combine G	Special slip	....	7,350–8,650*	4,160–4,200	7,470		1,600	Straight combining
21	25	Hammer mill H	Standard	9,030	17,500–20,150	4,145	7,270*		2,700	Grinding ear corn
22	35	Hammer mill H	Standard	6,130	....	3,740	14,900		2,140	Grinding ear corn
23	35	Hammer mill H	Special slip	8,230	....	....	6,920		4,210	Grinding ear corn
24	45	Hammer mill J	Standard	18,150	25,800	7,800	13,000		5,450	Grinding ear corn

* Safety clutch in PTO line slipped, limiting torsional load to this value.
SOURCE: Merlin Hansen, Loads Imposed on Power-take-off Shafts by Farm Implements, *Agr. Eng.*, 33:66–70, February, 1952.

8. Silver, E. A., and G. W. McCuen: A Study of Power Requirements and Efficiency of Threshing Machines, *Agr. Eng.*, 16:137–154, April, 1935.
9. McCuen, G. W.: unpublished data, Ohio State University, Columbus, Ohio, 1941.
10. Downing, C. G. W.: Horsepower Requirements of Power Take-off-driven Combines, *Agr. Eng.*, 23:47–50, February, 1942.
11. Clyde, A. W.: unpublished data, Pennsylvania State University, University Park, Pa., 1946.
12. Davidson, J. B.: "Agricultural Machinery," John Wiley & Sons, Inc., New York, 1931.
13. Duffee, F. W.: Efficiently Filling the Silo, *Agr. Eng.*, 6:4–12, January, 1925.
14. Randolph, J. W., I. F. Reed, and E. D. Gordon: Cotton Tillage Studies on Red Bay Sandy Loam, *USDA Circ.* 540, 1940.
15. Hansen, Merlin: Loads Imposed on Power-take-off Shafts by Farm Implements, *Agr. Eng.*, 33:67–70, February, 1952.
16. Larson, G. H., and G. E. Fairbanks: Draft Tests on a Killefer Chisel, *Kansas State Coll. Agr. Eng. Infor. Bull.* 7, 1952.
17. Burrough, D. E., and J. A. Graham: Power Characteristics of a Plunger-type Forage Baler, *Agr. Eng.*, 35:221–232, April, 1954.
18. Burrough, D. E.: Power Requirements of Combine Drives, *Agr. Eng.*, 35:15–18, January, 1954.
19. Blevins, F. Z.: "Some of the Component Power Requirements of Field Type Forage Harvesters," unpublished thesis, Purdue University, Lafayette, Ind., 1954.
20. Huntington, D. H.: unpublished thesis, Cornell University, Ithaca, N.Y., 1953.
21. Schwantes, A. J., M. J. Thompson, O. W. Swenson, and T. M. McCall: You Don't Gain with Deep Tillage, *Minn. Farm and Home Sci.*, May, 1952.
22. Elfes, L. E.: Design and Development of a High-speed Mower, *Agr. Eng.*, 35:147–153, March, 1954.
23. Raney, J. P., and J. B. Liljedahl: Impeller Blade Shape Affects Forage Blower Performance, *Agr. Eng.*, 38:722–725, October, 1957.
24. Richey, C. B., et al.: Corn Picker Features New Principle, *Agr. Eng.*, 37:93–97, February, 1956.

FRAMES

Frame design is particularly important in field machines because their frames must be as light as possible to reduce cost, soil compaction, and propelling power but yet strong enough to resist the shocks due to rough ground or obstacles. With lightweight construction there is often considerable deflection, and as a rule self-aligning bearings are necessary. Deflection is not a bar to satisfactory service if no permanent set results and points of stress concentration causing fatigue failure are avoided.

Before the advent of electric arc welding, farm implement frames consisted largely of rolled sections (angles, I beams, bars, etc.) bolted or riveted together or to castings which also carried functional parts. Pipe members were sometimes used, but it was difficult to take advantage of their full strength, particularly in torsion, because of the limitations of the bolted or clamp connections necessary.

Tubes or closed box sections are strongest for their weight, and arc welding of connecting members makes it possible to take full advantage of their strength in both torsion and bending. They serve particularly well as a central frame member to which various functional arms, axle brackets, hitch pivots, etc., can be attached. Standard pipe rather than tubing is commonly used in farm machinery where tolerances permit, and its structural properties are given in Table 2-3. Cold-drawn steel tubing is usually stronger for its size than pipe, even with the same-analysis steel, because of the effects of cold-working.

Arc welding has greatly accelerated the use of tapered cantilever arms which can be welded to a tube, thus developing the full strength of both. Such tapered arms are most economically fabricated from tapered flat blanks bent into tapered angles, channels, or Z sections on a press using general-purpose V dies. The leg of an angle should be put on the tension side to give greater length of weld grip where it is needed. For high loads, a tapered channel is preferable to an angle, not only because it is stronger, but also because it reduces the tendency to push into the pipe on the compression side. Buckling failure of thin sections (relative to cross-section size) may occur at much lower loads than those predicted by simple formulas.

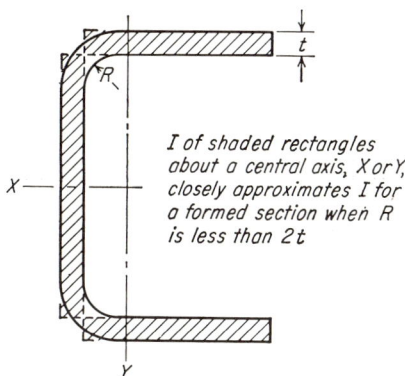

Fig. 2-1. Moment of inertia of formed sections.

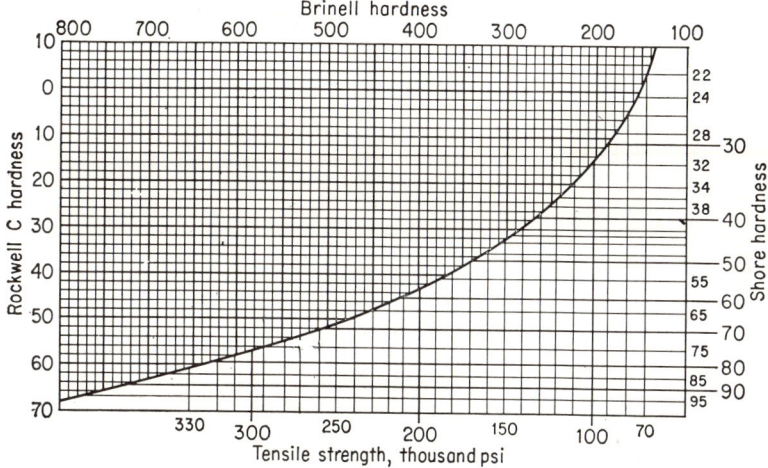

Fig. 2-2. Approximate relationships between Brinell, Rockwell, and Shore hardnesses and the tensile strength of steel. Conversions from one scale to another are made at the intercepts with the curve crossing the chart. For example, follow the vertical line representing 200 Brinell hardness to its intersection with the conversion curve. From this point follow horizontally to the right for equivalent Shore hardness value, 31, horizontally to the left for Rockwell C value, 14, and vertically downward for the tensile strength, 97,000 psi. (*International Nickel Co.*)

Since open sections are not strong torsionally, the load should be applied as near the flexural center as possible. Formulas for the location of the flexural center of several formed sections are shown in Table 2-4.

The calculation of moment of inertia and section modulus of formed sections can be simplified by assuming that a formed channel, for instance, is composed of three rectangles touching at the corners, as shown by the shaded areas in Fig. 2-1. This method gives a very close approximation when the inside bend radius is less than twice the material thickness.

TABLE 2-3. PROPERTIES OF STANDARD STEEL PIPE

Nominal diam, in.	Outside diam, in.	Inside diam, in.	Thickness, in.	Weight per ft, lb.	Properties			
					I, in.^{4}	A, in.^{2}	r, in.	Z, in.^{3}
1/8	0.405	0.269	0.068	0.24	0.001	0.072	0.12	0.005
1/4	0.540	0.364	0.088	0.42	0.003	0.125	0.16	0.012
3/8	0.675	0.493	0.091	0.57	0.007	0.167	0.21	0.022
1/2	0.840	0.622	0.109	0.85	0.017	0.250	0.26	0.041
3/4	1.050	0.824	0.113	1.13	0.037	0.333	0.33	0.070
1	1.315	1.049	0.133	1.68	0.087	0.494	0.42	0.133
1 1/4	1.660	1.380	0.140	2.27	0.195	0.669	0.54	0.234
1 1/2	1.900	1.610	0.145	2.72	0.310	0.799	0.62	0.326
2	2.375	2.067	0.154	3.65	0.666	1.075	0.79	0.560
2 1/2	2.875	2.469	0.203	5.79	1.530	1.704	0.95	1.062
3	3.500	3.068	0.216	7.58	3.017	2.228	1.16	1.72
3 1/2	4.000	3.548	0.226	9.11	4.788	2.680	1.34	2.39
4	4.500	4.026	0.237	10.79	7.233	3.174	1.51	3.21
5	5.563	5.047	0.258	14.62	15.16	4.300	1.88	5.44
6	6.625	6.065	0.280	18.97	28.14	5.581	2.25	8.48

TABLE 2-4. POSITION OF FLEXURAL CENTER Q FOR DIFFERENT VERTICALLY LOADED SECTIONS

Form of section	Position of Q
1. Sector of thin circular tube	$e = \dfrac{2R}{(\pi - \Theta) + \sin \Theta \cos \Theta}[(\pi - \Theta)\cos \Theta + \sin \Theta]$ For complete tube split along element ($\theta = 0$), $e = 2R$.
2. Angle	Leg 1 = rectangle $w_1 h_1$; leg 2 = rectangle $w_2 h_2$ I_1 = moment of inertia of leg 1 about Y_1 (central axis) I_2 = moment of inertia of leg 2 about Y_2 (central axis) $e_x = \dfrac{1}{2} h_2 \left(\dfrac{I_1}{I_1 + I_2}\right)$ $e_y = \dfrac{1}{2} h_1 \left(\dfrac{I_1}{I_1 + I_2}\right)$ If w_1 and w_2 are small, $e_x = e_y = 0$ (practically) and Q is at 0.
3. Channel	$e = h \left(\dfrac{H_{xy}}{I_x}\right)$ where H_{xy} = product of inertia of half section (above X) with respect to axes X and Y, and I_x = moment of inertia of whole section with respect to axis X. If t is uniform, $e = \dfrac{b^2 h^2 t}{4 I_x}$
4. Tee	$e = \dfrac{1}{2}(t_1 + t_2)\left(\dfrac{1}{1 + \dfrac{d_1^3 t_1}{d_2^3 t_2}}\right)$ For a T beam of ordinary proportions, Q may be assumed to be at 0.
5. I with unequal flanges and thin web	$e = b \left(\dfrac{I_2}{I_1 + I_2}\right)$ where I_1 and I_2, respectively, denote moments of inertia about X axis of flange 1 and flange 2.

SOURCE: R. J. Roark, "Formulas for Stress and Strain," McGraw-Hill Book Company, Inc., New York, 1954.

Tapered members are usually laid out to cut from standard stock widths in pairs. Amounts of stock to be allowed for 90° bends of various inside radii and stock thicknesses are shown in Table 2-5.

TABLE 2-5. AMOUNT OF STOCK REQUIRED TO MAKE A 90° BEND

Example: For 20-gage stock and $5/32$ inside radius, stock required for 90° bend = 0.270 in. (from table).

Stock thickness	Radius taken inside metal, in.												
	$1/16$	$3/32$	$1/8$	$5/32$	$3/16$	$7/32$	$1/4$	$9/32$	$5/16$	$3/8$	$1/2$	$5/8$	$3/4$
22 ga.*	0.116	0.171	0.219	0.268	0.318	0.367	0.416	0.464	0.515	0.613	0.808	1.005	1.202
20 ga.	0.118	0.173	0.224	0.270	0.323	0.372	0.421	0.469	0.519	0.616	0.812	1.010	1.208
18 ga.	0.124	0.175	0.232	0.282	0.333	0.381	0.430	0.478	0.529	0.626	0.823	1.02	1.217
16 ga.	0.130	0.179	0.234	0.289	0.341	0.391	0.439	0.488	0.538	0.640	0.831	1.028	1.225
14 ga.	0.138	0.188	0.235	0.293	0.344	0.404	0.447	0.498	0.550	0.649	0.842	1.04	1.236
13 ga.	0.146	0.195	0.244	0.295	0.348	0.409	0.452	0.512	0.561	0.659	0.855	1.052	1.248
12 ga.	0.154	0.206	0.251	0.305	0.353	0.416	0.461	0.523	0.572	0.671	0.868	1.063	1.26
10 ga.	0.169	0.219	0.267	0.315	0.366	0.420	0.470	0.529	0.596	0.694	0.890	1.088	1.284
9 ga.	0.177	0.226	0.274	0.324	0.374	0.425	0.481	0.535	0.608	0.706	0.902	1.098	1.295
$1/8$ in.	0.162	0.211	0.262	0.310	0.360	0.418	0.463	0.520	0.589	0.687	0.883	1.08	1.276
$3/16$ in.	0.196	0.245	0.294	0.344	0.392	0.442	0.491	0.540	0.614	0.712	0.932	1.128	1.325
$1/4$ in.			0.327		0.425		0.523	0.571	0.622	0.720	0.950	1.138	1.345
$5/16$ in.			0.360		0.458		0.556		0.658	0.770	0.960	1.158	1.160
$3/8$ in.			0.393		0.491		0.581		0.687	0.786	0.981	1.178	1.375
$7/16$ in.			0.425		0.522		0.621		0.719	0.819	1.014	1.211	1.408
$1/2$ in.			0.458		0.557		0.653		0.752	0.851	1.046	1.243	1.439
$5/8$ in.			0.522		0.621		0.720		0.818	0.917	1.11	1.258	1.50
$3/4$ in.			0.589		0.687		0.785		0.882	0.980	1.178	1.375	1.767

* Gages are U.S. Standard.
Metal bend allowance formula: $R = r + 1/3 T$
For radii over $2T$: $R = r + 1/2 T$
$L = 0.017453AR$; for 90° bends: $L = 1.5708R$

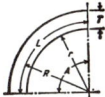

It is difficult to predict functional forces on frames in advance of field experience. Stresses due to the weight carried may be calculated with reasonable accuracy. Experience indicates that, with low- and medium-carbon annealed steels, the static stress should not exceed 40 per cent of the stress at the elastic limit in order to provide sufficient margin for dynamic loads encountered when rubber-tired machines travel over rough ground at tractor speeds. (See Chap. 30 for a discussion of forces acting on two-wheel trailed implement frames.)

Strain-gage equipment is extremely valuable for determining dynamic loads under operating conditions because it finds not only points of weakness but also areas of excess strength where weight and cost may be reduced (Chap. 5).

MECHANICAL DESIGN FORMULAS AND DATA

$$Hp = \frac{(\text{force, lb})(\text{velocity, fpm})}{33{,}000}$$

Torque, in.-lb = (force, lb)(radius of arm, in.)

$$Hp = \frac{(\text{torque, in.-lb})(\text{rpm})}{63{,}000}$$

$$\text{Hydraulic hp} = \frac{(\text{gal per min})(\text{pressure, psi})}{1714}$$

$$\text{Kinetic energy, ft-lb} = \frac{(\text{wt of body, lb})(\text{velocity, fps})^2}{64.32}$$

$$\text{Acceleration, fps/sec} = \frac{\text{velocity change, fps}}{\text{time, sec}}$$

$$\text{Linear acceleration force, lb} = \frac{(\text{acceleration, fps/sec})(\text{weight, lb})}{32.16}$$

$$\text{Radial acceleration force, lb} = \frac{(\text{weight, lb})(\text{radius, ft})(\text{rpm})^2}{2{,}934}$$

Moment of inertia = (area, sq in.^{2})($k^2 + x^2$)

where k = radius of gyration about neutral axis (through center of gravity), in.
  x = distance from neutral axis to a parallel axis, in.

$$\text{Bending stress, } S_b, \text{ psi} = \frac{Mc}{I} = \frac{M}{Z}$$

where M = bending moment, in.-lb
  c = distance from neutral axis to outermost fiber, in.
  I = moment of inertia about neutral axis, in.^{4}
  $Z = I/c$ = section modulus, in.^{3}

Combined stress, S_c, psi = $\sqrt{S_b^2 + S_s^2}$

Shearing Stress Due to Torsion

Round:

$$S_s \text{ max., psi} = \frac{\text{torque, in.-lb}}{0.2 \,(\text{diam, in.})^3}$$

Round tube:

$$S_s \text{ max., psi} = \frac{\text{torque, in.-lb}}{0.2 \left[\dfrac{(\text{diam}_o, \text{in.})^4 - (\text{diam}_i, \text{in.})^4}{\text{diam}_o, \text{in.}}\right]}$$

Rectangle:

$$S_s \text{ max., in middle of long side, psi} = \frac{(\text{torque, in.-lb})\left(3 + 1.8\dfrac{\text{height, in.}}{\text{breadth, in.}}\right)}{(\text{breadth, in.})(\text{height, in.})^2}$$

Thin open sections, straight, arc, channel, etc.:

$$S_s \text{ max., in middle of long side, psi} = \frac{3\,(\text{torque, in.-lb})}{(\text{thickness, in.})(\text{mean developed length of section, in.})}$$

TABLE 2-6. PROPERTIES OF VARIOUS MATERIALS OF CONSTRUCTION

Values within the range shown depend on composition, method of fabrication, and heat treatment.

Material	Specific weight, lb/cu ft	Tensile strength, 1,000 psi	Yield strength, 1,000 psi	Elongation in 2 in., %	Modulus of elasticity, 1 million psi	Brinell hardness
SAE 1010 steel................	489.6	47–53	26–44	28–20	30	95–105
SAE 1020 steel................	489.6	55–61	30–51	25–15	30	111–121
SAE 1045 steel................	489.6	82–91	45–71	16–12	30	163–179
SAE 1045 hardened.............	489.6	100–130	60–100	20–10	30	210–270
Low-alloy, high-strength steel....	489.6	65–90	40–90	30–15	30	150
SAE 1080 steel................	489.6	112–140	61–80	15–12	30	229
SAE 1080 hardened.............	489.6	125–200	100–150	20–12	30	270–400
Gray cast iron.................	442	18–60	8–40	0	13.5–21	100–300
Ductile cast iron...............	445	60–100	45–80	22–2	24	170–280
Malleable cast iron.............	457	50–53	32–35	18–10	25	110–145
Pearlitic malleable.............	457	60–90	43–70	10–3	25	180–240
Cast steel....................	487	65–100	35–70	24–10	28.5	131–207
Cast aluminum................	141	19–45	9–32	14–0.5	10	50–115
Wrought aluminum.............	141–146	15–80	5–70	35–8	10	23–150
Die-cast zinc alloy.............	412–418	41–52	......	10–7	......	82–100
Sintered metal powder..........	393–500	19–75	......	9–1		

TABLE 2-7. WEIGHTS OF FLAT ROLLED STEEL, LB/SQ FT (489.6 LB/CU FT)

Thickness, in.	Weight, lb/sq ft	Thickness, in.	Weight, lb/sq ft
0.0149 (28 ga.)*.....	0.625	0.1495 (9 ga.).......	6.25
0.0179 (26 ga.)......	0.75	0.1644 (8 ga.).......	6.875
0.0239 (24 ga.)......	1.00	0.1793 (7 ga.).......	7.50
0.0299 (22 ga.)......	1.25	3/16............	7.65
0.0359 (20 ga.)......	1.50	1/4.............	10.20
0.0478 (18 ga.)......	2.00	5/16............	12.75
0.0598 (16 ga.)......	2.50	3/8.............	15.30
0.0747 (14 ga.)......	3.125	7/16............	17.85
0.1046 (12 ga.)......	4.375	1/2.............	20.40
0.1196 (11 ga.)......	5.00	5/8.............	22.95
0.125................	5.10	3/4.............	30.60
0.1345 (10 ga.)......	5.625	7/8.............	35.70
		1...............	40.8

* Gages are U.S. Standard.

TABLE 2-8. UNIFIED SCREW THREADS (AMERICAN STANDARD B 1.1, 1949)

Size, in.	Coarse thread series, UNC and NC				Fine thread series, UNF and NF			
	Threads/in.	Minor diam (external threads), in.	Stress area, sq in.	Tap-drill size	Threads/in.	Minor diam (external threads), in.	Stress area, sq in.	Tap-drill size
1/4	20	0.1887	0.0317	No. 7	28	0.2062	0.0362	No. 3
5/16	18	0.2443	0.0522	F	24	0.2614	0.0579	I
3/8	16	0.2983	0.0773	5/16	24	0.3239	0.0876	Q
7/16	14	0.3499	0.1060	U	20	0.3762	0.1185	25/64
1/2	13	0.4056	0.1416	27/64	20	0.4387	0.1597	29/64
9/16	12	0.4603	0.1816	31/64	18	0.4943	0.2026	33/64
5/8	11	0.5135	0.2256	17/32	18	0.5568	0.2555	37/64
3/4	10	0.6273	0.3340	21/32	16	0.6733	0.3724	11/16
7/8	9	0.7387	0.4612	49/64	14	0.7874	0.5088	13/16
1	8	0.8466	0.6051	7/8	12	0.8978	0.6624	15/16
1 1/8	7	0.9497	0.7627	63/64	12	1.0228	0.8549	1 3/64
1 1/4	7	1.0747	0.9684	17/64	12	1.1478	1.0721	1 11/64
1 3/8	6	1.1705	1.1538	17/32	12	1.2728	1.3137	1 19/64
1 1/2	6	1.2955	1.4041	1 27/64	12	1.3978	1.5799	1 27/64

TABLE 2-9. DIMENSIONS OF HEXAGON BOLT HEADS AND NUTS (AMERICAN STANDARD B 18.2, 1952)

Bolt diam, in.	Heads							Nuts						
	Unfinished regular		Unfinished heavy		Cap screw		Finished regular				Unfinished heavy		
	Across flats, in.	Height, in.	Across flats, in.	Height, in.	Across flats, in.	Height, in.	Across flats, in.	Height, in.			Across flats, in.	Height, in.	
								Regular	Jam	Thick		Heavy	Jam
1/4	7/16	11/64			7/16	5/32	7/16	7/32	5/32	9/32	1/2	1/4	3/16
5/16	1/2	7/32			1/2	13/64	1/2	17/64	3/16	23/64	19/32	5/16	7/32
3/8	9/16	1/4			9/16	15/64	9/16	21/64	7/32	13/32	11/16	3/8	1/4
7/16	5/8	19/64			5/8	9/32	11/16	3/8	1/4	29/64	25/32	7/16	9/32
1/2	3/4	21/64	7/8	7/16	3/4	5/16	3/4	7/16	5/16	9/16	7/8	1/2	5/16
5/8	15/16	27/64	1 1/16	17/32	15/16	25/64	15/16	35/64	3/8	23/32	1 1/16	5/8	3/8
3/4	1 1/8	1/2	1 1/4	5/8	1 1/8	15/32	1 1/8	41/64	27/64	13/16	1 1/4	3/4	7/16
7/8	1 5/16	37/64	1 7/16	29/32	1 5/16	35/64	1 5/16	3/4	31/64	29/32	1 7/16	7/8	1/2
1	1 1/2	43/64	1 5/8	1 3/16	1 1/2	39/64	1 1/2	55/64	39/64	1	1 5/8	1	9/16

28

TABLE 2-10. PROPERTIES OF VARIOUS CROSS SECTIONS

(I = moment of inertia; I/c = section modulus; $r = \sqrt{I/A}$ = radius of gyration)

Section	Moment of inertia	Section modulus	Radius of gyration
$I = \dfrac{bh^3}{12}$ $\dfrac{I}{c} = \dfrac{bh^2}{6}$ $r = \dfrac{h}{\sqrt{12}} = 0.289h$	$\dfrac{bh^3}{3}$ $\dfrac{bh^2}{3}$ $\dfrac{h}{\sqrt{3}} = 0.577h$	$\dfrac{b^3h^3}{6(b^2+h^2)}$ $\dfrac{b^2h^2}{6\sqrt{b^2+h^2}}$ $\dfrac{bh}{\sqrt{6(b^2+h^2)}}$	$\dfrac{bh}{12}(h^2\cos^2 a + b^2\sin^2 a)$ $\dfrac{bh}{6}\left(\dfrac{h^2\cos^2 a + b^2\sin^2 a}{h\cos a + b\sin a}\right)$ $\sqrt{\dfrac{h^2\cos^2 a + b^2\sin^2 a}{12}}$
$I = \dfrac{b}{12}(H^3 - h^3)$ $\dfrac{I}{c} = \dfrac{b}{6}\dfrac{H^3 - h^3}{H}$ $r = \sqrt{\dfrac{H^3 - h^3}{12(H - h)}}$	$\dfrac{H^4 - h^4}{12}$ $\dfrac{1}{6}\dfrac{H^4 - h^4}{H}$ $\sqrt{\dfrac{H^2 + h^2}{12}}$	$\dfrac{H^4 - h^4}{12}$ $\dfrac{\sqrt{2}}{12}\dfrac{H^4 - h^4}{H}$ $\sqrt{\dfrac{H^2 + h^2}{12}}$	$\dfrac{bh^3}{36}; c = \dfrac{2}{3}h$ $\dfrac{bh^2}{24}$ $\dfrac{h}{\sqrt{18}}$
$I = \dfrac{bh^3}{12}$ $\dfrac{I}{c} = \dfrac{bh^2}{12}$ $r = \dfrac{h}{\sqrt{6}}$	$\dfrac{5\sqrt{3}}{16}R^4$ $\tfrac{5}{8}R^3$ $\sqrt{\dfrac{5}{24}}R$	$\dfrac{5\sqrt{3}}{16}R^4$ $\dfrac{5\sqrt{3}}{16}R^3$	$\dfrac{1 + 2\sqrt{2}}{6}R^4$ $0.6906R^3$ $0.475R$

Square, axis same as first rectangle, side = h; $I = h^4/12$; $I/c = h^3/6$; $r = 0.289h$.
Square, diagonal taken as axis: $I = h^4/12$; $I/c = 0.1179h^3$; $r = 0.289h$.

TABLE 2-10. PROPERTIES OF VARIOUS CROSS SECTIONS (*Continued*)

Section	Moment of inertia	Section modulus	Radius of gyration
Equilateral Polygon A = area, (see p. 1–39) R = rad circumscribed circle r = rad inscribed circle n = no. sides a = length of side Axis as in preceding section of octagon	$I = \dfrac{A}{24}(6R^2 - a^2)$ $= \dfrac{A}{48}(12r^2 + a^2)$ $= \dfrac{AR^2}{4}$ (approx)	$\dfrac{I}{c} = \dfrac{I}{r}$ $= \dfrac{I}{R \cos \dfrac{180°}{n}}$ $= \dfrac{AR}{4}$ (approx)	$\sqrt{\dfrac{6R^2 - a^2}{24}} \approx \dfrac{R}{2}$ $\sqrt{\dfrac{12r^2 + a^2}{48}}$
(trapezoid)	$I = \dfrac{6b^2 + 6bb_1 + b_1^2}{36(2b + b_1)} h^3$ $c = \dfrac{1}{3} \dfrac{3b + 2b_1}{2b + b_1} h$	$\dfrac{I}{c} = \dfrac{6b^2 + 6bb_1 + b_1^2}{12(3b + 2b_1)} h^2$	$\dfrac{h\sqrt{12b^2 + 12bb_1 + 2b_1^2}}{6(2b + b_1)}$
(cross/plus/H sections)	$I = \dfrac{BH^3 + bh^3}{12}$ $\dfrac{I}{c} = \dfrac{BH^3 + bh^3}{6H}$		$\sqrt{\dfrac{BH^3 + bh^3}{12(BH + bh)}}$
(hollow rectangle, I-beam, C)	$I = \dfrac{BH^3 - bh^3}{12}$ $\dfrac{I}{c} = \dfrac{BH^3 - bh^3}{6H}$		$\sqrt{\dfrac{BH^3 - bh^3}{12(BH - bh)}}$
(I-section unequal flanges)	$I = \tfrac{1}{3}(Bc_1^3 - B_1h^3 + bc_2^3 - b_1h_1^3)$ $c_1 = \dfrac{1}{2} \dfrac{aH^2 + B_1d^2 + b_1d_1(2H - d_1)}{aH + B_1d + b_1d_1}$		$\sqrt{\dfrac{I}{(Bd + bd_1) + a(h + h_1)}}$
(T/L sections)	$I = \tfrac{1}{3}(Bc_1^3 - bh^3 + ac_2^3)$ $c_1 = \dfrac{1}{2} \dfrac{aH^2 + bd^2}{aH + bd}$ $c_2 = H - c_1$		$r = \sqrt{\dfrac{I}{[Bd + a(H - d)]}}$
(circle)	$I = \dfrac{\pi d^4}{64} = \dfrac{\pi r^4}{4} = \dfrac{A}{4} r^2$ $= 0.05 d^4$ (approx)	$\dfrac{I}{c} = \dfrac{\pi d^3}{32} = \dfrac{\pi r^3}{4} = \dfrac{A}{4} r$ $= 0.1 d^3$ (approx)	$\dfrac{r}{2} = \dfrac{d}{4}$

TABLE 2-10. PROPERTIES OF VARIOUS CROSS SECTIONS (*Continued*)

Section	Moment of inertia	Section modulus	Radius of gyration
(hollow circle) $d_m = \frac{1}{2}(D+d)$ $s = \frac{1}{2}(D-d)$	$I = \frac{\pi}{64}(D^4 - d^4)$ $= \frac{\pi}{4}(R^4 - r^4)$ $= \frac{1}{4}A(R^2 + r^2)$ $= 0.05(D^4 - d^4)$ (approx)	$\frac{I}{c} = \frac{\pi}{32}\frac{D^4 - d^4}{D}$ $= \frac{\pi}{4}\frac{R^4 - r^4}{R}$ $= 0.8 d_m^2 s$ (approx) when $\frac{s}{d_m}$ is very small	$\frac{\sqrt{R^2 + r^2}}{2} = \frac{\sqrt{D^2 + d^2}}{4}$
(half circle)	$I = r^4\left(\frac{\pi}{8} - \frac{8}{9\pi}\right)$ $= 0.1098 r^4$	$\frac{I}{c_2} = 0.1908 r^3$ $\frac{I}{c_1} = 0.2587 r^3$ $c_1 = 0.4244 r$	$\frac{\sqrt{9\pi^2 - 64}}{6\pi} r = 0.264 r$
(half hollow circle)	$I = 0.1098(R^4 - r^4)$ $- \frac{0.283 R^2 r^2 (R - r)}{R + r}$ $= 0.3 t r_1^3$ (approx) when $\frac{t}{r_1}$ is very small	$c_1 = \frac{4}{3\pi}\frac{R^2 + Rr + r^2}{R + r}$ $c_2 = R - c_1$	$\sqrt{\frac{2I}{\pi(R^2 - r^2)}}$ $= 0.31 r_1$ (approx)
(ellipse)	$I = \frac{\pi a^3 b}{4} = 0.7854 a^3 b$	$\frac{I}{c} = \frac{\pi a^2 b}{4} = 0.7854 a^2 b$	$\frac{a}{2}$
(hollow ellipse)	$I = \frac{\pi}{4}(a^3 b - a_1^3 b_1)$ $= \frac{\pi}{4} a^2 (a + 3b) t$ (approx)	$\frac{I}{c} = \frac{\pi}{4} a(a + 3b) t$ (approx)	$\sqrt{\frac{I}{(\pi ab - a_1 b_1)}} =$ $\frac{a}{2}\sqrt{\frac{a + 3b}{a + b}}$ (approx)
(cross section)	$I = \frac{1}{12}\left[\frac{3\pi}{16} d^4 + b(h^3 - d^3) + b^3(h - d)\right]$ $\frac{I}{c} = \frac{1}{6h}\left[\frac{3\pi}{16} d^4 + b(h^3 + d^3) + b^3(h - d)\right]$		$\sqrt{\frac{I}{\pi\frac{d^2}{4} + 2b(h - d)}}$ (approx)
(U section) $h = H - \frac{1}{2}B$	$I = \frac{t}{4}\left(\frac{\pi B^3}{16} + B^2 h + \frac{\pi B h^2}{2} + \frac{2}{3} h^3\right)$ $\frac{I}{c} = \frac{2I}{H + t}$		$\sqrt{\frac{I}{2\left(\frac{\pi B}{4} + h\right) t}}$

Table 2-10. Properties of Various Cross Sections (*Continued*)

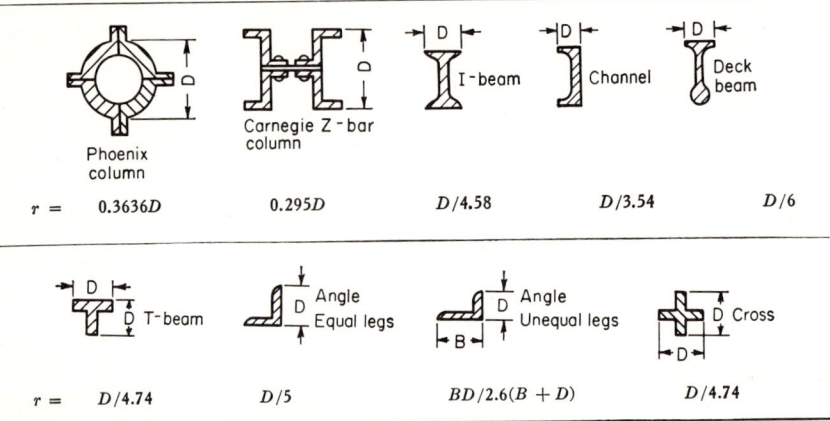

Section	Moment of inertia and section modulus	Radius of gyration
Corrugated sheet iron, parabolically curved	$I = \dfrac{64}{105}(b_1 h_1{}^3 - b_2 h_2{}^3)$, where $h_1 = \tfrac{1}{2}(H+t)$ $\;\;\; b_1 = \tfrac{1}{4}(B+2.6t)$ $h_2 = \tfrac{1}{2}(H-t)$ $\;\;\; b_2 = \tfrac{1}{4}(B-2.6t)$ $\dfrac{I}{c} = \dfrac{2I}{H+1}$	$r = \sqrt{\dfrac{3I}{t(2B+5.2H)}}$

Approximate values of *least* radius of gyration r

Phoenix column	Carnegie Z-bar column	I-beam	Channel	Deck beam
$r = 0.3636D$	$0.295D$	$D/4.58$	$D/3.54$	$D/6$

T-beam	Angle Equal legs	Angle Unequal legs	Cross
$r = D/4.74$	$D/5$	$BD/2.6(B+D)$	$D/4.74$

SOURCE: T. Baumeister (ed.), "Marks' Mechanical Engineers' Handbook," 6th ed., McGraw-Hill Book Company, Inc., New York, 1958.

This formula is for the **strength of beams**. For rectangular beams, $M = \tfrac{1}{6}Sbh^2$, where b = breadth, and h = depth; *i.e.*, the elastic **strength of beam sections** varies as follows: (1) for equal width, as the square of the depth; (2) for equal depth, directly as the width; (3) for equal depth and width, directly as the strength of the material; (4) if span varies, then for equal depth, width and material, inversely as the span.

If a beam is cut in halves horizontally, the two halves laid side by side will carry only one-half as much as the original beam.

The term **section modulus** is given to the value of I/c, where c is the distance to the fiber carrying greatest stress. Moment of inertia of cross section = I.

Tables 6 to 8 give the properties of various beam cross sections. For properties of structural-steel shapes, see Sec. 12.

Oblique Loading. It should be noted that Table 6 includes certain cases for which the horizontal axis is not a neutral axis, assuming the common case of vertical loading. The rectangular section with the diagonal as a horizontal axis (Table 9) is such a case. These cases must be handled by the principles of oblique loading.

Every section of a beam has two principal axes passing through the center of gravity, and these two axes are always at right angles to each other. The principal axes are axes with respect to which the moment of inertia is, respectively, a maximum and a minimum, and for which the product of inertia is zero. For symmetrical sections, axes of symmetry are always principal axes. For unsymmetrical sections, like a rolled angle section (Fig. 29), the inclination of the principal axis with the X-axis may be found from the formula $\tan 2\theta = 2I_{xy}/(I_y - I_x)$, in which θ = angle of inclination of the principal axis to the X-axis, I_{xy} = the product of inertia of the section with respect to the X- and Y-axes, I_y = moment of inertia of the section with respect to the Y-axis, I_x = moment of inertia of the section with respect to the X-axis. When this principal axis has been found, the other principal axis is at right angles to it.

Chapter 3

POWER-TRANSMISSION ELEMENTS

BEARINGS

Bearings for farm machines should meet the following operating requirements:

1. Self-alignment—often required because of the high deflections resulting from lightweight construction
2. Dirt resistance—achieved by wear-resistant materials or by adequate seals
3. Minimum maintenance—ideal when sealed and lubricated for season or life

The *types* of bearings commonly used in farm machines and some of their characteristics are listed in Table 3-1. Thrust loads can be carried only by plain bearings

TABLE 3-1. CHARACTERISTICS OF TYPES OF BEARINGS COMMONLY USED IN FARM MACHINES

Bearing	Relative cost	Range of coefficient of friction	Typical usage for bearing approx 1″ large on 1″-diam shaft		Lubrication period (not running in oil), hr	Expected life, % machine life
			Speed range, rpm	Radial load range, lb		
Wood......................	Low	0.05–0.15	25–100	50–25	5	25
Cast iron..................	Low	0.04–0.15	100–400	350–100	5	25
Bronze.....................	Low	0.02–0.12	200–1,000	400–80	5	35
Sintered metal..............	Medium	0.02–0.12	300–1,500	650–125	10	35
Nylon......................	Low	0.02–0.11	100–1,000	500–200	10	35
Nylon, sealed...............	Medium	0.02–0.11	100–1,000	500–200	50	50
Phenolic resin..............	Medium	0.06–0.15	100–500	300–50	10	35
Roller, soft on soft shaft....	Medium	0.02	100–1,000	225–105	10	25
Roller, hard on hard shaft, sealed	Medium	0.002–0.004	100–2,000	1,125–425	50	100
Needle roller, on hard shaft..	Medium	0.004–0.008	100–1,500	4,000–1,700	10	100
Tapered roller, sealed.......	High	0.002–0.005	500–3,000	1,400–800	200	100
Ball, unground, sealed.......	Medium	0.003–0.005	250–650	475–25	Life	50
Ball, ground, sealed.........	High	0.001–0.003	500–5,000	1,000–500	Life	100

made with end flanges. Ball and tapered roller bearings can carry thrust loads, but roller bearings require thrust washers. The use of one ball bearing and a pure radial-load bearing on the same shaft is frequently an economical method of carrying combined loads.

Plain bearings on open shafts are usually unsealed. Lubrication by grease gun serves to flush out dirt and maintain a grease seal at the bearing ends as well as to lubricate. Nylon is capable of running with very little lubricant, but when unsealed, lubrication is needed to flush out dirt and maintain a grease seal. This is also true of unsealed needle bearings. Some lightly loaded oil-impregnated or plastic bearings give satisfactory service without lubrication.

Self-aligning mountings vary from a plain cast-iron sleeve loosely carried by a hole in a plate to a spherical OD ball bearing held in stamped flanges. Plain and needle bearings are often mounted so as to be self-aligning. Tapered roller bearings are used exclusively in housings which maintain accurate alignment.

Bearing loads due to total chain or belt tension can be estimated by multiplying the net driving force by the following factors:[1]

Single chains	1.1
V belts	1.5
Single-ply flat leather belts	2.0
Double-ply flat leather belts	2.5
Triple-ply flat leather belts	3.0

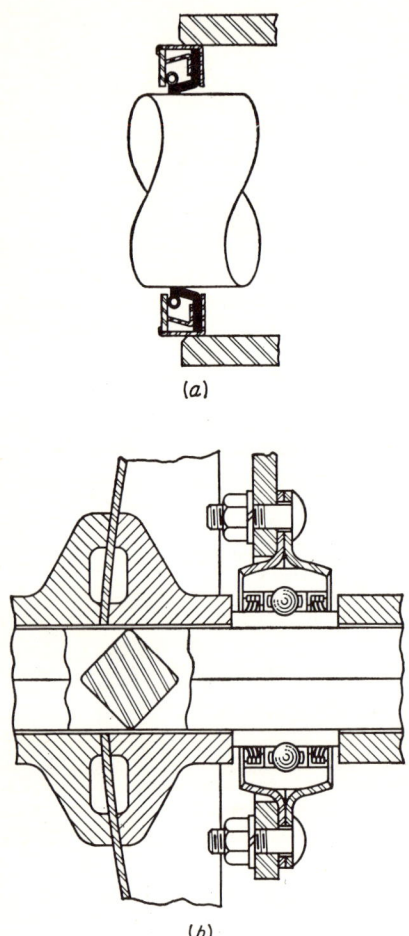

Fig. 3-1. Bearing seals: (*a*) seal with garter spring. (*Chicago Rawhide Co.*) (*b*) Multiple-lip-seal integral with bearing for extremely dirty conditions. (*New Departure Div., General Motors Corp.*)

Sealing is particularly important with precision-type antifriction bearings which are easily damaged by dirt because of their close internal fits. Sealing is usually effected by a sharp-cornered lip held against the rotating shaft by its own elasticity or by spring pressure. Leather, neoprene, or nylon is commonly used for seal lips. Felt is sometimes used for auxiliary protection but is no longer used for primary protection. Seal lips are usually turned in to retain oil in a gear case, but are otherwise turned out, particularly when they must relieve grease-gun pressure. Gear cases should be vented to prevent air pressure build-up while in use, and consequent oil leakage.

Most seals will not tolerate appreciable shaft eccentricity and for this reason work best with precision-type antifriction bearings. On shafts subject to wrapping with hay, etc., seal lips should be protected by an adjacent flange rotating with the shaft. When assembling seals over shafts with sharp shoulders, damage can be prevented by wrapping paper around the shaft. Seals must press into the housing tightly enough to prevent the passage of oil and dirt. Two types of seals are shown in Fig. 3-1.

Daily *lubrication* of farm machines is being eliminated as far as possible. "Sealed-for-life" antifriction bearings will run several years with normal usage, but should be replenished after 500 to 1,000 hr of use by injecting an oxidation-resistant oil through the seal, using a gun with a hollow needle.

Grease-gun lubrication of bearings in sealed housings is required at 50- to 200-hr intervals, only once a season in many cases. Plain bearings with little reservoir space require grease-gun lubrication at least twice a day to prevent excessive friction and wear. Few farmers today are willing to grease twice a day, and such bearings are being eliminated from new designs.

---

[1] "Engineering Handbook," General Motors Corp., Hyatt Bearing Division, Harrison, N.J., 1946.

GEARS

Gears are an essential part of all farm tractors and are also found in the majority of field machines. They vary from high-quality automotive-type gears to rather crude cast-iron gears, according to the requirements.

Gear design and manufacture is a complex art as well as a science. There is considerable difference of opinion concerning the best design methods and practices, even among men who have spent their lives specializing in gears. An attempt will be made here to point out only some of the principles and factors involved, and the references will have to be consulted for complete design information.

The function of gears is to transmit power at uniform angular velocity from one shaft to another, usually for one or more of the following purposes:
1. To change rotational velocity and torque
2. To change direction of rotation
3. To change direction of shafting, as with bevel gears

In order to meet the requirement of *uniform velocity,* either the epicycloidal or involute tooth form must be used. The involute form is used exclusively because it allows slight variations in center distance between gears without velocity fluctuation. The involute shape was first used in the early seventeenth century by Robert Willis, an English professor who also was the first to make practical application of the epicycloidal system.

In order to secure uniform velocity it is necessary that the force at the point of contact act at a constant angle to the line of centers and act through the pitch point. *Pressure angles* of $14\frac{1}{2}$ and $20°$ are in common use. The $14\frac{1}{2}°$ pressure-angle tooth has the advantage of a longer interval of driving contact, with two teeth carrying the load a greater percentage of the time and also less radial load and compressive stress. On the other hand, the $20°$ full-length and, particularly, the $20°$ stub systems have a shorter tooth with a broader base, which requires no undercutting to prevent interference with small numbers of teeth, unlike the $14\frac{1}{2}°$ system, and is thus stronger and also, because of the shorter tooth, has less bending moment. The $14\frac{1}{2}°$ system, however, is usually preferred with large numbers of teeth since no undercut is required with 32 teeth and over.

In the *Maag system* of gearing, developed in Switzerland, the teeth of a 12-tooth pinion meshing with a 30-tooth gear, for instance, have decreased dedendum, increased addendum, and special pressure angle to give the best running conditions. This pinion, however, will not mesh with another similar 12-tooth pinion. The system has the disadvantage that gears must be specially cut for each ratio but can secure strong pinion teeth and smooth running with as few as six teeth.

Since bevel gears are inherently noninterchangeable, the Gleason system for bevel gears uses the same approach, varying tooth height and pressure angle for each ratio to secure best results. Spiral bevel pinions with as few as five teeth have been used.

Gear-tooth failures can be outlined in part as follows:
1. Breakage or deformation due to loads exceeding the elastic limit
2. Breakage at root due to fatigue failures
3. Abnormal wear such as
    a. Scuffing due to heavy pressure and high rubbing velocities breaking the oil film, resulting in metal-to-metal seizure
    b. Abrasion due to foreign material
    c. Progressive pitting because the compressive fatigue limit of the surface is exceeded

Failures may be caused or aggravated by the following conditions:
1. Inaccuracies in tooth spacing, thickness, and profile
2. Incorrect center distances which cause bending if too close or increased leverage on the teeth if too far
3. Shaft misalignment or deflection which concentrates the load on one side of the tooth instead of applying it uniformly
4. Insufficient lubrication
5. Excessive loads

6. Excessive speeds
7. Vibration
8. Poor surface finish

In order to avoid failure, a gear must have teeth of adequate strength and wear resistance. Various formulas have been developed to evaluate these factors.

A formula for *tooth strength*, considering the tooth as a cantilever beam, was first developed by Wilfred Lewis in 1893. This formula contains a form factor Y which compensates for the variation in root area with the necessary undercut for small numbers of teeth as well as for the variation in tooth height and pressure angle. It assumes that the entire load is taken by a single tooth at its tip rather than in from the tip with less leverage, as is normally the case with the contact ratio greater than 1.0. This is considered to err on the safe side, however, and helps provide for the stress concentration at the fillet. In modern practice, the allowable stress is taken as being the endurance limit of the material, because of the repeated loading, and the strength calculated by the Lewis formula is called the endurance strength. The endurance limit of steel is approximately half the ultimate strength, or (250)(Brinell number), pounds per square inch, up to a limit of 100,000 psi. For case-hardened steels the endurance limit is usually taken as (300)(Brinell number of the core), pounds per square inch.

The actual or *dynamic load on gear teeth in operation* consists of the transmitted load plus additional forces due to inaccuracies in tooth manufacture, misalignment, velocity fluctuations, and tooth deflection. These factors are increasingly important as speed increases; hence the term dynamic load. On the basis of a great number of tests, Earle Buckingham developed an empirical formula for dynamic loading which adds to the transmitted load an additional load varying with velocity, face width (effect of misalignment), material, tooth form, and error in action due to manufacture and deflection. If the endurance strength, as calculated from the Lewis formula, is equal to the dynamic load, the tooth is considered to be sufficiently strong for normally smooth loads. Extra margin should be provided for shock loads, using double the dynamic load for extremely severe service.

In order to avoid damaging *wear* due to pitting, the compressive fatigue limit of the surface material must not be exceeded. On this basis Earle Buckingham has developed a formula for the limiting wear load which depends on gear ratio, velocity, pinion diameter, face width, surface endurance strength and modulus of elasticity of the material, and the pressure angle. Surface endurance strength of steel is approximately equal to (400)(Brinell number of surface) − 10,000 psi. The limiting wear load should exceed the dynamic load for long life.

Cast gears are extensively used in farm machines for low speeds and light loads where noise is not objectionable. The teeth are inherently inaccurate, but a rough estimate of the dynamic load F_d can be made by the following formula:[1]

$$F_d = \frac{(600 \, V_m) F_t}{400} \tag{3-1}$$

where F_t = transmitted load
V_m = pitch-line velocity, fpm

The endurance strength F_s can be approximated by the formula:[1]

$$F_s = 0.054 s_n b P_c \tag{3-2}$$

where 0.054 = a constant Lewis factor
s_n = endurance strength, 12,000–18,000 psi, depending on the grade of cast iron
b = tooth width, in.
P_c = circular pitch, in.

As with steel gears, the endurance strength should exceed the dynamic load.

The American Gear Manufacturers Association has adopted modifications of the above formulas. The Lewis formula has been modified by:

[1] V. M. Faires, "Design of Machine Elements," The Macmillan Company, New York, 1941.

1. Considering the load to be applied at the actual maximum height of single-tooth contact rather than at the tip
2. Adding a fillet stress-concentration factor
3. Adding a "combined face-width and inbuilt factor" for misalignment and deflection

This is then used with velocity and a velocity factor for dynamic load to give a direct horsepower-rating formula, rather than using endurance-strength and dynamic-load formulas.

A surface-durability horsepower-rating formula has also been developed. These ratings are for cut gearing having 8 to 10 hr daily service at uniform load, and service factors are given for other conditions. These rating formulas are available for spur, helical, and bevel gears, and information may be secured by writing the American Gear Manufacturers Association, Washington, D.C.

Heat-treatment of gears for maximum strength and wear is a specialized operation. Carburized gears are popular, and one reason for their high endurance strength is that proper quenching will result in a residual surface compressive stress (similar to that produced by shot peening) which reduces the starting of tension cracks leading to fatigue failure. Induction-hardened gears are being successfully substituted for carburized gears by some manufacturers.[1]

Most of the above formulas have been developed for industrial use where normal life is considered to be 5 years at 10 hr per day, or approximately 15,000 hr. The normal life of a farm tractor is about 12,000 hr, and most farm machines are used less than 2,500 hr. It can be seen that, in many cases, industrial standards as to wear and even as to endurance strength may not apply.[2]

The allowable load can be increased above the endurance limit where the desired life is less than 10 million loading cycles. The amount of increase is greatest for materials which are most vulnerable to fatigue failure because of surface roughness and low hardness.[3] Life factors for machined and hardened gears are as follows:[4]

Number of cycles	Spur and helical gears		Bevel gears
	250 to 450 BHN	Case carburized†	Case carburized†
1,000	3.0–4.0*	2.7	4.6
10,000	2.0–2.6*	2.0	3.1
100,000	1.6–1.8*	1.5	2.1
1 million	1.1–1.4*	1.1	1.4
10 million	1.0	1.0	1.0
100 million	1.0–0.9	1.0–0.9	1.0

* Use the higher values for higher hardness.
† 55–63R_c.

CHAIN DRIVES

Chain drives not only transmit power but are particularly useful in farm machines for material feeding, conveying, and elevating.

Chain velocity is not uniform as the links pass over a sprocket because of the variation in the effective radius. The change in velocity is a function of the number of sprocket teeth as shown in Fig. 3-2. This results in high induced tension loads as well as impact loads when small sprockets are used at high speed.

Small sprockets may also cause high wear because of the increased angle of hinge action. This results in internal hinge wear as well as wear of the chain against the

---

[1] H. B. Knowlton, Induction Hardening of Gears, *SAE Journal*, January, 1950.
[2] H. A. McAninch, Gear Design for Finite Life, *Agr. Eng.*, 39:396–399, July, 1958.
[3] Charles Lipson, G. C. Noll, and L. S. Clock, "Stress and Strength of Manufactured Parts," McGraw-Hill Book Company, Inc., New York, 1950.
[4] Extracted from AGMA Information Sheet, Strength of Spur, Helical, Herringbone and Bevel Gear Teeth (*AGMA* 225.01, October, 1959), with the permission of the publisher, The American Gear Manufacturers Association, Washington 5, D.C.

tooth, particularly with the rollerless types of chain. Cost and space considerations usually favor small sprockets. Less than 12 teeth should never be used on roller-chain sprockets or less than 8 with low-speed link chain, and more should be used if possible. Fewer teeth may be used on idler sprockets because they do not introduce velocity fluctuations.

Steel detachable link chain is often used with driving sprockets having a pitch slightly longer than the chain pitch and driven sprockets with slightly shorter pitch as shown in Fig. 3-3. This allows the chains to seat quietly, picking up the load gradually, and also counteracts the tendency of the chain to ride out as it elongates.[1] The releasing link, however, is always under full load. If the chain is operated with the hook leading, sprocket-tooth wear is minimized because the chain does not rotate against the tooth while under load. On the other hand, chain wear in the joint, and consequent elongation, increase because internal hinge action takes place under full load. Reversing the chain increases tooth wear and decreases hinge wear. Short chains may be operated with the hook trailing to prevent rapid hinge wear and elongation, while long chains usually have the hook leading to reduce sprocket wear. Conveyor chains are customarily run with the hook trailing to favor a driving sprocket with the same pitch as the chain, particularly where the driving sprocket is at the delivery end and the load is applied by the conveyed material.

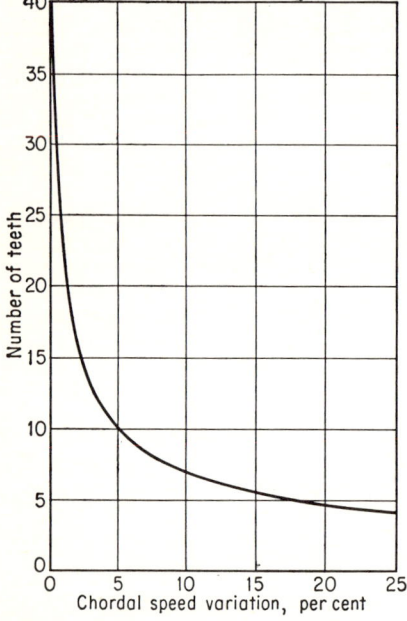

FIG. 3-2. Velocity fluctuations with number of teeth.

Steel detachable link chain is widely used in farm machines because of the low cost of chain and cast sprockets. It is not suitable for speeds exceeding 350 to 400 fpm and is usually subject to fatigue failure at loads exceeding 10 per cent of the ultimate strength, listed in Table 3-2. If a chain link breaks from the hook out

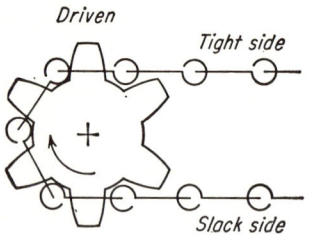
Driven pitch less than chain pitch

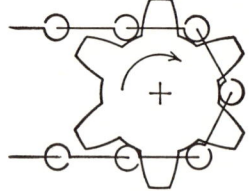
Driver pitch greater than chain pitch

FIG. 3-3. Sprocket pitch modifications.

through the side bars at 90° to the direction of travel, fatigue failure is indicated. If it breaks longitudinally at the side of the hook, tensile failure is indicated.

An improved type of detachable link chain, in which the hook is not formed from the material punched out from the center of the link, is claimed to have 33 per cent

---

[1] "Machinery's Handbook," The Industrial Press, New York, 1955.

Table 3-2. Steel-chain Dimensions and Ultimate Strengths

Chain number	Pitch, in.	Over-all width, in.	Max. sprocket width, in.	Weight per ft, lb	Av. ultimate strength, lb
Steel detachable chain					
25	0.904	$23/32$	$3/8$	1.9	950
32	1.157	$31/32$	$1/2$	3.3	1,650
34	1.402	$11/16$	$5/8$	3.8	1,750
42	1.375	$11/4$	$45/64$	5.1	2,300
45	1.630	$11/4$	$45/64$	5.0	2,100
50H	1.395	$15/16$	$23/32$	6.2	2,600
51	1.135	$11/8$	$5/8$	4.5	2,100
52	1.513	$13/8$	$49/64$	6.4	2,700
55	1.631	$15/16$	$23/32$	6.3	2,800
62	1.654	$19/16$	$7/8$	9.0	4,400
67	2.312	$115/16$	1	11.3	4,200
70	2.016	2	1	13.3	4,800
Steel roller chain					
41	$1/2$	0.536	$1/4$	0.277	2,000
40	$1/2$	0.628	$5/16$	0.41	3,700
50	$5/8$	0.796	$3/8$	0.64	6,100
60	$3/4$	0.978	$1/2$	1.0	8,500
80	1	1.23	$5/8$	1.68	14,500
Double-pitch steel roller chain					
840	1	0.628	$5/16$	0.34	3,300
1,050	$11/4$	0.796	$3/8$	0.44	6,100
1,260	$11/2$	1.104	$1/2$	0.79	8,500
1,680	2	1.356	$5/8$	1.25	14,500

SOURCE: Link-Belt Co. catalogue, Indianapolis, Ind., 1958.

more ultimate strength and an endurance limit 2½ times that of equivalent regular steel link chain.[1] It requires more blank material and is higher in price.

Steel roller chain is more efficient than link chain because the rollers reduce friction against sprocket teeth and because of more accurate machine-cut sprocket teeth and smoother finishes. Small sprockets are usually of hardened steel, but large ones may be of mild steel, cast steel, or cast iron. Allowable speeds may range up to 4,000 fpm for light chain on large sprockets, although the usual range is from 500 to 2,000 fpm.

The allowable working tension is influenced less by ultimate strength than by:

1. Pin bearing area
2. Loads due to impact and centrifugal force as affected by unit chain weight, sprocket size, and speed
3. Types of lubrication and dirt protection

Wear of chain hinges and sprocket teeth is usually the controlling factor. Manufacturers' ratings should be consulted. Sprockets having less than 15 teeth should be avoided if possible.

Roller-chain costs are roughly four time those of steel detachable link chains. *Double* or *extended-pitch steel roller chain* has half the joints of regular chain, costs less, and can be used with accurate cast sprockets. If used with the same-diameter sprockets (having half the teeth), velocity fluctuation and hinge action will increase in accordance with Fig. 3-2, resulting in additional stress and wear. Small sprockets

---

[1] J. H. Thuerman and E. A. Paul, Recent Agricultural Chain Developments, *Agr. Eng.*, 37:613-617, September, 1956.

should be avoided, and speeds kept below 600 fpm. It is desirable that the tight side be an even multiple of the pitch in order to reduce the effects of velocity fluctuation.

Double-pitch rollerless chain is again lower in cost. By eliminating the roller, the hinge bearing area is increased, thus increasing load capacity at low speed. It is claimed that the resulting additional sprocket-tooth wear is inconsequential in most cases. Clearances are opened up slightly to facilitate the use of cast sprockets and to allow more misalignment without damage.

The *length* of a chain in pitches is closely approximated by the following formula:

$$L = \frac{N_1 + N_2}{2} + \frac{2C}{P} + \frac{P(N_1 - N_2)^2}{39.5C} \tag{3-3}$$

where L = length of chain, pitches
 N = number of sprocket teeth
 C = center distance, in.
 P = pitch, in.

It is recommended that the driving sprocket have at least 135° of wrap to avoid jumping of the chain. Chain wear is inversely proportional to length, and this factor, plus the increase in elasticity of a longer tight side, favors long center distance. On the other hand, excessive length may cause whip and is more expensive. A *center distance* equal to the sum of the diameters normally represents a good compromise.

Chains of any type should ideally run with zero tension on the slack side, resulting in minimum bearing loads. Excessive slack, however, causes noise and wear. An adjustment for center distance is preferable to an idler adjustment for removing slack.

There is some question as to the desirability of lubricating detachable steel chain on farm machinery because of the tendency of the oil to collect and hold dirt. Where excessive quantities of abrasive dust are present, lubrication may do more harm than good.

Roller chain running in the open should be periodically removed and washed in a solvent and then soaked in oil. The excess oil should be drained and wiped off to reduce the tendency to collect dust.

V-BELT DRIVES

W. S. WORLEY

The use of V-belt drives in farm machines has increased considerably in recent years because of the following features:

1. A single belt can drive several units on a machine by passing over several sheaves.
2. Direction of shaft rotation can be reversed by crossing the belt.
3. The V belt can be so designed that slippage will occur at overloads and thus act as a safety clutch.
4. The V belt can be used as a clutch by controlling belt tension.
5. Speed changes are easily made, and speed variation is obtainable with adjustable sheaves.
6. The V belt is quiet-running.
7. No lubrication is required.
8. The V belt cushions shocks.
9. Alignment requirements are less critical than with other drives.
10. Belt failure does not damage other machine parts.

V belts are made of fabric and cords molded in rubber. Three types of V belts are used in the agricultural industry. The most frequently used is a belt of trapezoidal cross section having a ratio of top width to thickness of approximately 1.6. For use with adjustable sheaves, a trapezoidal cross section having a top width-thickness ratio of approximately 2¼ is used. For drives on which the direction of rotation of one or more wheels must be changed, a belt having a modified hexagonal form (known as double V belt) is used. The American Society of Agricultural Engineers (ASAE) has established a standard which covers cross sections, lengths, groove specifications, belt-length calculations, installation and take-up requirements, and belt-measuring specifications.[1]

As pointed out in the ASAE Standard: "The use of larger sheave diameters will result in lower bearing loads and can result in the use of smaller and less expensive belt cross sections." This is an important point. It is also important that idler diameters be chosen with some reference to the diameter of the loaded wheels. Although the ASAE Standard gives minimum diameters recommended for idlers, a better procedure is to choose an idler diameter somewhere between the diameters of the loaded wheels. The diameter of an idler which runs on the outside of the belt should be one-third larger than the diameter of an inside idler.

Mechanics of V-belt Drives. A V belt transmits power by virtue of a difference in tension in the belt between the point at which it enters and leaves a sheave. This difference in tension is developed through friction between the sidewalls of the belt and the sides of the groove in the sheave. The tensions in the belt cause it to wedge into the sheave groove and greatly increase the driving force available.

The designer is generally concerned with the service to be expected of a V-belt drive under given conditions of horsepower load, sheave diameters, and belt speed. In the following discussion, formulas used for determining tensions in belts are given. This is followed by an outline of the method used by one manufacturer to determine the service to be expeeted of a V-belt drive under given conditions.

The difference in tension required to carry the horsepower load at each wheel of a drive can be found by the following formula:

$$T_1 - T_2 = \frac{33{,}000 Hp}{S} \tag{3-4}$$

where Hp = horsepower load transmitted
T_1 = tight-side tension, lb
T_2 = slack-side tension, lb
S = belt speed, fpm = $\dfrac{(\text{pitch diam of wheel, in.})(\text{rpm})}{3.82}$

A design tension ratio, T_1/T_2, of 5.00 at 180° arc of contact has been found to be safe for V-belt-drive design, although V belts will operate easily at tension ratios as high as 8.00 or 9.00. With reduced arc of contact, both T_1 and T_2 must be increased to transmit the same power without slippage, resulting in a lower tension ratio.

It is customary in designing drives to calculate tensions on the basis of a horsepower load somewhat greater than the average load being transmitted. This provides for sufficient tension in the drive to handle overloads. For estimating drive service, the actual average load is increased to a value called the *design horsepower*, which, as a steady load, would be roughly equivalent in its effect on belt life to the fluctuating load which the drive will encounter. To find the design horsepower, the average load is multiplied by the service factor listed in Table 3-3 for the type of application. In the case of two or more driven sheaves, an appropriate service factor should be selected for each, and the design horsepower for the driver is the sum of the design horsepowers of the driven units.

TABLE 3-3. SERVICE FACTORS FOR COMPONENTS OF FARM IMPLEMENTS

Function of operating unit	Service factor
Cutting (sickle bars)	1.5
Cutting (sickle bars with counterweight)	1.3
Cutting (reels)	1.0
Pickup attachments for combines	1.0
Feeding (front cylinder beaters, feeder rolls, draper canvas, etc.)	1.3
Threshing, chopping, etc. (combine cylinders, corn-sheller cylinders, hammer-mill rotors, etc.)	1.5
Separation (rear cylinder beaters, straw walkers, etc.)	1.0
Cleaning (fans, cleaning shoes, sieves, etc.)	1.0
Expelling (straw spreaders, husk blowers, etc.)	1.3
Delivery (augers, elevators, etc.)	1.3
Traction for self-propelled machines	1.3
Hydraulic system oil pumps	1.3

[1] ASAE Standard: V-belt Drives for Farm Machines, *Agr. Eng. Yearbook*, 1958, pp. 49–55.

The following formulas for tensions after correction by the service factor are useful in estimating the life to be expected of the drive and in determining bearing loads:

$$T_1 = \frac{41{,}250 H p_d}{GS} \quad (3\text{-}5)$$

$$T_2 = T_1(1 - 0.8G) \quad (3\text{-}6)$$

where T_1 = tight-side tension, lb
T_2 = slack-side tension, lb
Hp_d = (average horsepower load)(F) = design horsepower
F = service factor from Table 3-3
G = arc-of-contact correction factor from Fig. 3-4 (corrects for extra tension needed at less than 180° arc of contact, and vice versa)
S = belt speed, fpm

Equations (3-5) and (3-6) are valid for drives with two grooved sheaves. Factor G is taken for the smaller wheel, since it has the smaller arc of contact. If the large wheel is a flat pulley, tension should be recalculated, using G for the arc of contact with the flat pulley. The larger value of T_1 is then used for the design.

For drives with more than two wheels, the analysis of tensions must be done in an accumulative manner. The tight- and slack-side tensions at the sheave at which the belt is most likely to slip, usually the driver, will be given by Eqs. (3-5) and (3-6). At the remaining wheels the tensions will be greater than those given by these equations and will instead be governed by Eq. (3-4). The procedure used is illustrated by the following analysis of the drive shown in Fig. 3-5, where

	Arc of contact	G	Av. Hp	F	Hp_d
A	145 − 15 = 130°	0.86	1.0	1.0	1.0
B	182 − 15 = 167°	0.97	2.0	1.3	2.6
C (flat)	172 − 15 = 157°	0.68	6.0	1.5	9.0
R	205 − 15 = 190°	1.02	9.0		12.6

$S = 5{,}000$ fpm

1. Find minimum required tensions at the driver R using Eqs. (3-5) and 3-6).

$$\text{Min. } T_{1R} = \frac{41{,}250 H p_d}{G_R S} = \frac{41{,}250(12.6)}{1.02(5{,}000)} = 102 \text{ lb}$$

$$\text{Min. } T_{2R} = T_{1R}(1 - 0.8 G_R) = 102[1 - (0.8 \times 1.02)] = 19 \text{ lb}$$

2. Find the tensions at wheel A with the design horsepower load at A, with T_{2R} as the slack-side tension, using Eq. (3-4).

$$T_{2A} = T_{2R} = 19 \text{ lb}$$

$$T_{1A} = T_{2A} + \frac{33{,}000 H p_d}{S} = 19 + \frac{33{,}000(1)}{5{,}000} = 26 \text{ lb}$$

3. This value of T_{1A} must be checked, using Eq. (3-5), to be sure that it will be sufficient to prevent slip at wheel A.

$$\text{Min. } T_{1A} = \frac{41{,}250(1)}{0.86(5{,}000)} = 10 \text{ lb}$$

This is less than the 26 lb found in step 2 above, so the 26 lb can be taken as the tight-side tension of wheel A. Had minimum T_{1A} been larger than 26 lb, the difference would be added to T_{1A} and T_{2A} as found in step 2. This condition would indicate that the belt is more likely to slip at wheel A than at the driver.

4. Find the tensions at wheel B using the design horsepower load at B.

$$T_{2B} = T_{1A} = 26 \text{ lb}$$

$$T_{1B} = T_{2B} + \frac{33{,}000(2.6)}{5{,}000} = 26 + 17 = 43 \text{ lb}$$

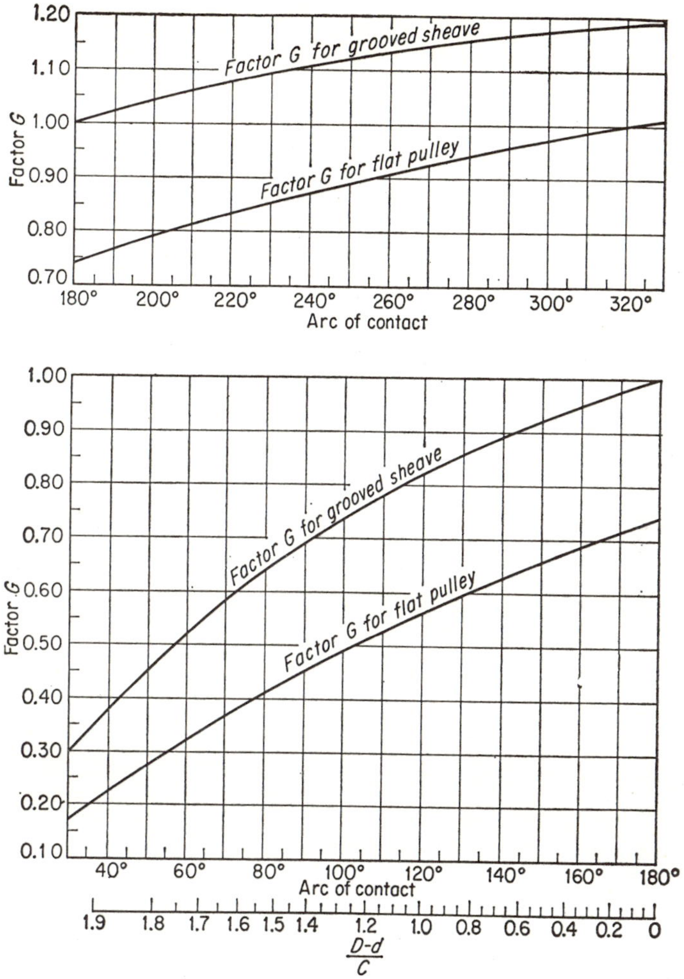

FIG. 3-4. Arc-of-contact correction, factor G. *Note:* for conventional V belts subtract 15° from arc of contact measured on layout [or found from $(D-d)/c$ on this page]. Use the result to find factor G. (*Gates Rubber Co., Denver, Colo.*)

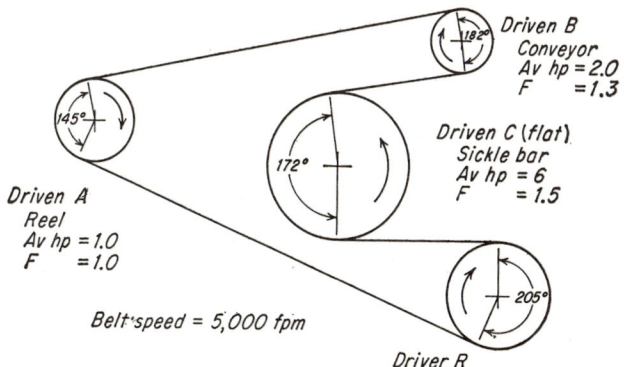

FIG. 3-5. Drive with data for tension analysis.

5. This value of T_{1B} must be checked, again using Eq. (3-5), to be sure it is greater than the minimum tension required to prevent slip at wheel B.

$$\text{Min. } T_{1B} = \frac{41{,}250(2.6)}{0.97(5{,}000)} = 22 \text{ lb}$$

6. Find the tensions at wheel C.

$$T_{2C} = T_{1B} = 43 \text{ lb}$$

$$T_{1C} = T_{2C} + \frac{33{,}000(9)}{5{,}000} = 43 + 59 = 102 \text{ lb}$$

7. Check T_{1C} for adequate tension to prevent slip.

$$\text{Min. } T_{1C} = \frac{41{,}250(9)}{0.68(5{,}000)} = 109 \text{ lb}$$

The 102 lb found in step 6 is not adequate and must be increased to 109 lb. The other tensions will increase by the same amount, 7 lb, giving

$$T_{1R} = 109 \text{ lb} \qquad T_{2R} = T_{2A} = 25 \text{ lb} \qquad T_{1A} = T_{2B} = 33 \text{ lb} \qquad T_{1B} = T_{2C} = 50 \text{ lb}$$

Note that $T_{1C} = T_{1R}$, giving a check on the calculations.

Service Life of a V-belt Drive. A V belt goes through a repeated cycle of stress each time it makes a complete revolution on the drive. It has been experimentally determined that the belt fails from fatigue under the repetition of this stress cycle. The rate at which it fails is determined by the peaks in the stress cycle as shown in Fig. 3-6. There is one stress peak at each wheel.

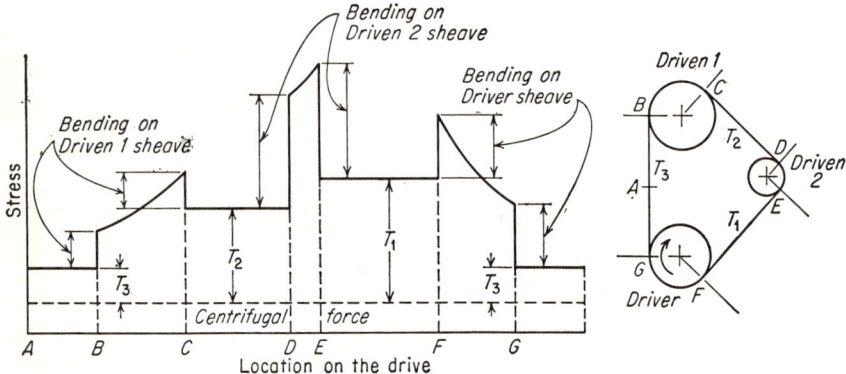

FIG. 3-6. Stress-cycle chart for V-belt drive. (*Gates Rubber Co., Denver, Colo.*)

As Fig. 3-6 shows, each peak stress is made up of stresses due to centrifugal force, bending, and tight-side tension. A typical curve showing the number of stress cycles resulting in fatigue failure of a belt at various peak stresses is given in Fig. 3-7.

V belts are manufactured in two qualities. One, known as the standard-quality belt, is the modern version of the V belt which was introduced on farm machines about 1937. The other, known as the premium-quality V belt, was first introduced in 1947 and has become generally available within the past few years. On a given drive, the premium-quality V belt will handle approximately 40 per cent higher horsepower loads than a standard-quality belt for the same service life. At the same horsepower load the premium-quality V belt will give much longer service life than a standard-quality belt. The ratio between the lives of premium- and standard-quality belts on the same drive depends on the details of the drive. It may, in extreme cases, exceed 10:1.

The *choice of the belt cross section* to be used on a drive depends on the tension in the tightest strand of the drive and the diameter of either the driver wheel or

POWER-TRANSMISSION ELEMENTS   45

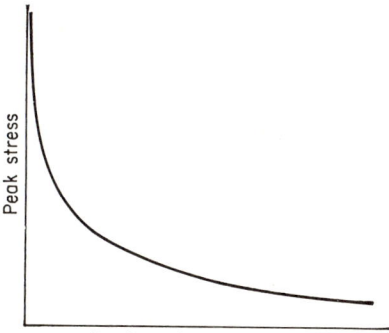

FIG. 3-7. Typical *S-N* curve for V belts.

the last driven wheel (wheel C of Fig. 3-5). Table 3-4 can be used to select a belt cross section for a check of service life. In some cases, sheave diameter will have to be increased in order to allow use of a belt cross section large enough to carry the load.

TABLE 3-4. CHOICE OF V-BELT CROSS SECTION ACCORDING TO VALUE OF T_1d
(where T_1 = tension in tightest strand of drive, pounds, and d = pitch diameter, inches, of driver or last driven wheel (e.g., wheel C of Fig. 3-5), whichever is smaller)

Cross section	Min. recommended d, in.	Range of T_1d			
		V belts, standard	Premium V belts	Standard double V belts	Premium double V belts
A	2	–190	–270	–250	–370
A or B		190–230	270–330	250–300	370–440
B	3	230–500	330–740	300–670	440–990
B or C		500–600	740–890	670–800	990–1,180
C	5	600–1,540	890–2,320	800–2,060	1,180–3,090
C or D		1,540–1,850	2,320–2,780	2,060–2,470	3,090–3,710
D	10	1,850–3,990	2,780–6,070	2,470–5,310	3,710–
D or E		3,990–4,780	6,070–7,280	5,310–6,380	
E	15	4,780–	7,280–	6,380–	

To estimate the average service life of a drive a method of computing the *fatigue rate* at each wheel of the drive has been developed. The fatigue rate is a measure of the number of stress peaks of a given magnitude which the belt will withstand. For computation purposes, fatigue rate is defined as the length of belt which will have a life of 1,000 hr if subjected to one peak stress of the indicated magnitude for each complete revolution of the belt at the actual belt speed. The fatigue rates for the various wheels of a drive are added together to find the total fatigue rate.

The fatigue rate for a particular wheel of the drive can be determined by the following formula:

$$\log Fr = 11.111 \log \left(T_1 + \frac{K_1}{d} + K_2 S_M{}^2 \right) + \log S_M - K_3 \qquad (3\text{-}7)$$

where Fr = fatigue rate, in./1,000 hr.
T_1 = tight-side tension, lb.
d = pitch diameter of wheel. (Use ¾ the actual pitch diameter for a flat wheel running on the back side of a V belt and for all wheels of a double-V-belt drive.)
$$S_M = \frac{S}{1,000} = \frac{\text{belt speed, fpm}}{1,000}$$

See accompanying table for constants K_1, K_2, and K_3.

Quality	Belt cross section	K_1	K_2	K_3
Standard...............	HA	156.8	0.5610	21.01
	HB	405.8	0.9653	23.63
	HC	1,112	1.716	26.41
	HD	3,873	3.498	29.85
	HE	7,332	5.041	31.61
Premium...............	HA-P	219.7	0.5610	22.56
	HB-P	575.9	0.9653	25.18
	HC-P	1,601	1.716	27.96
	HD-P	5,680	3.498	31.40
	HE-P	10,850	5.041	33.16

If an idler is used, its fatigue rate should also be calculated, using the tension in the section of the drive passing over the idler. If the idler runs on the back side of the belt, use three-quarters of its actual pitch diameter in calculating fatigue rate.

Belt life can then be computed as follows:

$$\text{Average belt life, thousands of hours} = \frac{\text{belt length, in.}}{\text{sum of fatigue rates}} \quad (3\text{-}8)$$

The life calculated from Eqs. (3-7) and (3-8) is a guide to the adequacy of the drive. If the horsepower loads have been correctly established and if the service factors used accurately reflect the effect of peak loads and load fluctuations on the drive, average belt life will be reasonably close to the life calculated above. If load and service factor have been overestimated, average belt life will be greater than calculated. Similarly, if load and service factor have been underestimated, average belt life will be less than calculated.

If the load to be transmitted is known to vary from time to time, the above method may be refined to get a more accurate estimate. Divide the usage into several load classifications and estimate the proportion of operating time for each. Determine the total fatigue rate for the load and speed conditions in each classification. Then multiply the total fatigue rate at each load by the proportion of operating time at that load and add up the results. Compute average belt life as follows:

$$\text{Average belt life, thousands of hours} = \frac{\text{belt length, in.}}{\text{sum of (fatigue rate)(proportion of time) for each load}} \quad (3\text{-}9)$$

Adjustable-speed V-belt drives are commonly used on farm implements as a means for providing a variable speed of rotation for one unit or group of units while maintaining a fixed speed for other driven units. The cylinder of a grain combine and propulsion drives for self-propelled combines and corn pickers are the most common applications of adjustable-speed drives.

An adjustable-speed drive uses a V belt somewhat thinner than usual, which generally operates in two adjustable sheaves. As shown in Fig. 3-9b, each sheave has one disk fixed in position on its shaft and one which can move axially with respect to the fixed disk. The sheaves are generally arranged so that the left-hand disk of one sheave moves in while the right-hand disk of the other moves out, and vice versa. This maintains belt alignment and tension while varying the speed ratio between the two sheaves. See ASAE Standard, V-belt Drives for Farm Machines, for sizes, lengths, grooves, etc., for adjustable-speed belts and sheaves.

In designing an adjustable-speed V-belt drive, the problem of providing the desired amount of *speed variation* (ratio of maximum to minimum output revolutions per minute) complicates the selection of belt and sheave sizes for satisfactory service life at minimum cost. The smaller the variation, the larger the sheaves may be, and vice versa. The use of two ASAE Standard minimum-diameter sheaves will give a speed variation slightly greater than 4.

The speed variation possible with a given minimum pitch diameter is greatest with equal-diameter sheaves and is reduced when the two sheaves differ in size.

Because of this problem, it is often best to use adjustable sheaves of equal diameter, one of which may be on a jackshaft to a conventional reduction to complete the power train. With equal-diameter sheaves, the speed of the driver sheave will be the *mean proportional* ($\sqrt{Nn}$) between the maximum and minimum driven speeds.

Selection of belt cross section depends on the expected load, which may vary widely. An analysis based on a *duty cycle* of three or four loads is recommended. Each assumed speed-load condition existing for 10 per cent or more of the time should be checked on the chart of Fig. 3-8, and the largest cross section used for the initial computation of belt life.

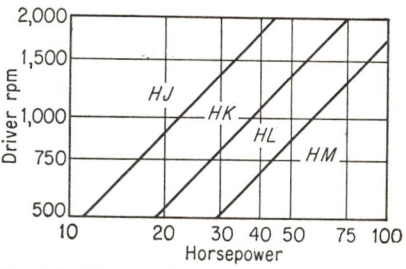

Fig. 3-8. Selection chart for belt cross-section in adjustable-speed V-belt drives.

The *change in pitch diameter p*, and the minimum allowable pitch diameter which may be used for design purposes are given in Table 3-5. If the take-up for belt stretch and wear is made by moving the shaft (as is preferable), p can be larger than where part of the sheave adjustment must be reserved for take-up.

A belt operates under the least tension for a given power at the highest speed possible up to the practical limit of 5,000 fpm. The following formulas are useful in

TABLE 3-5. CHANGE IN PITCH DIAMETERS FOR ADJUSTABLE-SPEED V BELTS

Cross section	p, change in pitch diameter, in.		ASAE Standard minimum allowable pitch diameter, in.
	Take-up by moving shafts	Take-up by narrowing sheaves	
HJ	4.23	3.6	4.2
HK	5.14	4.5	4.9
HL	6.02	5.5	6.0
HM	6.91	6.3	6.5

determining the largest-size sheaves which can be used for a given speed variation, within the belt speed limit of 5,000 fpm and the safe rim speed limit for the sheave material, usually taken as 12,000 fpm for carefully manufactured, dynamically balanced sheaves of steel or high-strength cast iron.

For equal-size adjustable sheaves:

$$N = r\sqrt{V} \tag{3-10}$$

$$V = \frac{N}{n} \tag{3-11}$$

$$\text{Max. belt speed, fpm} = \frac{\pi}{12} p \left(\frac{V}{V-1}\right)(n+r) \tag{3-12}$$

$$\text{Max. rim speed, fpm} = \frac{\pi}{12} p \left(\frac{V}{V-1}\right)(N+r) \tag{3-13}$$

where r = driver sheave, rpm
 N = maximum driven sheave, rpm
 n = minimum driven sheave, rpm
 V = N/n = speed-variation ratio

To find the maximum allowable driver speed for a given speed-variation ratio, find r by each of the following equations and use the lower value:

$$r = 12(V-1)\frac{(\text{max. allowable belt speed, fpm})}{\pi p(V + \sqrt{V})} \quad (3\text{-}14)$$

$$= 12\frac{(V-1)(\text{max. allowable rim speed, fpm})}{\pi p(V + \sqrt{V^3})} \quad (3\text{-}15)$$

After selecting r, N, and n for the p of the belt selected and the desired V, the following formula can be used to find the *minimum pitch diameter* for equal-size adjustable sheaves:

$$\text{Min. pitch diam, in.} = \frac{p(n+r)}{N-n} \quad (3\text{-}16)$$

If part of the sheave adjustment was reserved for take-up, use as the maximum pitch diameter for establishing belt length the minimum pitch diameter [Eq. (3-16)] plus the smaller value of p (Table 3-5). In any case, the actual maximum pitch diameter will be the minimum pitch diameter [Eq. (3-16)] plus the larger value of p (Table 3-5).

In the case of *unequal-diameter sheaves* the above formulas do not apply. Trial-and-error calculations are necessary to determine the speed variation which can be achieved with the cross section selected and within the allowable belt and rim speeds. Where both sheaves are mechanically shifted together, the change in belt-length requirement between settings for maximum and minimum speeds of the driver sheave must be taken up by a spring idler or by moving the shafts. Where only one sheave is adjusted mechanically and the other sheave is spring-loaded automatically to accommodate the length of belt available, some of the speed variation is lost in taking up the extra belt length. This must be taken into consideration in calculating the drive.[1]

To estimate belt life, the tight-side tension T_1 must be calculated for the various speed-load conditions in the duty cycle, using Eq. (3-5). Then the fatigue rate for each condition can be found by the use of Eq. (3-7), but with the use of the following K factors:

Cross section	K_1	K_2	K_3
HJ	3,560	2.768	30.28
HK	6,254	3.820	31.84
HL	10,070	5.020	33.15
HM	15,210	6.361	34.29

If an idler is used, its fatigue rate, as well as that of the driver and driven sheaves, should be determined. Use the tension in the section of the drive passing over the idler. If the idler runs on the back side of the belt, use three-quarters of its actual diameter in calculating fatigue rate. The total fatigue rate for each load condition is then the sum of the rates for driver, driven, and idler sheaves. Belt life can be determined by Eq. (3-9).

Speed controls for adjustable-speed drives may be arranged to allow speed changes while the drive is operating or may require stopping the drive to make an axial adjustment of the sheave disk (usually by spacers or by threads).

Common arrangements which permit changing speed while operating are as follows:

1. A double sheave with floating center flange mounted on a shiftable jackshaft, shown in Fig. 3-9d, is used for combine cylinder drives and also for traction drives.

[1] W. S. Worley, Designing Adjustable Speed V-belt Drives for Farm Implements, *SAE Trans.*, 63:321-333, 1955.

POWER-TRANSMISSION ELEMENTS 49

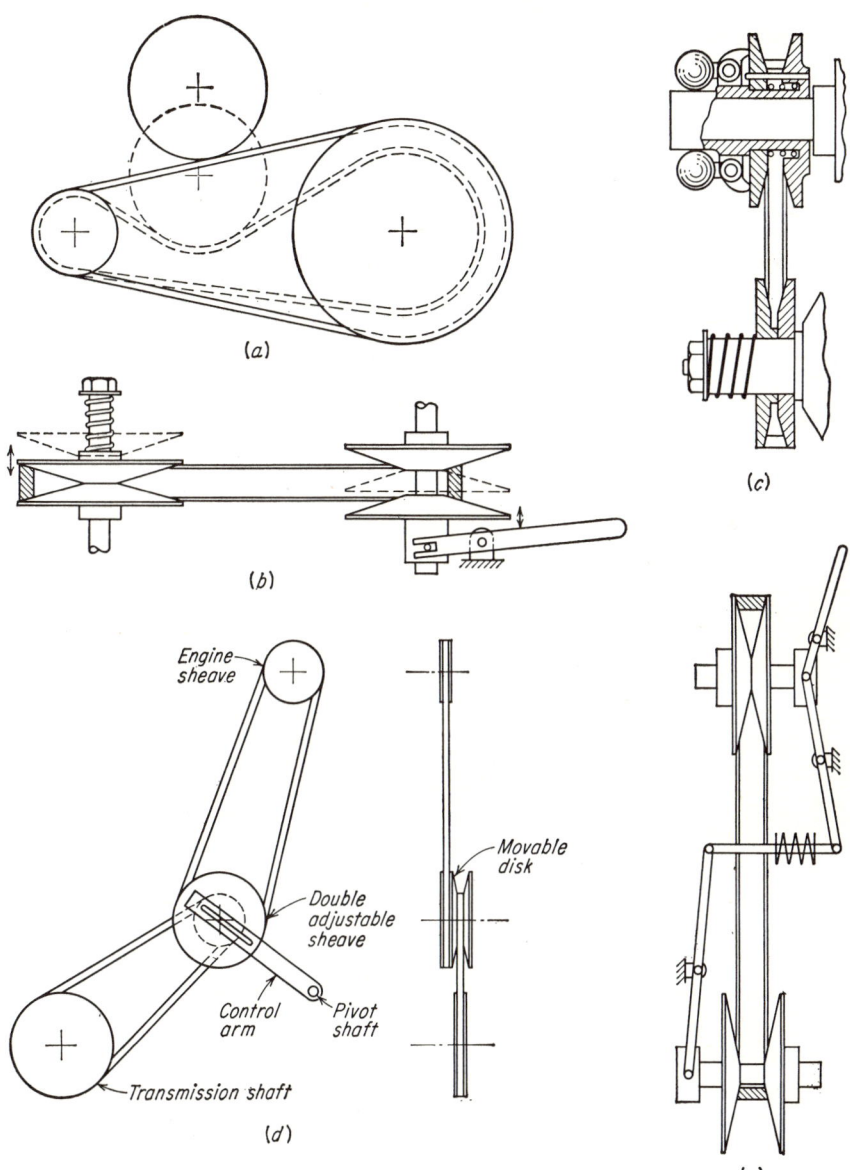

Fig. 3-9. Adjustable-speed V-belt drives: (*a*) one adjustable sheave with idler take-up; (*b*) manually controlled sheave and spring-loaded sheave; (*c*) governor-controlled sheave and spring-loaded sheave; (*d*) jackshaft sheave control; (*e*) dual shifting linkage.

The output speeds are the same as would be obtained by driving from the engine to a countershaft with the fixed-diameter sheaves and then using a conventional adjustable-speed drive with driver and driven the same diameter as the double-adjustable sheave. The following equations may be used to select sheaves for such a drive:

$$r^1 = \sqrt{Nn} \qquad (3\text{-}17)$$

where r^1 = countershaft rpm of an equivalent drive as explained above
N = max. output shaft rpm
n = min. output shaft rpm

$$d = \frac{p}{\sqrt{V-1}} \qquad (3\text{-}18)$$
$$D = d + p \qquad (3\text{-}19)$$

where d = min. pitch diameter of double-adjustable sheave, in.
D = max. pitch diameter of double-adjustable sheave, in.
p = change in pitch diameter, in.
$V = N/n$ = speed-variation ratio

$$D_R = \frac{dP_{max}}{r} \qquad (3\text{-}20)$$

where D_R = pitch diameter of driver sheave (on input shaft), in.
r = rpm of driver sheave (on input shaft)
$P_{max} = 19{,}200/D$ = max. permissible double sheave rpm for 5,000-fpm belt speed limit

$$D_N = D_R \frac{r}{r^1} \qquad (3\text{-}21)$$

where D_N = pitch diameter of driven sheave (on output shaft), in.

2. If only one adjustable sheave is used, as in Fig. 3-9a, a spring-idler sheave or movable shafts are required for take-up.

3. Mechanical shifting of both sheaves simultaneously by linked levers is shown in Fig. 3-9e. The spring shown is needed to provide belt tension in the intermediate positions.

4. Mechanical shifting of one sheave in combination with a spring-loaded sheave, as shown in Fig. 3-9b, is a popular method. It is usually best to spring-load the driven sheave and positively control the driver sheave in order to prevent decrease of speed under load.

5. Arrangement 4 above, with the driver sheave controlled by governor weights to close with speed increase, is shown in Fig. 3-9c. This has been used for propulsion drives in which a slowing down of the driven sheave with increase in torque is acceptable or can be counteracted by an increase in engine speed.

Mechanical shifting by control lever can be replaced by the use of a hydraulic positioning cylinder, although the action of such a cylinder should be slow, to avoid excessive side pressure on the belt.

Axial thrust on the movable disk of an adjustable sheave is given by the following formula:

$$F_s = Y(T_1 - T_2) \qquad (3\text{-}22)$$

where F_s = axial thrust, lb
T_1 = tight-side tension, lb
T_2 = slack-side tension, lb
Y = constant from Table 3-6

TABLE 3-6. VALUES OF Y FOR DETERMINING AXIAL THRUST BY EQ. (3-22)

	Arc of contact, deg	Y	
		Tension ratio = 5 at 180°	Tension ratio = 9 at 180°
Driver sheave............	180	2.83	2.54
	160	2.64	2.34
	140	2.46	2.15
	120	2.29	1.96
Driven sheave............	Does not depend on arc of contact	1.71	1.44

UNIVERSAL JOINTS

Universal joints are used to transmit power between two intersecting shafts, usually where the angle is variable. A single joint does not transmit uniform velocity to the driven shaft. Referring to Fig. 3-10a, it can be seen that if the driving shaft has an

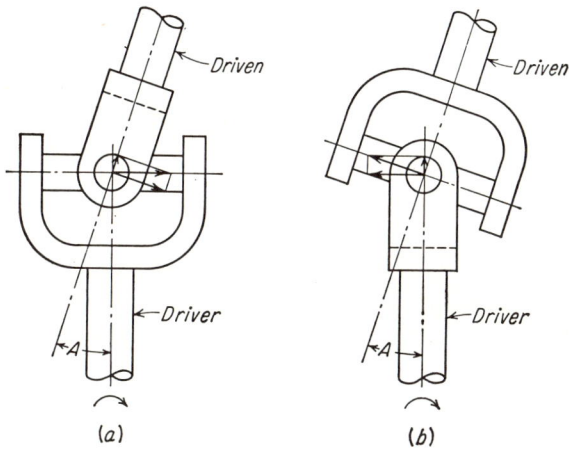

FIG. 3-10. Universal-joint force reactions exerted by cross (a) On driven fork, (b) On driving fork.

angular velocity ω_{driver}, the instantaneous angular velocity of the driven shaft ω_{driven} can be found as follows:

$$\omega_{\text{driven}} = \omega_{\text{driver}} \cos A \quad (3\text{-}23)$$

where A = angle between driven shaft and extension of operating shaft, deg

When the shafts have rotated 90°, as shown in Fig. 3-10b, the relationship is:

$$\omega_{\text{driven}} = \frac{\omega_{\text{driver}}}{\cos A} \quad (3\text{-}24)$$

Figure 3-11 shows the lead-and-lag cycle of the driven shaft with relation to the driving shaft for several values of shaft angularity, these curves being based on the formula

$$\tan Y = \frac{\tan X}{\cos A} \quad (3\text{-}25)$$

where X = angular displacement of driven shaft, deg
Y = angular displacement of driving shaft, deg

In order to impart *uniform velocity* to the driven shaft, universal joints are, wherever possible, used in pairs having equal angularity. If all three shafts lie in the same plane, as shown in Fig. 3-12, the driving yokes are turned 90° out of phase with each other to secure velocity compensation. If the intersection angles are equal but

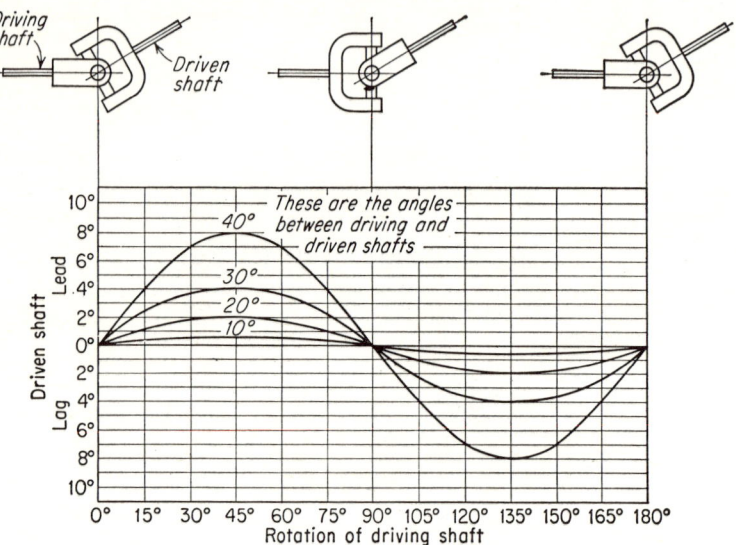

Fig. 3-11. Lead-and-lag cycle of driven shaft. (*Potgieter, Agr. Eng., January, 1952.*)

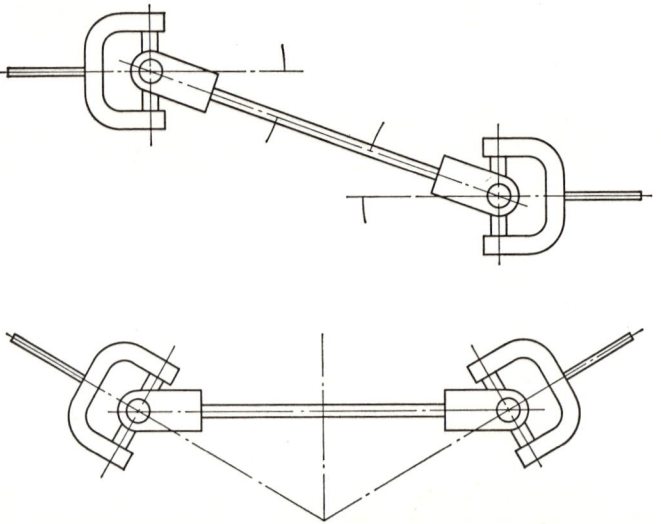

Fig. 3-12. Joints arranged for velocity compensation with shafts in one plane.

the shafts do not lie in a common plane, uniform velocity of the driven shaft can be secured by turning the driving yokes the proper amount out of phase. Where three or more joints are used in a train, the best combination of fork phasing can be most easily studied with a model setup having degree indicators at each joint.

Although uniform driven velocity may be secured by the use of two compensating joints, the shaft between them is subject to velocity fluctuation and its inertia will introduce fluctuations in torque. Because of this, its weight and polar moment of inertia should be kept as low as possible and a heavy protective slip clutch should not be located between joints if it can be placed elsewhere. The effect of two compensating joints as well as a single joint on torque fluctuation in a typical drive is shown in Fig. 3-13.

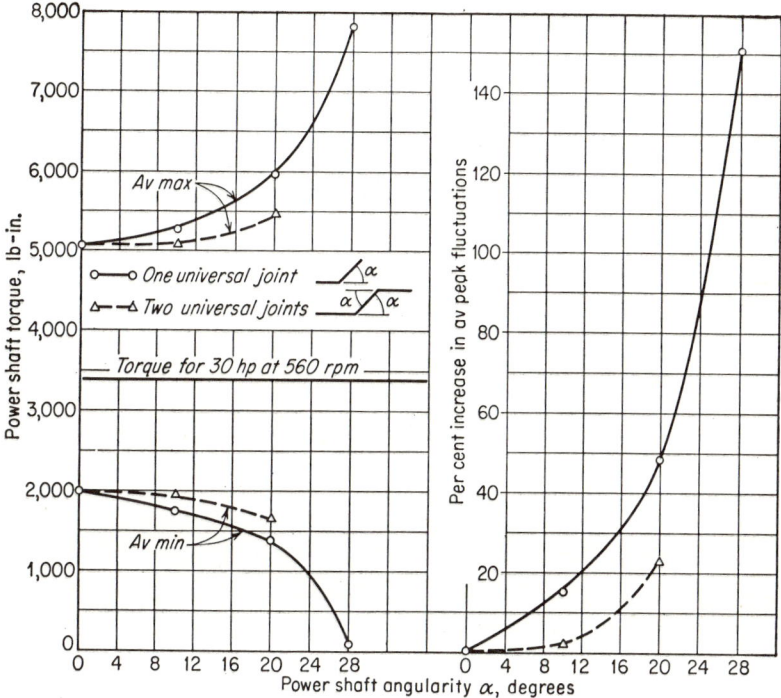

FIG. 3-13. Effect of power-shaft universal-joint angularity at 30 hp load. (*Hansen, Agr. Eng.*, February, 1952.)

It has been found that, in addition to universal-joint angularity, the "*torsional loads*" imposed on PTO drives are dependent on the following factors:

"1. The amount of kinetic energy stored in the rotating parts of the tractor.
"2. The moment of inertia of the rotating parts in the implement.
"3. The amount of resilience in the drive between the heavy rotating parts of the tractor and the rotating parts of the driven implement.
"4. The horsepower available at the PTO drive from the tractor.
"5. The horsepower required to operate the implement.

"Contrary to popular belief, items 1, 2 and 3 are far greater factors in influencing the magnitude of peak PTO torsional loads than are items 4 and 5." Peak loads have been found to range up to three times those calculated from the available engine horsepower.[1]

As can be seen in Fig. 3-14, *secondary couples* are introduced which tend to bend the shafts. The chart in Fig. 3-15 shows the values of these secondary couples for various degrees of angularity.

[1] Merlin Hansen, Loads Imposed on Power Take-off Shafts by Farm Implements, *Agr. Eng.*, 33:67–70, February, 1952.

This bending couple makes necessary a very firm connection between fork and shaft and also sufficient shaft stiffness to prevent whip. It is particularly important that the maximum speed be kept below the *critical speed* of the shaft as given by the formula

$$\text{Critical speed, rpm} = \frac{4{,}705{,}000(D^2 + d^2)^{1/2}}{L^2} \qquad (3\text{-}26)$$

where D = OD, in.
d = ID, in.
L = shaft length, in.

"For safe operation the speed of tubular shafts should be from 15 to 50% less than the critical speed. For speeds less than 2000 rpm and lengths less than 72 in., a

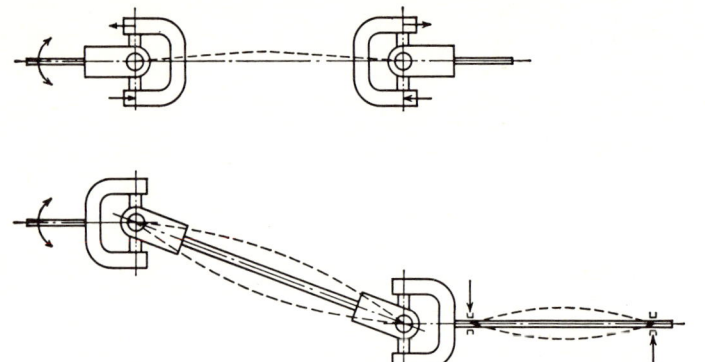

FIG. 3-14. Shaft bending and bearing loading resulting from secondary couples. (*Potgieter, Agr. Eng., January,* 1952.)

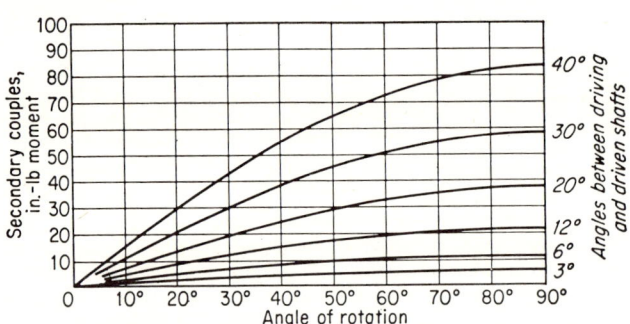

FIG. 3-15. Values of secondary couples for 100 in-lb. of input torque. (*Potgieter, Agr. Eng., January,* 1952.)

15% factor of safety may be used. For higher speeds or longer lengths a greater factor of safety should be used. These figures on maximum lengths and speeds should not be exceeded."[1]

The most important application of universal joints in farm machinery is in the *tractor power take-off drive* to trailed machines. Because of turning and variation in tilt due to uneven ground, the problem of arranging the joints for minimum velocity fluctuation is quite complex. Three joints are commonly used. Standards have been established for power take-off drives, making it possible to couple all makes of

[1] Fred M. Potgieter, Application of Universal Joints to Farm Machinery, *Agr. Eng.*, 33:21-27, January, 1952.

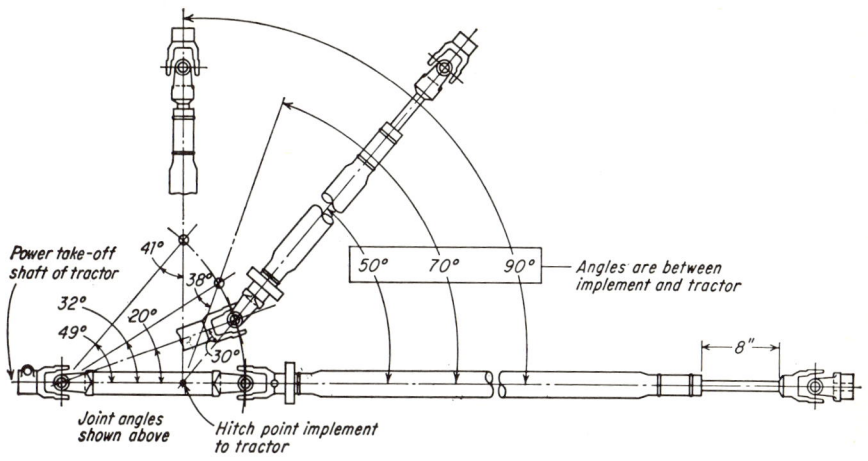

Fig. 3-16. "Straight-back" PTO drive. (*Potgieter, Agr. Eng.,* January, 1952.)

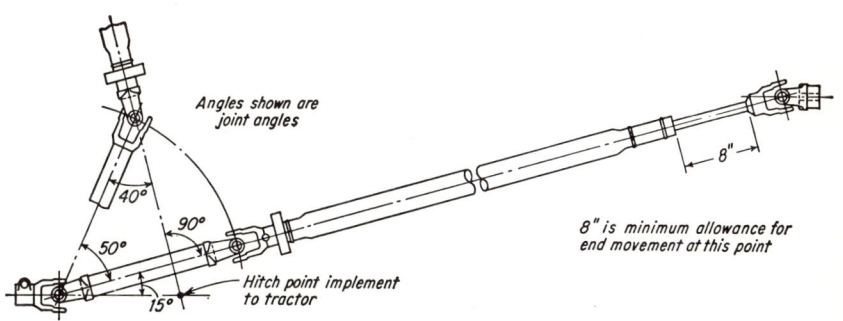

Fig. 3-17. Offset PTO drive. (*Potgieter, Agr. Eng.,* January, 1952.)

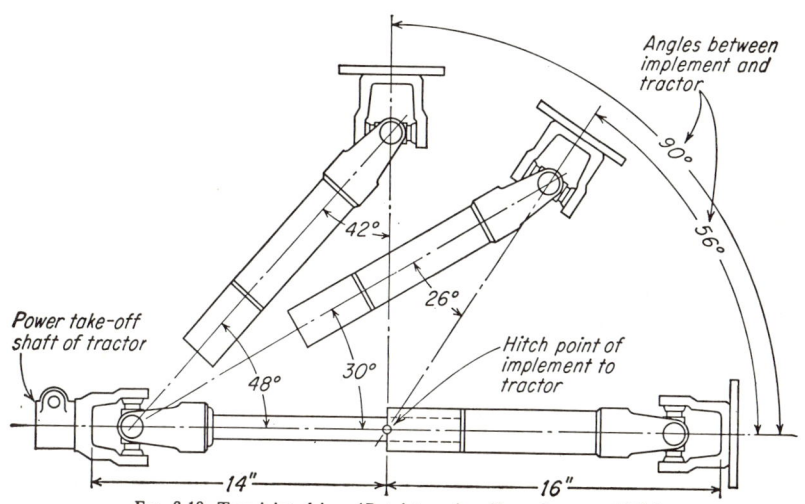

Fig. 3-18. Two-joint drive. (*Potgieter, Agr. Eng.,* January, 1952.)

tractors and implements conforming to the standards.[1] An example of a hitch with the joints all on the center line of the tractor PTO shaft is shown in Fig. 3-16 at various angles of turn. A drive with the rear driven shaft to the implement offset from the center line of the tractor PTO is shown in Fig. 3-17, and its inferiority to the drive in Fig. 3-16 can be seen by noting the additional joint angle required and inequality of the angles at extreme turns. Where room permits only a two-joint drive, it is best arranged as shown in Fig. 3-18 to fit the standard hitch point and still allow sufficient slip distance for extreme turns.

Constant-velocity universal joints have been developed, primarily for use in vehicle front-wheel drives, but cost has prohibited their application to farm machines.

Power-take-off drive shafts, where unshielded, have caused many serious injuries to farmers. Standard master *shields* for tractors will accept all implement shields following the standard, and it is essential that the implement shield be easy to attach to the

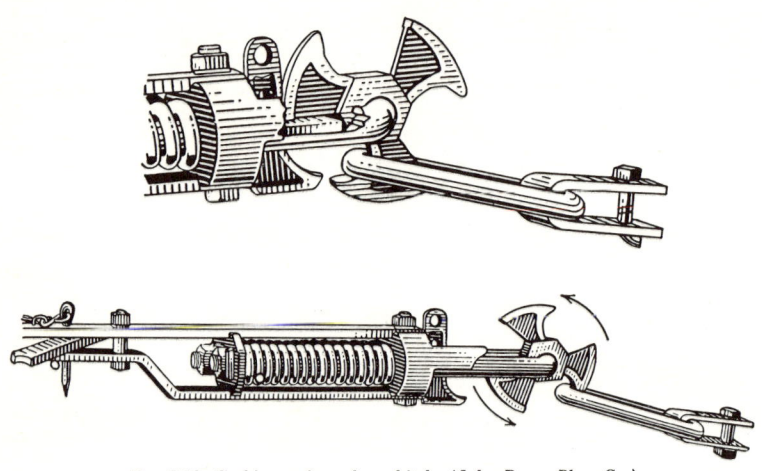

FIG. 3-19. Cushion-spring-release hitch. (*John Deere Plow Co.*)

tractor and difficult to detach from the implement in order to encourage the farmer to operate with it in place.

OVERLOAD PROTECTION DEVICES

Protective Hitches. The problem of preventing damage to an implement when it encounters an obstacle is an extremely difficult one. The basic relationship between the average stopping force at the drawbar and the distance to stop the tractor is expressed in the following formula:[2]

$$(H + R)d = 0.4WS^2 + Td + 0.01Wfd \tag{3-27}$$

where H = average force between implement and tractor, lb
R = rolling resistance of tractor, lb
d = total stopping distance, in.
W = weight of tractor and operator, lb
S = speed, mph
T = tractive effort following impact, lb
f = slope of ground, ft/100 ft

In the oldest and simplest protective device, the *pin-break hitch*, a wood or soft-steel shear pin carries the load and will break before the implement is damaged. Be-

---

[1] ASAE Standard: 540 rpm Power Take-off for Farm Tractors; ASAE Recommendation: Farm Tractor Auxiliary Power Take-off Drives; ASAE Recommendation: Operating Requirements for Power Take-off Drives; ASAE Standard: 1000 rpm Power Take-off for Farm Tractors, *Agr. Eng., Yearbook*, 1959.

[2] A. W. Clyde, Cushion Hitch Developments, *Agr. Eng.*, 30:169-171, April, 1949.

cause of the time required for replacement and the variability of the replacement pin which may be used by the farmer, this type of hitch is seldom used where frequent obstacles are encountered.

Friction breakbacks are sometimes used, but the action depends entirely on the tightness of clamping.

In a common protective hitch, the *cushion-spring-release type,* the load is carried by a cushion spring which, when deflected a predetermined amount, allows the hitch to release from the tractor. An example of this type of hitch is shown in Fig. 3-19. This type of hitch solves the problem of tractor inertia, but time is lost in rehitching when plowing in stony ground.

The alternative to complete release of the implement is controlled deceleration of the tractor with a maximum force which will not damage the implement. Soft cushion springs have the disadvantage of severe recoil. Declutching the tractor by cushion-spring actuation upon hitting an obstacle stops tractive effort but does not change inertia effects. The mechanical linkage required for declutching is usually cumbersome. Linkages which short the ignition to stop the engine are lighter and quicker-acting, but engine inertia is still effective. Neither of these expedients has been widely popular. Some work has been done on a declutching device which also applies the brakes to overcome inertia.

A *hydraulic release hitch* which opens a relief valve at the desired force and provides a uniform resistance while decelerating the tractor has been developed and can stop the tractor in much less distance than a spring without exceeding a given force.[1]

Clyde gives the following comparison of four types of hitches, based on the equation given above and the following assumptions:

H = 5,000 lb R = 200 lb T = 1,800 lb
S = 3½ mph f = 5% downgrade W = 4,000 lb
Plow elasticity and yielding of obstacle = 2 in.
Maximum spring stress of 110,000 psi and K factor of 1.22

Type of hitch	Extension needed, $d - 2$, in.	Approx. weight of spring, lb
Hydraulic with holding device....................................	4⅛	
Spring with 1,500-lb preload..	11½	73
Spring with 1,500-lb preload and declutcher..........................	4⅝	33
Spring with holding device, no preload, with declutcher...............	6⅝	33

Field tests indicate that a solid-steel share will stand a force of about 5,000 lb without damage, and a chilled-iron share will stand about half as much.[2]

It is apparent that in soil which requires a draft of 1,000 lb per bottom (not an uncommon condition), the total draft for a three-bottom plow will be 3,000 lb, or enough to break a cast-iron share if a rock is encountered. In the case of three-bottom and larger plows, individual frogs and beams must withstand the entire force of the tractor when one bottom hits an obstruction, and a point is soon reached where no type of protective hitch is adequate.

Breakback Mechanisms. The obvious answer to the problem of multiple-bottom plows is the same method which has long been in common use on cultivators—individual breakback devices on each beam. A plow beam with a breakback mechanism is shown in Fig. 3-20a. A cultivator spring trip shank is shown in Fig. 3-20b, and a field-tiller spring shank in Fig. 3-20c. A mower cutter-bar release mechanism which allows the cutter bar to be swung back by an obstacle is shown in Fig. 3-20d.

In general, such devices should hold with practically no movement until breakback

---

[1] *Ibid.*
[2] A. W. Clyde, Tractor Stop Hitches, *Agr. Eng.*, 23:5-8, January, 1942.

but require little force to push them on back, once motion has started.[1] This action may be secured by a combination of spring pressure and cam action, with the cam follower sliding, as in Fig. 3-20d, or having a roller.

Another method utilizes jack-knifing links with the actuating force depending on a combination of joint-angle adjustment and spring pressure as shown in Fig. 3-20b.

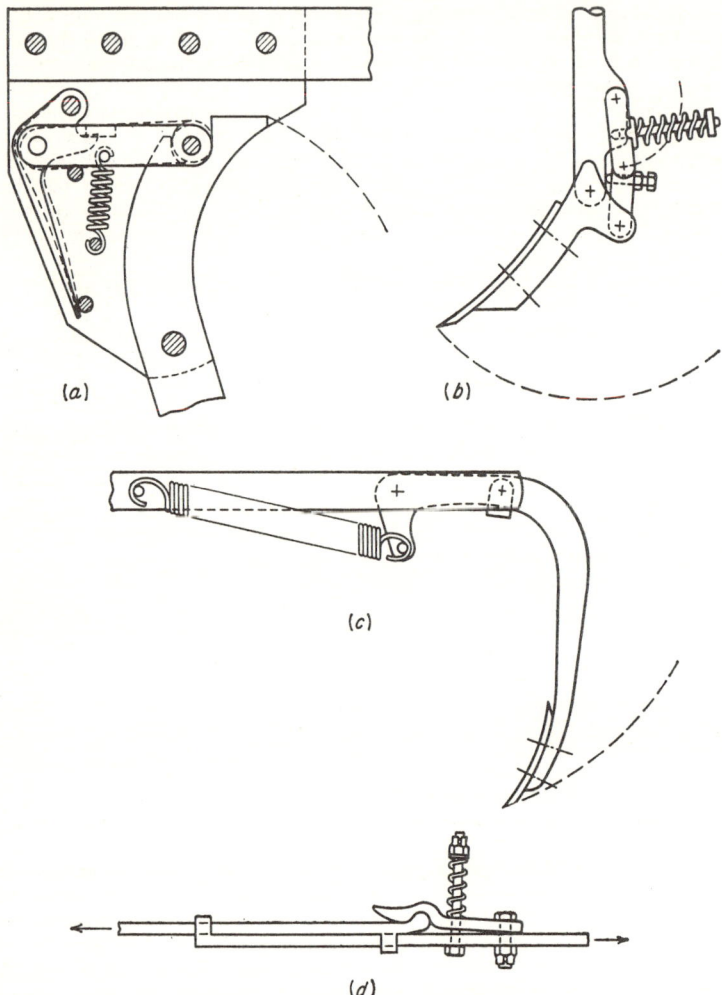

Fig. 3-20. Breakback mechanisms: (a) plow beam trip, using leaf spring to initiate movement of holding roller out of notch in top of beam (*Deere*); (b) cultivator spring trip; (c) field-tiller spring shank; (d) mower cutter-bar release. (*International Harvester Co.*)

Because of the large deflection after breakback, the holding load applied by the spring is often small. This results in primary dependence on joint angle with attendant variations in action due to pin friction effects.

Friction is an important factor in the breakout force required in all three mechanisms. This is particularly true in the sliding-cam arrangement where variations in

---

[1] A. W. Clyde, Spring Trip Cultivator Shanks, *Agr. Eng.*, 19:315–316, July, 1938. O. E. Johnson, Design Factors in a Spring Trip Beam Assembly for Moldboard Plows, *ASAE Paper* 58-59. R. W. Wilson, Discussion, *ASAE Paper* 58-59A. W. H. Silver, Discussion, *ASAE Paper* 58-59B, June, 1958.

smoothness of surfaces and degree of lubrication may make the required breakout force too variable and unpredictable for satisfactory protection. With the roller cam follower a reduction in friction is possible, as is true with the jack-knifing linkage. With both of these, a reduction in bearing pin friction through the use of lubrication, hardened surfaces, or antifriction bearings is beneficial in securing consistent action.

Slip Clutches. Overload protection is important in the driving of machine parts which are subject to clogging, e.g., corn-picker rolls and elevators. Cast-iron *jump clutches* (sometimes called "clatter" clutches) with rounded mating jaws held together by spring pressure, as shown in Fig. 3-21a, are commonly used for this purpose. They are set to slip before the other parts of the power train fail, and the noise made when they slip notifies the operator of clogging so he can immediately stop and remedy the situation. The friction factor is high in this type of clutch, and its action

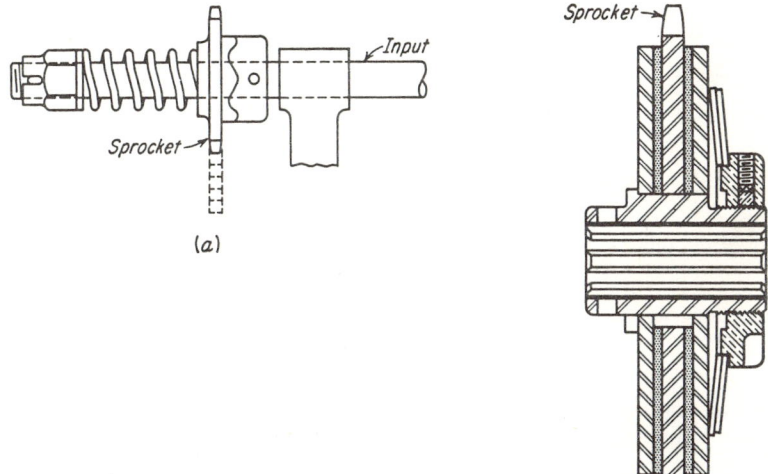

Fig. 3-21. Slip clutches: (a) jaw-type jump clutch; (b) friction type. (*Morse Chain Co.*)

is often inconsistent. Also, the hammering action as it slips may result in higher impact loads than the original breakout load.[1]

Conventional *friction-plate clutches*, such as shown in Fig. 2-21b, are sometimes used for overload protection with the spring pressure adjusted to drive normal loads and slip under abnormal loads. These have the advantage of more consistent breakaway torque with no damaging peaks during slippage. Tests have shown that the dynamic torque capacity under sudden load application is approximately twice the static value.[2] In addition torque may increase to twice normal or more after periods of idleness, because of corrosion of the plates. It is good insurance against breakage to free the clutch before starting a machine at the beginning of the season. Balers use this type of clutch to dampen severe torque fluctuations due to uneven universal-joint angularity in the PTO drive when making sharp turns.

[1] Sherman C. Heth, Development of Safety Clutch for Tractor PTO Drives, *SAE Preprint* 589, September, 1955.
[2] Martin A. Erickson, Strength Considerations in Agricultural Drive Lines, *SAE Preprint* 95U, September, 1959.

Chapter 4

WHEELS AND TIRES

The function of a transport wheel is to carry a load, with minimum energy expended to overcome rolling resistance. A driving wheel must also have adequate traction to exert the necessary propelling force.

ROLLING RESISTANCE OF RIGID WHEELS

The rolling resistance of a rigid wheel operating on a friable medium, such as soil, is due to the following conditions.

1. *Friction in axle bearing*, normally a minor item. One series of tests found a differential of 0.01 between the coefficients of rolling resistance for plain bearing and tapered roller-bearing wheels on a farm wagon.[1]

2. *Displacement of soil*, the major factor. A moving wheel will often sink deeper than a stationary wheel with the same load because of the loss of support area behind the lowest point. The angle of the resultant soil reaction, as shown in Fig. 4-1, varies according to the depth of compaction, the wheel diameter, and the angle of the soil resultant pressure against the rim.

3. *Friction between wheel and soil*, important only where side friction results from cutting in deeply.

4. *Adhesion*, which becomes important in sticky soil carried up by the wheel.

5. *Impact* of a rigid wheel against a projection. It tends to throw the wheel clear of the ground and to crush the projection. The draft of a farm wagon with steel wheels running on a gravel road increased 25 to 50 per cent when speed was increased from 2½ to 5 mph.[2] The height of free fall which will produce an equivalent impact force F is given by the following formula:[3]

$$H = 0.0334 S^2 \tan^2 \theta \sec \theta$$

where H = equivalent height of fall, ft
S = speed, mph
θ = angle of impact measured between a vertical radius and a radius through point of contact with obstacle, deg (Fig. 4-2)

Rolling-resistance Tests. McKibben ran an exhaustive series of tests of steel wheels of diameters ranging from 16 to 60 in. and widths from ¼ to 14 in. operating on various surfaces.

It was found that the rolling resistance tended to be inversely proportional to a soil's resistance to penetration or to its volume weight, as shown in Fig. 4-3. A transport wheel moves soil particles downward, ahead, and to each side in curved paths. It has negative slippage, its circumference being less than the actual distance traveled

---

[1] G. W. McCuen and E. A. Silver, Rubber-tired Equipment for Farm Machinery, *Ohio Agr. Expt. Sta. Bull.* 556, 1935.
[2] *Ibid.*
[3] E. G. McKibben, Some Kinetic and Dynamic Studies of Rigid Transport Wheels for Agricultural Equipment, *Iowa State Univ. Agr. Expt. Sta. Bull.* 231, 1938.

in one revolution. Increases in speed up to 5 mph on agricultural soils increased rolling resistance only slightly.

$$\text{Coefficient of rolling resistance} = \frac{\text{force to pull load, lb}}{\text{load, lb}}$$

The above coefficient tended to remain constant regardless of load, like coefficient of friction, only as long as penetration into the soil was slight by reason of large diameter, width, or compact soil. Coefficient of rolling resistance increased with load when penetration increased markedly with load, the normal condition.

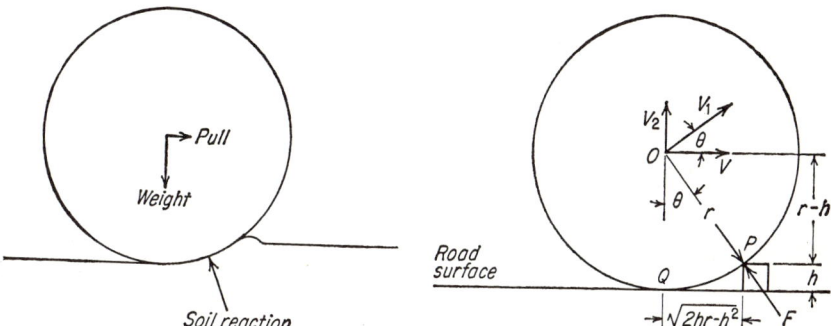

Fig. 4-1. Resistance to a wheel rolling on soil.

Fig. 4-2. Rigid wheel striking a solid obstruction. (*McKibben, Iowa State Univ. Agr. Expt. Sta. Bull. 231.*)

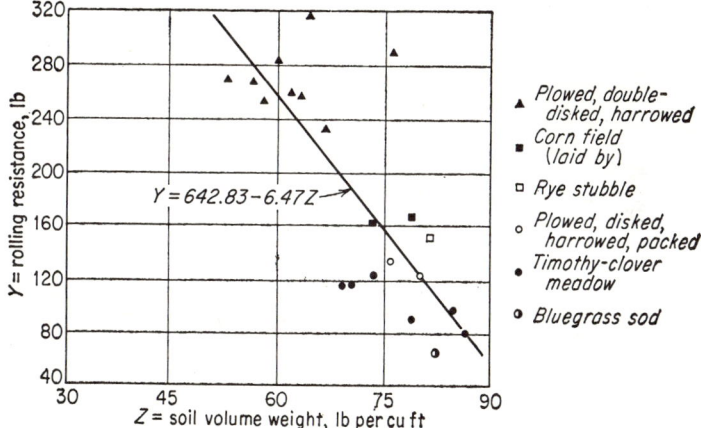

Fig. 4-3. Soil volume weight and rolling resistance for 36- by 2½-in. wheel and 1000-lb load on various surfaces. (*McKibben, Iowa State Univ. Agr. Expt. Sta. Bull. 231.*)

Increased *diameter* always reduced rolling resistance. Increased width usually decreased rolling resistance except when the wheel penetrated through a soft surface layer to be supported by a firm subsurface layer. The rolling resistance of various diameters and widths of wheels on *meadow* is shown in Fig. 4-4 and on *tilled soil* in Fig. 4-5.[1] Tests of three 4- by 36-in. steel wheels with flat, concave, and convex rims on various soils showed the concave rims to have 2 per cent more rolling resistance and the convex 9 per cent more than the flat rim.[2]

[1] *Ibid.*
[2] E. G. McKibben and J. B. Davidson, Effects of Steel Wheel Rim Shape and Pneumatic Tire Tread Design on Rolling Resistance, *Agr. Eng.*, 21:139–140, April, 1940.

62 CROP-PRODUCTION EQUIPMENT

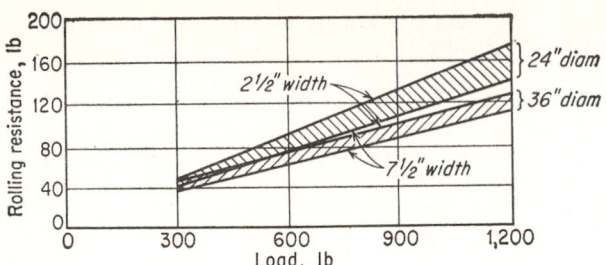

Fig. 4-4. Rolling resistance of 24- and 36-in.-diameter steel wheels, 2½ to 7½ in. wide, on meadow. (*McKibben, Iowa State Univ. Agr. Expt. Sta. Bull. 231.*)

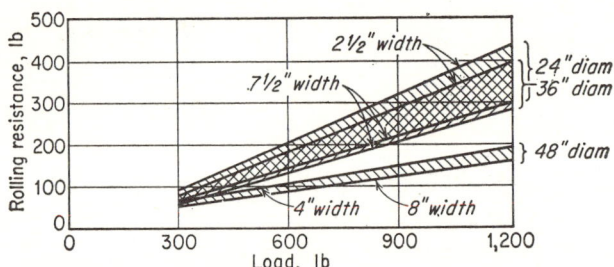

Fig. 4-5. Rolling resistance of 24-, 36-, and 48-in.-diameter steel wheels of various widths on plowed, disked, and harrowed loam. (*McKibben, Iowa State Univ. Agr. Expt. Sta. Bull. 231.*)

ROLLING RESISTANCE OF PNEUMATIC TIRES

The rolling resistance of pneumatic tires is caused by the same factors listed above for rigid wheels plus the factor of energy loss due to flexing of the tire during ground contact. Normally, rolling resistance is markedly lower than for steel wheels because:

1. Impact effects are reduced, since the tire accommodates itself to hard-surface irregularities rather than bouncing over them.
2. Soil displacement is reduced because tire deflection increases the support area, thus reducing penetration under most conditions.

Rolling-resistance Tests. Rolling-resistance coefficients for various pneumatic and steel wheels on concrete, bluegrass, tilled loam, and sand at loads near their recommended capacities are shown in Fig. 4-6.[1]

Reducing the *inflation pressure* increases the work of flexing the tire but decreases the energy expended in displacing the soil. Rolling resistance is least when the unit pressure exerted on the soil by the tire is equal to the soil's unit bearing strength, resulting in no soil displacement. High inflation pressure is most efficient on concrete, and low pressure on loose soils. Practically, reduction in inflation pressure is limited by the deflection allowable without damage to the tire, and a point is soon reached where a larger tire must be used to secure greater support area.

Increased *outside diameter* consistently reduces rolling resistance, as shown in Fig. 4-6, because of:

1. Less tire-wall deflection for equal supporting area
2. Less forward displacement of soil because of flatter angle at front edge of contact area
3. Narrower track for equal support area resulting in less work to compress soil

"Varying the cross-section diameter, however, has relatively little effect . . . except in the case of very narrow tires operated on soft loose surfaces."[2]

[1] E. G. McKibben and J. B. Davidson, Rolling Resistance of Individual Wheels, *Agr. Eng.*, 20:469–473, December, 1939.
[2] E. G. McKibben and J. B. Davidson, Effect of Outside and Cross-sectional Diameter on the Rolling Resistance of Pneumatic Implement Tires, *Agr. Eng.*, 21:57–58, February, 1940.

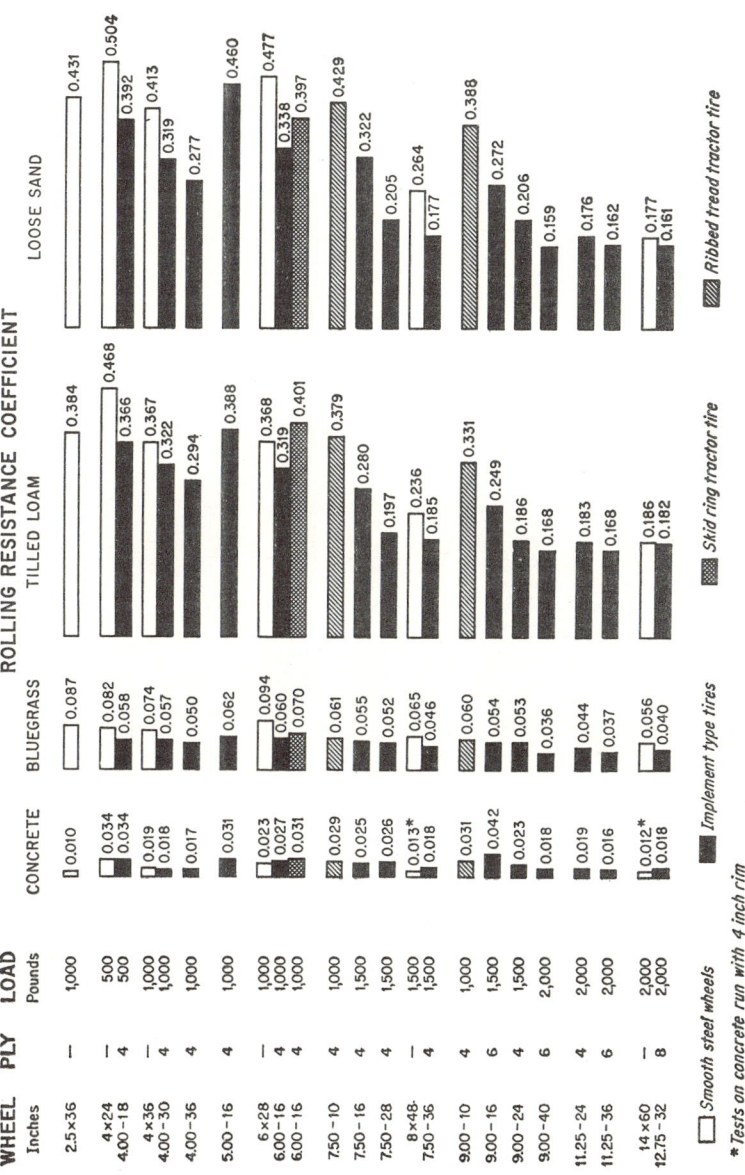

FIG. 4-6. Coefficients of rolling resistance for various-size wheels and tires. The loads selected are those nearest to the recommended capacities. It should be noted, however, that because of testing-equipment limitations these loads represent only one-third to one-half of the capacities of the 9.00-24 and larger tires. (*McKibben, Agr. Eng., December, 1939.*)

Tandem wheels had less rolling resistance than *duals* because the tandem arrangement allows the second wheel to run in the compacted track of the first. The reduction in rolling resistance for repeated runs in the same track is shown in Fig. 4-7.[1]

Compaction effects by steel wheels and rubber tires were studied by plotting the shift of small beads placed in molding sand. The steel wheel penetrated more deeply and caused more lateral displacement, as shown in Fig. 4-8. Forward displacement was also greater with the steel wheel.[2]

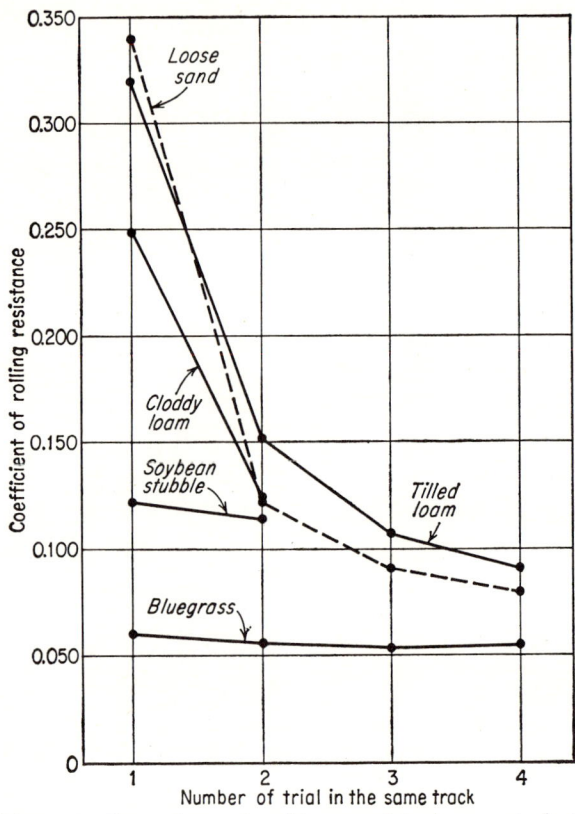

Fig. 4-7. Coefficients of rolling resistance for trials repeated in the same track. (*McKibben, Agr. Eng., March,* 1940.)

Slippage was found to be approximately proportional to the square of the rolling resistance, and the following formulas gave a reasonably close prediction:[3]

Slippage of steel wheels, % = 200 (coefficient of rolling resistance)^2
Slippage of rubber tires, % = 2 + 220 (coefficient of rolling resistance)^2

TRACTIVE EFFICIENCY OF WHEELS

A propelling wheel or track transforms the torque supplied by the engine into thrust against the soil. If thrust in excess of the rolling resistance is available, the vehicle

---
[1] E. G. McKibben and J. B. Davidson, Effect of Wheel Arrangement on Rolling Resistance, *Agr. Eng.*, 21:95–96, March, 1940.
[2] E. G. McKibben and R. L. Green, Relative Effects of Steel Wheels and Pneumatic Tires on Agricultural Soils, *Agr. Eng.*, 21:183–185, May, 1940.
[3] E. G. McKibben and J. B. Davidson, Effective Radius and Slippage, *Agr. Eng.*, 21:275–280, July, 1940.

WHEELS AND TIRES 65

can do useful work such as pulling a plow or pushing an attached cultivator through the soil.

The reaction of the soil to the weight and thrust of a wheel is complex and varies greatly with the soil type and condition. A smooth tire or track is limited to the

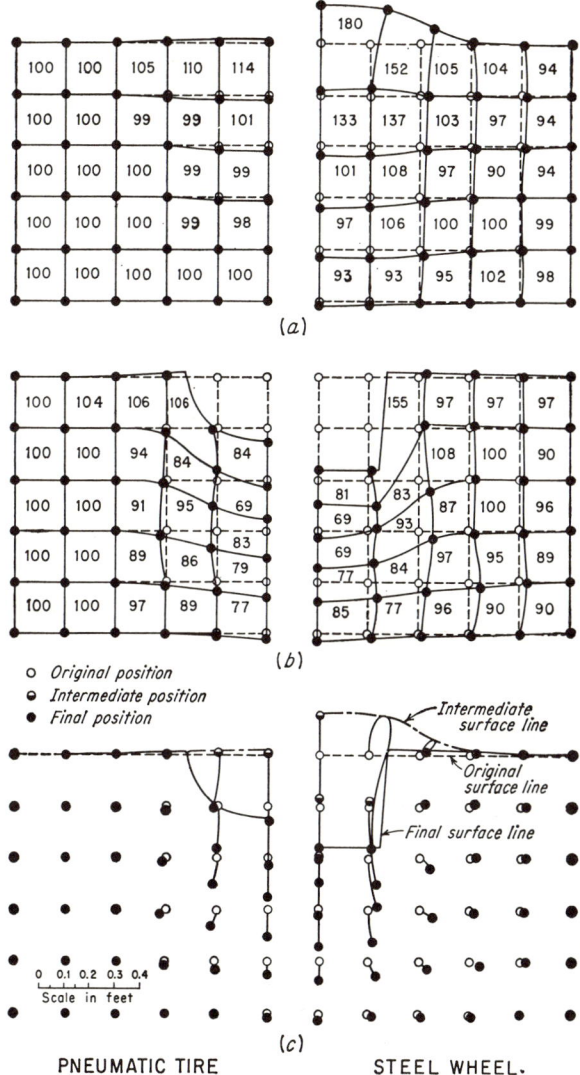

FIG. 4-8. Movement of soil particles in planes perpendicular to the direction of travel of a 6.00-16-in. pneumatic tire and a 6- by 28-in. steel wheel when carrying a 1000-lb load on Des Moines molding sand: (a) effects just ahead of wheels; numbers indicate per cent of original volume; (b) effects after wheels had passed; (c) paths taken by individual particles. (*McKibben, Agr. Eng., May,* 1940.)

thrust, which can be secured by friction alone, as is also the case with a cleated tire or track on a hard impenetrable surface. On agricultural soils, however, the cleats may penetrate the surface and take advantage of shearing strength as well as frictional resistance, although cleats do not increase traction if friction between the track or tire material and soil surface equals or exceeds the soil's shear strength.

The shearing strength of soil increases as it is pressed together either by vertical or horizontal pressure. The shearing characteristics of soil have been studied in a shear box such as shown in Fig. 4-9.[1] The effect of vertical loading on a particular soil is shown in Fig. 4-10. Soil moisture is also an important factor since shear strength increases with moisture content until the lower plastic limit is reached, after which it decreases rapidly. It has been found that, under a given vertical loading, the lateral shear strength of an agricultural soil builds up with horizontal displacement to a maximum, and then decreases as it shears loose, and resistance is due more to internal soil friction than to cohesiveness. Characteristic curves for several soils are

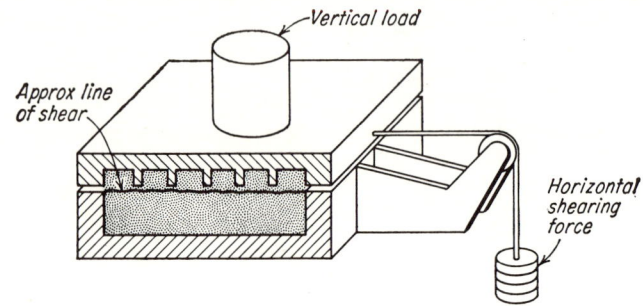

Fig. 4-9. Shear box for studying soil shear strength. (*Bekker*, "*Theory of Land Locomotion*," 1956.)

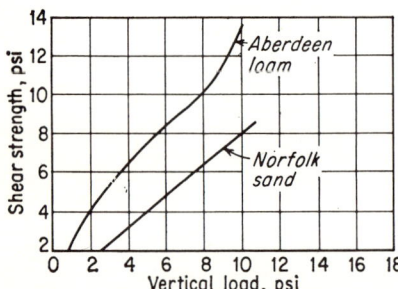

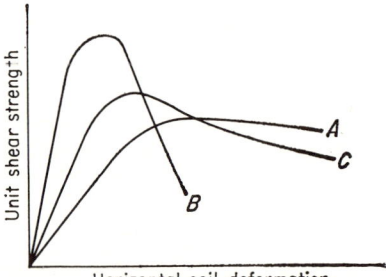

Fig. 4-10. Shear strength vs. vertical load. Shear strength in a nonplastic soil (Norfolk sand) varies as a straight-line function of the applied vertical load but not in a plastic soil (Aberdeen loam). (*Gross, SAE Journal, December,* 1946.)

Fig. 4-11. Typical curves for soil shear strength vs. horizontal deformation under a constant vertical load. (*Bekker,* "*Theory of Land Locomotion*", 1956.)

shown in Fig. 4-11. Curve *A* is characteristic of either loose frictional sands or wet plastic clays. The shearing strength builds up moderately during an initial period of deformation and, being primarily frictional, does not drop appreciably with additional displacement. Curve *B* is for a dry settled clay or firm silt in an undisturbed state. Less lateral deformation occurs before maximum shear strength is reached, but the strength drops rapidly after shearing off. Curve *C* is for a soil having intermediate properties.[2]

The action of cleats on a track is shown in Fig. 4-12. The horizontal soil displacement by an individual cleat varies from zero at the point of initial contact to a maximum, depending on the per cent slippage and the length of track, at the point of cleat withdrawal. It can be seen that the tractive effort exerted by a cleat varies throughout the period of its engagement with the soil according to the amount of

[1] W. A. Gross and E. D. Elliott, Soil Shear is Key to Army Vehicle Traction, *SAE Journal,* December, 1946, pp. 26–28.
[2] M. G. Bekker, "Theory of Land Locomotion," University of Michigan Press, Ann Arbor, Mich., 1956, p. 264.

soil deformation and the correspondig shear strength of the particular soil under this vertical loading. A firm brittle soil such as *B* in Fig. 4-11 may have so little strength after breaking that the tractive effort is largely concentrated at the front of the track.[1]

It follows that maximum traction per cleat for the amount of soil deformation is secured when the total deformation does not exceed that for peak shear strength. The per cent slip to give this deformation varies with the length of contact area. For instance, a maximum deformation of 2 in. will be secured with 2 per cent slip of a 100-in. track or 20 per cent slip of a rubber tire having a contact length of 10 in. Total traction will, of course, be proportional to the total contact area, assuming the same unit vertical loading. Since the work expended in displacing the soil is approximately the same in both cases, assuming equal widths, and the track has ten times the

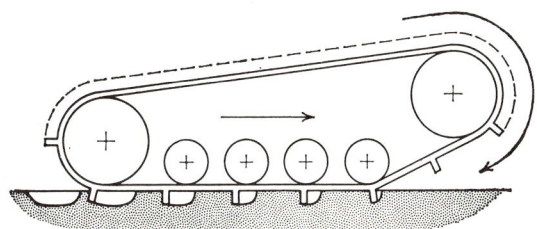

FIG. 4-12. Action of track-layer cleats in displacing soil. (*Bekker, "Theory of Land Locomotion," 1956.*)

traction, its tractive efficiency will be much higher. The advantage of a narrow large-diameter wheel over a wide small-diameter wheel applies to traction as well as to rolling resistance.

Sources of energy loss in doing drawbar work are:
1. Rolling resistance of tractor wheels
2. Work of compressing, shearing, and digging soil, because of thrust action of drive wheels
3. Deflection of tire sidewalls and lugs, with rubber tires
4. Simple slippage in the case of hard surfaces which are not displaced by the wheel thrust

The *force efficiency* of a traction wheel is defined as the ratio of thrust developed to the force input and can be found as follows:

$$\text{Force efficiency, \%} = \frac{(\text{wheel thrust, lb})(\text{loaded radius, in.})100}{\text{axle torque, in.-lb}}$$

Slippage or *travel reduction* is also a factor in energy loss.

$$\text{Travel reduction, \%} = (1 - \text{travel ratio})100$$

$$\text{Travel ratio} = \frac{\text{advance per wheel revolution } \textit{with} \text{ drawbar load}}{\text{advance per wheel revolution } \textit{without} \text{ drawbar load}}$$

Tractive efficiency of a traction wheel (also called power efficiency) is defined as the ratio of work output of the wheel thrust to the work input at the axle, or

$$\text{Tractive efficiency, \%} = (\text{force efficiency, \%})(\text{travel ratio})$$

In determining tractive efficiency of a complete tractor, parasitic rolling resistance of the front wheels is accounted for by using drawbar pull rather than drive wheel thrust. High efficiency depends on securing maximum thrust with least rolling resistance and slippage or soil working.

The ability of a wheel to develop tractive effort is expressed by

$$\text{Coefficient of traction} = \frac{\text{wheel thrust, lb}}{\text{weight on wheel, lb}}$$

[1] *Ibid.*

Tests of *steel wheels,* 38 to 58 in. in diameter with rims 12 in. wide, 4-in.-long lugs, and 1,945-lb static load showed coefficients of traction ranging from 0.185 to 0.283, travel reductions from 15 to 20 per cent, and maximum tractive efficiencies from 29 to 40 per cent in loose soil. In oat stubble on silty clay loam, coefficients of traction ranged from 0.58 to 0.62, travel reduction from 8 to 12 per cent, and tractive efficiencies from 58 to 69 per cent. The large diameters were most efficient.[1]

Progress has been made in developing formulae and simple field measurements of soil shear strength and sinkage values which can be used to predict wheel and track performance over a wide range of conditions.[2]

Pneumatic Rubber Tires. Rubber tires have almost completely displaced steel wheels on farm equipment because of their greater tractive efficiency, reduced rolling resistance, cushioning effect, and higher allowable transport speeds. Their greater trac-

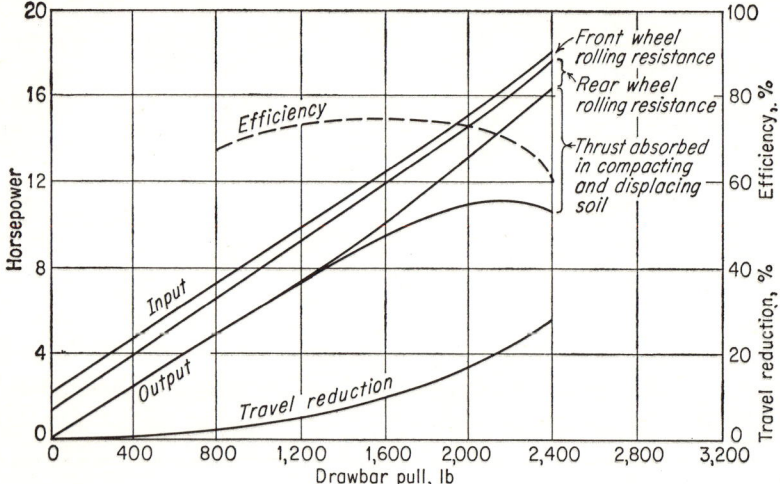

Fig. 4-13. Disposition of input horsepower of rubber-tired tractor on clay loam stubble: coefficient of traction (maximum drawbar horsepower), 0.63; coefficient of rolling resistance: front wheel, 0.108, rear wheel, 0.078. (*O'Harrow, SAE Quart. Trans., January, 1947.*)

tive efficiency is due primarily to the reduced rolling resistance. Over a large number of independent tests, normal loads could usually be pulled in the next higher gear than with steel wheels, resulting in an average saving of approximately 15 per cent in time and fuel.[3]

Since 4-in. lugs increase the rolling resistance of smooth steel wheels from 10 to 20 per cent in loose soil and over 100 per cent in firm soil, the difference between steel and rubber tractor drive wheels is much greater than indicated in Fig. 4-6 for smooth transport wheels.

The division of input horsepower between drawbar output, rolling resistance, and energy lost in soil displacement and slip at various loads on a clay loam stubble soil is shown in Fig. 4-13.[4]

The performance of 11-38 tractor tires on several types of soil and on concrete

---

[1] J. B. Davidson, E. V. Collins, and E. G. McKibben, Traction Efficiency of the Farm Tractor, *Iowa State Univ. Agr. Expt. Sta. Research Bull.* 189, 1935.

[2] M. G. Bekker, "Off-the-road Locomotion Research and Development," University of Michigan Press, Ann Arbor, Mich., 1960.

[3] G. W. McCuen, Ohio Tests of Rubber Tractor Tires, *Agr. Eng.*, 14:41–44, February, 1933. C. W. Smith and L. W. Hurlbut, A Comparative Study of Pneumatic Tires and Steel Wheels on Farm Tractors, *Agr. Eng.*, 15:35–48, February, 1934. R. H. Wileman, Effect of Tire Size on Tractor Efficiency, *Agr. Eng.*, 19:27–28, January, 1938.

[4] C. T. O'Harrow, Traction Efficiency, *SAE Quart. Trans.*, 1:71–75, January, 1947.

is shown in Fig. 4-14.[1] Work or tractive efficiency at maximum drawbar pull varied from 70 per cent on concrete to 30 per cent on loose sand.

The effects of *lug height* on tractor tire performance were investigated on freshly worked and packed soils in the soil bins of the USDA Tillage Laboratory, Auburn, Ala. Open angular lugs, ½, 1, and 1½ in. in height were tested on 11-38, 4-ply tires inflated to 12 psi with a static load of 2,185 lb. The short lugs had greater force efficiency, greater tractive efficiency, and less slippage. They exerted more drawbar pull in sand, but the high lugs exerted more drawbar pull on loam and clay.[2] It is generally recognized that, under adverse conditions such as wet or muddy ground

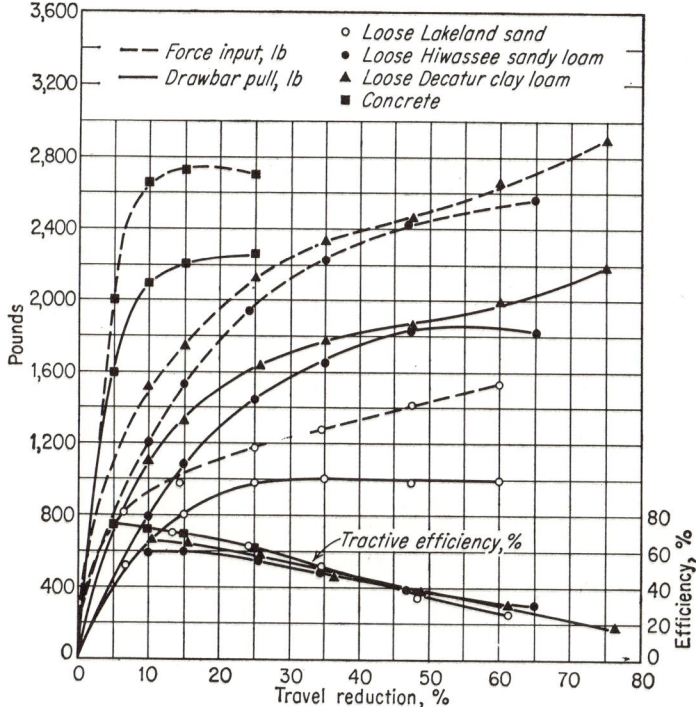

FIG. 4-14. Performance of 11-38 tractor tire, 12 psi air inflation with 2,325 lb static load, on various surfaces. (*Reed, Agr. Eng., June,* 1953.)

or green cover crops, shallow lugs will slip and high lugs tend to dig in until a firm soil layer is reached.

Tests indicate that the optimum *lug spacing* for best traction and highest efficiency is somewhat greater than is deemed acceptable from the standpoint of long wear and smooth riding on hard surfaces. A smooth tire without lugs is most efficient on dry hard surfaces or on dry loose soils where traction is due to friction rather than to shear strength of the soil. Traction of a rubber tire on rough concrete is influenced by the shear strength of the rubber and increases with the area of rubber contact, as evidenced by the increase in coefficient of traction with lowered inflation pressure shown in Table 4-2.

Tests of rubber tires with and without *chains* in loose soil showed a marked increase in shipping with chains.[3]

[1] I. F. Reed, C. A. Reaves, and J. W. Shields, Comparative Performance of Farm Tractor Tires Weighted with Liquid and Wheel Weights, *Agr. Eng.,* 34:391–399, June, 1953.
[2] I. F. Reed and J. W. Shields, The Effect of Lug Height and of Rim Width on the Performance of Farm Tractor Tires, *SAE Preprint* 504, September, 1950.
[3] Davidson, Collins, and McKibben, *op. cit.*

The coefficient of traction and tractive efficiency of the rear tires of a *four-wheel drive* is considerably higher than for the front tires, which do most of the work of compacting the soil. This corresponds to the results for tandem transport tires, discussed above. Two tractors assembled in tandem delivered up to 45 per cent more pull in dry cloddy soil than the total combined pull of the same tractors used conventionally and with up to 35 per cent higher tractive efficiency and consequent work output for the fuel consumed.[1]

Increased *weight* per tire reduces travel reduction for a given pull and increases the maximum pull unless flotation is inadequate.. In general, tests have indicated that for a given soil and per cent travel reduction, pull is approximately proportional to weight. The range of traction coefficients in numerous tests on various surfaces is shown in Table 4-1.

TABLE 4-1. AVERAGE TRACTION COEFFICIENTS OF RUBBER TIRES

Soil condition	Travel reduction, %	Traction coefficient at 12 psi		
		Low	High	Average
Concrete road	5	0.57	0.75	0.66
Dry clay	16	0.52	0.66	0.55
Sandy loam	16	0.45	0.58	0.50
Dry fine sand	16	0.29	0.42	0.36
Gravel road	5	0.32	0.41	0.36
Alfalfa	8	0.31	0.41	0.36

SOURCE: SAE Cooperative Tractor Tire Testing Committee, The Traction of Pneumatic Tractor Tires, *SAE Trans.*, 1938, pp. 13–26.

TABLE 4-2. EFFECT OF INFLATION PRESSURE ON TRACTION COEFFICIENTS

Soil condition	Average traction coefficients			Rating, %		
	8 psi	12 psi	16 psi	8 psi	12 psi	16 psi
Concrete road	0.71	0.66	0.61	108	100	92
Dry clay	0.56	0.55	0.53	102	100	96
Sandy loam	0.53	0.50	0.49	106	100	98
Dry fine sand	0.39	0.36	0.33	108	100	91
Gravel road	0.35	0.37	0.37	95	100	100
Alfalfa	0.39	0.36	0.35	108	100	97

SOURCE: SAE Cooperative Tractor Tire Testing Committee, The Traction of Pneumatic Tractor Tires, *SAE Trans.*, 1938, pp. 13–26.

Tests at the USDA Tillage Laboratory indicate that *liquid filling* of tires is as effective as an equivalent *metal wheel-weight* if the inflation pressure under load is the same. There is a tendency, however, for the pressure in the liquid-filled tires to build up more than in air-filled tires under the weight transfer resulting from pull. This increased pressure reduced performance slightly in sand.[2]

Extra weight results in higher rolling resistance and lowered efficiency at low drawbar pulls, as shown in Fig. 4-15. Ideally, weight should be adjustable to traction requirements, both for maximum efficiency and for minimum soil compaction.

Lowered *inflation pressure* increases support area and reduces vertical soil displacement. The average unit pressure of the tread on the soil is much greater than the tire inflation pressure, however, because the load is carried by the tire casing

---

[1] C. B. Richey, Traction Tests of a Tandem Tractor, *ASAE Trans.*, 21:16–17, 1959. I. F. Reed, A. W. Cooper, and C. A. Reaves, Effects of Two-wheel and Tandem Drives on Traction and Soil Compacting Stresses, *ASAE Trans.*, 21:22–25, 1959.

[2] Reed, Reaves, and Shields, *op. cit.*

walls as well as by the air pressure. Lowered inflation pressures reduce thrust absorption and rolling resistance enough on loose soils to give more drawbar pull for a given slippage. Higher inflation pressures reduce energy absorbed by tire deflection and give better lug penetration. Performance may be better on hard surfaces. Inflation pressures below 12 psi have the practical disadvantage of allowing excessive buckling of tire sidewalls, causing reduced tire life. Average traction coefficients for 8, 12, and 16 psi are shown in Table 4-2.[1]

Increased *section diameter* usually gives better flotation, reducing rolling resistance in loose soils, and may give a greater coefficient of traction where shear area engaged by lugs is a factor. From the practical standpoint, the extra capacity of the larger section for liquid weight or for added weight of any kind may be its greatest advantage in securing more traction.

Compaction Effects. There is increasing evidence that the soil-compaction effects of tractor tires are undesirable. Studies in California found a pronounced reduction in

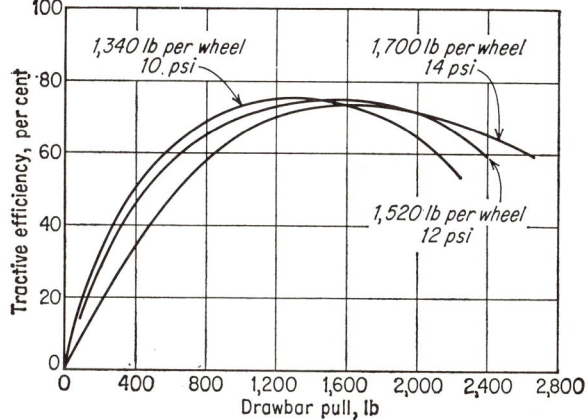

Fig. 4-15. Effect of tire load on tractive efficiency: 10-28 tires, clay loam stubble. (*O'Harrow, SAE Quart. Trans., January, 1947.*)

the water-infiltration rate of irrigated soil in the tracks of tractor wheels.[2] Compaction was reduced as the soil became drier. Successive passes in the same track increased compaction. The front wheels of a tricycle tractor reduced infiltration rate as much as the rear wheels, although they carried only half as much load.[2]

"The compactness and permeability of intensively cropped soil on Long Island within and below plow depth was shown to be determined by the traffic pattern imposed on the soil during the preceding three years. It was possible to detect and measure the compaction pattern even though all evidence of row location had been removed by leveling and planting a winter cover crop six months before."[3]

Measurements of compaction patterns under smooth rubber transport tires at the USDA Tillage Laboratory at Auburn, Ala., showed that a conventional 11-38 tire at 12 psi carrying 2,440 lb compacted a sandy soil slightly less than a 42-in.-diameter 60-in.-wide bag-type tire at 5.75 psi carrying 500 lb and having over five times the contact area.[4]

A strain-gage pressure cell has been developed to *measure soil pressures* at various depths due to a load applied to the surface. Pressures induced by a tractor tire at

---

[1] SAE Cooperative Tractor Tire Testing Committee, The Traction of Pneumatic Tractor Tires, *SAE Trans.*, 1938, pp. 13–26.
[2] L. D. Doneen and D. W. Henderson, Compaction of Irrigated Soils by Tractors, *Agr. Eng.*, 34:90–102, February, 1953. W. R. Gill, Soil Compaction by Traffic, *Agr. Eng.*, 40:392–402, July, 1959.
[3] George R. Free, Traffic Soles, *Agr. Eng.*, 34:528–531, August, 1953.
[4] William R. Gill and Carl A. Reave, Compaction Patterns of Smooth Rubber Tires, *Agr. Eng.*, 37:677–684, October, 1956.

various depths are shown in Fig. 4-16. This pressure curve was shaped like those calculated by Froehlich's formula for pressures under a circular plate:

$$P_d = P_s \left[1 - \left(\frac{d}{\sqrt{d^2 + r^2}} \right)^v \right]$$

where P_d = pressure under the load axis at depth d, psi
P_s = surface pressure applied, psi
d = depth under the surface, in.
r = radius of circular loading plate, in.
v = soil constant; $v = 4$ for hard dry soil, $v = 5$ for medium moisture and density, $v = 6$ for wet loose soil

This formula assumes that the load is supported by the frustrum of a cone of soil with a top area equal to the pressure plate. It indicates that induced pressures are more a function of total load than of surface area at depths below the plow sole. Con-

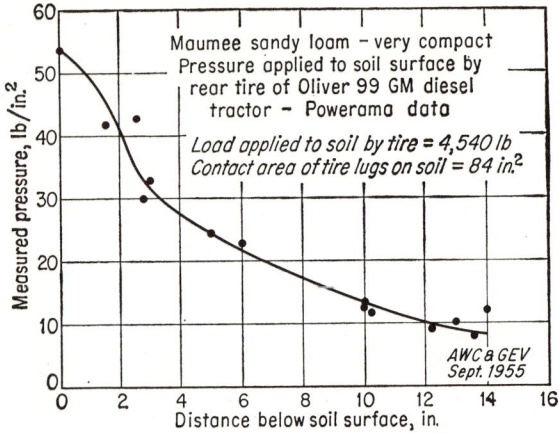

Fig. 4-16. Change with depth of soil pressure due to a tractor tire. (*Cooper, Agr. Eng., April, 1957.*)

sequently, larger tires cannot greatly reduce deep compaction by heavy loads.[1]

It is possible that part of the success of minimum tillage practices (Chap. 11) is due to a reduction in compaction by tractor tires.

Life. Tire life varies with service conditions, climate, and care, according to a study made by the USDA Bureau of Agricultural Economics.[2]

The years of life and average annual hours of use for various states gave an average life of approximately 7 years and 4,500 hr for rear tires with a high of 11 years and 5,500 hr in Washington and a low of about 4 years and 3,700 hr in the Mississippi Delta. Stumps, stones, under inflation, and high temperatures all contribute to lower tire life. New England and Wisconsin averaged 7 to 8 years and 3,500 hr, while the Corn and Wheat Belts were near the national average. Front tires on tricycle tractors had shorter lives than on four-wheel tractors and averaged about 70 per cent of the life of rear tires.

Selection of Traction Tires. A tractor tire is expected to transform full engine power into drawbar pull at normal plowing speed in normal soil without more than 15

---

[1] A. W. Cooper, et al., Strain Gage Cell Measures Soil Pressure, *Agr. Eng.*, 38:232–246, April, 1957. G. E. VandenBerg, Soil Pressure Distribution under Tractor and Implement Traffic, *Agr. Eng.*, 38:854–859, December, 1957. G. W. Trabbic, K. V. Lask, and W. F. Buchele, Measurement of Soil-Tire Interface Pressures, *Agr. Eng.*, 40:678–681. W. H. Soehne, W. J. Chancellor, and R. H. Schmidt, Soil Deformation and Compaction During Piston Sinkage, *ASAE Paper*, 59-100, June, 1959. G. E. VandenBerg and W. R. Gill, Pressure Distribution Between a Smooth Tire and the Soil, *ASAE Paper* 59-108, June, 1959. C. A. Reaves and A. W. Cooper, Stress Distribution in Soils Under Tractor Loads, *Agr. Eng.*, 41:20–31, January, 1960.

[2] Life of Tractor Tires, *Farm Implement News*, July 10, 1950, pp. 101, 146.

to 20 per cent slippage. In modern tractors the lower gears are furnished primarily for operation of PTO-driven harvesting machines under adverse conditions and tires are not expected to hold the full engine power at these speeds.

The following formula has been developed to determine the operating weight necessary on the two rear tires.:[1]

$$\text{Weight, lb} = \frac{3.75\ (\text{engine hp})(\text{tractive efficiency, \%})}{(\text{speed, mph})(\text{coefficient of traction})}$$

The figure obtained by this equation includes weight transfer from front to rear wheels due to drawbar pull, ranging from 15 to 25 per cent of drawbar pull, with an average of 20 per cent.

The tractive efficiency and coefficient of traction vary with the soil type and condition. The following figures have been used as representative by Shields:

	Sand	Sandy loam	Loam	Clay
Tractive efficiency, %	50	60	65	80
Coefficient of traction	0.30	0.40	0.50	0.65

The operating rear-wheel weight can be calculated for the desired speed and soil condition. Then tires are selected which will carry the required load at the desired inflation pressure, using the manufacturer's ratings.

TABLE 4-3. FARM TRACTOR AND IMPLEMENT DISK WHEELS

Diam, in.	Offset, in.	Pilot diam, in.	Bolts		Bolt-circle diam, in.	Series	Rim size	Max. tire size	Max. load, lb
			No.	Diam					
20	2¼	4.625	6	½	6	....	20 × 5.50F	7.50—20, 8-ply	2,670
20	1⅛	4.625	6	½	6	....	20 × 5.50F	7.50—20, 8-ply	2,670
18	2¼	4.625	6	½	6	....	18 × 5.50F	7.50—18, 8-ply	2,580
18	1⅛	4.625	6	½	6	....	18 × 5.50F	7.50—18, 8-ply	2,580
16	2¼	4.625	6	½	6	Light(L)	16 × 4.00E	6.00—16, 6-ply	1,520
							16 × 4.50E	6.50—16, 6-ply	1,740
16	2¼	4.625	6	½	6	Heavy(H)	16 × 4.50E	6.50—16, 8-ply	2,120
							16 × 6.00F	7.50—16, 8-ply	2,500
16	1⅛	4.625	6	½	6	Light(L)	16 × 4.00E	6.00—16, 6-ply	1,520
							16 × 4.50E	6.50—16, 6-ply	1,740
16	1⅛	4.625	6	½	6	Heavy(H)	16 × 4.50E	6.50—16, 8-ply	2,120
							16 × 5.50F	7.50—16, 8-ply	2,500
							16 × 6.00F	7.50—16, 8-ply	2,500
15	2¼	4.625	6	½	6	Light(L)	15 × 5.00K	7.10—15	1,765
							15 × 6.00L	7.60—15	1,765
15	2¼	4.625	6	½	6	Heavy(H)	15 × 6.00L	7.60—15	2,500
							15 × 6.50L	8.20—15	2,500
15	1⅛	4.625	6	½	6	Light(L)	15 × 3.00D	5.00—15, 4-ply	910
							15 × 5.00K	7.10—15	1,765
							15 × 6.00L	7.60—15	1,765
15	1⅛	4.625	6	½	6	Heavy(H)	15 × 6.00L	7.60—15	2,500
							15 × 6.50L	8.20—15	2,500
15	⅜	3.625	4	½	5	Light(L)	15 × 4J	5.90—15, 6-ply	1,250
							15 × 4½K	6.70—15, 6-ply	1,250
							15 × 4½KB	6.70—15, 6-ply	1,250
12	1¼	3.125	5	7/16	4½	Light(L)	12 × 3.00D	5.00—12, 2-ply	355
							12 × 3.00D	4.00—12, 4-ply	635
							12 × 4JA	5—12, 2-ply	315
							12 × 5JA	6—12, 2-ply	395

SOURCE: ASAE-SAE Recommendations, *Agr. Eng. Yearbook*, 1959.

[1] J. W. Shields, Selecting Rear Tires for Farm Tractors, *Agr. Eng.*, 33:485–486, August, 1952.

WHEEL STANDARDS

The following wheel standards have been approved:[1]

ASAE Recommendation: Wheel-mounting Elements for 8-bolt, 8-inch Bolt-circle Farm Implement Disk Wheels
ASAE Recommendation: Farm Tractor and Implement Disk Wheels
ASAE Recommendation: Preferred Drive Wheel Tire and Rear Sizes for General Purpose Farm Tractors
ASAE Recommendation: Interchangeability of Disk Halves for Farm Implement Press and Gage Wheels
ASAE Standard: Rim Contours for Agricultural Press and Gage Wheels
ASAE Standard: Agricultural Press and Gage Wheel Tires
ASAE Standard: Agricultural Planter Press Wheel Tires

The tractor and implement disk-wheel dimensions are summarized in Table 4-3. The preferred drive wheel tire and rim sizes are given in Table 4-4.

TABLE 4-4. PREFERRED DRIVE WHEEL TIRE AND RIM SIZES FOR GENERAL-PURPOSE FARM TRACTORS

Tire designation	Loaded radius, in.	Section width, in.	OD, in.	Bead seat diam, in.	Rim width, in.
Group 1:					
15.5–38	28.5	15.5	61.6	38.188	14
13.6–38	28.5	13.6	61.6	38.188	12
12.4–38	27.8	12.4	59.8	38.188	11
Group 2:					
13.9–36	26.8	13.9	57.8	36.188	12
12.4–36	26.8	12.4	57.8	36.188	11
11.2–36	26.0	11.2	55.3	36.188	10
Group 3:					
11.2–34	25.0	11.2	53.3	34.188	10
Group 4:					
13.6–28	23.5	13.6	51.6	28.188	12
14.9–28	24.5	14.9	54.0	28.188	13
Group 5:					
12.4–28	22.8	12.4	49.9	28.188	11

NOTE: The tire section dimensions shown are obtained with the rim widths as specified. All dimensions are subject to standard tolerances of tire and rim manufacturers.
SOURCE: ASAE-SAE Recommendations, *Agr. Eng. Yearbook*, 1959.

[1] *Agr. Eng. Yearbook*, 1959.

Chapter 5

STRESS MEASUREMENTS AND STRENGTH ANALYSIS

Martin A. Erickson

Modern experimental techniques make it possible to measure loads and stresses occurring in machine parts while they are in operation. The information is usually more complete and accurate than can be obtained by a theoretical strength analysis. The latter, however, is still essential for proportioning the members of the first prototype at the beginning of a design, and new design variations can often be accurately evaluated by calculations based on experimental results.

The steps used in the *experimental approach* are very similar to those used in the theoretical strength analysis, namely:

1. Evaluation of *loads* as to magnitude, direction, rate, and frequency of application.
2. Evaluation of *working stresses* occurring in the machine members as a result of the applied loads. The determination of these stresses as to magnitude, direction, and distribution is commonly referred to as *stress analysis*.
3. Evaluation of the *permissible* or *allowable stress* of the material of the part.
4. Comparison of the working stresses with the allowable stress to determine *service life*.

Stresses (force per unit area) obtained by the experimental method are calculated from measured strains (elongation per unit length).

TECHNIQUES AND EQUIPMENT FOR EXPERIMENTAL ANALYSIS

Brittle Coatings. The brittle-coating method of stress analysis is based upon the occurrence of visible cracks in a special lacquer or coating at predetermined values of tensile strain. When this coating is applied to machine parts it will register location, direction, and approximate magnitude of the strains resulting from applied loads. The brittle coating most widely used in experimental stress analysis is a lacquer known commercially as Stresscoat, manufactured and marketed by the Magnaflux Corp.

Brittle coatings are *very sensitive to temperature* and *humidity*. High ambient temperatures and high relative humidity tend to insensitize the lacquer, necessitating a higher load application to produce indications (cracks in the coating). Under these conditions the part coated may yield before indications are produced or cracks which appear under load may disappear when the load is removed. On the other hand, low ambient temperature and low relative humidity generally increase the sensitivity. Sudden cooling (4 to 5°F below the normal or design temperature) of the coating will often produce crazing, thus making the coating useless for further strain indications. Obviously, sudden temperature and humidity fluctuations must be avoided and testing must be done near the planned temperature and humidity for good results.

Stresscoat can be purchased in formulations designed for use at various temperatures ranging from 5 to 130°F combined with relative humidities varying from 0 to 100 per cent.

Before applying Stresscoat, *the parts must be carefully prepared* by removing from the surfaces all foreign matter such as paint, scale, weld spatter, grease, etc. Irregularities produced by blowholes, scale, poor welds, bad casting practice, etc., are generally removed by sand blasting, grinding, etc. Next, all surfaces are thoroughly degreased by means of a solvent such as trichlorethylene or an equivalent degreasing solvent. After masking all machine-part contact faces with tape to prevent deposits of coating, the parts are sprayed with an aluminum undercoating ST-840. This undercoating provides a background for detecting Stresscoat cracks. A Stresscoat lacquer is then selected for the anticipated temperature and relative humidity. This lacquer is applied by spraying or dipping, depending on the size and accessibility of the surfaces involved. Since the sensitivity of Stresscoat is also affected by coating thickness, the coating must be applied uniformly over the areas to be studied. Next the parts are assembled in the test machine or loading fixture and allowed to dry for a minimum of 6 hr at a temperature slightly above the anticipated test temperature.

The test procedure consists of applying loads to the parts under the planned coating-sensitivity conditions. Sensitivity, as indicated by tests of calibration bars, may usually be satisfactorily controlled by varying the test temperature. The load is usually applied in equal increments until the first crack is observed in the coating. The strain at which this crack occurs is evaluated by means of *calibration bars,* a *bending fixture,* and a *strain scale* furnished in the Stresscoat kit. The calibration bars, $\frac{1}{4}$ by 12 in. of steel or aluminum, are coated at the same time and with the same coating as the test part. At the time of the test, strain sensitivity of the coating is determined by loading the calibration bar as a cantilever beam, deflecting it a standard amount in the loading fixture. The bar is then placed in the strain scale, which indicates the strain in inches per inch applied throughout the length. The strain reading obtained in the region of the first crack at the loaded end is taken as the sensitivity of the coating on the part being tested.

Brittle coatings indicate tensile strains only. Compressive strains can be evaluated, however, by maintaining the test load on the part for 2 or more hours, giving the lacquer a chance to creep to a new neutral state, and then releasing the load. Compressive areas of strain will then be shown by virtue of a tensile fracture of the coating. Quantitative data are obtained by means of the calibration bar, which has been loaded in the loading fixture for the same period of time and then released, giving a pattern on the underside of the bar.

Stresscoat is also used to measure residual and assembly strains. Residual strains resulting from fabrication or heat-treatment are measured by drilling small holes in the area under consideration. Compressive strains appear as concentric circular cracks around the hole, while tensile strain appears as radial cracks emanating from the center of the hole.[1]

Compressive strains resulting from bolting parts together can be determined by disassembling a Stresscoated assembly. Tensile strains are indicated by cracks which appear in the coating when the Stresscoated parts are bolted together.

Stresscoat indications can usually be seen with proper light reflection. *Statoflux,* a powder also supplied by the Magnaflux Corp., is often applied to the test part to aid in the visual location of patterns. It does not affect the coating sensitivity and thus permits subsequent testing. Following a Stresscoat test, the patterns are generally dyed with a red dye ST-1300 A and photographed for record purposes. These photographs are very valuable guides for design modifications or for future redesigns.

Stresscoat, under controlled conditions, can usually indicate strains corresponding to stresses as low as 15,000 psi in steel, 7,000 psi in cast iron, 5,000 psi in aluminum, and 12,500 psi in malleable iron. The quantitative accuracy of any individual test is about ± 20 per cent. Greater accuracy can be obtained by averaging several tests. Brittle coatings, in general, are very valuable for measuring sharp stress concentrations.

The brittle-lacquer method of stress analysis is especially useful in locating areas of high strain and the direction of strain resulting from service loads. Thus the entire

---

[1] M. Hetényi, "Handbook of Experimental Stress Analysis," John Wiley & Sons, Inc., New York, 1950, pp. 655–658.

frame of a combine, corn picker, or tractor, for instance, can be coated and subjected to representative field service loads. Strain indications thus obtained are used as guides for the placement of strain gages for more accurate determination of service loads and stresses.

When it becomes necessary to use the brittle-lacquer method *out of doors*, as is generally the case when studying entire machines, extreme care must be exercised to prevent crazing of the coating. In cold weather the conditioning of the coating from the drying temperature down to operating temperature must be done gradually. In hot weather it becomes necessary to protect the coated surfaces from the direct rays of the sun by a cover of some kind. In outdoor tests under extreme conditions it is advisable to operate near a building so that a fast test run can start at the building, cover the test course, and end up in the building to avoid appreciable temperature changes in the coating.

To facilitate testing in variable environments, the Magnaflux Corp. has developed Stresscoat All-Temp, a ceramic strain-sensitive coating which is relatively insensitive

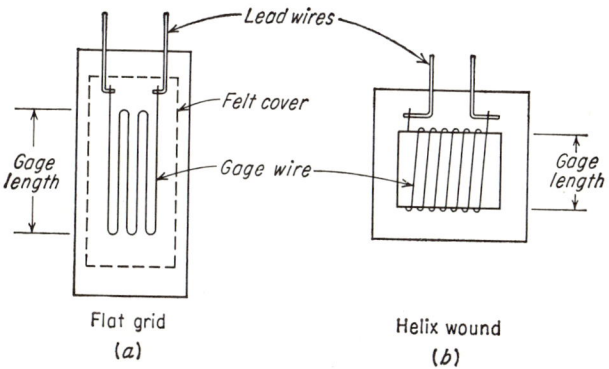

Fig. 5-1. SR-4 Strain gages: (*a*) flat grid; (*b*) helix-wound.

to temperature, humidity, and oils. It is applied by spraying and requires curing at 1000°F for 4 to 5 min. Its use is limited to materials which are not affected by the high curing temperature.

Strain Gages and Extensometers. These are mechanical, optical, or electrical elements used to measure *linear deformations* or strains over a specified length of material under load. Since quantities measured are usually very small, magnification is required. In the early devices, magnification was usually obtained by mechanical or optical means. Examples of these are the Huggenberger tensometer, Porter-Lipp strain gage, Whittemore fulcrum-plate strain gage, Tuckerman gage, Voce interferometer, G. M. photoelectric extensometer, and many others.[1]

Mechanical and *optical* strain gages have several disadvantages, namely:
1. They can only be used for static or steady applied load tests.
2. They are often difficult to apply to the part under study.
3. Gage lengths are too large for many applications.
4. Accuracy is generally low because of friction of joints, backlash of gears, and improper mounting of the gage.
5. Cost is high.

In recent years the *electrical* type of strain gage has been more highly favored than the mechanical or optical types. This is primarily due to its versatility of use, especially in *dynamic measurements*. It overcomes many of the disadvantages listed above and has enabled the designer to secure load data which were previously impossible to obtain.

[1] *Ibid.*, Chap. 3.

There are three types of electric strain gages, namely, the *resistance* type, the *inductance* type, and the *capacitance* type.

Resistance-type strain gages may be of the original SR-4 wire type or the more recent etched-foil type. SR-4 strain gages are of two basic designs, the *flat grid* for the larger size gages and the *helix-wound* for the smaller sizes, below ⅜ in. (Fig. 5-1). Gage wires may be made of Advance or constantan for stability (minimum drift) where static strains are involved or of isoelastic wire where high sensitivity and fluctuating strains are being studied. Table 5-1 shows properties of various wire materials used in strain-gage construction.

SR-4 strain gages may be made with nitrocellulose cement for average ambient temperature and humidity conditions or with phenol resin cements for temperatures up to 450°F, for high humidity or where greater permanence is required of the in-

TABLE 5-1. WIRE PROPERTIES*

Common name	Composition	Gage factor	Temp. coefficient of resistance†	Resistance, ohms/ft, in 1-mil diam	Stress equivalent to 10°C on steel, psi
Nichrome	Ni-0.80, Cr-0.20	2.0	300	638	2,000
Manganin	Ni-0.04, Mn-0.12, Cu-0.84	0.47	Nil	260	−400
Advance		2.1			−66
Copel	Ni-0.45, Cu-0.55	2.4	2	290	−200
Constantan‡		2.1			−60
Chromel-C	Ni-0.64, Fe-0.25, Cr-0.11	2.5		640	980
Isoelastic	Ni-0.36, Cr-0.08, Fe-0.52, Mo-0.005	3.5	175	680	5,000
Nickel		12.1	6,000	70	13,500§
Platinum		4.8	3,000	80	
Soft iron		4.2	5,000	68	
Carbon		20.0	−500	45,000	

* These data are not to be taken too literally, since most of the characteristics vary markedly with small changes in composition, with degree of cold-working, etc.
† Ohms per ohm per degree centigrade × 10⁶.
‡ Constantan is the name also applied to a 60–40 alloy with somewhat different properties.
§ Unstable.
SOURCE: C. C. Perry and H. R. Lissner, "The Strain Gage Primer," McGraw-Hill Book Company, Inc., New York, 1955, p. 17.

stallation. Adhesives most commonly used in making special gages or applying SR-4 gages are given in Table 5-2.

Etched foil gages, of the same materials used for wire gages, are finding wide usage because of their low transverse strain sensitivity, negligible hysteresis under cycling stresses, low creep during sustained deflections, and flexibility of application. They are made by using printed circuit techniques to produce a configuration similar to the flat grid gage shown in Fig. 5-1a. They are available in a variety of gage lengths, configurations, backings, and temperature adaptations.[1]

Good results require extreme care in surface preparation, in removal of oily films and moisture, in cleanliness, and in adhesive and gage application.[2]

There are many different SR-4 strain gages available commercially.[3] For the average laboratory operation an adequate selection of gages for almost any type

[1] SR-4 Etched Foil Strain Gages, *Bull.* 4320, Baldwin-Lima-Hamilton Corp., Waltham, Mass., February, 1959. T.M.S. Epoxy Back Metalfilm Strain Gages, *Bull.* BN-6001, BN-6001-P, Tatnall Measuring Systems Co., Phoenixville, Pa., December, 1958.
[2] C. C. Perry and H. R. Lissner, "The Strain Gage Primer," McGraw-Hill Book Company, Inc., New York, 1955, p. 31. How to Apply SR-4 Strain Gages. *Bull.* 279-B, Baldwin-Lima-Hamilton Corp., 1949.
[3] "SR-4 Strain Gage Instruments and Accessories: Specification and Price List 4310," Baldwin-Lima-Hamilton Corp., Waltham, Mass., Feb. 1, 1959.

of stress-analysis problem can be had from the 19 varieties shown in Table 5-3 compiled by W. T. Bean. Special high-temperature gages are available for tests that must be run at elevated temperatures. Other special-type gages include temperature-compensated gages and postyield gages.

In selecting a strain gage for an application, the following factors must be considered:

1. *Gage length.* Since the grid of the gage measures the average strain over its length, it becomes necessary to select a gage length compatible with the size of the area to be measured.

TABLE 5-2. STRAIN-GAGE CEMENTS

Cement	For use with gage types	Drying or curing time			Bonding pressure	Max. operating temperature, °F	Source
		Notes	Air-dry	Heat-dry			
Duco	A-1, C-1 A-51, C-5 A-19, C-19	a b c	24 hr 12–18 hr 48 hr	8 hr 5 hr 12 hr	1 lb	160	Department store
SR-4	A-1, C-1 A-5-1, C-5 A-19, C-19	a b c	12 hr 6 hr 24 hr	4 hr 3 hr 6 hr	1 lb	160	Baldwin-Lima-Hamilton Corp.
Bakelite BC-6035	AB, CB	5 hr with heat-curing			50–200 psi	450	Baldwin-Lima-Hamilton Corp.
Armstrong A-1	A, AB, C, CB	1 hr at 150–165° F			1 lb or less	200	Armstrong Products Co., Warsaw, Ind.
deKhotinsky	AB, CB	Fusible cement; hardens when cool, and gage can be used immediately			1 lb or less	125	Central Scientific Co., Chicago

a Regular-paper gages.
b Thin-paper gages.
c Small gages.
SOURCE: C. C. Perry and H. R. Lissner, "The Strain Gage Primer," McGraw-Hill Book Company, Inc., New York, 1955, p. 30.

2. *Gage factor.* The ratio of the change in unit resistance of a strain gage, as it is loaded by the surface to which it is attached, to the change in its unit length is known as the *gage factor,* or *strain sensitivity.* It is expressed as follows:

$$K = \frac{\Delta R/R}{\Delta L/L}$$

where K = gage factor
ΔR = change in gage resistance, ohms
R = gage resistance, ohms
ΔL = change in gage length, in.
L = gage length, in.

It is usually desirable to select for the highest gage factor in order to obtain maximum accuracy of measurement. This may not always be possible, however, because of other considerations, such as temperature drift, impedance match of indicating instrumentation, etc.

3. *Gage output.* Gages should be selected to give maximum output in order to obtain maximum accuracy. For example, if only the fluctuating component of strain is required, a type C gage with a gage factor of 3.5 would be used.

4. *Temperature characteristics.* The temperature conditions at which the test is to be run will determine the type of gage and bonding material required. For temperatures up to 450°F, the bakelite type AB gage is generally the best. Cellulose cement gages are reliable up to approximately 175°F (Table 5-3).

5. Creep characteristics. Creep (change in unloaded resistance evidenced by drift of zero) must be reckoned with where accuracy is required. Factors affecting drift include gage length, gage material, grid configuration, strain level, cement, bonding technique, operating temperature, time, excitation current, lead wire attachments, environment, and humidity.[1] Type ABD gages (dual leads) using bakelite or phenolic resin cement appear to give the best results as far as creep is concerned.

TABLE 5-3. BILL BEAN'S SR-4 SELECTION CHART

Gage type	Dimensions, in.			Gage factor	Resistance, ohms	Output,* mv per gage	Temperature, max.† °F		Remarks
	Length		Width, min.				Static	Dynamic	
	Gage	Paper							
A-19	0.062	0.56	0.12	1.7	60	2.5	±175	±200	For stress analysis of small fillets, etc.
A-18	0.125	0.75	0.18	1.8	120	5.4	±175	±200	Small, easy to apply
AD-7	0.250	0.87	0.18	1.9	120	5.7	±175	±200	All-purpose, good endurance
A-5-1	0.500	1.50	0.31	2.0	120	6.0	±175	±200	Thin paper, all-purpose
AX-5	0.500	1.75	0.75	2.0	120	6.0	±175	±200	Two 90° gages for torque; etc.
A-12-2	1.625	2.50	0.09	2.1	60	3.1	±175	±200	Single-strand, narrow gage
A-15	0.375	0.87	0.53	2.0	750	37.5	±175	±200	High output with zero stability
A-23	0.468	1.12	0.53	2.0	1,500	75.0	±175	±200	Highest output of any A gage
AB-19	0.062	0.56	0.12	1.7	60	2.5	±350	±500	Small gage for elevated temperatures
ABD-11	0.125	0.75	0.25	1.8	120	5.4	±350	±500	Small gage, good endurance
ABD-7	0.250	0.87	0.31	2.0	120	6.0	±350	±500	All purpose, good endurance
AB-20	0.500	1.12	0.62	1.9	1,000	47.5	±350	±500	Highest output of any AB gage
C-19	0.062	0.56	0.12	2.7	100	6.7	.....	±175	Small, high dynamic signal
CD-7	0.250	0.87	0.28	3.3	500	41.3	.....	±175	Small, all-purpose, good endurance
CD-3	1.000	2.50	0.53	3.3	500	41.3	.....	±175	Thin paper, good endurance
C-14	0.375	1.00	0.53	3.2	2,000	160.0	.....	±175	Highest output of any C gage
CB-19	0.062	0.56	0.12	2.9	100	7.2	.....	±250	Small, high output
CBD-7	0.250	1.00	0.28	3.3	500	41.3	.....	±250	Small, all-purpose, good endurance
CBD-10	0.312	1.00	0.50	3.2	1,000	80.0	.....	±250	Highest output of any CB gage

* Bridge output in millivolts per active gage is based on 50 ma current per gage and 1,000 microin./in. of strain. This current may be used only when the heat generated by the gage can be dissipated in the structure or by external cooling. The bridge-output formula is $E = KVN$ (four active legs), where E is output in microvolts, K is gage factor, V is supply voltage in peak volts, and N is strain in microinches per inch. This equation becomes more universal when K is expressed in microvolts per volt per unit of physical phenomena measured, namely, pressure, torque, weight, temperature, acceleration, displacement, etc., and N equals the number of units.

† While most SR-4 gages exhibit individual temperature characteristics, the approximate maximum operating temperature (in which the gage factor is maintained within ±5 per cent) is listed above. These gages do not compensate for the apparent strains due to temperature changes above or below ambient. Therefore, in static testing, correction must be made by establishing a curve of apparent strain vs. temperature for each gage type or by compensating with dummy gages of identical type mounted on identical material and maintained at the same operating temperature as the active gage.

NOTES: The above gages were selected from a group of 91 commercial gages on the basis of size, performance, fatigue resistance, adaptability, *and not cost*. They were selected because of the wide range of jobs they will do and not to match specific instrument requirements. It is hoped that this list will simplify gage selection by the beginner, and cut gage inventories.

SOURCE: C. C. Perry and H. R. Lissner, "The Strain Gage Primer," McGraw-Hill Book Company, Inc., New York, 1955, p. 39.

6. Endurance. Where strain gages are to be used over long periods of time at relatively high cyclic strains it becomes necessary to consider the fatigue strength of the gage and leads. Type ABD gages were developed for high fatigue resistance.

7. Cost. It is desirable to select a gage that can fulfill all the requirements and yet be low in cost. It would be poor economy, for example, to use the type A-19 gage ($\frac{1}{16}$-in. gage length) on a straight member where the cheaper type A-5-1 would be adequate.

[1] L. H. Schoenleber, Strain Gages and Stresscoat in Machinery Design, *Agr. Eng.*, 36:309–317, May, 1955.

TABLE 5-4. RELATIONS BETWEEN STRAIN ROSETTE READINGS AND PRINCIPAL STRESSES

Rosette types / Required solution	Two-gage	Rectangular	Delta	T-Delta
Maximum normal stress, σ max.	$\dfrac{E}{1-\mu^2}(e_1 + \mu e_2)$	$\dfrac{E}{2}\left\{\dfrac{e_1+e_3}{1-\mu} + \dfrac{1}{1+\mu}\sqrt{(e_1-e_3)^2 + [2e_2-(e_1+e_3)]^2}\right\}$	$E\left[\dfrac{e_1+e_2+e_3}{3(1-\mu)} + \dfrac{1}{1+\mu}\sqrt{\left(e_1 - \dfrac{e_1+e_2+e_3}{3}\right)^2 + \left(\dfrac{e_2-e_3}{\sqrt{3}}\right)^2}\right]$	$\dfrac{E}{2}\left[\dfrac{e_1-e_4}{1-\mu} + \dfrac{1}{1+\mu}\sqrt{(e_1-e_4)^2 + \dfrac{4}{3}(e_2-e_3)^2}\right]$
Minimum normal stress, σ min.	$\dfrac{E}{1-\mu^2}(e_2 + \mu e_1)$	$\dfrac{E}{2}\left\{\dfrac{e_1+e_3}{1-\mu} - \dfrac{1}{1+\mu}\sqrt{(e_1-e_3)^2 + [2e_2-(e_1+e_3)]^2}\right\}$	$E\left[\dfrac{e_1+e_2+e_3}{3(1-\mu)} - \dfrac{1}{1+\mu}\sqrt{\left(e_1 - \dfrac{e_1+e_2+e_3}{3}\right)^2 + \left(\dfrac{e_2-e_3}{\sqrt{3}}\right)^2}\right]$	$\dfrac{E}{2}\left[\dfrac{e_1+e_4}{1-\mu} - \dfrac{1}{1+\mu}\sqrt{(e_1-e_4)^2 + \dfrac{4}{3}(e_2-e_3)^2}\right]$
Maximum shearing stress, τ max.	$\dfrac{E}{2(1+\mu)}(e_1 - e_2)$	$\dfrac{E}{2(1+\mu)}\sqrt{(e_1-e_3)^2 + [2e_2-(e_1+e_3)]^2}$	$\dfrac{E}{1+\mu}\sqrt{\left(e_1 - \dfrac{e_1+e_2+e_3}{3}\right)^2 + \left(\dfrac{e_2-e_3}{\sqrt{3}}\right)^2}$	$\dfrac{E}{2(1+\mu)}\sqrt{(e_1-e_4)^2 + \dfrac{4}{3}(e_2-e_3)^2}$
Angle from gage 1 axis to maximum normal stress axis, φ_p	0	$\dfrac{1}{2}\tan^{-1}\left[\dfrac{2e_2-(e_1+e_3)}{e_1-e_3}\right]$	$\dfrac{1}{2}\tan^{-1}\left[\dfrac{\dfrac{1}{\sqrt{3}}(e_2-e_3)}{e_1 - \dfrac{e_1+e_2+e_3}{3}}\right]$	$\dfrac{1}{2}\tan^{-1}\dfrac{2(e_2-e_3)}{\sqrt{3}(e_1-e_4)}$

SOURCE: C. C. Perry and H. R. Lissner, "The Strain Gage Primer," McGraw-Hill Book Company, Inc., New York, 1955.

Single SR-4 strain gages placed at right angles to the Stresscoat indications are sometimes used to evaluate stress. This can be done with relatively small error under uniaxial loading on simple sections such as a bar in tension.

Under more complex conditions such as generally occur in castings, forgings, and fabricated members, it is necessary, for accuracy, to use a two-element cross gage, type AX, or two single-element gages, with one element placed at right angles to the Stresscoat indications and the other parallel to the indications. The two perpendicular planes containing the strain-gage elements in the above are called planes of *principal strain,* and the resulting stresses are known as the *principal stresses*. It will be found that the maximum and minimum stresses occur in these planes. They are evaluated as follows:

$$\sigma_x = \frac{E}{1 - \mu^2}(e_x + \mu e_y)$$

and

$$\sigma_y = \frac{E}{1 - \mu^2}(e_y + \mu e_x)$$

where σ_x = maximum principal stress, psi
σ_y = minimum principal stress, psi
e_x = maximum principal strain, in./in.
e_y = minimum principal strain, in./in.
μ = Poisson's ratio
E = modulus of elasticity, psi

For cases where the stress condition of a point is desired and the principal stress planes are not known, a strain-gage rosette is required in order to accurately evaluate the stress as to magnitude and direction. Four typical rosette configurations and their solution are given in Table 5-4.

The *wire* strain gage is generally used as a leg of a Wheatstone bridge in the *electrical circuit* as shown in Fig. 5-2. This type of circuit gives maximum temperature compensation and sensitivity. The output voltage of this circuit can be expressed as

$$E_b = KVe$$

where E_b = output voltage, μv
K = gage factor
V = bridge excitation, volts
e = unit strain, microin./in.

In terms of current

$$I_b = \frac{E_b}{R + R_g} = \frac{KVe}{R + R_g}$$

where I_b = bridge output, μamp
R = gage resistance, ohms
R_g = galvanometer resistance, ohms

Gages are usually arranged with (1) one active and three inactive, (2) two active and two inactive, and (3) four active.

The following applications are most common:

1. Measurement of strain at a point where the direction of principal strain is known and the effect of the minor principal strain is neglected as shown in Fig. 5-2a.

2. Measurement of strain to determine magnitude and direction of major and minor principal strains at a point, either by the use of two gages at right angles to each other where the direction of strain is known or by using a rosette where direction of strain is unknown

3. Measurement of strain to obtain average direct tension or compression in a bar in which a bending moment may or may not be present as in Fig. 5-2b

4. Measurement of the bending strain in a bar subjected to combined or pure bending moment as in Fig. 5-2c.

STRESS MEASUREMENTS AND STRENGTH ANALYSIS 83

5. Measurement of shear strain in a shaft subjected to static torsional or combined loads as in Fig. 5-2d

6. Measurement of strain in a shaft subjected to rotational loads by means of rotating contacts

The two types of rotating contacts commonly used are the slip ring with sliding contacts and the disk rotating in mercury. The slip-ring type, when made with coined

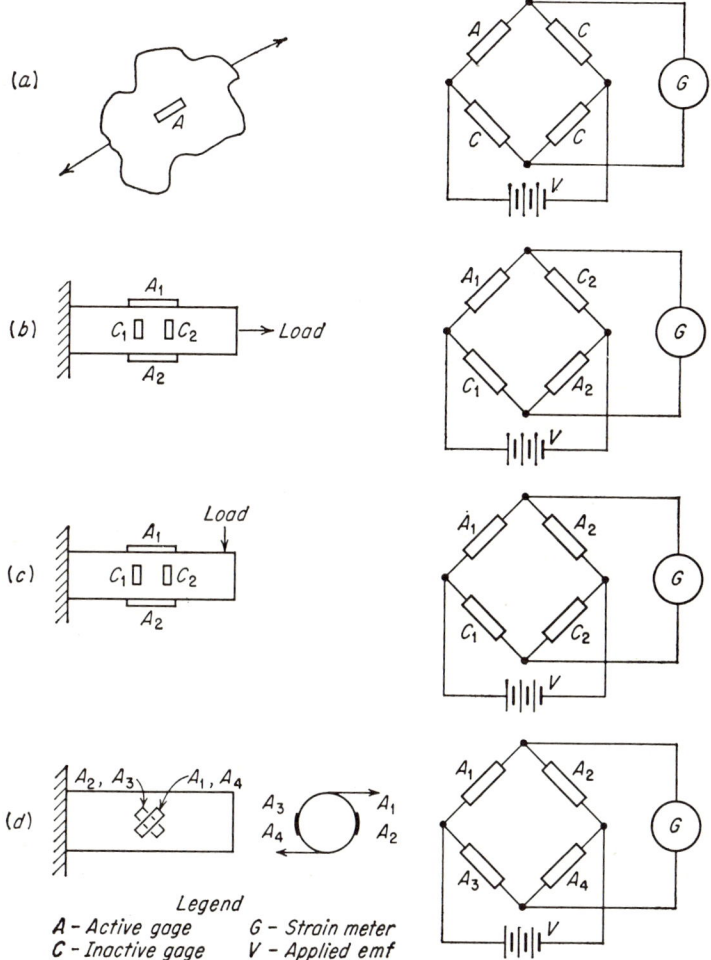

Fig. 5-2. Strain-gage bridge circuits: (a) single-gage-strain circuit; (b) double-gage direct-stress circuit; (c) double-gage bending circuit; (d) four-gage torsion circuit.

silver rings and silver graphite brushes operating at 20 psi pressure, gives excellent results up to 2,000 fpm. The mercury type is preferred in some cases because of low contact resistance and low drag on the rotating shaft.

One of the most useful applications for the wire strain gage is its use in a *transducer*[1] for measurement of such quantities as force, pressure, acceleration, displace-

[1] R. J. Fyffe and A. Arobone, Strain Gage Transducers for Measurement and Control, *Prod. Eng.*, November, 1952, pp. 121–148.

ment, temperature, and torque. Typical examples of these transducers are shown in Fig. 5-3.

Loads can also be measured in the actual machine members through the use of strain gages. For example, the rear-axle housing of a tractor can be gaged so as to register total tractive effort and also weight transfer to the rear wheels resulting from a draft load; the front axle can be gaged to measure weight transfer; or a PTO shaft to measure torque, etc.

Indicating and *recording instruments* used with strain gages are of the electronic type, employing a galvanometer as the strain-indicating element. The galvanometer deflection, which is proportional to strain, is either read directly or recorded as the ordinate length on a direct written or photographically filmed oscillogram. Magnetic tape recording systems may also be used. Although most strain-gage instrumentation

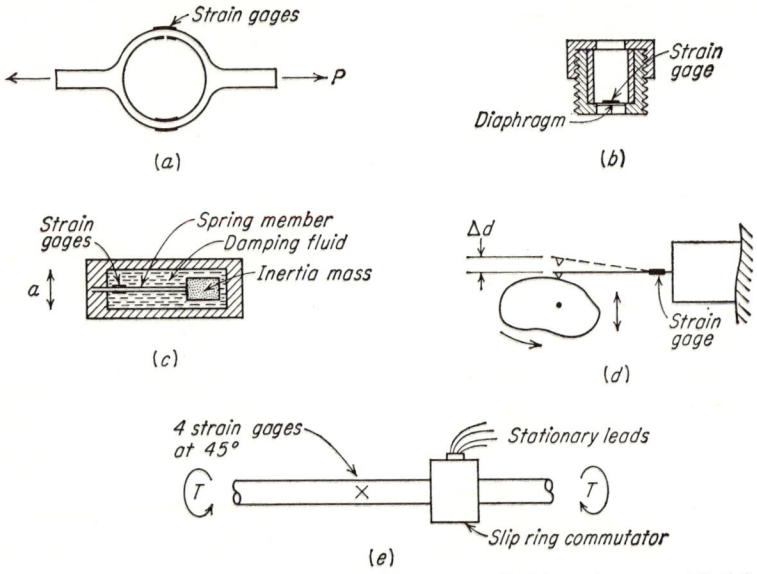

FIG. 5-3. Strain-gage transducers: (*a*) load ring; (*b*) pressure cell; (*c*) accelerometer; (*d*) deflection gage; (*e*) torquemeter.

is complex, the stress analyst need not be an expert in electronics. His concern is only with the scope and limitations of his instrumentation. Conversely, he should be able to select instrumentation required for a particular job. Instruments used for stain evaluation and their functions are shown in Fig. 5-4.[1]

Selection of instrumentation for a particular stress problem involves the following factors:

1. *Sensitivity of measurement* is usually a function of electronic amplification.
2. The *frequency response* of the instrumentation should be high enough to give an accurate record of all the strains occurring during the test. Frequencies may range from 0 in a static test to 5,000 cycles per sec, under some dynamic tests on farm machinery.
3. *Recording speed* must be high enough to accurately measure the time interval of the phenomenon under test.
4. The *inherent shielding* and *design* of the instrumentation must be capable of reducing outside interferences, such as occur from power lines, transformers, lights, arc welders, etc., to a minimum.

[1] Perry and Lissner, *op. cit.*, p. 92.

5. The *data record* may consist of numerical readings, an ink or optical trace on paper tape, or a photograph of an optical trace, depending on the instrumentation used.

6. *Recordings* may be made from a single gage or several gages simultaneously. Multichannel instrumentation is often required, not only for saving time, but also for the determination of phase relationships.

7. *Power* for instrumentation is usually secured from 110-volt a-c outlets for laboratory testing. In field tests, however, it is usually necessary to use batteries or an engine-powered generator unit.

8. *Stability* and *accuracy* of instrumentation must be considered. Drift and other inaccuracies may result from temperature changes.

9. The *handling* of instrumentation under test must also receive consideration. Some types are adaptable to stationary tests only, while others are more rugged and lend themselves to being transported while in operation.

When testing farm machinery under service conditions, the strain-gage instrumentation is usually transported by:

1. A self-propelled vehicle such as a truck, station wagon, or car
2. A tractor-drawn trailer
3. A platform built on the test machine

Strain gages are connected to the recording instruments by means of wire cables. Suitable holding brackets and flexible cord suspensions are usually required where a separate transporting vehicle is used. For multichannel operation, much time is saved by using coded connections and standard lengths of cable.

Telemetering or wireless transmission of strain signals from the moving test machine to stationary recording equipment enables the test machine to operate without the inconvenience of suspended lead wires to an accompanying instrument carrier and is also easier on delicate recording equipment.[1]

Photoelasticity. This is the science of evaluating strains in certain *doubly refractive* transparent materials under load by means of polarized light. As polarized light passes through this loaded material it is divided into two components, one vibrating in the direction of one of the principal strains, the other vibrating perpendicular to the first or in the direction of the other principal strain. The difference in the speed of these two components as they pass through the doubly refractive transparent material creates *interference bands* or *fringes*, which increase in number as the load is increased. The magnitude of stress at the boundary of the model is obtained as follows:

$$S = \frac{NF}{t}$$

where S = stress, psi
N = number of bands
F = fringe constant, psi/in. of model thickness
t = thickness of model, in.

The photoelastic model used can be either full size or a scale model of the prototype. Materials generally used include bakelite BT-61-893, Marblette, Catalin, Columbia Resin No. 39, Fosterite, etc.

Photoelasticity is used mainly in the evaluation of *stress concentrations* produced by holes, fillets, and notches.

Photoelastic stress analysis has the following limitations:

1. It is often difficult to duplicate a three-dimensional problem by means of a two-dimensional analysis.
2. Three-dimensional analysis is approximate at best because of the difficulty in curing model material uniformly.
3. Dynamic and impact studies are difficult, and results are subject to many sources of error.

---

[1] Albrecht Gerlach, Field Measurement of Tractor Transmission Forces, *ASAE Paper* 58-53, June, 1958.

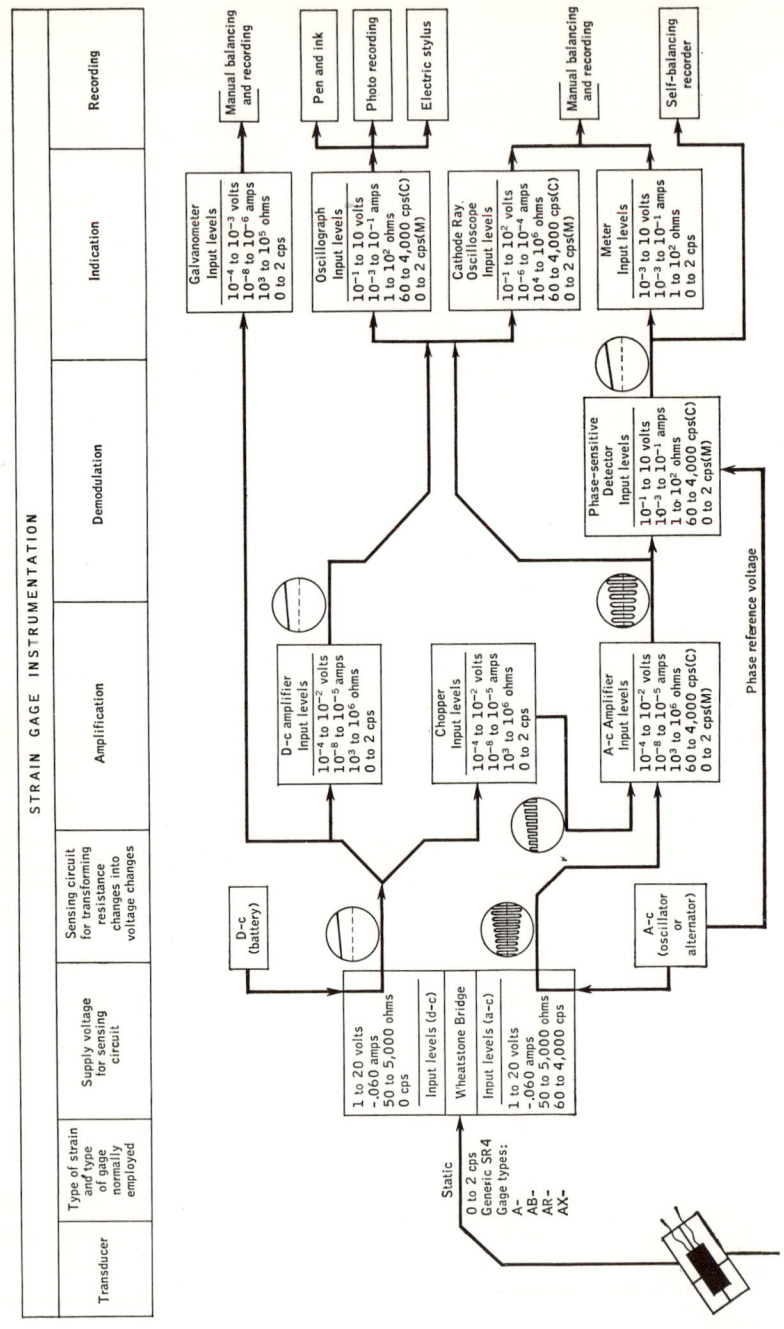

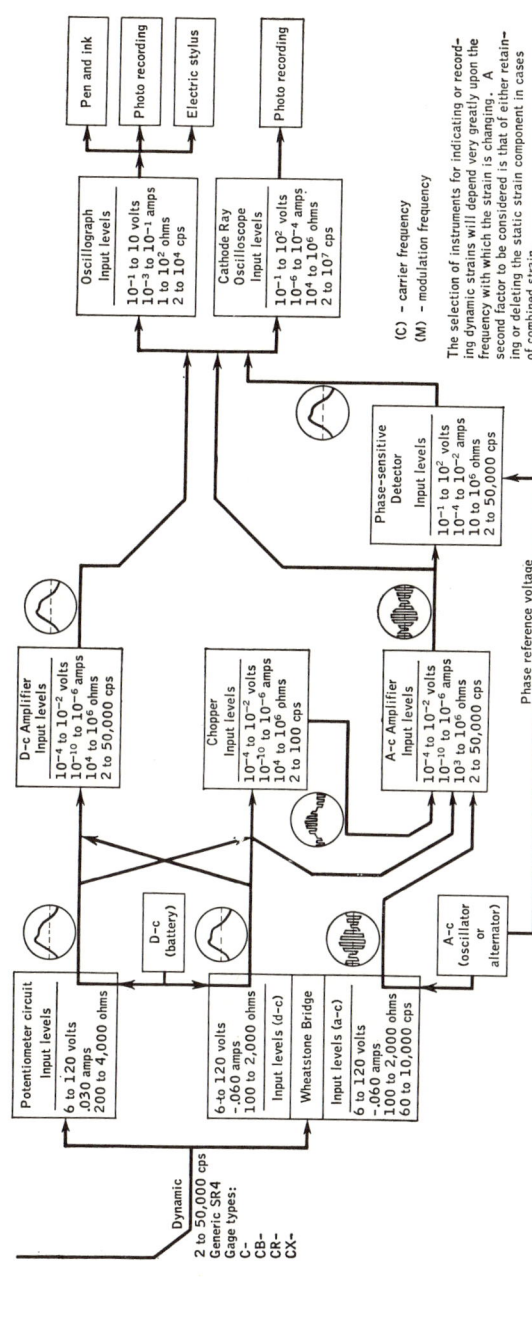

FIG. 5-4. Strain-gage instrumentation diagram. (Perry and Lissner, "Strain Gage Primer," 1955.)

Birefringent coatings and *photoelastic strain gages* have been developed. By using these techniques, actual parts can be tested rather than photoelastic models. Birefringent, or doubly refractive, coatings are available in sheet or liquid form and permit the evaluation of the stress field by a reflective-type polariscope.[1]

A photoelastic strain gage consists of a small, thin plate of birefringement material. This plate is bonded to the part under study and subjected to load. The stress pattern is again evaluated by a reflective-type polariscope.[2] Photoelastic strain gages which can be read directly in normal light are also available.[3]

Plastic models of parts in the design stage may be used for brittle-lacquer or strain-gage studies. These models are often made of Durez resin cast in wood or plaster-of-paris molds. Such procedures become very useful where expensive intricate parts are under study and where time is short. *Rubber models* can also be used advantageously

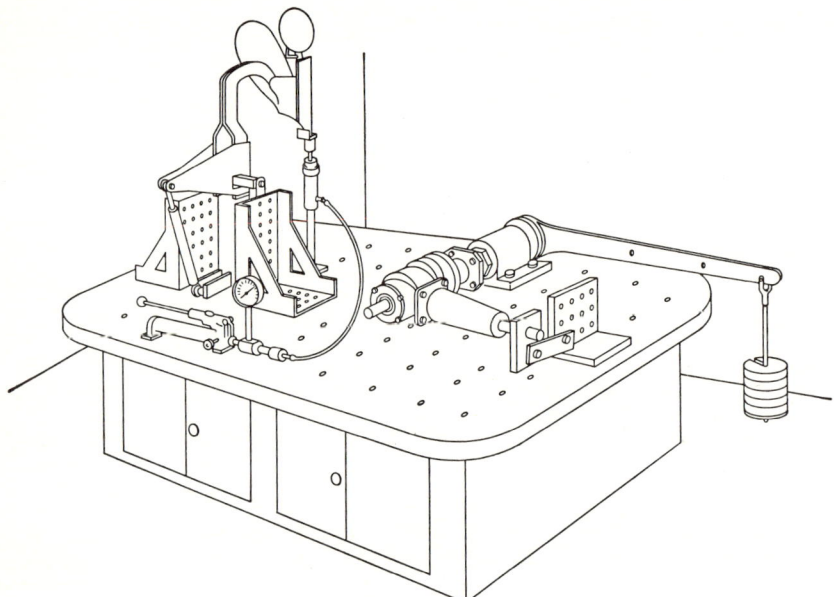

Fig. 5-5. Laboratory static-loading devices.

to indicate strain by laying out a grid pattern in white ink on black rubber prior to application of the load. When the load is applied, stress concentrations are located qualitatively by observing distortions produced in the grid pattern.

LABORATORY LOADING EQUIPMENT

Both *static-* and *dynamic-load* application equipment are required for a strength analysis of farm machinery. Static tension or compression loads are applied by machines such as the universal test machine. A torsion machine is used for torque applications, and various types of presses, screws, and air or hydraulically operated rams may also be used. Many types of static-load applications may be made by using a steel base plate with various combinations of hydraulic rams, angle brackets, torque fixtures, levers, and weights. Much fixture machining and laboratory time and expense are saved by the use of multipurpose brackets and fixtures, as shown in Fig. 5-5.

[1] Felix Zandman, "Photo Stress—Principles and Applications," Tatnall Measuring Systems Co., Phoenixville, Pa., March, 1959.
[2] G. U. Oppel, Photoelastic Strain Gages, *Soc. Exptl. Stress Analysis Paper* 574, October, 1959.
[3] Felix Zandman, *op. cit.*

Dynamic- or *fatigue-load* applications are made by various commercial machines: the Sonntag, Ivy, etc., which use an unbalanced weight exciter, the Shenck universal fatigue machine, the Baldwin Southwork stroking machine, the G. M. hydraulically actuated fatigue machine, and many others. In many cases it is advantageous to construct fatigue-test fixtures designed for the part and mounted on steel or cast-iron bedplates. Fluctuating loads can be applied to the part by such devices as unbalanced weights, cranks, pulsating air or hydraulic cylinders, push-pull electric vibrators, solenoids, air jets, etc.[1]

EXPERIMENTAL STRENGTH-ANALYSIS PROCEDURE

A typical strength analysis consists of the following steps:
1. Select and prepare the test area.
2. Make a pretest shakedown run over the test area.
3. Make a run over the test area using the Stresscoated parts under study, followed by a study of the Stresscoat patterns to determine location and direction of strains, as well as nature and direction of the applied loads.
4. Load-strain calibrations are then made using SR-4 strain gages cemented to the critical areas. Loads are applied in previously determined directions up to a magnitude not exceeding the yield point of any strain-gaged area.
5. From previous calibration tests, strain gages are selected that are sensitive to a particular direction of loading. Field tests are then run, measuring the strains on the selected gages by means of a multichannel oscillograph.
6. Convert the measured strains to load, and record load direction, magnitudes, and frequency of application. This is known as *load analysis*.
7. Apply known loads to the test part using laboratory fixtures. Measure strains in all critical areas by means of strain gages, extensometers, etc. These strains are converted to stress by applying the proper formula and material relationships. The resulting stresses are known as the working stresses. This process is known as *stress analysis*.
8. Determine the allowable stress of the material by means of fatigue tests on the part or data from published texts. This is known as *material analysis*.
9. Compare the working stress to the allowable stress to evaluate the factor of strength of the design. This completes the *strength analysis* of the design.[2]

In *applying published material-strength data* to design problems the following general rules should be observed:
A. Where loads are static in nature, i.e., where steady loads are applied less than 1,000 times during the life of the part, the following points apply:
   1. The elastic limit (the stress at which the stress-strain relationship departs from a straight line) of the material is the criterion for strength.
   2. Sharp fillets, notches, holes, and other stress concentrations produce very little if any weakening effect.
   3. Creep phenomenon may exist and should be guarded against.
   4. Residual stresses (internal stresses produced by manufacture or assembly) are often dangerous.
   5. Loading below the transition temperature of the material (the temperature below which the fracture changes from brittle to ductile) should be avoided.
B. Where loads are suddenly applied (*impact*), the following points are relevant:
   1. Sharp notches and other stress concentrations decrease load capacity.
   2. Effective elastic limit may be up to twice the static elastic limit.
   3. Residual stresses are often dangerous.
   4. Ductility has a major effect on load capacity.
   5. Loading below the transition temperature of the material should be avoided.
C. Where loads are *fluctuating*, with from 1,000 to an infinite number of applications during the life of the machine, making fatigue strength the primary criterion, the following points apply:

---

[1] A. B. Skromme, Accelerated Testing of Farm Machinery, *ASAE Paper* 59-632. December, 1959.
[2] R. J. Miller and M. A. Erickson, New Experimental Techniques Accelerate Farm Equipment Design Analysis, *Agr. Eng.*, 37:321–324, May, 1956.

1. The *allowable stress* for a particular material depends on the relationship between the maximum stress and the average, or mean, stress. When the mean and maximum are the same (the static-load condition) the allowable stress is the elastic limit. Where the mean stress is zero, as with completely reversed loading, the allowable stress for infinite life is the endurance limit. The modified Goodman diagram, shown in Fig. 5-6, provides a graphical method for determining the maximum allowable working stress for various mean working stresses. The endurance limit and the ultimate strength must be known in order to construct the diagram.[1]

    a. *Fatigue stress* is the alternating stress magnitude that can be withstood by the material for a specified number of cycles. The highest alternating stress

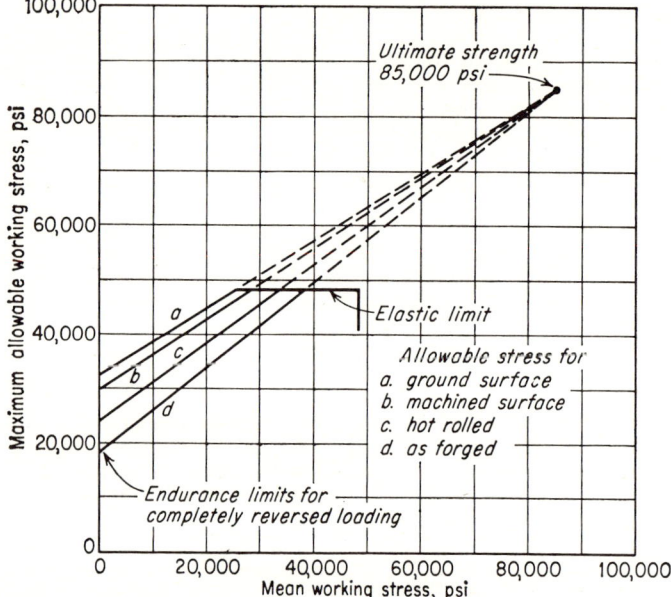

Fig. 5-6. Modified Goodman diagram showing allowable working stresses for infinite life under fluctuating load with various mean stresses. Curves shown are for SAE 1045 steel (Bh 170) with various surface finishes.

that can be withstood for an infinite number of cycles is defined as the *endurance limit* [see *S-N* curve, Fig. 5-7].

    b. *Mean stress* is the average value of the fluctuating stress under the service application.

2. Surface imperfections and irregularities lower life.[2]
3. Stress concentrations caused by notches, fillets, holes, etc., lower life.[3]
4. Endurance limit is not always proportional to hardness (Fig. 5-8).
5. Notch sensitivity affects strength through its relationship to the fatigue-stress-concentration factor[4] (Figs. 5-9 and 5-10).

[1] G. C. Noll and M. A. Erickson, "Allowable Stresses for Steel Members of Finite Life," *Proc. Soc. Exptl. Stress Anal.*, 5:132–143, Addison-Wesley Publishing Company, Reading, Mass., 1948.

[2] M. F. Garwood, H. H. Zurburg, and M. A. Erickson, Correlation of Laboratory Tests and Service Performance, in "Interpretation of Tests and Correlation with Service," American Society for Metals, Cleveland, Ohio, 1951, pp. 1–77.

[3] R. E. Peterson and A. M. Wahl, Two and Three Dimensional Cases of Stress Concentration and Comparison with Fatigue Tests, *Trans. ASME*, 58:A-15–22, 1936.

[4] R. E. Peterson, "Stress Concentration Design Factors," John Wiley & Sons, Inc., New York, 1953. R. J. Roark, "Formulas for Stress and Strain," 3d ed., McGraw-Hill Book Company, Inc., New York, 1954.

6. Heat-treatments involving carburizing, induction hardening, flame hardening, and nitriding usually produce beneficial residual compressive stresses in the surface. This condition reduces the maximum tensile stress at the surface and thus improves fatigue life.
7. Mechanical surface working done by hammer peening, shot peening, burnishing, and tumbling produces beneficial residual compressive stresses and improves surface finish, each of which increases service life.

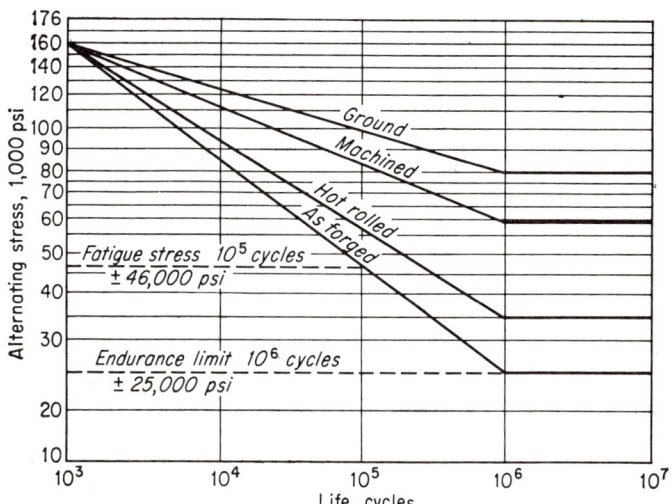

Fig. 5-7. S–N (stress vs. number of cycles) curve for steel of hardness Bh 375–401.

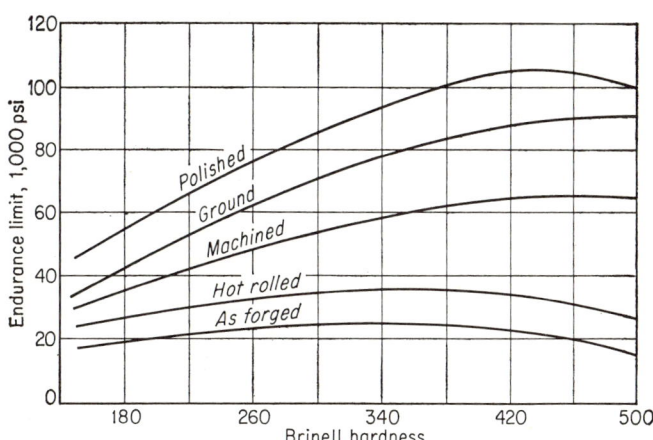

Fig. 5-8. Effect of hardness on the endurance limit for various surface conditions.

8. Corrosive atmosphere generally decreases service life.
9. Plating as a rule decreases service life because of hydrogen embrittlement.[1]
10. Severe grinding introduces harmful residual tensile stresses which reduce service life.
11. Decarburization, produced during hot-forming and heat-treat operations in an uncontrolled atmosphere, reduces service life.

[1] Plating Affects Parts Adversely, *Machine Design,* August, 1950, p. 124.

Occasionally, problems arise where adequate data are not available in published literature. They may include:

1. Endurance and wear data on assemblies such as gear drives, hydraulic systems, clutches, etc.
2. Allowable stresses in manufactured parts where such unknown factors as residual stresses, decarburization, surface roughness, etc., are introduced by the forming, heat-treat, and machining operations
3. Allowable stresses in case-hardened parts

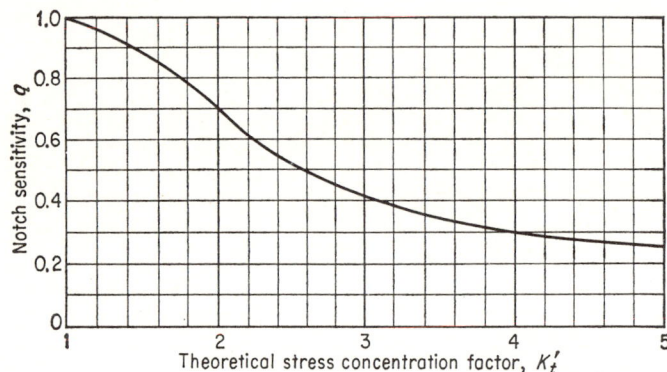

FIG. 5-9. Notch sensitivity of annealed steels, data from 0.5 to 5.25 in. circular specimens tested in reversed bending.

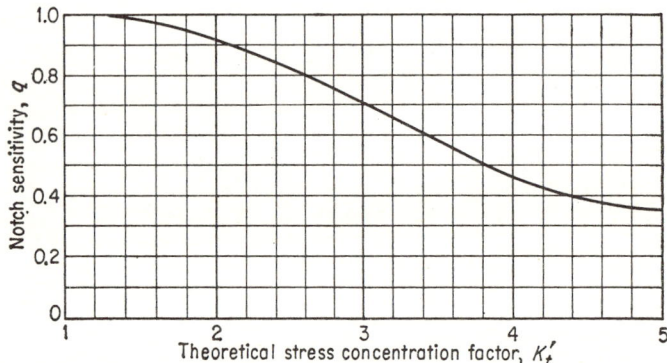

FIG. 5-10. Notch sensitivity of quenched and drawn steels, data from 0.5 to 2.0 in. circular specimens tested in reversed bending.

In the above cases it is necessary to conduct laboratory fatigue or life tests. This may consist of operating entire machines in such cases as:

1. Cycling of a tractor hydraulic system loaded to simulate field conditions
2. Cycling of a loader hydraulic system with a capacity load in the bucket
3. Cycling of a PTO-driven baler using an air cylinder to react against the ram instead of hay
4. Driving a combine or a tractor with a mounted implement over a calibrated obstacle course on a test track to study field transport loads

In other cases it may be possible to fatigue-test machine components on commercially available fatigue machines or laboratory-built fixtures.

The analysis of fracture patterns[1] occurring in failed parts usually gives valuable information regarding service loadings and material behavior characteristics. Fracture patterns of smooth bars may give the following indications:

[1] Hetényi, "Handbook of Experimental Stress Analysis," pp. 593–635.

1. *Tension loading.* Mild steel usually necks down prior to failure, resulting in a cuplike fracture. For harder steels the necked-down diameter is less noticeable and the fracture is usually cupped much less. With a brittle material such as cast iron, certain plastics, etc., no necking down occurs and the fracture is flat across the bar.

2. *Torsion loading.* A ductile material will fracture across the bar or parallel to the longitudinal axis of the bar in the planes of maximum shear. A brittle material will usually fracture at 45° to the longitudinal axis along the planes of maximum direct stress.

3. *Impact loading.* For suddenly applied or impact loads the behavior is similar to static loads in brittle materials. Thus in a mild-steel or cast-iron part in tension, fracture would appear across the bar with no necking down while torsion failure would occur on planes 45° from the longitudinal axis.

4. *Fluctuating loading.* Fracture of parts subjected to fluctuating loads are particularly informative in steel or ductile materials. Fractures of the type shown in Fig. 5-11b indicate failure by reversed bending, often occurring in such parts as steering knuckles, stub shafts, etc. Fractures of the type shown in Fig. 5-11a indicate failure by rotating bending such as occurs in rotating shafts. Fractures of the type

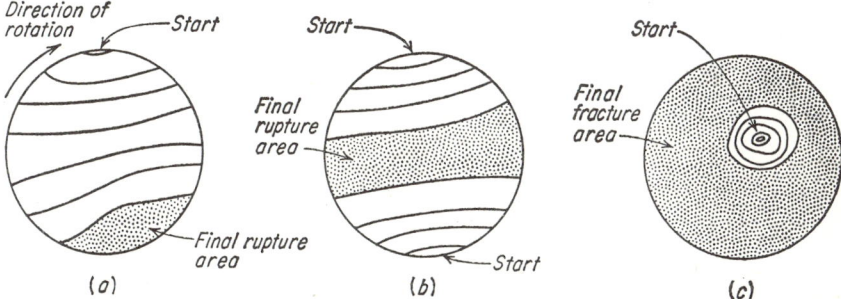

FIG. 5-11. Shaft fatigue failure patterns: (*a*) rotating bending; (*b*) reversed bending; (*c*) alternating direct load.

shown in Fig. 5-11c usually occur with fluctuating direct stress as in push rods, tension bars, etc. Stress concentrators such as fillets, holes, notches, and surface irregularities will usually modify the fracture pattern characteristic of smooth bars, especially in the case of fluctuating loads. Brittle materials show no characteristic fracture under fluctuating loads, and hence very little load data are gained from a brittle-fracture-pattern study.

The complete strength evaluation of a machine part, made by comparing the working stresses with the allowable stress of the material, may require a large number and variety of tests. Working-stress and allowable-stress studies on experimental parts often result in modifications, both in the design and in material composition and processing.

Before the final design is released for mass production, it is wise to make a complete stress study on the production prototype. At the same time, field tests are usually in progress on several prototype machines under a wide variety of service conditions. These final tests generally yield much additional information on production processes (such as welding) and design variations required by production, as well as unusual loading conditions that were not uncovered in the original service load measurements. Weaknesses encountered in these tests can usually be taken care of by modifications before mass production begins.

It is desirable to follow the design not only from its conception through development stages to production, but also after production machines have been put into use by customers. A history of service failures is invaluable in guiding the designer in new projects of a similar nature. It is only when field failures have been brought down to an acceptable minimum that the designer is assured that the allowable stress of the materials in the machine is sufficient to withstand the imposed working stresses and that the strength of the design is adequate.

Chapter 6

TRACTOR FORCE REACTIONS

R. L. ERWIN

The forces reacting on a tractor as a result of its loading and motion have important effects on its stability, traction, and steering. This chapter outlines methods of calculating these effects. For more detailed or technical discussions, the references listed at the end of the chapter may be used.

The common practice is to reduce the dynamic forces acting upon the tractor into static values for purposes of calculating stability, weight transfer, and loading. However, for a complete understanding it should be realized that the tractor is

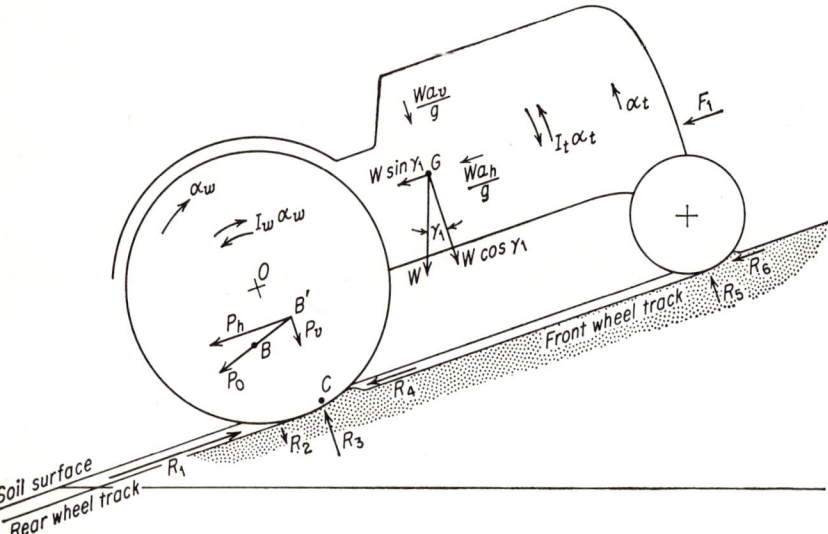

FIG. 6-1. External forces on a tractor acting in a plane perpendicular to the rear axle. (*McKibben, Agr. Eng., May,* 1927.)

usually moving and all the forces are dynamic. For simplicity of calculation these forces can be separated into those acting in a plane perpendicular to the ground surface and those horizontal to the ground surface.

FORCES IN VERTICAL PLANE

Basic Forces. The tractor at rest without a load has three basic forces acting on it: gravity, or the weight of the tractor, the reaction of the soil on the front wheels, and the reaction of the soil on the rear wheels. As the tractor starts to move, other forces are set up as shown in Fig. 6-1. These forces are explained as follows:

W = weight of tractor acting through G, center of gravity.
γ_1 = angle of tractor to horizontal.
$W \sin \gamma_1$ = component of W acting parallel to line of motion of tractor.
$W \cos \gamma_1$ = component of W acting perpendicular to line of motion of tractor.
R_1 = resultant tractive reaction of soil against rear wheels. Except when tractor is decelerating or moving downhill, this is only forward force. As shown in Fig. 6-2, this force is applied between instantaneous rolling radius OQ_1 and lowest point on the wheel and is usually considered as acting through C.
R_2 = resultant reaction of soil against lifting action of rear tire lugs, small and usually neglected.
R_3 = supporting action of soil on rear wheel, acting through C.
R_4 = resultant of forces opposing forward motion of rear wheel, acting through C.
C = intersection of resultant of tractive and supporting forces. Under many conditions when tractor stability is a problem, C is below the ground surface, forward of OQ_1Q_4 and above the lowest point of the wheel. Moments can be conveniently taken about C for force calculations.[1]
R_5 = resultant of supporting forces on front wheels.
R_6 = resultant of forces opposing forward motion of front wheels.
P_0 = external force applied to tractor by trailing load.
B = point on tractor frame to which load P_0 is attached.
P_h = component of load P_0 parallel to direction of motion considered to be acting through B', a point of intersection of the line of pull with a plane through C perpendicular to direction of motion of tractor.
P_v = component of load P_0 perpendicular to plane of motion of rear axle.
G = center of gravity.

The following forces are small in a tractor and are not usually calculated:

$\dfrac{Wa_h}{g}$ = inertia force parallel to direction of motion, where a_h is acceleration of G

$\dfrac{Wa_v}{g}$ = inertia force perpendicular to direction of motion (usually small)

I_t = moment of inertia of tractor about G
$I_t\alpha_t$ = inertia couple due to angular acceleration, α_t, of entire tractor about G
$I_w\alpha_w$ = inertia couple of rear wheels and all attached parts with respect to tractor frame
F_1 = air resistance due to forward motion

Weight Transfer. In the following discussion, weight transfer is defined as the reduction in weight on the front wheels due to the application of external forces and loads, comparing the static weight on the front wheels of the tractor alone with the dynamic weight on the front wheels when operating with an implement.

On pavement or hard ground the effective tire thrust times the rolling radius is approximately equal to the torque input to the rear axle. Under these conditions the point of resistance C can be easily determined and the weight transfer can be found by considering the tractor as a free body acted on by external forces only. However, on soft ground where a much greater proportion of the torque is expended in overcoming rolling resistance of the rear wheels and in work on the soil, weight transfer is more accurately calculated by considering that the tractor chassis is rotating about the rear axles. Thus the primary factors affecting weight transfer are the torque input to the rear wheels, modified by the load and pull on the tractor, the tractor weight, and the rolling resistance of the front wheels.

When the tractor is at rest on level ground, the forces acting on it are R_3, R_5, and W. The point of soil reaction on the rear wheel is directly under the center of the

---

[1] Data for approximate locations of C are given in Ref. 13. On firm level ground when traction conditions were good, C was found to be approximately below the center of the axle and at the ground surface.

axle. As the tractor moves forward, the center of resistance C moves forward as shown in Fig. 6-2.

Neglecting several of the minor forces, the basic formula for weight transferred from the front wheels as a result of forward motion can be developed as follows (Fig. 6-3):

$$R_{5,\text{static}} = \frac{Wa}{b}$$

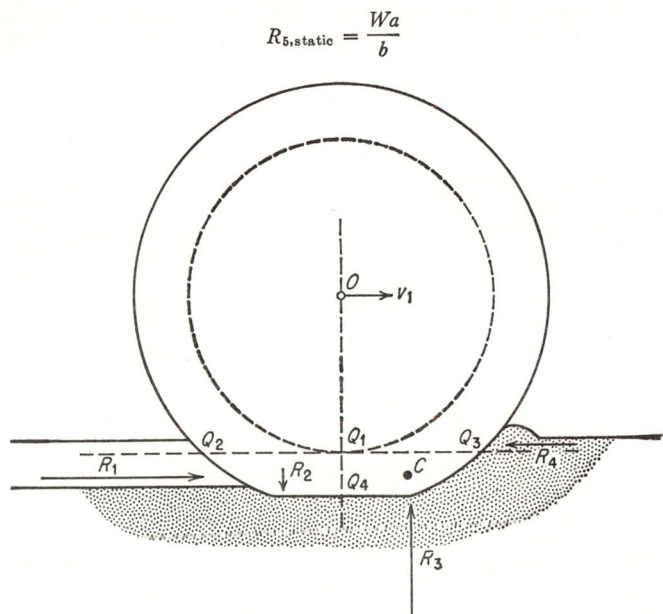

Fig. 6-2. Soil reactions on a rear tractor wheel.

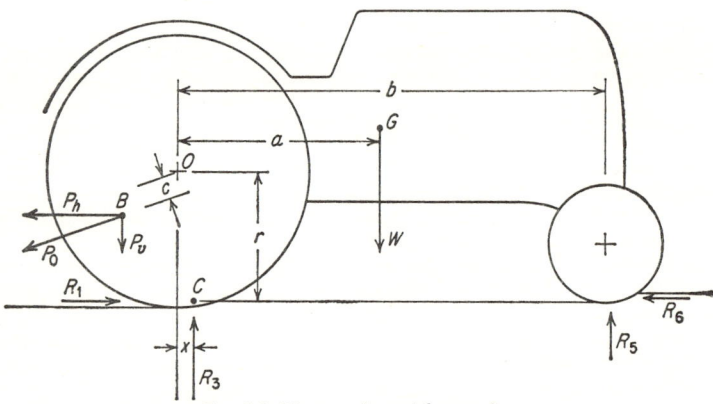

Fig. 6-3. Moments for weight transfer.

Taking moments about the rear axle for the dynamic condition,

$$R_{5,\text{dyn}}b + T_i = Wa + P_0c + R_6r$$

$$R_{5,\text{dyn}} = \frac{Wa + P_0c + R_6r - T_i}{b}$$

$$\text{Weight transfer} = R_{5,\text{static}} - R_{5,\text{dyn}} = \frac{T_i - P_0c - R_6r}{b} \qquad (6\text{-}1)$$

TRACTOR FORCE REACTIONS

When T_i = torque input to rear axle, in.-lb.
  c = distance from O to line of action of P_0, in.
  a = distance from O to vertical plane through G, in.
  b = wheel base, neglecting forward shift of R_5 with motion, in.
  r = rolling radius or distance OC, in.

The dynamic weight on the rear wheels can be found as follows:

$$R_{3,\text{dyn}} = R_{3,\text{static}} + P_v + \text{wt. transfer} \tag{6-2}$$

where
$$R_{3,\text{static}} = W\left(1 - \frac{a}{b}\right)$$

The forward shift of R_3 is related to the torque expended in working the soil as follows:

$$T_i = R_1 r + R_{3,\text{dyn}} x$$

and
$$x = \frac{T_i - R_1 r}{R_{3,\text{dyn}}} \tag{6-3}$$

where c = forward shift of R_3.

If the tractor is on a grade as shown in Fig. 6-1, R_3 and R_5 must be calculated on the basis of $W \cos\gamma$.

The principles shown can be applied to any tractor loading problem regardless of type of hitch, etc.

The *rigid drawbar with a pulled load,* as shown in Fig. 6-4, affects weight transfer as shown in Eq. (6-1). It can be seen that the load P_0 reduces the weight on the

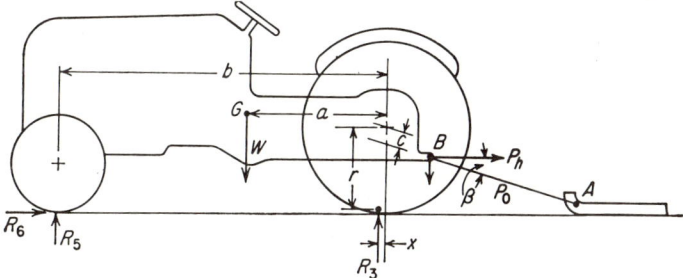

FIG. 6-4. Rigid drawbar with pulled load.

front wheels, even if the pull is horizontal. Increasing the slope of the line of pull AB by moving the load closer to the tractor, or by raising the hitch point B increases the weight shift from the front to the rear. In soft ground, the greater torque input T_i to the rear wheels for a given pull also increases weight transfer.

The common *three-point hitch* is illustrated by a tractor and mounted-plow combination as shown in Fig. 6-5. Usually the bottom links are supported by the tractor hydraulic system and the plow is for all practical purposes a part of the tractor chassis.

Taking moments about the rear axle for the dynamic condition,

$$R_{5,\text{dyn}} b + T_i + P_v d = Wa + R_6 r + Le$$

$$R_{5,\text{dyn}} = \frac{Wa + R_6 r + Le - T_i - P_v d}{b}$$

$$\text{Weight transfer} = R_{5,\text{static}} - R_{5,\text{dyn}} = \frac{T_i + P_v d - R_6 r - Le}{b} \tag{6-4}$$

or
$$= \frac{T_i - P_0 c - R_6 r}{b} \tag{6-5}$$

where L = horizontal resolution of complex soil forces acting on plow moldboards.
V = vertical resolution of soil forces on plow. For some implements, such as a disk plow, this may be negative.
W_p = weight of plow.
$P_v = W_p + V$ = total downward component.
E = center of resistance of plow.
P_0 = resultant of all forces acting on plow through E.

The above three-point hitch becomes a *floating hitch* if the tractor hydraulic system does not support the lower links. In this case the lower links no longer hold the plow up, so the resultant of all the forces on the plow, P_0, extension of line EE_1, must pass through point K (Fig. 6-6), the instantaneous intersection of the lower and upper links. With this type of hitch the plow will seek a depth to satisfy the above con-

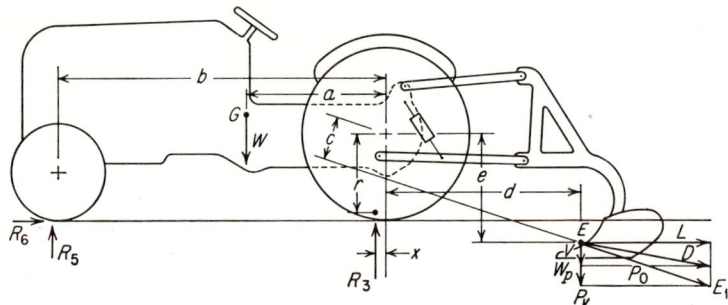

Fig. 6-5. Three-point rigid hitch.

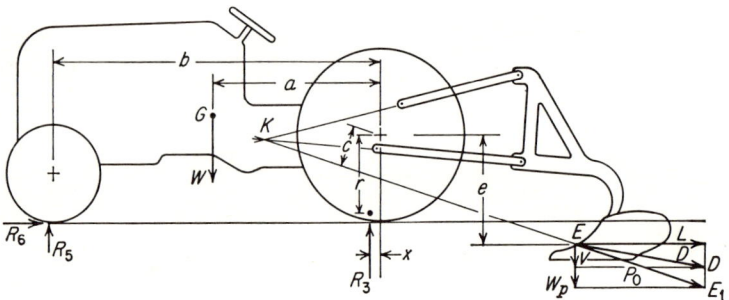

Fig. 6-6. Three-point floating hitch.

ditions. The depth can be adjusted within limits determined by the hitch geometry by varying the length of the links or by changing the pitch of the plow. Weight transfer can be determined by using Eq. (6-5).

A *semimounted implement* supported by caster wheels at the rear and by a three-point floating hitch at the front, with the center of gravity of the implement between the implement wheels and the tractor, generally causes weight transfer from the rear tractor wheels to the front tractor wheels. This can be modified on those tractors that are "draft-sensitive" by causing the hydraulic system to support all or part of the weight of the front of the implement. This type hitch is shown in Fig. 6-7.

If the rolling resistance of the wheels is neglected, the weight transfer can be calculated statically. To find R_i, the upward reaction on the implement wheels, take moments about K, the point of link convergence in the vertical plane. With the lower links level,

$$R_i = \frac{W_i f}{f + g} \tag{6-6}$$

where W_i = weight of implement
f = horizontal distance from K to G_i, center of gravity of implement
g = horizontal distance from R_i to G_i

By taking moments about C

$$R_5 = \frac{Wa - F_l h - F_{uv} k + F_{uh} m}{b} \qquad (6\text{-}7)$$

where F_l, tension in lower links $= \dfrac{R_i r}{p} - \dfrac{W_i q}{p}$.

(For simplicity F_l is shown as horizontal since it is nearly so in most hitches of this type.)

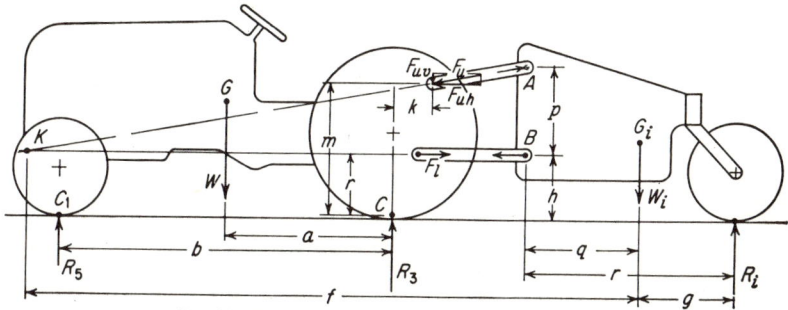

FIG. 6-7. Floating three-point hitch with gage wheels.

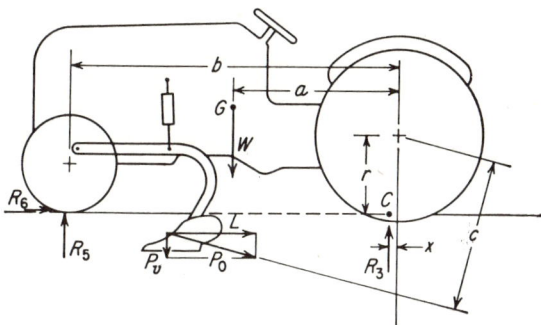

FIG. 6-8. Front-mounted implement.

By taking moments about C_1

$$R_3 = \frac{W(b - a) + F_l h + F_{uv}(b + k) - F_{uh} m}{b} \qquad (6\text{-}8)$$

assuming C_1 in the same plane as C.

Midmounted soil working implements have a reaction on the tractor much different from rear-mounted or pull-type implements. In general, they tend to reduce the weight transferred to the rear wheels, resulting in less traction. The implement may be supported by gage wheels or by the tractor. Figure 6-8 shows a front-mounted middlebuster supported by the tractor hydraulic system. Weight transfer may again be found by Eq. (6-5), taking P_0 as the resultant of all forces acting on the implement, including its weight.

Implements such as *loaders* may be *mounted ahead of the front wheels* of the tractor. The load on the front wheels is consequently increased with a decrease in rear-wheel weight. In Fig. 6-9, the solid lines show R_5 and R_3 in a static position. The weight on the front wheels can be calculated by taking moments about C:

$$R_{5,\text{static}} = \frac{Wa + W_L(d + b)}{b}$$

where W_L = load on bucket or fork
d = distance center of gravity of load is ahead of front axle

When the tractor is moving forward to load, C and C_1 will move forward as noted previously. Additional forward force is needed to push the fork or bucket into the material and overcome rolling resistance of the wheels. This condition tends to

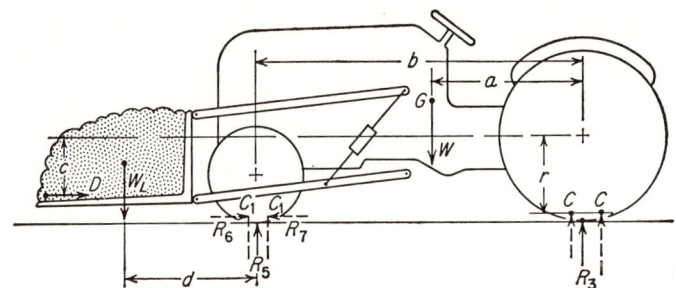

FIG. 6-9. Front-mounted loader.

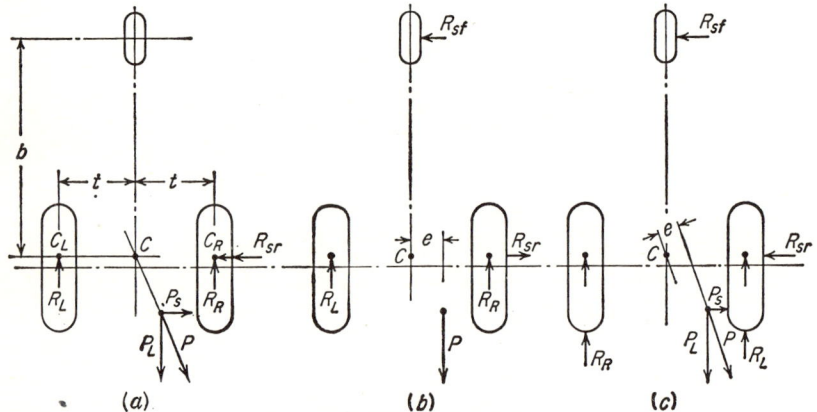

FIG. 6-10. Basic horizontal loading.

reduce the weight on the front wheels and add to the rear unless the loader is being pushed into the pile and lifted simultaneously. If it is not being lifted,

$$\text{Weight transfer} = \frac{T_i - Dc - R_6 r}{b} \qquad (6\text{-}9)$$

where D = resultant of forces necessary to push bucket into material.

If the tractor is backed, D is removed or may become negative and the rear-axle torque T_i is reversed, giving

$$\text{Weight transfer} = \frac{R_7 r - T_i}{b} \qquad (6\text{-}10)$$

Force R_5, the front-wheel load, gives an indication of *stability* with respect to overturning backward and also traction for steering. As soon as this force approaches zero or becomes negative, the tractor becomes unstable and may overturn backward. Conditions that reduce stability are:

1. Steep grade where the moment arm of W to vertical line through rear axle is reduced
2. High hitching where c is decreased or becomes negative
3. High axle torque T_i, with low pulling moment $P_o d$, such as with the rear wheels in a deep hole or with a log or rail ahead of the rear wheels

HORIZONTAL FORCES

Steering. The basic horizontal forces acting on the tractor or the tractor-and-implement combination are most apparent as they affect the steering of the tractor and implement. Figure 6-10 illustrates three basic effects of implement hitches on the tractor. The center of the horizontal plane of the reaction of the soil on the rear wheels is shown by C. If P passes through C as in Fig. 6-10a, the moment of P

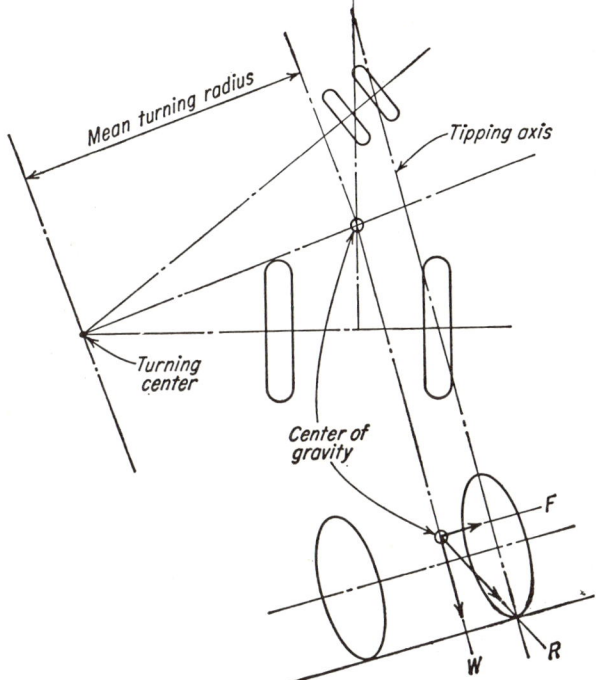

FIG. 6-11. Forces resulting when the tractor is turning. W is the weight of tractor acting downward; F is the centrifugal force acting horizontally; and R is the resultant. (*Worthington, Agr. Eng., April, 1949.*)

about C becomes zero, there is no appreciable effect on the steering of the tractor, and R_L should equal R_R. However, the more common conditions are shown in Fig. 6-10b and c where there is a steering force R_{sf} necessary on the steering wheels. Moments can be taken about C as follows:

$$R_L t - R_R t + Pe - R_{sf} b = 0$$

where R_L = tractive force on left wheel
R_R = tractive force on right wheel
P = pull of load
e = offset of pull from C
t = ½ tractor tread
b = tractor wheel base
R_{sf} = side force on front wheels
R_{sr} = side force on rear wheels

Since $R_R = R_L$ if tractive conditions and weight are the same,

$$Pe - R_{sf}b = 0$$

and $$P_s = R_{sr} + R_{sf}$$

Linkage Hitches. In linkage-type hitches, where the implement tends to become a part of the tractor, the hitch itself must be analyzed and the soil forces acting on the implement must be measured to calculate the horizontal forces.

Lateral Instability. Turning at high speeds creates this problem in modern rubber-tired tractors with high road speeds. Figure 6-11 shows the major forces acting on a tricycle tractor during turning. The tractor remains stable only so long as R, the resultant of W and F, remains inside the line of contact between the outside front and rear wheels. Turning stability may be further reduced by such factors as shifting tire fluid level, tire deflection, braking the wrong wheel, and rolling over obstructions.

REFERENCES

1. Clyde, A. W.: Mechanics of Plow and Tractor Hitches, *Agr. Eng.*, 15:388–390, November, 1934.
2. Clyde, A. W.: Vertical Hitching of Farm Implements, *Agr. Eng.*, 16:358–364, September, 1935.
3. Clyde, A. W.: Measurement of Forces on Soil Tillage Tools, *Agr. Eng.*, 18:5–9, January, 1936.
4. Clyde, A. W.: Load Studies on Tillage Tools, *Agr. Eng.*, 18:117–121, March, 1937.
5. Clyde, A. W.: Mounted Plows and Their Effects on the Tractor, *Agr. Eng.*, 21:167–170, May, 1940.
6. Clyde, A. W.: Shifting the Center of Pull of a Tractor, *Agr. Eng.*, 28:11–13, January, 1947.
7. Clyde, A. W.: Mounted Plows and 3-point Linkage, *Farm Implement News*, 73:34–78, June 25, 1952.
8. Heitshu, D. C.: Kinematics of Hitches for Tractor Mounted Farm Implements, *Agr. Eng.*, 33:343–356, June, 1952.
9. Johannsen, B. B.: Tractor Hitches and Hydraulic Systems, *Agr. Eng.*, 35:789–800, November, 1954.
10. LaConte, J. W.: Graphic Analysis as Applied to the Study of Wheel Type Tractors, *Agr. Eng.*, 8:190–191, July, 1927.
11. McKibben, E. G.: The Kinematics and Dynamics of the Wheel Type Farm Tractor, *Agr. Eng.*, 8:15–16, January, 1927; 8:39–43, February, 1927; 8:58–60, March, 1927; 8:90–93, April, 1927; 8:119–122, May, 1927; 8:155–160, June, 1927; 8:187–189, July, 1927.
12. McKibben, E. G.: Effect of Drawbar Pull upon the Effective Weight on Front and Rear Wheels of Farm Tractors, *Agr. Eng.*, 9:243–245, August, 1928.
13. Worthington, Wayne H.: Evaluation of Factors Affecting the Operating Stability of Wheel Tractors, *Agr. Eng.*, 30:119–123, March, 1949; 30:179–183, April, 1949.
14. Sack, Hans W.: Longitudinal Stability of Tractors, *Agr. Eng.*, 37:328–333, May, 1956.
15. Richey, C. B.: Traction Tests of a Tandem Tractor, *ASAE Trans.*, 2:16–17, 1959.

Chapter 7

TRACTOR DESIGN OBJECTIVES

The function of the modern farm tractor is to pull, push, carry, supply rotative power through the power take-off or belt pulley, and supply power for controlling implements. Originally the tractor was regarded only as a substitute for the horse and the steam engine. The introduction of the power take-off, or PTO, in 1918 to supply rotative power to drawn machines in place of ground drive wheels was a milestone in broadening the usefulness of the tractor. The introduction of the tricycle or cultivating-type tractor in 1924, the introduction of low-pressure rubber tires in 1932, and the introduction of hydraulic lift equipment in 1935 are other important milestones.[1]

Functional factors affecting the value of a tractor to the user, in addition to its being of an economic power size and of a type adapted to perform the necessary operations, are as follows:

A. Efficiency in generation and application of power as indicated by
  1. Engine economy at various speeds and loads
  2. Controllability of power flow to points of application by
     a. Engine governor
     b. Engine clutch
     c. Adequate transmission ratios to rear wheels
     d. Independent PTO drive while forward motion is stopped and started
  3. Power-transmission efficiency (minimum friction losses) to points of application
  4. Tractive efficiency (ratio of power available at drawbar to that delivered to the wheels) and adequate traction with the weight and weight distribution of the various tractor and implement combinations
  5. Arrangement of components and provision of mounting points to efficiently accommodate mounted implements of as many types as possible

B. Efficient use of operator's time by
  1. Adequate rate of performing operations
  2. Minimum of nonoperating time for
     a. Routine maintenance
     b. Implement attachment and detachment
     c. Travel and implement transport
     d. Adjustments (principally wheel tread)
     e. Bogging down and breakdowns
  3. Minimum of human fatigue assured by[2]
     a. Ease of steering
     b. Ease of control
     c. Good visibility

---

[1] Roy B. Gray and E. M. Dieffenbach, Fifty Years of Tractor Development in the USA, *Agr. Eng.*, 38:388–397, June, 1957. Roy B. Gray, Development of the Agricultural Tractor in the United States, *USDA ARS AERB Infor. Ser.*, no. 107, 1954.

[2] Heinrich Dupuis, Effect of Tractor Operation on Human Stresses, *Agr. Eng.*, 40:510–525, September, 1959. J. B. Liljedahl, R. Gluck, and M. E. Schroeder, Steering Force Requirements of Wheel Tractors, *Agr. Eng.*, 40:522–525, September, 1959. W. H. M. Morris, J. B. Liljedahl, and J. E. Wiebers, Heat Stresses in Tractor Operation, *Agr. Eng.*, 40:672–683, November, 1959.

        d. Comfortable ride
        e. Quietness and minimum vibration
        f. Freedom from dust and exhaust fumes
        g. Ease of stepping on and off
        h. Protection from weather
C. Safety as to
        a. Adequate stability under operating conditions
        b. Shielding of moving parts
        c. Minimum temptation for tractor to be operated from other than the regular position

Durability factors affecting the value of a tractor to the user are:

A. Freedom from breakage secured by
   1. Parts of adequate strength
   2. Adequate overload protection by
        a. Hydraulic pressure-relief valves
        b. Friction slip clutch in PTO power line, where required
B. Resistance to mechanical wear from friction by
   1. Wear-resistant materials for wearing surfaces, from cylinder walls to tires
   2. Adequate lubrication with a minimum of attention
   3. Bearings, gears, etc., of adequate capacity
   4. Exclusion of dirt
C. Resistance to chemical deterioration by using
   1. Heat-resistant materials where needed and adequate cooling
   2. Crankcase ventilation and oil filtering to remove combustion products and prevent acid formation
   3. Protective coatings or treatments for vulnerable surfaces, interior and exterior
D. Low repair costs, secured by
   1. Design to permit easy replacement of wearing parts
   2. Exchange provisions for worn assemblies
   3. Adjustments to compensate for wear where practical
   4. Simplicity of construction

TYPES OF TRACTORS

The major components of a tractor are the engine with radiator and fuel tank, transmission, driving axle and wheels, steering wheels, operator's seat, and controls. In the early days of tractor development many different arrangements were tried but the primary requirement of pulling trailed implements led to the general adoption of the *four-wheel type* with rear drive wheels, front steering, and engine in front so that its weight balanced the overturning couple caused by tractive effort and drawbar pull. This arrangement resulted in the greatest practical proportion of the total weight on the drive wheels for traction at high drawbar loads, as analyzed in Chap. 6.

The *tricycle* tractor substituted a single or dual front wheel for the wide front axle in order to allow the attachment of front-mounted cultivators. The body and rear axle were raised for cultivator gang clearance as well as for high clearance over the rows. The center of the tractor was narrowed to reduce the obstruction to clear vision of the cultivating operation. Stability was unavoidably reduced, the high center of gravity caused loss of traction by the drive wheel on the land-wheel side when plowing, and the plow furrow no longer guided the front wheels. The necessity for row-crop cultivation, however, made this type of tractor preferred where corn, cotton, and other row crops were grown.

Particularly adapted for cultivation of vegetables is a small *four-wheel tractor with the engine over the rear axle* and operator's seat just ahead of the engine. Cultivating vision is obstructed only by the front frame.

The *operations* involved in crop production can be classified according to tractor requirements as shown in Table 1-7. It can be seen that conventional tillage, such as plowing, disking, subsoiling, etc., is the only classification which requires high tractive effort. It is in this area that more powerful tractors are most useful. Traction tire size is limited in many areas by the necessity of running between normal row widths and in open plow furrows. Such tires are currently inadequate for more than about 75-hp engines. The *four-wheel-drive* principle appears increasingly attractive for higher-powered tractors for drawbar work, both from the standpoint of tire-size limitations and higher tractive efficiency, as discussed in Chap. 4.

Conventional low-clearance tractors, driving with the two rear wheels, are reasonably efficient for rotary tillage, planting and seeding, mowing and crushing, raking, baling, chopping, blowing, manure loading, manure spreading, feed transport, and feed grinding. With increased clearance and tricycle front wheels, they are also efficient for cultivating, corn picking, and cotton picking when modified.

Combining is best done by large *self-propelled* machines. Other harvesting operations such as baling, chopping, corn picking, and cotton picking can advantageously be done by self-propelled machines. The cost of individual self-propelled machines is

TABLE 7-1. TRACTOR REQUIREMENTS FOR VARIOUS CROP-PRODUCTION OPERATIONS

Operation	Type and amount of power						Best method of transporting machine		Under-axle clearance		
	Tractive			PTO							
	Low	Medium	High	Low	Medium	High	Trailed	Carried	Low	Medium	High
Conventional tillage........		x					x		x		
Rotary tillage.............	x				x		x	x	x		
Planting and seeding.......	x						x	x	x		
Cultivating................		x						x		x	
Spraying and dusting.......	x			x				x			x
Mowing and crushing.......	x			x				x	x		
Raking....................	x			x			x	x	x		
Baling....................		x			x		x	x	x		
Chopping.................		x				x	x	x	x		
Blowing (stationary).......						x					
Windrowing...............	x				x			x		x	
Combining................		x				x	x	x		x	
Cotton picking............	x			x				x		x	
Corn topping.............	x			x				x			x
Corn picking.............		x		x				x		x	
Manure loading...........		x		x				x	x		
Manure spreading.........	x				x		x		x		
Feed transport............	x			x			x		x		
Feed grinding (stationary)...					x						

high, and a chassis specially adapted for carrying various harvesting machines offers the possibility of lower total cost.

A *high-clearance self-propelled chassis* is needed for spraying and dusting tall crops. Power and traction requirements are low for these operations, and the machines are light in weight.

Only the very large farm will have enough use for each of the above types of tractors to benefit from their increased efficiency on specialized jobs.

It is probable that the above requirements must be met by not more than two different types of tractors if the average-size farm as well as the large farm is to benefit. The primary variables involved are weight, engine size, and configuration.

The ideal tractor would have only enough *weight* to do the job at hand, since weight compacts the soil and requires extra fuel to transport when the weight is not required for traction. Soil compaction is probably the most important factor. The ideal tractor would be basically lightweight but would have provision for doubling its weight without manual effort.

An *engine* is most efficient when operating at three-fourths to full load. Modern engines have a wide range of operating speeds and can usually be operated at low speed for light loads if teamed with adequate power-transmission ratios. In this way, reasonable fuel economy can be obtained with light loads, and high power is still available at high engine speeds.

The development of tractor *configurations* which have a broader range of efficient use will require thorough analysis of implement needs and perhaps new implement mounting systems. In view of the past history of farm-equipment development, it is reasonable to assume that such designs will be forthcoming.

Chapter 8

TRACTOR PERFORMANCE TESTS

Tractor performance tests are usually concerned with engine output (measured by belt or power-take-off dynamometers) and fuel economy and with drawbar pull, drawbar horsepower, and drawbar fuel company.

DYNAMOMETERS

Engine dynamometers must absorb and measure power. If operated by belt, they do not account for belt transmission losses. There are many types of dynamometers.

A *Prony brake* consists of a large wheel with a block or band brake and attached arm. The torque on the arm is measured by scales. The band can be adjusted to vary torque. Water cooling is necessary. These brakes are simple and useful for short runs, although the load cannot be controlled as well as with the other types.

A *water brake* consists of a wheel or disk rotating in water, with the resulting fluid friction absorbing the power. The water is allowed to circulate through the wheel housing to carry away the heat generated. Torque is measured by the reaction of the cradle-mounted housing against a scale. The Clayton water dynamometer utilizes a closed fluid circuit passing through a heat exchanger. The rotor and housing vanes function somewhat like a fluid coupling, and torque is changed by varying the amount of water in the homogeneous water-air mixture resulting from the vigorous agitation.

A *hydraulic pump* forcing oil through a restricting orifice may be used for power absorption. Torque on the pump may be determined by using a cradle mounting. The oil must be cooled to remove heat.

Electric generators with resistance coils are popular for precise test work. The generator may be cradle-mounted for torque measurement, or electrical output may be used as an indicator of power output after a calibration curve has been established. The resistance coils must be cooled by air circulation.

Magnetic-drag or *eddy-current brakes* are cradle-mounted for torque measurement. They may be air- or water-cooled.

Drawbar dynamometers measure drawbar pull where power absorption is provided by a trailing implement, a trailing tractor with the engine being motored against compression, and/or a specially built vehicle having wheels which drive one of the power-absorption devices listed above under engine dynamometers.

A *spring scale* is the simplest drawbar dynamometer, but it is difficut to dampen the vibrations enough to permit accurate readings.

Hydraulic cylinders or pressure cells are popular because the oil pressure indicates pounds pull and an orifice restriction can provide adequate dampening.

Electric strain gages are very simple to apply if instrumentation is available. Recording instruments give very accurate records of draft peaks as well as average draft throughout a test. An *electrical integrating unit* eliminates the necessity of planimeter measurements to determine the average reading where the reading is proportional to electric-current flow, as in the case of strain-gage recording of draft. The electrical integrator is a small d-c electric motor in which the speed of rotation is proportional to current flow. It is geared to a dial which, in effect, records the total current that has passed through the motor. The dial reading divided by the time gives the average current value, or the average recorded force.

PERFORMANCE CRITERIA

One horsepower in the foot-pound-second system is developed when working at the rate of 550 ft-lb per sec. A metric horsepower is 75 kg-m per sec and is equivalent to 0.9863 hp in the fps system.

Observed belt horsepower is that delivered by the belt to the dynamometer. For a dynamometer having a torque arm

$$\text{Hp} = \frac{2\pi LFN}{33,000}$$

where L = length of torque arm, ft
F = force at end of torque arm, lb
N = dynamometer speed, rpm

Commercial cradle-type dynamometers are made with the torque arm of such a length that the constant $(2\pi L/33,000)$ has some convenient value such as 0.001 or 0.0005 to facilitate calculations.

Corrected belt horsepower is the observed horsepower corrected for standard conditions. The formula is as follows:

$$\text{Corrected hp} = Hp_o \left(\frac{P_s}{P_o - P_w}\right) \sqrt{\frac{T_o + 460}{T_s + 460}}$$

where Hp_o = observed horsepower
P_s = standard barometric pressure, in. Hg
P_o = observed barometric pressure, in. Hg
P_w = water vapor pressure in air, in. Hg
T_s = standard temperature, °F
T_o = observed temperature, °F

Brake mean effective pressure, or bmep (end of belt), per square inch of piston area is a measure of the power output of an engine for its size and speed. The formula for a four-stroke-cycle engine is

$$\text{bmep, psi} = \frac{792,000 \text{ (belt hp)}}{(\text{engine displacement, cu in.})(\text{engine rpm})}$$

The formula for a two-stroke-cycle engine is

$$\text{bmep, psi} = \frac{396,000 \text{ (belt hp)}}{(\text{engine displacement, cu in.})(\text{engine rpm})}$$

Bmep is proportional to engine torque and peaks at the same speed as torque.

One *horsepower-hour* is the amount of work performed by working at the rate of one horsepower for one hour and is equivalent to 1,980,000 ft-lb.

Horsepower-hours per gallon of fuel consumed is a measure of engine efficiency, as is pounds of fuel per horsepower hour. The latter is also called *specific fuel consumption,* commonly used by engineers the world over. The former term, abbreviated hp-hr per gal, is more convenient for the user who buys his fuel by the gallon. Conversion of one figure to the other can be made by using the weights per gallon listed in Table 9-1.

Drawbar pull is the horizontal component of the force exerted at the drawbar.

$$\text{Drawbar horsepower} = \frac{(\text{drawbar pull, lb})(\text{speed, fpm})}{33,000}$$

Drawbar horsepower-hours per gallon is a measure of engine and tractive efficiency, as is pounds of fuel per drawbar horsepower-hour.

Since tractive efficiency varies greatly with the type of surface (Chap. 4), comparisons between tractors must be made on similar surfaces. A hard weatherproof surface, such as concrete or asphalt, is favored because the change in condition is at a minimum. Unfortunately, test results on such a surface are greatly different from those which would be obtained on agricultural soils. A comparative advantage on the hard surface, however, usually holds true on other surfaces.

An approximation of the drawbar pull on a particular soil surface can be made by multiplying the pull observed on the test surface by the ratio of the coefficient of traction known or estimated for the soil to that observed on the test surface and then subtracting the increase in front-wheel rolling resistance. In the same way drawbar horsepower can be approximated by multiplying the test horsepower by the ratio of tractive efficiencies. Tractive efficiencies and coefficients of traction for several soils are listed in Chap. 4.

NEBRASKA TRACTOR TESTS

The state of Nebraska requires that any tractor model to be sold in the state undergo tests by the Nebraska Tractor Testing Board at the University of Nebraska.

TABLE 8-1. PERFORMANCE FACTORS FOR 62 TWO-WHEEL-DRIVE FARM TRACTORS, BASED ON 1958 AND 1959 NEBRASKA TRACTOR TESTS

	Gasoline			Diesel		
	Min.	Av.	Max.	Min.	Av.	Max.
Bmep at full load...........................	92	101	110	66	90	103
Belt hp-hr per gal at full load	9.8	11.4	12.8	12.6	14.3	17.8
Av. belt hp-hr per gal for varying load tests.......	7.7	9.1	9.9	9.8	12.1	15.2
Drawbar hp-hr per gal at full load..............	8.4	10.0	11.2	10.5	12.8	15.2

	Gasoline and diesel		
	Min.	Av.	Max.
Max. drawbar hp in per cent of max. belt hp at same engine speed.......................	80.1	89.0	94.5
Drawbar pull in per cent of total weight.......	66.1	74.7	84.7

Maximum belt or power take-off horsepower and fuel consumption over a 2-hr period are first determined, followed by varying load and fuel-consumption tests. Drawbar tests include a 10-hr test at 75 per cent of the pull exerted at maximum drawbar horsepower. Any repairs or adjustments required during the test are also reported. Detailed information on these tests and copies of test reports are available from the University of Nebraska, Agricultural Engineering Department, Lincoln, Neb. In general, these tests comply with the SAE and ASAE Standard, Agricultural Wheel Type Tractor Test Code.[1]

The range and average of several basic performance factors for tractors are listed in Table 8-1. These values are based on Nebraska Tractor Test results.

[1] *Agr. Eng. Yearbook,* 1959, pp. 65–71. L. W. Hurlbut, L. F. Larsen, G. W. Steinbrugge, and J. J. Sulek, Nebraska Tractor Tests in Relation to the SAE-ASAE Agricultural Tractor Test Code, *ASAE Paper* 59-624, December, 1959.

Chapter 9

FUELS AND COMBUSTION

WAYNE W. WORTHINGTON

The operating cycle of all internal-combustion engines includes the following steps:
1. *Introduction of fuel into the combustion air* may occur
   a. Before the combustion air enters the engine cylinder: The mechanical means for metering the fuel and air and mixing them to form a homogeneous mixture may employ a carburetor or injectors which spray the fuel into the air stream immediately ahead of the intake valves. In either, the fuel-air charge is ignited with an electric spark timed for most efficient combustion.
   b. Directly into the combustion chamber as the piston nears the end of its compression stroke. This method employs high-pressure pumps (1,800 to 4,200 psi) and injectors which function to finely atomize the fuel and propel it into the air, as is necessary to ensure homogeneous mixing and efficient combustion. This procedure characterizes all diesel engines.
2. *Compression* of the fuel-air charge (method 1a) or of the air alone (1b) to the highest point consistent with controlled combustion.
3. *Ignition and combustion* of fuel. In the case of spark-ignition engines, because of the rapid velocity of the flame front throughout the mixture, combustion is said to take place at constant volume. In diesel engines, because of the relatively long duration of both the injection and burning of the fuel, combustion is said to take place at constant pressure.
4. *Expansion* of heated gases, at which time the heat of combustion is converted into work.

When but two strokes of the engine piston, as effected by a single revolution of the crankshaft, are necessary to include all the above events (including the introduction of air into the cylinder and scavenging of gases resulting from combustion), the engine is said to operate on the *two-stroke-cycle* principle (commonly termed *two-cycle*). Similarly, when four strokes of the engine piston, involving two revolutions of the crankshaft, are necessary to effect the same cycle of events, the engine is known as *four-cycle*.

The approximation most commonly used in analyzing the compression and expansion strokes of the combustion engine is that both involve "perfect gases," i.e., those which observe the laws of Boyle and Charles and have a constant specific heat. Since the actual medium used in combustion engines is largely air, its characteristics are likewise generally used:

Density.......................... 0.0765 lb/cu ft at 14.7 psi and 60°F
Specific heat:
   At constant pressure........... 0.240 Btu/lb
   At constant volume............. 0.171 Btu/lb

When compression or expansion occurs without loss or addition of heat (constant enthalphy), the process is known as *adiabatic*. It can be demonstrated that the relationships of the changing pressures and volumes (P_1, P_2, P_3, etc., and V_1, V_2, V_3, etc.) and *air standard efficiency* (η) may be represented by the following equations:

$$P_1 V_1^{1.4} = P_2 V_2^{1.4}$$

and

$$\eta = 1 - \left(\frac{1}{r}\right)^{0.4}$$

where

$$r = \frac{P_2}{P_1} = \frac{V_1}{V_2} = \text{expansion ratio}$$

Under *actual conditions of operation,* liquid fuel within the cylinder takes its heat of vaporization from the combustion air. Heat passes from the combustion-chamber surfaces into the gas charge, and vice versa, by both radiation and convection and at varying rates. As a result, both compression and expansion are *polytropic*, rather than either isentropic or adiabatic. It has been found that in diesel engines, when the "practical" limit of compression ratio is reached, compression is more nearly *isothermal* (at constant temperature) than adiabatic. In a polytropic process, the ratios of pressure and volume tend to change constantly throughout either compression or expansion, although the ratio of pressure and volume at the beginning and end, respectively, of either process may be expressed by the equation

$$P_1 V_1^n = P_2 V_2^n$$

The exponent n falls within a range between 1.2 and 1.45 and is not the same during compression as during expansion. However, both the *mean effective pressure* (specific output) and *thermodynamic efficiency* within the engine cylinder remain some function of the expansion ratio. This characteristic accounts for the continuing increase in compression (expansion) ratios of spark-ignition engines, as means have been developed to control the detonation characteristics of the fuels.

FUELS

Tractor engines of today operate upon the following generally available fuels:
1. Gasoline with controlled octane number.
2. "Tractor" fuels. These usually are mixtures of fuels having varying volatility characteristics to effect easy starting, but with low octane numbers.
3. Liquefied petroleum gases (LPG).
4. Diesel fuels.

The first three are used in spark-ignition engines, and the latter in diesel engines. A few spark-ignition tractors, intended for use principally in areas where the only liquid fuels available are those with uncontrolled volatility and combustion characteristics, employ highly heated intake manifolds to effect fuel vaporization and low compression ratios to hold detonation to a tolerable, although undesirable, level. Both contribute to low specific output and high specific-fuel consumption.

Largely for reasons of economy, the use of LPG fuels has rapidly increased in geographical areas where these fuels are freely available, usually at "distress" prices.

In addition to pure propane and butane, commercial LP gas frequently contains other hydrocarbons, including isobutane, pentane, propylene, and butylene, in varying proportions. These all have low boiling points and high vapor pressures and must be handled in closed containers. Pressure reducers, heating (vaporizing) elements, and mixers are used in place of conventional carburetors.

Characteristics of distilled tractor fuels, which are carefully controlled by the refiners, are as follows:

1. *Volatility and distillation range.* Since gasoline is the most generally used motor fuel, its volatility and distillation range are carefully controlled and changed with the seasons to effect easy starting and prevent dilution of crankcase oil. To a lesser extent, this is true of "tractor" and "power" fuels, which in summer permit starting without the use of a more volatile starting fuel. In the case of both "tractor" and "power" fuels, the end point is largely determined by the requirements necessary to the operation of burners used in domestic heating furnaces. The end point and re-

covery of diesel fuels are controlled to maintain a low smoke level and limit dilution of crankcase lubricating oils. The distillation characteristics of commercial tractor fuels are shown in Fig. 9-1.

2. *Initial boiling point (IBP) and vapor pressure.* These are carefully tailored, as required for easy starting and minimum evaporation loss during storage, by the use of "light fractions," which are usually available in great quantities. At summer temperatures, the vapor pressure must be kept sufficiently low to prevent vapor lock.

3. *Sulfur content.* The presence of sulfur tends to form corrosive compounds which are destructive of parts within the crankcase, particularly where moisture is present and electrolytic action can be established between adjacent surfaces. Sulfur also causes rapid and excessive wear of upper-cylinder surfaces and piston rings. Highly com-

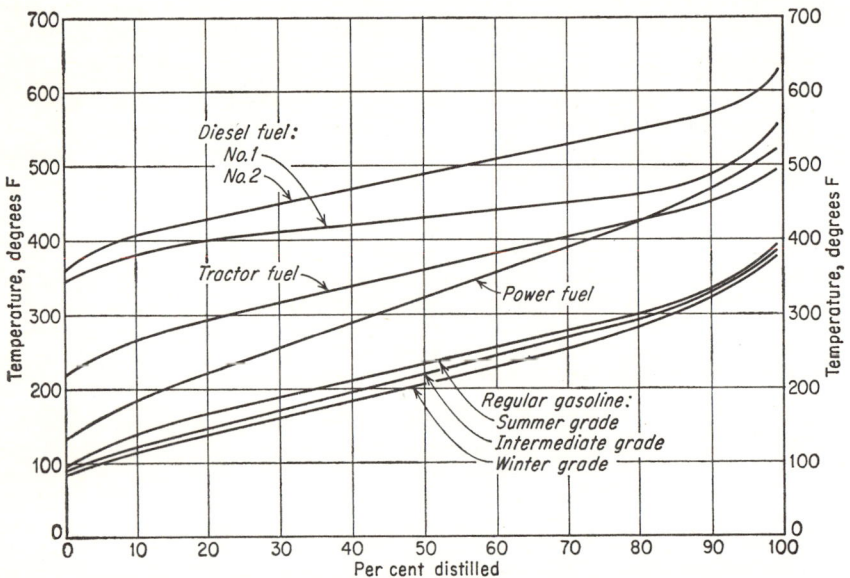

Fig. 9-1. Distillation characteristics of tractor fuels. (*Standard Oil of Indiana.*)

pounded oils, which limit and reduce the harmful effects of sulfur, have become increasingly available and are coming into more general use. Federal and many state specifications on gasoline usually limit maximum sulfur content to 0.1 per cent. ASTM Tentative Specifications limit sulfur content as follows:

	Per cent
No. 1 fuel (furnace)	0.5
Diesel fuel:	
No. 1	0.5
No. 2	1.0

4. *Carbon residue.* The commonly used Conradson carbon-residue test measures the residue remaining upon evaporation under specified conditions. Although it does not correlate closely with operating results, it is an indication of carbon-forming qualities under many engine conditions.

5. *Gum content.* Many hydrocarbon fuels, particularly cracked fuels containing unstable unsaturated compounds, tend to form viscous liquids or solids which clog carburetor float valves, screens, and jets, as well as causing "sticky" engine inlet valves. With distillates and diesel fuels, gum is rarely a serious problem.

FUELS AND COMBUSTION

6. *Flash test.* The flashpoint of a fuel is the temperature of the fuel sample at which inflammable vapors are given off. This test is frequently used as a control test for diesel fuel.

7. *Detonation and ignition characteristics.* With fuels for spark-ignition engines, the detonating characteristics are expressed as octane numbers. Two methods of testing are employed. The *research method* gives reproducible laboratory results and is most commonly used for process control. The other is the *motor method,* which correlates closely with road and field operation. The *octane number* of a fuel is the percentage of iso-octane in the mixture of a reference fuel, composed of iso-octane and heptane, which has the same intensity of "knock" as the fuel being tested. Some fuels exceed iso-octane in antiknock quality, and the octane scale has been extended above 100 in order to rate them.

For smooth knock-free operation with diesel fuels, it is important that ignition take place with a minimum of delay following the injection of the fuel into the com-

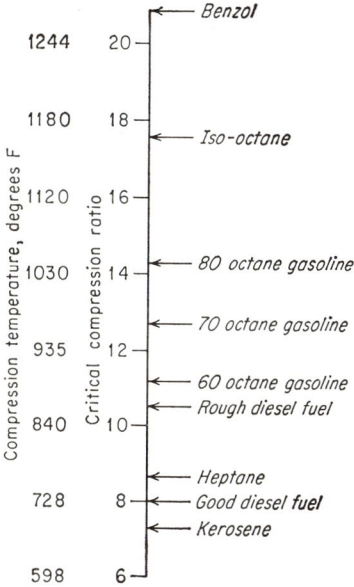

Fig. 9-2. Compression temperatures and critical compression ratios for various fuels which will ignite by compression-ignition in a CFR engine. (*Waukesha Motor Co.*)

bustion chamber. The measure of the "ignition characteristic" of diesel fuel is its *cetane number,* which is the percentage of normal cetane in a reference fuel composed of normal cetane and alpha-methylnaphthalene, which has the same ignition delay as the fuel being tested. The American Society for Testing Materials has established and published standardized procedures for making all these tests, with the exception of determining octane ratings higher than 100.

The characteristics of commercially available LPG and liquid fuels are shown in Table 9-1.

Figure 9-2 shows the critical compression ratios and compression temperatures of various fuels which will ignite by compression ignition in a CFR (Cooperative Fuel Research) single-cylinder variable-compression-ratio engine.

COMBUSTION

All commercial internal-combustion-engine fuels are composed chiefly of carbon, hydrogen, and oxygen, and the substance with which they react is always air. The *stoichiometric ratio* of air to fuel defines the proportion of fuel and air necessary to effect a complete and final chemical combination of all the carbon and hydrogen in the fuel with all the oxygen in the air. Since the burning of carbon to CO liberates more heat than the burning of CO to CO_2, *maximum specific power output* (per pound of air burned) is obtained when incomplete combustion occurs and all the carbon in the fuel burns to CO. *Maximum economy* results when the ratio of available fuel to air is such that all carbon is burned to CO_2.

The various impurities in the fuels, and the rarer gases in the air, are generally present in such small quantities that their effects upon combustion may be neglected. It follows that the products of combustion are made up of various compounds of oxygen, hydrogen, carbon, and nitrogen, including CO_2, H_2O, N_2, O_2, CO, H_2, OH, NO, N_2O, and NO_3. Changes in chemical composition occur continually throughout combustion, and the composition of the combustion products in equilibrium after com-

TABLE 9-1. CHARACTERISTICS OF COMMERCIALLY AVAILABLE TRACTOR FUELS

Fuel	Weight, lb per gal	Heating value		Average octane number
		Btu/lb	Btu/gal	
Propane.................	4.25	21,680	92,140	115
Butane..................	4.80	21,300	102,200	100
Regular-grade gasoline......	6.10	20,300	124,100	85
Power fuel...............	6.50	20,020	130,100	40–50
Kerosene.................	6.76	19,830	134,100	0–10
Tractor fuel..............	6.85	19,800	135,630	25–40
Diesel fuel:				
No. 1.................	6.90	19,600	135,250	45–55 cetane
No. 2.................	7.10	19,500	138,750	40–50 cetane

SOURCE: Standard Oil Co. of Indiana, 1957.

bustion is a function of combustion temperature and pressure. Considering the theoretically complete reaction of iso-octane and air,

$$C_8H_{18} + 12\tfrac{1}{2}O_2 + 47N_2 \rightarrow 8CO_2 + 9H_2O + 47N_2$$

In terms of weight,

1 lb iso-octane + 15 lb air → 3.1 lb carbon dioxide + 1.4 lb water + 11.5 lb nitrogen

Detonation in the Spark-ignition Engine. *Detonation* is that phenomenon commonly causing the "pinging," "spark knock," "preignition," and "carbon knock" familiar to all tractor and automobile drivers. Laboratory investigations have found that this characteristic sound is accompanied by the following combustion phenomena:

1. Rapid increase in flame speed near the end of the flame travel.
2. An increase in rate of pressure rise and maximum pressure as the last portion of the change burns.
3. High-pressure waves within the cylinder gases, starting near the end of combustion.
4. A great increase in the amount of heat radiated from the flame front near the end of combustion. This is frequently accompanied by overheating and boiling of the cooling water.

Even with slight detonation, the pressure waves are of sufficient intensity to set up audible vibrations in the cylinder walls and throughout the engine. In general, *detonation is the factor which determines the maximum practicable compression ratio for any given engine.*

On the part of the fuel refiners, there has been a consistent trend over the years to increase the octane number of motor fuels, both by fuel composition and by the addition of tetraethyl lead, an antiknock agent. Tractor manufacturers have been quick to take advantage of fuel improvement by increasing the compression ratios of their engines.

Design features introduced by manufacturers of spark-ignition engines to *reduce the tendency to detonate* include the following:

1. Use of more compact combustion chambers, thus reducing the length of flame-front travel.
2. Increase of surface to volume ratio. This characterizes engines with large bore-to-stroke ratios.
3. Introduction of *quench areas* at the more remote distances from the spark plug. This has the duel effect of cooling the fuel air charge in the remote areas and creating rapid turbulence by producing violent inward movement of the air, termed *squish*.

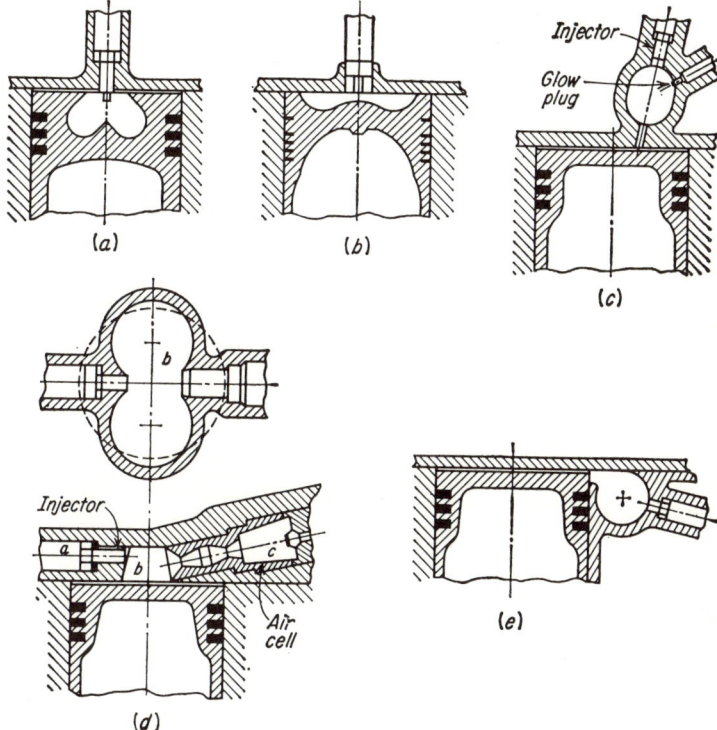

Fig. 9-3. Diesel combustion chambers: (*a*) toroidal; (*b*) open; (*c*) precombustion; (*d*) **Lanova** air-cell; (*e*) swirl.

4. Reduction in temperature of "hot-spot" areas not subject to direct cooling, by means of improved water circulation around exhaust-valve seats and spark-plug bosses.
5. Locating the spark plug as close as possible to the hottest spot in combustion-chamber wall consistent with minimum flame travel. This is usually close to the exhaust valve.
6. Water-jacketing all cylinder-wall areas exposed to combustion gases.
7. Higher rotative speeds.
8. Introduction of high turbulence, through the use of directional intake ports to induce swirl.

In general, design features which improve specific output effect a similar improvement in efficiency.

Combustion in Diesel Engines. A wide variety of combustion-chamber configurations have been developed for diesel engines to obtain the high air utilization and effective combustion necessary for (1) high specific output and low specific-fuel consumption, (2) low detonation (particularly at light loads when thermal conditions are poor), (3) minimum exhaust smoke (particularly when lugging), (4) easy starting, and (5) control of the explosion forces acting on the piston head.

The principal *styles of combustion chambers* used in the tractor engines of today are shown in Fig. 9-3. The characteristics of these systems, as determined from published official tests and impartial investigations, may be summarized as follows:

1. *Toroidal chamber* in piston (Fig. 9-3a). High turbulence, due to high swirl rate resulting from use of directional ports and squish. Excellent fuel consumption and cold-weather starting, due to minimum surface exposed to flame. High explosion pressures, effective directly upon piston heads.

2. *Open chamber* with recessed piston (Fig. 9-3b). Same as for toroidal.

3. *Precombustion chamber* (Fig. 9-3c). Satisfactory operation with simple pintle-type injector nozzles. High ratio of surface area to combustion-chamber volume makes for poor cold-weather-starting characteristics and comparatively high specific-fuel consumption. Moderate pressures effective on piston heads.

4. *Air cell* (Lanova, Fig. 9-3d). Regularly used with multiple-orifice fuel injectors. Ratio of surface area to chamber volume and refrigerating action of air emerging from air cell makes for poor starting characteristics and high specific-fuel consumption, as compared with open and toroidal chamber. Moderate pressures effective on piston heads.

5. *Swirl chamber* (Fig. 9-3e). Most prevailing British and Continental type. Insensitive to minor changes in configuration. At least two such commercial engines are highly insensitive to fuel with low cetane values and have been developed for emergency gasoline operation. Other characteristics are generally comparable with air-cell type.

Chapter 10

POWER CONTROLS FOR IMPLEMENTS

The first use of the tractor was to replace the horse as a source of tractive effort. Eventually the power take-off evolved as a more efficient means of supplying rotative power than through implement ground wheels. With the advent of implements carried by the tractor, the problem of applying mechanical power to lifting and lowering was attacked.

Mechanical power lifts were developed first and consisted of a half-revolution dog clutch (similar to that used on trailing plows) usually powered from the tractor PTO shaft. A lift shaft and arm were actuated to lift an attached implement clear of the ground or to lower it into working position. The depth was varied by hand levers. A power lift of this type is shown in Fig. 10-1.

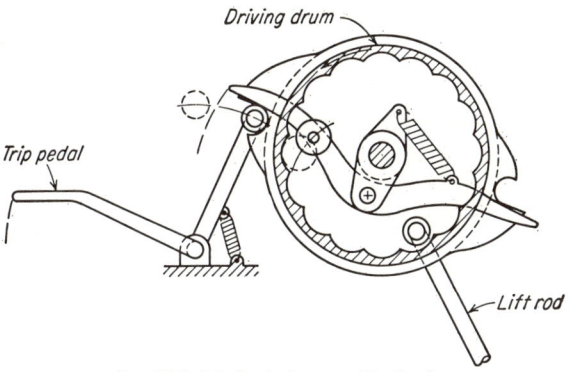

FIG. 10-1. Mechanical power lift clutch.

This control gave welcome relief from hand lifting of heavy cultivator gangs but presented some mechanical problems because of the shock loading inherent in this type of clutch.

Cable controls powered by winch drums are extensively used with heavy earth-moving equipment. The first tractor loaders were lifted by wire cables. Some earth-moving machines are using an electric generator driven by the tractor engine to power electric-motor-driven winches mounted on the trailed section.

A *pneumatic cylinder* powered from the tractor exhaust was used to lift and lower implements on small tractors but, because of space and power limitations, is no longer used.

The first *hydraulic power lift,* introduced in 1935 by Deere, was interchangeable in function with a mechanical power lift and had only two positions, up and down. It did, however, cushion the application of lifting power and control the rate of drop. This was followed by other hydraulic controls having additional features, until the scope has extended far beyond mere control and into application of working power as in the case of hydraulic loaders, stackers, back hoes, etc.

In view of the applications of hydraulic power which are used with farm tractors, a tractor *hydraulic system* should have the following:

1. At least sufficient power output to lift the heaviest implement used in 1½ to 2 sec.[1] If loaders or stackers are to be operated, additional capacity may be needed.
2. Direct engine-driven pump, particularly important in applications with, say, loaders, where it is desirable to interrupt forward motion by clutching while continuing to lift.
3. Built-in linkage lift for attached lift-type implements.
4. Provision for connecting one or more remote cylinders and controlling them independently.
5. Position-responsive control (setting of cylinder position by control-lever position) of either built-in cylinders or remote cylinders.
6. Changeover to draft control (automatic adjustment to give constant draft) for either built-in cylinders or remote cylinders.
7. Delayed lift and drop for rear cultivator gangs in order to stop and start at same place as front gangs.
8. Selective lift for either side of a two- or four-row cultivator for "point" rows.
9. Provision for operating a power-steering system.

The evolution and development of hydraulic implement controls is still proceeding at a rapid pace, and there is much variation in approach and execution. Space does not permit a complete description and explanation of the various systems, but the common components and problems are covered briefly below.

Pumps. The pump work capacity required for implement control is governed by the heaviest implement to be lifted, usually a trailing type. As a result of cooperative tests by various manufacturers, recommended thrust capacities of standard ASAE-SAE remote cylinders for the various sizes of tractors, based on manufacturer's ratings, are shown in Table 10-1.

TABLE 10-1. MINIMUM RECOMMENDED THRUST CAPACITY* WITH STANDARD ASAE-SAE REMOTE CYLINDER

Size tractor (manufacturer's rating)	Thrust, lb	Lifting hp†
2-plow	5,625	3.4
3-plow	7,500	4.5
4-plow	9,375	5.7
5-plow	12,500	7.6

* Capacity at relief-valve opening pressure.
† Lifting horsepower based on standard 8-in. stroke in 2 sec.
SOURCE: Wayne H. Worthington and J. Waldo Seiple, Hydraulic Capacity Requirements for Farm Implements, *Agr. Eng.*, 33:273–278, May, 1952.

The hydraulic horsepower delivered by a pump may be found by the following formula:

$$\text{Hydraulic hp} = (\text{output, gpm})(\text{delivery pressure, psi})/1{,}714$$

The figure obtained, however, must be corrected for valve leakage and pressure losses in valves and lines and the efficiencies of the respective hydraulic cylinders or motors used in order to find the actual lifting horsepower available.

A positive-displacement type of pump is required for tractor hydraulic systems, and piston pumps or constant-delivery rotary pumps such as the gear, internal rotor, and vane types are preferred. Cross sections of typical pumps are shown in Fig. 10-2.

The *piston pump* is generally considered to be best adapted to pressures of 2,000 psi or more, but it is necessary to use several pistons to avoid excessive pulsation. Constant-delivery rotary pumps are simpler and perform well at their recommended

[1] ASAE Standard: Application of Hydraulic Remote Control to Farm Tractors and Trailing-type Farm Implements, *Agr. Eng. Yearbook*, 1959, pp. 51–56.

pressures, usually below 1,500 psi. The pump is driven direct from the engine or from a constant-running power take-off in the newer designs in order to retain hydraulic control when the clutch to the drive wheels is disengaged.

Valves. Provision must be made for stopping or diverting the flow of oil when a constant-running positive displacement type of pump supplies power intermittently.

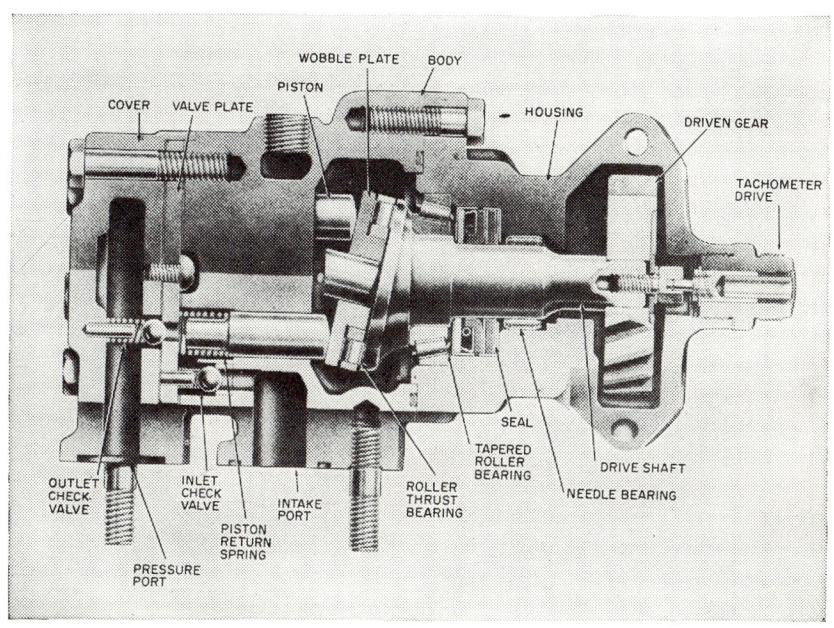

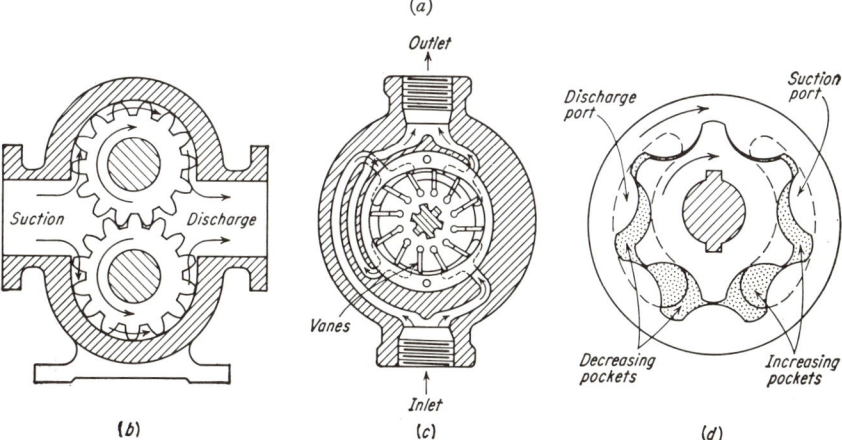

Fig. 10-2. Hydraulic pumps: (a) axial piston (*Tractor and Implement Div., Ford Motor Co.*); (b) gear; (c) balanced vane (*Vickers, Inc., Detroit, Mich.*); (d) internal rotor (*Eaton Mfg. Co., Pump Div., Detroit, Mich.*).

This may be done by throttling the intake, as in the Ferguson piston pump, by diverting the flow through the control valve, as in the Allis-Chalmers system (described later), or by an *unloading valve*, which bypasses oil to the reservoir under low pressure when it is not needed. Unloading valves are usually hydraulically actuated from the control valve as shown in Fig. 10-3b.

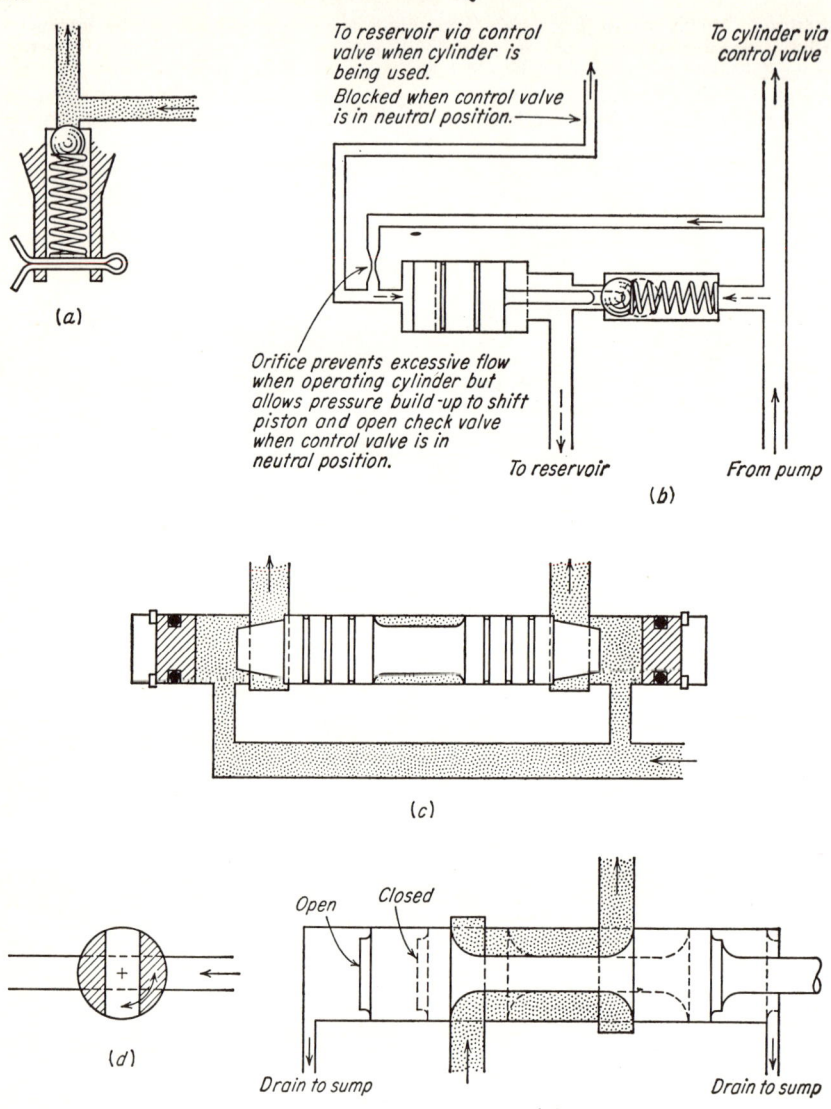

Fig. 10-3. Hydraulic valves: (a) ball-check relief valve; (b) differential-pressure unloading valve (*International Harvester Co., Chicago, Ill.*); (c) flow-dividing valve (*Deere & Co.*); (d) rotary valve in closed position; (e) spool valve.

A safety or *relief valve* is included in all systems to prevent damage to the system or to the implement due to excessive pressure or cylinder thrust. Such valves are designed to open at a particular pressure and should be capable of discharging the pump's maximum output without appreciably exceeding the opening pressure and also of reseating at a minimum drop below the opening pressure.[1] A simple ball-type relief valve is shown in Fig. 10-3a.

[1] H. A. Lehmann, Hydraulic Relief Valve Problems and Design, *ASAE Paper* 59-626, December, 1959.

Check valves, often used to hold the piston in working position, are similar to the relief valve shown in Fig. 10-3a but with a light spring which allows oil to pass at low pressure in one direction and prevents flow in the other direction unless the valve is mechanically pushed from its seat.

An *automatic-flow dividing valve* is desirable when two cylinders must operate simultaneously from one pump, regardless of differences in load. The valve shown in Fig. 10-3c utilizes Bernoulli's principle, the spool shifting to restrict flow on the side having greater flow and consequent lower static pressure. Several cylinders can also be operated simultaneously by a *series circuit*.

Control valves are usually of the spool type because they can be hydraulically balanced more easily than rotary valves. An unbalanced control valve has excessive actuating friction at high pressures. As shown in Fig. 10-3d, a simple unbalanced rotary

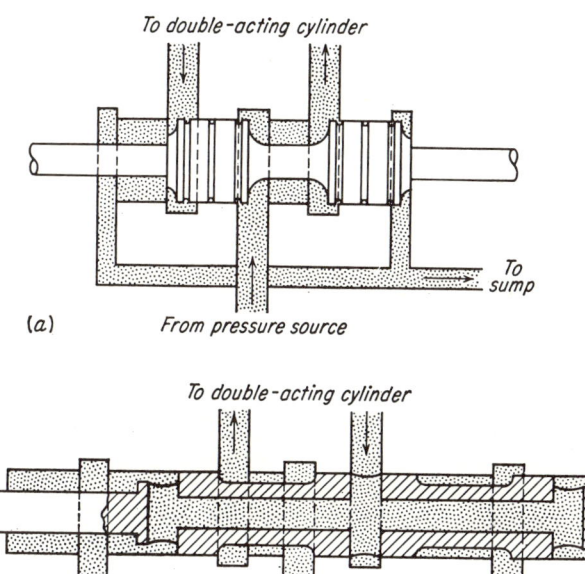

Fig. 10-4. Four-way valves: (a) closed center; (b) open center.

valve is forced against the opposite side of a journal when it blocks a port, whereas the spool valve in Fig. 10-3e has a uniform pressure loading around it at all times and at each section, regardless of whether a port is opened or closed. Annular ports are used to maintain balanced pressure around the valve. Annular grooves are often cut in the pistons of a spool valve to collect leakage and prevent unbalanced pressure. Figure 10-4a shows a closed-center four-way valve suitable for applications which do not require low-pressure discharge of oil when the cylinder ports are both closed. Figure 10-4b shows an open-center spool valve which allows free-flow return when in neutral position.

Such valves may be spring-centered (returned to neutral by spring pressure when the control lever is released) and held open manually or, more often, held open by a cam or detent which is released by the increased oil pressure resulting when the operating piston reaches the end of its stroke.

Control Systems. *Position-responsive control* is achieved by the use of a follow-up or servo system, whereby the operating piston follows the movement of the control lever and stops in a corresponding position. This is achieved by having both the hand-control lever and the operating piston act on the control valve through a lever arrangement. When the hand-control lever is shifted, the valve is opened and shifts the operating piston until its movement is sufficient to bring the valve to neutral again, thus stabilizing the piston in a new position corresponding to the new control-lever setting. The principal of position-responsive control is illustrated in Fig. 10-5.

Draft control is achieved by having a draft-responsive member act on the control valve in cooperation with the hand-control lever. If draft increases, the valve is shifted and moves the operating piston in the direction which will reduce draft to the equilibrium value, returning the valve to neutral. A plow, for instance, will be

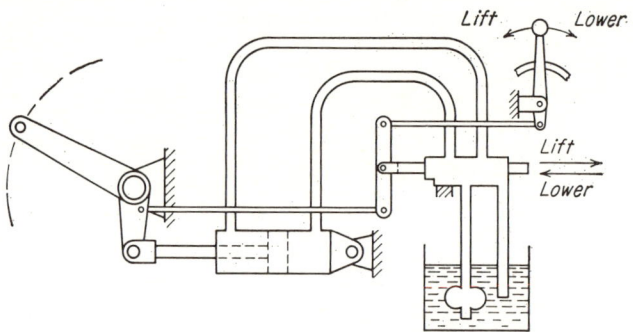

Fig. 10-5. Position-responsive control by follow-up system.

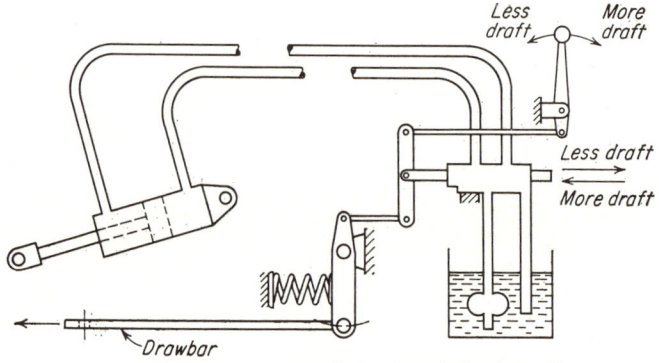

Fig. 10-6. Draft control of cylinder through drawbar pull.

lifted to reduce draft. In Ford tractors the top-link movement, as it pushes against a spring, actuates the control valve. In Massey-Ferguson tractors, the control valve is controlled by top-link tension as well as compression. In the Allis-Chalmers system, drawbar movement resulting from drawbar pull against a control spring actuates the control valve. With this system, a remote cylinder on a trailing machine, as well as the cylinder lifting a mounted implement, can be draft-controlled. The principle of draft control is illustrated in Fig. 10-6.

Automatic delayed lift of rear cultivator gangs can be secured, where front and rear gangs are operated by separate cylinders, by incorporating a "delayed-action" check valve in the line to the rear cylinder. Sufficient pressure to open the check valve is not built up until the front gangs are completely lifted against a stop, thus delaying the lifting of the rear gangs. Delayed drop of the rear gangs is usually secured by moving the control lever halfway to drop the front gangs, the check valve holding the rear cylinder, and the remaining control-lever movement mechanically lifts the check valve from its seat to drop the rear gangs. The delay time should be adjusted according to the speed of travel. Independent manual control of front and rear gangs is often highly desirable.

Selective lift of right-hand or left-hand cultivator gangs can be secured only where independent cylinders are used for each side, with separate control valves manually operated.

Remote cylinders and hoses for use on trailing machines usually comply with the ASAE Standard[1] as to dimensions, allowing any make of cylinder to be used with any implement complying to the standard. Double-acting cylinders are more versatile than single-acting. The diameter of the cylinder will vary as required to give the necessary operating force with the pressure available in the hydraulic system of the tractor for which the cylinder was designed. Accurate positioning of remote cylinders to give the desired depth, etc., is a problem, and several approaches are in use.

One method is to set a mechanical stop for the maximum desired action on the implement and provide slack in the connection to allow full movement of the hydraulic cylinder. Maximum plow depth can be set by a lever and shallower depths secured by stopping the cylinder before full depth has been reached.

A second method is to provide mechanical stops on the cylinder itself which limit its movement and "blow" the relief valve until the control valve is manually shifted to neutral or until pressure builds up and automatically returns a pressure-centered control valve to neutral.

A third method is to provide an adjustable stop on the cylinder which actuates an oil-blocking valve stopping return flow from the low-pressure side of the piston, thus stopping piston movement and "blowing" the relief valve until the control valve is shifted to neutral.

In the Oliver Hydra-lectric system, the control valve is electrically actuated by a solenoid. A stop collar on the remote piston makes electrical contact at the end of the stroke and thus returns the control valve to neutral. The position of the stop collar can be shifted from the tractor seat by an alternative control-switch position which causes the stop collar to be held with the cylinder by electromagnet and moved along the piston rod when the piston is actuated until it reaches the desired position where it is released. The two independent cylinders furnished with this system can be used on a trailing implement or mounted one on each side of the tractor, to control cultivator gangs, simultaneously or independently as may be desired.

A fifth method, used in the Massey-Harris Depth-O-Matic system, utilizes the slave-cylinder principle to control the remote cylinder. Here the remote cylinder is connected "in series" with the built-in cylinder and moves in unison with it. In the Massey-Harris system the built-in cylinder has adjustable stops on a rod attached to and moving with the piston. These stops contact the control-valve stem at either end of the stroke and automatically return the control valve to the center or neutral position. The linkage system normally actuated by the built-in cylinder can be disconnected or removed when the remote cylinder is used, since the built-in cylinder serves only to position the remote cylinder. The problem of synchronization and drift of the remote cylinder is solved by check valves in the piston which are opened by contact with the cylinder end each time the implement is fully lifted, allowing free passage of oil through the piston.

The slave-cylinder principle can be used to secure true position-responsive control of a remote cylinder by connecting it in series with a built-in cylinder controlled by the follow-up principle illustrated in Fig. 10-5.

When both lines to a double-acting cylinder are closed to hold the cylinder in a fixed position, a "thermal safety valve" is required to prevent possible cylinder breakage from oil expansion due to an increase in temperature.

Couplings. Where remote cylinders are used on trailing plows having spring-release hitches for overload protection, it is necessary that breakaway quick-disconnect hydraulic couplings be used in the hose lines to the remote cylinder mounted on such implements. Desirable characteristics for such couplings are as follows:

1. Breakaway with a low enough pull to prevent damage to the hoses or other parts of the system

[1] *Ibid.*

2. Automatic self-sealing on both sides without appreciable loss of oil when uncoupled
3. Flush surfaces on sealed ends to facilitate wiping off dirt before recoupling
4. Manual recoupling under working pressure permitted
5. Manual disconnecting under pressure permitted

Such couplings, in addition to breakaway ability, allow quick unhitching of the tractor from the implement when the remote cylinder is temporarily left on the implement. A support is necessary to keep the loose ends from dropping into the dirt or being damaged after uncoupling. A spring connection to the support is desirable to allow operating flexibility.

The principles involved in hydraulic control systems are basically simple, but in practice, systems which attempt to meet most of the ideal requirements become quite complex. As is usually the case in the early stages of a development, there has been much variation in details and in approach. Pressures in use vary from 610 to 3,500 psi, and controls from simple lift-lower systems to optional draft control or positioning control. Mounted implements may be controlled by built-in cylinders or by detachable cylinders which double as remote cylinders, and the cylinders may be single-acting ram type or double-acting piston type. Space does not permit a detailed description of the various systems, but they are usually composed of combinations and variations of the basic elements which have been briefly discussed here.[1]

[1] Henry A. Ferguson, Hydraulic Control Systems for Farm Tractor Implements, *SAE Quart. Trans.*, 5:259–272, April, 1951.

Chapter 11

TILLAGE OBJECTIVES

The basic purpose of tillage is to provide a favorable soil environment for the germination and growth of a particular crop.

Seedbed requirements for satisfactory germination are (1) sufficient warmth, (2) sufficient air, (3) adequate moisture, and (4) a layer of soil between the seed and the surface which can be penetrated by the sprout of the particular seed. Seeds vary widely in their soil-moisture requirement for germinating and also in the energy of their sprouts. Corn and potatoes have a low moisture requirement and also high sprout energy; hence they can be planted fairly deep and in loose soil. Grass seeds must be planted close to the surface because they have low sprout energy, but they also require moisture. If planted deep enough to stay moist, they may not be able to penetrate to the surface. A light mulch is valuable in securing a stand of grass because it retains moisture but does not offer resistance to the sprouts. In general, a fine firm seedbed favors germination but may crust over excessively and is often not the most desirable rootbed.

Good soil *tilth* is the most important *rootbed* factor, tilth usually being defined as the degree of aggregation of the soil. Ideal tilth is secured when the soil is aggregated into crumblike particles that allow free access of air and water and are not easily broken down by rainfall.[1]

Adequate *air circulation* in a soil is needed for the following purposes:

1. To supply oxygen for plant root respiration and absorption of nutrients. (There is evidence that potassium absorption is especially dependent on aeration.)

2. To supply oxygen for activity of the soil bacteria which transform organic matter into nitrates available to the plant. (The depressing effect of mixing large amounts of green manure or other organic matter into the soil may be partly due to the competition between the microorganisms and the growing plants for a limited supply of oxygen.)

3. To supply nitrogen for nitrogen fixation by the bacteria associating with legume roots.

A high degree of *aggregation* in a soil is also desirable for:

1. Freedom of root development and penetration
2. Maximum water-storage capacity to supply plant needs
3. Resistance to breakdown of surface particles under rainfall, preventing sealing over and allowing maximum water intake, thus reducing erosion due to runoff and to breakdown of aggregates into fine particles which are transported by water
4. Resistance to compaction by wheels or tracks of tractors and field machines and to compaction by the action of tillage tools.[2]

The importance of proper soil structure has recently been emphasized by attempts to produce record-breaking corn yields where plenty of water and nutrients were present but inadequate aeration was a limiting factor.

Inasmuch as water-stable soil aggregates are normally secured by the binding action of grass roots and soil organic matter, the exhaustion of organic matter by poor

---

[1] John A. Slipher, Soil Preparation for Meadow Crops, *Agr. Eng.*, 37:681–684, October, 1956.
[2] M. L. Nichols and C. A. Reaves, Soil Structure and Consistency in Tillage Implement Design, *Agr. Eng.*, 36:517–522, August, 1955.

farming practices is accompanied by a deterioration in soil structure, accelerating the decrease in soil productivity.[1] Recent experiments with chemicals which act as a binder for water-stable soil aggregates point up the importance of good soil structure. These materials do not serve as plant food but only modify soil structure. Nevertheless, increases in germination, water absorption, and yields have been observed, even where only the surface 2 or 3 in. was treated to secure water-stable aggregates.[2]

Deep tillage by deep plowing, chiseling, or subsoiling is done in an effort to provide a deeper rootbed with more moisture and plant nutrient capacity. It is most effective when it shatters a hardpan, a thin impervious layer at plow-sole depth consisting of fine soil particles carried down by percolating water over a long period of time. Subsurface drainage is improved for a time, but plant roots do not feed significantly deeper unless the loosened subsoil is naturally fertile or is fertilized as it is tilled. In many cases, deep tillage without deep fertilization has not repaid the cost.[3] Deep-rooted legumes such as sweet clover are favored for deepening the rootbed in many soils.

In addition to securing a favorable seedbed and rootbed, tillage is also depended upon for:

1. Killing competitive plant growth, as in turning under a sod or cultivating row crops
2. Turning under trash for pest control or for easier cultivation

Tillage has been undergoing a process of reevaluation during recent years, primarily because of the erosion resulting from clean-cultivated row crops. It has been found that a surface mulch of crop residues prevents surface sealing, thus increasing water intake and reducing runoff and erosion. Good results have been secured in the Wheat Belt where this practice was developed, but yields in the Corn Belt have been lower than with conventional tillage.

The traditional practice of plowing followed by much disking, rolling, and harrowing for a fine firm seedbed results in good germination and maximum temporary oxidation of organic matter, but:

1. The breakdown of surface particles results in quick sealing under rainfall with increased runoff and erosion as well as reduced aeration.
2. The packing action of disks and rollers reduces porosity below the surface.
3. The compaction of the soil by a number of passages of the tractor reduces porosity.

A vicious circle is initiated by depletion of organic matter and the accompanying loss in soil aggregation. Large hard clods are turned up by the plow, and the extra working required to secure a satisfactory seedbed results in undesirable compaction and mechanical breakdown of aggregates.

Experiments in Michigan,[4] Ohio,[5] and New York[6] comparing *different tillage treatments* have shown that, in some soils, preparation of sod ground by a plow, trailing a section of packer or reversed rotary hoe for smoothing, with no subsequent tillage has given as good yields of corn, oats, beans, and beets as conventional preparation and saved considerable cost of ground preparation. Apparently the gain in porosity and maintenance of surface aggregation have offset any loss due to lessened germination or oxidation of organic matter. Surprisingly, weed-seed germination has been less on these plots than on the conventionally prepared plots, possibly because the coarse surface structure does not favor germination of small weed seeds. This method of

---

[1] G. M. Browning, Principles of Soil Physics in Relation to Tillage, *Agr. Eng.*, 31:341–344, July, 1950.

[2] G. S. Taylor and W. P. Martin, Effect of Soil-aggregating Chemicals on Soils, *Agr. Eng.*, 34:550–554, August, 1953.

[3] A. J. Schwantes et al., You Don't Gain with Deep Tillage, *Univ. Minn. Farm and Home Sci.*, May, 1952. I. L. Saveson and Z. F. Lund, Deep Tillage for Crop Production, *ASAE Trans.*, 1:40–42, 1958.

[4] R. L. Cook and F. W. Peikert, A Comparison of Tillage Implements, *Agr. Eng.*, 31:211–214, May, 1950.

[5] C. J. Willard, G. S. Taylor, and W. H. Johnson, Tillage Principles in Preparing Land for Corn, *Ohio Agr. Expt. Sta. Research Circ.* 30, March, 1956.

[6] R. B. Musgrave, P. J. Zwerman, and S. R. Aldrich, Plow-planting of Corn, *Agr. Eng.*, 36:593–594, September, 1955.

preparation, combined with planting in the tracks of a narrow-tread tractor, is a growing farm practice, particularly in Michigan.[1]

Both in Ohio[2] and Iowa experiments,[3] tillage to plow depth with subsurface-type sweeps gave results markedly inferior to plowing, although the ground seemed as thoroughly loosened at the time of planting. This has been at variance with results from the Wheat Belt and has not been adequately explained. It may be possible that plowing turns up well-aggregated soil, particularly with a good sod, which has much better resistance to sealing by rainfall than the old top layer which has been broken down by cultivation, raindrop action, and oxidation of organic matter. Disking and particularly subsurface sweep tillage leave this old layer of soil at the top, and unless considerable residue is present, rainfall may quickly wash the fines into the open pores, thus causing sealing and reduced aeration. Such sealing may not be as pronounced with the lighter soils and lower rainfall of the Wheat Belt as it is in the Corn Belt.

In many of these experiments difficulty was encountered in killing a vigorous sod by any method except plowing. Even rotary tillage was often comparatively ineffective in securing an acceptable grass kill.

A *mulch planter* which may be used in undisturbed soil to till, plant, and fertilize in a once-over operation has been developed.[4] The machine consists of a front sweep running 2½ in. deep, followed by a sweep running 7 in. deep with fertilizer deposited at this depth, then four rotary hoe wheels preparing an 8-in.-wide seedbed, and finally a stub runner planter followed by a shallow fertilizer application. This machine has given acceptable control of sod crops, and good yields have resulted where additional fertilizer has compensated for the difference in availability of nutrients compared with plowing. The surface residue has reduced sealing and erosion.

Planting soybeans directly in grain stubble has been successful in South Carolina experiments.[5]

Corn has been grown on *contoured ridges* formed by moldboard plows in an effort to secure a favorable soil environment for intertilled crops with negligible long-term soil and water losses. In extensive experiments in Iowa the following advantages have been noted:

1. Soil erosion is negligible.
2. Less power is required for soil preparation.
3. Earlier planting date is allowable because of better drainage and higher temperature of the soil in the ridges.
4. There is less damage from standing water.

Corn yields were equal to those obtained by conventional practices.[6]

Many aspects of the effects of tillage on the interaction between soil and plants are as yet imperfectly understood, but a period of change of emphasis in tillage objectives may be at hand.

---

[1] Wendell Bowers and H. P. Bateman, Minimum Tillage in Illinois, *ASAE Paper* 58-619, December, 1958. C. M. Hansen, L. S. Robertson, and B. H. Grigsby, Plow-plant Equipment Designed for Corn Production, *ASAE Trans.*, 2:65–67, 1959. W. H. Johnson and G. S. Taylor, Tillage Treatments for Corn on Lakebed Clay, *ASAE Paper* 58-618, December, 1958. A. A. Swamy Rao, R. C. Hay, and H. P. Bateman, Minimum Tillage—Its Effect on Soil Physical Properties and Crop Responses, *ASAE Paper* 58-500, December, 1958.

[2] Willard et al., *op. cit.*

[3] R. A. Norton, E. V. Collins, and G. M. Browning, Present Status of the Plow as a Tillage Implement, *Agr. Eng.*, 25:7–10, January, 1944.

[4] R. R. Poynor, An Experimental Mulch Planter, *Agr. Eng.*, 31:509–510, October, 1950.

[5] J. T. McAlister, Mulch Farming Gains Favor in Southeast, *Agr. Eng.*, 38:312–315, May, 1957.

[6] W. F. Buchele, E. V. Collins, and W. G. Lovely, Ridge Farming for Soil and Water Control. *Agr. Eng.*, 36:324–331, May, 1955.

Chapter 12

MOLDBOARD PLOWS

The plow is the primary tillage implement, and plowing requires more horsepower-hours than any other single crop-production operation. The plow loosens and pulverizes the soil with efficient use of power for the depth tilled, and it has the unique ability to invert the soil, burying surface residues, killing surface growth, and bringing up a new surface layer which often has a better aggregated structure for resisting sealing over by rainfall than the weathered surface layer which is turned under.

Many attempts have been made to describe the surface of a plow by mathematical formulas, among them those of E. A. White[1] and of M. L. Nichols with T. H. Kummer[2] and with Ralph D. Doner.[3] In general, all moldboards were found to be functionally divided into three sections: (1) the lower or share portion, forming a wedge for breaking the soil loose, (2) a central pulverization area, and (3) a turning

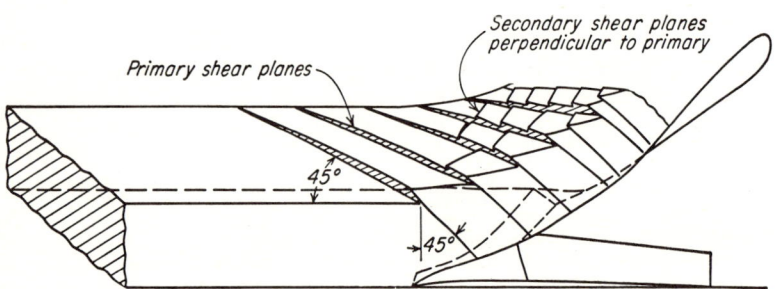

Fig. 12-1. Pulverization of furrow slice by shear plane formation. (*Nichols, Agr. Eng., June,* 1934.)

and inversion area on the upper part of the moldboard. The furrow slice is pulverized in primary shear planes as the slice is wedged loose and in secondary shear planes because of the turning action as shown in Fig. 12-1. A gentle turning action, such as that of a *sod and clay bottom,* is needed for stiff soils because of the furrow slice's resistance to secondary shear. Steep sharply turned moldboards, used in *stubble bottoms,* are desirable for sand and loose soil in order to lift the furrow slice while it is being supported by the soil in front, preventing the furrow slice from collapsing.

Stubble bottoms also give maximum pulverization. *General-purpose bottoms* have a compromise shape for acceptable performance under a wide range of conditions. The relocation of points in a cross section of the furrow slice after plowing is shown in Fig. 12-2.[4]

[1] E. A. White, A Study of the Shape of the Plow Bottom, *Trans. ASAE,* vol. 12, 1918.
[2] M. L. Nichols and T. H. Kummer, A Method of Analysis of Plow Moldboard Design Based upon Dynamic Properties of Soil, *Agr. Eng.,* 13:279–285, November, 1932.
[3] Ralph D. Doner and M. L. Nichols, Dynamics of Soil on Plow Moldboard Surfaces Related to Scouring, *Agr. Eng.,* 15:9–13, January, 1934.
[4] M. L. Nichols and I. F. Reed, Physical Reactions of Soils to Moldboard Surfaces, *Agr. Eng.,* 15:187–190, June, 1934.

Scouring. This is defined as the soil sliding freely over the surface of the plow bottom. Soils which give scouring trouble are usually sticky and too loose to develop enough sliding pressure to overcome the friction. Slat moldboards aid scouring by reducing friction drag and increasing the sliding pressure on the surface in contact. It has been found that plow shapes which give a constant soil pressure normal to the surface scour the best and that curvatures low at the share and high at the wing tend to equalize and reduce all forces.[1] Blackland moldboards are shaped in this way and are as small as will turn the furrow.

An investigation into the nature of adhesion between soil and metal surfaces indicated that it is caused by competition between the soil and plow surface for film

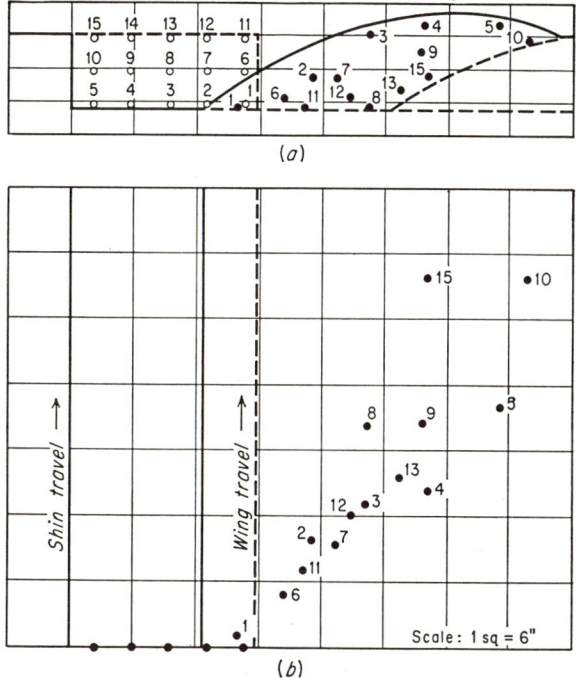

FIG. 12-2. Diagrams showing the relocation of various points in the furrow slice by plowing as found by Ashby and associates, USDA. (*Nichols, Agr. Eng., June,* 1934.)

moisture. Correlation was found between soil adhesion and "wettability" of the metal surface by the soil solution as indicated by the angle of meniscus.[2] Oil- or paraffin-impregnated wood was found to have the least adhesion of any material tried, and a moldboard of such wooden slats was found to scour considerably better than a conventional steel slat moldboard.[3] Low-friction plastics such as nylon and Teflon show very good scouring qualities.[4]

Draft. Plow draft may vary from 3 to 20 psi of furrow section and is affected primarily by soil conditions but also by operating conditions and plow-bottom characteristics. An analysis of manufacturer's tractor ratings (two-plow, three-plow, etc.), compared with Nebraska Test operating maximum drawbar pull in the gear

---

[1] Doner and Nichols, *op. cit.*
[2] F. A. Kummer and M. L. Nichols, A Study of the Nature of Physical Forces Governing the Adhesion between Soil and Metal Surfaces, *Agr. Eng.*, 19:73–78, February, 1938.
[3] F. A. Kummer, The Effect of Certain Experimental Plow Shapes and Materials in Heavy Clay Soils, *Agr. Eng.*, 20:111–114, March, 1939.
[4] Roy T. Tribble, The Teflon-covered Moldboard Plow, *ASAE Paper* 58-615, December, 1958.

used for the 10-hr drawbar test H, usually an efficient plowing gear, showed an average pull of 937 lb per rated bottom. On the basis of a 14-in. bottom operating 6 in. deep, this indicates that the average tractor is rated to plow soil having a draft of 11 psi of furrow section, assuming very favorable tractive-efficiency conditions.[1]

Some of the *soil factors affecting draft* are as follows:

1. Shear strength—increases with percentage of clay, compaction, lack of organic matter, dryness, and rate of shear. (Studies at Iowa State University indicate that the major cause of the increase in draft with speed, from 25 to 80 per cent when speed is increased from 3 to 6 mph, is due to an increase in the shear strength of soil with rate of shear, particularly in clay. Since the increase in acceleration of the soil with speed is a lesser factor, redesigning the plow bottom for reduced throwing action at high speeds may give only limited gains.[2])

2. Density—usually varies with shear strength

3. Coefficient of friction—increases with percentage of clay and with moisture content up to the lubrication phase

4. Cutting resistance of roots

Plow draft, the longitudinal force needed to move the plow through the soil, must overcome the following *sources of resistance:*

1. Vertical cutting and shearing by coulter and share
2. Cutting roots
3. Wedging furrow slice loose
4. Lifting furrow slice against gravity
5. Bending furrow slice[3]
6. Pushing and throwing furrow slice sideways
7. Frictional drag of furrow slice sliding over share and moldboard
8. Frictional drag of underside of share, landside, and wheels

It is apparent from the above list of elements that plow draft is a complex subject involving many opposing elements. For instance, dry soil may have a higher unit draft than a moist soil but the per cent of increase in draft with speed may be lower because of the lower coefficient of friction.[4] Also, it would appear that a prairie-breaker type of bottom might move through the soil with a minimum of resistance because of its gentle action on the furrow slice. In many soils, however, the high frictional drag resulting from its greater area and the greater distance the furrow is shifted make its draft higher than stubble bottoms. Significant differences in draft have been found between bottoms designed for the same function, indicating that compliance with the principles involved is to be desired.[5]

Forces on a Plow Bottom. The forces exerted on the face of a plow bottom have been measured by attaching it to a carrier frame running on metal tracks and with pressure cells arranged to measure vertical, lateral, and longitudinal forces as well as couples.[6] It has been found that the longitudinal force L and the lateral side force S usually lie in approximately the same plane and can be combined into a horizontal resultant RH. The vertical force V is usually offset to the land, introducing a couple, as shown in Fig. 12-3.

Measurements indicate that the *side force S* may range from 25 to 50 per cent of the longitudinal force L.[7] The force S tends to be less affected by soil variations than L but increases with speed, as would be expected. Balancing the side force S is usually accomplished by furrow-wall pressure against the landside, which, of course,

---

[1] Wayne H. Worthington and J. Waldo Seiple, Hydraulic Capacity Requirements for Control of Farm Implements, *Agr. Eng.*, 33:273–278, May, 1952.

[2] R. J. Rowe and K. K. Barnes, The Influence of Speed on Draft of a Simple Tillage Tool from Model Studies, *ASAE Paper* 59-630, December, 1959.

[3] I. F. Reed, Effect of Shape on the Draft of 14-inch Moldboard Plow Bottoms, *Agr. Eng.*, 22:101–104, March, 1941.

[4] O. J. J. Rogers and J. C. Hawkins, "Soil Loads on Plough Bodies," Parts 1 and 2, National Institute of Agricultural Engineering, Silsoe, Bedfordshire, England, April, 1956.

[5] Reed, *op. cit.*

[6] A. W. Clyde and R. J. McCall, Tillage Tools, *Penn. State Univ. Agr. Expt. Sta. Bull.* 465, 1944. I. F. Reed, Test Equipment and Procedure for Moldboard Plows, *Agr. Eng.*, 18:111–115, March, 1937.

[7] John W. Randolph and I. F. Reed, Effects of Several Factors on the Reactions of Fourteen-inch Moldboard Plows, *Agr. Eng.*, 19:29–33, January, 1938.

introduces parasitic drag. Clyde has found the coefficient of landside friction to be around 0.3, indicating a parasitic drag ranging from 7½ to 15 per cent of L.[1] Some reduction in draft is possible by using a rolling wheel instead of a landslide, tests by Clyde indicating an average of 4 per cent.

The *vertical force* V due to furrow-slice reaction is very difficult to isolate because of the parasitic upward force on the cutting edge, which varies greatly with share sharpness and soil hardness. Measurements have shown the net downward force V to range from 70 per cent of L in sandy soil to 23 per cent with sharp shares in clay soil. These tests were made with a coulter preceding the plow, but the coulter forces

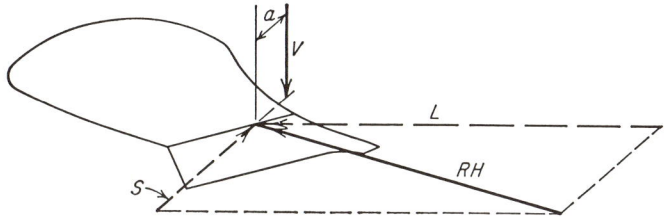

Fig. 12-3. Forces exerted by the furrow slice on the face of a plow bottom, showing couple Va. (*Clyde, Penn. State Univ. Agr. Expt. Sta. Bull.* 465, Part 2, 1944.)

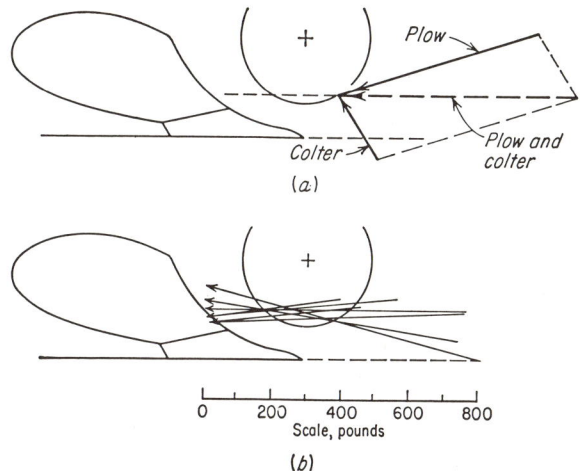

Fig. 12-4. (*a*) Division of the resultant forces on plow and coulter in the vertical plane for moist sod, spring plowing. (*b*) Typical resultants for a wide range of conditions showing that in hard ground the increased coulter force causes an upward resultant. (*Clyde, Penn. State Univ. Agr. Expt. Sta. Bull.* 465, Part 2, 1944.)

were not included. The soil exerts an upward reaction on the coulter, depending on soil hardness, and the net reaction of plow bottom and coulter is the pertinent factor. Measurements for the net V ranged from 130 lb downward in moist sod to 200 lb upward in dry hard sod. It is obvious that the plow must be kept in the ground by weight when the latter condition is encountered and that shallow set coulters, or jointers alone, will be advantageous. Typical net resultant forces in the vertical plane are shown in Fig. 12-4, as measured by Clyde.[2]

When the net V is downward, it is balanced by the upward angle of pull and the upward pressure against the landside heel or the rear wheels. It can be seen that the

[1] Clyde and McCall, *op. cit.*
[2] *Ibid.*

substitution of rolling friction for sliding friction in balancing the vertical force is often of little benefit, particularly under high draft conditions.

The *average forces,* in pounds per square inch of furrow section, on a typical stubble bottom, a general-purpose bottom, and a disk plow are compared at various speeds in Fig. 12-5. The plow bottoms tested are shown in Fig. 12-6 by top views with 1-in. contour lines.

Penetration under difficult conditions is greatly affected by *share suction.* Proper share conformation to secure land suction and down suction consists of relief behind

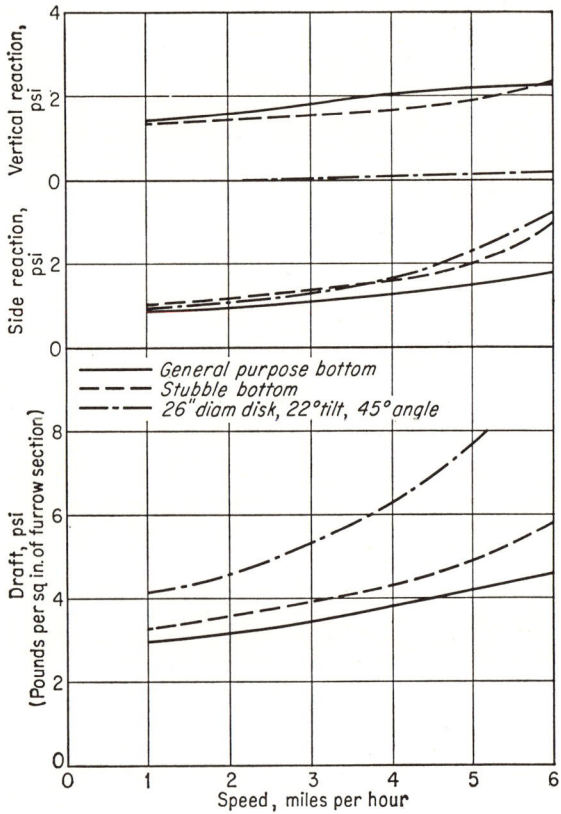

FIG. 12-5. Soil forces on typical stubble bottom, general-purpose bottom, and disk plow (average of Norfolk sand, Davidson loam, and Decatur clay at USDA Tillage Laboratory). (*Unpublished data from Tractor and Implement Div., Ford Motor Co.*)

the share edge underneath and at the side in order to avoid as much upward and side reaction as possible.

Hitch Adjustment. Proper hitching of a pull-type plow to a tractor is important because an improper hitch may cause:
1. Poor penetration if hitched too low
2. Improper furrow turning if hitched too high and plow is running on its nose
3. Improper furrow turning if plow beams are not parallel to furrow wall
4. Uneven furrows if front bottom is not cutting the proper width
5. Extra draft if plow is being pulled against the furrow wall by an excessive angle of pull
6. Extra effort to steer the tractor if the line of pull is offset on the tractor

To avoid a twisting effect on the tractor, the *line of draft* should pass through the tractor's *center of pull,* a point halfway between the driving wheels and slightly ahead of the axle. Swinging drawbars are usually pivoted near this point and, if free to swing, automatically meet this requirement.

The *center of draft* of a single plow bottom is usually from 2 to 4 in. in from the landside near the top edge of the share. For two bottoms it is halfway between the centers of the individual bottoms or at the center of the middle bottom for a three-bottom plow. This center, however, may shift with soil conditions and also with the

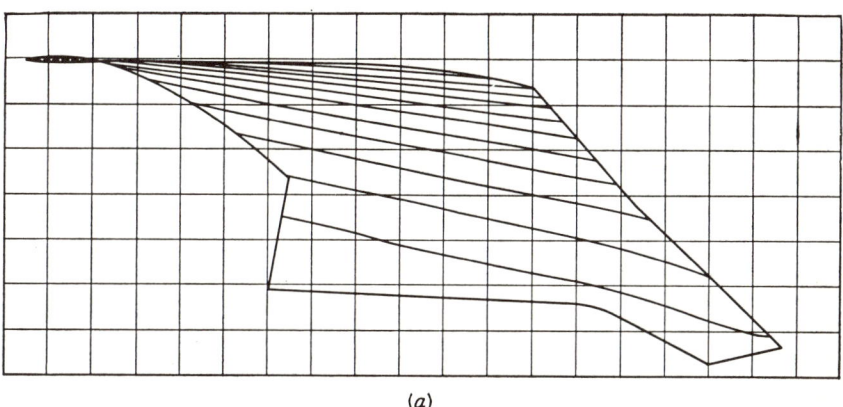

(a)

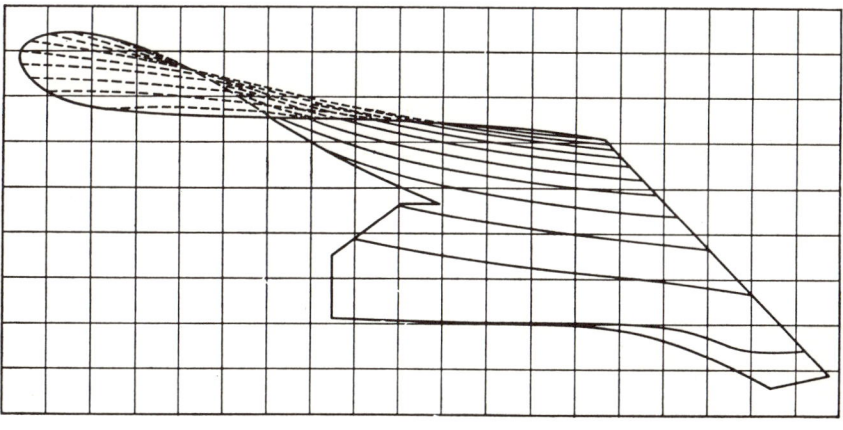

(b)

FIG. 12-6. Top views of plow bottoms with 1-in. contour lines: (a) stubble bottom; (b) general-purpose bottom.

amount of landside force, as affected by the angle of pull, and also with the landside length.[1] The *line of pull* should pass through the plow's center of draft if the plow is to run true. If the plow is pulled by a chain or free-swinging link, it will automatically assume an equilibrium position, with the chain or link pointing toward the plow's center of draft in the horizontal plane. This is easily done by pulling the plow with the drawbar swinging. If the plow does not cut the proper width, the plow hitch bars can be adjusted until the proper width is cut with the drawbar free to

[1] *Ibid.*

swing. Then the requirement that the line of pull pass from the center of draft of the plow through the center of pull of the tractor has been met and the tractor drawbar can be pinned in this position without causing side draft.

The *vertical line of pull* is indicated by the angle of the plow hitch bars, which should point toward the center of draft of the plow. A view of the hitch from the side, as shown in Fig. 12-7, will give an indication of its adjustment, and a practical check can be made by comparing the force needed to manually hold and slide the rear furrow wheel with that required for the front wheels. These forces should be approximately equal, indicating equal weight, on a well-adjusted plow. A high hitch on the tractor gives best traction but also tends to reduce plow penetrating ability in hard ground.

Mounted Plows. Mounted plows, carried by the tractor, have become increasingly popular because of their convenience for transport and turning as well as their economy of manufacture. The Ferguson three-point linkage system has been most widely adopted and consists of two lower draft links and a top compression link, all connected with ball joints and free to swing sideways, except when constrained. The lower links converge forwardly in the horizontal plane, thus pointing the plow back to center when it swings over. The lower links are hitched to a cross-bar extending

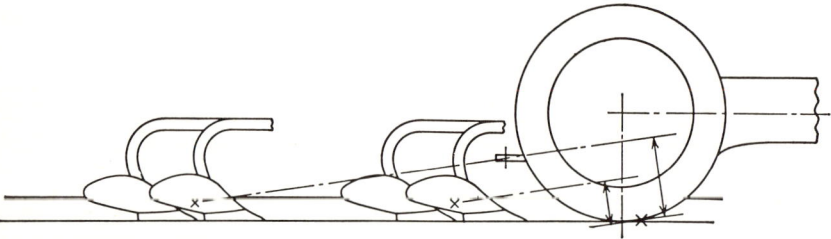

Fig. 12-7. Effect of positions of trailed and mounted plows on length of moment arm which causes weight transfer from front to rear tractor wheels.

across the plow, and the width of cut can be controlled by an adjustment changing the horizontal angle of the crossbar with respect to the plow beam assembly. Level of the plow is controlled by raising or lowering one of the lower links with respect to the other. Suction is controlled by the top-link-length adjustment, and the lifting action is approximately parallel throughout the normal plowing range.

With the Ferguson type of linkage, the lower links may be automatically actuated to vary plowing depth so as to give approximately uniform draft, or they may be set at a given height to give uniform depth where the ground is level.

Three-point linkages with increased convergence in the vertical plane accentuate the change in level of the plow as it moves up and down. The top-link length can be adjusted to make the plow level out or float at the desired plowing depth, and the plow allowed to find its way through the soil without constraint from the hydraulic lift. In this case the plow is supported by the upward component of the pull, the upward reaction on the cutting edge, and the supporting pressure on the rear wheel or landside heel. (See Chap. 6 for discussion of tractor force reactions to both types of three-point linkage systems.)

Plows attached by three-point linkages may also have a depth-gaging wheel which runs on the unplowed ground, thus maintaining a uniform depth.

Another type of mounted plow is essentially a trailed plow pulled from a point under the center of the tractor. Like a walking plow, depth is controlled by raising or lowering the front end of the beam, and the hydraulic lift connections behind the tractor are normally loose when plowing, serving only to lift the plow for turning or transport. A modification of this type of hitch uses articulated lift links to move the virtual hitch point back to conventional drawbar position for improved entry and flexibility.[1]

[1] E. W. Tanguary and A. W. Clyde, New Principle in Tractor Hitch Design, *Agr. Eng.*, 38:88–95, February, 1957.

Plowing Traction. With modern tractors having a high power-weight ratio, getting sufficient weight on the rear wheels to provide traction for plowing is somtimes a problem. As shown in Fig. 12-7, a trailed plow has the advantage of a greater moment arm from the tractor's center of pull compared with a mounted plow being pulled with a resultant force having the same vertical angle, simply because the trailed plow is farther back. This additional moment arm results in increased weight transfer from the tractor front wheels to the rear wheels, giving more traction as the pull increases. This, of course, is desirable as long as there is no danger of the front wheels leaving the ground.

Traction can be increased by the mounted plow if the vertical angle of pull is increased. This, however, has practical limitations because, when plowing hard dry

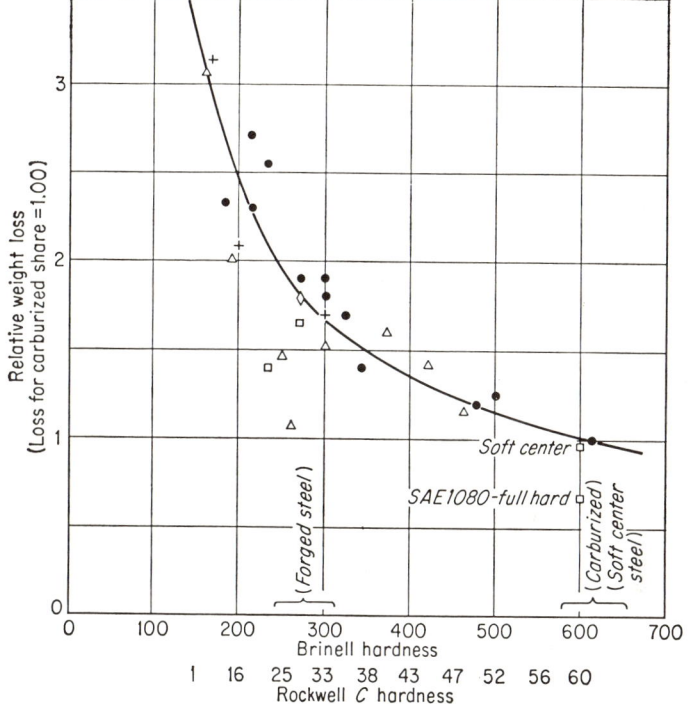

FIG. 12-8. Comparative rate of weight loss by shares of various hardnesses. (*Unpublished data from Tractor and Implement Div., Ford Motor Co.*)

ground which normally has the greatest draft, the net vertical soil force on plow and coulter may be upward, requiring some of the weight of the plow to keep it in the ground. Under these conditions a steep angle of pull, needed for good traction with the mounted plow, causes difficulty in keeping the plow in the ground. Obviously, wheel weights and added weight on the plow may be required. In easy plowing conditions where the net vertical soil force on plow and coulter is downward, some traction is gained with a mounted plow because it is being carried by the tractor instead of on its own wheels.

Trash Coverage. Complete burial of trash is important where row crops are to be cultivated or where corn borer control is desired. Jointers, cover boards, weed hooks, trash wires, or trash shields[1] can be added for this purpose. Notched coulters may be used for better cutting of trash, and moldboard extensions increase furrow turn. It is

[1] R. H. Wileman, Trash Shields for Plows, *Agr. Eng.*, 16:260, 286, July, 1935.

possible to almost completely cover cornstalks or growing vegetation if the proper equipment is used.

Materials Used in Plow Bottoms. Share wear is extremely difficult to overcome since the soil acts as a grinding compound, carrying particles of sand and rock which are extremely hard and abrasive. The iron carbide constituents of hardened high carbon steel and chilled cast iron rank near the top in abrasive resistance but tend to be brittle and break when obstructions are hit. The comparative wear resistance of various steels (as measured by weight loss in a number of different tests) has been plotted against hardness in Fig. 12-8. Wear-test abrasion machines for testing specimens of various materials have given results showing good correlation with field measurements of weight loss.[1] Laboratory wear tests of steels varying from 0.45 to

TABLE 12-1. TYPICAL MATERIALS USED FOR PLOW-BOTTOM PARTS

Usage	Material	Heat-treatment	Comparative advantages
Rolled-section plow share	C-1095	Heat, quench, and draw to Brinell 415-477	Low cost with high wear resistance
Forged plowshare	C-1024	Carburize, quench, and draw to Rockwell C58-62	Homogeneous material for good strength with good wear resistance
	Soft-center, C-1095 over C-1010 with C-1080 welded patch on point	Heat, quench, and draw to Rockwell C58-62	Full hardness to greater depth on wearing surface for high wear resistance
Crucible	C-1080	Heat and air-cool to Brinell 240-305	Low cost and good resistance to breakage
Cast-steel plowshare	Alloy steel	Heat, quench and, draw to Brinell 375-450	High strength and good wear resistance with lower manufacturing cost for small-volume requirements
Cast-iron plowshare	Gray iron with chilled cutting edge	None	Low cost and high wear qualities for obstruction free soils
Plow moldboard	C-1024	Carburize, quench, and draw to Rockwell C60-65	Homogeneous material for greater strength and uniform hardness and good wear resistance
	Soft-center, C-1095 over C-1010	Heat, quench, and draw to Rockwell C60-65	Full hardness to greater depth on wearing surface for high wear resistance and scouring ability
Plow landside	C-1050	Heat, quench, and draw to Brinell 281-400	High wear resistance with forging qualities

0.95 per cent carbon content with samples of each hardened to various degrees indicated that if two steels have the same relative hardness, the one with the higher carbon content will offer more resistance to abrasion.[2]

Typical materials for plow bottom parts along with heat-treatment and comparative advantages are shown in Table 12-1. Coulters are made of the same steels used for disks, as shown in Table 13-1.

Listers and Middlebusters. These are double plows, throwing the dirt both ways and leaving a trench. In the light soils of the western Corn Belt, where rainfall is normally less than it is farther east, corn is planted in moist soil at the bottom of this furrow and the furrow filled in by later cultivations. The lister is normally used twice, first making furrows in the flat land and then breaking out the ridges as the

---

[1] I. F. Reed and E. D. Gordon, Determining the Relative Wear Resistance of Metals, *Agr. Eng.*, 32:98–100, February, 1951. Nuri Mohsenin et al., Wear Tests of Plowshare Materials, *Agr. Eng.*, 37: 816–820, December, 1956. F. G. Lechner and H. F. McColly, Abrasive Wear Resistance of Hardfacing Materials Used on Agricultural Tillage Tools, *ASAE Trans.*, 2:55–57, 1959.
[2] I. F. Reed and W. F. McCreery, Effects of Methods of Manufacture and Steel Specifications on the Service of Disks, *Agr. Eng.*, 35:91–97, February, 1954.

corn is planted. In this way most of the ground is worked. In the South this tool is called a middlebuster and is used for cotton as well as corn. In blackland and other sections where drainage is a problem, the crop is bedded, being planted on top of the ridge in warmer and better-aerated soil than that in the furrow bottoms.

In general, lister moldboards are similar in conformation to those on stubble plow bottoms and are preferred in the light western soils, while middlebuster moldboards are more gradual in slope, like those on a blackland bottom, in order to better handle the stiff high-draft blackland soils of the South. Lister bottoms are usually available in 12- and 14-in. sizes, as measured across the share wings, and middlebuster bottoms in 10-, 12-, and 14-in. sizes.

In flat busting (first operation), the furrow slice is not thrown in a previously opened furrow as in plowing, but must be thrown up on the flat land. This results in more draft per square inch of actual cross section worked than plowing. In rebusting (second operation), the previously opened furrows are filled but there is still 7 in. of unworked plow sole in the case of 14-in. bottoms and 42-in. rows. Draft for the second operation is not greatly lowered in spite of the open furrows because of the greater mass of soil to be moved. Draft per row of a 14-in. lister planter operating 4 in. deep in heavy Webster soil in Iowa was 725 lb or 11.9 psi of theoretical cross section.[1] Unpublished Kansas data by Promersberger list an average draft of 550 lb per bottom, and Montana data give a range of 350 to 733 lb per bottom, with an average of 570 lb.[2] The total horsepower-hours for the two operations, flat busting and rebusting, is usually less than for plowing, and, of course, there is a considerable saving in eliminating secondary tillage operations and in combining planting with the second operation.

Lister and middlebuster bottoms are furnished with or without runners (sliding shoes which help support the downward soil force similarly to the underside of the landside on a walking plow). The lateral soil forces on the two sides of the bottom oppose each other. A coulter may be used to cut trash and vines ahead of the beams and to split the furrow slice.

[1] C. K. Shedd, J. B. Davidson, and E. V. Collins, Machinery for Growing Corn, *USDA Circ.* 592, 1940.
[2] H. E. Murdock, Mechanical Tests on Tractor Farming Equipment, *Montana State Coll. Agr. Expt. Sta. Bull.* 243, 1931.

Chapter 13

DISK TOOLS

Disk plows, tillers, and harrows rank next in importance to moldboard plows for seedbed preparation. Disk tools are of comparatively recent origin, concave disk blades having been patented in 1877.

The action of a disk blade is somewhat like that of a plow in that it tends to granulate the soil, using lifting action with a minimum of compression. It has, however, the following characteristics, which differentiate it from the plow:

1. Does not completely invert the soil (a disadvantage where vegetation is to be killed or complete trash coverage is desired, but an advantage where some trash should be left exposed to check wind erosion and surface sealing by rainfall)
2. Can cut through crop residues making subsequent tillage easier
3. Can roll over roots and rocks, reducing likelihood of damage
4. Can operate in nonscouring soils by using scrapers
5. Can be used in harder ground than a plow because the cutting edge is thinner, although extra weight is required
6. Can cut large clods, particularly with closely spaced disks as in a harrow
7. Has a packing effect due to soil support of the edge and back of the blade

Disk blades are commonly made of the steels listed in Table 13-1. Plain disks are heat-treated to a hardness of Rockwell C38-45, while cutaway disks (used for better penetration in trash or hard ground) are slightly softer and tougher, Rockwell C35-40.

In laboratory impact and fatigue tests at the USDA Tillage Laboratory, Auburn, Ala., it was found that cross-rolled steel gave better results than strip-rolled steel. It was also found that disks with beveled-edge notches formed by angle shearing had higher impact strength than disks with ground notches, because of residual surface tensile stresses resulting from grinding.[1]

Disks used for disk harrows and disk tillers range from 16 to 26 in. in diameter, while plow disks commonly range from 24 to 32 in. in diameter. Proportions are shown in Table 13-2. The spherical radius of curvature ranges from 22 to 28 in.

The *forces on a disk blade* are even more variable than those on a plow bottom, since they are affected by disk size, concavity, angle, and inclination, as well as by depth and soil variation. Results of a series of tests by Clyde on a 24-in.-diameter disk tiller blade having a 21⅝-in. radius of curvature are shown in Fig. 13-1.[2] The vertical force V is negative, the values shown being the weights necessary to keep the blade in the ground. Since a disk exerts a side thrust on the soil at its edge and it is held at its axis, there is always a couple present. The center of side thrust was found to be below the surface from 25 to 45 per cent of the depth of cut. In the top view of the disk shown in Fig. 13-1, this couple has been indicated by displacing V the distance a, Va giving the value of the couple.

The S curve in Fig. 13-1 for a disk angle of 30° with the direction of travel shows side support on the back of the blade increasing with depth, thus reducing the net S and also increasing V, the weight necessary for penetration. The curves for 38 and

---

[1] I. F. Reed and W. F. McCreery, Effects of Manufacture and Steel Specifications on the Service of Disks, *Agr. Eng.*, 35:91–97, February, 1954.

[2] A. W. Clyde, Improvement of Disk Tools, *Agr. Eng.*, 20:215–221, June, 1939.

TABLE 13-1. CHEMICAL COMPOSITION OF TYPICAL DISK STEELS

Element, percentage by weight

Type	Carbon		Manganese		Silicon		Molybdenum		Nickel		Chromium		Sulfur, max.	Phosphorus, max.
	Min.	Max.	Min.	Max.	Min.	Max.	Min.	Max.	Min.	Max.	Min.	Max.		
Carbon steel..................	.75	.90	.70	1.00	.10	.40	...	...	...	.30*	...	.30*	.05	.05
Chrome molybdenum...........	.55	.65	.60	.90	.20	.35	.25	.40	...	.30*	1.15	1.45	.05	.05
Nickel chrome molybdenum steel..	.55	.65	.75	1.00	.20	.35	.15	.25	.40	.70	.40	.60	.05	.05
Nickel chrome steel............	.55	.65	.70	.90	.20	.35	...	...	1.00	1.50	.60	.90	.05	.05

* Included as an impurity.
SOURCE: Disk Subcommittee, Farm Equipment Institute, Chicago, Ill., 1958.

45° do not show this tendency, indicating that these angles were sufficient to provide clearance behind the leading edge. This characteristic of disk blades is used to control penetration where ground-gaging wheels are not used.[1]

In general, the *draft* of disk tools in undisturbed soil is as much or more than with plows for the cross-section area worked (Fig. 12-5). The substitution of rolling friction for sliding friction is not complete enough for an appreciable advantage. Tests by Gordon at the USDA Tillage Laboratory, Auburn, Ala., found unit draft ranging from 1.36 psi in sand to 11.3 psi in clay.[2] In these tests (mostly with a 26-in.-diameter disk with 4⅛-in. concavity, 22³⁄₁₆-in. radius of curvature, cutting 7 in. wide and 6 in. deep), S, the side reaction, was from 50 to 70 per cent of L, as was the case in Clyde's tests of 22- and 24-in.-diameter disks. In Clyde's tests of 18-in.-diameter disk harrow blades with 1¾-in. concavity, however, S ranged from 120 to 150 per cent of L.[3]

TABLE 13-2. TYPICAL DISK SIZES AND DIMENSIONS

Diameter, in.	Thickness, in.	Concavity, in.
Plain and cutaway* harrow disks		
16	0.125, 0.148	1½
18	0.125, 0.148	1¾
20	0.148, 0.187	2
Plain and cutaway* tiller disks		
22	⁵⁄₃₂, ³⁄₁₆	2½
24	⁵⁄₃₂, ³⁄₁₆, ¼	3½
26	³⁄₁₆, ¼	4
Plow disks		
24	³⁄₁₆, ¼, ⁵⁄₁₆	3½
26	³⁄₁₆, ¼, ⁵⁄₁₆	3¾
28	³⁄₁₆, ¼, ⁵⁄₁₆	4
30	¼, ⁵⁄₁₆, ⅜	4¾
32	¼, ⁵⁄₁₆, ⅜	5¼

* Cutaway disks commonly have a notch for each inch of radius.
SOURCE: Disk Subcommittee, Farm Equipment Institute, Chicago, Ill., 1958.

Disks with deep concavity lift the soil more, and the corresponding increase in downward soil force helps them to penetrate better than disks with shallow concavity. Draft is higher, but the side force S is lower. They cannot operate through as wide a range of angle since the extra curvature gives more soil support on the back of the blade at small angles.[4]

In Gordon's tests of 26-in.-diameter blades, the lowest unit drafts were secured at a depth of 6 in., indicating that optimum depth for the conventional disk blade is near one-fourth its diameter.

When a disk is inclined back from the vertical, as in a disk plow, the soil is lifted higher, increasing penetrating ability, but is not thrown sideways as much, thus reducing S. In general, the unit draft was slightly higher. The clearance in back of the blade is reduced by inclination, and the range of angle possible without the soil bearing on the back of the blade is greatly reduced.

Speed of travel is an important factor since it increases L and S because it throws the dirt farther. With the disk vertical, penetrating ability is lost as speed is in-

[1] W. F. McCreery and M. L. Nichols, The Geometry of Disks and Soil Relationships, *Agr. Eng.*, 37:808–820, December, 1956.

[2] E. D. Gordon, Physical Reactions of Soils on Plow Disks, *Agr. Eng.*, 22:205–208, June, 1941.

[3] Clyde, *op. cit.*

[4] W. F. McCreery, Effect of Design Factors of Disks on Soil Reactions, *ASAE Paper* 59-622, December, 1959.

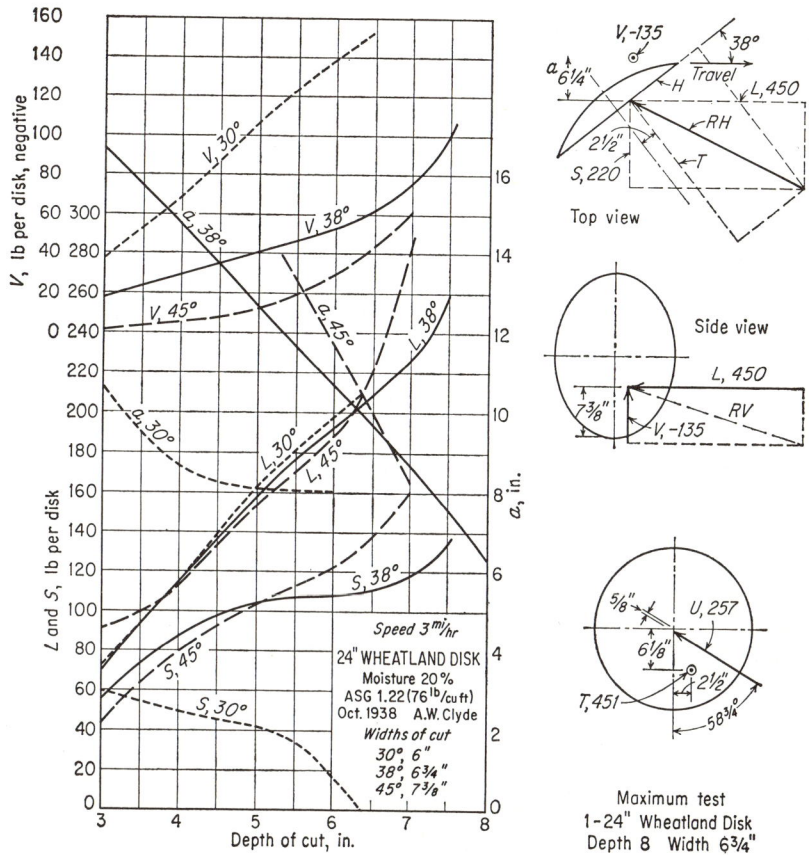

Fig. 13-1. Forces acting on a disk blade. (*Clyde, Agr. Eng., June, 1939.*)

creased, and it is a common observation that disk harrows run shallower with the same angle setting when speed is increased. With the disk inclined, Gordon found an increase in penetrating ability with speed, due to the increase in the downward soil force.[1]

DISK HARROWS

The vertical and side-thrust forces acting on a gang of disk blades are shown in Fig. 13-2. It can be seen that the weight must act off center to balance the thrust couple, but since weight and thrust both are proportional to the number of disks, the offset b is independent of gang width.

Single-disk harrows, throwing out from the center, tend to cut deep at the outside and shallow at the center. Provision is made for applying the frame weight near the center to overcome this tendency.

Tandem-disk harrows may balance the couples of the front and rear gangs on each side by cross-connections, or they may use frame weights or springs at the inner ends of the front gangs and springs, truss connections, or a continuous bar on the rear gangs to hold them level.

Older types of tandem-disk harrow have the *rear gangs pulled from the front gangs by a hinge joint,* and Clyde's *V-* and *L*-force diagram for this arrangement is shown

---

[1] Gordon, *op. cit.*

in Fig. 13-3. It is apparent that pulling the rear gang transfers weight to the front gang, giving it more penetrating ability than the rear gang, particularly if the hinge point is high. Weights may be required for the rear gangs in order to secure adequate penetration.

Some tandem-disk harrows have a *rigid frame connecting front and rear gangs* and use a hinged drawbar. Clyde's L- and V-force diagram for this type of frame is

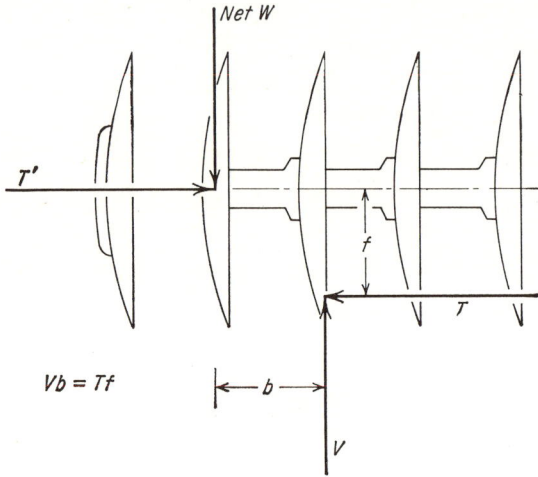

Fig. 13-2. Vertical and thrust forces acting on a gang of disk blades. (*Clyde, Penn. State Univ. Agr. Expt. Sta. Bull.* 465, 1944.)

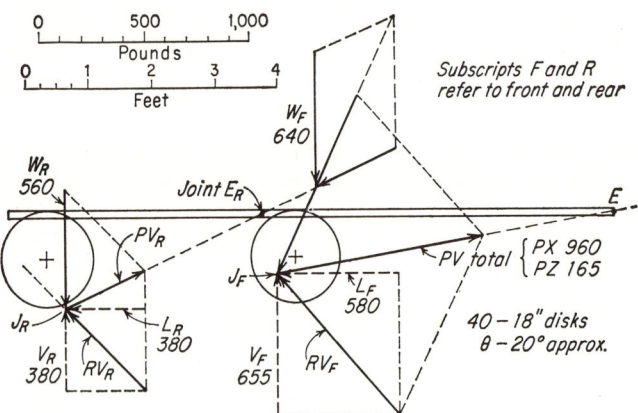

Fig. 13-3. Analysis of V and L forces for front and rear gangs of tandem disk with jointed frame. (*Clyde, Agr. Eng., June,* 1941.)

shown in Fig. 13-4. Here weight transfer between gangs can be controlled by the line of draft as affected by drawbar angle and hitch height on the disk. One harrow of this type is made with adjustable angle, and the gangs are hinged at the inner ends for flexibility in following the ground.

A second type has a *rigid frame which holds all gangs level* and at a fixed, nonadjustable angle. In this case, depth is not controlled by angle but by combination depth-gaging and transport wheels in the pulled type and by the tractor in the mounted type. These disks are not supported by the soil reaction except at maximum penetration and consequently can be used for light surface working without a packing effect. They are also more effective for their weight in cutting trash or high spots than flexible disks because the individual gangs cannot float up. Since they are always operated at full angle, there is more side thrust at shallow depths but also greater soil pulverization. Impact loads on disk blades are higher when rocks are encountered, and some flexibility, particularly in the hitch, is desirable for rocky conditions.[1]

The operation of a *mounted* type of rigid-frame disk with a three-point linkage system, such as shown in Fig. 13-5, is greatly benefited by draft-control action in

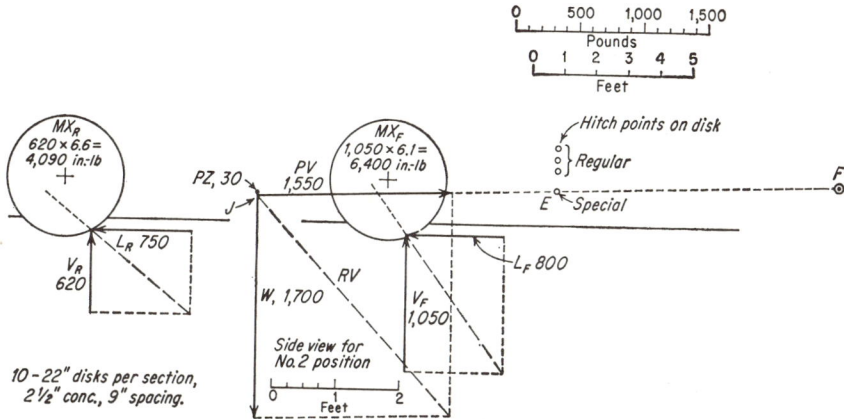

Fig. 13-4. Analysis of V and L forces for rigid-frame tandem or offset-disk harrow. (*Clyde, Agr. Eng., June*, 1941.)

loose plowed ground because the disk is lifted before the tractor wheels spin and dig in. With a fixed depth setting, tractor wheel slippage and digging in lower the disk, making it pull harder.

Disk-harrow bearings are subject to heavy radial loads as well as thrust and usually have dirt pouring over them. Oil-soaked-wood bearing inserts operating against chilled cast-iron journals are widely used because they are cheaply replaceable, partially self-lubricating, and quiet. Chilled-iron bearings and journals are more durable but squeak when not lubricated. In some abrasive soils durability is not noticeably improved by lubrication, since the mixture of grease and soil makes a good grinding compound. With some soils it is important that the bearing be designed so that the dirt which unavoidably enters may fall out at the bottom rather than fill up the clearance and cause binding. Success in improving bearings has been had in gang applications where the disk blades are mounted on a tube which has plain bearings inside it running in oil on a center shaft

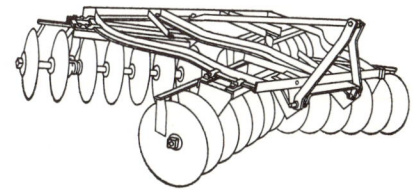

Fig. 13-5. Lift-type disk harrow with floating hitch. (*Tractor and Implement Div., Ford Motor Co.*)

and protected by seals in the ends of the tube. Ball bearings with multiple-lip seals, such as shown in Fig. 3-1b, are being used in disk harrows. These bearings are usually lubricated for life at the factory. Bearing friction is a small proportion of the total draft of disk harrows in general.[2]

Single-disk harrows are made from 5 to 28 ft wide, and tandem-disk harrows from 5 to 21 ft wide. Blades 16 and 18 in. in diameter are usually spaced about 7 in. apart, while 20- and 22-in. blades are spaced about 9 in. apart. Penetration under

[1] A. W. Clyde, Disk Harrow Design Improvements, *Agr. Eng.*, 37:173–176, March, 1956.
[2] Gordon, *op. cit.*

difficult conditions is largely a function of the weight per blade, and extra frame weight and strength are provided with the larger blades.

OFFSET-DISK HARROWS

The offset-disk harrow was developed in California around 1920, and its use has spread steadily. It is usually made with 20-, 22-, or 24-in. disks spaced 9 in. apart and in widths from 4 to 13 ft. The principal advantages over tandem-disk harrows are as follows:

1. A heavy-duty tool can be built with fewer parts and bearings because of the use of two gangs instead of four.
2. The ground is left level.
3. All the ground is worked, since a center strip is not left.
4. It can be offset from the center of the tractor, (important for orchard work).

The last feature is a peculiarity resulting from the side thrusts of the two gangs. Since these cause a couple about the center of resistance, this couple must be opposed by offsetting the line of pull from the center of resistance. The angle assumed by the tool tends to produce equal side forces in the two gangs. If the front gang is cutting deeper or in firmer soil than the rear gang, it will take less angle and the rear more, in order to develop equal side thrust in both gangs. The side component of an angled pull is also a factor. As shown in Fig. 13-6, changing the amount of offset of the tool with respect to the tractor, and the resulting change in angle of pull cause the tool to assume a new angle of attack in maintaining equilibrium. Differences in side-thrust characteristics of different-size blades also affect the equilibrium angle. The forces in the vertical plane are similar to the rigid-frame tandem-disk harrow as shown in Fig. 13-4. The rotational couples causing the gangs to try to dig deeper at the concave ends counteract each other through the rigid frame, although the condition shown in Fig. 13-6d, where the front gang is cutting deepest, because of the hitch, may throw the entire assembly out of level.[1]

Trail-type offset harrows without transport wheels have adjustable angling for transport and depth control. They also have automatic straightening of the gangs during right-hand turns, in the direction of dirt thrown by the front gangs. Such a turn is almost impossible with the gangs angled, although the opposite turn is easily made. Trail-type offset harrows with hydraulically actuated wheels, are of fixed-angle design since the wheels control depth and lift the tool for turning and transport. The advent of the ASAE-SAE standard for hydraulic remote controls has facilitated the use of this type.

Lift-type offset harrows are also of fixed-angle design since they are carried by the tractor for transport and the tractor controls the depth. When mounted on links with lateral freedom, the level must be adjusted to give equal side thrust of front and rear gangs in order that the tool may stay centered behind the tractor. Where the fore-and-aft level of the disk cannot change to follow uneven ground, a high spot will cause the front gangs to cut deep and swing to one side, followed by a swing to the other side as the rear gangs cut deep.

DISK TILLERS

These tools are also called one-way disk plows, harrow plows, and Wheatland disk plows. They consist of a single gang of disk blades, 20 to 26 in. in diameter, on a common axle, all throwing the soil the same way. The common axle differentiates these tools from regular disk plows which have the disk blades tipped back, necessitating separate axles and bearings for each blade.

Disk tillers are made as small as two-blade models cutting 15 in. wide up to models cutting 20 ft wide. Wide-trailed models, such as shown in Fig. 13-7, are particularly popular in the Wheat Belt because they leave vegetation and stubble partially exposed but tied down so that it is effective in preventing wind erosion. They are also economical of power in the light soils of the area. Seeding boxes are available for some models, enabling seedbed preparation and wheat seeding to be done in one operation (see Chap. 17).

[1] R. W. Kramer, Offset Disk Harrow Design, *Agr. Eng.*, 36:587–590, September, 1955.

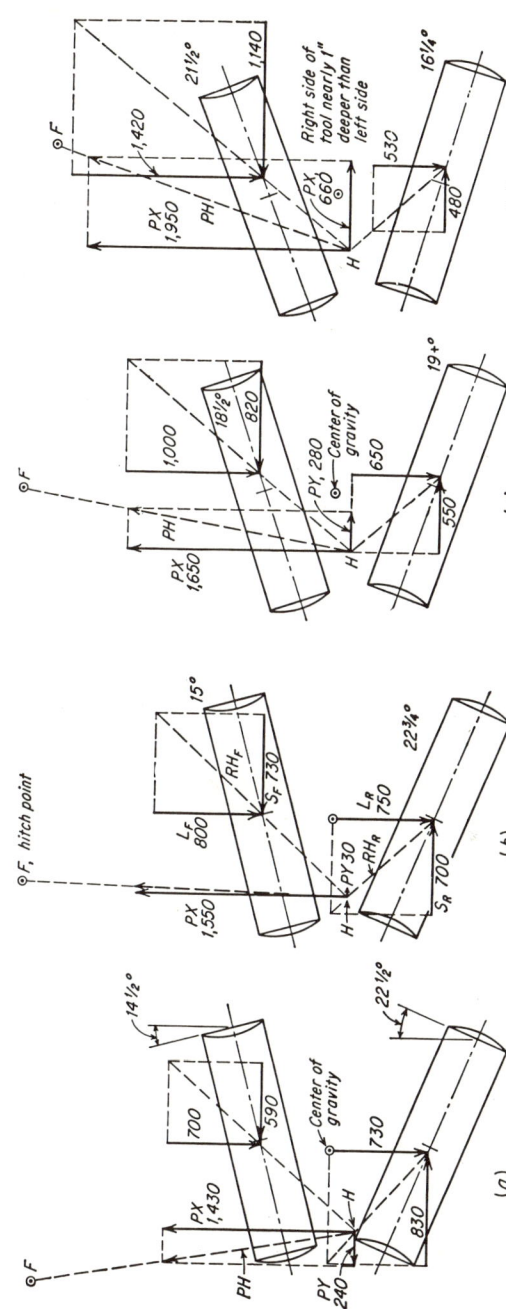

FIG. 13-6. Effects of angle of line of draft to tractor hitch on equilibrium position of offset-disk harrow. (a) Right offset with respect to tractor; (b) normal; (c) trailing; (d) left offset. (Clyde, Agr. Eng., June, 1939.)

The *forces* acting on a disk tiller are similar to those acting on a plow except that the resultant lengthwise soil force on the blades is usually upward, requiring extra weight for penetration, and the side force S is from 50 to 70 per cent of the lengthwise force F instead of from 25 to 50 per cent as with a moldboard plow. An analysis made by Clyde of the forces on this tool is shown in Fig. 13-8. Balancing S is a prob-

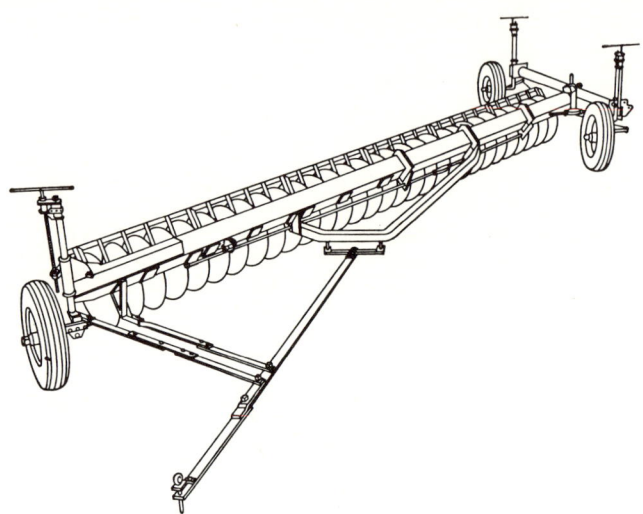

Fig. 13-7. Disk tiller. (*Case Co.*)

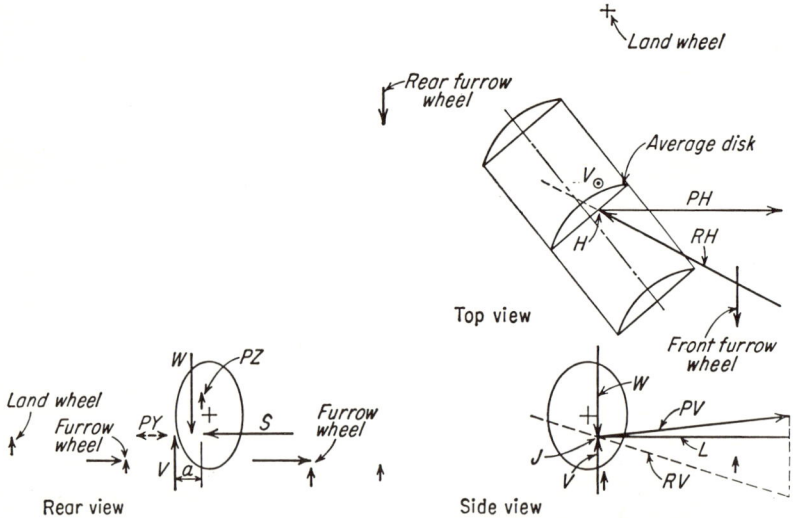

Fig. 13-8. Analysis of forces on a disk tiller. (*Clyde, Agr. Eng., June,* 1939.)

lem, not only because it is proportionately larger than with a plow, but also because there are no individual landsides and clean-cut vertical furrow walls for the furrow wheels to run against.

Trail-type disk tillers usually permit the angle of attack of the disk gang to be adjusted from 40 to 55° by steps. In hard high-draft soil, penetration can be improved and width of cut decreased by increasing the angle, while maximum width can be taken with less angle in loose easy-working soils. This adjustment is secured by changing the angle of the land wheel and the furrow wheels with respect to the disk gang and may require hitch adjustment also.

Hitching principles are similar to those for moldboard plows except that the side-thrust problem is accentuated if the line of pull angles to the left. The center of draft is at the center disk blade slightly below the surface of the ground. Fortunately, disk tillers are usually wide enough so that the line of draft can be straight ahead or angled to the right.

Disk tillers are necessarily built heavier per foot of cutting width than moldboard plows, and wheel weights can be added for penetration in hard soil. Weight is best utilized at the rear because of the force couples, and rear-furrow-wheel weights are most popular.

DISK PLOWS

Disk plows have the blades inclined backward for maximum penetration without excessive side throw. They are used where the following features are important:
 1. Ability to operate where moldboard plows will not scour
 2. Ability to roll over stumps, roots, and some rocks without damage
 3. Ability to penetrate hard dry soil if of sufficient weight and strength

Blades of 26 in. in diameter are most popular, and they are individually mounted, usually on well-sealed tapered roller bearings which are not exposed to as much dirt as on a disk gang because the single sealing surface is close in behind the blade.

The *forces* involved are similar to those in a disk tiller except that the side force S is proportionately smaller because of the pitch of the blades. With individual blade mounting, more fore-and-aft clearance is available for trash clearance than with a disk tiller and the over-all length is greater for the same width of cut. The width of cut per blade is commonly from 7 to 11 in. compared with 6 to $7\frac{1}{2}$ in. for disk tillers, and maximum depth is 10 to 12 in. compared with 7 in. for most disk tillers.

In trailed models the land wheel is mounted with the rear furrow wheel in a common truck frame adjustable for angle with the beam assembly, thus changing the angle of attack and the width of cut. This construction allows two-wheel trailer-type transport and turning, eliminating the need for a steering connection to the rear wheel. The front furrow wheel steers from the hitch in the conventional manner.

Proper hitching may be more difficult than with the disk tiller because the narrower cut often results in a line of draft angling to the left, thus adding to the soil side force S. The same principles still apply, however.

Chapter 14

SHOVEL AND SWEEP TOOLS

These tillage tools, like the plow, are descendants of the primitive forked stick. They are used primarily for loosening and stirring the soil and for killing weeds. Like the plow and disk, they lift the soil, but they do not usually have an inverting effect.

The variety of shapes available is almost infinite, ranging from double-pointed reversible alfalfa points, 1½ in. wide, to 60-in.-wide sweeps. Shovels are more effective in stirring the soil, while sweeps cut off all weeds and loosen the soil. Shovels penetrate more easily in hard ground because there is less upward soil reaction on a point than on the horizontal sweep blades.

Draft of shovels is probably in the same range as of plows and disk tools for the cross-section area worked and in soil of similar compaction. Cultivator shovels usually work in soils previously loosened by seedbed preparation, and unpublished data by Clyde give the draft for a 5-in.-wide spearpoint shovel of 22, 45, and 95 lb at 2-, 3-, and 4-in. depths, respectively, or an estimated 4½ to 6 psi of shovel cross section in a silt loam seedbed. In the same soil a 10-in.-wide sweep running 2 in. deep required a 60-lb pull, or approximately 3 psi, indicating that sweeps may be somewhat lower in draft for the area worked because they stir the soil less, although shovels may loosen a wider cross section than the actual shovel width. After compaction by heavy rains, draft may double. Draft of wide sweeps used for bindweed control in summer fallow in Nebraska ranged from 3½ psi at 3-in. depth to 3 psi at 7-in. depth,[1] while duckfoot sweeps in Montana in plowed summer fallow ranged from 3 psi at 3 in. deep to 2 psi at 5 in. deep.[2]

Shovels are operated at the *angle* which gives maximum penetrating ability, usually near 45°. A steeper angle reduces the lifting effect and the resulting downward soil force which aids penetration, while a flatter angle may reduce point and edge clearance, giving a sled-runner effect. Small sweeps are usually set at a slight angle, while large sweeps run almost flat.

In shovels, *sharpness* at the point is important in securing penetration. A rounded sled-runner edge on a sweep reduces root-cutting ability and causes extra drag on the underside as well as additional upward soil reaction, which reduces penetrating ability.

The shanks which carry shovels or sweeps must either be very rugged to resist obstructions or allowed to yield. Deep chisels are an example of the former, while pin-break and spring-trip shanks or spring teeth are examples of yieldable construction.

Materials used in teeth, shovels, and sweeps are listed in Table 14-1.

SUBSOILERS

Subsoilers have a chisel-shaped blade on a strong narrow shank capable of being pulled from 12 to 24 in. deep or more for the purpose of shattering the subsoil to admit air and water (see Deep Tillage in Chap. 11). In recent experiments at Purdue University, chopped vegetative residue has been blown down behind the subsoiler

---

[1] L. F. Larsen and E. C. Joy, A Subsurface Row-crop Cultivator, *Agr. Eng.*, 24:123, April, 1943.
[2] H. E. Murdock, A Study of the Operation of Tractors and Implements under Farm Conditions, *Montana State Coll. Agr. Expt. Sta. Bull.* 344, 1937.

shank so as to fill a narrow trench. This has prevented sealing over and greatly increased water intake in many cases.

Subsoilers are high-draft tools, a single shank with a chisel point running 15 in. deep being roughly equivalent to a two-bottom plow. The shank should be as narrow as possible, with a sharp leading edge in order to part the soil with a minimum of resistance. It must, however, be strong and rigid to avoid deflection in operation and to resist the impact of rocks and roots. It must be capable of stopping and holding the tractor without damage unless a shear pin or friction breakback is used.

Table 14-1. Typical Materials Used in Teeth, Shovels, and Sweeps

Part	Material	Heat-treatment
Harrow spring tooth, flat	C-1095	Heat, quench, and draw to Rockwell C38-46
Tiller spring tooth, square	SAE 5050	Heat, quench, and draw to Bh 265-340
Narrow shovel for field tiller	C-1080	Heat, quench, and draw to Bh 341-401
Cultivator sweep	C-1080	Heat, quench, and draw to Bh 341-401; sometimes air-hardened to Bh 228

The point is commonly 2 to 3 in. wide and set near an angle of 15° for maximum lifting effect, the downward soil force often being more than one-half of the longitudinal resistance.[1] This characteristic makes the subsoiler exceptionally satisfactory as a tractor-mounted tool because it increases rear-wheel traction enough to avoid slippage, and with hydraulic depth control the tractor can be kept from stalling.

CHISELS

Chisels, originally developed as orchard cultivators, consist of several shanks with narrow double-pointed shovels and are strong enough to operate up to 12 in. deep without breakback protection. The shanks are often clamped to a square tool bar and spaced according to conditions.

The chisel is most commonly used in the irrigated sections of the West where the soil is very hard and tight when dry. It loosens the soil, and the resulting large chunks are worked down with offset-disk harrows. It is also used for orchard cultivation and ripping up old roads.

The forces on a chisel shank and point are similar to those on a subsoiler.

FIELD TILLERS

Field tillers include tools which are also called field cultivators, subsurface tillers, and mulch tillers. They can usually be equipped with either shovels or sweeps and are used for tillage rather than row-crop cultivation. They vary widely in construction and duty.

Stubble-mulch tillage has received increasing acceptance in the Great Plains Wheat Belt for soil and water conservation because leaving the stubble on the surface prevents wind erosion and reduces sealing from rainfall, thus increasing water intake and reducing runoff erosion. The mulch also helps reduce evaporation (Chap. 6). Sweep tillage from 2½ to 5 in. below the surface is used to control weeds and leave the stubble or other residue on the surface. A subsurface tiller is shown in Fig. 14-1.

Field tillers are also especially adapted for *control of noxious weeds* such as bindweed, thistles, and quack grass. With overlapping sweeps, all weeds can be cut off, or with narrow shovels on spring shanks, quack grass roots can be brought to the surface.

The tillage action with spring teeth is similar to that of a spring-tooth harrow, loosening the ground and bringing clods and weed roots to the surface. The teeth are largely immune to damage from stones.

Design considerations for field tillers are as follows:

1. *Trash clearance* is very important in either stubble-mulch- or weed-control work. Shanks must be long enough so that heavy trash can pass under the frame.

[1] D. C. Heitshu, The Kinematics of Tractor Hitches, *Agr. Eng.*, 33:343–356, June, 1952.

Plenty of fore-and-aft clearance between rows of shanks is needed so that the trash can zigzag through.

2. *Minimum disturbance of stubble and trash* is important in stubble-mulch tillage. Maximum width per sweep allows fewer shanks to disturb the cover, but with sweeps wider than 24 in., overload protection of sweeps and shanks becomes difficult and they must be built much heavier, resulting in higher cost. In stubble-mulch work, shanks should be as narrow as possible with coulters ahead to cut trash, preventing bare grooves which may allow erosion to start.[1]

3. *Sweep-blade pitch, angle,* and *shape* are very important in mulch tillage. If too steep, the soil and trash will not flow over uniformly and uneven unprotected spots may be left. If too flat, penetration may be poor and the soil not loosened enough to kill the weeds.[2] Small sweeps can have pitch adjustable within a limited range, but wide sweeps cut too shallow at the wings if given pitch. The included angle of the V must be such that the blade sheds roots and trash, usually from 100 to 65°.

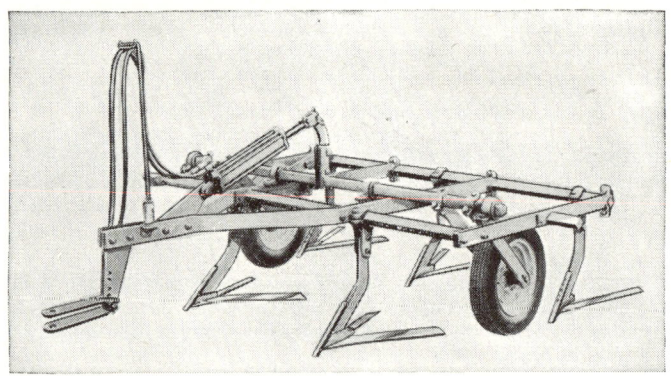

Fig. 14-1. Subsurface tiller. (*Massey-Ferguson, Inc.*)

Some wide sweeps are given an upwardly convex cross section for smoother soil flow and minimum mulch disturbance with satisfactory scouring and penetrating ability.

Sweep blades must have sufficient overlap to cut all weeds, but excessive overlap may cause nonscouring because the outer wings of the rear sweeps run in loosened soil. An overlap of 2 to 2½ in. has been found best for stubble-mulch work.[3]

4. *Overload protection* is commonly obtained by spring trips for stiff shanks with sweeps and by using spring teeth for narrow shovels and points. Spring-trip shanks are discussed in Chap. 3 under Overload Protection Devices.

Spring teeth vary in stiffness according to the service expected, but double flat-leaf construction is most common. Teeth of 1- or 1¼-in. square steel bars with a double coil at the top for adequate deflection have become popular for use with tool bars because they are strong enough to allow extended teeth for a rear row. These teeth are better adapted for carrying sweeps than flat spring teeth because they have more torsional rigidity to prevent sweeps from dodging tough taproots. The stiff shank, however, still has the advantage of holding sweep pitch uniform instead of constantly varying, as is the case with spring teeth. Spring teeth are usually attached by clamps to avoid bolt holes, which cause stress concentration and fatigue failure.

SPRING-TOOTH HARROWS

This was the first tool to use spring teeth and came into use in the 1860s in the eastern states for the purpose of working over and around roots and stones without

---

[1] F. L. Duley and J. C. Russell, Machinery Requirements for Farming through Crop Residues, *Agr. Eng.*, 23:39–42, February, 1942.
[2] L. W. Chase, A Study of Subsurface Tiller Blades, *Agr. Eng.*, 23:43–50, February, 1942.
[3] *Ibid.*

being damaged. It is also useful for loosening partially prepared seedbeds which have been compacted by heavy rains and for control of quack grass because of the ability of the vibrating teeth to bring the roots to the surface. It does not have the packing effect of a disk harrow.

Trailed spring-tooth harrows are commonly built in sections 3 to 5 ft in width, such as shown in Fig. 14-2, and pulled by a common drawbar. Longitudinal frame members carry the tooth bars and act as skid runners, gaging depth and carrying the teeth for transport. Depth of penetration is controlled by rotating the tooth bars, either by levers or by a sliding drawbar connection actuated by the tractor.

Clearing collected trash which will not pass through is occasionally necessary with all spring-tooth harrows, and the tractor-mounted type can easily be lifted for this purpose. With trailed tractor models having trip-rope tooth-bar control, the teeth can be raised as the harrow is pulled forward, leaving the trash, and then be reset by

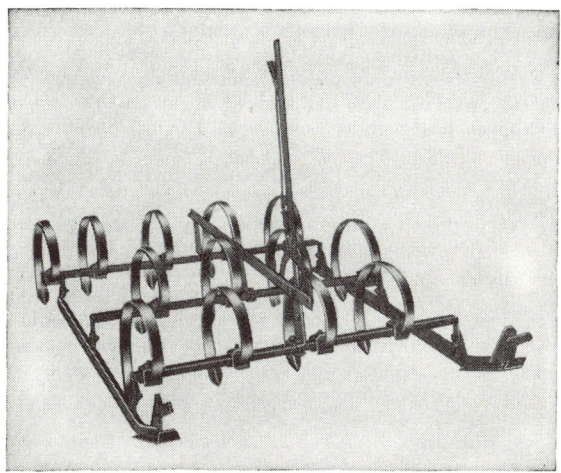

Fig. 14-2. Spring-tooth-harrow section. (*Deere & Co.*)

backing. One model features roll-over tooth-bar action, trip-rope-controlled, for clearing trash.

Spring teeth are commonly pointed in four styles: (1) arrow-shaped for digging, (2) pointed and rounded for lifting quack grass roots without breaking them, (3) narrow and rounded for alfalfa cultivation, and (4) with holes for bolting on a detachable double-pointed shovel. Clyde has found draft per tooth at 3- to 3½-in. depth to be from 30 lb in freshly plowed soil to 60 lb in moderately firm ground. With a tooth spacing of 5 in. this would give from 1.9 to 3.7 psi of cross section worked. The soil reaction was approximately horizontal, indicating that the weight of the tool was required to oppose the upward component of the hitch line.[1]

ROW-CROP CULTIVATORS

The primary purpose of row-crop cultivation is to control weeds, although it is also desirable to stir some soils for aeration and better rainfall intake.

Cultivation may control weeds in three ways:[2]

1. Uprooting the weed, separating its roots from the soil, and exposing them to die
2. Cutting off the stem just below the ground surface, as is usually done with deep-rooted weeds
3. Smothering the aerial parts of the weed by covering with soil

[1] A. W. Clyde and R. J. McCall, Tillage Tools, *Penn. State Univ. Agr. Expt. Sta. Bull.* 465, 1944.
[2] C. K. Shedd and E. V. Collins, Equipment for Cultivating Corn, *Agr. Eng.*, 22:5–6, January, 1941.

Any of these three methods may be used in the space between the rows, but only the third can be used to destroy the weeds in the row, except that weed seedlings growing in a crust can be killed by breaking and stirring the crust with a rotary hoe, finger weeder, or spike-tooth harrow when the crop is deeply enough rooted and yet small enough so that it is not damaged by this treatment.

Some of the *design requirements* for tractor and cultivator combinations, and methods used to meet them, are as follows:

1. *Clearance* of 2 ft under axles and cultivator parts is desirable for individual gang cultivators for corn and cotton. Beet and bean cultivators, also used for vegetables, use a continuous tool frame extending across under the tractor. Clearance is about 1 ft, but tools can be set for very narrow rows and the rigid lateral spacing permits very precise work.

2. *Speed* ranges from ½ mph for precise vegetable cultivation to 5 mph for the final cultivation of corn.

3. *Steering* should be accurate and easy, and the cultivator should be tied to the tractor as rigidly as possible. The gangs are usually located toward the front wheels for quick response. The cultivator gang bearings for vertical movement usually have provisions for take-up of wear to avoid lateral lost motion.

4. *Visibility* of the working area is improved by making the tractor as narrow as possible through the central portion. Some small tractors used for vegetable cultivation have the engine offset to one side and the operator's seat to the other side for maximum visibility, while one model has the engine over the rear axle with an open frame to the front axle.

5. Accurate *depth gaging* is required, particularly for small plants and vegetables. Two-row corn and cotton cultivators are commonly gaged from the tractor, with the gangs held to depth by spring pressure rods. Four-row cultivators, however, have small gage wheels on all gangs, except the two or four adjacent to the tractor. These wheels are set the desired distance above the shovels, and enough spring action is provided in the spring pressure rods to let the gang float with respect to the tractor. Almost all cultivator gangs are carried by parallel-action linkages in order to maintain the relative depth of front and rear shovels on a gang.

6. Quick, easy *lifting* for turning and transport is secured by the use of hydraulic power. Where gage wheels on the gangs are not used, the hydraulic lifts can also be used to adjust depth. Selective and delayed lifts are available for some cultivators. Selective lift allows either side to be raised independently for cultivating point rows resulting from angling fences or contoured rows between terraces. Delayed lift delays lifting and lowering of the rear gangs to stop and start cultivating at the same spot as the front gangs.

7. *Row-width adjustments* are made by loosening clamps and sliding the gangs and lift arms sideways into the desired position and reclamping. Square tubing is often used for the main frame members carrying the gangs and square bars for the lift arms. Corn and cotton cultivators are usually adjustable from 28 to 42 in. in the two-row models and from 36 to 42 in. in the four-row models. Vegetable cultivators may be set for rows as close as the tractor tires will allow.

8. *Gang interchangeability* is necessary because of the wide variety in types of shovels, sweeps, etc., demanded by different areas. A common parallel-linkage assembly is used for all types of gangs and also utilized for planter attachments in many cases.

9. Quick, easy *attachment to the tractor* is necessary because the tractor is needed for much other work during the cultivating season. Supporting legs, guide plates, locating pins, hook bolts, "buttonhole" slots, and self-locking pins are used to expedite attachment. Several models have a hinged frame which closes in against the sides of the tractor as the tractor pushes it forward and is then attached by nuts on each side plus pins for the lift pipes.

10. Simple rugged *construction* with a minimum of wearing parts is desirable. Breakback shovels are a necessity, except where spring teeth are used, and the frame and gangs must resist shock loads. Lateral deflection reduces accuracy and allows the shovels to dodge tough weeds. Clamps must be positive to maintain position, yet easily loosened for adjustment.

SHOVEL AND SWEEP TOOLS 153

Shovel equipment for cultivating is extremely varied. Sweeps and spear-point shovels are most popular, but many others are available, including knife weeders, various blade hillers, and disk hillers. Some of the shovels and sweeps available are shown in Fig. 14-3. High-speed sweeps, with flatter pitch and lower crown than regular sweeps, reduce ridging at high speeds. Standards have been developed for cultivator sweep and shovel mountings.[1]

Shields are used to prevent small plants from being covered, and they may be of the blade, tent, or rotary type. Rotary hoe-wheel attachments, consisting of three or

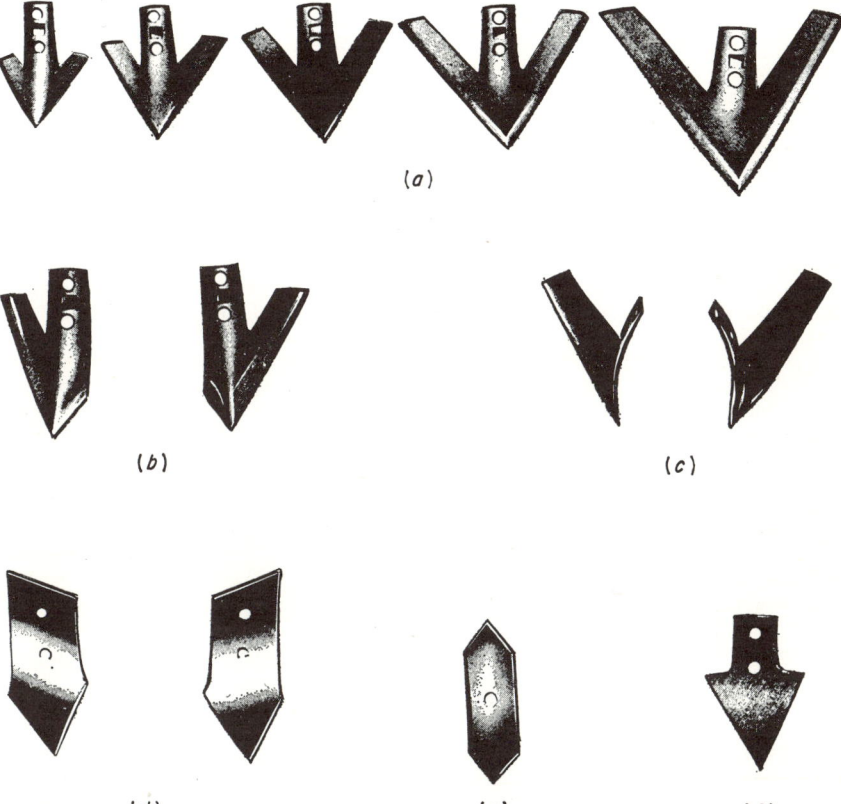

FIG. 14-3. Cultivator shovels and sweeps: (*a*) 6- to 16-in. sweeps; (*b*) half sweeps; (*c*) knife weeders; (*d–f*) single-pointed, double-pointed, and spearpoint shovels.

four wheels running over the row, may be used to break the crust around the plants and also to allow fine dirt thrown by the shovels to cover weeds in the row without covering the crop. These attachments sometimes double the allowable speed of travel.

Rear-gang equipment is primarily for the purpose of loosening the tractor-wheel tracks. Spring teeth are quite popular on rear gangs. Finger-weeder sections are sometimes used on the rear gangs because they smooth out the tractor tracks for cross-cultivation and also drag out weeds to die which might otherwise be only transplanted.[2]

Lister cultivators are used for the first two cultivations of corn planted in the bottoms of furrows made by a lister, as is common practice in the western Corn Belt.

[1] ASAE Standard: Cultivator Sweep and Shovel Mounting, *Agr. Eng. Yearbook,* 1959, pp. 85, 86.
[2] Shedd and Collins, *op. cit.*

These cultivators have long tent-type shields with disk hillers throwing away from the row, followed by shovels throwing in. The disk hillers are set to throw in for the second cultivation, almost leveling the field, so that a conventional cultivator can be used for the third cultivation. Regular trailing lister cultivators give the units for each row lateral freedom so that they can automatically follow the trench, being guided by a pair of furrow wheels. This is particularly needed where a skip-row lister has been used, since the row spacing may not be constant.

Power-driven *rotary-tiller units* with suitable shields are sometimes used for close cultivation of very small plants. They are effective in killing weeds without digging deeply enough to disturb the plants. Their use has been largely limited to vegetable crops grown on organic soils but is being extended to other crops on mineral soils.

Some fruits such as grapes, raspberries, and blueberries are spaced several feet apart in wide rows. Cultivation in the row is desirable to control weeds, and horse- or tractor-drawn *grape hoes* are often used. They are manually guided in and out of the row to kill the weeds and miss the plants. A hydraulically controlled tractor hoe was shifted out of the row by plant actuation of a feeler rod, which in turn operated a hydraulic control valve and cylinder.[1]

Flame cultivators direct a flame into the row near the ground from each side to kill small weeds. The theory of flaming is not to burn the weeds but to apply enough heat so that the liquid in the weed cells expands and ruptures the cell walls, causing the plant to gradually wilt and die. By careful control of the flame, reduction in cotton yield was avoided in spite of several flamings and control of small weeds and grasses in the row was aided.[2] Flame cultivators are generally used at the rear of the tractor in conjunction with a conventional front-mounted cultivator. From 2 to 4 gal of LP gas are usually required per acre.[3] A special fuel tank is required along with the necessary gages and valves, a vaporizer to change the fuel from liquid to gas, a pressure-regulator control valve, special fuel lines and fittings, and special burners with flat-spray nozzles.[4]

Weed-controlling chemicals such as 2-4-D are valuable adjuncts to conventional cultivation, and preemergence spraying of the row has been helpful in controlling weeds in the row. Spraying for weed control is discussed in Chap. 18.

Mulch tillage of corn requires that the residues be left on the surface, and single 30-in. sweeps have been used between the rows with disk hillers next to the row throwing in or out as required.[5]

Considerable experimentation is in progress on *wide-row planting* of corn in order to grow legume cover crops in between for erosion control and increased nitrogen and organic matter. Preliminary results indicate that corn yields are satisfactory with up to 60-in. widths when normal plant population is maintained and that soil can be protected and fertility maintained with a corn-corn-clover rotation. Wide-row spacing also permits planting winter wheat with tractor equipment. It is possible that these practices will require modification of existing cultivating practices and equipment.

---

[1] W. M. Carleton and Dwight Kampe, An Experimental, Hydraulically Manipulated Blueberry Weeder, *Mich. State Univ. Agr. Expt. Sta. Quart. Bull.*, May, 1954, pp. 426–434.

[2] J. R. Tavernetti and H. F. Miller, Jr., Studies on Mechanization of Cotton Farming in California, *Univ. Calif. Agr. Expt. Sta. Bull.* 747, 1954.

[3] Thomas L. Baggette, Farm Engineering in Flame Cultivation, *Agr. Eng.*, 28:548–550, December, 1947.

[4] L. M. Carter, Rex F. Colwick, and J. R. Tavernetti, Evaluation of Flame Burner Design for Weed Control in Cotton, *ASAE Paper* 59-139, June, 1959.

[5] Larsen and Joy, *op. cit.*

Chapter 15

MINOR TILLAGE TOOLS

ROTARY TILLERS

Rotary tillers utilize mechanically powered rotary motion of soil-working parts in combination with forward motion in order to secure more pulverization in one operation than is possible with pull-type tillage tools. A number of different approaches have been tried, one of the earliest being a series of small plows or teeth attached to the underside of a circular revolving frame which was carried forward by the tractor. Another approach, introduced about 1929, was a *combined plow and pulverizer*, shown in Fig. 15-1, where a vertical PTO-driven beater pulverized and turned the

FIG. 15-1. Combined plow and PTO-driven pulverizer. (*Pulverator Plow Co.*)

furrow slice as it left a stub moldboard. This machine was capable of burying surface trash, unlike many rotary tillers, and its power consumption was not excessive.[1] Its chances of acceptance were greatly reduced by the adverse economic situation following its introduction.

Rotary tillers now on the market consist of a series of pointed spring teeth or L-shaped blades mounted on a horizontal shaft and rotated so as to cut down against the soil as the tool advances. Soil is thrown up behind and is usually confined by a hood unless it is desired to allow noxious weed roots to settle on the surface to dry up and die. Because of the soil reaction against the teeth, the drive wheels of the tractor often act to brake and control forward motion rather than to propel.

Garden tractor sizes of rotary tillers are most popular because of the convenience of once-over preparation of small garden plots and in nursery and greenhouse work.

[1] J. B. Davidson and E. V. Collins, The Direct Application of Mechanical Power to Soil Tillage, *Agr. Eng.*, 10:165–168, May, 1929.

The larger tractor-powered models are little used for general farming because of the high-power consumption compared with conventional soil preparation. They are used, however, for quack grass control and grub control and where it is desired to thoroughly incorporate a heavy cover crop or even light brush into the soil. Heavy-duty models equipped with thin L-shaped blades, widely spaced, shown in Fig. 15-2, are popular in tropical areas for chopping up and burying thick layers of sugar cane and pineapple residues for quick decomposition, eliminating the fertility loss of burning them.

Heavy-duty models are also extensively used in constructing roads, airport runways, etc., where a stabilized base must be mixed in place.[1]

Design problems encountered in rotary tillers, and current solutions are as follows:

1. Excessive *power consumption* is a basic problem in rotary tillage because soil requires more force for pulverization by impact than by lifting or crushing as is done

Fig. 15-2. Rotary tiller with widely spaced L-shaped blades. (*Howard Rotavator Co., Inc.*)

by conventional tillage tools. In general, any expedient which reduces the kinetic energy applied to the soil, such as reducing peripheral speed, frontal area of springs, etc., increasing slice thickness or forward travel per revolution, and decreasing the number of teeth, has reduced power requirements.[2] The ideal objective is to produce only the required degree of pulverization and to do it by shearing and lifting rather than impact of teeth against soil and soil against hood. Slow-speed L-shaped blades which give a minimum number of vertical cuts seem to be most economical of power.[3] The time-circle diameter should be about 1½ times the depth for greatest efficiency but is usually larger because of mechanical requirements.[4] Power requirements are extremely variable, but a 36-in. width is made for 25-hp tractors with auxiliary transmissions giving a speed down to ½ mph.

2. *Working the soil in front of the drive* is a problem because the rotor shaft operates at ground level.[4] An enclosed chain drive to one end of the rotor shaft, as shown in Fig. 15-2, offers a narrow frontal area and can be run in previously loosened soil after the first round.

---

[1] John M. Greene, Some Rotary Tillage Applications, *Agr. Eng.*, 27:175–176, April, 1946.
[2] C. W. Kelsey, Rotary Soil Tillage, *Agr. Eng.*, 27:171–182, April, 1946.
[3] W. J. Adams, Jr., and Donn B. Furlong, Rotary Tiller in Soil Preparation, *Agr. Eng.*, 40:600–607, October, 1959.
[4] *Ibid.*

3. *Protection from stones and obstacles* is very important. Spring-mounted teeth have been most popular, although careful design and the highest-quality material are needed to reduce fatigue failure.[1] Flexible rubber drives and slip clutches are also used. One machine has the teeth mounted on a series of plates which are stacked on the shaft between spacers and friction plates. Any tooth plate can slip individually, the force being determined by the pressure adjustment at the end of the shaft.

4. *Prevention of excessive wear* of drive parts and bearings requires the utmost attention to sealing of the portions operating in the soil. Soil-working parts move faster and farther than on plows, cultivators, etc., and special alloys and hard surfacing are often required to resist rapid wear.

Rotary tillage has been compared with conventional and various other tillage methods in preparing ground for corn and other field crops by experiments in Ohio and Michigan.[2] Little difference in yield resulted when the surface was free of vegetation, but yields were significantly lower with rotary tillage where a legume or grass sod had to be worked up. Weed and grass control was more difficult because the residues interfered with cultivation, not being completely buried as with a plow, and weed and grass germination was stimulated because of the fine pulverization of the surface. In these tests, plowing followed by only a treader or reversed-rotary-hoe section gave least weed germination, apparently because of the resulting coarse surface structure. There has also been some evidence of reduced nutrient availability following rotary tillage of sods, possibly because the thorough mixing of the residues temporarily ties up nitrogen during its decomposition.

SPIKE-TOOTH HARROWS

Spike-tooth harrows engage the ground with short stiff pointed teeth mounted on a series of crossbars and are extensively used for a final smoothing and surface pulverization of the ground before planting.

The spike-tooth harrow has a crushing and stirring effect with no lifting effect, since the teeth are usually angled back. Modern all-steel designs are usually built in sections about 5 ft wide and having five tooth bars with teeth spaced so as to cut from 1½ to 2 in. apart. The harrow is a comparatively lightweight tool, weighing about 25 lb per ft of width. It does not ordinarily penetrate more than 2 in. deep and is most effective when the surface is mellow, following plowing or a rain.

Draft of a spike-tooth harrow is light, ranging from 30 to 60 lb per ft of width. As much as a 30-ft width may be pulled in large-scale farming operations. In the Corn Belt two sections are commonly trailed behind a tandem-disk harrow or one section behind a plow.

Transport of wide spike-tooth harrows is a problem. Usually sections are detached from the drawbar and hauled in a wagon. Some four-section harrows are made with folding drawbars so that the outer sections can be folded onto the center sections for passing through gates. Lift-type harrows with folding outer sections are convenient for transport, but the lifting frame adds considerably to the cost. Another method uses a wheel-supported drawbar which can be folded into a transport trailer after the sections have been tipped up and fastened in a vertical position.

Flexible spike-tooth harrows have flexible connections between tooth bars, enabling them to follow uneven ground closely. These harrows penetrate more uniformly than rigid-frame harrows. Usually only two tooth angle settings are provided, one straight for pulverizing and the other slanted for smoothing, determined by the end from which the section is pulled. Some flexible harrows have lever adjustments. Individual sections can be rolled up for transport or storage when they are detached from the drawbar.

Chain-type spike-tooth harrows are popular in Europe for cultivating young wheat plants and also for scattering animal droppings on pastures. Round steel rods are

---

[1] *Ibid.*
[2] R. L. Cook and F. W. Peikert, A Comparison of Tillage Implements, *Agr. Eng.*, 31:211–214, May, 1950. C. J. Willard, G. S. Taylor, and W. H. Johnson, Tillage Principles in Preparing Land for Corn, *Ohio Agr. Expt. Sta. Research Circ.* 30, March, 1956.

pointed and bent to form one-piece spike teeth with connecting links. The teeth are thus free to deflect laterally as well as backward.

KNIFE-TOOTH HARROWS

The best known example of this tool is the *acme harrow,* used primarily in truck-crop areas. It consists of a row of twisted blades spaced about 6 in. apart which cut and stir the surface for pulverization and destruction of small weeds.

A recently developed tool is a *combination smoothing float and knife-tooth harrow.* Each section consists of a steel pan about 4 ft square with four rows of U-shaped steel knives attached underneath and pivoted to facilitate turning at the end of the field. The knives cut about 4 in. deep and 1 in. apart.

FINGER WEEDERS

The finger weeder has closely spaced light spring teeth about 18 in. long terminating in a "pencil" end and is effective in destroying weed sprouts in a light crust without damaging corn or other deeper-rooted crops. It can be used on larger corn (12 to 15 in.) than can the spike-tooth harrow because of its clearance and the ability of the teeth to deflect around plants without damaging them. Draft is light, being from 25 to 35 lb per ft of width. Finger-weeder sections are also used for rear cultivator gangs.

ROD WEEDERS

The rod weeder is a unique tool used for controlling weeds on summer fallow ground in the light soils of the semiarid West. The operating element is a rotating rod about 1 in. thick extending across the machine and drawn under the surface to pull out weeds. The rod is supported by several shanks and is rotated to clear the weed roots, being driven from the wheels. The soil is normally loosened at the beginning of the season by plowing or sweep tillage, and weeds controlled thereafter by periodic use of the rod weeder. Draft is from 80 to 110 lb per ft of width, and total widths up to 36 ft are handled by the large tractors popular for the extensive type of farming practiced in these areas.

ROTARY HOES

This tool is operated over row crops to control weeds in much the same manner as the spike-tooth harrow and the finger weeder, but is able to break up a heavier crust and can be used until the crop plants are much larger. It consists of two rows of rimless wheels with curved pointed spokes which enter the soil straight and tear out a small section on leaving. The wheels usually range from 12 to 20 in. in diameter with 10 to 16 spokes and are set about 4 in. apart in each row, thus working every 2 in. The wheels in the two rows usually overlap enough to clean trash from each other.

Current rotary hoes have two to four sections about 3½ ft wide hinged together for flexibility in following the ground. Wide models sometimes allow the outer gangs to be folded for transport. These hoes have fabricated steel wheels which are not easily bent. The sections can be pulled from either end, the curved spokes leading for hoeing and trailing for transport or for treading. A single section makes an efficient treader to trail behind a plow since it is light in draft and very effective in settling the freshly plowed soil and in pulverizing large lumps.

The *draft* of a rotary hoe is from 30 to 60 lb per ft of width, and a four-row model can be pulled 6 to 8 mph by a two-plow tractor. The rotary hoe is most effective at high speed, and as much as 80 acres per day can be covered.

A recent variation on the rotary hoe, used for tillage rather than cultivation, is the *skew treader,* which consists of two rows of rotary-hoe teeth set an an angle with each other so as to exert a side thrust on the soil. The curved spokes trail for this use, and the hoe wheels are somewhat smaller and stronger than on the conventional hoe. The soil-stirring effect is greatly increased by the skew action, and the tool is effective in killing weeds in stubble mulches without burial of the residues. Provision must, of course, be made for carrying the end thrust resulting from the skew. Some models can be adjusted to run straight and reversed for use as a rotary hoe. A small tool of this type with small four-spoke wheels is designed to be trailed behind a plow for pulverizing and compacting the soil.

MECHANICAL THINNERS

Some row crops, such as cotton, beets, and vegetables, may vary widely in percentage of emergence from one season to another, and it is common practice to plant at a high rate and thin out the surplus plants. This may be done by hand hoeing, cross-blocking, or by oscillating or rotating down-the-row thinners.

Hoeing is best in that the healthiest plants can be selected, but labor costs are high. Mechanical thinning is satisfactory where the stand is uniformly distributed.

Cross-blocking (cross-cultivation with sweeps spaced to leave the desired number of plants) is acceptable on flat planted fields.

Where the rows are on beds or the ground is uneven, down-the-row thinners are best. The rotary type (Fig. 15-3) is most popular and may be driven by ground con-

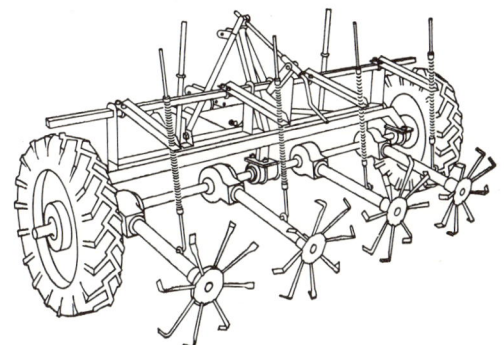

Fig. 15-3. Rotary thinner. (*Silver Engineering Works, Denver, Colo.*)

tact or by power. The size and spacing of the gaps left by the rotating element are controlled by the openings between cutting elements and the speed of rotation.

ROLLERS

Rollers are used to pulverize clods by crushing and to compact the soil. Smooth one-piece rollers are not used except on grassland because bare soil is left so smooth that it crusts excessively after a rain. A series of cast-iron wheels stacked on a shaft is the most popular construction. The shaft turns in bearings on the ends, but the individual wheels can slip on the shaft for ease in turning.

Double-gang rollers usually have wheels with V-shaped rims, the rear row splitting the ridges formed by the front row. This action gives maximum pulverization of the surface and is also very efficient in covering and compacting grass seed dropped between the rollers.

Single-gang rollers usually alternate V-rim wheels with toothed wheels which roughen the surface slightly to reduce the tendency for the fine soil to blow or crust after a rain. The toothed wheel has a large center hole, allowing it to lag with respect to the V-rim wheel. This gives a cleaning effect and also allows it to follow irregular ground more closely. Some western machines have very large center openings in all wheels, which give them extreme ability to conform to irregular ground. They can be used to compact bedded rows in irrigated soils.

Subsurface packers have wheels spaced apart several inches with V rims or crowfoot rims so as to penetrate more deeply and pack underneath rather than pulverize the surface. Short sections of this type are made for trailing behind plows.

Draft of rollers is fairly light, ranging from 30 to 60 lb per ft of width.

Chapter 16

FERTILIZING AND LIMING MACHINES

In order to maintain soil fertility it is necessary to replace the plant food removed by the crops. Nitrogen can be partially replaced by bacterial fixation of nitrogen from the air when legumes are grown, but other elements must be completely replaced. When the manure is returned, livestock farming does not remove fertility as fast as cash cropping, but even so, a deficit is incurred. The use of commercial fertilizer, consisting of mined or synthesized compounds which furnish the required plant nutrients, is the only way to make up the deficit. The function of the machines considered here is to measure out fertilizer at the desired rate and apply it at the desired location. (See Chap. 56 for a discussion of fertility.)

TABLE 16-1. HYGROSCOPIC POINTS OF VARIOUS FERTILIZER SALTS

Fertilizer salt	Per cent relative humidity at:	
	68° F	86° F
Calcium nitrate	54.8	46.5
Ammonium nitrate	63.3	59.4
Sodium nitrate	74.5	73.7
Urea	80.7	75.2
Ammonium chloride	79.2	77.5
Ammonium sulfate	81.0	81.1
Diammonium phosphate	83.2	82.8
Potassium chloride	85.3	84.4
Monoammonium phosphate	93.1	92.9
Monopotassium phosphate	93.2	93.0
Potassium nitrate	94.5	93.3
Potassium sulfate	97.0	96.5

SOURCE: A. L. Mehring and G. A. Cumings, Factors Affecting Mechanical Application of Fertilizers to the Soil *USDA Tech. Bull.* 182, 1930.

There are many different types of metering devices and also many methods of application. Some of the *application practices* are as follows:
1. Broadcast on the surface
2. Deposited on the bottom of the plow furrow and covered by the succeeding furrow slice
3. Applied at planting with the seed, as by grain drills
4. Applied at planting near the seed as by corn, cotton, potato planters, etc.
5. Applied at cultivation under the surface near the plant
6. Applied under the surface before planting with subsoiler or chisel openers
7. Drilled into established pastures with special equipment
8. Applied in irrigation water

Chemical fertilizers must be water-soluble in order to be available to plants, and consequently they can absorb moisture from the atmosphere, with an ensuing change in physical properties, usually becoming more difficult to spread. Mehring and Cumings found the drillability of fertilizers to be affected by the following physical properties:[1]

1. *Hygroscopicity,* or tendency to absorb moisture from the atmosphere. When the relative humidity exceeds the hygroscopic point of the fertilizer salt it absorbs moisture rapidly and tends to dissolve. Table 16-1 gives hygroscopic points for various common fertilizer salts.

2. *Particle size and shape* is a major factor since coarse particles flow more uniformly than fine particles and because of their smaller total surface area, are much less affected by high humidity. Smooth spherical particles flow easily, but irregular particles tend to lock together. Granular and pelleted fertilizers have much better drillability than common pulverized fertilizers.

3. *Heterogeneous* fertilizers, consisting of widely different sizes and types of particles may separate in the hopper, with the finest and heaviest settling to the bottom, where they are distributed first. In a mixed fertilizer the result may be much variation in the composition and amount of plant food spread throughout the field. Such materials, when broadcast by throwing a spinner spreader, show much segregation over the width of spread.

4. *Specific gravity* affects weight delivered because distributors work on a volume basis and should logically be calibrated on that basis. The specific gravity of common fertilizer materials varies from 0.63 for urea to 1.06 for nitrate of soda.

The *kinetic angle of repose* of a fertilizer material gives a rough indication of drillability since it affects (1) the rate of delivery, (2) the size of gate opening through which the fertilizer will escape when the distributor is not operating, (3) delivery-rate variations with depth change in the hopper, and (4) uniformity of discharge. Common fertilizer distributors give best results with materials having an angle of repose of not more than 40°.

Commercial fertilizers commonly contain conditioners, relatively inert materials which can absorb moisture without losing drillability or can reduce the hygroscopicity of the mixture. Animal tankage, fish scrap, cottonseed meal, peat, lime, ground limestone, and other materials are often used. High-analysis fertilizers necessarily contain less of this inert material and are more affected by humidity, with consequent reduced drillability.

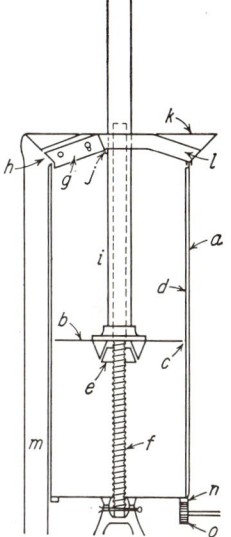

Fig. 16-1. Top-delivery fertilizer distributor with rotating hopper for experimental plot work: (a) cylinder; (b) diaphragm; (c) diaphragm slot; (d) diaphragm drive flange; (e) tapered split nut; (f) stationary threaded rod; (g) fertilizer delivery blade; (h) delivery opening; (i) shield; (j) shield cleaner; (k) top carriage; (l) spider brace; (m) delivery tube; (n) cylinder ring gear; (o) drive pinion. (*Mehring and Cumings, USDA Tech. Bull. 182,* 1930.)

METERING MECHANISMS

Metering mechanisms of many different types have been used for distributing fertilizer. The simplest is a plain gravity feed through a hole, with the rate of flow controlled by the size of the hole. This of course, works only for a free-flowing material such as dry sand, and the rate of application varies with ground speed. The next step is some sort of agitator over the hole to keep the material flowing, preferably in direct ratio to speed of ground travel so as to give a uniform rate of application. The delivery rate for the same feed opening and agitator speed will vary widely according to the flow characteristics of the fertilizer, however, depending on the degree of dependence on gravity flow. The ideal metering device, of course, would discharge a uniform predictable set volume of fertilizer by a positive displacement feed unaffected by gravity flow. In

---

[1] A. L. Mehring and G. A. Cumings, Factors Affecting Mechanical Application of Fertilizers to the Soil, *USDA Tech. Bull.* 182, 1930.

addition, it should be easily adjustable, simple, rugged, inexpensive, capable of pulverizing lumps, corrosion-resistant, and easily cleaned.

Existing metering devices may be classified as top-delivery, side-delivery, or bottom-delivery. The *top-delivery type* feeds from the top surface of the fertilizer, either by lifting it into contact with blades which scrape it into delivery tubes as shown in Fig.

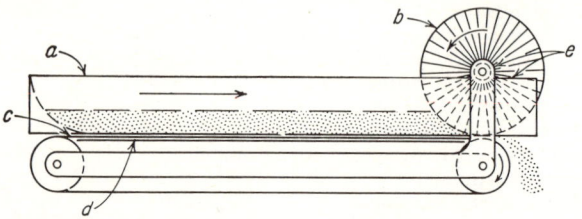

FIG. 16-2. Side-delivery fertilizer feed for experimental plots: (*a*) three-sided hopper; (*b*) rotating discharge brush; (*c*) endless-belt bottom; (*d*) bottom support slide; (*e*) pinion and rack for moving hopper. (*Fairbank, Agr. Eng., November,* 1950.)

16-1 or by descending scraping blades. This method has the advantage of delivering a predictable set volume, but because of settling of the fertilizer after filling the hopper, it is difficult to start feeding promptly and at the desired rate. This type is more complex and costly than bottom-delivery types and has not been successful in the United States except for experimental plot work, although it is in commercial production in Europe.[1]

The *side-delivery type* was introduced in Germany in 1927 and is currently used in the United States for close control of fertilizer applications to experimental plots. One of the models developed for plot use consists of a three-sided bottomless box riding on an endless belt.[2] The box is filled evenly to the desired level with fertilizer, and the open end is advanced toward the rear roller of the endless belt. Box and belt move together, and the fertilizer drops over the rear roller, being removed evenly by a stationary revolving brush. The relationship between box travel, ground travel, and depth of fertilizer in the box determines the rate of application. A cross section of this type of applicator is shown in Fig. 16-2. With it, neither gravity flow nor settling affects feeding, but the mechanical complications, large hopper area, and care required for leveling the fertilizer in the hopper have prevented the principle from being adopted for commercial production.

The *bottom-delivery types* depend on gravity and agitation to ensure settling of the fertilizer to the bottom of the hopper where a metering device of more or less positive action controls the rate of delivery. These metering devices may be classified as agitator, revolving plate, star wheel, chain finger, paddlewheel, reciprocating

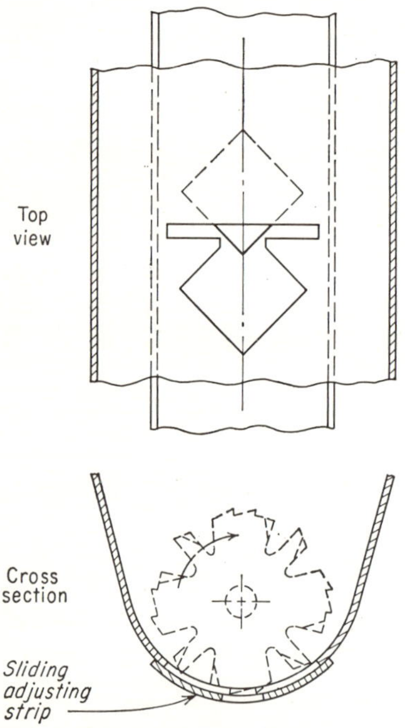

FIG. 16-3. Agitator feed with agitator disk running in upper feed opening. (*Ezee Flow-New Idea.*)

[1] *Ibid.*
[2] G. E. Fairbank, Fertilizer Placement Machine for Experimental Plot Work, *Agr. Eng.*, 31:556–564, November, 1930.

knocker, endless belt, roller, and screw feeds. The first three are most popular in the United States.

The *agitator feed* usually consists of a constant-speed rotary or oscillating agitator operating over a discharge hole, the rate being controlled by the size of the hole.

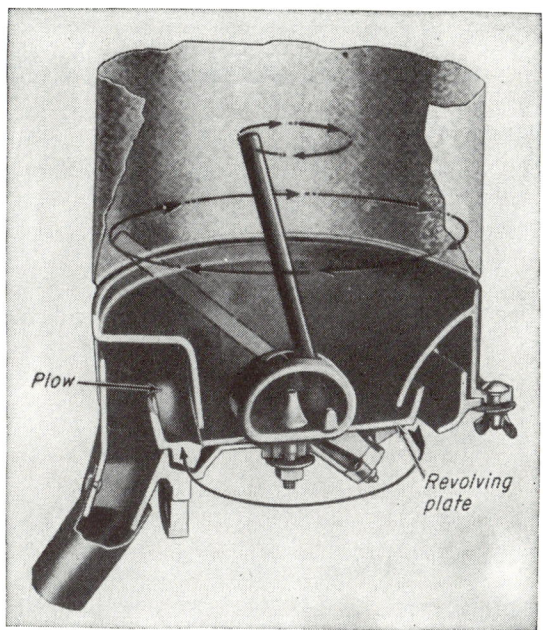

FIG. 16-4. Revolving-plate feed. (*J. I. Case Co.*)

Triangular or diamond-shaped holes are commonly used in both the hopper bottom and the adjusting strip. Only the corners overlap for small openings. Feed disks with twisted and L-shaped blades have good ability to grind up lumps, but the hopper and agitator must be properly designed to prevent bridging of damp material. Difficulty with holes clogging at the smaller openings has been successfully overcome in recent machines by slotting at the bottom of the hopper at the corner of the hole used for light feeds and allowing a narrow notched portion of the feed disk to pass into this slot, running next to the adjusting strip hole under the hopper, thus forcing material through the opening. This construction is shown in Fig. 16-3.

The *revolving-plate feed* is used with vertical cylindrical hoppers for planting and side dressing. Accurate reproducible adjustment of the feed opening is difficult, and different plate speeds may be used to help control rate. A cross section of this feed is shown in Fig. 16-4. These feeds give fairly uniform delivery with free-flowing fertilizer, although the rate is influenced somewhat by agitator position as well as by inaccuracies in the revolving plate outer wall. They do not crush lumps as well as some other feeds, and lumpy fertilizers cause irregular delivery. They are affected by the fertilizer condition and by depth in the

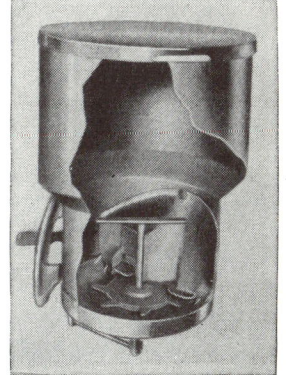

FIG. 16-5. Star-wheel feed for row-crop fertilizer attachment. (*International Harvester Co.*)

hopper because of the dependence on gravity flow. Minor inclinations of the hopper, as on hillsides, affect delivery rates up to 10 per cent for 10° of tilt.[1]

The *star-wheel feed* is commonly used in grain drills and some broadcasting machines as well as in some planter and side-dressing attachments. This feed is somewhat similar to the revolving-plate feed except that the plate has a notched edge which carries the fertilizer out under a gate. A knocker may be used to dislodge sticky fertilizer from between the teeth, and agitators may be used above the star wheel. Wheel speed as well as gate height is adjustable to control feeding rate. A corn-planter star-wheel feed is shown in Fig. 16-5. The feed rate is less affected by gravity

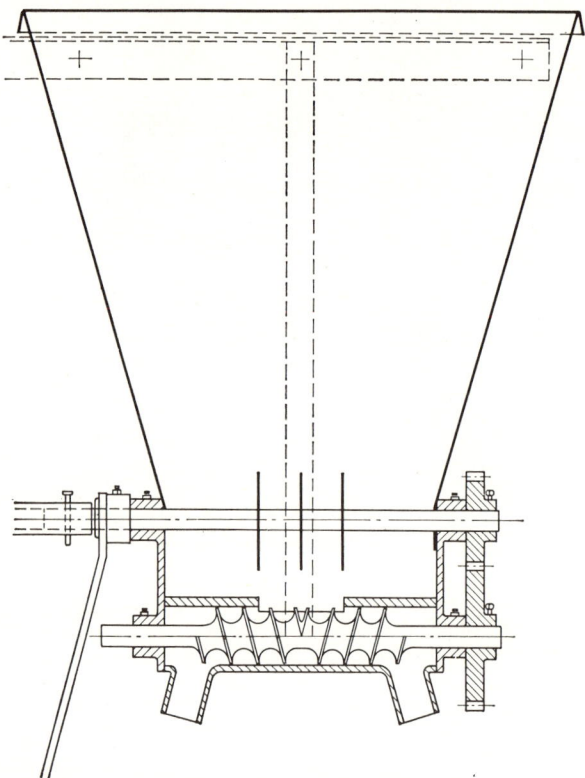

Fig. 16-6. Auger feed. (*L. H. Schultz Mfg. Co.*)

flow than with the revolving plate, but the teeth tend to cause some variations in delivery rate. Inclination toward or away from the discharge opening increases or decreases the rate. Star wheels with different numbers of teeth are available for some makes in order to extend the range of delivery rate.

An *auger feed* is shown in Fig. 16-6, and an *endless-belt feed* in Fig. 16-7.

Calibration of any fertilizer distributor for rate of application per acre may be performed by jacking up the drive wheel and turning it 100 revolutions at approximate operating speed while collecting the discharged material. Then

$$\text{Application rate, lb/acre} = \frac{(\text{amount collected, lb})(1{,}664)}{(\text{wheel diam, in.})(\text{effective width, ft})}$$

This formula also applies to grain drills, planters, or any other type of distributor.

[1] Mehring and Cumings, *op. cit.*

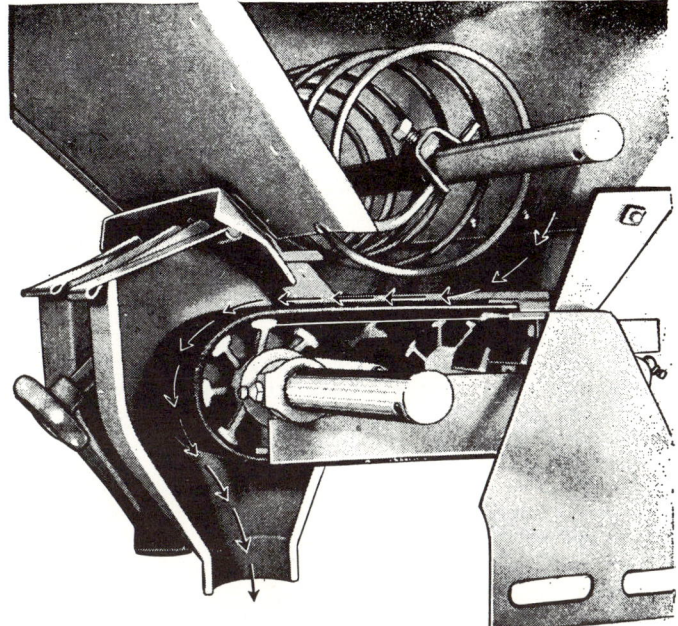

Fig. 16-7. Endless-belt feed. (*Deere & Co.*)

WIDE-HOPPER FERTILIZER AND LIME SPREADERS

These machines are used to broadcast material on the surface. A large trough-shaped hopper, usually from 8 to 12 ft wide, is carried between two wheels. Hopper capacity ranges from 1 to 2 cu ft per ft of width. An axle-mounted agitator feed is almost exclusively used, although some machines capable of light applications of fertilizer use the star-wheel feed. Advertised delivery rates ordinarily range from 100 to 8,000 lb per acre, although many machines are not capable of accuracy at the lower settings with sticky fertilizer.

Because of the corrosiveness of fertilizer it is important that corrosion-resistant steel be used and that the feeding parts can be easily cleaned. Many machines are designed for easy removal of metering parts for cleaning. After cleaning, parts should be given a protective coating of oil.

CENTRAL-HOPPER FERTILIZER AND LIME SPREADERS

This type of machine discharges the material from a central hopper on to spinning plates or other broadcasting means. The small two-wheeled ground-driven type is trailed behind a wagon or truck, and the material shoveled or dumped into the hopper while spreading. Material drops through adjustable openings onto horizontal spinner plates, which give about a 15-ft effective spread. Ordinarily two counterrotating spinner plates with vanes are used. The distribution pattern can be modified by shifting the point of discharge upon the spinner plate but is also greatly affected by materal characteristics and by wind. Overlapping is necessary because distribution tapers out at the edges without a sharp cutoff. With a heterogeneous material, the large particles are thrown farther than the small, but the finest particles will be blown more by the wind. These machines are not ordinarily used for anything but heavy applications of ground limestone, where the unevenness of distribution is not as serious as with fertilizer.

Much limestone is now applied by the *trucks* which haul it to the farm (an obvious economy), and special truck bodies are built for this purpose. A V body with conveyer delivery to the rear is commonly used, with adjustable conveyer speeds and gate openings to control delivery rate. Broadcasting is usually done by spinners because of their simplicity and compactness, although distribution is frequently poor. One truck spreader attachment uses foldable lateral chain conveyors extending 10 ft out from the center of the truck at each side. Uniform distribution is secured by using a tapered conveyor floor which allows material to fall over the open side as it is conveyed out. A canvas skirt can be attached to the conveyor to shield the falling material from the wind. This type of distributor should be particularly valuable with rock phosphate, which, if ground fine enough to become quickly available, is easily blown but hard to throw with a spinner.

GRAIN-DRILL FERTILIZER ATTACHMENTS

Fertilizer attachments for grain drills were originally offered as optional equipment, but the increase in use of fertilizer has led to the fertilizer grain drill having the fertilizer feeds and hopper built as an integral part of the machine. The star-wheel feed has been most popular in grain drills. By gate and speed adjustments, and also optional star wheels, rates of 50 to 800 lb per acre are usually available. Agitators to prevent bridging are sold as attachments where they are needed.

Fertilizer can usually be deposited in direct contact with the seed without damage because the concentration of fertilizer salts is quite low and small-grain seeds have some measure of resistance to fertilizer injury. For susceptible crops such as beans and peas, special tubes are used to place the fertilizer above and separated from the seed. For heavy applications on susceptible crops, it is customary to either drill the fertilizer in a separate application or to broadcast it on the surface.

ROW-CROP FERTILIZER UNITS FOR CORN AND COTTON PLANTERS AND CULTIVATORS

These attachments are usually of the cylindrical hopper type with one or two delivery tubes per hopper. The revolving-plate feed is most common, although star-wheel, belt, and auger feeds are also used. Late designs for tractor use have large hoppers with a capacity of 60 lb or more, often made of corrosion-resistant epoxy resin reinforced with glass fiber. Hoppers and feeds are designed for easy dumping and disassembly to allow thorough cleaning. They are also kept as low as practical for easy filling.

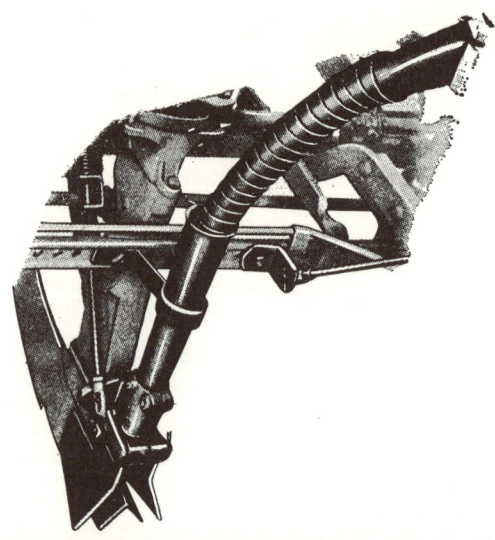

Fig. 16-8. Split-boot for corn-planter attachment. (*J. I. Case Co.*)

Placement of starter fertilizer for corn is commonly accomplished by a split boot, shown in Fig. 16-8, which, at low speeds, places the fertilizer in a band, separated from the seed by ½ to 1 in. of soil at a depth slightly above the seed, but often does not achieve separation at modern tractor speeds. When checkrowing or hill dropping, a valve is closed in the split boot to accumulate the fertilizer between hills and opened simultaneously with the corn valves to place bands 6 to 8 in. long on each side of the hill. Continuous bands are used for drilled corn, and a heavier application can be made without seed damage.

It has been found that it is somewhat better to place the starter fertilizer slightly below the level of the seed by a small fertilizer shoe preceding the planter shoe or by a single-disk opener spaced to one side or slightly to the rear of the seed opener. This arrangement places a continuous or interrupted (valved) band of fertilizer approximately 3 in. to one side and 1 in. deeper than the seed.

A *dual-level applicator* is shown in Fig. 16-9. The coulter opener has a streamlined, saber-shaped boot fitting against each side, one depositing fertilizer slightly below the

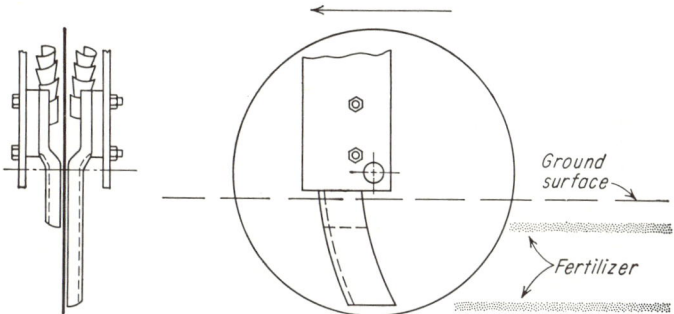

Fig. 16-9. Dual-level fertilizer applicator. (*Farmcraft Mfg. Co., Ft. Wayne, Ind.*)

seed and the other at plow-sole depth to eliminate the need for side dressing when the corn becomes larger.[1] The two sides can be fed from different halves of the fertilizer hopper to enable different analyses to be used for the two depths.

Cottonseed is more sensitive to fertilizer than corn, and the fertilizer should be placed approximately 2½ in. to either one or both sides of the row and 2 to 2½ in. below the level of the seed.[2] This requires that the fertilizer be placed by a special shoe or shovel running ahead of the cotton-planting shoe.

Side-dressing attachments for corn and cotton cultivators are quite similar to planter attachments and use the regular hoppers, with the delivery tubes feeding behind a regular cultivator shovel or a special deep applicator shovel. For cotton it should be applied behind the inside shovel, and for corn, 2 to 3 in. deep and 14 to 16 in. from the row on one or both sides.

Where it is desirable to use heavier applications of fertilizer than can be placed near the seeds, it must be either broadcast on the surface and plowed under or placed in bands deep enough to prevent injury to the plants and remain moist for availability. When fertilizer is mixed throughout the soil, fixation of phosphates is increased, reducing the amount of nutrients immediately available for plants. Deep-band application minimizes fixation and also is out of reach of shallow-rooted weeds.

LIQUID-FERTILIZER DISTRIBUTORS

The use of *anhydrous ammonia* as a nitrogen fertilizer began in California in the early 1930s, where it was added to irrigation water. Direct application to the soil began about ten years later. In modern synthetic-ammonia plants, hydrogen from natural gas and steam is combined with nitrogen from the air to produce anhydrous

---

[1] C. M. Hansen and R. E. Lucas, A Rolling Coulter Furrow Opener for Deep Placement of Fertilizer, *Mich. State Univ. Agr. Expt. Sta. Quart. Bull.* 36(3):310–317, February, 1954.
[2] National Joint Committee on Fertilizer Application, "Methods of Applying Fertilizer," *National Fertilizer Association*, Washington, D.C., 1948.

ammonia, which can be converted to ammonium nitrate for use as solid fertilizer. The conversion process is costly, and in addition, about 7 per cent of the nitrogen is lost. Thus a pound of nitrogen in the form of anhydrous ammonia is less expensive than in the form of ammonium nitrate.[1] This economic advantage has encouraged the use of anhydrous ammonia in spite of the handicap of requiring high-cost equipment for transport, storage, and application.

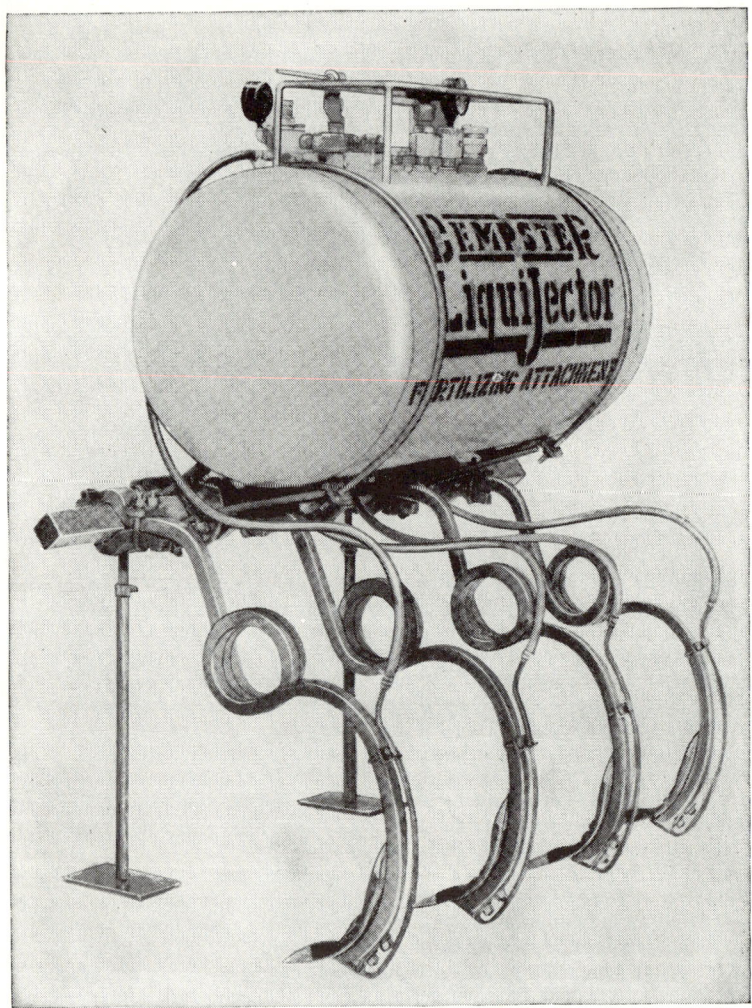

FIG. 16-10. Applicator for anhydrous ammonia. (*Dempster Mill Mfg. Co., Beatrice, Nebr.*)

Anhydrous ammonia is a gas at normal temperatures and atmospheric pressure, and it must be handled as a liquid under pressure. At 50°F it has a vapor pressure of 75 psi, and at 100°F the vapor pressure is 197 psi. It must be injected into the soil and sealed immediately to prevent loss by evaporation, but it is rapidly absorbed by the clay in the soil, and subsequent cultivation does not cause a loss. It weighs about 5 lb per gal in the liquid state and is 82 per cent nitrogen by weight.

[1] W. B. Andrews, F. E. Edwards, and J. G. Hammond, Ammonia as a Source of Nitrogen, *Mississippi State Coll. Agr. Expt. Sta. Bull.* 451, 1949.

FERTILIZING AND LIMING MACHINES

Liquid-fertilizer distributors for anhydrous ammonia ordinarily consist of a metering device, hoses, and applicator blades supported by a conventional tool bar or frame. This equipment may be tractor-mounted, as shown in Fig. 16-10, or trailed and is often owned by a custom operator.

The tank, fittings, and hoses must allow a working pressure of at least 250 psi. Steel or iron fittings are immune to attack by ammonia and must be used rather than brass, which will corrode rapidly with commercial grades of anhydrous ammonia

TABLE 16-2. DENSITIES OF ANHYDROUS AMMONIA LIQUID AND AMMONIA VAPOR AT VARIOUS TEMPERATURES

Temp., °F	Liquid anhydrous ammonia			Ammonia vapor, lb/100 gal vapor capacity at saturation
	Gage pressure	Weight, lb/gal	Correction factor to 60°F	
30	45.0	5.3419	1.0379	2.77
32	47.6	5.3293	1.0355	2.88
34	50.2	5.3167	1.0330	3.00
36	52.9	5.3041	1.0306	3.11
38	55.7	5.2916	1.0282	3.24
40	58.6	5.2790	1.0257	3.37
42	61.6	5.2659	1.0232	3.50
44	64.7	5.2527	1.0206	3.63
46	67.9	5.2398	1.0181	3.77
48	71.1	5.2265	1.0155	3.91
50	74.5	5.2136	1.0130	4.06
52	78.0	5.2002	1.0104	4.22
54	81.5	5.1868	1.0078	4.36
56	85.2	5.1735	1.0052	4.53
58	89.0	5.1601	1.0026	4.69
60	92.9	5.1467	1.0000	4.86
62	96.9	5.1333	0.9974	5.03
64	101.0	5.1199	0.9948	5.22
66	105.3	5.1066	0.9922	5.40
68	109.6	5.0932	0.9896	5.59
70	114.1	5.0797	0.9870	5.78
72	118.7	5.0659	0.9843	5.98
74	123.4	5.0520	0.9816	6.19
76	128.3	5.0381	0.9789	6.41
78	133.2	5.0242	0.9762	6.62
80	138.3	5.0103	0.9735	6.84
82	143.6	4.9964	0.9708	7.07
84	149.0	4.9820	0.9680	7.31
86	154.5	4.9676	0.9652	7.56
88	160.1	4.9531	0.9624	7.80
90	165.9	4.9393	0.9597	8.05
92	171.9	4.9249	0.9569	8.31
94	178.0	4.9101	0.9540	8.58
96	184.2	4.8955	0.9512	8.85
98	190.6	4.8806	0.9483	9.13
100	197.2	4.8657	0.9454	9.42

which contain up to 1 per cent water. Tanks are made of steel and must not be filled above the 85 per cent level. A 125-gal tank will hold 500 lb ammonia which contains 410 lb of nitrogen, the equivalent of 1,242 lb of ammonium nitrate.

Metering can be accomplished by a pressure-regulator valve which gives a constant discharge flow or by a positive displacement pump. The pressure-regulator valve requires flow calibration for various temperatures and maintenance of a constant ground speed for uniform delivery rate. The effect of temperature on the vapor pressure and density of anhydrous ammonia is shown in Table 16-2. The positive displacement pump can be driven from a ground wheel so that speed does not affect delivery rate and the flow stops when the tractor stops. One machine uses a pump with adjustable effective stroke which will deliver from 2 to 250 lb of ammonia per acre, with the pump setting giving a fairly predictable rate.

Applicator blades have a sharp leading edge and are just wide enough to shield the delivery tube, which discharges just behind the point for maximum depth. They must be operated from 4 to 6 in. deep, or more, to prevent loss of ammonia vapor. The blades must be strong and wear-resistant because of their depth and speed of operation. Spring-trip or spring-tine mountings are needed in rocky ground. For row-crop side dressing, one applicator is used per row, and for grain or pasture the blades are spaced from 14 to 20 in. apart.

Since the evaporation of ammonia has a refrigerating effect, sometimes causing ice formation, the design of control valves, pumps, and applicator blades must take account of this factor. A heat exchanger is combined with the pump in one case.

Because of the lesser weight to be handled and the trouble-free discharge, a liquid-fertilizer tractor outfit can cover about twice as many acres as a solid-fertilizer outfit and with one man instead of two.

Storage tanks for anhydrous ammonia must withstand 250 psi if unprotected from the sun. The weight of anhydrous ammonia and ammonia vapor in a storage tank can be determined by using Table 16-2 to find liquid and vapor density at a given temperature and gage pressure. Special precautions are necessary when transferring the liquid into or from the storage tanks, because it or the gas is dangerously irritating to the human body. Tight-fitting goggles should be worn, and gas masks and rubber gloves should be available, as well as water for washing in case of contamination.

It is possible to get a mixture of ammonia and air which will ignite at high temperatures, and welding should not be done on tanks which contain ammonia. When a mixture of ammonia and butane or propane is burned, poisonous hydrocyanic acid is formed. Consequently, tanks which have held ammonia should be rinsed out with water before using them for LP gas.

Aqua ammonia is made by adding enough water to anhydrous ammonia to reduce the vapor pressure to atmospheric. It weighs about 7½ lb per gal and contains 20 to 25 per cent nitrogen by weight, requiring over three times the weight of anhydrous ammonia for the same nitrogen. It will develop a vapor pressure of 10 to 20 psi at 100°F, requiring caution in handling. Like anhydrous, it is corrosive to brass and aluminum but not to iron and steel. It must be covered with dirt immediately after application, but does not require deep sealing like anhydrous. Gravity feed can be used when the hoses discharge on the surface, with covering by shovels, but gravity will not supply enough pressure to keep the openings clear when operating under the surface behind an applicator blade. Metering pumps may be used if gravity-fed to avoid vapor lock. A PTO-driven air compressor with pressure regulator may be used to force the liquid out of the tank at an even rate and is also efficient in transferring the liquid from nurse tank to applicator tank.

Aqua ammonia is slightly more expensive than anhydrous as a source of nitrogen but can be handled with the same tanks, fittings, and pumps used for mixed liquid fertilizers (see below) or weed sprays.

Nitrogen solutions contain urea, ammonium nitrate, and/or anhydrous ammonia dissolved in water. The proportion of nitrogen varies from 20 to 41 per cent, depending on the composition. They are termed low-pressure if they contain free ammonia and nonpressure if not. The low-pressure solution should be covered with soil, but nonpressure solutions can be applied to the surface without loss.

These solutions are corrosive to iron or brass and require aluminum or stainless-steel tanks and fittings. Pumps must be resistant to corrosive action and may be positive-displacement ground-driven or used with a pressure regulator and bypass. In the latter case, ground speed should be held constant to maintain an even rate of distribution. As with aqua ammonia, air pressure is often favored.

Mixed liquid fertilizers contain nitrogen, potash, and phosphorus carried in a solution which is near chemical neutrality and comparatively noncorrosive. The concentration is limited by the tendency to "salt out" in cool weather. Analyses of

10-10-10, 16-8-8, and 5-20-5 are at the present upper limit of concentration for Corn Belt temperatures. They weigh about 10 lb per gal.

These solutions are usually nonpressure and can be sprayed or dribbled on small grain and pastures. They permit heavy applications of fertilizer while planting, with much saving in time and effort compared with solid fertilizer.

Metering may be by gravity flow, pump, or air pressure. The gravity-flow systems frequently use an airtight tank with an air vent near the bottom to maintain a constant head as the tank empties. The trend is toward tractor-mounted tanks and metering devices which can be used with aqua ammonia, complete liquid fertilizer, or weed sprays, since none of these are corrosive to steel or iron. Fertilizer can then be applied with any trailed or mounted implement.

Pesticides or weed sprays with compatible chemicals are sometimes mixed with liquid fertilizer and applied simultaneously. They are also available mixed with solid fertilizers.

Nonpressure types of liquid fertilizer may be sprayed from airplanes. This permits fertilization when fields are muddy or crops would be damaged by ground equipment.

Chapter 17

SEEDING AND PLANTING MACHINES

The function of seeding and planting machines is to meter out seeds without damage at the desired rate and place them in moist soil, uniformly distributed or in hills, at the desired depth, with the covering soil compacted as desired. The major problems are metering and placement.

Some of the *physical characteristics of seeds* which affect planting are:
1. *Size*, which varies from corn and beans to fine grass seeds.
2. *Uniformity* of shape and size, particularly important in single-seed metering.
3. *Shape* as it affects cell design and also flow characteristics of seed for feeding to the metering mechanism.
4. *Density* per unit volume, which affects flow and is a function of shape and actual seed density. Brome grass seed has a very low density because it is long and thin and has a high proportion of hull to seed. It is sometimes necessary to mix it with sand to get it to flow through a seeder.
5. *Surface smoothness*, which affects flow and cell filling. Smooth round clover seed and fuzzy cottonseed are examples of the extremes.
6. *Resistance to injury* from compression and abrasion by the metering mechanism. Peanuts, beans, cotton, and sugar-beet seed are easily damaged, with resulting poor germination.

METERING MECHANISMS

Many different methods are used to meter seeds for planting, depending on the characteristics of the seed and the spacing desired. Metering mechanisms in common use are as follows.

Agitator with Adjustable Hole. This feed is used primarily with small seeds which cannot be selected singly. A wide hopper seeder for small grass seeds uses a reciprocating rope or cord in the bottom of the trough-shaped hopper as an agitator. Seeders using spinner-plate spreading for broadcasting may use a rotary agitator on a vertical shaft. Some vegetable planters use a wavy-edged disk on a horizontal drive shaft, with the disk edge wiping back and forth across the hole. The hopper bottom plate has a circle of holes of different sizes, any one of which can be fixed in delivery position to vary the rate.

Fluted or External Force Feed. This feed is one of the oldest and most widely used for bulk-seed metering, as in grain drills. Jethro Tull is said to have built a grain drill with a cylinder feed in 1733, although the refinements of the present fluted feed were worked out in the United States between 1850 and 1877. It consists of a fluted cylinder rotating about a horizontal axis. The seeds are carried under the cylinder and discharged about 45° past bottom center into a feed tube leading to the furrow opener, as shown in Fig. 17-1a. The primary adjustment for seeding rate is by sliding the entire feed shaft sideways to vary the length of fluted cylinder exposed to the seed. The delivery-gate opening can be adjusted to several positions according to seed size, and in some cases the speed of rotation can be varied by changing sprockets.

SEEDING AND PLANTING MACHINES 173

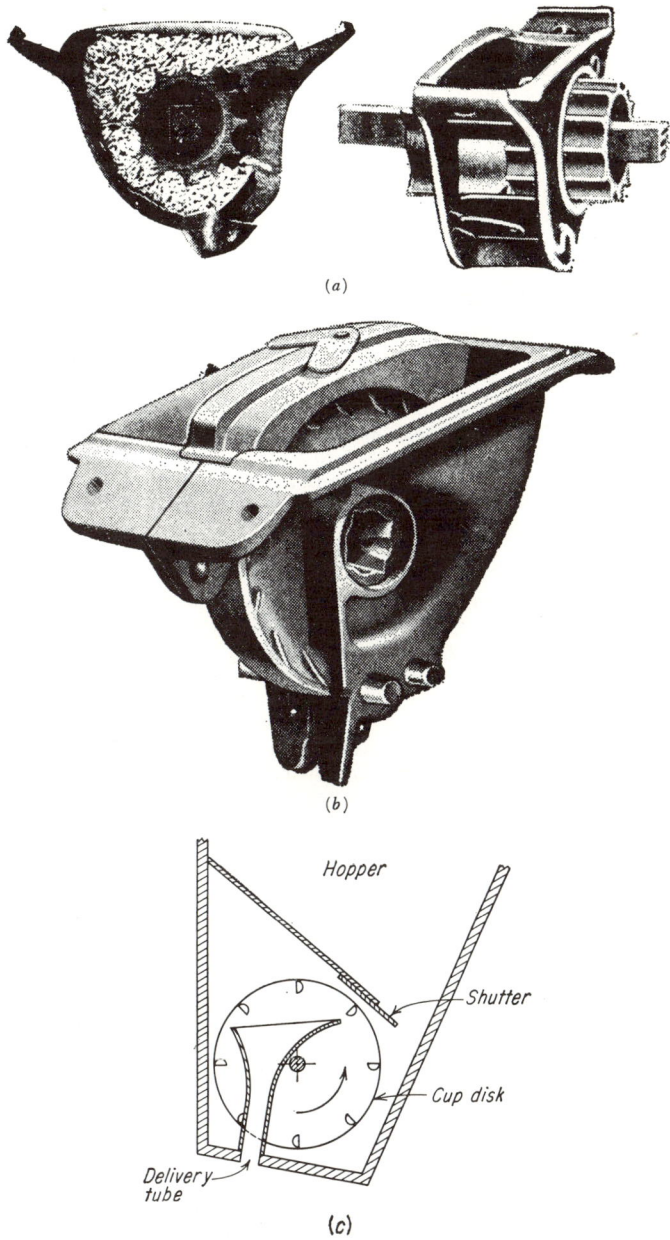

Fig. 17-1. Grain-drill feeds: (a) fluted; (b) double-run; (c) cup.

The fluted feed is made in small sizes for grass drills as well as in the standard size, which can be used for all small grains. It is simple and easily adjusted with acceptable accuracy and uniformity of distribution for small grains.

Double-run or Internal Force Feed. This feed for grain drills uses a wide-rim disk rotating on a horizontal axis, the feeding surface being the corrugated inner surface of the rim which carries the seed out of the hopper into the delivery tube as shown in Fig. 17-1b. The two inner surfaces of the rim on opposite sides of the disk are made unequal in width, and one side is covered over while the other side is used for feeding, the side selected depending on the seed size. This characteristic gave rise to the name *double-run*. The primary rate adjustment is by (1) a speed-changing mechanism driving the feed wheels, using a selection of gear and sprocket sizes, (2) the use of a plate gear with a pinion sliding in or out to vary speed, or (3) a cone gear with sliding pinion. (The latter two drives require unorthodox tooth shapes in the cast gears used but serve reasonably well.) The range of seeding rates is extended by adjustable gates or by reducers inserted to narrow the feed outlets.

The internal force feed gave less variation in feeding rate because of hopper tilt forward and backward than the external feed in Iowa State University tests,[1] and it is sometimes claimed to be less easily clogged by trash at light rates because of the larger wheel area exposed. The drive is more complex, however, and the choice between the two feeds seems to be one of individual preference. Most manufacturers offer grain drills with either type of feed.

A recent design of internal force feed includes a slide adjustment similar to that of a fluted feed to control the amount of opening. This gives more exact and convenient control than the conventional double-run design.

Cup Feed. This feed is popular for grain drills in England but is not used in United States grain drills. A series of cups on the rim of a vertical rotating wheel dip into a shallow pool of seed, lifting a few at a time and carrying them over the top, where they drop into a delivery funnel, as shown in Fig. 17-1c. Rate of seeding is controlled by wheel speed. There is no restricted gate area through which the seed must pass, thus eliminating a possible source of seed cracking and damage. The accuracy of this type of feed is adversely affected by tilt of the hopper and jarring over rough ground. Leveling adjustments for hoppers are often provided.

Vertical-plate Feed. This feed with single-selection cells has been used in recent years for planting segmented beet seed and vegetable seeds. A vertical plate containing cells large enough for single seeds has the upper portion of the rim passing through the bottom of the hopper, where the cells fill with seed and pass out under a positive cutoff. The seeds are ejected from the cells at the bottom of the wheel by a positive knockout. One type, shown in Fig. 17-2a, has a circumferential groove passing through the cells which allows a thin ejector plate to run in the groove at the discharge point.[2]

Inclined-plate Planter. This is a variation of the cup feed. An inclined seed plate with indented cells in the edge dips into a well of seed fed under a baffle plate from the hopper, lifts the seeds, and drops them into a delivery tube as shown in Fig. 17-2b. This type of feed, like the cup feed, requires no cutoff, which sometimes may cause seed injury, but it may be affected by hopper tilt and rough ground. Since the seed is not carried in a tight-fitting cell, a knockout is not normally required. This type of feed is used in some beet and vegetable planters as well as in a few combination cotton and corn planters. Speed of rotation must be slow to avoid centrifugal force which will throw the seeds from the cells prematurely. Some planters use two plates dropping alternately into one seed tube for more capacity.

Horizontal Seed Plate. This is by far the most popular feed for planting where accuracy requirements dictate single-seed selection. A horizontal seed plate contains shaped cells indented in the edge or a circle of round-hole cells. The cells fill with one seed each (if the seed is uniform and the cell of the proper size) and pass under a cutoff which stops additional seeds. The cell passes over the discharge spout, and a positive knockout is used to ensure the seed dropping out of the cell. The general arrangement is shown in Fig. 17-2c. For hill-dropping, the desired number of kernels for a hill are accumulated by a valve which is then actuated to drop the group of kernels in a hill. Major problems encountered in horizontal-seed-plate planters and current solutions are as follows:

Cell fill is facilitated by (1) low pressure in the filling area which allows seeds to turn as required to enter the cell rather than bridging over, (2) low lineal cell speed,

---

[1] J. B. Davidson, "Agricultural Machinery," John Wiley & Sons, Inc., New York, 1931.
[2] Roy Bainer, Precision Planting Equipment, *Agr. Eng.*, 28:49–54, February, 1947.

and (3) long interval of cell exposure. Seed plates and other hopper bottom parts are shaped to facilitate flow of seeds to the cell area for better filling and complete emptying of the hopper. The horizontal-plate design usually allows more time for filling than vertical or inclined plates.

Cutoffs cause seed damage if they are too positive, hence they are usually spring-mounted so as to deflect upwardly rather than damage a second seed partially in a

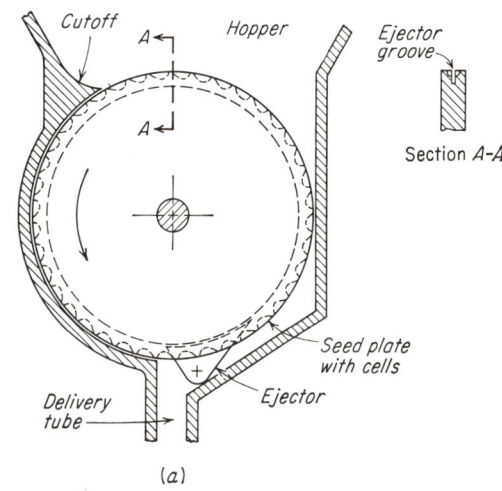

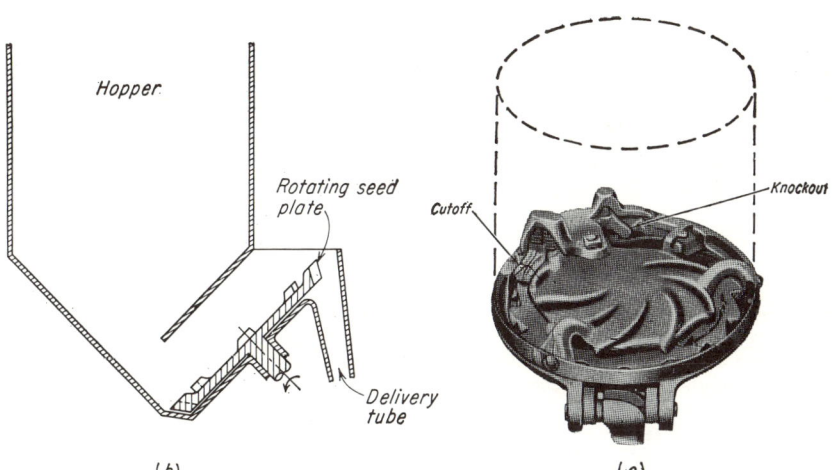

FIG. 17-2. Seed plate hoppers: (*a*) Vertical plate. (*Cobbley, manufactured by Edwards Mfg. Co., Yakima, Wash.*) (*b*) Inclined plate. (*c*) Horizontal plate. (*International Harvester Co.*)

cell when accuracy is less important than seed damage. For easily damaged seeds, such as beans, brush cutoffs are often used.

Knockouts are usually rounded points, rollers, or star wheels which are forced into the cells by spring pressure as they pass underneath and thus eject seeds which have not fallen out.

Proper cell shape and size as well as accurate fit of plates and other hopper bottom parts are necessary for accurate selection and avoidance of damage to seeds. These parts are usually made of unmachined cast iron and demand the utmost of the foundryman's art to maintain the required dimensional accuracy. Edge-drop cornplanter plates are made for each of the various sizes of grades of hybrid seed corn, some of which vary by only $\frac{1}{32}$ in. in one dimension. Round holes, usually taper-reamed, are used where the shape of the seed permits.

Rate of planting is changed by plate travel in relation to ground travel and by number of cells per plate where necessary. With hill-drop plates, the size of cell controls the number of kernels per hill.

Easy changing of plates and also emptying and cleaning of the hopper are facilitated by hinged hoppers which can be tipped upside down when a latch is released. The driving connection to the seed plate must automatically disengage when this is done.

Single-cell Belt Feed. This feed has been used experimentally for high-speed planting of easily damaged peanuts.[1] It carries seeds up an incline out of a well by holes

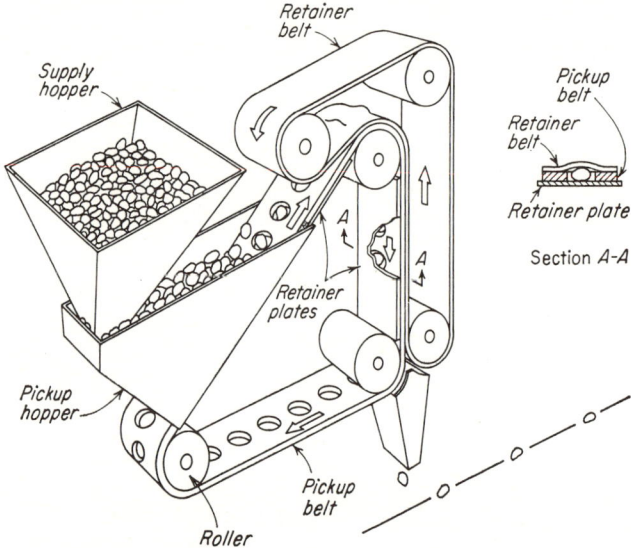

FIG. 17-3. Belt feed planter. (*Futral and Allen, Agr. Eng., April,* 1951.)

in an endless belt, with a retainer belt used to hold the seeds in place until they are discharged downward into the furrow as shown in Fig. 17-3. This arrangement has the gentle-handling characteristics of the inclined plate without being affected by jarring or high speed, plus the advantage of positive downward throwing of the seed into the furrow. A belt-feed planter, the Stanhay, is being made in England.

Calibration of seeding and planting machines for rate per acre can be facilitated by the formula given in Chap. 16.

DELIVERY PASSAGES

All planting and seeding mechanisms require a delivery passage leading from the hopper to the furrow opener. Where the opener is lifted independently of the seed box, the passage must be of telescoping or ribbon construction. In planting to a stand to avoid or reduce hand thinning, it is important that the seed be evenly spaced. Work with sugar-beet planters at the University of California, Davis, indicates that seed spacings vary from the theoretical spacing, based on seed-plate delivery intervals,

[1] J. G. Futral and R. L. Allen, Development of a High-speed Planter, *Agr. Eng.*, 32:215–216, April, 1951.

because of variations in the time for the seed to fall since some seeds bounce back and forth across the delivery passage as they drop.[1] Factors favoring minimum dispersion of seed from the average spacing are:

1. Positive knockout action imparting the same dropping velocity to each seed.
2. Smooth walls in the delivery tube to reduce bounce. (Spiral ribbon tubes encouraged excessive bounce.) Square cross sections reduce spiral bouncing.
3. Straight tubes of minimum length and diameter.

Transparent delivery tubes or windows in the side of the passages are sometimes used to enable the operator to see whether the seed is feeding properly. They must be of a material which is static-free.

FURROW OPENERS

Furrow openers must place the seed at the desired depth with minimum dispersion. Many types are used, some of which are as follows:

1. *Hoe* or *shovel openers* are probably the oldest type, and they consist of a narrow pointed shovel followed by a delivery boot. The dirt falling in from the sides behind the boot covers the seed. Hoe openers equipped with spring-trip shanks are used on grain drills in rocky soils. They are also used on cotton planters.

2. *Single-disk openers* are very popular for grain drills since they cut trash and penetrate well. A seed delivery boot drops the seed just behind the concave disk, where it is covered by moist soil. A scraper for the convex portion of the disk is attached to the disk. A scraper may also be used on the concave side of the disk, placed so as to reduce excessive throwing of soil at high speeds as well as to prevent soil build-up. A grain-drill single-disk opener is shown in Fig. 17-4a.

3. *Double-disk openers* are used for grain drills and also to a limited extent for corn and cotton planters. For grain-drill use the seed is usually discharged between the disks just below their center, as shown in Fig. 17-4b. The disk blades are straight rather than concave as with single-disk openers. For deeper planting of corn and cotton the delivery boot extends to the ground at the rear of the disks.

For precise depth control, depth-gaging bands or rims may be fitted to double-disk openers, particularly on beet drills.

The double-disk opener does not cut trash as well as the single-disk, but has a more positive furrow-opening action in rough seedbeds and, because of the use of flat instead of concave disks, does not throw excessive amounts of dirt at high speeds.

4. *Shoe openers,* also called runner or sword openers, are the prevalent type for corn planters and vegetable planters. They are also popular for cotton planters and are sometimes used on grain drills. A wedge-shaped blade opens the soil enough for the boot at the rear to deposit the seed, as shown in Fig. 17-5. The sharp blade cuts through clods and sod, although it does not cut trash as well as a disk opener. It opens the soil for receiving the seed with a minimum of disturbance and draft. The notch or skag just behind the bottom portion of the sword allows moist soil to cover the seed before the dry upper soil closes over it.

5. An experimental *wheel opener* with a V-shaped rim which pressed a groove in the soil gave improved emergence of sugar beets.[2]

Gage shoes or skids are sometimes used on each side of a runner opener to control depth of planting when the runners are allowed to float to follow uneven ground. They also smooth and compact the soil when planting in rough seedbeds.

Where it is necessary to go deeper than the seed should be covered, in order to plant in moist soil, several expedients are used. Lister planters for corn or cotton use a lister bottom to throw the soil to both sides and plant in the trench with a hoe opener followed by covering blades.

The single-disk deep-furrow grain-drill opener operates on the same principle, using wide spacing between rows and distributing the seed in a wide trench made by a plowlike blade on the opener. Where seeds are planted on beds in the South, a wide sweep may be used to scrape off the top of the ridge to expose moist soil,

---

[1] Bainer, *op. cit.*
[2] R. D. Barmington and S. W. McBirney, Mechanizing the Production of Sugar Beets, *Colo. Agr. Expt. Sta. Bull.* 420A, 1952.

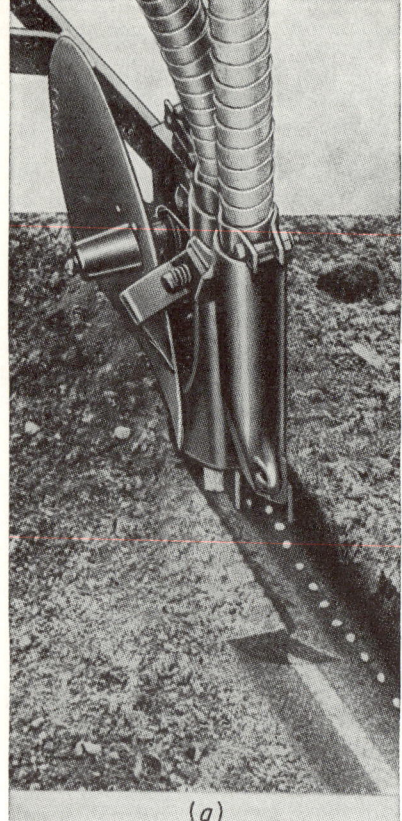 

Fig. 17-4. Grain-drill furrow openers: (a) Single disk with twin-boot placement of seed and fertilizer. (*Allis-Chalmers Mfg. Co.*) (b) Double disk with near disk removed. (*International Harvester Co.*)

with a hoe opener following. This may be accomplished by vertical bed-leveler wings attached to a long runner opener.

Another method tried with sugar beets was to cover deeper than required with a ridge of dirt which was scraped off after the seedlings had just germinated. This resulted in additional moisture for germination and removed any surface crust.[1]

COVERING DEVICES

For proper germination, seed must be in contact with moist soil and covered with a layer of soil which the sprout can penetrate. Insufficient moisture will delay germination, while too thick a covering layer or a hard-crusted layer can cause the sprout to die.[2] Seeds vary greatly in their requirements, and the soil type and climate also affect covering practices. Some of the devices used to provide the proper seed environment are as follows:

1. *Drag chains* are often used behind grain-drill disk openers to help fill the furrow and cover the seed with fine soil, although they may not be needed in loose soil and humid climates, particularly with single-disk openers.

[1] S. W. McBirney, The Relation of Planter Development to Sugar Beet Seedling Emergence, *Agr. Eng.*, 29:533-536, December, 1948.
[2] C. T. Morton and W. F. Buchele, Basic Factors Affecting the Emergence Energy of Seedlings, *ASAE Paper* 59-104, June, 1959.

2. *Covering blades* may be used to deflect loose soil over the row behind the openers.

3. *Covering shovels* running beside the row will push dirt over the row for covering.

4. *Disk coverers* similar to disk hillers are also used to throw dirt over the row.

5. *Press wheels* compact the soil around the seed, facilitating the rise of capillary moisture, and also push some soil over the furrow for covering, particularly if the rim is concave. They may also act as gage wheels to hold the openers at a constant depth over uneven ground. In sugar-beet planting experiments, a toothed center wheel between press-wheel flanges improved emergence.[1]

It has been found that a seed press wheel with a rubber-covered rim about 1 in. wide and 8 in. in diameter is an aid to germination, particularly for cotton,[2] in some areas where soil moisture is low. This wheel runs behind the planting shoe or disk and presses the seed into firm moist soil in the bottom of the furrow. It is followed by a regular concave press wheel. Seed press wheels are furnished as optional equipment on some planters, as shown in Fig. 17-5.

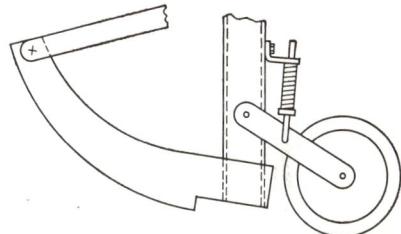

FIG. 17-5. Shoe or runner opener with seed press wheel.

BROADCASTERS

Broadcasting of seed distributes it on the surface of the soil where it is covered by further tillage or by raindrop-splash action. As a method of seeding, broadcasting is not as efficient as drilling because of the variable depth of coverage, some seeds being too deep and others too shallow.

Broadcasting, or sowing by hand, dates back to prehistoric times, and grass seed is still sometimes sown on foot by swinging a tube from a sack of seed or by cranking a small spinner spreader.

Endgate seeders use spinner distributors and are designed to be mounted on the rear of a wagon box. The hopper is kept filled from a supply of seed carried in the wagon. Driving power is taken from a sprocket bolted to the wagon wheel. Twin horizontal spinner disks with vertical vanes are used in combination with two fluted feeds or other metering mechanisms. These machines are used primarily for oat seeding in cornstalks and for grass seeding. Small grass seeders of this type can be mounted on the rear of a tractor and driven from the PTO, avoiding the need for pulling a wagon.

Airplane seeding is extensively used for rice, and its use is growing for reseeding rangeland, for seeding wheat or rye in standing corn, and for spring seeding of oats or grass seed on land too muddy to drive over. The metering mechanism is commonly driven by a small propeller and feeds into an air stream, which is spread by deflecting vanes. A strip 45 ft wide can be covered with small-grain seed. Airplanes are also used to broadcast fertilizer on rice, small grains, and hilly pasture land. (See Chap. 18 for a discussion of airplane dusting and spraying.)

Wide-hopper broadcast seeders vary from light grass seeders with agitator feeds to regular grain-drill hoppers and feeds but without openers. A successful type of grass seeder combines a double-gang corrugated roller with a full-width hopper having small fluted feeds which drop seed between the two roller gangs so that it is covered by the action of the rear roller. This machine covers the seed with a shallow layer of fine well-compacted soil which encourages good germination.[3]

GRAIN DRILLS

Grain drills are available with a great many combinations of feed and opener equipment and in row widths from 6 to 14 in. Grain-drill sizes are designated by

---
[1] McBirney, *op. cit.*
[2] James R. Tavernetti and H. F. Miller, Jr., Studies on Mechanization of Cotton Farming in California, *Univ. Calif. Agr. Ext. Bull.* 747, 1954.
[3] F. W. Duffee, A Soil Packer Grass Seeder, *Agr. Eng.*, 21:21–22, January, 1940.

numbers such as 16 × 7, meaning 16 openers spaced 7 in. apart. They are often available with either fluted or double-run feeds and with single-disk, double-disk, hoe, or shoe-type openers. Models with fertilizer hoppers are called fertilizer drills. Small grass-seed hoppers with special fluted feeds are available as attachments. In addition, chain-drag coverers may be replaced by depth-gaging press wheels or by gang press wheels.

Tractor-mounted grain drills are necessarily limited in size but are reported to cover more ground per day than a trailed drill of the same width because of the convenience of operation.[1]

Press drills are popular in dry farming areas when it is desirable to compact the soil over the seed for better germination and also to reduce blowing. The frame is carried between a pair of forecarriage wheels and a row of press wheels, one to an opener. These wheels are usually grouped in several gangs with a frame bearing in the middle of the gang, leaving it free to rock in following uneven ground. Wide-width press drills with extra-heavy frames are most popular for use on the large acreages of the semiarid West.

Adjustable plow press drills are made to pull behind plows, usually with a roller or packer in between. The distance between openers can often be adjusted so that the drill for a four-bottom plow, for instance, can be set to fit either 14- or 16-in. bottoms.

One-horse drills are made to drill between rows of standing corn and usually have five disk openers adjustable in spacing to fit the normal range of corn row widths. Because it is slow and arduous, this seeding practice is being dropped in favor of early picking of corn, made possible by artificial drying, followed by conventional ground preparation and drilling.

Disk tiller seeders use conventional drill hoppers and feeds mounted on a disk tiller and with seed-delivery tubes dropping seed between the disks at a location where it will be covered to the desired depth. This seeding method saves considerable labor and power and has been growing in popularity. The lift system of the tiller must have sufficient capacity to handle the added weight of the seeder attachment.

Pasture-renovating drills of very rugged construction and spacings up to 20 in. are designed to open the sod in rundown pastures for placement of fertilizer and grass seed. A coulter to cut the sod precedes a narrow hoe opener for the fertilizer, followed by a seed drop tube and, in some cases, a press wheel. The coulter also serves to lift the assembly over rocks and other obstructions, although safety trip openers are also desirable.

Drills have a large number of seed and fertilizer feed parts which can be easily damaged if not operating freely. Because of this it is very important to clean the hopper and feeds thoroughly at the end of the season, to coat with rust preventative, and to turn carefully by hand before starting the next season. The drill should also be checked for bent frame or axles, proper opener spacing, worn disk bearings, disk-scraper adjustment, and loose bolts. Because of the nature of a grain drill's mechanism and its cost, extra care is worthwhile.

CORN PLANTERS

Two-row corn planters are usually adjustable by 2-in. steps from at least 32 to 42 in., most corn being planted in 38- to 42-in. rows. A high degree of precision in planting rate is possible because the size and nature of the corn kernel make possible accurate single-seed selection of graded kernels with properly fitted seed plates.

Corn is *checkrowed* in many areas, being planted in hills which line up crossways for cross-cultivation. This practice facilitates weed control, particularly in very wet years when weeds may get as high as the corn before the first cultivation is possible. There has been a trend away from checkrowing to drilling because of the time saved and because chemical weed control has reduced dependence on cultivation.

Modern tractor planters may be operated up to 5 mph, particularly when drilling. Large fertilizer boxes and seed hoppers holding up to $\frac{1}{3}$ bu each reduce filling time.

Corn planters are usually built in two-row units, two of which may be combined

[1] A. G. Buhr, Mounted Grain Drill Development, *Agr. Eng.*, 36:649–653, October, 1955.

for a four-row machine, or in one-row units which are combined as desired. The *two-row units*, stemming from horse-drawn models, support the openers (usually runner type) between the press wheels and the tractor drawbar. Wheels follow the openers as closely as possible for accurate depth gaging and act as transport wheels when the hinged frame lifts the openers. The frame has some vertical flexibility to allow the runners to follow uneven ground.

Two, four, or six *one-row units* are carried by a rigid frame on transport wheels or by a tractor-mounted tool bar. Hoppers and small press wheels are mounted in a

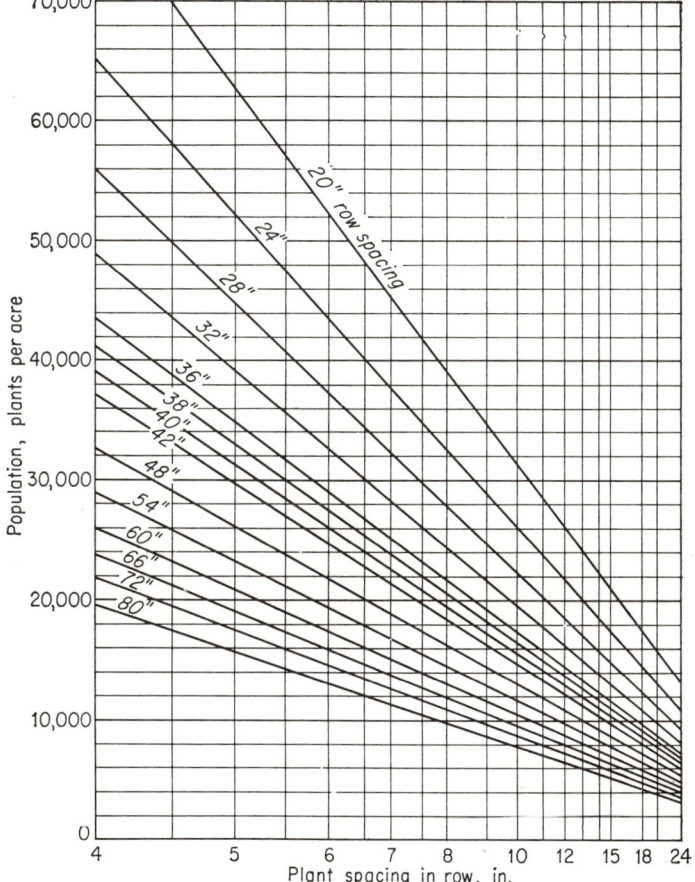

Fig. 17-6. Plant populations for various row and plant spacings.

unit with the openers, and the complete assembly is connected to the main frame, usually by a parallel linkage. Planting depth is gaged by the press wheel, and the complete unit is lifted for transport.

Press wheels are of a concave open-center design which presses the soil in from the sides without inducing surface crusting over the hill. Zero-pressure rubber press-wheel tires give approximately the same compaction characteristics and shed sticky dirt better.

Planting rates for corn usually vary from 6,000 to 20,000 plants per acre, with 15,000 considered optimum for conditions capable of producing 100 bu per acre. Corn planters almost all use horizontal-plate seed boxes and single-kernel selection.

Planting rates depend on the number of cells per plate and the drive rates. Checkrow planters traditionally drop two, three, or four kernels per hill, the plates being turned the proper distance when the check wire button trips a single-revolution dog clutch at the same time it operates the valves to drop the previously accumulated hill. Some recent designs dispense with the clutch and plant to a given population, similar to a drill planter. Research in recent years has shown that corn plants have the ability to reach out and use nutrients from a considerable distance and that, with reasonably uniform distribution, plant population per acre is the pertinent factor. In fact, 60-in. row widths have yielded almost as well as 40-in. with equal population. Figure 17-6 shows plant populations for various row and plant spacings.

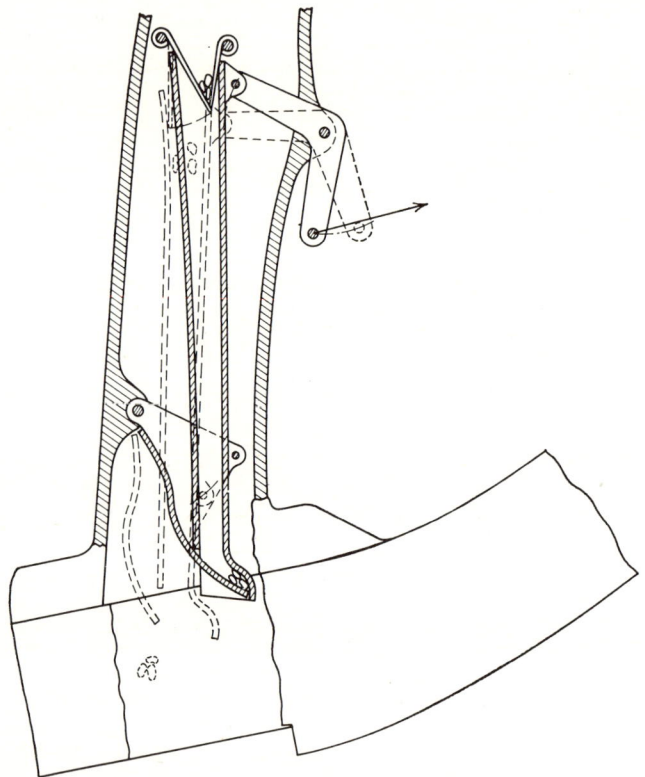

Fig. 17-7. High-speed checkrow valves. (*Deere & Co.*)

Accurate *checkrow hill placement and grouping* at high speed has been the subject of considerable study.[1] One of the major causes of hill scattering at high speed was found to be bouncing of the kernels after they hit the lower valve, which, at high speeds, opened before the kernels had settled together. This was remedied by using a curved entrance into a V-shaped valve pocket which smothered the bouncing tendency and threw the kernels down the back in a group. A V-shaped upper valve pocket also helped drop the kernels in a more compact group than when they were pushed off a flat ledge. A modern boot-and-valve design is shown in Fig. 17-7.

Straight crossrows in checked corn, without offsets and zigzags, require proper planter adjustment. There is bound to be a slight unevenness at the extreme ends of the field because of changes in wire angle and cumulative stretch. Several hills should be uncovered near the center of the field to see if they are in line. It will be found

[1] A. C. Sandmark, Check-row Planting at Higher Speeds, *Agr. Eng.*, 25:386, October, 1944.

that the hills must be placed slightly behind the wire buttons to compensate for wire stretch and angle. If the hills are not in line, they can be shifted with respect to the wire button by changing the level of the front frame of the planter, thus throwing the heel of the opener forward or back with respect to the check fork. Some planters incorporate an adjustment to shift the check-fork assembly forward or back on the front frame for the same purpose. These adjustments serve to align pairs of hills dropped together.

If the two hills dropped together are not in line, it is due to a mechanical imperfection in the planter, such as:
1. Twisted front frame or bent runner throwing one heel back
2. Check shaft twisted between boots
3. Check fork and valves loose on the check shaft

Proper handling of the check wire is also necessary for a good cross-check. It must be pulled to the same tension every time to give uniform stretch from one round to the next. About 50 lb tension is recommended, more tension causing excess stretch, due to the angle when approaching the ends, and possibly pulling out the stakes. This problem is accenuated with a four-row planter, one aid being pay-out stakes which unlock and pay out the wire under spring or slip-clutch tension when the wire exceeds a certain angle as the planter approaches the end.[1] Stakes with spring scales for setting the wire to uniform tension are also available.

The check wire is wound up on a reel for transport and storage. The reel is detachably mounted on the planter frame and driven by a friction clutch which maintains the proper tension when winding up the wire, in spite of the change in effective diameter of the reel as the wire accumulates.

COTTON PLANTERS

Cotton planters range from the one-horse walking type to six-row tractor-drawn or mounted models. In general, it is required that the cotton planter also plant the other row crops common in the South such as corn, sorghum, peas, beans, and peanuts.

Cotton is traditionally drilled thicker than necessary and thinned by hand hoeing. Mechanical thinning by cotton choppers using intermittent rotating blades or by cross-cultivation has supplanted hand thinning to a large extent, and much cotton is drilled or hill-dropped to a stand (see Chap. 15). It has been found that thick stands tend to inhibit low fruiting without a penalty as to yield and quality, a considerable advantage in machine harvesting because of the difficulty of picking bolls within 6 in. of the ground without taking dirt and trash.[2]

For successful *single-cell selection* of cottonseed, the seeds must be delinted mechanically or by acid. When this has been done, a conventional horizontal-plate feed is most popular, using edge-drop plates and a gentle pressure spring to force seed into the cell. Cutoff, shed, and knockout must be designed for gentle action as cottonseed is very easily cracked. This type of feed is generally used for planting to a stand.

The *picker-wheel feed* is most common for handling fuzzy seed which is planted thickly for subsequent thinning. It consists of a horizontal spider plate with raised radial fingers at its periphery which carry the cottonseed over a small vertical-toothed picker wheel that turns opposite the spider plate and picks seed downward into the seed tube as shown in Fig. 17-8. The rate of seeding is regulated by the height of a gate bar above the spider plate at the picker-wheel location, which controls the thickness of the layer of cottonseed exposed to the picker wheel. This type of feed takes trash through without clogging and is little affected by the depth of cotton in the seed box. This type of hopper bottom is made interchangeable with conventional plate-feed bottoms for the other crops.

Shoe openers are most popular for cotton. Cotton is sometimes planted in hills because of the advantage of several seeds together in breaking through a crust. Hill-drop plates may be used although the seeds scatter too much for a compact hill unless the delivery tube is short, smooth, and straight. Better hills are usually secured by a rotary valve in the shoe which collects several seeds and drops them together.

---

[1] C. K. Shedd, Check-wire with a Four-row Planter, *Agr. Eng.*, 15:18–20, January, 1934.
[2] Tavernetti and Miller, *op. cit.*

Experiments in Oklahoma indicate substantial protection from the usual crusting and water drainage following heavy rains by the use of a pair of disk hillers set to leave a small furrow on each side of the row as shown in Fig. 17-9. Satisfactory stands were secured in several cases where replanting was necessary with conventional shallow-furrow planting.[1]

Cotton is universally fertilized, and cotton planters usually have *fertilizer attachments*. Since it is recommended that the fertilizer be placed below and to one side of the seed, a separate fertilizer shoe preceding the seed opener is required.

Because of the practice of planting beans or peas with corn in the South, *auxiliary pea-and-bean seed boxes* are required. These are small horizontal plate hoppers and may be mounted above the corn hopper, in the corn hopper, or independently at one side, but in all cases feeding into the same delivery tube with the corn.

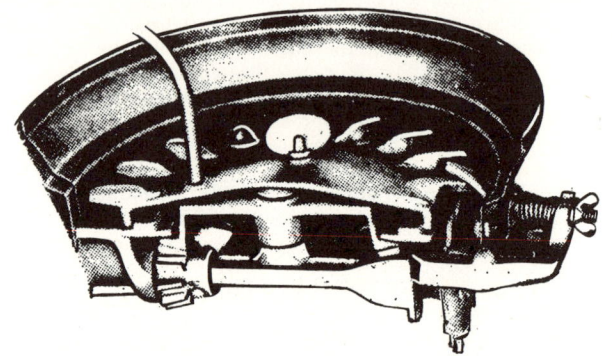

FIG. 17-8. Cotton-planter picker-wheel feed. (*Deere & Co.*)

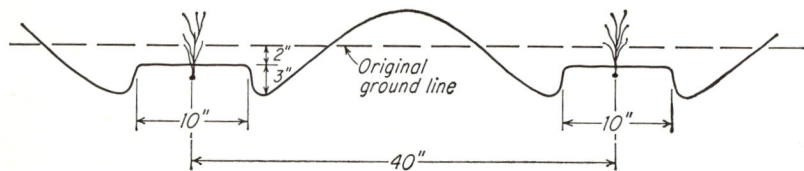

FIG. 17-9. Seedbed designed to reduce crusting and water damage. (*Porterfield and Smith, Oklahoma Agr. Expt. Sta. Bull.* B-449, 1956.)

The use of *planter attachments* for tractor-mounted cultivators is increasing in the South. The seed and fertilizer boxes are mounted on the side of the tractor, and the cultivator gangs are fitted with the required opener equipment. By using flexible delivery tubes, the openers can be lifted without lifting the heavy hoppers. The same fertilizer hoppers and delivery tubes can later be used for side dressing during cultivation. Hoppers for solid fertilizer commonly have a capacity of 100 lb per row unit. Liquid-fertilizer attachments provide 250 to 350 lb tank capacity per row unit.

POTATO PLANTERS

Potato-seed pieces are entirely different in character from other seeds, and the seed-selection mechanisms differ accordingly. Because of the irregularity of the size and shape of cut seed, conventional cell-type selection is not successful. The automatic mechanisms used are somewhat like the English cupfeed for small grain in that the rotating selection elements pass up through a well of seed. One type of selector arm

---

[1] J. Porterfield and E. M. Smith, Development and Test Performance of a New Seedbed for Cotton, *Oklahoma Agr. Expt. Sta. Bull.* B-449, 1956. J. G. Porterfield, E. W. Schroeder, and D. G. Batchelder, *Plateau Profile Planter, ASAE Paper* 59-105, June, 1959.

has cam-actuated jaws which close to grasp a seed piece while passing up through the well and then release it into the boot on the opposite side. Another type of selector arm has cam-actuated picker points which are projected through the picker arm face as it passes up through the seed well, thus spearing a seed piece and carrying it over to the opposite side, where the points are withdrawn, allowing the seed to drop in the furrow made by the opener. Seed flows sideways from the hopper into the seed well, the rate being regulated by the operator. This type of picker wheel is shown in Fig. 17-10. Seed spacing is controlled by speed of rotation of the picker wheel, varying from approximately 7 to 18 in.

Potatoes are planted deep, and a wide V opener is required to open a deep enough furrow for the seed. Disk coverers without press wheels are most common.

Large amounts of fertilizer are often used with potatoes, and to avoid damage the fertilizer should be placed in bands at each side and slightly below the seed. A pair of disks open furrows for the fertilizer bands. The seed opener then splits the soil between the bands, thus covering the fertilizer and preventing contact with the seed.

Fig. 17-10. Potato-planter picker-wheel feed. (*Deere & Co.*)

Fertilizer hoppers have belt-type or horizontal plate feeds capable of applying up to 3,000 lb per acre.

Potato planters are made as two-wheel trail-type implements only, since the weight and bulk of the seed and fertilizer required make tractor mounting impractical. One- and two-row machines are available. An operator is required on the planter to see that the seed flows into the well properly, and he is also relied upon to lift and adjust the furrow opener and to actuate the feed clutch and in some cases the row marker.

The weight carried and the draft make it necessary for the frame, axle, and wheels to be sturdy and well designed. Seed and fertilizer hoppers should be placed as low as possible to ease the labor of filling hoppers.

TRANSPLANTERS

The operation of transplanting has not been completely mechanized, but transplanters do open a furrow for the plants and press the soil around them after they are placed in the furrow and also apply water and fertilizer if desired. Two operators usually ride the machine for each row being planted, and they place the plant in the opened furrow at the desired spacing or into a mechanical setting device in the case of the most recent designs. Transplanters may be either tractor-mounted or trail type.

Furrow openers are of the conventional V type, similar to potato planter openers, and leave a wide enough furrow to avoid cramping the plant roots.

Where the plants are placed in the furrow by the operators, the seats are as low and close beside the furrow as possible. With mechanical setting, higher seats are possible. Plants are spaced by a bell signal, by an automatic water release, or in one case by a reciprocating plant-setting gage against which the operator places his hand while holding the plant in the open furrow, the action of the gage moving his hand back so that the closing soil will take the plant at the proper instant.

The soil is closed and packed around the plants by *press wheels*, or *press plates* in the case of loose sandy soils. Press wheels are used in pairs, pitched and gathered for most efficient action. In some cases, these wheels have tilted rims to gather more soil around the plants. The furrow-opener depth may be gaged by the press wheel or plate setting, with the entire assembly floating as a unit and lifted for turning. Other designs control opener depth from the transport wheels and use spring pressure to control the degree of press-wheel packing.

With *mechanical setting devices* the operator places the plant in fingers or flexible wheel flanges which grasp it and carry it down into the open furrow where it is released to the closing dirt just in front of the press wheels. One type of setting device

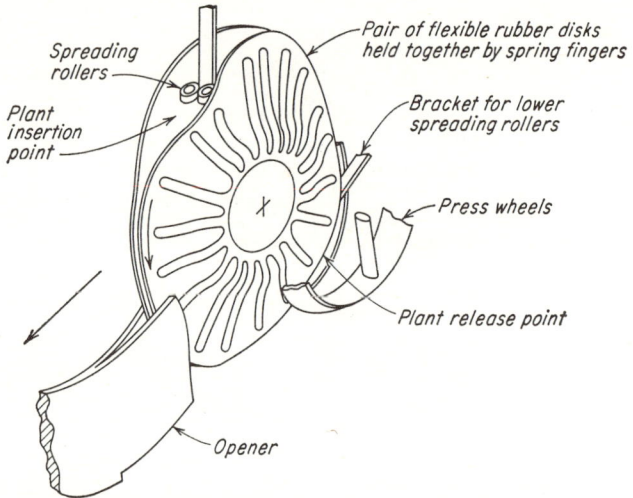

Fig. 17-11. Planting wheel for transplanter. (*Deere & Co.*)

uses cam-operated fingers carried by a chain. Another design, shown in Fig. 17-11, uses a pair of rubber disks held together by spring fingers and opened by a pair of small rollers at the top to allow insertion of the plant top and another pair at the bottom for release into the furrow. It is claimed that these devices speed planting considerably.

Fertilizer attachments use shoe or disk openers ahead of the planting opener to place bands on each side of the row at the desired depth.

Many plants require extra moisture at the time of transplanting, and a large water tank with an automatically actuated valve in the opener releases water at the desired spacing. In some cases, fertilizer and water application are combined by dissolving fertilizer in the water, particularly where anhydrous ammonia is used.

A *tree transplanter* for use in unprepared sod or brush ground has a middlebuster plow bottom to clear a wide strip, in combination with a wide chisel to open a trench 3½ in. wide into which the seedling is inserted. Covering blades and press wheels complete the operation.[1]

[1] H. D. Bruhn and F. B. Trenk, Fundamentals of Tree Planter Design and Performance, *Agr. Eng.*, 28:387–396, September, 1947.

Chapter 18

SPRAYERS AND DUSTERS

BY W. W. GUNKEL AND C. B. RICHEY

Sprayers and dusters are used to control undesirable insects, fungi, and vegetation through the application of suitable toxic materials.

The basic problem is to apply the necessary quantity of toxic materials at the desired points with a minimum of wastage. The primary uses in agriculture are to protect field crops or fruit trees from insect and disease pests and, by application of chemicals, to control weeds and brush and to defoliate crops or accelerate their maturity. This variety of uses results in many special requirements.

Metering is usually best accomplished by an adjustable, positive feed mechanism delivering pesticide material at a constant time rate, although for uniform-growing row crops the discharge should be proportional to ground speed. Dust is even more diffi-

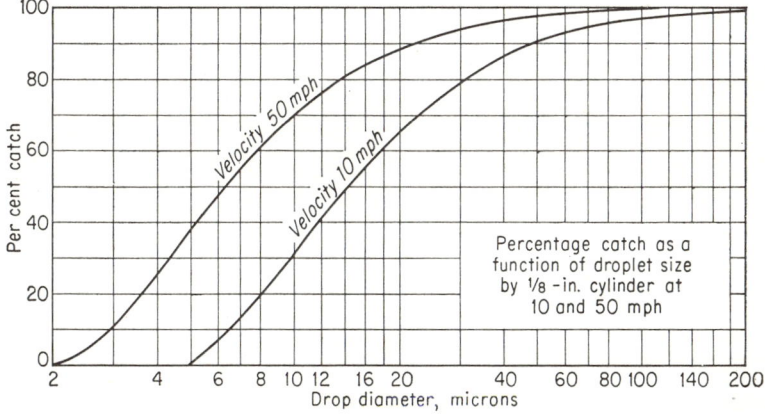

FIG. 18-1. Percentage catch of droplets in an approaching ⅛-in.-wide stream by a ⅛-in.-diameter stem, as affected by droplet size. (*Brooks Agr. Eng., June,* 1947.)

cult to meter accurately than liquid spray. Inert material is used in dusts and water in sprays to increase metering accuracy and coverage.

Choice of *droplet size* of sprays is a compromise. Within limits the smallest droplets give the most uniform coverage, while the larger droplets give better penetration and greater range of controlled "throw" with less drift. The same factors enter into the choice of dust particle sizes. The behavior of various-size water droplets falling through air is shown in Table 18-1.[1] Large particles are more likely to be intercepted by a stem or leaf than small particles because their greater inertia enables them to penetrate the boundary layer of the stem or leaf rather than "slide around" with the air stream. Figure 18-1 shows the effect of droplet size on the percentage of droplets

[1] F. A. Brooks, The Drifting of Poisonous Dusts Applied by Airplanes and Land Rigs, *Agr. Eng.*, 28:233–244, June, 1947.

which are intercepted by a ⅛-in.-diameter stem when the droplets approach in a ⅛-in.-wide spray stream. Particle velocity is also an important factor since the fast-moving particles are less easily deflected. In general, the impingement of the particles on the leaves is a function of their kinetic energy.

Particles ranging from 1 to 80 microns in diameter are so small that, in settling by gravity, the air they force aside flows around them in a viscous manner, in accordance with Stoke's law.[1] From Table 18-1 it can be seen that drift control is virtually impossible with particles smaller than 10 microns in diameter. Particles under 20 microns in diameter are called aerosols (air solutions) and are easily air-borne.

Coverage of the plants may differ considerably, depending upon the purpose of the chemicals. Materials which poison chewing insects or which are absorbed into the

TABLE 18-1. TERMINAL VELOCITY OF WATER DROPLETS FALLING IN AIR AND TIME FOR FALLING 10 FT

Equiv. diam of particle, microns	Description of size	Terminal velocity, ft/min	Time to fall 10 ft from rest, min	Distance carried in 3 mph uniform drift
½	Brownian max.	0.0015	6750	388. miles
1	....	0.0059	1686	84. miles
2	....	0.024	420	21. miles
3.5	....	0.072	139	7.0 miles
5	Sea fog	0.148	67.5	3.4 miles
10	....	0.593	16.9	4,452. ft
20	....	2.371	4.20	1,109. ft
33	Cloud	6.46	1.55	409. ft
50	....	14.82	0.675	178. ft
80	....	37.9	0.264	70. ft
100	Mist	55.0	0.1818	48. ft
170*	Atomized 1,000 lb/sq in.	120.	0.0833	22. ft
200	Drizzle	142.	0.0704	18.6 ft
290*	Atomized 400 lb/sq in.	229.	0.0437	11.5 ft
380*	Atomized 200 lb/sq in.	313.	0.0320	8.4 ft
500	Light rain	421.	0.0265	7.0 ft
1,000	Moderate rain	763.	0.0179	4.7 ft
2,000	Excessive rain	1,041.	0.0152	4.0 ft
3,000	Cloudburst	1,172.	0.0145	3.8 ft
4,000	....	1,237.	0.0141	3.7 ft
6,000	Max. drop size	1,307.	0.0138	3.6 ft
Free fall..........	....	1,524.	0.0131	3.5 ft

1 micron = 1/1,000 mm, 1,600 microns = 1/16 in.

* O. C. French, Spraying Equipment for Pest Control, *Univ. Calif. Agr. Exp. Sta. Bull.* 666, May, 1942, fig. 8, p. 11.

SOURCE: F. A. Brooks, The Drifting of Poisonous Dusts Applied by Airplanes and Land Rigs, *Agr. Eng.*, 28:236, June, 1947.

plant, like 2-4-D, need not cover the plant uniformly and completely and can be applied in coarse, widely separated drops or particles. Materials which attack fungi or insects on the plant by direct contact should cover the plant as completely and uniformly as possible.

Application of the material may be made directly into the plant foliage by nozzles adjacent to the row or, at the other extreme, by drift-spraying or dusting with a cloud of particles of as near 80 microns as possible so that they will be carried into the plants or trees by the wind and dropped without excessive drift.[2]

Dusting is usually a faster and less expensive operation than spraying because less bulk is handled and the application equipment is simpler, although less of the material is deposited on the plant and more may drift. Best dust adherence is obtained when the plant surfaces are wet and the air is still. In many cases only about 10 per cent of the dust material is actually intercepted by the plant. Electrostatic charging

[1] *Ibid.*
[2] E. R. Hoare, The World Spraying Problem, *Farm Mechanization*, 4:280–284, May, 1952.

of dust particles, causing them to be attracted to the plant, offers hope of increased dusting efficiency.[1]

Spraying is more independent of adverse weather conditions and permits more hours of operation per day. The percentage of deposition of spray material is greater than with dust.

Spray mixtures vary greatly as to characteristics which affect the spraying equipment. They may carry abrasive solids in suspension or be in the form of emulsions or solutions. They may be acid, causing corrosion, or contain solvents which attack rubber. Sprayers should be able to perform satisfactorily with as many of these mixtures as possible.

SPRAYER COMPONENTS

Nozzles. The liquid spray material must be projected with adequate velocity and directed to the desired area. It must also be broken up into droplets of the desired size. Nozzles are used to direct the liquid in a thin film or stream which will break up into droplets. In free space, surface tension acts on the liquid and forms it into spherical droplets since a sphere has minimum surface area for its volume.

Droplet size is dependent upon:
1. Velocity differential between liquid and air
2. Surface tension
3. Liquid density

Velocity differential may be secured by a pressurized discharge of the liquid into still air through a nozzle, the velocity varying with the pressure. It may also be secured by low-pressure discharge of the liquid into a high-velocity air stream which breaks the liquid into drops and carries it to the treated surface.

The liquid in the *hollow-cone-type nozzle* is forced through a series of sloping ports inside the nozzle. This gives the liquid a rotating action, and on being discharged, the liquid is moving radially under the impulse of a high tangential velocity. All the emerging liquid receives this impulse, and in consequence a circular hollow cone of liquid is discharged from the nozzle in the form of a thin film, as shown in Fig. 18-2a. This film in turn breaks up into droplets in two directions: one in the forward direction because of surface tension, and the other at right angles because of the increase in circumference as the cone leaves the nozzle.

With *flat fan-type nozzles,* two jets of liquid, directed against each other, impinge in such a manner that the liquid forms a flat fan-shaped film, the width increasing with distance from the nozzle, as shown in Fig. 18-2b.

In the *internal-type flat-fan nozzle,* the distance between the two jets is reduced to the extent that they form one nozzle. The inside of this nozzle is formed to have a long slot as shown in Fig. 18-2c.

In the *shear-plate nozzle,* a stream of liquid is directed against a stationary plate of disk. The forward motion of the liquid is converted into radial velocity and results in the formation of a large thin fan-shaped liquid film. The included angle between the face of the shear plate and liquid stream determines the direction of the thin liquid fan.

There are several other methods of obtaining small droplets from liquids, such as liquid dropped on a rotating disk, liquid passed through aerofoils in an air stream, and liquid directed against a rotating propeller in an air stream. These methods are not widely used, however.

Nozzle wear occurs when any abrasive material is discharged through the nozzle. The resulting abrasion rounds off the sharp edges of the nozzle orifice. This in turn increases the orifice coefficient of discharge and causes a nonuniform spray pattern.

Pumps. Pressure for discharging the liquid through the nozzles may be supplied by air or by direct pumping. The former method uses a piston-type pump to compress air in a leakproof tank. Spray solutions may carry abrasive solids, corrosive liquids, or solvents which act on rubber. Choice of a pump depends on resistance to these materials as well as on the pressure and capacity desired.

---

[1] Henry D. Bowen et al., Application of Electrostatic Charging to the Deposition of Insecticides and Fungicides on Plant Surfaces, *Agr. Eng.,* 33:347–350, May, 1952. R. W. Brittain and W. M Carleton, How Surfaces Affect Pesticidal Dust Deposition, *Agr. Eng.,* 38:22–31, January, 1957.

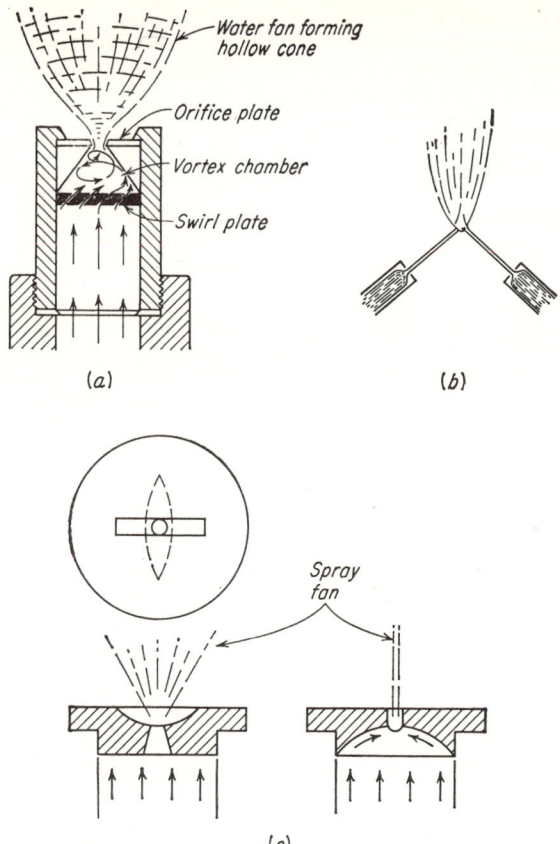

Fig. 18-2. Sprayer nozzles: (a) hollow-cone type; (b) fan formation by impinging jets; (c) flat fan.

Centrifugal pumps are simple and not readily damaged by abrasive material but must be driven at a high speed to develop adequate pressure. The pressure developed by a given pump varies with the square of the impeller rpm, and a pressure-regulating valve is usually required. A diagram of a centrifugal pump is shown in Fig. 18-3.

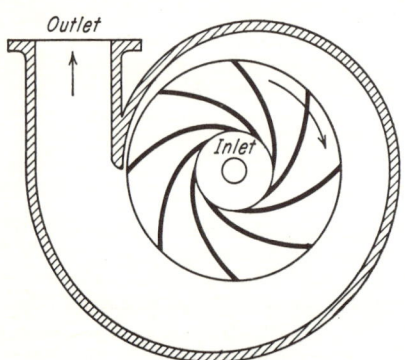

Fig. 18-3. Centrifugal pump.

Capacity per working stroke of positive displacement pumps (piston, diaphragm, gear, vane, etc.) equals the volume swept by the pumping element on the working stroke minus leakage and lack of fill in the pumping chamber. These two factors are covered by *volumetric efficiency,* the ratio of the actual volume pumped to the swept volume.

Piston pumps are most popular for high-pressure sprayers of large capacity. In order to resist the abrasive and corrosive action of spray liquids the steel cylinders are lined with porcelain and a molded-rubber expanding-type packing or plunger cup is expanded against the cylinder walls by the pressure to prevent leakage.

Corrosion-resistant ball or flat-type valves and valve seats, usually of hardened stainless steel, are used. The valve area should[1] equal 50 per cent of the piston area for average piston speeds of 100 ft per min, and 150 per cent of the piston area where piston speed is 300 ft per min.

$$\text{\textit{Capacity} of piston pump, gpm} = \frac{NnD^2L}{294} \text{ (volume efficiency)}$$

where N = number of cylinders
n = rpm
D = cylinder diam, in.
L = length of stroke, in.

A *plunger-type pump* usually has stainless-steel plungers and stationary internal packings.

Diaphragm pumps are not easily affected by abrasion since natural- or synthetic-rubber diaphragms and valves are the only moving parts in contact with the liquid.

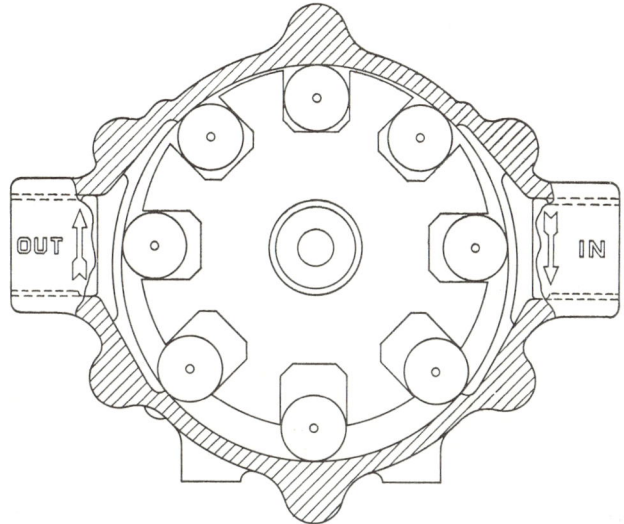

FIG. 18-4. Roller-vane-type pump. (*Hypro Engineering Inc., Minneapolis, Minn.*)

Sprays containing solvents, such as xylene, damage the rubber and should not be used with a diaphragm pump unless a special xylene-resistant diaphragm is used.

Diaphragm, piston, and plunger pumps require a *pressure dome* to cushion the discharge-pressure fluctuations and induce a uniform flow. A *suction dome* helps prevent liquid from pulling apart and causing water hammer. The size of these domes depends upon the number of cylinders, capacity, and operating pressure, varying from four to six times the piston displacement. Air in this chamber is absorbed by the liquid, and some means has to be provided to replace it.

Gear pumps are commonly used for handling spray emulsions and solutions in low-volume spraying. Sprays containing solids in suspension, however, cause abrasive wear of the gear teeth and should not be used. For the principle of operation of external and internal gear pumps, *see* Fig. 10-2b.

Vane-type pumps may also be used for low-volume spraying. For their principle of operation see Fig. 10-2c. Roller vane pumps (Fig. 18-4) with nylon rollers and nickel-chrome ferrous-alloy casings are widely used for pumping emulsions and low concentrations of wettable powders. Radial movement of the roller vanes compensates for any peripheral housing wear, but there is not comparable compensation for

---

[1] M. Bogema, "Pumps and Turbines," unpublished lecture notes, Cornell University, Ithaca, N.Y., 1950.

housing and plate wear. Therefore installation of new vanes in a worn pump may not improve its performance.

Single-screw pumps with a stainless-steel rotor and a rubber stator are capable of pumping abrasive spray suspensions with little damage. The rotor, which revolves within the stator, is a true helical screw offset from the rotor axis by eccentricity e. Its action in the double internal helical thread of the stator tends to trap the fluid and positively carry it through in pockets.

The capacities, speeds, and operating pressures of the various types of sprayer pumps are shown in Table 18-2.

TABLE 18-2. CAPACITIES, SPEEDS, AND PRESSURE RANGES OF SOME SPRAYER PUMPS

Type of pump	Capacity, gpm	Speed, rpm	Pressure, psi
High-volume, medium-pressure, single-stage centrifugal	100–150	1,700–2,500	50–100
Low-volume, medium-pressure, single-stage centrifugal	5–20	1,700–2,500	50–100
High-volume, high-pressure, single-stage centrifugal	50–100	7,000	500–800
Diaphragm	5–30	200–550	50–100
Piston	3–60	80–550	300–800
Plunger	1½–60	80–550	200–800
Gear	2–15	200–1,800	50–100
Vane	2–15	200–1,000	50–250
Single-screw	3–10	1,500–2,000	50–100

The *pump capacity* required for a sprayer is the maximum rate needed for spraying plus an allowance for normal wear, slower than rated operating speeds and, in some cases, hydraulic agitation. This allowance varies but should be at least 20 per cent greater than the nozzle discharge.

Power for spray pumps, neglecting suction lift head, is given by the formula

$$\text{Hp} = \frac{(\text{gpm})(\text{pressure, psi})}{1,714 \; (\text{over-all efficiency})}$$

Over-all efficiency =
 (mechanical efficiency of drive and pump)(hydraulic efficiency, based on friction losses in pump and pipes)(volumetric efficiency, based on pump leakage and lack of fill)

Pump Protection and Pressure Regulation. Positive displacement pumps require a pressure-relief valve to protect the pump and sprayer parts from being damaged by extreme pressures when lines clog or the nozzles are shut off. In addition, intermittent operation of the relief valve may be required for pressure regulation at low delivery rates.

A simple ball-check relief valve is shown in Fig. 10-3a. This type of valve closes with only a slight drop in pressure, and more power is required with the nozzles closed than open. In addition, operation is noisy and the ball and seat are subject to wear and erosion.[1]

Low-pressure relief valves permit the flow from the pump to be bypassed at low pressure when it is not needed to maintain pressure to the nozzles. A piston-type pressure regulator is shown in Fig. 18-5. When the pressure on the piston exceeds the holding force of the spring, it moves and lifts the ball-type overflow valve, which allows the pump flow to bypass the nozzles. A check valve between the pressure-regulating piston and the pump closes when the bypass valve opens, and thus maintains the pressure on the piston to hold the bypass valve open. The resulting bypass-pressure and pumping-power requirement is only a fraction of that required for spraying. A diaphragm can be used instead of a piston, and a sliding sleeve valve and port instead of a ball-type valve.[2]

[1] K. R. Frost, The Unloading Characteristics of Orchard Sprayer Pressure Regulators, *Agr. Eng.*, 15:191–197, June, 1934.

[2] K. R. Frost, The Design of a Double-piston Pressure Regulator for Spray Pumps, *Agr. Eng.*, 16:227–228, June, 1935.

Tanks. Steel tanks from 50 to 600 gal capacity are commonly used on power sprayers. Some tanks are galvanized on the inside, while others are enameled or coated with a plastic material to resist corrosion and abrasion. Once these coatings have been penetrated by chemical or mechanical action, however, they are subject to cracking and flaking, which in turn may result in clogged nozzles or strainers. Tanks should be cleaned and flushed after use and before storage. To prevent corrosion, bare tanks

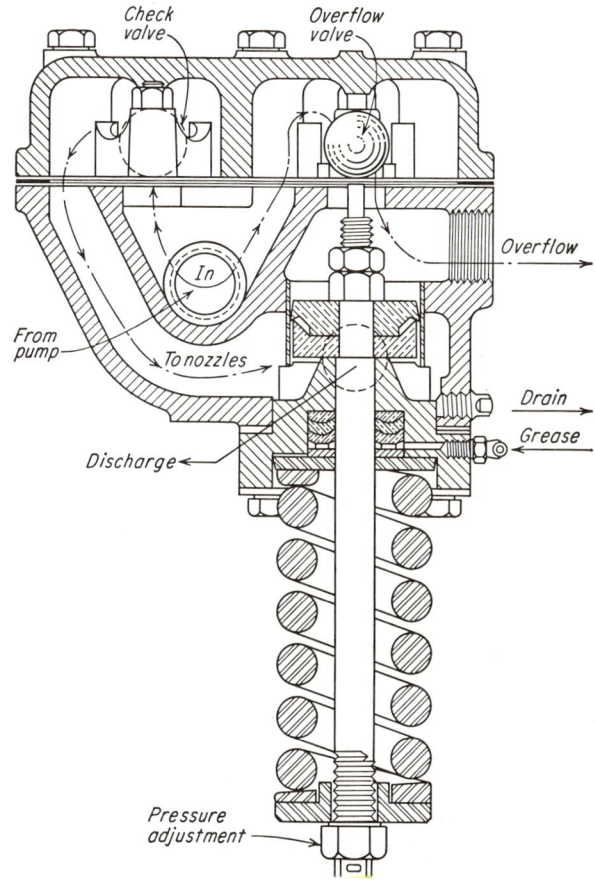

Fig. 18-5. Piston-type pressure regulator. (*Friend Mfg. Corp., Gasport, N.Y.*)

should be coated before storage with a rust-preventative compound which will not clog nozzles when the tank is used again.

Plastic tanks of clear epoxy resin reinforced by glass fiber are coming into use. This material is corrosion-proof and transparent enough to reveal the liquid level.

All power and hand sprayers used for supplying mixed sprays require some method of *agitation* to keep the spray concentration uniform. Agitation requirements vary with different types of spray mixtures. Suspensions of solid materials require positive or mechanical agitation, while hydraulic agitation, using a portion of the pump delivery, is satisfactory for some emulsions and all solutions.

Either flat or propeller-type blades may be used on *mechanical agitators*. Power requirements vary with the shape of the tank bottom.[1]

[1] K. R. Frost, Tests of Flat Steel Blades, *Agr. Eng.*, 16:443–445, November, 1935.

Hydraulic agitation is simpler than mechanical and satisfactory if sufficient liquid recirculation is utilized. With this method, excess liquid at boom pressure is recirculated to the tank and forced out through numerous small holes in a pipe placed along the tank bottom. These openings are frequently fitted with replaceable nozzles that can be adjusted for variations in liquid flow and changed for wear. The nozzles or orifices should be arranged to prevent the discharge from the jets striking the tank walls or bottom through less than 1 ft of liquid. Otherwise the continuous action of the liquid will abrade the tank linings and in time may wear through the tank.

No accurate data are available as to the amount of hydraulic agitation flow required for various spray mixes and tank shapes. However, experience indicates that approximately 30 gpm at 100 psi should be provided for a 3-ft-diameter round-bottom 250-gal tank. Flat-bottom tanks should have 30 to 40 per cent more agitation. Power requirements range from 3 to 4 hp for the above-size tanks.[1]

Hose. Hose design depends on the operating pressure and capacity requirements. Ordinary garden hose may be suitable for low-volume low-pressure spraying where the spray solution does not attack rubber.

The *pressure loss* in sprayer hose is caused by friction of the moving liquid against the inside surface of the hose. This loss increases with hose length, inside roughness, and fluid velocity and decreases with larger diameters and higher temperatures reducing viscosity, but is not affected by pressure. The following formula gives the pressure loss for 50-ft lengths of high-pressure spray hose and fittings:[1]

$$\text{Pressure loss, psi/50 ft} = K(\text{gpm})^2$$

where $K = 3.6$ for $\frac{3}{8}$-in. ID
$= 1.39$ for $\frac{7}{16}$-in. ID
$= 0.69$ for $\frac{1}{2}$-in. ID
$= 0.22$ for $\frac{5}{8}$-in. ID
$= 0.096$ for $\frac{3}{4}$-in. ID

Strainers. A system of strainers is required to prevent clogging of the lines and nozzles and to reduce wear on the pump, pressure regulator, and nozzles. A complete system consists of:

1. A tank strainer located at the filler opening to prevent foreign material from entering the tank. A 50-mesh screen is recommended.

2. A pump suction-line strainer located between the tank and inlet side of the pump. This strainer should be 50-mesh-screen, the area of which should be at least ten times the area of the suction line to prevent "starving the pump." This strainer should be easily accessible for cleaning.

3. A discharge-line strainer located between the pressure regulator and boom. This strainer mesh size varies with different types of sprayers. A 100-mesh screen is frequently used.

4. The nozzle strainer is part of the nozzle assembly, and the mesh size should be such as to prevent nozzle-orifice clogging.

FIELD SPRAYERS

Field sprayers are used primarily for pest control, weed control, and to accelerate maturity. The advent of 2-4-D for weed control has greatly increased the use of this type of sprayer. A general-purpose field sprayer must handle many types of spray materials at rates ranging from 5 to 10 gal of 2-4-D per acre at 20 to 40 psi up to 100 gal per acre for other materials. Many field sprayers are suited only for low-volume low-pressure usage. Most field sprayers are mounted directly on the tractor and powered from the power take-off. Some are mounted on a separate carriage with their own engine and drawn by a tractor. For tall mature row crops, special self-propelled high-clearance sprayers are built.

---

[1] Norman B. Akesson and W. A. Harvey, Chemical Weed Control Equipment, *Univ. Calif. Agr. Expt. Sta. Circ.* 389, 1948.
[2] This formula is derived from O. C. French, Spraying Equipment for Pest Control, *Univ. Calif. Agr. Expt. Sta. Bull.* 666, 1942.

Booms 15 to 25 ft wide are the distinctive feature of hydraulic field sprayers. These booms fold for transport and can hinge backward against spring pressure if an obstruction is encountered when spraying.

Boom height must be adjustable to clear plants and to allow the spray fans to meet at the desired level. *Quick shutoff* of booms is desirable to prevent damage from an overdosage when turning at the ends. Dripping is prevented by automatic valves which close at pressures below 5 psi or by a reverse-flow valve system which sucks the liquid back up from the booms and nozzles.

Nozzles producing a flat fan-shaped spray are considered to give the most uniform coverage and strongest drive at low pressures (40 psi and lower). Nozzles producing a solid or hollow cone-shaped discharge give a more uniform coverage of foliage at higher pressures.

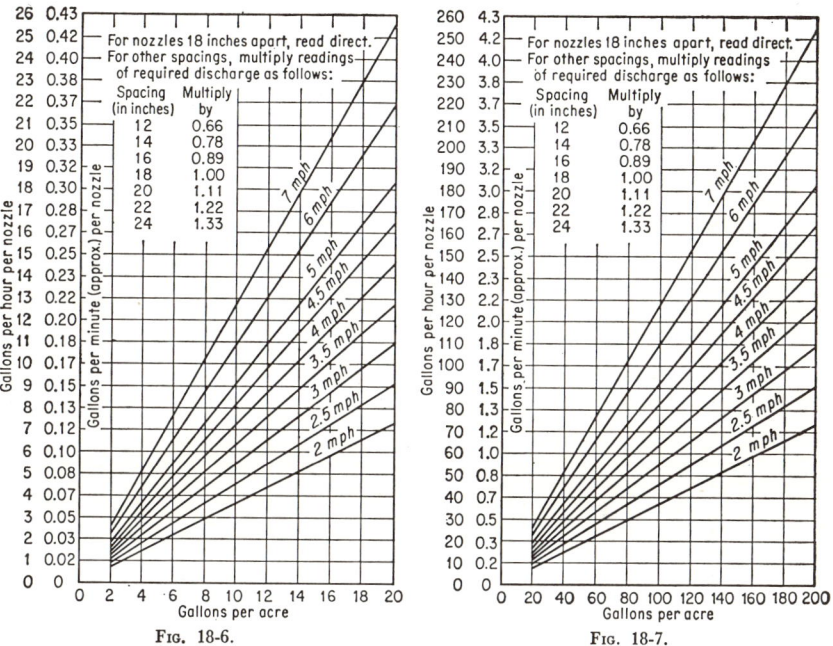

Fig. 18-6. Discharge per nozzle to give 2 to 20 gal per acre at various field speeds. (*Akesson and Harvey, Agr. Eng., September,* 1948.)

Fig. 18-7. Discharge per nozzle to give 20 to 200 gal per acre at various field speeds. (*Akesson and Harvey, Agr. Eng., September,* 1948.)

"Higher pressures, 75 to 125 psi, are generally used with water-base sprays, general contact spraying, heavy weed growth, and if fogging and drift will cause no harm. Lower pressures, 15 to 50 psi, will be used with oil sprays, 2-4-D or other translocated sprays, short weed growth, and in any case where fogging and drift might cause damage."

"*Nozzle spacing* on the boom depends upon the type of crop to be sprayed. When the outlets for the nozzles are welded into the boom, it may be necessary, for specific row or bed widths, to have extra outlets (which may be plugged to obtain other widths). Several manufacturers are using hose connections between the nozzles or from the boom to the nozzles to carry the spray liquid. Special nozzle holders which clamp on the boom may be moved to adjust spacing. This system is particularly adapted to rows and beds where plantings are at different widths. For spraying row crops, it is customary to have one nozzle directly over each row, and the boom height is adjusted to allow the spray fans to meet between the rows. On bed plantings, the nozzles will be spaced to fit the rows on the beds. Nozzle spacing should be uniform for open field work, such as grain and alfalfa spraying, and may vary from 12" to 24".

"The nozzle fan width and the spacing on the boom will determine how high the boom will have to be above the weed growth to have the fans overlap one-fourth of their width at the tops of the weeds, as is necessary for uniform coverage. *Double coverage,* where each strip of ground is covered by two nozzles, will give the most uniform and thorough application. This is particularly true in heavy weed growth, when using contact herbicides, or when operating on rough ground where skipping may result when the boom height changes. Nozzles for double coverage are frequently arranged with alternate nozzles on opposite sides of the boom, and each row may be tilted slightly toward the other to get a different angle of attack. The *fan angle* is altered by pressure. That is, if a 0.15 gpm nozzle at 10 psi will give a 60° fan; at 80 psi, the fan may be 70°; and at 100 psi, 80°. When ordering nozzles, the manufacturer's catalog on the nozzles used should be consulted for fan angle, as well as discharge volume for a given pressure."[1] Table 18-3 indicates the relationship of nozzle spacing, fan angle, and nozzle height above uniform-coverage level.

Boomless field sprayers have a central single nozzle or cluster of nozzles which are designed to cover up to a 30-ft swath. They are less expensive and more convenient

TABLE 18-3. NOZZLE HEIGHTS FOR UNIFORM COVERAGE

Nozzle spacing, in.	Height of nozzles* above single-coverage level,† in.			
	60° fan	80° fan	90° fan	110° fan
12	13	9	8	5
16	17	12	10	7
18	20	13	11	8
20	22	15	13	9
24	26	18	15	10

* Flat fan-type nozzles spaced apart three-fourths of fan width at single coverage level.
† For double coverage, double the given height.
SOURCE: N. B. Akesson, R. G. Curley, and W. E. Yates, When You Buy or Build a Field Sprayer, *Univ. Calif. Agr. Ext. Serv.,* March, 1956.

than boom-type sprayers but at a rate of 10 gal per acre gave less even distribution than boom-type sprayers and were more affected by wind in California tests.[2]

Calibration of a sprayer for the desired rate of application is important for maximum effectiveness and economy. For field sprayers, the rate of application depends on ground speed, nozzle spacing, and flow rate per nozzle. Nozzle flow rates for a particular pressure and orifice size are given by the manufacturer. The charts shown in Figs. 18-6 and 18-7[3] can be used to determine the desired nozzle discharge rate for a particular nozzle spacing and ground speed. Then the manufacturer's catalogue can be used to select the proper nozzle for the pressure to be used, and the listed fan angle can be applied to Table 18-3 to set the boom at the proper height for the desired coverage.[4]

Nozzle discharge rates can be checked by collecting the discharge of a nozzle for a timed interval in a calibration jar and calculating the rate. Calibration can also be accomplished by measuring the spray used for one round, calculating the acreage covered, and readjusting the rate if necessary. The ground speed must be accurately controlled for uniform application. An accurate low-range speedometer helps greatly.

---

[1] N. B. Akesson and W. A. Harvey, Equipment for the Application of Herbicides, *Agr. Eng.,* 29:384–389, September, 1948.
[2] J. E. Dibble, R. G. Curley, and N. B. Akesson, Boom and Broadcast Sprayers, *Calif. Agr.,* January, 1958.
[3] Akesson and Harvey, *op. cit.*
[4] E. L. Barger et al., Problems in the Design of Chemical Weed-control Equipment for Row-crops, *Agr. Eng.,* 29:381–389, September, 1948.

ORCHARD SPRAYERS

These sprayers are of two general types, high-pressure sprayers, which impart enough hydraulic energy to a large volume of dilute spray to secure coverage of trees, and blower sprayers, which use an air stream to carry more concentrated spray solutions into the trees.

High-pressure sprayers discharge from 20 to 80 gpm and may apply from 500 to 2,500 gal per acre. Trees and foliage are wetted until there is a slight excess runoff. Application may be by hand guns or by various booms or masts for automatic spraying. Piston pumps are most common, and a pressure regulator with a low-pressure bypass, such as shown in Fig. 18-5, is required. Comparatively large droplets, 200 to 400 microns in diameter,[1] are used in order to secure good carrying distance.

Blower sprayers, also called air-carrier sprayers, are occasionally used for bulk applications with as great a volume of liquid as high-pressure sprayers but are commonly used for more concentrated sprays. *Semiconcentrate* applications use one-fourth to four-fifths the volume of bulk applications and wet the trees only to the point of drip. Concentrate applications have one-sixth to one-eighth the volume of bulk applications (up to 250 gal per acre), and there should be no drip or excess runoff. Concentrate and semiconcentrate sprays both require about the same amount of active spray material and may save up to 20 per cent compared with the bulk applications.

Semiconcentrate spraying usually requires high-volume blower sprayers delivering up to 30,000 cfm for two-sided delivery or 15,000 cfm for one-sided delivery with 8 to 18 gpm of liquid.[2] Air volume must be adequate to fill the tree area with spray-laden air, displacing the original air. Droplets do not have to carry through air and can be 75 to 100 microns for maximum coverage. The pattern of the air and spray discharge and the speed of travel are adjusted to give coverage in one pass as the sprayer is driven by. Mechanically oscillated deflector vanes are sometimes used. A typical blower sprayer has a 65-hp engine powering an axial fan which delivers 25,000 cfm at 90 to 100 mph and a centrifugal pump capable of delivering up to 120 gpm at 55 psi. The complete machine weighs about 3,200 lb empty and 5,700 lb with the 300-gal tank filled.

Concentrate spraying requires less air volume than semiconcentrate, and liquid delivery may be up to 8 to 10 gpm from a piston-type pump. Pressures up to 500 psi are used to obtain atomization. The amount of water handled is reduced, along with labor and time. Semiconcentrate spraying, however, often gives more reliable control because of better coverage, particularly in marginal wind conditions, and there is less danger of foliage and fruit damage.[3]

DUSTERS

Dusters use air to carry dry toxic material and deposit it on the plant. Less weight and volume of material are required than with sprays, the usual range being 20 to 50 lb per acre.

Precise metering of the dust into the air stream is a difficult problem. Some agitation of the dust is necessary and is accomplished by mounting revolving or reciprocating brushes, scrapers, or rods inside of the hopper, usually over the discharge port. This port is generally located near the bottom of the hopper. Since rates of feed are dependent upon the head of dust above the port opening, wide variation in rates of application occurs for any one setting with this arrangement.

An even-feed dust hopper (Fig. 18-8), with a vertical agitator or lift that provides a constant dust head to an adjustable feed port, with the excess overflowing back into the hopper, has shown advantages over a standard hopper. The feed rate is fairly constant over the entire range from a full to an empty hopper. The outlet from the dust hopper is connected to the fan intake, and some fan suction is used to draw the dust from the hopper into the fan.[4]

---

[1] French, *op. cit.*
[2] Arthur D. Borden, Spray Chemical Concentrations, *Calif. Agr.*, January, 1952.
[3] E. H. Glass et al., How Shall We Apply Orchard Sprays? *N.Y. State Agr. Expt. Sta. Farm Research*, April, 1953.
[4] A. H. Glaves, A New Dust Feed Mechanism for Crop Dusters, *Agr. Eng.*, 28:551–552, December, 1947.

Centrifugal and axial-flow *fans* are used for dust application. Some dusters use a volute-type housing and "feather-feed" centrifugal fan, in which all the dust is fed through one hollow fan blade which distributes the dust uniformly in the various tubes. Volute fan casings with one outlet and tapered booms require internal baffles to control the uniformity of discharge through the individual nozzles.

Light fan-shaped metal *nozzles* with the rear sides cut back to prevent clogging are commonly used for row-crop applications. Special-design bean nozzles that direct dust up into the undersurface of the plant foliage are also used.

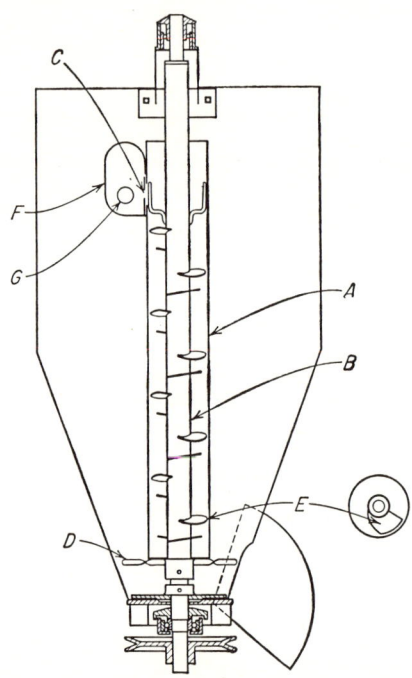

Fig. 18-8. Dust-feed mechanism developed by USDA: A cross section through the hopper and vertical impeller tube, showing the relative position of the vertical tube A, the impeller unit B, and the dust port C. The pair of dual-pitched impeller blades are shown at D. The shape of an individual elevator blade is shown in the plan view at the right and indicated by E. As the dust is fed through the dust port C, it is delivered to the fan through the dust delivery tube F, which also surrounds the fan shaft G. (*Glaves, Agr. Eng., December,* 1947.)

Granular insecticides have been found very effective for corn-borer control. The granules are dropped on the corn plant and funneled by the leaves to the leaf junction with the stalk, the point where the young borers feed before entering the stalk. These materials are applied at the rate of 10 to 20 lb per acre by grass-seeder-type hoppers and metering mechanisms on high-clearance axles or by airplane.[1]

AIRPLANE DUSTING AND SPRAYING

In many areas, dusting and spraying by airplane is an accepted and essential agricultural practice. The United States has over 6,000 planes doing the work.[2] The major advantages over ground equipment are speed of application and independence of ground or crop conditions. Major problems are thorough coverage of thick crops, cost on small acreages, and drift of small particles. Because of drift, less wind can be tolerated than with ground equipment (Table 18-1). As much as 70 per cent of dust

[1] H. A. Myers and Walter G. Lovely, Granular Insecticide Applicators for Control of European Corn Borer, *Agr. Eng.*, 38:298–319, May, 1957.
[2] How to Spray the Aircraft Way, *USDA Farmers' Bull.* 2062, 1954.

containing fine particles may drift out of the treatment area,[1] possibly damaging neighboring crops. Airplane dusting is confined to early-morning and late-evening hours when the air is quiet.

Planes used should be capable of carrying 800 to 1,200 lb of dust or spray, taking off from unprepared fields, flying within 10 ft of the ground at 100 mph under precise control, and turning quickly and easily at the ends of the field.[2] They should also be of simple and rugged construction, with the greatest possible protection for the pilot in case of a crash.

Dust is commonly metered by an agitator over an adjustable opening and fed into *venturi* spreader tubes located behind the plane propeller. The width of swath depends on weather conditions and particle size, ranging from 50 to 200 ft. A quick effective dust shutoff is needed to hold the dust while turning over neighboring fields. Felt seals are effective, although dust which has built up in the venturi may continue to discharge.[3]

Spraying by airplane has been growing in favor. Spray particles usually range from 100 to 200 microns, drift less than dusts, and stick to the crop better than dust in case

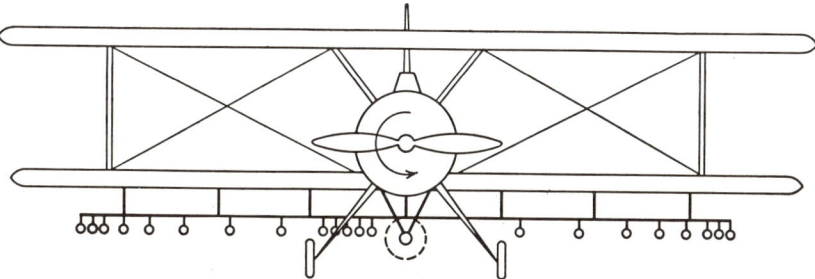

FIG. 18-9. Nozzle arrangement to compensate for wingtip and propeller distortion of spray pattern. (*USDA Farmers' Bull.* 2062, 1954.)

of rain. Dusts, however, may penetrate dense crops better, and some insecticides are more effective in the dust form.

In addition to the usual nozzle and pressure factors, *spray particle size* is affected by air speed and by direction of nozzle discharge into the air stream. Discharge backward gives coarsest particle size, across stream gives medium size, and 45° from the direction of motion gives finest size.[4]

Tests of coarse, medium, and fine atomization (300, 150, and 80 microns mmd) sprays released 50 ft above the ground for forest spraying indicated that medium atomization was most efficient, giving better distribution than the coarse and less loss than the fine spray.[5]

Pumps are usually gear or centrifugal and are driven by a small propeller or by a hydraulic motor. Centrifugal pumps can handle abrasive materials and require no relief valve. The required speed of 3,000 to 4,000 rpm is easily secured, but centrifugal pumps are not self-priming and must be located under the tank.

Nozzle locations on the booms must take into account the tendency of the wing tips to spread the swath and the tendency of the propeller slip stream to shift the center material to one side. A typical nozzle arrangement is shown in Fig. 18-9. As with dusts, quick shutoff at the ends is required. Individual nozzle check valves in combination with a suck-back arrangement at shutoff have been found effective.[6]

[1] Brooks, *op. cit.*
[2] Fred E. Weick, Development of an Agricultural Airplane, *Agr. Eng.*, 33:361–364, June, 1952.
[3] O. C. French, Use of the Airplane for Pest Control, *Agr. Eng.*, 28:240–242, June, 1947.
[4] How to Spray the Aircraft Way.
[5] D. A. Isler and D. G. Thornton, Effect of Atomization on Airplane Spray Patterns, *Agr. Eng.*, 36:600–604, September, 1955.
[6] Roy Bainer, R. A. Kepner, and E. L. Barger, "Principles of Farm Machinery," John Wiley & Sons, Inc., New York, 1955.

Chapter 19

MOWERS AND CRUSHERS

The function of a mower is to sever the stems of standing vegetation, primarily in cutting hay. The mower cutter bar must be capable of cutting from approximately 1¼ to 4 in. above the ground. It must be able to cut cleanly and without clogging almost anything from bluegrass sod to light brush. The cutting parts must be protected from rocks and be capable of cutting through occasional ridges of dirt without damage.

Many attempts have been made to replace the reciprocating knife (and resultant vibration) by chain or band knives, but functional, mechanical, and cost handicaps have not yet been overcome.

TABLE 19-1. FRICTION ANGLES OF STALKS OF WHEAT AND QUACK GRASS AGAINST VARIOUS TYPES OF CUTTING EDGES

Cutting edge	Angles of pinching, deg, at cutting heights of:				
	2 in.	4 in.	5.9 in.	7.9 in.	Stalks suspended
Stalks of wheat straw (⅛-in. diam)					
Two blades with smooth edges..................	49	41	34	32	27–34
Two blades with upper serrations..............	98	82	63	59	34–40
Two blades with under serrations..............	141	131	123	112	66–84
Smooth blade and serrated ledge plate.........	65	55	45	42	29–35
Blades with dulled upper serrations............	...	...	...	...	26–32
Blades with dulled under serrations...........	...	...	...	...	49–55
Stalks of quack grass					
Two blades with smooth edges.................	61	58	49	42	32–40
Two blades with upper serrations..............	122	110	106	98	65–78
Two blades with under serrations..............	144	141	125	116	84–90
Smooth blade and serrated ledger plate........	66	60	53	47	40–45
Blades with dulled upper serrations............	...	...	...	...	35–40
Blades with dulled under serrations...........	...	...	...	...	66–75

NOTES: (1) Information in this table is from test results reported by Bosoi. (2) Values given are maximum included angles between cutting edges at which stalks would be cut rather than pushed ahead. Values with stalks suspended (corresponding to infinite cutting height) represent double friction angles.

SOURCE: R. A. Kepner, Analysis of the Cutting Action of a Mower, *Agr. Eng.*, November, 1952.

Mower Cutting Action. A mower cuts by shearing the stems of the vegetation between the moving knife sections and the stationary guards. The included angle between the two cutting edges is usually about 38°. Tests in Russia by Bosoi on the included angle which would pinch and cut stems rather than push them ahead are shown in Table 19-1. Wheat straw required a smaller angle than quack grass, and serrated edges increased the allowable angle.[1]

[1] Robert A. Kepner, Analysis of the Cutting Action of a Mower, *Agr. Eng.*, 33:693–704, November, 1952.

A pendulum knife was used to measure *energy requirements for cutting* single stalks, with the results shown in Table 19-2. A dulled knife with a 0.010-in. flat leading

TABLE 19-2. ENERGY REQUIRED TO SHEAR SINGLE STALKS OF TYPICAL FORAGE PLANTS
(Knife velocity = 113 in. per sec.)

Grass	Energy required (E = energy, in.-lb, D = diam, in.)	Diam range, in.	Age, days	Deviation from equation, in.-lb	Average moisture content, %
Alfalfa	$E = 0.0895 - 2.56D + 24.7D^2$	0.06–0.14	28	±0.02	80
Alfalfa	$E = -0.26 + 2.58D + 11.5D^2$	0.10–0.30	35–37	±0.15	78
Alfalfa	$E = 0.00211 - 0.633D + 26.6D^2$	0.08–0.13	55	±0.04	73
Alfalfa (second cutting)	$E = -0.0412 + 0.316D + 21.6D^2$	0.06–0.14	35–42	±0.04	72
Blue grass	$E = 0.0066 + 0.372D + 6.53D^2$	0.02–0.06	35–40	±0.01	70
Ladino clover	$E = 0.0366 - 0.560D + 6.79D^2$	0.04–0.14	40–42	±0.01	87
Orchard grass	$E = -0.0379 + 1.24D + 5.64D^2$	0.06–0.11	30	±0.02	74
Orchard grass	$E = 0.215 - 3.65D + 33.3D^2$	0.09–0.13	45	±0.04	68
Reed canary	$E = 0.106 - 0.210D + 20.4D^2$	0.15–0.27	35	±0.15	76
Timothy	$E = -0.0522 + 1.65D + 8.96D^2$	0.06–0.12	35–40	±0.03	78

SOURCE: O. A. Fisher, J. J. Kolega, and W. C. Wheeler, An Evaluation of the Energy Requirements to Cut Forage, Grasses, and Legumes, *Univ. Conn. Storrs Agr. Expt. Sta. Progr. Rept.* 17, 1957.

edge required, compared with the sharp knife used above, from 2.5 times the energy for 0.10-in.-diam alfalfa to 3.5 times for 0.06-in.-diam. A knife with a 0.005-in. flat required from two to three times the energy.[1]

Energy was measured for cutting groups of stems at velocities comparable with mower knives and found to range from 0.4 to 2.4 hp-hr per ton of dry matter to cut into ½-in. lengths.[2]

Knife Speed vs. Ground Travel. A study of a cutting diagram, Fig. 19-1, will show that, before being cut, almost all stems are bent sideways, forward, or both. It is apparent that between strokes there is an accumulation of stems which are bent forward and must be cut off at the beginning of the succeeding stroke. Too slow a relative knife speed results in excessive forward bend of the stems, causing uneven stubble. It also results in the collection of a large group of stems which must be cut at the beginning of each stroke, and the excessive force required may cause clogging. An analysis of the cutting diagram for a conventional cutter bar traveling forward 3 in. per stroke indicated that 25 per cent of the area cut per stroke was crowded forward in this manner.

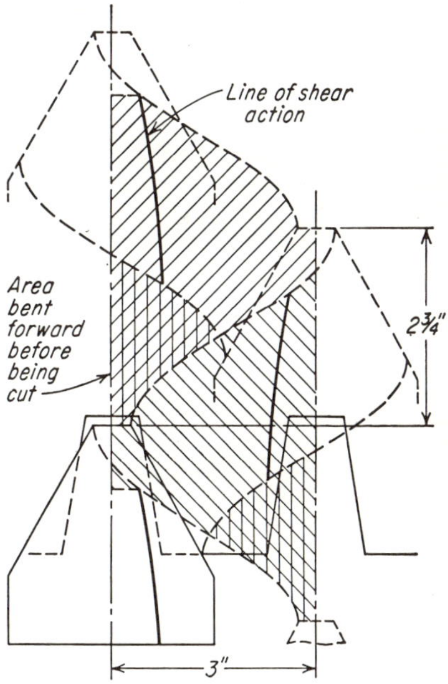

Tractor mowers have much more power and momentum available for this peak load than horse-drawn mowers and are commonly designated to travel forward about 3 in. per stroke where horse-drawn mowers travel about 1¾ in. Scything action (impact cutting) comes into effect at

FIG. 19-1. Mower cutting diagram.

[1] O. A. Fisher, J. J. Kolega, and W. C. Wheeler, An Evaluation of the Energy Requirements to Cut Forage, Grasses, and Legumes, *Univ. Conn. Storrs Agr. Expt. Sta. Progr. Rept.* 17, 1957.
[2] W. J. Chancellor, Energy Requirements for Cutting Forage, *Agr. Eng.*, 39:633–636, October, 1958.

speeds above 6 mph in easily cut crops, such as irrigated alfalfa. In such cases, forward travel up to 5 in. per stroke may be satisfactory with a sharp knife.

It will be noted from the diagram that the stroke can be somewhat shorter than the 3-in. spacing between guards without appreciably lowering cutting efficiency, reducing inertia loads correspondingly. Some mowers use 2.6 in., but precise register is necessary.

A theoretical analysis indicated that a double-knife mower with 3-in spacing of sections and 3-in. stroke, making two cuts per stroke, would accumulate only 6 per cent of the area at the beginning of a cut with 3-in. forward travel per stroke, rather than the 25 per cent accumulated by a conventional mower.[1] The two knives serve to counterbalance each other, reducing vibration at high speeds, and the double number of cuts results in less side bending of stems, allowing more forward travel per stroke. To avoid interference with gather, any guards must be placed on the lines of cut and be narrow, with enough clearance above and below the knife to allow the stems to be gathered to the center and cut.

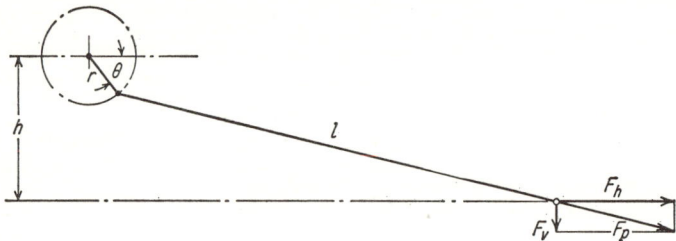

FIG. 19-2. Schematic mower pitman drive.

Forces Due to Knife Inertia. Tractor mowers are commonly considered to require a maximum of 1 hp per ft of cutter-bar width. Thus the load is light and the only limitations on the speed of operation are the roughness of the ground and the capabilities of the cutting mechanism. Most tractor mowers are geared for about 900 rpm. With a forward travel of 3 in. per stroke, this permits a speed of approximately 5.1 mph with acceptably clean cutting under normal conditions.

The following formula closely approximates the knife inertia force:[2]

$$F_h = 2.84 W r N^2 \left(\cos \theta + \frac{r}{l} \cos 2\theta + \frac{h}{l} \sin \theta \right) (10^{-5})$$

where F_h = accelerating force, lb
W = weight of knife plus knife end of pitman, lb
r = crank radius, in.
N = crank, rpm
θ = angle between crank and horizontal, deg (Fig. 19-2)
l = pitman length, in.
h = offset of crank center above pitman connection to knife, in. (Fig. 19-2)

A typical 7-ft mower knife weighs 11½ lb. A typical pitman weighs 6½ lb, with 3 lb resting on the knife end when supported at both ends. Then, in the above formula, $W = 14.5$ lb. If we assume $r = 1.45$ in., $N = 900$ rpm, $l = 45$ in., and $h = 12$ in., then θ is 15° at the outer end of the stroke, and the solution gives $F_h = 515$ lb. If the speed is increased to 1,200 rpm, F_h is increased to 916 lb.

Strain-gage equipment was used to determine actual knife loads on a mower having reciprocating knife parts weighing 9.15 lb and with the drive shown in Fig. 19-3.

---

[1] Kepner, *op. cit.*
[2] T. Baumeister (ed.), "Marks' Mechanical Engineers Handbook," 6th ed., McGraw-Hill Book Company, Inc., New York, 1958, p. 8-93.

The results are shown in Table 19-3. The maximum cutting loads were determined while mowing moderately heavy mixed hay (1¼ to 1½ tons per acre) in third gear (4.8 mph at 1,500 engine rpm).[1]

Counterbalance and Vibration Control. A mower is similar to a single-cylinder gasoline engine in its balancing problem, with the added complication of crank offset, and there is no simple method of securing perfect balance. The accepted practice is to counterbalance the flywheel, using trial-and-error procedure until the best results are obtained. Although the flywheel counterbalance weight reduces horizontal vibration, it adds a vertical vibration component, so ordinarily counterbalance weight is added until the horizontal and vertical amplitudes of vibration at the flywheel are approximately equal.

Vibration stresses almost every part in a mower and results in many possibilities of fatigue failure. As a result, the development of a durable and trouble-free mower is an exacting and tedious job, requiring many hundreds of hours of testing.

In the case of a rigidly attached tractor mower, the vibration is transmitted into the tractor, causing radiator leaks and fatigue failure of other loosely attached parts. One remedy is a flexible mounting attachment which will allow the mower itself to

TABLE 19-3. MOWER KNIFE LOADS

	Tractor engine, rpm	Knife speed, rpm	Av. max. connecting-rod load, lb	Equivalent knife load, lb	PTO horsepower av.
Running empty...............	1,500	942	920	410	1.70
	2,000	1,255	1,420	605	2.75
Cutting.......................	1,500	942	1,280	570	2.55
	2,000	1,255	1,760	780	3.41
Jamming (shearing section rivets)........	1,500	942	7,280	3,200	

SOURCE: L. E. Elfes, Design and Development of a High Speed Mower, *Agr. Eng.*, March, 1954.

freely vibrate sideways, enough to counterbalance the knife (usually about 5/32 in.) but still hold the mower to its work. When this is properly done, little vibration is transmitted into the tractor even at open throttle. With a flexible mounting, the reaction from the inertia force of the knife is transmitted into the frame and in turn into the cutter bar. In effect, the counterbalancing is similar to the case of a double-knife mower, but here the frame and cutter bar are much heavier than a second knife and move a correspondingly shorter distance. High stresses are placed in the frame, and careful design is required throughout, particularly in the connections between frame and cutter bar.

Another remedy is to introduce an opposing reciprocating weight as nearly as possible on the line of action of the knife, as shown in Fig. 19-3. The knife is attached to a pitman-driven lever. A second crank throw, 180° from that driving the knife lever, drives a second lever, with counterweights at the bottom which move opposite the knife.[2] Another design uses the Lanchester balancer principle, consisting of twin horizontal counterrotating flywheels with short pitman drives to a bar across the knifehead. The double flywheels balance each other in the fore-and-aft direction and balance the knife in the lateral direction.

Some highway mowers are being built with *hydraulic drives*. A hydraulic motor, powered from a pump in the tractor, drives the knife through a flywheel and a short pitman. The major advantages are flexibility, due to the elimination of a mechanical drive connection, and the use of a pressure-relief valve to protect the knife from damage when jammed by foreign objects. The latter feature is particularly valuable

[1] L. E. Elfes, Design and Development of a High Speed Mower, *Agr. Eng.*, 35:147–153, March, 1954.
[2] *Op. cit.*

for highway mowing.[1] The cost of hydraulic drives has prevented their general adoption. Direct hydraulic piston drives to the knife have been attempted, but sealing of the piston and cushioning the drive are major problems.

Hay Crushers. The function of a hay crusher or conditioner is to break the skin of the hay stems in order to speed drying in the fields. Splitting, bruising, or scraping

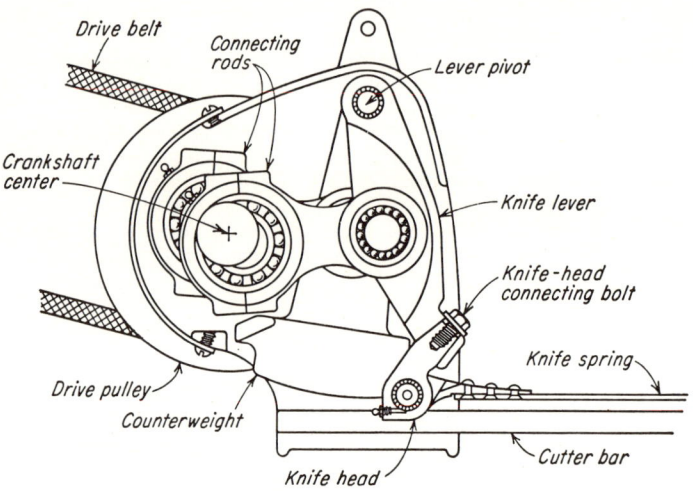

FIG. 19-3. Mower drive with swinging counterweight. (*Massey-Ferguson Inc.*)

the skin is effective, although splitting by crushing or bending between rolls is most easily accomplished. The leaves of uncrushed hay are usually dry enough to shatter by the time the stems are dry enough to bring the average moisture content low enough for safe storage (20 to 25 per cent for long hay). Quicker drying of the stems permits an increase in allowable leaf moisture and a consequent reduction in leaf-shatter-

TABLE 19-4. TYPICAL MATERIALS USED FOR MOWER CUTTER-BAR PARTS

Usage	Material	Hardness
Knife section	SAE 1080	Edges R_c52-57, center R_c25-35
Knife back	SAE 1040	Bh 179-225 (unhardened)
Knife head	SAE 1020	Carbonitride 0.025-0.035 deep, R_c58 min.
Guard	Malleable cast iron	
	SAE 1020 forging	Bh 116-128 (unhardened)
Guard plate	SAE 1080	Edges R_c52-57, center R_c20-30
Wearing plate	SAE 1055	Edges R_c45-55
Knife clip	SAE 1020	Bh 111-121 (unhardened)
	SAE 1045	Bh 163-179 (unhardened)
Sole	SAE 1045	Bh 163-179 (unhardened)
Cutter bar	SAE 1045 stress-relieved special-process cold-rolled	Bh 229-270 (unhardened)

ing losses.[2] Typical drying curves for crushed and uncrushed alfalfa in good and average drying weather in Wisconsin are shown in Fig. 19-4.[3]

Crushing effectiveness, the percentage of stems which have been split open, is increased by high roll pressure and by feeding in a thin uniform layer. High roll periph-

[1] E. P. Morgan, The Hydraulic Sickle Drive, *Agr. Eng.*, 36:736–738, November, 1955.
[2] T. T. Pedersen and W. F. Buchele, Drying Rate of Alfalfa Hay, *Agr. Eng.*, 41:86–108, February, 1960.
[3] H. D. Bruhn, Status of Hay Crusher Development, *Agr. Eng.*, 36:165–170, March, 1955.

eral speed, up to over three times ground speed, contributes to a thin layer if feeding is uniform. Excessively high pressures, however, may clip and smear leaves and break stems into short pieces, causing excessive losses. A pressure of 30 lb per lineal inch of roll was found to be near the optimum for smooth rolls in Wisconsin studies. Drying time was further reduced in the Wisconsin studies by passing the hay through two sets of rolls, the second set running slower than the first set. This tended

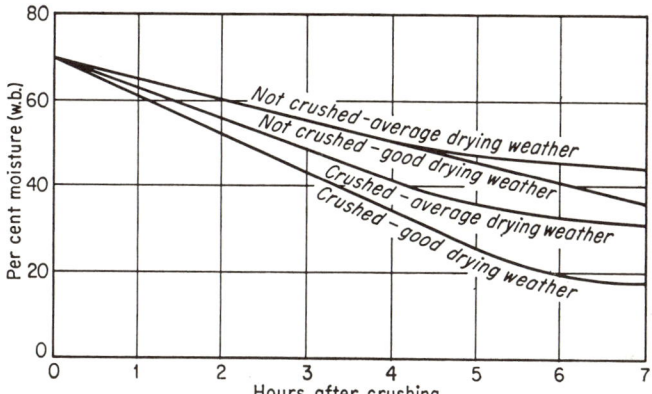

Fig. 19-4. Effect of weather conditions on the comparative drying rates of crushed and uncrushed alfalfa. (*Bruhn, Agr. Eng., March,* 1955.)

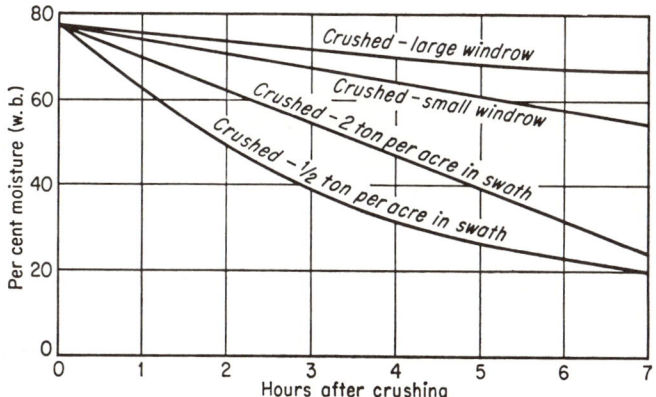

Fig. 19-5. Effect of density of the crop on the drying rate of crushed alfalfa and clover. (*Bruhn, Agr. Eng., March,* 1955.)

to buckle and split the stems as they passed from the first set of rolls to the second. Drying time was reduced by about one-fourth compared with single-roll crushing.[1]

Clogging of smooth rolls by coarse-stemmed weeds which turn crossways is a problem. A thick mat of forage helps carry these stems through, and slower roll speeds may be less troublesome under these conditions. Some crushers provide a triprope release of roll pressure as an aid to clearing clogged rolls. Roll diameter determines the pinching angle on a given diameter stem. A theoretical analysis, assuming a coefficient of friction of 0.3, indicated that 8.5-in.-diam rolls would take crossways stems up to 0.38-in.-diam while 6.5-in.-diam rolls would take only up to 0.29-in.[2]

[1] *Ibid.*
[2] *Ibid.*

Long pressure springs with a low rate are desirable to pass thick layers of hay without clogging.

Maximum *exposure to air and sun* is essential for rapid drying of crushed hay. As can be seen from Fig. 19-5, drying rate was greatly reduced when crushed hay was raked into a windrow before drying.

In tests at Michigan State University, the use of black polyethylene plastic film under the cut hay to seal off ground moisture approximately halved the time required to dry crushed alfalfa.[1]

In recent designs, the lower roll is fluted or grooved to enable it to pick up the mower swath, thus eliminating a conventional hay pickup cylinder. Roll combinations

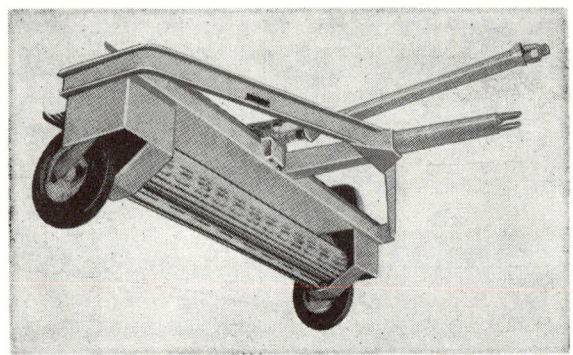

FIG. 19-6. Underneath view of hay conditioner with intermeshing fluted rolls. (*Ray Cunningham and Sons, Inc., La Crosse, Wisc.*)

include two fluted metal rolls (often called *crimpers,* Fig. 19-6), two grooved rubber rolls, and combinations of these rolls with each other and with smooth steel rolls.

Severe crushing of stems for maximum drying rate is usually accompanied by increased *field losses* due to clipping off leaf clusters. Crimpers have slightly higher losses than smooth rolls when operated to give equal drying rate.[2] On the other hand they are less subject to clogging. Hay cut with a flail-type forage harvester had much higher losses for its drying rate than mown and conditioned hay.[3]

[1] T. T. Pedersen and W. F. Buchele, Hay-in-a-Day Harvesting, *Agr. Eng.,* 41:172–175, March, 1960.
[2] R. A. Kepner, J. R. Goss, J. H. Meyer, and L. G. Jones, Curing Rates, Field Losses and Feeding Response with Crimped, Rolled and Untreated Alfalfa Hay, *ASAE Paper* 59-132, June, 1959.
[3] H. M. Boyd, Hay Conditioning Methods Compared, *Agr. Eng.,* 40:664–667, November, 1959. H. D. Bruhn, Performance of Forage Conditioning Equipment, *Agr. Eng.,* 40:667–670, November, 1959.

Chapter 20

RAKES

The side-delivery rake was developed to sweep cut vegetation sideways into a continuous windrow, which is more easily picked up by a hay loader, field baler, or chopper than the windrow left by a dump rake. The ideal windrow has the leaves on the inside with the stems outside and is rolled loose enough for good drying but still has a tendency to hang together, thus facilitating feeding into a machine.

The first side-delivery rake was tried in 1893 and used a double conveyer chain to drag attached tooth bars across the direction of travel. In 1914 the left-hand side-

Fig. 20-1. Lift-type side-delivery rake with cylindroid or roller bar reel. (*Massey-Ferguson Inc.*)

delivery rake with cylindrical reel appeared and has been unchanged in principle since. The side-stroke or roller-bar type of reel was patented in England by Martin in 1905, but is comparatively recent in this country. A mounted rake of this type is shown in Fig. 20-1. In 1946 the wheel rake, consisting of a series of floating overlapping wheels with spring teeth, appeared in California and has since been sold nationally. A trailed model of this type is shown in Fig. 20-2.[1]

[1] Roy Bainer, New Concepts in Side Delivery Rakes, *Agr. Eng.*, 32:266–268, May, 1951.

Raking Action. Reel-type side-delivery rakes feather the teeth to facilitate their withdrawal from the hay. Feathering action is also commonly used on windrow pickup attachments for feeding swathed grain into combines and hay windrows into balers or forage harvesters, as well as on the pickup-type reels used on cutter-bar attachments for forage harvesters and on combines for handling down grain. A few rakes and hay pickups have used stripper bars or straps to effect tooth withdrawal without feathering action, but they are in the minority.

Feathering action of tooth bars may be obtained by the following mechanisms:

1. Tooth-bar cranks linked to spokes of a control spider offset from the reel spider as shown in Fig. 20-3a. The eccentric mounting of the control spider requires that the bearing be large enough for the reel drive shaft to pass through it. Tooth angle can be changed by rotating the control-spider center about the reel center. This mechanism has been very popular on side-delivery rakes.

2. Tooth-bar crank ends controlled by a cam, as shown in Fig. 20-3b. The crank ends are fitted with a cam roller run in a cam track or controlled during the work-

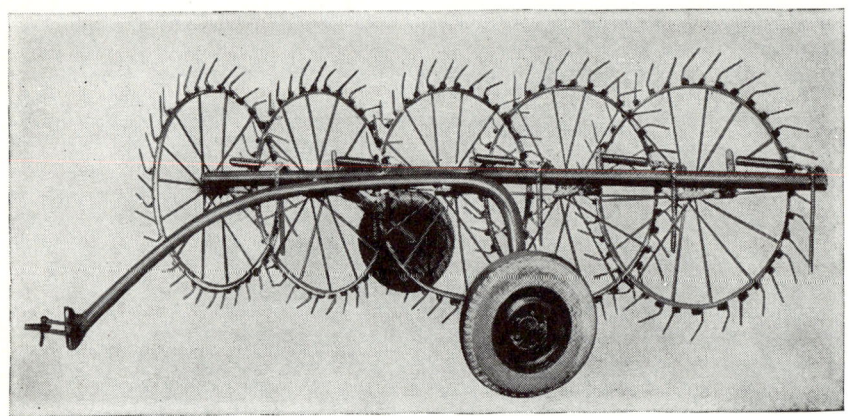

Fig. 20-2. Trailed wheel rake. (*The Farmhand Co., Hopkins, Minn.*)

ing stroke by an internal or external cam which maintains the teeth in the proper position. The teeth are not necessarily kept parallel, as with the other mechanisms, but can be varied as desired throughout the cycle by the cam contour. This mechanism is widely used on hay pickups and pickup reels. It has also been used on side-delivery rakes.

3. A system of planetary gears to the tooth bars as shown in Fig. 20-3c. Tooth angle is controlled by the position in which the sun gear is locked. A system of chains and sprockets may also be used.

4. Oblique reel ends as used in the side-stroke or roller-bar type of rake, shown in Fig. 20-1. This design is not suited to pickup reels or cylinders because of the oblique tooth circle.

Raking with rotating teeth is accomplished by means of a series of strokes whereby each stroke picks up and passes on the hay raked by the previous stroke. This action is illustrated in Fig. 20-4. It is apparent that cleanness of raking is influenced by x, the height of the take-over point above the bottom of the tooth circle, and by the distance between the raking paths of adjacent teeth.[1]

The successive raking paths of two teeth similarly located on adjacent reel bars are shown in Fig. 20-6. An explanation of the symbols used in Figs. 20-5 and 20-6 follows:

[1] C. B. Richey, An Analysis of the Raking Action of a Side Delivery Hay Rake, *Agr. Eng.*, 24:330–331, October, 1943.

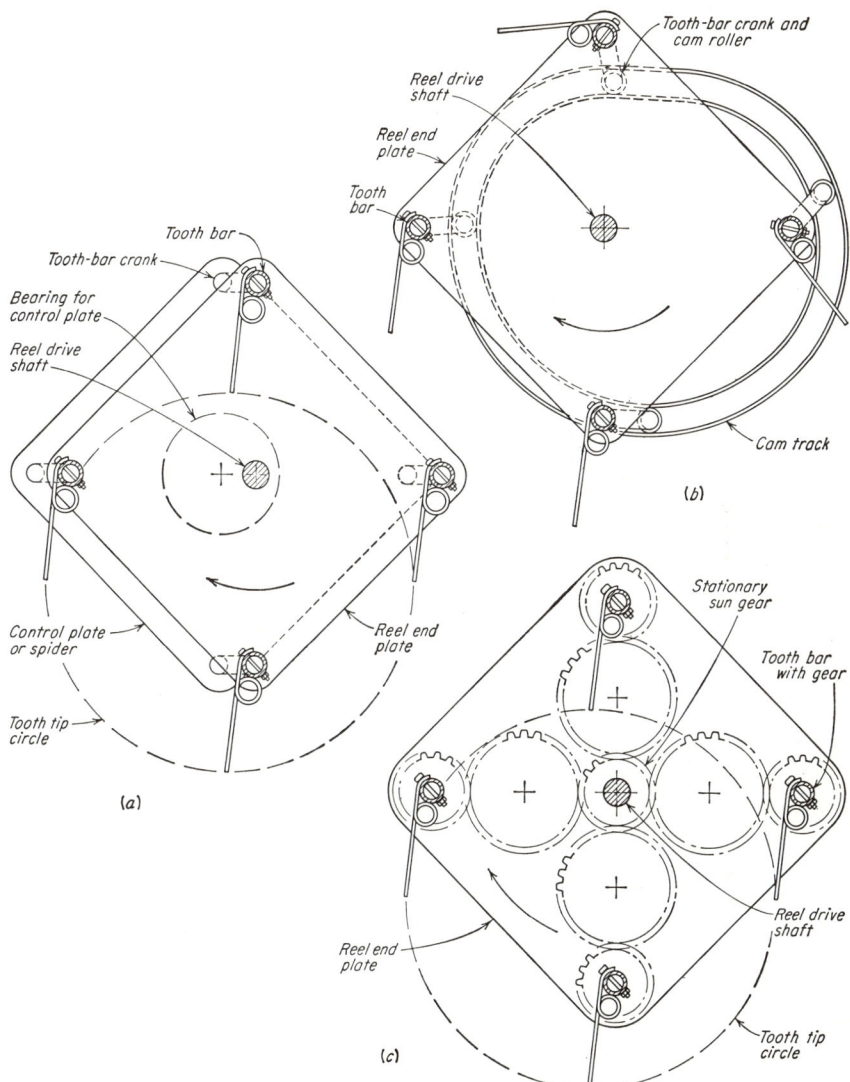

Fig. 20-3. Feathering actions used in raking and pickup reels: (*a*) eccentric spider control; (*b*) cam control; (*c*) planetary gear control.

α = angle turned by reel during effective stroke, deg
β = angle between tooth bars, deg
l = length of chord of tooth circle during effective stroke, in.
x = variation in height from ground during effective stroke, in.
φ = angle between plane of tooth circle and direction of forward travel, deg
θ = angle between tooth bar and perpendicular to forward travel, deg
σ = angle between path of effective stroke and direction of forward travel, deg
R = radius of reel, in.
T = forward travel per reel revolution, in.

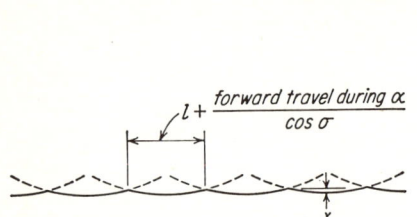

Fig. 20-4. Side view of successive raking strokes.

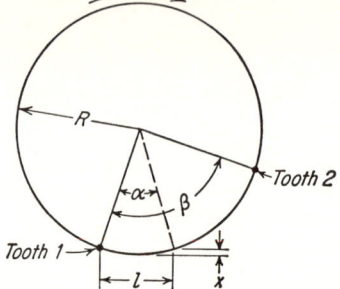

Fig. 20-5. Reel rotation during raking stroke.

The relationships of the above factors are expressed in the following formulas:

$$x = R - \sqrt{R^2 - \left(\frac{l}{2}\right)^2}$$

$$T = 360l \left(\frac{\cos \varphi + \sin \varphi \tan \theta}{\beta - \alpha}\right)$$

$$\tan \sigma = \frac{l \sin \varphi}{l \cos \varphi + T\alpha/360}$$

Where T is known and x, l, and α are to be found, it is easiest to try various values of x and check the resulting T against the known value.

With the wheel rake, the wheels are turned by the contact of the teeth with the ground and the hay. The path of the effective stroke is parallel to the wheel axle.

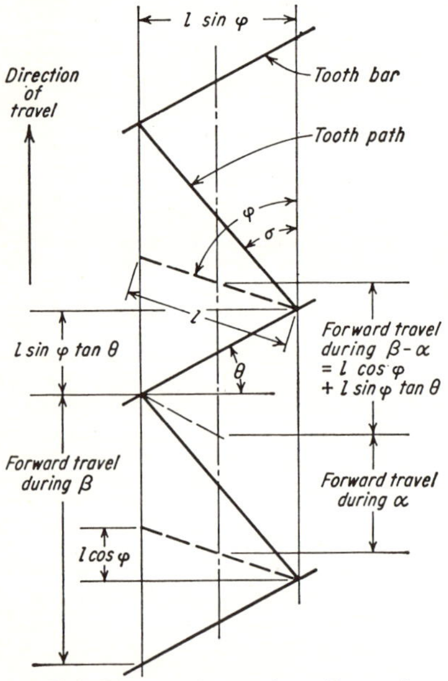

Theoretically, the path of hay travel should be in the same direction as the path of the effective tooth stroke. Actually, the hay is often carried forward more than this by the reel rake because of tooth slippage along the roll of hay and contact of the hay with the stationary stripper rods. Inasmuch as hay should be raked with a minimum of agitation to conserve the leaves, it is considered desirable that the distance it is moved from swath to windrow be a minimum and that the resultant maximum tooth velocity during the effective stroke be as low as possible. Minimum change in hay velocity between raking strokes is also desirable in order to reduce leaf shatter.

A factor influencing the tightness of the windrow is the amount the roll of hay in front of the rake must stretch or draw down as it leaves the rake. With more stretch, the windrow is tighter and more liable to vary in size when the rake changes direction slightly. If the tooth-bar length equals the forward travel necessary to move hay across the swath into the windrow, there will be no change in length as the roll leaves the rake.

Fig. 20-6. Diagram of successive raking strokes.

An analysis of several typical rakes by means of the above formulas gives the results itemized in Table 20-1.

Speed of Operation. The operating speed of a side-delivery rake is limited by leaf shatter of hay, excessively wide windrows, and the durability of the machine rather than by power requirements. As mentioned above, a low ratio of maximum resultant tooth velocity to travel velocity, with the accompanying lessened acceleration and deceleration of the hay during raking, permits higher speeds.

In the case of a side-stroke reel, where the tooth bars are supported at the ends only, tooth-bar whip due to centrifugal force is a definite speed-limiting factor, although excessively wide windrows usually occur first.

Teeth. Because of the constant flexing from hay and ground contact, side-delivery-rake teeth are subject to fatigue failure. Proper coil design and support to prevent stress concentration are essential, and turning the coil in the direction of the tooth path has been found helpful. The practice of passing the tooth bar through the coil has been dropped to facilitate replacement of broken teeth. Double teeth are common since they reduce the number of bolts and mounting clips.

Tooth spacing is a factor in cleanness of raking, and current rakes have spacings which give approximately 3 and 8 in. between tooth scratches on the ground.

Table 20-1. Theoretical Analysis of Various Rakes

		Type of reel				
		3-bar cylinder	4-bar cylinder	4-bar side-stroke	5-bar side-stroke	6-bar side-stroke
	β	120°	90°	90°	72°	60°
	φ	45°	45°	72°	90°	100°
	θ	45°	45°	27°	25°	35°
	R	11 in.	11 in.	12 in.	12 in.	12 in.
	T	44.5 in.	60.2 in.	51.5 in.	69.6 in.	65.2 in.
	l	7.16 in.	6.57 in.	8.3 in.	9.8 in.	7.8 in.
	x	0.6 in.	0.5 in.	0.8 in.	1.05 in.	0.65 in.
Theoretical σ		27⅓°	24°	41½°	46½°	54°
Length of tooth path across 7-ft swath		15.2 ft	17.2 ft	10.5 ft	9.7 ft	8.7 ft
Ratio of max resultant tooth velocity to forward-travel velocity		2.36	1.97	2.01	1.5	1.39
Stretch ratio as roll of hay leaves rake		2.08	2.30	1.46	1.23	1.16

Tooth Bars and Bearings. Tubular tooth bars are commonly used because they give maximum torsional and bending strength for their weight. If the tooth characteristics are known, a maximum tooth deflection and loading can be assumed, along with a maximum operating speed, giving a basis for strength and fatigue calculations. Location of tooth-bolt holes is an important factor in fatigue strength.

Tooth-bar bearings are more heavily loaded in side-stroke reels than in cylinder reels because they must provide torsional stability for the tooth bar. In the side-stroke rake it is important that reel-end alignment and tooth-bar length be held within close limits to avoid excessive bearing preloads. Side-stroke rakes often use a rubber mounting for the rear tooth-bar bearing in order to reduce bearing loads due to manufacturing variations and tooth-bar whip.

Stripping. Side-delivery rakes require stripping means to prevent hay from being carried up and back by the tooth bars. Conventionally, the teeth pass down between stationary stripper rods during their raking stroke. As they pass upward they withdraw through the stripper rods, which prevent hay from being carried up. Because of the oblique plane of the tooth circle, it is difficult with side-stroke rakes to space the teeth close enough for clean raking and still clear the stripper rods if the teeth become slightly bent. One such rake uses a rotating bladed cylinder as a stripper rather than stripper rods. English rakes of the side-stroke type use a stationary board for a stripper, aided by long teeth and shields above the tooth bars to reduce the tendency to carry hay up. The necessary height from tooth tip to the top of the shield is secured by using a large tooth-bar circle and not more than four tooth bars.

Flexibility for Following Ground Contour. This is a very important factor in rake design. Some type of spring suspension is often used to reduce damage when a high spot is encountered. In many rakes, the reel is suspended from the frame by springs. Gaging is improved if the rake wheels are located as close as possible to the raking line, and by this means, lightweight lift-type side-stroke rakes with comparatively short teeth dispense with spring support of the reel.

The wheel-type rake with individual spring-floated wheels can follow uneven ground more closely than any rake with a continuous tooth bar.[1]

Reel Drive. Most rakes are driven from ground wheels in order to maintain the proper relationship between reel speed and forward travel. The reel drive must accommodate up-and-down movements if the reel is of the floating type, and universal joints at this point are common.

Many rakes with cylinder-type reels incorporate a reverse gear for tedding (loosening the hay in the swath or windrow for drying after a rain).

In PTO-driven rakes, a speed-changing device is desirable to allow the tractor to be used in more than one gear, unless the tractor has a ground-speed PTO geared to the rear wheels.

Attachment to Tractor. Uniformity and straightness of windrows are extremely important when field balers and choppers are working near capacity. The trailed type of rake supported entirely on its own wheels has some tendency to snake and in doing so also changes the size of the windrows. Rakes which are supported and trailed from the tractor drawbar and have noncastering rear wheels are much more stable, as are semimounted lift-type rakes, and the recent trend has been to these types.

A *dump rake* collects cut vegetation by means of a row of hooked spring teeth dragging along the surface of the ground. These teeth are periodically raised to discharge the collected hay. The popularity of the dump rake has greatly diminished in recent years because its windrow is not as well suited for hay loaders, field balers, and choppers as that left by the side-delivery rake. It is still popular, however, in western areas where hay is collected by sweep rake and stacked.

[1] G. W. Giles and C. A. Routh, The Finger Wheel Rake, *Agr. Eng.*, 32:537–544, October, 1951.

Chapter 21

HAY-HANDLING EQUIPMENT

Hay harvesting involves one of the most difficult material-handling problems on the farm. Up to 2 tons per acre of dry hay or 8 tons of green hay for silage may be harvested, transported to the farmstead, and stored. Hay is very bulky for its weight and requires transport equipment of high-volume capacity. It is not a high-value crop, but at present high-cost equipment is required for complete mechanization.

Methods of hay handling have changed rapidly during the past few years, as shown by Table 21-1. The new methods of baling and chopping have had the advantage of easing hand labor for feeding out of storage as well as for harvesting and storage. Storage space is also reduced by baling or chopping.[1]

TABLE 21-1. METHODS OF HANDLING HAY, 1918–1953

Year	Stationary baler, %	Pickup baler, %	Total baled, %	Chopped, %	Loose, %
1918	24.3	....	....	...	75.0
1939	12.0	2.5	14.5	...	85.5
1944	13.2	13.6	26.8	1.7	71.5
1948	10.0	37.5	47.5	5.6	46.9
1951	5.0	56.7	61.7	7.5	30.8
1953	4.0	66.0	70.1	8.0	22.0

SOURCE: L. E. Campbell, G. R. Mowry, and C. H. Gordon, Machinery and Labor Requirements for Forage Harvesting, unpublished paper, *USDA ARS*, 1954.

The relative labor required and cost from windrow to storage in Michigan by various methods is shown in Table 21-2. The costs for each method are calculated for the usual annual tonnage of farms using that method. The cost per ton is greatly affected by the annual tonnage, particularly with the high-investment methods.

Nutrient losses due to wet weather during curing are very serious with dry hay.[2] Crushing and barn drying help but have met with limited acceptance thus far. The problem of efficient hay harvesting is far from solved, and there will undoubtedly be further rapid changes in the future.

HANDLING EQUIPMENT FOR LONG HAY

Long hay is handled in more different ways and with more different types of equipment than almost any other crop. It may be loaded onto a wagon for transport to storage by pitchfork or hay loader and unloaded into the mow by pitchfork, sling, or hay fork. A sweep rake, also called a buck rake, may be used to collect the hay directly from the windrow and transport it to the barn, where it may be elevated to the mow by sling, hay fork, or long-hay blower. Where hay is stacked in the field, sweep rakes are commonly used to collect the hay and bring it to the stack, although wagons may be used to bring hay to large stacks at the farmstead.

[1] R. E. Marx and J. W. Birkhead, Hay Harvesting Methods and Costs, *USDA Circ.* 868, June, 1951.
[2] J. B. Shepard et al., Experiments in Harvesting and Preserving Alfalfa for Dairy Cattle Feed, *USDA Tech. Bull.* 1079, February, 1954.

Stacking may be done by overshot stacker, slide stacker, derrick stacker, cable stacking outfit, or by combination sweep rake and stacker, also called buck-stacker. The latter is gaining in popularity in many areas because its ability to distribute the hay over the stack reduces stacking labor, but the slide stacker with portable backboard for forming the stack is still preferred in many wild-hay areas because of its simplicity and efficiency.

Hay loaders have been popular in the Middle West since 1900, but the labor connected with wagon transport of long hay is causing this method of hay handling to lose ground. Although hay loaders are a relatively simple machine, there are several different types. The hay may be picked up by a pickup cylinder with cam-operated

Table 21-2. Comparison of Hay-harvesting Methods, 1953*

Method	Average investment in equipment	Estimated number of tons of hay harvested annually	Tons of hay moved from field to mow in:		Average		Estimated average cost per ton of hay from windrow into storage
			1 hr	8 hr	No. of men	Man-hours per ton	
Hay loader (long hay)	$ 871	50	1.4	11	3	2.3	$4.52
Buck rake (long hay)	468	40	1.6	13	3	1.5	4.55
Baler (automatic)	3,618	300	2.2	18	3–4.5	2.2	5.71
Hay crusher and baler	4,522	175	2.2	18	3–4	2.2	6.71
Field choppers:							
Small	2,993	150	1.5	12	3–4	1.5	5.12
Large	4,319	300	3.0	24	3–4	1.6	4.50
Barn driers:							
Small chopper	3,958	...	1.5	12	3–4	1.5	8.37
Large chopper	5,284	...	3.0	24	3–4	1.6	7.75
Grass silage:†							
Small chopper	2,993	250	3.3	26	3–4	1.7	4.07
Large chopper	4,319	375	5.0	40	3–4	1.6	3.52

* Cost estimated on the basis of 1953 cost of equipment and 1946 to 1950 performance records.

† Basis of 68 per cent moisture; to place costs on an approximate comparable basis as other hays at 20 per cent moisture, multiply grass silage costs by 2.6.

source: Karl A. Vary, Hay Harvesting Methods and Costs, *Mich. State Univ. Agr. Expt. Sta. Spec. Bull.* 392, May, 1954.

tooth bars or by oscillating rakes, and it may be elevated by conveyer or by oscillating rake bars. Various combinations of these elements are used, the rake-bar pickup and elevating mechanism being the simplest. All are ground-driven and trailed behind the wagon.

Sweep, or buck, rakes date back to the 1880s and have the unique ability to collect, transport, and unload long hay efficiently in spite of their simplicity. Those primarily used with field stackers push rather than carry the hay and are 12 ft wide with teeth approximately 8 ft long. Commercial tractor models are front-mounted, but it is common in areas of heavy usage for farmers to mount them on the rear and reverse the direction of travel of the tractor. Buck rakes developed by farmers in the eastern Corn Belt during the late 1930s to transport hay from field to barn are commonly 10 ft wide and have 12-ft teeth lifting by power to a 30° angle or more in order to avoid dragging and dropping of hay. These rakes are often mounted at the rear of old automobiles and trucks and are loaded in reverse gear.[1] They carry up to 1,000 lb of hay per load, and the man-hours per ton for handling long hay compares favorably with much more elaborate methods.[2]

[1] C. B. Richey and R. D. Barden, The Automotive Buck Rake in Ohio, *Agr. Eng.*, 23:196–198, June, 1942.

[2] Ellis W. Lanborn and L. B. Adkinson, Costs and Labor Used to Handle Hay by Different Methods in New York, *Cornell Univ. Agr. Expt. Sta. Bull.* 569, August, 1946.

In England small strongly built rear-mounted tractor buck rakes are popular for bringing long green hay to trench silos for grass silage. The tractor packs as it is driven in for unloading. The removal of this silage without excessive hand labor has been facilitated by the use of a power chain saw in at least one instance in the United States.

Combination buck-stackers are usually tractor-mounted and lifted by hydraulic cylinders. Horse-operated models, lifted by cables from a ground-wheel drive, date back to the 1920s. Tractor power was applied during the 1930s, and the widespread adoption of hydraulic power lifts in the 1940s accelerated the use of full-mounted types, which can also be used as manure loaders when a bucket replaces the sweep rake head.

The rake head is somewhat stronger with longer teeth than the western type of sweep rake because of the necessity for lifting the hay. Lifting capacity varies from

Fig. 21-1. Tractor buck-stacker with push-off discharge. (*The Farmhand Co., Hopkins, Minn.*)

500 to 1,000 lb, although the problem of stability and excessive front-wheel loading often requires weight boxes at the rear of a tricycle tractor.

Discharge onto the stack may be by dumping the head or by a push-off mechanism which discharges the hay up over the ends of the elevated teeth, as shown on the machine in Fig. 21-1. The latter is popular on stacker attachments for loaders because it allows the use of the regular loader frame and arms and also reduces the overhang of the load. A separately controlled hydraulic cylinder powers the push-off. Delivery heights vary from 15 to 21 ft.

In western areas of high usage the hay is often brought to the stack by sweep rakes and then placed on the stack by the buck-stacker. The high-lift dump type is often mounted on a truck chassis for better stability.

The field-stacking method of hay storage is very efficient as regards labor, and in areas where spoilage is not serious, it will no doubt be continued. This hay is fed to livestock during the winter, and tractor power is being utilized in various ways to reduce hand labor in this connection. Tractor-mounted booms with cable-operated forks are being used in some sections to take hay off the stack and scatter it for feeding.

Chapter 22

BALERS

J. H. BORNZIN

The baler, or baling press, is a machine used to compress hay or straw into bales for the purpose of facilitating handling and saving storage space. Rectangular bales are produced by ramming successive charges of hay into a chamber of 14 by 16 in. up to 17 by 22 in. or more in cross section. Round bales are made by rolling a windrow into a cylinder. Bale lengths vary from 24 to 48 in., and densities from 7 to 18 lb per cu ft, with total weight varying from 35 to 125 lb. A PTO-driven pickup baler is shown in Fig. 22-1.

FIG. 22-1. Pickup baler, PTO-driven. (*International Harvester Co.*)

Pickups. The windrow is usually lifted from the ground by a pickup reel having spring teeth. The teeth may be feathered by cams for easy withdrawal from the lifted hay, or they may be on fixed tooth bars and withdrawn perpendicularly through stationary stripper plates if a large-diameter drum is used. The simplest pickup device, used on a machine for round bales, is a double-chain conveyer with sawtooth crossbars which pass close enough to the ground to get under and lift the windrow and which are stripped by a beater at the top.

A power-driven pickup allows the baler to be stopped while the pickup tines continue to operate and untangle a bunch of hay. A ground drive gives best control where ground speed varies considerably and is also preferable in heavy viny hay which will not break and stop feeding with a power drive when forward motion is stopped. A combination drive may be used in which either type can be selected or in which the ground drive takes over from the power drive above a certain ground speed.

Feeding the Bale Chamber. The feeding mechanism pushes individual charges of hay into the bale chamber during the intervals when the ram is withdrawn. In addition to providing for accumulation of hay between feeding strokes, it must meet the following requirements:

1. Have some packing or precompressing effect to achieve maximum capacity
2. Distribute the hay evenly in the chamber to avoid bales which are more dense on one side than the other and curve after discharge
3. Have positive action, with no dead spots, capable of handling moist hay or slugs without plugging
4. Be free from winding or wrapping
5. Minimize shatter loss of dry leaves

With *top feeding,* the hay must be conveyed up to the level of the top of the baling chamber and across to the packer, or wadboard, which pushes the hay down into the chamber while the ram is retracted. The packer head is linkage-mounted for approximately vertical action and is driven by a connection to the ram or to the connecting

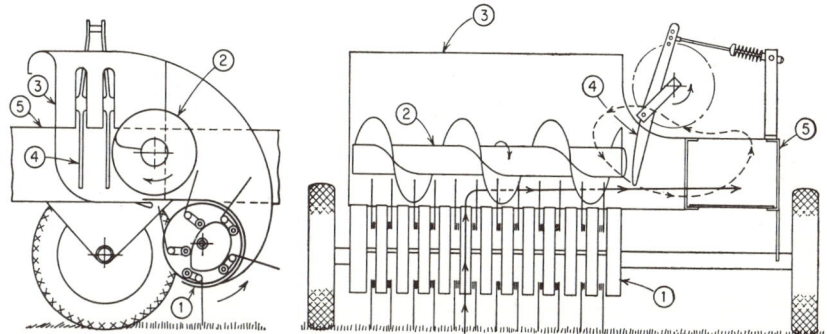

Fig. 22-2. Flow of material from windrow to bale chamber: (1) pickup; (2) feed auger; (3) cross conveyer; (4) packer fingers; (5) bale chamber. (*International Harvester Co.*)

rod. A quick return motion is common in order to get maximum packing leverage in the time interval available.

The *side-feed* principle has come into use because of its simplicity and compactness. The pickup reel usually delivers hay directly to a cross-auger, which pushes the hay into the path of packer fingers that sweep the hay sideways into the bale chamber and retract through the top. Such a feed is shown in Fig. 22-2. Some balers have replaced the auger by additional packer fingers in tandem or mounted on a bar moving in strawwalker fashion.

Compression. The ram, or plunger, of an automatic baler has the following functions:

1. Compresses the hay or straw within the bale chamber
2. Separates the charges of hay by shearing
3. Protects the needles which place the binding means around the bale

The force exerted by the ram must overcome:

1. The resistance of the new charge to being packed against the previously compressed material and its friction against the chamber sides as it is compressed.
2. The resistance due to shearing the unseparated portion of the new charge.
3. The frictional resistance of all the compressed material in the chamber as it is displaced by the new charge. [This resistance depends on (*a*) the coefficient of friction between material and chamber surfaces; (*b*) modulus of elasticity of the material as it affects lateral expansion under ram pressure and resistance to compression by side convergence; (*c*) length of chamber; and (*d*) number of hay retaining wedges at

front, removable wedges, and *chamber convergence,* the latter two being controllable. It appears that both factors *a* and *b* vary with the type of material and the moisture content.]

An actual force-displacement curve for a particular baler at maximum load is shown in Fig. 22-3.[1] The hump at 23 in. displacement represents the shearing peak, the magnitude of which depends on the compression force at the shearing point and the effectiveness of the feed in separating the material between charges. The peak shear load is estimated to be about one-fourth the compression peak with most balers,[2] although this may vary according to design details. (One baler has a shearing knife actuated by a separate connecting rod on a crank throw about 30° ahead of the throw for the ram connecting rod. Shearing is done early in the compression stroke, reducing the peak load, and the knife is partially withdrawn for better needle clearance when the ram is at the outer end of its stroke.) The compression peak occurs just before the previously compressed material breaks loose and slides in the

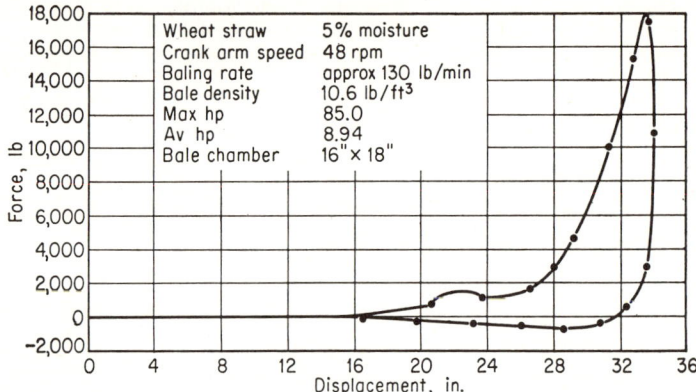

Fig. 22-3. Plunger force vs. displacement for maximum load cycle. (*Burrough and Graham, Agr. Eng., April,* 1954.)

chamber, usually at a crank angle near 160° from forward dead center. The negative force results from forward expansion of the baled material as the ram withdraws. Crankshaft torque, the primary design factor, can be calculated at various points in the cycle by also taking into account ram acceleration, ram friction, and connecting-rod angularity effects.

The compression force and the resulting density increase rapidly with an increase in moisture content of the material being baled, primarily because of the increased coefficient of friction. The effect of moisture on alfalfa bale density with constant tension setting is shown in Fig. 22-4, and it can be seen that dry-matter density increased as well as the density as baled. Twine-tied bales cannot be made as dense as wire-tied because the twine has less tensile strength than wire and may be broken by expansion of the bale after it leaves the chamber. High densities require much more baling energy, as is shown by Fig. 22-10.

Density is most simply *controlled* by adjusting the convergence of the bale chamber sides. Removable wedges may also be used. Bales with short chambers require more side convergence for adequate resistance than bales having long chambers. Frictional resistance varies with moisture content, and automatic control of side-plate pressure to give constant dry density would be highly desirable. Various expedients are being used to approach this ideal, such as hydraulic means to secure constant chamber side pressure or pressure plates which swing out against spring pressure as friction

---

[1] D. E. Burrough and J. A. Graham, Power Characteristics of a Plunger-type Forage Baler, *Agr. Eng.,* 35:221–232, April, 1954.
[2] *Ibid.*

increases. A direct solution to this problem gages density by the penetration into the bale of a pressure-held star wheel, which in turn controls hydraulic pressure in a cylinder that adjusts bale chamber convergence.

The ram, or plunger, carries a *shearing knife* on one side. Shearing action results in high longitudinal forces on the knife, as well as side thrust resulting from the tendency of the material to wedge between the knife and the shear plate. The plunger must provide adequate support for the knife as well as bearing area to take the side thrust. Knife adjustment must be provided to compensate for knife wear, often rapid when the hay contains gritty dust.

Plungers may be fitted with wooden wearing pieces, usually hard-maple, steel, or plastic runners, or rollers. In any case, the wearing members opposite the shear knife should have the greater bearing surfaces.

The binding strands are placed around the bale by *needles* when it has reached the desired length. The needles are usually timed with the plunger so as to pass through slots in the plunger while it is holding the bale compressed. This makes

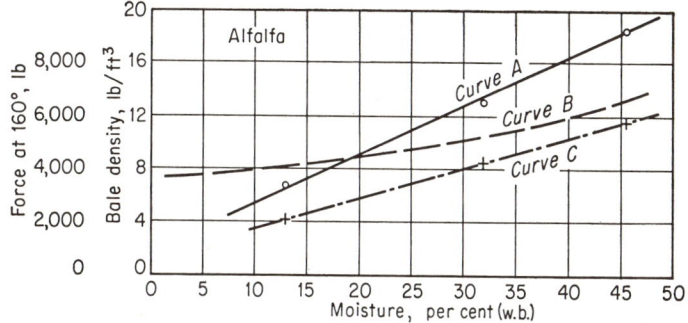

FIG. 22-4. Effect of moisture on bale density and plunger force at a constant tension setting: curve A, bale density as baled; curve B, bale density on basis of 20 per cent moisture (w.b.); curve C, force to bale at 104 lb dry matter per minute at the bale densities given by curve A. (Burrough and Graham, *Agr. Eng.*, April, 1954.)

certain of bale separation and allows the plunger to protect the needles from foreign obstructions and hay.

The cycle of the needles may take place during one or more cycles of the plunger crank, depending on the design. In the one-cycle method, it is necessary to complete tying and withdraw the needles from the bale chamber before the plunger compresses the next charge of material against them. When more than one plunger-crank cycle is consumed for needle and knotter functioning, a stop holds the plunger in its extended position, keeping the material compressed, while a spring-loaded telescopic connection in the plunger pitman or connecting rod allows the plunger crank to continue revolving. After tying is completed, the stop is withdrawn and the crank again drives the plunger. Greater baling capacity as well as less mechanism is claimed for tying without missing a plunger stroke. The advantages claimed for stopping the plunger and consequently losing the feed stroke are (1) more time for the tying cycle and (2) binding the bale while it is under compression from the plunger.

Tying or Binding. The bale is held together under compression by wire or twine. Two ties are most common, although three may be used for very large heavy bales. The needle-and-tying mechanism is engaged when the bale has reached a particular length as measured by a pronged metering wheel which is turned by the passage of the bale.

Bale length is controlled by the circumference and consequent rotation of the metering wheel and is usually varied by one of the following methods:
 1. Changing metering wheels to different diameters
 2. Using an adjustable-diameter metering wheel

3. Changing gears or sprockets in the drive between the metering wheel and the trip mechanism

Wire and *twine* are the common materials used for binding bales. Generally, twine is preferred by the farmer who feeds his stock from hay he himself has grown and baled, one reason being that it is less injurious than wire if eaten by livestock. Wire is preferred, however, when bales must be handled several times before feeding. This is particularly true when bales must be shipped long distances.

Wire is *fastened* by twisting, while twine is tied by a double overhand knot. Mechanisms for handling either material must function under extremely adverse conditions. Dirt, chaff, rough fields, and oxidation from weather must not affect their efficiency. Adjustments should be eliminated whenever possible in order to prevent the operator from tampering, but where necessary they should be generous. Timing of re-

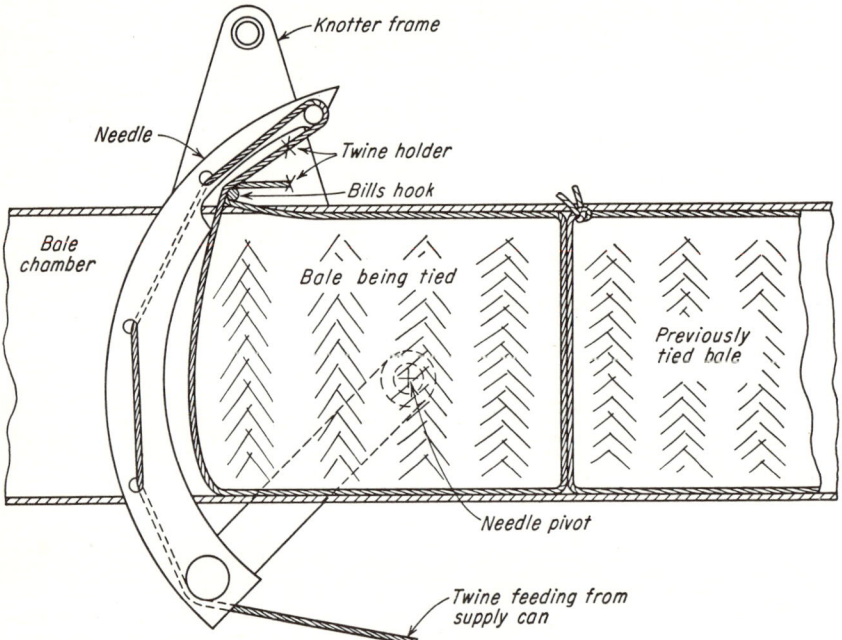

Fig. 22-5. Diagram of binding elements of a twine-tie baler.

lated parts is critical because of the many functions performed. At no time during a binding cycle should the end of the binding strand be released. It must be passed from one unit of the mechanism to the other without being turned loose; otherwise it may be lost by the mechanism and cause a "miss." Wire mechanisms must be built stronger than twine mechanisms because of the greater stiffness of wire and its greater wearing action on contacted parts.

Twine for bales is usually made of sisal fiber. A popular grade has a tensile strength of 325 lb and usually runs about 231 ft per lb compared with a strength of 90 lb and 500 ft per lb for binder twine. It should be uniform in strength and size, having no thick spots which may cause malfunctioning of the tying mechanism.

Wire for bales is 0.076 ± 0.002 in. in diameter with 50,000 to 70,000 psi tensile strength and 12 per cent minimum elongation in 10 in. length.[1]

Both twine and single-coil wire-tying *mechanisms* require the following elements:
1. Twine or wire *holder* to hold the end while the bale is being formed
2. *Needle* to complete the encirclement of the bale and place the twine or wire in position for the fastening mechanism

[1] ASAE Standard: Baling Wire for Automatic Balers, *Agr. Eng. Yearbook*, 1959, p. 94.

3. *Knife* to cut off the piece encircling the bale
4. *Knotter bills* for twine or *twister* for wire to tie the ends together
5. *Stripper* to remove the tied ends from the tying mechanism

The general arrangement of these elements in a twine-tie baler is diagramed in Fig. 22-5.

Twine or wire is *held* by kinking or compressing or a combination of both. Kinking cannot be too severe or it will result in fractures, thereby reducing the tensile strength of the binding strand. Compressing of twine tends to crush fibers and has proved unsatisfactory unless no pull is exerted upon the twine during bale formation. To accomplish this would involve a means for releasing slack twine into the system. Compressing of wire has proved satisfactory, and usually compression members have notches or teeth to prevent slippage of wire. However, wire is held by kinking means also. A type of holder using both kinking and compression and adaptable to either wire or twine is shown in Fig. 22-6.

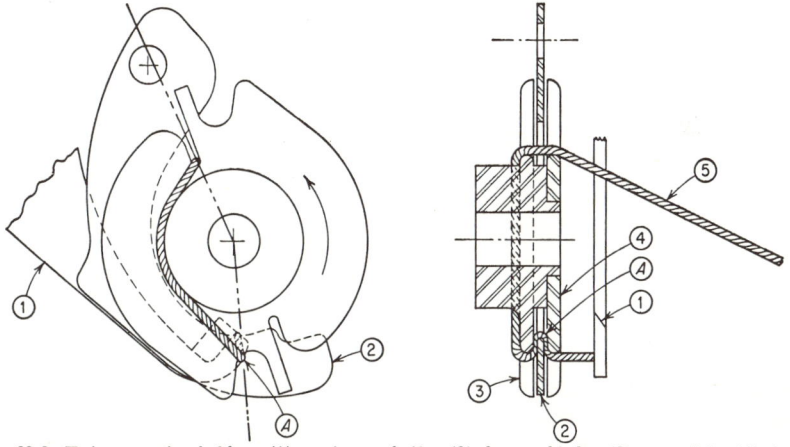

Fig. 22-6. Twine or wire holder: (1) stationary knife; (2) keeper blade; (3) rear disk; (4) front disk; (5) twine end to knotter bill hook.

Needles may be classified as closed-end (tubular), as shown in Fig. 22-5, through which twine or wire feeds out from the supply; or they may be open-end, picking up the strand as they start their stroke. Open-end needles are more effective in rethreading the knotter in some cases of strand breakage, but closed-end needles do not have to "pick up" the strand since it is confined at all times. Needle action is usually fast, and the needles must be strong, particularly for wire. A careful study of the actuating cam or crank is required to secure as gradual acceleration and deceleration as possible within the timing limits.

The hardened-steel *knife* may be moving, or stationary with the strand forced against it. Wire must be sheared and requires a better grade of knife steel than twine.

Twine ends are *tied* by a knotter bill hook, using the principle invented by Appleby in 1869 for the grain binder. They have also been fastened by steel clips, etc., but tying has proved most practical. The action of the knotter bills in tying a double overhand knot is shown in Fig. 22-7. Causes of malfunctioning of twine knotters can often be deduced from the appearance and location of the faulty knot and are usually well covered in the instruction manual furnished with the baler.

Wire has been fastened by twisting, knotting, or welding. Since wire is not flexible, it does not lend itself readily to knotting, and welding near hay is a fire hazard. Twisting is generally accepted and has been accomplished either by hook or by twister gears. When the wire is fastened by the twister hook, shown in Fig. 22-8b, the system is similar to that shown in Fig. 22-5 for twine.

A *twister gear* twists two strands of wire together on both sides of the gear, as shown in Fig. 22-8a, and in some cases cuts the wire between the twists. This leaves both pairs of ends fastened, one pair holding a completed bale and the other joining wires to receive charges for a new bale. Thus each wire around a bale has two ties and is formed from two coils, as shown in Fig. 22-9. Systems which do not cut the wire between twists handle the wire much like twine, and each wire around a bale has only one tie.

Simple twists tend to unwind under tension and may develop only about one-third the strength of the wire. Some machines kink the twisted ends so they will lock the twist as they start to unwind and thus develop additional strength.

After the twine or wire has been fastened, it must be *released* from the tying mechanism. The twine knot is pulled or stripped from the knotter bills either by bale motion or by a stripper arm, and in this process the ends are released last, thus pull-

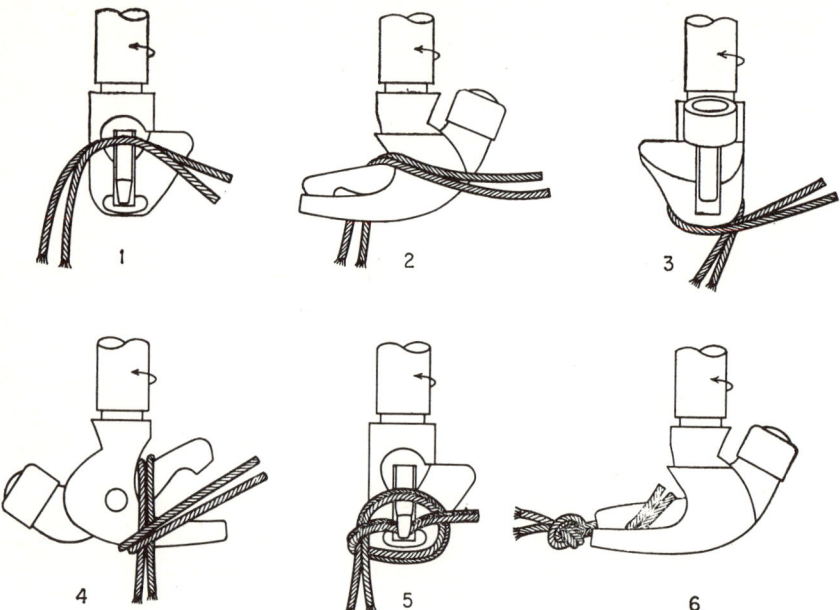

Fig. 22-7. Action of knotter bill hook in tying a knot.

ing them through to complete the knot, as shown in Fig. 22-7. Wires fastened by twister gears may be released by being sheared between the twists or by ejecting means.

Timing of the tying parts must be such that the binding strand is positively positioned in the proper place at the proper time, so that it can be passed from one unit of the binding mechanism to the other without being released. This is particularly true of twine because it is more flexible than wire and will not support itself.

All parts of the binding mechanism which contact the strand must be smooth and polished, otherwise fracture of twine fibers or scoring of wire will weaken the binding strand.

In forming a twine knot by knotter hook or a wire twist by a twister, *additional binding material must be released* to the fastening mechanism. Otherwise breakage will occur because the material is stretched beyond its elastic limit. Some machines allow the cut end to pull through the holder and the other end to pull through the needle. Another method is to move the knotter or twister hook toward the bale during the tying cycle and thereby get sufficient binding material from the original

BALERS

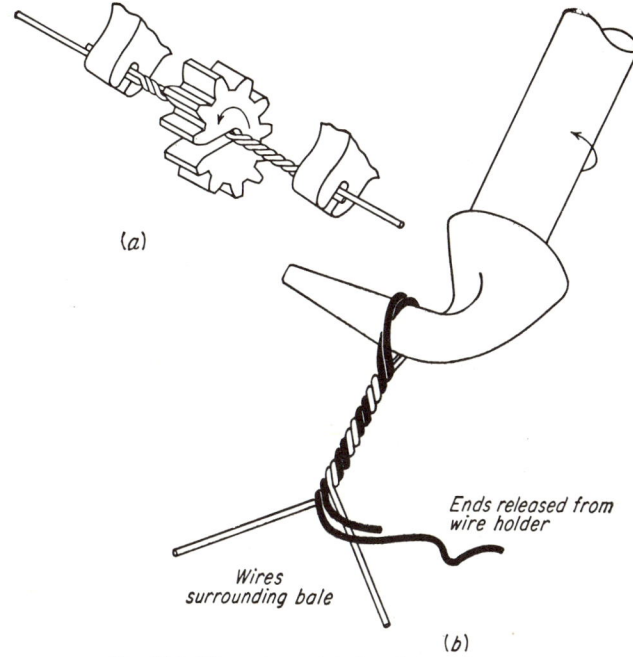

FIG. 22-8. Wire twisters: (a) slotted gear; (b) hook.

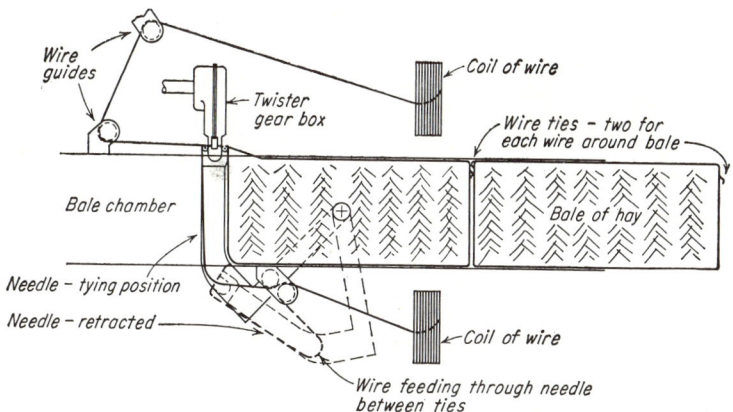

FIG. 22-9. Diagram of binding elements in a wire tie baler using twister gears. (*Oliver Corp.*)

span between the hook and the bale. Excessive slack must be prevented because the binding material will flex out of position and be missed by the parts which should pick it up.

Protective Devices. The feeding fork (packer fingers or wadboard) should be released if it encounters a slug of hay too large to force into the chamber. Protection is furnished by a shear pin in the drive or by a relief spring in the linkage.

A *ram stop* may be used to keep the ram from pushing a fresh charge of hay against the needles if, for any reason, they fail to withdraw. The stop is linked with the needles so that it is always in a blocking position when the needles are in the chamber.

224   CROP-PRODUCTION EQUIPMENT

A *needle-drive shear pin* or release device is usually used to protect the needles in case an obstruction is encountered as they enter the bale chamber.

The *hay pickup* is commonly protected by a slip clutch, and the *knotter head* by a shear pin in the drive line.

A heavy flywheel is essential in the drive line to the ram to maintain speed with not more than 20 per cent slowdown during the great increase in force requirement throughout the compression cycle. A friction slip clutch or shear pin of consistent and accurate breakaway torque must be placed between the flywheel and the crankshaft in order to protect the ram and its drive from overloads or ram stop action. In the case of PTO-driven balers, a friction slip clutch is also desirable between the

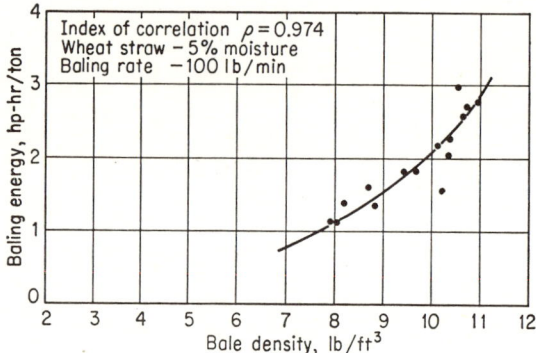

FIG. 22-10. Effect of bale density upon baling energy requirements. (*Burrough and Graham, Agr. Eng., April,* 1954.)

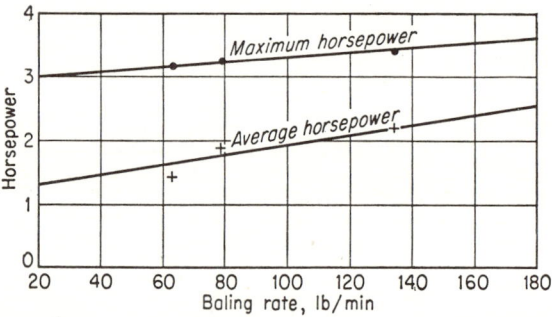

FIG. 22-11. Power requirements for auxiliary drives in wheat straw. (*Burrough and Graham, Agr. Eng., April,* 1954.)

PTO and the flywheel for protection in case of sudden tractor clutch engagement or in case of velocity fluctuations due to universal joint angle when making sharp turns. Some balers use an overrunning clutch to avoid torque reversals in the power line. It is recommended that "when the frequency of the instantaneous load does not exceed 10 cycles per hr, implements imposing loads greater than 7,500 lb-in for the 1⅜ diameter shaft or 12,000 lb-in for the 1¾ diameter shaft should have a power-line protective device which does not exceed a maximum instantaneous slip value of 13,000 lb-in for the 1⅜ diameter shaft or 26,000 lb-in for the 1¾ diameter shaft."[1]

Power Requirements. Baler power requirements vary widely according to the type of crop, density of baling, and rate of baling. Strain-gage tests indicate that total

[1] ASAE Recommendation: Operating Requirements for Power Take-off Drives, *Agr. Eng. Yearbook*, 1959, pp. 63–64.

power requirements at the compression peak range from 56 to 93 hp, with averages throughout the cycle from 12 to 20 hp.[1] Engines rated from 15 to 25 hp are used on engine-driven models.

Bale density was found to be the most important factor in baling energy requirements, and the relationship for a particular baler is shown in Fig. 22-10.[2] Power requirements for auxiliary drives were also determined and are shown in Fig. 22-11.

Round Bales. One baler forms cylindrical bales of 14 to 22 in. in diameter and 36 in. long by rolling up the windrow like a roll of cloth. Belts are arranged as shown in Fig. 22-12 to roll and compress the hay. When the desired diameter is reached, the feed conveyer stops and binder twine is fed into the bale chamber as the bale con-

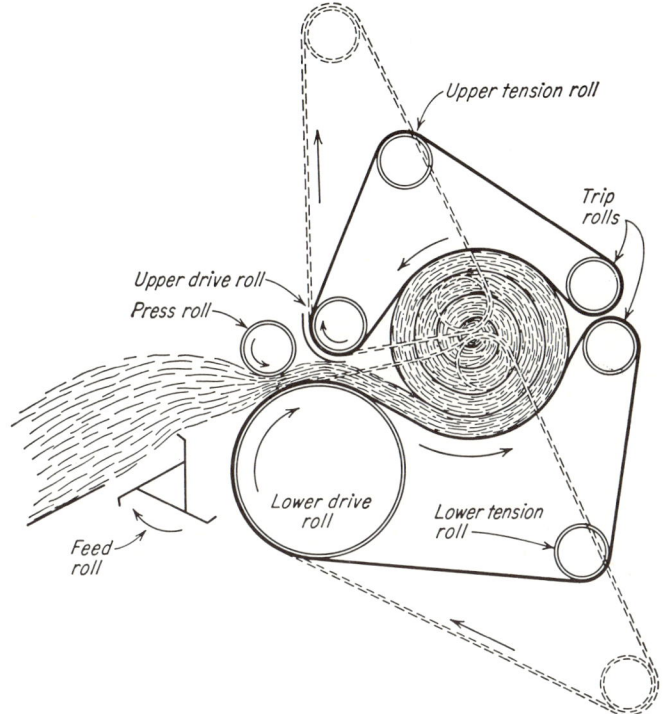

Fig. 22-12. Diagram of forming belts for round bales. (*Allis-Chalmers Mfg. Co.*)

tinues to revolve, resulting in several turns of twine being spirally wrapped around the cylinder of hay. The bale is then discharged, and the feed conveyer is automatically reengaged. Large double windrows are desirable to feed a layer of hay as wide as the bale chamber. The bales do not tend to expand like rectangular bales, and the wrapped twine normally keeps them from unrolling in spite of the loose twine ends.

Hay Wafers. Experiments at the University of Wisconsin indicate that hay of 30 per cent or less moisture can be compressed into stable cylindrical wafers or pellets, 2 in. in diameter by 1 in. thick, with a density of 30 to 40 lb per cu ft (about four times that of baled hay), by a pressure of around 4,000 psi. These wafers were easily

---

[1] Merlin Hansen, Loads Imposed on Power Take-off Shafts by Farm Implements, *Agr. Eng.*, 33: 67–70, January, 1952.
[2] Burrough and Graham, *op. cit.*

eaten by cattle and would appear to offer great handling and storage advantages if they can be made by an economical field machine.[1]

Bale Handling. Balers package the hay, but it must still be collected, loaded, transported, and stored. Unless lifting onto the wagon or truck and up to the mow or stack is mechanized, a great deal of arduous hand labor remains. Where bales are to be hauled short distances, wagons are trailed behind the baler and a chute at the end of the bale chamber places the bales within reach of one or more men who stack the bales on the wagon.

Mechanical bale loaders are often used to load trucks. They are usually ground-driven and attach to the side of the truck, picking up the bales and lifting them to be stacked by hand on the truck as it is driven through the field. Bale loaders are of the conveyer, wheel, or fork types.

Bales may be stacked on a sled pulled behind the baler and slid off in piles, convenient for truck loading. Bales may be lifted up to mow or stack by elevator or by a grapple fork which can lift several at once. Stacking in storage is arduous labor, and some farmers allow bales to drop at random since many barns cannot support a mow full of stacked bales.

Some balers make short bales and have an attachment to throw them back into a wagon with high sides. No loading labor is needed, and if the wagon is power-unloading, a minimum of labor is needed to unload the bales into an elevator and conveyer, which bulk-piles them in the barn.[2]

[1] H. D. Bruhn, Engineering Problems in Pelletized Feeds, *Agr. Eng.*, 38:522–525, July, 1957. H. D. Bruhn, A. Zimmerman, and R. P. Niedermeier, Developments in Pelleting Forage Crops, *Agr. Eng.*, 40:204–207, April, 1959. J. L. Butler and H. F. McColly, Factors Affecting the Pelleting of Hay, *Agr. Eng.*, 40:442–446, August, 1959. John B. Dobie, An Engineering Appraisal of Pelleting, *Agr. Eng.*, 40:76–93, February, 1959. John B. Dobie, Progress Report on Hay Wafering, *ASAE Paper* 59-617, December, 1959. J. R. McCalmont and D. T. Black, Research Problems and Progress in Handling Field Wafered Forage, *ASAE Paper* 59-814, December, 1959.

[2] M. W. Forth and C. S. Morrison, One-man Baled Hay Harvesting, *ASAE Paper* 57-521, December, 1958.

Chapter 23

FORAGE HARVESTERS AND BLOWERS

FORAGE HARVESTERS

The forage harvester is a comparatively recent machine developed to chop standing or windrowed forage crops and deliver them into a wagon or truck. Row crops such as corn or sorgo are stored as silage, while hay crops may be chopped green or after partial or complete field drying.

The *functional elements* of a forage harvester are as follows:

1. A hay pickup, hay cutter bar, or row-crop cutting knife, as the case may be, with associated parts to collect and move the material back to the feed rolls

2. Feed rolls which force the material into the cutting knives, compressing and holding it firmly while it is chopped

3. Cutting blades *cylindrically* arranged or *radially* arranged like spokes on a wheel, coacting with a stationary shear bar to chop the material into short lengths

4. An impeller blower, separate or combined with the knives, to deliver the material into a wagon or truck

The *feeding problem* differs from that with balers in several important respects: (1) green or wilted hay has more friction and weight than dry hay; (2) the windrow must usually be narrowed to enter a 14- to 20-in.-wide throat; and (3) with short cuts the windrow is usually fed much more slowly than ground speed, and it must be thickened accordingly.

The windrow may be narrowed to throat width by tapered sides in combination with the action of a web conveyer and floating hold-down wheel as shown in Fig. 23-1. It may also be centered and fed to a web conveyer by a cross-auger as shown in Fig. 23-2. Rake bars, spring-tooth conveyers, and other devices are also used in feeding.

Cutter-bar Attachments. These are increasing in popularity for hay to be dehydrated, for grass silage, and for green-feeding, since the mowing and raking operations are eliminated. They also reduce the possibility of picking up stones, broken rake teeth, and other tramp metal, which frequently causes extensive breakage.

Mower guards and knife sections are used on the *cutter bar*. The knife drive must be compact, but ground speeds and the corresponding knife speeds and loads are lower than for mowers. Short pitman drives, bell cranks, and wobble bearing drives are all used.

Reels, usually of the feathering type (see Fig. 20-3 for mechanisms), are necessary to sweep the mown material back on the feed table. Gathering the material into the center is even more of a problem than with windrow pickups. Some machines with feathering reels have a table contoured to the sweep of the reel so that the hay is delivered high enough at the outer edges for gravity flow into the web conveyer.

A T-type header, similar to that shown in Fig. 24-1 for large combines, is shown in Fig. 23-3. The reel delivers the mown material to a cross-auger which gathers it onto a web conveyer at the center.

Row-crop Attachments. These must cut the stalks close to the ground and carry them back, butt first, into position to enter the feed rolls, preferably spread uniformly

228 CROP-PRODUCTION EQUIPMENT

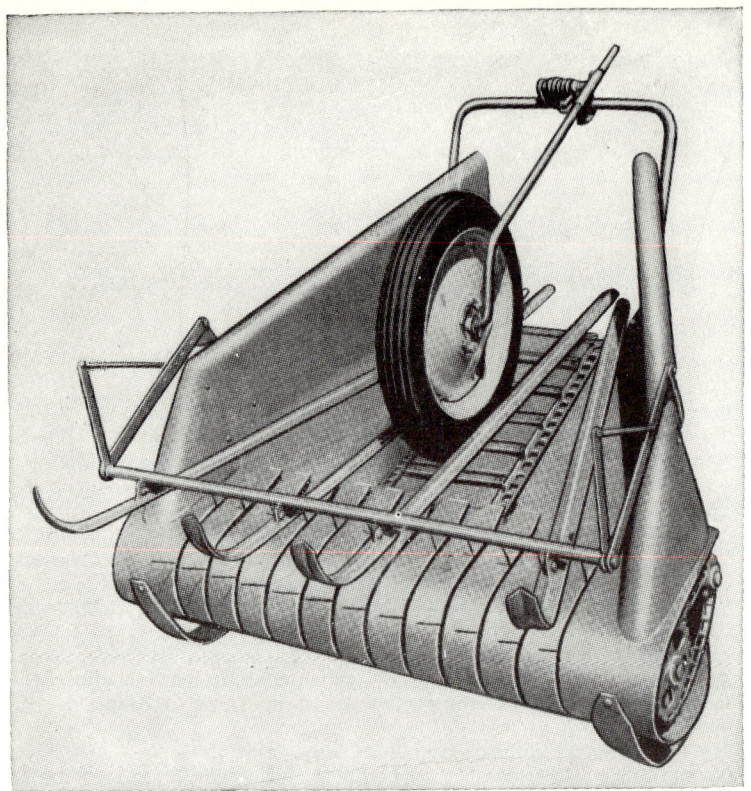

Fig. 23-1. Hay pickup with floating press wheel. (*Fox River Tractor Co., Appleton, Wisc.*)

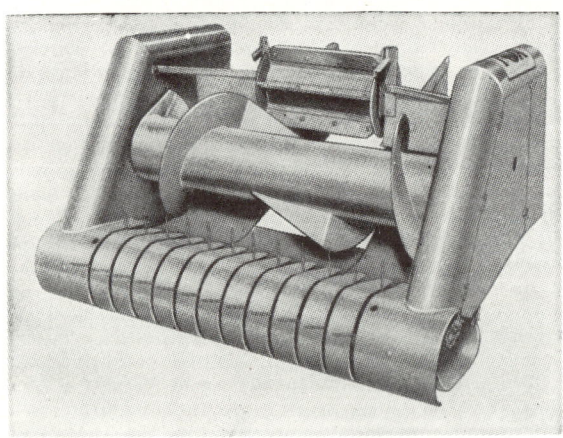

Fig. 23-2. Hay pickup with cross-auger. (*Fox River Tractor Co., Appleton, Wisc.*)

across the feed opening. Tall thick corn grown especially for silage may be harvested at rates of 25 tons per hr or more, requiring considerable strength and capacity in the gathering unit.

The *cutting mechanism* is similar to that on a corn binder and consists of a single large knife section reciprocating across the V formed by two stationary knives.

The gathering chain fingers must be properly stripped at the top to avoid clogging. They are often allowed to carry stalks around the corner at the top in order to spread them over the full width of the feed rolls.

Stripping of chain fingers is accomplished by the following methods on corn pickers and binders as well as forage harvesters:

1. Uniform gradual withdrawal through a straight slot before the fingers start to swing around the upper sprocket
2. Uniform gradual withdrawal through a slot in a cylindrical shield while the finger is passing around a sprocket

Fig. 23-3. Cutter-bar attachment. (*Fox River Tractor Co., Appleton, Wisc.*)

3. Uniform gradual deflection of the stalk away from the fingers by ledges, either straight or curving around a sprocket
4. Deflection of the stalks away from the fingers by a disk turning with the sprocket

In the first three cases the included angle between finger and stripping surface should not be less than 75° for trouble-free operation.

Feed Rolls. A typical feed-roll arrangement is shown in Fig. 23-4. Peripheral speed should be the same for all rolls and is usually varied by changing sprockets or gears to control the theoretical length of cut. Actual length of cut is the same as the theoretical only when the stems feed in straight, and it may average twice the theoretical in hay. Although the power required by feed rolls is usually 3 hp or less, the overloads encountered, particularly when plugging occurs, make rugged construction necessary.

A reversing gear is usually supplied to clear the feeder when it plugs, and it must allow engaging without stopping the machine. Because of this and other plugging loads it is desirable that the roll drive be protected by an overload clutch or shear bolt. Where a spring-tooth pickup reel is used in combination with a reversing gear, a one-way clutch is necessary in the pickup drive to avoid turning it backward and bending teeth.

Cutting. *Flywheel-type* machines use from one to six knives, and the relationship of flywheel speed, number of knives, and feeder speed can usually be varied to give at least a range of $\frac{1}{2}$ to 3 in. theoretical length of cut. Capacity, assuming that adequate power is available, is largely governed by throat and shear-bar width, which in turn depend on knife length, a function of wheel diameter. Few machines use a throat width of less than 14 in. or a flywheel impeller tip diameter under 42 in. Maximum throat height ranges from 5 to 8 in., depending on the feed-roll lift provided.

Major considerations in the design of flywheel-type cutterheads are:

1. Minimum adjustment of knives after sharpening
2. Freedom from knife interference with the incoming material during the cut, as may occur at the inner end of the knife with long cuts
3. Free flow of the cut material out to the paddles without long-cut material catching on knife supports
4. Minimum damage if a rock or piece of steel is encountered
5. Minimum deflection from cutting loads to avoid excessive clearance or pulling into the shear bar

Stationary ensilage cutters were used primarily to cut corn, and a flat-mounted knife was evolved for the purpose. This type of mounting is used in several present-

FIG. 23-4. Feed roll arrangement. (*Gehl Bros. Mfg. Co., West Bend, Wisc.*)

day forage harvesters. The mounting spools are simple, and since the unground side of the knife works against the shear bar, minimum adjustment is required.

Studies at the University of Wisconsin[1] indicated, however, that this type of knife mounting resulted in hay interference with the inner end of the knives at long cuts. A new mounting was developed which reduced interference by carrying the knife at an angle of 30° with the plane of rotation. Free flow of cut material was facilitated by moving the flywheel plate approximately 6 in. back of the shear bar and carrying the knife by a cup-shaped bracket which had no projections or attaching points to interfere with outward radial movement of the cut material. A cutterhead utilizing these principles is shown in Fig. 23-5a. In the course of the Wisconsin studies it was observed that the angled knife did not pass through the cut material but accelerated it up to the speed of the knife. A study of the scour marks on the flywheel plate indicated that the material tended to be thrown approximately in the direction it

[1] Orrin I. Berge, Design and Performance Characteristics of the Flywheel-type Forage Harvester Cutterhead, *Agr. Eng.*, 32:85–91, February, 1951.

was cut until it encountered the outer housing, where it was picked up by the fan blades.

Shear bars are made of hardened steel and have square corners. Usually they can be turned to use all four corners before being sharpened or discarded.

Knives may be made of mild steel with hardened high-carbon inserts at the cutting edge, of hardened carburized steel, or of an alloy steel such as SAE 4150 with the cutting edge hardened to R_c50-54.

(a)

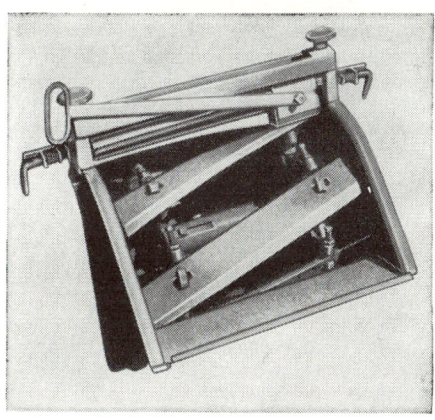

(b)

Fig. 23-5. Cutterheads: (a) flywheel (*Deere & Co.*); (b) cylinder with sharpener (*Fox River Tractor Co., Appleton, Wisc.*)

An *impeller*, or paddle blade, is usually mounted behind each knife, although three paddles are often used with six knives. These paddles must be shaped and located to avoid interference with the incoming material. Some machines use narrow blades which have considerable slip. Flywheel inertia can be most efficiently secured by using heavy paddle blades. For a discussion of the delivery of material from impeller-blower blades, see the latter portion of this chapter dealing with blowers.

Cylinder-type cutterheads are composed of several spiral knives bolted to spiders to form a cutting cylinder, as shown in Fig. 23-5b. Four knives are commonly used for short lengths of cut, with two removed for long cuts. Cylinder lengths and throat widths are usually wider than with flywheel cut machines, and some machines use cylinders approximately 3 ft long. Cylinder diameters are usually about 14 in.

Cylinder-type machines are usually considered to have the following advantages over flywheel machines:

1. Less cutterhead inertia, allowing some measure of protection from obstacles by means of shear pins or slip clutches in the cylinder drive

2. Knife sharpening in place by passing a built-in sharpening stone along the length of the cylinder while it is rotating

3. Less kinetic energy imparted to cut material because of lower cutting speeds

Disadvantages of conventional cylinder-cut machines are:

1. More expensive knives

2. Some tendency to carry long-cut material around within the knives, resulting in a different discharge path than with short-cut material

3. Necessity for a separate impeller blower to elevate and throw the material into a wagon

Some cylinder-cut machines use knives which also act as impellers to throw the material back into the wagon. These knives keep the material from going into the center of the cylinder, thus avoiding a discharge problem, and they also make a separate blower unnecessary, reducing power requirements.

Capacity. The theoretical capacity of a forage harvester can be calculated on the basis of the volume of material fed per minute and the density of the material. The following formula[1] can be used to determine the theoretical maximum capacity:

$$C = 1.738 \times 10^{-5} DHWLNR$$

where C = theoretical maximum capacity, tons/hr
D = density, lb/cu ft
H = maximum throat height, in.
W = throat width, in.
L = theoretical length of cut, in.
N = number of knives
R = cutterhead speed, rpm

In forage-harvester runs made to determine actual maximum capacity it was not found possible to maintain the maximum thickness of feed. Effective densities, based on the above formula, were calculated to be 3.22 lb per cu ft with dry hay and about 24 lb per cu ft in silage corn.[2]

The capacity of current forage harvesters is more likely to be limited by the power available than by ability to take in material.

Power Requirements. The power required for picking up, conveying, and compressing hay in the feed rolls is usually less than 5 hp.[3] Direct cutting will require more power, as will heavy row crops. In any case clogging loads are high, and a slip clutch should be used for protection against breakage.

The number of horsepower-hours per ton required for cutting and delivering material into the wagon is influenced by the following factors:

1. Shearing energy, depending on length of cut and shearing resistance of material
2. Kinetic energy imparted to material by knives and paddles
3. Frictional resistance of the material as it passes through the housing
4. Air and bearing friction

There is little information available concerning *shearing energy*. It cannot be directly measured in a chopper because it is impossible to shear at normal speeds without also imparting kinetic energy to the material. Shearing resistance varies with the type of material (corn, grass, or legumes) and may be proportional to the dry fiber content for a particular material. It is logical to assume that shearing energy per ton is inversely proportional to theoretical length of cut. If power comparisons

---

[1] F. W. Duffee, Efficiently Filling the Silo, *Agr. Eng.*, 6:4–12, January, 1925. Berge, *op. cit.*
[2] G. P. Barrington, O. I. Berge, and F. W. Duffee, "Cutting Corn for Silage," unpublished report, University of Wisconsin, Agricultural Engineering Department, Project 406, 1953.
[3] F. Z. Blevins and H. J. Hansen, Analysis of Forage Harvester Design, *Agr. Eng.* 37:21–29, January, 1956.

are available for the same material cut at two different lengths but at the same knife speed, tons per hour and thickness of feed, it is possible to estimate shearing energy from the difference in energy per ton by the use of the following formula:

$$\text{Shearing, hp-hr/ton, at one cut per inch} = \frac{W_1 - W_2}{1/c_1 - 1/c_2}$$

where W_1 = hp-hr/ton to cut in lengths c_1
W_2 = hp-hr/ton to cut in lengths c_2
c = theoretical length of cut, in.

Results reported by Duffee for cutting silage corn at two different lengths with ensilage cutters give from 0.12 to 0.26 hp-hr per ton at one cut per inch.[1] There is an increase with peripheral speed, indicating that length of cut may influence power consumption through friction as well as shearing energy. An analysis of various tests in green alfalfa with 70 to 75 per cent moisture at theoretical lengths of cut of ½ and 1 in. gave 0.20 to 0.30 hp-hr per ton at one cut per inch for shearing alone.[2]

Kinetic energy can be calculated from the peripheral speed of the paddle tips since the material is discharged approximately at this speed. This energy is proportional to the square of the speed and can be calculated as follows:

$$\text{Kinetic energy, hp-hr/ton} = \frac{(\text{peripheral velocity, fps})^2}{63,680}$$

About 6,000 fpm is the practical minimum for satisfactory delivery of all materials to the rear of the wagon.

Air horsepower depends on the efficiency of the fan and upon the volume and velocity of the air moved. Paddle area, number, peripheral speed, and air-intake opening are all factors. It has been found that the material is thrown from the paddles faster than the velocity of the air discharge and that the air helps convey the material only after it has lost considerable velocity.[3] The air may help delivery into high silos but is not needed in field forage harvesters. The action is that of an impeller rather than of a fan, with the air discharge an incidental effect. The no-load power for a typical 42-in.-OD flywheel-type cutterhead with four knives and paddles is shown in Fig. 23-6. The power required for the air moved during cutting may possibly be less if the feeding of material blocks part of the air-intake opening.

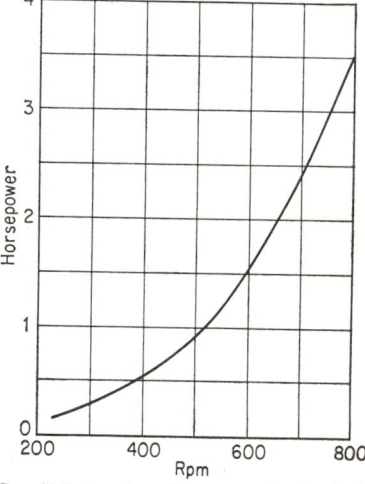

FIG. 23-6. Speed vs. horsepower for flywheel-type cutterhead with four knives and four fan blades.

Friction losses result from the sliding of the cut material over knife, flywheel, and paddle surfaces and, principally, on the inner periphery of the housing. Theoretical friction losses can be calculated by the following formula:[4]

$$\text{Friction, hp} = \frac{\mu V^2 \theta R}{1,016,000}$$

where μ = coefficient of friction
V = peripheral velocity, fps
θ = arc of contact, deg
R = rate of feed, lb/sec

[1] Duffee, *op. cit.*
[2] C. B. Richey, Discussion of Energy Requirements for Cutting Forage, *Agr. Eng.*, 39:636, 637, October, 1958.
[3] G. Segler, Calculation and Design of Cutterhead and Silo Blower, *Agr. Eng.*, 32:661–663, December, 1951.
[4] Barrington, Berge, and Duffee, *op. cit.*

The effect of moisture content on coefficient of friction of alfalfa on clean stainless steel is shown in Fig. 23-7.[1]

The power required by a flywheel forage harvester cutting green alfalfa at various rates is shown in Fig. 23-8.[2]

The available information concerning power requirements of conventional *cylinder-cut* machines indicates little difference from flywheel-cut machines. Lower peripheral cutting speeds reduce the kinetic energy imparted in cutting, but additional kinetic energy is imparted by a separate impeller blower which also has friction losses. A

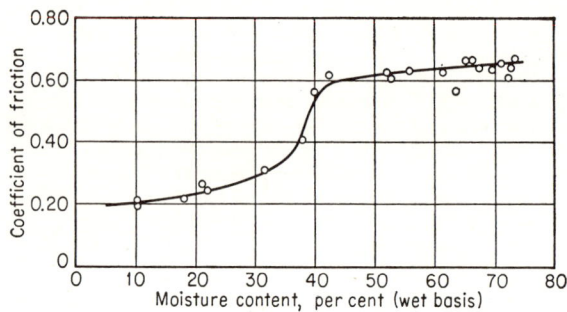

FIG. 23-7. Coefficient of friction vs. moisture content for alfalfa on clean stainless steel. (*Richter, Agr. Eng. Yearbook*, 1959, pp. 125–126.)

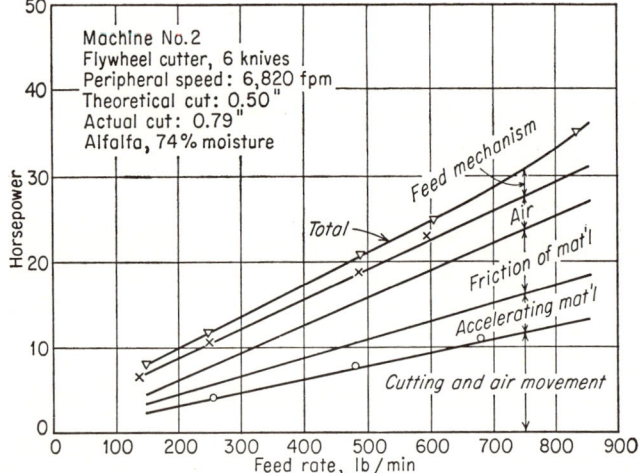

FIG. 23-8. Power required by flywheel forage harvester to cut green alfalfa. (*Blevins and Hansen, Agr. Eng., January*, 1956.)

cylinder-cut machine with blades acting as impellers may have an advantage in power requirement because rubbing friction and windage should be reduced to a minimum. Peripheral speed must, of course, be sufficient to assure delivery into the wagon.

The flail-type forage harvester, shown in Fig. 23-9, has evolved from the hammer-mill-type stalk shredder (Chap. 29). Swinging hook-shaped knives with a chisel-like leading edge rotate upward in front with a peripheral speed of approximately 8,500 fpm. They cut off the standing vegetation in 2- to 10-in. lengths by impact, carry it

---

[1] ASAE Data: Friction Coefficients of Chopped Forages, *Agr. Eng. Yearbook*, 1959, pp. 125, 126.
[2] Blevins and Hansen, *op. cit.*

up into the hood, and throw it back through the delivery spout into the wagon. A sharpened edge at the entrance to the hood aids shearing action. Standing crops can be cut off cleanly and reasonably close on level ground. Windrowed hay can also be picked up and cut since the hammers have a pronounced suction and lifting action. Other machines of this general type collect the cut material in a lateral auger trough inside the hood at the rear, and an auger feeds the material to a blower impeller. This type of machine took more power than conventional forage harvesters, but the direct-throw type is in the same range.[1] Because of their simplicity and low cost, these machines are particularly well adapted for green-feeding.

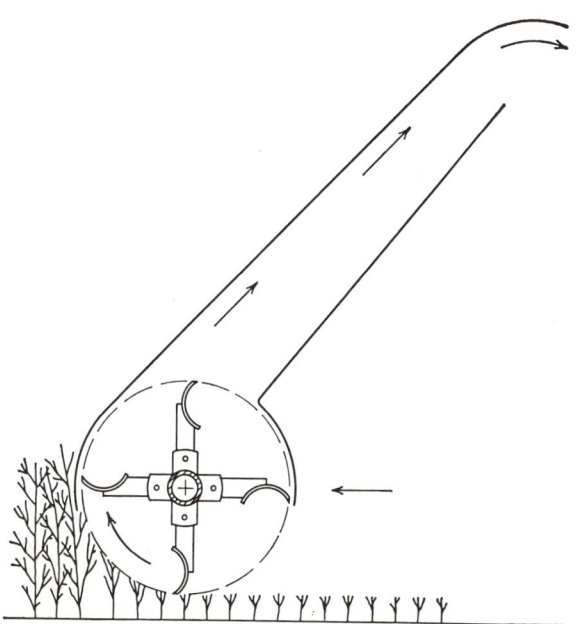

FIG. 23-9. Adaptation of hammer-mill type of stalk shredder as a direct-cut forage harvester. (*Lundell Mfg. Co., Cherokee, Iowa.*)

A horizontal rotary-knife shredder-mower has also been adapted for direct harvesting of hay crops by attaching impeller blades to the swinging knives and providing an S-shaped delivery pipe.

BLOWERS

Chopped forage is commonly elevated into the silo or barn by an impeller blower which throws the material up a delivery pipe and also supplies enough air to help carry it after it loses its initial velocity. The material is dropped from the back of a wagon or truck into a feed trough with slat or auger conveyer which delivers it to the impeller blower.

Recommended *peripheral velocities* range from 7,000 to 10,000 fpm. Kinetic energy amounts to 0.35 hp-hr per ton for 9,000 fpm. Tests at Purdue University indicated power requirements of 1.1 to 2.0 hp-hr per ton for green alfalfa-timothy, ½-in. cut.[2] Wilted hay may require up to double this power. Since only 0.04 hp-hr is theoretically required to elevate 1 ton 40 ft, it is apparent that blowers are quite inefficient. Much energy is lost by friction of the material on the paddle blades and on the inside

---

[1] C. W. Bockhop and Kenneth K. Barnes, Power Distribution and Requirement of a Flail-type Forage Harvester, *Agr. Eng.*, 36:453–457, July, 1955.
[2] J. P. Raney and J. B. Liljedahl, Impeller Blade Shape Affects Forage Blower Performance, *Agr. Eng.*, 38:722–725, October, 1957.

of the housing. In many cases, at least half the material makes at least one unnecessary revolution in the housing.[1]

The *unloading action of impeller blades* has been analyzed by Raney,[2] and he has developed the following formula for the maximum wing length which can be unloaded while passing a given angular discharge opening (Fig. 23-10):

$$L = R\left(1 - \frac{1}{\cosh \theta}\right)$$

where L = wing length, in.
R = radius to wing tip, in.
θ = outlet angle, radians

Speed of rotation is not a factor and does not appear in the equation. This formula is of value in determining the angular rotation required for a blade to unload a given amount of material and the corresponding size of discharge opening required. The

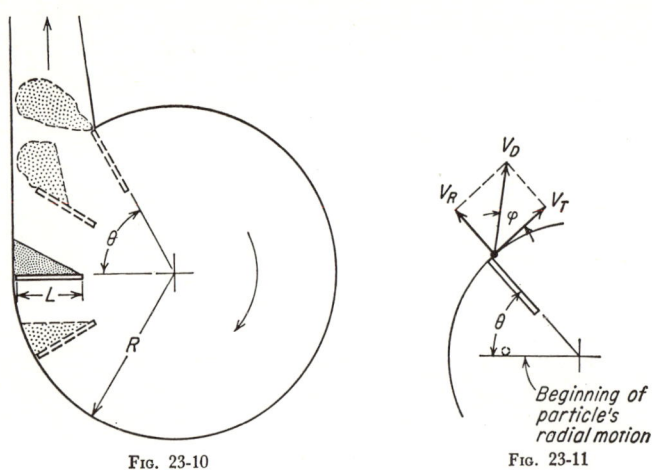

Fig. 23-10.
Fig. 23-11.

Fig. 23-10. Unloading action of impeller blower blades.
Fig. 23-11. Discharge direction and velocity of impeller blower blades.

quantity of material per blade can be calculated, but the length of blade occupied is variable because the angle of repose is influenced by centrifugal force and housing friction resulting from the material bearing against the inner periphery of the housing and being pushed by the paddle.

Raney also gives the following formulas for direction and velocity of discharge of a particle (Fig. 23-11):

$$\tan \varphi = \frac{\sinh \theta}{\cosh \theta}$$

where φ = angle of discharge from tangential
θ = angle of rotation past beginning of discharge opening (start of particle's radial motion), radians

and

$$V_D = \frac{V_T}{\cos \varphi}$$

where V_D = discharge velocity
V_T = tangential velocity

[1] William J. Chancellor, Influence of Particle Movement on Energy Losses in an Impeller Blower, *Agr. Eng.*, 41:92–94, February, 1960. Relations Between Air and Solid Particles Moving Upward in a Vertical Pipe, *Agr. Eng.* 41:168–176, March, 1960.
[2] Russel R. Raney, "The Free Throw Theory of Blower Discharge," unpublished paper, International Harvester Company, Chicago, Ill., 1946.

The relationship between peripheral speed and both theoretical and practical height of conveying corn silage with a 10-in.-diam blower pipe is shown in Fig. 23-12.[1]

The *feed opening* of a blower ideally should be located so that the material slides to the end of the blades just as the discharge opening is reached in order to achieve minimum housing friction. No material should, of course, be carried past the discharge opening. In practice, the feed opening is approximately half the diameter of the impeller, may be adjustable in size, and is centered slightly below the impeller hub. Because of gravity, most of the material is introduced near the bottom of the opening.

Some machines with conveyer *feed tables* have retarder beaters above the feed opening which turn against the material and rake back the upper portion of slugs in order to prevent clogging. Auger feed tables do not use such beaters because of the metering control of the augers. Conveyers are commonly 16 in. wide, and augers 9 to 12 in. in diameter. Table feed rate can often be varied by different drive sprocket sizes in order to match the feed rate of a particular material to the capacity of the blower and power source. Feed tables are usually 8 ft or more in length in order to accommodate rear unloading from 7- or 8-ft-wide wagons.

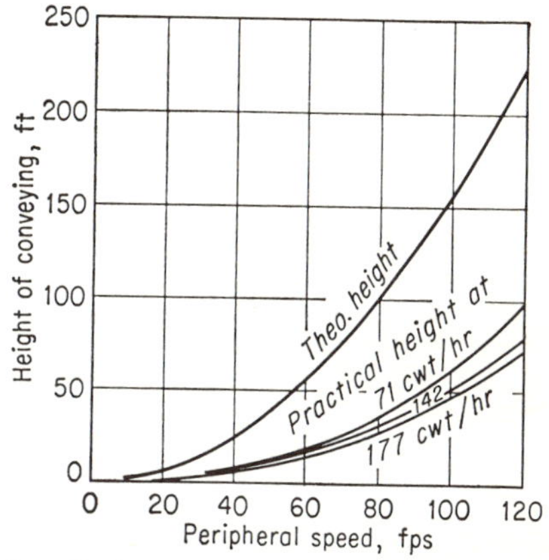

Feed tables are usually folded up to allow the wagon to pass by and then swung down into operating position. The feed-table drive must, of course, accommodate this action, and long balance springs are provided to ease manual lifting and lowering.

Some blowers have, in place of a feed table, a hopper adapted to be fed from power-unloading wagon boxes with side delivery at the front or rear for unloading into feed bunks.

Fig. 23-12. Peripheral speed vs. theoretical and practical height of conveying. (*Segler, Agr. Eng., December,* 1951.)

Blowers are usually *driven* by belt but are also being driven directly from the tractor PTO. This type of drive is quicker to connect than a belt and does not require staking to resist belt pull. It also eliminates the problem of belt slippage at peak loads and thus reduces clogging. With a direct PTO drive, the speed is limited to about 540 rpm with most tractors, requiring greater diameter to secure the same peripheral speed as the faster-turning belt-driven impellers.

Delivery pipes vary in diameter from 6 to 9 in. Hay at 35 to 45 per cent moisture, to be dried in the barn, is commonly regarded as being the material most liable to clog. It has been theorized that a bunch leaves the paddle partially compressed and expands as it travels up the pipe. Large delivery pipes have been found advantageous in handling this material. The deflector elbow at the top has been found to initiate clogging in many cases.[2]

Power-drive attachments for actuating *wagon unloading* endgates or canvases are available for many blowers. These drive the roller shaft at the rear of the wagon box through a quick-detachable coupling and a pair of universal joints. A ratchet drive, similar to that used on manure-spreader conveyers, allows several feed rates to be selected. Power-unloading wagons are a necessary link in the complete mechanization of forage harvesting.

---

[1] Segler, *op. cit.*
[2] William J. Chancellor and Gordon E. Laduke, Analysis of Forage Flow in a Deflector Elbow, *Agr. Eng.,* 41:234–240, April, 1960.

Chapter 24

COMBINES

THOMAS CARROLL

The function of a combine harvester is to cut, thresh, winnow, and clean grain or seed of any kind. With a pickup attachment, a combine can be used to handle grain that has been swathed.

The crop being harvested is fed from the cutting table to a threshing drum, where the grain is knocked out of the heads as it passes between a revolving cylinder and a stationary concave. If the concave is of the open type, from 20 to 80 per cent of the grain may fall directly to the sieves. The grain in the straw and chaff is separated by straw-rack agitation, and after passing over agitated sieves where cleaning is completed by fan blast, the grain is delivered to the bulk container or to the bags, according to the system of marketing in the locality. A cross section of a large self-propelled combine is shown in Fig. 24-1.

The idea of a combine, a single machine for all grain-harvesting operations, is almost as old as that of the reaper. The first combine was built in Michigan in 1836 by Moore and Hascall. It was not successful in Michigan, but was later shipped to California and used successfully there in 1854.[1] Commercial production of combines started in California in the 1880s. Combines developed independently in Australia about the same time, stemming from Ridley's stripper built in 1843.

FUNCTIONAL ELEMENTS OF COMBINES

Headers. The function of the header is to cut and gather the grain and deliver it to the threshing cylinder. The reel pushes the straw back on the platform as it is being cut by the sickle.

Small combines often use scoop-type headers with a canvas conveyer to take the grain straight back to a wide cylinder. Some gather may be achieved by flared side sheets. Large combines use T-type headers with auger tables which bring the grain to a conveyer feeding to the cylinder. Auger tables have largely replaced the canvas tables used earlier.

Reels have four to six bats and range from 40 to 60 in. in diameter. Power drive is favored over ground drive, although various sprockets are furnished to vary the speed. The following reel speeds for 42-in.-diam reels are suggested as average, but the reel must be fast enough at any ground speed to bring the crop back to the cutter bar.

Ground speed, mph	Reel speed, rpm
½–¾	14
1–2	20
2¼–3½	30
4 and over	35–40

Reels are adjustable up and down, and in or out; the operator sets the reel in the position best suited to the crop he is harvesting, and there is usually little need to make changes in the adjustment. In sections where the crop is cut close to the

[1] Chris Nyberg, Highlights in the Development of the Combine, *Agr. Eng.*, 38:526–535, July, 1957.

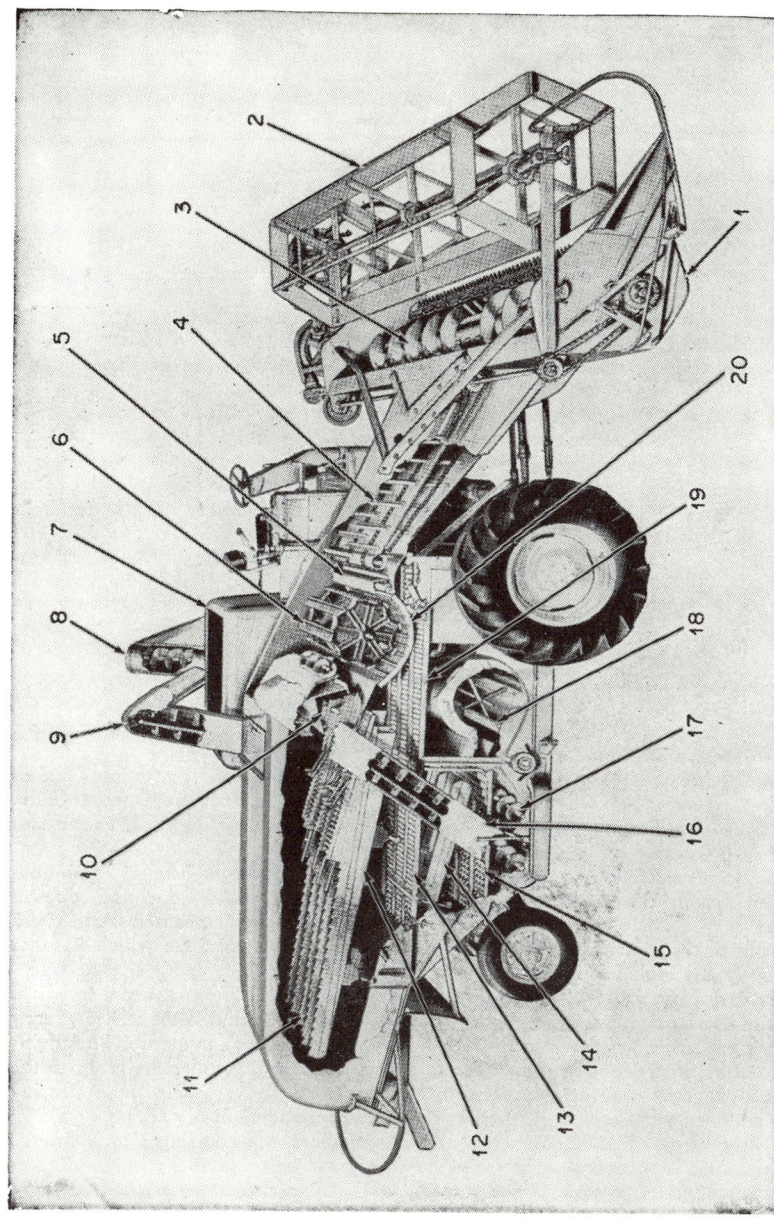

FIG. 24-1. Large self-propelled combine: (1) header; (2) reel; (3) auger; (4) undershot conveyer; (5) stripper-beater; (6) cylinder; (7) grain tank; (8) unloading auger; (9) grain elevator; (10) beater behind cylinder; (11) strawwalkers; (12) walker return pan; (13) chaffer sieve; (14) shoe return pan; (15) shoe sieve; (16) tailings return elevator; (17) grain auger; (18) fan; (19) grain pan under cylinder; (20) concave. (*Massey-Ferguson, Inc.*)

ground, it is desirable to provide, especially for the up-and-down range, reel adjustments that can be controlled from the operator's platform.

Feathering reels (see Fig. 20-3 for mechanisms) which keep the slats vertical during rotation may be used for difficult conditions. The pickup type with spring teeth on the slats was developed to pick up down and tangled crops, particularly rice, which is usually wet and heavy. Another type has flaps on the bottoms of the slats and is used to sweep short crops back from the cutter bar into the table conveyer. Feathering reels are usually not standard equipment, but the additional cost is quickly recovered by savings in down or short grain.

The *dividers* at the ends of the header are very important in down, tangled, and weedy conditions. In the United States, closed-end reels, with divider hoops to hold down the grain so that the cutter bar can cut it for separation, are usually satisfactory. When extreme down conditions and damp grain are met, such as in rice fields, a vertical knife divider seems to be the only solution. This shears through the tangled crops, giving a clean division, allowing the material to be readily handled by the machine.

The reciprocating knife (mower-type) *cutter bar* is commonly used on combine harvesters. As in mowers, the knife is made up of 3-in.-wide sections. More forward travel per stroke is permissible than with mowers because of the greater height of cut and thinner material. Travel ranges up to 6 in. with 3-in. stroke (one guard space) and 7½ in. with 6-in. stroke (two guard spaces). As with mowers, the stroke need only complete its cutting action and can be reduced from 6 to 5¼ in., for instance. Knife drive speeds are much lower than with mowers, around 430 rpm for a nominal 3-in. stroke and 300 rpm for 6-in. stroke. Mower speeds would result in very high loads on 10- to 14-ft headers.

While ball bearings have been commonly used on oscillating crank and sway-bar types of drives, rubber bushings are coming into use.

Coarse-stemmed crops, such as sunflowers and grain sorghums, require special equipment, particularly with auger tables, to cut and feed them properly, although these grains are easy to thresh and save. Inclined cutter bars, special fingers, and, for sunflowers, special finger trays have been developed.

For raising down crops off the ground to be engaged by the cutter bar, *pickup guards* are available. Flexibly mounted guards, 12 to 18 in. long and spaced 12 in. apart, help in down crops when dry, but in wet conditions they are likely to increase clogging of the cutter bar.

Canvas *table conveyers* are used with narrow-cut scoop-type headers, and both auger and canvas conveyers are used for wide-cut narrow-body machines. Canvas conveyers may have rubber-coated canvas and integral rubber-coated slats. Back feeding under the canvas is a problem of the canvas-table machine which has not been entirely eliminated in any design up to the present.

Augers vary from 16 to 20 in. in diameter and pitch, with flights about 4 in. wide. The grain is swept underneath the augers and conveyed behind them where gravity and a stripper ledge, as shown in Fig. 24-2, keep it from being carried up. The flights are often straightened out at the center to act as paddles in feeding the grain to the undershot conveyer web which delivers the grain to the cylinder. Disappearing fingers or hinged paddles may be used at the center instead of fixed paddles.

Wide headers, particularly on self-propelled machines, are controlled by hydraulic or electric *lifts*. It is desirable to spring-balance the header, allowing only sufficient weight to hold it down to working position so that it can float up without damage when uneven ground is encountered. Electric lifts usually have less power available than hydraulic, and spring-balancing over the entire range of action is required.

In many areas, grain is cut with a *swather* and left on the stubble to mature, later to be picked up and threshed by a combine. This method is used where weedy conditions are encountered or where weather or other conditions are generally unfavorable for straight combining. In many cases, it is better to have the crop on the ground in the swath rather than standing in the field. The swather's width of cut should not exceed the capacity of the combine at its minimum ground speed.

The eccentric-type rotary *pickup* with spring teeth is probably most commonly used for grain. For the handling of fine grass seeds and other specialty crops, the draper or canvas-type pickup with attached light spring fingers is used, as its design prevents grass seed falling to the ground.

The cylinder is *fed* by a canvas conveyer on scoop-type headers. As the slope increases, more difficulty is experienced in securing an even feed. If the grain hesitates and then feeds in a double thickness, clogging may result. A slope of 40° is about the practical maximum, and less slope is desirable. On wide-cut narrow-body machines, the undershot slatted chain conveyer is favored, as shown in Fig. 24-2. This feed is independent of gravity and, in combination with the auger table, gives good control and even feeding under most conditions. In some machines the bottom shaft and sprocket assembly is allowed to accommodate various crop thicknesses by floating.

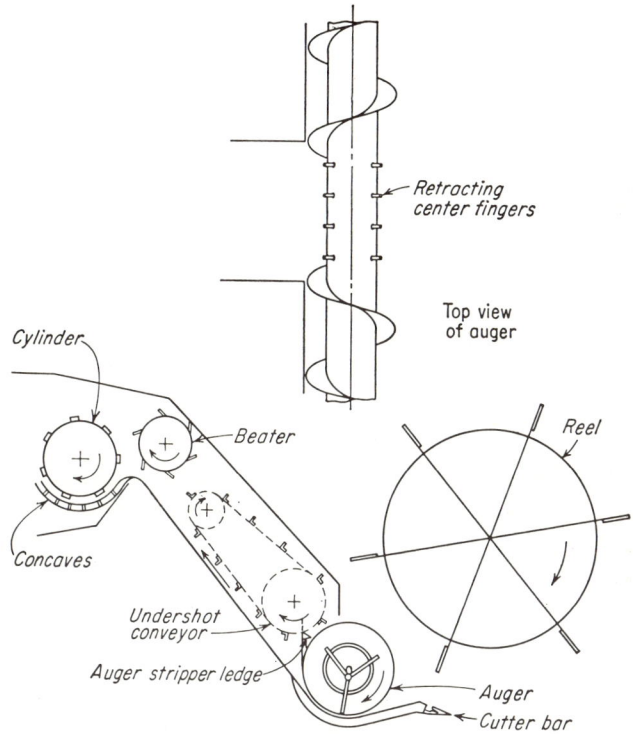

Fig. 24-2. Cross section of auger header and undershot feed to cylinder.

A *beater* in front of the cylinder and rotating in the same direction is used on most machines with undershot conveyers in order to aid in stripping the crop from the conveyer and feeding it to the cylinder. A beater or a floating upper canvas is used to aid feeding of scoop-type machines.

Threshing. The seeds are shattered out of the heads by the impact of blows from the cylinder and the concaves. It appears that the intensity and number of blows received by a head of grain as it passes between the cylinder and concaves depends on cylinder speed, the number of concaves, and the slowing effect of the concaves as determined by the clearance in relation to seed and straw size.

Cylinders may be of the rasp-bar or angle-bar type, originally developed in Europe, or the spike-tooth type which originated in the United States. The rasp-bar cylinder has come into more general use on combines in America as well as in Europe. Rasp-bar cylinders, as used in nearly all scoop-type combines, range from 15 to 17½ in. in diameter and 48 to 60 in. in width. On narrow-body, wide-cut machines, they range from 18 to 24 in. in diameter and 22 to 42 in. in width. Spike-tooth cylinders are usually found only on narrow-body machines and follow the general dimensions of width and diameter of rasp-bar cylinders used on similar installations. It is generally conceded that spike-tooth cylinders have more capacity than equal-size rasp-bar cylinders, but they are not as versatile in the variety of crops combined today. Rasp bars break up straw and weeds less and are easier to adjust and less expensive. Corrugated rasp bars are most common, but one manufacturer uses rubber-faced angle bars working against rubber-faced concaves. Spring-tooth cylinders and concaves have been developed to reduce crackage of edible beans.[1]

An experimental machine, using an impeller-type blower (see Chap. 23) for impact threshing with rotary drum separation utilizing a combination of centrifugal force, air blast, and mechanical agitation, has shown high capacity for its size in small grains, although it may not be as versatile as the conventional combine.[2] Laboratory experiments with individual heads of wheat have shown that centrifugal force can be used to extract the kernels from the head without removing the chaff.[3]

For successful results, a cylinder and concave must do a first-class job of threshing the grain from the heads the first time through, without cracking the grain and with as little chopping of the straw into chaff as is possible. High-efficiency threshing with a low volume of chaff gives fastest combining and maximum saving of grain. Provision is made in combines for the return of unthreshed heads from the sieves to the cylinder, but the objective of every operator is to keep these tailings returns as low as possible.

Cracking is a problem in certain varieties of wheat, barley, some grass seeds, and edible beans. It has been found that cracking is due to cylinder action, and not to chains and other operating parts. It is possible to encounter conditions which make it imperative to compromise and accept some loss of grain and/or a small percentage of cracked grain to obtain adequate threshing.

In threshing crimson clover, which requires very high cylinder speeds, rubber-faced or plain angle bars threshed more seed and damaged less than rasp-bar cylinders.[4] See Chap. 28 for a more detailed discussion of problems encountered in combining forage seeds.

The modern combine has a wide range of *cylinder speeds* to suit various crops and conditions. The usual range of regular peripheral speeds is from approximately 2,000 to 7,000 fpm. This range may ordinarily be extended by optional sheaves, etc. The same crops vary widely, but average recommendations for rasp-bar cylinders as to peripheral speed and clearance are listed in Table 24-1. Because of design differences between makes, the instruction book for the particular machine should always be followed.

Since cylinder speed and clearance may require *adjustment* as the crop dries through the day, these adjustments should be easy to make. Cylinder speeds are usually changed by change-speed sprockets or by adjustable V-belt sheaves. These sheaves may be adjusted by threads, shims, or more conveniently by hand crank.

Flax rolls, consisting of a rubber-covered roll and a steel roll ahead of the cylinder, squeeze open the pods and control the feed to the cylinder. Peripheral speeds of the two rolls may differ slightly to give a rubbing action as well. Flax rolls have been found advantageous in harvesting legume seeds in California.[5]

The cylinder threshes the grain against the bars or teeth of the *concave*. The open-type concave, the most popular type, is made of crossbars and, in some cases, ¼-in.-diam wires on ½-in. centers, leaving spaces of ¼ in. between wires. The wires are removable, and for beans, oats, or barley, every other wire may be removed.

---

[1] H. F. McColly, Harvesting Edible Beans in Michigan, *ASAE Trans.*, 1958, pp. 68–75.
[2] W. L. Zink and G. W. McCuen, New Principles in Combining: A Progress Report, *ASAE Paper* 58-623, December, 1958.
[3] B. J. Lamp, Jr., and W. F. Buchele, Centrifugal Threshing of Small Grains, *ASAE Paper* 59-600, December, 1959.
[4] Joseph K. Park, Harvesting Small Grass and Legume Seed, *Agr. Eng.*, 35:562–564, August, 1954.
[5] Philip R. Bunnelle, L. G. Jones, and J. R. Goss, Combine Harvesting of Small-seed Legumes, *Agr. Eng.*, 35:554–558, August, 1954.

Tests with *open concaves* of various arrangements have shown that 20 to 80 per cent of the grain, depending upon kind and condition of the crop, can be separated at this point. Maximum separation decreased in direct proportion to moisture content or to the amount of weeds in the crop. The greater the separation at the concave, the less work there is to be done by the separating mechanism, and designers try to have concaves as large as possible in relation to intake and discharge by the cylinder. Larger concaves are practical with the European rasp-bar stationary threshers with overshot intake than is possible with the front, or undershot, intake.

TABLE 24-1. AVERAGE RANGES OF CLEARANCE AND PERIPHERAL-SPEED RECOMMENDATIONS FOR RASP-BAR CYLINDERS

Crop	Peripheral speed, fpm	Concave clearance, in.
Crimson clover	5,500–6,500	$\frac{1}{16}$–$\frac{3}{16}$
Alfalfa, alsike clover, Ladino clover, red clover, and white clover	5,200–6,200	$\frac{1}{16}$–$\frac{1}{4}$
Flax (use flax rolls) and canary grass	5,000–6,000	$\frac{1}{16}$–$\frac{3}{16}$
Bermuda grass, blue grass, timothy, and sweet clover	5,000–6,000	$\frac{1}{8}$–$\frac{3}{8}$
Crested wheat grass, English rye grass, Johnson grass, orchard grass, and redtop	4,500–5,500	$\frac{1}{16}$–$\frac{1}{4}$
Barley and oats	4,000–5,500	$\frac{3}{16}$–$\frac{1}{2}$
Wheat and rye	4,000–6,000	$\frac{1}{8}$–$\frac{3}{8}$
Rice, common	3,500–5,000	$\frac{1}{8}$–$\frac{3}{8}$
Buckwheat and grain sorghum	3,000–4,500	$\frac{3}{16}$–$\frac{1}{2}$
Lespedeza	2,500–4,500	$\frac{1}{8}$–$\frac{5}{16}$
Soybeans	2,000–3,000	$\frac{1}{4}$–$\frac{5}{8}$
Edible beans	1,500–2,500	$\frac{3}{8}$–$\frac{3}{4}$
Corn	2,500–3,000	$1\frac{1}{4}$ at front, $\frac{5}{8}$ at rear

Closed-type concaves, in which no separation takes place, are used on many combines, both large and small. Some concaves, and the cylinders used with them, are equipped with rubber-faced bars. Closed concaves are particularly effective in threshing clover seeds, which are extremely difficult to rub out without cracking the seed and, with effective walker separation, are very suitable in many threshing conditions.

Both the open- and closed-type concaves are provided with adjustments at the front and rear to change the relationship of the concave to the cylinder. The space between the concave and cylinder at the front can usually be adjusted from $\frac{1}{16}$ to $1\frac{1}{4}$ in. and at the rear from $\frac{1}{16}$ to $\frac{3}{4}$ in. With closed-type concaves, the normal setting of the concave is closer to the cylinder at the front than at the rear. With open-type concaves, the concave normally is closer to the cylinder at the rear than at the front, experience having shown that this setting allows more separation to take place through the open concave. Threshing action can be modified by number of concaves as well as by cylinder speed and clearance. Some open concave machines have inserts to close the grates for tough-threshing crops.

As with cylinder-speed adjustments, correct concave clearance may vary as conditions change through the day, and adjustment should be quick and convenient. A good operator will make use of the adjustments to do clean threshing without cracking the grain and also to keep chaff to the minimum, so that separation can be made on strawwalkers and sieves without loss.

The wear on cylinder bars taken from machines should be even over the *full width;* too much wear at any point indicates uneven feeding of the crop, with consequent poor threshing results and excessive cracking. With the advent of auger-type headers on the self-propelled combine, it has been easier to employ the full width of the cylinder than with large-capacity pull-behind machines having the crop flowing from one end of the table only. It might be stated, as a general principle, that 36 in. is about the maximum cylinder width over which effective, even feeding can be made where wide cuts are gathered with an auger-type header.

In straight combining, the wide cylinders on scoop machines receive the crop in a thin, even flow across their full width and do a good job of threshing as well as delivering the crop in a manner that enables the separating mechanism to do its work effectively. However, these machines are at a disadvantage in threshing crops that have been swathed. Until some method is devised of splitting up a swath and delivering it to the full width of the cylinder, these machines will never be so successful as narrow-body machines in swathed crops.

Separation of Grain from Straw. The threshed material is shaken and tossed back by the straw rack so that the grain sifts out and falls through openings in the rack onto the cleaning shoe while the straw is discharged at the rear. The *time interval required for separation* depends upon (1) the size of the seed in relation to the size

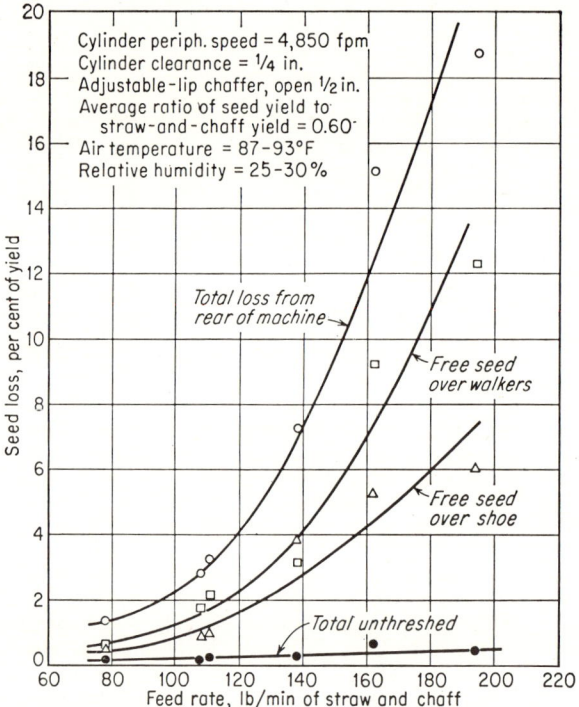

FIG. 24-3. Effect of feed rate upon seed losses for 12-ft self-propelled combine No. S-1 in barley (1955 tests). (*Goss, Kepner, and Jones, Agr. Eng., November,* 1958.)

of the openings between straws, (2) the thickness of the straw layer, (3) the coefficient of friction between the seed and the straw, and (4) the density of the seed in relation to the straw. Increased agitation increases the average size of the openings between straws. Rate of travel over the racks controls the thickness of the straw layer and the time available for separation. Chopped straw resulting from overthreshing slows down separation. Green leafy weeds greatly impede separation, because of blanketing and greater friction, particularly if finely chopped. The amount of straw retained on the rack varies widely from crop to crop. In wheat 90 to 95 per cent of the straw and chaff is retained on the rack, two-thirds in barley, about one-half in alfalfa, and less than one-third in windrowed red clover.[1] The rapid increase in grain lost from the rear of the machine as the separating and cleaning capacity is exceeded is shown in Fig. 24-3.[2]

Modern combines have much greater separating capacity per unit area than older machines as a result of refinements in rack shape, openings, and motion.

[1] Bunnelle, Jones, and Goss, *op. cit.*
[2] J. R. Goss, R. A. Kepner, and L. G. Jones, Performance Characteristics of the Grain Combine in Barley, *Agr. Eng.*, 39:697–711, November, 1958.

Lower cylinder speeds than used for other crops were found to be necessary for the threshing of beans, including soybeans, without excessive cracking. Repeating, or wrapping around the cylinder, resulted from the slow cylinder speeds, and to overcome this undesirable condition, a *beater* was installed *behind the cylinder* to control the movement of the threshed crop from the cylinder to the strawwalkers. These beaters are 10 to 15 in. in diameter, operate from 600 to 900 rpm, and are located so that the trailing-type nonwinding blades have a stripping action. A well-designed beater, properly located in relation to cylinder and grate and used with deflectors or curtains, has contributed greatly to effective grain separation in a smaller area than was thought possible some years ago.

Various combinations are used for grain separation, but the three main types in general use are the one-piece rack, strawwalkers, and the conveyer raddle. With the cylinder and concave set to thresh the grain out of the heads with a minimum creation of chaff, the work of the separating mechanism is made easier. By maintaining a constant even flow of material in the most open manner possible, good separation is obtained without overloading the sieves with excessive amounts of chaff. Tests show that the bulk of separation takes place at the front end of the separator, where there is the maximum amount of grain and the least amount of chaff.

A low-density baling-press attachment has been developed and is in fairly common use on combines in Europe today, baling the straw as it comes off the strawwalkers. These presses mount directly on the combine and, driven from the combine motor, make a loose bale of straw up to 20 lb in weight, about 4 ft by 24 in. by 10 in. in size, and use 2 to 4 hp according to size. They tie with binder twine, and both single and double knotter attachments are available. The principal advantage of the loose bale over the high-density bale made by the pickup baler is that damp straw or straw mixed with green weeds or undergrowth can dry in the loose bale.

One-piece racks oscillate on swinging hangers through an arc which moves the rack back and up, then forward and down. This action tends to toss the straw backward, and aided by the fishbacks, it moves toward the rear. This type of separator is most common on small combines with scoop-type headers and wide cylinders and bodies. From 4 to 6 sq ft of separating surface is usually provided per foot of cutter-bar width. One-piece racks are operated at speeds from 200 to 270 cycles per min.

Strawwalkers, or sectional racks, have more separating effect for their area than one-piece racks and are favored for wide-cut narrow-body machines. From 2 to 3 sq ft of separating surface are provided per foot of cut. Strawwalkers are operated somewhat slower than one-piece racks, ranging from 170 to 215 rpm. They are usually mounted on two crankshafts with throws of approximately 2-in. radius. Some lower-cost walkers use one crankshaft in combination with a hanger at the other end.

Conveyer raddles are not popular except on large prairie and hillside machines used in the far west. The area provided is somewhat greater than with strawwalkers. Beaters combined with blowers are used over the conveyer to agitate the straw and aid separation.

The grain separated from the straw must be dropped on the front of the cleaning shoe. That separated at the concaves may be transported back by an oscillating deck or by a chain conveyer. Grain falling through one-piece racks at the rear is usually moved forward by a chain conveyer, although inclined pans moving with the rack or an oscillating deck may also be used. Strawwalkers commonly have inclined pans attached underneath to drop the grain forward as desired.

Small machines with wide separators and narrower shoes must prevent overloading of the outside edges of the shoe due to the inward gather of the material falling from the racks.

Straw spreaders and *cutters* may be used to facilitate plowing where the straw is not saved. Where the crop is cut close to the ground to save straw, a smaller header should be used to avoid overloading the machine.

Centrifugal separation has been tried, but without agitation, the straw is pressed into a mat which will not let the grain through. Air separation has also been extensively investigated, but capacity and versatility have been inadequate to date.

Cleaning. Grain is separated from the chaff and other plant residue falling through the rack by a combination of agitation and air separation. The cleaning mechanism usually consists of two sieves and a fan. The material first falls on an upper, or *chaffer, sieve* which has air blowing up through it and across under it. The sieve is oscillated so as to toss the material and move it rearward. The grain falls through, and the finer chaff which also falls is blown away. The chaffer-sieve openings should let as much air through as possible to float chaff away without blowing out grain, but coarse heavy material should be retained and discharged at the rear. The air coming through is also controlled in volume and direction. The air passing between the sieves must not blow out grain. The process is repeated on a lower, or *shoe, sieve,* with smaller openings which let the clean grain through but retain larger materials. This may be adjustable for grains, but round-hole sieves are often needed for small-seeded crops. The cleaned crop is collected by an auger and transported to the grain

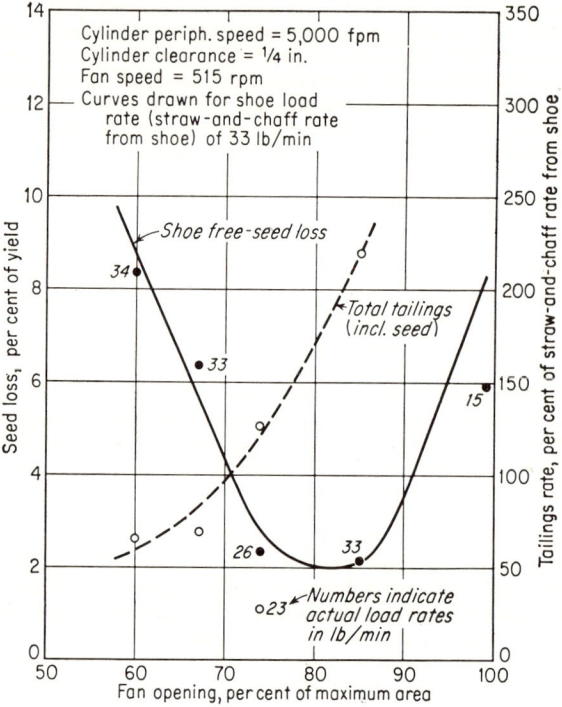

FIG. 24-4. Effect of fan opening upon shoe performance in barley (1954 tests, riffle chaffer with finger extension). (*Goss, Kepner, and Jones, Agr. Eng., November,* 1958.)

tank or bagging platform, as the case may be. The material passing over the shoe sieve drops in the tailings return for rethreshing, as do unthreshed heads and other material which drop through the coarse openings of the *chaffer extension.*

Cleaning shoes, like straw racks, have a sharply defined capacity for a particular crop and condition, as shown in Fig. 24-3. The time interval required for the seed to sift down through a sieve depends on (1) the size of the seed in relationship to the average size of the openings between the chaff and other material on the sieve, (2) the thickness of the chaff layer, (3) the coefficient of friction between seed and chaff, and (4) the terminal settling velocity of the seed in relation to the rate of air flow up through the material on the chaffer. Where the air can keep the material loose without blowing out grain, friction may be unimportant. Where there is little difference in size and density between seed and chaff or weed particles, thorough cleaning is impossible without seed loss. The sharp increase in loss when air velocity exceeds the settling velocity of the seed is shown in Fig. 24-4.

Shoe action is similar to that of a one-piece rack, oscillating on the arc of the supporting hanger. A common action is 1¼-in. travel with 5/16 uplift at a speed of 300 to 320 cycles per min. Ball bearings and rubber bushings acting as bearings through torsional deflection have increased the durability of shaker devices. European stationary threshers use a pendulum action which is effective with large sieves.

An upper recleaner shoe, located on top of the machine, was formerly used but is no longer popular. It usually employed a "shiver" shake, still used in many stationary fanning mills.

Shoe adjustments are extremely important in cleaning the crop without loss. Sieve size, amount and direction of air, and tailboard height are the common adjustments. The air not only carries chaff away, but is relied on to keep the material on the chaffer "alive" so that the grain can shake out. Much attention has been given to keeping adjustments simple for satisfactory operation by unskilled operators.

Paddlewheel *fans* are used, and the amount of air is regulated by shutters on the intake, by outlet restriction, or by fan speed. Direction of air may be regulated on some machines by adjustable deflectors called windboards. It is very important to have a uniform well-controlled flow of air through the shoe.

Some European machines use a suction fan to draw air through the sieves and also act as a blower to deliver the chaff to a container.

Rotary screen attachments or recleaners may be used to separate weed seeds, fine dust, and cracked kernels from whole grain. They are usually mounted over the grain tank.

Combines specially built for use on *hillsides* have long been popular in many parts of Washington, Oregon, Idaho, and other similar areas. These machines are equipped with automatically controlled mechanical devices for keeping the threshing and separating mechanisms level, independent of the position of the table. Thus an even flow of material is maintained over the full width of the machine for most efficient operation.[1]

Standard-type combines, with auger-type table and undershot elevator, give full-width delivery from cylinder to separator area regardless of slope, and fins, or dividers, running lengthwise in the sieves keep the grain from accumulating on one side. Such prairie-type combines are able to operate much more efficiently on sloping land than older machines.

Grain Collection. The grain is moved horizontally by augers and elevated by paddles attached at intervals to a chain running in an enclosed passage to give positive action regardless of slope. Metal paddles are being replaced by rubber tire–carcass paddles which are quieter and crack less seed.

The grain is collected in a tank except for areas where the grain is sacked and a bagging platform is supplied. Tank capacity averages about 4 bu per ft of cut, and the unloading auger should empty a tank in 1 to 1½ min. Large self-propelled machines require long unloading augers which can unload into a moving truck as they cut. These augers must be folded in for road transport.

POWER REQUIREMENTS

Combines are of three general types with respect to power:
1. Trailed with auxiliary motor
2. Trailed with power take-off
3. Self-propelled

The scoop-type combines with 5- to 7-ft cut require an engine of 20 to 25 hp, usually running 1,500 to 2,000 rpm, or a two- to three-plow tractor to pull and operate the combine by power take-off.

The narrow-body wide-cut machines of 12-ft cut or more require a 35- to 40-hp engine, usually running 1,500 to 2,000 rpm, or a four- to five-plow tractor if PTO-operated. The tractor should have an independent PTO and a low ground speed of approximately 1 mph for best results in adverse conditions.

[1] Homer D. Witzel and B. F. Vogelaar, Engineering the Hillside Combine, *Agr. Eng.*, 36:523–528, August, 1955. S. D. Pool, Controls for Full-leveling Hillside Combine, *Agr. Eng.*, 37:245–248, April, 1956.

Self-propelled combines of 12- to 16-ft cut require engines of 48 to 68 hp. It is vitally important to be able to vary or halt the ground speed while the combine mechanism runs at constant speed. This permits slowing for down grain and easing across rough spots, usually achieved by a variable-speed V-belt drive from the engine to the transmission. The newer machines use hydraulic actuation of the sheave-shifting mechanism. A fluid coupling with controlled variable slip has also been used.

Combine engines should have a governor allowing not more than 50-rpm change between idle and full load with very quick response to overloads.

Power distribution for a 7-ft-cut combine with a wide rasp-bar cylinder in wheat is shown in Fig. 24-5.[1] Since the cut wheat crop is usually about half grain and half straw in humid areas, a straw rate of 60 lb per min would approximately represent, for instance, a 30 bu per acre crop cut at 2⅓ mph with a 7-ft machine. It can be

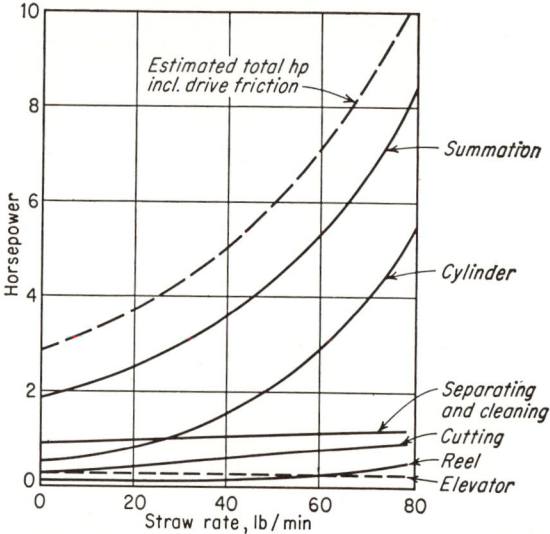

FIG. 24-5. Power distribution for a 7-ft combine in wheat. (*Burrough, Agr. Eng., January, 1954.*)

seen that the cylinder power increases rapidly, but the cutting, separating, cleaning, and elevating power requirements increase only a little with the straw rate. The curves in Fig. 24-5 are for steady uniform feeding and do not allow for uneven feeding, weeds, and other adverse factors normally encountered. Much more power must be available to take care of peak loads.

In Saskatchewan, Canada, tests of a self-propelled combine in windrowed hollow-stemmed wheat, the cylinder required 5 hp at a rate of 100 bu per hr, 6¼ hp at 140 bu per hr, and 8 hp at 177 bu per hr.[2]

Maximum power and force measurements for the various parts of the above combine are listed in Table 24-2. These figures do not allow for friction losses in the power train.

Also, in the same study, traction horsepower was measured for a 6,600-lb self-propelled combine with 5,000 lb on the 10-24 traction tires and 1,600 lb on the 6.00-16 steering tires. On level soybean stubble, approximately 12 hp was required at 2½ mph.

Traction power for self-propelled combines may be considerably higher for hills or adverse conditions such as rice fields. Special high-cleat rubber tires or tracks are

---

[1] D. E. Burrough, Power Requirements of Combine Drives, *Agr. Eng.*, 35:15–18, January, 1954.
[2] F. W. Bigsby, Power Requirements for Combining Solid and Hollow-stemmed Wheat, *Agr. Eng.*, 40:453–455, August, 1959.

available for use in harvesting rice. Ground speeds down to ½ mph are desirable for rice, although speeds up to 6 mph are required for light wheat in the Great Plains.

COMBINE EFFICIENCY

Grain losses are usually classified as follows:
1. *Shatter loss* of overripe grain prior to harvest
2. *Cutter-bar loss* consisting of missed and dropped heads and grain shattered on the ground in cutting
3. *Cylinder loss* of unthreshed heads passing out in the straw
4. *Rack loss* of loose grain passing out over the rack with the straw
5. *Shoe loss* of loose grain blown or carried out with the chaff

Cutter-bar loss is determined by picking up the grain on sample areas where the straw and chaff have not fallen. Prior shatter loss must be negligible or picked up previously for accurate determinations of loss chargeable to the machine. For minimum cutter-bar loss, the grain must be cleanly cut low enough to get the heads and the reel must be running at the proper speed in relation to ground speed and positioned so as to lay the crop on the platform. Where the reel tends to catch and throw

TABLE 24-2. MAXIMUM POWER AND FORCE MEASUREMENTS FOR VARIOUS PARTS OF A 7-FT COMBINE WITH WIDE RASP-BAR CYLINDER

	Max horsepower over 1-sec interval	Max force or torque	Crop
Cutter bar, 360 rpm	1.6	394 lb	Grass
Reel	0.35	283 lb-in.	Soybeans
Cylinder	10.0	....	Wheat
Straw rack	0.9	280 lb-in.	Soybeans
Fan	0.3	....	No load

SOURCE: D. E. Burrough, Power Requirements of Combine Drives, *Agr. Eng.*, 35:15–18, January, 1954.

drooping heads, extra wide bats may be used. Canvas flaps may be needed to sweep short grain off the cutter bar. Where a machine does not have the capacity to separate the volume of material resulting from low cutting, it is best to cut higher and accept a slight increase in cutter-bar loss or to run more slowly if traction speed will permit.

Cylinder loss is determined accurately by catching the straw coming off the rack over a measured area and, after the loose grain is shaken out, rethreshing to get any grain left in heads. This loss is minimized by using as much threshing action as possible without cracking grain. Overthreshing, however, chops straw and weeds into short lengths which drop through the racks and overload the sieves, causing excessive loss there. It is often better to accept a small cylinder loss in order to lower rack and shoe losses.

Rack loss is the loose grain shaken out of the straw collected over a measured area. Incorrect rack speed or overthreshed straw and green weeds may cause rack loss, although plain overloading due to cutting too low or traveling too fast is the most common cause. "Where walker losses tend to be high and threshing is not difficult, total seed losses will be minimized by having the concave grate open as much as possible so that maximum separation will be achieved at this point."[1]

Shoe loss is determined by collecting the material coming from the shoe over a measured area. Losses result primarily from overloading, excessive air, and too little sieve opening in attempting to clean the grain.

From tests conducted as above, the loss per acre can be calculated. If the threshed grain delivered to the tank is collected simultaneously with the straw and chaff, an accurate percentage of loss can be calculated where the crop is uniform enough so that the lag in grain delivery is not important. Cylinder, rack, and shoe losses should total less than 2 per cent when harvesting wheat under normal conditions with a

[1] Goss, Kepner, and Jones, *op. cit.*

properly adjusted and operated machine. Losses from the various sources at progressive harvest dates are shown in Fig. 24-6.[2] Harvesting at 20 per cent moisture gave the highest net yield, although artificial drying was required.

A great deal can be determined about a combine's adjustment and efficiency of operation by suddenly stopping the machine for inspection when operating under uniform conditions. If operating properly, the following conditions should be met:

1. Straw should be spread evenly over the straw racks with no grain on the rear 12 in.

2. Grain dropped from the racks and lying on the pans or conveyer, carrying it forward to the shoe, should taper off sharply toward the rear of the rack if rack loss is at a minimum.

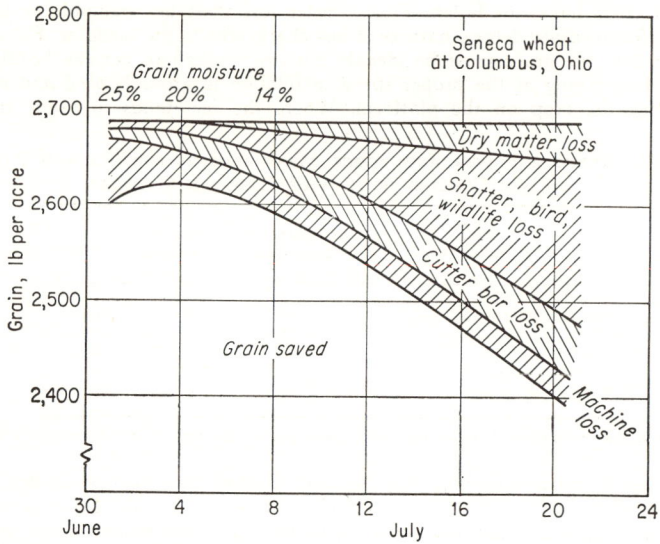

FIG. 24-6. Sources of loss of grain as the harvest date progresses. (*Johnson, Agr. Eng., January,* 1959.)

3. The upper sieve should be loaded uniformly across its width, and only coarse material should remain on it at the rear with no free seeds discernible.

4. The lower sieve should be clean except for coarse particles passing over into the tailings auger. The smallest opening which meets this requirement will give the best cleaning.

Excessive losses most often result when a small PTO machine is pulled by a large tractor which has enough power to put through much more material than the combine was designed to handle. The horsepower provided with engine-powered and self-propelled combines is usually in keeping with the over-all capacity of the machine, and losses from overloading are minimized.

[2] W. H. Johnson, Machine and Method Efficiency in Combining Wheat, *Agr. Eng.*, 40:16–29, January, 1959.

Chapter 25

CORN-HARVESTING MACHINES

Corn occupies a greater acreage than wheat, its closest competitor in the United States, and has approximately twice the value. It also yields by far the greatest tonnage per acre of any grain crop, average yields of ear corn weighing approximately 2½ times as much as average yields of wheat.

CORN PICKERS

Gathering. In order to miss as few of the ears as possible, a corn picker must be able to pick up and feed low or down ears, avoid shaking off ears, and catch ears which do drop. (Under some conditions many ears may drop to the ground before picking, but as yet no mechanism for recovering them has been successful.) The gathering and picking parts of a two-row mounted picker can be seen in Fig. 25-1. The floating gathering points, or snouts, guide standing corn into the picker and get under down stalks, lifting them enough for the gathering chains to catch them.

Flare sheets funnel the corn into the snapping rolls. They should blend into the gathering points and be well rounded to allow down stalks to pull over without breaking. They are commonly set at an angle 30 to 45° with the horizontal and should be high enough to catch ears which may shake loose as the stalk is pulled into the snapping rolls or those which bounce when snapped.

Gathering chains having finger links 8 to 10 in. apart lift low ears and stalks and feed all stalks back to the snapping rolls. They also prevent loose ears from sliding down forward and being lost. It is important that the upper chains overlap as close to the ground as possible in order to lift low ears which may otherwise pass under or through the lower ends of the snapping rolls and be lost.

For maximum effectiveness in down corn, the *lineal chain speed* should be somewhat faster than ground speed. Often the chains are geared to run at, or slightly less than, ground speed in the normal picking gear of the tractor, usually second, with the desired extra relative speed for picking up down corn being secured by running in low gear. It is essential that the chain speed match the speed at which the lower flutes on the snapping rolls feed the stalks back. Otherwise stalk breakage will be aggravated in dry brittle corn.

Snapping. The snapping rolls straddle the row and pull the stalks down and through, pinching off the ears. They are commonly made of cast iron, about 3½ in. in diameter, and have a tapered point to facilitate stalk entrance. Peripheral speed is usually about 600 fpm. A typical pair of snapping rolls is shown in Fig. 25-2. It will be noted that the spiral flute is smooth at the lower end but the upper portion has notches on the leading edge which aid in pulling stalks through. Some pickers have adjustable snapping-roll shields which can be set to prevent ears from touching the rolls but allow the stalks to pass through. In this case, severe straight-fluted rolls may be used.

One picker, shown in Fig. 25-3, skews the snapping rolls so that the rear end of the upper roll is almost directly above the other. The stalks are bent sideways by the gathering chains as they progress back into the snapping point. The snapped ears drop away from the rolls immediately, reducing the shelling which occurs when a loose ear is pushed into the rolls by an incoming stalk and ear.[1]

---
[1] C. B. Richey, J. F. O'Donnell, J. T. Ashton, and R. J. Groves, Corn Picker Features New Principle, *Agr. Eng.*, 37:93–97, February, 1956.

252 CROP-PRODUCTION EQUIPMENT

The *space* between snapping rolls is *adjustable* at the lower end and should be kept as narrow as possible without causing clogging because any extra width increases the tendency of ears to pull into the rolls and shell. Tests in Illinois indicated a loss of 1⅓ bu per acre for each ⅛ in. of roll clearance over the minimum.[1]

The rolls are usually adjusted to snap most of the ears just back of the gathering chains, thus leaving upper roll space for stubborn stalks and trash. Rolls can usually

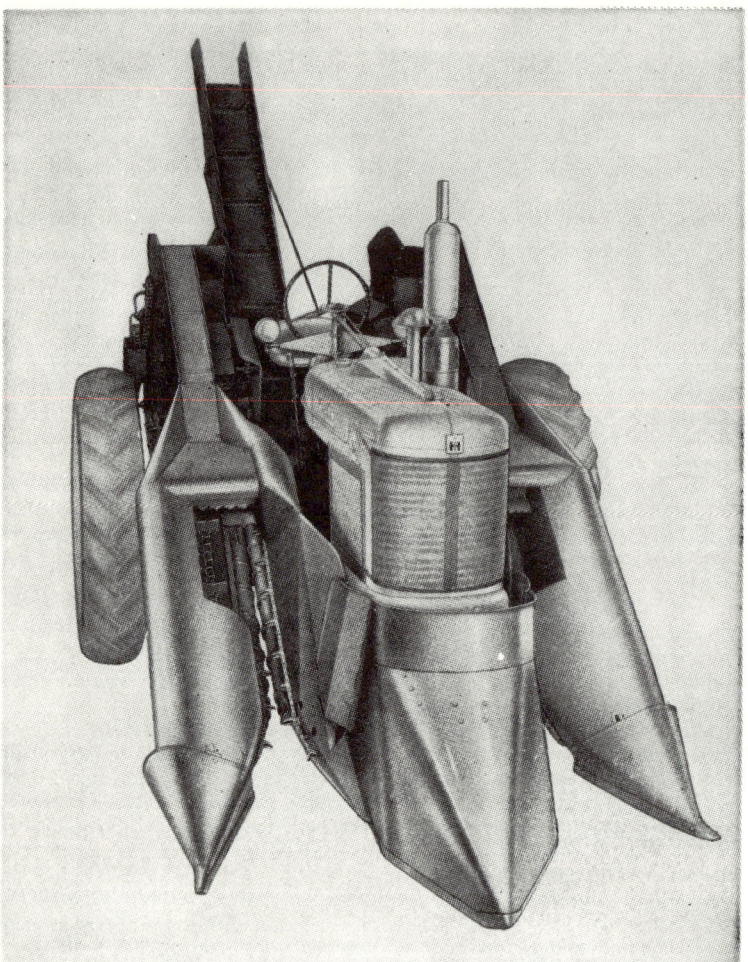

Fig. 25-1. Two-row mounted corn picker. (*International Harvester Co.*)

be set close for minimum shelling when the stalks are damp and tough, but must be opened to prevent excessive stalk breakage as the corn becomes dry and brittle. Many pickers have a roll-adjusting lever which can be changed by the operator while picking, a feature which helps clear clogged rolls without danger to the operator.

Snapping rolls should be held by rigid enough *framing* so that they do not spread appreciably, causing shelling, when several stalks pass through at once. Tests by the author of adequate frames showed the rolls spreading ¼ in. near the bottom when a spreading force of 1,000 lb was applied near the center of the rolls.

[1] H. P. Bateman, G. E. Pickard, and Wendell Bowers, Corn Picker Operation, *Univ. Illinois Coll. Agr. Ext. Serv. Circ.* 697, 1952.

In dry brittle corn it is often necessary to increase the *aggressiveness* of the snapping rolls in order to pull through broken stalks and trash. One expedient is to bolt on cleats or to insert pegs. This may cause more corn to be shelled and lost through the rolls, although quantities of loose trash also increase shelling because the ears are not free to get off the rolls after being snapped. The use of rubber at the upper end of one snapping roll is very helpful in taking through dry stalks and trash as well as in removing more husks when the ear is snapped, but shelling losses may increase. Arc-welded beads on the edges of the notches or on smooth portions of the rolls are also used to increase aggressiveness. One picker uses a $\frac{1}{8}$-in.-thick coating of crushed

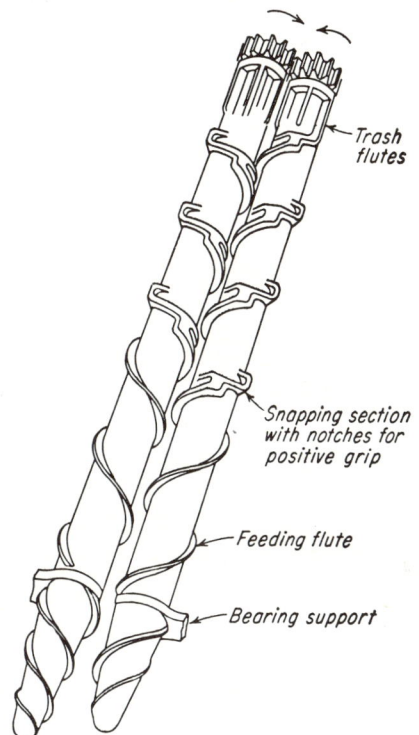

Fig. 25-2. Notch-type snapping rolls. (*International Harvester Co.*)

quartz bonded to the upper snapping roll by epoxy resin in order to secure a favorable surface for gripping stalks with minimum shelling.

A *trash beater* with flexible flaps, rotating partially over the upper ends of the rolls, may be used to either feed loose trash through the rolls or start it up the elevator, thus preventing its accumulation and eventual clogging.

Some pickers use *stalk-ejector rolls* at the top of the snapped-corn elevator to eject broken stalks and snap off ears which may be on them. A deflector hood is necessary to guide the stalks into the ejector rolls properly. The rolls are usually held together by spring pressure so as to spread rather than clog when several stalks go through at once.

Husking. Some machines, called snappers, do not use a husking bed. In the South, husks are often left on for protection from weevils and for extra feeding value. Husks do, however, impede air flow through a crib of corn and thus reduce the allowable moisture content for safe storage. Corn varieties differ in the proportion of husks to ears by weight, ranging from 5 to 8 per cent.

254 CROP-PRODUCTION EQUIPMENT

Husks are removed by pairs of husking rolls which grasp and pull them down through the rolls, usually taking all the husks at once when one is caught. A typical husking bed is shown in Fig. 25-4.

The coefficient of friction betwen metal and corn husks is greatly affected by the moisture content of the husks. Consequently, cleanness of husking is extremely variable. It is not uncommon for the snapping rolls to remove more husks when corn is damp than the husking bed can remove when the corn is dry.

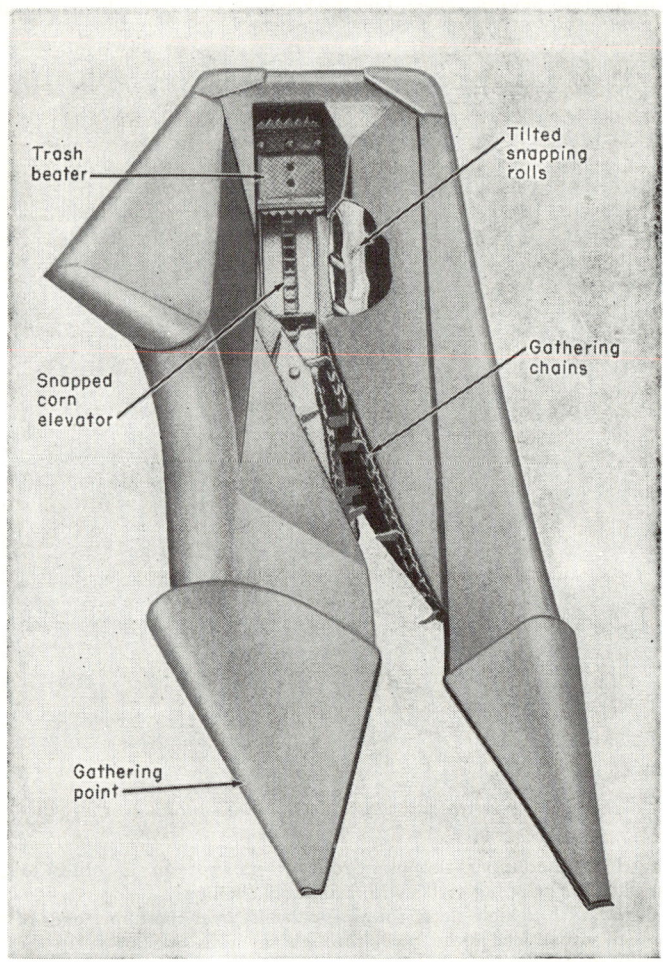

FIG. 25-3. Snapping unit with tilted snapping rolls. (*Ford Motor Co., Tractor and Implement Div.*)

Many combinations of *roll materials* are in use, the most popular being metal against metal, metal against rubber, and wood against rubber. Many combinations of smooth or serrated ribs and spiral grooves are used. The use of rolls composed of rubber-fabric washers punched from used automobile-tire casings and stacked on a center shaft has become popular because of their effectiveness, durability, and reasonable cost. One of these rolls is used against a metal roll.

Ear retarders, or forwarders, may be of the belt, pressure-wheel, or pressure-plate types. The ear retarder aids husking by:

1. Straightening ears so that they are parallel to the rolls
2. Assuring a constant rate of flow over the rolls in spite of change of ground slope
3. Forcing ears against the rolls for cleaner husking when necessary
4. Aiding rotation of ears with some designs[1]

A chain *husk conveyer,* which also acts as a shelled-corn saver, commonly runs underneath the husking bed. Husks are conveyed away over a screen which allows the shelled corn to drop to a second level where the returning conveyer slats carry it to the wagon elevator. Considerable corn can be lost with the husks if the following requirements are violated:

1. Ample hole area in the screen. (Transverse louvers rather than holes provide more area and are not as easily clogged.)
2. Adequate separating distance beyond the upper ends of the rolls since shelled corn discharged from the snapped-corn elevator will go through the rolls at the first opportunity. Husk agitators are sometimes used to aid separation.

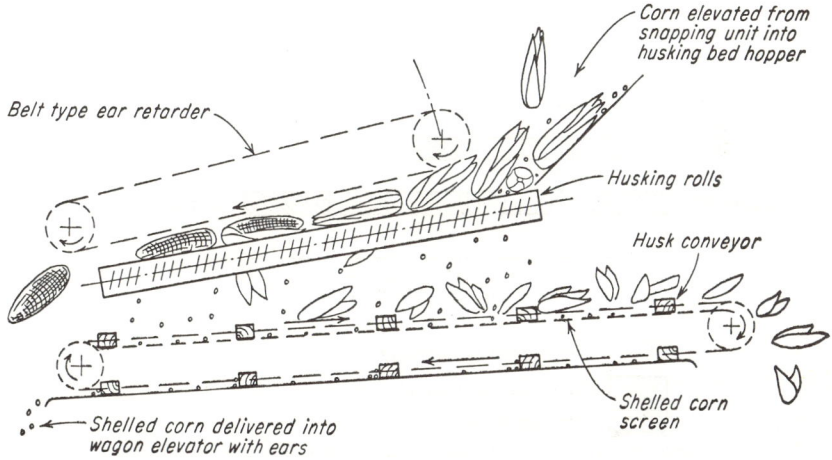

FIG. 25-4. Husking bed.

Some pickers, instead of dropping the snapped ears off the snapping rolls onto an elevator, use *combination snapping-husking rolls,* over which a side-paddle chain conveyer elevates the ears back over a husking section at the upper ends of the snapping rolls. It is difficult to prevent excessive shelling in the snapping area with this arrangement because many of the snapped ears are not conveyed away before the next stalk and ear come through. Thus the loose ears are forced against the rolls and shelled.

Another approach is the use of a snapped-corn elevator wide enough to accommodate a four-roll husking bed. This saves space, and the rolls can be omitted to produce the snapper model. Aggressive rubber husking rolls are used to help compensate for the tendency of the ears to be turned sideways and lifted off the rolls as they are conveyed uphill.

Cleaning fans may be used in several ways. The purpose of the fan is to remove silks and other light trash which may increase spoilage. One very effective system is to attach the fan to the bottom of the wagon elevator and, by means of a duct running up underneath the elevator, blow through the corn as it drops from the elevator. This removes the light trash after the corn has had a maximum of mechanical agitation. The air blast may also be supplied where the corn drops from the husking bed into the elevator hopper, up through the shelled corn dropping from the husk

[1] E. V. Collins, J. M. Trummel, and C. K. Shedd, Results of a Corn Husking Mechanism Study, *Agr. Eng.,* 21:425–428, November, 1940.

rake, or where the corn drops from the snapped-corn elevator. The latter location often keeps much trash off the husking bed and thus aids shelled-corn separation under the beds but does not remove silks and trash loosened in husking. Some pickers have the fan blast directed against the husking bed to aid the rolls in grasping husks and trash. It is desirable, in any case, to use as much air as possible without blowing out shelled corn.

Corn-picker *elevators* are of the chain-conveyer type. Tire-fabric paddles have gained favor because they are quiet and shell less corn than steel paddles. Elevators which handle two rows of corn are 8 in. or more in width so that the ears can lie crosswise. These elevators can operate at angles of 50° and more, depending on the amount of cup in the paddles, whereas elevators which push the ear endways have excessive tumble above 45°. Elevator speeds range from 180 to 250 fpm.

Pickers are *classified* as one-row pull, one-row mounted, two-row pull, two-row mounted, and two-row self-propelled. The two-row mounted and self-propelled are often preferred because of the ease of control, better traction and flotation in soft wet ground, and the ability to open up fields without breaking down an unpicked row.

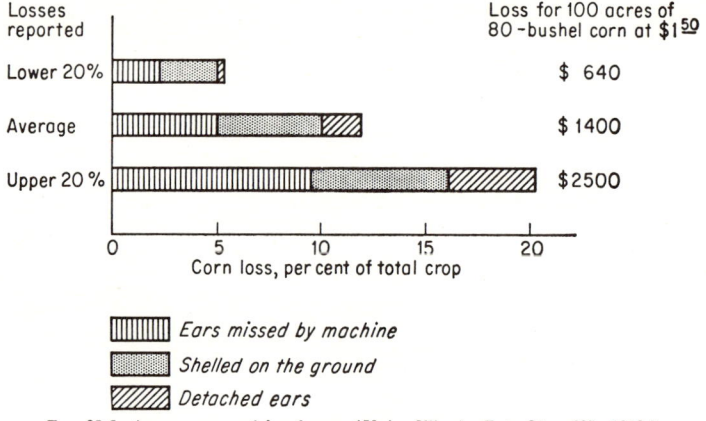

Fig. 25-5. Average corn-picker losses. (*Univ. Illinois, Ext. Circ.* 697, 1952.)

The self-propelled with driving wheels in front is somewhat handicapped in pulling a heavily loaded wagon around a short turn, but it affords better operator safety, comfort, and visibility than the two-row mounted. The center gathering point is more efficient because it is not as far ahead of the gathering chains as on two-row pickers mounted on tricycle tractors.

Pull-type pickers do not monopolize a tractor for the entire picking season as do mounted pickers, but are harder to keep on the row and require a wide headland for turning.

Some one-row pickers with 20-bu tanks to carry the corn were built around 1930, but trailing a wagon proved more practical, particularly after the advent of rubber-tired wagons.

Losses. Corn-picker losses are normally much greater than is considered acceptable for combines. The losses reported for a number of experiments and contests ranged from 5 to 20 per cent, with an average of 10 per cent, excluding previously detached ears. As shown in Fig. 25-5, approximately half of these losses were from missed ears and half from shelled corn lost through the snapping rolls.[1] Results of efficiency checks made around 1930, when the corn picker was just coming into general use, do not differ greatly from those made recently.[2]

---

[1] Bateman, Pickard, and Bowers, *op. cit.*

[2] L. G. Hobson and R. H. Wileman, Mechanical Corn Pickers in Indiana, *Ind. Agr. Expt. Sta. Bull.* 362, 1932. C. W. Smith, W. E. Lyness, and T. A. Kisselbach, Factors Affecting the Efficiency of the Mechanical Corn Picker, *Nebr. Univ. Agr. Expt. Sta. Bull.* 394, 1949. A. L. Young, Present Status of Mechanical Corn Picking, *Agr. Eng.*, 12:267, July, 1931.

Careful operation in standing corn should result in very little ear loss by most pickers. At the 1952 Illinois State Corn Picker Contest held in Bloomington, Ill., 23 machines averaged an ear loss of only ⅓ bu per acre in corn yielding 70 bu per acre.

Shelled-corn losses may occur at the snapping rolls, with the husks, or through leaks in elevator hoppers, etc. With a properly designed shelled-corn saver under the husking rolls and tight elevator connections, little corn should be lost at these points. Snapping-roll losses, however, are extremely difficult to avoid under many conditions. The 1951 Iowa State Corn Picker Contest was held near Audubon in 90-bu-per-acre corn containing about 25 per cent moisture and damp from rains the previous day. Shelled-corn losses for eight contestants ranged from 4.5 to 15 bu per acre, with an average of 11 bu per acre.

Design factors which reduce shelled-corn losses at the snapping rolls are:

1. Rolls which are aggressive in taking through stalks and trash but present a relatively smooth surface to the ear. (This is difficult to secure, particularly under dry trashy conditions.)
2. Small-diameter rolls which have less tendency to pinch ears, as long as trash-handling is not impaired.
3. Adequate room in the snapping area for ears to turn butt down before being snapped.
4. Maximum opportunity for ears to drop away quickly after being snapped.
5. Aggressive sections at tops of rolls or trash beaters to prevent accumulation of trash and consequent holding of ears on the rolls.
6. Sheet metal around rolls arranged to catch as much shelled corn as possible.

Shelling losses are often increased by high ground speeds which keep the rolls crowded and do not give the snapped ears enough chance to fall away.

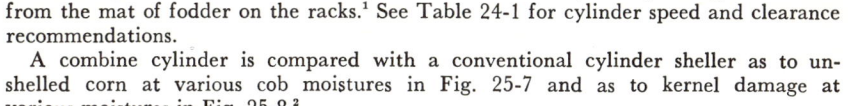

Fig. 25-6. Stripper plates and traction rolls on snapping unit for combine. (*Deere & Co.*)

Ear-corn losses can be estimated by counting the number of average size (0.54 lb) ears lost per 100 ft. One ear per 100 ft will represent 1 bu per acre with 40-in. rows. Several checks should be made, and the average used. Shelled-corn loss should be checked over a 15- to 25-ft length of row because of its variability. If a short length is taken, the result may depend on whether an ear was nipped by the snapping rolls in the area or not. An average of six kernels per lineal foot of 40-in. row usually represents a loss of 1 bu per acre.

Combines are being used to harvest shelled corn. One approach is to use a special row-crop header attachment to cut off the stalks and feed them into the combine where the corn is shelled, separated, and cleaned. The other approach is to use conventional corn-picker snapping units which deliver only ears and husks to the combine. Since husks need not be removed, shelling at the snapping rolls can be reduced as much as 50 per cent by using protective stripper plates over severe traction rolls, as shown in Fig. 25-6. Tests in California, summarized in Table 25-1, indicated no advantage for the cutoff machine because its reduction in gathering unit loss was balanced by the increase in rack loss due to the difficulty of separating the kernels from the mat of fodder on the racks.[1] See Table 24-1 for cylinder speed and clearance recommendations.

A combine cylinder is compared with a conventional cylinder sheller as to unshelled corn at various cob moistures in Fig. 25-7 and as to kernel damage at various moistures in Fig. 25-8.[2]

[1] Roy Bainer et al., Combine Used in Corn, *Calif. Agr.*, July, 1955.
[2] C. S. Morrison, Attachments for Combining Corn, *Agr. Eng.*, 36:796–799, December, 1955.

TABLE 25-1. CORN-HARVESTING LOSSES BY COMBINES TESTED IN CALIFORNIA

	Average loss, % of total yield	
	Cutoff type	Snapper type
Gathering-unit loss:		
Ear corn	2.9	2.0
Shelled corn	0.3	3.2
Total	3.2	5.2
Combine loss, exclusive of gathering unit:		
Unshelled corn	0.4	0.6
Shelled corn	2.8	0.4
Total	3.2	1.0
Total harvesting loss	6.4	6.2

NOTE: Kernel moisture content varied from 9.9 to 20.6 per cent. Yield varied from 2,700 to 7,000 lb of shelled corn per acre.

SOURCE: Roy Bainer et al., Combine Used in Corn, *Calif. Agr.*, July, 1955.

PTO power requirements for corn picking vary greatly with amount of picker mechanism, types of bearings, and amount and toughness of cornstalks. Average requirements of 4.75[1] and 6.3 hp[2] have been reported for two-row pickers.

The results of power tests by the author of a one-row mounted picker, using the snapping unit shown in Fig. 25-4, are given in Table 25-2. The tests were made in corn planted in 56-in.-wide rows and yielding 107 bu per acre in the area checked, a row yield equivalent to 150 bu per acre in 40-in. rows. Tests were made on two

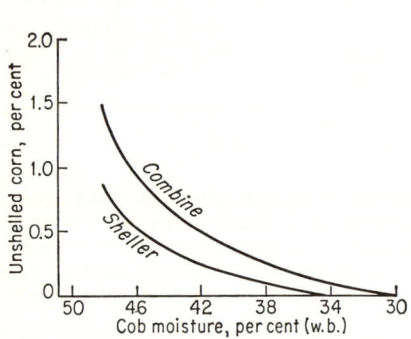

FIG. 25-7. Shelling loss by combine vs. conventional sheller. (*Morrison, Agr. Eng.,* December, 1955.)

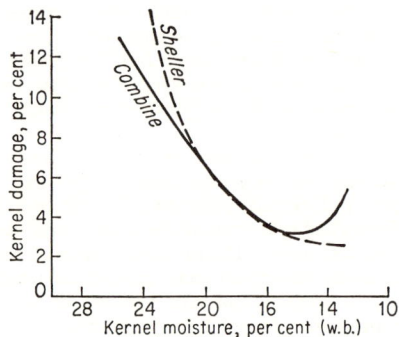

FIG. 25-8. Kernel damage by combine vs. conventional sheller. (*Morrison, Agr. Eng.,* December, 1945.)

successive days, one dry and one damp. The ears contained about 22 per cent moisture, and many stalks were slightly green.

Power requirements for snapping corn vary greatly with conditions. Early picking of green corn requires much more power than picking after stalks have been killed by frost, when ears snap more easily. Some tall heavy-stalked southern corn requires much more power than any in the Corn Belt. Power requirements for propelling tractor, picker, and wagon are extremely variable but occasionally use the remaining available engine capacity.

---

[1] K. Shedd, E. V. Collins, and J. B. Davidson, Labor, Power and Machinery in Corn Production, *Iowa State Univ. Agr. Expt. Sta. Bull.* 365, 1937.

[2] Merlin Hansen, Loads Imposed on Power Take-off Shafts by Farm Implements, *Agr. Eng.*, 33: 67–70, February, 1952.

TABLE 25-2. POWER REQUIREMENTS FOR ONE-ROW CORN PICKER IN HEAVY CORN

	No. runs	Av. snapping unit hp	Av. husking bed and elevator hp	Av. fan hp	Av. total hp
No load....................	5	1.5	1.1	0.7	3.3
Dry conditions:					
First gear, 1.9 mph..........	3	4.3	1.3	0.7	6.3
Second gear, 3 mph..........	3	5.0	1.6	0.7	7.3
Damp conditions:					
First gear, 1.9 mph..........	4	6.2	1.4	0.7	8.3
Second gear, 3 mph..........	4	6.6	1.5	0.7	8.8
Snap unit clogged...........	2	13.8	1.2	0.7	15.7

SOURCE: C. B. Richey et al., Corn Picker Features New Principle, *Agr. Eng.*, 37:93–97, February, 1956.

Safety is a major consideration in corn-picker operation since more accidents occur with pickers than with any other field machine. Most of these accidents occur when the operator tries to unclog the machine while it is running.[1]

CORN BINDERS

Corn binders are used to cut and tie standing corn into bundles which may be hauled to a stationary ensilage cutter or shocked in the field. The advent of field choppers has almost eliminated the part of the corn binder in silage making. The practice of shocking corn is disappearing because of the labor involved.

The corn binder uses gathering chains similar to those on a corn picker, but more of them are required. A reciprocating knife cuts the stalk close to the ground after it has been firmly grasped by the gathering chains, and the chains then feed the stalks into an accumulating chamber, where they are tied into bundles by a twine-tying mechanism similar to that on a grain binder or hay baler. The bundles may be dropped directly on the ground, onto a bundle carrier which periodically drops groups of bundles, or onto an elevator which carries them up to a wagon drawn alongside.

It is estimated that the PTO horsepower requirement for a one-row corn binder is less than for a corn picker, averaging 1½ to 2 hp. Three horses were commonly used on the one-row horse-drawn models.

HUSKER-SHREDDERS

The husker-shredder, a companion machine to the corn binder, is also being eliminated by newer methods.

It can be described functionally as a stationary hand-fed corn picker with a cylinder cutter for shredding the stalks and a blower to transport them into the barn mow. The snapping rolls are horizontal when separate, and the ears fall directly onto a husking bed similar to those used in corn pickers. One make utilizes combination snapping-husking rolls whereby the stalks are fed through the upper portion and the snapped ears are husked as they flow to the lower end. The snapping rolls feed the stalks into a horizontal shredding cylinder, and the shredded stalks fall onto a shaker rack with the husks, where they are carried to the blower and discharged through the stacker pipe. Shelled corn is separated out by the shaker rack, cleaned by the screen, and discharged into a bag.

Size of husker-shredders is designated by the number of husking rolls, ranging from 2 to 10. Four-roll models usually have a capacity of about 50 bu per hr and require 15 to 20 hp.

CORN SHELLERS

Shelled corn occupies less storage space on the farm, but because of greater resistance to air circulation, must not be above 13 per cent moisture for safekeeping, compared with 20 per cent for ear corn. Field shelling in combination with picking offers savings in transport and storage space but often requires artificial drying.

[1] C. J. Scranton, Safety and the Mechanical Corn Picker, *Agr. Eng.*, 33:140–142, March, 1952.

When an ear of corn is shelled by hand it is necessary to start at an end. The *spring sheller* works on this same principle, feeding the ears endways into an opening bounded by a rotating fluted cylinder, a rotating toothed disk, and a spring pressure plate. The cylinder shells the ears as the disk revolves them. The cob passes through undamaged after the kernels are removed.

The *cylinder sheller* has largely replaced the spring sheller because of its greater capacity, its ability to shell snapped corn, and its simplicity, although more power is required because more cob is broken and crushed. The cylinder sheller consists of a cylinder with spiral flutes or paddles which turns inside a cage with longitudinal bars as shown in Fig. 25-9. The cylinder clears the bars by approximately 2 in., and the bars are spaced apart enough to let kernels fall through but retain cobs. The ears are fed into an opening at one end of the cage. The spiral flutes feed them through the cage and at the same time shell the ears by rolling and crushing action against the cage and each other. In some machines the cage bars are slightly flexible to reduce crushing of cobs and kernels. An adjustable cob gate serves to control the flow of corn through the cylinder, and it is adjusted to hold the corn in the cylinder long enough to be completely shelled.

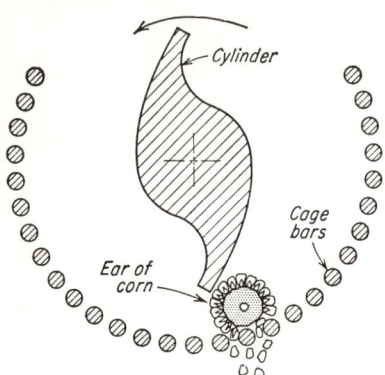

FIG. 25-9. Cylinder sheller.

The cobs pass over an oscillating cob rack and cleaning sieve to remove kernels and then into a cob-stacker elevator. A cleaning fan similar to that on a combine blows air up through the shelled corn and cobs to remove light trash and husks. A paddle-type blower near the end of the cob rack augments the cleaning-fan action and blows the husks and trash out the blower pipes for discharge into a separate pile. The shelled corn is elevated and delivered into the truck or wagon by a swinging spout.

Ear corn is fed into the feeder elevator by a drag conveyer which can be swung 180° laterally. It can have extension sections attached as needed for greatest convenience in getting the corn from the crib.

Sheller *cylinder cages* are usually from 11 to 15 in. in diameter, and the cylinders have a peripheral speed of from 1,500 to 2,000 fpm. Large-cylinder shellers capable of shelling 1,000 bu per hr deliver about 25 bu per hp-hr and have a maximum capacity of from 1½ to 2 bu per hr for each square inch of projected cage area.[1]

Ability to *shell high-moisture corn* is a major consideration in cylinder-sheller attachments for corn pickers. Tests in Indiana showed sheller losses of 5 to 10 per cent with kernel moisture of 29.6 per cent and cob moisture of 56.5 per cent. Part of the loss was due to unshelled corn or crushed-cob sections, and part to difficulty in separating the damp kernels from the cobs.[2] The percentage of kernels left on the cobs and the percentage of kernels damaged by shelling were almost directly proportional to the moisture content of the kernel, as indicated in Figs. 25-7 and 25-8. One manufacturer states that the rate of shelling of 26 to 30 per cent moisture corn is only one-half that of 20 per cent corn. Another approach to field shelling of corn is a separate sheller with its own engine trailed behind the corn picker and delivering the shelled corn into its own tank or into a trailed wagon. Thus the corn can be picked with or without shelling, depending on its moisture content.

[1] "Red Book," Implement and Tractor Publications, Inc., Kansas City, Mo., 1959.
[2] D. E. Burrough and R. P. Harbage, Performance of a Corn Picker–Sheller, *Agr. Eng.*, 34:21-22, January, 1953. R. L. Beldin and A. B. Skromme, Shelling Attachment for Mounted Corn Picker, *Agr. Eng.*, 40:87–91, February, 1959.

Chapter 26

COTTON-HARVESTING MACHINES

Cotton resisted mechanical harvesting the longest of any of our major crops. The bolls do not ripen uniformly in most areas, and as many as three hand pickings may be desirable. The human hand is able to reach into the plant and pick the cotton from an open boll without getting leaves or branches or injuring the rest of the plant, but this is very difficult to accomplish mechanically.[1]

Current cotton harvesters are of two general types:

1. *Pickers,* with rotating spindles to engage the open cotton bolls without damaging the foliage or unopened bolls.

2. *Strippers,* which comb the bolls from the plant.

Stripping has been largely confined to the semiarid regions of Texas and Oklahoma where the plants are small, the cotton can be left in the field until all the bolls are open without serious damage from rain, and the leaves are usually removed by frost in the fall, although the use of chemical defoliants has considerably extended this area. Mechanical harvesting gathers more foreign material than hand picking, as indicated by Table 26-1.

TABLE 26-1. AVERAGE GIN TURNOUTS IN COTTON BELT, 1950–1951

	Per cent turnout	Pounds of seed cotton harvested for 500-lb bale of lint
Hand-picked	37	1,350
Hand-snapped	26	1,920
Machine-stripped	23	2,170
Machine-picked	36	1,390

SOURCE: Leonard J. Watson, "The Effect of Mechanical Harvesting on Quality," 1951 Cotton Mechanization Conference, National Cotton Council, Memphis, Tenn.

"Since many references will be made to loss in grade and consequent loss in dollar value, it seems advisable to explain how the grading and marketing system affects the prices received by the farmer. The price of cotton depends upon color, grade, and staple length. The market pattern is based on white color, Middling grade, and $15/16$-inch staple, with premiums (points on) for better grades and longer staple lengths, and discounts (points off) for lower grades and shorter staples, and additional points on or off for differences in color. Each point represents one one-hundredth of a cent per pound or 5 cents per 500-pound bale. Figure 26-1 shows the price pattern on the Memphis market on May 17, 1945, and is representative only, since the number of points on or off for given grades or staple lengths vary with the supply and demand. This plate brings out clearly the sharp drop in prices obtained when the grade drops from Middling to Strict Low Middling and Low Middling, especially when the staple lengths are more than $15/16$ inch. Because of the short staple length and the comparative harshness of the fiber grown in those areas where strip-

---

[1] H. P. Smith et al., The Mechanical Harvesting of Cotton, *Texas Agr. Expt. Sta. Bull.* 452, 1932.

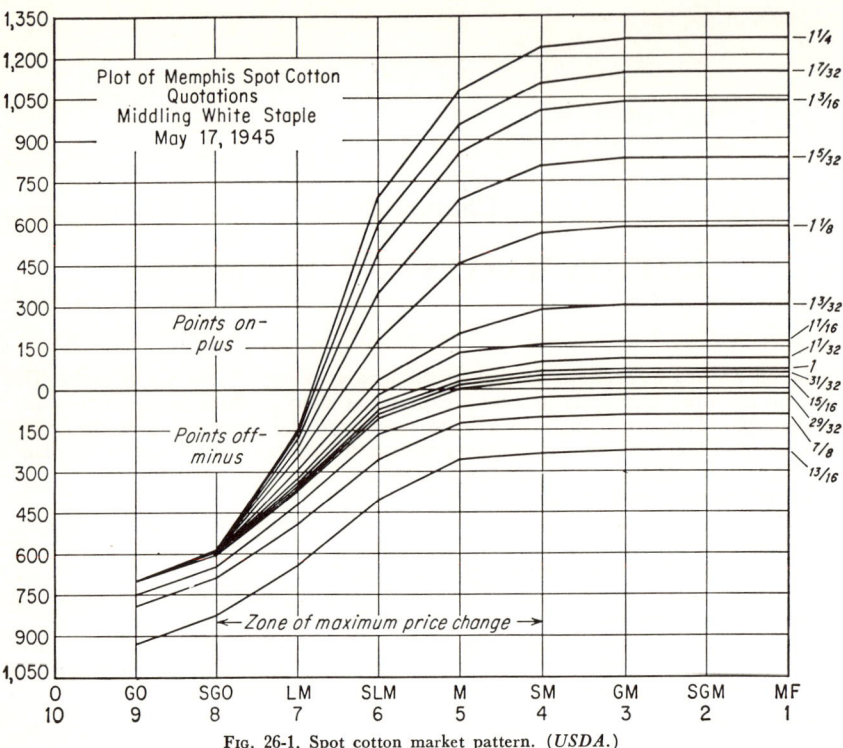

Fig. 26-1. Spot cotton market pattern. (*USDA*.)

ping is most favored, cleaning is easier than in other sections. Then, too, the difference in price differentials is not so great in these grades as with longer staple cotton, so the grower's losses are considerably less than the savings in picking costs."[1]

SPINDLE PICKERS

The *principle of the spindle picker* is shown in Fig. 26-2. The cotton plant is compressed while passing through the picking zone, and rotating spindles, spaced about 1½ in. apart to prevent damage to unopened bolls, are inserted into the plant and withdrawn without forward travel for minimum plant disturbance. When cotton lint contacts the rotating spindle, it is wound up and pulled out of the boll. After the spindle is withdrawn from the picking zone, the cotton lint must be removed and transported to the storage basket.

Spindle bars may be carried on a drum, cam-actuated to avoid raking action, as shown in Fig. 26-2, or on a belt with gradual entry as shown in Fig. 26-3. Reciprocating action has been tried whereby alternate banks of spindles are thrust into the plant, but the mechanical complications have been greater than with the drum or belt types.[2]

The major functional problem is to secure a spindle which is aggressive in catching cotton lint, permits doffing (removal) of the lint, and is nonaggressive to leaves and trash. With the drum-type arrangement shown in Fig. 26-2, the spindles rotate continuously and the cotton is doffed by rubber doffing disks. The tapered spindle permits doffing by this method, and barbs provide aggressiveness. These spindles are

[1] George R. Boyd, C. A. Bennett, and W. E. Meek, Mechanization of Cotton, USDA BPISAE, October, 1947.
[2] C. R. Hagen, Twenty-five Years of Cotton Picker Development, *Agr. Eng.*, 32:593–599, November, 1951.

chromium-plated for wear and corrosion resistance. An individual spindle with integral bevel drive gear is shown in Fig. 26-4a.

With the belt-type arrangement shown in Fig. 26-3, the spindles rotate only while in the picking zone, being driven by friction of the roller drive ends against rubber-

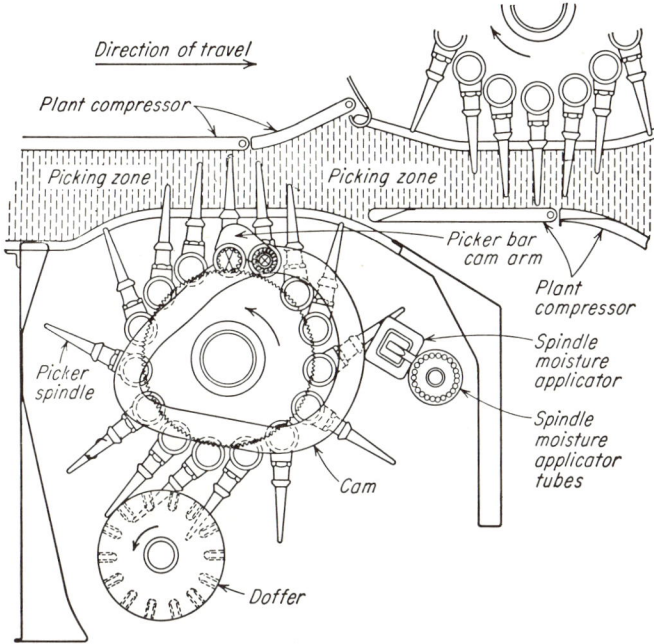

Fig. 26-2. Drum-type spindle picker. (*International Harvester Co.*)

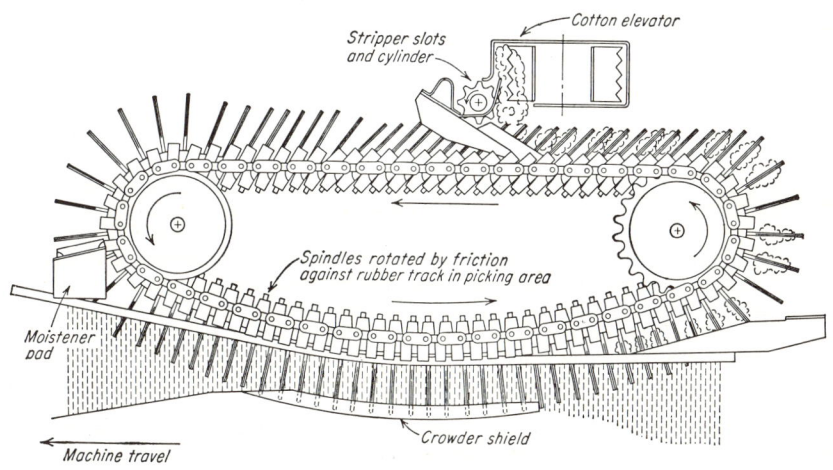

Fig. 26-3. Belt-type spindle picker. (*Allis-Chalmers Mfg. Co.*)

faced tracks. Doffing is somewhat easier because the spindles are not rotating and long straight spindles, as shown in Fig. 26-4b, can be used. They are doffed as they pass between stationary spring-held stripper fingers. The spindles may be roughened by contact with abrasive disks as often as necessary to maintain aggressiveness. Be-

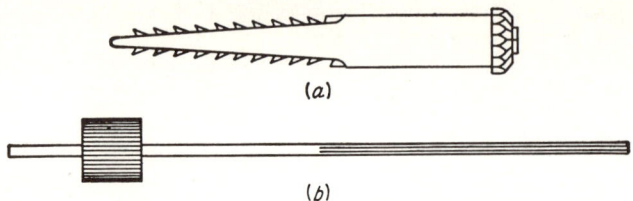

Fig. 26-4. Cotton-picker spindles: (a) tapered gear-driven (*International Harvester Co.*); (b) straight friction-driven. (*Allis-Chalmers Mfg. Co.*)

cause of the more numerous and longer spindles and the longer period of contact, one picking head is sometimes sufficient with the belt arrangement.

Selective aggressiveness is aided by *wetting* the spindles just before they enter the cotton plant. Water is accurately metered to rubber wiping pads which contact the entire spindle. The water also helps prevent the build-up of plant sap, dirt, and cotton fiber on the spindles.[1]

The addition of a chemical wetting agent is sometimes effective in reducing the amount of water required.

The cotton may be *conveyed* to the storage basket by air or mechanically. With air, a paddle-type fan pulls the cotton up from the picker head and blows it into the basket. The intake into the fan usually allows the cotton to bypass the blades, as shown in Fig. 26-5, so as to reduce cracking out of the cotton seeds. Seed cracking leaves locks of cotton attached to small bits of seed hull which cannot be ginned out like whole seeds. These bits of hull lower the grade because they interfere with the spinning process. Mechanical conveyers are usually of the paddle type.

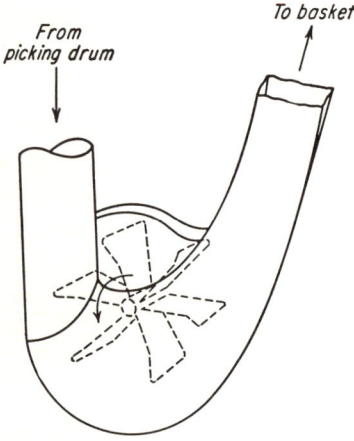

Fig. 26-5. Bypass-type cotton fan. (*International Harvester Co.*)

The number of picker spindles per row varies from 420 to 2,560, depending upon the type of machine and the height of cotton it is designed to pick. Picker-head height varies from 24 to 34 in. Spindle speeds vary from 1,250 to 2,900 rpm to allow picking speeds of 2 to 3 mph. It can be seen that the multiplicity of fast-moving parts provides problems unique in the field of farm machinery. Wear is a serious problem, many spindles requiring replacement after one or two seasons.

Power requirements for operating the picking mechanisms are not high, but the weight of the complete machine and tractor, over 10,000 lb for some one-row machines, requires considerable power for propulsion in muddy fields. A three-plow tractor has been used for carrying these one-row machines, although another two-row machine of approximately the same total weight is also carried by a three-plow tractor.

Picking rates from studies in the Yazoo-Mississippi Delta and in California are summarized in Table 26-2.

Cotton-picker efficiency, in terms of cotton harvested, exclusive of that previously dropped to the ground, ranges from 95 per cent or over in very favorable conditions down to 80 per cent or less.[2] In the latter case, of course, hand picking has an economic advantage if labor is available.

---

[1] *Ibid.*
[2] J. R. Tavernetti and H. F. Miller, Jr., Studies on Mechanization of Cotton Farming in California, *Univ. Calif. Agr. Expt. Sta. Bull.* 747, 1954.

TABLE 26-2. PICKING RATES WITH ONE-ROW SPINDLE PICKERS

	Yazoo-Mississippi Delta*			Louisiana,† 1948	California,‡ 1949
	1945	1946	1947		
No. of machines....................	27	20	26	38	63
Av. bales picked per machine.......	111	87	109	181	229
Av. bales per hour.................	0.4	0.2	0.35	0.73	0.57
First picking:					
Bales per hour..................	...	...	...	....	0.75
Acres per hour..................	...	...	....	....	0.62
Second picking:					
Bales per hour..................	...	...	....	....	0.29
Acres per hour..................	...	...	....	....	0.85

* Grady B. Crowe, Mechanical Cotton Picker Operation in the Yazoo-Mississippi Delta, *Mississippi State Coll. Agr. Expt. Sta. Bull.* 465, 1949.
† Frank D. Barlow, Jr., unpublished data on the performance of 38 mechanical cotton pickers, *Univ. Louisiana Agricultural Experiment Station*, 1948.
‡ T. R. Hedges and W. R. Bailey, Economics of Mechanical Cotton Harvesting, *Univ. Calif. Agr. Expt. Sta. Bull.* 743, 1954.

Sources of loss in mechanical picking compared with hand picking are:
1. Unpicked low bolls 6 in. or less from the ground, which cannot be reached without damage to the spindles from dirt and clods. This loss was 41 per cent of the total loss in California tests.[1]
2. Unpicked bolls above 6 in. which were not engaged by the spindles, amounting to 17 per cent of the total loss in the above tests.
3. Shatter (bolls and locks dropped and locks left clinging on the plant) due to agitation of the plant prior to spindle engagement and also to locks being stripped from the spindles as they retract through heavy growth. This loss averaged 42 per cent of the total in the above tests.
4. Knocking off or damaging unopened bolls.
5. Grade loss due to:
    a. Admixture of leaf trash, hulls, grass, or other material which cannot be completely cleaned out at the gin.
    b. Stain from green leaves or grass.
    c. Graying or mildewing due to moisture from spindles if drying and ginning is delayed.
    d. Twisting and matting the lint due to spindle action.
    e. Discoloration from machine oil or grease.

Studies in California in 1948, 1949, and 1950 found the *total cost* of machine picking to be distributed as follows:[2]

   Machine operation and overhead............ 48%
   Field loss over hand picking................. 30%
   Grade reduction under hand picking......... 22%

The total cost of machine picking averaged two-thirds that for hand picking.

In general, machine-picked cotton averages one grade lower than hand-picked cotton, the difference being greatest at the beginning of the season and least at the end when the cotton has weathered.[3]

Efficient *limb lifters* help get low bolls by lifting the lower limbs and bolls up into contact with the spindles. They are made of rods so as to help sift out any dirt which may be lifted.

[1] J. P. Fairbank and K. O. Smith, Cotton Mechanization in California, *Agr. Eng.*, 32:219–222, May, 1950.
[2] Tavernetti and Miller, *op. cit.*
[3] Grady B. Crowe, Mechanical Cotton Picker Operation in the Yazoo-Mississippi Delta, *Mississippi State Coll. Agr. Expt. Sta. Bull.* 465, 1949.

Shatter loss during picking is reduced by a minimum of agitation prior to spindle engagement. Cotton plants often spread wide, and it has been found desirable for the picking head to contact the plants in advance of wheel shields, etc. With several pickers, the tractor is operated in the reverse direction to permit this.

The remaining boll and shatter losses depend on the efficiency of the spindles in contacting the cotton, pulling it from the boll, and holding it as they retract through the plant. The mechanical factors involved are "type of spindle; number, shape and length of spindles; size; barbs on spindle and their height, sharpness and angle on spindle; dampness and cleanliness of spindle; r.p.m. of spindles; spacing of spindles in picking zone; synchronized movement of the spindle or r.p.m. of picker drum with the forward travel of the machine; width of throat adjacent to spindles; tension or pressure given the plate holding the plants into the spindles; fingers attached to the pressure plate; use of picking spindles on one side or both of the row; and the thoroughness of removal of the cotton from the spindles by the doffing device."[1]

The *varietal characteristics* of the cotton plant affect losses from mechanical picking more than the acre yield.

"In general, a variety of cotton for the mechanical picker should have plants of medium size with a relatively narrow spread and not too tall. It should have a wide-opening boll with fluffy locks and a staple length long enough to wrap well around the spindle. There should be enough storm resistance to prevent the locks from falling out of the bur or being blown out by moderate winds, and from being beaten out by rains."[2]

Cultural practices for efficient mechanical picking differ somewhat from the traditional practices used with hand picking. For instance, it has been found that plant spacings of 2 to 6 in., rather than the traditional spacings of 9 to 15 in., are quite effective in reducing the number of bolls within 6 in. of the ground. This spacing also helps keep branch limbs short, thus reducing the loss of bolls folded tightly into the center of the plant when it is picked.[3] High fruiting can also reduce maintenance costs since the picker drums can be operated higher, thereby reducing the amount of sand and dirt collected with the cotton.

Weather conditions during the growing season have marked effects on the efficiency of mechanical pickers. Drought may cause stunted bolls which do not open enough to allow adequate spindle engagement of the fibers. Under these conditions, picker efficiency may drop below the acceptable level. If a storm occurs after the bolls have opened, the open-boll varieties preferred for machine picking may drop many locks of cotton to the ground. Moderately dry weather may favor machine picking by reducing branch and leaf growth. Wet weather may cause excessive growth or undesirable regrowth after defoliation.

COTTON STRIPPERS

Cotton strippers are of two general types: the finger stripper, with several sloping steel fingers which comb the cotton plant, and the roll stripper, which has a single slot formed by a pair of rolls or a roll and a stationary bar.

Finger strippers require a cleaning reel above the rear ends of the fingers with blades which pass through the slots and push back the bolls. Approximately ¼ in. of space is left between the fingers, which are slightly flexible sideways to accommodate varied stalk sizes. The cotton is then passed back by cleaning beaters over a screen to the wagon elevators.

The finger stripper will not handle as large plants as the roll stripper or as short because the plant must carry the bolls back. Since it does not need to be precisely centered on the row, a pull-type machine can be operated at speeds up to 6 mph.

The construction of a popular *single-roll stripper* is shown in Fig. 26-6. The cotton pulls down through an opening between a stationary bar and an upwardly rotating metal roll, which throws the stripped bolls into a trough where beaters convey it back

---

[1] H. P. Smith and D. L. Jones, Mechanized Production of Cotton in Texas, *Texas A & M Coll. Agr. Expt. Sta. Bull.* 704, 1948.

[2] *Ibid.*

[3] J. R. Tavernetti and B. B. Ewing, Cotton Mechanization Studies in California, *Agr. Eng.*, 32:489–492, September, 1951.

to the wagon elevator over a screen. Double-roll strippers require a conveyer on each side, and the extra width complicates the construction of a two-row mounted machine for tricycle tractors.[1]

Experimentation with double-roll strippers having brush rolls and rubber flap rolls indicates that much larger plants can be handled than with solid rolls, offering the possibility of stripper-harvesting economy for a much greater portion of the cotton crop.[2]

Power requirements for stripping are low, the friction horsepower of the stripper normally exceeding the additional load due to the cotton. It is estimated that 5 hp is normally adequate for a two-row stripper.

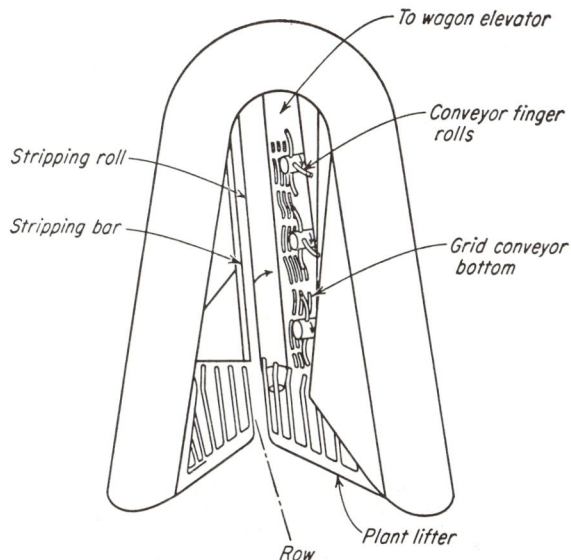

FIG. 26-6. Single-roll stripper. (*Deere & Co.*)

Losses in stripping are usually compared with those for hand snapping, since this is the normal practice in areas where stripping is practical. Sources of loss in stripping are:

1. Unpicked low bolls, although the stripper is more effective in this respect than the picker because it can lift low bolls out of loose dirt without damage to the machine.

2. Unpicked bolls in the upper part of the plant, a rarity except with large bushy plants not normally occurring in stripping areas.

3. Lodged plants.

4. Shatter loss, small with stormproof varieties, although open-boll types may shatter before getting into the machine and open locks may hang to the plant or be pulled around and dropped by the rolls.

5. Green bolls stripped, a loss compared with two hand snappings or one hand snapping followed by stripping.

6. Grade loss over hand snapping, due primarily to weeds, leaves, and sticks. Grass leaves and moisture from green leaves and bolls are very objectionable. During the period 1939 to 1945, tests on various varieties at College Station and Lubbock,

---

[1] H. P. Smith et al., Factors Affecting the Performance of Mechanical Cotton Harvesters (Stripper Type), *Texas A & M Coll. Agr. Expt. Sta. Bull.* 686, 1946.
[2] W. J. Oates, R. H. Witt, and W. S. Wood, The Development of a Brush Type Cotton Harvester, *Agr. Eng.*, 33:135–142, March, 1952. H. P. Smith, *Texas A & M Coll. Agr. Expt. Sta. Progr. Rept.* 1838, 1956.

Tex., showed stripped cotton averaging 0.2 grade lower than hand-snapped and about one grade lower than hand-picked.[1]

In the dry areas to which they are adapted, strippers have the following advantages over the spindle pickers:

1. Faster operating speeds.
2. Picking cost per acre is much less than with spindle pickers, allowing profitable use on yields down to one-fifth bale per acre.
3. Greater percentage of cotton harvested.
4. Less moistening and matting of the lint.

Stripper tests of several varieties in Texas for the period 1939 to 1945 found efficiencies to average 89 per cent in the College Station area and 96 per cent at Lubbock (irrigated).[2]

CULTURAL PRACTICES FOR MECHANICAL HARVESTING

Proper cultural practices are extremely important for efficient mechanical harvesting. Recommended practices are as follows:

1. Plant in 40-in. rows, particularly for two-row machines designed for this width.
2. Plant storm-resistant or stormproof cotton for conventional strippers and open-boll cotton for spindle pickers.
3. Space plants close enough to keep stalk size down. The desired spacing may vary from two to six plants per foot, depending on the soil fertility and moisture condition of the area.[3]
4. Space plants as evenly in the row as possible and do not hill-drop for stripping.
5. Set sweeps during late cultivations so that the middle will be lower than the ridge of dirt at the base of the plants. This will give the plant lifters a better chance to get low bolls without collecting defoliated leaves.
6. Apply defoliants so that cotton can be picked 7 to 15 days after application. For stripping, time defoliation so that as few as possible green bolls will be left unopened when cotton is ready to strip.
7. Use tractor-wheel shields to prevent damage to plants and bolls during late insect control and defoliation. Use airplanes in rank cotton.[4]

Weeds and cotton-plant leaves are the two greatest obstacles to successful mechanical harvesting. Both increase trash and stain as well as causing clogging of strippers and preventing spindles from contacting open locks.

Preemergence and postemergence sprays, flame cultivation, and late cultivation are being tried and show some promise for better control of weeds.[5]

Defoliation may occur naturally because of frost or be artificially induced by several chemicals, of which calcium cyanimide has been most used. The latter compound is applied as a dust and requires a little moisture, such as dew, for efficient action. Rain immediately following application washes it off and instead of defoliating it fertilizes the soil, encouraging new leaf and weed growth.

The defoliant should be applied when the bolls are 90 to 95 per cent open, and 7 to 15 days are required for the leaves to drop. In the humid sections, rain and muddy fields may prevent harvesting when defoliation is complete and delay it until regrowth of leaves has occurred. This is serious because the succulent new leaves are a great handicap to machine harvesting and are very difficult to defoliate.

Considerable research is under way to find chemicals which are more effective defoliants, and success in this endeavor will remove one of the last barriers to mechanical cotton harvesting.

[1] H. P. Smith et al., *op. cit.*
[2] *Ibid.*
[3] J. G. Porterfield, O. G. Batchelder, and W. E. Taylor, Plant Population for Stripper Harvested Cotton, *Oklahoma State Univ. Agr. Expt. Sta. Bull.* B-514, 1958.
[4] H. F. Miller et al., Mechanical Harvesting of Cotton in Texas, *Texas A & M Coll. Agr. Expt. Sta. Progr. Rept.* 1337, 1951.
[5] Tavernetti and Miller, *op. cit.*

Chapter 27

ROOT-HARVESTING EQUIPMENT

Harvesting of root crops involves lifting and separating the crop from the soil. Thus the variables of soil type and condition are added to the usual crop variables encountered in aboveground harvesting. Because of this, root-harvesting machines are often adapted only to the area where they have been developed.

POTATO DIGGERS AND HARVESTERS

Potato diggers may lift and handle as much as 5 tons of dirt and potatoes per minute per row. Draft and PTO power requirements are both high, necessitating extremely rugged construction. Conventional diggers deposit the potatoes in a row on the surface of the ground to be picked up, sacked, and loaded by hand. Diggers

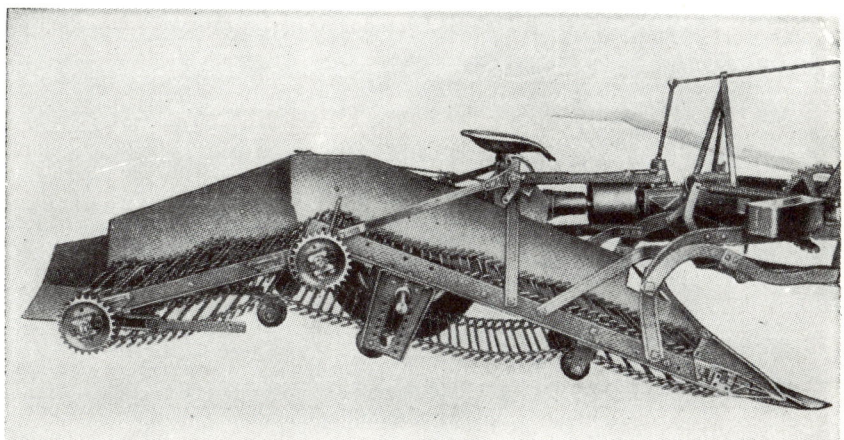

FIG. 27-1. One-row potato digger with right-hand side removed. (*International Harvester Co.*)

which deliver the potatoes directly into sacks or into trucks have been developed and are meeting increasing acceptance by commercial growers. This market, however, consists of several widely separated areas with varying requirements, and such machines tend to be produced in and for a particular area. A one-row conventional potato digger is shown in Fig. 27-1.

Lifting *shovels* for potato diggers must lift enough dirt to get all the potatoes and also cut and shed roots and trash. They are as wide as the elevators, usually 22, 24, or 26 in., and are available in several shapes to fit various degrees of ridging, weed-root conditions, and soil characteristics. Shovels are made of high-carbon heat-treated steel to wear and scour well and must be firmly held to resist shocks. Shovel draft per square inch of cross section is roughly equivalent to plow draft in comparable soil.

Weeds and trash may cause clogging, particularly with two-row diggers, unless coulters or disk blades are used to divide the trash so that it can be taken up the elevator. One two-row digger uses a continuous blade which takes up all trash in the middle instead of trying to divide it. The practice of shredding top growth before digging, to reduce clogging and separation difficulties, has become common.

Separation of potatoes from the dirt is normally performed by a potato chain conveyer web consisting of a series of crossrods having the ends bent into interlocking chain hooks. The crossrods are spaced about 2 in. apart with the central portion bent up on one rod and down on the next to facilitate carrying the potatoes up the incline. Oval idler sprockets agitate the web to assist separation. A roller is sometimes placed under the web in light loose soils to make a firm smooth bed for easier picking up of potatoes.

Potato harvesters ordinarily use pairs of rubber rolls or belts for the mechanical elimination of vines. Considerable separation of stones and clods from the potatoes is achieved by adjustable tilt of a conveyer belt or chain. The remaining foreign material is removed manually by transported workers before the potatoes are discharged into sacks or truck, as the case may be.[1]

In Europe *potato spinners* are popular because they are simpler than the conveyer web machines. They consist of a loosening share and a series of rotating fingers, or spokes, which pass sideways through the loosened soil. These spokes lift out and throw the potatoes onto a substantially horizontal rotating spoked disk which sieves out the remaining dirt and leaves the potatoes on top of the ground.

Bruising of potatoes leads to spoilage and reduced marketability.[2] The web-conveyer digger is much better than the spinner in this respect but still leaves much to be desired. Experiments at the University of Idaho resulted in the following recommendations for reducing injury to a minimum.

"1. Separate the soil from the potatoes on rubber rolls after they have traveled up a rubber-coated digger chain.

2. Keep the potatoes on rubber from soil to sack.

3. Use rubber padding to cushion the potatoes.

4. Use rubber tubing cemented to the digger links.

5. Allow the potatoes no free fall over 6 inches.

6. Reduce the operating speed and carry a cushion of soil up the digger chain."

Machines built in accordance with these principles, using rubber-roll separation, could be used under muddier conditions and gave much less bruising than conventional machines. A two-row bulk-handling digger of this type for elevating into a truck required approximately 15 hp to drive digger chains separating rolls and elevator.[3]

BEET HARVESTERS

Successful mechanical beet harvesters are a comparatively recent development, and some of the machines currently available are adapted only to the beet-growing area in which they were developed. A one-row tractor-mounted beet harvester with a trailed power-unloading collecting trailer is shown in Fig. 27-2.

Topping. Beet topping is necessary to prevent growth in storage and also to facilitate the sugar-extraction processes. With hand topping, the crown is ideally cut just below the lowest leaf, a distance below the crown which will vary with the diameter of the beet. Powers[4] found a positive correlation between beet diameter and height of crown above the ground surface, indicating the desirability of a variable-cut topping mechanism which varies crown-cut thickness according to beet crown height. This differential gaging principle is used in most machines which top the beet in the

---

[1] H. D. Bartlett and D. H. Huntington, Mechanical Potato Harvesting, *Univ. Maine Agr. Expt. Sta. Bull.* 549, 1956.

[2] R. Bruce Hopkins, Effect of Potato Digger Design on Tuber Injury, *Agr. Eng.*, 37:109–111, February, 1956.

[3] J. W. Martin and E. N. Humphrey, Development of Idaho Potato Harvesters, *Agr. Eng.*, 32:261–269, May, 1951; The Idaho Potato Harvester, *Univ. Idaho Agr. Expt. Sta. Bull.* 283, 1951.

[4] John B. Powers, The Development of a New Sugar Beet Harvester, *Agr. Eng.*, 29:347–354, August, 1948.

ground. In such machines, a power-driven finder wheel or web is often used to prevent pushing the beets over, as shown in Fig. 27-3. The cutting knife or disk should be light in weight for quick, easy gaging, and it should also cut with a minimum of tipping force.

Disk toppers cut well and shed trash better than knife toppers but are more complicated and have more inertia, making quick accurate positioning more difficult. One

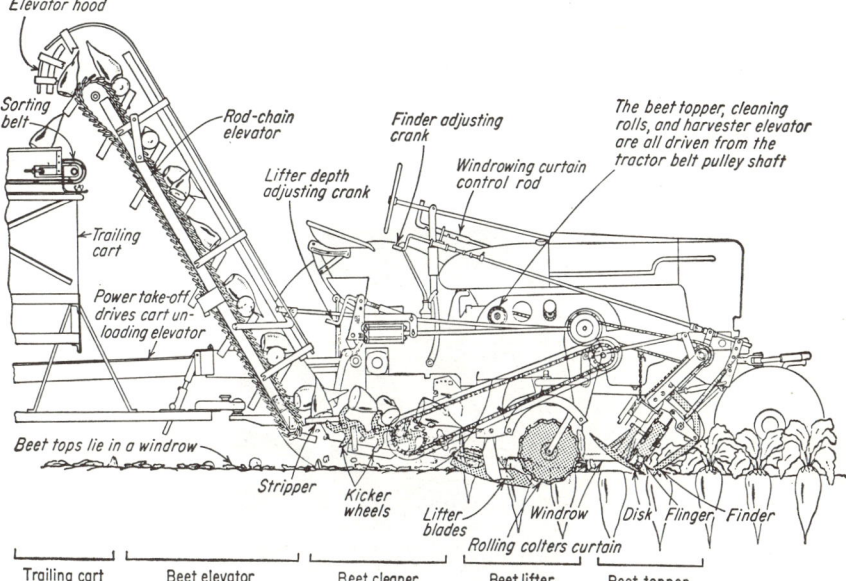

Fig. 27-2. One-row tractor-mounted beet harvester with collecting trailer. (*International Harvester Co.*)

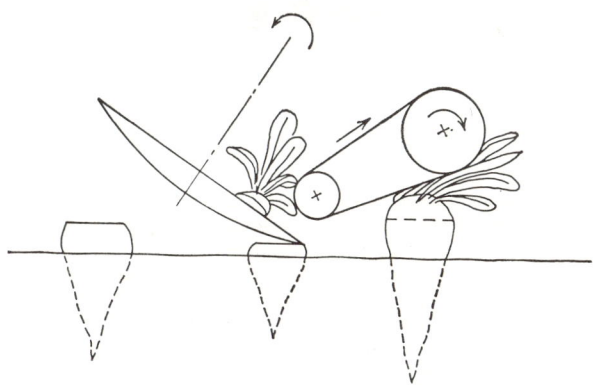

Fig. 27-3. Disk topper and power-driven finder. (*International Harvester Co.*)

disk topper has flinger fingers which throw the tops aside. Another machine picks up the tops with a spiked wheel and dumps them in a windrow for collection.[1] Beet tops are a valuable livestock feed, ranking high in nutritive value per pound.

A recent development in topping is the use of a *rubber flail shredder* which beats off the tops without damaging the beets. A scalping operation is still required to remove the terminal bud, but this can be done by a floating knife-and-finder assembly

[1] Claude W. Walz, Mechanization of Sugar Beet Harvesting, *Agr. Eng.*, 27:549–550, December, 1946.

cutting at a constant thickness. Less of the beet is wasted by this method, but the tops are lost.

Machines which top the beet after picking it up, such as shown in Fig. 27-6, have an advantage in collecting or windrowing the tops but tend to top at a constant thickness.

Lifting. Lifters may use single or double shares. Single shares must run deeper than double shares, which straddle the beets and exert upward soil pressure on each side of the beet to lift and loosen it. The shape, angle, and spacing of the lifter shares should be such as to lift the beet without breaking the taproot. They must also produce maximum pulverization of the soil around the beet in order to avoid lifting large clods, which may be difficult to separate in the machine. Powers found a helical shape to be most effective.[1]

A pair of tilted rimless *lifter wheels* with chisel-shaped spokes, as shown in Fig. 27-4, may be used for lifting. The wheels loosen and lift the beets, crushing the soil around

FIG. 27-4. Beet-lifting wheels. (*Deere & Co.*)

them. Paddles running between the wheels push the beets onto a separating or elevating unit. This arrangement is comparatively simple and works very well where the ground is not too hard to prevent penetration.

Separation. Separation of beets from dirt is aided by a Reink's screen in many machines. This consists of several rows of overlapping star wheels, as shown in Fig. 27-5, which pass the beets back to an elevator and also tend to crush clods and separate dirt without clogging with trash. Potato chain elevator webs and elevators with grill floors are also used to separate dirt. Another expedient used by Powers for breaking clods and loosening dirt from the beets was tumbling in the final elevator, secured by half-width elevator flights on alternate sides of the conveyer crossbars.

Rocks and clods which cannot be broken without damaging the beets are sometimes picked up. Some harvesters dump the beets on a sorting belt, where transported workers pick off foreign material.[2]

Some machines achieve separation by selectively lifting the beets out of the dirt and topping them afterwards. One type grasps the tops between two belts as the beets are loosened by a lifter blade and elevates the beets back to topping disks that sever the tops and allow the beets to fall into an elevator, which dumps them into a truck.[3] The tops are dropped in a windrow in good shape for collection. This principle of

[1] *Ibid.*
[2] C. E. Guelle, A New Sugar Beet Harvester, *Agr. Eng.*, 27:552–553, December, 1946.
[3] W. E. Urschel, A Sugar Beet Combine, *Agr. Eng.*, 27:551–553, December, 1946.

operation gives good dirt separation. Topping is less precise than is possible before the beets are lifted from the ground. Two adjustable gaging rods grasp the tops adjacent to the crown and guide the beet into the topping disks as shown in Fig. 27-6. This machine requires healthy tops by which the beets can be elevated and has been most successful in the Ohio-Michigan area. It has also been adapted to conditions in Ireland.[1]

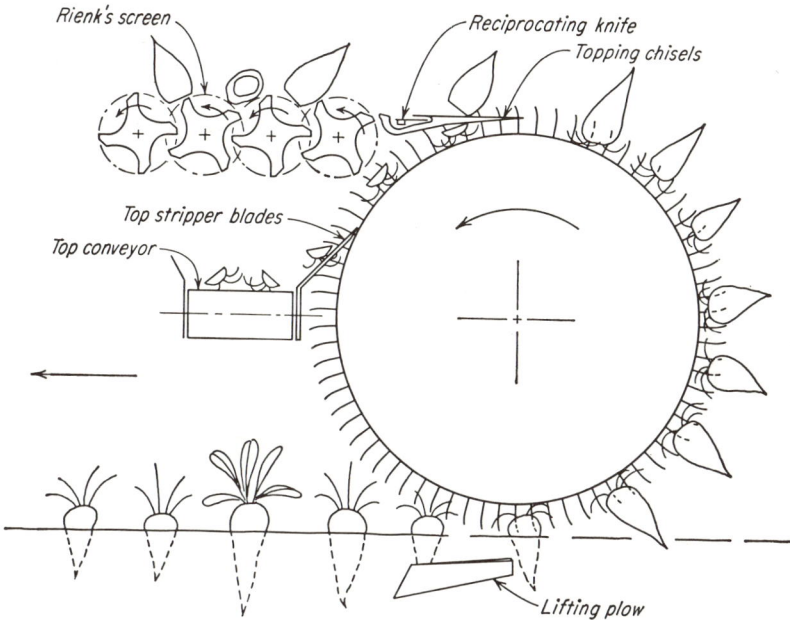

Fig. 27-5. Spiked-wheel beet harvester—Marbeet. (*Blackwelder Mfg. Co., Rio Vista, Calif.*)

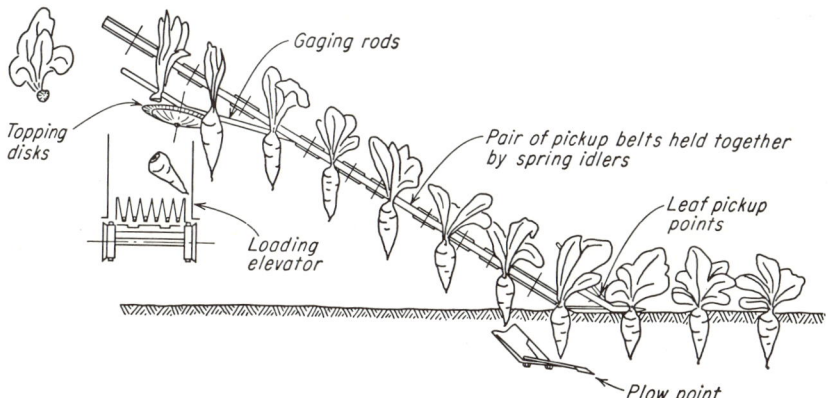

Fig. 27-6. Beet harvester which lifts beets by tops. (*Scott Viner Co., Columbus, Ohio.*)

Another type of machine uses a large spiked wheel, 30 in. to 6 ft in diameter, to lift the beets out of the soil.[2] Lifter blades loosen the beets as the wheel spikes penetrate the beet crowns, as shown in Fig. 27-5. The beets are carried up and into top-

---

[1] Austin A. Armer, A Harvester for Ireland's Sugar Beets, *Agr. Eng.*, 34:312–313, May, 1953.
[2] E. F. Blackwelder, Development of the Marbeet Beet Harvester, *Agr. Eng.*, 29:481, November, 1948.

ping chisels between the rows of spikes which, aided by a reciprocating knife, sever the tops. The beets pass over a Reink's screen and are elevated into a truck. This machine is well adapted to California conditions and harvests most of the crop there. A somewhat similar machine modified for muddy conditions has been successful in the Red River Valley of North Dakota and Minnesota.

Some differential topping action is secured with these machines since the larger beets grow higher and are spiked somewhat deeper than small beets. Topping in place in the ground followed by spiking has reportedly been unsuccessful because the beets do not all stay on the spikes unless the spikes pass through the tough crown.

Harvesting losses are due to missed or dropped beets, broken taproots, overtopping, or tare penalty at the mill due to undertopping and dirt. Beets are a high-value crop, and losses are correspondingly serious. Modern beet harvesters will harvest as much as 98 per cent of the beets under favorable conditions. Overtopping loss and undertopping tare may run as low as 2½ per cent each, and dirt tare as low as 5 per cent. These figures represent performance equal to average hand labor. Under adverse conditions performance is much impaired, particularly in beets missed and dirt tare.

Beet harvester *power requirements* vary greatly with soil condition and type as well as with the mechanism used. The draft for a conventional one-row lifter is roughly equivalent to the draft of a single-bottom plow. A full three-plow tractor is normally required to operate complete one-row harvesters which top, lift, separate dirt, and elevate the beets into a truck. Very high lifter draft is encountered in some of the heavy California irrigated soils when they have been allowed to dry out for harvesting.

Future developments in beet harvesters may emphasize more efficient collection of tops, the utilization of which is necessary in some areas to make beets an economic crop. Since the tops can be most efficiently collected as they are removed, one possibility may be to divide beet harvesting into two separate operations, top harvesting and root harvesting.[1]

PEANUT-HARVESTING EQUIPMENT

Current peanut-harvesting practices require a drying period between digging and threshing. Consequently, harvesting cannot be done in a single operation although the curing period and consequent weather hazard can be shortened by artificial drying.[2]

Peanuts are ordinarily lifted by digger blades resembling large half sweeps which pass under the nuts, loosening them and severing the lower roots.

Separation of dirt from the peanuts and vines is traditionally done by hand with a pitchfork as part of the process of stacking the vines and nuts on poles for curing. This operation is very tedious and, in combination with hand feeding of the stacks to portable peanut threshers, requires as much as 30 man-hours per acre for harvesting. The current trend in areas growing peanuts primarily for oil is toward drying in a windrow followed by combining. Where the digger-shaker leaves the peanuts in a suitable windrow, drying has been satisfactory and the labor requirement for harvesting is reduced to about 4 man-hours per acre.[3] Edible peanuts are still stacked in order to obtain the highest quality possible.

The ideal peanut *digger-shaker* should lift all the nuts, shake off all the dirt, and lay the vines in a windrow upside down with the nuts on top for rapid drying and prevention of spoilage. Many machines which function satisfactorily in light sandy soils are unsatisfactory in heavy sticky soils. Current digger-shakers use conventional digger blades with several methods of dirt separation and windrow forming, the most

---

[1] R. D. Barmington, P. N. Davis, and H. S. Wilgus, Conservation of Sugar Beet Tops by Dehydration, *Colo. A & M. Coll. Agr. Expt. Sta. Bull.* 47, 1952.

[2] J. L. Butt and F. A. Kummer, Artificial Curing of Peanuts, *Agr. Eng.*, 32:27–32, January, 1951. Vernon H. Baker, B. M. Cannon, and J. M. Stanley, A Continuous Drying Process for Peanuts, *Agr. Eng.*, 33:351–356, June, 1952. Norman C. Teter, Peanut Curing as Related to Mechanization, *Agr. Eng.*, 35:568–573, August, 1954. N. C. Teter and R. L. Givens, Technical Progress Report on Curing Virginia Type Peanuts, *USDA ARS* 42-12, July, 1957.

[3] C. M. Stokes and I. F. Reed, Mechanization of Peanut Harvesting in Alabama, *Agr. Eng.*, 31: 175–177, April, 1950.

popular being a pickup cylinder and elevator web to lift the vines and nuts 4 ft or more before dropping them. Deflector rods are used to form a single partially inverted windrow from the two rows dug at once. Side-delivery rakes are also used to shake out dirt and form windrows, although most rakes form too tight a windrow for good drying.[1]

Peanut threshers use spring-tooth cylinders and concaves to comb the nuts from the vines. Cylinder speed is low, from 300 to 400 rpm, since it is not desired to shell the nuts.[2] Separation of the nuts from the vines, dirt, and trash is similar to that in conventional threshers.

Considerable development work has been done on a peanut harvester with a pickup conveyer similar to that used on digger-shakers; two slow-speed intermeshing spike-tooth picking cylinders; and three spring-tooth separation cylinders which carry the vines back over expanded metal concaves.[3]

Small-grain *combines* are being successfully modified to pick up and thresh peanuts from the windrow. Spring-tooth cylinders and concaves, larger straw-rack openings, and other minor changes are necessary, but the labor saved makes this method very attractive. The large amount of abrasive dirt and sand passed through the combine can cause excessive wear of interior parts if they are not adequately designed. In some areas peanut threshers have been equipped with pickup attachments and used as combines. Picking up a peanut windrow requires more precise header adjustment than a grain windrow because there is no stubble to hold it off the ground.

SWEET-POTATO HARVESTING EQUIPMENT

Sweet potatoes can be dug with regular potato diggers equipped with dividing coulters, but the rank vine growth greatly hampers the operation. The vines have considerable feed value and can be harvested for silage before harvesting the potatoes. An experimental vine harvester uses half sweeps to sever the vines just below the surface of the ground, followed by a pickup cylinder and web conveyor to load the green vines into a wagon or truck.[4] The use of a field chopper with pickup attachment should offer a method of making silage from the severed vines.

Rod lifters behind a lister-type digger increased the number of sweet potatoes exposed on the surface in South Carolina studies.[5]

ONION HARVESTERS

A mechanical onion harvester which performs the functions of digging, lifting, topping, and sacking the onions has been developed at the University of California.[6] The principles of operation are somewhat similar to those used in the beet harvester shown in Fig. 27-6.

A commercial onion harvester lifts the onions onto a chain conveyer with a pair of disks and top-grasping belts. An air blast under the conveyer removes dirt and orients the tops for removal by a sickle-type knife. The topped onions are then conveyed into crates.

---

[1] I. F. Reed and O. A. Brown, Developments in Peanut Harvesting Equipment, *Agr. Eng.*, 25:125-128, April, 1944. G. B. Duke, Progress Report on Harvesting Virginia Type Peanuts, *USDA ARS* 42-11, July, 1957.

[2] J. W. Sorenson, Mechanization of Peanut Production in Texas, *Agr. Eng.*, 33:561-562, September, 1952.

[3] Charles E. Rice and James H. Ford, The Georgia USDA Peanut Harvester, *Agr. Eng.*, 35:168-170, March, 1954.

[4] O. A. Brown, A Machine for Harvesting Sweet Potato Vines, *Agr. Eng.*, 27:303-304, July, 1946.

[5] Joseph K. Park et al., Machinery for Growing and Harvesting Sweet Potatoes, *S. Carolina (Clemson Coll.) Agr. Expt. Sta. Bull.* 404, 1953.

[6] C. Lorenzen, Jr., The Development of a Mechanical Onion Harvester, *Agr. Eng.*, 31:13-15, January, 1950.

Chapter 28

SPECIALIZED HARVESTING EQUIPMENT

Mechanization of the harvesting of minor crops is proceeding at a rapid rate, with much of the development work being done by growers or local shops. Some of this harvesting equipment is briefly described and available references are listed in the following discussion.

FORAGE-SEED HARVESTING

With the increased emphasis on high-producing pasture and on the inclusion of legume and grass crops in rotations, the demand for forage seed has made its production profitable in adapted areas.

Forage seeds vary greatly in physical and growth characteristics, necessitating several different types of harvesting equipment.

Strippers. Strippers are used primarily for harvesting bluegrass seed. The most popular type uses an 18-in.-diam beater drum studded with steel spikes 2½ in. long to beat the seeds out of the heads and into a collecting box or onto a cross-conveyer which delivers the seed into sacks for transport to the curing yard.

Comb strippers, similar in principle to the finger-type cotton stripper, have also been used for bluegrass. The collected seed and stems were originally cut and raked back off the comb by hand, although this is done mechanically on later models.[1] The beater type of machine harvests fewer hulls and stems than the comb type, facilitating subsequent cleaning of the seed, although there is more shatter loss.

Bluegrass seed is harvested before it is dry enough to store in order to reduce shatter loss. It is dried in large windrows on outdoor curing yards and must be turned frequently to prevent heating and consequent poor germination of the seed. Side-delivery rakes are used to turn the windrows, and power sweepers may be used to help collect shattered seed after the windrows have been picked up.

Combines. Combining is the most popular method of harvesting grass and legume seed, but the varying requirements of the different seeds require special adjustments, techniques, and accessories. Many forage crops are mowed and raked into windrows for curing before combining. Proper combine pickup action is essential in easy-shattering crops, and the belt-type pickup, ground-driven at a peripheral speed about 10 per cent faster than the speed of travel, is most efficient.[2]

Many seeds are easily cracked, and excessive *cylinder speed* is very damaging. On the other hand, crimson clover is very difficult to thresh out of the head, and special high-speed cylinder-drive sheaves are available for some combines, a peripheral speed of approximately 7,000 fpm being desirable. In alfalfa studies in California, proper cylinder adjustment increased germination 10 per cent and reduced losses by one-half.[3]

It has been found that in alfalfa, for instance, the shoe must handle several times as much material as when operating in grain, because of the great amount of fine material settling through the straw racks and recirculating several times as tailings. Therefore, proper adjustment or choice of chaffer sieves is extremely important since

[1] J. B. Kelly, Machinery for Harvesting Bluegrass Seed, *Agr. Eng.*, 22:353–354, October, 1941.
[2] L. G. Jones, R. A. Kepner, Roy Bainer, and J. P. Fairbank, Alfalfa Seed Harvesting Equipment, *Calif. Agr.*, 8:9–16, August, 1950.
[3] *Ibid.*

any more tailings than necessary cause extra recirculation and overload the shoe, resulting in heavy losses. In the same way, too little wind can increase tailings and losses, although the wind velocity must be less than the settling velocity of the particular seed.[1] (See Chap. 24.)

Typical alfalfa and red clover seed losses found in California studies are shown in Table 28-1. Losses of various forage seed, found in Oregon studies for windrowing and combining compared with direct combining, are shown in Table 28-2.[2]

Large producers of high-value hard-to-save seed crops have, in some cases, built their own self-propelled combines, using high-capacity thresher bodies and often mounting recleaners on the threshers. In addition, the threshed straw may be blown

TABLE 28-1. TYPICAL SEED LOSSES UNDER AVERAGE AND HEAVY LOADS IN IRRIGATED ALFALFA AND RED CLOVER STANDS WITH SEED YIELDS OF 800 TO 900 LB/ACRE IN CALIFORNIA

Crop	Swath width, ft	Speed, mph	Material rate, lb/min	Seed losses, % of total yield		
				Free	Unthreshed	Damaged
Alfalfa	9	0.6	60	0.01	0.3–1.2	2 and up
	9–10	1.2–1.5	120	0.2–1.0	0.8–1.5	0–3
Red clover	8½	0.4–0.5	40	0.2–0.3	2–5	3–4
	8½	0.8–0.9	80	0.3–0.5	4–11	1–3

SOURCE: P. R. Bunnelle et al., Small-seed Legume Harvesting, *Calif. Agr.*, 6:11–12, September, 1952.

TABLE 28-2. LOSSES OF FORAGE SEEDS BY WINDROW COMBINING AND DIRECT COMBINING

Losses, %	Crimson clover		Alta fescue		Hairy vetch		Subterranean clover
	Windrow combine	Direct combine	Windrow combine	Direct combine	Windrow combine	Direct combine	
Field germination	3	....	8	4.4	12.7	16.0	3.0
Shatter:	32	25	25.2	29.8	25.0	35.0	68.0
Preharvest	5	....	2.8	19.5	7.2	22.2	} 48.4
Windrowing	23	....	14.6	....	12.2	....	
Windrow pickup	4	....	7.8	....	5.6	....	19.7
Cutter bar	..	....	.....	10.3	.....	12.8	
Combine:	32	29	1.2	4.0	9.3	19.5	6.3
Cylinder (unthreshed)	23	12.7	0.3	1.0	2.6	1.4	3.2
Rack	2	7.2	0.5	1.0	3.3	2.5	0.8
Shoe	1	1.9	2.7	0.4	1.2	7.0	0.6
Broken seed	2	1.9	0.0	0.0	2.4	8.7	1.2
Machine germination	4	5.3	−2.3	1.6	−0.2	−0.1	0.5
Total loss	67	63	35	38	47	71	77

SOURCE: J. E. Harmond, Seed Harvesting Research, *ASAE Paper* 57-518, December, 1957.

into a trailer for rethreshing, and shattered seed gleaned from the ground by a vacuum harvester.

Direct combining of legume seed crops is made possible by spraying the plants with a general contact weed-killing chemical such as dinitro-general type in oil. This kills the plant growth above the ground and causes it to dry rapidly, permitting direct combining within 1 to 5 days. Under favorable conditions the cost of spraying is offset by the saving of seed normally lost by shattering or during mowing, raking and, picking up.

[1] P. R. Bunnelle, L. G. Jones, and J. R. Goss, Combine Harvesting of Small-seed Legumes, *Agr. Eng.*, 35:554–558, August, 1954.
[2] Jesse E. Harmond, Seed Harvesting Research, *ASAE Paper* 57-518, December, 1957.

In addition, the higher-moisture seed is less easily cracked by the cylinder, the higher-moisture straw breaks up less and reduces the shoe load, and the direct-combined material feeds into the combine more uniformly than a windrow. All these factors reduce machine losses.[1]

The *efficiency* of small combines in harvesting native grass seed in the Great Plains area is reportedly increased by the following modifications:[2]

"(a) Extending the platform about 9", thus moving the sickle forward and tilting the bar so the sickle can run flat on the ground.

(b) Bridging the gap between the platform and sickle bar with tin.

(c) Lengthening the pitman rod and resetting the bell crank to fit the new sickle placement.

(d) Equipping reel bats with brushes of broom to sweep seed material up on the canvas."

Table 28-3 lists the most common native grasses, their seed characteristics, and cylinder-speed and clearance recommendations.

TABLE 28-3. NATIVE-GRASS-SEED CHARACTERISTICS AND COMBINE-CYLINDER ADJUSTMENTS

Grass	Fill			Purity		Average germination, %	Length of seed harvest, days	No. of seeds per pound pure seed	Cylinder speed,* rpm	Concave clearance,* in.
	Average, %	High, %	Minimum for harvest, %	Thresher run, %	Re-cleaned, %					
Crested wheat grass	55	80	35	65	94	88	10–15	165,000–200,000	1,400	3/8
Western wheat grass	50	80	35	55	88	80	15–21	100,000–125,000	1,500	3/8
Slender wheat grass	70	88	35	70	95	92	10–15	140,000–160,000	1,500	3/8
Big bluestem	21	70	20	24	40	60	5–25	140,000–170,000	900	1/2
Little bluestem	35	80	30	35	55	60	5–25	254,000–263,000	900	1/2
Sand bluestem	13	60	20	20	40	60	5–25	105,000–122,000	900	1/2
Side-oats grama	10	50	20	12	30	65	5–20	500,000	1,100	3/8
Smooth brome	70	85	40	70	90	90	7–10	170,000–200,000	1,000	3/8
Buffalo grass	151 (burr)	250 (burr)	100 (burr)	65	85	50	330	40,000–55,000	1,400	3/8
Canada wild rye	75	90	40	65	94	90	15–21	110,000–120,000	1,100	3/8
Sand love grass	40	80	25	50	98	45	5–30	1,300,000	1,500	1/8
Switch grass	46	75	40	50	95	45	5–20	370,000–420,000	1,500	1/4
Green needle grass	60	80	40	70	98	5–60	7–10	80,000–100,000	1,100	3/8

* These settings are intended as a guide in starting the harvest; they will need some changes in the field.
SOURCE: A. D. Stoesz, *Crops and Soils*, Vol. 10, April–May, 1953.

Vacuum Harvesters. Vacuum harvesters are able to collect small easily shattered seeds growing close to the ground more effectively than any other machine. They are most commonly used for collecting the shattered seed left on the ground after the windrow has been combined, particularly in Ladino clover.

This type of harvester was pioneered in Kansas for harvesting the extremely low-growing buffalo grass seed. A machine was developed, shown in Fig. 28-1, which was

[1] L. G. Jones and P. R. Bunnelle, 'Direct Combining of Small-seeded Legumes,' University of California, Agronomy Dept., January, 1953, mimeo. J. K. Park and B. K. Webb, Seed Harvesting Research in the Southeast, *ASAE Paper* 57-519, December, 1957.
[2] A. D. Stoesz, Native Grasses—Better Seed Harvesting, *Crops and Soils*, 5:9–11, 1953.

able to get from 50 to 90 per cent of the seed where no practical harvesting method had existed before.[1]

Development was continued in a machine for gleaning fallen peppermint leaves in Indiana. It was found that a vacuum of 0.95 in. of water and an air velocity of 2,300 fpm at the mouth of a 3-in.-wide nozzle was most effective for this purpose.[2]

One modification uses the vacuum to suck seed and stems off a mower cutter bar and blows them into a trailer for hauling to a central thresher.

Another type of vacuum machine harvests seed from tall standing plants by straddling a row with a pair of vertical nozzles, one of which blows across into the other to collect ripe seed without damaging the plant and immature seed. This type of vacuum machine permits several harvestings of grasses which bear seed over a prolonged period.

In Oregon studies of harvesting subterranean clover, a vacuum attachment on a combine, located behind the header, reduced losses from 77 to 8 per cent.[3]

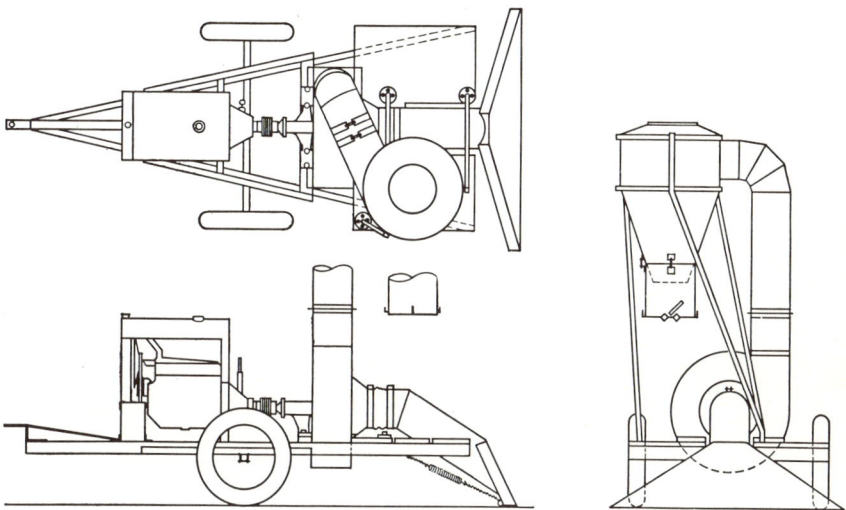

FIG. 28-1. Experimental vacuum harvester for buffalo grass seed. (*Zink, Agr. Eng., May,* 1936.)

Alfalfa Pollinating Machine. Alfalfa blossoms must be tripped to expose the pollen for pollination. The tripping action delivers a blow to the bee which seeks the pollen, and consequently bees prefer other blossoms. As a result, alfalfa seed yields are often low because of inadequate pollination. An experimental machine trips the blossoms by passing them gently between foam-rubber rollers and then blows warm air through the blossoms to spread the pollen dust.[4]

Shatter Prevention by Baling. Bird's-foot trefoil is very difficult to harvest because the ripe pods tend to burst open and drop the seed. A successful technique for reducing losses has been developed by H. D. Hughes of Iowa State University. A crop is mowed and windrowed when fairly green and, after partial curing, baled into round bales which shed rain fairly well. The pods ripen inside the bale, where the seed is not lost even though the pods pop. The bales are then cut open and fed into a combine or thresher and the straw saved for hay.

---

[1] Frank J. Zink, Design of a Machine for Harvesting Buffalo Grass Seed, *Agr. Eng.*, 17:197-198, May, 1936.
[2] R. H. Wileman and N. K. Ellis, A Machine for Collecting Fallen Peppermint Leaves, *Agr. Eng.*, 24:237-238, July, 1943.
[3] Harmond, *op. cit.*
[4] Al Goff, Now—A Mechanical Bee, *Farm Journal*, Vol. 36, December, 1952.

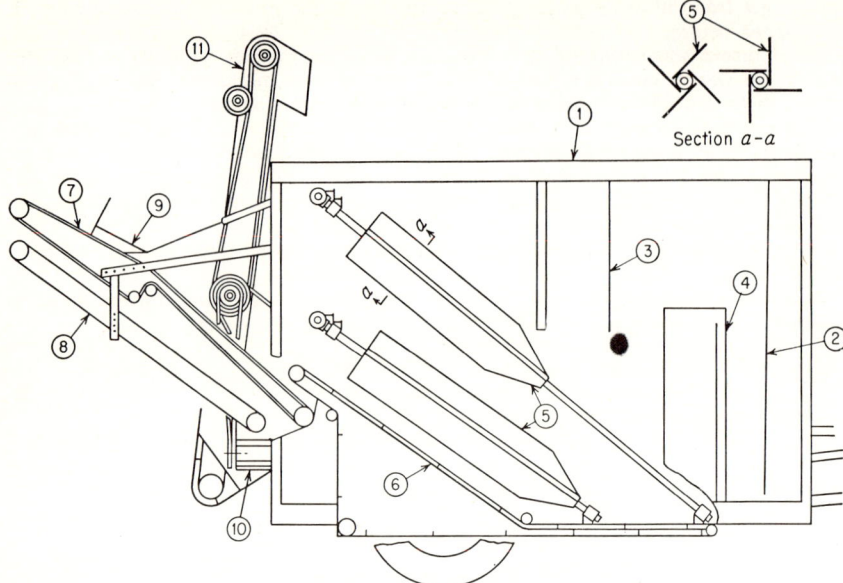

Fig. 28-2. Experimental castor bean harvester: (1) stripping compartment; (2) canvas curtain; (3) canvas curtain; (4) rigid baffle; (5) stripper paddles; (6) conveyer to separator; (7) separator raddle; (8) endless apron cleaner; (9) floating blocks; (10) conveyer to elevator; (11) elevator. (*Arms and Hurlbut, Agr. Eng., December,* 1952.)

CASTOR BEAN HARVESTING

Oil from castor beans is an important ingredient in paint, lubricants, and many other products. Because of our increasing requirements, the USDA has been developing adapted varieties of castor beans along with harvesting and hulling machines in an effort to make them an economic crop in the United States.

The major harvesting problem is shatter loss, since the ripe seed clusters drop with very little agitation. Also, the seed ripens over a long period, the early-ripened spikes shattering before the late ones are mature, although plant breeders are developing varieties which ripen more uniformly.

Dwarf varieties of castor beans can be combined, although a special header must be used to reduce shatter losses and the coarse woody stalks are hard on conventional combines.

Because of the ease of shattering, stripping appeared practical, and the initial development work on a castor bean stripper was done at the University of Nebraska.[1] The beans were stripped by two pairs of upwardly rotating rubberized fabric beaters. The stripping chamber was sealed to retain the loose beans. Two pairs of beaters were used, one above the other, to better handle tall bushy plants. The upper pair stripped high beans which would otherwise be folded in by the lower branches and not stripped if a single pair of beaters were used. A cross section of this harvester is shown in Fig. 28-2. The development of the machine has been continued in Oklahoma, and a commercial model produced.[2]

Castor beans must be shelled and well cleaned before oil extraction, and a rubber-disk type of huller which cracks a minimum of beans has been developed.[3]

---

[1] Milo F. Arms and Lloyd W. Hurlbut, An Experimental Harvester for Castor Seed, *Agr. Eng.*, 33:784–786, December, 1952.

[2] E. W. Schroeder and I. F. Reed, Developing Tractor-mounted Castor Bean Harvesters, *Agr. Eng.*, 33:775–779, December, 1952.

[3] J. G. Porterfield and F. J. Oppel, Jr., An Experimental Castor Bean Huller, *Agr. Eng.*, 33:713–716, November, 1952. L. G. Schoenleber and W. E. Taylor, Development of a Portable Castor Bean Huller, *Agr. Eng.*, 35:327–336, May, 1954.

SUGAR CANE HARVESTERS

The harvesting of sugar cane involves the handling of from 15 to 25 tons of cane stalks per acre in Louisiana, with annual harvesting, to a maximum of 65 to 130 tons per acre in Hawaii with harvesting after 1½ to 2 years of growth.

With hand harvesting, the stalk is cut off close to the ground, the green watery top severed, and the leaves knocked off, all with a cane knife, or machete. The stalks of cane are then loaded on carts for transport to the sugar mill.

A machine which used rotating circular blades to cut off the stalks and trim the tops was developed in Louisiana in the early 1940s by the Thomson Machinery Co., Inc., of Thibodeau, La. Each of these one-row machines replaced about 70 hand harvesters. The same concern has developed a new machine which also strips the leaves and loads the stalks into a wagon.[1] USDA engineers have continued with the development of this type of machine.[2]

Another new cane-harvesting machine has been developed, primarily to meet Caribbean requirements where yields run from 30 to 60 tons per acre. It cuts the stalks into 14-in. lengths for easier loading and handling and in this process also knocks loose and separates the leaves.[3]

Sugar cane in Hawaii provides an even more difficult problem since the stalks are from 1 to 2 in. in diameter and average 15 ft long with a maximum of 28 ft, most of them recumbent. A harvester has been developed there which cuts the cane, elevates it into a cross-conveyer collector, and dumps it into a windrow. Heavy cranes with grab forks are then used to load the cane for transport to the mill.[4]

TREE-CROP HARVESTING

Mechanical aids are rapidly being developed for tree crops, and harvesting methods are changing rapidly.

Portable, hydraulically lifted platforms for pruning and picking are coming into use. They vary from a parallel-lift one-man platform fitted to a tractor loader to a trailer equipped with several independent hydraulically actuated boom-carried platforms.

A one-man self-propelled unit, controllable from the hydraulically lifted platform, is reported to multiply output from 1½ to 2 times compared with ladder equipment.[5]

Power lifting of picked fruit onto trucks is a great labor saver in large orchards. Pallet handling of fruit boxes by tractors equipped with fork lift attachments is popular. For bulk handling of fruits such as juice oranges, loader attachments have been developed which can pick up a special box, empty the contents into a truck, and relocate the box for refilling.[6]

Knocking, or shaking of the tree, is used to dislodge nuts and fruits which are not damaged by dropping. This may be done manually with poles, by a tractor carrying an eccentric shaft and bearing which actuates a boom attached to the tree, or by pneumatic shakers. The last are effective and economical for large orchards which already have air compressors for operating pneumatic pruning shears.

Canvas sheets may be used to collect the falling fruit or nuts, particularly where freedom from bruising and dirt is important. They should be spring-supported from light metal frames designed for easy setup, take-down, and shifting.[7]

---

[1] B. C. Thomson, The Mechanical Harvesting and Loading of Clean, Fresh Sugar Cane in Louisiana, *SAE Journal*, 60:63, 1952.

[2] R. M. Ramp, Development of a Sugar Cane Harvester, *Agr. Eng.*, 37:821–824, December, 1956.

[3] J. L. Hipple, "Sugar Cane Harvester," unpublished paper, International Harvester Co., Chicago, Ill., 1953.

[4] Richard A. Duncan, Development of a Sugar Cane Harvester, *Agr. Eng.*, 31:65–70, February, 1950.

[5] Fred L. Hill and R. W. Brazelton, The Steel Squirrel, *Agr. Eng.*, 36:17–19, January, 1955.

[6] J. H. Levin and H. P. Gaston, Equipment Used by Deciduous Fruit Growers in Handling Bulk Boxes, *USDA ARS* 42-20, Beltsville, Md., August, 1958. S. W. McBirney and A. Van Doren, Pallet Bins for Harvesting and Handling Apples, *State Coll. Wash. Agr. Expt. Stas. Circ.* 355, April, 1959.

[7] P. A. Adrian and R. B. Fridley, New Concept of Fruit-catching Apparatus Tested, *ASAE Trans.*, 2:30–31, 1959. P. A. Adrian et al., Low Profile Catching Equipment for Fruit, *ASAE Paper* 59-601, December, 1959. R. B. Pridley and P. A. Adrian, Some Aspects of Vibratory Fruit Harvesting, *Agr. Eng.*, 41:28–31, January, 1960.

Pickup machines have been developed to collect nuts or fruit from the ground. One such machine, made by the A. D. Goodwin Nut Harvester Co. of Manteca, Calif., uses rotating fingers to sweep the nuts into the machine, where sieves and air are used to clean out the dirt. These machines work best where the ground has previously been made firm and smooth without clods. They have been successful on almonds, walnuts, filberts, pecans, and tung nuts and have also been adapted for prunes.[1]

Vacuum pickup machines have been successfully used to pick up filberts and almonds. Suction nozzles 2 in. wide riding 1½ in. above the ground are used, and an air speed of 4,500 fpm produced by a vacuum of 2½ in. of water at the nozzle is required to lift the nuts.[2] Dust is a major problem because it wears the fan and ducts as well as bothering the operator. Smoothing and rolling the soil before harvesting helps control dust, particularly if the smoothing is followed by flooding.

TRUCK-CROP HARVESTING

Hand labor is the rule in truck-crop harvesting, although progress is being made. An *onion harvester* (Chap. 27) and an *asparagus harvester*[3] have been developed at the University of California. The asparagus harvester uses a high-speed (5,000 fpm) band saw to cut off the asparagus spears just below the soil surface. Pairs of flexible disks grip the spears just before they are cut off and then lift and drop them into a hopper, using an action somewhat the reverse of the transplanter shown in Fig. 17-20. In tests this machine missed 3 to 10 per cent and damaged 4 to 8 per cent for an efficiency of 80 to 90 per cent compared with hand harvesting. It indicated a capability of replacing about 12 men.

Narrow *conveyer-belt booms* extending 20 ft or more from each side of a truck or trailer are being used to collect lettuce, celery, tomatoes, etc., from the pickers and thus speed up picking by eliminating baskets and boxes.

A conveyer plus slings to carry the pickers in a horizontal position is being used for low-growing crops easily damaged by tramping.

Raspberry-picking machines are being used in the Pacific Northwest. The berries are shaken from the canes onto the machine apron, where they are collected and cleaned by screens and air. The smaller machines work on one side of the row, and shaking may be done by hand or mechanically. These machines are most successful when the rows of canes are tied to an offset wire so that they lean over, permitting more efficient collection of the berries on the apron. The larger machines straddle the row, use mechanical shaking, and have aprons on both sides.

The fruit picked by these machines is reported to be well cleaned and of good quality because of uniform maturity, and hence it is superior to hand-picked berries.[4]

Blueberries are being successfully harvested in Michigan by hand-held vibrators applied to individual branches and a vibrator-type harvesting machine is being developed at Michigan State University.

A harvesting machine for *raisin grapes* has been developed at the University of California, Davis, Calif. The grapevines are carried on horizontal wires supported by crossbars and offset posts in such a way that the clusters of grapes hang below the wires. A short cutter bar is carried at the side of the tractor high enough to clip off the clusters of grapes. The grapes are conveyed under the tractor and spread for drying on a strip of paper laid down from a roll carried at the front of the tractor. The dried grapes are also picked up and loaded mechanically.[5]

A mechanical *cucumber harvester* has been developed at Michigan State University. Vines trained to extend to one side of the row are picked up and shaken over a

---

[1] Louis E. Davis et al., Improving Prune Harvesting, *Calif. Agr.*, 6:3-14, April, 1952. R. B. Fridley and P. A. Adrian, Development of a Fruit and Nut Harvester, *Agr. Eng.*, 40:386-391, July, 1959.

[2] R. R. Parks and J. P. Fairbanks, Suction Machines for Harvesting Almonds, *Agr. Eng.*, 29:305-306, July, 1948.

[3] Robert A. Kepner, Harvester for Green Asparagus, *Calif. Agr.*, 6:7-9, October, 1952.

[4] Dorothy E. Campbell, Raspberries Yield to Mechanical Harvesting, *Implement Record*, November, 1951, pp. 28-29.

[5] L. H. Lamouria et al., Designing a Grape Harvester, *Agr. Eng.*, 39:218-236, April, 1958.

collecting conveyor to obtain the cucumbers. Since the vines are not damaged by this operation, they can be picked as many times as required.[1]

Mechanical *tomato harvesting* has been successful with varieties on which all the tomatoes ripen simultaneously. A machine developed at Michigan State University cuts the plants just under the ground surface and elevates the vines to a bed which removes the tomatoes by lateral shaking.[2]

---

[1] G. W. Bingley et al., Design and Development of a Mechanical Cucumber Harvester, *ASAE Paper* 59-604, December, 1959.

[2] B. A. Stout and S. K. Ries, Mechanical Harvesting of Tomatoes, *ASAE Paper* 59-603, December, 1959.

Chapter 29

STALK AND BRUSH SHREDDERS

The impact-type shredder and cutter for use on standing vegetation became popular in the 1940s. The invasion of the western Corn Belt by the corn borer resulted in many types of stalk shredders being placed on the market. Subsequent tests indicated that this method of corn borer control was incomplete since 10 to 50 per cent often survived. Shredders, however, proved to be a very versatile tool and are being used for the following purposes:

1. Cotton-stalk chopping to facilitate subsequent tillage[1] and for pink bollworm control in southern Texas
2. Cornstalk shredding for winter mulching and easier plowing
3. Potato-vine shredding to facilitate digging
4. Beet-top shredding to facilitate subsequent topping and harvesting, made possible by the use of rubber flails
5. Weed and heavy cover-crop shredding to facilitate plowing and accelerate decay
6. Roadside, pasture, and lawn mowing
7. Brush clearing
8. Chopping of standing or windrowed hay when designed to throw the chopped material back into the wagon (Chap. 23)

TYPES OF SHREDDERS

The *cylinder*, or hammer-mill type of shredder, shown in Fig. 29-1, has multiple blades, usually hinged, turning about a horizontal axis transverse to the direction of

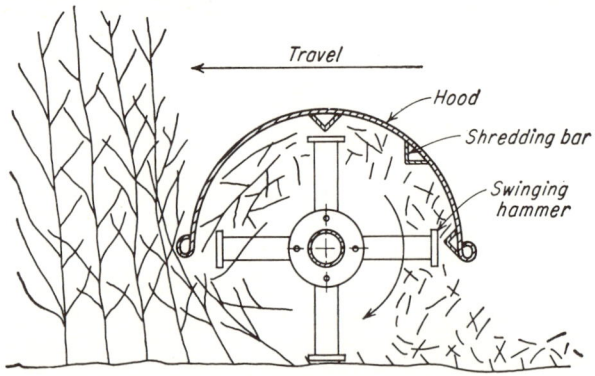

Fig. 29-1. Shredding action of cylinder-type shredder.

travel. The hammers rotate up in front so as to pick up the material and throw it back over the top. The degree of disintegration varies with hammer peripheral speed,

---

[1] Harris P. Smith, Stalk Disposal and Land Preparation for Cotton, *Agr. Eng.*, 30:485–488, October, 1949.

spacing and width of hammer ends, and the amount of retardation of the material under the hood.

Tests in Indiana of an experimental cylinder-type shredding attachment for corn pickers indicated that a hammer peripheral speed of 9,000 fpm killed 95 per cent of the corn borers.[1] The blades shredded the stalks as they came through the snapping rolls and, in standing corn, did a very thorough job.

The *cross-row* type of shredder has blades turning across the direction of travel on longitudinal axes. The stalks are bent over, and one or more sets of blades rotating on an axis over the row cut across the stalks. These shredders work well on rows, but do not give complete coverage for brush and pasture work.

The *horizontal-blade* shredder with blades on one or more vertical spindles is simple and rugged and gives complete coverage. A single-spindle model is shown in Fig. 29-2. It does not shred as finely as the cylinder type, but requires less power and does

Fig. 29-2. Lift-type single-horizontal-blade shredder. (*Ford.*)

an acceptable job of shredding for most purposes. It is very effective as a pasture mower, although there is some windrowing effect.

DESIGN PROBLEMS AND SOLUTIONS

Blade wear and breakage is a problem, since most machines operate at peripheral speeds from 10,000 to 14,000 fpm and dirt, rocks and stumps are often encountered. Swinging hammers, chain flails, or rubber flails give some protection and are used on cylinder-type machines. Because of their greater weight, hinged blades do not give as much protection on horizontal-blade machines but are often used. Shear bolts are also used. Blade dullness does not decrease efficiency under most conditions because impact rather than cutting does most of the work. Blades are usually made of a steel with high impact strength and low notch sensitivity, such as SAE 5150 through-hardened to $R_c 38 - 43$.

Vibration due to unbalance resulting from blade wear is a problem. As in a mower, the best procedure is to isolate the shredder vibration from the tractor by using a flexible attachment on a mounted machine or by using a pull-type machine built to with-

---

[1] R. H. Wileman, Progress in Controlling the European Corn Borer, *Agr. Eng.*, 25:419–20, November, 1944.

stand the vibration. In some cases, rubber mounting of the rotor reduces vibration due to unbalance.

Accurate *ground gaging* is particularly important in pest control where close cutting is essential. Some cylinder machines vary the length of the hammers across the width so as to better conform to row contours. Adjustable skids or caster wheels are used to gage mounted horizontal-blade models, since they are too far behind the tractor to be satisfactorily gaged by direct suspension.

The *final drive* to the rotor is often by V belts because of their quietness, shock absorption, and overload protection by slipping. The rotor speeds permit high enough belt speeds to provide adequate power. A rugged bevel gear box is required on cylinder and horizontal-blade machines unless a "corner" belt drive is used. A friction slip clutch is desirable for overload protection when the blades are positively driven by gears or chains.

The *hood* must be substantial and well braced to withstand vibration and protect the operator from flying stones, sticks, etc. It may need belting flaps near the ground to reduce the dust thrown out under dry conditions.

Power requirement depends on the amount of material shredded, the disintegrating work done per unit of material, and the power required for windage. Windage power is highest with cylinder-type machines, particularly where wide rubber flails are used, but it can be reduced by T-end or U-shaped hammers, which give complete coverage with fewer hammers. The disintegrating work done depends on the resistance of the material, blade speed, and number of blows. A 5-ft-wide machine may require less than 5 hp in light cutting, but up to 30 hp for thorough chopping of heavy cover crops or brush. Horizontal-blade machines have less windage and less disintegrating effect than cylinder type machines.

Rugged construction throughout is a necessity because almost every shredder will sooner or later encounter a job which requires full tractor power. In brush-clearing work it is common practice to drive forward until the tractor stalls and then back up and try again. A tractor with a constant-running PTO drive has a great advantage for such work.

Chapter 30

FARM TRANSPORT EQUIPMENT

Drawn wheeled vehicles for transport purposes date back to the origin of the wheel in antiquity. The wooden wagons of the American pioneers were highly developed vehicles which played an important part in the settling of the West.

TWO-WHEELED TRAILERS

The two-wheeled trailer is a very simple form of transport equipment. An analysis of the static and dynamic forces acting on the two-wheeled trailer is also applicable to the wheels and framing of many drawn field machines.

Static forces acting on a trailer are illustrated in Fig. 30-1a. Additional dynamic forces are introduced when the trailer is in motion.

Rolling resistance introduces a couple which transfers some of the load from the trailer wheels to the hitch connection, as shown in Fig. 30-1b.

When *one wheel hits a bump,* side forces at the hitch and at the wheels are introduced, as well as a severe backward and upward force on the wheel, as shown in Fig. 30-1c.

When the trailer is *accelerated,* the resulting couple transfers weight from the hitch to the wheels. When the trailer is *decelerated by the towing vehicle,* the reverse couple transfers weight to the hitch from the wheels. When the trailer is *decelerated by brakes* on the trailer wheels, the greater couple arm results in a more severe weight transfer from the wheels to the hitch than when decelerated by the towing vehicle.

When the trailer is *turning sharply,* an overturning couple is introduced in addition to a side reaction on the hitch if the center of gravity is not directly over the axle.

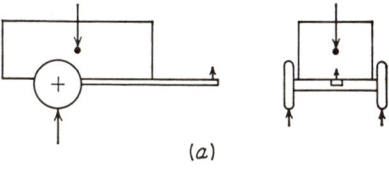

(a)

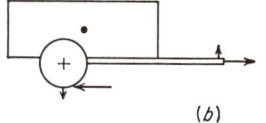

(b)

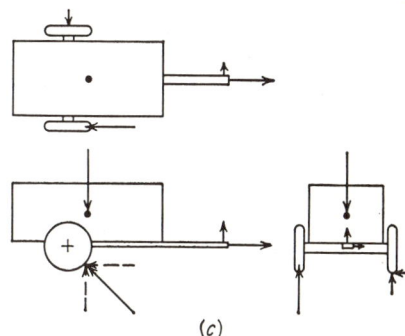

(c)

FIG. 30-1. Major static and dynamic forces on a two-wheel trailer or implement-carrying frame: (a) forces due to static load; (b) forces due to rolling resistance; (c) forces due to right-hand wheel hitting a bump.

Each part of the trailer must, of course, be strong enough to resist the greatest load imposed on it. The axle and wheel spindles are most severely stressed when both wheels hit a bump. The hitch frame from axle to towing vehicle is most highly stressed in the horizontal plane when one wheel hits a bump and in the vertical plane when the trailer is decelerated by its own brakes.

Farm trailers vary from small sizes of 500 lb capacity for use behind automobiles to large 8- by 14-ft dual-wheel tilt-bed trailers of 8,000 lb capacity for use behind trucks and tractors. Many farm trailers are locally made from used auto and truck parts.

Trailing characteristics are important for road work. Short beds or concentrated loads having a minimum radius of gyration about the center of the axle exert least side force on the hitch and have least tendency to whip. A wide tread resists overturning but exerts more side force on the hitch when one wheel hits a bump. Springs cushion bumps and reduce the shock on the towing vehicle but may increase sway and whip. The weight and rigidity of the towing vehicle are also important factors in controlling *whip*.[1]

FOUR-WHEEL WAGON GEARS

The modern four-wheel steel wagon gear on rubber tires is a versatile transport unit with light draft in soft fields and high-speed capability on the highway. The

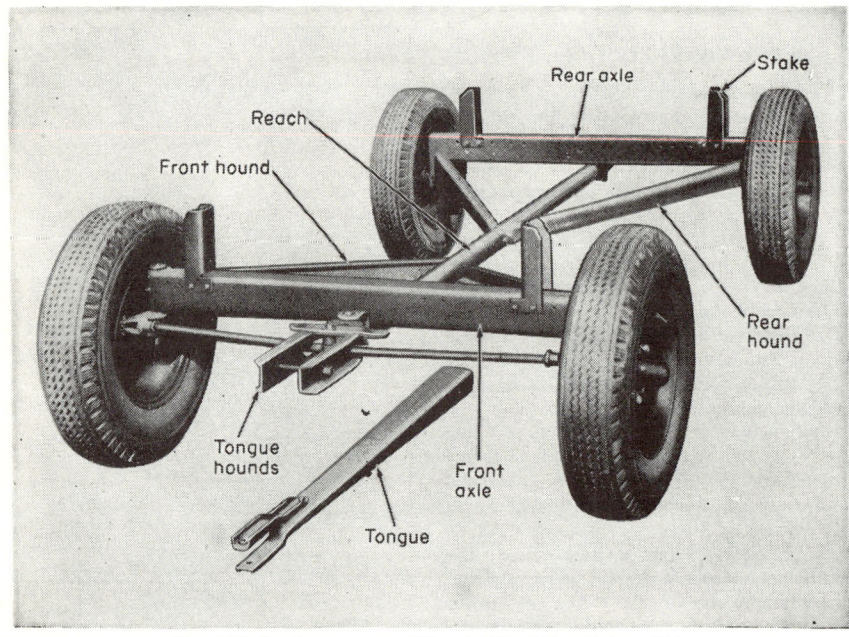

FIG. 30-2. Steel wagon gear. (*Electric Wheel Co., Quincy, Ill.*)

highly developed wooden wagon gear has become virtually obsolete in the past thirty years. A typical steel wagon gear with names of the various parts is shown in Fig. 30-2.

The *front axle* may be of fifth-wheel or autosteer construction. The fifth-wheel axle pivots in the center and is usually stabilized by a turning circle or ring. With autosteer construction the front wheels pivot individually on vertical spindles at the ends of the front axle. The front-axle spindle and pivot may be of L- or T-type construction (Fig. 30-3.) The L-type construction is simplest and most common, with the T-type used for low-clearance construction.

An *Ackerman steering diagram* for autosteer wagons is shown in Fig. 30-4. With ideal steering geometry, the projection lines of the front-wheel spindles would intersect

---

[1] Henry H. DeLong, Homemade Rubber-tired Wagons and Trailers, *S. Dakota State Coll. Agr. Expt. Sta. Bull.* 349, 1941.

on the rear axle line at any angle of turn, allowing turning with no skidding. With conventional steering linkages, the geometry is not perfect but is reasonably close. For normal turning radii, best results are secured if the projections of the front-wheel steering arms intersect at the middle of the rear axle when the front wheels are straight. With short-turn wagons this layout may give excessive skid at full turn, requiring a trial-and-error procedure to find the setting giving the lowest average departure from the ideal over the full range. The wheels can be made to turn slower or faster than the tongue, as desired. Changing the wheel-base setting of a wagon

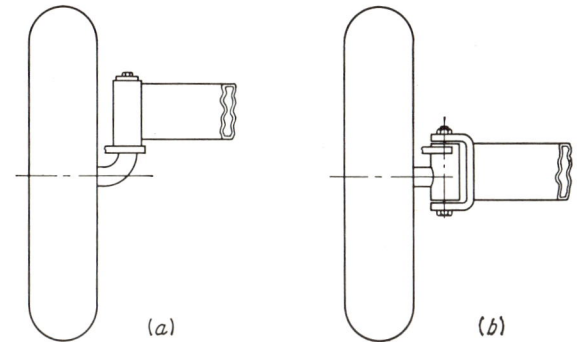

FIG. 30-3. Front axle spindles for autosteer wagons: (a) L type; (b) T type.

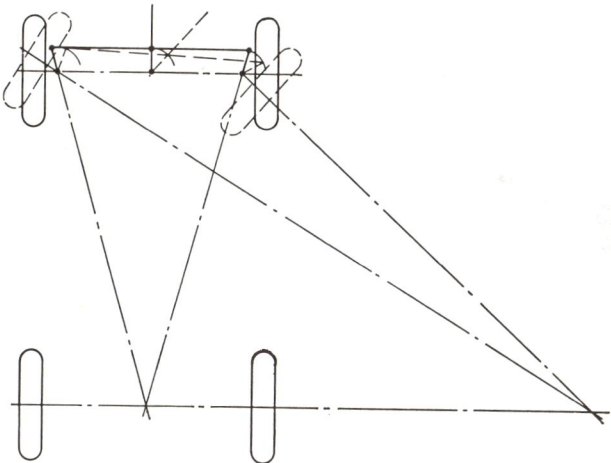

FIG. 30-4. Diagram of Ackerman steering for autosteer wagons.

obviously affects the steering geometry, but little damage is done by slow-speed skidding on soil.

A *short turning radius* is important in close quarters and for use behind field choppers and corn pickers. Stops are provided to prevent the wheels from cutting into the lower corner of the box. The fifth-wheel type of front axle can turn very short when the bolsters carry the box high enough for the tires to pass under, although the lateral stability of the front axle is reduced during sharp turns. Short-turn autosteer wagons must carry the box high or have a wider-than-normal front axle so that the inside front wheel can turn almost 90°. In addition, the front hounds must be set narrow and a special steering linkage may be required. The minimum turning radius to the outside front wheel ranges from 10 ft on short-turn wagons to 14 to 16 ft on regular models when set at 7-ft wheel base.

The *axle clearance* required depends on the rocks, stumps, and other obstacles where the wagon will be used and on row crops and windrows which must be cleared. Few wagons have less than 13 or more than 21 in. axle clearance, with the average near 16 in. Where low loading height is a major consideration, low axles are an advantage.

Wagons must have *flexibility* to travel over uneven ground without damage. This is usually secured by allowing the rear axle to pivot about the reach pipe. The pivoting action twists the box, and a rocking front bolster may be supplied to reduce the twisting effect, although it may accentuate any tendency to whip at high speeds. In an alternative method of construction, the front and rear axles are connected by longitudinally formed steel channel beams at each side. When properly designed, these channels flex without damaging stress concentration as the wagon passes over uneven ground.

Whipping results when steering connections become loose or worn. Wear-resistant connections, or provision for take-up of wear, are highly important. Fifth-wheel wagons have fewer wearing parts than the autosteer type.

Wheel bearings are most commonly the antifriction tapered-roller type, packed with grease and sealed.

In the *design and testing of rubber-tired wagon gears,* it has been noted that parts stressed higher than one-third of the material's yield strength by the static load tend to take permanent set during rough treatment at 15 mph.

Wagon *draft* can be predicted from the rolling-resistance coefficients shown in Fig. 4-6 for various-size wheels and surfaces. For instance, a loaded wagon with a gross weight of 6,000 lb pulled at 2 mph on 6.00-16 tires inflated to 40 psi would have a draft of approximately 125 lb on concrete, 360 lb on bluegrass pasture, 2,125 lb on freshly tilled loam, and 2,500 lb on loose sand.

Brakes are required by law in many states for wagons used on the highway. They are a necessity where a heavily loaded wagon is being hauled by an auto or light tractor on hilly roads. The rim brakes formerly used with wooden and steel wheels are not practical with rubber tires, and automotive-type internal expanding brakes are most common. They may be manually controlled from the tractor seat by a rope or automatically applied by the action of a telescoping tongue when the towing vehicle holds back the wagon. Electric brakes controlled by a rheostat lever and powered from the towing vehicle's battery are very convenient where the amount of use justifies their cost. Hydraulic brakes actuated by a telescoping hitch or by a hose connection to the brake system of the towing vehicle are also popular for heavy-duty hauling service. A wagon with rear brakes needs a stronger reach than one without because of the additional downward bending movement induced.

A *telescoping tongue* facilitates hitching where two or more wagons are hauled in a train. The first wagon need not be backed into exact position, but can be pulled past the second wagon and stopped. A telescoping tongue on the second wagon is swung over, extended as required, and hitched to the first wagon. Both wagons are then pulled forward until the rear wagon is straight, after which the tractor and front wagon are backed until the second wagon's tongue latches in its original short position.[1]

WAGON BOXES

Faster rates of harvesting and the use of tractor power for towing have resulted in the obsolescence of the old 38-in.-wide straight-sided box. A level full capacity of 100 bu (125 cu ft) is popular in a flare box or flat deck with sides.

Flare boxes may be made of wood, steel, or steel sides with a wood floor. They are usually 38 in. wide at the bottom, 10½ ft long, about 5 ft wide at the top, and 3 ft or more deep. Wood boxes are somewhat lighter than steel and resist corrosion, although they are subject to wear and rotting out. Steel boxes must either be able to

[1] C. K. Shedd, and E. V. Collins, A Telescoping Wagon Tongue, *Agr. Eng.*, 17:343–345, August, 1936.

flex without fatigue failure or be stiff enough to resist flexing when the wagon gear conforms to uneven ground. Since a wagon box is an open channel, it is most practical to allow it to flex. Front-corner fatigue cracks can be prevented by retaining the upper portion of the front end between loose-fitting flanges attached to the sides, which are held in by a tie rod.

Flat-deck platforms approximately 7 ft wide by 14 ft long with detachable grain sides or rack sides are popular because of their versatility. With hay ladders, long or baled hay can be hauled. Low solid sides are used for grain, and slat sideboards are added for ear corn or chopped forage. It is important that the solid sides for grain fit well and are pulled down tightly to prevent leakage.

These platforms are usually carried by a conventional wagon gear, although wagon front and rear axles may be attached directly to the main sills. In the latter case, the axle mountings must accommodate flexing over rough ground without eventual fatigue failure.

Platform decks may be made as large as 8 by 16 ft, primarily for hay, or down to 6 by 12 ft with higher sides, primarily for ear corn and grain. In picking corn planted on the contour, a wide box is necessary to catch the corn when the wagon trails around curves.

The location of a hitch point on corn pickers and field choppers for trailing wagons is a problem because of the variation in wagon tongue lengths, box heights from the ground, and in some cases box widths. The front edge of the box must, of course, clear the delivery elevator or blower pipe over rough ground and around turns.[1]

WAGON UNLOADING

Power unloading of wagons may be performed in several ways. The oldest and most common method for grain and corn is by gravity dump. Hoist attachments for lifting the front wheels of the wagon are often used in conjunction with elevators. Built-in hydraulic or cable hoists, powered from the tractor, may also be used to lift the front of the box for dumping.

It is difficult to secure an even feed of chopped forage into a forage blower by dumping, and conveyer-type unloading is favored for this purpose. By using a suitable reduction, the load can be moved back uniformly at a rate near the capacity of the blower. An attendant is needed to rake down the material and stop the motor intermittently if the feed is too fast. The work, however, is very easy compared with hand unloading.

Conveyer unloading may be performed by an unloading attachment adaptable to a regular wagon box or by a box with a built-in conveyer bottom. A popular type of unloading attachment uses a roller extending across the rear of the wagon to pull cables attached to a *sliding-front* endgate. Such a roller may also be used to pull a *canvas apron* extending the length of the floor and up the front end.

Force requirements for pulling off a load vary with the depth and kind of material and the smoothness of the inside of the box. Sliding endgates require more power than canvas unloaders because the end compression causes side expansion and extra drag against the sides. Canvas is sometimes laid down with one or two folded sections so as to spread out the load and avoid starting it all at once. Comparative tests with a load of 2½ tons of corn silage indicated a maximum *coefficient of friction* of 0.9, with an average of 0.65 for the sliding endgate compared with 0.85 and 0.6 for canvas.[2] Grass silage has a higher coefficient of friction than corn, and it can be seen that, with 4- to 5-ton loads, forces up to 10,000 lb may be required.

Power requirements for unloading attachments vary from a ⅓-hp electric motor with a well-made gear reducer with antifriction bearings to ¾ hp for an intermittent ratchet drive similar to that used on manure spreaders.

PTO-driven power unloading boxes with built-in conveyer save much labor during harvest and even more in large-scale feeding operations when equipped with a cross-

---
[1] Sherman C. Heth, A Proposed Standard Short Tongue Farm Wagon, *Agr. Eng.*, 31:230–231, May, 1950. ASAE Recommendation: Hitch and Box Dimensions for Farm Grain Wagons, *Agr. Eng. Yearbook*, 1959, p. 82.
[2] E. L. Barger et al., Results of Tests of Mechanical Unloading Devices for Chopped Forages, *Agr. Eng.*, 30:221–225, May, 1949.

conveyer to unload directly into feed bunks. A three-chain conveyer is often used to reduce the bending loads on the conveyer bars. A rake-down mechanism, consisting of several beaters arranged vertically or a vertical chain conveyer, is needed in conjunction with a side-delivery cross-conveyer. Some PTO-driven manure spreaders have attachments for converting them into power unloading wagons.

V-type boxes with auger bottom and swinging delivery auger, as shown in Fig. 30-5, are simple and well adapted to handle grain or ground feed.

Fig. 30-5. Power unloading V box with augers. (*Dodgen Industries, Ft. Dodge, Iowa.*)

MANURE SPREADERS

The function of a manure spreader is to transport the load to the field and then distribute it evenly over the ground. The treatment received during tractor loading, plus the stresses resulting from the use of tractor power to tear apart and unload large loader forkfuls, requires very rugged construction. A two-wheel PTO-driven spreader is shown in Fig. 30-6.

The usual *distributing mechanism* consists of an upper beater which throws off the upper portion of the load, a lower beater which throws out the lower portion, and a rear widespread beater with angled paddles or large auger flights. The upper beater must be located far enough ahead of the lower beater to avoid interference with the delivery of the lower beater. Ideally, the location and peripheral speed of the upper and lower beaters should be such that they deliver most of the material directly into the widespread beater for additional shredding and wider distribution. Because of variations in manure and operating speed, the widespread beater often intercepts only a portion of the material and the effective width of spread is usually about the width of the wheel tread.

Since it is desirable to spread the manure thinly and uniformly, other methods of increasing width of spread have been tried. One spreader, no longer in production, used a divided lower beater, the two sections forming a V so as to throw the manure outward. A recent design uses conventional upper and lower beaters at the front of the box feeding forward under a hood to a pair of vertical spinner disks, which are claimed to double the effective width of spread.

The *beater cylinders* have steel teeth riveted to steel tooth bars. Lower beaters usually have six tooth bars, range from 16 to 20 in. in diameter, and are geared from $\frac{1}{2}$ to $\frac{3}{4}$ revolution per foot of forward travel. Upper beaters usually have three bars, range from 13 to 15 in. in diameter, and make from $\frac{5}{8}$ to 1 revolution per foot of forward travel.

Shielding rings extending over or inside the ends of the tooth bars prevent twine and stems from wrapping on the beater shafts next to the sides.

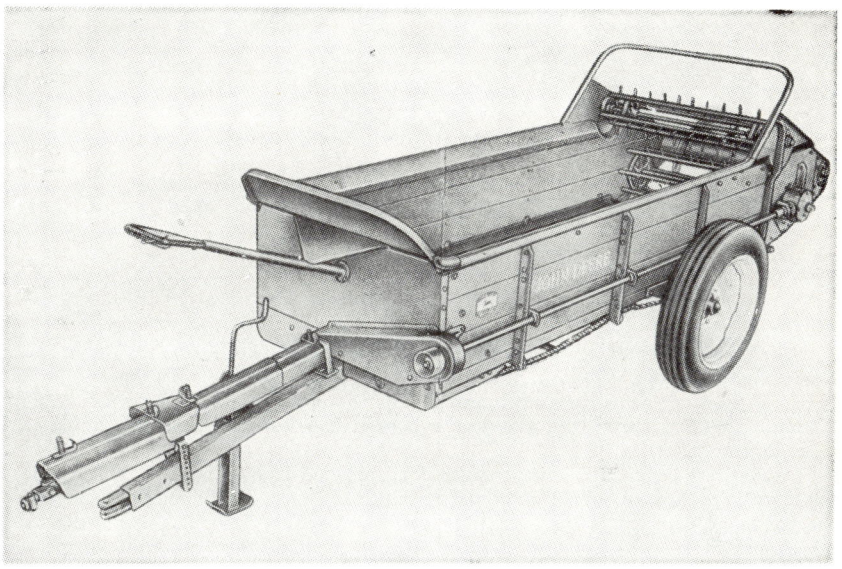

Fig. 30-6. Two-wheel PTO-driven spreader. (*Deere & Co.*)

Widespread beaters are slightly larger in diameter than lower beaters and are geared for $1\frac{1}{2}$ to 2 revolutions per foot of forward travel.

The load is *fed* to the beaters by a conveyer web using two steel chains with steel-angle crossbars. Several rates of feed are desirable, the slowest ranging down to $\frac{1}{10}$ in. per foot of forward travel. On ground-driven spreaders, a three- or four-lobe cam on the rear axle actuates a lever with a roller cam follower, which in turn drives a ratchet wheel on the conveyer drive shaft as shown in Fig. 30-7. By regulating the distance the cam roller drops into the cam, from one to four or five ratchet teeth can be moved per stroke. The resulting intermittent action is not objectionable in this application. Because of the resistance of dry heavy manure, or worse, manure frozen to the box, the conveyer drive must be very strong. Locating the holding pawl near the driving pawl reduces fatigue failure of the conveyer drive shaft in contrast to locating the holding pawl opposite the driving pawl, which reverses the load and deflection at every stroke. Spreader boxes are usually 1 to 2 in. wider at the back than at the front to reduce side friction as the load moves back.

PTO-driven spreaders need fewer feed speeds because the various ground speeds of the tractor can also be used to vary the rate of feed. Some spreaders do not vary the conveyer feeding rate, relying entirely on the tractor speeds.

Thorough *shielding* is functionally important on a manure spreader. The feed becomes inoperative if the ratchet-wheel notches fill with foreign material, and chains may be stretched or broken if straw gets on the sprockets. Large rounded shields over the various drives improve appearance as well as function.

Spreader *boxes* may be made of wood or steel or both. Wood should be creosoted to prevent rotting, and steel must be corrosion-resistant or well coated to prevent rapid rusting. Most current spreaders use wood floors because the conveyer soon wears through a protective coating over steel. Wood is most satisfactory when it is used as a box liner with a steel frame to carry the loads. All operating parts should be tied to the steel framing, if possible, and the top edges of the sides should be well protected by steel to resist bumping by the loader fork.

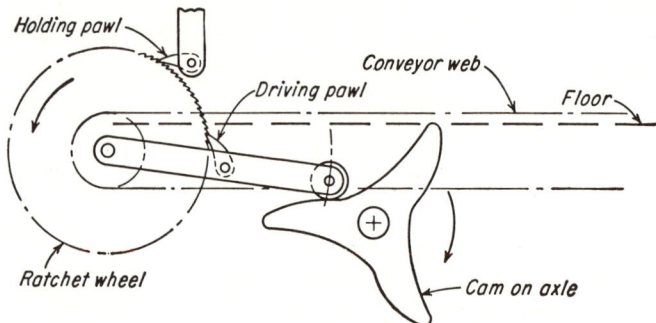

FIG. 30-7. Cam and ratchet drive for conveyer web.

The use of power gutter cleaners loading directly into a spreader results in much *liquid manure* being placed in the spreader. To prevent its loss on the way to the field, one spreader has a tight steel bottom sloping from each end to the center to form a sump. The feed apron conforms to the bottom and feeds the solid and liquid portions together.

Spreader-box *capacity* is usually given for a heaped load in bushels. The standard for calculating heaped capacity assumes an angle of repose of 60° and a height above the box of 60 per cent of the average box width at the top.[1] Popular sizes range from 60 bu, handling up to 2 tons, to 150 bu. PTO-driven spreaders are built for greatest capacity, particularly where the beaters are detachable and sideboards can be added for use as a self-unloading wagon. Some spreaders are available with a mixer-feeder attachment with mixing beaters, a hood, and a cross-auger spout to deliver feed into feed bunks.

[1] ASAE Standard: Volumetric Capacities of Manure Spreaders, *Agr. Eng. Yearbook*, 1959, p. 64.

Chapter 31

TRACTOR LOADERS

Tractor loaders have permitted the application of tractor power to many arduous lifting and material-handling jobs such as loading manure or gravel; loading or feeding silage from trench silos; lifting and transporting hog houses, feed bunks, etc.; pulling fence posts; piling logs; bulldozing earth or snow; and serving as a scaffold for picking fruit or painting.

The *lift height* required depends upon the use. Manure spreaders are low, and a lift giving 4½ ft of clearance under the tips of the teeth when dumped is usually adequate. Most loaders, however, are sometimes used for loading trucks with sides up to 7 ft high, requiring dumped clearance of this much or more, as shown in Fig. 31-1.

Bucket overhang beyond the front of the tractor should be enough to dump the load near the middle of the vehicle body. An overhang of 1½ ft, from front of tractor to the tooth tips when dumped, may be satisfactory for loading spreaders, but up to 3 ft may be desirable for loading trucks.

Tractor stability with lifted loads of 1,000 to 2,000 lb is a problem, particularly with tricycle tractors which can tip sideways if the rear wheels leave the ground. Ideally, the bucket should be as close in as possible for loading and transport, extending out only for dumping. This type of motion has been secured by using telescoping arms and a curved track to guide the bucket as it is lifted. Solid pivoted lift arms are preferred, however, even if rear weights must be added for stability. Aside from stability, enough weight must be kept on the rear wheels for traction. Some industrial loaders use a reverse arrangement, with driving wheels in front to get additional weight for traction and less weight on steering wheels for easier steering.

Lift arms, or *booms*, are usually of tubular trussed construction or fabricated tapered-box sections. Farm loaders used primarily for manure may have the lift arms pivoting near the rear axle for minimum overhang and maximum stability without adding rear weights. Industrial loaders used for loading trucks usually pivot above the hood line near the center of the tractor for more reach at high lifts. They require rear weights because of the extra overhang when the lift arms are level. Buckets may have a parallel linkage for level lift or be locked rigidly to the lift arms. The latter construction is most common on farm loaders because of its simplicity and strength.

When one side of the fork or bucket catches under an obstacle, excessive twist of the lift arms can result if they are not adequately stabilized. Where a hydraulic cylinder is used on each side, a flow divider can be used to meter equal amounts of oil to each side, regardless of load. Usually, however, stability is secured by an adequate tubular crosstie between the lift arms.

Bucket pivots are usually located so that the loaded bucket will dump when released but return by gravity when empty. A spring-held latch is commonly used to hold the bucket in loading position. Overcenter linkages are also used and allow the bucket to be manually returned when sticky material does not empty completely. Hydraulically controlled dumping by a third cylinder is preferred for industrial loaders because it allows dumping at a controlled rate, gives a positive return, and can be used to control digging pitch while loading.

Optimum distribution of lifting force throughout the lifting cycle varies according to the work. In loading matted manure, breakaway loads may be more than twice the weight of the material actually lifted free, and it is desirable to have extra lifting capacity for the first 2 or 3 ft of lift. On the other hand, extra breakaway force may start a load of sand or gravel which cannot be fully lifted.

FIG. 31-1. Heavy-duty tractor loader. (*Ford Motor Co., Tractor and Implement Div.*)

Hydraulic pumps for loaders are preferably driven direct from the engine to allow the forward motion of the tractor to be halted by depressing the clutch while the bucket is being loaded and lifted. The loader may be powered from the tractor hydraulic system, although most do not have adequate output for industrial loaders which should be capable of lifting up to 2,500 lb in not more than 6 sec. This requires approximately 20 hydraulic hp or 17 gpm at 2,000 psi, for example.

Loaders often operate under dirty dusty conditions, and effective cylinder seals are required to resist wear. Piston rods should be chrome-plated or covered by boots to prevent corrosion during periods of idleness.

Control valves are usually of the four-way spool type (Fig. 10-4), with a second independent valve where a dumping cylinder is used. It is important that the valve design allow feathering for slow lifting and lowering movements when desired. Self-centering control valves, which are returned to neutral by pressure build-up when the cylinder completes its stroke, are sometimes used.

A conventional ball or poppet relief valve (Fig. 10-3a) is used to prevent extreme pressures when the control-valve spool momentarily blocks flow or when the rated lifting capacity is exceeded.

Rear-mounted loaders, actuated from the tractor hydraulic linkage arms aided by a booster cylinder, are often effective for manure-loading work. Height of lift is adequate for loading spreaders, rear-wheel traction is increased, close-quarter maneuverability is often superior to front-mounted loaders, and they can be quickly attached and detached.

Stacker attachments broaden the use of loaders in hay-stacking areas (Chap. 21).

Loaders, like earth-moving equipment, must be of rugged construction because full power will often be used, both for pushing and for lifting. In addition to withstanding bucking and lifting loads, the lift arms and frame must resist sideway of the loaded bucket when the tractor goes over uneven ground or is turned quickly. Some loaders for light tractors have a subframe bridging from rear axle to front axle to avoid loading of transmission housing and motor-block castings.

Rear-mounted hydraulically powered *back hoes* have proved flexible and efficient for digging trenches and small excavations where mobility and compactness are desirable.

The tractor is stationary when digging, and the operator faces the rear. Support pads are hydraulically lowered to stabilize the tractor. The boom-and-bucket assembly is actuated by four hydraulic control valves, which (1) tilt bucket, (2) swing bucket arm in and out, (3) raise and lower main boom, and (4) swing main boom sideways.

A typical back hoe of this type can dig 12 ft deep and load into a truck with sides 10 ft above the ground. The boom can be swung sideways through an angle of 185°.

Chapter 32

FEED GRINDERS

Advantages often cited for grinding of feeds are:
1. Increased digestibility
2. Increased palatability
3. Easier mixing of feeds, particularly when roughage is included
4. Less waste and rejection of coarse stems, etc.[1]

The economic advantage of grinding varies with the feeds and animals involved but is often considerable. Medium and coarse grinding are usually preferred to fine grinding, which reduces palatability and requires extra power. Roughage with low nutritive value may not be worth grinding.

BURR MILLS

Burr mills grind by crushing and shearing between two chilled cast-iron *plates* such as shown in Fig. 32-1a. The feed is introduced through a hole in the middle of the

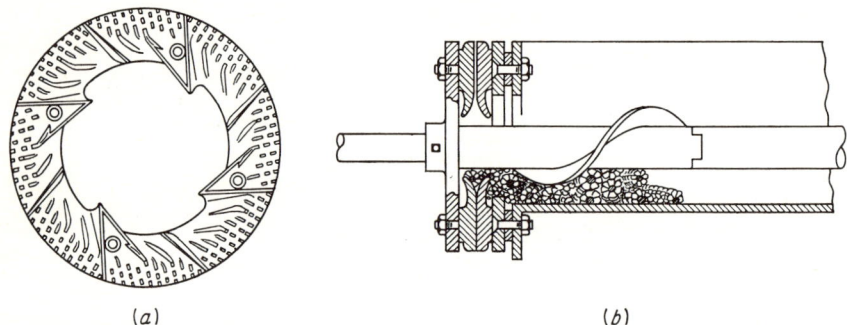

FIG. 32-1. Burr mill: (a) grinding plate; (b) cross section showing plates and feed screw.

rotating plate and is moved outward and ground by spiral ribs on the plates. The rib design may vary according to type of grain and fineness of grinding. Conical grinding burrs are also used. Plates are usually held together by spring pressure, adjustable for fineness, and should not be run empty unless the pressure is released or the plates are positively held apart. Ear corn must be crushed before it can be fed into the plates, and a common design uses a single shaft to carry spiral crushing and feeding blades as well as the moving grinding plate as shown in Fig. 32-1b.

University of Wisconsin tests of a small burr mill operating at high speed indicated that uniform feeding and fixed burr clearance with an overload release were desirable features for efficient operation at high speed. In these tests, 5½-in.-diam plates operating at 1,800 rpm required about 0.3 hp-hr per cwt of oats ground to a fineness modulus of 3.0, or 2 hp to grind 670 lb per hr.[2]

[1] F. C. Fenton and C. A. Logan, Farm Grinding of Grain and Forage, *Kansas State Coll. Eng. Expt. Sta. Bull.* 27, 1931.
[2] H. D. Bruhn, Burr Mill Design and Performance, *Agr. Eng.*, 17:101–107, March, 1936.

Fineness modulus is a measure of the fineness of grinding and follows the same principle used in grading materials for concrete. *Modulus of uniformity* indicates uniformity and fineness.[1]

A popular size of burr mill has 10½-in.-diam plates operating from 650 to 750 rpm and requiring from 20 to 30 hp, with a capacity of 2 to 5 tons of coarsely ground ear corn per hour. Burr mills are driven by belt or by tractor PTO in the case of portable models.

The relationship between capacity, speed, and energy requirement for an 8-in. burr mill is shown in Fig. 32-2.[2]

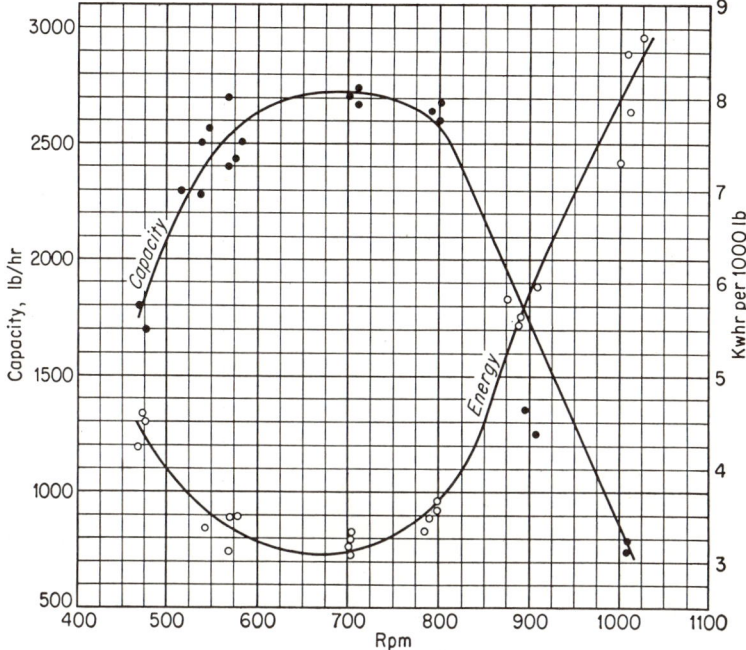

FIG. 32-2. The effect of speed on capacity and energy requirements for an 8-in. burr mill grinding shelled corn. (*Kans. Eng. Expt. Sta. Bull.* 27, 1931.)

Curves showing the effects of moisture content and fineness of grinding on power requirements are shown in Figs. 32-7 and 32-8.

HAMMER MILLS

A typical *hammer mill* is shown in Fig. 32-3. Hammer-tip speeds usually range from 12,000 to 17,000 fpm. Feed is fed onto the tips of the hammers, which, because of their speed, present an almost solid surface. Brittle kernels may be shattered by impact, but many feeds are carried around in a thin layer between the hammer tips and the cylindrical screen and are reduced in size by the tips plowing through this layer. The screen retains large particles until they have been ground small enough to drop through. Fineness of grinding is primarily controlled by screen size, although hammer speed and number are also factors.

Hammers are made of tough steel with wear-resistant hardened tips, since square sharp corners are required for efficient shearing of fibrous materials such as oat hulls.

---

[1] ASAE Recommendation: Method of Determining Modulus of Uniformity and Modulus of Fineness of Ground Feed, *Agr. Eng. Yearbook*, 1959, p. 127. R. C. Nicholas and C. W. Hall, Particle Size Distribution Analysis, *ASAE Paper* 57-594, December, 1957.

[2] Fenton and Logan, *op. cit.*

They are usually reversible for extra wear. Either swinging or stiff hammers may be used since little difference in function has been observed. Considerable variation exists as to number and thickness of hammer tips for a given width of mill, but total tip thickness often averages about half the actual width.

Because of the high speeds, it is important that the rotor assembly be accurately balanced to prevent vibration.

Hammer mills are usually *driven* by a flat-belt pulley on the rotor shaft. A 16-in.-diam rotor operating at 3,400 rpm is common, but the standard 3,100-fpm belt speed requires a 3½-in.-diam pulley. This is smaller than recommended and requires a thin flexible belt especially designed for hammer-mill work. Jackshaft speed step-ups are

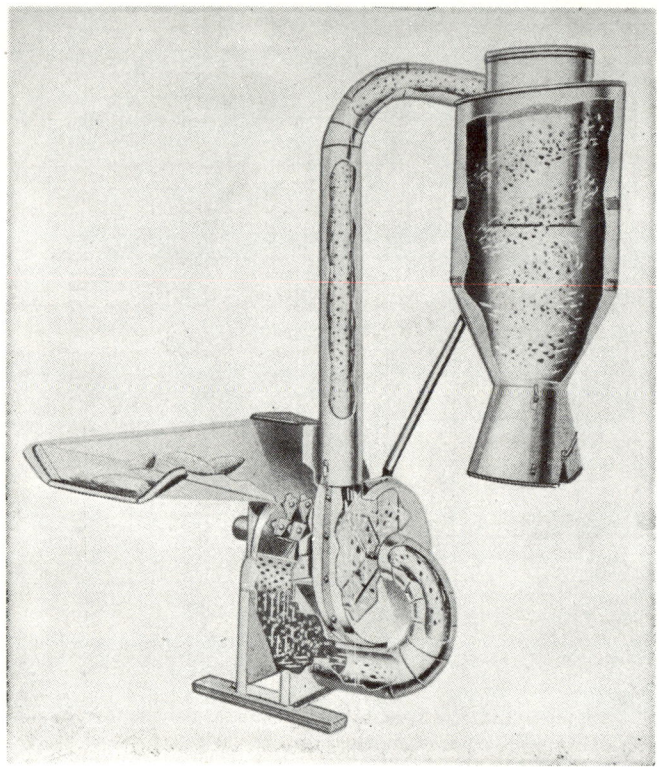

Fig. 32-3. Hammer mill with cyclone feed collector. (*Case.*)

seldom used on farm hammer mills because it is less expensive to obtain optimum tip speed by increasing rotor diameter.

PTO-driven mills have a severe speed problem in that the standard 540-rpm PTO speed must be stepped up to 2,000 to 3,000 rpm. Multiple V-belt drives are commonly used. The new 1,000-rpm PTO will alleviate this problem.

Screens are subject to wear as the grain is ground against them. They are made of high-carbon steel and can often be reversed to distribute wear. One or more breaker bars are often used to take the initial impact of the material being ground. Quickly changed screens with openings ranging from $\frac{1}{16}$ to $1\frac{1}{2}$ in. in diameter are usually available.

A *fan* is used to deliver the ground feed to sack, wagon, or bin, as the case may be. The suction of the fan may help draw ground material through fine screens. For the sake of simplicity, the fan is usually mounted on the rotor shaft with a suction duct leading to the fan from the hopper under the screen. One mill uses a screw conveyer under the screen, ending in an impeller which aids suction in getting the feed up to the fan. Large mills may use a screw-conveyer section to pull the material from under the screen into a smaller fan on the same shaft, requiring less power than a fan, which has to lift the material by suction. In either case it is essential that the removal capacity equal the maximum grinding capacity; otherwise clogging will result.

Finely ground oats bridge easily, and the sides of the *hopper* under the screen should have a slope of 60° to the horizontal, be as smooth as possible, and clear the screen by 1½ in. or more. Fine material blankets the inside of the screen and reduces the amount of air which the fan can pull through. Because of this, *additional air intakes* are provided, usually adjustable to meet various conditions.[1]

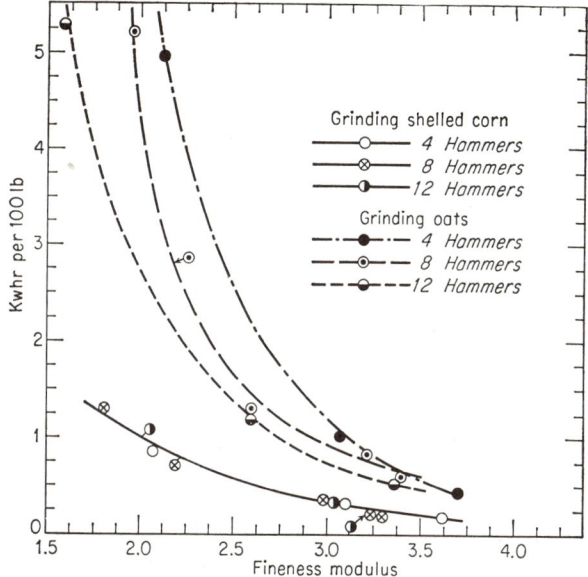

FIG. 32-4. Effect of number of hammers on mill efficiency. (*Hendrix, Agr. Eng., October,* 1937.)

In experiments with a small mill for automatic grinding, "no difficulty was experienced with screens becoming clogged, due to absence of a fan, when the mill was constructed so that the ground material was free to drop away from the screen into a sack or feed bin."[2]

A *cyclone feed collector* is used to separate the feed from the air. As shown in Fig. 32-3, air is introduced tangentially into the top of an inverted cone. The feed is thrown outward and settles downward through a delivery spout while the air reverses halfway down to go upward through a central discharge stack.[3]

Power requirements of hammer mills depend greatly on the type of material being ground and the fineness of grinding. No-load power to run mill and fan at rated speed varied from one-eighth to one-fourth of the total maximum power requirement in tests at Ohio State University[4] Tests on a small 10-in.-diam mill suitable for automatic operation indicated a no-load power requirement for mill and fan of about 50 per cent of the grinding load when operating at 3,750 rpm, the optimum speed.[5] The *effect of number of hammers* on the efficiency of this mill is shown in Fig. 32-4.

[1] F. W. Duffee, The Design and Performance of Small Hammer-type Feed Mills, *Agr. Eng.,* 11: 171–176, May, 1930.
[2] Andy T. Hendrix, Design and Performance of a Small Automatic Hammer Mill, *Agr. Eng.,* 18: 445–450, October, 1937.
[3] Olin M. Geer, A Small Diameter Feed Collector, *Agr. Eng.,* 19:109–110, March, 1938.
[4] E. A. Silver, Feed Grinder Investigation, *Ohio State Univ. Agr. Expt. Sta. Research Bull.* 490, 1931.
[5] Hendrix, *op. cit.*

The effect of speed on fineness of grinding is shown in Fig. 32-5, and the relation of screen size to fineness of grinding is shown in Fig. 32-6.

Moisture content of grain has a pronounced effect on grinding-power requirements, and Fig. 32-7 shows the results of the Ohio tests on burr and hammer mills. The

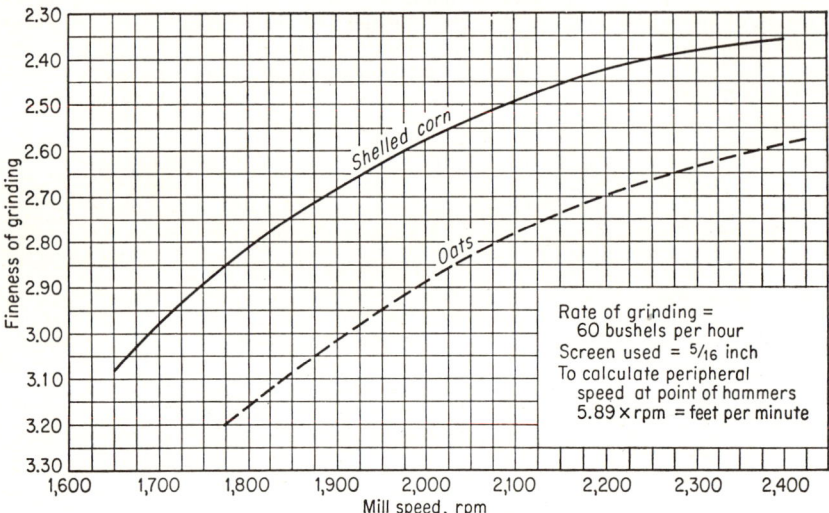

Fig. 32-5. Effect of speed on fineness of grinding. (*Ohio State Univ. Agr. Expt. Sta. Bull.* 490, 1931.)

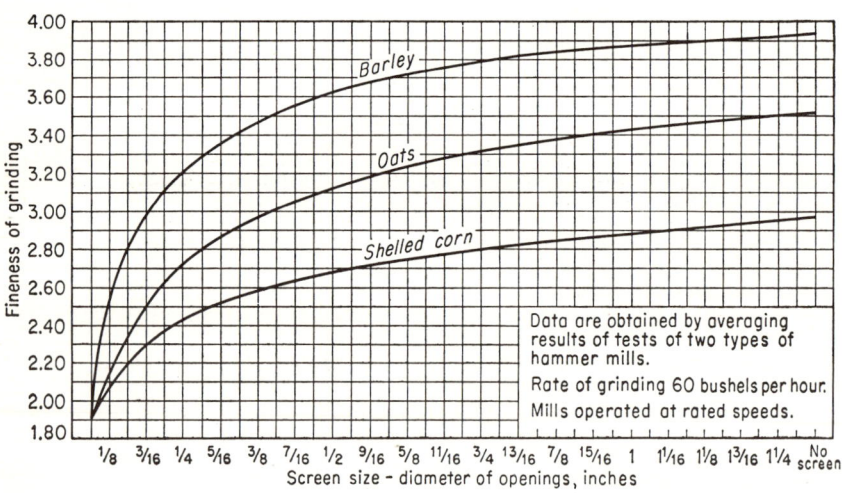

Fig. 32-6. Effect of screen size on fineness of grinding. (*Ohio State Univ. Agr. Expt. Sta. Bull.* 490, 1931.)

effect of fineness of grinding of shelled corn on power requirement is shown in Fig. 32-8 for both burr and hammer mills.[1] Distribution of particle sizes in samples of ear corn ground by hammer, burr, and knife mills is shown in Fig. 32-9. Feeding trials of beef cattle disclosed no significant differences due to the type of feed mill used.[2]

[1] Silver, *op. cit.*
[2] R. W. Kleis, and A. L. Neumann, Cattle Neutral on Feed Grinding Methods, *Agr. Eng.*, 37:544–547, August, 1956.

COMBINATION, OR ROUGHAGE, MILLS

Either burr or hammer mills may be provided with a conveyer feed table and precutting knives in order to grind hay, cornstalks, kafir corn, etc. A helical-blade cutting cylinder is used with burr mills and some hammer mills.

In general, grinding roughage requires more power than grain, often twice the horsepower-hours per ton. Overfeeding causes clogging, so to secure maximum capac-

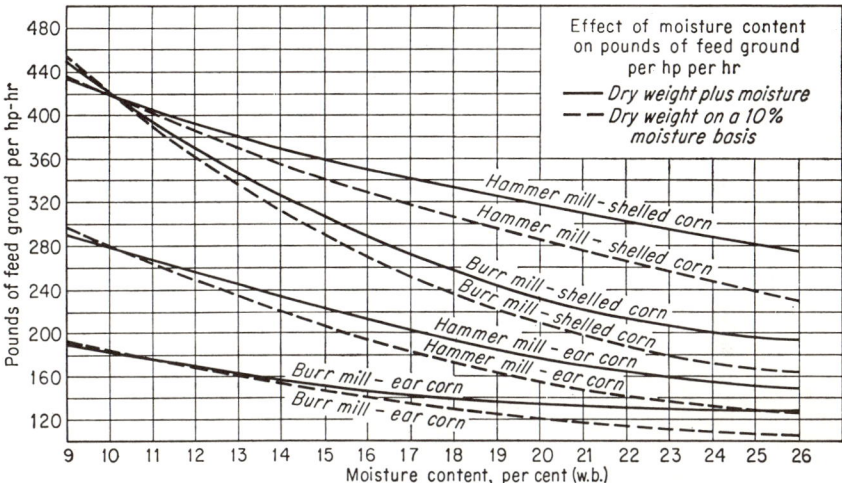

Fig. 32-7. Effect of moisture content on energy requirement. (*Ohio State Univ. Agr. Expt. Sta. Bull.* 490, 1931.)

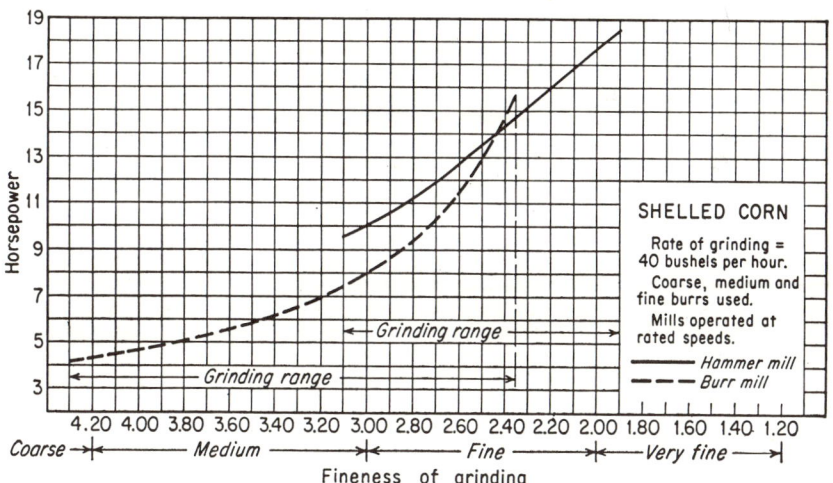

Fig. 32-8. Effect of fineness of grinding on power requirements. (*Ohio State Univ. Agr. Expt. Sta. Bull.* 490, 1931.)

ity, a uniform rate of feeding is desirable. A feed-table governor is often used to stop the feed table when the rotor speed falls below normal. A volume governor may also be used to stop the feed table in anticipation of an overload when an excessive feed thickness raises the upper feed roll.[1]

[1] W. Vutz, Some Observations on Hammer Type Feed Grinders, *Agr. Eng.*, 12:271–274, July, 1931.

Sharpness of precutting knives has a marked effect on power requirements for grinding roughage. High moisture content greatly increases power requirement, as does increased fineness. Roughage containing grain should be ground only fine enough to crack all the grain.[1]

OTHER GRINDERS

Many variaitions of the above grinding methods have been used. A bar or bladed cylinder working against bar concaves has been used, particularly for ear corn, but power requirements are excessive.

Another grinder uses a *cylinder of helical knives* cutting against a shear plate and running close to a screen, the grain being cut and crushed by the knives until it

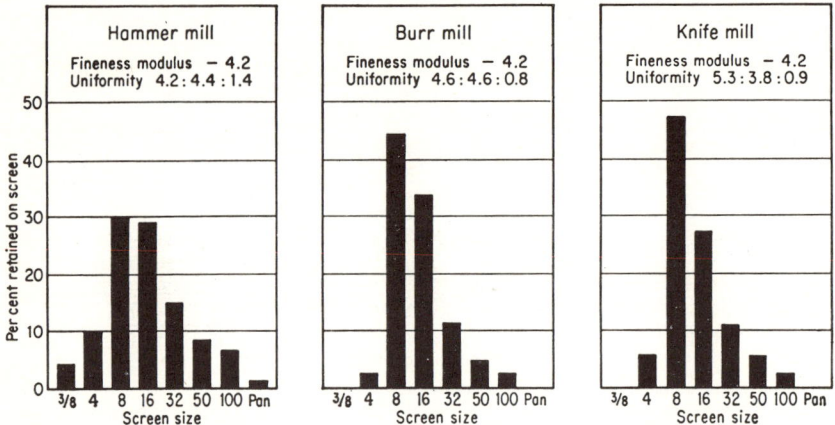

FIG. 32-9. Distribution of particle sizes in samples of ear corn ground by hammer, burr, and knife mills. (*Kleis and Newmann, Agr. Eng., August,* 1956.)

passes through the screen. It will handle ear corn and roughage and produces a minimum of dust, although the knives require sharpening.

Roller mills pass the grain between two grooved rollers about 10 in. in diameter which crack or crush and crimp the grain, depending on the pressure adjustment. Power consumption is low, and little dust is produced, although roughage or grain mixtures cannot be handled.

AUTOMATIC GRINDERS

In connection with the mechanization of farmstead operations, considerable study is being devoted to automatic systems for metering the various components of a ration directly into the grinder. This type of system eliminates the need for a separate batch-mixing operation.[2]

[1] Fenton and Logan, *op. cit.*
[2] W. R. Peterson, A Coordinated Design of Farmstead Structures for Improved Efficiency, *Agr. Eng.,* 27:399–402, September, 1946. W. F. Millier, Batch Weighing and Processing of the Dairy Ration, *Agr. Eng.,* 32:547–549, October, 1951. M. W. Forth et al., Automatic Feed Grinding and Handling, *Agr. Eng.,* 32:601–605, November, 1951. M. W. Forth and E. W. Lehmann, Performance and Electrical Load Characteristics of Automatic 5-horsepower Grinders and Motors, *Univ. Illinois Agr. Expt. Sta. Bull.* 581, 1954. R. W. Mowery, Electrical Control System for Automatic Feed Grinding, *Univ. Illinois Agr. Expt. Sta. Bull.* 555, 1952. R. P. Prince and E. F. Olver, A Continuous Flow Automatic Feed Grinder and Mixer, *Penn. State Coll. Agr. Expt. Sta. Progr. Rept.* 164, 1957.

Chapter 33

FEED ELEVATING AND CONVEYING EQUIPMENT

In general, farmstead operations have lagged behind field operations in mechanization, but the pitchfork and scoop shovel are being replaced by power elevating and conveying equipment. Grain, ear corn, silage, ground feed, chopped hay, and bales of hay are being handled mechanically at a great saving in time and human exertion.[1]

In general, feed is mechanically transported by being carried by elevators or conveyers, thrown and blown by impellers, or blown by air.

ELEVATORS

Flight conveyer elevators have chain-drawn flights or crossbars which slide material up an inclined flat-bottomed trough. The maximum permissible angle depends on the height and shape of the flights and the tumbling characteristics of the material. Ears of corn traveling endways in a narrow trough tumble much more than in a wide trough where the ears can lay crossways. The maximum recommended operating angle for grain is normally 45°, with capacities increasing with less angle.

Elevators up to 65 ft in length may be used to elevate silage into vertical silos. They are operated at angles up to 70° and require much less power than forage blowers (Chap. 23), handling silage at 20 tons per hour with a 3-hp electric motor in many cases.

In order to handle square-cornered hay bales of standard size, the *trough* must be 14 in. or more in width or designed to take the bale cornerwise. A general-purpose elevator with a trough large enough to handle bales must have enough power and capacity to handle full troughs of grain. Some single-chain elevators have folding trough sides which can be adjusted for the material to be elevated.

The *chains* are usually driven from the bottom, requiring that they be kept fairly tight to prevent slack and breakage just above the driving sprocket. Double chains often run on shelves at the side and are attached above the bottom of the flight to equalize the load.

A *horizontal receiving hopper* with conveyer can be tipped up to allow the wagon to pass and then lowered to receive the load. The wagon is usually dumped by lifting the front wheels with an overhead hoist powered from the elevator.

The *transport truck* may be a four-wheel or two-wheel trailer type. A *lifting frame,* extending back of and pivoting on the rear axle, is swung up by cables and a hand-operated winch to lift the elevator into operating position as shown in Fig. 33-1. This lifting frame usually slides along the bottom of the elevator as it swings up or down.

Portable elevators, because of their weight and size, can cause serious injuries if they overturn, upend, or collapse. Design features should include a wide track for lateral stability, constraint and limited travel of lifting frame, and a self-locking winch to prevent loss of control under load.[2]

---

[1] R. W. Kleis and D. E. Wiant, Material Handling: Methods and Labor Requirements, *ASAE Paper* 57-595, December, 1957.
[2] B. J. Lamp and K. A. Harkness, Recommendations for Improving the Design of Portable Farm Elevators, *ASAE Paper* 59-914, December, 1959.

Portable elevators may be *powered* by electric motor, gasoline engine, or tractor PTO. A formula for the *chain pull* of a *horizontal conveyer* is as follows:

$$\text{Chain pull, lb} = 2WLF + W_1LF_1$$

where W = weight per ft of element, lb
L = length of conveyer, ft
F = coefficient of friction of element, 0.33 for sliding, nonlubricated conditions
W_1 = weight of material per foot of conveyer, lb
F_1 = coefficient of friction of material on steel plate, 0.30 to 0.40 for grains (see Fig. 23-7 for forages)

The power requirement for an *inclined conveyer* is the sum of that required for the horizontal run plus that required to lift the load. The formula for chain pull at the pitch line of the upper sprocket is as follows:

$$\text{Chain pull, lb} = WL(F\cos\theta + \sin\theta) + W_1L(F_1\cos\theta + \sin\theta) + WL(F\cos\theta - \sin\theta)$$

where θ = angle between conveyer and horizontal, deg

The third term may be a minus quantity, indicating that the return run then is assisting the turning effort.

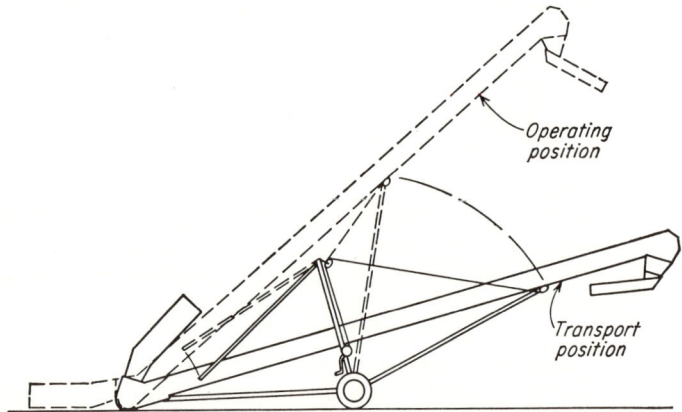

Fig. 33-1. Two-wheel portable elevator. (*Deere.*)

"For the motor horsepower, add 10 per cent for friction in the head bearings, plus 10 per cent for loss in the speed-reduction gearing, plus 10 per cent for starting or surge loads."[1]

The capacity of an inclined conveyer compared with its level capacity is 90 per cent at 20°, 80 per cent at 25°, 70 per cent at 30°, and 60 per cent at 35°.[1]

Because of the friction involved, the actual power required is about three times the theoretical lifting power at the steeper angles and more at flatter angles.

It is desirable to have an *overload slip clutch* or V-belt drive in the power line to prevent breakage when clogging occurs, particularly with a tractor PTO drive.

Auger loaders, or elevators, are popular for handling small grain because of their simplicity and light weight. They consist of a long screw conveyer running in a tube but projecting beyond at the lower end to take grain out of the pile or bin. Some models have an open cage which supports a bottom bearing for the screw. The auger flights are often flanged at the bottom to help pull in grain.

The screw is *driven* from the upper end by a large V-belt sheave, bevel gears, or worm gears. An air-cooled single-cylinder gasoline engine is usually mounted about 5 ft above ground level and drives to the top of the tube by V belt. Engines of 2½ to 8 hp are used, depending on tube length, which ranges from 14 to 40 ft.

[1] W. G. Hudson, Hoisting and Conveying, in T. Baumeister (ed.), "Mechanical Engineers' Handbook," 6th ed., McGraw-Hill Book Company, Inc., New York, 1958.

In studies at Purdue University, the capacity of a 16-in.-diam screw conveyer running 600 rpm or less was found to vary directly with screw speed. At 500 rpm, shelled corn was conveyed at a rate of 22 bu per hr at 0° inclination, 18 bu per hr at 45° inclination, and 10 bu per hr at 90° or vertical. Capacities and power requirements of a 4-in.-diam screw soybean conveyer at various speeds and angles are shown in Fig. 33-2. The power efficiency in various tests at 85° inclination and 700 rpm ranged from 16 per cent for oats to 9 per cent for barley.[1]

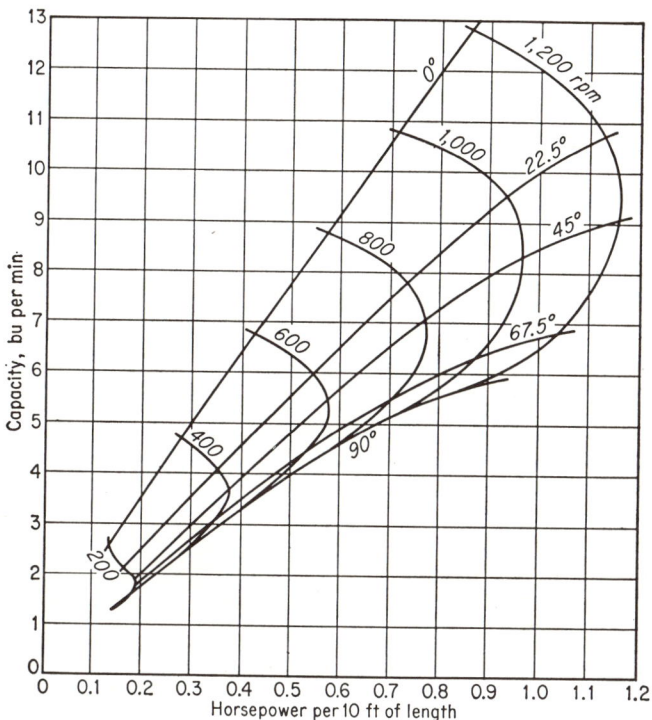

Fig. 33-2. Characteristic curves for a 4-in.-diam screw conveyer handling soybeans. The characteristics of this conveyer are as follows: 13 per cent moisture, wet basis tube ID 4.00 in., screw OD 3.37 in., and screw shaft OD 0.844 in. (*Ross and Isaacs, ASAE Paper* 59-915, *December,* 1959.)

Short models are popular for loading trucks and may be supported by the side of the truck. Larger models have transport trucks and means for lifting them into operating position by hand crank.

Horizontal auger unloaders, sweeping around a center delivery point, are capable of efficiently unloading flat-bottom grain bins.[2]

Vertical elevators in permanent grain-storage buildings are of the bucket type. Large buckets which will hold several ears are necessary for efficient elevation of ear corn.[3]

---

[1] I. J. Ross and G. W. Isaacs, Theory of Operation and Construction of Enclosed Screw Conveyers, *ASAE Paper* 59-915, December, 1959. W. M. Regan and S. M. Henderson, Performance Characteristics of Inclined Screw Conveyors, *Agr. Eng.,* 40:450–452, August, 1959. W. F. Millier, Bucket Elevators and Auger Conveyors for Handling Free-flowing Materials, *Agr. Eng.,* 39:552–555, September, 1958. Floyd L. Herum, Performance of Screw Conveyors at Rates Less than 2,500 Pounds per Hour, *ASAE Paper* 59-916, December, 1959.

[2] B. A. McKenzie and I. J. Ross, A Mechanical Unloader for Round, Flat-bottom Grain Bins, *ASAE Paper* 57-600, December, 1957. H. B. Puckett, A New Automatic Unloader for Flat-bottom Bins, *Agr. Eng.,* 40:388–391, July, 1959.

[3] W. F. Millier, *op. cit.*

IMPELLER BLOWERS FOR GRAIN

The *impeller blower* is a simple machine useful for elevating grain vertically. The grain is primarily thrown rather than blown, and there is not enough air to convey it horizontally. The grain is introduced near the center of an impeller rotor having two to six paddles and leaves the blade tips at near their peripheral speed.

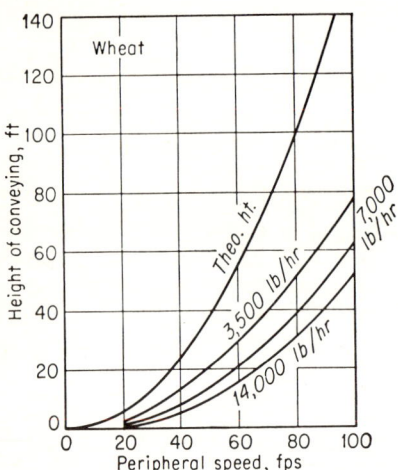

Fig. 33-3. Relationship between impeller peripheral speed and conveying height for wheat. (Segler, "Pneumatic Grain Conveying," 1951.)

Tests indicate that wheat containing 12½ per cent moisture is not appreciably *damaged* as to cracking or germination at impeller peripheral speeds below 70 fps. Power consumption is at a minimum when peripheral speed is only enough to elevate to the desired height.

The theoretical *heights* and actual heights for different delivery rates are shown in Fig. 33-3.[1] The actual delivery height falls short of the theoretical height to an increasing extent as internal friction increases with high rates of grain flow.

It can be seen from the chart that higher tip velocity is required to deliver greater rates to a given height and that 14,000 lb per hr can be lifted only 20 ft without exceeding 70 fps.

The following formula gives the power required to elevate wheat at rates of 3,500 to 14,000 lb per hr to heights of 5 to 30 ft:[2]

$$\text{Hp} = \frac{V^{2/3} W^{1/3}}{460}$$

where V = tip velocity, fps

W = weight of grain per hour, cwt

Radial paddles gave the best results. Grain damage occurred at the initial impact with the blade, and it is important that the grain be introduced at the inner circle of the blades where the velocity is the lowest. It was found possible to increase the allowable peripheral speed to 100 fps when the paddles were covered with rubber to reduce crackage.[2]

A typical American impeller blower has a six-blade 21-in.-diam rotor with 6-in.-diam delivery pipe and is rated at 450 to 1,000 bu per hr. Many forage blowers can be satisfactorily used as impeller blowers for grain. For minimum crackage, peripheral speed should be kept below 70 fps.

PNEUMATIC GRAIN CONVEYING

Pneumatic grain conveying is not commonly used on farms but is widely used in large grain-storage and processing plants. The power requirement is much greater than for a mechanical conveyer of equal capacity, but the duct may be led along almost any path and distance. The vacuum-cleaner action provides dustless operation when unloading from boxcars or trucks with a flexible nozzle, and there are no dangerous moving parts.

The grain moves through a horizontal pipe in a series of jumps, and air velocity must be great enough to lift an individual kernel from the bottom of the tube in order

---

[1] G. Segler, "Pneumatic Grain Conveying," *National Institute of Agricultural Engineering*, Silsoe, Bedfordshire, England, 1951.

[2] *Ibid.*

to avoid blockage. The total *resistance coefficient* for the mixture of grain and air is due to (1) friction of air on pipe, (2) friction between air and grain, (3) friction of grain on pipe, and (4) friction between grains, as illustrated in Fig. 33-4.[1] In vertical conveying, the air velocity must be somewhat greater in order to lift the kernels directly upward.

Tests of *damage* to various whole grains at different moisture contents indicate that the upper limit for air velocity is from about 73 fps at low throughputs to 79 fps at high throughouts with pipes which contain curves. As shown in Fig. 33-5, blockage

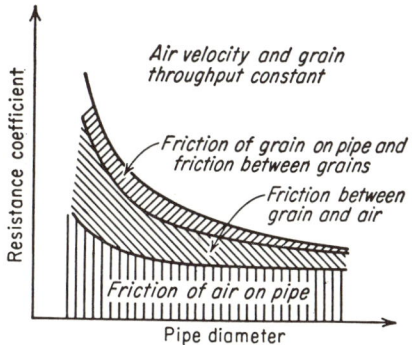

FIG. 33-4. Distribution of friction in pneumatic conveying of solids. (Segler, "Pneumatic Grain Conveying," 1951.)

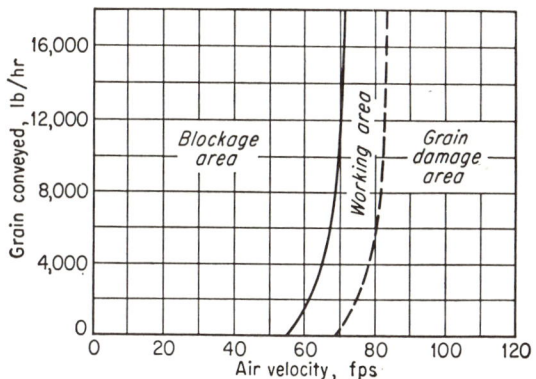

FIG. 33-5. Safe air velocities to avoid blockage and grain damage. (Segler, "Pneumatic Grain Conveying," 1951.)

may occur at air velocities about 12 fps less than these, which indicates that the satisfactory working range is comparatively narrow, requiring careful engineering of pneumatic conveying systems. Higher velocities can, of course, be used with ground feed with no damage other than further size reduction.

As can be seen from Fig. 33-6, small-pipe diameters require less power for a given capacity than large-pipe.[2] On the other hand, much higher pressures must be used to carry the denser mixture of air and grain, requiring compressors instead of fans and heavier, tighter pipelines. Large pipes, 8 to 12 in. in diameter, and lower pressures are often more practical for smaller-scale installations because they are better adapted to the use of simple injector feeders than to bucket-wheel or screw feeders.

[1] *Ibid.* Jack W. Crane and W. M. Carleton, Predicting Pressure Drop in Pneumatic Conveying of Grains, *Agr. Eng.*, 38:168–180, March, 1957.
[2] Segler, *op. cit.*

The feeder is commonly located on the discharge side of the fan to avoid damage to the grain from the fan blades. An efficient injector design developed by the British National Institute of Agricultural Engineering is shown in Fig. 33-7.[1]

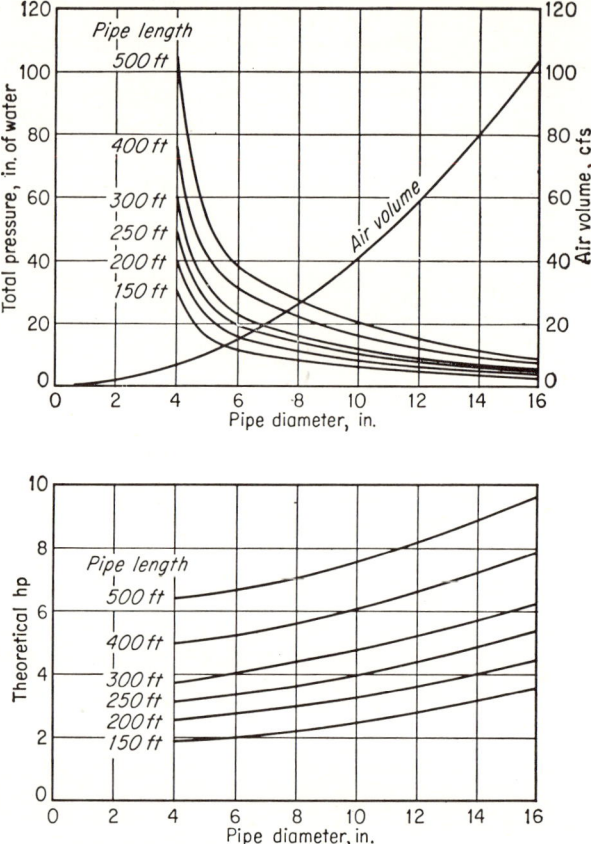

FIG. 33-6. Air volume, total pressure, and power requirement for pneumatic grain conveying in various pipe diameters at 16,000 lb per hr and 75 fps. (Segler, "Pneumatic Grain Conveying," 1951.)

Blowing has proved to be a simple, convenient method of transporting ground and mixed feed from storage to the feeding area. The following points are helpful in planning a system:[2]

1. An air velocity of 4,000 fpm is necessary for satisfactory operation.
2. Horsepower requirements to maintain 4,000 fpm air velocity without grain in the system are: 6-in. pipe, 1½ hp for each 100 ft; 5-in. pipe, 1¼ hp for each 100 ft; 4-in. pipe, 1 hp for each 100 ft. Add ⅓ hp for each 1,000 lb of grain per hour.
3. Maximum grain-conveying rate is: 6-in. pipe, 5,500 lb per hr; 5-in. pipe, 4,500 lb per hr; 4-in. pipe, 3,500 lb per hr.
4. Use a 4-in. pipe for grinders driven by a 7½-hp electric motor or smaller. Grinders driven by tractors will usually require a 6-in. blower pipe.
5. Type of grain and fineness of grind have no significant effect on power requirements or size of pipe needed.

[1] *Ibid.*
[2] R. W. Kleis, Blowing Grain from Storage to Feed Area, *Univ. Illinois, Dept. Agr. Eng.*, RE-6, 1954; Operating Characteristics of Pneumatic Grain Conveyers, *Univ. Illinois Agr. Expt. Sta. Bull.* **594,** 1955.

6. For greatest operating efficiency use elbows with a radius of curvature equal to at least four pipe diameters.

7. Do not place an elbow closer than 8 ft from the dust collector.

8. Slope the blower pipe toward the discharge end to drain off water caused by condensation or leakage.

9. The air intake of the blower must not be restricted in any way, because a large quantity of air is vital to satisfactory operation.

10. It is desirable to vent the top of the dust collector to the outside of the building.

Pneumatic conveying of sorghum grain was studied at Texas A & M College. At a rate of 7 tons per hour through a 6-in. duct with an air velocity of 77 fps, a static pressure of 3.1 in. of water was required for grain entrance and acceleration, 0.6 in. of water for each 10 ft of duct length for horizontal conveying, plus 1 to 4 in. of additional water for each elbow.[1]

The principle of *fluidization*, originally applied to cement and pulverized coal, has been experimentally applied to grain conveying. In these tests, grain was conveyed through 75 ft of 1-in.-ID tube at rates of 1 to 2 tons per hr by applying air

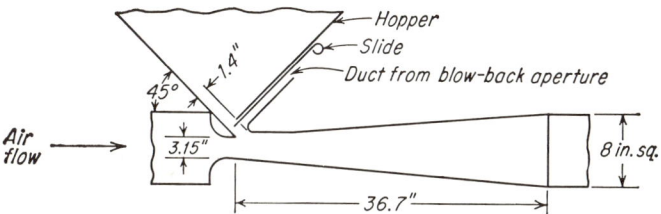

Fig. 33-7. Injector design developed at National Institute of Agricultural Engineering, Silsoe, Bedfordshire, England. (Segler, "Pneumatic Grain Conveying," 1951.)

pressures of 5 to 24 psi requiring 0.35 to 1.7 hp. About 20 lb of grain was moved per pound of air.[2] In Illinois tests of an installation of this type, ground feed was conveyed through 250 ft of 1-in.-ID tube at a rate of 1,231 lb per hr, with a pressure of 6.2 psi and a pumping power of 2 kw, requiring 0.0656 kwhr/(cwt) (hr) (100 ft).[3]

AUTOMATIC FEEDING

The laborsaving possibilities of automatic feeding by conveyers are being explored, particularly in large-scale poultry enterprises. Various types of simple conveyers are being used to carry feed into the feeding trough, including an oscillating hopper which walks the feed along like a combine shoe.[4]

Auger bunk feeders for cattle distribute silage to a line of feed bunks. The open-bottom type fills the bunks progressively as it loads up underneath. The tube type has openings gradually spiraling downward so as to give simultaneous distribution over the full length.

[1] N. K. Person, Jr., and J. W. Sorenson, Jr., Pneumatic Handling of Grain, *ASAE Paper* 59-912, December, 1959.
[2] A. D. Longhouse et al., The Application of Fluidization to Conveying Grain, *Agr. Eng.*, 31:349–352, July, 1950.
[3] H. B. Puckett, Performance of Medium-pressure Pneumatic Conveying System, *ASAE Paper* 59-911, December, 1959.
[4] R. R. Parks, Mechanizing Poultry Feeding, *Agr. Eng.*, 31:23–25, January, 1950; Mechanical Feeding with Worksavers, *Agr. Eng.*, 32:554–558, October, 1951. D. C. Sprague et al., Automatic Poultry Feeders, *Agr. Eng.*, 31:21–25, January, 1950. John B. Dobie, Mechanical Feeding of Livestock, *Agr. Eng.*, 36:458–466, July, 1955.

SECTION II
SOIL AND WATER CONSERVATION
BY PAUL JACOBSON, *Subeditor*

Chapter 34

PRINCIPLES OF AGRICULTURAL HYDROLOGY
LLOYD L. HARROLD

Hydraulics is the science dealing with flowing water or other fluids and their artificial conveyance through machines, pipes, or channels. For ordinary work of the agricultural engineer, the Manning formula, Bernoulli's theorem, and the orifice formula or derivations of these are all that are needed for solving problems. The use of these formulas is adequately covered in handbooks[1] which are readily available. Some formulas as they pertain to various phases of the flow of water in irrigation, in drainage, and through structures will be covered in the individual chapters.

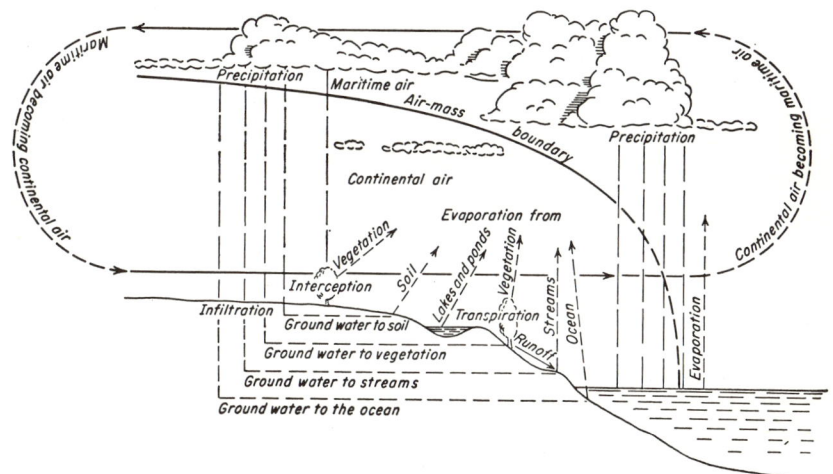

Fig. 34-1. The hydrologic cycle according to modern meteorologic evidence. (*"Climate and Man,"* USDA Yearbook of Agriculture, 1941.)

Hydrology is the science treating water in its entire cycle from the atmosphere to the earth's surface, to the sea, and back to the atmosphere as shown in Fig. 34-1. The major phases of hydrology treated in this chapter are (1) precipitation, (2) infiltration, (3) transpiration, (4) evaporation, (5) ground water, and (6) runoff. Each of these subjects will be treated separately.

PRECIPITATION

Precipitation is the general term used to embrace any or all forms of moisture falling from the clouds as rain, snow, hail, or sleet. The common unit of measurement

[1] Horace W. King and Ernest F. Brater, "Handbook of Hydraulics," 4th ed., McGraw-Hill Book Company, Inc., New York, 1954.

is depth of water in inches and hundredths of inches. Data on solid precipitation, such as snow, are given in equivalent depth of water. Kincer indicated that the moisture equivalent of snow ranged from 5 to 6 in. of very wet snow to an inch of water to 15 in. or more of fluffy snow to an inch of water.[1] Where the actual water equivalent is not determined, the arbitrary rule for average snow density is 10 in. of snow equivalent to 1 in. of water.

Sources of Precipitation Data. The U.S. Weather Bureau is the principal source of basic information concerning the amount and area distribution of precipitation

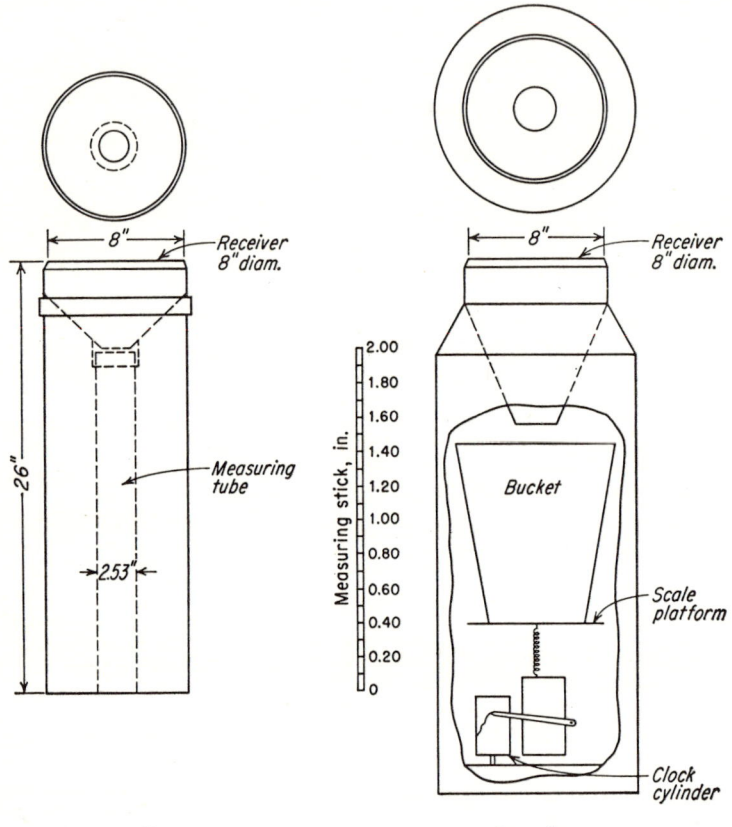

Fig. 34-2. Precipitation gages.

throughout the United States. Much of its data is published by the U.S. Department of Commerce. Many other Federal and state agencies have gathered a great volume of precipitation data, very little of which has been published. These agencies usually make their records available for public examination on location. The Natural Resources Planning Board reported on the availability of published and unpublished data.[2]

Equipment for Measurement. Nonrecording and recording gages are illustrated in Fig. 34-2. The nonrecording gage merely collects total precipitation. The recording

---

[1] J. B. Kincer, "Climate and Weather Data for the United States," USDA Yearbook of Agriculture, 1941.
[2] Natural Resources Planning Board, Principal Federal Sources of Hydrologic Data, *Tech. Paper* 10, 1943.

gage records the depth of precipitation at a given time on a chart operated by a clock-work mechanism.

Measurements of snow and ice cover are frequently made prior to the melt period. Equipment usually consists of a metal cutting tube to obtain a small-diameter vertical core from the entire depth of snow cover. The sample is weighed, and the equivalent depth of water is computed. This procedure is quite accurate for snow depths of several feet or greater. For shallow snow depths it is more accurate to obtain larger samples and either weigh them or melt them and calculate the water depth. (See Chap. 42 for additional information on snow measurements.)

Where records of snowfall are desired, either the recording or nonrecording gages can be used. In such cases, the gage should be shielded so as to reduce the effect of wind movement to a minimum and antifreeze should be added to prevent damage to the bucket by freezing of the melted snow.

The Alter shield has been adopted as a standard by the U.S. Weather Bureau[1] to reduce the effect of wind on the gage catch of rain or snow.

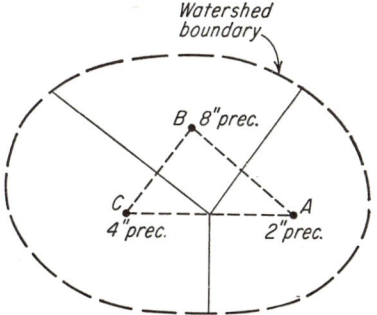

FIG. 34-3. Thiessen method of averaging rainfall. A, B, and C are rain-gage sites. Thiessen weights of relative areas for each gage: $A = 30\%$, $B = 30\%$, $C = 40\%$. Average rain on watershed:

$A = 2'' \times 0.30 = 0.60''$
$B = 8'' \times 0.30 = 2.40''$
$C = 4'' \times 0.40 = 1.60''$
$\overline{4.60''}$

Compilation of Data. The total storm precipitation or totals for specific periods of the storm for entire watersheds can be obtained from:

1. *Arithmetic average* of values from all gages in the watershed. (This is satisfactory if the gages are uniformly spaced over the watershed area.)

2. *Thiessen method.*[2] Construct perpendicular bisectors to lines connecting adjoining gages on a map of the watershed (Fig. 34-3). The area of each polygon thus formed is determined and expressed as a per cent of the total watershed area. This factor is then multiplied by the rainfall measured at that station. The total of all such values represents the average depth of precipitation for the entire area. (The value is considered more accurate than that obtained by the simple-arithmetic process, where there is nonuniform areal distribution of the gage.)

3. *Isohyetal method.* On the watershed map, list at each gage location the total precipitation for the storm, day, or any selected period. Contour lines of equal precipitation (isohyets) are then drawn on the map at their proper position between the gage locations as shown in Fig. 34-4. The sum of values obtained by multiplying the areas between adjacent contour lines within the watershed by the average of the two contour values is divided by the watershed area to determine the average watershed precipitation.

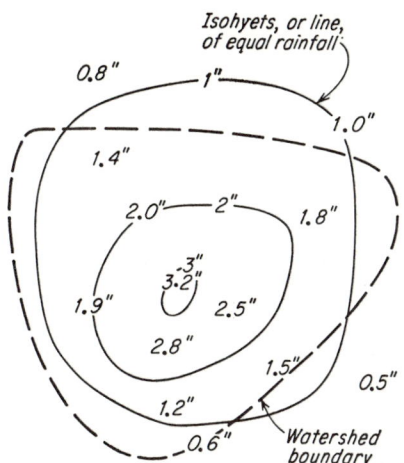

FIG. 34-4. Isohyetal method of showing rainfall distribution over an area. Decimal point of each precipitation value designates the location of the gage.

[1] J. C. Alter, Shielded Storage Precipitation Gages, *Monthly Weather Rev.*, 65:262–265, 1937.
[2] A. H. Thiessen, Precipitation Averages for Large Areas, *Monthly Weather Rev.*, 39:1082–1084, 1911.

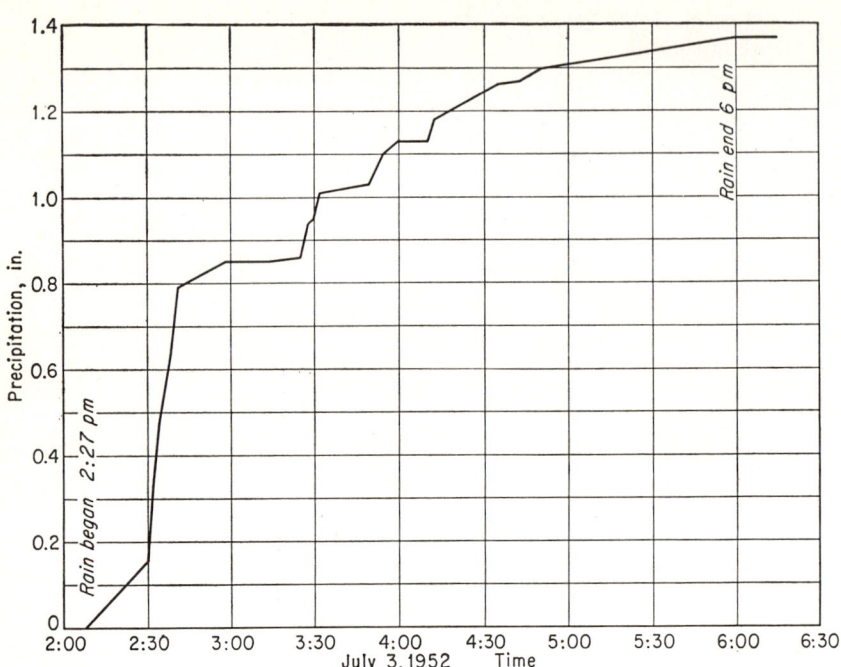

Fig. 34-5. Time vs. precipitation—accumulation graph. See Table 34-1 for data.

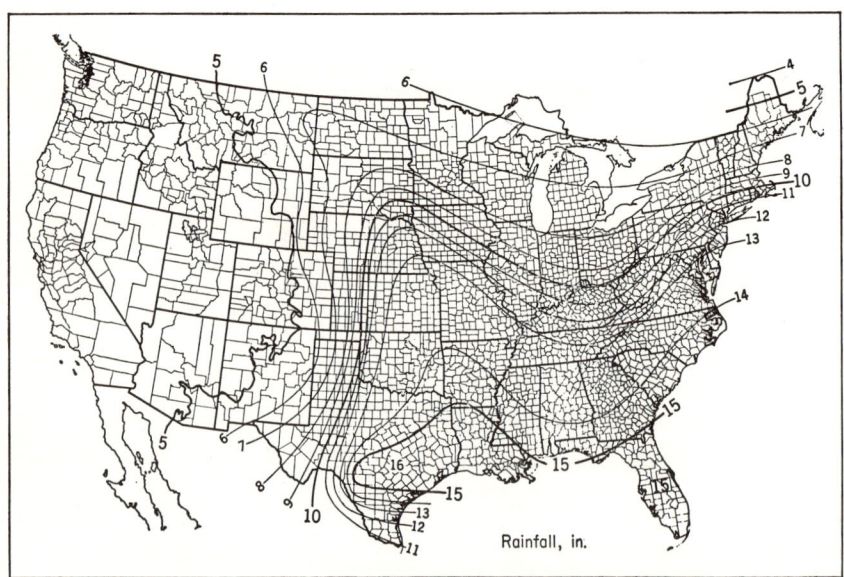

Fig. 34-6. Six-hour-point rainfall, United States, except West. For details west of the 5-in. line, see Fig. 34-7. (*USDA SCS, Eng. Div., Central Tech. Unit*, 1955.)

The recording type of precipitation gage is the principal source of data for studies involving rainfall intensities. The most common record is the mass-precipitation curve (Fig. 34-5). Detailed tabulation of this record appears in Table 34-1. The data are thus placed in a form for almost every kind of use.

TABLE 34-1. SAMPLE FORM FOR RAINFALL-INTENSITY TABULATION

Project Coshocton, Ohio
Station Y101
Record of Rainfall Intensity   Date July 3, 1952

Date	Time, hr and min	Time interval, min	Accumulated depth, in.	Depth for each time interval, in.	Intensity for each time interval, in./hr	Remarks
(1)	(2)	(3)	(4)	(5)	(6)	(7)
1952 July 3	2:27 P.M.		0			Begin
	:30	3	.15	0.15	3.00	
	:32	2	.33	.18	5.40	
	:34	2	.48	.15	4.50	
	:38	4	.63	.15	2.25	
	:42	4	.79	.16	2.40	
	:58	16	.85	.06	.23	
	3:14	16	.85	0	0	
	:25	11	.86	.01	.05	
	:28	3	.94	.08	1.60	
	:30	2	.95	.01	.30	
	:32	2	1.01	.06	1.80	
	:49	17	1.03	.02	.07	
	:54	5	1.10	.07	.84	
	4:00	6	1.13	.03	.30	
	:10	10	1.13	0	0	
	:12	2	1.18	.05	1.50	
	:34	22	1.26	.08	.22	
	:42	8	1.27	.01	.08	
	:50	8	1.30	.03	.23	
	6:00	70	1.37	.07	.06	End

Maximum depth and intensity for selected time intervals

Duration	2 min	5 min	10 min	15 min	20 min	30 min	1 hr	2 hr	4 hr	6 hr	12 hr
Depth, in.	0.18	0.38	0.59	0.79	0.81	0.85	0.91				
Intensity, in./hr	5.40	4.56	3.54	3.16	2.43	1.70	0.91				

Tabulated by _____   Computed by _____   Checked by _____
           (Date)   (Date)   (Date)

Sheet _____ of _____ sheets

Many engineering problems involve the use of *rainfall intensity-frequency data*. Sometimes, where extreme loss of life and property would result from failure or overtopping of a structure, it may be necessary to consider in the design the maximum possible precipitation, as has been done by the Corps of Engineers of the Department of the Army in cooperation with the U.S. Weather Bureau in their Hydrometeorological Reports for various river basins. Figures 34-6 and 34-7 show the maximum 6-hr-point rainfall (occurring at a single point) for the United States, and Table 34-2 gives constants used by the U.S. Soil Conservation Service for extending the 6-hr rainfall to storms of greater duration.

On small agricultural structures, where there is little hazard downstream from a

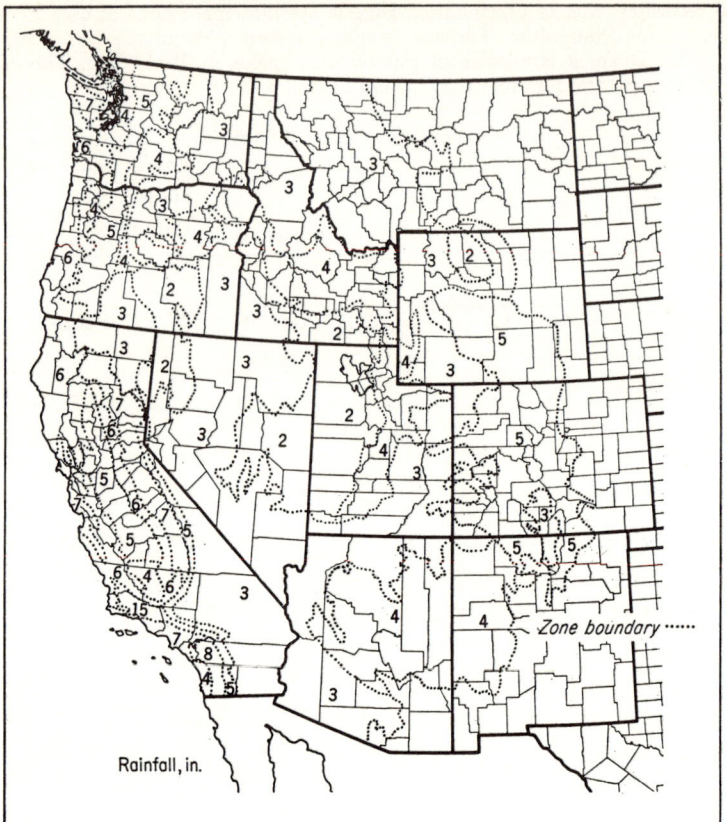

Fig. 34-7. Six-hour-point rainfall, western United States. (*USDA SCS, Eng. Div., Central Tech. Unit*, 1955.)

failure which might occur, designs are ordinarily not made to provide for a maximum probable storm. In these cases designs are based on storm return periods of 5, 10, 25, 50, and 100 years. The return period used should be based on size of structure and the permanence desired.

TABLE 34-2. CONSTANTS FOR EXTENSION OF DESIGN STORM

Duration, hr	Constant	Duration, hr	Constant
7	1.026	18	1.552
8	1.121	20	1.619
9	1.176	22	1.682
10	1.227	24	1.741
11	1.274	28	1.852
12	1.320	32	1.953
14	1.402	36	2.048
16	1.481		

SOURCE: USDA Soil Conservation Service.

Example: Given a 6-hr storm rainfall of 11.0 in., what is the equivalent maximum amount for a duration of 12 hr?

From the table, at 12 hrs the constant is 1.320. Then

$$1.320 \, (11.0) = 14.52 \text{ in.}$$

which is rounded to 14.5 in.

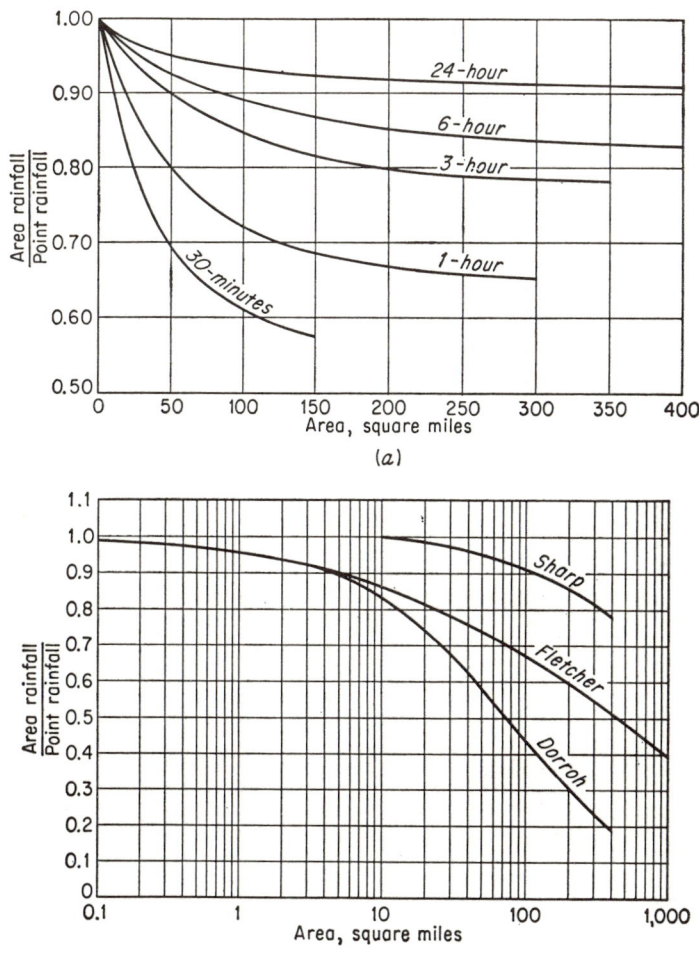

Fig. 34-8. Generalized relationship between point rainfall and area rainfall: (a) This curve is for use with gages whose catch is less than the maximum rainfall depth of the storm. It is quite often being used by field engineers as a basis for design. If the gage has caught the maximum, use the curves given in (b). (U.S. Weather Bureau Tech. Paper 29, pt. 1.) (b) The *Sharp* curve applies to the West Coast, west of the Cascade and Sierra Nevada ranges. The *Dorroh* curve applies to the intermountain area between the Cascade and Sierra Nevada ranges and the Rocky Mountain Divide. The *Fletcher* curve (adapted from an equation of Robert D. Fletcher, Trans. Am. Geophys. Union, June, 1950) applies to the area east of the Rocky Mountain Divide.

Yarnell,[1] in 1935, compiled for the United States a *summary of rainfall-intensity data*. In December, 1955, the Weather Bureau published rainfall intensity-duration-frequency curves[2] which include 20 years more record than Yarnell's material.

To use this material it is merely necessary to know the location, the estimated return period in question, and the duration of the storm. Thus for Fresno, Calif., the average rainfall intensity for a 6-hr duration with a return period of 10 years would be about 0.19 per hour and the total rainfall for the 6-hr period would be 1.14 in.

Rainfall determined as above will be point rainfall. It will be necessary to adjust these to reflect watershed size. Figure 34-8 gives a series of curves which can be used

[1] David L. Yarnell, Rainfall Intensity-Frequency Data, *USDA Misc. Publ.* 204, 1935.
[2] Rainfall Intensity-Duration-Frequency Curves, *U.S. Weather Bur. Tech. Paper* 25, 1955.

to adjust point rainfall to area rainfall on a watershed. These are generalized curves for the areas indicated under the figure. The U.S. Weather Bureau is now working on the refinement of these curves, which will give assistance in application.

Often it will be necessary to distribute a total precipitation according to time increments. Figure 34-9 gives a breakdown of percentage of rainfall that can be expected for a proportionate duration of the storm. Curve A on this figure generally applies where the Dorroh curve (Fig. 34-8) applies and is for the advanced type of storm in which the greatest intensity occurs in the early part of the storm. Curve B is for highest intensity in the middle of the storm (Fletcher-curve area). Curve C is for retarded or high intensity late in the storm (Sharp-curve area). The last usually produces the greatest runoff.

For a particular problem facing the engineer, there may be available in the vicinity an actual long-time record of precipitation from which he can develop a specific rainfall-frequency relationship for the problem. There is considerable controversy regarding suitable methods of fitting frequency curves to such data. An article by Chow[1] gives an excellent discussion of the subject and an extensive literature review. The Chow method of fitting frequency curves, though somewhat cumbersome to apply, is very general and should cover most cases encountered in actual practice. A much simpler but less exact method of fitting is given by Mockus.[2]

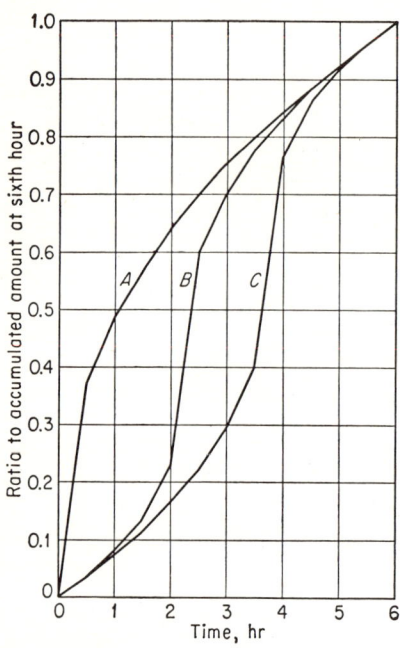

FIG. 34-9. Generalized accumulated rainfall curves for A (advanced), B (intermediate), and C (retarded) types of storms. (*USDA SCS, Eng. Div., Central Tech. Unit*, 1955.)

In cases where the *theory of extreme values* is applicable, a simple procedure is available. The data are ranked in decreasing order and plotted on extreme-probability paper as shown on Fig. 34-10. A line of best fit through the plotted points may be drawn by eye or computed.[3] The line of best fit can be extended to establish estimates of rainfall intensities for 25- and 100-year frequencies. Caution is needed in using values obtained from such extended lines based on very short records since considerable error may be introduced.

The accuracy of frequency relationships of this type depends largely on the length of record—the longer the record, the more reliable is the result.

Annual precipitation records summarized from daily observations over a long time period show a definite pattern over the United States (Fig. 57-26). The lines of equal precipitation are termed isohyets, and the map on which the lines are shown is termed an isohyetal map.

Antecedent rainfall has a major effect on runoff. Normally, rainfall from 1 to 30 days previous is considered as having an effect on the runoff. In general, the greater the antecedent rainfall and the closer the time it occurs to rain being studied, the more will be the runoff from a storm. In setting up a method of estimating runoff it is very difficult to determine antecedent storm conditions from data normally avail-

---

[1] Ven Te Chow, The Log-probability Law and Its Engineering Applications, *Proc. ASCE*, vol. 80, no. 536, September, 1954.
[2] W. L. Cowan, Victor Mockus et al., "Hydrology," *USDA*, 1957, suppl. A, sec. 4.
[3] Emil J. Gumbel, Statistical Theory of Extreme Values and Some Practical Applications, *Natl. Bur. Standards (U.S.), Appl. Math. Ser.*, no. 33, 1954. W. D. Potter, Simplification of the Gumbel Method for Computing Probability Curves, *USDA* SCS-TP-78, 1949.

able. Therefore the Soil Conservation Service,[1] in a newly developed runoff procedure, has set up three conditions:

Condition I: The optimum condition with soil dry but not to the wilting point, allowing satisfactory plowing or cultivation
Condition II: The average for annual floods
Condition III: When heavy rainfall or light rainfall and low temperatures have occurred during the 5 days previous to the given storm

The 5-day period is considered minimum for estimating antecedent conditions. However, it should be recognized that the additional work required to analyze longer

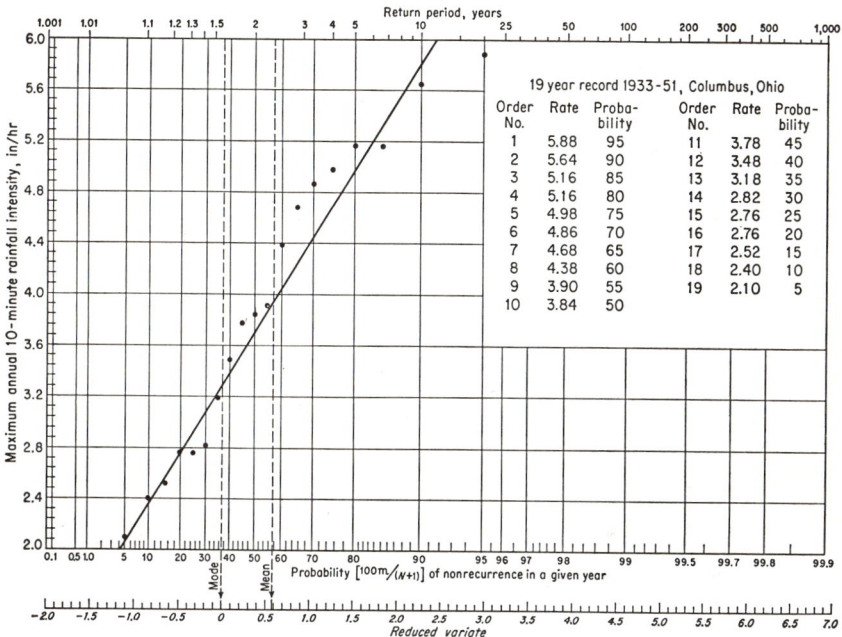

FIG. 34-10. Ten-minute rainfall intensity frequency.

periods does not always produce additional accuracy.

Table 34-3 shows the antecedent rainfall used to classify each condition.

Hartman and others[2] developed a procedure for estimating antecedent soil moisture in the Blacklands of Texas.

TABLE 34-3. RAINFALL GROUPS FOR ESTIMATING ANTECEDENT RAINFALL CONDITION
(Five-day antecedent rainfall, in.)

Condition	Dormant season	Growing season	Annual basis
I	Less than 0.5	Less than 1.4	Less than 0.5
II	0.5 to 1.1	1.4 to 2.1	0.5 to 1.5
III	More than 1.1	More than 2.1	More than 1.5

[1] Cowan, Mockus, et al., *op. cit.*
[2] M. A. Hartman, R. W. Baird, J. B. Pope, and W. G. Knisel, Antecedent Soil Moisture Index from Daily Precipitation, *USDA, ARS, SWC Research Report* 312, July, 1958.

INFILTRATION

Infiltration is the downward movement of water into the earth's surface, and the rate is generally expressed in inches per hour. Infiltration rates are needed for runoff determination and application of water to the land in irrigation. In most states in the United States irrigation guides for both sprinkler and surface irrigation have been developed by cooperative efforts of the Soil Conservation Service, Agricultural Research Service, state experiment stations, and other interested agencies. In these guides, which are generally available by writing to the Soil Conservation Service or state college experiment stations, intake rates for various soils are given and are a valuable aid in developing specific irrigation recommendations for a given site.

Intake rates for a soil may be determined by the following methods:
1. Artificial water application
   a. Two *rings* are driven into the ground and filled to a given height with water. The inner ring is usually 8 or 10 in. in diameter, and the outer ring is of a larger dimension. Measurements are made on the inner ring. The outer ring is merely to provide a larger wetted area to reduce the effect of lateral flow. Several locations must be used since a root or wormhole will materially influence the intake rate.
   b. Sprinkled plot[1]
2. Natural rainfall
   a. Hydrograph analyses of watershed runoff.[2] (A hydrograph is developed by plotting runoff cubic feet per second as ordinate against time as abscissa. The area under this curve represents total volume of runoff. The difference between the runoff and rainfall consists of surface storage, evaporation, and infiltration.)

In runoff studies the engineer is interested not in infiltration as a separate item, but in the difference between rainfall and runoff. With this thought in mind, the hydrologists of the U.S. Soil Conservation Service, using all available research, developed a method of determining runoff when the rainfall is known. The method uses two variables, antecedent rainfall (see under Precipitation) and the hydrological-soil-cover complex.

Soils were divided into four major groups based on their intake rates at the end of long-duration storms occurring after previous wetting and without the protective effects of vegetation:

Group A. Soil with lowest runoff potential. Includes deep sands with little silt and clay, also deep rapidly permeable loess.

Group B. Sandy and loess soils less deep or less aggregated than those in group A but having above-average infiltration after thorough wetting.

Group C. Compromise shallow soils and soils containing clay and colloid, though less than group D. The group has below-average infiltration after saturation.

Group D. Soils with highest runoff potential. Includes mostly clays of high swelling per cent, but the group also includes some shallow soils with nearly impermeable subhorizons near the surface.

The vegetative cover on the soil and physical barriers such as contour rows and terraces have a major effect on reducing runoff. Tables 34-4 and 34-5 and Fig. 34-11 have been developed by the U.S. Soil Conservation Service and by the U.S. Forest Service for determining a runoff number for a watershed. These tables assume the initial abstraction I_a (the amount of rainfall absorbed before runoff starts) to be $0.2S$, the maximum potential infiltration and retention of moisture by the soil at the time the storm begins. A study of the tables will indicate that the higher the number

---

[1] Robert E. Horton, Analysis of Simulated Rainfall Experiments, *USDA* SCS-TP-18, June, 1938. Don Johnstone and William P. Cross, "Elements of Applied Hydrology," The Ronald Press Company, New York, 1949. L. D. Meyer and D. L. McCune, Rainfall Simulator for Runoff Plots, *Agr. Eng.*, 39:644–648, October, 1958.

[2] A. L. Sharp and H. N. Holton, Extension of Graphic Methods of Sprinkled-plot Hydrographs to the Analysis of Control Plots and Small Homogeneous Watersheds, *Trans. Am. Geophys. Union*, pt. II, 23:578–593, 1942. H. N. Holton, Time Condensation in Hydrograph Analysis, *Trans. Am. Geophys. Union*, 26(3):407–413, 1945. Leonard Schiff and F. R. Dreibelbis, Movement of Water within the Soil and Surface Runoff with Reference to Land Use and Soil Properties, *Trans. Am. Geophys. Union*, 30:401–411, 1949.

TABLE 34-4. RUNOFF CURVE NUMBERS FOR HYDROLOGIC SOIL-COVER COMPLEXES
(For Watershed Antecedent Condition II and $I_a = 0.2S$)

Land use or cover	Treatment or practice	Hydrologic condition of soil cover	Hydrologic soil group			
			A	B	C	D
Fallow...................	Straight row	....	77	86	91	94
Row crops................	Straight row	Poor	72	81	88	91
	Straight row	Good	67	78	85	89
	Contoured	Poor	70	79	84	88
	Contoured	Good	65	75	82	86
	C and T*	Poor	66	74	80	82
	C and T*	Good	62	71	78	81
Small grains..............	Straight row	Poor	65	76	84	88
	Straight row	Good	63	75	83	87
	Contoured	Poor	63	74	82	85
	Contoured	Good	61	73	81	84
	C and T*	Poor	61	72	79	82
	C and T*	Good	59	70	78	81
Legumes† or rotation meadow......	Straight row	Poor	66	77	85	89
	Straight row	Good	58	72	81	85
	Contoured	Poor	64	75	83	85
	Contoured	Good	55	69	78	83
	C and T*	Poor	63	73	80	83
	C and T*	Good	51	67	76	80
Native pasture or range............	....	Poor	68	79	86	89
	....	Fair	49	69	79	84
	....	Good	39	61	74	80
	Contoured	Poor	47	67	81	88
	Contoured	Fair	25	59	75	83
	Contoured	Good	6	35	70	79
Meadow, permanent...............	....	Good	30	58	71	78
Woods, farm woodlots.............	....	Poor	45	66	77	83
	....	Fair	36	60	73	79
	....	Good	25	55	70	77
Farmsteads....	....	....	59	74	82	86
Roads:						
Dirt‡..........................	....	....	72	82	87	89
Hard-surface‡.................	....	....	74	84	90	92

* Contoured and terraced.
† Close-drilled or broadcast.
‡ Including right of way.
SOURCE: "Hydrology Guide," USDA SCS, 1957.

the greater will be the runoff. As was discussed under precipitation, antecedent rainfall also has a major effect on runoff. Table 34-6 gives conversion factors for converting watershed numbers for different antecedent-rainfall conditions.

Next an envelope of curves was developed using the equation

$$Q = \frac{(P - 0.2S)^2}{P + 0.8S} \qquad (34\text{-}1)$$

where Q = direct runoff, in.
P = storm rainfall, in.
S = maximum potential difference between P and Q (mostly infiltration), in., at time of storm beginning

To use the curves in Figs. 34-12 and 34-13 it is merely necessary to know the rainfall and compute the watershed-conditions number. The curves give a direct rela-

TABLE 34-5. RUNOFF CURVE NUMBERS FOR HYDROLOGIC SOIL-COVER COMPLEXES

Part 1. Commercial or national forest, for watershed antecedent condition II and $I_a = 0.2S$

Hydrologic soil-cover condition class	Hydrologic soil group			
	A	B	C	D
I (Poorest)	56	75	86	91
II (Poor)	46	68	78	84
III (Medium)	36	60	70	76
IV (Good)	26	52	62	69
V (Best)	15	44	54	61

Part 2. Forest-range areas in western United States, for watershed antecedent condition III and $I_a = 0.2S$

Cover	Condition	Soil groups			
		A	B	C	D
Herbaceous	Poor	...	90	94	97
	Fair	...	84	92	95
	Good	...	77	86	93
Sagebrush	Poor	...	81	90	
	Fair	...	66	83	
	Good	...	55	66	
Oak-aspen	Poor	...	80	86	
	Fair	...	60	73	
	Good	...	50	60	
Juniper	Poor	...	87	93	
	Fair	...	73	85	
	Good	...	60	77	

SOURCE: Forest Service, June, 1956.

tionship between rainfall and runoff. Thus, as an example: Given a watershed with soils in group B with 56.2 per cent of area in straight-row crop with a good rotation, 37.5 per cent in legumes contoured with a good rotation, and 6.3 per cent in permanent meadow. Find the runoff for a 3-in. rain.

Crop	No. from Table 34-4	Part of area	Index number (col. 2 × col. 3)
Row crop, good rotation	78	0.562	43.84
Legumes contoured, good rotation	69	0.375	25.88
Meadow	58	0.063	3.65
Total	...	.....	73.37

Thus 73.37 in the table is a weighted number.
Runoff for 3-in. rain, from curve number 73, Fig. 34-12, is 0.85 in.

TABLE 34-6. CONVERSIONS AND CONSTANTS FOR VARIOUS ANTECEDENT CONDITIONS
(For the case $I_a = 0.2S$)

Curve number for condition II	Corresponding curve numbers for:		S values*	Curve* originates where $P =$
	Condition I	Condition III		
(1)	(2)	(3)	(4)	(5)
100	100	100	0	0
95	87	99	0.526	0.10
90	78	98	1.11	0.22
85	70	97	1.76	0.35
80	63	94	2.50	0.50
75	57	91	3.33	0.67
70	51	87	4.29	0.86
65	45	83	5.38	1.08
60	40	79	6.67	1.33
55	35	75	8.18	1.64
50	31	70	10.00	2.00
45	27	65	12.2	2.44
40	23	60	15.0	3.00
35	19	55	18.6	3.72
30	15	50	23.3	4.66
25	12	45	30.0	6.00
20	9	39	40.0	8.00
15	7	33	56.7	11.34
10	4	26	90.0	18.00
5	2	17	190.0	38.00
0	0	0	Infinity	Infinity

* For curve number in column 1.

Example: Weighted II curve number using Table 34-4 is 64.
The weighted III curve number is 82 by interpolation:

$$79 + \frac{(64 - 60)(83 - 79)}{(65 - 60)} = 82.2$$

Which can be rounded off to 82.

TRANSPIRATION

Transpiration refers to the passage of soil water through the plant system and its vaporization into the atmosphere. Both transpiration and evaporation deplete soil moisture. The combined process of evaporation from the soil and transpiration by plants will be referred to as *evapo-transpiration,* or ET. It will be used in an evaluation of soil-moisture depletion.

The variation of evapo-transpiration values for various crops throughout different parts of the growing season are important to the modern irrigation engineer. Water consumption by crops—seasonal, monthly, and even daily—affects the operation of irrigation systems, the method of water application, the cropping pattern, and the farm-labor requirement.

Soil moisture affects infiltration rates and resultant runoff. Evapo-transpiration affects soil moisture. Therefore evapo-transpiration data are of considerable value in runoff problems.

The evapo-transpiration process in climatic areas like that in Ohio has been summarized and evaluated in detail by Harrold and Dreibelbis.[1] Tanner[2] used the heat budget or energy balance method to evaluate evapo-transpiration.

[1] L. L. Harrold and F. R. Dreibelbis, Agricultural Hydrology by Monolith Lysimeters, *USDA Tech. Bull.* 1179, 1959.
[2] C. B. Tanner, Energy Balance Approach to Evapo-transpiration from Crops, *Proc. Soil Sc. Soc. Am.,* 24:1–9, January, 1960.

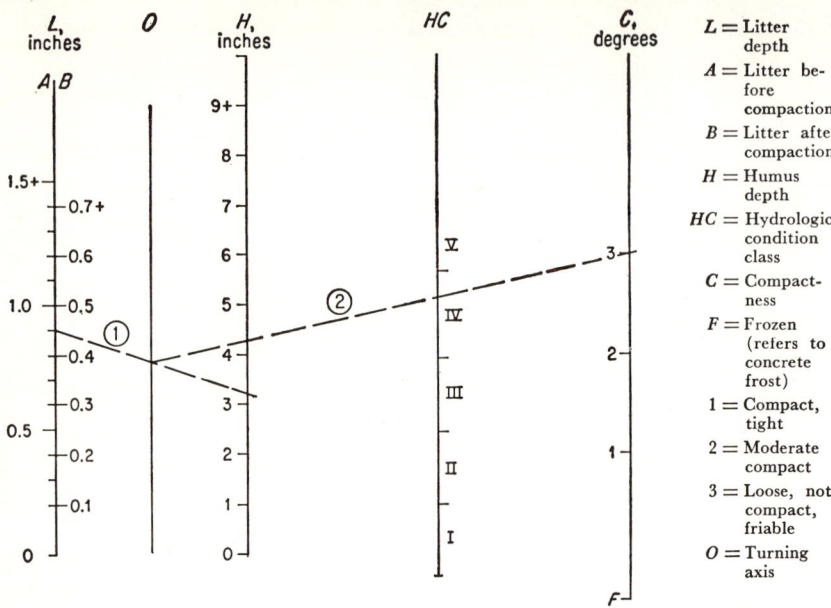

Fig. 34-11. Chart for determining hydrologic condition of forest and woodland. (*Morey, USDA Forest Service*, 1955.)

Example: $L = 0.9$ in. (before compaction); $H = 3.2$ in.; $C = 3$. Draw line ①, connecting $L = 0.9$ in. and $H = 3.2$ in.; Draw line ② connecting O intersection and $C = 3$. Read $HC = $ IV.

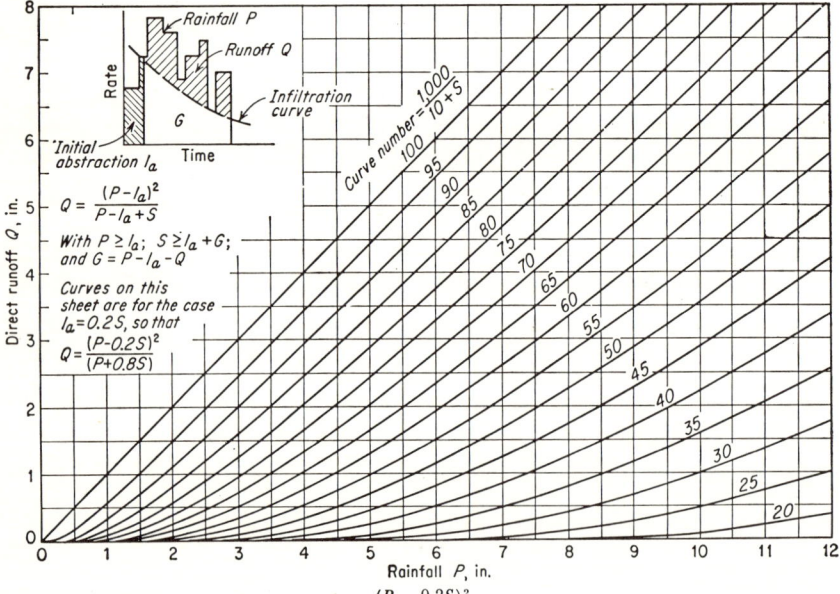

Fig. 34-12. Solution of runoff equation $Q = \dfrac{(P - 0.2S)^2}{P + 0.8S}$. $P = 0$ to 12 in., $Q = 0$ to 8 in. (*Mockus, Estimating Direct Runoff Amounts from Storm Rainfall, USDA SCS, Central Tech. Unit, October*, 1955.)

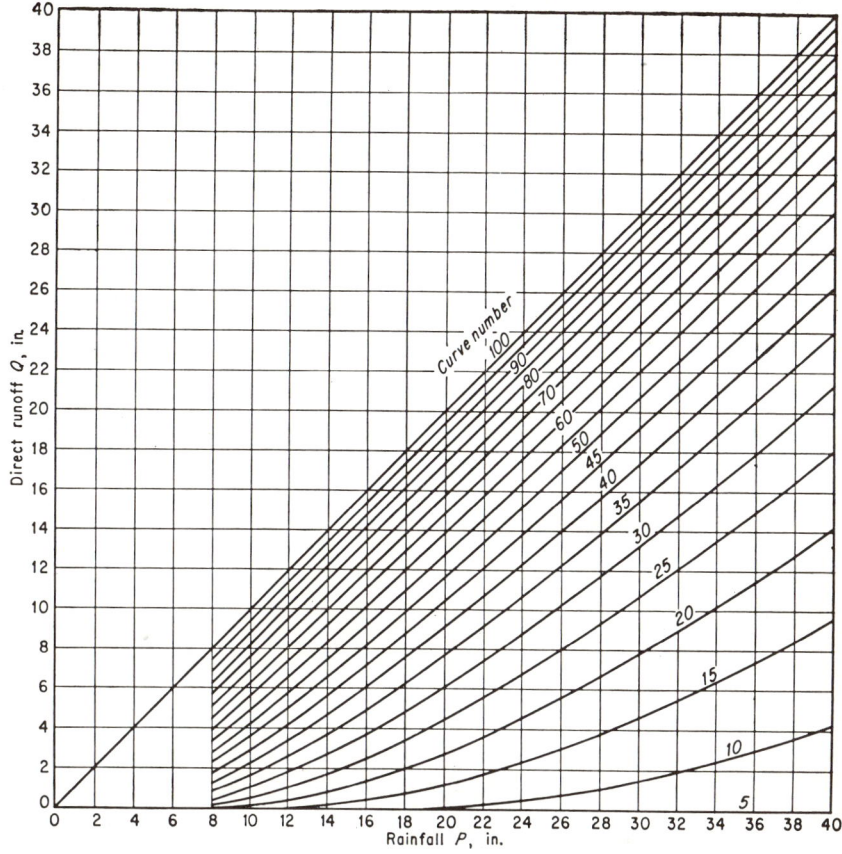

Fig. 34-13. Solution of runoff equation $Q = \dfrac{(P - 0.2S)^2}{P + 0.8S}$. $P = $ 8 to 40 in., $Q = 0$ to 40 in. (*Mockus, Estimating Direct Runoff Amounts from Storm Rainfall, USDA SCS, Central Tech. Unit, October, 1955.*)

Evapo-transpiration data for cropland in the Ohio area are summarized in Table 34-7. Average crop yields per acre in these years were corn, 94 bu; wheat, 38 bu; meadow, 2.2 and 2.5 tons.

The evapo-transpiration ratio in pounds of soil water per pound of dry vegetal matter of various crops and plants produced in such nonirrigated areas is as follows:

	Pounds of water
Corn, May–Sept	430
Wheat, Apr.–June	632
Meadow:	
Apr.–Aug	1,090
Apr.–Aug	1,170

These data are often referred to in irrigation studies as *consumptive-use values*. Blaney and Criddle compiled the results of various studies of irrigated crops (Table 34-8) and devised the method of computing evapo-transpiration values presented below.

Method of Estimating Crop Water Requirements. The measured and generally

TABLE 34-7. EVAPO-TRANSPIRATION DATA FOR UNIRRIGATED CORN, WHEAT, AND MEADOW CROPLAND, AVERAGE VALUES 1941-1951, COSHOCTON, OHIO
(Inches of Water)

Item	Corn		Wheat		Meadow*		Meadow†	
	Muskingum‡	Keene §	Muskingum	Keene	Muskingum	Keene	Muskingum	Keene
April	3.07	3.37	3.31	3.34	2.01	2.31	2.72	3.30
May	3.55	4.37	5.23	5.55	4.78	4.92	5.55	5.17
June	3.91	4.67	5.29	5.91	5.60	5.84	5.58	6.22
July	5.86	7.14	4.17	5.06	6.15	5.94	4.89	5.13
August	4.28	4.67	4.31	4.98	3.09	3.17	3.65	3.42
September	2.59	2.20	2.99	3.17	4.03	3.85	2.88	2.76
Total	23.26	26.42	25.30	28.01	25.66	26.03	25.27	26.00

* First-year meadow of alfalfa, red clover, and timothy.
† Second-year meadow of alfalfa and timothy.
‡ Muskingum silt loam soil, well drained.
§ Keene silt loam soil, slowly permeable.

TABLE 34-8. SEASONAL CONSUMPTIVE USE OF WATER FOR DIFFERENT CROPS MEASURED BY STATE AND FEDERAL AGENCIES IN THE WESTERN UNITED STATES

Crop	Range of consumptive use per season			
	Low		High	
	Depth, in.	Location	Depth, in.	Location
Alfalfa	21.6	Gooding, Idaho	52.5	Mesa, Ariz.
Beans	12.8	Davis, Calif.	18.0	Davis, Calif.
Corn	19.4	Vernal, Utah	29.3	Bonners Ferry, Idaho
Cotton	23.6	Los Banos, Calif.	31.0	Mesa, Ariz.
Flax	....	....	34.0	Mesa, Ariz.
Grain, sorghum	....	....	12.4	Mesa, Ariz.
Grains, small	12.0	Davis, Calif.	18.0	Prosser, Wash.
Orchard:				
Oranges	18.1	Azusa, Calif.	32.4	Mesa, Ariz.
Grapefruit	....	....	40.2	Mesa, Ariz.
Deciduous	19.5	Albuquerque, N. Mex.	28.4	Ontario, Calif.
Walnuts	26.3	Tustin, Calif.	27.4	Tustin, Calif.
Pasture	19.0	Redmond, Ore.	25.0	Vernal, Utah
Potatoes	15.0	Logan, Utah	23.0	Prosser, Wash.
Soybeans	....	....	22.3	Mesa, Ariz.
Sugar beets	22.8	Spanish Fork, Utah	26.3	Davis, Calif.
Tomatoes	17.0	Mercedes, Tex.	22.8	Davis, Calif.
Truck	21.4	Stockton, Calif.	24.6	Stockton, Calif.

SOURCE: Harry F. Blaney and Wayne D. Criddle, Determining Water Requirements in Irrigated Areas from Climatological and Irrigation Data, *USDA* SCS-TP-96, 1950.

available climatic factors affecting evapo-transpiration are air temperature and length of daytime hours. These factors are used to predict consumptive use as follows:

$$u = kf \qquad (34\text{-}2)$$

where u = month's consumptive use of water (evapo-transpiration), in.
k = monthly consumptive-use coefficient (Table 34-9)
$f = tp/100$ = monthly consumptive-use factor
t = mean monthly temperature, °F
p = monthly daytime hours as percentage of annual total daytime hours (Fig. 34-14) from sunshine tables[1]

[1] Sunshine Tables, *U.S. Weather Bur. Bull.* 805, 1905 (reprinted 1944).

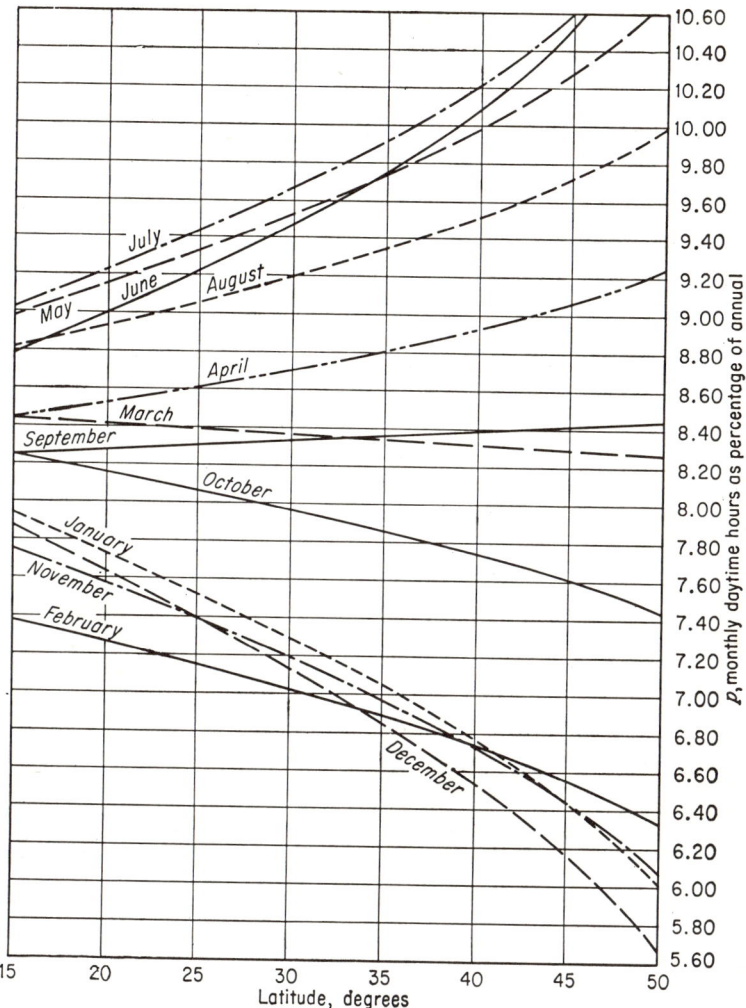

Fig. 34-14. Monthly daytime hours as percentage of annual total. (*Blaney and Criddle*, USDA SCS-TP-96, 1950.)

On a seasonal basis

$$U = KF \qquad (34\text{-}3)$$

where U = consumptive use of water by crop for the season, in.
  K = empirical consumptive-use coefficient for the season (Table 34-9)
  F = sum of monthly consumptive-use factors, f

The seasonal consumptive-use coefficients K in Table 34-9 are generally assumed to be the same for each month of the season, although there is some evidence that the k values for the individual months are different from those for the entire season.

In the design of farm irrigation systems, particularly the more expensive, the *peak consumptive-use rate* is of major importance. A nomograph (Fig. 34-15) has been developed for the solution of the consumptive-use formula $u = kf$. As an example, assume that it is desired to know what the July consumptive use of water by sugar beets might be in an area of latitude 36°, whose mean temperature during the month was 85°F.

TABLE 34-9. CONSUMPTIVE-USE COEFFICIENTS (K) IN IRRIGATED AREAS WEST OF THE 100TH MERIDIAN

Crop	Normal K for use in western U.S.	Factor	
		Average	Range
Alfalfa............	0.85	0.84	0.78–0.90
Beans.............	0.65	0.57	0.40–0.62
Corn..............	0.75	0.89	0.70–0.96
Cotton............	0.70	0.62	0.58–0.63
Small grain........	0.75	0.80	0.68–0.91
Orchard:			
Oranges.........	0.55	0.51	0.47–0.56
Walnuts.........	0.65	0.70	0.69–0.71
Peaches.........	0.65	0.61	0.51–0.75
Pasture...........	0.75	0.74	0.64–0.91
Peas..............	0.42	0.42	
Potatoes..........	0.70	0.71	0.52–0.86
Soybeans..........	0.60	0.60	
Sugar beets.......	0.70	0.69	0.64–0.73
Tomatoes.........	0.72	0.72	0.70–0.75

SOURCE: Harry F. Blaney and Wayne D. Criddle, Determining Water Requirements in Irrigated Areas from Climatological and Irrigation Data, *USDA* SCS-TP-96, 1950.

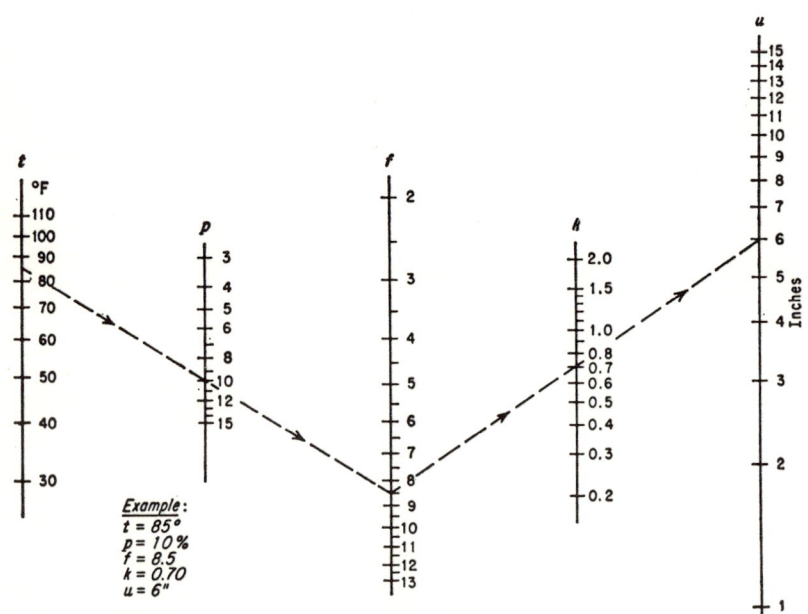

Fig. 34-15. Nomograph for solution of consumptive-use formula:

$$u = kf = k(tp)$$

where: u = monthly consumptive use (evapo-transpiration), in.
k = empirical coefficient for crop
t = mean monthly temperature, °F
p = monthly per cent of daytime hours of year

From Table 34-9 the consumptive-use coefficient for sugar beets is 0.70. From Fig. 34-15, p is 10 per cent. Entering the nomograph with the above values of t and p, we find that $f = 8.5$. With a k of 0.70, the use of water by sugar beets during July will be about 6 in. Had the crop been alfalfa with a k of 0.85, the normal July use would be about 7.2 in. This requirement may be met from precipitation, soil-moisture reserves from the previous month, ground water, and/or irrigation. In the hot dry western areas, most, if not all, of this requirement must be met by irrigation.

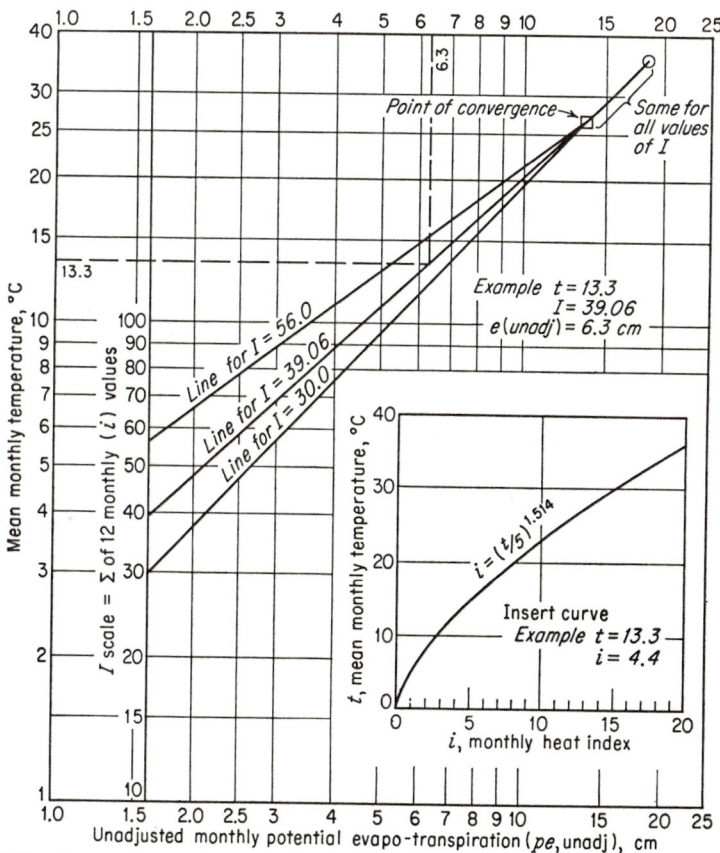

Fig. 34-16. Nomograph for determining unadjusted monthly potential evapo-transpiration from mean monthly temperature (t) and heat index (I) for the year. (*After Thornthwaite.*)

Maximum daily evapo-transpiration (ET maximum) that prevails for a period of several days is often a critical factor in irrigation schedules. ET maximum is not expected to be much greater than 150 per cent of ET average. The following is recommended:

$$\text{ET max.} = 1.4 \text{ ET av.} \tag{34-4}$$

Thornthwaite indicates that the mathematical development of evapo-transpiration data from these empirical formulas is far from satisfactory and recommends the use of a heat index i related to mean monthly temperature as shown in the insert in Fig. 34-16.[1] The sum of the 12 monthly values of i gives the annual heat index I for the

[1] C. W. Thornthwaite, An Approach toward a Rational Classification of Climate, *Geograph. Rev.*, 38:55–94, 1948.

location. (See Chap. 57.) With the value of I for a particular location, a line for determining the unadjusted potential evapo-transpiration is drawn on the nomograph in Fig. 34-16. The monthly unadjusted potential evapo-transpiration value can then be read for the monthly temperature at the particular location under study. These monthly potential evapo-transpiration values are adjusted for duration of daylight and number of days in the months to the particular location under study by multiplying potential evapo-transpiration (PE) by the factor obtained from Table 34-10. A sample set of data for Banner Elk, N.C. (latitude 36°N) is given in Table 34-11.

TABLE 34-10. MEAN POSSIBLE DURATION OF SUNLIGHT IN THE NORTHERN HEMISPHERE EXPRESSED IN UNITS OF 30 DAYS OF 12 HOURS EACH

N. lat.	Jan.	Feb.	Mar.	Apr.	May	June	July	Aug.	Sept.	Oct.	Nov.	Dec.
10	1.00	0.91	1.03	1.03	1.08	1.06	1.08	1.07	1.02	1.02	0.98	0.99
15	0.97	0.91	1.03	1.04	1.11	1.08	1.12	1.08	1.02	1.01	0.95	0.97
20	0.95	0.90	1.03	1.05	1.13	1.11	1.14	1.11	1.02	1.00	0.93	0.94
25	0.93	0.89	1.03	1.06	1.15	1.14	1.17	1.12	1.02	0.99	0.91	0.91
30	0.90	0.87	1.03	1.08	1.18	1.17	1.20	1.14	1.03	0.98	0.89	0.88
35	0.87	0.85	1.03	1.09	1.21	1.21	1.23	1.16	1.03	0.97	0.86	0.85
40	0.84	0.83	1.03	1.11	1.24	1.25	1.27	1.18	1.04	0.96	0.83	0.81
45	0.80	0.81	1.02	1.13	1.28	1.29	1.31	1.21	1.04	0.94	0.79	0.75
50	0.74	0.78	1.02	1.15	1.33	1.36	1.37	1.25	1.06	0.92	0.76	0.70

SOURCE: C. W. Thornthwaite, An Approach toward a Rational Classification of Climate, *Geograph. Rev.*, 38:55–94, 1948.

TABLE 34-11. POTENTIAL EVAPO-TRANSPIRATION VALUES (PE) FOR BANNER ELK, N.C., LATITUDE 36°N

	Jan.	Feb.	Mar.	Apr.	May	June	July	Aug.	Sept.	Oct.	Nov.	Dec.	Year
Temperature: °C	0.9	1.1	4.7	9.2	13.3	17.3	18.8	18.4	16.1	10.3	4.7	1.2	9.7
i	0.07	0.10	0.91	2.52	4.40	6.55	7.43	7.19	5.87	2.99	0.91	0.12	
Unadjusted PE, cm/month	0.2	0.4	2.0	4.2	6.3	8.4	9.2	9.0	7.2	4.7	1.9	0.4	39.06
Adjusted PE, cm/month	0.2	0.3	2.1	4.6	7.6	10.2	11.4	10.4	7.4	4.6	1.6	0.3	60.7

SOURCE: C. W. Thornthwaite, An Approach toward a Rational Classification of Climate, *Geograph. Rev.*, 38:55–94, 1948.

Considerable evidence has been collected by Thornthwaite and others to show that the actual and the calculated evapo-transpiration values are in reasonably close agreement.[1] Evapo-transpiration values computed by the Thornthwaite method are potential values, i.e., evapo-transpiration that would take place if ample moisture were available. Such may be the case in irrigated areas.

Van Bavel and Wilson and Krimgold have illustrated how such estimates apply to supplemental irrigation.[2] The method is a simple bookkeeping procedure involving daily accretion and depletion of soil-moisture content in the crop root zone. Supplemental irrigation can be scheduled to prevent the soil moisture in the root zone from depleting below any desired limit. Two major steps in this bookkeeping procedure are (1) evaluating the daily evapo-transpiration and (2) estimating that portion of the rainfall that is effective in replenishing soil-moisture supplies in the root zone.

[1] C. W. Thornthwaite, Report of the Committee on Transpiration and Evaporation, 1943–1944, *Trans. Am. Geophys. Union*, 25:686–693, 1944.
[2] C. H. M. van Bavel and T. V. Wilson, Evaporation Estimates as Criteria for Determining Time of Irrigation, *Agr. Eng.*, 33:417–418, July, 1952, and *USDA ARS* 41-11, August, 1956. D. B. Krimgold, Determining Time and Amount of Irrigation, *Agr. Eng.*, 33:705–706, November, 1952.

EVAPORATION

Evaporation as discussed in this section pertains to evaporation from free-water surfaces, i.e., water changed from liquid form into vapor form and transferred from the body of water into the atmosphere.

Measurements of Evaporation. The most common means of measuring evaporation from a water surface are (1) the exposed land pan, (2) the buried pan, and (3) the floating pan. Rohwer has assembled data comparing values from the various types of pans and reservoirs and has computed the coefficients necessary to convert pan evaporation data.[1] These coefficients appear to vary with the season and the location. In its final report, the subcommittee on evaporation of the Special Committee on Irrigation Hydraulics of the American Society of Civil Engineers in 1934 adopted 0.70 as the value of the coefficient to reduce annual evaporation values from U.S. Weather Bureau Standard Class A evaporation pan to reservoir values. Blaney and Muchel[2] described various evaporation pans and gave us coefficients for each.

The *Class A pan* is used as a standard type of pan principally because (1) more data are available from this type of pan than any other type; (2) geographical location has little effect on the coefficient; and (3) it is reasonably free from drifting dirt and debris. The unpainted galvanized steel pan is 4 ft in diameter and 10 in. deep, and water depth is maintained between 7 and 8 in. The pan is 6 in. above the ground surface, on a level spot, and exposed to the maximum possible sunshine. Water depth is observed periodically with a hook gage in a stilling well.

Numerous evaporation formulas have been developed over a period of years. Some are based on theoretical considerations of energy exchange, and others are empirical relations involving experimental evidence and weather data. Since it is difficult to evaluate actual evaporation from natural water surfaces, it is not possible to rate the reliability of the evaporation formulas. Linsley and others have compiled a number of such formulas and described their functions.[3] The *Meyer evaporation formula* given below has produced evaporation data that correlate satisfactorily with observed data.[4]

$$E = C(V_w - V_a)\left(1 + \frac{W}{10}\right) \qquad (34\text{-}5)$$

where E = evaporation depth, in.

V_w = maximum vapor pressure corresponding to mean temperature of surface of water of given lake or reservoir, in. Hg (temperature measurements made about 1 ft below water surface)

V_a = actual vapor pressure in atmosphere about 25 ft above water surface or above surface of surrounding land area, in. Hg

W = wind velocity measured about 25 ft above surface of water or above surrounding cleared land or tops of trees or buildings, mph

C = an empirical constant having the following values:

C = 15 for monthly evaporation from pans or small puddles of water

C = 11 for monthly evaporation from small lakes and reservoirs when V_a is obtained from daily maximum and minimum temperatures and mean of morning and evening relative humidity measured at a height of 25 ft

C = 10 for monthly evaporation from small lakes and reservoirs when V_a is obtained from mean, morning and evening, or more frequent determinations of actual vapor pressure at a height of 25 ft

Meyer has prepared separate United States maps showing mean annual evaporation, mean summer evaporation, and mean monthly evaporation from shallow lakes and reservoirs. Maps for the year and for summer appear as Figs. 34-17 and 34-18, re-

---

[1] Carl Rohwer, Evaporation from Different Types of Pans, *Trans. ASCE*, 60:673–703, 1934.
[2] H. F. Blaney and Dean C. Muchel, Evaporation from Water Surfaces in California, *Cal. Dept. Water Resources, Bull.* 73, 1959.
[3] Ray K. Linsley, Jr., Max A. Kohler, and Joseph L. H. Paulhus, "Applied Hydrology," McGraw-Hill Book Company, Inc., New York, 1949.
[4] Adolph F. Meyer, "Evaporation from Lakes and Reservoirs," Minnesota Resources Commission, St. Paul, Minn., June, 1942.

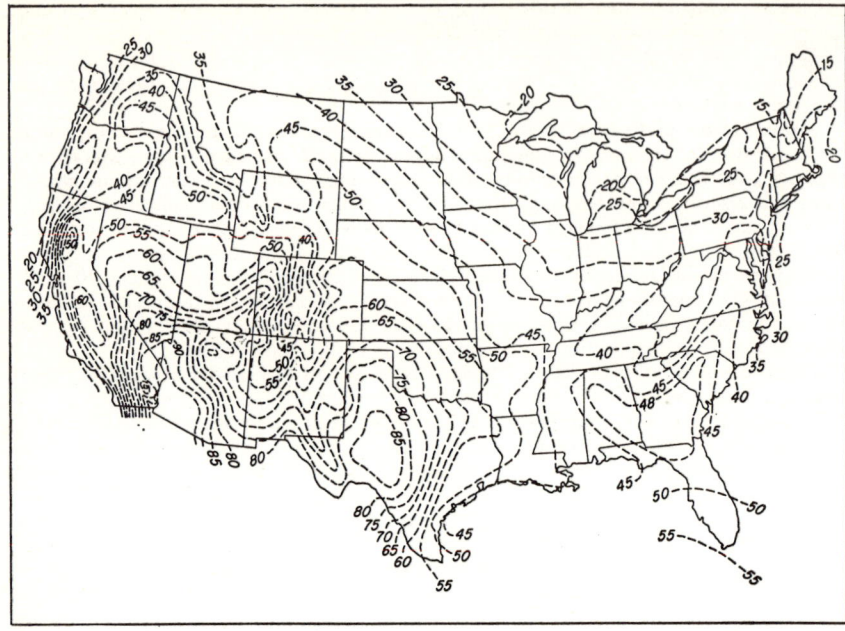

FIG. 34-17. Mean annual evaporation from shallow lakes and reservoirs, inches depth. (*After Meyer.*)

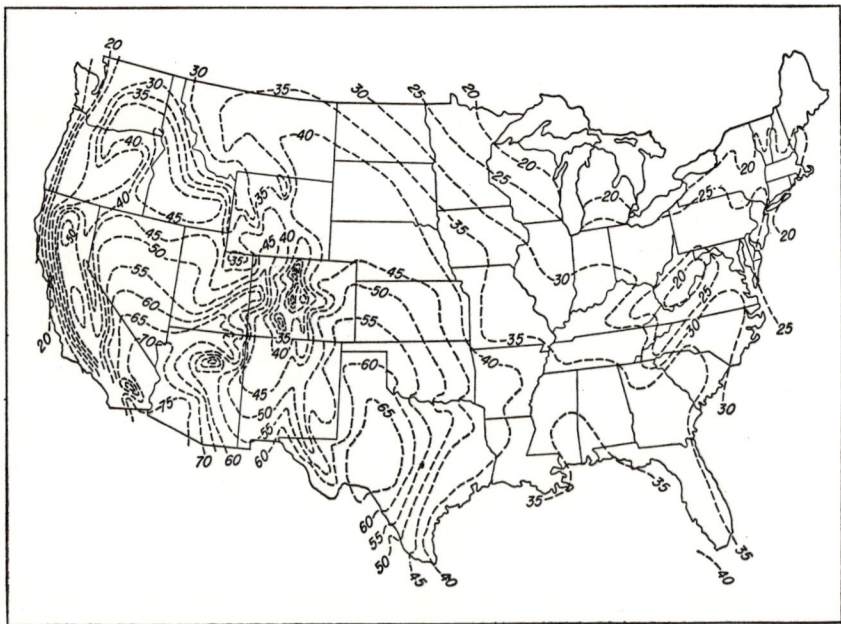

FIG. 34-18. Mean summer, April–October, evaporation from shallow lakes and reservoirs, inches depth. (*After Meyer.*)

spectively. Adequate pond depth and ample inflow are necessary to offset evaporation losses. This will be discussed in greater detail under Runoff.

GROUND-WATER MOVEMENT

Ground-water movement is generally considered to be lateral flow through pores and openings in the zone of saturation below the earth's surface under the influence of hydraulic head. As the geologic water-bearing formations vary from place to place, the ground-water movement correlations over the world are extremely complex. A brief summary of these ground-water conditions in the United States was compiled by Meinzer.[1]

It is commonly accepted that most of the water in the zone of saturation comes from precipitation that infiltrates into the soil and drains by gravity through the soil zone of aeration. Such recharge to the ground-water reservoir occurs mostly in the wet season.

Effects of Soil Characteristics. The more important physical characteristics of soils which influence the passage of water down to the water table are the pore-size distribution and the stability of soil aggregates. The pores between the soil particles

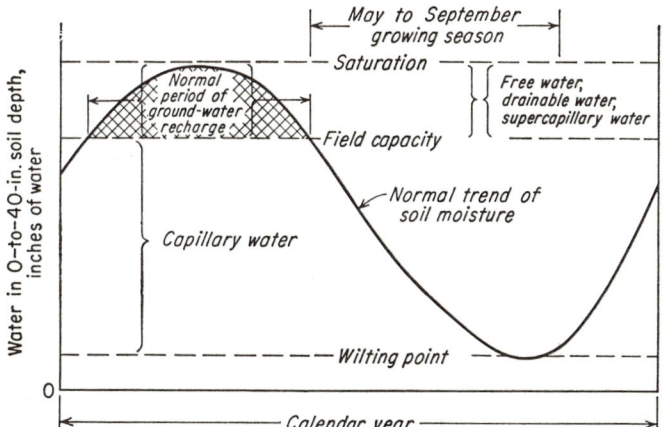

Fig. 34-19. Basic principles of soil-water movement and ground-water recharge, humid region.

range from the capillary, or small size, from which water does not drain but from which water is released for plant needs, up to the supercapillary, or large size, from which water drains by gravity.[2]

A soil having a high percentage of its pores of the supercapillary size drains soil water rapidly, thus recharging the ground-water reservoir more readily than a soil having a high percentage of capillary pores, but often retains insufficient water for crop needs. Poorly drained soil profiles are undesirable for both ground-water recharge and for crops. These physical conditions of the soil, at least near the surface, are not permanent and land-use practices have a profound effect upon them.

The physical soil properties pertaining to normal soil-water fluctuations in the humid region and resultant normal period of ground-water recharge are illustrated in Fig. 34-19 and Table 34-12.[3]

Natural Recharge. Beneath the ground surface at any given location there may be several separate zones of saturation. Whenever the downward movement of percola-

---

[1] Oscar E. Meinzer, Ground Water in the United States, *USGS Water Supply Paper* 836-D, 1939.
[2] Oscar E. Meinzer, Outline of Ground-water Hydrology, *USGS Water Supply Paper* 494, 1923.
[3] F. R. Dreibelbis, A Summary of Soil-moisture Data Useful in Soil- and Water-conservation Investigations, *Trans. Am. Geophys. Union*, 26:1041–1047, 1944.

tion is greatly reduced because of an impermeable layer, a zone of saturation is formed. It may be permanent throughout the year or only temporary. At some level beneath both this zone of saturation and its supporting impervious layer there may be one or more saturated zones. Situations like this (Fig. 34-20) are encountered at

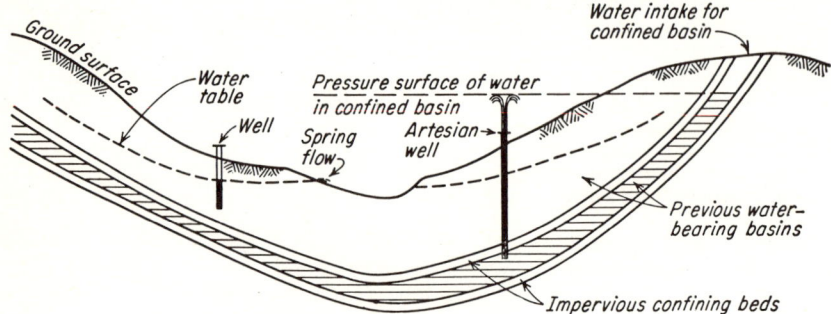

Fig. 34-20. Sketch of a section of a ground-water artesian basin.

numerous locations. If the water-bearing basin is confined by beds of low permeability, the water is under artesian pressure at points where the water-pressure surface is above the ground surface.

Table 34-12. Sample of Water Characteristics of 0- to 40-in. Depth of Soil Near Coshocton, Ohio

Soil	Land use	Inches of water in 40-inch soil depth at:				
		Saturation	Field capacity	Hygroscopic	Max. supercapillary water	Max. capillary water
Well-drained (Muskingum loam)..........	Woodland	16.5	7.7	1.4	8.8	6.3
Well-drained (Muskingum silt loam)......	Crop rotation	16.8	12.2	2.6	4.6	9.6
Slowly permeable (Keene silt loam).......	Crop rotation	17.1	15.7	4.6	1.4	11.1

note: See Fig. 34-20 for graphic illustration of this hydrologic evaluation of soil.

Other natural sources of ground-water recharge are flowing streams or bodies of stored water on land surfaces. Seepage from streams in flood or from lakes may be expected to extend back into the ground away from the body of water to various distances and raise water-table levels to a limited degree, depending on local conditions.

Artificial Recharge. In many areas where ground-water supplies are being depleted by increased pumpage, various means for artificial recharge are under investigation. The value of this practice is a local problem, depending on the cost of water, the demand for water, the permeability of the soil over which the water is spread, and the quality and quantity of water for spreading. Of the various possible methods of recharging water by spreading, the basin method appears to offer the most promise.

The basin method is the impounding of water diverted from the river channel in a series of small basins formed by dikes or banks. Percolation rates generally decrease during the life of the basin with continued use. Special treatment of the land surface of these basins improves its ability to absorb water at good rates for longer periods, as illustrated in Table 34-13. See Chap. 42 for additional information on recharging of ground water.

TABLE 34-13. TYPICAL INFILTRATION RATES IN SMALL SPREADING BASINS, BAKERSFIELD, CALIF., 1950

Treatment	Infiltration rates, feet per day, at end of:				
	1 day	5 days	20 days	30 days	80 days
None	4.0	2.3	3.6	2.7	0.4
Cotton-gin trash	4.5	7.0	11.4	8.6	3.8
Bermuda grass	4.3	5.4	10.2	11.4	8.9
Para grass	4.5	6.2	13.4	9.7	4.6

Natural Depletion. Natural depletion of ground water from the zone of saturation takes place because of vegetal plants and through effluent seepage or spring flow where the water table comes to the land surface along the sides of hills, rivers, or ditches (Fig. 34-20). Water flows laterally to the outlets under hydrostatic head.

Artificial Depletion. Pumpage removal of ground water may be for the purpose of (1) lowering water table so land operation can be improved, and (2) providing water supplies for agricultural, industrial, or human needs.

Where pumpage of ground water is planned, investigations are usually necessary to determine the quantity of water available and the rate at which it can be withdrawn without seriously affecting the supply. The Thiem method as described by Wenzel has been found useful in making estimates of permeability.[1] The pumping method, outlined by Meinzer, is useful for estimating specific yield as well as permeability.[2]

Specific yield is defined as the ratio of water gravity-drained from a saturated water-bearing formation to the volume of the aquifer; that is, if from 100 cu ft of saturated formation, there drained 20 cu ft of water, the specific yield would be 20 per cent. Specific-yield values can be determined from laboratory tests or from pumpage. The latter method involves pumping water from a well and observing the fluctuations of water-table surfaces in nearby observation wells throughout the area of the cone of water-table depression.

RUNOFF

Runoff is that portion of the precipitation upon a drainage basin which is discharged from the area in stream channels. Surface runoff results from excess rainfall or snow melt—that water which the land cannot absorb or retain on the surface. The various terms of runoff and convenient equivalents are shown in Table 34-14.

Runoff during periods of rainfall may consist only or principally of surface-water flow. During other periods, runoff may result only from base or ground-water flow. It is not possible to divide stream flow accurately into these two components, but approximate methods are summarized by Linsley, Kohler, and Paulhus.[3]

Runoff Totals. Total annual runoff over the United States (Fig. 34-21) varies widely from year to year.[4] For large streams the runoff trend in general appears to follow the precipitation and temperature patterns, according to Langbein.[5] For small basins, the effect of storm rainfall intensity, the seasonal occurrence of precipitation, vegetal cover, land treatment, and soil types are more influential in precipitation-runoff relationships than the annual precipitation totals alone.

Monthly distribution of runoff is influenced by precipitation, soil moisture, and temperature. In areas where snow accumulations and snow melt are predominant in causing runoff, the greatest total normally occurs in June, with May and July amounts

---
[1] Leland K. Wenzel, Theim Method for Determining Permeability of Water-bearing Materials, *USGS Water Supply Paper* 679A, 1936.
[2] Oscar E. Meinzer, Outline of Methods for Estimating Ground-water Supplies, *USGS Water Supply Paper* 638-C, 1931.
[3] *Op. cit.*
[4] W. G. Hoyt et al., Studies of Relation of Rainfall and Runoff in the United States, *USGS Water Supply Paper* 772, 1936.
[5] Walter B. Langbein et al., Annual Runoff in the United States, *USGS Circ.* 52, 1949.

TABLE 34-14. CONVENIENT HYDRAULIC EQUIVALENTS

1 United States gallon of water weighs 8.34 pounds avoirdupois.
1 cubic foot of water weighs 62.5 pounds avoirdupois.
1 second-foot = 7.48 United States gallons per second = 448.8 United States gallons per minute = 26,928.9 United States gallons per hour = 646,317 United States gallons per day.
1 second-foot = 60 cubic feet per minute = 3,600 cubic feet per hour = 86,400 cubic feet per day = 31,536,000 cubic feet per year = 0.000214 cubic mile per year.
1 second-foot = 0.9917 acre-inch per hour = 1.983471 acre-feet per day = 723.966942 acre-feet per year.
1 second-foot = 50 miner's inches in Idaho, Kansas, Nebraska, New Mexico, North Dakota, and South Dakota = 40 miner's inches in Arizona, California, Montana, and Oregon = 38.4 miner's inches in Colorado.
1 second-foot = 0.028317 cubic meter per second = 1.699 cubic meters per minute = 101.941 cubic meters per hour = 2,446.58 cubic meters per day.
1 cubic meter per minute = 0.5886 second-foot = 4.403 United States gallons per second = 1.1674 acre-feet per day.
1 million gallons per day = 1.55 second-feet = 3.07 acre-feet per day = 2.629 cubic meters per minute.
1 second-foot falling 8.81 feet = 1 horsepower.
1 second-foot falling 10 feet = 1.135 horsepower.
1 second-foot falling 11 feet = 1 horsepower, 80 per cent efficiency.
1 second-foot for 1 year (365 days) will cover 1 square mile 1.1312 feet or 13.5744 inches deep.
1 inch deep on 1 square mile = 2,323,200 cubic feet = 0.0737 second-foot for 1 year.
1,000,000,000 (1 United States billion) cubic feet = 11,570 second-feet for one day = 413 second-feet for one 28-day month = 399 second-feet for one 29-day month = 386 second-feet for one 30-day month = 373 second-feet for one 31-day month.
100 California miner's inches = 18.7 United States gallons per second = 4.96 acre-feet in one day.
100 Colorado miner's inches = 2.60 second-feet = 19.5 United States gallons per second = 5.17 acre-feet in one day.
100 United States gallons per minute = 0.223 second-foot = 0.442 acre-foot in one day.
1 foot deep (head of 1 foot) = 0.434 pound pressure on 1 square inch.

next in order. Methods of estimating runoff from melting snow, particularly where used for irrigation, have been briefly summarized by Linsley, Kohler, and Paulhus, who also give references for more detailed study.[1]

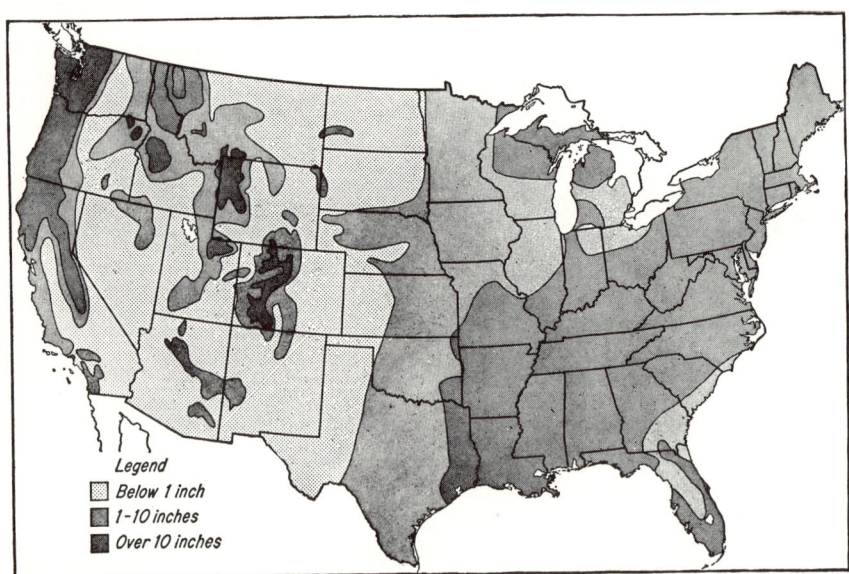

FIG. 34-21. Average annual runoff for the United States (*U.S. Dept. of the Interior, Geological Survey.*)

Information concerning the amount of runoff water for public or private water supplies can in many areas be obtained from actual measurements. Such data have, for many years, been collected and published by the U.S. Department of the Interior,

[1] Linsley, Kohler, and Paulhus, *op. cit.*

Geological Survey, in its *Water Supply Paper* series. References to other sources of runoff and other hydrologic data were compiled by the Natural Resources Planning Board.[1]

Estimates of water supply for areas where no data are available can, in most cases, be developed from values gathered for similar nearby areas.

The *yield of runoff* water from a drainage basin is an important factor in the location of impounding structures for water supplies, especially on the small watersheds. Drainage areas must be large enough to supply the reservoir, but the area must be kept to a minimum so that the expected flood rates are not excessive. The Soil Conservation Service, and later the Agricultural Research Service, in a number of areas, have compiled small-watershed runoff data and recommended watershed areas required for farm ponds of various sizes.[2]

Rainfall, evaporation, and runoff values for critical drought periods are important factors in these recommendations. Dry ponds indicate failure to give proper consideration to the probable watershed runoff, to reservoir losses, and to water usage.

Maximum Rates of Runoff. In the hydraulic design of channels, ditches, pipes, or other structures through which water flows, it is necessary to consider the maximum rate of water flow, usually in cubic feet per second. Where overtopping or flooding is not permissible under any circumstances, the maximum instantaneous peak rate of runoff is determined. In other cases, like drainage-ditch design, where temporary overtopping and flooding are permissible, a flow rate somewhat less than the peak flow rate is used, such as the flow rate which would remove a given depth of water from the basin in 24 hr (see Chap. 36 on Agricultural Drainage). The Agricultural Research Service[3] has compiled maximum flow data for a number of small watersheds.

Runoff measurements furnish the best basis for determination of design peak flow rates. They also supply the data for the development of flood-flow formulas. Jarvis and others classified various methods of estimating flood peaks somewhat as follows: (1) extreme-flood method, (2) flood-frequency methods by formulas or by statistical or probability methods, and (3) hydrographs.[4] A brief description of each of these methods is given below.

Extreme-flood methods are used mostly where failure of the structure cannot be allowed because of possible extreme flooding damage downstream. The value of river discharge for the expected extreme flood may be obtained from (1) a study of maximum-flood-discharge records of this and other countries or (2) a value derived from a study of the maximum possible precipitation (Figs. 34-6 and 34-7) on the drainage basin and calculations of the expected resultant runoff.

The method based on maximum-flood records is illustrated by a compilation made by Jarvis in 1926.[5] The maximum-possible-precipitation method has been used extensively by the Corps of Engineers of the Civil Works Department. The Ohio River Report is a typical example.[6]

Flood-frequency methods have a wide application where it is deemed uneconomical to build the structure to handle the rare maximum-possible flood. For example, if a terrace, gully-control structure, or highway embankment overtops, there may be no

---

[1] Principal Federal Sources of Hydrologic Data, *Natl. Resources Planning Bd. Tech. Paper* 10, 1943.

[2] L. L. Harrold, D. G. Krimgold, and L. A. Westby, Preliminary Report of Watershed Studies near Waco and Garland, Texas, *USDA* SCS-TP-53, 1944. D. B. Krimgold and N. E. Minshall, Hydrologic Design of Farm Ponds and Rates of Runoff for Design of Conservation Structures in the Claypan Prairies, *USDA* SCS-TP-56, 1945. W. D. Potter and Maurice Cox, Hydrologic Design of Small Farm Ponds in the Rolling Plains, Central Prairies, and West Cross Timber Areas of Kansas, Oklahoma, and Texas, *USDA* SCS-TP-57, 1946. W. D. Potter, Hydrologic Design of Small Farm Ponds in the Forested Interior West Gulf Coastal Plains Areas of Arkansas, Louisiana and Texas, *USDA* SCS-TP-59, 1946. L. L. Harrold, Hydrologic Design of Farm Ponds and Rates of Runoff for Design of Conservation Structures in the North Appalachian Region, *USDA* SCS-TP-46, 1947; Minimum Water Yield from Small Agricultural Watersheds, *Trans. Am. Geophys. Union*, 38(2):201–208, April, 1957; Monthly Precipitation and Runoff for Small Agricultural Watersheds in the U.S., *USDA, ARS, SWC,* 1957.

[3] Annual Maximum Flows for Small Watersheds in the U.S., *USDA, ARS, SWC,* June, 1958.

[4] C. S. Jarvis et al., Floods in the United States, *USGS Water Supply Paper* 771, 1936.

[5] C. S. Jarvis, Flood Characteristics, *Trans. ASCE*, 89:985–1032, 1926.

[6] C. S. Jarvis, Maximum Possible Precipitation, Ohio River Basin above Pittsburgh, *U.S. War Dept., Corps of Engineers, and U.S. Weather Bur. Hydrometeorological Rept.* 2, Waterways Experiment Station, Vicksburg, Miss., 1941.

serious loss downstream. It may be more economical, therefore, to design these and other similar engineering structures on the basis of safely handling floods up to 5-, 10-, 25-, or 50-year recurrence expectancy. In agricultural-engineering practice on small watersheds, the rational formula has had wide usage:

$$Q = CIA \qquad (34\text{-}6)$$

where Q = flood peak, cfs.
A = drainage area, acres.
I = point rainfall for watershed time of concentration and for frequency desired in design, in./hr. (Time of concentration is considered to be that duration of rainfall which results in all parts of drainage basin contributing runoff water to flood peak simultaneously; therefore it is time of water travel from most remote portion of watershed to outlet.)
C = coefficient indicating that part of rainfall rate which is expected to contribute to flood-peak flow. Values are about as follows:

	Cultivated	Pasture	Timber
Hilly land (10 to 30 per cent slope)............	0.72	0.42	0.21
Rolling land (5 to 10 per cent slope)..........	0.60	0.36	0.18

Values of C are generally increased to 0.9 for the 100-year storm.

In the hands of an engineer experienced in the field of hydrology, the rational method is a useful tool for estimating peak flows from small areas of agricultural or urban areas. Since (1) time of concentration, (2) average rainfall for the drainage area, and (3) a coefficient must often be estimated, a wide range of results is likely to be obtained by different individuals. Improvements on the use of the rational method were made by Gregory and Arnold in 1931.[1]

Variations in the selection of values based on individual judgment can be further reduced by use of refinements made by Bernard in 1939.[2]

It has been found desirable and practical in some cases for the experienced engineer to use the rational method to derive flood-peak values for various frequencies and to relate them to watershed characteristics that can be readily evaluated by the less experienced individual. Ramser's work aided in establishing the value of several of these watershed factors.[3]

Hamilton and Jepson[4] developed a method of figuring runoff dependent upon characteristics of a watershed. The Soil Conservation Service[5] refined this method and set up values for the characteristics.

Figure 34-22 illustrates the form in which the procedure was developed and how it can be used to estimate flood-peak values for various frequencies.

It should be realized that data from which these procedures were developed are very meager and that the results obtained therefrom are not as reliable as flood-peak values derived from actual records collected over a period of time. However, since records are generally not available on small watersheds, these methods and the development of hydrographs are the best available at this time for quick estimates of peak runoff for waterway or small-structure design.

Statistical analysis of observational data provides a more reliable method of estimating flood-peak values for various frequencies. The usual problem in this case is to obtain estimates of infrequent flood peaks from a record of at least a few years duration. The solution involves the development of a procedure for plotting the short-term

---

[1] R. L. Gregory and C. E. Arnold, Runoff: Rational Runoff Formulas, *Trans. ASCE*, 96:1038–1099, 1932.

[2] "Low Dams: A Manual of Design for Small Water Storage Projects," National Resources Committee, Washington, D.C., 1939; prepared by Subcommittee on Small Water Storage Projects, Perry A. Fellows, Chairman.

[3] C. E. Ramser, Runoff from Small Agricultural Areas, *J. Agr. Research*, 34:798–823, 1927.

[4] C. L. Hamilton and Hans G. Jepson, Stock Water Developments, *USDA Farmers' Bull.* 1859, 1940.

[5] Farm Planners' Engineering Handbook, *USDA SCS Farmers' Publ.* 57, 1953.

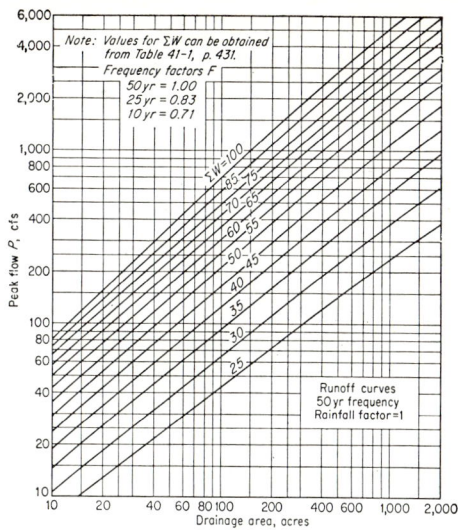

Example: Given a watershed with 4% slope, having brown silt loam upland prairie soil 95% cultivated to a rotation corn-corn-oats-hay, with a well-defined drainage system in lower one-half of watershed with no ponds or marshes, and a drainage area of 344 acres in Woodford County, Ill., what is the runoff for 25-year frequency?

Watershed characteristics	Runoff characteristics "W" from page 431
Relief 8	
Soil infiltration 10	
Vegetal cover 13	
Surface storage 12	
	$\Sigma W = \overline{43}$

$Q = PRF$
$= (430)(0.88)(0.83)$
$= 314 \text{ cfs}$

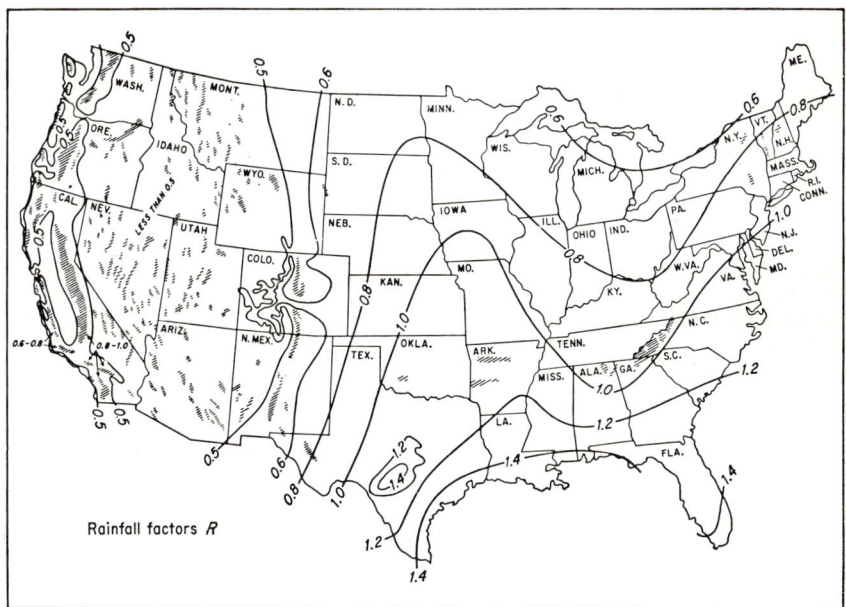

FIG. 34-22. Peak runoff determinations. The map shows distribution of rainfall factors to be used in adjusting runoff rates. The curves in the graph are based on a rainfall factor of 1. In crosshatched areas on the map, the rainfall is more intense than in the surrounding country. If the range of rainfall factors for these areas is not indicated on the map, use a factor of 0.1 to 0.2 higher than the value applicable to the surrounding area. Boundaries of the areas are only roughly indicated.

record of flood peaks against recurrence interval in years, so that the relationship thus defined can be extended by a straight line or mathematically to recurrence intervals of 25, 50, or 100 years. Potter and others found that the Gumbel method satisfied the straight-line-extension criterion especially well for runoff peaks which do not follow the statistical law of normal or log-normal distribution.[1]

Hydrographs of Runoff. A hydrograph is a graph of water flow vs. time. It is required for reservoir-storage and outflow-peak analysis, and since observed hydrographs are seldom available, it is often necessary to derive a hydrograph by approximate methods. (See Chap. 41 for construction of inflow and outflow hydrographs for a storage reservoir.)

A *distribution graph* or *unit hydrograph* as described by Bernard has been used mostly for large (over 500 miles) river-flow studies.[2] These graphs represent the

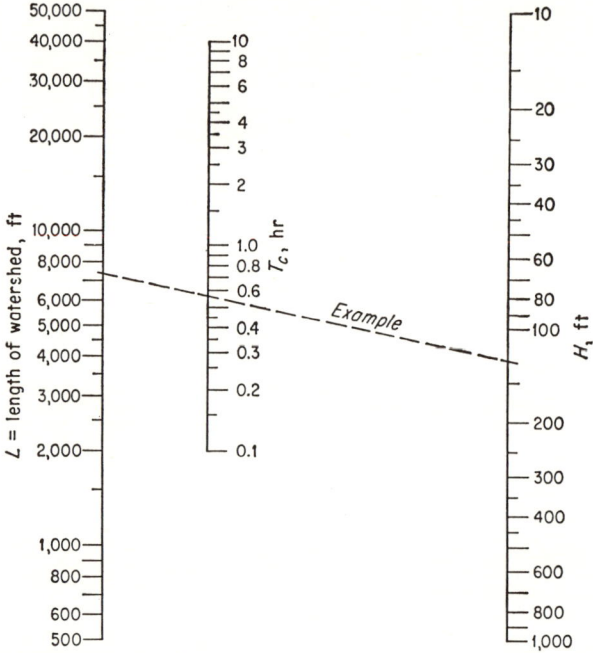

Fig. 34-23. Nomograph for solution of time of concentration. H = difference in elevation from outlet under consideration and top of watershed less the height of any overfalls. *Example:* $L = 7,250$ ft, and $H = 130$ ft. Then $T_c = 0.57$ hr.

flow-assembly characteristics of the particular drainage basin and can be derived from the observed runoff records.

A method of preparing a *synthetic hydrograph* has been developed by the Soil Conservation Service. In this method it is necessary from either nomograph (Fig. 34-23)[3] or waterway velocities to estimate the time of concentration for the watershed. In a number of observations, the relation of concentration to the time to peak has been found to be

$$T_p = \sqrt{T_c} + 0.6 T_c \tag{34-7}$$

where T_p = time to peak
T_c = time of concentration

[1] E. J. Gumbel, Simplified Plotting of Statistical Observations, *Trans. Am. Geophys. Union*, 26:69–82, 1945. W. D. Potter, Simplification of the Gumbel Method for Computing Probability Curves, *USDA* SCS-TP-78, 1949.
[2] Merrill M. Bernard, An Approach to Determine Stream Flow, *Trans. ASCE*, 100:347, 1935.
[3] P. Z. Kirpich, Time of Concentration of Small Agricultural Watersheds, *Civil Engr.*, June, 1940.

In the usual case where no hydrographs are available (1) divide watershed into channel reaches which have similar slope and cross section; (2) find the channel bankful discharge for each reach (a 2-year frequency discharge, if available, should be used where the gullies are so deep that overflow does not occur); (3) compute the average velocity for bankful or 2-year frequency discharge if available for each reach; (4) compute travel time for each reach based on length of reach and velocity of the water; (5) add travel times for various reaches to get time of concentration.

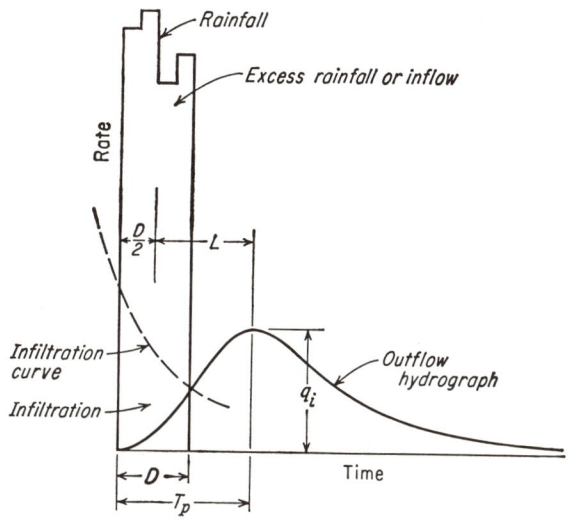

Fig. 34-24. Example of terms used in a natural hydrograph.

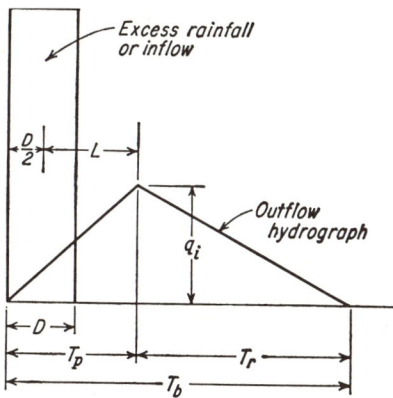

Fig. 34-25. Same hydrograph as Fig. 34-24 shown in triangular form.

Figure 34-24 shows a simple hydrograph of runoff from rainfall. Figure 34-25 shows the same hydrograph with average rates rather than actual rainfall, and the hydrograph of runoff is drawn in triangular shape. For the preparation of hydrographs for increments of rainfall this is much simpler and is almost as accurate.

In analyzing a large number of natural hydrographs, it was found that

$$T_r = 1.67 T_p \tag{34-8}$$

where T_r = time of recession on hydrograph in Fig. 34-25
T_p = time to peak

$$T_b = 2.67 T_p \tag{34-9}$$

where T_b = time of base, hr

and

$$T_p = \frac{D}{2} + 0.6 T_c \tag{34-10}$$

where D = duration of rain, hr
T_c = time of concentration, hr

It was also found that

$$L = 0.6 T_c \tag{34-11}$$

where L = time of lag from $D/2$ to T_p, hr

From this relationship, the following equation was developed:

$$q = \frac{484AQ}{D/2 + 0.6T_c} = \frac{484AQ}{T_p} \tag{34-12}$$

where q = peak discharge, cfs, for hydrograph
A = drainage area, sq miles
Q = runoff as previously determined, in.
D = time of duration, hr
T_c = time of concentration, hr

Figure 34-26 shows a nomograph for solving Eq. (34-12).

In developing the hydrograph by this method, storm duration is taken as 6 hr or the time of concentration, whichever is greater. The storm is divided into increments of time and subhydrographs figured for each increment. The total rate of runoff at any given time is the sum of subhydrograph rates in effect at that time, as shown in Fig. 34-27. The increments (d = duration of increment) suggested by the USDA SCS are 0.5 hr where $T_c < 3$; 1 hr where $3 < T_c < 6$; ⅓ T_c where $T_c > 6$.

Example:
Watershed area = 1.86 sq miles
Rainfall = 13.0 in.
T_c, time of concentration = 1.25 hr
Modified runoff for 1.86 sq miles from Fig. 34-8, Fletcher curve = 12.2 in.
From above recommendations, $d = 0.5$ hr
Watershed runoff number for condition III is 82
Thus, from Eq. (34-9),

$$T_b = 2.67 \text{ hrs}$$

from Eq. (34-10),

$$T_p = 1.0 \text{ hr (using duration } d \text{ of 0.5 hr)}$$

and from Eq. (34-12),

$$q \text{ for 1-in. runoff} = \frac{(484)(1.86)(1)}{1} = 898 \text{ cfs}$$

Computation of subhydrograph peaks, basing Accumulative P on Fig. 34-9, curve B, and Accumulative Q on Figs 34-13 and 34-14:

Time, hr	Acc. P, in.	Acc. Q, in.	ΔQ, in.	$q = (\Delta Q)898$, cfs
0	0	0		
0.5	0.43	0	0	0
1.0	0.95	0.09	0.09	81
1.5	1.65	0.40	0.31	278
2.0	2.80	1.21	0.81	727
2.5	7.32	5.18	3.97	3,560
3.0	8.60	6.44	1.26	1,130
3.5	9.50	7.28	0.84	755
4.0	10.20	7.96	0.68	611
4.5	10.80	8.55	0.59	530
5.0	11.30	9.05	0.50	449
5.5	11.80	9.53	0.48	431
6.0	12.20	9.93	0.40	359

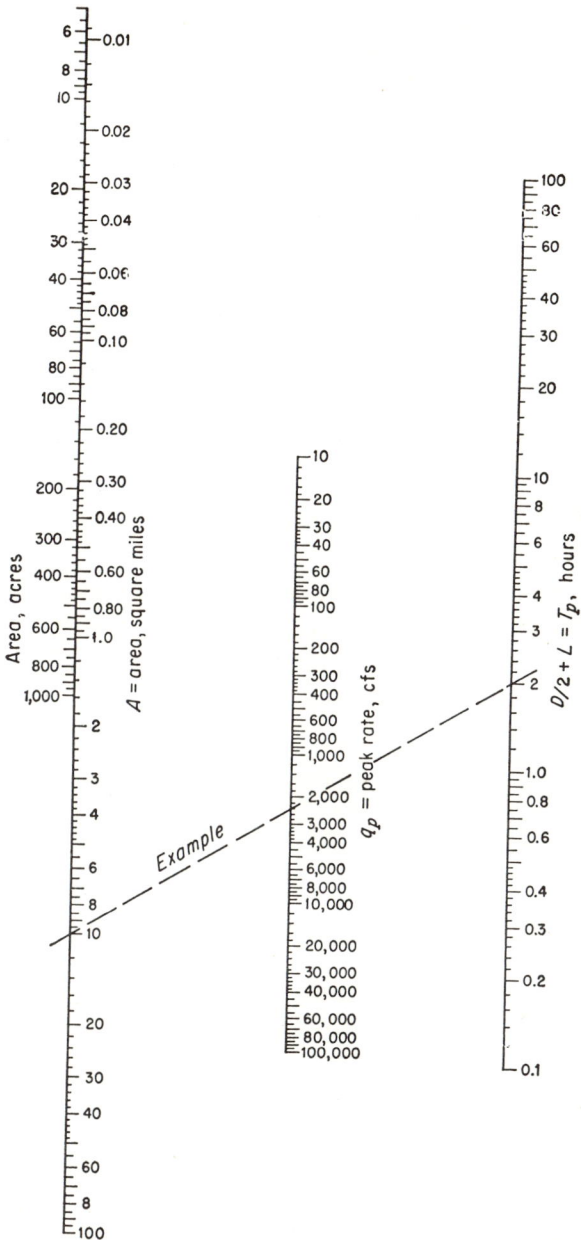

Fig. 34-26. Peak runoff rates when $Q = 1$ in. Example: $T_p = 2.0$ hr, $A = 10.0$ sq miles. Then $q_p = 2,400$ cfs. (Mockus, USDA SCS, Eng. Div., Central Tech. Unit, 1957.)

Figure 34-27 shows the plotting of the example given. Note that each triangular subhydrograph is plotted with the base starting at the corresponding time interval. Thus the hydrograph for peak of 81 cfs is plotted with base starting at 0.5 hr. The base of the hydrograph as given by formula is 2.67 hr. The composite hydrograph for the watershed is prepared by adding the subhydrographs as indicated in Fig. 34-27.

If a watershed is divided into two major separate branches, individual hydrographs can be developed by the method outlined above for the branches and a composite

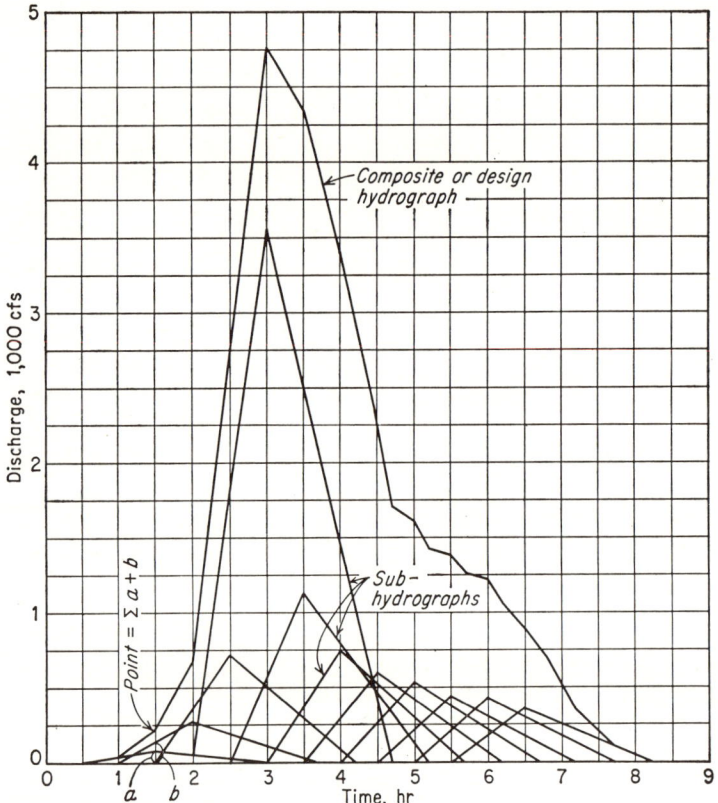

FIG. 34-27. Composite hydrograph development.

hydrograph developed by adding together the two hydrographs, displaced to allow for any pertinent time lag.

The Soil Conservation Service has developed[1] for operational use a set of semi-dimensionless hydrographs. These hydrographs are divided into families based on storm distribution pattern, 6-hour rainfall, and runoff curve number. After the hydrograph family is determined, the selection of the correct unit hydrograph is based on the ratio of the time of excess rainfall to the time to peak. If a large number of hydrographs are to be developed, it would be well to secure this or similar information, since this is a great timesaver. However, for developing a limited number of hydrographs, the method outlined in the handbook will give reliable results, as it was used to develop the unit hydrographs which were set up in families.

[1] Cowan, Mockus et al., "Hydrology," chap. 3.21.

Chapter 35

LAND LEVELING AND GRADING

John T. Phelan

The term *land leveling* is commonly used in the field of irrigation, and *land grading*, in drainage, to refer to a grading or smoothing operation which modifies the land surface to meet the slope criteria for the irrigation or drainage practice to be applied. Usually the land is not graded to a truly level surface since, to meet drainage requirements in both irrigated and nonirrigated fields, some minimum grade is usually desirable.

In some areas the terms floating, smoothing, or planing are also common, but generally their use is restricted to a fine grading operation, where the soil is moved only a short distance to fill minor depressions and eliminate small ridges. Equipment to do this type of work is commercially available and is usually automatic in operation. In irrigated areas it is common for the farmer to float, or plane, his land as a cultural practice.

Soils that are shallow have limitations in the depth of cut that is allowable. A knowledge of subsoil conditions is essential to prevent damage to the field. Exposures or serious reductions in depth to rock, hardpans, sands or gravels, or caliche must be avoided if the agricultural productivity of the area is to be maintained. Undulating topography when coupled with a shallow soil makes land leveling especially difficult.

The *rainfall characteristics* of an area are very important in determining the maximum and minimum grades allowable for a field. Minimum grades must meet drainage requirements which are in part determined by the quantity, intensity, and seasonal occurrence of rainfall. Likewise the minimum permissible grade for a given soil and crop is often determined by the maximum grade which is nonerosive under the prevailing climatic conditions.

Each *method of irrigation or drainage* has its own limitations as to the permissible "cross" slope and "downfield" slope. Thus the leveling or grading criteria must specify slopes within these limits. When several crops are to be grown on a field and several methods of irrigation are to be used, the most restrictive limitations must be the basis for design. In general, surface flooding is the most restrictive as far as cross slope is concerned and furrow methods are the most restrictive for length-of-run and downfield slope.

A consideration of the *crops to be grown* is important in selecting the irrigation or drainage method and the resulting land grading or leveling criteria. It is also important in that a high-value crop or a crop with a high labor requirement may justify a higher degree of leveling to reduce labor or production costs. It is readily apparent that the same standard of land leveling suitable for production of truck crops might not be justified for the production of a hay crop.

The *location and elevation of the water source* will influence the choice of slopes for surface irrigation.

The *location of the drainage outlet* for the disposal of irrigation waste water as well as natural runoff, including that to be expected from adjoining land, is a factor that must be provided for in the design.

Field subdivision based on natural topographic boundaries should be considered for the entire farm, even though the leveling of only one field may be under consideration. In those areas where slopes are excessive and must be reduced in order to meet drainage or irrigation criteria, it is often possible to subdivide the farm into relatively narrow strips on the approximate contour to reduce the slope to acceptable limits. This results in a series of benches down the slope, each bench separated from the adjoining ones by a narrow escarpment area. Each strip is considered a separate field for purposes of land-leveling design. The location of the irrigation supply system and the water-disposal outlets is especially critical in this instance.

Plans should be made for the use of the *waste soil* from drains and *borrow* for construction of fills in the water-supply system during the leveling operation. Best results can be obtained if these works are constructed at the same time as the adjacent areas are leveled. Farms that are leveled field by field without consideration of an integrated pattern can readily lead to eventual difficulties involving expensive reworking, makeshift improvisations, or an inefficient installation.

It is quite common for a landowner to consider the *optimum leveling* job to be too expensive or beyond his capabilities at the moment, and he is willing to accept higher crop-production costs and lower irrigation or drainage efficiencies in order to reduce his investment in leveling. Changing economic conditions, labor or water shortages, or other factors may later cause him to reconsider and to perform additional work to meet a higher standard. In some areas fields have been leveled and releveled several times over a period of years, each time more nearly approaching the optimum condition. Since it is obviously more efficient to perform the construction work in a single operation, it is desirable to select criteria which are as high as economic conditions and the landowner will permit.

Land grading for *surface drainage* has a distinct advantage over bedding or other surface-drainage methods covered in Chap. 36 in that a positive grade is provided for each row and an adequate outlet ditch is provided at a given length of row. The limitation to its more general use in drainage is the cost of land grading. Costs of grading to a true plane or profile usually run from $50 to $100 per acre. This limits the use of land grading for surface drainage to lands on which fairly high value crops are to be grown or to areas where supplemental irrigation is needed along with drainage.

Additional details for length of run for drainage or irrigation are available in many states in surface irrigation or drainage guides of the Soil Conservation Service or state experiment stations.

SURVEYS AND STAKING

Prior to making the survey, it is advantageous to remove heavy vegetative growth from the land and, if the surface is very rough, to perform a smoothing operation to remove furrows, ridges, or other irregularities. It is desirable that the surface soil be in a fairly compact condition so that the construction equipment can make the required depth of cut accurately.

On new lands, clearing should be done before making the survey. Vegetative matter should be removed or burned. It is usually found desirable to level native sod areas before plowing.

With the field boundaries considered and established, the next step is to *locate reference points* in the field which can be used as a basis for design and construction. General practice is to establish a coordinate, or grid, system over the field and *set stakes* at 100-ft intervals as shown in Fig. 35-1. This grid system need not be precisely located. It is usually satisfactory to establish accurately two or more base lines in each direction and then to sight in the rest of the stakes. Thus in Fig. 35-1, line B might first be established by line and tape, a right angle turned, and lines 4 and 5 likewise located. Line C should also be established by measurement. Other stakes can then be located by sighting. In this manner stake D-3 can be located at the intersection of the line of sight over D-4 and D-5 and B-3 and C-3.

Where surface irregularities are such that a 100-ft spacing does not accurately reflect field conditions, additional stakes should be set on the "highs" or "lows" or a closer grid spacing used. Other key points such as the point of entry of the irrigation water or the drainage outlets should also be located on the coordinate system.

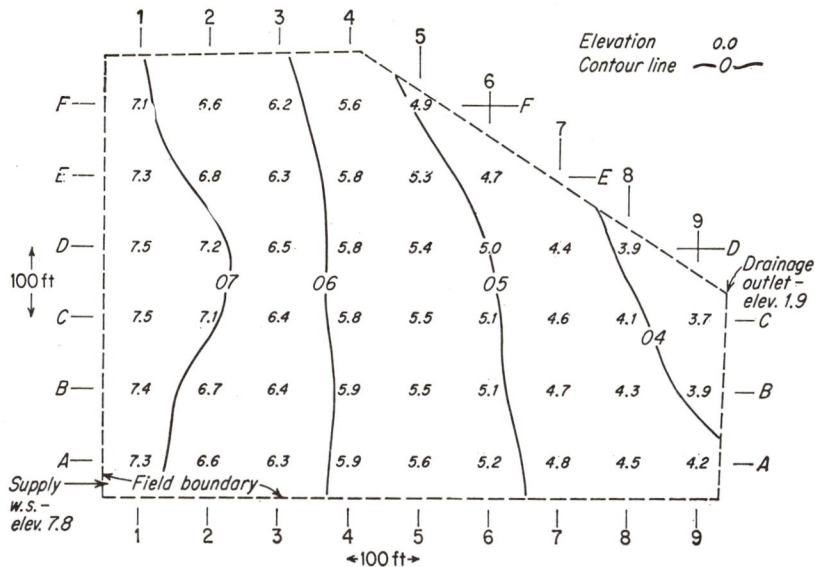

Fig. 35-1. Survey for land leveling.

A *level survey* is then made to determine the ground elevation to the nearest 0.1 ft at each stake and the water-surface elevation of the water supply and drainage outlet.

METHODS OF LEVELING

There are four basic design methods in common use for grading or leveling. Each has its advantages and disadvantages and is best adapted to specific site conditions:
1. The contour adjustment method
2. The plan inspection method
3. The profile method
4. The plane method

Modifications of each of these methods are common.

The Contour Adjustment Method. This method is especially adapted to the smoothing of steep lands that are to be irrigated. Some workers also find it useful on moderate slopes. The survey data are plotted in plan, and contours drawn. The contour interval in feet should not exceed the slope of the land in per cent. By examination of the contour lines, it can be determined where it is feasible to remove ridges and fill swales. *Proposed contours* to meet land-leveling criteria are then drawn on the map as shown in Fig. 35-2. The uniformity of the downfield slope can be controlled by the uniformity of the horizontal spacing between contours, and the cross slopes can be examined by scaling the distance between contours at right angles to the direction of irrigation. A balance between cut and fill can be approximated by maintaining the proposed contour in an average position with relation to the original contour at the same elevation. The design elevation of grid points can be determined by interpolation between the design contours.

A still closer approximation can be made by *summing the design cuts and fills* from the staked points and comparing the totals. This method ordinarily gives a satisfactory approximation on steep slopes. On moderate slopes, the earthwork quantities should be computed and the proper balance between cut and fill attained by adjustment of the design elevations. On steep slopes, particular attention must be given to the maximum permissible depth of cut that the soil will permit, and on moderate slopes, the length of haul between the cut-and-fill area is often the point which requires closest attention.

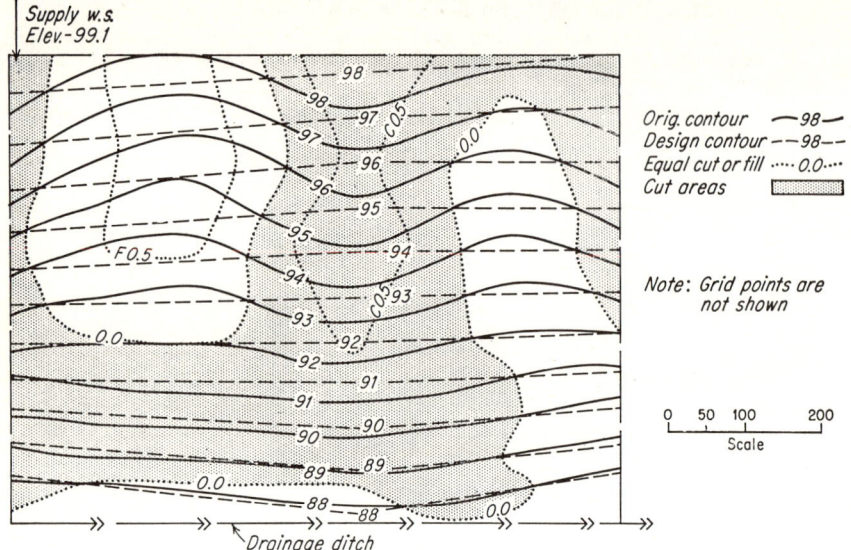

Fig. 35-2. The contour adjustment method of land leveling.

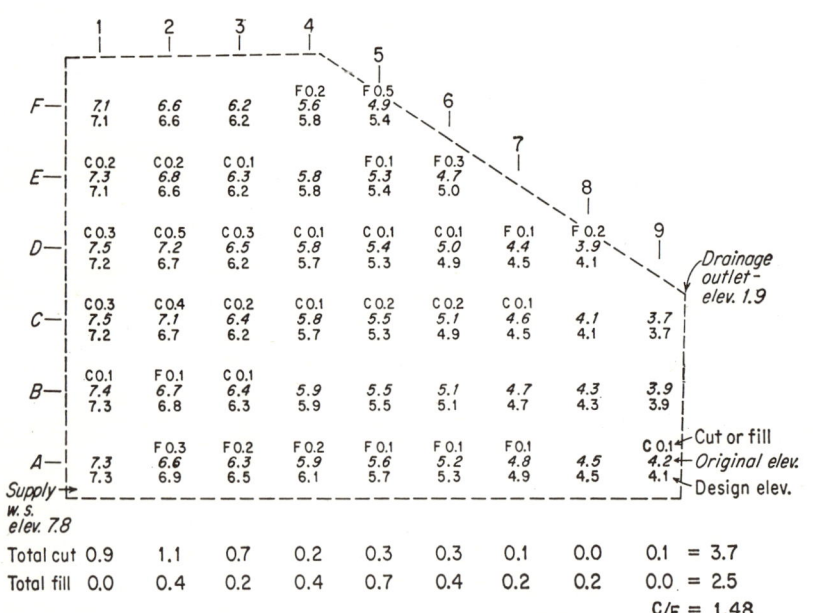

Fig. 35-3. The plan inspection method of land leveling.

Plan Inspection Method. This method is especially adapted for moderate to flat slopes when the irrigation criteria provide for some variation in cross slope and downfield grade. It is best adapted to the use of experienced personnel but is rapid and satisfactory

In this method, survey data are plotted in plan and *determination of the desired grid-point elevations made by inspection* as shown in Fig. 35-3. Usually the grid map is carefully studied and then trial elevations assumed for one or two lines of stakes. It is necessary to keep in mind cross slope and downfield grade limitations simultaneously so that neither will be exceeded. Earthwork balance is approximated by comparing the sum of the cuts with the sum of the fills. Cuts or fills at the border of the area are discounted in totaling in proportion to the area they represent. The grades are frequently adjusted, and the adjacent lines of stakes are worked out until the field is covered. Earthwork quantities are then computed and minor revisions made to provide the proper earthwork balance.

Thus by trial and error a set of elevations are found that satisfy leveling criteria. It should be noted that this method does not necessarily provide minimum quantities of excavation or the shortest length of haul. The quality of the design depends on the success of the designer in weighing all factors as he selects the grades. Because of its rapidity, it is the method most often used in the areas to which it is adapted.

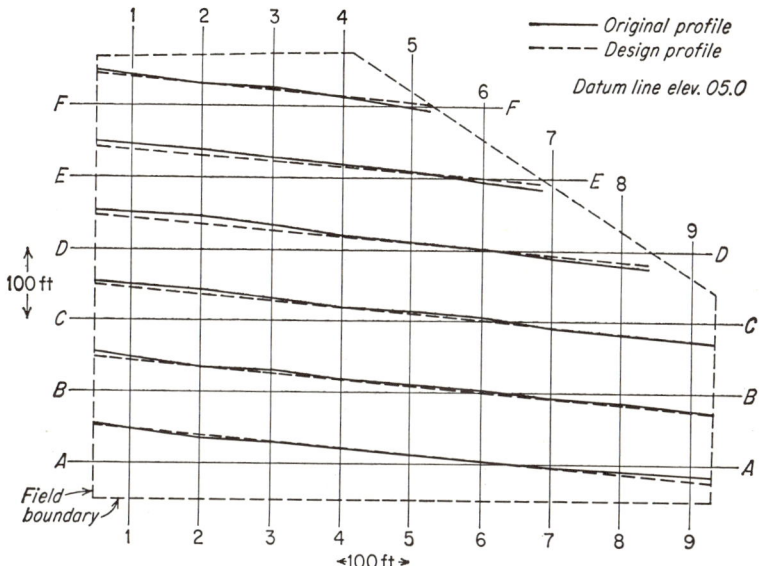

Fig. 35-4. Profile method of land leveling.

Profile Method. Several variations of the profile method are in common use. They are usually best adapted to very flat or undulating lands that have many variations in slope between contour lines. Basically the method consists of plotting profiles of the grid lines and then *laying the desired grade on the profiles*. Some designers prefer to plot the profiles at approximate right angles to the direction of irrigation or drainage; others plot them downfield.

On flat lands it is sometimes possible to plot profiles immediately adjacent to each other so that adjacent profiles are approximately the same distance apart as the horizontal scale on the individual profile would indicate. For example, the profile of the C line on a horizontal scale of 1 in. equals 100 ft could be plotted with its datum line 1 in. above the datum for the B line. Figure 35-4 is an example of this method of plotting. It has the advantage that design grades of one profile as compared with another are easily determined.

By inspection of the profiles a grade can be established which will provide an approximate balance of cut to fill and will also keep the haul distance within reasonable limits. If desired, profiles at right angles to the original plot can also be drawn. The elevations of the proposed grade can be simultaneously adjusted on both sets of profiles until cross-slope and downfield grade criteria are met.

The cuts or fills are then taken from the profiles and the earthwork quantities computed. Minor adjustments are usually necessary to attain the desired balance.

Plane Method. The use of this method is restricted to those fields where it is feasible to grade the field to a true plane. In this method the average elevation of the field is determined and this elevation applied to the centroid of the area. The elevations of the other points of the plane are then *computed from the centroid* by applying the desired downfield grade and cross slope.

The location of the centroid must first be found. On rectangular areas it is simply the center of the area, but on irregular fields, the area must be divided into rectangles and right triangles and the centroid located by computing moments around one corner of the area. In ordinary practice it is satisfactory to assume that each grid stake represents the same area. This was done in the calculation in Fig. 35-5.

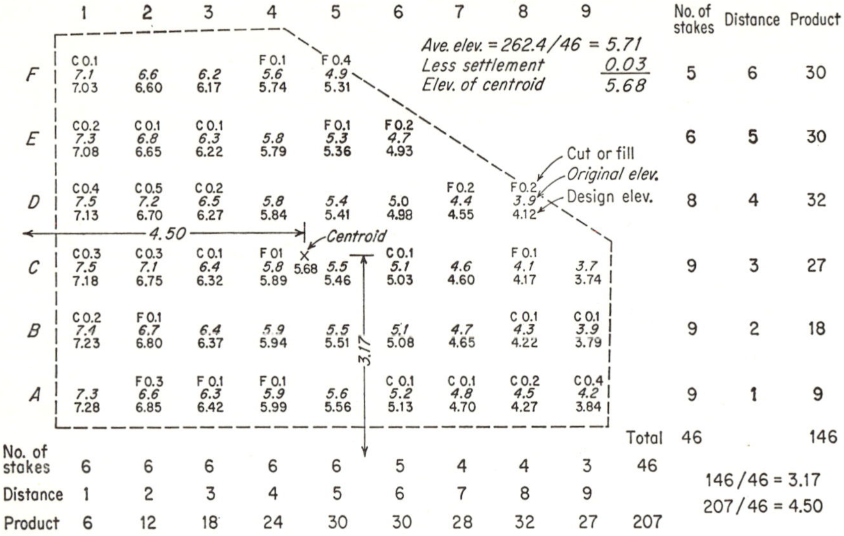

	1	2	3	4	5	6	7	8	9	No. of stakes	Distance	Product
F	C 0.1 7.1 7.03	6.6 6.60	6.2 6.17	F 0.1 5.6 5.74	F 0.4 4.9 5.31	Ave. elev. = 262.4/46 = 5.71 Less settlement 0.03 Elev. of centroid 5.68				5	6	30
E	C 0.2 7.3 7.08	C 0.1 6.8 6.65	C 0.1 6.3 6.22	5.8 5.79	F 0.1 5.3 5.36	F 0.2 4.7 4.93				6	5	30
D	C 0.4 7.5 7.13	C 0.5 7.2 6.70	C 0.2 6.5 6.27	5.8 5.84	5.4 5.41	5.0 4.98	F 0.2 4.4 4.55	F 0.2 3.9 4.12		8	4	32
C	C 0.3 7.5 7.18	C 0.3 7.1 6.75	C 0.1 6.4 6.32	F 01 5.8 5.89	5.68 5.5 5.46	C 0.1 5.1 5.03	4.6 4.60	F 0.1 4.1 4.17	3.7 3.74	9	3	27
B	C 0.2 7.1 7.23	F 0.1 6.7 6.80	6.4 6.37	5.9 5.94	5.5 5.51	5.1 5.08	4.7 4.65	C 0.1 4.3 4.22	C 0.1 3.9 3.79	9	2	18
A	7.3 7.28	F 0.3 6.6 6.85	F 0.1 6.3 6.42	F 0.1 5.9 5.99	5.6 5.56	C 0.1 5.2 5.13	C 0.1 4.8 4.70	C 0.2 4.5 4.27	C 0.4 4.2 3.84	9	1	9
									Total	46		146
No. of stakes	6	6	6	6	6	5	4	4	3	46		
Distance	1	2	3	4	5	6	7	8	9		146/46 = 3.17	
Product	6	12	18	24	30	30	28	32	27	207	207/46 = 4.50	

(Annotations on figure: "Cut or fill / Original elev. / Design elev."; "←—— 4.50 ——→ Centroid"; vertical "3.17")

FIG. 35-5. The plane method of land leveling.

A weighted average of the grid-point elevations can be made to determine the elevation of any plane through the centroid.

If the field is rectangular in area, the slope of any line on the plane which will best fit the natural ground surface can be determined by the least-squares method. This can be represented as follows:

$$S = \frac{\Sigma(DH) - (\Sigma D)(\Sigma H)/n}{\Sigma(D)^2 - (\Sigma D)^2/n}$$

where S = slope of line on plane
D = distance from reference axis to point
H = elevation of point
n = number of points

The slope of the plane can be determined on both the x and y axes, and the elevation of any point calculated from the elevation at the centroid.

Many fields are irregular in shape, and this method is not practical in such instances. However, it is often satisfactory to enclose most of the field within an arbitrary rectangular boundary and extend the slopes determined for this area to cover the entire field.

The grades so determined must be checked to see if they fall within the limitations selected for cross slope and downfield grade. Should they exceed these values, it will be necessary to use the nearest values that satisfy the criteria.

Since passing a plane through the centroid at an elevation which is the average for the field will theoretically provide exactly the same volumes of excavation and fill, it will be necessary to *lower the entire plane* somewhat to provide the proper balance factor. In instances where borrow is to be taken from the field for other construction or where waste from drains or ditches is to be utilized in making land-leveling fills, a like adjustment must be computed. This can be done by dividing the cubic feet of waste or borrow by the area of the field in square feet. The adjustment of the elevation of the plane can be made at the centroid by subtracting the value for borrow or shrinkage and adding the value for waste.

The elevation of other points is then computed, and the cut or fill at each point determined. While the elevations are computed to hundredths, cuts and fills are ordinarily marked to the nearest tenth since construction tolerances do not warrant a higher degree of precision.

EARTHWORK QUANTITIES

Any of the conventional methods of computing earthwork volumes can be used if a proper prismoidal correction is made. The prismoidal formula is as accurate as the survey data used, but it is time-consuming and therefore not commonly used. Approximate methods are usually satisfactory when construction is performed on an hourly rate basis or when the method of determining the earthwork volumes is explained in the contract.

The *end area* method contains a prismoidal error but is commonly used. The areas of cuts and fills on the profiles or lines are calculated, and the volume between the adjacent profiles on line computed by

$$V = \frac{L}{27} \frac{A_1 + A_2}{2}$$

where V = volume of cut or fill, cu yd
L = distance between profiles or lines, ft
A_1 = area of cut or fill in first profile or line, sq ft
A_2 = area of cut or fill in second profile or line, sq ft

A modification of this method may be expressed

$$V = \frac{A}{27} \frac{C_1 + C_2 + \cdots + C_n}{n}$$

where V = volume of cut (or fill) in area A, cu yd
A = surface area, sq ft
$C_1 + C_2 + \cdots + C_n$ = sum of cuts (or fills) at corners of area, ft
n = number of corners of area

With this formula, n may be 4 for a square or 3 for a triangular area. Grid squares that have cuts (or fills) at one, two, or three corners only should be further subdivided into smaller squares or triangles, and the value of each computed.

Another formula which has been useful in determining approximate land-leveling quantities is

$$V_c = \frac{1}{108} \frac{L^2 H_c^2}{H_c + H_f}$$

$$V_f = \frac{1}{108} \frac{L^2 H_c^2}{H_c + H_f}$$

where V_c = volume of cut, cu yd
V_f = volume of fill, cu yd
L = grid spacing, ft
H_c = sum of corners of one grid in cut, ft
H_f = sum of corners of one grid in fill, ft

As with other types of construction involving earthwork, it will rarely be found in practice that the volume of cut equals the volume of fill. Experience has shown that an *additional volume* of cut must be allowed to provide for the required volume of fill. The reasons for this are usually attributed to the compaction of the field surface in both the cut and fill areas by the heavy construction equipment so that the yield from the cut areas is reduced and the requirement for the fill areas is increased. It has also been claimed[1] that a tendency to crown the ground surface between stakes exists.

Some engineers make the required adjustment by increasing the volume of cut over fill by an *estimated percentage* as determined by experience in the area. This percentage increase is usually higher for work where the cuts and fills are small and lower for large cuts and fills. It is also appraised by soil texture and the density of the surface soil at the time of leveling. The ratio of cut volume to fill volume is usually about 1.3:1.6, although values as low as 1.1 and as high as 2.0 have been found. It is common practice to make adjustments in the grade design after construction is under way to provide the proper balance factor.

Other engineers prefer to estimate the volume of fill over cut by assuming a *uniform settlement* due to compaction over the entire field surface. If they consider the settlement to be 0.05 ft on a field 50 acres in size, they would provide an equivalent volume of $(1/27)(0.05 \times 50 \times 43{,}560)$, or 403 cu yd for this settlement. The settlement is estimated according to the type of soil material and its density. Usual figures are as low as 0.01 ft for dry compact sands to as high as 0.10 ft for loose clays.

CONSTRUCTION AND INSPECTION

After the design is complete, the desired cuts or fills are marked on the grid stakes. The most common method of moving the earth is by heavy equipment operated by contractors who specialize in such work.

Usually carrier-type scrapers powered by crawler or rubber-tired power units are used for the heavy earthwork, and large floats or planes are used to attain the desired finish. A wheel scraper with a PTO-driven loading chain, such as shown in Fig. 35-6, which permits loading the scraper without a pusher tractor, is generally pulled with a high-speed rubber-tired farm tractor.

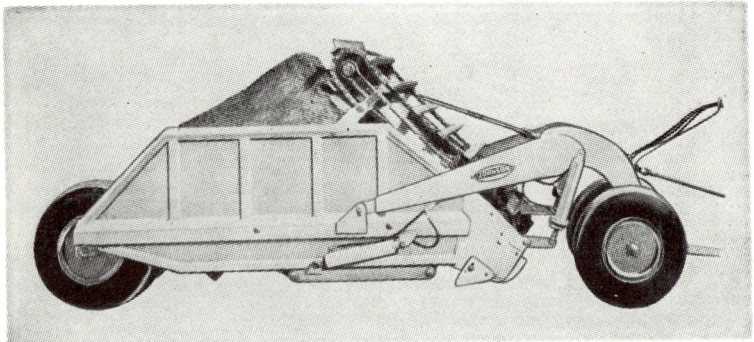

FIG. 35-6. Scraper with PTO-driven loading chain. (*Johnson Mfg. Co., Lubbock, Tex.*)

The normal tolerance required in construction will somewhat depend upon the slope of the land, the methods of irrigation to be used, and crops to be grown. Commonly a tolerance of 0.1 ft from the design grade is permitted as long as no reverse grade occurs in the direction of irrigation or drainage and cross-slope criteria are not exceeded. On very flat lands or with specialized crops, a smaller tolerance may be specified.

The cuts and fills are made in lanes between the grid stakes, with care being taken not to disturb the stakes. When all the earth within the lanes has been moved, the

[1] James C. Marr, Grading Land for Surface Irrigation, *Calif. Agr. Expt. Sta. Circ.* 438, 1954.

construction is checked, additional cuts or fills made as necessary, the stakes removed, and the small area surrounding them leveled to match the adjacent area. Heavy floating is accomplished after the grading is completed.

MAINTENANCE OF LEVELED FIELDS

After a field has been leveled, it is not uncommon for unequal settlement or subsidence to occur, necessitating additional work at a later date. In most areas, these minor irregularities can be removed a year or two later with a heavy floating operation. In this floating or smoothing operation three trips with a leveler are generally needed. The first should be diagonally across field, the second oppositely diagonal, and the last in the same direction as the flow of irrigation water.

In some areas of the West considerable subsidence is experienced after irrigation water is applied to new lands. This subsidence may range from $\frac{1}{10}$ ft to several feet. All the factors contributing to this subsidence are not evaluated, and experience in an area seems to be the best guide at present. In locations where this danger exists, it will usually be best to perform only minimum leveling operations for a year or two and then do the heavy leveling after a period of irrigation.

BIBLIOGRAPHY

Bamesberger, John G.: Land Leveling for Irrigation, *USDA Leaflet* 371, 1954.

――――: "Preparing Land for Efficient Irrigation," USDA Yearbook of Agriculture, 1955.

"Engineering Handbook, Region 5, Soil Conservation Service," USDA, 1946.

"Engineering Handbook, Region 6, Soil Conservation Service," USDA, 1947.

"Engineering Handbook, Region 7, Soil Conservation Service," USDA, 1947.

"Land Leveling," National Engineering Handbook, USDA SCS, 1959, sec. 15, chap. 12.

Marr, James C.: Grading Land for Surface Irrigation, *Calif. Agr. Expt. Sta. Circ.* 438, 1954.

Phelan, John T.: Bench Leveling for Surface Irrigation and Erosion Control, *Soil and Water Trans. ASAE*, 1960.

――――, and Wayne D. Criddle: "Surface Irrigation Methods," USDA Yearbook of Agriculture, 1955.

Stewart, K. V., Jr., and I. L. Stevens: "Systems for Draining the Surface," USDA Yearbook of Agriculture, 1955.

Chapter 36

AGRICULTURAL DRAINAGE

John G. Sutton

Agricultural drainage is the removal of excess water from agricultural land by means of open or covered drains. Shallow surface drains, bedding, and land grading or smoothing are measures used to collect and remove surface water from fields. Tile drains are used to lower the ground-water table and to provide the soil drainage and aeration needed for proper plant growth. Levees, dikes, and diversion ditches are used to prevent overflow. Pumping plants are installed to lift drainage waters where gravity flow is not obtainable. Controlled drainage means regulation of drainage water in open or closed drains to subirrigate crops. Any or all of these practices which are needed should be combined by the engineer into an effective and economical drainage system.

Economic considerations are crucial in determining the feasibility of a drainage system. If the work is to be worthwhile, the value of the benefits during the life of the project should exceed the costs of the work. If the project is not feasible because of too high costs or other factors, it is well to obtain that information early so as to avoid unnecessary costs.

The principal *benefits* from drainage include increased land values resulting from increased crop yields and from better land uses, protection of structures, roads, railroads, and other utilities, and improved public health conditions. The cost of drainage, including amounts for organization, construction, and maintenance, should be estimated and compared with the value of benefits.

A good drainage system should provide an adequate outlet for each farm. The outlet should be large enough to provide drainage for at least 10 to 15 years, after which a major cleanout is to be expected. Drainage systems which would fill with *sediment* rapidly should be protected as required by erosion-control measures.

Maintenance[1] is facilitated by provisions for drop structures for lowering side or lateral drainage, side slopes to permit mowing, planted side slopes, gates and fences for pasturing, roads along berms for use by heavy equipment and chemical-spray outfits, and extra capacity for sediment deposits.

Group Drainage Enterprises. Drainage work that involves more than one landowner may be done by a drainage enterprise organized under state laws, by landowners working under some voluntary agreement, or by a government agency.

Provisions for the granting and recording of right-of-way easements and for maintenance should be made where more than one landowner depends upon a drainage outlet.

INVESTIGATIONS AND SURVEYS

For planning new systems, the usual first step is a reconnaissance survey to determine the general feasibility of draining the land, kind of system needed, and approximate costs. If the results are favorable, a preliminary topographical survey is made and a preliminary report prepared which includes designs and cost estimates.

For construction, it is necessary to make additional surveys having the accuracy

[1] John G. Sutton, Maintaining Drainage Systems, *USDA Farmers' Bull.* 2047, 1952.

required for location of right of way with respect to property boundaries and for carrying out the construction contracts.

Where only the rehabilitation of an existing drainage system is involved, all the surveys outlined above may not be required. If existing drains are located properly and need only deepening and enlarging, the surveys may be limited primarily to profiles and cross sections of existing ditches and to determining watershed areas and other essential data.

The principal information needed on a *drainage survey* is as follows:

1. A *map* of the area showing watershed location, areas, slope, and vegetative cover of area tributary to each ditch. This map should also show ditch locations, roads, railroads, and other physical features which will affect the design of the proposed drainage system.

2. A *profile* on or near the proposed ditch center line with elevations and information on existing channels, ditches, tile outlets, bridges, culverts, utilities, pipelines, levees, gates, buildings, and other structures which will affect the drainage design.

3. Necessary *watershed topographical information* which will be needed for design. Elevation of low points which need drainage is of especial importance. Generally this information can best be obtained by running cross sections approximately at right angles to the profile at intervals of ¼ to 1 mile across the land needing drainage. Land elevations should be read to the nearest 0.1 ft.

4. *Outlet* conditions: adequacy of capacity, high-water elevations, frequency of floods, and authority for use of outlet.

5. *Capability* of land to be drained, agricultural values, yields of adaptable crops, present and recommended land use, value of timber, and costs of clearing.

6. *Soil borings* along drains and levees and at structures to determine character of excavation and foundations should be made to sufficient depth to determine suitability of soil for excavation or footings.

7. *Types* and estimated *costs* of farm drainage required.

8. Conditions which affect the *costs of construction,* such as kinds and amounts of clearing of right of way, accessibility, kinds of equipment needed, and availability of contractors.

9. Township, section, and property *lines* as related to location of drainage system.

10. *Benchmarks* should be established along the proposed system at ½- to 2-mile intervals on permanent structures and should be referenced so they can be relocated. Sea-level elevations should be used where possible. The error in closure of benchmarks should not exceed $0.05 \text{ (traverse length, miles)}^{\frac{1}{2}}$ ft.

DESIGN OF OPEN DITCHES

This section covers the location and design of large outlet ditches and ditch systems which are generally constructed by drainage enterprises, but does not include smaller farm ditches (see Surface Drainage).

Location and Adequacy of Outlet. The first step, and one on which the success of the project rests, is locating the outlet and determining the water levels at which the ditch system will operate. Ordinarily the use of natural outlets will obviate the difficulties encountered in diverting water from one watershed to another. Alternative locations of point of outlet are sometimes available, and careful study is necessary to select the best. Often it is necessary to survey a natural channel and provide for cleanout and straightening to a point well downstream from the land drained in order to find an adequate outlet for the drainage system.

Where a ditch enters areas subject to flooding, such as a large stream, lake, or other body of water, a study of the frequency, duration, height, and time of flooding should be made in order that the success of the drainage improvement may not be impaired by flooding out of the outlet.

Location of Ditches. The location of main ditches and laterals is generally fixed by the natural drainage and the topography. Usually these ditches will be located in the lowest areas of the watershed. Drainage systems covering large flat areas are frequently designed to provide laterals 1 mile apart. An outlet ditch or tile main should be provided for each farm. Soils data should be studied to secure the most favorable location. For example, deep cuts in sand should be avoided, but underdrainage may be improved if the bottom of the ditch merely intercepts the sand which underlies a large area.

Watershed Areas. The watershed area tributary to each ditch at each design point needs to be determined. Ordinarily, design points are at the junctions of mains, submains, large lateral ditches, changes in grade, and at culverts.

Drainage or Runoff Coefficients. Agricultural drainage is usually designed to remove surface water within 24 hr, and some overflow can be expected. This is especially true for the smaller areas. It is more economical to permit occasional flooding than to construct ditches large enough to give complete protection from overflow. A higher degree of protection may be required for some crops which may be damaged by flooding of short duration. This can be secured by using a design curve

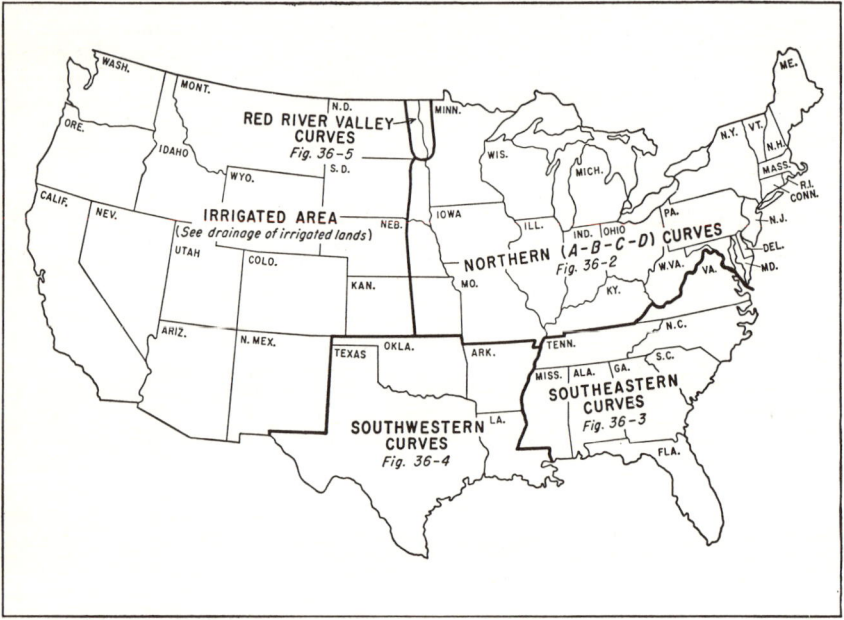

FIG. 36-1. Key map showing drainage coefficients for use in drainage design. ("*Drainage Engineering Handbook,*" USDA SCS, 1955.)

with higher runoff. Ditches designed in accordance with the regular curves have extra capacity to provide for normal sedimentation so as to provide good drainage for 10 to 15 years. On watersheds with high soil losses, erosion should be controlled or, under some conditions, the drainage or runoff coefficient should be increased by an arbitrary percentage to provide additional reserve capacity for sedimentation.

The drainage-curve groups recommended by the Soil Conservation Service for use in the various areas of the United States are shown in Fig. 36-1. The actual runoff curves are shown by groups in Figs. 36-2 to 36-5.

Runoff Computations. Runoff design points are selected just above or below large laterals or at culverts or bridges. The watershed area and proper curve are determined for each design section. On watersheds where there are no major laterals it is necessary to know the drainage area to obtain runoff from the curves.

It will be noted from the runoff curves that the peak discharge per square mile decreases with increase in area. This is because of the greater spread in the time of concentration (arrival of the water) at a particular point.

The *size of ditch needed below the junction of two branches* is somewhat smaller

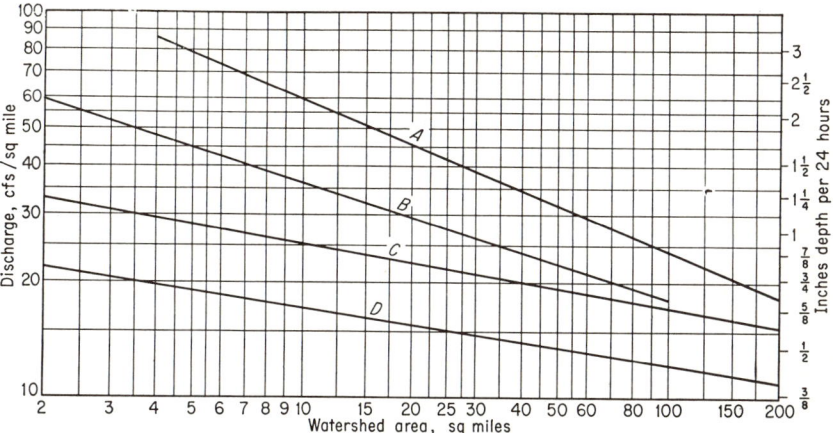

FIG. 36-2. Northern (*ABCD*) curves: *A* for good protection from overflow (not maximum flood runoff); *B* for excellent drainage; *C* for very good agricultural drainage in Ohio, Indiana, Illinois, Iowa, and northern Missouri; for good agricultural drainage in Kentucky and southern Missouri; *D* for fair agricultural drainage in Ohio, Indiana, Illinois, Iowa, and northern Missouri. Watershed area to be determined above each section of a ditch for which capacity is to be computed. Applicable only to flat watershed areas having average slope less than 25 ft per mile. (*Sutton, Hydraulics of Open Ditches, Agr. Eng.*, 20:175-180, May, 1939.)

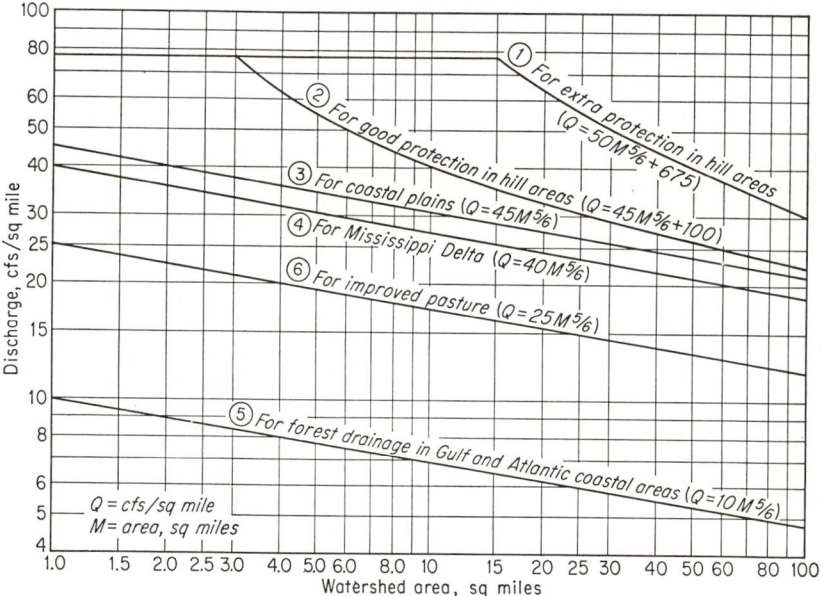

FIG. 36-3. Southeastern curves. (*Curves 1, 2, 3, and 4 developed by Lewis A. Jones, curve 5 developed by E. A. Schlaudt, curve 6 developed by John G. Sutton, USDA SCS, Eng. Div., 1954.*)

than the combined capacity of the two if their time of concentration differs. If the two branches are *nearly equal* in size, the main ditch should have a capacity equal to their sum. For example, if two ditches join, each draining 10 sq miles, *C* curve (Fig. 36-2) gives 250 cfs design volume for each ditch, or a total of 500 cfs. Under this condition the main ditch should be designed for 500 cfs rather than for 446 cfs, as would be obtained from a single area of 20 sq miles. The same ditch cross-

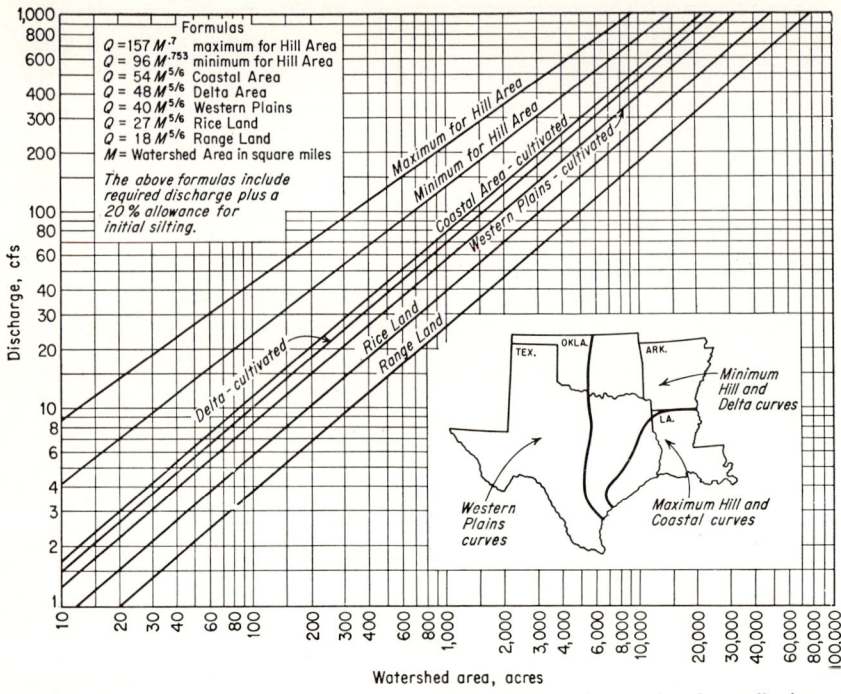

FIG. 36-4. Southwestern curves. The map insert shows the general areas for the applications of runoff curves. (*Prepared by USDA SCS, Western Gulf Region, 1951.*)

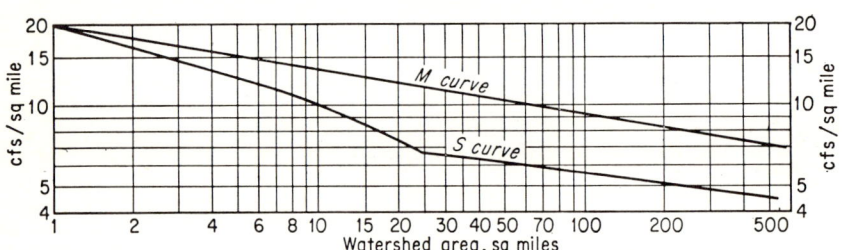

FIG. 36-5. Red River Valley curves. *M* curve should be used for conditions producing high rates of runoff including steep slopes, impervious soils, crops requiring better drainage, and areas having intensely developed systems of farm drainage. *S* curve should be used for conditions producing small rates of runoff which may include flat slopes, pervious soils, crops requiring ordinary drainage, and areas with least development of farm drainage systems. (*USDA SCS, Eng. Div., 1946.*)

section can be extended downstream without change in size until it drains about 23 sq miles.

For cases where one of the ditches above a junction has a flow *between* 60 *and* 70 *per cent of the total flow*, it is recommended that the design figure be taken as the average of the results of the two methods. For ditches draining 7 and 13 sq miles, respectively, the computations would be as follows, using the *C* curve:

Q for 7 sq miles: $7 \times 26.6 = 186$ cfs
Q for 13 sq miles: $13 \times 24 = 312$ cfs
Total Q of the ditches: 498 cfs
Q for 20 sq miles: $20 \times 22.3 = 446$ cfs
Average result of 2 methods: 472 cfs

The figure 472 cfs may be used for design for the ditch below the junction.

Where the largest branch at a junction has *more than 70 per cent of the total flow,* the capacity of the main ditch may be based on the total flow, 446 cfs for 20 sq miles, in the case of branches draining 5 and 15 sq miles.

Some drainage engineers prefer to use the method of adding quantities in all cases, since it gives higher values for the ditch flow below junctions.

Where three or more ditches join, the same principles apply. The percentage is computed from the flow in the largest tributary ditch and the total flow of all tributaries above the junction.

If an area in a drainage system has a different curve from the rest of the area, its equivalent tributary area is used instead of its actual area. The equivalent area is found and used as follows:

Assume that a lateral draining 8 sq miles, designed on the *A* curve, drains into a main being designed on the *C* curve. Trial computations based on Fig. 36-2 show that 8 sq miles on the *A* curve gives the same total discharge as 24 sq miles on the *C* curve. Downstream from this junction the equivalent area of 24 sq miles is added to other tributary areas to take account of the flow from the lateral having the high

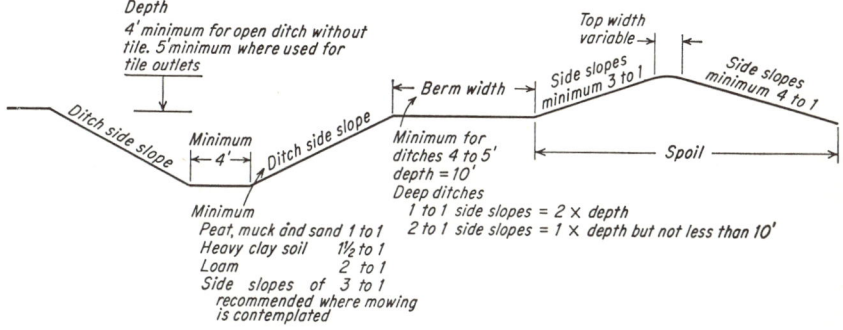

Fig. 36-6. Drainage-ditch specifications.

rate of discharge based on the *A* curve. Transition curves may be developed to simplify such computations.

Under some conditions, such as when designing important channels to provide flood protection, it is not permissible to use drainage coefficients to determine the capacity. It then becomes necessary to estimate the probable flow for a given frequency. Methods of estimating runoff in these cases can be obtained by reference to runoff discussions in Chap. 34, or in hydrology textbooks,[1] or on the basis of measured stream discharges.

The *minimum depth, berm widths,* and *ditch side slopes* which are usually recommended are shown in Fig. 36-6.

Maximum Velocity. The maximum safe velocity for open drains depends upon the soil conditions and on the age and state of vegetative growth. A study of irrigation canals by Fortier and Scobey[2] provides principles applicable to drainage ditches. Permissible canal velocities quoted from this article are shown in Table 36-1. Since ditch banks generally become vegetated and since most drainage waters in humid areas carry colloidal silts, the higher values shown in the table may be used safely. Severe damage to drainage systems sometimes occurs during floods shortly after construction. Such damage can be lessened by providing time for dangerous sections to stabilize with vegetation before excavating all tributary ditches. In some cases grade-control structures are required to keep down the maximum velocity. In others bank

---

[1] R. K. Linsley, M. A. Kohler, and J. L. H. Paulhus, "Applied Hydrology," McGraw-Hill Book Company, Inc., New York, 1949. Daniel W. Mead, "Hydrology," McGraw-Hill Book Company, Inc., New York, 1950.

[2] Samuel Fortier and Fred C. Scobey, Permissible Canal Velocities, *Proc. ASCE,* 51:1397, December, 1925.

TABLE 36-1. PERMISSIBLE CANAL VELOCITIES

Original material excavated for canal	Permissible velocity, feet per second, after aging, of canals carrying:		
	Clear water, no detritus	Water-bearing colloidal silts	Water-bearing noncolloidal silts, sands, gravels, or rock fragments
Fine sand, noncolloidal	1.50	2.50	1.50
Sandy loam, noncolloidal	1.75	2.50	2.00
Fine loam, noncolloidal	2.00	3.00	2.00
Alluvial silts when noncolloidal	2.00	3.50	2.00
Ordinary firm loam	2.50	3.50	2.25
Volcanic ash	2.50	3.50	2.00
Fine gravel	2.50	5.00	3.75
Stiff clay, very colloidal	3.75	5.00	3.00
Graded, loam to cobbles, when noncolloidal	3.75	5.00	5.00
Alluvial silts when colloidal	3.75	5.00	3.00
Graded, silt to cobbles, when colloidal	4.00	5.50	5.00
Coarse gravel, noncolloidal	4.00	6.00	6.50
Cobbles and shingles	5.00	5.50	6.50
Shales and hardpans	6.00	6.00	5.00

SOURCE: Samuel Fortier and Fred C. Scobey, Permissible Canal Velocities, *Proc. ASCE*, 51:1397, December, 1925.

protection, revetment, or jetties are needed to prevent serious bank erosion. In some locations, sections of ditches are shingled over with gravel and no serious erosion occurs under conditions which might otherwise be dangerous.

Roughness Coefficients. The same value of n is used in both Manning's and Kutter's formulas for drainage-ditch design. Manning's formula is the most widely used for computing flow in open ditches, but some authorities prefer the Kutter-Chezy formula for flat slopes. Tables and charts to solve these formulas are available in publications on hydraulics.[1]

The value of n varies greatly during the year with the growth of vegetation and during the life of the ditch as silt and obstructions accumulate. The actual values of n have been found to be larger for small ditches than for large ditches, because of the greater influence of vegetation, irregularities, and debris. The following range in values is recommended:

Hydraulic radius	Value of n
Below 2.5	0.040–0.045
2.5–4.0	0.035–0.040
4.0–5.5	0.030–0.035
Over 5.5	0.025–0.035

These values are considered good average values over the life of a well-maintained ditch. They will be smaller for newly constructed channels; consequently, initial velocities will exceed the design velocities. In some channels this results in erosive velocities for which bank protection or other preventative treatment may be needed.

Hydraulic Grade Line. The hydraulic grade line represents the surface of the water in the ditch for the designed flow conditions. Control points to consider in establishing a hydraulic grade line are (1) low areas to be drained, (2) elevations of tile outlets, and (3) elevation of bridges, culverts, and other structures. After the ground line and control points are plotted, a trial hydraulic gradient should be drawn about 1 ft below the average elevation of the low ground. It is not always feasible or economical to hold the hydraulic grade line below all low areas. This is a decision which will depend mainly on the judgment of the design engineer.

[1] H. W. King and E. F. Brater, "Handbook of Hydraulics," McGraw-Hill Book Company, Inc., New York, 1954, pp. 7–44. Horace W. King, "Manning Formula Tables," vol. 2, "Flow in Open Channels," McGraw-Hill Book Company, Inc., New York, 1939. U.S. Department of the Interior, Bureau of Reclamation, "Hydraulic and Excavation Tables," 11th ed., 1951. U.S. Department of the Army, Corps of Engineers, "Hydraulic Tables," 2d ed., 1944.

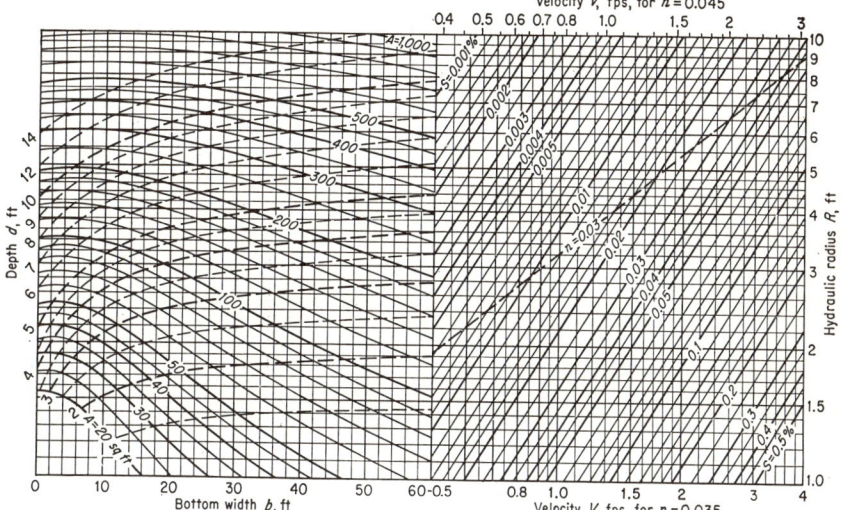

FIG. 36-7. Graphic solution of drainage-channel dimensions, by Manning formula. Trapezoidal cross section with 1.5 : 1 side slopes. (*USDA SCS, Milwaukee, Wis.*, 1940.)

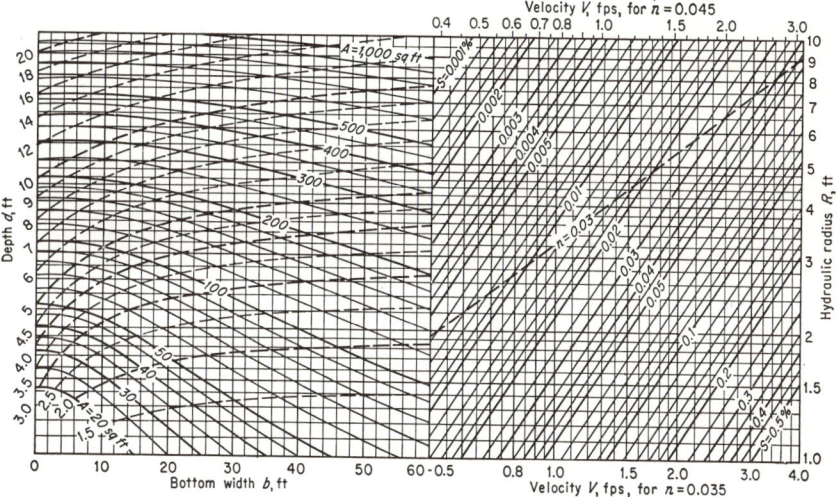

FIG. 36-8. Graphic solution of drainage-channel dimensions, by Manning formula. Trapezoidal cross section with 2 : 1 side slopes. (*USDA SCS, Milwaukee, Wis.*, 1940.)

The hydraulic grade line is determined by use of a flow formula. As indicated previously, the Manning formula is most widely used.

$$Q = \frac{1.486 A r^{2/3} s^{1/2}}{n} \qquad (36\text{-}1)$$

where Q = capacity, cfs
A = area of channel, sq ft
r = hydraulic radius, which is the area divided by wetted perimeter, ft
s = slope of channel, ft/ft
n = coefficient of roughness (see tabulation above for values)

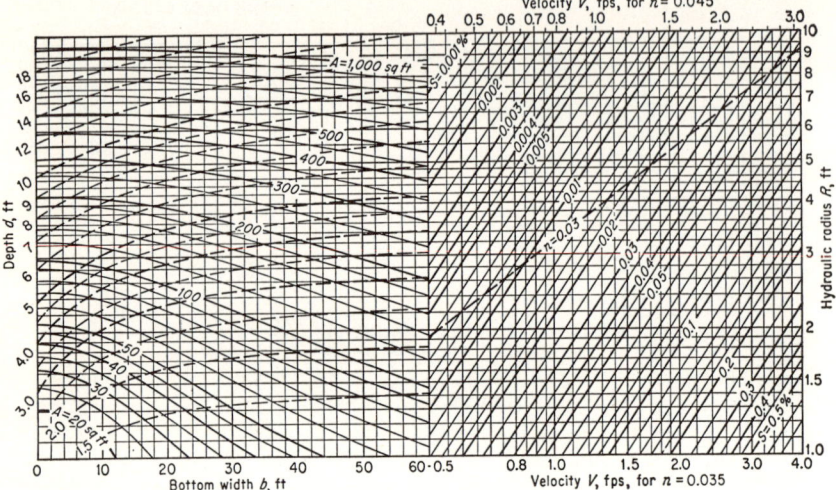

FIG. 36-9. Graphic solution of drainage-channel dimensions, by Manning formula. Trapezoidal cross section with 2.5 : 1 side slopes. (*USDA SCS, Milwaukee, Wis.*, 1940.)

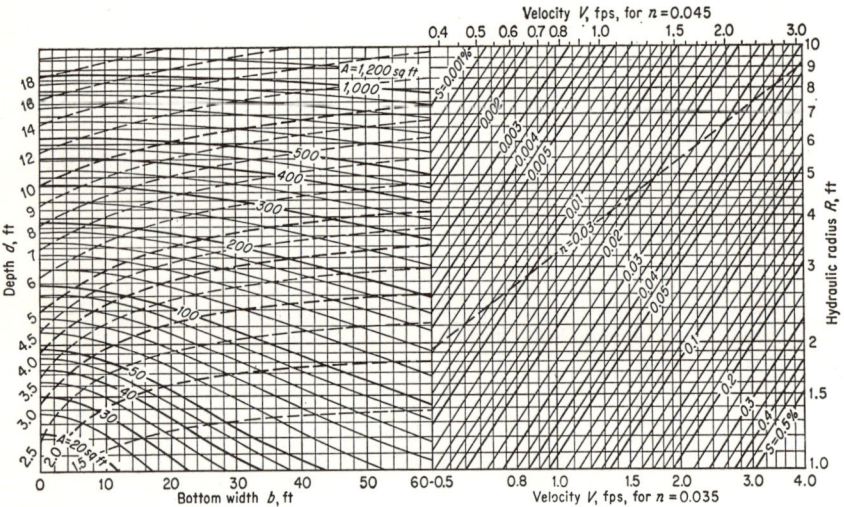

FIG. 36-10. Graphic solution of drainage-channel dimensions, by Manning formula. Trapezoidal cross section with 3 : 1 side slopes. (*USDA SCS, Milwaukee, Wis.*, 1940.)

Figures 36-7 to 36-10 are graphic solutions for determination of capacities of trapezoidal ditches with side slopes ranging from 1.5:1 to 3:1.

The capacity of culverts will be needed in the solution of most drainage problems. Tables 36-2 to 36-6 can be used to solve most culvert design problems. Additional details for culvert design are found in hydraulics handbooks, which are readily available and have been previously cited.

Figure 36-11 shows the calculation of culvert and ditch sizes for a given hydraulic grade.

Auxiliary Structures. Drop spillways, chutes, flumes, pipe drops, and other structures are used to drop the water from the ground surface or from a lateral into the

TABLE 36-2. CAPACITY OF PIPE WITH WATER SURFACE AT INLET THE SAME ELEVATION AS THE TOP OF PIPE, AND FREE OUTLET

Corrugated metal pipe—free outlet
(Capacity, cfs)

Diam of pipe, in.	Slope, per cent												
	.1	.2	.3	.4	.5	.6	.8	1.0	1.2	1.4	1.6	1.8	2.0	2.2	2.4
12	0.6	1.0	1.3	1.5	1.7	1.9	2.1	2.3	2.4	2.5	**2.6**	2.6	2.6	2.6	2.6
15	1.3	2.0	2.4	2.8	3.0	3.3	3.7	4.0	4.3	4.4	4.5	**4.6**	4.6	4.6	4.6
18	2.1	3.1	3.9	4.4	4.9	5.4	6.1	6.5	6.8	7.0	**7.1**	7.1	7.1	7.1	7.1
21	3.3	4.7	5.9	6.8	7.5	8.1	9.0	9.6	10.	10.	10.	**11.**	11.	11.	11.
24	4.7	6.8	8.3	9.5	10.	11.	13.	14.	14.	**15.**	15.	15.	15.	15.	15.
30	8	12	15	17	19	21	23	24	25	25	**26**	26	26	26	26
36	12	19	25	28	31	33	37	39	**40**	40	40	40	40	40	40
42	20	30	37	42	46	50	55	57	**59**	59	59	59	59	59	59
48	25	42	53	62	68	72	77	80	82	**83**	83	83	83	83	83
54	36	57	72	83	90	97	100	**110**	110	110	110	110	110	110	110
60	47	77	97	110	120	130	140	140	**150**	150	150	150	150	150	150

2   4   6   8   10

Smooth pipe—free outlet
(Capacity, cfs)

Diam of pipe, in.	Slope, per cent												
	.1	.2	.3	.4	.5	.6	.8	1.0	1.2	1.4	1.6	1.8	2.0	2.2	2.4
12	1.2	1.8	2.2	2.3	2.4	2.5	**2.6**	2.6	2.6	2.6	2.6	2.6	2.6	2.6	2.6
15	2.1	3.2	3.8	4.1	4.4	4.5	**4.6**	4.6	4.6	4.6	4.6	4.6	4.6	4.6	4.6
18	3.5	5.0	5.9	6.5	6.8	7.0	**7.1**	7.1	7.1	7.1	7.1	7.1	7.1	7.1	7.1
21	5.4	7.6	8.9	9.7	10.	10.	**11.**	11.	11.	11.	11.	11.	11.	11.	11.
24	7.5	10.8	12.8	13.9	14.4	**15.**	15.	15.	15.	15.	15.	15.	15.	15.	15.
30	14	19	23	24	25	**26**	26	26	26	26	26	26	26	26	26
36	22	31	37	39	**40**	40	40	40	40	40	40	40	40	40	40
42	34	47	54	58	**59**	59	59	59	59	59	59	59	59	59	59
48	50	67	77	**83**	83	83	83	83	83	83	83	83	83	83	83
54	70	95	104	108	**110**	110	110	110	110	110	110	110	110	110	110
60	90	125	137	142	145	**150**	150	150	150	150	150	150	150	150	150

2   4   6   8

NOTE: The values in bold face type indicate discharge of the approximate "critical slope" beyond which the discharge remains constant for any given size culvert. The "stairs" of heavy horizontal lines beginning at the upper left and reading downward and to the right indicate approximate velocities of 2, 4, 6, 8, and 10 feet per second.

SOURCE: "Ready Reference Tables," USDA SCS, rev., 1959.

Critical velocity in feet per sec = $V = 4.475 \sqrt{D}$
Critical slope in per cent = $S = \frac{1.04}{D^{1/3}}$ (smooth pipe)
Maximum discharge in cfs = $Q = 2.58 D^{5/2}$
D = diameter of pipe in feet
$n = 0.015$ smooth pipe, $n = 0.021$ corrugated pipe

TABLE 36-3. CAPACITY OF CONCRETE CULVERTS 20 TO 40 FT IN LENGTH WITH SUBMERGED OUTLET OR LAID ON LESS THAN CRITICAL SLOPE

Square-cornered-entrance capacity, cfs

Diam of pipe, in.	Area of pipe, sq ft	Head on pipe, ft																																		
		0.1	0.2	0.3	0.4	0.5	0.6	0.7	0.8	0.9	1.0	1.2	1.4	1.6	1.8	2.0	2.2	2.4	2.6	2.8	3.0	3.2	3.4	3.5
12	0.79	1.4	1.9	2.4	2.7	3.1	3.3	3.6	3.8	4.1	4.3	4.7	5.1	5.4	5.8	6.1	6.4	6.7	7.0	7.2	7.5	7.7	7.9	8.1
15	1.23	2.3	3.2	3.9	4.5	5.0	5.5	6.0	6.3	6.7	7.1	7.8	8.4	8.9	9.5	10	11	11	11	12	12	13	13	13
18	1.77	3.3	4.7	5.7	6.5	7.4	8.0	8.7	9.2	9.8	10	11	12	13	14	15	15	16	17	17	18	19	19	19
21	2.40	4.6	6.4	7.9	9.0	10	11	12	13	14	14	16	17	18	19	20	21	22	23	24	25	26	26	27
24	3.14	6.1	8.5	10	12	13	15	16	17	18	19	21	22	24	25	27	28	29	30	32	33	34	35	35
30	4.91	9.6	13	16	19	21	23	25	27	28	30	33	35	38	40	42	44	46	48	50	52	54	55	56
36	7.07	14	19	24	27	31	33	36	38	41	43	47	51	54	58	61	64	67	69	72	75	77	79	81
42	9.62	19	26	32	37	42	45	49	52	56	59	65	69	74	79	83	87	91	94	98	101	105	108	110
48	12.57	24	34	42	48	54	58	64	67	72	76	83	89	95	101	107	112	117	122	126	131	135	139	141
54	15.90	31	43	53	60	68	74	80	85	91	96	105	113	121	128	135	141	148	154	160	165	171	176	179
60	19.63	37	52	64	73	83	90	98	103	110	116	128	137	147	156	164	172	180	187	194	202	208	214	218

Velocities: 2 | 3 | 4 | 5 | 6 | 8 | 10 ft/s

Beveled-lip-entrance upstream capacity, cfs

Diam of pipe, in.	Area of pipe, sq ft	Head on pipe, ft																																		
		0.1	0.2	0.3	0.4	0.5	0.6	0.7	0.8	0.9	1.0	1.2	1.4	1.6	1.8	2.0	2.2	2.4	2.6	2.8	3.0	3.2	3.4	3.5
12	0.79	1.5	2.1	2.5	2.9	3.2	3.5	3.8	4.1	4.3	4.6	5.0	5.4	5.8	6.1	6.4	6.8	7.1	7.4	7.6	7.9	8.2	8.4	8.5
15	1.23	2.4	3.4	4.2	4.8	5.4	5.8	6.4	6.8	7.2	7.6	8.3	9.0	9.6	10	11	11	12	12	13	13	14	14	14
18	1.77	3.6	5.1	6.2	7.1	8.0	8.7	9.5	10	11	11	13	13	14	15	16	17	18	18	19	20	20	21	21
21	2.40	5.1	7.1	8.7	10	11	12	13	14	15	16	17	19	20	21	22	23	25	26	26	27	28	29	30
24	3.14	6.7	9.4	12	13	15	16	18	19	20	21	23	25	26	28	29	31	32	34	35	36	37	38	39
30	4.91	11	15	19	21	24	26	28	30	32	34	37	40	43	45	48	50	52	55	57	59	61	62	63
36	7.07	16	22	27	31	35	38	41	44	47	49	54	58	62	66	70	73	76	79	82	85	88	91	92
42	9.62	22	31	38	43	49	53	58	61	65	69	76	81	87	92	97	102	106	111	115	119	123	126	128
48	12.57	29	40	49	57	64	69	75	80	85	90	99	106	113	120	126	133	139	144	150	155	160	165	168
54	15.90	37	52	63	72	81	88	96	102	109	115	126	135	145	154	162	170	178	185	192	198	206	212	215
60	19.63	45	64	78	89	101	109	119	126	135	142	156	167	179	190	200	210	220	228	237	245	254	261	265

Velocities: 3 | 4 | 5 | 6 | 8 | 10 ft/s

NOTE: Based on Yarnell, Nagler, Woodward experiments. University of Iowa.
30 foot length of pipe used in computations.
See Table 36-2 for critical slopes.
Dark lines across chart indicate velocities of 2-3-4-5-6-8-10 feet per second.

TABLE 36-4. CAPACITY OF CORRUGATED METAL PIPE AND VITRIFIED CLAY PIPE 20 TO 40 FT IN LENGTH WITH SUBMERGED OUTLET OR LAID ON LESS THAN CRITICAL SLOPE

Corrugated-metal-pipe capacity, cfs

Diam of pipe, in.	Area of pipe, sq ft	Head on pipe, ft																																	
		0.1	0.2	0.3	0.4	0.5	0.6	0.7	0.8	0.9	1.0	1.2	1.4	1.6	1.8	2.0	2.2	2.4	2.6	2.8	3.0	3.2	3.4	3.5
12	0.79	1.0	1.4	1.7	1.9	2.1	2.3	2.5	2.7	2.9	3.0	3.3	3.5	3.8	4.0	4.2	4.4	4.6	4.8	5.0	5.2	5.4	5.5	5.6
15	1.23	1.7	2.3	2.9	3.3	3.7	4.0	4.4	4.6	5.0	5.2	5.7	6.2	6.6	7.0	7.4	7.7	8.1	8.4	8.7	9.0	9.4	9.6	9.8
18	1.77	2.5	3.6	4.4	5.0	5.6	6.1	6.7	7.1	7.5	7.9	8.7	9.4	10	11	11	12	12	13	13	14	14	15	15
21	2.40	3.6	5.1	6.3	7.2	8.1	8.8	9.6	10	11	11	12	13	14	15	16	17	18	18	19	20	20	21	21
24	3.14	5.0	7.0	8.6	9.8	11	12	13	14	15	16	17	18	20	21	22	23	24	25	26	27	28	29	29
30	4.91	8.2	12	14	16	18	20	22	23	24	26	28	30	32	34	36	38	40	41	43	44	46	47	48
36	7.07	12	17	21	24	27	30	32	34	37	39	42	46	49	52	54	57	60	62.64	67	69	71	72
42	9.62	17	24	30	34	38	42	45	48	51	54	59	64	68	72	76	80	84	87	90	93	97	99	101
48	12.57	23	32	39	45	51	55	60	64	68	71	79	84	90	96	101	106	111	115	119	124	128	132	134
54	15.90	29	41	51	58	65	71	77	82	87	92	101	108	116	123	129	136	142	148	153	159	164	169	172
60	19.63	37	52	63	72	82	88	96	102	109	115	126	135	145	154	162	170	178	185	192	199	206	211	215

Dark lines across chart indicate velocities of 2-3-4-5-6-8-10 feet per second.

Capacity of vitrified clay pipe with bell upstream, cfs

Diam of pipe, in.	Area of pipe, sq ft	Head on pipe, ft																						
		0.1	0.2	0.3	0.4	0.5	0.6	0.7	0.8	0.9	1.0	1.2	1.4	1.6	1.8	2.0	2.2	2.4	2.6	2.8	3.0	3.2	3.4	3.5
12	0.79	1.6	2.2	2.6	3.0	3.4	3.7	4.0	4.3	4.6	4.8	5.3	5.7	6.1	6.5	6.8	7.1	7.4	7.7	8.0	8.3	8.6	8.9	9.0
15	1.23	2.5	3.6	4.3	5.0	5.6	6.1	6.6	7.0	7.5	7.9	8.7	9.3	9.9	11	11	12	12	13	13	14	14	15	15
18	1.77	3.7	5.2	6.4	7.3	8.3	9.0	9.8	10	11	12	13	14	15	16	16	17	18	19	19	20	21	21	22
21	2.40	5.1	7.2	8.8	10	11	12	13	14	15	16	18	19	20	22	23	24	25	26	27	28	29	30	30
24	3.14	6.8	9.6	12	13	15	16	18	19	20	21	23	25	27	28	30	31	33	34	35	37	38	39	40
30	4.91	11	15	18	21	23	25	28	29	31	33	36	39	42	44	47	49	51	53	55	57	59	61	62
36	7.07	15	21	26	30	34	37	40	42	45	48	52	56	60	64	67	71	74	77	80	83	85	88	89
42	9.62	21	29	35	40	45	49	54	57	61	64	70	76	81	86	90	95	99	103	107	111	115	118	120
48	12.57	26	37	45	52	59	64	69	74	78	83	91	98	104	111	116	122	126	133	138	143	148	152	154
54	15.90	33	46	56	64	72	79	86	91	97	102	112	120	128	137	144	151	158	164	170	177	183	188	191
60	19.63	39	55	68	77	87	95	103	109	117	123	135	145	155	165	173	182	191	198	205	212	220	226	230

NOTE: Based on Yarnell, Nagler, Woodward experiments. University of Iowa.
30 foot length of pipe used in computations.
See Table 36-2 for critical slopes.
Dark lines across chart indicate velocities of 2-3-4-5-6-8-10 feet per second.
SOURCE: "Ready Reference Tables," USDA SCS, rev., 1959.

TABLE 36-5. CAPACITY OF CONCRETE CULVERTS 40 TO 60 FT IN LENGTH WITH SUBMERGED OUTLET OR LAID ON LESS THAN CRITICAL SLOPE

Square-cornered-entrance capacity, cfs

Diam of pipe, in.	Area of pipe, sq ft	Head on pipe, ft																							
		0.1	0.2	0.3	0.4	0.5	0.6	0.7	0.8	0.9	1.0	1.2	1.4	1.6	1.8	2.0	2.2	2.4	2.6	2.8	3.0	3.2	3.4	3.5
12	0.79	1.3	1.8	2.2	2.5	2.8	3.0	3.3	3.5	3.7	3.9	4.3	4.6	4.9	5.2	5.5	5.8	6.1	6.3	6.6	6.8	7.0	7.2	7.3
15	1.23	2.1	2.9	3.5	4.0	4.5	4.9	5.4	5.7	6.1	6.4	7.1	7.6	8.1	8.6	9.0	9.5	9.9	10	11	11	11	12	12
18	1.77	3.1	4.3	5.3	6.1	6.8	7.4	8.1	8.6	9.2	9.7	11	11	12	13	14	14	15	15	16	17	17	18	18
21	2.40	4.3	6.0	7.3	8.4	9.5	10	11	12	13	13	15	16	17	18	19	20	21	21	22	23	24	25	25
24	3.14	5.7	8.1	9.8	11	13	14	15	16	17	18	20	21	23	24	25	26	28	29	30	31	32	33	33
30	4.91	9.1	13	16	18	20	22	24	25	27	28	31	33	36	38	40	42	44	46	47	49	51	52	53
36	7.07	13	19	23	26	29	32	35	37	39	41	45	49	52	55	58	61	64	67	69	72	74	76	77
42	9.62	18	25	31	35	40	43	47	50	53	56	62	66	71	75	79	83	87	91	94	97	101	104	105
48	12.57	24	33	40	46	52	57	62	66	70	74	81	87	93	99	104	109	114	118	123	127	132	135	138
54	15.90	30	42	51	59	66	72	78	83	88	93	102	110	117	124	131	137	144	150	155	161	166	171	174
60	19.63	37	52	63	72	82	88	97	102	109	115	126	135	145	154	162	170	178	185	192	199	206	211	215

Velocity markers: 3, 4, 5, 6, 8, 10 ft/sec

Beveled-lip-entrance upstream capacity, cfs

Diam of pipe, in.	Area of pipe, sq ft	Head on pipe, ft																							
		0.1	0.2	0.3	0.4	0.5	0.6	0.7	0.8	0.9	1.0	1.2	1.4	1.6	1.8	2.0	2.2	2.4	2.6	2.8	3.0	3.2	3.4	3.5
12	0.79	1.3	1.8	2.3	2.6	2.9	3.2	3.5	3.7	3.9	4.1	4.5	4.9	5.2	5.5	5.8	6.1	6.4	6.6	6.9	7.1	7.4	7.6	7.7
15	1.23	2.2	3.1	3.7	4.3	4.8	5.2	5.7	6.1	6.5	6.8	7.5	8.0	8.6	9.1	9.6	10	11	11	11	12	12	13	13
18	1.77	3.3	4.7	5.7	6.5	7.4	8.0	8.7	9.2	9.8	10	11	12	13	14	15	15	16	17	17	18	19	19	19
21	2.40	4.6	6.5	8.0	9.1	10	11	12	13	14	15	16	17	18	19	20	21	22	23	24	25	26	27	27
24	3.14	6.2	8.7	11	12	14	15	16	17	18	19	21	23	24	26	27	29	30	31	32	34	35	36	36
30	4.91	10	14	17	20	23	25	27	28	30	32	35	38	40	43	45	47	49	51	53	55	57	59	60
36	7.07	15	21	26	30	33	36	39	42	45	47	52	56	59	63	66	70	73	76	79	81	84	87	88
42	9.62	21	29	36	41	46	51	55	58	62	66	72	77	83	88	93	97	101	106	109	113	117	121	122
48	12.57	28	39	48	55	62	67	73	77	82	87	95	102	109	116	122	128	134	139	145	150	155	159	162
54	15.90	36	50	61	70	79	86	93	99	105	111	122	131	140	149	156	164	172	179	185	192	199	204	206
60	19.63	43	62	76	87	98	107	116	123	131	138	152	163	174	185	195	204	214	223	232	239	248	254	259

Velocity markers: 3, 4, 5, 6, 8, 10 ft/sec

NOTE: Based on Yarnell, Nagler, Woodward experiments. University of Iowa.
50 foot length of pipe used in computations.
See Table 36-2 for critical slopes.
Dark lines across chart indicate velocities of 3-4-5-6-8-10 feet per second.

SOURCE: "Ready Reference Tables," USDA SCS, rev., 1959.

TABLE 36-6. CAPACITY OF CORRUGATED METAL PIPE AND VITRIFIED CLAY PIPE 40 TO 60 FT IN LENGTH WITH SUBMERGED OUTLET OR LAID ON LESS THAN CRITICAL SLOPE

Corrugated-metal-pipe capacity, cfs

Diam of pipe, in.	Area of pipe, sq ft	Head on pipe, ft																									
		0.1	0.2	0.3	0.4	0.5	0.6	0.7	0.8	0.9	1.0	1.2	1.4	1.6	1.8	2.0	2.2	2.4	2.6	2.8	3.0	3.2	3.4	3.5
12	0.79	0.8	1.1	1.4	1.6	1.7	1.9	2.1	2.2	2.3	2.5	2.7	2.9	3.1	3.3	3.5	3.7	3.8	4.0	4.1	4.3	4.4	4.5	4.6
15	1.23	1.4	2.0	2.4	2.7	3.1	3.3	3.6	3.9	4.1	4.3	4.8	5.1	5.5	5.8	6.1	6.4	6.7	7.0	7.2	7.5	7.7	8.0	8.2
18	1.77	2.1	3.0	3.7	4.2	4.7	5.1	5.5	5.9	6.3	6.7	7.3	7.9	8.4	8.9	9.4	9.9	10	11	11	12	12	12	13
21	2.40	3.1	4.4	5.3	6.1	6.9	7.5	8.1	8.6	9.2	9.7	11	11	12	13	14	14	15	16	16	17	17	18	18
24	3.14	4.3	6.0	7.4	8.4	9.5	10	11	12	13	13	15	16	17	18	19	20	21	22	22	23	24	25	25
30	4.91	7.2	10	12	14	16	17	19	20	21	22	25	26	28	30	32	33	35	36	37	39	40	41	42
36	7.07	11	16	19	22	25	27	29	31	33	35	38	41	44	46	49	51	54	56	58	60	62	64	65
42	9.62	16	22	27	31	35	37	41	43	46	49	53	57	61	65	69	72	75	78	81	84	87	89	91
48	12.57	21	30	36	41	46	50	55	58	62	66	72	77	83	88	93	97	101	105	109	113	117	121	123
54	15.90	27	38	46	53	60	65	70	75	80	84	92	99	106	113	118	124	130	135	140	145	150	154	157
60	19.63	34	48	59	67	76	82	90	95	102	107	118	126	135	143	151	158	166	172	179	185	191	197	200

Capacity of vitrified clay pipe with bell upstream, cfs

Diam of pipe, in.	Area of pipe, sq ft	Head on pipe, ft																									
		0.1	0.2	0.3	0.4	0.5	0.6	0.7	0.8	0.9	1.0	1.2	1.4	1.6	1.8	2.0	2.2	2.4	2.6	2.8	3.0	3.2	3.4	3.5
12	0.79	1.4	2.0	2.4	2.7	3.1	3.4	3.7	3.9	4.1	4.4	4.8	5.2	5.5	5.9	6.2	6.5	6.8	7.0	7.3	7.6	7.8	8.0	8.2
15	1.23	2.3	3.2	3.9	4.5	5.0	5.5	6.0	6.3	6.7	7.1	7.8	8.4	9.0	9.5	10	11	11	11	12	12	13	13	13
18	1.77	3.4	4.8	5.9	6.7	7.6	8.2	8.9	9.5	10	11	12	13	13	14	15	16	17	18	18	19	20	20	
21	2.40	4.8	6.7	8.2	9.4	11	12	13	13	14	15	16	18	19	20	21	22	23	24	25	26	27	27	28
24	3.14	6.3	8.8	11	12	14	15	16	17	19	20	22	23	25	26	28	29	30	32	33	34	35	36	37
30	4.91	10	14	17	20	22	24	26	28	30	31	34	37	40	42	44	46	49	51	52	54	56	58	59
36	7.07	15	20	25	29	32	35	38	40	43	45	50	53	57	61	64	67	70	73	76	78	81	83	85
42	9.62	20	28	34	39	44	47	52	55	58	62	68	73	78	83	87	91	96	99	103	107	110	113	115
48	12.57	25	36	44	50	56	61	67	71	76	80	88	94	100	106	112	118	123	128	133	138	142	146	149
54	15.90	32	45	55	63	71	77	84	89	95	100	109	117	125	133	140	147	154	160	166	172	178	183	186
60	19.63	38	54	66	75	85	92	100	106	113	120	132	141	151	160	169	177	185	192	200	207	214	220	224

NOTE: Based on Yarnell, Nagler, Woodward experiments. University of Iowa.
50 foot length of pipe used in computations.
See Table 36-2 for critical slopes.
Dark lines across chart indicate velocities of 3-4-5-6-8-10 feet per second.

SOURCE: "Ready Reference Tables," USDA SCS, rev., 1959.

main ditch. In many instances structural work can be eliminated by excavating laterals back from the main ditch on a gentle grade. In some localities a sod flume can be used to admit surface water into drainage ditches. Structures which admit water into a ditch are ordinarily designed to take care of drainage flow usually allowed for small

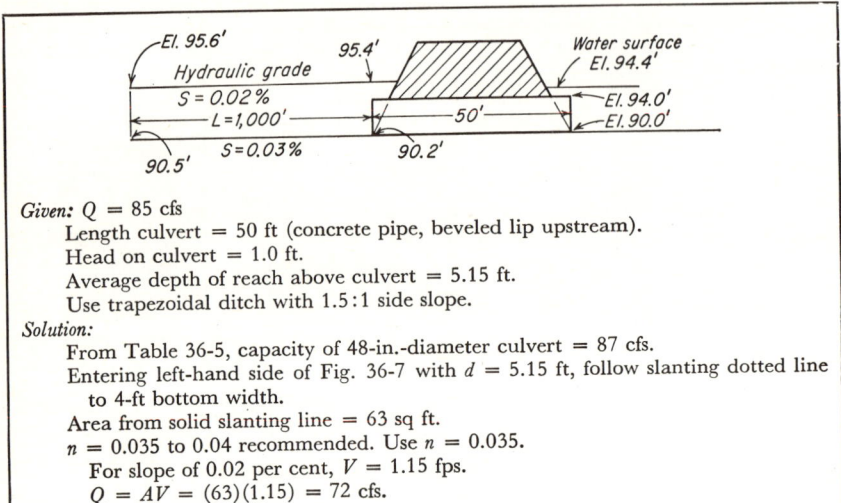

Given: $Q = 85$ cfs
    Length culvert = 50 ft (concrete pipe, beveled lip upstream).
    Head on culvert = 1.0 ft.
    Average depth of reach above culvert = 5.15 ft.
    Use trapezoidal ditch with 1.5:1 side slope.
Solution:
    From Table 36-5, capacity of 48-in.-diameter culvert = 87 cfs.
    Entering left-hand side of Fig. 36-7 with $d = 5.15$ ft, follow slanting dotted line to 4-ft bottom width.
    Area from solid slanting line = 63 sq ft.
    $n = 0.035$ to 0.04 recommended. Use $n = 0.035$.
    For slope of 0.02 per cent, $V = 1.15$ fps.
    $Q = AV = (63)(1.15) = 72$ cfs.

Fig. 36-11. Sample problem on ditch and culvert size using culvert tables and graphic charts.

areas but are not large enough to take care of the maximum runoff. Auxiliary spillways are provided through the spoil bank so that such structures will not wash out when the runoff is more than the design flow. This is called the *island-type design*. A difference of 12 to 24 in. in elevation from the bottom of the auxiliary spillway to the top of the wing walls is usually allowed.

Corrugated and other pipes are used to drop water into outlet ditches, using construction similar to that described for tile drains (Fig. 36-12).

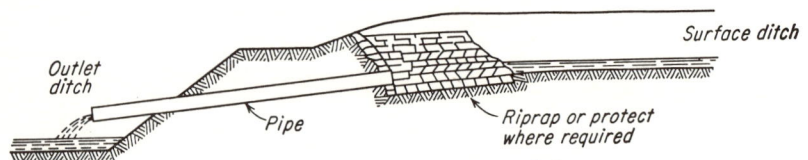

Fig. 36-12. Outlet pipe for surface ditch.

Drop spillways and similar structures are used to control the grade in drainage ditches by concentrating any excess fall at one or more locations. This construction is required frequently in drainage of sloping irrigated lands. In such cases the drop should be designed for the drainage flow and the wing walls should be built high enough to protect the structure when the ditch overflows.

Bridges and Culverts. Permanent structures of steel, concrete, or treated timber generally conform to the requirements of the railroad, highway, or road system for which they are built. Farm bridges are usually the responsibility of the landowner but should be designed to support the heaviest tractor or truck likely to use it. A minimum loading of 5 tons is recommended for a single-lane and 10-ton loading for a double-lane farm bridge. A sign stating the load for which the structure is designed should be erected.

For most drainage jobs bridges are preferred to culverts because they are less likely to cause overflow and damage to lands upstream. The cross-section area of the opening under the bridge should be at least as large as the designed cross-section area of the ditch. The bridge should be set high enough for the stringers to be above high water. Openings should be left in spoil banks above and below such crossings, and the bridge should be higher than nearby road elevations so that flood waters may flow around bridges or culverts when overflows occur.

Where possible, bridge piers or pile trestles should be avoided in the perimeter of the ditch. In wide ditches where more than one span is required, an odd number of spans should be used so that the center section of the ditch may be left unobstructed. Yarnell investigated and published information on the hydraulic losses of bridge piers[1] and pile trestle[2] as channel obstructions.

If culverts need to be considered as well as bridges, the flood flows of 5- to 25-year frequency should be studied as well as the design drainage flow. A culvert introduces an obstruction in the ditch which may cause flooding a distance upstream; this may be determined by backwater computations. It may be advisable to enlarge a ditch where part of the available fall is used by a culvert, leaving less fall available.

If there are no culverts, a flow in excess of the design flow causes a rise in the ditch above the design stage but the water recedes rapidly if there are no obstructions. Such rises generally cause little damage. Stoppage by debris and sediment is an added hazard in the use of culverts, and they require frequent maintenance.

Where culverts are to be used, capacities can be obtained from Tables 36-2 to 36-6. Talbot's formula[3] or other methods acceptable to highway officials are sometimes used for computing the required area of runoff for culvert discharge. Hydraulic losses, when not covered by the above tables, may be computed by procedures developed by Mavis,[4] by Yarnell, Nagler, and Woodward,[5] and as covered in King and Brater's "Handbook of Hydraulics"[6] and other handbooks. Usually it has been found desirable that the design drainage flow not exceed 4 to 5 fps passing through a culvert and that the culvert area be not less than half the required area of the drainage channel. The criteria which give the most conservative size should be used. Culverts may prove economical near the upper end of a drainage system where the ditch may have excess capacity because of minimum size and depth requirements or where the ditch gradients are steep.

CONSTRUCTION OF OPEN DITCHES

The plans, specifications, and contract should be clear and complete and should cover the following information: location of each item of work, amount of each type of work, working conditions for clearing and excavation, time limits for starting and completion of work, and penalties if specified. Provisions for setting stakes, inspection, tolerances to be required, and methods of payment should be covered.

Staking Ditches. An experienced contractor equipped with a good level needs few stakes. He may do satisfactory work if provided with a center-line stake giving the station number, grade stakes every 300 to 500 ft, and a schedule showing station number, cut, top width and bottom width, berm width, spoil-bank heights, and location and dimensions of openings in the spoil bank. Usually, however, *grade* and *cut stakes* are set every station or two to mark the depth of cut and *slope stakes* are set to mark the points where the side slopes intersect the natural ground surface. Grade stakes should be set at 25- to 50-ft intervals on short-radius curves. Stakes to mark the berm width and height of spoil bank may be provided at intervals of about 300 to

---

[1] D. L. Yarnell, Bridge Piers as Channel Obstructions, *USDA Tech. Bull.* 442, 1934.
[2] D. L. Yarnell, Pile Trestles as Channel Obstructions, *USDA Tech. Bull.* 429, 1934.
[3] "Handbook of Drainage and Construction Products," Armco Drainage & Metal Products, Inc., 1955.
[4] F. T. Mavis, The Hydraulics of Culverts, *Penn State Univ. Bull.* 56, 1943.
[5] D. L. Yarnell, Floyd A. Nagler, and Sherman M. Woodward, Flow of Water through Culverts, *Iowa Univ. Bull.* 1, 1926.
[6] *Op. cit.*

600 ft. Side inlets and openings should be staked so that they will not be overlooked during construction. Local custom should be followed in marking stakes.

The contractor should have the dragline operator set stakes to mark both sides of the ditch bottom as a section is completed. Since the machine backs away from the excavated section, these stakes are excellent guides to enable the operator to keep on line and to secure straight tangents and uniform side slopes.

It is well to provide more than the minimum number of stakes at the start of a job until the operator is familiar with the requirements and permissible tolerances on the job.

Allowable Tolerances in Construction. Since drainage work does not have to meet the close tolerances required for some other types of work, low unit prices may be expected if conditions are fully understood by the bidder. Even so, the permissible tolerances should be covered in the contract and not left to chance. For larger outlet ditches over 4 ft deep with 2:1 or steeper side slopes, any overcut which deepens the ditch up to 1 ft more than the depth specified is usually permissible. It is easier for the machine operator to overcut than to stop the machine frequently and measure the grade. To avoid moving the machine back for minor deviations above the designed grade, a tolerance of 0.1 or 0.2 ft for 100 to 300 ft is ordinarily permitted above the designed grade. Side slopes may be above grade by a similar allowance. To ensure that the tolerance will be used as intended, a fair penalty providing, for example, that triple the yardage remaining above grade will be moved free, or another suitable penalty clause may be worked out.

On shallow drains which may be mowed or pastured, cutting below grade should be limited to 0.3 to 0.6 ft and the ditch should be widened to maintain the designed side slopes if permissible limits are exceeded.

Inspection. Satisfactory inspection of drainage-ditch work consists of taking levels of the completed bottom grade and measuring cross sections. These levels may be taken at intervals of 100 to 1,000 ft, depending upon the type of job and the experience of the contractor and inspector. Such measurements should be checked against plans as soon as possible so that errors may be corrected with a minimum of cost.

Clearing. The right of way should be cleared of trees, brush, and logs to allow efficient use of equipment and planned disposal of spoil. Logs or pulpwood should be salvaged where economical. Brush and limbs from trees are ordinarily piled and burned. On many jobs it is permissible to cut trees on the berm and spoil-bank area to a height of from 12 to 18 in., then bankfill with spoil over the berm area, to a height about 4 to 10 in. above tops of stumps. This provides a road along the berm and avoids the necessity of removing stumps from the berm or cutting them flush with the ground. Stumps of trees 8 in. or more in diameter may be removed by draglines from the area to be excavated. Stumps which are too large for the dragline to handle economically should be blasted.

Large tractors equipped with bulldozers or treedozer attachments are economical for clearing work. Stump and debris should be disposed of so as not to interfere with spoil-bank leveling or cultivation of adjoining fields.

Excavation. *Draglines* are standard construction equipment for ditches over 3 ft deep and are used for shallow ditches under wet conditions. By using mats they can be operated on soft ground. Where soils are too unstable for use of mats alone, a corduroy road made of transverse logs and backfilled with mineral soil about 1 to 2 ft over the logs will provide a stable foundation for the mats. Such a road is satisfactory for practically all bad conditions, including deep organic soils on which trees are growing. Under swamp conditions it may pay to blast a shallow pilot ditch ahead of the dragline to drain surface water and facilitate clearing.

The *tractor and grader* and *motor patrol* are well adapted to excavating shallow ditches under 3 ft deep in regular topography. They are not efficient for making cuts which vary greatly in depth along the ditch line or for working in wet unstable soils. The *bulldozer* is used to excavate shallow ditches on some jobs. The *carryall scraper* is efficient in cutting wide-bottom ditches where the soil is stable enough for them to operate. It can deposit excavated soil in field depressions to improve drainage

Special equipment which moves dirt with small buckets on a chain is used on some work. Blasting may be used, but it is difficult to construct ditches to accurate sections or grade with dynamite; therefore this practice is generally limited to initially opening up an area or to constructing a pilot ditch, as was previously suggested.

Spoil-bank Leveling. In cultivated fields it is desirable to level the spoil banks for cropping. For pasturing and hay crops the spoil-bank side slopes should not exceed 4:1, the slope on which a tricycle type of farm tractor may operate. For row crops, slopes from 6:1 to 10:1 are preferred.

Spoil banks in woodland and other areas where farming is not practical may be left in a rough state and planted to shrubs which furnish food and cover to wildlife.[1] In some locations the spoil is built up to serve as a levee.

Construction Costs. Construction costs may be estimated in accordance with the principles and details discussed in available texts.[2] Costs of drainage work vary so greatly with size and location of job that it is always desirable to secure local current prices to estimate costs.

MAINTAINING OPEN DRAINAGE SYSTEMS

Regular and effective maintenance must be carried out if a drainage system is to remain effective. When outlet ditches are neglected, the lower lands are damaged first. Conflicts in interest may arise between owners whose lands are damaged and those whose lands are high enough to get good drainage. However, as ditches fill with vegetation and sediment, the higher lands also are damaged. It is better for all owners to carry out maintenance as needed in order to keep the system operating at high efficiency.

Many ditches commonly lose half their capacity in 2 to 4 years because of vegetative growth. Vegetation on a ditch may be controlled by pasturing, mowing, chemical treatment, burning, hand cutting, and mechanical equipment.

More detailed information on methods of maintaining drainage systems has been prepared by the U.S. Department of Agriculture, Soil Conservation Service.[3]

Controlling Vegetative Growth. Cutting by hand is a practical way of keeping vegetation under control on some deep outlet ditches having steep side slopes. In most of the humid areas the cutting of weeds and brush once or twice a year gives native grasses a chance to come in.

One of the principal objectives is to secure a good stand of close-growing grasses adapted to the soil and climate on the ditch banks and the berm. Such grasses include bluegrass, redtop, tall fescues, and love grass, where adapted in the northern states. In the southern states fescues, Dallis grass, Bermuda grass, and Bahia grass will stabilize ditch banks. Hay crops of adapted species are often grown and harvested along berms and spoil banks.

Pasturing is effective in controlling vegetation in some locations where the ditch can become part of a well-managed pasture. Proper fencing to permit good management, water gates which will not obstruct ditches, and ramps for easy access to ditches will aid a pasturing program. Good grasses should be planted where they do not come in naturally.

Mowing is becoming more important as a maintenance method, especially for the shallow drains. For a tricycle-type farm tractor to operate with reasonable safety, the side slope of the ditch should be $3\frac{1}{2}:1$ or flatter. However, track-type tractors may operate on 3:1 side slopes. When not used for hay, clippings may be used as a mulch or burned.

Chemical control of ditch-bank weeds by selective chemicals, such as 2,4-D and 2,4,5-T, kills the broadleaf plants but does not permanently injure the desirable grasses. These chemicals are dangerous to use near cotton, tomatoes, legumes, and many flowering ornamentals, brush fruits, and truck crops. The proper selection of chemicals requires consideration of effects on crops and ditch weeds and of the costs.

---

[1] W. L. Anderson, Making Land Produce Useful Wildlife, *USDA Farmers' Bull.* 2035, 1951.
[2] H. E. Pulver, "Construction Estimates and Costs," McGraw-Hill Book Company, Inc., New York, 1947.
[3] John G. Sutton, Maintaining Drainage Systems, *USDA Farmers' Bull.* 2047, 1952.

Recommendations of state experiment stations or chemical companies should be secured.

A road along the ditch with crossings at laterals and gates in fences is needed for efficient spray-rig operation. A *spray rig* with one or more tanks for chemicals, a pump, an engine, and a flexible hose or pipe and nozzles to direct the spray may be mounted on a truck or on a tractor-drawn trailer. Trial use of a hand spray may be advisable for a year or two to determine the effectiveness and cost of treatment by chemicals. Chemical control is usually cheaper than hand cutting if floating water weeds and those which grow under water can be controlled effectively. For example, in Michigan[1] the application of 6.8 to 13.6 lb of Dalapon per 100 gal of water gave seasonal control of cattail. In some cases, water hyacinths have been brought under control only to be succeeded by other water weeds just as bad. It is well to find by trials what will happen under the changed environmental conditions produced by chemicals.

Mechanical ditch-maintenance equipment includes such items as the Ruth dredger with buckets mounted on a chain, rakes attached to dragline excavators, chains drawn by tractors or horses, and boats to cut and remove vegetation. A track-type tractor equipped with an angle dozer, operating on a 2½:1 slope, has proved effective in removing brush and small willows when the ditch was dry.

Maintenance is facilitated by leveled spoil banks and a road along the berm which may be used by heavy equipment.

Reducing Sediment Deposits. It is essential to locate the sources of excessive sediment deposits and apply conservation measures to reduce the rate of ditch siltation. Effective stream-bank control should be established where needed. Where the beneficial effects of such measures will be delayed, it may be desirable to construct desilting or sedimentation basins to collect excess silt entering a ditch.

Locations where shallow drains enter outlet ditches should be examined to determine the need for additional drop spillways, pipe drops, and sod flumes. Such structures may be installed as are discussed in previous sections.

LEVEES AND DIKES

Levees as defined in this handbook are embankments of earth used along small streams to prevent flooding during periods of relatively high runoff. They provide for extra capacity over ordinary drainage design by raising the hydraulic grade line above the surrounding land, increasing the area and velocity of the ditch.

A stream with levees provides a narrow high-velocity channel for small flows, tending to prevent sediment deposition.

Since levees may have the effect of reducing flood storage, the damage of increased flow downstream needs to be studied before levees are recommended.

Runoff for Levees. The runoff to be used in the design of levees differs from ordinary drainage in that capacity is provided to take care of a storm of 10-, 25-, or 50-year recurrence interval. The actual capacity provided will depend on the economics of the situation and the desires of landowners. For computation of maximum flows for a given recurrence interval, see Chap. 34.

Design of Levees. The surveys and computations of hydraulic grade lines for levees will be the same as explained previously for ditch drainage except that ordinarily composite ditch sections will be encountered rather than simple trapezoidal sections. Figure 36-13 shows an example of such computation.

Design Criteria. The side slope's minimum bottom width and width of berm will be the same as for ditch design. In levee design the spoil is used to build the levee. Figure 36-14 shows a typical levee section and gives side slope, top width, and banquette recommendations for levees.

Culverts are placed through levees where it is necessary for drainage water to drain from the land side into the stream channel during low stages. Usually a floodgate is placed on the end of the culvert so that it will swing open when water is higher on the land of the levee and will close when water is higher in the ditch. Sometimes a manually operated gate is preferred.

---

[1] W. A. Cutler et al., Observations on Control of Cattails, Typha Spp., by Chemical Sprays, art. 37–46, *Mich. State Univ. Quart. Bull.*, February, 1955.

Such culverts should be designed to remove the drainage flow rapidly enough to minimize crop losses after stages on the water side of the levee recede. Such determinations should be made on the basis of a study of hydrographs of water stages inside and outside the levee under design conditions.

Selection and Compaction of Soils. The principles which apply to the construction of earth dams[1] apply to the selection of soils and compaction of levee fills. A well-graded soil of high density, with fine material to prevent seepage and with sharp

Given:
  Channel cross section as shown with 1.5:1 side slopes
  $S = 0.00055$ ft per ft or 0.055 per cent

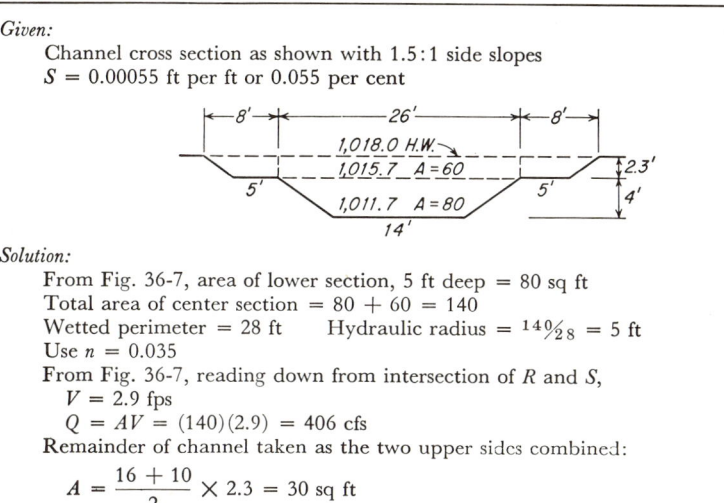

Solution:
  From Fig. 36-7, area of lower section, 5 ft deep = 80 sq ft
  Total area of center section = $80 + 60 = 140$
  Wetted perimeter = 28 ft   Hydraulic radius = $140/28 = 5$ ft
  Use $n = 0.035$
  From Fig. 36-7, reading down from intersection of R and S,
  $V = 2.9$ fps
  $Q = AV = (140)(2.9) = 406$ cfs
  Remainder of channel taken as the two upper sides combined:

  $$A = \frac{16 + 10}{2} \times 2.3 = 30 \text{ sq ft}$$

  $P = 17.6$ ft $R = \dfrac{30}{17.6} = 1.7$ ft

  Use $n = 0.045$
  $V = 1.1$ fps $Q = AV = (30)(1.1) = 33$ cfs
  Total $Q = 406 + 33 = 439$ cfs

FIG. 36-13. Use of graphic solution charts to solve composite channel section.

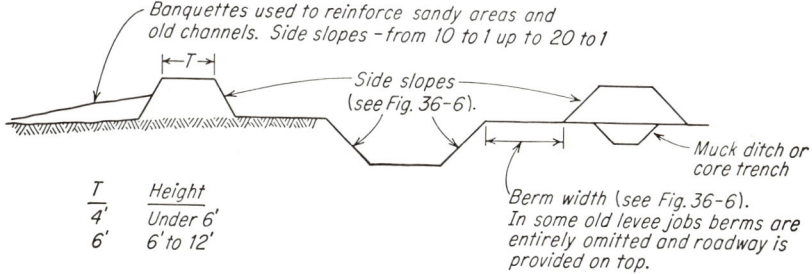

FIG. 36-14. Typical levee section.

particles of sand to provide stability, is desirable. Soil should be placed in layers 3 to 6 in. thick and compacted. The proper moisture content should be secured by irrigation of the borrow pit or sprinkling of the fill, if necessary.

Construction of Levees. The area on which the levee will rest should be cleared of trees and brush and the topsoil removed. Stumps and roots should be grubbed. The

---

[1] "Low Dams: A Manual of Design for Small Water Storage Projects," National Resources Committee, Washington, D.C., 1939; prepared by Subcommittee on Small Water Storage Projects, Perry A. Fellows, Chairman.

foundation area should be plowed or scarified to provide a good bond between foundation and levee. The muck ditch is then cut if one is specified. All large roots, debris, and soft spots revealed by the muck ditch or during clearing should be dug out and holes filled and compacted with good material.

In the humid area it is nearly always desirable to *maintain* levees by pasturing. A good sod should be established and maintained in accordance with the recommendations of the local agricultural agencies.

SURFACE DRAINAGE

Keith H. Beauchamp

Surface drainage is the removal of surplus water from the surface of agricultural land as quickly as possible, allowing the wind and sun to start their job of drying the soil as soon as possible after a rain and thus preventing damage to crops which occurs from the exclusion of essential air. The runoff water should move off the field as rapidly as possible, without causing erosion of the land surface or the ditches.

The most complete surface drainage system is produced by land leveling or grading as discussed in Chap. 35. Other methods which may be used where economic conditions will not justify the more complete job are:

1. Bedding, or crowning, system
2. Random ditch system
3. Cross-slope ditch system
4. Parallel ditch system

Additional details of spacing and capacity for various soil types are given in the state drainage guides generally available from the Soil Conservation Service or state experiment stations.

Adequate drainage for many fields requires a complete surface drainage system. The system should provide for (1) the correct length, slope, and direction of crop rows and (2) sufficient and strategically placed open field ditches to collect and remove runoff water from the crop rows and to drain impounded areas.

Surface Drainage Systems. The *bedding system* of surface drainage is generally used on fields that are practically flat, slope from 0 to 1.5 per cent, where the soils are slowly permeable, and where tile drainage is not justified by value of crops grown or an outlet for tile drainage is not available. The system is designed, constructed, and maintained so that excess surface water drains laterally from crowned beds similar to plowlands into dead furrows, thence into the collection ditches, and finally into an outlet ditch, as shown in Fig. 36-15. The beds should be laid out with the dead furrow running in the direction of greatest slope.

Widths of the beds, based on drainage characteristics of the soil, are shown in Table 36-7.

TABLE 36-7. WIDTH OF BEDS FOR GENERAL FIELD CROPS

Degree of internal drainage of soil	Width of bed in feet center to center of dead furrows	No. of 3½' corn rows with 2' allowed per dead furrow	No. of rounds using 2-14" plows
Very slow............	23	6	5
	30	8	6½
	37	10	8
Slow................	44	12	9½
	51	14	11
Fair................	58	16	12½
	65	18	14
	72	20	15½
	79	22	17
	86	24	18½
	93	26	20

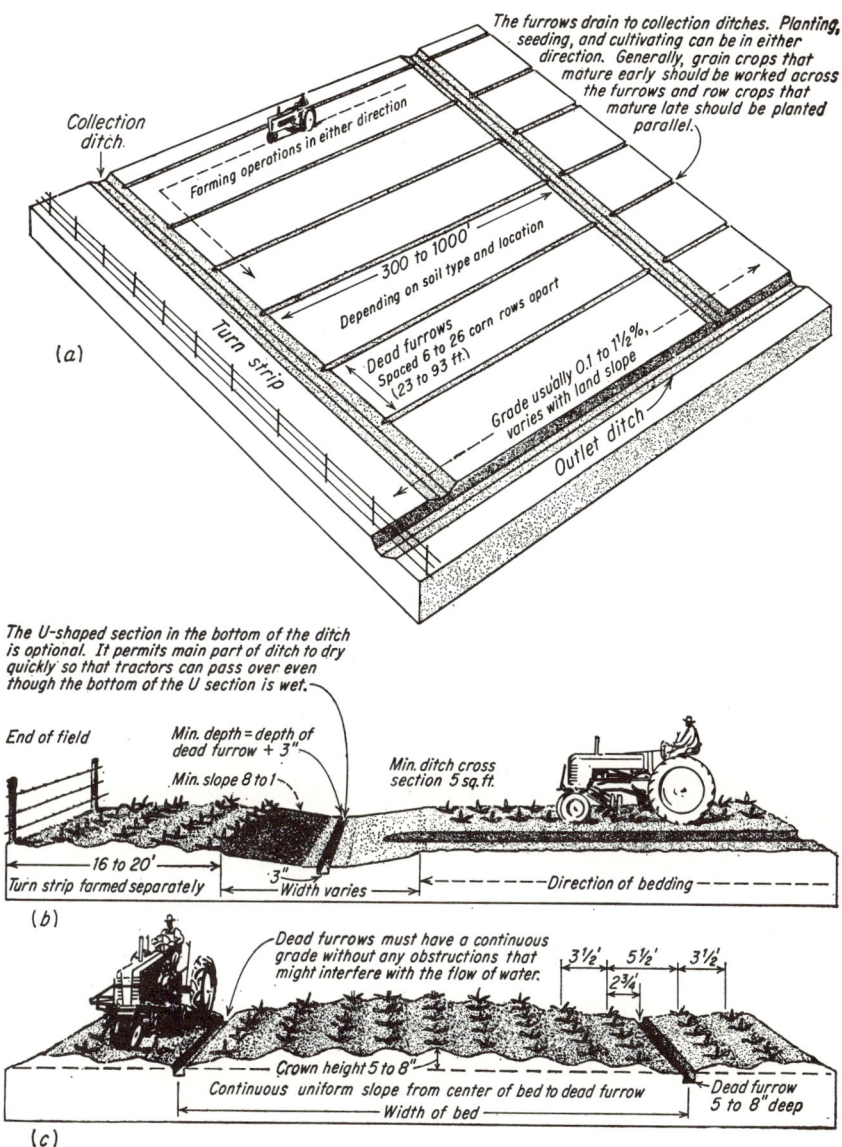

Fig. 36-15. Bedding system for surface drainage: (a) Outlet ditch should be at least 0.5 to 1.0 ft deeper than collection ditch. This will provide complete drainage of collection ditch so that it can be crossed with farm machinery. In soils subject to severe erosion the overfall should be graded back on a nonerosive grade. (b) Cross section at end of field showing collection ditch and turn strip. (c) Cross section of bed showing crown effect and proper spacing of corn rows. (*USDA SCS, Milwaukee, Wis., December,* 1951.)

For the bedding system to function properly all dead furrows must have a continuous grade with no obstructions or low points that would interfere with the complete drainage of the furrow. The collection ditches should be spaced at regular intervals across the slope of the field, depending upon the amount of slope and the permeability of the soil. Where the land is nearly flat and the permeability of the soil very slow, a spacing of about 300 ft is generally required. This can be varied up to 1,000 ft for more sloping soil of fair permeability. Where collection ditches are placed at the end of the field, the ditch should be placed about 16 to 20 ft in from the end to provide a turn strip so that drainage from the rows into the ditch will not be obstructed.

Some general recommendations on bed width and layout details are given in Fig. 36-15. The beds are developed by back-furrowing to the center of the bed. This will leave a dead furrow between beds which acts as a drain. Several plowings of the field and some filling of low places may be required before beds are adequate. To maintain beds after sufficient height is developed, a two-way plow can be used, throwing the furrows all one way, or a one-way plow can be used by laying out a land on two beds throwing dirt toward the previous dead furrow on each side of the land and away from the dead furrow between the two beds. The next year plowing is reversed and furrows plowed toward the center of the land.

The *random ditch system* is adapted to pothole-type topography having wet pockets or depressions too deep or large to fill by land smoothing. The random surface ditch should meander from one low spot or depression to another so that the water flows toward the lateral, or outlet, ditch.

To secure the highest degree of drainage the entire field should be smoothed or graded to fill all the minor depressions and thus allow the surface water to flow to the ditched-out depressions or directly to the ditch. On flat, very slowly permeable soils it may be necessary to combine this system with the bedding system to do an adequate job of surface drainage. On tilable land having large and deep depressions, this system should be used in conjunction with tile drainage. General recommendations are given in Fig. 36-16.

The *cross-slope ditch system* resembles terracing in that the drainage ditches are constructed around the slope on a uniform grade according to the lay of the land. This method of surface drainage is adapted to sloping wet fields of 4 per cent slope or less, where internal drainage is poor from the plow sole downward and where many shallow depressions hold water after rains. The surface water which collects in these low spots cannot infiltrate into such soil. The depressions are too numerous and the slope too great for successful bedding, and generally tiling is not practicable or feasible.

The cross-slope ditch differs from the regular terrace in that little or no ridge is permitted on the downslope side of the ditch. This provides for ease in crossing the ditches and reduces the damage caused from overflow. The excavated material from the ditches should be placed in the low areas between the ditches. Any excavated material not used in this operation should be spread out on the downhill side of the ditch so that the ridge is not over 3 in. above the actual ground surface. After the excavated material from the ditches has been placed in the depressions, the area between the ditches should be smoothed or graded to eliminate all minor depressions and humps which might obstruct the free flow of surface water to the ditches. The key to the success of this system of surface drainage is the elimination of the depression between ditches so that no runoff will be permitted to collect and stand on the surface of the ground. Recommendations on location, size, and spacing are given in Fig. 36-17.

The *parallel ditch system* is adapted to flat, poorly drained soils in which there are numerous small shallow depressions. In the past, bedding has often been used on this type of soil and slope. On many soils, bedding can be eliminated by smoothing or grading the land to eliminate all minor depressions and to provide uninterrupted crop-row drainage to the parallel field ditches.

The shallow field ditches should be parallel, but not necessarily equally spaced. The success of the system depends largely upon the spacing of the parallel ditches and the smoothing or grading between ditches. The land should be prepared so that every row throughout its entire length will drain to a ditch. The ditches must be spaced so that the length of row drainage from a high point to a ditch is such that the row can safely carry the runoff without much overflow or erosion damage in the row. Another factor which influences ditch spacing is the amount of earth and the distance it will have to be moved to provide complete row drainage.

The original installation of this system may be a little more costly than the bedding system; however, it provides for ease in farming operations and fits in well with the use of large farm machinery. It is particularly adaptable to large fields, since crops can be planted from one end of the field to the other over the ditches, thus providing

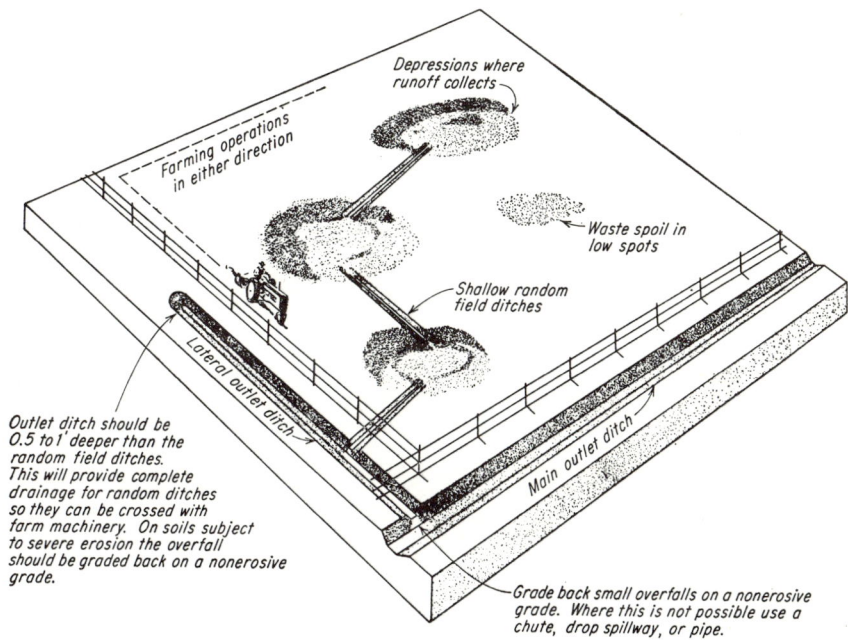

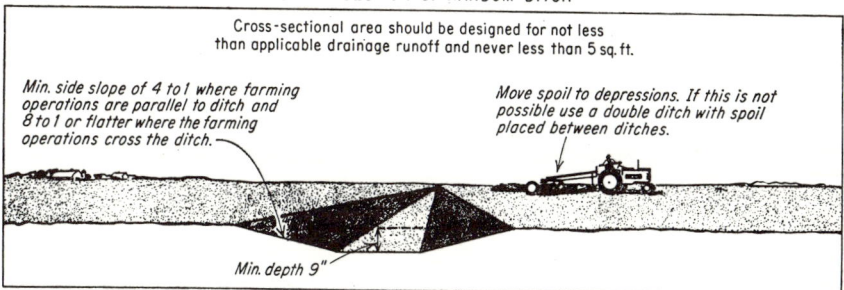

Fig. 36-16. Random ditch system for surface drainage. Remove minor depressions by land smoothing with land plane or leveler. Smooth area so that land will drain to the large depressions or random ditches. (*a*) Cross section of field. (*b*) Cross section of random ditch. (*USDA SCS, Milwaukee, Wis., December,* 1951.)

the maximum length of crop rows. Suggestions for layout and design are shown in Fig. 36-18.

In some areas parallel ditches are used for both surface-water removal and water-table control. In this case the ditches are deeper, with steeper side slopes which prohibit crossing with farm machinery. For mineral soils except sands, the ditches are generally spaced from 100 to 300 ft apart and constructed about 2.5 ft deep, with a 12- to 18-in. bottom and side slopes of 1 or 1¼:1. Such a ditch can easily be construted and maintained with an agricultural ditcher that straddles the ditch and has a dragline bucket shaped to the ditch section. In very sandy soils the ditches are spaced approximately 660 ft apart and have a minimum cross section of 4 ft deep, 4 ft bottom, and side slopes of 1:1.

Field Ditches for Surface Drainage. An important part of any surface drainage system is the field ditch that must carry away the excess surface water. The two types

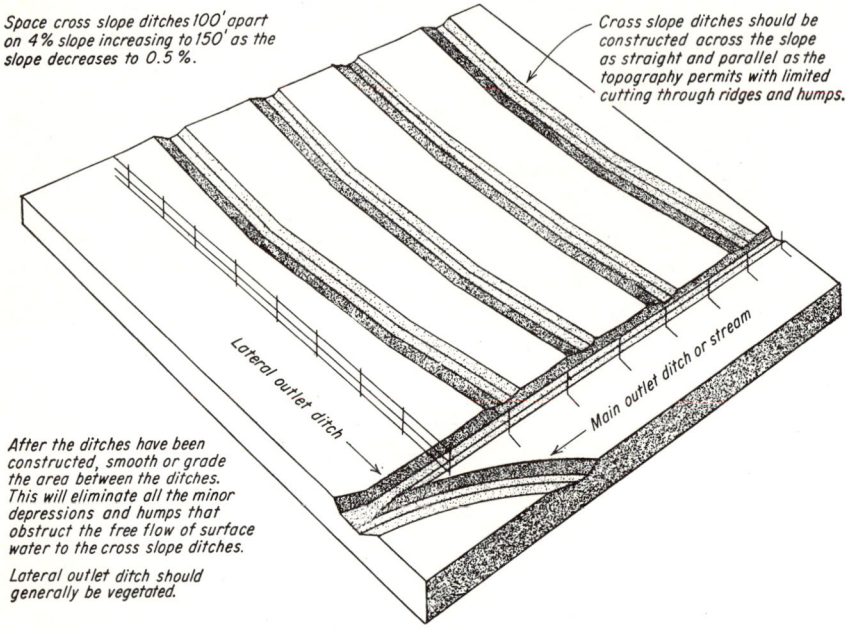

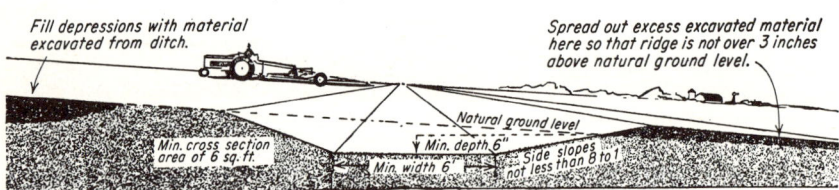

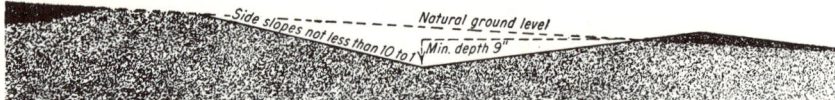

Fig. 36-17. Cross-slope ditch system for surface drainage (terrace-type drainage). (*USDA SCS, Milwaukee, Wis., December,* 1951.)

of field ditches commonly used in connection with surface drainage are the single ditch and the double ditch, the latter sometimes called W ditch, or twin ditches. Their cross section may be trapezoidal, parabolic, or flat V-shaped, depending upon location, use, and available construction equipment. For ease in farming and maintaining, the side slopes should not be steeper than 8:1 where farming operations are across the ditch. When a ditch is to be crossed only occasionally, the side slopes may be 4:1. If the ditch will not be crossed with farm machinery, side slopes of 2:1 are satisfactory. The ditches should be large enough to provide the degree of drainage required by the crops that are to be grown. Figure 36-19 gives capacities for trapezoidal ditches with 10:1 side slopes.

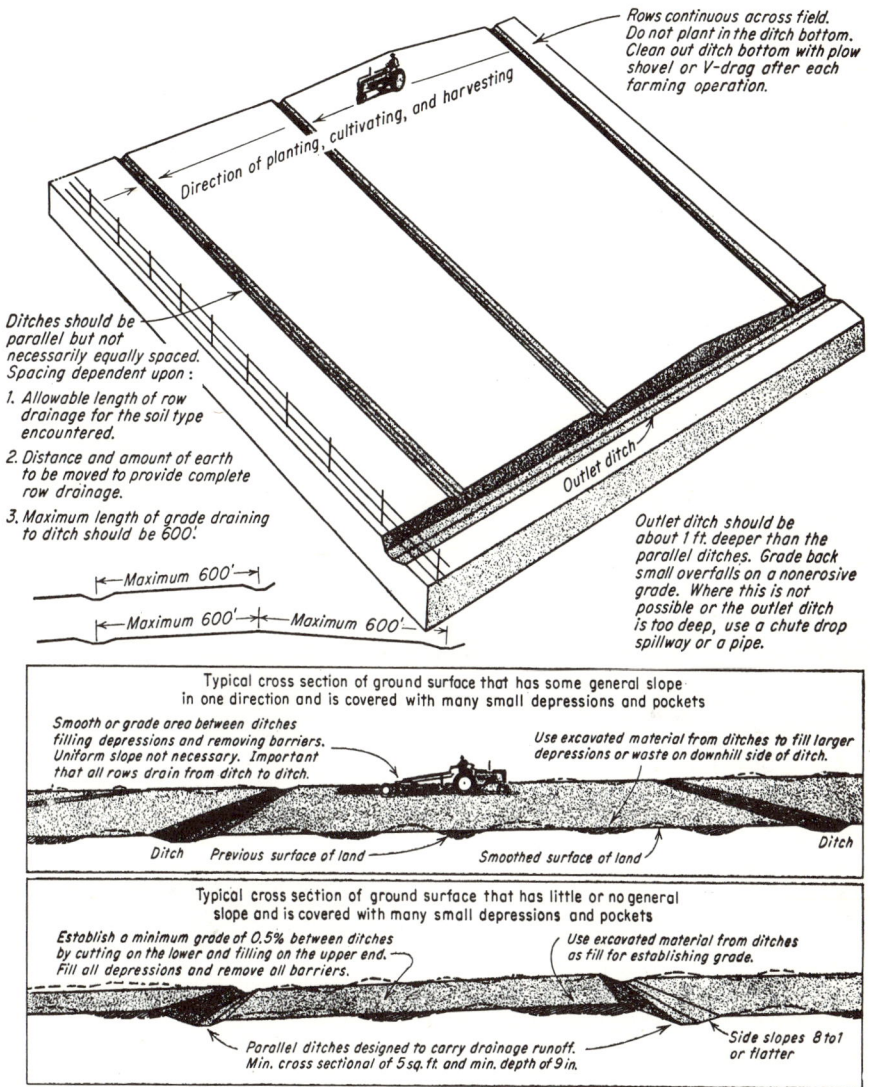

Fig. 36-18. Parallel ditch system for surface drainage. Parallel ditches to intercept and rapidly remove surface water from the field and reduce the length of row drainage. (*USDA SCS, Milwaukee, Wis., December,* 1951.)

The *single ditch* should be used when the excavated material is needed and can be placed in depressions adjacent to the ditch or where putting the soil on one side of the ditch will not interfere with or obstruct surface-water flow into the ditch.

The *double*, or *W, ditch* is actually two parallel ditches spaced a short distance apart with the excavated material placed between the ditches. In this way surface water can enter the double ditch from each side without being obstructed. It is particularly adaptable to plow construction. This ditch has application (1) in depressional areas or where the land drains toward the ditch from both sides, (2) where the land is very flat and row drainage enters from both sides, and (3) where excavated material will not be needed for filling depressions. Farming can be across the ditch or the area between the ditches planted to some other field crops, maintained as a hay strip, or used as a field road. The channels of the two ditches should be about 35 ft apart as a minimum for a channel about 9 in. deep. As the depth of channel increases, the width between ditches should be increased so that a side slope of about 8:1 or flatter can be maintained. This is extremely important where the

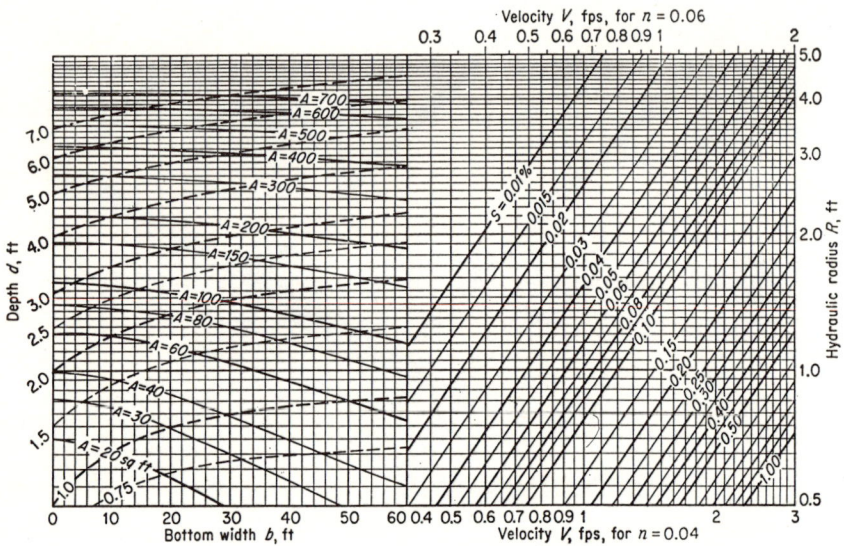

FIG. 36-19. Graphic solution of drainage-channel dimensions by Manning formula. Trapezoidal cross section with 10:1 side slopes. (*USDA SCS, Milwaukee, Wis.*, 1958.)

ditches are to be farmed across. A simple formula that can be used is $L = 50D$, where $L =$ distance between drains in feet, and $D =$ depth of drain in feet.

TILE DRAINAGE

Tile drainage systems designed to provide economical drainage for agricultural lands are considered in this section. Similar systems are used for flood protection of airports, urban areas, and special installation, but are not discussed. Most systems use clay or concrete tile. Wood box, pipe of bitumen-impregnated fiber, metal, and other materials are used in some systems.

Tile drainage is used for soils which are permeable enough for economical spacing of tile and which are fertile enough to justify the initial investment. Soils of low permeability are drained by surface drains.

Benefits of Tile Drainage. High crop yields and profitable farming ordinarily result from successful tile drainage, since the removal of free water allows good soil aeration, thus promoting soil bacterial action which releases plant food and permits deep rooting of crops. Since plants have a deeper root zone, they have more plant food available and are better able to withstand drought periods.

Tile-drained soil warms up earlier in the spring and can be cultivated sooner, extending the growing season in some northern areas. Tile drainage facilitates cultivation because tractors can operate in fields soon after rains, and soil dries uniformly, leaving no wet spots that might be left uncultivated. Seed germination is high, and stands of crops are uniform. Fertilizers and lime are utilized effectively by plants, particularly deep-rooted legumes, only in a well-drained soil. Only excess gravity water is removed through tile, leaving the full supply of capillary moisture to be utilized by plant roots.

The best drainage is frequently obtained by a combination of tile, surface drains or diversions, land grading, and proper row arrangement. Fields to be tile-drained should be carefully inspected for needed surface drainage. In many cases it is more economical to remove excess surface water from fields by open ditches rather than by installing large tile.

Surveys needed for tile drainage are generally the same as indicated at the start of this chapter. For planning the most economical design on very flat areas, however, a topographical survey may also be needed and contour lines sketched in as shown in Fig. 36-20.

Tile Trenching Equipment. Tile trench is most commonly dug with wheel-type diggers especially designed for accurate control of grade. Endless-chain diggers are

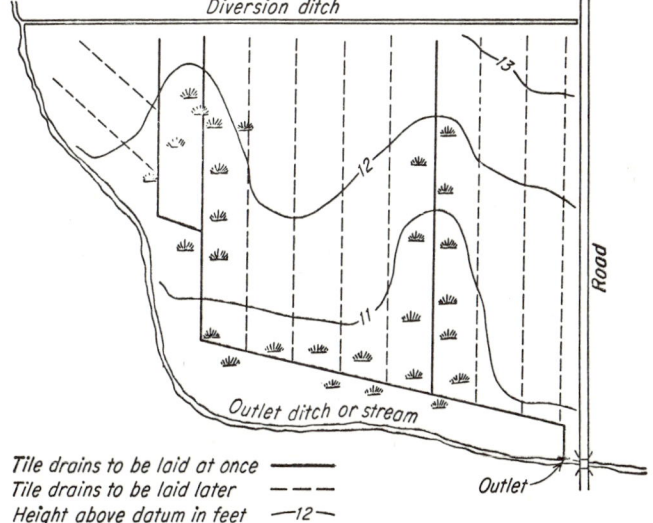

Fig. 36-20. Tile-layout plan.

also used. Average digging rates for several machines in various soil conditions in Iowa ranged from 430 ft per hr for a 3-ft depth to 200 ft per hr for a 5-ft depth.[1] Because of the many variables involved, deviations from the average were large.

Mole drainage consists of opening a drain in the subsoil by a bullet-shaped opener on a subsoiler beam. This tool leaves a slit in the soil, with a widened portion at the bottom. It is usually effective as a drain only for a year or so, since it fills in rapidly.

The effectiveness of mole drainage has been greatly increased in research trials at Cornell University by feeding a slitted plastic strip through the mole into the drain passage in such a way as to form an arched lining.[2]

A vibration-type machine designed to lay long unbroken tubes of permeable concrete has been used to line mole drains in England.[3]

Types of Tiling Systems. A tile system having parallel laterals is usually best to drain flat fields with uniform soil. A *gridiron system* such as illustrated in Fig. 36-20 with a main drain and laterals extending from one side is frequently used. Where there are differences in natural drainage, it may pay to install some tile laterals before others, as shown.

[1] G. O. Schwab, R. K. Frevert, and L. L. DeVries, Tile-trenching Machine Performance, *Agr. Eng.*, 37:469–472, July, 1956.
[2] Charles D. Busch, Low Cost Subsurface Drainage, *Agr. Eng.*, 39:92–103, February, 1958.
[3] A. N. Ede, Continuously Formed Concrete Tube for Drainage, *Agr. Eng.*, 38:864–866, December, 1957.

A *random system* of tile is used where wet spots are due to undulating topography or varying soil conditions. In this system (Fig. 36-21) the mains follow the swales and laterals branch off to wet areas.

Many areas are wet because of *upward seepage* from permeable layers of gravel, sand, or silt lying under tight layers of soil. Such soil strata containing water under hydrostatic pressure can be detected by boring with an auger or by installing piezometers or observation wells. Water from such strata rises in a well according to the pressure. Many areas of this type can be drained by a tile line laid as deeply as possible through the permeable layer. For best location of these tile drains, numerous soil borings are required to determine the source and path of the seepage water.

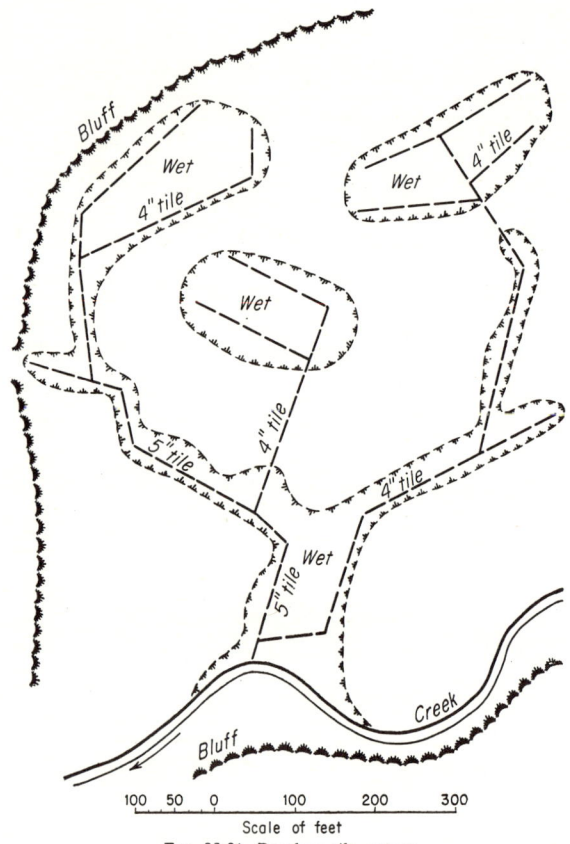

FIG. 36-21. Random tile system.

A tile drain is frequently used to *intercept seepage water* from hillsides. On many hillsides, wet areas are due to seepage water moving horizontally through permeable layers lying over an impermeable layer. This condition is indicated by seepage along a horizontal plane near the foot of the slope or at a break in grade. The impervious layer should be located by soil borings or by trenching uphill from the seepy area. The interception drain should be laid along the bottom of the permeable layer in order to intercept the seepage causing the damage.

Changes in Alignment. In installing tile lines a change in horizontal direction is made by one of the following methods:

1. A gradual curve of the tile trench on a radius that the trenching machine can dig and still maintain grade

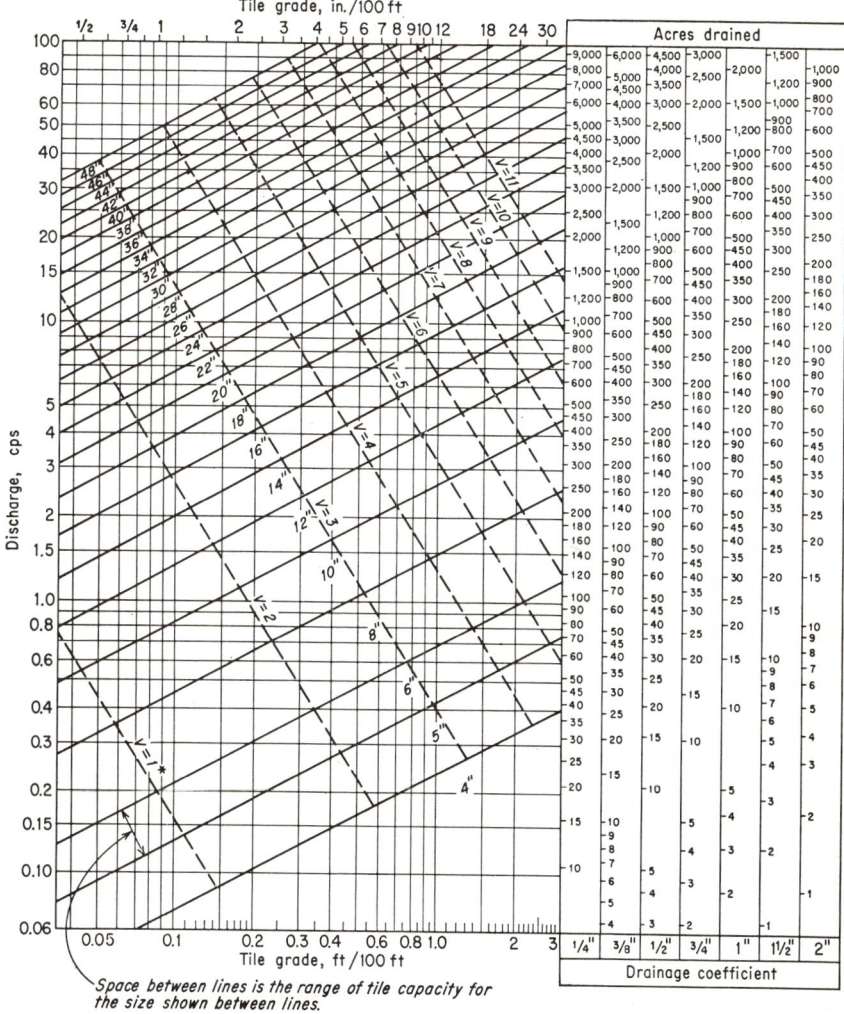

FIG. 36-22. Tile drainage chart. Acres drained by various sizes of tile. (*USDA SCS, Milwaukee, Wis.*, 1955.)

2. A curve of shorter radius made by shaping the inner side of the curve and chipping the tile, the radius of curvature in no case less than 5 ft
3. The use of manufactured bends or fittings
4. The use of junction boxes and manholes

Capacity of Tile Drains. A convenient chart for determining the required capacity and size of drain tile is shown in Fig. 36-22. The chart is based on the Yarnell-Woodward formula[1]

$$V = 138 r^{2/3} s^{1/2} \tag{36-2}$$

where V = velocity, fps
r = hydraulic radius, ft
s = slope, ft/ft

[1] D. L. Yarnell and S. M. Woodward, The Flow of Water in Drain Tile, *USDA Bull.* 854, 1920.

The chart is used by extending a horizontal line from the area drained to intersect the proper slope line. For example, a 6-in.-ID tile on a slope of 0.2 ft per 100 ft will drain 19 acres using a drainage coefficient R of $\frac{3}{8}$ in. per 24 hr.

Tile Drainage Coefficients. The rate of removal expressed as depth in inches over the watershed area in a 24-hr period is called the drainage coefficient.

In the humid areas of the United States a drainage coefficient ranging from $\frac{1}{4}$ to 1 in. has been found to give good results where surface runoff is not handled by the tile system. For best results the open drainage system should be capable of removing from 1 to 3 in. in 24 hr from the tiled field in addition to the amount removed by the tile drains. The recommended tile drainage coefficient for a locality varies with soil, crop, rainfall, topography, and condition of surface drainage as shown in Table 36-8. Experienced drainage engineers should be consulted on large and important jobs and where unusual conditions are encountered.

TABLE 36-8. DRAINAGE COEFFICIENTS, INCHES DEPTH PER 24 HR FROM WATERSHED AREA, GENERALLY RECOMMENDED BY THE SOIL CONSERVATION SERVICE

Surface drainage condition	Field crops	Truck crops
Northern climatic region including midwestern and northeastern states		
Fields having good surface drainage or a surface drainage system:		
Silt loam and clay loam soils....................................	$\frac{3}{8}$	$\frac{1}{2}$
Soils underlain by sand or deeper permeable soils.............	$\frac{1}{4}$	$\frac{3}{8}$
Peat and muck..	$\frac{3}{8}-\frac{3}{4}$	$\frac{1}{2}-1\frac{1}{2}$
Fields where tile removes surface water:		
Silt loam and clay loam soils....................................	$\frac{1}{2}-1$	$\frac{3}{4}-1\frac{1}{2}$
Peat and muck..	$\frac{3}{4}-1\frac{1}{2}$	$1\frac{1}{2}-3$
Southern climatic region including states south of the Ohio and Potomac Rivers for field crops		
Soils with surface drainage.......................................	$\frac{1}{2}$	*
Soils with limited surface runoff..................................	1 or more	*

* Truck crops follow local practice.

A 4-in. line laid on a 1 to 2 per cent grade is adequate to drain most hillside seepy spots. However, if ground waters are concentrated into a well-defined spring, a 6-in. or larger tile is needed to handle the maximum flow.

Depth and Spacing of Tile. For many soils, tile laterals of uniform depth and spacing are needed for satisfactory drainage. Inadequate drainage between laterals spaced too far apart is indicated by uneven height and yield of crops such as corn. The determination of depth and spacing is the most difficult and crucial decision in designing a complete tile system.

The best depth and spacing depend upon the permeability of soil layers through which ground water or a perched water table moves in reaching the tile. In a tiled field the ground-water elevation exists on a curved line, the lowest point being over the tile and the highest midway between tile lines. After rains the ground water at mid-points between tile should be lowered at least a foot in 24 hr to prevent crop damage.

Depth and spacing are interrelated because, in soils with uniform permeability, deep tile is effective for a greater distance than shallow tile. The trend is toward deeper tile as equipment and construction methods are improved. Besides permitting wider spacing, deep tile provides a larger water-storage capacity in the soil profile.

The determination of depth and spacing has been based chiefly on experience and on studies of ground-water levels in similar soils. The depth of outlet ditch and the location of sand and slowly permeable soil layers are also controlling factors.

Tiles should be laid below normal frost depth to avoid damage by freezing and thawing. They should be covered with not less than 24 in. of earth to prevent breakage by equipment.

In several states the state agriculture college and the Soil Conservation Service have developed joint recommendations for depth and spacing of tile for different soil types. The usual range in these recommendations is given in Table 36-9, applicable

TABLE 36-9. RECOMMENDED DEPTH AND SPACING OF DRAIN TILE

Texture of soil	Tile spacing, ft	Tile depth, ft
Clay	33–60	2½–3
Clay loam	50–80	2½–3½
Silt loam	60–120	3–4
Sandy loam	80–300	3–4
Peat and muck	50–300	3½–6

in the humid sections of the country. It should be noted that these recommendations are for drainage only. If the tile system is to be used for subirrigation, closer spacings may be required. A spacing closer than 50 ft is usually not economical.

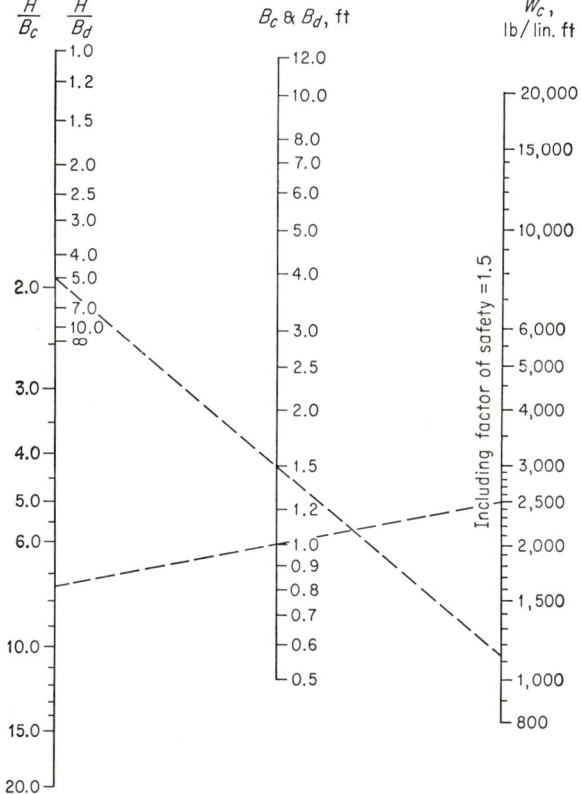

Based on $w = 120$ lb/cu ft, $K\mu'= 0.130$, $K\mu = 0.1924$, $r_{sd}p = 0.75$.

FIG. 36-23. Nomograph for loads on tile installed in thoroughly wet clay: H = depth above top of tile; B_c = outside diameter of tile; B_d = width of trench above tile; W_c = load on tile, lb/cu ft; w = weight of soil, lb/cu ft. (*Schilfgaarde, Frevert, and Schlick, Agr. Eng., July,* 1951.)

The table applies only under humid conditions. In irrigated lands where salts may rise to the soil surface, tile is often spaced from 200 to 600 ft apart and 5½ to 8 ft deep.

In recent years considerable research work has been done toward the development of a scientific basis for determining depth and spacing of drain tile. Donnan, Aronovici, and Blaney[1] reported use of a tile-spacing formula based on soil permeability and the application of Darcy's law to flow of ground water. This formula was used to compute depth and spacing of some tile systems in the Imperial Valley, Calif.

The *maximum depths for safe loading of tile* can be obtained by nomographs (Figs. 36-23 and 36-24) which were prepared from the Marston formula.[2] Both the *projecting conduit basis* (shallow bedding and covered with an embankment) and *ditch*

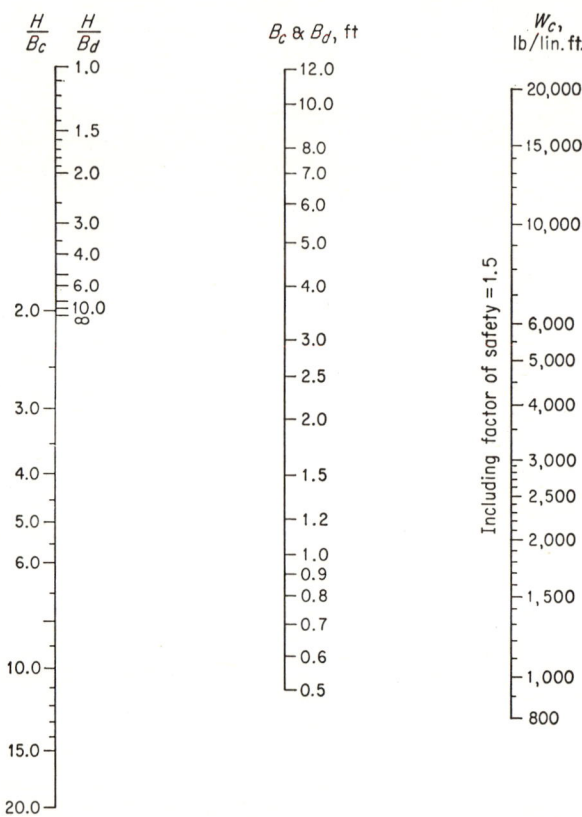

Based on $w = 120$ lb/cu ft, $K\mu^1 = 0.165$, $K\mu = 0.1924$, $r_{sd}p = 0.75$.

FIG. 36-24. Nomograph for loads on tile installed in saturated sand. (*Schilfgaarde, Frevert, and Schlick, Agr. Eng., July,* 1951.)

conduit basis (in a ditch two to three times wider than the tile) are included in the nomograph, and both conditions should be checked for each tile loading. For the projecting conduit the outside diameter of the tile is used for computation in the nomograph, and on ditch conduit the width of trench above the top of tile is used. The lower value of loading obtained from the nomograph should be used in design as shown in following examples:

Example 1. Given: 10-in.-ID tile installed in 18-in. trench 8½ ft in depth, in thoroughly wet clay.

[1] W. W. Donnan, V. S. Aronovici, and Harry F. Blaney, Report on Drainage Investigations in Irrigated Areas of Imperial Valley, California, *USDA SCS Mimeo. Rept.* 1947, p. xviii.
[2] Tentative ASAE Recommendation: Design and Construction of Tile Drains in Humid Areas, *Agr. Eng. Yearbook,* 1957.

Solution:
Load, projecting conduit basis.
Depth above top of tile, $H = 8.5 - 1.0 = 7.5$ ft.

$$\frac{H}{\text{Tile OD, } B_c} = \frac{7.5}{1}, \text{ or } 7.5$$

Then from Fig. 36-23, $W_c = 2{,}470$ lb per lineal ft.
Load, ditch conduit basis.

$$\frac{H}{\text{Trench width, } B_d} = \frac{7.5}{1.5} = 5.0$$

Then from Fig. 36-23, $W_c = 1{,}130$ lb per lineal ft.
Use lowest loading, 1,130 lb per lineal ft.

Example 2. Given: 10-in. standard drain tile, strength 1,200 lb per lineal ft, to be installed in an 18-in. trench in thoroughly wet clay.
Find: Maximum depth of trench allowable.
Solution:
Projecting conduit basis.
Read on left line, $W_c = 1{,}200$.
Center line, 1.0.

$$\frac{H}{B_c} = 3.8 \qquad H = 3.8 B_c = 3.8$$

Depth of trench $= 3.8 + 1$, or 4.8 ft.
Ditch conduit basis.
Read on left line, $W_c = 1{,}200$.
Center line, 1.5.

$$\frac{H}{B_d} = 5.5$$
$$H = 5.5 \times 1.5 = 8.2 \text{ ft}$$

Depth of trench $= 8.2 + 1.0 = 9.2$ ft.
Since this problem is the reverse of Example 1, the larger value, or 9.2 ft depth, is the maximum permitted.

The loads which tile will carry are also influenced by the type of bedding. Figure 36-25 gives load factors for various types of bedding. The nomographs were prepared based on ordinary bedding. As shown in Fig. 36-25, strength may be reduced 30 per cent by the impermissible trench or increased 30 to 100 per cent by other types of trench bottom.

Minimum Size of Tile. Tile sizes are designated by inside diameter. The wall thickness is usually about $\frac{1}{12}$ ID. A 4-in. tile is generally the smallest size recommended for clay and clay loam soils. In soils of single-grain structure containing fine sands and for peat and muck soils, 5-in. is the minimum, with 6-in. in 2-ft lengths considered good practice.

Limits of Length and Grade. Where possible, laterals should be laid in the direction of the greatest slope in the field. Mains may be on a flatter grade. When the soil becomes saturated, many tile lines flow under hydrostatic pressure. Higher velocities and discharges may result than provided in design. Jones[1] recommended that limits of grade be established in accordance with the following specifications:

Size of tile, in.	Minimum allowable grade, ft/100 ft
4	0.10
5	0.07
6 and up	0.05

[1] Lewis A. Jones, Farm Drainage, *USDA Farmers' Bull.* 2046, 1952.

These limits apply only if the capacity, based on watershed area and seepage flow, is adequate and if soil does not wash into the tile. For soils in the southeastern states containing considerable amounts of silt or sand, minimum grades to provide a velocity of 1.4 fps are recommended, 0.3 ft for 4-in. tile, and less for larger sizes. The maximum length depends on area drained as determined from Fig. 26-23, based on tile size, grade, and runoff coefficient.

The *maximum permissible grades* depend primarily on stability and resistance to erosion of the soil in which the tile is laid. For doubtful cases successful local experience is the best guide. Where tile is laid on steep grades, good construction is especially important, although the control of the grade may be less precise.

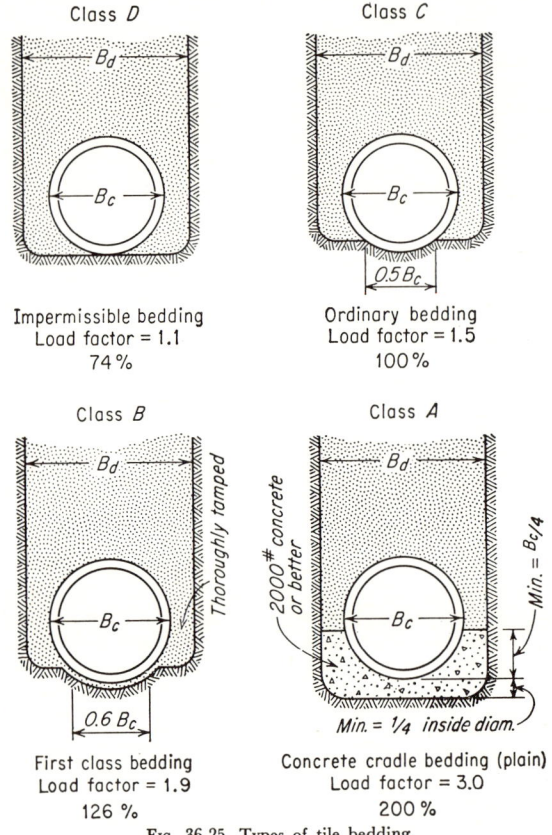

Fig. 36-25. Types of tile bedding.

In *sandy soils* which flow readily, a tight or continuous pipe or gravel filter should be used for grades over 1 per cent. For flatter grades, see *joint protection* below.

For *silts,* single-grain soils and soils containing a mixture of sand and silts which erode readily, the maximum grade for tile should be 2 per cent. Continuous pipes with small holes or sealed bell or tongue-and-groove tile with gravel filters may be used up to 4 per cent. For such soils on grades from ½ to 2 per cent, the tile should be laid to a close fit and fine-textured soil should be tamped firmly under and about the tiles.

For erosion-resistant soils including *clay,* clay loams, and silty clays, the maximum grade may be 6 per cent for main drains running near capacity and 8 per cent for laterals, which will seldom operate at more than two-thirds capacity. In New York

State, laterals carrying limited flow have been laid successfully in silty clay soils on grades up to 15 per cent. However, tile lines have on flatter grades failed where construction methods were not the best. On such soils on grades between 1 and 6 per cent the tile should be laid to a close fit and a fine-textured soil should be tamped firmly under and about the tile.

Gap between Tile. The tile should be laid so that the maximum gap between tile at bottom grade will conform as nearly as possible to the following specifications:
 1. Sandy soils, tight fit
 2. Silt and silt loams, about $\frac{1}{16}$ in.
 3. Silty clay and clay loams, about $\frac{1}{8}$ in.
 4. Clay, about $\frac{1}{8}$ in. unless local experience has shown a wider gap is desirable
 5. Peat and muck, $\frac{1}{4}$ to $\frac{3}{8}$ in.

Where the gap exceeds the amount specified above, the joint should be covered in an acceptable manner such as by broken pieces of tile. Where the ends of the joints are irregular, each tile should be turned so that a close fit is secured at the top of the joint and prescribed gaps apply at the bottom of the joint.

Protecting Tile Joints, Blinding, and Backfill. The usual ways of *protecting tile laid in sands,* silts, and sandy clay soils to prevent soil from entering tile lines include (1) obtaining tight fit of tile and wrapping the upper three-quarters of the circumference of the joint with building paper impregnated with tar, burlap, or other acceptable material; (2) encasing the tile in a French-drain type of construction with well-graded pea-sized gravel as filter material; (3) backfilling with layer of straw, hay, corncobs, or vegetal material; (4) tamping silt or clay loam topsoil around tile; (5) laying section of continuous pipe with small openings or tight pipe.

Methods of *protection for tile laid in clay,* clay loam, or silt clay soils to improve permeability around the tile include (1) blinding (initial covering) with topsoil if it contains humus material and little fine sand, (2) blinding with straw or other vegetal material if clay may seal the joints, (3) using gravel backfill. Tile laid in peat or muck soil should be blinded with straw or vegetal material if the peat or muck has a tendency to seal joints.

Blinding in clay, clay loam, and silty clay soils is usually done by shaving off topsoil or sod from the sides of the trench. This holds the tile in place during backfilling. Backfilling may be completed by moving the spoil into the trench with a farm tractor.

To simplify construction, a continuous strip of tar-impregnated paper may be laid along the top of the tile back of the tile-trenching machine instead of wrapping each joint. A gravel filter may be laid from two hoppers attached to the tile-trenching machine. One can be used to feed gravel into the trench below the tile, and the other to cover the tile after it has been laid.

Tile Types and Specifications. Manufactured connections or branches for joining two tile lines should be used where available. If such connections are not available, the junction should be chipped and fitted and the connection sealed with mortar. Laterals should be connected to the main tile line at the approximate mid-point of the main tile.

Tile used in drainage work should meet the latest ASTM specifications. Many states now have an organized testing procedure and maintain a list of tile manufacturers who meet specifications. If soil sulphates or acids are present, clay tile or extra-quality steam-cured, concrete tile made from resistant cements should be used.[1]

Staking the Tile and Checking Grade. The location of tile mains should be marked with grade stakes at 100-ft intervals or at more frequent intervals if required. Grade stakes should be set at a convenient distance from the center line so that they will not be disturbed during construction. The station number should be marked on a suitable guard stake, which serves to locate the grade stake.

The *line-and-gage method* is ordinarily used for hand-dug trenches. Batter boards are set at the gage distance (usually about 6 ft) above the bottom grade of the trench. A taut line is strung along the batter boards. The bottom grade is then dug to grade by measuring with the gage stick from the line.

[1] Philip Manson, Durability of Concrete and Mortars in Acid Soils with Particular Reference to Drain Tile, *Univ. Minn. Agr. Expt. Sta. Tech. Bull.* 180, 1948; Longtime Tests of Concrete and Mortars Exposed to Sulphate Waters, *Univ. Minn. Agr. Expt. Sta. Tech. Bull.* 194, 1951.

Targets are set up above the grade stakes for tile trenchers. Targets are set so that the proposed trench bottom is parallel to the target grade. The trenching-machine operator sights over the sight bar attached to the machine, attempting to keep his sight bar in line with the top of the targets so that the trench bottom has the same grade as the targets.

Permissible Tolerance in Laying. Checking and inspection should be done before backfilling is complete. A *practical working tolerance* for agricultural tile, attainable with the average tile trenching machine, is a deviation of 0.1 ft, which may be expected from true grade, but not over 0.03 ft of this should occur in any 50-ft station. A suitable penalty for minor deviations or rejection of the work for major deviations should be provided on contract jobs.

Outlets. Outlet protection is required at the ends of tile lines where they empty into open ditches or natural channels. For most installations a watertight pipe with 8 to 20 ft embedded in the bank is satisfactory. The length embedded and the desirability of using antiseep collars around the pipe to prevent washouts depend upon the soil and flow conditions. Corrugated pipe is widely used for outlet protection. Sewer

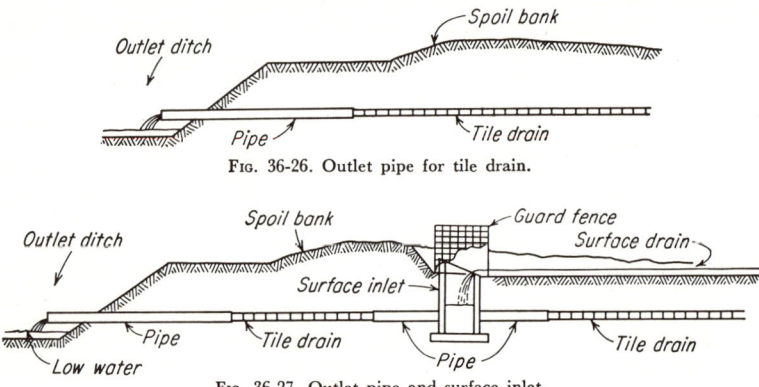

FIG. 36-26. Outlet pipe for tile drain.

FIG. 36-27. Outlet pipe and surface inlet.

and tile pipe with sealed joints and other permanent materials are also used for tile-outlet protection. Sewer or tile should be cradled on a slab of concrete or other permanent material to prevent washouts.

An *overhang* of pipe is desirable to prevent serious bank erosion (Figs. 36-26 and 36-27). In most cases there is little erosion, because a water cushion exists in the ditch where the tile is flowing. In unstable soils it often may be advisable to protect pipe outlets with riprap. Where ice may cause damage, the pipe should be recessed into the ditch bank. A swinging gate or screen should be used on a tile outlet to exclude rodents and small animals.

Drop-spillway structures, often called *headwalls,* are recommended where a large quantity of surface water discharges at the tile outlet. A surface inlet (Fig. 36-27) may be used to drop small quantities of surface water into the tile pipe near the tile outlet.

Auxiliary Structures. *Open surface inlets* should be used to drain depressed areas only where it is not feasible to use blind inlets or to construct surface ditches to remove runoff. Open surface inlets permit sediment and debris to enter the tile system unless they are well constructed and maintained. A simple type consists of sewer-pipe joints from the ground surface and connected into the tile drain with a T connection. A sediment trap should be provided downstream in the line where sediment is a problem. This may consist of a well of larger-diameter tile through which the tile water flows and from which silt may be removed. The surface inlet is often offset from the tile line. A beehive or other grate should be used to prevent the entrance of trash.

Blind inlets are especially useful and are constructed by backfilling a section of the trench with gravel, stone, corncobs, or other permeable material. The upper foot of the trench may be backfilled with topsoil so that cultivation can be carried on over the tile. Some inlets of this kind are effective for a long time, but others made of corncobs or large stone may fill rapidly in soils containing silts and fine sand. An effective filter can be designed by mixing graded sands and gravel.

Breathers, or *air vents,* consist of pipes of small diameter extending from the tile to the ground surface to admit or discharge air. These are valuable at the beginning of steep sections to admit air into the lines, resulting in a more steady hydraulic flow. Air vents relieve hydraulic pressures at the end of steep sections or where several branches join. Some engineers use them at ¼- to ½-mile intervals along tile mains for inspection holes and to admit air into the main, but other engineers prefer not to have such openings in tile lines unless required.

Junction boxes or wells include manholes or wells of large-size tile to connect three or more lines where there is a drop in grade. They may be designed to collect and remove sediment. They are often capped and covered with 18 in. or more of soil to avoid interference with cultivation.

Tree Removal. All lowland-type trees, such as willows, elm, and cottonwood, should be removed from an area 100 ft each side of a tile line. Other trees should be removed for a distance of 50 ft. An exception is where tile is used to drain orchards.

DRAINAGE-PUMPING PLANTS

In recent years, numerous electrically powered small farm drainage-pumping plants have been installed to drain fields which are too low to be drained economically by ditches. There are a number of such plants in the lake plains bordering the Great Lakes in Ohio and Michigan. Small pumping plants are also installed at tile outlets where the outlet ditches are not deep enough.

For these small pumping plants in the north central part of the United States the pump capacity[1] can be determined from the following information:

Tile system for subsurface drainage only, 7 gpm per acre (⅜ in. per 24 hr)
Surface drainage by ditch or tile field crops, 10 gpm per acre (½ in. per 24 hr)
Surface drainage by ditch or tile truck crops, 15 gpm per acre (¾ in. per 24 hr)

In a number of pumping-plant installations of this type it was found that, with automatic operation, the electric motor should not start and stop more than five times per hour. If a gasoline engine is used, the pump capacity should be increased 20 per cent and the sump should be designed so that the engine will only need to be started twice a day. This type of installation should be provided with an automatic shutoff when the sump is pumped out.

For automatic operation a sump can be used on areas up to 100 acres. On larger areas an open ditch should be provided. To provide the five cycles per hour, using the pump capacity recommended above, the capacity of the sump in cubic feet can be determined by dividing computed pump capacity in gallons per minute by 2.5, and the area of sump in square feet is determined by dividing this figure by the depth of storage. A circular sump is recommended since it does not need extensive reinforcing.

For nonautomatic operation, to provide the suggested pumping cycles it will be necessary to multiply pump capacity in gallons per minute by 25 to obtain the cubic feet of storage required in the sump.

For the detailed design and operation of large drainage-pumping plants, it is advisable to refer to U.S. Department of Agriculture publications on the subject.[2] Some of the material in this section has been abstracted from *USDA Technical Bulletin* 1008.

---

[1] Curtis Larson, "Pumping Outlets for Farm Drainage," University of Minnesota, Agricultural Engineering Dept., 1950.

[2] Carl Rohwer, Putting Down and Developing Wells for Irrigation, *USDA Circ.* 546, 1941. John G. Sutton, Cost of Pumping for Drainage in the Upper Mississippi Valley, *USDA Tech. Bull.* 327, 1932; Design and Operation of Drainage Pumping Plants, *USDA Tech. Bull.* 1008, 1950. Ivan D. Wood, Pumping for Irrigation, *USDA* SCS-TP-87, 1950.

Determination of Maximum Runoff to Be Pumped. The pumping capacity should provide for surface runoff and the seepage which exists in a large pumping district. The maximum pumping-plant capacity is usually only a fraction of the maximum runoff from the land.

These relationships may be expressed as follows:

Maximum plant capacity = pumping district coefficient
(gravity drainage runoff + seepage runoff)

For the Upper Mississippi Valley conditions the following empirical formula was found applicable:

$$C = K_1(G + K_2 r) \tag{36-3}$$

where C = plant capacity at maximum lift runoff per 24 hr, in.
K_1 = pumping district coefficient; for small areas it should approach unity, but for areas of 15,000 acres it was found to equal about 0.33
G = applicable runoff coefficient used in ditch design
K_2 = coefficient to convert annual runoff to seepage
r = av. annual runoff, which was found to vary as follows:
  5–12 in./year for districts having considerable gravity drainage
  13–16 in./year for nonseeped districts pumping all runoff
  16–33 in./year for heavily seeped districts

A summary of major-pumping-plant-capacity recommendations for the United States is shown in Table 36-10.

TABLE 36-10. RECOMMENDATIONS FOR PUMPING-PLANT CAPACITIES

Area	Drainage area	Runoff coefficient, in./24 hr for pumping-plant design
Upper Mississippi Valley................	5,000 acres or more	0.33–0.55
For field crops.....................	Small farm areas	¾–1¼
For truck crops....................	Small farm areas	1½–2½
Coastal marsh areas, Texas and Louisiana...	All areas	0.93–2.55*
Texas:		
Sugar cane.......................	Under 3,000 acres	3.0†
Rice or pasture...................	Under 3,000 acres	2.0†
Florida Everglades (1927):		
By A. Marston....................	16 sq miles	1.0‡
By S. H. McCrary.................	43 sq miles	¾‡
By George B. Hills................	322 sq miles	½‡
Florida Everglades, SCS, for truck crops....	1 sq mile	3.0
	2–3 sq miles	2.0
	4–9 sq miles	1.4
	10–16 sq miles	1.0

* C. W. Okey, The Wet Lands of Southern Louisiana and Their Drainage, *USDA Bull.* 652, 1918.

† Represents both storage and pumping required. T. C. Anderson and R. B. Moore, unpublished report, USDA SCS, 1947.

‡ Formula $Q = 69.1/M + 9.6$, where Q = runoff, cfs/sq mile, and M = drainage area, sq miles. Report of the Engineering Board of Review to Board of Commissioners of the Everglade Drainage District.

Static Lift. The static lift is the difference in elevation between the water levels at the point of discharge and the suction bays. The stages of the discharge and suction bays should be studied to determine the maximum, minimum, and average static lifts. The maximum static lift for design conditions must be accurately determined in order to obtain a pump which can operate satisfactorily.

Total Dynamic Head. The total dynamic head on the pump is equal to the total energy in the water at the discharge flange minus the total energy at the suction flange of the pump. It is expressed by the formula

$$H_t = \left(H_d + \frac{V_d^2}{2g} + d_1\right) - \left(H_s + \frac{V_s^2}{2g} + d_2\right) \tag{36-4}$$

where H_t = net total head, ft of water
  H_d = discharge pressure head, ft of water, measured near discharge flange of pump (positive if pipe is under pressure and negative if under vacuum, at point of measurement)
  V_d = average velocity, fps, in pipe at point where H_d is measured
  d_1 = elevation of gage measuring H_d, ft above some reference plane (positive if gage is above reference plane, negative if below)
  H_s = suction pressure head, ft, measured near the suction flange of pump (nearly always negative, except for submerged pumps, since suction pipe is usually under vacuum)
  V_s = average velocity, fps, in pipe at point where H_s is measured
  d_2 = elevation of gage measuring H_s, ft above same reference plane from which d_1 is measured
  g = acceleration due to gravity, approximately 32.16 fps/sec

The expressions $V_d^2/2g$ and $V_s^2/2g$ are the velocity heads in the discharge and suction pipes, respectively. The total head is equal to the static lift plus losses in suction and discharge pipes.

Pump efficiency is computed by the following formula:

$$E = \frac{H_t \text{ gpm}}{3{,}960 \text{ bhp}} \qquad (36\text{-}5)$$

where E = pump efficiency, per cent
  gpm = gallons per minute
  H_t = total head on pump, ft
  bhp = brake-horsepower input into pump shaft

Selection and Operation of Pumps. The type of pump, size and number, operating speeds, arrangement of units, required efficiency, and power used are determined on

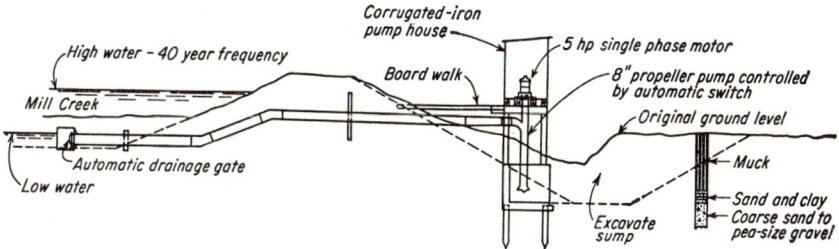

Fig. 36-28. Typical pumping-plant layout.

the basis of the pumping requirements of the individual plant. The pump manufacturer will usually assist the user in selection of pumps which generally operate at 70 to 80 per cent efficiency. The pumping requirements must be anticipated with precision if a satisfactory and economical pumping unit is to be obtained. Centrifugal pumps are most popular where the maximum total head exceeds 18 ft. The mixed-flow pump is adapted to medium heads. Tests on one unit gave efficiencies above 70 per cent at all heads from 7 to 26 ft. The axial-flow, or propeller, pump is highly efficient at heads below 10 ft. This type of pump, also called the screw pump, is widely used for vertical submerged pumps. Its capacity decreases rapidly if the operating head exceeds the designed head, and it is seldom used where the maximum head exceeds 12 ft. All the above references are to total dynamic head on the pump, and not to the static lift.

In designing *small plants* which drain up to a few hundred acres, emphasis is on simplicity and low first cost rather than high operating efficiency. Small low-lift plants usually consist of a submerged vertical-type propeller pump driven at constant speed and designed for the maximum lift (Fig. 36-28). These arrangements are well adapted to automatic operation if electric power is available because special priming equipment is not needed and a float switch or electrodes can control the pump operation.

For *large plants* a high operating efficiency is desirable. Because drainage plants usually operate at a wide range of static lifts, a change in pump speed is often required to secure best efficiency. This may be provided by changing pulley size where the pump is belt-driven by an electric motor. Two synchronous motors mounted on the same shaft but operating at different speeds provide a simple way of changing pump speeds. Diesel engines can be slowed down enough so that pumps can operate efficiently at low heads. One pump adapted for efficient pumping against low heads is sometimes provided in multipump plants. In large plants, two or three smaller pumps rather than one large one allow the plant to operate even when one unit is down for repairs or maintenance.

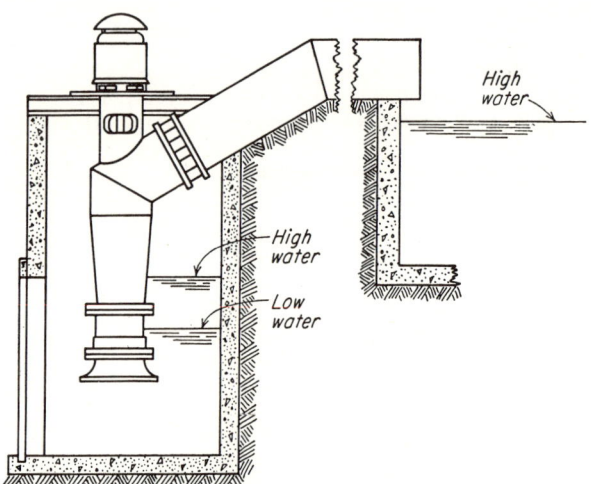

Fig. 36-29. Typical pump installation.

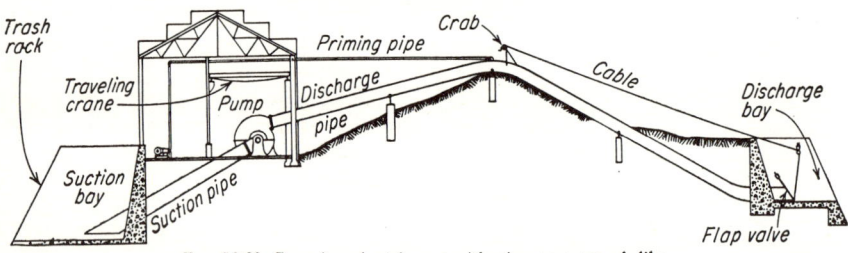

Fig. 36-30. Pumping-plant layout with pipe over top of dike.

As a general rule, pumps in most drainage plants have a water velocity of 8 to 10 fps through the pump against the maximum lift. Lower velocities through the pumping unit result in higher operating efficiencies because the hydraulic losses vary as the square of the velocity. However, using a larger pump to handle the same quantity of water adds to the initial investment.

If the pumps are not submerged, priming equipment is needed. Wet-vacuum and dry-vacuum pumps are used for priming. Pipes should be arranged so water cannot enter a dry-vacuum pump.

Suction and Discharge Pipes. If the water is discharged at the high point in the discharge line, the static lift is the difference in elevation between the high point and the suction bay as shown in Fig. 36-29. If the outlet pipe is discharged under water with no air in it, advantage is taken of the siphon action, and the static lift is the difference in elevation between the discharge and the suction bays as shown in Fig. 36-30.

The head loss in suction and discharge pipes, based on Scobey's formula, may be estimated from Fig. 36-31. The velocity head and the discharges of pipes of various sizes are given in this figure. Expansion of the suction pipe is recommended to reduce the entrance loss. For good results, the entrance velocity should be about 3 fps. The suction pipe should be cut so that the lower end is a horizontal plane. The discharge pipe is usually expanded from 2 to 6 in. at the pump discharge to reduce friction losses. An automatic flap gate on a submerged discharge pipe is needed to prevent backflow through the pump.

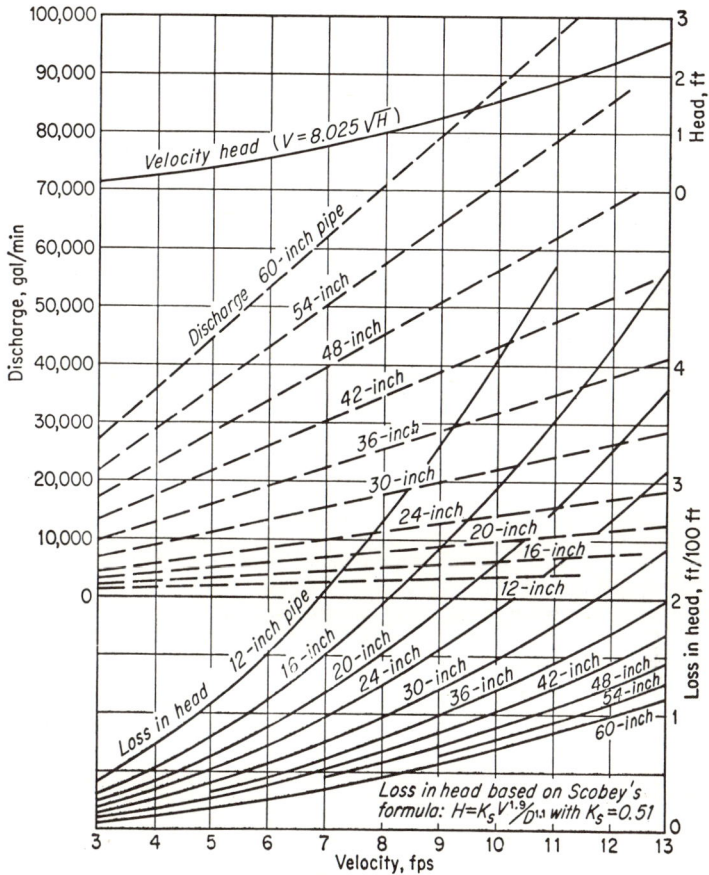

Fig. 36-31. Head losses in pumping layouts.

Type of Power. Electric power is used in most drainage plants, because of the simplicity of operation and the low power rates which may often be obtained. Electric plants are easily adapted to automatic operation. In some locations electric power is not available at a competitive rate and diesel engines are more economical. Factors affecting the cost of pumping for drainage are discussed in *USDA Technical Bulletin* 327.[1]

Tests of drainage pumping plants should be made by a qualified engineer before acceptance of new equipment and are discussed in *USDA Technical Bulletin* 1008.[2]

[1] Sutton, Cost of Pumping for Drainage in the Upper Mississippi Valley.
[2] Sutton, Design and Operation of Drainage Pumping Plants.

A good *foundation* for large and medium-sized units is usually secured by driving piling. A reinforced-concrete suction bay supported on piling is usually needed for the larger plants in order to prevent undermining. Screens should be provided to prevent trash from entering the pumps. A walkway should be constructed to permit easy cleaning of the screens.

A *gravity sluiceway* may be located near a plant to allow drainage by gravity when the discharge bay is at low stage. However, if such a sluiceway is through a levee, it should be located several hundred feet away from the plant to avoid the danger of the plant being washed out in the event of a levee failure at the sluiceway.

DRAINAGE OF IRRIGATED LANDS

Drainage systems for irrigated lands need to be planned to take care of (1) storm runoff, (2) excess irrigation water, and (3) subsurface water. Storm runoff is ordinarily handled through natural channels. It is usually essential to prevent such waters from flowing into irrigation systems. The disposal of excess irrigation water should be planned along with the land-leveling and the irrigation system.

After an irrigation system operates for some time, seepage from canals, from higher irrigated areas, and from the application of excess irrigation water creates subsurface drainage problems. The seepage travels through permeable strata and results in a high water table, often accompanied by a rise of saline and alkali salts to the ground surface.

Ordinarily a subsurface drainage system is not planned until high-water-table conditions and the rise in alkali salts become severe enough to encourage remedial action. Thorough investigations are required to determine the source of the damaging water, its path, the proper location of drains, and feasibility of leaching out salts. The methods of making the investigations are described by Donnan and Bradshaw.[1]

Such investigations include deep soil borings to determine the texture and permeability of the soil and to locate any stratum through which the damaging water is moving. Piezometers are used to measure hydrostatic pressures and locate strata through which ground water is moving. The recording of ground-water levels in shallow wells during an irrigation season is desirable to show the rise and fall of the water table. Such wells may be set on a grid system at intervals of $\frac{1}{4}$ to 1 mile to determine ground-water contours during the irrigation season. Drains are located in so far as possible in permeable strata to intercept the ground-water flow and to tap strata under the greatest hydrostatic pressure and along the lines where the ground-water table is nearest the ground surface.

Large areas of irrigated lands have been abandoned because the soils have become saline. The practicability of restoring such lands to productivity depends upon the feasibility of drainage, soil characteristics and permeability, and the character of the salts. Hilgard, an early California investigator, classified the salty soils as "white alkali" and "black alkali"; now they are more generally called saline and alkali, respectively.

Saline ("white alkali") soils are often distinguished by a white crust on the surface in which sodium predominates and is absorbed on the surface of clay particles. This results in a poor soil structure, and such soils "puddle" readily, take water slowly, and are difficult and expensive to drain, treat, leach, and reclaim. *Alkali soils* can be improved by the addition of gypsum, which causes a base exchange of calcium and sodium salts, and the soluble sodium salts may then be leached. The reclamation of alkali land is expensive and should be attempted only under supervision of experienced technicians.

Systems for subsurface drainage of irrigated lands are generally an integral part of reclamation or improvement of lands damaged by high-water-table or saline or alkali conditions. Such systems include open drains, tile drains, and pumps. Tile drains should be at least $4\frac{1}{2}$ ft deep, and open drains at least 5 ft. They are often 10 to 12 ft deep and sometimes deeper. They are usually quite effective if laid in gravel near the high side of a field to intercept ground flow from a hill, leaking canal,

---

[1] W. W. Donnan and G. B. Bradshaw, Drainage Investigation Methods for Irrigated Areas in Western United States, *USDA Tech. Bull.* 1065, 1952.

or reservoir. Such drains are called *interceptors,* and drains on the low side, parallel to ground-water flow and laid out in a gridiron system, are usually called *relief* drains.

Drainage wells with pumps have been used with success in the Salt River Valley, Ariz., near Fresno, Calif., and in other locations. Such wells produce a drawdown and lower the ground water for distances up to half a mile. They are most likely to prove feasible when they can pump from permeable gravels which lie too deep to drain by gravity and are more likely to be economical if the water pumped from the wells is of suitable quality for irrigation and has value for such use.

DRAINAGE LAWS

Among the first drainage statutes enacted in many states were those which provided relief for any landowner who needed to construct an outlet ditch across the land of another owner. These statutes made it possible to obtain right of way after an investigation and appraisal by a county official or neutral appraiser. The knowledge that such laws are in force frequently encourages landowners to reach a compromise without legal action.

There are two distinct rules relating to surface waters. One rule is known as the civil-law rule, and the other one is known as the common-law, or common-enemy, rule.

Civil-law Rule. According to Etcheverry[1] the civil-law rule is followed in 13 states: Alabama, California, Georgia, Illinois, Iowa, Kentucky, Louisiana, Maryland, Michigan, North Carolina, Ohio, Pennsylvania, and Texas. This rule provides that the owner of the higher land, known as the dominant owner, is entitled to the natural advantage which the elevation of his land gives him, and he has the right to have his surface waters flow naturally from his land upon the land of the lower owner. The owner of the lower land (servient owner) does not have the right to interfere with the natural flow of these surface waters. In most of the states which follow the civil-law rule, the rule has been modified to permit the dominant owner to construct ditches and drains on his own land so long as they empty into natural channels or depressions.

Common-law Rule. Etcheverry states that the English common-law, or common-enemy, rule is followed in 18 states: Arkansas, Connecticut, Indiana, Kansas, Maine, Massachusetts, Minnesota, Missouri, Nebraska, New Hampshire, New Jersey, New Mexico, New York, Oklahoma, South Carolina, Virginia, Washington, and Wisconsin. Under this rule the lower landowner may lawfully protect his land from surface waters flowing on it from higher land, and such waters are considered a "common enemy." In states following this rule the owner of higher land is responsible for damages caused by increases in the volume or velocity of the flow due to the construction of artificial channels on his land. This rule has been modified in some states to permit construction of drains on higher lands provided no appreciable injury is done. However, under this rule the lower owner has the right to obstruct the flow of surface waters or divert them from his land without incurring a liability to the upper owner for such obstruction or diversion. This rule does not give him the right to obstruct natural channels, which are covered by a different rule. When constructing drains the upper owner may protect himself against damages by securing an easement from the lower owner.

In states which have adopted the common-law rule, drainage districts may assess higher lands for outlet benefits by providing such lands with a lawful outlet. In some cases all lands in a watershed are assessed.

ORGANIZED DRAINAGE ENTERPRISES

Nearly all states have enacted statutes covering the organization, financing, construction, and maintenance of drainage works including drainage districts which have their own governing bodies and county drains which are administered by the governing body of the county. The organization of a drainage enterprise makes it possible for the majority of the landowners of an area in need of drainage to construct the required drains, to assess the lands benefited, and to maintain and recon-

---

[1] B. A. Etcheverry, "Land Drainage and Flood Protection," McGraw-Hill Book Company, Inc., New York, 1931, p. 284.

struct the drainage improvements. Such enterprises have permitted the drainage and improvement of large areas of wet lands in this country.

State laws covering the procedures for organizing drainage enterprises vary greatly between states. It is nearly always essential to employ an attorney to carry out the procedures because all provisions of the state laws must be met. Otherwise the acts of the enterprise or the bonds issued may be held invalid. The engineer engaged in this work should be familiar with the provisions of the state laws where he is employed. An abstract of state laws covering drainage enterprises is given in the Drainage Census.[1]

The laws relating to maintenance, rehabilitation, and operation of drainage improvements vary greatly among states. The importance of adequate provisions for maintenance is evident because at least one-fourth of our cultivated lands are within some kind of drainage enterprise and their productivity depends upon maintenance or improvement of their outlet drains. Many drainage enterprises are well maintained, but others receive little or no maintenance. The adequacy of the provisions of the state laws relating to maintenance and operation has an important bearing on the quality of maintenance even though it is recognized that the kind of management, technical supervision, and economic circumstances are likewise of great importance.

[1] Seventeenth Census of the United States, 1950, Drainage of Agricultural Lands, U.S. Department of Commerce, Bureau of the Census.

Chapter 37

MECHANICS OF WATER EROSION

PAUL JACOBSON

Early investigators in erosion control considered that water erosion was caused by soil moving downslope with flowing surface water, but in the last few years, research men have concluded that raindrop splash is of major importance as a contributor to erosion. Borst and Woodburn, working with mulch supported 1 in. above the soil, came to the conclusion that the reduction of raindrop impact rather than a reduction in overland flow of water is the main contribution of the mulch in reducing erosion.[1]

Raindrop splash affects soil in a number of ways. One result is to gradually remove the fine material from the soil and leave the less fertile sands and gravels. Ellison noted that after a splash test the particles left in the splash dishes consisted of the coarser material. In checking, it was found that the soil samples had changed from 25.3 per cent sand and gravel at the start of the test to 31.2 per cent sand and gravel at the end. A check of the surface flow showed that it consisted of 95 per cent silt and clay and only 5 per cent sand and gravel. This demonstrates the grading effect of the erosion caused by raindrops on the size of particles.

Another effect of splash erosion is to detach soil and break down soil aggregation. Since finer particles are more easily transported, erosion is thus accelerated. Ellison's experiments show that on a 10 per cent slope slightly more than 70 per cent of the soil splashed is moved downhill. This means that raindrop splash is causing a gradual downward movement on a hillside.

Raindrop splash tends to seal the surface through puddling of the soil. This causes higher surface runoff. The higher runoff, in turn, is able to transport more material. Thus a hard, beating rain brings about a vicious cycle, multiplying erosion in many ways.

The major effect in surface flow of water is to carry off soil loosened by splash erosion. As the depth and concentration of this overland flow increases, we go to microchannels and then to small rivulets or gullies. When this point is reached, the flowing water is not only transporting soil but is actually picking up added soil from these small gullies. A good portion of the soil which is picked up in this manner is moved along the bottom of the stream as bed load.

Thus, in the final analysis, water erosion is a product of both splash erosion and erosion caused by water flowing over the surface. Zingg[2] developed a formula for the sum of this erosion. This formula is

$$X = CS^m L^n$$

where X = total soil loss, weight units
C = a constant of variation
S = degree of land slope, per cent
L = horizontal length of land slope, ft
m = exponent of degree of land slope
n = exponent of horizontal length of land slope

[1] W. D. Ellison, Studies of Raindrop Erosion, *Agr. Eng.*, 25:131–136, April, 1944; 25:181–182, May, 1944.

[2] A. W. Zingg, Degree and Length of Land Slope as It Affects Soil Loss in Runoff, *Agr. Eng.*, 21:59–64, February, 1940; An Analysis of Degree and Length of Slope Data as Applied to Terracing, *Agr. Eng.*, 21:99–101, March, 1940.

TABLE 37-1. BASE TABLE OF ESTIMATED SOIL LOSS, TONS/(ACRE)(YEAR)
[Using a 3-year rotation of R-O-M on 9 per cent slope, 72.6 ft in length as a base = 8 tons/(acre)(year). Soil factor 1.0; $L^{0.5}$ and S values from $a + bS + cS^2$ relationship.]

Slope length, ft	Per cent slope										
	2%	4%	6%	8%	10%	12%	14%	16%	18%	20%	24%	30%

Up- and downhill

Slope length, ft	2%	4%	6%	8%	10%	12%	14%	16%	18%	20%	24%	30%
40	1.1	2.1	3.4	5.0	6.9	9.2	11.7	14.5	17.7	21.2	29.0	43.1
60	1.3	2.5	4.1	6.1	8.5	11.2	14.3	17.8	21.7	25.9	35.6	52.8
80	1.5	2.9	4.8	7.1	9.8	12.9	16.5	20.6	25.0	29.9	41.1	61.0
100	1.7	3.3	5.3	7.9	10.9	14.5	18.5	23.0	28.0	33.5	45.9	68.1
150	2.1	4.0	6.5	9.8	13.4	17.7	22.6	28.2	34.3	41.0	56.2	83.6
200	2.4	4.6	7.6	11.2	15.5	20.5	26.1	32.5	39.6	47.3	64.9	96.5
250	2.7	5.2	8.4	12.5	17.3	22.9	29.2	36.4	44.3	52.9	72.6	107.9
300	2.9	5.7	9.3	13.7	18.9	25.1	32.0	39.8	48.5	58.0	79.5	118.2
350	3.2	6.1	10.0	14.8	20.5	27.1	34.6	43.0	52.4	62.6	85.9	127.6
400	3.4	6.6	10.7	15.8	21.9	28.9	37.0	46.0	56.0	67.0	91.8	136.4
450	3.6	7.0	11.3	16.8	23.2	30.7	39.2	48.8	59.4	71.0	97.4	144.7
500	3.8	7.3	11.9	17.7	24.5	32.4	41.3	51.4	62.6	74.9	102.6	152.5

Contouring

Slope length, ft	2%	4%	6%	8%	10%	12%	14%	16%	18%	20%	24%	30%
40	0.7	1.1	1.7	3.0	4.1	5.5	9.4	11.6	14.2	19.1	26.1	43.1
60	0.8	1.3	2.1	3.7	5.1	6.7	11.4	14.2	17.4	23.3	32.0	52.8
80	0.9	1.5	2.4	4.3	5.9	7.7	13.2	16.5	20.0	26.9	37.0	61.0
100	1.0	1.7	2.7	4.7	6.5	8.7	14.8	18.4	22.4	30.2	41.3	68.1
150	1.3	2.0	3.3	5.8	8.0	10.6	18.1	22.6	27.4	36.9	50.6	83.6
200	1.4	2.3	3.8	6.7	9.3	12.3	20.9	26.0	31.7	42.6	58.4	96.5
250	1.6	2.6	4.2	7.5	10.4	13.7	23.4	29.1	35.4	47.6	65.3	107.9
300	1.7	2.9	4.7	8.2	11.3	15.1	25.6	31.8	38.8	52.2	71.6	118.2
350	1.9	3.1	5.0	8.9	12.3	16.3	27.7	34.4	41.9	56.3	77.3	127.6
400	2.0	3.3	5.4	9.5	13.1	17.3	29.6	36.8	44.8	60.3	82.6	136.4

Strip cropping

Slope length, ft	2%	4%	6%	8%	10%	12%	14%	16%	18%	20%	24%	30%
80	0.5	0.7	1.2	2.1	2.9	3.9	5.0	6.2	7.5	9.0	12.3	18.3
100	0.5	0.8	1.3	2.4	3.3	4.4	5.6	6.9	8.4	10.1	13.8	20.4
150	0.6	1.0	1.6	2.9	4.0	5.3	6.8	8.5	10.3	12.3	16.9	25.0
200	0.7	1.2	1.9	3.4	4.7	6.2	7.8	9.8	11.8	14.2	19.5	29.0
250	0.8	1.3	2.1	3.8	5.2	6.9	8.8	10.9	13.3	15.9	21.8	32.0
300	0.9	1.4	2.3	4.1	5.7	7.5	9.6	11.9	14.6	17.4	23.9	35.0
350	1.0	1.5	2.5	4.4	6.2	8.1	10.4	12.9	15.7	18.8	25.8	38.0
400	1.0	1.7	2.7	4.7	6.6	8.7	11.1	13.8	16.8	20.1	27.5	41.0

Terracing (usually not practical on slopes above 14%)

	2%	4%	6%	8%	10%	12%	14%	16%	18%	20%	24%	30%
All slope lengths...	0.6	0.8	1.2	2.0	2.7	3.5	5.9	7.3	8.8	11.7	15.8	

$$\frac{\text{Value from base tables (above)}}{\text{Rotation index (Table 37-2)}} = \text{estimated soil loss, tons/acre}$$

SOURCE: "Ready Reference Tables," USDA SCS, rev. 1959.

TABLE 37-2. ROTATION INDEXES FOR SOILS WITH ERODIBILITY AS SHOWN

Rotation	Soil-erodibility factor											
	0.50		0.75		1.00		1.25		1.50		1.75	
	Rotation index											
	Without mgmt.	With mgmt.	Without mgmt.	With mgmt.	Without mgmt.	With mgmt.	Without mgmt.	With mgmt.	Without mgmt.	With mgmt.	Without mgmt.	With mgmt.
R-R-R-O$_x$	0.62	0.74	0.41	0.49	0.31	0.37	0.25	0.30	0.20	0.25	0.18	0.21
R-R-O$_x$	0.70	0.82	0.46	0.55	0.35	0.41	0.28	0.33	0.23	0.27	0.20	0.23
R-O$_x$	0.90	0.98	0.60	0.65	0.45	0.49	0.36	0.34	0.30	0.33	0.26	0.28
R-R-R-O-M	0.92	1.12	0.61	0.75	0.46	0.56	0.37	0.45	0.30	0.37	0.27	0.32
R-O-O-W	1.12	1.14	0.75	0.76	0.56	0.57	0.45	0.46	0.37	0.38	0.32	0.35
R-R-O-O-M	1.16	1.40	0.77	0.93	0.58	0.70	0.46	0.56	0.38	0.47	0.34	0.40
R-R-O-M	1.22	1.46	0.81	0.97	0.61	0.73	0.49	0.58	0.40	0.49	0.35	0.42
R-R-O-W-M	1.28	1.48	0.86	0.99	0.64	0.74	0.52	0.59	0.43	0.49	0.37	0.42
R-R-O-M-M	1.74	2.06	1.15	1.34	0.87	1.03	0.70	0.82	0.58	0.69	0.50	0.50
R-O-W-M	2.02	2.16	1.34	1.44	1.01	1.08	0.81	0.86	0.67	0.72	0.58	0.62
W-O-O-O-M	2.04	2.04	1.36	1.36	1.02	1.02	0.82	0.82	0.67	0.67	0.58	0.58
R-O-M	2.00	2.24	1.33	1.39	1.00	1.12	0.80	0.90	0.66	0.73	0.57	0.64
R-R-O-M-M-M	2.08	2.48	1.38	1.65	1.04	1.24	0.83	0.99	0.69	0.86	0.59	0.71
R-W-M	2.30	2.30	1.50	1.50	1.15	1.15	0.92	0.92	0.75	0.75	0.65	0.65
R-O-M-M	3.08	3.44	2.04	2.29	1.54	1.72	1.23	1.38	1.02	1.15	0.88	0.98
O-O-M	3.12	3.12	2.07	2.07	1.56	1.56	1.25	1.25	1.03	1.03	0.90	0.90
R-O-M-M-M	3.84	4.26	2.55	2.85	1.92	2.13	1.54	1.70	1.27	1.42	1.09	1.22
R-W-M-M-M	4.34	4.34	2.88	2.88	2.17	2.17	1.74	1.74	1.43	1.43	1.24	1.24
R-O-M-M-M-M	4.54	5.12	3.03	3.41	2.27	2.27	1.81	2.05	1.51	1.71	1.30	1.46
W-O-M-M	5.72	5.72	3.80	3.80	2.86	2.86	2.29	2.29	1.89	1.89	1.63	1.63
W-O-M-M-M	7.68	7.68	5.11	5.11	3.84	3.84	3.07	3.07	2.53	2.53	1.99	1.99
W-M	8.32	8.32	5.55	5.55	4.16	4.16	3.33	3.33	2.77	2.77	2.38	2.38
O-M-M-M	16.00	16.00	10.64	10.64	8.00	8.00	6.40	6.40	5.28	5.28	4.56	4.56
W-M-M-M	21.04	21.04	13.99	13.99	10.52	10.52	8.42	8.42	6.94	6.94	5.60	5.60

Legend: R—row crop; O—spring grain; W—winter grain; x—catch crop; M—meadow.
"With management (mgmt.)" includes cornstalks carried over to spring and spring-plowed for row crop following row crop and cornstalks carried over to spring and disked for oats following row crop.
SOURCE: "Ready Reference Tables," USDA SCS, rev. 1959.

In a series of experiments at Bethany, Mo., on Shelby soil, cultivated to a 3-in. depth, he found this formula to be

$$X = 0.026 S^{1.37} L^{1.60}$$

where X = soil loss for runoff plot, lb

The same year, Smith[1] expanded this material and modified the equation to

$$Y = CS^{1.4} L^{0.6}$$

where Y = soil loss, tons/acre

Smith also developed some factors for the reduction of soil loss due to contouring, strip cropping, and terracing. Later Browning[2] further refined the factors for soil-conservation practices and introduced a factor for the different soil types. He set his material up as a base table for Marshall soil. At a meeting held at Purdue University in the summer of 1956, it was agreed that the exponent of 1.4 for the S value still

[1] Dwight D. Smith, "Interpretation of Soil Conservation Data for Field Use," unpublished report, USDA SCS, 1946.
[2] G. M. Browning, "A Method for Estimating Soil Management Requirements," unpublished paper, Iowa State University Agricultural Experiment Station, 1947.

generally holds, but the analysis of the present data by Wischmeier indicated that a parabola-type equation fits the data better than other equations that have been used or suggested. Based on this, the expression $0.520 + 0.363S + 0.52S^2$ is substituted for S in the basic equation for computing soil loss. At the meeting it was also agreed, for simplicity, to use an exponent of 0.5 for L rather than 0.6. This tended to liberal-

TABLE 37-3. RELATIVE SOIL LOSS FROM CROPS UNDER VARIOUS CROPPING SYSTEMS IN PERCENTAGE OF LOSS FOR CONTINUOUS ROW CROPS*

Crop	Relative soil loss
Continuous row crop	100
Row crop after 1 year meadow	40
Row crop after 2 or more years meadow	35
Row crop after small grain	90
Row crop after row crop after small grain	100
Row crop after row crop (or grain) after 1 year meadow	80
Row crop after row crop (or grain) after 2 or more years meadow	70
Third-year row crop after (1 year) meadow	95
Fourth-year (or more) row crop after (1 year) meadow	100
Spring grain after 1 or 2 years row crop (or grain) after catch crop	35
Spring grain after row crop (or grain) after 1 year meadow	30
Spring grain after row crop (or grain) after 2 or more years meadow	25
Spring grain after second-year row crop (or grain) after 1 year meadow	35
Spring grain after second-year row crop (or grain) after 2 or more years meadow	30
Spring grain after 3 or more years row crop (or grain)	40
Spring grain after 1 year meadow	15
Spring grain after 2 or more years meadow	10
First-year meadow	1.0
Second-year and succeeding meadow	0.5
Lespedeza hay	4.0

Soil loss from winter grain is the same as that from spring grain in Kentucky, southern Illinois, Indiana, and Missouri; elsewhere it is 0.7 that of spring grain.

* Fertility treatments sufficient to produce good yields based on experimental results and farm planning recommendations are assumed in these values.

The relative soil loss value for each crop is multiplied by the factor for the management practice used on it, such as:

	Factor
Residues (2 tons/acre or more) left on surface throughout year	0.50
Residues (moderately grazed) left on surface throughout year	0.75
Residues (2 tons/acre or more) left on surface until planting time	0.80
Winter cover crops plowed at planting time	0.80
Mulch (manure 6 tons or more) applied immediately after planting	0.60
Plow planting—a minimum of cultivation and packing	0.60
Use of field cultivator on meadow instead of plowing	0.50

To determine rotation indexes:

$$\frac{\text{All crop losses} \times \text{mgmt. factors}}{\text{Years of rotation}} \div \frac{\text{crop losses (R-O-M)}}{3} = \text{rotation factor (ratio to R-O-M)}$$

$$\frac{1}{\text{Rotation factor} \times \text{soil-erodibility factor (Table 37-4)}} = \text{rotation index (Table 37-2)}$$

SOURCE: "Ready Reference Tables," USDA SCS, rev. 1959.

ize the results slightly on the steeper slopes. The basic equation was then used by the Soil Conservation Service to prepare tables for field use in the Corn Belt area. Technicians in other parts of the country are working on similar developments for estimating soil loss, but to date they are only in the tentative state. The revised material for soil loss is shown in Table 37-1. This table is considered as a base table for erosion control and can be adapted for other humid areas. Tables 37-2 and 37-3 give factors which adapt it to other soils and crop rotations. The soil-erodibility factor, a measure

TABLE 37-4. SOIL-ERODIBILITY FACTORS AND PERMISSIBLE SOIL LOSS FOR THE CORN BELT STATES

Soil type	Soil-erodibility factor	Permissible soil loss, tons/(acre)(yr)	Soil type	Soil-erodibility factor	Permissible soil loss, tons/(acre)(yr)	Soil type	Soil-erodibility factor	Permissible soil loss, tons/(acre)(yr)
Alford	1.25	4	Flanagan	1	4	Nicholson	1.5	3
Almena	1.5	3	Flandreau	1	3	Nicollett	1	5
Ayrshire	1.5	3	Fox	1.25	3	Norden	1.25	3
Barnes	1	4	Fox (gl)	1.25	3	Norris	1.25	3
Baxter	1	3	Frederick	1.25	3	Oaktown	0.75	4
Beasley	1.5	3	Freer	1.5	3	Octagon	1	3
Bedford	1.25	3	Froberg	1.5	3	Onaway	1.25	3
Bedford (Ky.)	1.5	3	Gale	1.25	3	Ontonagon	1.25	3
Bertrand	1.25	4	Gibson	1.5	3	Oshkosh	1.25	3
Bewleyville	1.25	3	Gogebic	1.25	3	Oshtemo	1	4
Blount	1.5	3	Grenada	1.5	3	Ostrander	1	4
Blue Lake	0.75	4	Grundy	1.25	4	Oswego	1	3
Bohemian	1.5	4	Grundy (Mo.)	1.25	3	Ottawa	0.75	4
Boone (Minn., Wis.)	0.75	4	Hagerstown	1.25	4	Otterholt	1.25	4
Boyer	1	3	Hampshire	1.5	3	Otway	1.75	1
Bratton	1.25	3	Harrison	1.25	4	Parr	1	3
Calamus	1.25	4	Hartsells	1.5	3	Parsons	1	3
Canfield	1.25	3	Hayden	1.25	3	Pembroke	1.25	4
Captina	1.5	3	Hersey	1.5	3	Plainfield	0.75	4
Carrington (heavy till)	1.25	4	Hiawatha	0.75	4	Princeton	1.25	4
			Hillsdale	1.5	3	Putnam	1	3
Carrington (reg.)	1	5	Hiwood	0.75	4	Rockton	1	3
Casco	1	2	Hosmer	1.75	3	Russell	1.25	3
Celina	1.25	3	Hubbard	0.75	4	Russellville	1.25	4
Christian	1.25	3	Ida	1.5	5	Sango	1.5	3
Cincinnati	1.5	3	Isabella	1.25	3	Santiago	1.25	3
Clarence	1.5	2	Kalamazoo	1.25	3	Saybrook	1	4
Clarion	1	5	Kalkaska	1	4	Seaton	1.25	3
Clarksville, cherty	0.75	2	Keamah	1.25	4	Seymour	1.25	3
Clary	1.5	3	Kennan	1.25	3	Sharpsburg	1	4
Clinton	1.25	4	Kent	1.5	2	Shelby	1	4
Cloquet	1.5	3	Kewaunee	1.25	3	Shelbyville	1.25	4
Coggen	1.25	3	Keweenaw	0.75	4	Sidell	1	4
Coloma	0.75	4	Knox	1.25	4	Sisson	1.25	3
Cookeville	1.25	3	Lakeville	1	3	Sparta	0.75	4
Craig	1.25	3	Lebanon	1.5	3	Spinks	0.75	4
Crider	1.25	4	Lester	1	4	St. Clair	1.75	3
Crosby	1.25	3	Lindley	1.25	3	Superior	1.5	2
Culleoka	0.75	3	Loring	1.75	4	Sverdrup	0.75	4
Curran	1.5	3	Lowell	1.25	3	Tama	1	5
Cushing	1.25	3	Mahaska	1.25	4	Tama (Ill.)	1	4
Dakota	0.75	4	Mahoning	1.75	3	Taylor	1.25	3
Dane	1.25	3	Manitou (Ky.)	1.75	3	Tilsit	1.5	3
Decatur	1.25	4	Marathon	1.25	3	Tracy	1	3
Derinda (sp)	1.25	2	Marenisco	0.75	4	Trenary	1.25	3
Dexter	1.5	4	Marquette	0.75	2	Union	1.25	3
Dickinson	0.75	4	Marshall	1	5	Vigo	1.5	3
Dixon	1.5	3	Maury	1.25	4	Vilas	0.75	5
Dodgeville (dp)	1	4	Medary	1.5	3	Wakefield	1.25	3
Dodgeville (sp)	1	3	Memphis	1.5	4	Warsaw	1.25	3
Downs	1	4	Menfro	1.25	4	Watton	1.5	3
Dubuque (dp)	1.25	4	Menominee	1	4	Waukon	1	4
Dubuque (sp)	1.25	3	Mercer	1.5	3	Weldon	1.25	3
Eden	1.5	5	Metes	1	4	Weller	1.25	3
Eldon	1.25	3	Mexico	1	3	Wellston	1.25	3
Elk	1.25	4	Miami	1.25	3	Wellston (Ohio)	1	4
Elkinsville	1.25	3	Milaca	1.25	3	Westmoreland (Ky.)	1.25	3
Elliott	1.25	4	Monona	1.25	5			
Ellsworth	1.5	3	Monongahela	1.5	3	Westmoreland (Ohio)	1	4
Emmet	1.25	3	Moody	1	5	Wheeling	1.25	3
Estherville	1	2	Morley	1.5	3	Winfield	1.25	3
Fairmount (Ky. and Ind.)	1.25	3	Mountview	1.5	3	Withee	1.5	3
			Muscatine	1	5	Wooster	1	3
Fairmount (Ohio)	1	4	Muskingum (Ky.)	1	3	Worthen	1	5
Fawcett	1.5	2	Muskingum (Ohio)	1	4	Zanesville	1.5	3
Fayette	1.25	4	Nebish	1.25	4	Zimmerman	0.75	4
			Nester	1.25	3			

NOTE: dp = deep phase
sp = shallow phase
gl = glacial phase

SOURCE: "Ready Reference Tables," USDA SCS, rev. 1959.

of the relative susceptibility to erosion is given at the top of Table 37-2. Marshall and comparable soils have a factor of 1. The factor for sandy soils is generally less than 1 and is greater than 1 for heavy soils. Factors for a few soils in the Midwest are given in Table 37-4. The estimated annual soil loss can be determined from the soil-erodibility factor, the crop rotation index, and the per cent and length of slope by using the following formula:

Estimated soil loss, tons/(acre)(year)
$$= \frac{\text{(base soil loss from Table 37-1)(soil-erodibility factor)}}{\text{rotation index}}$$

Rotation index can be taken from Table 37-2 for ordinary management or calculated from the data in Table 37-3 for various rotation and conservation practices. Thus, for R-R-O-M (row crop–row crop–oats–meadow) rotation, 2 per cent slope, 200 ft long, on a soil with erodibility factor of 1.00, from Table 37-2 the rotation index = 0.61; from Table 37-1 for up- and downhill, soil loss = 2.4 tons/(acre)(year).

$$\text{Estimated soil loss} = \frac{2.4(1)}{0.61} = 4 \text{ tons/(acre)(year)}$$

For shallow soils, authorities generally agree that 3 tons/(acre)(year) is about all that should be allowed. On deep soils, annual loss of 4 or 5 tons per acre is considered tolerable. This presents another way in which the base tables may be used. Select from Table 37-1 the soil loss based on length and per cent of slope. This is then divided by the allowable soil loss from Table 37-4, and the figure obtained can be used in selecting the crop rotation as shown in the following example.

An 8 per cent slope, 200 ft long, on a soil with erodibility factor of 1.00 has an allowable soil loss of 4 tons/(acre)(year). Table 37-1 shows the base loss for contouring as 6.7 tons/(acre)(year). To reduce this to 4 tons/(acre)(year), a rotation index of 1.67 or more is required. From Table 37-2, R-O-M-M-M without management or R-O-M-M with management should be satisfactory (management is defined in the table.

In the northeastern part of the United States, the Soil Conservation Service has used a similar system with a slightly different formula.[1] The formula is

$$X = CS^{1.35}L^{0.35}P^{1.75}$$

where P = maximum rainfall in 30 min to be expected in 2 years, in.

This formula includes the addition of a factor for local rainfall intensity. Values in Table 37-1 can be corrected for the northeastern area by multiplying by a slope factor from Table 37-5 and then by $P^{1.75}$.

TABLE 37-5. CONVERSION TABLE FOR CONVERTING BASE TABLE 37-1 TO NORTHEASTERN UNITED STATES CONDITIONS

Slope length, ft....	40	60	80	100	150	200	250	300	350	400	450	500
Factor............	0.58	0.54	0.52	0.50	0.47	0.45	0.44	0.43	0.42	0.41	0.40	0.39

Gully Erosion. As the water from surface flow accumulates into channels, gully erosion sometimes occurs. Generally a steep gradient or an overfall is present in gully erosion. The control of this overfall or steep gradient requires the development of grassed waterways and/or gully-control structures. The need for these treatments and some points on location are covered in Chaps. 40 and 41.

[1] Gail W. Eley and Charles H. Lloyd, "Soil Loss in SCS Region 1," unpublished material, USDA, SCS, 1948.

Chapter 38

TERRACES AND DIVERSIONS

PAUL JACOBSON

TERRACES

Effect of Terraces in Controlling Erosion. According to the accepted formula described in Chap. 37, Mechanics of Water Erosion, soil loss by erosion is proportional to the length of slope to the 0.5 power. Doubling the length of slope increases the erosion 1.4 times. By shortening the length of slope, terraces contribute greatly to reducing soil losses. They are effective throughout the entire year, and since 70 per cent of the earth disturbed by splash erosion is moved downhill, terraces produce a barrier to partially stop this downhill movement of soil. A great part of the splashed soil is caught in the terrace channel. This deposition in terrace channels can be largely offset by always plowing uphill with a two-way plow. Baird[1] found that erosion between terraces amounted to 0.794 cu ft for each foot of terrace length. A 16-in. two-way plow plowing 7 in. deep will move 0.778 cu ft per foot of length. If this method of plowing is used, erosion between terraces can largely be offset by this farming operation.

Planning the Terrace System. Terraces are a very important part of a water-disposal system. Planning a water-disposal system[2] requires the consideration of (1) land use, (2) outlets, (3) terrace location, (4) terrace direction, (5) farm road location, (6) fence location or field boundary, and (7) crop rotation.

It is important to consider *land use* in terrace layouts. Ordinarily, unless needed for gully control, terraces do not prove economical on grassland. Also, if the topography is too irregular to permit satisfactory contouring, a long crop rotation with the land in grass most of the time should be used for erosion control rather than terraces.

In the South, and to a lesser extent in other parts of the United States, a number of small steep areas commonly occur in a field. These areas should first be outlined and planned for permanent vegetation before terraces are planned on a field.

Outlets are the next important feature that must be considered in planning a terrace system. On some highly pervious soils, terraces are made level and outlets are not needed, but in most areas outlets are needed. In the past, many constructed terrace outlets were planned for terrace systems. They are still advocated in some areas to eliminate waterways which must be crossed. *Constructed outlets* located at the edge of the field were generally used in lieu of natural waterways. For the most part, this practice now meets with disfavor among engineers. A *waterway* is much less of a hindrance to farm operation now that tractor equipment has hydraulic controls. Therefore the present tendency is to use every suitable waterway for an outlet. This has a number of advantages, such as: (1) Terrace lengths are greatly reduced, thus permitting smaller terraces. (2) Terraces are straightened, permitting easier contour farming. (3) In many systems some parts or all of the terraces can be made

---

[1] Ralph W. Baird, Requirements of Farm Machinery for Terraced Land, *Agr. Eng.*, 16:97–102, March, 1935.
[2] L. D. Worley, Planning Terrace Systems to Facilitate Farm Machinery Operation, *Trans. ASAE*, pp. 68–70, 1959.

parallel, thus eliminating point rows. (4) Individual row drainage is provided since each row drains to the waterway. (5) Each outlet has to carry only small amounts of water, with the result that fewer outlet difficulties will be experienced.

No matter what type of outlet is finally selected, its location will have a major effect on the final layout of the terrace system. Each outlet should have adequate capacity, as shown in the tables in Chap. 40, Waterways for Erosion Control.

After it has been decided that a field should be terraced, the location of the outlets having been determined, the next step is to *locate the terraces*. In some fields there may be an area farther upslope which cannot be terraced because the slope is too steep or because the land belongs to an owner who does not wish to terrace. In these cases, a diversion of adequate capacity (see below) having a suitable outlet or outlets must be provided at the upper edge of the area that is to be terraced.

In early terrace layouts, the procedure was to locate the top terrace from the top of the hill by use of the interval table and to space the terraces downslope using the intervals shown in the table for the average slope of the land. In modern, well-planned systems, more care is taken and a *key terrace* is selected. Factors affecting the selection of a key terrace are (1) slope, (2) row drainage, (3) saddles (low portions on the ridge) which will affect over-all terrace systems, and (4) possibilities of keeping the terraces parallel. Quite often the key terrace providing the best alignment will be one or two intervals below the top of the ridge. Experience in terrace layout is the only sure guide for proper selection of this terrace.

Terrace direction is important in terrace layouts. If at all practical, terraces should drain from the ridges to the waterways. With terraces laid out in this manner, since rows are parallel to the terrace, each row will drain from the ridge to the waterway. This reduces to a minimum the silting caused by water breaking over and flowing across rows to the terrace channels. Thus benching due to silting of channels will be held to a minimum since there will be fewer breakovers under these conditions.

Farm road layouts are very important in terrace planning. Farmers must have access to the field without driving over the terraces. Where terraces drain from the ridges to the waterways, the divide on the ridge can be left open and easy access into the field is provided. Another common road location is on the contour with the terraces. Either the road can be between terraces, or the ridge of a terrace can be widened out and used for a road.

The *location of field boundaries* or fences is an important consideration in terrace layout, and it is often beneficial to change field boundaries to facilitate farming operations in connection with terraces. Field boundaries may be along a waterway or ridge or parallel with the terraces. The approximate location of the field boundaries to provide needed field units should be determined before the terraces are laid out.

Rotation of crops has a distinct effect on soil loss, as can be determined from the tables in Chap. 37, Mechanics of Water Erosion. Therefore, if it is known that terraced land will be farmed heavily with row crops, terrace spacings should probably be reduced somewhat. On the other hand, if the farm is to be used largely for grass, spacings can be increased slightly. Since the farmer may wish later to change his cropping program, caution should be exercised in making deviations because of the present cropping system.

Selecting the Proper Type of Terrace. *Bench* terraces are used in densely populated areas of foreign countries which are intensively cultivated. Two kinds of terraces commonly known as (1) the ridge type and (2) the channel type have been developed in the United States.

The *ridge-type* terrace is largely used where water holding is the principal function. This type of terrace is largely limited to the Great Plains area and sandy areas of coastal plains of the Southeast.[1] They are also used on some soils that have an extremely high infiltration rate, such as the loess soils of western Iowa and some southern states. Since the spreading of water over the maximum area is of prime concern, the ridge is the important item and the channel is only incidental to producing the ridge. As can be noted in Fig. 38-1, this terrace is usually built from both sides.

The *channel-type* terrace is mostly used on soils where controlled removal of water

[1] C. L. Hamilton, Terracing for Soil and Water Conservation, *USDA Farmers' Bull.* 1789, 1938.

is of prime importance. It is largely built from one side, and the important feature is to dig a channel for carrying the water. The ridge resulting from this type of construction is low and offers the least hindrance to farming operations.

A number of methods for building this type of terrace have been proposed during the last few years. The use of wheel-scraper type of equipment has been advocated by Zeasman and Wojta[1] of Wisconsin. The method is to move the dirt out of the channel to small depressions in the field. The resulting terrace consists mainly of a channel. This method has considerable merit when consideration is given to the long-time use that will be made of well-planned terrace systems. Although the first cost may be high, the ease of operation and the possibility of making the terraces parallel by this method may more than offset the first cost.

Terrace Spacing. Terrace spacing in the northern states can be greater than in the southern states, since the ground surface is frozen during the winter and is not subject to erosion. Also, the land is protected by grass and legumes more of the time in the North because of the heavier livestock program. However, farmers in the South have been moving toward a livestock program and increased grass acreages.

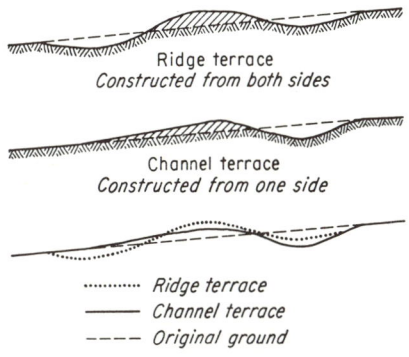

FIG. 38-1. Comparison of ridge and channel terraces.

Recently, there has been considerable tendency to increase terrace spacings. The latest research and observations have indicated that this is feasible. Figure 38-2 gives the spacing formula presently being used by the Soil Conservation Service in the United States.

Terrace Grades. Too much emphasis has been placed in the past on a variable-grade terrace. The idea was to increase the grade at about 300- to 400-ft intervals from the start of the terrace to the outlet. This has considerable merit on long terraces, and if long terraces are used, varying the grade will reduce the cross section required at the outlet. However, the present tendency is to use every suitable natural waterway for an outlet. Under these conditions, a uniform grade can be used or the grade varied within certain limits to permit terraces to be constructed parallel. Each soil and cropping system has a maximum channel grade which is permissible. A. W. Zingg,[2] in a summary of tests run at Bethany, Mo., states that a grade of 6 in. per 100 ft may be used safely on the Shelby soil of that station. The maximum permissible grade should be determined by designed runoff and permissible velocity for a given soil and rotation. It should be checked by local observation.

By constructing terraces as nearly *parallel* as possible Smith showed that an overall saving of 25 per cent of farming time on row crops resulted. To construct terraces parallel without large amounts of cuts and fills, variation in grades is often desirable. As stated previously, terrace grades can be varied within limits to permit the construc-

---

[1] O. R. Zeasman and A. J. Wojta, Terraces, *Univ. Wisconsin Ext. Circ.* 386, 1950.
[2] A. W. Zingg, Terrace Grades on Shelby Soil as They Affect Soil and Water Losses, *Agr. Eng.*, 23:159–162, May, 1942.

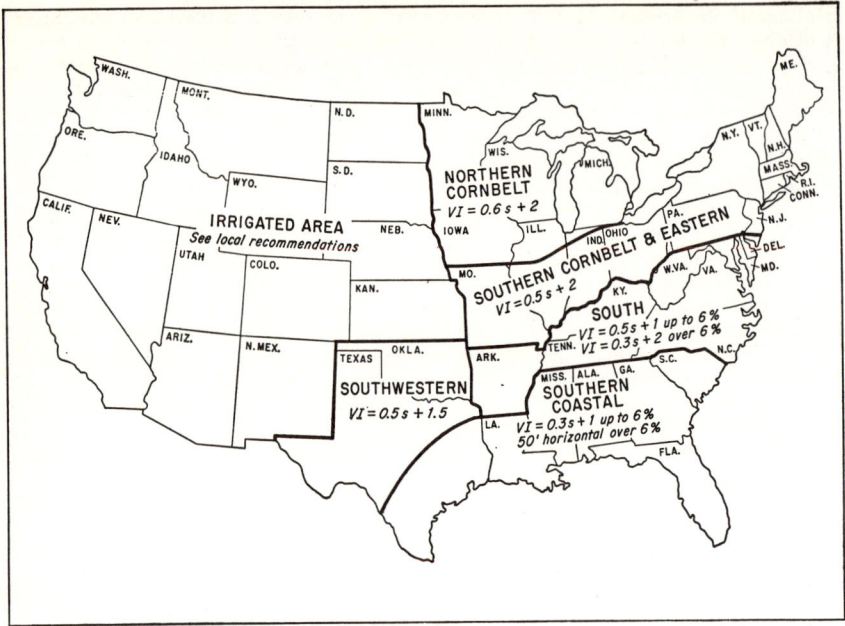

Fig. 38-2. Terrace-spacing formula in use in the United States. VI = vertical interval, ft; s = slope, per cent. (*USDA SCS, Milwaukee, Wis.*, 1953.)

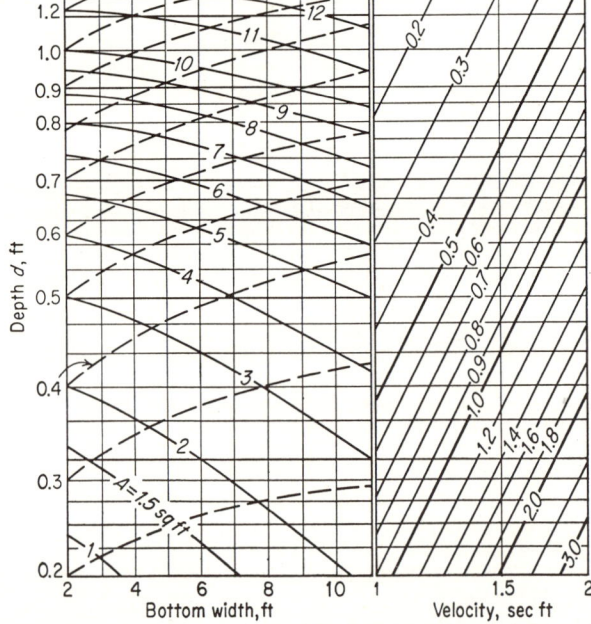

Fig. 38-3. Terrace velocity check chart. (*Design notes:* Provide grades ranging from 0.3 to 0.6 ft per 100 ft of length. For short distances terrace grades can be increased to improve alignment. The chart permits a check of channel velocity.) This chart is based upon 8:1 slopes and a roughness coefficient n of 0.04. Runoff can be figured by multiplying area drained in acres by 4 cfs on the North and 6 cfs in the South. Add at least 0.3 ft to the design depth to provide for silting and settlement.

tion of parallel terraces on slopes which are fairly uniform.[1] Velocity establishes the limits of the slope permitted. *Safe velocity* for bare soil is usually set at 1.5 fps on sandy soils and 2 fps on other soils. The only point of caution in grade variation is that over-all average grade for a soil must be adhered to and a steep section must be compensated for by a flat section. Using the bottom width and cross section planned for terrace construction, Fig. 38-3 can be used in determining the maximum variation allowable in a short section of channel.

For parallel terraces, the *key terrace* is first selected, as discussed above, and then laid out. Usually it should be laid on a uniform grade of 0.4 or 0.5 per cent, depending on erodibility of soil. The remaining terraces are laid out equidistant from the key terrace with a chain. The vertical interval for the average slope obtained from the formula in Fig. 38-2 can be converted to horizontal interval by dividing by the slope expressed as a decimal. Some adjustment in location of the key terrace may be required to provide drainage in the terraces staked above and below the key line. In some cases it may be more desirable to make cuts and fills in these terraces than to relocate the key terrace.

Staking Terraces. Stakes are usually set at 50- to 100-ft intervals, depending on the uniformity of the topography. They are first set on the true desired grade. After a line is staked, some adjustment of the stakes is made (moving a few feet laterally

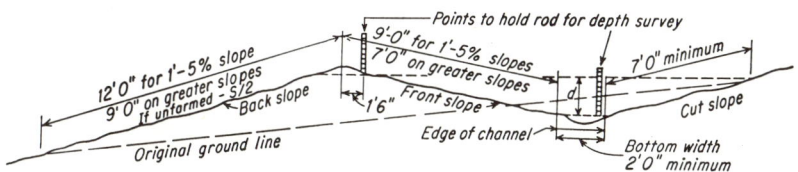

FIG 38-4. Minimum terrace cross section (graded or level). Corn Belt standards for terraces on cropland. Individual dimensions may be decreased 1 ft 0 in. provided over-all dimensions are not decreased more than 2 ft 0 in.

uphill or downhill) to permit rounding out of curves for efficient operation of farm equipment.

Where a key terrace is used, it is staked first and the adjustments made. If topography permits and it is desired to have the terraces parallel, the other terraces are measured off with a chain. It is necessary to set the stakes at uniform distances on these terraces, which are laid out with a chain. Grades should then be determined with an instrument to be sure that the terraces will drain and that the grades are not too steep. Adjustments of the key terrace may be necessary after the grades are determined in order to prevent backfall or too steep grades. In some cases where backfall is minor (6 in. or less), it may be more desirable to make cuts in the terrace channels rather than to adjust the stakes. This is a point that must be determined by the over-all effect of these shifts on terrace grades and row drainage.

Terrace Cross Sections. Figure 38-4 gives general Soil Conservation Service recommendation for terrace cross sections in the Corn Belt area. The dimensions may need to be increased to permit the use of four-row equipment. Required heights of graded terraces may be determined based on length and runoff by use of Fig. 38-3. It is desirable to furnish a safety factor of about 1.5 to provide some capacity for sediment. This is a guide, and terrace cross sections should be kept as wide and as convenient to farm as conditions require. Where single-row equipment is used, a wide section is not as important as where two- or four-row equipment is used.

Level-terrace Specification. Generally, level terraces can be spaced much the same as graded terraces. Enough storage must be provided to handle the expected runoff. In normal practice, capacity is provided to take care of runoff from rains of about a 10-year recurrence interval. In an unpublished report on level-terrace observations in western Iowa, overtopping seldom occurred where storage capacity for 1½ in. of

---

[1] Paul Jacobson, Are Point Rows Necessary on Terraced Land? *Soil and Water Conserv.*, November, 1951.

runoff was provided. The specifications for this area call for terraces which will hold 2 in. of runoff to provide a safety factor for irregularities in construction. The proper level-terrace capacity for an area can be obtained by figuring a watershed index number and referring to figures, as in Chap. 34, for runoff for a given rainfall. The maximum inches of runoff will be the necessary storage needed in the channel. The cross-section area required, neglecting infiltration in the terrace (usually small), can then be computed by the following formula:

$$\text{Cross-section area, sq ft} = \frac{(\text{runoff, in.})(\text{terrace horizontal interval, ft})}{12}$$

Building Terraces. In some localities, many of the early terraces were constructed by plow. With the present tendency to build much larger terraces which can be farmed easily with two- or four-row equipment, construction of terraces by plow requires many more rounds and is prohibitively slow. Many terraces are now being built by contractors. The bulldozer is popular for terrace construction since it can build the wide sections now in demand and can be used in making cuts and fills to improve alignment or permit terraces to be constructed parallel. The patrol grader is popular and well adapted where cuts and fills are not too much of an item.

Other machines adapted to building terraces are disk plows, carryall scrapers, whirlwind terracers, and various types of elevating terracing machines. Instructions for using these machines to build terraces usually can be obtained from the manufacturers, the U.S. Soil Conservation Service, or the Agricultural Extension Service.

Farming Terraced Land. In general, all soil-tilling operations should be performed parallel with the terraces, since it is extremely important that farm implements do not destroy the terraces. All row crops should be planted parallel to the terraces. This reduces erosion to the minimum and is a much more efficient way of farming terraced land.[1]

If permanent agriculture is the goal, a two-way plow is a necessity for plowing terraced land. Any other method of plowing will leave dead furrows between terraces which, over a long period, will gradually bench the land. Another advantage of the two-way plow is that the dead furrow can be placed in the channel, and as was pointed out previously, soil deposition in the channel will be largely offset by uphill plowing.

Some operations, such as cutting grain or hay, can be done by crossing the terraces on the more gentle slopes. On steep land, however, these operations must be parallel to the terraces because of the difficulty in crossing them.

Maintenance of Terraces. If a two-way plow is used in plowing the field, no further maintenance will ordinarily be required. Maintenance by the conventional one-way plow requires replowing the front slope of the terrace two or three times. The method used can be much the same as that used for the construction of terraces with a moldboard plow.

DIVERSIONS

Diversions differ from terraces in that they consist of individually designed channels across a hillside. They may be used to protect bottomland from hillside runoff, or they may be needed above a terrace system for protection against runoff from an unterraced area. They may also divert water out of active gullies, protect farm buildings from runoff, reduce the number of waterways, and in some cases they are used in connection with strip cropping to shorten the length of slope so that the strips can effectively control erosion.

Planning the Diversion System. Deposition of sediment is the most common cause of diversion failure. Therefore the area above a diversion should be maintained in permanent grass or trees. If the area is to be planted to row crops, it should be strip-cropped or terraced. Another method of protecting a diversion from excessive sedimentation is to use a long crop rotation with a high proportion of grass. If these precautions are not taken, overtopping may occur and the area below may be damaged.

[1] Paul Jacobson and Walter Weiss, Farming Terraced Land, *USDA Leaflet* 335, 1952.

The location of an adequate *outlet* will affect the location of the diversion. A field of permanent pasture or meadow is sometimes used for an outlet. Unless the vegetation is uniformly heavy, it is necessary to plan the outlet to spread the water rather thin. This generally requires placing the outlet on a ridge. Constructed outlets along a field boundary must be planned and designed as grassed waterways as shown in Chap. 39. Natural waterways probably furnish the best and most trouble-free outlets for diversions. They should be checked for cross-section area to be assured of sufficient width to permit safe velocity for the present or expected vegetation. In a few instances permanent structures may be required for diversion outlets. For details, see Chap. 41, Erosion-control Structures.

Design of the Diversion. Velocity of water flow is of major importance in the design of a diversion. To avoid sediment deposition, the velocity should be kept as high as will be safe for the channel vegetation. Safe velocities are as follows:

	fps
Bare channel:	
Sand	1.5
Other soils	2.0
Poor channel vegetation	3.0
Fair channel vegetation	4.0
Good channel vegetation	5.0

Velocities for bare channels should be used for designing all diversions subject to cultivation. Normally, poor to fair vegetation will improve except on extremely poor soil sites. Therefore diversion velocities generally should be low enough to permit cultivation or be based on good channel vegetation.

The watershed area must be determined from field survey or from aerial or other maps as the first step in diversion design. In short diversions, only the area at the outlet is necessary. For long diversions, it is well to break the diversion into several reaches and determine the drainage area of each reach. This procedure will permit reducing the size of the diversion as the watershed area decreases. After the area has been measured, the runoff should be determined (see Chap. 34 for details). If a vegetated outlet is to be used, runoff from a storm of 10-year recurrence interval is usually sufficient for design. If a diversion is used in connection with a permanent structure, it should be designed on the same storm runoff as the structure.

It is a simple procedure, after the runoff is determined, to select the proper cross section and slope from Fig. 40-3, 40-4, or 40-5, depending on the cross section desired. As can be noted by studying the figures, considerable flexibility in grade is permitted by varying the height or bottom width. If the required runoff exceeds those given in the figures, Manning's formula may also be used.

Staking, Constructing, and Maintaining Diversions. Diversions are staked in the same way as terraces. Since the cross sections are similar to terrace cross sections, the same method of construction can be used. Where some diversions are larger than terraces, additional rounds will be needed during construction to provide for the increased height.

If the diversion velocity was planned for vegetation, seeding practices should be similar to those recommended for waterways in Chap. 40. Heavy fertilization in accordance with local practice generally will be required, especially where diversions are constructed on infertile soils.

Where diversions are farmed, the same maintenance methods that apply to terraces will apply to diversions. Where the diversion is left in permanent grass, trees and rodents may become problems. Trees can be kept off the diversion by mowing, and rodent holes must be filled to prevent ridge breaks.

Chapter 39

FIELD PRACTICES FOR EROSION CONTROL

B. D. BLAKELY, WALTER WEISS, AND F. L. DULEY

Contour Farming. The practice of planting across the slope on a line perpendicular to the slope of the land is called contour farming. The small ridges and plant stems in the contoured rows hold water and thus prevent runoff and soil erosion. The practice is most effective on row crops, but the water-holding ability of the ridges and plant stems carries over to grain and meadow crops.

On average hillsides the first contour line should ordinarily be laid about $4\frac{1}{2}$ or 5 ft vertically below the top of the hill. Laying a contour line merely consists of running a level line across the hillside with a surveyor's level or hand level. In general, on soils which are slow to take up water, the line need not be exactly level but should have a 0.5 per cent slope toward the waterways. Rows should then be planted parallel to the contour line on both sides. Whenever the slope of a parallel row exceeds 2 or 3 per cent for a distance in excess of 50 to 100 ft, a new contour line should be laid out. Exact values of this allowable deviation and the effectiveness of contouring vs. up- and downhill farming can be determined by studying Table 37-1. When used without terraces, contouring should be limited to the length of slope indicated by the stepped line on the contouring part of the table. With contouring, it is desirable to leave headlands or turn strips of grass at the ends of the rows as well as grassed waterways. This practice allows machinery to turn at the ends of the row without planting turn rows up- and downhill, which is very erosive.

Contour Strip Cropping. This is the system by which ordinary farm crops are produced in relatively narrow strips of variable or even width on which close-growing meadow crops alternate with clean-tilled crops such as corn or grain. The strips are placed crosswise of the line of slope approximately on the contour. Such strips are also effective in reducing wind erosion. It is difficult to state rules that will apply to all conditions in laying out a strip-cropping system. There are three ways of using the base line (contour line). It may form either the top or the bottom boundary of the strip, or it may be used for the center line. The width of the strip depends upon per cent of slope and equipment to be used. As a rule, the steeper the slope the narrower the strip width. As a general guide, strips should rarely, if ever, be more than 150 ft in width and in order to be practical should seldom be less than 50 ft in width.

The exact width of strip needed to obtain adequate control of erosion will depend not only upon the soil type and per cent of slope, but also upon the length of slope and the rotation used. A study of the material in Table 37-1 shows that strip cropping reduces the soil loss over contour farming by a factor of $\frac{1}{2}$ or more. Observations at the Upper Mississippi Valley Conservation Experiment Station, near La Crosse, Wis., indicated that in a corn-oats-meadow-meadow rotation on a moderately eroded Fayette silt loam surface-planted to corn on slopes up to 300 ft long, the most effective and practical strip widths were 70 ft for 2 to 5 per cent slope and 50 ft for 14 per cent slope and over. In laying out strips, these widths should be adjusted to provide an even number of rounds when using two- or four-row planters at the 42- or 40-in. row spacing (Table 39-1).

TABLE 39-1. GENERAL RECOMMENDATIONS FOR STRIP WIDTH FOR A GIVEN SLOPE ADJUSTED FOR EVEN ROWS WITH TWO- OR FOUR-ROW EQUIPMENT

Slope, per cent	42-in. row spacing				40-in. row spacing			
	2-row equipment		4-row equipment		2-row equipment		4-row equipment	
	Strip width, ft	No. of rows	Strip width, ft	No. of rows	Strip width, ft	No. of rows	Strip width, ft	No. of rows
2–5	98–112	28–32	112	32	93–106	28–32	106	32
6–9	84	24	84	24	80	24	80	24
10–14	70	20	56–84	16–24	67	20	53–80	16–24
14	56	16	56	16	53	16	53	16

Layout Methods for Strips

1. For areas where topography is such that uniform-width strips cannot be used, the upper and lower boundaries of the strip should be laid on the contour. This will result in each strip being irregular in width.
2. For areas where topography is such that every other strip can be of uniform width:
    a. Lay out the top contour (base) line; then measure off equal distance above and below the contour line to obtain the width of strip desired.
    b. Location for running the second contour line on strips over 70 ft in width is determined by going to the steepest part of the slope and measuring down from the top-strip boundary one-half the strip width plus 28 ft (if rows are to be 42 in. wide). Thus for an 84-ft width, the width at the narrowest point would be 42 plus 28, or 70 ft, as shown in Fig. 39-1. The lower line of the next strip (which is uniform width) is then laid 84 ft below the contour line. On strips less than 70 ft in width the distance should be one-half the strip width plus 14 ft. If more strips are needed on the hillside, the above process is repeated as many times as needed. Thus the bottom of each variable-width strip will always be on the contour.
3. For areas where topography is such that more than one uniform strip may be used:
    a. Lay out upper contour (base) line.
    b. Lay out one uniform strip above and as many strips below as topography will permit. If the deviation from the contour is not more than 2 or 3 per cent in short distances (50 to 100 ft in length), additional strips can be laid out. If the deviation does not stay within this limitation, the starting point for running the second contour line is normally one-half strip width plus 28 ft below the bottom line of the last strip established at the steepest point in the field, as explained above in 2*b*.
4. For areas where topography is relatively uniform, a contour (base) line may be laid about the middle of the slope or field. Then lay strip boundaries above and below the contour line until the deviation from the contour is greater than the limits set in 3*b*.
5. For area where topography is too irregular or undulating for practical contour strips, field strip cropping may be used. Strips should be made uniform in width and across the general slope, but should not curve to conform to the contour. This is a poor substitute for contour strip cropping and should only be used on very irregular slopes.
6. For areas with level topography, where wind erosion is the major problem, the strips may be placed crosswise of the prevailing wind without regard to the con-

tour of the land. This method is used in the production of regular farm crops in relatively narrow, straight, parallel strips.

Green-manure and Cover Crops. Greater use of commercial fertilizer, new kinds of farm machinery, and other improvements in methods of producing crops have made it possible for the farmer to farm his land more intensively than in the past. Row crops are grown more frequently, often subjecting the soil to more erosion, destroying soil tilth, and lowering the per cent of organic matter. Cover crops and green-manure crops therefore have an increasingly important place in the present-day cropping systems.

The term *green-manure crop* applies to any crop of which the total growth is worked into the soil—green or mature. It may be considered as part of the crop ro-

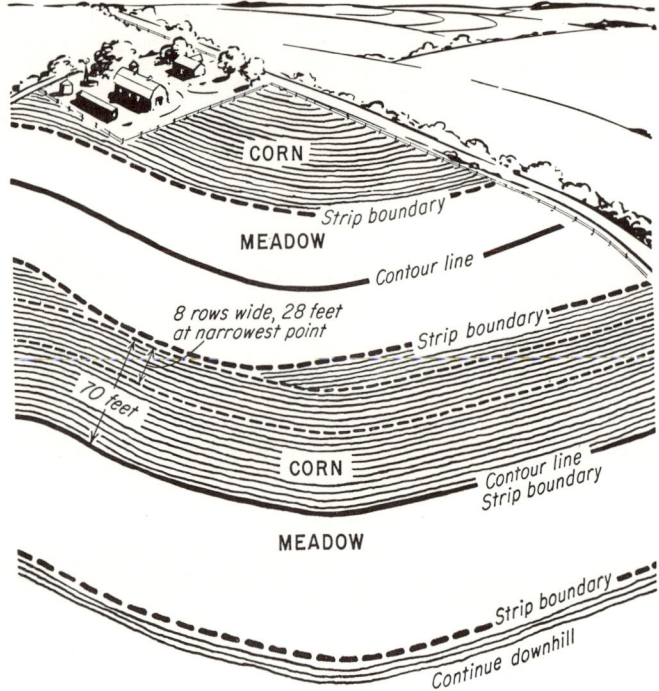

FIG. 39-1. Contour-strip-cropping layout.

tation, used to provide organic material, and if leguminous, it will add nitrogen. It also serves as a cover to protect the soil from excessive erosion and runoff.

The term *cover crop* applies to crops that are planted between the regular cropping periods for the purpose of protecting the soil from erosion and to prevent leaching of valuable plant food such as nitrogen. Such crops also add organic material to the soil and improve tilth, as do green-manure crops. Cover crops are often plowed under for green manure.

In selecting the type of crop, the following points should be considered: (1) adaptation of the crop to the soil and climate, (2) quantity of organic material produced in the time available for growth, (3) character of the root growth, (4) ease of working the growth into the soil, (5) time of seeding, and (6) whether the crop is to be turned under while green or as dead material.

Some of the principal crops used for cover and green manures are:[1]

[1] "Soils and Men," USDA Yearbook of Agriculture, 1938; Cover Crops for Soil Conservation, *USDA Farmers' Bull.* 1758, 1936.

Vetch. Two varieties, the smooth-stem type adapted to southern United States, where the winters are not severe and where the temperatures during the growing season are moderately cool, and the hairy-stem, adapted, because of its winter hardiness, to northern United States. Vetch requires 30 to 35 lb of seed per acre alone or 20 to 30 lb when seeded with a winter grain.

Austrian winter field peas. Adapted to the same general area as smooth-stemmed vetch. This crop should not be fall-seeded north of Tennessee and Maryland. Seeding rates are from 35 to 40 lb per acre.

Crimson clover. A good winter cover and green-manure crop for the Atlantic Seaboard from New Jersey to northern Georgia and for the northern part of the Cotton Belt where similar conditions prevail. Its use is limited by the difficulty of obtaining good stands. The usual seeding rate is 20 to 25 lb per acre.

Crotalaria. Used in the South, especially on sandy soils or soils low in fertility. Seed can be produced easily and at a low cost.

Sweet clover. Perhaps the most satisfactory green manure for the Corn Belt, since it fits in well with the cropping system. It can be used over much of the northern and western part of the United States on soils that are well supplied with lime and phosphate. Insect damage during later years has limited the use of sweet clover; however, spraying is providing effective control.

Medium red and mammoth red clover and alfalfa. Also in use.

Rye. Seeded at 1 to 1½ bu per acre, rye is one of the best all-around cover crops in the Corn Belt or Cotton Belt. It is easy to grow and has a wide range of adaptability. On fertile soils it makes a rapid dense growth and will produce a fairly satisfactory cover on thin and droughty soils.

Winter oats and barley. Excellent cover crops in areas where winter temperatures are not severe.

Cover or green-manure crops should be turned under while there is ample moisture and before growth reaches the point of maturity where it will be slow to decay. Usually this is about 2 weeks before corn planting or 3 weeks before cotton planting. In orchards the cover should be turned under or cut prior to the start of the spring growth.

Mulch Tillage. The use of crop residues to protect the land against excessive runoff erosion and wind erosion came into use in a limited way during the 1930s as very effective protection against wind erosion. About this time it was shown that any protective residue cover on the surface of the soil kept the pores of the soil open and prevented sealing of the surface during a rain. This condition resulted in a much greater amount of water being absorbed, with consequent reduced runoff and erosion. The presence of the residue reduced the splash of soil caused by the raindrops striking the soil, slowed down the rate of flow of water over the surface, and thus reduced erosion in greater proportion than the reduction in runoff.[1]

Where a good covering of mulch is maintained in the Great Plains, wind erosion can be greatly reduced. As much stubble as possible should be left anchored to the soil as long as possible during the process of preparing the seedbed. It has long been known that a mulch tends to reduce the evaporation from the soil.

In order to take advantage of the protective effects of mulches, it was necessary to develop a system of farming whereby the soil can be thoroughly tilled without necessarily inverting it or burying the residue. To do this, subsurface tillers were devised (see Chaps. 14 and 15). These use the principle of running a sweep or rod under the soil, lifting and pulverizing the soil, and at the same time allowing it to fall back without burying the residue. Some disk tools can be used to a limited extent, but they are likely to bury so much residue that not enough is left to protect the soil.

Subsurface tillage takes the place of plowing in the conventional system of farming. As with plowing, the surface must be finely pulverized and packed uniformly in order to put the soil in condition for planting. On plowed land these surface-working operations are usually done with disks and harrows, which tend to bury residue or collect it in piles. To avoid this, a *treader* has been devised. It is essentially a rotary

---

[1] F. L. Duley and J. C. Russell, Machinery Requirements for Farming through Crop Residues, *Agr. Eng.*, 23:39–42, February, 1942.

hoe with the fingered wheels reversed to avoid clogging. If each gang of wheels is set at opposite angles like an offset disk, an excellent job of pulverizing the surface soil results. When set in this way the implement is called a *skew treader*. The fingered wheels of the front gang penetrate the soil and push it sidewise, while those on the rear gang push the soil in the opposite direction. They tend to break and shatter the soil more than a disk. Treaders also do considerable packing and thus leave the soil in proper condition for planting.

Small grain can be drilled with standard equipment after land with residue on the surface has been prepared as indicated above. Narrow spaced drills cause some difficulty from clogging, and for this reason the semideep furrow drill with 10-in. spacing has some advantage. This is particularly true in regions of limited rainfall. The important thing is to plant the seed in a furrow clean and free of residue, but with sufficient straw left anchored between the rows to reduce runoff and erosion and to control wind erosion where it is a problem.

Row crops may be planted through light mulches. This is best accomplished by the use of a press-wheel drill which makes a furrow in which the seed is planted. The furrows may be made with either narrow shovels or, better, with disk-furrow openers.

Row crops should be cultivated with sweeps which do not bury the remaining surface residue but kill weeds by undercutting. Disk hillers which may be adjusted to throw soil either toward the row or away from it help materially in eradicating weeds in the rows.

The use of mulches is proving very effective in *reducing wind erosion* on sandy soils. By providing sufficient plant nutrients on such soils, either through the use of legumes or by application of commercial fertilizers, crops large enough to supply abundant residue can be produced. If these residues are retained on the surface as a mulch by the use of subsurface tillage implements, wind erosion can be effectively controlled during the interval between growing crops.

In areas of greater rainfall, as indicated above, mulch farming greatly reduces erosion from runoff water. However, yields have been somewhat reduced because of lower soil temperature and consequent slower nitrification. In recent experiments it has been found that this can be largely overcome by the addition of nitrogen fertilizer.

Chapter 40

WATERWAYS FOR EROSION CONTROL

PAUL JACOBSON

Almost every field on sloping land needs grassed waterways to carry runoff to a stabilized drainageway. Waterways are the foundation of water management for all other erosion-control practices except level terraces.

The cost and feasibility of blading in gullies and establishing waterways should be investigated before structures are considered for gully control. Quite often such waterways will be the most permanent and economical type of gully control.

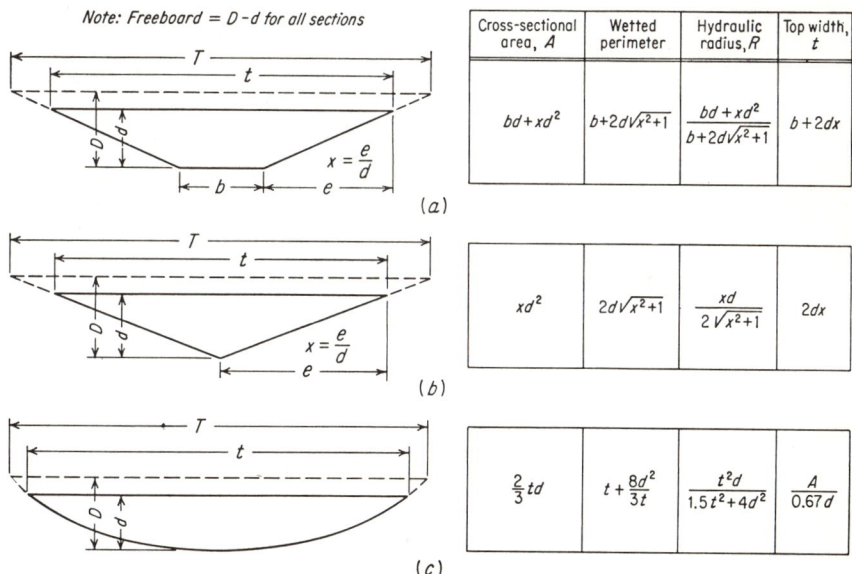

Fig. 40-1. Channel cross sections, notations, and formulas: (a) trapezoidal cross section; (b) triangular or flat V cross section; (c) parabolic cross section. (*USDA SCS-TP*-61, 1947.)

Waterway Shape. There are three common cross sections considered in waterway design, namely, the parabolic, trapezoidal, and the flat V as shown in Fig. 40-1. Manning's formula (Chap. 34) is generally used to compute the velocity and quantity of water handled. The value of n varies greatly as shown by Ree and Palmer[1] and Smith.[2] Table 40-1 provides a classification of various types of grasses

[1] W. O. Ree and V. J. Palmer, Flow of Water in Small Channels Protected with Vegetative Linings, *USDA Tech. Bull.* 976, 1949.
[2] Dwight D. Smith, Blue Grass Terrace Outlet Channels, *Agr. Eng.*, 24:333–342, October, 1943.

into retardance classes.[1] Figure 40-2 sets up values of n based on these retardance classes in relationship to the velocity times the hydraulic radius. As depth of flow increases, increasing velocity, grass is laid over and n will decrease. With the range of velocities ordinarily recommended and the usual type of grasses, a value of n of 0.04 will generally fit in the northern part of the United States and a value of 0.06 in the southern part. Before the vegetation is laid over, high retardance may cause the water to overflow the designed channel. To overcome this, about 0.5 ft of freeboard is added to the vegetated channel.

TABLE 40-1. CLASSIFICATION OF VEGETAL COVERS BY DEGREE OF RETARDANCE

Retardance	Cover	Condition
A	Weeping love grass	Excellent stand, tall (average 30 in.)
	Yellow bluestem Ischaemum	Excellent stand, tall (average 36 in.)
B	Kudzu	Very dense growth, uncut
	Bermuda grass	Good stand, tall (average 12 in.)
	Native grass mixture (little bluestem, blue grama, and other long and short middle-west grasses)	Good stand, unmowed
	Weeping love grass	Good stand, tall (average 24 in.)
	Lespedeza sericea	Good stand, not woody, tall (average 19 in.)
	Alfalfa	Good stand, uncut (average 11 in.)
	Weeping love grass	Good stand, mowed (average 13 in.)
	Kudzu	Dense growth, uncut
	Blue grama	Good stand, uncut (average 13 in.)
C	Crab grass	Fair stand, uncut (10 to 48 in.)
	Bermuda grass	Good stand, mowed (average 6 in.)
	Common Lespedeza	Good stand, uncut (average 11 in.)
	Grass-legume mixture—summer (orchard grass, redtop, Italian rye grass, and common Lespedeza)	Good stand, uncut (6 to 8 in.)
	Centipede grass	Very dense cover (average 6 in.)
	Kentucky bluegrass	Good stand, headed (6 to 12 in.)
D	Bermuda grass	Good stand, cut to 2.5-in. height
	Common Lespedeza	Excellent stand, uncut (average 4.5 in.)
	Buffalo grass	Good stand, uncut (3 to 6 in.)
	Grass-legume mixture—fall, spring (orchard grass, redtop, Italian rye grass, and common Lespedeza)	Good stand, uncut (4 to 5 in.)
	Lespedeza sericea	After cutting to 2-in. height; very good stand before cutting
E	Bermuda grass	Good stand, cut to 1.5 in. height
	Bermuda grass	Burned stubble

NOTE: Covers classified have been tested in experimental channels. Covers were green and generally uniform.

The parabolic channel is generally recommended for waterway design because it approaches more closely the cross sections found in natural waterways. Since the center is low, small flows will be carried with less meandering than in a trapezoidal channel. The parabolic-shaped channel can be standardized in chart form over a wide range.

Waterway Location. Waterways should generally be located in natural drainageways where the water can drain in from all sides. Also, moisture conditions and soil fertility are usually the best for establishment of the needed vegetation. It may be necessary on terraced fields, where the natural outlet is on an adjoining farm, to construct a waterway along a fence line. Some engineers recommend locating terrace outlets only along field boundaries, but these cases require placing a waterway in an unnatural location. Under all conditions a parabolic channel can be used.

[1] V. J. Palmer, William P. Law, and W. O. Ree, Handbook of Channel Design for Soil and Water Conservation, *USDA Tech. Bull.* 61, 1954.

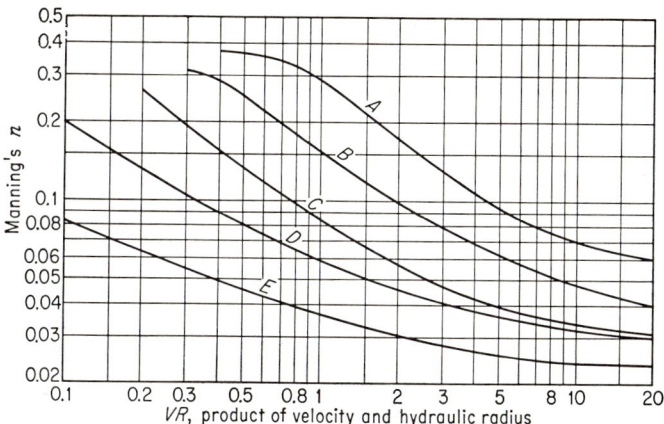

Fig. 40-2. Degrees of vegetal retardance for solution of Manning formula.

Drainage of Waterways. Waterway failures are often a result of poor drainage. If natural drainage is so poor that the waterway will be saturated for periods of several days or weeks, tile drainage should be provided. The tile should be laid on one side of the waterway. This will prevent the danger of surface water washing out the tile. On extremely wide waterways, tile lines should be laid on both sides. The depth of these tile should be based on local jobs which have been installed and are working satisfactorily. See Chap. 36 for additional details regarding random drainage systems.

Waterway Design. Two physical factors are considered in the design of a waterway. These factors are watershed area and slope of the waterway. For design purposes, it is best to break long waterways into reaches and determine the watershed area and the corresponding slope for each reach. With the watershed area known, the runoff can be determined from Chap. 34. Most waterways are designed for a 10-year-frequency runoff.

Permissible Velocity. Table 40-2 gives the permissible velocity based on slope of land, type of vegetation, and the erosiveness of soils. If velocities are in the higher

TABLE 40-2. PERMISSIBLE VELOCITIES FOR CHANNELS LINED WITH VEGETATION*
(Values apply to average, uniform stands of each type of cover.)

Cover	Slope range,† %	Permissible velocity	
		Erosion-resistant soils, fps	Easily eroded soils, fps
Bermuda grass...	0–5	8	6
	5–10	7	5
	Over 10	6	4
Buffalo grass, Kentucky bluegrass, smooth brome, blue grama...	0–5	7	5
	5–10	6	4
	Over 10	5	3
Grass mixture..	0–5†	5	4
	5–10	4	3
Lespedeza sericea, weeping love grass, yellow bluestem, kudzu, alfalfa, crab grass..	0–5‡	3.5	2.5
Common Lespedeza, Sudan grass§.........................	0–5¶	3.5	2.5

* Use velocities exceeding 5 fps only where good covers and proper maintenance can be obtained.
† Do not use on slopes steeper than 10 per cent, except for side slopes in a combination channel.
‡ Do not use on slopes steeper than 5 per cent, except for side slopes in a combination channel.
§ Annuals—used on mild slopes or as temporary protection until permanent covers are established.
¶ Use on slopes steeper than 5 per cent is not recommended.

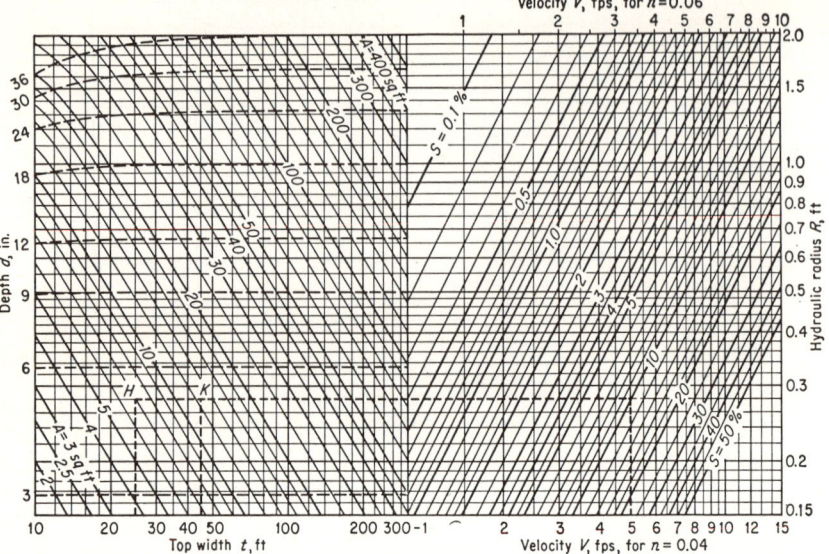

FIG. 40-3. Graphic solution of vegetated channel dimensions by Manning formula for parabolic cross section. (*USDA SCS*, 1947.)

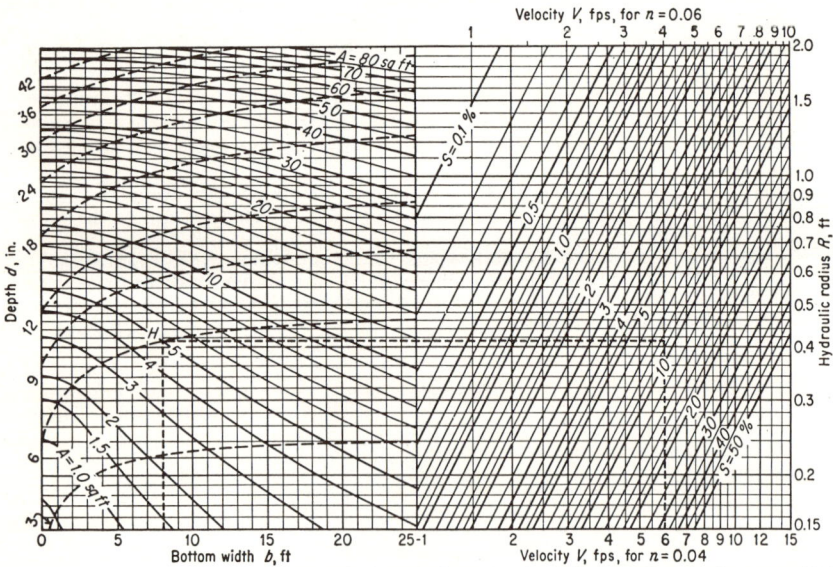

FIG. 40-4. Graphic solution of vegetated channel dimensions by Manning formula for trapezoidal cross section with 4:1 side slopes. (*USDA SCS*, 1947.)

range, it may be necessary to protect the channel by diversions while a seeding is being established.

Selection of Waterway Dimensions. After the runoff, slope of the waterway, and the design velocity have been determined, the next step is to decide on the waterway dimensions. Figure 40-3 gives the top width and depth of parabolic waterways. The shape of a parabolic waterway will approach a channel having a top width and

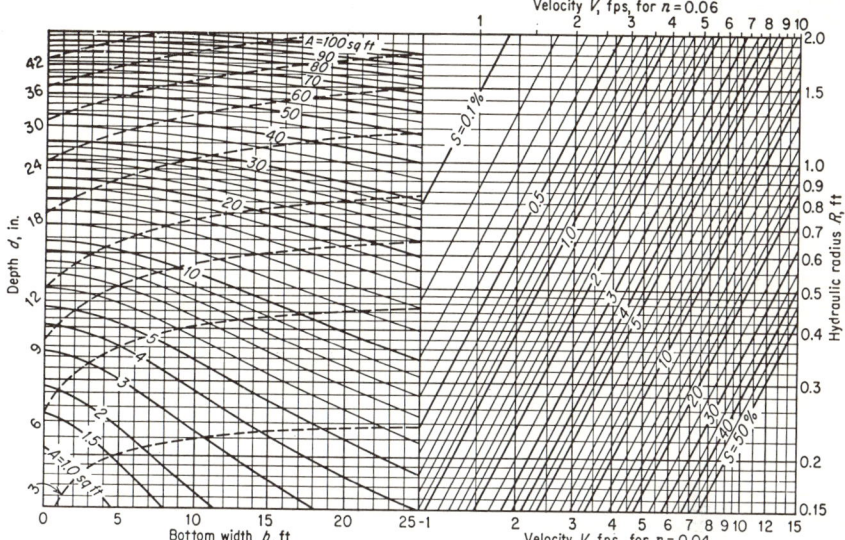

Fig. 40-5. Graphic solution of vegetated channel dimensions by Manning formula for trapezoidal cross section with 6:1 side slopes. (*USDA SCS*, 1947.)

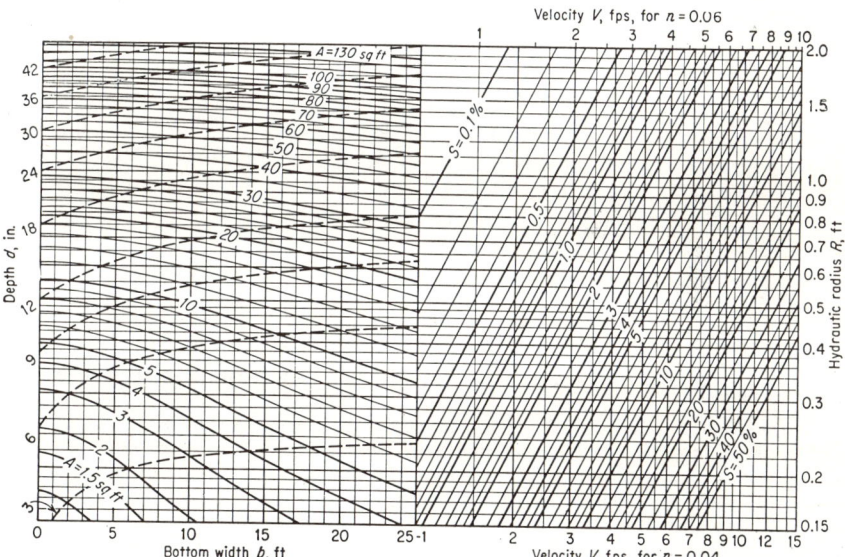

Fig. 40-6. Graphic solution of vegetated channel dimensions by Manning formula for trapezoidal cross section with 8:1 side slopes. (*USDA SCS*, 1947.)

depth as given and 4:1 side slopes, with the flat center section dished about $\frac{1}{10}$ ft to prevent meandering. About 0.5 ft of freeboard should be provided to take care of build-up by sediment and variations in n value.

Sod Chutes. Where a grassed waterway is used for gully control, it is often necessary to use a steeper section on the lower end to conduct the water from the bladed-in waterway to the bottom of the existing outlet channel. In these cases a high-velocity sod chute of trapezoidal section is often the correct solution. Careful mowing of the sod and constant maintenance will be required where these high velocities are used. Velocity recommendations are given in Table 40-2. The dimensions for sod chutes of various cross sections can be determined from Figs. 40-4 to 40-6. Manning's formula may be used to design chutes requiring greater capacities, but generally chutes wider than shown are not feasible. In these cases some type of stabilizing structure may be required.

Toewalls. Waterways very often develop an erosion problem at the lower end. In other cases, rock outcrops or wet areas may prevent a good growth of sod. Under these conditions a low toewall as indicated in Fig. 40-7 is needed. Such a toewall provides a means whereby the sod can be protected from unraveling.

Constructing Waterways. It is difficult to construct a waterway where a gully has developed or where the waterway is too wide and flat to permit adequate drainage.

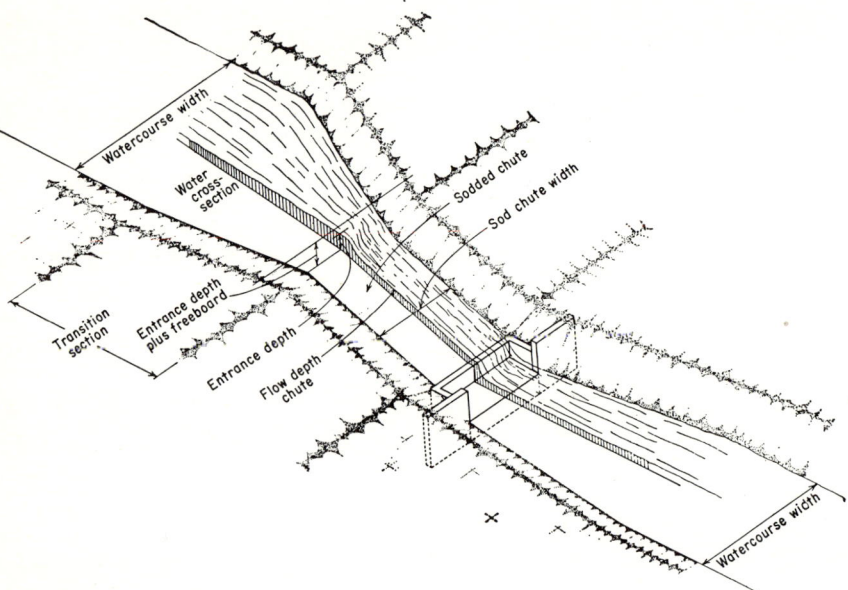

Fig. 40-7. Sod chute and toe wall.

The latter condition may occur in a wide flat natural waterway, or it may be the result of silting.

To construct a waterway *where a gully exists* it is necessary to move the topsoil to one side so that it can later be used as a top-dressing to assure the establishment of vegetation. A bulldozer is efficient for working in large gullies. Other types of equipment such as blades or plows may be used, but more time will be required to fill the gully. After the major portion of the gully is pushed in, it is well to stake the center line. Curves can be straightened out, and alignment of the finished waterway improved.

To construct a suitable waterway *where the drainageway is too flat and wide,* it is necessary to push the dirt out from the center to provide the designed cross section. If a bulldozer or a wheel scraper is available, the earth can be feathered out on each side so that a single channel can be made. It is often more economical to make a *W ditch* if a plow or grader is used. This type of waterway consists of two constructed channels with the dirt pushed into the center, allowing free access for the water from each side. The chief objection to building a single channel with light equipment is the tendency to produce a ridge on both sides of the waterway, which prevents water from getting into it. In a W waterway, the ridge in the center produces a divide and each channel should be designed on the basis of its individual watershed.

Checking the Constructed Waterway. A constructed waterway should be checked at frequent intervals to determine whether the designed depth and width have been met in construction. Any deviation should be marked on the centerline stakes, indicating whether cuts or fills are required. This checking is extremely important, since it is of little use to plan a waterway if the design is not achieved in construction. Usually, only a hand level and tape are necessary for checking.

Seeding the Waterway. After the proper cross section is obtained, a good firm seedbed should be prepared. Liberal applications of manure and fertilizer should be made. Information about the best-adapted type of grass, nurse crop, and amounts of seed to use can be obtained from the local agricultural extension or Soil Conservation Service representative. Studies are now being made on protecting waterways during the critical seeding period. The use of diversions to divert water during this period seems to be especially effective. Another method being tried is the use of asphalt spray to stabilize soil until vegetation is present.

Waterway Sprigging. Some plants such as reed canary grass, Bermuda grass, and kudzu are best propagated by breaking the sod into individual plants, which are then set out individually or placed on the waterway and worked into the ground. This process is called sprigging and is a quick way to get good cover on a waterway with these plants.

Sodding a Waterway. Where it is desired to cut and lay sod, a sod cutter is needed. If sod is used, it is best to lay it in the spring or fall when moisture conditions are good. If laid during the dry period, it should be watered. Sod strips should be laid across the waterway with the joints staggered. Loose dirt should be placed between the joints. The sod should be rolled with a rubber-tired tractor or roller for best results.

Chapter 41

EROSION-CONTROL STRUCTURES

Fred W. Blaisdell and Arthur F. Moratz

The subject of this chapter is the investigation, design, and construction of pipe conduits, drop spillways, head spillways, and chute spillways and the related earth embankments in common use in soil- and water-conservation work. The detailed analysis of the more complex installations will require additional study, using the information contained in this section as a guide. The Bibliography and other reference material should be the basis of such study.

A well-planned program of soil- and water-control structures will necessarily need supporting conservation practices on the watershed. Vegetated channels and simple conservation practices alone often prove inadequate in handling concentrations of runoff or flows of long duration. Permanent installations of steel, concrete, and masonry are required in some cases to supplement the conservation practices. This is particularly true where permanence and a high degree of safety are required.

There are no pattern solutions or formulas that can be applied to all the problems encountered in this field. Any design should include an analysis of all factors affecting the work. The experienced engineer will study the possibilities of each problem and make the final design on the basis of including minor improvements that are economically justified. Only after a thorough study of all the factors (i.e., economic, control, utility, etc.) should he develop the final designs.

The general classification of structures includes those required for grade stabilization, head control, storage reservoirs for irrigation, measuring devices, silt-detention basins, flood storage, water impounding for ground-water supply or flood control, and ponds for stock water supply, fire protection, spraying, and recreation.

PRELIMINARY INVESTIGATIONS

Preliminary investigations in the form of an analysis of available data and outline surveys are an important prerequisite in the over-all planning of the work. The proper use of this material will enable the designer to study alternative site locations and determine the economic justification of all or part of the project. When adequate basic data are not available, a safe design requires conservative assumptions. The more dependable the data, the smaller the margin of overdesign, resulting in a greater saving in the final cost of the installation. Where a large number of small structures having similar characteristics are involved, a few typical samples can be carefully studied and the results projected to other structures, thus saving a great deal of time in preliminary investigations.

The survey plan in the form of a preliminary route survey includes establishing temporary and permanent benchmarks, sufficient channel shots to establish grades, rough cross sections at possible structure sites, and regular and unusual channel sections. Cross sections should be taken at locations where high-water marks have been established. Elevations of any present controls (i.e., structures, roads, wells, springs, buildings, etc.) should be a part of this survey. Linear measurements at this stage are made by pacing or transit. Stations are established at permanent fence lines and other landmarks. This information is plotted and used as a basis for the preliminary design. Stations originally paced and other horizontal controls are verified by large-scale aerial photographs or planimetric maps. These are usually available locally through the Soil Conservation Service, Bureau of Reclamation, Corps of Engineers, or the Production and Marketing Administration. The aerial photographs will enable the designer to check the general pattern of the land use in the area and determine watershed boundaries. The soils and slopes can be determined by reference to soil-survey maps of the Soil Conservation Service, state college soil reports, or if in the vicinity of recently surveyed state or Federal highways, from the state highway department. Small-scale contour maps may be available from U.S. Geological Survey quadrangles or Corps of Engineers or Bureau of Reclamation maps. Flood-survey reports may also be available from these agencies. Water-survey and geology reports have been prepared for many areas and can be secured from the respective state departments. All the above are useful as guides in preparing the preliminary plans. These plans determine the approximate locations and control elevations of structure sites. Preliminary borings of sufficient depth and coverage should be made to determine stability of foundations and the availability of adequate fill material. The borings are often the most important consideration in the final selection of structure sites. At this stage in the work, a report is prepared compiling all the known information, cost estimates of proposed installations, approximate benefits, land treatment, and the proposed conservation practices.

FINAL SURVEYS

Final surveys are correlated and expanded from the preliminary survey work. The required detail will depend on the complexity of the problem and the type of structure(s). For a control or a series of controls, profiles of channels should extend to stable grade below the limits of the work area and above to a point beyond the effects of the controls. This should be a minimum of 500 ft in each direction for the single-grade-control types of structures. Cross sections at each structure site should give sufficient detail to enable the designer to adequately orient the various parts of the structure and accurately determine fill quantities. Three cross sections are the minimum requirements, and as many as ten or twelve may be needed on the larger fill jobs. Additional cross sections are needed at average and unusual channel sections to determine velocities and depths of flow and to plan construction improvements. Where reservoir volumes or temporary storage are important factors in the design, complete coverage of flooded area by cross sections gives the most accurate detail. These cross sections are spaced equally along a traverse or base line, with sufficient readings to determine the required contours. A stadia survey of the impounded area is acceptable provided it results in adequate coverage of the site. Some jobs may require a traverse or base line for their entire length, with the channel and bank profile tied to the base line by cross sections. On all jobs a base line should be established through the structure area(s) to provide accurate design and layout work. Cross sections can be taken at even stations except where major improvements are required. Watershed maps are usually taken from aerial photographs. Preliminary boundaries are plotted on the photographs from available contour or soil maps and checked in the field. Soil borings are taken along established cross sections and plotted along these cross sections at their true elevations. Where numerous borings are required, other than on established cross sections, they are usually plotted in groups (log), but should be shown at their relative elevations as an aid in the design. A legend covering the various strata encountered should be shown adjacent to the borings.

PREPARATION OF PLANS

The plans should completely and clearly show all works of improvement and indicate sufficient details so that accurate quantities can be determined. For record purposes the plans should include all the survey information which was used in computing the hydrologic, hydraulic, structural, foundation, and embankment designs. The scale of the drawings should be governed by the amount of detail needed for layout and construction. Large-scale drawings are required at all structure sites and at locations of all major works of improvement. Work areas and structure outlines are indicated on all drawings where they appear. A minimum set of plans should contain profiles, cross sections, watershed and site maps, and detailed profiles and cross sections at all works of improvement. Contour maps of the impounded area, detailed location maps, and longitudinal sections may be required on the more complex jobs. The detailed structural plans and specifications should be a part of the plans.

RUNOFF DETERMINATIONS

Design runoff rates and flood volumes are important decisions for the engineer. The relationship between the design flows and spillway capacities requires careful study and sound judgment. Chapter 34, on Principles of Agricultural Hydrology, gives methods of determining the design flows and should be carefully studied before the spillway design is started. The design flow is usually given in terms of an expected storm recurrence interval and is based on the damage that might result to the structure and the possible danger to life and property in case of failure.

DEFINITIONS AND GENERAL DESIGN INFORMATION

The Principal Spillway. This spillway, commonly termed the mechanical or structural spillway, is designed to carry the shorter recurrence-interval discharges from the headwater pool or entrance channel. This spillway is usually constructed of the more permanent materials. On small-farm pond structures where the vegetative spillway conveys the major part or total discharge, a mechanical spillway consisting of a 6- or 8-in. pipe may be needed as a drawdown or trickle tube to prevent small prolonged flow which might damage the vegetative spillway. Table 41-1 gives details for capacity of these small installations based on pond surface area, watershed area, and runoff characteristics.

The Auxiliary Spillway. This may be of either permanent structural material or vegetation and is designed to discharge only a part of the higher-frequency flows. The elevation of the crest of the auxiliary spillway is located so that the required discharge of the principal spillway is satisfied before the auxiliary spillway begins to function. The design will likely require both spillways to discharge simultaneously at the design stage.

The Emergency Spillway. Usually this spillway is of the vegetative type and is designed to discharge flood flows in excess of the design storm or as a relief for the other spillway(s) in case of clogging or damage. The crest of the emergency spillway is set at the design water level of the entrance channel or reservoir or higher. The emergency spillway is not required to carry discharges at nonerosive velocities, but it is required to convey this flow beyond the structure so as to prevent damage to any major part of the dam.

Freeboard. Freeboard is the difference in elevation between the design water surface in the entrance channel or reservoir and the top of the settled embankment. Where emergency spillways are part of the design, freeboard is usually the difference in elevation between the crest of the emergency spillway and the top of the settled fill. The freeboard on small-farm pond and grade-control structures is usually 1 ft. For the larger types of structures and the small irrigation reservoirs and other water-impounding structures 2 ft is needed. Greater freeboard will be required for the more costly installations and where life or high property loss might be involved. The freeboard required for the larger reservoir spillways is usually determined by the Stephenson formula for wave action,

$$H = 1.5(D)^{1/2} + 2.5 - D^{1/4} \qquad (41\text{-}1)$$

where H = height of wave from trough to crest, ft

D = fetch or straight length, miles, across water surface with prevailing wind

Two times the wave height is allowed above high-stage water surface for freeboard. This height is required to allow for the run of the wave up to the face of the embankment. A 2-ft 0-in. freeboard is the minimum recommended for spillways of this type.

Critical Depth. The critical depth (d_c) for a given flow of water in a channel (Fig. 41-27) is that depth[1] at which the discharge is a maximum for a given total head (depth + velocity head).

Critical Flow. Critical flow will occur where the channel slope is just sufficient to maintain flow at critical depth. It is a very unstable condition.

Structure Spillway Capacities. A structure used as a grade or head control in a terrace outlet or upland waterway has the spillway section designed on the basis of the peak rate of flow from the contributing watershed. An auxiliary overflow section should also be provided to carry the expected difference in the peak rate between the water-channel design and the maximum flow expected during the life of the structure. Site conditions may require a structure to carry in excess of the commonly accepted design flow or in some cases to eliminate the need for an auxiliary spillway. Then the spillway should be designed on the basis of the life of the structure.

A guide to the life of common construction materials is given in the following table:

	Years
Plain to treated wood plank	10–25
Concrete block, corrugated-metal sheets, and pipe	25–50
Heavy sheet steel, reinforced concrete	50–100

The life of the materials may be reduced by high-velocity flows with erosive bed loads and when the foundation or embankment contains high-acid or high-alkali soils.

Spillways used as major controls in small water-disposal systems are generally designed to carry the maximum storm expected once in 50 years.

An *island type* of design is used where site conditions limit the size of weir opening or allow periodic flooding of the surrounding land. It is most commonly used at the head end of drainage ditches, where it may also serve as a tile outlet. The principal spillway is designed to fill the downstream channel bank full before any flow is allowed to pass over the auxiliary spillway(s). Some grading of the outlet of the auxiliary spillway(s) into the downstream channel can be provided to reduce the discharge requirements of the principal spillway. Auxiliary spillways may be provided around both ends of the principal spillway. The auxiliary spillway(s) should discharge the difference between the design storm and the full discharge capacity of the principal spillway. Submergence is often a factor in the discharge capacities of both the principal and auxiliary spillways. An approximate check of the effect of submergence should be made to determine the need for additional study. This is done by backwater projections of downstream tailwater. In the more involved cases combined stage discharge curves of both the spillways and downstream channel are necessary to properly design the island-type structure. Figure 41-1 shows an island-type structure using a box-inlet drop spillway.

Flood-detention Structures. Flood-detention structures are dependent more on the total volume of the flood runoff than on the peak rate of the design storm. This general group may include a wide range of structures, from the small-farm pond to the large upland watershed flood-prevention type of installation. It is not the intent of this chapter to attempt the design of the major stream or the multiple-purpose type of structure. On all detention structures additional storage should also be provided in the permanent pool for the loss of capacity from sediment. An estimate should be made of the soil losses from the watershed, the amount of this sediment transported to the pool, and the trap efficiency of the reservoir. The determination of the volume of sediment contribution from various types of watersheds is a special study, and additional texts on the subject should be consulted.[2]

Inlet Protection. Guardrails and trash racks are required on many spillways. Guardrails are desirable on all types except the drop spillways to protect both man and animal from injury due to high overfall at the crest. The rails circumvent the weir

---
[1] Calvin V. Davis, "Handbook of Applied Hydraulics," 2d ed., McGraw-Hill Book Company, Inc., New York, 1952.
[2] Henry M. Eakin and Carl B. Brown, Silting of Reservoirs, *USDA Tech. Bull.* 524, rev. August, 1939. Carl B. Brown, The Control of Reservoir Silting, *USDA Misc. Publ.* 521, 1943.

TABLE 41-1. STAGE REQUIREMENTS FOR FARM PONDS*

Pond area, acres / Stage, ft

Acres and watershed	ΣW	0.2	0.25	0.3	0.4	0.5	0.6	0.7	0.8	0.9	1.0	1.25	1.50	1.75	2.0	2.25	2.5	2.75	3.0
30	70										2.75	2.5	2.0	1.75	1.5	1.25	1.25	1.25	1.0
	60										2.50	2.0	1.75	1.5	1.25	1.25	1.0		
	50										2.25	2.0	1.5	1.25	1.25	1.0			
	40										1.75	1.5	1.25	1.0	1.0				
25	70									3.0	2.25	2.0	1.5	1.5	1.25	1.0			
	60								3.0	2.75	2.0	1.75	1.5	1.25	1.0				
	50							3.0	2.5	2.25	2.0	1.75	1.25	1.0					
	40	2.5			3.0	3.0	2.75	2.5	2.0	2.0	1.25	1.0	1.0						
20	70	3.0		2.75				3.0	2.75	2.5	1.75	1.5	1.25	1.0					
	60	2.75	3.0		3.0	3.0	3.0	2.75	2.5	2.25	1.5	1.25	1.0						
	50	2.25			2.25	2.5	2.75	2.5	2.25	2.0	1.25	1.0							
	40	1.25					2.0	1.75	1.75	1.5	1.0	1.0							
15	70		3.0	3.0	3.0	3.0	2.75	2.5	2.25	2.0	1.25	1.0							
	60		2.25	2.75	2.5	3.0	2.5	2.25	2.0	1.75	1.25	1.0							
	50			1.75	2.0	2.5	2.0	1.75	1.75	1.5	1.0								
	40				1.5	1.75	1.5	1.25	1.25	1.0	1.0								
10	70		2.75	2.5	2.0	2.5	2.25	1.75	1.75	1.5	1.25								
	60		2.5	2.0	1.75	2.25	1.75	1.5	1.5	1.25	1.25								
	50		1.75	1.5	1.25	1.75	1.5	1.25	1.25	1.0	1.0								
	40		1.0	1.0	1.0	1.25	1.0	1.0	1.0		1.0								
5	70	1.5	1.25	1.25	1.0	1.5	1.25	1.25	1.0										
	60	1.0	1.0	1.0		1.25	1.25	1.0											
	50					1.0	1.0												
	40																		

DIAGRAMMATIC SKETCH

Freeboard — stage — Crest of mechanical spillway

Crest of vegetative spillway

Horizontal cross section area of riser shall be at least two times the area of the conduit. Minimum size for jointed clay, concrete or corrugated metal pipe shall be 8". For smooth continuous steel pipe 6" diameter can be used.

Runoff-producing Characteristics of Watersheds with Corresponding Weights W

Designation of watershed characteristics	Runoff-producing characteristics				
	100 extreme	75 high	50 normal	25 low	
Relief	(40) Steep, rugged terrain, with average slopes generally above 30%	(30) Hilly, with average slopes of 10 to 30%	(20) Rolling, with average slopes of 5 to 10%	(10) Relatively flat land, with average slopes of 0 to 5%	
Soil infiltration	(20) No effective soil cover; either rock or thin soil mantle of negligible infiltration capacity	(15) Slow to take up water; clay or other soil of low infiltration capacity, such as heavy gumbo	(10) Normal; deep loam with infiltration about equal to that of typical prairie soil	(5) High; deep sand or other soil that takes up water readily and rapidly	
Vegetal cover	(20) No effective plant cover; bare or very sparse cover	(15) Poor to fair; clean cultivated crops or poor natural cover; less than 10% of drainage area under good cover	(10) Fair to good; about 50% of drainage area in good grassland, woodland, or equivalent cover; not more than 50% of area in clean-cultivated crops	(5) Good to excellent; about 90% of drainage area in good grassland, woodland, or equivalent cover	
Surface storage	(20) Negligible; surface depressions few and shallow; drainageways steep and small; no ponds or marshes	(15) Low, well-defined system of small drainageways; no ponds or marshes	(10) Normal; considerable surface-depression storage; drainage system similar to that of typical prairie lands; lakes, ponds, and marshes less than 2% of drainage area	(5) High; surface-depression storage high; drainage system not sharply defined; large flood-plain storage or a large number of lakes, ponds or marshes	

Problem

Find required stage. Drainage area 20 acres. Pond area 0.7 acres.

Runoff-producing characteristics   *Weighted values*

Relief: rolling land average slope 8% 16
Infiltration upland prairie soil brown silt loam 10
Vegetal cover: 100% pasture and woodland 5
Surface storage: well defined system of small drainage ways. 13
$\Sigma W = 44$

Solution from table above, for

$\Sigma W 50 \quad s = 1.75$
$\Sigma W 40 \quad s = 1.25$
difference 0.50

$\Sigma W = 44 \therefore 0.4 \times 0.5 = 0.20$
$\underline{1.25}$
1.45

Always increase to the next quarter foot; therefore,

$s = 1.50$

Design storm = Approx. Q_5

* Maximum stage for this chart is 3.0 ft. Higher values require design using information given in this chapter. The minimum stage given by this chart is 1.0 ft. If the determined stage falls below this value use 1.0 ft.
SOURCE: *USDA SCS Ag. Publ. 57*, 1953.

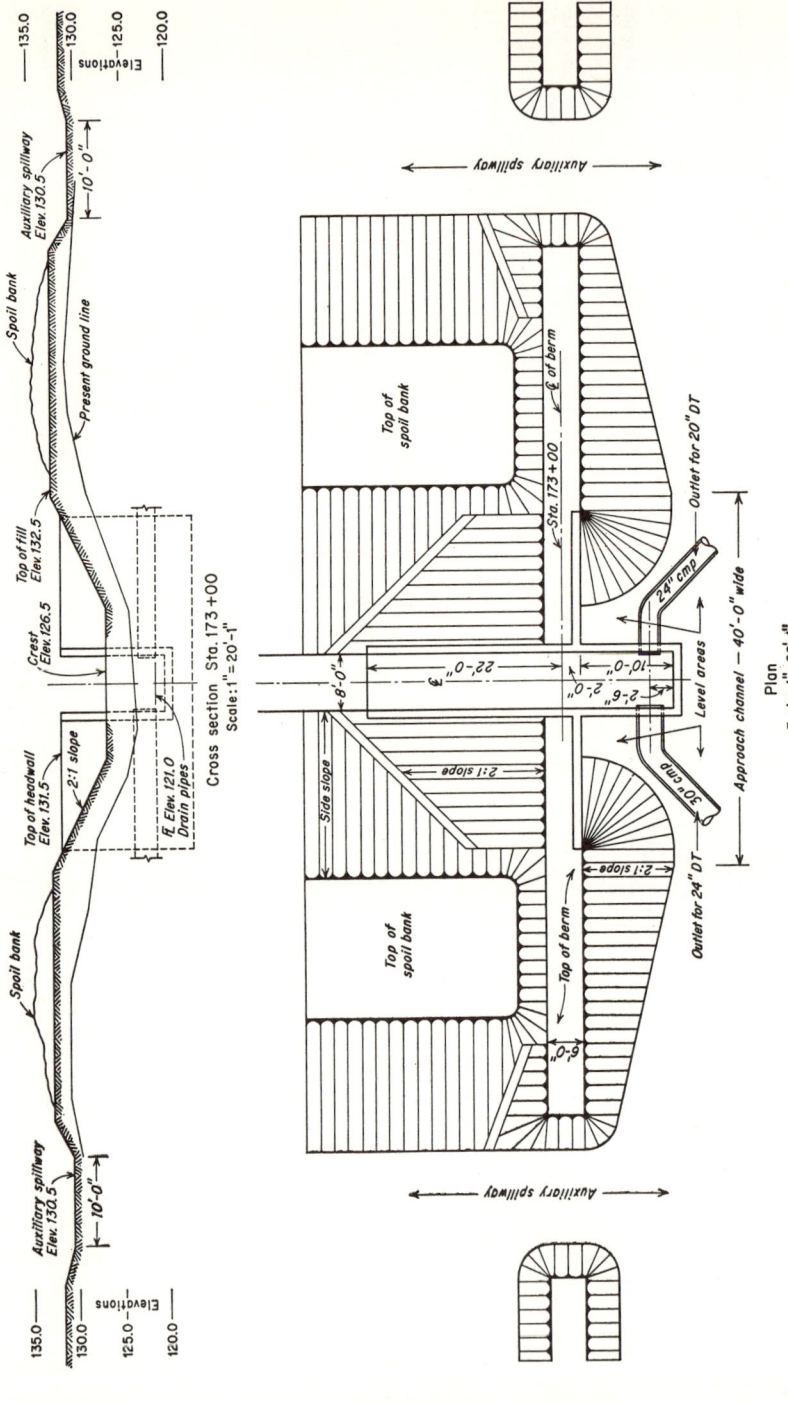

Fig. 41-1. Structure layout for island type of dam. (*USDA SCS, Eng. Div., Central Tech. Unit*, 1954.)

crest (Fig. 41-2) and protect the spillway from obstructions that would alter the flow conditions. Trash racks are usually installed on drop inlets with pipe conduits to protect the spillway from clogging. They are important when the watershed is timbered and are required where small conduits need protection from stalks of cane, corn, hay, or larger debris. The openings in the rack are smaller than in guardrails and are governed by the size of the debris than can safely pass through the spillway. Trash racks completely cover the inlet as shown in Fig. 41-3. The rack is designed so that excessive flow losses are not created at the inlet. Generally the vertical dimension of the opening is approximately twice the horizontal dimension. Both guardrails and trash racks may be made of reinforced concrete as well as of steel or aluminum pipe. If used, they must be cleaned periodically, and after large storms; otherwise they reduce the spillway capacity.

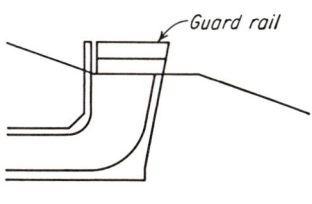

Fig. 41-2. Structure guardrail.

Antiseep Collars. These collars provide the necessary bond between the spillway and embankment and extend the creep distance for internal hydrostatic pressures (Fig. 41-4). The length, depth, and height of the collars are governed by the hydraulic head and the kind and density of the embankment materials. They should provide an increase in creep distance of at least 10 per cent for flood-detention structures

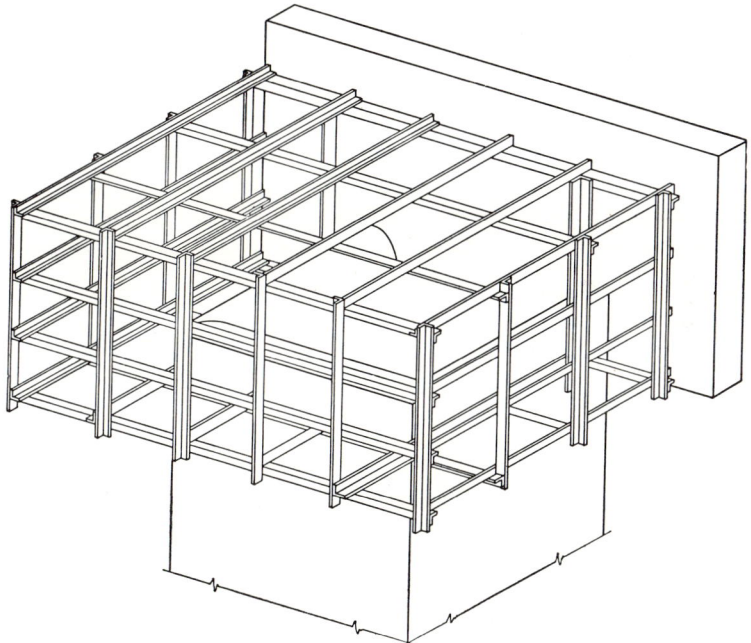

Fig. 41-3. Heavy-duty trash rack.

and 15 per cent for retention structures. The creep distance is measured horizontally from the entrance headwall to the base of the downstream slope or to a toe drain, where one is used. The increase in creep distance is the extra distance water would travel because of the collar, as shown in Fig. 41-4.

Toewalls, or Cutoffs. Toewalls, or cutoffs, as shown in Fig. 41-5 should extend into the foundation below the frost level or a minimum of 3 ft for the smaller installations. In unstable foundations of shallow depth an extension of the cutoff down to solid material is required. In concrete spillways the concrete walls may be given additional depth, or where the unstable conditions exceed 5 to 6 ft it is good design to drive sheet piling of corrugated metal, steel, or wood into the stable base. Additional support for the apron under these conditions is provided by a transverse series

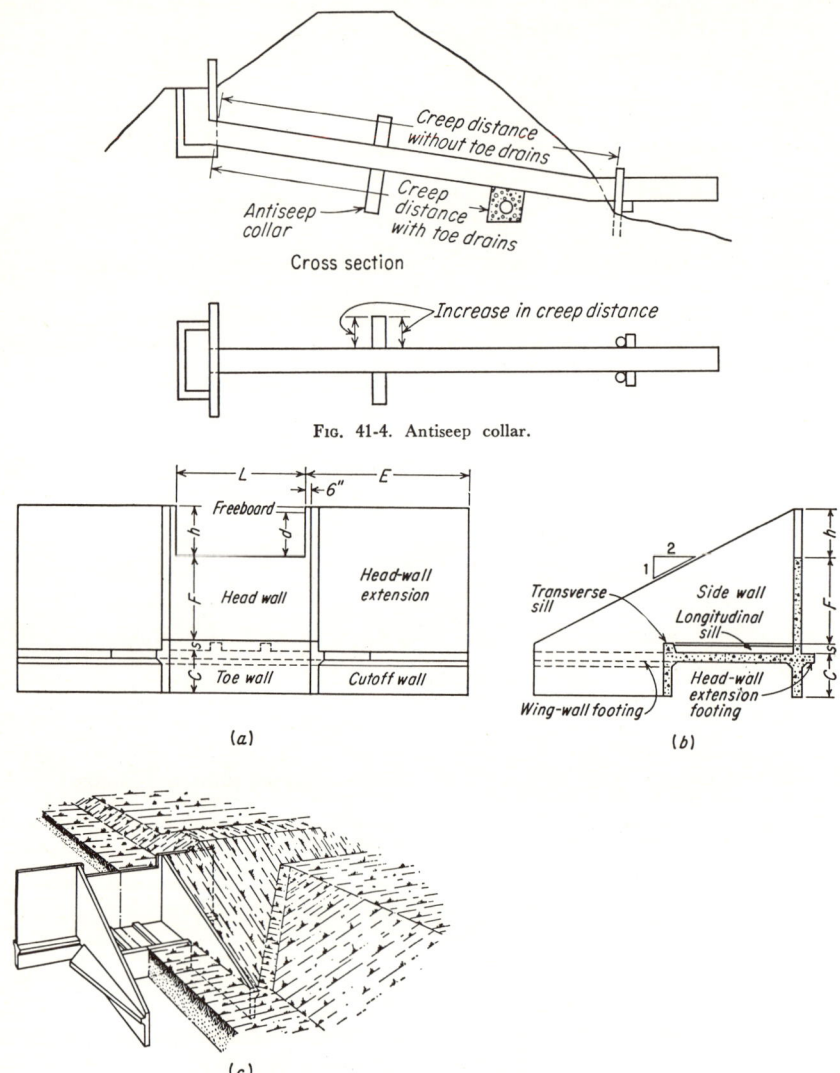

Fig. 41-4. Antiseep collar.

Fig. 41-5. Toewalls and cutoff walls. (*USDA SCS, Eng. Div., Central Tech. Unit*, 1955.)

of piles. Pile caps are formed in concrete spillways immediately below the bottom of the apron.

Headwall Extension. Headwall extension for the low-drop spillways should be extended from the edge of the weir as shown in Fig. 41-5 a minimum of $3d + 4$ ft, where d is the depth of the weir. Special entrance design may require a greater distance. The larger spillways will require a study of these dimensions based on embankment materials, construction, and drainage.

EROSION-CONTROL STRUCTURES 435

Expansion Joints. Expansion joints are provided in many structures to allow for the expansion and contraction of the construction materials. Joints in flat slabs are usually provided for each 400 sq ft of slab. Joints are used to separate members that have stress in opposite directions or different types of members. Joints are also incorporated in the design of some structures to provide for settlement of the foundation and embankment, particularly in drop inlets and chutes. In these cases some rotation should be allowed in the joint to permit closure when settlement takes place. Where either monolithic or pipe drop inlets are involved, the total fall in the conduit must exceed the planned settlement. The total fall Z is distributed over the length of the conduit to provide an approximate equal opening at all joints, as shown in Fig. 41-6. Where expansion occurs, a minimum of two joints spaced on either side

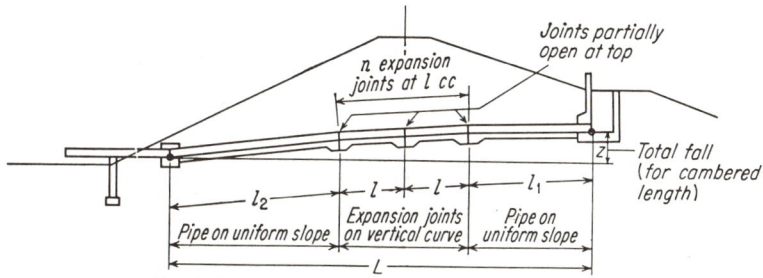

FIG. 41-6. Pipe laid with camber to allow for settlement.

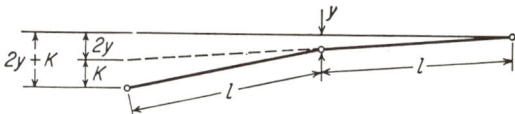

FIG. 41-7. Diagram showing l, y, and k used in calculating camber.

of the center line of the fill should be used. In conduits where the length is less than 50 ft and anticipated settlement is insignificant, the joints may be omitted. If the outlet is some type of headwall, then a joint should be formed by covering the contact faces between the conduit and headwall, with premolded expansion-joint material. The procedure using a vertical curve through the expansion joints is one method of providing camber.

The general formula for determining invert elevations for cambered conduit is

$$z - y\left(\frac{L}{l}\right) = K\left[\left(\frac{n_2 - n}{2}\right) + \frac{l_2}{l}n\right] \qquad (41\text{-}2)$$

where y = proportional amount of total fall less fall due to vertical curve for length (Fig. 41-7).
  l = basic length between expansion joints, ft (Fig. 41-6).
  L = total length of pipe to be cambered, ft (Fig. 41-6).
  K = difference in fall between two adjacent expansion joints, ft. This value will determine amount of opening at top of pipe at expansion joint (Fig. 41-7). For reinforced-concrete pipe up to 48 in. diam., use $K = 0.4l/24$ maximum.
  z = total fall in conduit, ft.
  n = number of expansion joints in vertical curve section (Fig. 41-6).
  l_2 = length of pipe, ft, on uniform slope between downstream end of pipe to be cambered and last expansion joint of vertical curve (Fig. 41-6).
  l_1 = length of pipe from inlet to first expansion joint, ft.

The procedure for determining invert elevations at expansion joints is as follows:
1. Determine the following: l, L, n, l_2, and total fall.
2. Solve general formula for y. (y must be positive. Reduce K as necessary to obtain positive y.)

3. Determine l_1 (Fig. 41-6).
4. Determine elevation at invert of first expansion joint on vertical curve as shown in Fig. 41-8b.

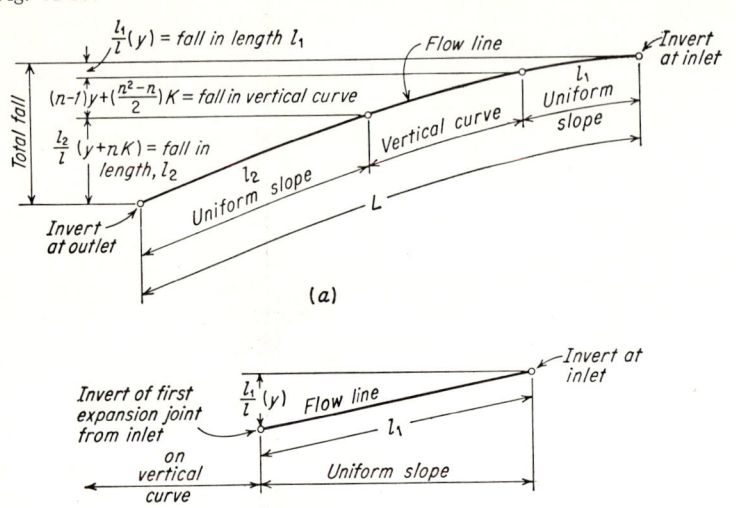

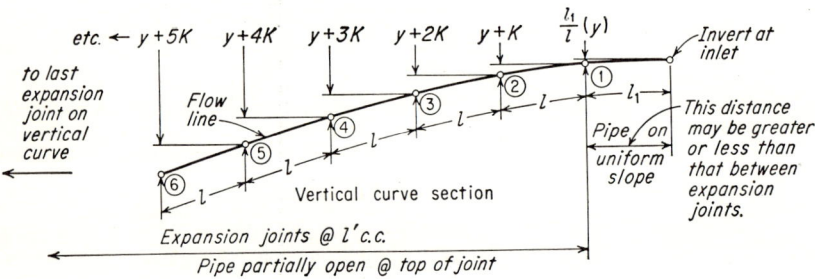

Fig. 41-8. Diagram showing formulas for vertical curve values for cambering pipe outlet.

5. Determine invert elevations of remainder of expansion joints in vertical-curve section.

Example: Given $l = 24$ ft, $n = 2$, $l_2 = 36$ ft, $L = 84$, $l_1 = 24$ ft; total fall 5.3 ft, inlet elevation 966.0 ft, outlet elevation 960.7 ft.

Determine camber elevation at inverts of expansion joints.
Solution: (see K above). Try

$$K = \frac{l_1}{24}(0.4') = \tfrac{24}{24}(0.4) = 0.4$$

Now from Eq. (41-2) solve for y:

$$5.3 - y(84/24) = 0.4\left[\left(\frac{2^2-2}{2}\right) + \tfrac{36}{24}(2)\right]$$

$y = 1.057$ ∴ value of K used is correct for this case

From Fig. 41-8c,

$$\text{Invert elevation at first expansion joint} = 966.00 - \frac{l_1}{l}(y)$$

$$= 966 - \tfrac{24}{24}(1.057) = 964.94$$

from Fig. 41-8c

$$\text{Elevation second expansion joint} = 964.94 - (y + k)$$
$$= 964.94 - (1.057 + 0.4) = 963.48$$

and from Fig. 41-8a,

$$\text{Elevation of outlet} = 963.48 - \frac{l_2}{l}(y + nK) = {}^{36}\!\!/_{24}[1.057 + 2(0.4)]$$

$$= 963.48 - 2.78 = 960.7, \text{ or given elevation of outlet}$$

SPILLWAY CAPACITIES

Straight-drop Spillway. The straight-drop spillway is a simple type of structure that is useful in stabilizing gullies, drainage channels, irrigation ditches, etc., and as a tile outlet structure over which surface runoff can be carried. The drop ordinarily ranges from 2 to 15 ft. Heads over the crest range from 2 to 6 ft. Crests may be any length adaptable to the site.

There are no completely reliable data to define the discharge coefficient C in the equation

$$Q = CLH^{3/2} \tag{41-3}$$

where Q = discharge, cfs
 L = crest length, ft
 H = head on spillway, ft

Kessler suggests a C value of 3.6 for a weir without end contractions.[1] However, end contractions exist on most straight-drop spillways.

When the approach channel is horizontal, a value of C based on critical depth may be computed theoretically. For this condition $C = 3.1$, and this value is suggested for design purposes until more reliable coefficients have been determined. In this case H is the depth of flow plus the velocity head. Recent experiments indicate that C will be between 3.1 and 3.6 and that the exact value depends on criteria which are presently unknown.

The water passing over straight-drop spillways is frequently allowed to fall onto a plain horizontal apron. This may be satisfactory for a time if the flows are intermittent and not large, if the soil is resistant to erosion, and if the apron is set below stream-bed level so that the water always falls into a pool.

It should be remembered that the nappe hits the floor at a considerable distance from the overfall edge. The locus of the upper nappe surface is given by[2]

$$\frac{x}{d_c} = -0.406 + \sqrt{3.195 - 4.386 \frac{y}{d_c}} \tag{41-4}$$

where y = distance above (+) or below (−) the overfall crest, ft
 x = distance from overfall edge, ft
 d_c = critical depth, ft

This equation is solved graphically in Fig. 41-9. There should be a considerable horizontal length of apron beyond the point where the nappe strikes to ensure that the stream will leave the apron horizontally. An end sill to direct the stream upward is preferred because it will form a horizontal roller under the deflected stream that will provide some protection to the stream bed at the end of the apron. An end-sill height equal to one-six the tailwater depth over the apron is suggested.[3] If the spillway is partly submerged by high tailwater, the nappe tends to float and this requires an additional length of apron to ensure that the nappe will not strike the stream bed beyond its end.

---

[1] L. H. Kessler, Experimental Investigation of the Hydraulics of Drop Inlets and Spillways for Erosion Control Structures, *Wisconsin Eng. Expt. Sta. Bull.* 80, 1934.
[2] C. A. Donnelly and F. W. Blaisdell, Straight Drop Spillway Stilling Basin, *Univ. Minn. St. Anthony Falls Hydraulic Lab. Tech. Paper* 151B, 1954.
[3] F. W. Blaisdell and C. A. Donnelly, Hydraulic Design of the Box Inlet Drop Spillway, *USDA SCS* 1-TP-1106, 1950.

The trajectory of the submerged nappe and the point at which the submerged nappe strikes the apron can be determined from Fig. 41-9. The distance of the tailwater above (+) or below (−) the crest is y_t, and the distance from the crest to the upper surface of the submerged nappe for design purposes as determined experimentally is x_a.

A straight-drop spillway stilling basin that completely prevents scour in the downstream channel should be used if the downstream channel is readily erodible or if flows approaching the design value can be expected to occur frequently and for long

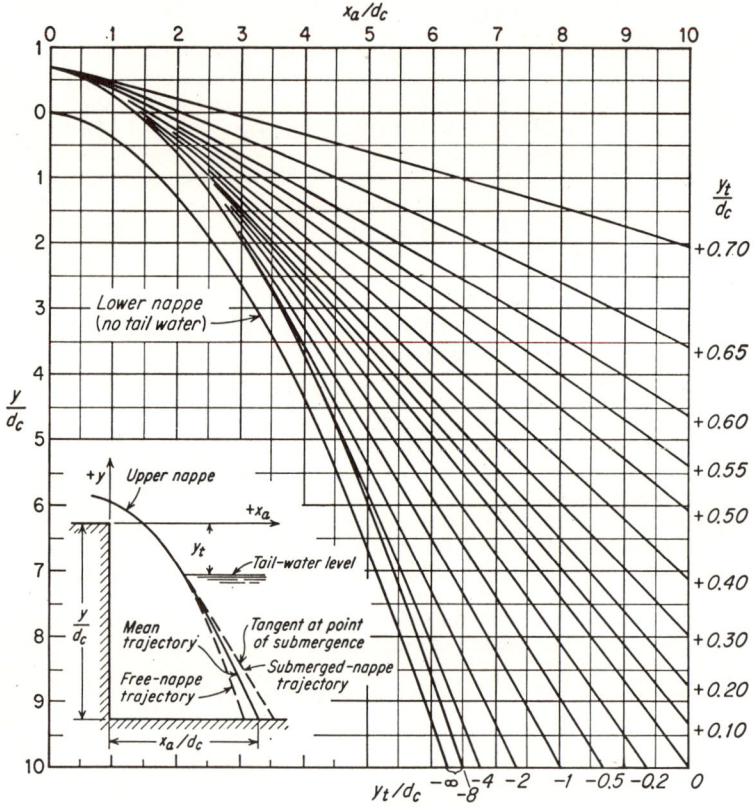

FIG. 41-9. Design chart for determination of x_a.

periods. Drops in irrigation ditches or downstream of floodwater detention reservoirs especially require adequate stilling basins. Such a basin is shown in Fig. 41-10. The rules for the design of this basin are:

1. The minimum length of the stilling basin L_B is

$$L_B = x_a + x_b + x_c = x_a + 2.55d_c \qquad (41\text{-}5)$$

  a. The distance from the headwall to the point where the surface of the upper nappe strikes the stilling basin floor x_a is determined from Fig. 41-9.

  b. The distance from the point at which the surface of the upper nappe strikes the stilling basin floor to the upstream face of the floor blocks x_b is

$$x_b = 0.8d_c \qquad (41\text{-}6)$$

c. The distance between the upstream face of the floor blocks and the end of the stilling basin x_c is

$$x_c = 1.75d_c \qquad (41\text{-}7)$$

2. The floor blocks are proportioned as follows:
   a. The height of the floor blocks is $0.8d_c$.
   b. The width and spacing of the floor blocks should be approximately $0.4d_c$, but a variation of $\pm 0.15d_c$ from this limit is permissible.

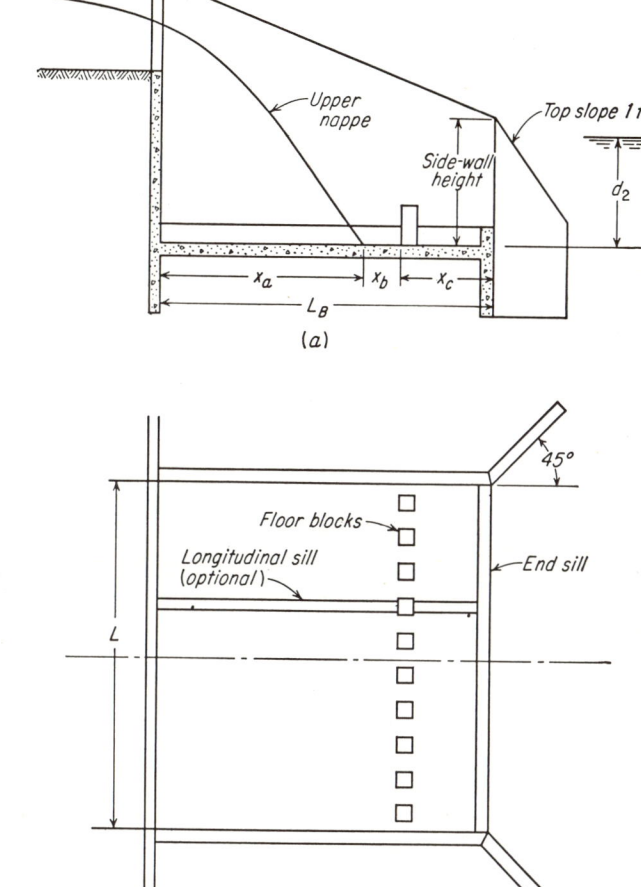

Fig. 41-10. Straight-drop spillway stilling basin.

   c. The floor blocks should be square in plan.
   d. The floor blocks should occupy between 50 and 60 per cent of the stilling-basin width.
3. The height of the end sill is $0.4d_c$.
4. Longitudinal sills may be used for structural purposes. They are neither beneficial nor harmful hydraulically. If used, they should pass through, not between, the floor blocks.

440   SOIL AND WATER CONSERVATION

5. The sidewall height above the tailwater level should be $0.85d_c$.
6. The wingwalls should be located at an angle of 45° with the outlet center line and should have a top slope of 1:1.
7. The minimum height of the tailwater surface above the floor of the stilling basin d_2 is $2.15d_c$.
8. The approach channel:
    a. Should be level with the crest of the spillway.
    b. Should have the toe of the dike or the toe of the side slope intersect the approach channel floor at the ends of the spillway notch; the approach channel

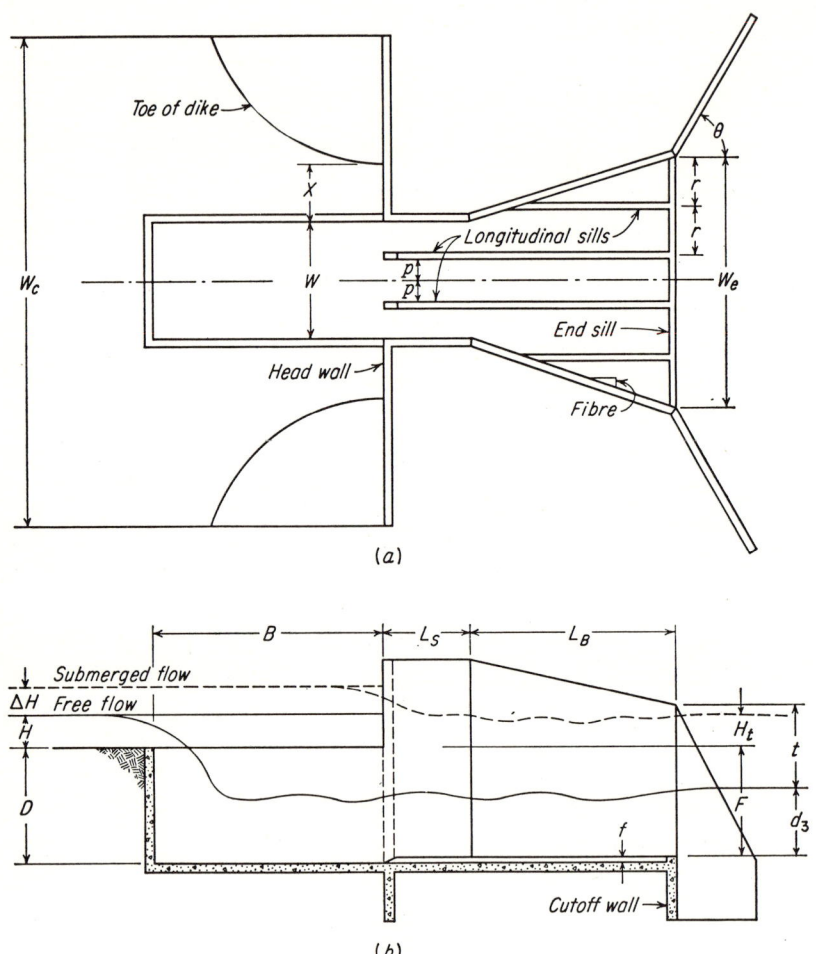

Fig. 41-11. Box-inlet drop spillway.

at the headwall should have a bottom width equal to that of the spillway notch.
    c. Should be protected by riprap or paving for a distance upstream from the headwall equal to two times the notch depth.
9. No special provision for aeration of the space beneath the nappe is required if the approach channel is shaped as recommended here.

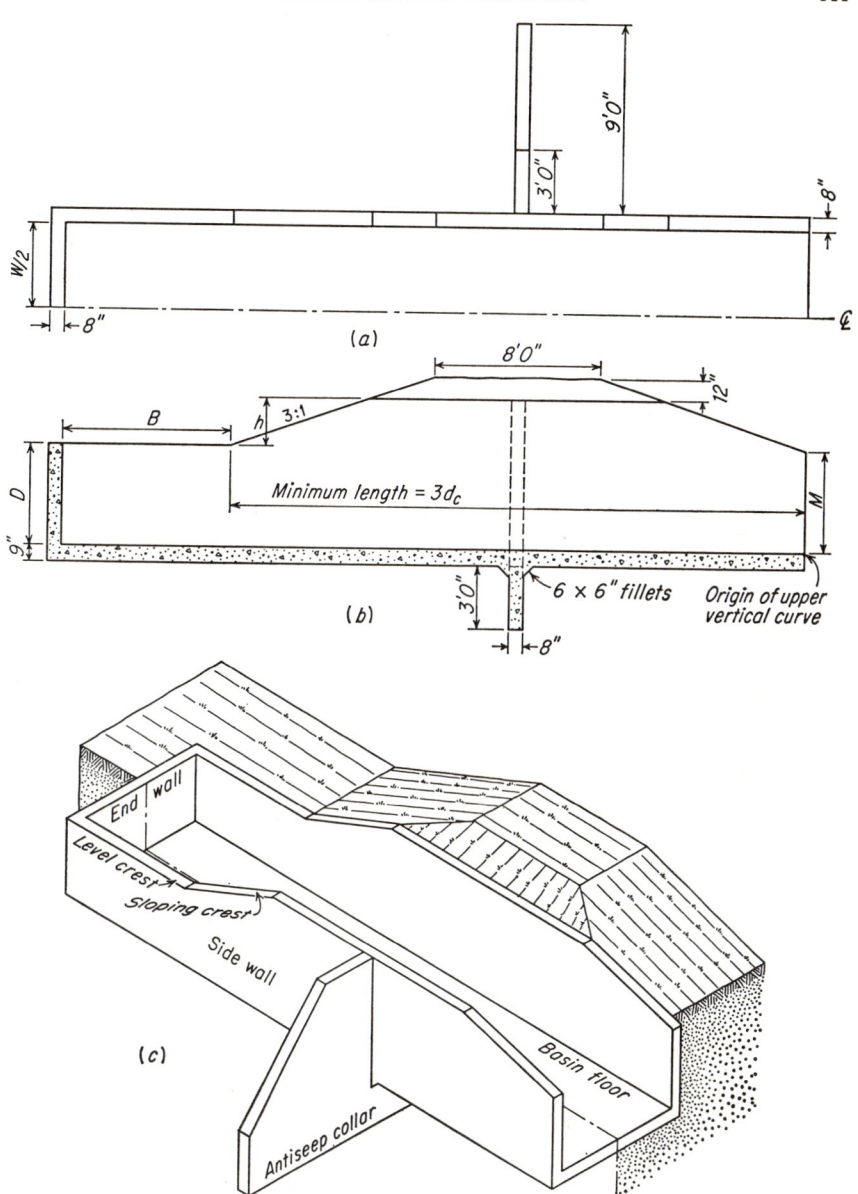

Fig. 41-12. Trapezoidal weir box inlet. (*USDA SCS, Eng. Div., Central Tech. Unit,* 1955.)

Box-inlet Drop Spillway. The box-inlet drop spillway[1,2] shown in Fig. 41-11 is particularly useful where a long crest length is required to keep the head on the crest low and where the downstream channel is so narrow that a straight-drop spillway of equivalent crest length cannot be used. As an erosion-control structure, it is used to

---
[1] Blaisdell and Donnelly, Hydraulic Design of the Box Inlet Drop Spillway.
[2] F. W. Blaisdell and C. A. Donnelly, The Box Inlet Drop Spillway and Its Outlet, *Trans. ASCE*, Vol. 121, 1955, 1956.

control drops of from 2 to 12 ft. It is also useful in drainage work, serving as a tile outlet structure and permitting surface water to enter the drainage ditch without eroding the head end or the adjacent banks.

The box-inlet crest controls the head-discharge relationship at low flows. When the *approach channel is level with the crest,* the following formula applies:

$$Q = 3.43LH^{3/2} \tag{41-8}$$

where Q = discharge, cfs
L = crest length, ft = $2B + W$ (Fig. 41-11)
H = head on crest, ft

Huff[1] and Kessler[2] indicate that a deep approach channel will increase the capacity by 10 per cent. The discharge found by Eq. (41-8) must be multiplied by the corrections for head, box-inlet shape, approach-channel width, and dike proximity given in Tables 41-2 to 41-5 to obtain the true capacity.

The opening through the headwall controls the head-discharge relationship at the *higher flows.* The general formula is

$$Q = C_2 W(H - H_{02})^{3/2} \tag{41-9}$$

where W = headwall opening, ft
C_2 = coefficient of discharge from Table 41-6
H_{02} = head correction obtained from Table 41-7 (can be computed by multiplying by D)

The discharge should be computed by both Eqs. (41-8) and (41-9). The lesser discharge is the actual discharge.

The *trapezoidal weir box inlet* shown in Fig. 41-12 does not require the long headwall used for the rectangular weir box inlet and makes it possible to run the dike up to the box inlet. The box-inlet sidewalls are carried up the 3:1 fill slopes and provide

TABLE 41-2. CORRECTION FOR HEAD (CONTROL AT BOX-INLET CREST)

H/W	0.00	0.01	0.02	0.03	0.04	0.05	0.06	0.07	0.08	0.09
0.0	....	....	....	....	....	0.76	0.80	0.82	0.84	0.86
0.1	0.87	0.88	0.89	0.90	0.91	0.91	0.92	0.92	0.93	0.93
0.2	0.93	0.94	0.94	0.95	0.95	0.95	0.95	0.96	0.96	0.96
0.3	0.97	0.97	0.97	0.97	0.98	0.98	0.98	0.98	0.98	0.98
0.4	0.99	0.99	0.99	0.99	0.99	0.99	0.99	0.99	0.99	1.00
0.5	1.00	1.00	1.00	1.00	1.00	1.00	1.00	1.00	1.00	1.00
0.6	1.00									

Correction is 1.00 when H/W exceeds 0.6.
SOURCE: *USDA* SCS-TP-106, 1950.

TABLE 41-3. CORRECTION FOR BOX-INLET SHAPE (CONTROL AT BOX-INLET CREST)

B/W	0.0	0.1	0.2	0.3	0.4	0.5	0.6	0.7	0.8	0.9
0	0.98	1.01	1.03	1.03	1.04	1.04	1.03	1.02	1.01	1.01
1	1.00	0.99	0.99	0.98	0.98	0.98	0.97	0.97	0.96	0.96
2	0.96	0.96	0.95	0.95	0.95	0.95	0.95	0.95	0.94	0.94
3	0.94	0.94	0.94	0.94	0.94	0.94	0.94	0.94	0.93	0.93
4	0.93									

SOURCE: *USDA* SCS-TP-106, 1950.

[1] A. N. Huff, The Hydraulic Design of Rectangular Spillways, *USDA* SCS-TP-71, 1944.
[2] Kessler, *op. cit.*

TABLE 41-4. CORRECTION FOR APPROACH-CHANNEL WIDTH
(CONTROL AT BOX-INLET CREST)

W_c/L	0.0	0.1	0.2	0.3	0.4	0.5	0.6	0.7	0.8	0.9
0	0.00	0.09	0.18	0.27	0.35	0.44	0.53	0.62	0.71	0.80
1	0.84	0.87	0.90	0.92	0.93	0.94	0.95	0.96	0.97	0.97
2	0.98	0.98	0.99	0.99	0.99	0.99	1.00	1.00	1.00	1.00
3	1.00									

Correction is 1.00 when W_c/L exceeds 3.0.
SOURCE: *USDA* SCS-TP-106, 1950.

TABLE 41-5. CORRECTION FOR DIKE EFFECT (CONTROL AT BOX-INLET CREST)

B/W	X/H					
	0.0	0.7	1.4	2.9	5.7	∞
2.0 (actual)	0.84	0.85	0.93	0.97	0.99	1.00
1.0 (estimated)	0.73	0.75	0.88	0.95	0.98	1.00
0.5 (estimated)	0.60	0.62	0.82	0.92	0.98	1.00

SOURCE: *USDA* SCS-TP-106, 1950.

TABLE 41-6. COEFFICIENT OF DISCHARGE (CONTROL AT HEADWALL OPENING)

D/W	0.0	0.1	0.2	0.3	0.4	0.5	0.6	0.7	0.8	0.9	1.0
C_2	2.76	2.76	2.77	2.78	2.81	2.85	2.90	2.99	3.10	3.22	3.43

SOURCE: *USDA* SCS-TP-106, 1950.

TABLE 41-7. HEAD CORRECTION H_{02}/D FOR $D/W \geq \frac{1}{4}$
(CONTROL AT HEADWALL OPENING)
(H_{02}/D is negative.)

B/D	0.0	0.1	0.2	0.3	0.4	0.5	0.6	0.7	0.8	0.9
0	0.00	0.07	0.13	0.20	0.25	0.30	0.35	0.39	0.42	0.46
1	0.49	0.52	0.54	0.56	0.59	0.61	0.63	0.65	0.67	0.68
2	0.70	0.71	0.72	0.74	0.75	0.76	0.77	0.79	0.80	0.81
3	0.82	0.83	0.84	0.85	0.86	0.87	0.87	0.88	0.89	0.90
4	0.90	0.91	0.91	0.92	0.92	0.92	0.92	0.92	0.92	0.92
5	0.92	0.92	0.92	0.92	0.93	0.93	0.93	0.93	0.93	0.93
6	0.93	0.93	0.93	0.93	0.93	0.93	0.93	0.93	0.93	0.93
7	0.93	0.93	0.93	0.93	0.93	0.93	0.93	0.93	0.93	0.93
8	0.93	0.93	0.93	0.93	0.93	0.93	0.93	0.93	0.93	0.93

SOURCE: *USDA* SCS-TP-106, 1950.

additional crest length as the head increases. Minshall gives the formula[1]

$$Q = 3.5LH^{3/2} + 9H^{5/2} \tag{41-10}$$

for values of H/W between 0.2 and 0.5 and states that the only correction required is that for B/W given in Table 41-3.

Submergence corrections to the discharge are necessary if the tailwater depth is close to or above the crest of the box inlet. These corrections are given in the above references by Blaisdell and Donnelly.

[1] N. E. Minshall, Discussion of the Box Inlet Drop Spillway and Its Outlet, *Proc. ASCE*, sep. no. 719, June, 1955.

TABLE 41-8. CRITICAL DEPTHS, IN FEET, FOR VARIOUS DISCHARGES PER FOOT OF WIDTH

Q/W, cfs/ft	0.0	0.1	0.2	0.3	0.4	0.5	0.6	0.7	0.8	0.9
0	0.000	0.068	0.107	0.141	0.171	0.198	0.224	0.248	0.271	0.293
1	0.314	0.335	0.355	0.374	0.393	0.412	0.430	0.448	0.465	0 482
2	0.499	0.515	0.532	0.548	0.563	0.579	0.594	0.609	0.624	0.639
3	0.654	0.668	0.683	0.697	0.711	0.725	0.738	0.752	0.765	0.779
4	0.792	0.805	0.818	0.831	0.844	0.857	0.869	0.882	0.894	0.907
5	0.919	0.931	0.943	0.955	0.967	0.979	0.991	1.003	1.015	1.026
6	1.038	1.049	1.061	1.072	1.084	1.095	1.106	1.117	1.128	1.139
7	1.150	1.161	1.172	1.183	1.194	1.204	1.215	1.226	1.236	1.247
8	1.257	1.268	1.278	1.289	1.299	1.309	1.319	1.330	1.340	1.350
9	1.360	1.370	1.380	1.390	1.400	1.410	1.420	1.430	1.439	1.449
10	1.459	1.469	1.478	1.488	1.498	1.507	1.517	1.526	1.536	1.545
11	1.555	1.564	1.573	1.583	1.592	1.601	1.611	1.620	1.629	1.638
12	1.648	1.657	1.666	1.675	1.684	1.693	1.702	1.711	1.720	1.729
13	1.738	1.747	1.756	1.764	1.773	1.782	1.791	1.800	1.808	1.817
14	1.826	1.835	1.843	1.852	1.860	1.869	1.878	1.886	1.895	1.903
15	1.912	1.920	1.929	1.937	1.946	1.954	1.962	1.971	1.979	1.988
16	1.996	2.004	2.012	2.021	2.029	2.037	2.045	2.054	2.062	2.070
17	2.078	2.086	2.094	2.103	2.111	2.119	2.127	2.135	2.143	2.151
18	2.159	2.167	2.175	2.183	2.191	2.199	2.207	2.215	2.222	2.230
19	2.238	2.346	2.254	2.262	2.269	2.277	2.285	2.293	2.301	2.308
20	2.316	2.324	2.331	2.339	2.347	2.354	2.362	2.370	2.377	2.385
21	2.393	2.400	2.408	2.415	2.423	2.430	2.438	2.445	2.453	2.460
22	2.468	2.475	2.483	2.490	2.498	2.505	2.513	2.520	2.527	2.535
23	2.542	2.549	2.557	2.564	2.572	2.579	2.586	2.593	2.601	2.608
24	2.615	2.623	2.630	2.637	2.644	2.652	2.659	2.666	2.673	2.680
25	2.687	2.695	2.702	2.709	2.716	2.723	2.730	2.737	2.744	2.752
26	2.759	2.766	2.773	2.780	2.787	2.794	2.801	2.808	2.815	2.822
27	2.829	2.836	2.843	2.850	2.857	2.864	2.871	2.878	2.885	2.891
28	2.898	2.905	2.912	2.919	2.926	2.933	2.940	2.946	2.953	2.960
29	2.967	2.974	2.981	2.987	2.994	3.001	3.008	3.015	3.021	3.028
30	3.035	3.042	3.048	3.055	3.062	3.068	3.075	3.082	3.089	3.095
31	3.102	3.109	3.115	3.122	3.129	3.135	3.142	3.148	3.155	3.162
32	3.168	3.175	3.181	3.188	3.195	3.201	3.208	3.214	3.221	3.227
33	3.234	3.240	3.247	3.253	3.260	3.266	3.273	3.279	3.286	3.292
34	3.299	3.305	3.312	3.318	3.325	3.331	3.338	3.344	3.350	3.357
35	3.363	3.370	3.376	3.383	3.389	3.396	3.402	3.408	3.414	3.421
36	3.427	3.433	3.440	3.446	3.452	3.459	3.465	3.471	3.478	3.484
37	3.490	3.496	3.503	3.509	3.515	3.522	3.528	3.534	3.540	3.547
38	3.553	3.559	3.565	3.571	3.578	3.584	3.590	3.596	3.603	3.609
39	3.615	3.621	3.627	3.633	3.640	3.646	3.652	3.658	3.664	3.670
40	3.676	3.683	3.689	3.695	3.701	3.707	3.713	3.719	3.725	3.731
41	3.737	3.743	3.750	3.756	3.762	3.768	3.774	3.780	3.786	3.792
42	3.798	3.804	3.810	3.816	3.822	3.828	3.834	3.840	3.846	3.852
43	3.858	3.864	3.870	3.876	3.882	3.888	3.894	3.900	3.906	3.912
44	3.918	3.924	3.929	3.935	3.941	3.947	3.953	3.959	3.965	3.971
45	3.977	3.983	3.988	3.994	4.000	4.006	4.012	4.018	4.024	4.030
46	4.035	4.041	4.047	4.053	4.059	4.065	4.070	4.076	4.082	4.088
47	4.094	4.099	4.105	4.111	4.117	4.123	4.128	4.134	4.140	4.146
48	4.152	4.157	4.163	4.169	4.175	4.180	4.186	4.192	4.198	4.203
49	4.209	4.215	4.220	4.226	4.232	4.238	4.243	4.249	4.255	4.260
50	4.266									

The outlet design of a box-inlet drop spillway is based on the *critical depth of flow* in the straight section and at the end sill. These depths are, respectively,

$$d_c = \sqrt{\frac{(Q/W)^2}{g}} \qquad (41\text{-}11)$$

$$d_{ce} = \sqrt{\frac{(Q/W_e)^2}{g}} \qquad (41\text{-}12)$$

where d_{ce} = critical depth at end sill exit, ft.

The linear dimensions are defined in Fig. 41-11. Critical depths may be found in Table 41-8. The minimum length of the outlet is

$$L_S = d_c \left(\frac{0.2}{B/W} + 1 \right) \qquad (41\text{-}13)$$

for values of $B/W \geq 0.25$. Greater lengths, even sufficient to provide a highway crossing, may be used. The sidewalls may flare from 2:1 to ∞:1.

The *minimum length* of the stilling basin is

$$L_B = \frac{L}{2B/W} \qquad (41\text{-}14)$$

for values of $B/W \geq 0.25$. Longer lengths of stilling basin may be used, but it will require less material if the straight section is lengthened to secure the same over-all outlet length.

When the stilling basin is less than $11.5 d_{ce}$ wide at its exit, the *minimum tailwater depth* over the basin floor is

$$d_2 = 1.6 d_{ce} \qquad (41\text{-}15)$$

When the stilling basin is more than $11.5 d_{ce}$ wide at its exit, the minimum tailwater depth over the basin floor is

$$d_2 = d_{ce} + 0.052 W_e \qquad (41\text{-}16)$$

However, stilling basins as wide as $11.5 d_{ce}$ may make inefficient use of the outlet.

The *height of the end sill* is

$$f = \frac{d_2}{6} \qquad (41\text{-}17)$$

Longitudinal sills are required to improve the flow distribution. They should be located as follows: (1) When the stilling basin sidewalls are parallel, the longitudinal sills may be omitted. (2) The center pair of longitudinal sills should start at the exit of the box inlet and extend through the straight section and stilling basin to the end sill. (3) When W_e is less than $2.5W$, only two sills are needed, which should be located at a distance p of $W/6$ to $W/4$ each side of the center line. (4) When W_e exceeds $2.5W$, two additional sills are required which should be located parallel to the outlet center line and midway between the center sills and the sidewalls at the exit of the stilling basin. (5) The height of the longitudinal sills is the same as the height of the end sill.

The *minimum height of the sidewalls* above the water surface at the exit of the stilling basin should be

$$t = \frac{d_2}{3} \qquad (41\text{-}18)$$

or greater. The sidewalls should extend above the tailwater surface under all conditions of flow. The wingwalls should be triangular in elevation and have a top slope of 45° with the horizontal. Top slopes as flat as 30° are permissible. Wingwalls should flare in plan at an angle of 60° with the outlet center line. Flare angles of 45° are permissible. Wingwalls parallel to the outlet center line are not recommended.

Closed-conduit Spillway. This type of spillway is made of concrete or tile pipe cradled in concrete, corrugated- or plain-metal pipe, or concrete poured in place to form a box culvert. It is used ordinarily as the principal spillway for earth dams in connection with farm ponds, gully control, floodwater retardation, permanent storage, and similar soil- and water-conservation structures. Highway culverts are one form of closed conduit spillway. The conduit sizes may range from small pipes to the large morning glory spillways used for major flood-control, water-power, and irrigation reservoirs. The fall through the spillway will vary from 6 to 40 ft for the ordinary farm and soil-conservation installation. The slope of the barrel portion of the conduit may range from horizontal to as steep as it is practical to install.

The *inlet* may take many forms. Reentrant inlets (conduit projects in past headwall) and simple square-edged inlets in headwalls will ensure that the conduct flows full only if the barrel slope is less than the friction slope of the barrel flowing full and the barrel is hydraulically long. If the barrel is hydraulically short or is on a steep slope, a well-formed or special inlet is required to ensure that the barrel fills. Culvert barrels do not necessarily flow full when the barrel is on a slope flatter than the friction slope.[1] The contraction caused by square-edged and reentrant inlets creates a higher-than-normal velocity near the inlet. If the barrel slope is so flat that this velocity cannot be maintained, the flow depth increases with distance along the barrel. The barrel will fill only if the water surface touches the barrel crown. The barrel is "long" if the water surface touches its crown and the barrel fills. The barrel is "short" if the water surface is below the barrel crown at the spillway exit. The water-surface profile is computed, using accepted methods, and beginning at the barrel entrance with the assumption that the flow area is equal to the entrance area times the entrance contraction coefficient. The contraction coefficient is 0.5 for a thin-walled, reentrant inlet and 0.61 for a square-edged inlet in a headwall. The contraction coefficient for a well-rounded inlet is 1.0, signifying that the barrel fills at the entrance. Properly proportioned well-rounded inlets, hood inlets, and drop inlets are examples of inlets that will ensure full conduits for any barrel slope. The radius of rounding for a well-rounded inlet set in a headwall perpendicular to the conduit center line is 0.14 pipe diameter. For a reentrant entrance the corresponding radius of rounding is 0.21 pipe diameter.

On *pipe with inlets designed to flow full,* the following two equations are required to set up stage discharge curves as shown in Fig. 41-13a.

$$Q = CLH^{3/2} \qquad \text{(weir flow)} \qquad (41\text{-}19)$$

$$Q = A\sqrt{\frac{2gH_t}{K_e + K_b + K_o + K_p l \text{ (or } K_c l)}} \qquad \text{(pipe flow)} \qquad (41\text{-}20)$$

where Q = discharge, cfs
L = length of weir, ft
H = head over inlet, ft
C = inlet coefficient ft$^{1/2}$/sec
A = area of pipe, sq ft
g = acceleration of gravity = 32.16 ft/sec^2
H_t = total acting head, ft, as shown in Fig. 41-13b (on unsubmerged pipe normally taken as difference in elevation of water at inlet pool and the point where the hydraulic grade line pierces the plane of the pipe exit)[2]
K_e = head loss at entrance
K_b = head loss at bend
K_o = head loss at outlet, generally assumed to be 1.0
K_p (or K_c) = f/D = friction head loss coefficient per ft of pipe diameter
l = length of pipe, ft
f = Darcy-Weisbach friction coefficient

The actual flow is the lesser value obtained from the two formulas.

[1] L. G. Straub et al., Importance of Inlet Design on Culvert Capacity, *Univ. Minn. St. Anthony Falls Hydraulic Lab. Tech. Paper* 13-B, 1953.
[2] L. G. Straub, et al., Hydraulic Tests of Large Concrete Pipes, *Univ. Minn. St. Anthony Falls Hydraulic Lab. Tech. Paper* 22B, 1960.

TABLE 41-9. HEAD-LOSS COEFFICIENT K_P FOR CIRCULAR PIPE, FLOWING FULL

Pipe diam, in.	Flow area, sq ft	Manning's coefficient of roughness n														
		0.010	0.011	0.012	.013	.014	.015	.016	.017	.018	.019	.020	.021	.022	.023	.024	.025
6	0.196	0.0467	0.0565	.067	.079	.091	.105	.12	.13	.15	.17	.19	.21	.23	.25	.27	.29
8	0.349	.0318	.0385	.046	.054	.062	.072	.081	.092	.103	.115	.13	.14	.15	.17	.18	.20
10	0.545	.0236	.0286	.034	.040	.046	.053	.060	.068	.077	.085	.094	.104	.114	.12	.136	.148
12	0.785	.0185	.0224	.027	.031	.036	.042	.047	.054	.060	.067	.074	.082	.090	.098	.107	.116
14	1.069	.0151	.0182	.022	.025	.030	.034	.039	.044	.049	.054	.060	.066	.073	.080	.087	.094
15	1.23	.0138	.0166	.020	.023	.027	.031	.035	.040	.045	.050	.055	.061	.067	.073	.079	.086
16	1.40	.0126	.0159	.018	.021	.025	.028	.032	.036	.041	.046	.050	.056	.061	.067	.073	.079
18	1.77	.01078	.0130	.016	.018	.021	.024	.028	.031	.035	.039	.043	.048	.052	.057	.062	.067
21	2.41	.00878	.01062	.013	.015	.017	.020	.022	.025	.028	.032	.035	.039	.042	.046	.051	.055
24	3.14	.00735	.00889	.0106	.012	.014	.017	.019	.021	.024	.027	.029	.032	.036	.039	.042	.046
27	3.98	.00628	.00760	.0090	.0106	.012	.014	.016	.018	.020	.023	.025	.028	.030	.033	.036	.039
30	4.91	.00546	.00660	.0079	.0092	.0107	.012	.014	.016	.018	.020	.022	.024	.026	.029	.031	.034
36	7.07	.00428	.00518	.0062	.0072	.0084	.0096	.0110	.012	.014	.015	.017	.019	.021	.023	.025	.027
42	9.62	.00348	.00422	.0050	.0059	.0068	.0078	.0089	.0101	.0113	.013	.014	.015	.017	.018	.020	.022
48	12.57	.00292	.00353	.0042	.0049	.0057	.0066	.0075	.0084	.0095	.0105	.012	.013	.014	.015	.017	.018
54	15.90	.00249	.00302	.0036	.0042	.0049	.0056	.0064	.072	.0081	.0090	.0100	.0110	.012	.013	.014	.016
60	19.63	.00217	.00262	.0031	.0037	.0042	.0049	.0055	.0063	.0070	.0078	.0087	.0096	.0105	.0115	.0125	.0135

SOURCE: USDA SCS, National Design Division, 1947.

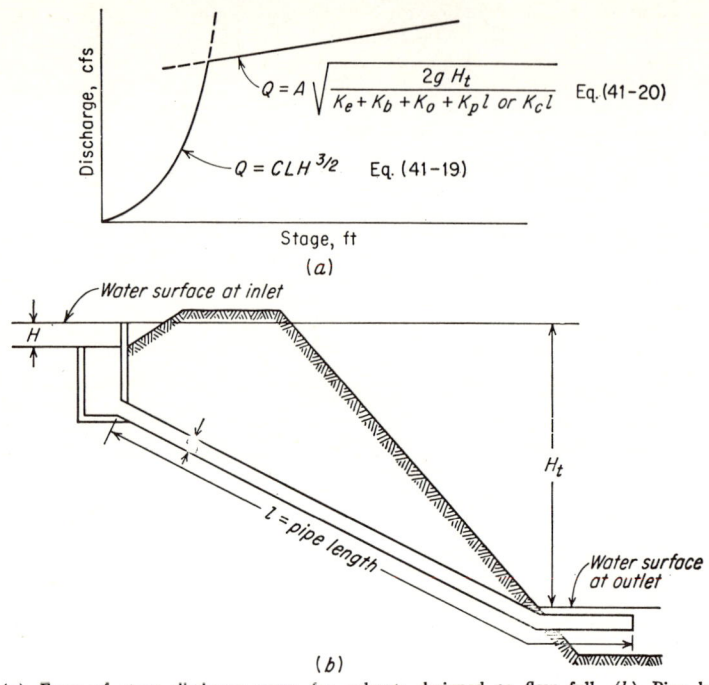

Fig. 41-13. (a) Form of stage discharge curve for culverts designed to flow full. (b) Pipe layout showing H_t, H, L, and d.

Table 41-10. Head-loss Coefficient K_c for Square Conduit* Flowing Full

Conduit size,* ft	Flow area,* sq ft	Manning's coefficient of roughness n				
		0.012	0.013	0.014	0.015	0.016
2½ × 2½	5.75	0.0074	0.0087	0.0101	0.0116	0.013
3 × 3	8.50	0.0058	0.0068	0.0079	0.0091	0.0103
3½ × 3½	11.75	0.0047	0.0055	0.0064	0.0074	0.0084
4 × 4	15.50	0.0040	0.0046	0.0054	0.0062	0.0070
4½ × 4½	19.75	0.0034	0.0040	0.0046	0.0053	0.0060
5 × 5	24.50	0.0030	0.0035	0.0040	0.0046	0.0053
5½ × 5½	29.75	0.0026	0.0031	0.0036	0.0041	0.0046
6 × 6	35.50	0.0023	0.0027	0.0032	0.0036	0.0041
6½ × 6½	41.75	0.0021	0.0025	0.0029	0.0033	0.0037
7 × 7	48.50	0.0019	0.0022	0.0026	0.0030	0.0034
7½ × 7½	55.75	0.0017	0.0020	0.0024	0.0027	0.0031
8 × 8	63.50	0.0016	0.0019	0.0022	0.0025	0.0028
8½ × 8½	71.75	0.0015	0.0017	0.0020	0.0023	0.0026
9 × 9	80.50	0.0014	0.0016	0.0019	0.0021	0.0024
9½ × 9½	89.75	0.0013	0.0015	0.0017	0.0020	0.0023
10 × 10	99.50	0.0012	0.0014	0.0016	0.0019	0.0021

* Flow area based on the use of 6-in. fillets on all four corners of conduit.
SOURCE: USDA SCS, National Design Division, 1947.

To aid in solving this equation for circular pipe, Table 41-9 has been set up. Values from this table for K_p can be substituted in the above formulas. Table 41-10 gives comparable figures for a square conduit. Values for solving the two basic formulas, relationships and design of inlets, and antivortex devices for common layouts designed to flow full are shown in Fig. 41-13c.

Antivortex walls for drop inlets may be tangent to one side of the drop inlet, may cross (split) the center of the drop inlet, or may be a cover supported over the drop inlet. The latter two arrangements are most efficient. Splitter antivortex walls should be at least one pipe diameter high and extend one pipe diameter outside the drop inlet. Little is known about the dimensions of cover antivortex devices, but it is suggested that they extend well beyond the edges of the drop inlet and that they be supported at a height above the drop-inlet crest equal to the head over the crest at which full conduit flow first occurs.

The form of head-discharge curve, Fig. 41-13a, makes the closed conduit spillway an excellent flood control device. At low flows, the capacity of the spillway increases rapidly as the head increases. This is characteristic of weir control, Eq. (41-19). The rate of rise of the reservoir level and the resulting depletion of the available reservoir storage capacity can be controlled if the weir length can be adjusted. After a pipe spillway flows completely full, the capacity of the spillway increases slowly as the reservoir level rises. Any inflow to the reservoir in excess of the spillway capacity must be stored until after the peak of the inflow hydrograph has been passed. The maximum capacity of the spillway can be adjusted to fit the capacity of the downstream channel by choosing the proper conduit size. This type of design requires that sufficient capacity be available to store that part of the runoff which exceeds the spillway capacity as explained later in this chapter. An emergency spillway, perhaps vegetated, should always be provided whenever a closed-conduit spillway is designed to flow full.

(Reinforced Concrete) Monolithic Drop Inlets with Curved Elbow

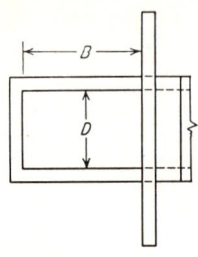

Minimum inlet dimensions:

Area of inlet = 1.5 × area of conduit
Depth of inlet (Z) = $2D$

Weir flow formula:

$$Q = 3.4(L + 4R - \text{any obstruction})H^{3/2}$$

where
Q = discharge, cfs
$L = 2B + D$
R = radius of curvature of lip, ft
H = head over weir, ft
any obstruction = width of guard rail posts, etc., ft

Conduit flow formula:

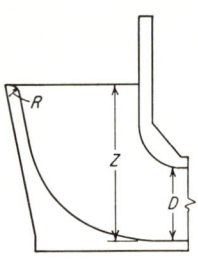

$$Q = A \sqrt{\frac{2gH_t}{1 + K_e + K_b + K_c l}}$$

where Q = discharge, cfs
A = area of conduit = D^2—fillets, sq ft
H_t = effective head causing conduit discharge, ft
$K_e + K_b$ = head-loss coefficient at entrance plus head-loss coefficient at bend
  = 0.15
K_c = conduit-friction head-loss coefficient
n = 0.012
l = length of conduit, ft
g = acceleration of gravity
  = 32.2 ft/sec²

Standard R/C Drop Inlets with Tongue and Groove R/C Pipe

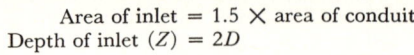

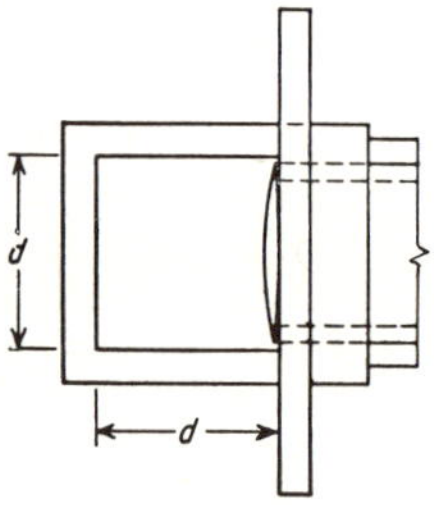

Minimum inlet dimensions:

Area of inlet = 1.5 × area of conduit
Depth of inlet (Z) = $2D$

Weir flow formula:

$$Q = 3.4(L + 4R - \text{any obstruction})H^{3/2}$$

where
Q = discharge, cfs
$L = 3d$
R = radius of curvature of lip, ft
H = head over weir, ft
any obstruction = width of guard rail posts, etc., ft

Pipe flow formula:

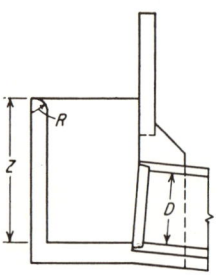

$$Q = A \sqrt{\frac{2gH_t}{1 + K_e + K_b + K_p l}}$$

where Q = discharge, cfs
A = area of pipe, sq ft
H_t = effective head causing pipe discharge, ft
$K_e + K_b$ = head-loss coefficient at entrance plus head-loss coefficients at bend
  = 0.65
K_p = pipe-friction head-loss coefficient
n = 0.012
l = length of pipe, ft
g = acceleration of gravity
  = 32.2 ft/sec²

Female pipe joint upstream is accepted as a rounded edge entrance

FIG. 41-13c. Dimensions, specifications, and flow formulas for pipe inlet designed to flow full.

Hood Inlets on Pipe Conduits with Square-edged Entrance[1]

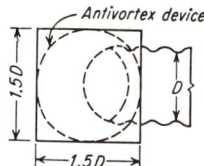

Weir flow formula:

$$\frac{Q}{D^{5/2}} = \left(1.83 S^{1/15} + 0.60 \frac{H}{D}\right) \frac{a}{A} \sqrt{\frac{H}{D}}$$

Values of $a/A \sqrt{H/D}$

H/D	.00	.01	.02	.03	.04	.05	.06	.07	.08	.09
0.0	.0000	.0002	.0007	.0015	.0027	.0042	.0060	.0081	.0106	.0134
0.1	.0164	.0199	.0236	.0275	.0318	.0364	.0413	.0465	.0519	.0577
0.2	.0637	.0700	.0765	.0833	.0904	.0978	.1053	.1132	.1213	.1296
0.3	.1382	.1470	.1561	.1653	.1748	.1845	.1945	.2046	.2149	.2255
0.4	.2362	.2472	.2583	.2696	.2811	.2928	.3046	.3166	.3287	.3411
0.5	.3536	.3661	.3789	.3918	.4048	.4179	.4312	.4445	.4580	.4716
0.6	.4843	.4990	.5128	.5267	.5407	.5547	.5688	.5829	.5971	.6114
0.7	.6256	.6398	.6540	.7783	.6825	.6967	.7110	.7251	.7392	.7531
0.8	.7670	.7809	.7947	.8083	.8218	.8352	.8485	.8614	.8742	.8869
0.9	.8994	.9115	.9232	.9347	.9457	.9565	.9667	.9763	.9852	.9933
1.0	1.0000									

Source: *Univ. Minn. St. Anthony Falls Tech. Paper* 20B, 1958.

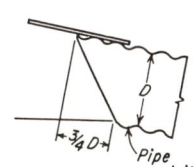

Slug flow formula (before full steady flow starts):

$$\frac{H}{D} = 1.05 + 0.025 \frac{Q}{D^{5/2}}$$

where Q = discharge, cfs
D = pipe diameter, ft
S = pipe slope (sine of angle)
H = head over invert of inlet
a = wetted pipe area caused by h
A = area of pipe, sq ft

Pipe flow formula:

$$Q = A \sqrt{\frac{2gH_t}{1 + K_e + K_p l}}$$

where Q = discharge, cfs
A = area of pipe, sq ft
H_t = effective head causing pipe flow
l = length of pipe, ft
g = acceleration of gravity
 = 32.2 ft/sec²

K_e = head-loss coefficient at entrance
 = 1.08 for CMP (corrugated metal pipe) and smooth steel pipe
K_p = pipe-friction head-loss coefficient
n = 0.025 for CMP
 = 0.010 for smooth pipe
 = 0.023 for helical CMP

Fig. 41-13c. (*Continued*)

[1] F. W. Blaisdell and C. A. Donnelly, Hydraulics of Closed Conduit Spillways, part x, The Hood Inlet, *Univ. Minn. St. Anthony Falls Hydraulic Lab. Tech. Paper* 20B, 1958.

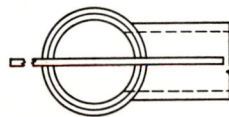

Plotting coordinates for weir and slug flow formula

Pipe diam	b		c		d		e	
	h	Q	h	Q	h	Q	h	Q
12	0.50	0.7	1.00	2.2	1.05	0	1.3	10
15	0.62	1.2	1.25	3.8	1.31	0	1.6	20
18	0.75	1.8	1.50	6.0	1.57	0	2.0	30
21	0.87	2.7	1.75	8.8	1.84	0	2.3	40
24	1.00	3.7	2.00	12.3	2.10	0	2.6	60
30	1.25	6.5	2.50	21.5	2.62	0	3.2	90
36	1.50	10.3	3.00	33.8	3.15	0	3.8	140
42	1.75	15.1	3.50	49.5	3.68	0	4.4	200
48	2.00	21.1	4.00	69.4	4.20	0	5.1	280

(Weir flow based on $S = 0.10$)

R/C Pipe Drop Inlet with Rounded Lip R/C Pipe Conduit

Minimum inlet dimensions:

Area of inlet = 1.5 × area of conduit
Depth of inlet $(Z) = 2D$

Weir flow formula:

$$Q = 3.4(\pi d_1 - \text{any obstruction})H^{3/2}$$

where Q = discharge, cfs
d_1 = inside diameter of pipe bell or groove
H = head over weir, ft
any obstruction = 2 × width of antivortex device, etc.

Pipe flow formula:

$$Q = A\sqrt{\frac{2gH}{1 + K_e + K_b + K_p l}}$$

where Q = discharge, cfs
A = area of pipe D, sq ft
H_t = effective head causing pipe discharge, ft
$K_e + K_b$ = head-loss coefficient at entrance plus head-loss coefficient at bend
       = 0.65
K_p = pipe-friction head-loss coefficient
n = 0.012
l = length of pipe, ft
g = acceleration of gravity
   = 32.2 ft/sec^2

Female pipe joint used as the crest of riser is accepted as a rounded-edge entrance.

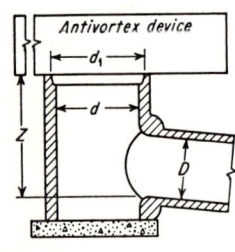

Inlet proportions		Antivortex device	
Pipe D conduit, in.	Pipe d riser, in.	Length, ft (min.)	Height,* ft
8–12	18	5	2
15	24	6	2
18	30	7	2
21	30	7	2
24	36	8	2.5
30	42	9	2.5
36	48	10	3
42	54	11	3.5
48	60	12	4

* Height may be reduced to maximum depth of water.

Fig. 41-13c. (*Continued*)

CMP (Corrugated Metal Pipe) Drop Inlets with CMP Conduit

Minimum inlet dimensions:

CMP is considered square entrance
Area of inlet = 1.5 × area of conduit
Depth of inlet $(Z) = 5D$

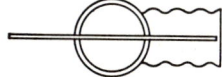

Weir flow formula:

$$Q = 3.4(\pi d - \text{any obstruction})H^{3/2}$$

where Q = discharge, cfs
d = diameter of inlet pipe, ft
H = head over weir, ft
any obstruction = 2 × width of antivortex device plus any other obstruction

Pipe flow formula:

$$Q = A\sqrt{\frac{2gH_t}{1 + K_e + K_b + K_p l}}$$

where Q = discharge, cfs
A = area of pipe, D, sq ft
H_t = effective head causing pipe discharge, ft
$K_e + K_b$ = head-loss coefficient at entrance plus head-loss coefficient at bend
= 1.0
K_p = pipe-friction head loss
n = 0.023 helical CMP
= 0.025 CMP
g = acceleration of gravity
= 32.2 ft/sec²
l = length of pipe, ft

Inlet proportions		Antivortex device		
Pipe D, conduit-in.	Pipe d, riser-in.	Length, ft-min	Height,* ft	Gage corrugated metal sheets
8–12	18	5	2	16
15	24	6	2	16
18	30	7	2	16
21	30	7	2	16
24	36	8	2.5	14
30	42	9	2.5	14
36	48	10	3	12
42	54	11	3.5	12
48	60	12	4	10

* Height may be reduced to maximum depth of water.

FIG. 41-13c. *(Continued)*

The pressures within the spillway should never be permitted to approach absolute zero. The maximum vacuum permitted should probably never exceed 25 ft of water. The pressure may be computed from the equation

$$h_p = \left(\frac{h_n}{h_{vp}}\right) h_{vp} - z - h_f \tag{41-21}$$

where h_p = pressure in spillway above or below atmospheric, ft.

$\frac{h_n}{h_{vp}}$ = local pressure constant. (This is zero unless local disturbances such as the drop inlet, the barrel entrance, bends or other disturbances affect the normal pattern of flow. See Fig. 41-14 for values with various drop-inlet proportions.)

h_n = difference between the friction and hydraulic grade lines, ft.
h_{vp} = velocity head in barrel, ft.
z = elevation of point above hydraulic grade line at conduit exit, ft.
h_f = friction head loss between point under consideration and outlet of conduit, ft, and is equal to $K_p l$ as discussed in Eq. (41-20).

Minimum values of the local pressure constant for a number of entrances are given in Fig. 41-14.

Z_1	D_1	S	h_n/h_{vp} minimum
5.0 D	1.25 D	0.025	-1.3
5.0 D	1.25 D	0.05	-1.2
5.0 D	1.25 D	0.10	-1.2
5.0 D	1.25 D	0.20	-1.1
5.0 D	1.25 D	0.30	-1.1
4.0 D	1.25 D	0.30	-1.0
2.0 D	1.25 D	0.30	-1.3
3.5 D	2.0 D	0.30	-0.8
3.5 D	1.5 D	0.30	-0.9
3.5 D	1.0 D	0.30	-1.4

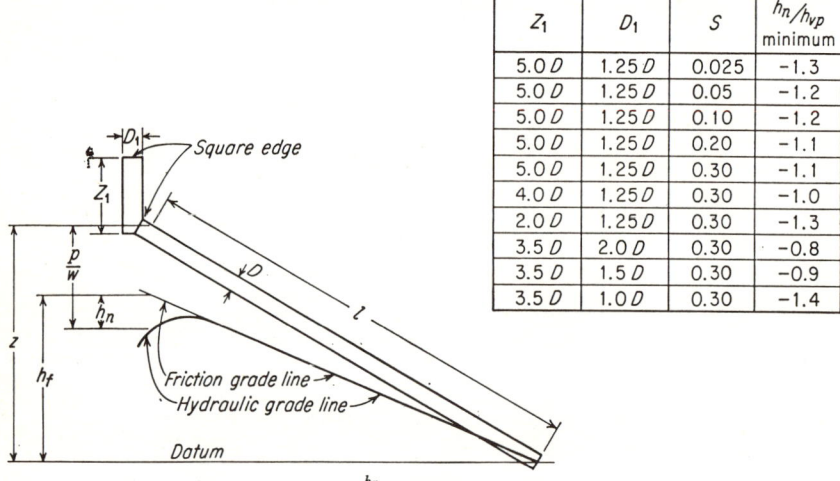

FIG. 41-14. Values of $\frac{h_n}{h_{vp}}$ for various drop inlet proportions.

Pressures within the spillway may *fluctuate* at some discharges. The alternating pressure and vacuum will draw the fines out of the soil around the pipe and lead to ultimate failure if the joints are not absolutely watertight. Special jointing material, such as rubber gaskets, should always be used for concrete pipe, and reliance should not be placed on ordinary mortared joints. Where high pressures are encountered, corrugated-metal pipe for closed-conduit spillways should be watertight.

Chute Spillway. A chute is used as a spillway when the drop through the structure is greater than can be economically handled with a straight-drop spillway or a box-inlet drop spillway. It can replace a drop-inlet spillway when large discharges are required.

A *level-floor chute entrance* is shown in Fig. 41-15. Either the rounded or the flared entrance walls may be used. The discharge[1] may be estimated from the equation

$$Q = 3.4 \, WH^{3/2} \tag{41-22}$$

[1] F. W. Blaisdell and A. N. Huff, Report on Tests Made on Three Types of Flume Entrance, *USDA* SCS-TP-70, 1948.

where Q = discharge, cfs
H = head on entrance, ft
W = chute width, ft

The water-surface profile at the upper end of the chute should be determined by referring to the original paper.[1]

A typical box-inlet chute-spillway entrance is shown in Fig. 41-16. The box-inlet proportions and capacity may be computed as for a box-inlet drop spillway. The

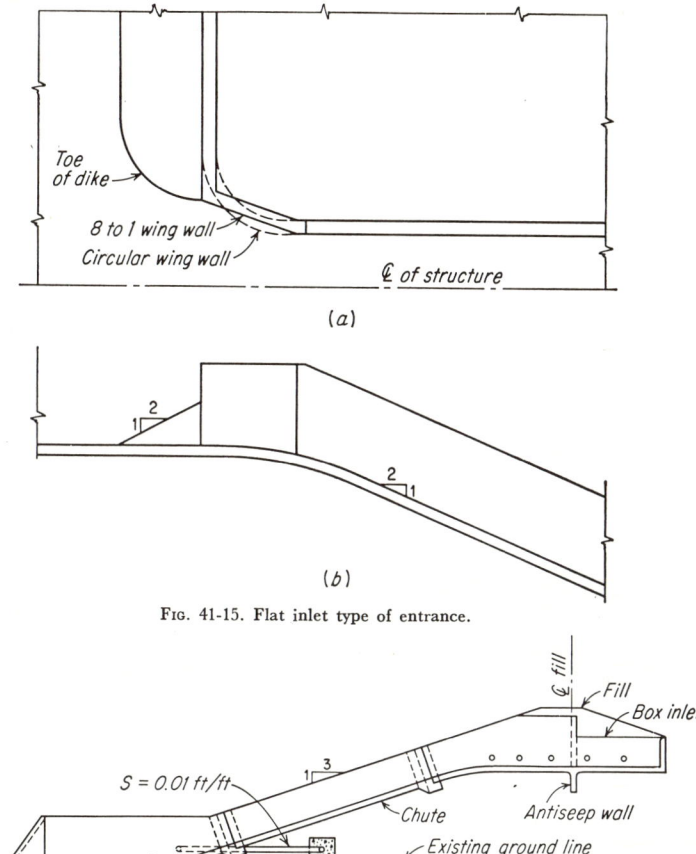

FIG. 41-15. Flat inlet type of entrance.

FIG. 41-16. Box-inlet chute spillway.

minimum distance between the headwall opening and the beginning of the floor curve should be L_s (Eq. 41-13) to permit some smoothing of the very rough water surface before entering the chute. For purposes of computing the water-surface profile and sidewall height in the chute, critical depth may be assumed to occur at the beginning of the floor curve.

OUTLETS

Outlets for closed-conduit and chute spillways may be simple aprons, a cantilevered extension of the closed conduit or chute, or some type of energy dissipator.

[1] *Ibid.*

A simple *horizontal apron* has proved generally satisfactory for the smaller structures having only occasional flows at the design capacity. In the case of pipe spillways, the upstream end of the apron provides a support for the end of the pipe, the sidewalls direct the flow into the downstream channel, and the floor prevents scour close to the end of the pipe. A typical apron outlet is shown in Fig. 41-17. The apron should be long enough so that the jet from the pipe lands well upstream from the end of the apron. An equation for the path of the jet is

$$y = x \tan \theta + \frac{16.1}{V^2 \cos^2 \theta} x^2 \tag{41-23}$$

The symbols are defined in Fig. 41-17. In the case of a chute spillway the apron serves to turn the stream into a horizontal direction. In either case the apron should

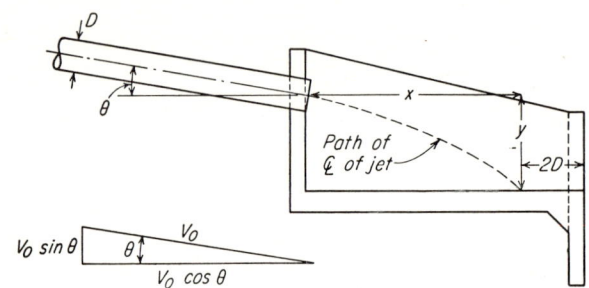

Fig. 41-17. Apron outlet for a pipe.

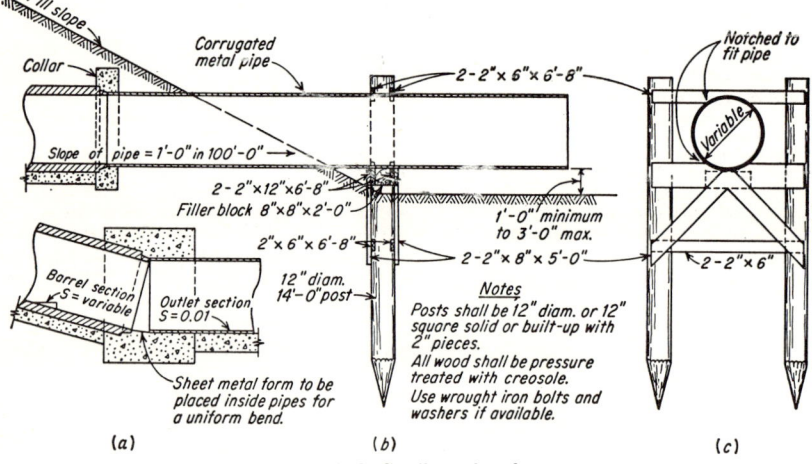

Fig. 41-18. Cantilevered outlet.

be at stable stream-bed elevation or below and the bed material should be only slowly erodible.

The *cantilevered outlet* (Fig. 41-18) is being used rather extensively. This outlet is formed by making the end of the spillway horizontal and cantilevering it beyond a supporting bent. At present, cantilevered outlets should be used with full realization that strongly erosive side whirls may develop that will greatly widen the channel and may eventually reach upstream to undermine the pipe or the chute.[1]

[1] F. W. Blaisdell and C. A. Donnelly, Hydraulic Model Studies for Whiting Field Naval Air Station, *Univ. Minn. St. Anthony Falls Hydraulic Lab. Project Rept.* 23, 1950.

The *impact type of energy dissipator*[1] shown in Fig. 41-19 may be used when it is necessary to completely dissipate the energy at pipe outlets. The procedures and rules for the design of this basin are:

1. Use of this stilling basin is limited to cases where the velocity at the entrance to the stilling basin is about 30 fps or less.
2. From the maximum expected discharge, determine the stilling-basin dimensions as shown in Fig. 41-19, using Table 41-11, columns 3 to 13. The use of multiple units side by side may prove economical in some cases.
3. Compute the necessary pipe area from the velocity and discharge. The values in Table 41-11, columns 1 and 2, are suggested sizes based on a velocity of 12 fps and so that the pipe may run full at the discharge given in column 3. Regardless of

TABLE 41-11. STILLING-BASIN DIMENSIONS
(Impact-type Energy Dissipator)

Suggested pipe size*		Maximum discharge	Feet and inches										Inches				
Diam, in.	Area, sq ft	Q	W	H	L	a	b	c	d	e	f	g	t_w	t_f	t_b	t_p	K
(1)	(2)	(3)	(4)	(5)	(6)	(7)	(8)	(9)	(10)	(11)	(12)	(13)	(14)	(15)	(16)	(17)	(18)
18	1.7672	21†	5-6	4-3	7-4	3-3	4-1	2-4	0-11	0-6	1-6	2-1	6	6½	6	6	3
24	3.1416	38	6-9	5-3	9-0	3-11	5-1	2-10	1-2	0-6	2-0	2-6	6	6½	6	6	3
30	4.9087	59	8-0	6-3	10-8	4-7	6-1	3-4	1-4	0-8	2-6	3-0	6	6½	7	7	3
36	7.0686	85	9-3	7-3	12-4	5-3	7-1	3-10	1-7	0-8	3-0	3-6	7	7½	8	8	3
42	9.6211	115	10-6	8-0	14-0	6-0	8-0	4-5	1-9	0-10	3-0	3-11	8	8½	9	8	4
48	12.5664	151	11-9	9-0	15-8	6-9	8-11	4-11	2-0	0-10	3-0	4-5	9	9½	10	8	4
54	15.9043	191	13-0	9-9	17-4	7-4	10-0	5-5	2-2	1-0	3-0	4-11	10	10½	10	8	4
60	19.6350	236	14-3	10-9	19-0	8-0	11-0	5-11	2-5	1-0	3-0	5-4	11	11½	11	8	6
72	28.2743	339	16-6	12-3	22-0	9-3	12-9	6-11	2-9	1-3	3-0	6-2	12	12½	12	8	6

* Suggested pipe will run full when velocity is 12 fps or half full when velocity is 24 fps. Size may be modified for other velocities by $Q = AV$, but relation between Q and basin dimensions shown must be maintained.
† For discharges less than 21 sec-ft, obtain basin width from curve of Fig. 41-19. Other dimensions proportional to W; $H = 3W/4$, $L = 4W/3$, $d = W/6$, etc.

SOURCE: *U.S. Bur. Reclamation, Hyd. Lab. Rept.* 399, Denver, Colo., 1955.

the pipe size chosen, maintain the relation between discharge and basin size given in the table. An open-channel entrance may be used in place of a pipe. The approach channel should be narrower than the basin, with invert elevation the same as the pipe.

4. Although tailwater is not necessary for successful operation, a moderate depth of tailwater will improve the performance. For best performance set the basin so that maximum tailwater does not exceed $d + g/2$ (Fig. 41-19).
5. The thickness of various parts of the basin as used in the Commissioner's Office, U.S. Bureau of Reclamation, Denver, Colo., is given in Table 41-11, columns 14 to 18.
6. The entrance pipe or channel may be tilted downward about 15° without affecting performance adversely. For greater slopes use a horizontal or sloping pipe (up to 15°) two or more diameters long just upstream from the stilling basin. Maintain proper elevation of invert at entrance as shown on the drawing.
7. If a hydraulic jump is expected to form in the downstream end of the pipe and the pipe entrance is sealed by incoming flow, install a vent about one-sixth the pipe diameter at any convenient location upstream from the jump.
8. For best possible operation of basin, use alternate end sill and 45° wall design shown on Fig. 41-19 to reduce the erosion tendencies.

A stilling basin should be used for all moderate- and large-size spillways and for detention storage flood-control structures, where the design flow can be expected to occur at frequent intervals and persist for long periods of time. There are a number

[1] J. N. Bradley and A. J. Peterka, Progress Report II, Research Study on Stilling Basins, Energy Dissipators and Associated Appurtenances, *U.S. Bur. Reclamation, Hyd. Lab. Rept.* 399, Denver, 1955.

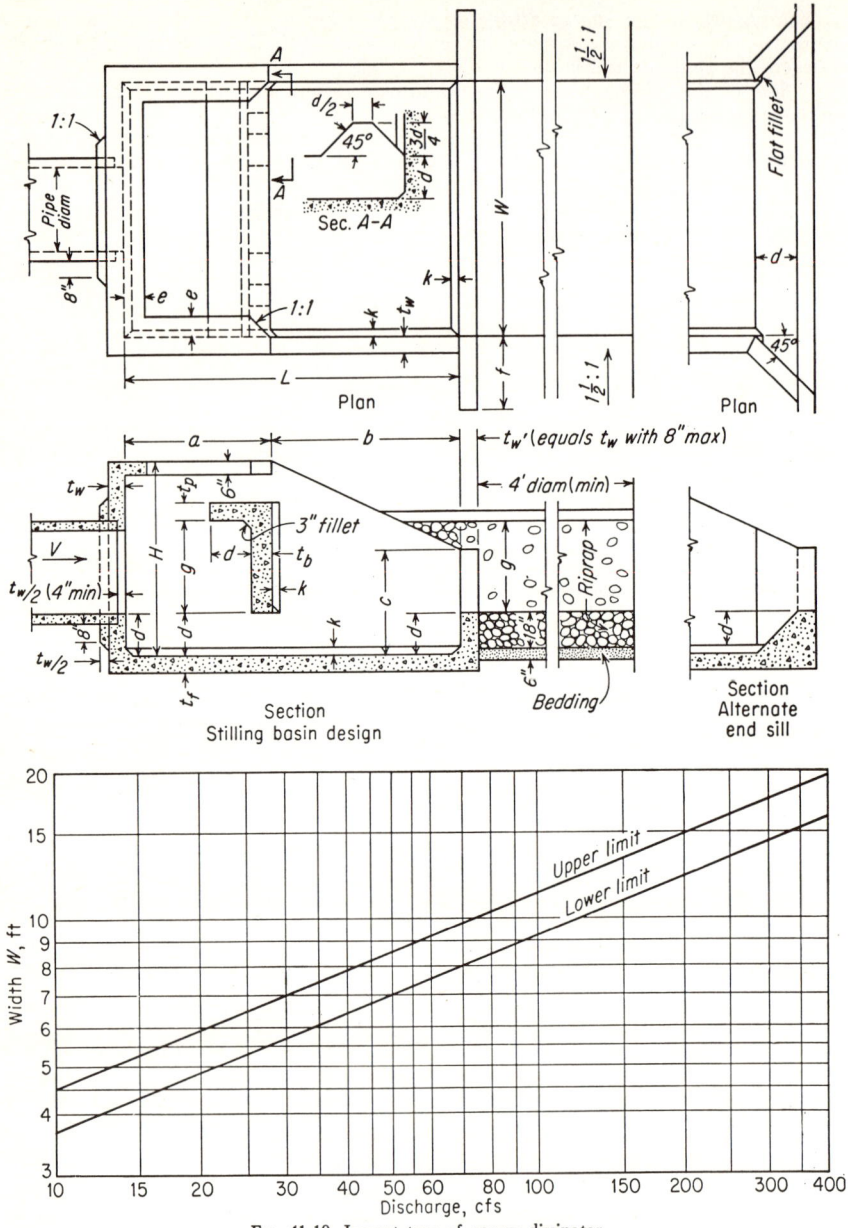

Fig. 41-19. Impact type of energy dissipator.

of stilling basins suitable for this purpose. The hydraulic jump basin[1] is simple, but it is also quite large. The rectangular stilling basin[2] developed by the U.S. Bureau of Reclamation is somewhat smaller and is an excellent performer. Probably the

[1] B. A. Bakhmeteff and A. E. Matzke, The Hydraulic Jump in Terms of Dynamic Similarity, *Trans. ASCE*, 101:630, 1936. Bradley and Peterka, *op. cit.*
[2] J. E. Warnock, Spillways and Energy Dissipators (Proceedings of Hydraulics Conference), *Univ. Iowa Studies in Eng. Bull.* 20, p. 142, 1940.

smallest and most economical energy dissipator is the SAF stilling basin[1] shown in Fig. 41-20. The equations for computing the proportions of the various elements of the SAF stilling basin are

$$\mathbf{F} = \frac{V_1^2}{gd_1} \tag{41-24}$$

$$d_2 = \frac{d_1}{2}(-1 + \sqrt{8\mathbf{F}+1}) \tag{41-25}$$

$$d_2' = 1.4 d_1 \mathbf{F}^{0.45} \tag{41-26}$$

$$L_B = \frac{4.5 d_2}{\mathbf{F}^{0.38}} \tag{41-27}$$

$$z = \frac{d_2}{3} \tag{41-28}$$

$$c = 0.07 d_2 \tag{41-29}$$

where, in addition to the symbols defined in Fig. 41-20, d_2 = downstream depth computed by the momentum equation for the hydraulic jump in feet, $\mathbf{F}$ = Froude number, g = acceleration due to gravity in feet per second per second, and V_1 = velocity at the entrance to the stilling basin in feet per second.

The SAF stilling basin will operate satisfactorily with tailwater levels higher than the design value if the sidewalls are high enough so that they are not overtopped. However, the sloping apron stilling basin[2] is suggested for high tailwater levels at the exit of chutes or for chutes which discharge into gullies which have not eroded to stable grade and where the tailwater level may change during the life of the structure. For the latter case, the chute is extended below the stream bed to an elevation corresponding to the minimum anticipated tailwater elevation minus the tailwater depth required for the sloping apron stilling basin. As the bed erodes, the hydraulic jump moves down the chute until it ultimately reaches the design location. The tailwater depth may be computed with satisfactory accuracy from the equation

$$d_2 = \frac{d_1}{2 \cos \phi}\left(-1 + \sqrt{\frac{8\mathbf{F}\cos^3 \phi}{1 - 2K \tan \phi} + 1}\right) \tag{41-30}$$

where K is given by the curve of Fig. 41-21 and ϕ is the angle of inclination of the chute. The sidewalls of the chute must be raised so that they are not overtopped by the tailwater.

Transitions are required at the entrance to many of these stilling basins to change the flow cross section from circular or trapezoidal to rectangular.

A transition from circular to rectangular used at the entrance to an SAF stilling basin is shown in Fig. 41-22. The sides, bottom, and top are plane triangular surfaces. The corners are quadrants of oblique cones having a base diameter equal to that of the pipe. The moderate flare increases the width and the Froude number and decreases the depth of flow to provide a more economical over-all installation. The velocity along the transition is assumed to be equal to that in the pipe. The top plane surface is raised to ensure that the flow does not cling to the top of the transition and cause an unsymmetrical and nonuniform depth of flow at the transition exit.

Stilling basins that are trapezoidal in cross section are not recommended. A transition from a trapezoidal chute to rectangular is shown in Fig. 41-23, where it is used in connection with a cantilevered stilling basin. The sloping chute sidewalls are continued until they intersect the horizontal stilling-basin floor. The vertical stilling-basin sidewall spacing is slightly greater than the width of the water surface in the trapezoidal channel, and the walls extend upstream to a chute elevation equal to the maximum water-surface height in the stilling basin.

---

[1] F. W. Blaisdell, The SAF Stilling Basin, *USDA Agr. Handbook* 156, 1959.
[2] Bradley and Peterka, *op. cit.*

460 SOIL AND WATER CONSERVATION

The Froude number at the exit of a closed-conduit spillway is frequently so low that it is economical to use a transition between the outlet and the stilling basin. Transition design for flow at supercritical velocities is entirely different from that for flow at subcritical velocities. The properties of the flow in a transition having straight

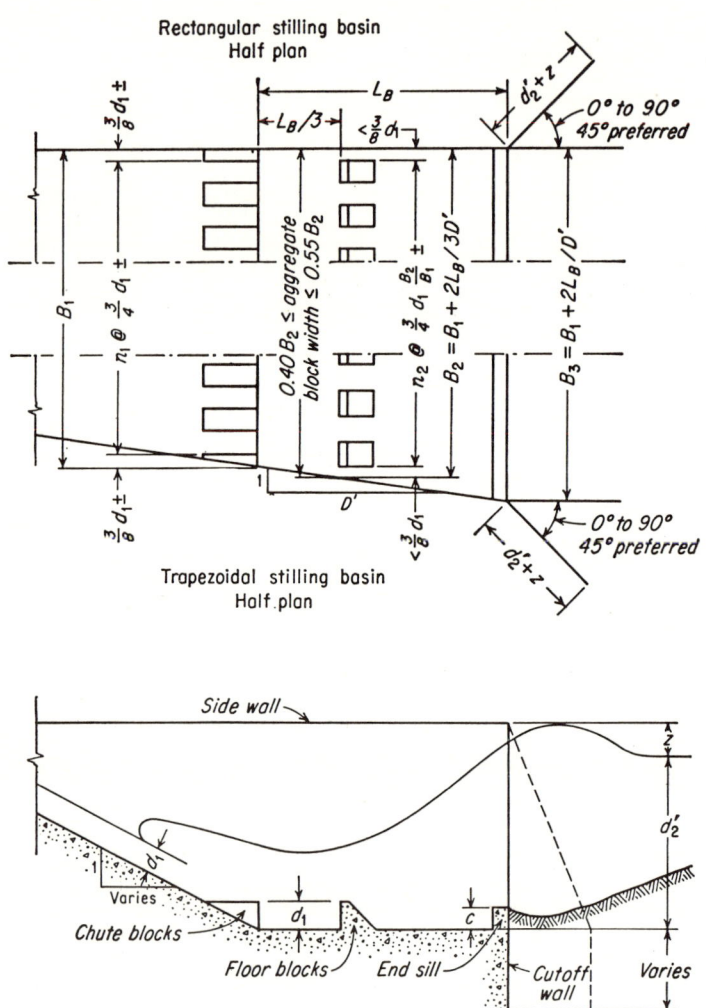

Fig. 41-20. Proportions of the SAF (St. Anthony Falls) stilling basin. (*Blaisdell*, *USDA Agr. Handbook* 156, 1959.)

flaring sidewalls and a rectangular cross section and laid on a floor having a 1 per cent slope are given in Fig. 41-24. The sidewall flare is one transverse to $3\sqrt{F}$ longitudinal. In Fig. 41-24, A is the cross-section area, b is the width, d is the depth of flow, and F is the Froude number. The subscript 1 denotes conditions at the entrance to the transition, and lack of a subscript denotes points along the transition at distances x from its entrance.

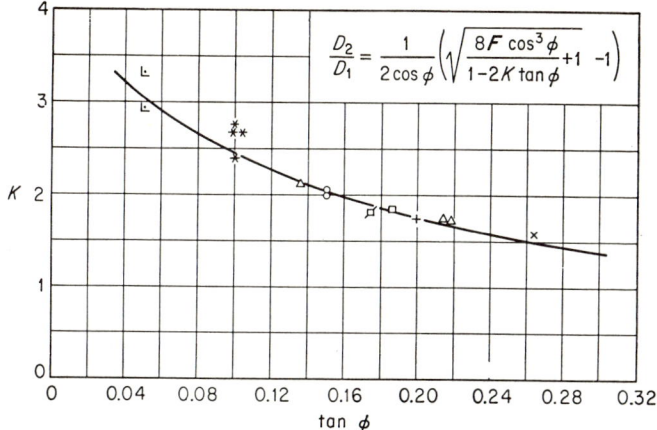

Fig. 41-21. K values for Eq. (41-30).

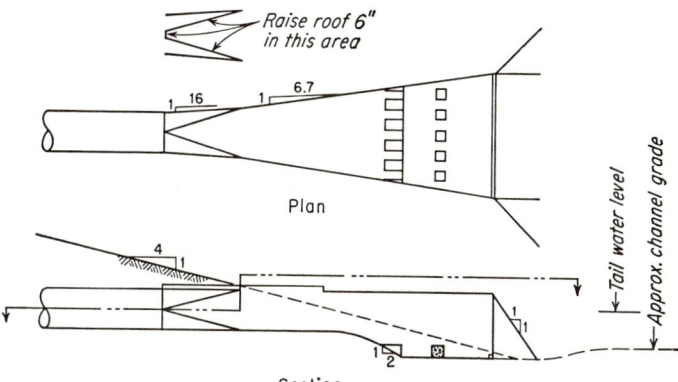

Fig. 41-22. Transition from circular to rectangular section with SAF stilling basin.

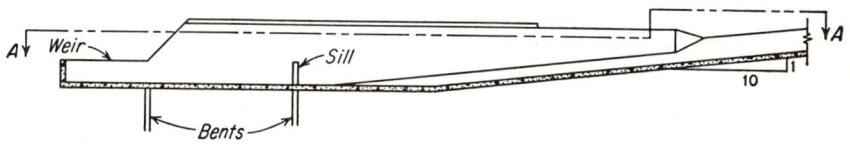

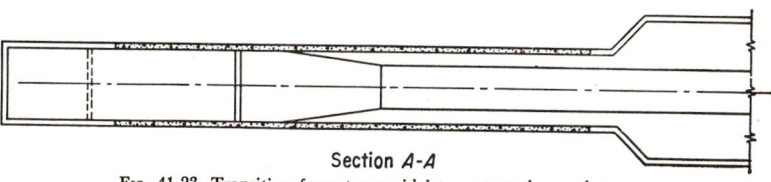

Fig. 41-23. Transition from trapezoidal to rectangular section.

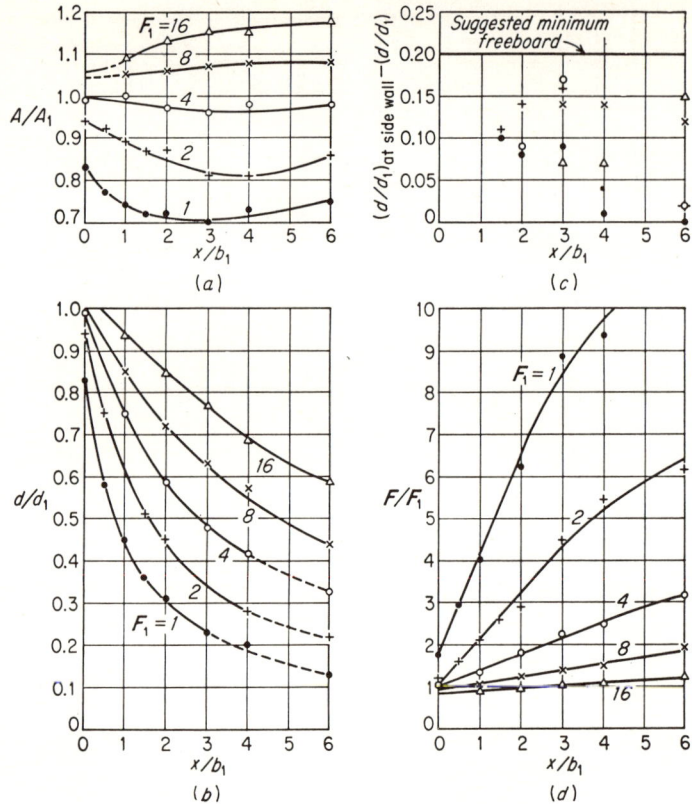

Fig. 41-24. Properties of transitions.

EMERGENCY OR AUXILIARY SPILLWAYS

Emergency or auxiliary spillways are generally constructed in earth and located at one or both ends of the embankment. Where a rock channel is available within a reasonable increase in cost, it should be utilized. Where only part of the channel is rock, the design should be based on the allowable velocities in the earth section, or the reach of the earth channel should be paved with material of life comparable with that of the rock channel. Emergency spillways sometimes are constructed of more permanent material and designed as chutes, flumes, or drop-inlet spillways, and their hydraulic characteristics and capacities are based on these spillway designs. The hydraulic design of earth spillways or chutes may be based on channel-flow criteria if the entrance is designed to provide sufficient flow to fill the crest or control section. This requires flaring the crest entrance to permit a minimum of 1.5 increase in area. The slope in the inlet channel from the headwater pool to the control section may be less than the friction slope. The best design has the channel reach a maximum elevation at the control section, with a slope toward the reservoir pool. Depending on site conditions, this slope can be small for drainage only or can be steep or excavated to reduce the head loss in the entrance channel. The position of the control section will also greatly depend on site conditions. Ordinarily the best location is as far upstream in the emergency spillway as possible. In this way the head loss in the entrance channel is held to a minimum. At many sites, the control section will be on the approximate center line of the embankment. The location of the emergency spillway through low areas or saddles immediately upstream from the structure site should be carefully considered. Saddles generally afford wider spillways than are possible at the ends of the embankment. A wider spillway will have a lower velocity through the control section and, in many cases, a lower embankment. Of most importance the saddle location provides a possibility of diverting the flow into another subchannel that will outlet a considerable distance below the embankment. Spillways at one or both ends of the embankment must be carefully designed as shown in Figs. 41-25 and 41-26, so that the discharge will not be permitted to overflow against the toe of the embankment except as backwater in the downstream channel. The exit channel below the control section must be straight for as great a distance as possible

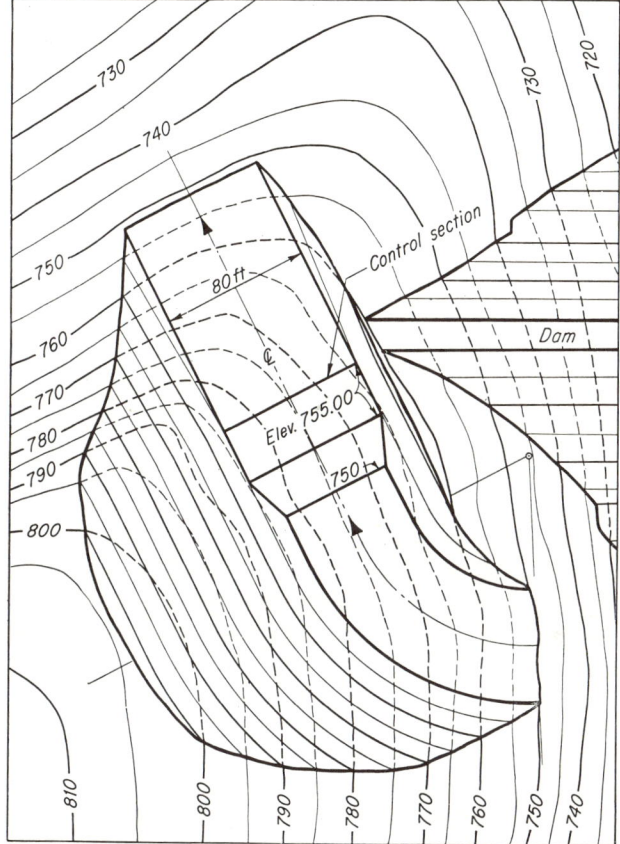

FIG. 41-25. Typical layout and grading plan for earth spillway. (*USDA SCS, Natl. Eng. Div., Design Sect.*, 1955.)

to provide this protection and to avoid nonuniform flow distribution across the channel. At many locations where a saddle is used, it is possible to locate the control section near the entrance pool or reservoir and in this way keep the entrance head loss as low as possible. Excavation of the outlet channel below the control section should be at sufficient grade to provide a suitable transition through the downstream channel and also to ensure supercritical flow below the control section. In most cases the excavated material required in this construction can be utilized in the embankment. It is also practical to allow higher velocities in the saddle spillways because serious damage to this spillway, in most cases, would not endanger the structure. Velocities in this type of emergency spillways might vary from 10 to 20 fps, depending on the thickness of the saddle and other site conditions, the soil, cover, and the estimated frequency of flow. Under the same conditions, the allowable velocities at the control section or emergencies at the end of the embankment should probably be limited to a range of 6 to 12 fps. Emergency spillways through rock can exceed 20 fps, depending on the type, density, uniformity, and other characteristics of the channel surface.

In the hydraulic design, the control section remains fixed for all significant discharges with subcritical flow in the inlet channel and supercritical flow in the exit channel. The normal depth of flow in the exit channel must not exceed the critical

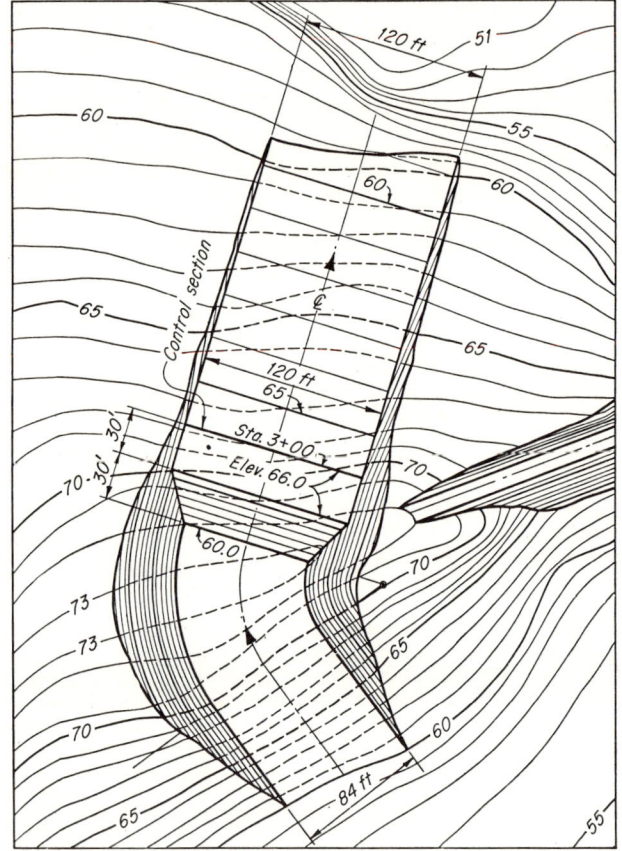

FIG. 41-26. Typical layout and grading plan for earth spillway. (*USDA SCS, Natl. Eng. Div., Design Sect.*, 1955.)

depth, if the control section is to remain fixed. Because of this, the minimum permissible slope is the critical slope, that which will just maintain steady, uniform flow at critical depth. As the flow passes beyond the control section there should be at least a slight acceleration to ensure this condition.

For the design of small dams with low failure hazard and for the preliminary investigation and design of the larger earth spillways, Culp[1] has developed the following approximate procedure, based on three assumptions: (1) the control section is a rectangle equivalent to the design shape and is without side friction; (2) the exit channel is straight grade alignment and the same cross section as the control

[1] M. M. Culp, Earth Spillways, USDA SCS Technical Release No. 5, 1955.

section; (3) a conservative estimate of friction head loss in the inlet channel is acceptable. Figure 41-27 shows the nomenclature used in the design.

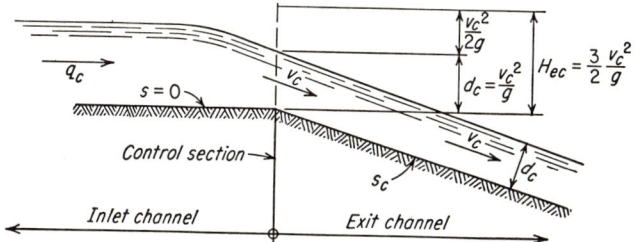

Fig. 41-27. Diagram of spillway with rectangular cross section for design calculations.

For any channel and control section of the same cross section, there exists a different and unique critical depth d_c and critical slope s_c for each different discharge q. The symbols used in this procedure are as follows:

Q = total discharge, cfs
w = width of equivalent rectangular section, ft
q_c = critical discharge per foot of width, cfs
H_{ec} = specific energy head at control section, ft
d_c = critical depth at control section, ft
v_{em} = maximum allowable velocity in exit channel, fps
v_c = critical velocity at control section, fps
g = acceleration of gravity = 32.16 ft/sec^2
s_c = critical slope, i.e., the slope which will just sustain the critical discharge q_c at critical depth d_c, ft/ft
α = correction factor
$s_{o,max}$ = maximum allowable slope in exit channel, ft/ft
H_P = difference in elevation between crest at control section and water surface in reservoir, ft
s_o = bottom slope of constructed channel, ft/ft
L = length of inlet channel, ft
b = bottom width of trapezoidal section, ft
z = side slope ratio (horizontal to vertical), ft/ft
n = Manning's friction factor

First select an allowable velocity v_c at the control section, depending on site conditions; then entering Fig. 41-28 with a known v_c, mark this value on the horizontal scale at the bottom and project vertically to the v_c curve. From this point, project horizontally to the H_{ec} curve and, continuing horizontally, determine the value of q_c on the vertical scale; also, from the point on the H_{ec} curve, project down vertically to the horizontal scale to determine the value of H_{ec}. For example, if

$$v_c = 8.0 \text{ fps}$$

then $\qquad q_c = 16 \text{ cfs} \qquad$ and $\qquad H_{ec} = 3.0 \text{ ft}$

Then entering Fig. 41-29 with Q, the design discharge for the emergency spillway, and the selected value of v_c, determine w, the width of the equivalent rectangular channel. For example, if

$$Q = 1{,}000 \text{ cfs} \qquad \text{and} \qquad v_c = 8.0 \text{ fps}$$

then $\qquad w = 63 \text{ ft}$

Bottom width b can then be found by the formula

$$b = w - zH_{ec} \tag{41-31}$$

With the side slopes of the trapezoidal channel at 3:1 and $w = 63$ and $H_{ec} = 3$ (as found above), $b = 54$ ft.

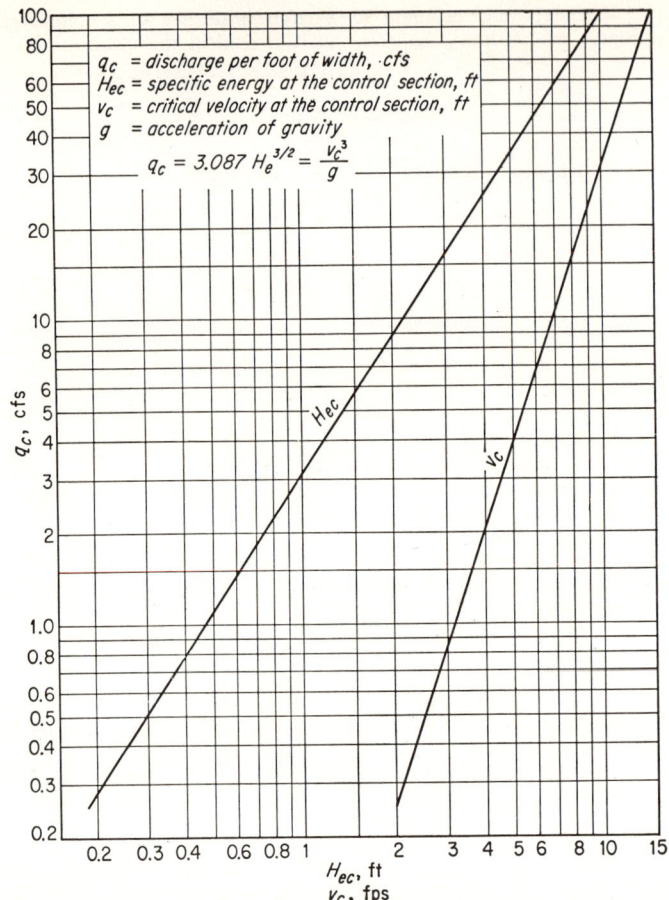

Fig. 41-28. Earth spillways: plot of q_c versus H_{ec} and q_c versus v_c. (*USDA SCS, Natl. Eng. Div., Design Sect.*, 1955.)

We should next determine the minimum slope possible in the exit channel without causing a backwater effect on the control section, defined as the critical slope for the given discharge. It should be understood that greater slopes will be acceptable up to the point of exceeding the permissible maximum velocity in the exit channel. With known values of q_c and n, we can make a direct determination of s_c. Since the critical slope is directly proportional to the square of Manning's roughness coefficient n, good design requires the use of the highest reasonable value of n for determining s_c for the exit channel below the control section. For example, with an estimated value of

$$n = 0.04 \quad \text{and} \quad q_c = 16 \text{ cfs}$$

as previously determined, we find from Fig. 41-30 that

$$s_c = 1.85 \times 10^{-2} = 0.0185 \tag{41-32}$$

To determine the maximum allowable slope in the exit channel, with a maximum allowable velocity in this channel v_{em}, use the formula

$$s_{o,\max} = s_c(v_{em} + v_c)^{10/3} \tag{41-33}$$

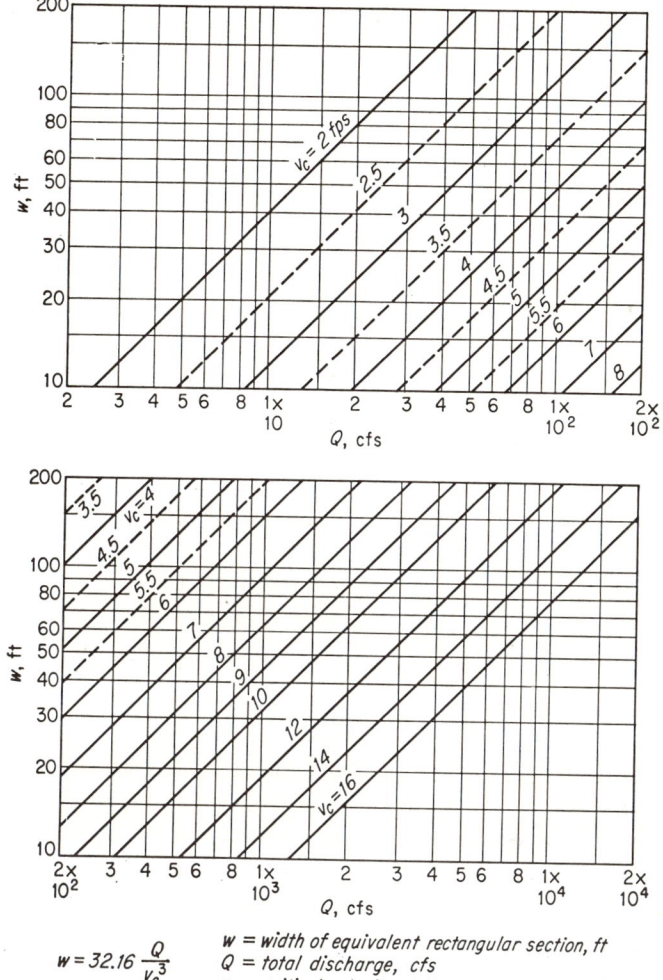

$$w = 32.16 \frac{Q}{v_c^3}$$

w = width of equivalent rectangular section, ft
Q = total discharge, cfs
v_c = critical velocity at the control section, fps

FIG. 41-29. Earth spillways: plot of relation between w and Q for various values of v_c. (USDA SCS, Natl. Eng. Div., Design Sect., 1955.)

Another important estimate is the determination of head loss in the inlet channel from the reservoir pool to the control section. This can be approximated by entering Fig. 41-31 with known values of n and H_{ec} and finding α, the correction factor for the head loss. For example, if

$$H_{ec} = 3.0 \text{ ft} \quad \text{and} \quad n = 0.04$$

then
$$\alpha = 1.6 \times 10^{-3}$$

The head loss H_p can be found by

$$H_p = H_{ec}(1 + \alpha L) \tag{41-34}$$

if
$$L = 50 \text{ ft}$$
then
$$H_p = 3.29 \text{ ft}$$

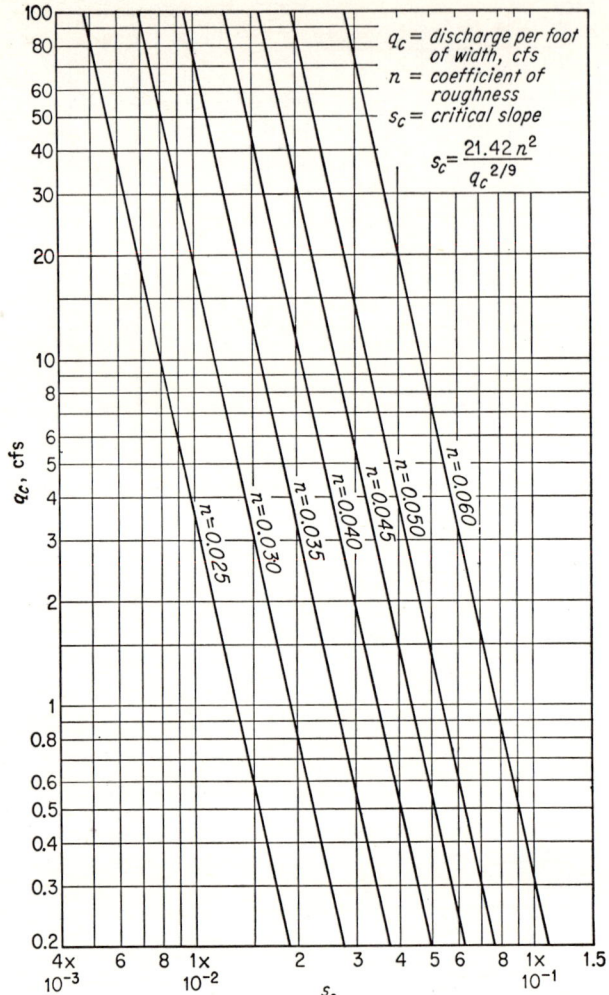

FIG. 41-30. Earth spillways: plot of s_c versus q_c. (USDA SCS, Natl. Eng. Div., Design Sect., 1955.)

A more accurate value of H_p can be made by determining a value H_{ec} at the control section and then plotting the energy gradient back to the reservoir pool. Similarly, a more accurate determination of downstream channel depths and velocities can be made by plotting a dropdown curve from the control section downstream.

MATERIALS OF CONSTRUCTION

All material used in construction should conform to specifications issued by the American Society for Testing Materials, Philadelphia, Pa. All concrete mixtures should be in accordance with recommendations of the Portland Cement Assoc., Chicago.

FLOOD ROUTING

Flood routing analyzes the disposition of the water from the design storm and is accomplished by various methods, using either algebraic or graphical solutions. This analysis should be included in the design whenever temporary storage is available.

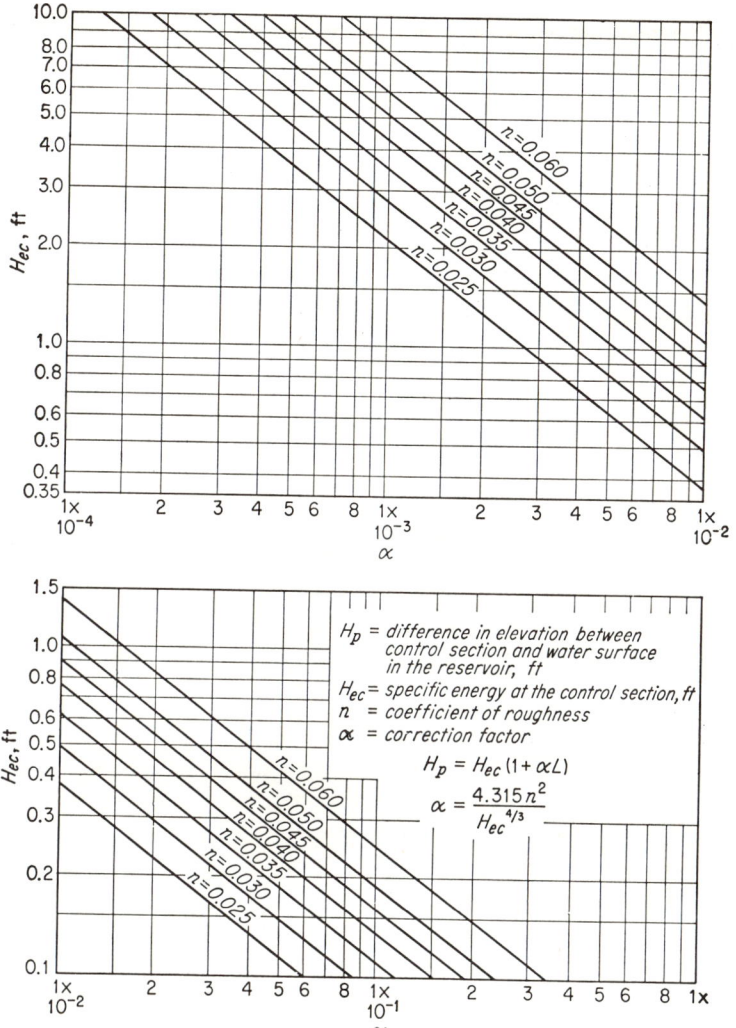

FIG. 41-31. Earth spillways: plot of relation between α and H_{ec} for given values of n. (*USDA SCS, Natl. Eng. Div., Design Sect., 1955.*)

Temporary storage is that water stored in the reservoir or other flooded land areas between the elevation of the normal pool level (crest of the low-stage spillway) and the elevation of the water surface for the design storm. The *required storage capacity* depends on the available volume above the crest of the low-stage spillway, the rate of inflow into the reservoir, and the discharge capacity of the spillway(s) at various stages. Utilizing temporary storage in the design reduces the size of the spillways, the peak rate of flow in the downstream channel, and in most cases the cost of the structure. The spillway(s) size and cost at succeeding fill heights should be compared with embankment costs to provide the most economical installation. Where a considerable volume of temporary storage is available, closed-conduit spillways are generally less expensive than open spillways.

The basic premise used in flood routing is that the total inflow into the reservoir equals the total outflow plus the change in storage. This is true for the entire period or for any part of it. Then

$$t\frac{(i_1 + i_2)}{2} = t\frac{(o_1 + o_2)}{2} + s \qquad (41\text{-}35)$$

where i_1 = inflow rate at beginning of period, cfs
i_2 = inflow rate at end of period, cfs
o_1 = outflow rate at beginning of period, cfs
o_2 = outflow rate at end of period, cfs
s = increase in storage during period, cu ft
t = length of period, sec

To be mathematically correct, the above equation would have to be stated in terms of differentials, but for all practical purposes, it is sufficiently accurate if reasonable values are chosen for T, the conversion time interval (see below).

The following graphical solution of this equation is one of several methods that have been devised. It is most easily explained by the use of an example (Fig. 41-32).

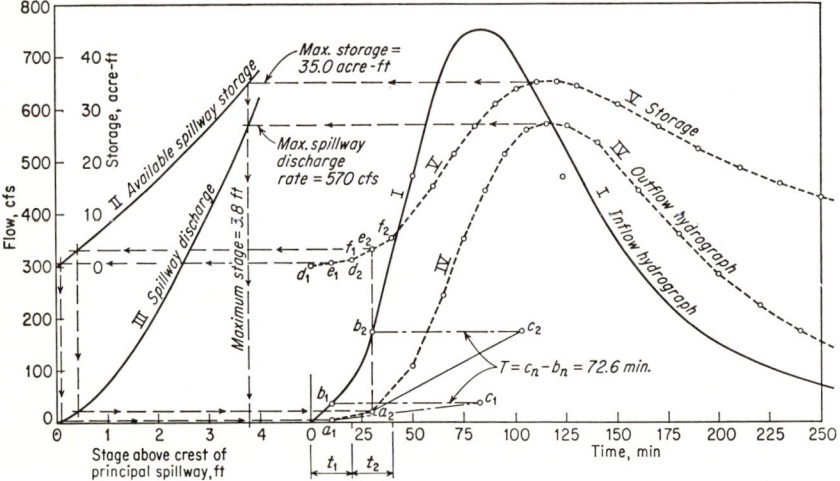

FIG. 41-32. Example of graphical flood routing solution to determine maximum storage requirements.

Three separate curves are needed to solve the storage equation, numbered as follows:

I. Inflow hydrograph. This curve gives the rate of inflow as a function of time. For development of this curve see Chap. 34, Principles of Agricultural Hydrology.

II. Available spillway storage. This curve shows the volume of storage in the reservoir for various stages of the water surface above the lowest outlet elevation. This curve must be plotted from calculations based on the topography of the reservoir.

III. Spillway discharge. This curve gives the relationship between the rate of discharge over the spillway and the stage of the water surface in the reservoir. It is computed from the hydraulic characteristics of the spillway being investigated and will vary with the size and type of spillway being used.

From these three curves and the storage equation, two additional curves can be derived:

IV. Outflow hydrograph. This curve gives the rate of outflow (discharge over the spillway or spillways) as a function of time.

V. Storage. This curve gives the volume of spillway storage existing in the reservoir as a function of time.

The procedure for derivation of curves IV and V is outlined as follows:
1. Compute and plot curves numbered I, II, and III, as indicated in the attached example. It will be found convenient to plot curve I with the ordinates measuring rate of flow in cubic feet per second and abscissas measuring time in minutes. Curve II should be plotted with ordinates of storage in acre-feet and abscissas as stage above spillway crest in feet. Curve III must have the same scale of ordinates as curve I and the same scale of abscissas as curve II. The scale of abscissas of curves II and III should be chosen so as to make curves II and III fairly steep; this cannot be accomplished for curve III when the spillway is of the pressure-conduit or orifice type. The scales for these three curves should be so located that the curves do not overlap.
2. Compute the conversion time interval T. The conversion time interval is the time required for a flow as measured by one unit (say, 1 in.) of ordinate on the flow scale to accumulate to the same unit (1 in.) of storage on the storage scale. In computing T, a factor to make the units of the equation consistent must be included.

$$T, \text{min} = \frac{\text{cu ft represented by 1 in. of ordinate}}{\text{cfm represented by 1 in. of ordinate}} \qquad (41\text{-}36)$$

The two ordinate scales and the time scale should be so chosen that T will plot on the time scale to a length of from 2 to 6 in.
3. Construct the derived curves numbered IV and V. This procedure (trial and error) is broken up into the following steps:
   a. Select a time interval t_1; assume an average rate of outflow for that time interval and plot it as point a_1 at the mid-point of the time interval. The shorter the time interval selected, the more accurate will be the graphical construction; however, it is not ordinarily necessary to select time intervals of less than about 0.025 times the total runoff period; and in some parts of the analysis, these time limits may be doubled. The shorter time intervals should be used where there are sharp breaks or rapid changes of slope of any of the five curves.
   b. From point b_1, which is on curve I directly above a_1 measure the distance T horizontally to the right, thereby locating point c_1. Point b_1 represents the average rate of inflow into the reservoir during the time interval t_1.
   c. The slope of the line a_1c_1 represents the average rate of change of storage for the time interval t_1; in other words, a flow equal to $b_1 - a_1$ measured on the flow scale will in time T accumulate an amount of spillway storage equal to $b_1 - a_1$ measured on the storage scale.
   d. Locate point d_1 with an abscissa of time $= o$ and an ordinate of storage $= o$.
   e. From point d_1 draw a line parallel to line a_1c_1. Locate point e_1 at the mid-point of the time interval on this line.
   f. Now check the accuracy of the assumption as to the value of a_1 as follows: From e_1 project horizontally to the left to curve II, then down to curve III, and then horizontally to the right to a vertical line through point a_1. If this last projection intersects the vertical line through a_1 at a_1 then the assumption of the value of a_1 was correct. If it does not intersect at point a_1, then a new trial value of a_1 must be selected and the process repeated until the graphical construction checks the trial value of a_1.
   g. After the location of a_1 has been checked, locate point f_1 (which corresponds to point d_1 in the first time interval) by drawing a line from d_1 parallel to line a_1c_1 to an intersection with an ordinate through the division point between time intervals t_1 and t_2.
   h. Select a new time interval and repeat steps a to g above and continue this process until the outflow hydrograph has intersected the inflow hydrograph.

The point at which the outflow and inflow hydrograph intersect will be the maximum value of the outflow hydrograph, and directly above this point the storage curve will also have zero slope and reach a maximum. From this time on, the outflow rate will exceed the inflow rate and the water level in the reservoir will be dropping.

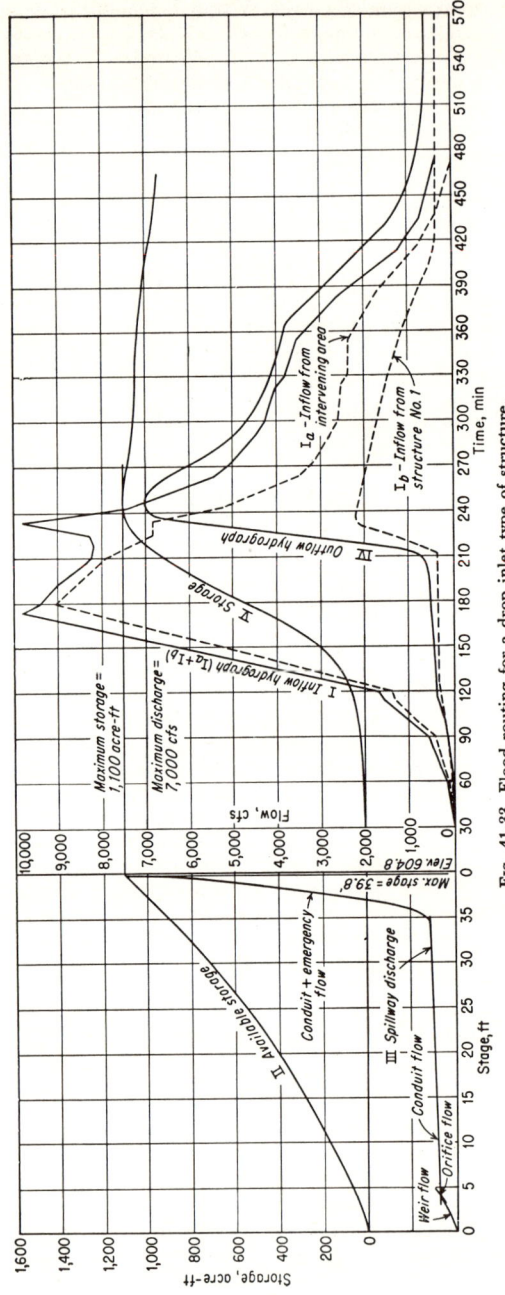

Fig. 41-33. Flood routing for a drop inlet type of structure.

It should be pointed out that true points on the outflow hydrograph fall at the mid-points of the selected time intervals and the true points on the storage curve occur at the end of the time intervals. The accuracy of the graphical construction can be checked by the fact that the area between curves I and IV, up to the point of intersection, when converted into volume, must equal the maximum spillway storage (storage above inlet) reached in the reservoir.

The spillway in the above example was a chute type, and the spillway discharge curve III was based on weir flow above the crest. Any type or combination of flows for the various types of spillways can be computed and inserted in curve III. Figure 41-33 illustrates a *drop-inlet type* of spillway with both weir and full-conduit flow. It also has flow through an emergency spillway added to the control flow in the principal spillway at higher stages in the temporary storage pool. The hydrograph for this structure is developed from an outflow hydrograph of structure 1 plus the hydrograph from the intervening watershed between structures 1 and 2.

Before proceeding with the detailed graphical flood routing, it is often necessary to make a preliminary estimate of the size and capacity of the spillway(s). The estimate can be made by use of one of the short-cut algebraic forms of flood routing. This method may be used as the final design where the ratio of temporary storage to the total flood volume is small. The following method, developed by Culp,[1] is a good example. Here the inflow hydrograph is assumed to be a triangle, as shown in Fig. 41-34, and the outflow built up with triangles and a rectangle.

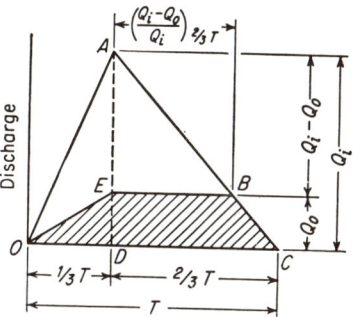

Fig. 41-34. Assumed triangular inflow hydrograph used in Culp equations.

It is these geometric relationships that provide the following rather simple equations:

$$V = ad + \frac{0.85d^2 \sqrt{a}}{S} \qquad (41\text{-}37)$$

$$\beta = \frac{Q_o}{Q_i} = 1.25 \pm \sqrt{\frac{18V}{AD} + 0.06} \qquad (41\text{-}38)$$

$$d = -\frac{S\sqrt{a}}{1.70} \pm \sqrt{\frac{DAS}{30.6\sqrt{a}}(1-\beta)(3-2\beta) \frac{aS^2}{2.89}} \qquad (41\text{-}39)$$

where A = area of watershed, acres
a = area of reservoir at elevation of crest of principal spillway (normal pool elevation), acres
d = stage above crest of principal spillway, ft
D = average depth of runoff, in.
Q_i = peak rate of inflow, cfs
Q_o = rate of outflow when the principal spillway first flows full as a pipe (under pressure), cfs
S = average slope of banks of reservoir, through range of stage d, per cent
T = total duration of inflow, min
V = available storage, acre-ft

The above equations should be used only for drop-inlet spillways that start to flow full at a comparatively early time period and do not have the combined flows of auxiliary or emergency spillways.

An additional equation developed by Culp is

$$\frac{V_s}{V_f} = (1-\beta)^2 \qquad (41\text{-}40)$$

[1] M. M. Culp, The Effect of Spillway Storage on the Design of Upstream Reservoirs, *Agr. Eng.*, 29:344–346, August, 1948.

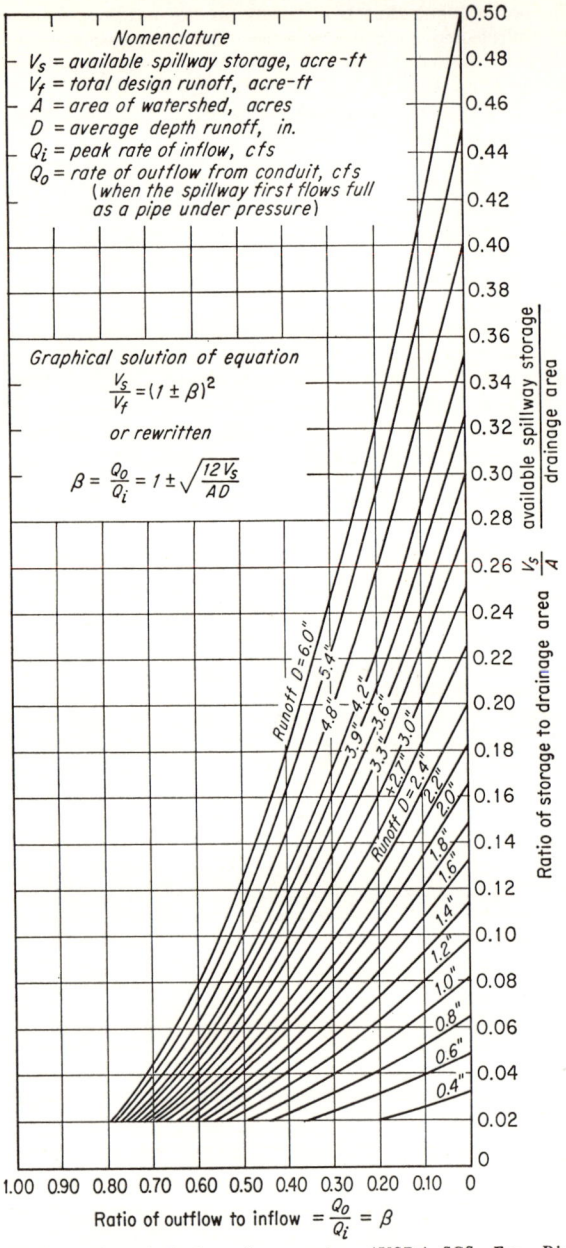

Fig. 41-35. Graphical solution of flood routing equation. (*USDA SCS, Eng. Div., Central Tech. Unit*, 1956.)

in which the nomenclature is given on Fig. 41-35. These curves enable the designer to make an estimate of either the spillway size or the required spillway storage.

FOUNDATIONS AND EMBANKMENTS

Mechanical analysis of construction materials is necessary in the design of foundations, abutments, and embankments for earth and rock-fill dams. An earth dam can be installed in almost any location, provided that the design can overcome weaknesses of the site. At some locations the cost of the improvement may be prohibitive and an alternative site is more economical. In many cases this can only be determined after the foundation and embankment materials have been tested and analyzed. Blankets, grouting, plating, drainage, and zoned fills are modern tools that the engineer can utilize to adapt earth dams to sites that had been rejected in the past. The embankment is usually constructed of selected, proportioned, and well-compacted materials, and the foundation is of unconsolidated and sometimes stratified materials that must be analyzed. In some cases, the variation in acceptable stratified material is severe. The boundaries of the unsatisfactory materials should be determined. If this is not possible and it is required in the design, then the design should be based on the poorest material found. The analysis and design of the foundation material should be studied in the light of future loading by the embankment weight and the possible hydrostatic pressures of impounded water. An interpretation of the possible rearrangement of the soil particles is important in the design of the foundation and embankment.

Some knowledge of the foundation is required for all structures. The coverage and intensity of investigation depends on the importance, size, and the degree of hazard of the installation. The degree of hazard, which is one item in the required safety factor of the structure, covers a wide range and, assuming an immediate failure of the embankment, may involve (1) only a small amount of damage to the agricultural land immediately below the structure, (2) damage to more extensive land and to fences and other farm property, including residential, without endangering life, (3) in addition to the above, some residential and highway damage that might possibly endanger life, (4) damage to residential areas, highways, crops, livestock, etc., and (5) in addition, damage to heavily residential areas and urban property of high value.

The small structures that do not seriously endanger life or property will only require borings that log the foundation materials. As an example, these could be drop spillways with less than 6-ft overfall and small-farm ponds. Here, visual inspection of the *boring log* may be sufficient to establish the stability and permeability of the foundation. As a guide to recognizing unstable and porous foundation materials, the following materials should be carefully studied: (1) gravel and coarse sand of uniform size, (2) high percentages of sand and silts alone or in combination, (3) gypsum, marls, or chalky soils, (4) mucks, peat, and other high-organic soil, and (5) weathered rock surfaces that have become separated from the bedrock, including those weathered surfaces that may be several feet below ground level.

On larger jobs, a mechanical analysis and a few simple tests of selected samples may be sufficient to develop a foundation design. The *mechanical analysis* will more accurately enable the designer to determine the permeability and volume weight of various soil strata without more detailed laboratory analysis. A knowledge of the combination of materials that produces not only a dense and impermeable soil but also the other limits of combinations that produce a porous and low-volume-weight soil is required in the proper interpretation of the mechanical analysis. The mechanical analysis has as its objective the determination of particle-size distribution. The gradation of grain-size distribution must show not only the predominant size, but also the relative amount of the various sizes. The results of the mechanical analysis may be presented in several different ways; however, the two chief methods are (1) listing percentages of the soil falling into the different textural groups and (2) developing a distribution curve, which is obtained by plotting grain size against per cent of fines on a graph. Figure 41-36 illustrates a mechanical analysis of representative soils, showing the per cent of the different size groups. As a guide, proportions of gravel, sand, silt, and clay that deviate considerably from that producing the greatest density should be studied. Generally, not over 20 per cent gravel, with sand from 20 to 50 per cent, silt not to exceed 30 per cent, and clay from 15 to 25 per cent will produce a satisfactory combination of materials with maximum density.

Porous zones will induce excessive percolation and hydrostatic pressure that may result in piping or direct flow through the foundation and uplift, sliding, or flotation

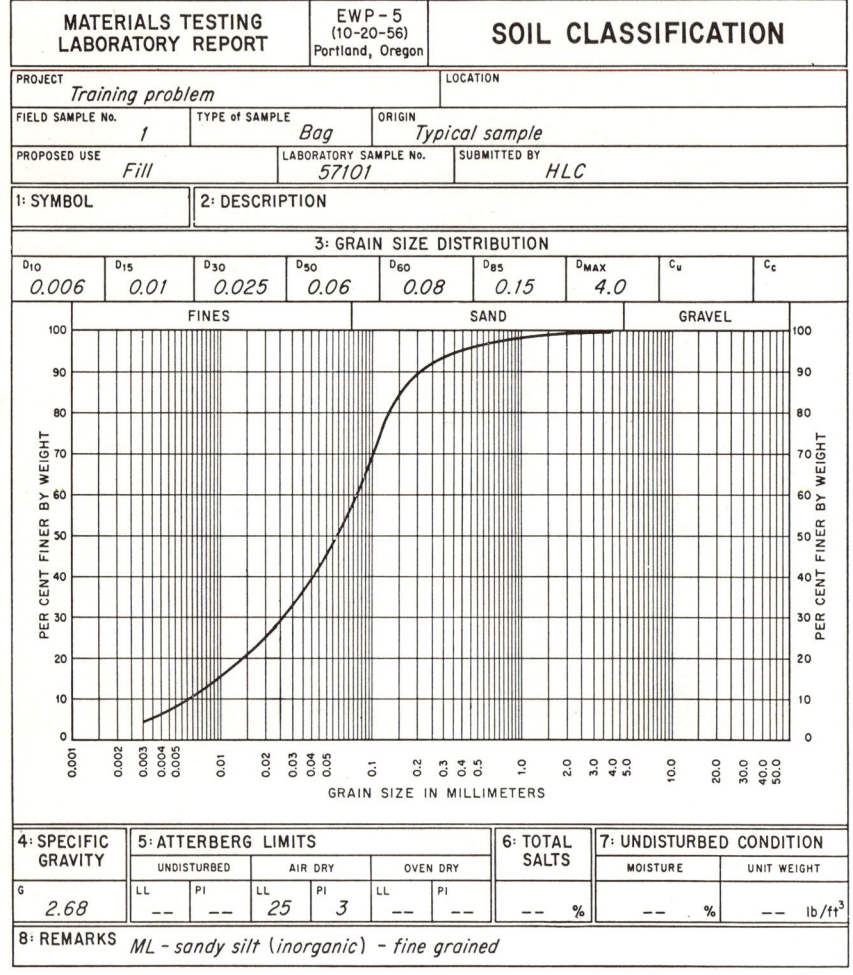

Fig. 41-36. Laboratory report showing grain size distribution.

of the base of the embankment or spillway. It could also cause rapid depletion of water in the reservoir, which would be a serious loss if this is an important design feature. Low density or volume weight is often found in porous zones and should be carefully studied to determine the possibility of consolidation and settlement of the foundation. While many of these conditions are evident on visual inspection of the boring logs and mechanical-analysis charts and satisfactory in the design of the smaller structures, a *laboratory analysis* will be required on the larger jobs. Any evidence of gypsum or salty or dispersed soils should be investigated at this stage.

When the soil contains more than 0.5 per cent total soluble salts or gypsum, treatment must be considered in the design of the abutment or foundation. These soils are also often low in volume weight and subject to the formation of solution channels. Dispersion ratio is that portion of clay (0.005 mm) which is easily slaked by water (Chap. 56). Dispersion often results in unstable foundation conditions by causing plastic flow, piping, and settlement. Those soils high in organic matter are also usually unstable, having a low density and bearing capacity and high permeability. When they are encountered in this intermediate-size structure, their effects on the foundation design should be considered.

As the height of the structure and the degree of hazard increase, a *more detailed analysis* is necessary. These tests and analyses could include bearing capacity, permeability, consolidation, shear, and triaxial shear. Investigation of geologic conditions also increases in detail as the corresponding size and hazards increase. An important consideration in the investigation is the amount and kind of weathering on rock surfaces. Although bedrock is reached in borings or test pits, there may still be cracks, faults, and crevices on the surface of the rock that will require treatment. The dip and stratification of the rock material should also be studied. In limestone, solution channels and sinkholes may be an important consideration. The stability of shales (soapstone) and sandstone, particularly after exposure to water or air, should be studied. The elevation of the water table, particularly the location of springs, should be noted at the same time that borings are made. These elevations should be indicated in the boring log for use in design. A detailed account of the geologic conditions should be contained in a report prepared by competent geologists. The extent of the investigation and the type and detail of the report will depend on size and type of structure and complexity of geologic materials. For the latter, a more complete boring coverage of the base area will be required. The number of borings will vary directly with the importance of the installation and the variation in the materials encountered.

The smallest class of job will require only two or three borings along the center line of the embankment and an additional one to three borings along the center line of the spillway. The intermediate class of structures will usually require twice this number at both locations. These borings are taken at different elevations to give a complete coverage of the site's material. The greater number are usually made at the lower elevations to examine the heavily loaded areas in more detail. An effort should also be made to take them at regular horizontal intervals so that large areas of the poorer foundation material do not escape the attention of the design engineer. Abrupt changes in slope are sometimes used as a guide to locate borings.

The larger structures may require a *complete boring grid* of the foundation. Often it is necessary to take additional borings other than those at planned locations. These are taken to verify unusual foundation conditions or to determine the extent of unstable material. Their position is determined in the field at the time the other borings are made. Depth of the borings will vary from one-half to twice the height of the embankment. The intermediate-size structure will require depths of from one-half to three-quarters the height. The weight of the structure, the nature of the material encountered, and the design of the foundation will govern the depth. If unstable material is found at a predetermined depth, it will usually be necessary to extend the boring into dependable material at a reasonable depth. Where bedrock is encountered, boring operation can usually be terminated except for that class of structure that needs to be cored or anchored to the rock. Where the surface of the rock is important in the design, test pits covering a considerable area will need to be excavated to the rock's surface.

A good design of foundation and embankment requires a careful analysis of the foundation materials and careful selection of soil materials for construction. This evaluation and design is based on the physical properties of the soil materials, which are determined by means of well-established laboratory and field-test methods. The methods and procedures for laboratory testing are not described here. Most of the reference texts on soil mechanics given at the end of this chapter will describe these tests in detail. The results of these tests are important as a means of determining the physical properties of the soil materials. The engineer is further aided in the evaluation of properties of soils by the use of some standard *soil-classification system*. There are many good classification systems in use today, and in general, the system selected is based on the specific use and coverage of the range of conditions for which the soil materials are to be used. The unified soil-classification system, developed jointly by the U.S. Army, Corps of Engineers, and the U.S. Department of the Interior, Bureau of Reclamation, is the system most directly applicable to the selection of soil materials for foundations and earth-dam embankments. It is designed to apply generally to the identification of soils regardless of the intended engineering uses. A condensed tabulation of the soil groups established by this system is presented in Table 41-12. The soil-plasticity chart is shown on Fig. 41-37. As in any classification system, the grain size or distribution of grain size plays an important part here. In addition to the grain-size classification indicated in Table 41-12, the amounts of clean gravel and sands are determined when the sample contains less than 5 per cent fines (material

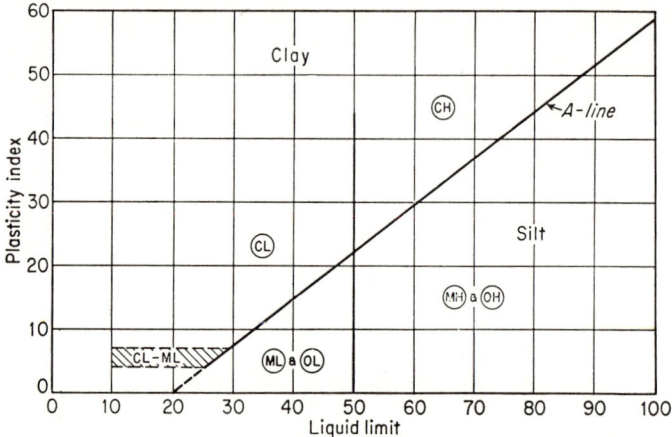

FIG. 41-37. Plasticity chart for Unified Soil Classification. (*USDA SCS, West Coast Area*, 1954.)

passing through a 200-mesh sieve). It should also be noted that the liquid limit and plasticity index (Chap. 56) are used as a division of the silt and clays, and a liquid limit greater than 50 indicates high plasticity and one less than 50 low.

A further breakdown of the unified classification system has been made in Table 41-13 by H. L. Cappleman, Jr., design engineer for the Soil Conservation Service, Portland, Ore. The additional breakdown includes the division of the silty sands into the coarse silty sands, SM-1, and the fine silty sands, SM-2. The division is based on calling those silty sands coarse in which more than 50 per cent of the grain sizes are larger than 0.15 mm and fine where 50 per cent of the grain sizes are less than 0.15 mm. It should be kept in mind that the designations d_{10}, d_{30}, d_{50}, etc., refer to the percentage of a given grain size contained in an average unit of the soil material.

An additional breakdown is given as CL-1 and CL-2. The division here is based on clays of low plasticity where the plasticity index (Chap. 56) is less than 15. CL-2, or clays of medium plasticity, are those with a plasticity index greater than 15.

In order to separate the well and poorly graded sands and gravels, a coefficient of uniformity C_u and a coefficient of curvature C_c were used. The formulas for these coefficients are given in column 6 under the Laboratory Classification Criteria. The soil material must satisfy both of these conditions to fall in the well-graded classification. The most important part of Table 41-13 is the behavior classification given in column 7. This specification will be used in the analysis and design of the embankments and foundations.

Embankment Materials. An earth-dam embankment must be designed to be stable for any force condition or combination of forces which may reasonably develop during the lifetime of the structure. Other than overtopping, caused by inadequate spillway capacity, the three most critical conditions to be considered in the design are:

1. Development of *shearing stresses* within the embankment due to the weight of the fill. If the magnitude of the shearing stresses exceeds the strength of the fill material, sliding of the embankment slopes and displacement of large portions of the embankment may occur. The same condition may occur in the foundation, the displacement of which would have a detrimental effect on the stability of the embankment.

2. The development of *differential settlement* within the embankment or its foundation is caused by variation of materials, variation of height of the embankment above the foundation, or compression of the underlying strata. This condition may cause the foundation to crack through the embankment approximately parallel to the abutments.

TABLE 41-12. UNIFIED SOIL-CLASSIFICATION SYSTEM (OUTLINE)

Major divisions	Group symbols	Typical description
Coarse-grained soils, d_{50}* > No. 200 mesh		
Gravels, more than half of coarse fraction > No. 4 mesh:		
Clean gravels................	GW	Well-graded gravels, gravel-sand mixtures, little or no fines
	GP	Poorly graded gravels, gravel-sand mixtures, little or no fines
Gravels with fines............	GM	Silty gravels, gravel-sand-silt mixtures
	GC	Clayey gravels, gravel-sand-clay mixtures
Sands, more than half of coarse fraction < No. 4 mesh:		
Clean sands..................	SW	Well-graded sands, gravelly sands, little or no fines
	SP	Poorly graded sands, gravelly sands, little or no fines
Sands with fines..............	SM	Silty sands, sand-silt mixtures
	SC	Clayey sands, sand-clay mixtures
Fine-grained soils, d_{50}* < No. 200 mesh		
Silts and clays:		
Liquid limit < 50...........	ML	Inorganic silts and very fine sands, rock flour, silty or clayey fine sands, or clayey silts with low plasticity
	CL	Inorganic clays of low to medium plasticity, gravelly clays, sandy clays, silty clays, lean clays
	OL	Organic silts and organic silty clays of low plasticity
Liquid limit > 50...........	MH	Inorganic silts, micaceous or diatomaceous fine sandy or silty soils, elastic silts
	CH	Inorganic clays of high plasticity, fat clays
	OH	Organic clays of medium to high plasticity, organic silts
Highly organic soils............	Pt	Peat and other highly organic soils

* d_{50} = diameter of 50 per cent of sample. See Fig. 41-36.
SOURCE: USDA SCS, Portland, Ore., 1954.

These cracks could encourage concentration of normal seepage through the dam and subsequent failure by piping.

3. The development of *seepage* through the embankment and foundation. This condition may cause piping or progressive internal erosion to occur within the embankment or foundation. Another noticeable effect of seepage is the continuing softening and sloughing of the toe of the slope.

The following are characteristics inherent in the embankment material itself:

1. The *shear strength* of the material influences the choice of the slope of the embankment. This can be a very important economic item on the larger embankments, so that a selection based on shear strength may have great importance.

2. The *permeability* of the soil material may have an effect on the shear strength. Permeability should also be considered in the zoning of bank materials and in the need and type of the drainage design.

TABLE 41-13. WORKING CLASSIFICATION OF SOILS

Major divisions	Symbols	Typical name	Laboratory classification criteria			Behavior
Coarse-grained soils [d_{50}* > No. 200 mesh (0.074 mm)]: Sands and gravels: Clean.........	GW	Well graded gravel; sandy gravels	Determine percentages of gravel and sand. Depending on percentage of fines (200 mesh), coarse-grained soils are classified as follows:	$C_u > 4$; $1 < C_c < 3$	Must satisfy both $C_u = \dfrac{d_{60}}{d_{10}}$ $C_c = \dfrac{d_{30}^2}{d_{10} \cdot d_{60}}$	I
	GP	Poorly graded gravels; sandy gravels		$C_u < 4$ or $1 > C_c > 3$		
	SW	Well graded sands; gravelly sands		$C_u > 6$; $1 < C_c < 3$		
	SP	Poorly graded sands; gravelly sands		$C_u < 6$ or $1 > C_c > 3$		
With clay fines..	GC	Clayey gravels; sandy, clayey gravels	Less than 5%— GW, GP, SW, SP More than 12% —GC, GM, SC, SM 5% to 12%— Borderline cases	Atterberg limits above A-line		II
	SC	Clayey sands; gravelly clayey sands		Atterberg limits above A-line		
With silt fines...	GM	Silty gravels; sandy, silty gravels		Atterberg limits below A-line		III
	SM-1	Coarse silty sands		$d_{50} > 0.15$mm	Atterberg limits below A-line	
	SM-2	Fine silty sands		$d_{50} < 0.15$mm		
Fine-grained sands [d_{50}* < No. 200 mesh (0.074 mm)]: Silts and clays: LL < 50.......	ML	Silts; rock flour; ash; very fine sandy silts; sandy, clayey silts				IV
	CL-1	Clays of low plasticity; silty clays; sandy gravelly clays; PI < 15				V
	CL-2	Clays of medium plasticity; silty; sandy, or gravelly clays; PI > 15				VI
LL > 50.......	CH	Clays of high plasticity; fat clay				VII
	MH	Elastic silts, micaceous and diatomaceous silt				VIII
Organic.......	OL	Organic silts and clays	Same as ML	LL (oven-dry soil) / LL (air-dry soil) < 0.7		IX
	OH	Organic silts and clays	Same as MH			

* d_{50} = diameter of 50 per cent of sample. See Fig. 41-36.

3. The *compressibility* of the embankment may cause excessive settlement, which could lead to overtopping or cracking of the fill.

All these characteristics of the embanking materials, according to their behavior group, have been evaluated by Cappelman in Table 41-14. Any one of these characteristics may have considerably more weight than any or all of the others, depending on the size, height, and the use of the embankment.

In the selection of *embankment materials* one should also keep in mind the possibility of either zoning or mixing of the soil materials. The design under these conditions will be based on the kind and amount of materials that are available from the borrow areas. The use of blankets and drains should also be part of the design process in the selection and placement of fill material. The type of compaction control required to produce a fill of adequate density depends upon the physical properties of the soil. Soils which indicate susceptibility to piping and cracking must be placed at carefully controlled moisture content and compacted to a density determined by laboratory compaction tests. The most effective types of compaction equipment for various soil types are indicated in Table 41-14.

TABLE 41-14. CHARACTERISTICS OF COMPACTED FILL MATERIALS

Behavior group	Relative resistance to failure (1) greatest to (6) least			Relative characteristics		
	Shearing	Piping	Cracking	Permeability	Compressibility	Compaction
(1)	(2)	(3)	(4)	(5)	(6)	(7)
I	1	...	...	High	Very slight	Good; crawler tractor; steel-wheeled roller
II	3	3	4	Low	Slight	Fair; sheep's-foot roller; rubber-tired roller
III	2	5	3	Medium	Slight	Good; rubber-tired roller; sheep's-foot roller
IV	3	6	6	Medium	Slight to medium	Good to poor; sheep's-foot roller; close control essential
V	4	4	5	Low	Medium	Good to fair; sheep's-foot roller; close control essential
VI	5	2	2	Low	Medium to high	Good to fair; sheep's-foot roller; rubber-tired roller
VII	6	1	1	Low	High	Fair to poor; sheep's-foot roller
VIII	6	Variable	Variable	Medium low	Very high	Poor to very poor; sheep's-foot roller
IX	6	Variable	Variable	Medium low	Very high	Very poor; not suitable for embankments

SOURCE: USDA SCS, Portland, Ore., 1954.

Several designs for *zoned embankments* are shown in Fig. 41-38 as types 4, 5, and 6. They vary primarily in the percentage relationship of the pervious and impervious soil materials. In many cases, this relationship will be governed by the amount of the materials available at the site or adjacent borrow areas. At some locations, the quantity available for the impervious zone will be even less than that illustrated in these examples. Type 4 in Fig. 41-38 illustrates an embankment composed of coarse-grained pervious soil in approximately the upstream and downstream third of the dam. The *downstream pervious shell* serves to lower the line of saturation in the same manner as does a toe drain, as shown in Fig. 41-38. It increases the total shear resistance of the embankment. The *upstream shell* protects the embankment from failure due to stresses caused by rapid drawdown of the reservoir. Also, this type of zoning usually results in the placement of the more shear-resistant materials in the outer zones where such resistance is required.

The *line of seepage,* or saturation (see under Selection of Embankment Design Type), in a homogeneous earth dam intersects the downstream slope at a point above the toe of the embankment. The use of proper materials and construction methods can reduce the quantity of seepage at the toe to a point where it has little effect on the stability of the embankment. In other cases, the saturation of the toe will cause sloughing or serious reduction in the shear strength. It is always desirable to include a *toe drain* in the design of homogeneous embankments. Likewise, dams founded on pervious foundations or constructed of materials which exhibit susceptibility to piping and cracking should always be protected by an adequate drainage system. Toe drains may be constructed of sand, gravel, or rock or a mixture of these materials, depending upon the nature of the fill materials. Whenever a rock toe drain is installed, a graded filter blanket should be placed between the fill and the drain as illustrated in types 2 and 3 of Fig. 41-38. Transition filters that function in a similar manner are shown as types 4 and 5. A minimum 12-in. layer of well-graded stream-run gravel placed either horizontally or vertically within the embankment will satisfy the drainage condition on some of the smaller structures. A *trench-type drain,* using similar filter materials, has also proved very satisfactory. This drain is usually located in the downstream third of the embankment and can be so constructed that it will control the saturation line in the embankment and relieve some of the drainage requirements of

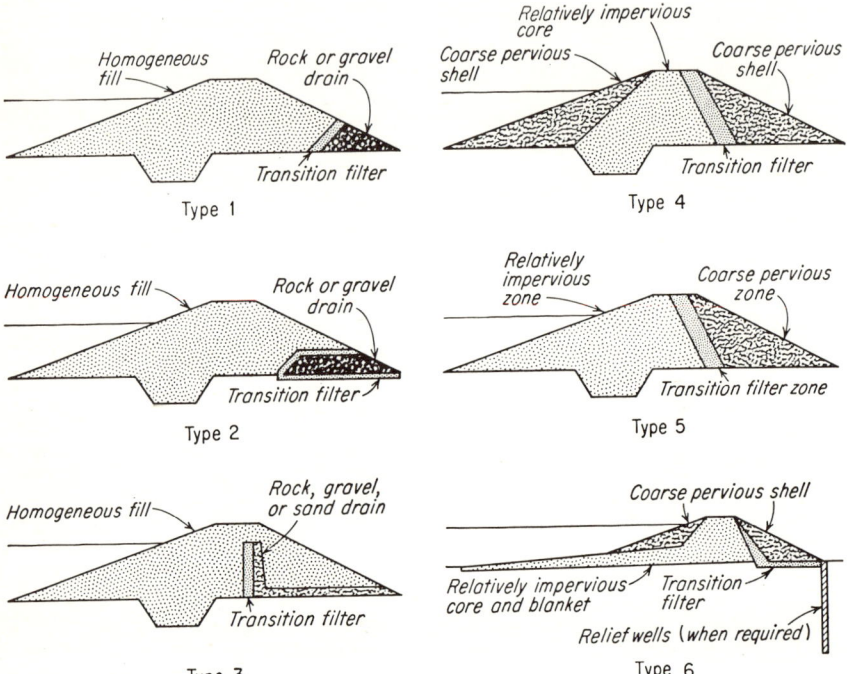

FIG. 41-38. Basic design types—rolled earth fill dams. (*USDA SCS, West Coast Area,* 1954.)

the foundation. The trench-type drains are usually 10 to 25 sq ft or more in cross-section areas.

Where more positive drainage is required, a *perforated conduit* is installed at the approximate center of the drain, as shown in Fig. 41-39. The drainage flows into a nonperforated conduit and is led to some safe outlet beyond the toe of the embankment. Many other embankment designs and toe-drainage systems have been developed.[1]

The criteria for *gravel filters* as presented by Terzaghi and others are based upon the size of the smallest 15 per cent of the filter material, the 15 per cent size of the foundation material, and the 85 per cent size of the foundation material (the smallest size of the coarsest 15 per cent of the foundation). The filter material should conform to the ratio

$$\frac{15\% \text{ size of filter}}{85\% \text{ size of foundation}} \leq 4:5 \leq \frac{15\% \text{ size of filter}}{15\% \text{ size of foundation}}$$

[1] W. P. Creager et al., "Engineering for Dams," vols. I to III, John Wiley & Sons, Inc., New York, 1945.

The openings in screens with or without filters and perforations with filters should follow the ratio

$$\frac{85\% \text{ size of foundation}}{\text{Perforated opening}} = 1$$

The screen-size openings of wells or well points without filter materials should be 0.175 mm. The size and shape of the slots or perforations are not important if filter materials surround the well. Well screens without filters should have at least 100 perforations with an open area of 3 sq in. per square foot of screen.

Foundation Materials. The physical properties and distribution of the materials occurring in the foundation must also be considered from the standpoint of shear strength, compressibility, and permeability. The shear strength of the foundation, in

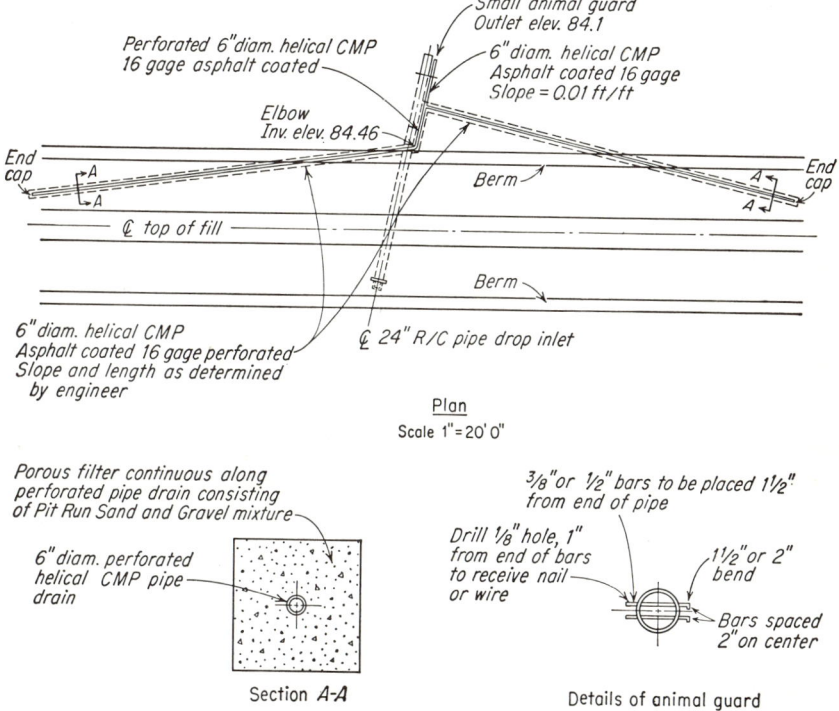

Fig. 41-39. Embankment drainage layout. (*USDA SCS*, 1954.)

many cases, governs the choice of embankment slopes. The foundation may contain compressible or plastic soils which may settle under the weight of the embankment even through their shear strength may be satisfactory. When such settlement occurs in the foundation, the embankment will also settle. Such settlement is rarely of uniform dimension over the entire base of the embankment, and fill materials used on such sites must be sufficiently plastic to deform without cracking. Permeability and stratification of the foundation materials will govern the drainage system and possibly influence the choice of the zoning plan.

The relative characteristics of the various soil groups pertinent to foundation evaluations are presented by Cappleman in Table 41-15. This table should be used as an aid to the evaluation of specific foundation conditions, and it is not presented as a general solution to all foundation problems. A foundation composed of one homogeneous soil type would be simple to evaluate; however, such a condition is rarely found in natural soil deposits. The more usual condition is a stratified deposit composed of layers of several soil types. The number of possible combinations of materials and arrangements of materials within such a deposit are so great as to render a general solution impossible of achievement. The geologic history of the site, the extent of stratification, and the order in which the materials occur within the stratification are of great significance in determining the suitability of the foundation. Complex stratified foundations containing plastic or compressible soil materials should be evaluated by engineers or geologists experienced in such conditions.

Table 41-15 enables the designer to make a classification of the foundation type based on the compressiblity and permeability of this material. Table 41-16 also presents the type of site classification based on the shape and proportion of the valley cross section at the embankment. The site classifications are a means of evaluating the differential settlement between adjacent sections of the embankment and of selecting the proper embankment materials to compensate for this problem.

TABLE 41-15. CHARACTERISTICS OF FOUNDATION MATERIALS

Group symbol	Characteristics influencing embankment design				Seepage-control requirements
	Shear strength	Sensitivity to shock	Compressibility	Permeability	
GW	High	None	Very slight	High	Positive cutoff
GP	High	None	Very slight	High	Positive cutoff
GM	High	None	Very slight	Medium-low	Toe trench to none
GC	High	None	Slight	Low	None
SW	High	None	Very slight	High	U.S. blanket and toe drainage; cutoff
SP	Usually high*	High for loose fine sand	Very slight	High	U.S. blanket and toe drainage; cutoff
SM	Usually high*	High for loose fine sand	Very slight	Medium	U.S. blanket and toe drainage; cutoff
SC	High	None	Slight	Low	None
ML	Medium	High for loose silts	Medium	Medium	Toe trench
CL	Medium	None	Medium	Medium-low	None
OL	Low	None	High	Low	None
MH	Low	High for loose silts	Very high	Low	None
CH	Medium to low†	None	Usually very high†	Low	None
OH	Low	None	Very high	Low	None
Pt	Very low	None	Very high	Very high	Remove from foundation

* Depends on density and drainage condition.
† Depends on consistency in place.
SOURCE: USDA SCS, Portland, Ore., 1954.

Selection of Embankment Design Type. The foundation classification and site classification as determined from Table 41-16 are carried over to Table 41-17. The previously determined characteristics of the available fill materials as determined from Table 41-13 are also inserted in Table 41-17. Here the materials group is the same classification as the behavior group indicated in Table 41-13. The variation in the soil materials that are available will then guide the engineer in the determination of the type of section. In this case there are only two classifications, that of an entirely homogeneous section or the zone section. There will be many primarily homogeneous sections that could have minor zone parts, in which case, the homogeneous design should be used. With the four items evaluated, Table 41-17 will give the design type of embankment as shown in Fig. 41-38. In the zone section, it should be noted that the materials group has been divided to allow for a selection of material for both the shell and the core of the embankment. The preceding qualitative evaluation of soil characteristics, site topography, and foundation conditions should be used only as a guide for making a structurally sound preliminary embankment design for the larger structures. The validity of conclusions reached by this method will depend on the competency of the on-site investigation, the adequacy of the testing program, and the

EROSION-CONTROL STRUCTURES

TABLE 41-16. EVALUATION OF SITE AND FOUNDATION CONDITIONS

Site classification		Foundation classification		
Site configuration*	Site type	General characteristics†		Foundation type
		Compressibility	Permeability‡	
1. Relatively broad gap 2. Gently sloping abutments	Type A	None to very slight...	Impervious Pervious	1a 1b
		Slight...............	Impervious Pervious	2a 2b
		Medium.............	Impervious Pervious	3a 3b
		High................	....	4
1. Relatively narrow gap 2. Steep abutments (Slope 1:1 or steeper)	Type B			

* Determined from topographic survey.
† Determined from report of geological investigation and laboratory testing.
‡ Relative to that of the embankment.
SOURCE: USDA SCS, Portland, Ore., 1954.

TABLE 41-17. SELECTION OF BASIC DESIGN TYPE

Site type	Foundation type	Type of embankment design required				
		Homogeneous section		Zoned section		
		Materials* group	Design type	Materials group*		Design type
				Shell	Core	
A	1a	II–VII	1	I, III, IV	II–VII	4 or 5
	1b	II, V, VI, VII	2	I, III, IV	II, V, VI, VII	4, 5, or 6
	2a	II, VI, VII	1	I, III	II, VI, VII	4 or 5
		V	3	I	V	4
	2b	II, VI, VII	2	I	VI, VII	4, 5, or 6
	3a	VI, VII	1	I, III, IV	VI, VII	4
	3b	VII	2	I	VII	4 or 6
	4	Evaluate on basis of test data from undisturbed samples only				
B	1a	II, VI, VII	1	I, III, IV	II–VII	4 or 5
		III, IV, V	3			
	1b	II, VI, VII	2	I	II, VI, VII	4, 5, or 6
	2a	II, VI, VII	1	I, III	VI, VII	4 or 5
	2b	II, VI, VII	2	I	VI, VII	4, 5, or 6
	3a	VI, VII	1	I, III	VI, VII	4
	3b	VII	2	I	VII	4 or 6
	4	Evaluate on basis of test data from undisturbed samples only				

* Behavior group, Tables 41-13 to 41-16 inclusive.
SOURCE: USDA SCS, Portland, Ore., 1954.

soundness of the designer's judgment. Design practices based on quantitative soil tests such as slope stability, seepage, and settlement analysis are not discussed here. However, it should be noted that they are necessary for the design of large dams and that such analyses would logically follow the selection of a preliminary design. For small earth dams with low hazard potential which do not justify extensive investigation or testing programs, a design based on this procedure may constitute the final design.

A more accurate analysis and design of the foundation can be made from laboratory tests of consolidation, permeability, and shear. *Consolidation* tests provide the means of determining the amount of settlement required in the design of large embankments. The tests, presented either graphically or numerically, give a reduction in thickness of sample for a given time. This may be based on the reduction of the void ratio of the sample, giving the settlement directly from the formula

$$S = h \left(\frac{e_1 - e_2}{1 - e_1} \right) \tag{41-41}$$

where S = settlement, in.
h = sample thickness, in.
e_1 = initial void ratio
e_2 = final void ratio

Laboratory tests for *permeability* may be made on disturbed samples for fill material or undisturbed for the foundation. Permeability has the dimensions of velocity and is usually expressed in either inches per hour or feet per day. The computation for permeability is based on the formula

$$k = \frac{QL}{hA} \tag{41-42}$$

where k = permeability
Q = quantity of water/time
h = total head
L = length of sample
A = area of soil sample

Shear tests are used to determine the stability of the embankment. Shear tests give the angle of internal friction from which the tangent or the coefficient of internal friction is obtained. The tests also yield the shearing force C of the sample. The shear tests are either direct or triaxial. Direct shear tests are made on cohesive soils, while the triaxial test may also be run on the cohesionless soils.

The shearing strength of the soil is divided into two components, one caused by cohesion between the individual particles and the other by the frictional resistance. The following equation based on Coulomb's work is used to express the relationship between the two component parts of the shearing strength:

$$S = C + N \tan \rho \tag{41-43}$$

where S = shear strength
C = cohesion
N = normal stress
ρ = angle of internal friction

For cohesionless soils, C approximates zero and the formula reduces to $S = N \tan \rho$. The angle of repose of cohesionless sand in a loose state is just slightly smaller than the angle of internal friction of the sand in a loose dry condition. A completely saturated sand has an angle of internal friction that is about 1 or 2° less than the angle of internal friction for the same sand in a dry state but at the same relative density. Shear strength of cohesive soils results from a combination of frictional resistance between the soil particles and cohesion between the particles. Values for the angle ρ generally vary between 10 and 25°, with the more clayey samples yielding the lowest angles. Values of cohesion for cohesive soils will range from 100 to 1,000 psf or more, and as a general rule cohesion will increase with the increase in clay content.

One of the most widely used tests for determination of the bearing capacity of the foundation material is that developed by the California State Highway Department and called the California Bearing Ratio Test. The test is used as a comparative measure of the bearing capacity as a relationship to a standard sample of compacted crushed stone, and results are presented as percentages of the load intensities for this sample.

Laboratory tests for embankment design should include, in addition to the above (of which the shear tests are the most important), moisture-density tests and those tests necessary to determine the treatment of the foundation or bottom of the reservoir to prevent excessive seepage losses. The compaction of moist cohesive soils increases the density, decreases the void space and consolidation, reduces permeability, and increases the shear strength. The size and grading of the soil particles in the material also affect the density of the material. Materials that contain large amounts of sand and gravel are heavier than materials made up primarily of clay or silt. Sandy materials that contain sand particles of all sizes are heavier than sandy materials with particles of uniform size.

Every cohesive material has an *optimum moisture content* at which the maximum density is obtained. The amount of moisture necessary to obtain maximum density is much higher for clay soils than for sandy soils. The optimum moisture content can be reduced by increasing the compactive energy. The compactive effort applied in standard compaction tests corresponds to that produced by a 200-psi sheep's-foot roller. The compactive effort applied by modified Proctor or heavy compaction tests corresponds to that produced by a 400- to 500-psi roller. A moisture-density report from most laboratories includes a graph for each sample showing a wet curve, dry curve, and penetration-resistance curve. Designations as "wet" and "dry" refer only to the wet weight and dry weight. Both compaction curves refer to the wet-rolled densities.

The *wet curve* shows the moist densities, or weights per cubic foot of soil plus water at different moisture contents, on a dry basis, right after compaction. The curve shows the moist densities and corresponding moisture contents as determined in a laboratory compaction test. Figure 41-40 is an example of a graph prepared by the laboratory showing a wet and dry curve. The peak of the wet curve illustrates optimum moisture. However, in most embankment designs, a reduction in the maximum density is permitted. This reduction is usually expressed in a percentage factor which will give a range of 2 to 4 per cent in the moisture requirement. The wet curve is supplied to the field to assist in control of compaction operations. Moisture percentage and density values obtained from penetration and volume-weight tests of the compacted fill during construction can be compared directly with wet-weight laboratory values as shown on the wet curve for that particular borrow material.

The *dry curve* shows oven-dry densities corresponding to the determined moist densities. The dry curve therefore illustrates weights per cubic foot of oven-dry soil obtainable when soil is compacted at the specified moisture contents. Dry-density data are obtained by correcting the moist densities for per cent moisture according to the equation

$$d = \frac{ww}{1 + \text{per cent moisture in sample}} \quad (41\text{-}44)$$

where d = dry weight, lb
 ww = wet weight, lb

It is important to remember that the moisture contents plotted on the density curves represent the percentages of moisture that must be added to oven-dry soil to get the compacted weights shown by the wet curve.

The data represented by the dry curve are the important moisture-density relationships. The oven-dry weight of any soil material is the only constant-weight value obtainable. Laboratory reports and computations pertaining to maximum density and optimum moisture for compaction refer to dry density and per cent moisture on the oven-dry basis as shown by the high point of the dry curve.

The *line of seepage* which represents an actual water table or saturation through the embankment is also an important consideration on the larger dams. A determination of the line of seepage has been the subject of much theoretical analysis. Casagrande's analysis[1] is contained in most soil-mechanics textbooks. His analysis presents a method for plotting the *phreatic line* for several different possibilities regarding the point of exit or intersection point of the seepage line and the downstream toe.

Justin has developed a rough approximation indicating that the seepage line through a homogeneous fill founded on impervious materials will intersect the toe of the dam at a vertical height equal to one-third of the head of water behind the dam

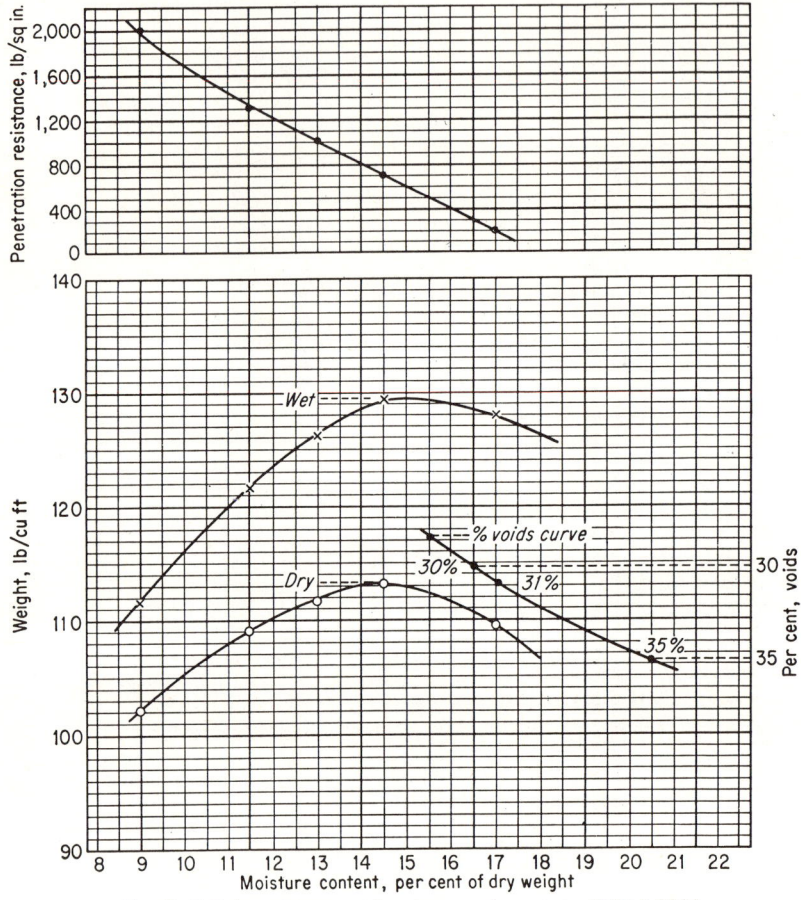

FIG. 41-40. Laboratory report of optimum moisture tests. (*USDA SCS.*)

above the impervious base.[1] The length of the line of seepage from the point of entrance to the exit point can thus be estimated from

$$l = \left(1.3h + 2y - \frac{h}{6}\right) \cot \alpha + W \qquad (41\text{-}45)$$

where l = length of seepage path, ft
 h = max. height of water behind dam, ft
 y = freeboard, ft
 α = angle of downstream slope from horizontal
 W = crest width, ft

[1] A. Casagrande, Seepage through Dams, *J. New Engl. Water Works Assoc.*, June, 1937.

An estimate of seepage losses through the dam can then be obtained by

$$q = \frac{4kh^2}{9l} \qquad (41\text{-}46)$$

where q = discharge, cfs
k = permeability of fill

All the seepage-line determinations presented above consider impervious foundations. Justin says, however, that the method shown by this equation can also be used for pervious foundations when it is assumed that the foundation is still a boundary, with seepage through the pervious material below this boundary down to an impervious layer. He has shown that the line of saturation through a dam composed of an impervious center section and pervious upstream and downstream sections is the same for both pervious and impervious materials.

Allowable seepage losses from reservoirs or ponds will depend to a great extent on the purpose of the structure and the value and availability of the water. Those reservoirs planned for long-duration storage cannot stand a high rate of water loss. However, the reservoir of a flood-retention structure could be more pervious and have a higher seepage loss. In arid or semiarid localities the maximum allowable water loss by seepage will be affected by the evaporation. In this case it is necessary to maintain a permanent and effective storage balance against the high evaporation loss. When laboratory tests disclose that soil-permeability rates for reservoir and canal construction are too high, then the various methods of reducing seepage losses must be investigated. The most common tests involve a treatment with either salt or bentonite or the use of local clays.

Effective *sealing treatment with salt* depends upon the amount of clay. Significant reductions can be effected when soils contain at least 20 per cent 5-micron clay. Soils with high clay content and base-exchange capacity will take more salt than soils with low clay and low exchange capacity. Soils containing gypsum and lime are not adapted to sealing with salt.

Reservoir losses in amounts exceeding 1 ft of water per day have been reduced to 0.08 ft of depth per day or less by applications of salt at rates of from 1 to ¼ lb salt per square foot. Permeability rates have been reduced from more than 2 down to 0.0001 cu ft/(sq ft)(day) with like treatments.

Bentonite linings are applied in two ways: as pure bentonite diaphragms and as mixtures with the surface soil. Bentonite diaphragms must be protected from rupture, freezing, and drying. Bentonite is a clay which swells to several times its original volume when moistened. Soil-bentonite mixtures are generally favored. Soil-bentonite admixtures work better when soils are sandy. Such mixtures are subject to less cracking when frozen or dried. The use of bentonite is not recommended as an admixture with soils high in clay.

The amount of bentonite required depends upon the quality of bentonite, the amount of original clay binder in the soil, and the amount of gypsum. Rates of bentonite applications are expressed as pounds per unit area or as per cent of soil volume. Applications of ½ to 2 psf have significantly reduced permeability and seepage losses.

Treatment with Locally Occurring Clays. Dispersed clays occur locally in many areas. Such materials have been successfully used as blankets for canals and reservoirs. Recommendations generally call for at least 1 ft of clay as a blanket. Wet compaction of natural material in canals and reservoirs has been effective in reducing seepage. Thickness of a bottom blanket depends on the head of water as well as the permeability of the blanket itself. Experimentation and actual field practice indicate that the soil mixture for sealing the pond should consist of 50 to 75 per cent of sharp graduated sand, 10 to 30 per cent of heavy clay, and enough silt to provide a good gradation of particle sizes from coarse sand to clay. Sheet piling may also be used to reduce excessive permeability on large ponds. The economic justification for the use of this material should be carefully analyzed.

The use of a *cutoff trench* dug into the foundation and filled with an impervious *core* of dense soil material is standard procedure in the design of earth dams. Core trenches are necessary to effect an adequate bond between the embankment and the foundation, to reduce seepage through the foundation, and to provide a stable center section for the embankment. The depth of cutoff depends upon the thickness of the pervious or unconsolidated foundation material and to some extent on the use of the structure. Core trenches under water-impounding structures constructed on pervious material should completely intercept the pervious strata where practical. Partial cutoffs may be considered if the proportional reductions in seepage will affect the stability or provide economical water storage. Core trenches are usually located along the center line of the embankment. Cutoff walls extending into rock foundations may be of compacted earth if the surface of the rock is sufficiently indented to form a good bond with the embankment; otherwise a low concrete wall might be required to control seepage. Core trenches constructed of compacted fill have the minimum bottom width controlled by the operating width of the equipment to be used in construction. In farm ponds, this may be as low as 4 ft. On the larger structures, the minimum is usually 10 to 12 ft. From the bottom of the core trench to the natural ground, the

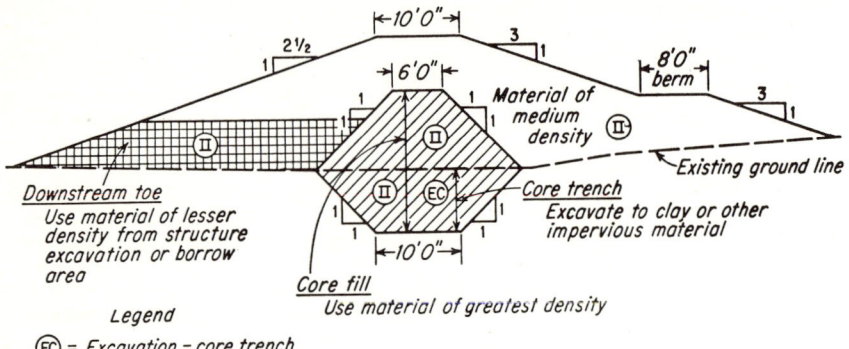

Fig. 41-41. Earth embankment cross section showing core trench and disposition of various earth fill materials.

sides are usually sloped 1:1 or flatter. An impervious core built in the center of the embankment is usually of the same material and the same degree of compaction as the core trench in the base of the dam. Figure 41-41 illustrates this type of construction and other proportions of the embankment.

Small water-storage structures on the order of farm ponds have 2:1 downstream and 3:1 upstream *slopes*. Steeper slopes and slope design of the larger structures should be based on laboratory analysis. It is common practice on the larger structures to use flatter slopes at the lower elevations, based on the maximum requirements of the design, and steeper at the higher elevations. This break in slope usually occurs at berms constructed to intercept the flow of runoff down long slopes. The berms should be provided at least every 40 ft in elevation and be spaced so as to provide approximate uniform drainage areas between berms. Berm widths vary from a minimum of 4 ft to over 12 ft. The type of equipment used on construction will usually govern the minimum size because it is necessary to grade the berms toward the embankment to provide a drainage channel. The berm is then graded from the center of the embankment toward both ends so that the channel will drain to sodded gutters that are formed between the slope intersection of the embankment and abutments. On the larger structures, these gutters are paved.

Berms are also provided at the crest of the overflow spillway on retention structures to provide control against *wave action*. Water-tolerant vegetation, floating rafts, loose rock, or other horizontal controls may be used where the exposed reach against the prevailing winds is small. For large reaches, protection of the embankment slope with either rock riprap or paving is required at points where serious wave damage could occur. A berm is provided to act as a base at an elevation established as the maximum drawdown of the reservoir. The drawdown elevation should include estimated seepage, evaporation loss, and usage for an extended period. The protection should extend up the slope to the maximum wave height plus freeboard, as determined by Eq. (41-1).

Rock riprap for wave protection should have an average minimum thickness of 12 in. if hand-placed or 18 in. if dumped, with a permissible variation of 3 in. It should be dense, hard, durable stone, with 50 per cent weighing at least 75 lb each. The riprap is placed on a filter bed of graded sand and gravel, concrete-aggregate, or pit-run gravel, a minimum of 6 in. in thickness for the graded aggregates to 8 in. for the pit-run material. The top width of the embankment varies from a minimum of 6 ft for simple terrace outlet structures, 8 ft for farm ponds, to that for large structures determined by the equation

$$W = \frac{H + 35}{5} \tag{41-47}$$

where H = maximum height of embankment, ft
  W = top width of embankment, ft

Where the top of the fill is used as a roadway, the top width is usually 12 ft for farm lanes and a minimum of 18 ft for township roads. Greater widths are required to meet the criteria of local highway specifications.

BIBLIOGRAPHY

1. King, H. W., and Ernest F. Brater: "Handbook of Hydraulics," 4th ed., McGraw-Hill Book Company, Inc., New York, 1954.
2. Donnelly, C. A., and F. W. Blaisdell: Straight Drop Spillway Stilling Basin, *Univ. Minn. St. Anthony Falls Hydraulic Lab. Tech. Paper* 15-B, 1954.
3. Stanley, C. M.: Study of Stilling Basin Design, *Trans. ASCE*, 99:490, 1934.
4. Prepared by Subcommittee on Small Water Storage Projects, Perry A. Fellows, chairman, National Resources Committee: "Low Dams," Washington, D.C., 1938.
5. Ree, W. O.: "The Capacities of Rectangular Notches," USDA SCS, 1938, mimeo.
6. Tschebotarioff, Gregory P.: "Soil Mechanics Foundations and Earth Structures," McGraw-Hill Book Company, Inc., New York, 1951.
7. Taylor, D. W.: "Fundamentals of Soil Mechanics," John Wiley & Sons, Inc., New York, 1948.
8. Bennett, P. T.: The Effect of Blankets on Seepage through Pervious Foundations, *Proc. ASCE,* June, 1944.
9. Middlebrok, T. A.: Compaction of Soils, *Proc. Soil Sci. Soc.,* 1943.
10. Turnbull, W. J.: Field Compaction Tests, *Second Intern. Conf. on Soil Mechanics,* 1948.
11. Making the Fill at Dorena Dam, *Eng. News-Record,* Nov. 3, 1949.

Chapter 42

WATER MANAGEMENT, CONSERVATION USE, AND LEGAL ASPECTS

HOWARD W. MATSON

Water conservation in a broad sense includes all phases of the protection, control, and beneficial use of water resouces. It requires an acquaintance with geology, soils, and plants and an understanding of their relationship to water resources; a knowledge of the basic principles of meteorology, hydrology, and hydraulics; and an appreciation of the legal aspects of water management. Water conservation lies largely in the field of the agricultural engineer because of the intimate relationship of agriculture to the engineering problems involved. The purpose here is to present basic elements and principles of the major phases of water conservation in concise form, with references to authoritative sources of data and detailed discussions of engineering procedures involved in each phase.

FARM PONDS

The major use of farm ponds is for livestock water supply. Most ponds are also used for fish production and some for irrigation, orchard spraying, and other purposes. Water for fire protection is an important side line, particularly in the eastern states. The requirements for each prospective use must be considered in the design and construction of the pond.

Design. Dependability of the water supply is the primary consideration. In determining dependability, average values of precipitation, runoff, and evaporation are valueless. For 95 per cent dependability of supply (desirable), it is necessary to determine the values of precipitation and runoff which may be expected to be equaled or exceeded 95 per cent of the time and the value of evaporation which may be equaled but not exceeded 95 per cent of the time. By relating these values to the dimensions of the pond and the water-use requirements it is possible to estimate how often a particular pond may fail to supply the required amount of water.[1]

Water requirements for livestock and poultry are listed in Table 55-9.

Evaporation exceeds *precipitation* during critical periods of low or no runoff. For convenience in design, seasonal values of expectancies of evaporation minus precipitation should be determined. Records of monthly and daily precipitation for many years are available in most localities, but long-time evaporation records are few. In many cases it may be necessary to use seasonal expectancies of evaporation computed by empirical formulas, as given in Chap. 34. The expected total value of evaporation minus precipitation during the selected design length of critical period is directly proportional to the required pond depth.

Runoff is determined by precipitation, temperature, and the physical characteristics of the watershed, including the type and density of vegetative cover. Many short-term

---

[1] D. B. Krimgold and N. E. Minshall, Hydrologic Design of Farm Ponds and Rates of Runoff for Design of Conservation Structures in the Claypan Prairies, *USDA* SCS-TP-56, May, 1945. W. D. Potter and D. B. Krimgold, Area Relationships That Simplify the Hydrologic Design of Small Farm Ponds, *Agr. Eng.*, 27:269, 1946.

records of runoff from small watersheds have been collected by the Soil Conservation Service, Agricultural Research Service, the Forest Service, and various state experiment stations, but few long-time records exist. From available precipitation and runoff records the Soil Conservation Service has prepared a series of technical publications on the hydrologic design of farm ponds for various physiographic areas. For areas in which these data are not directly applicable, generalized information is available in other Federal and state publications.[1]

To avoid excessive spillway discharge and capacity the ratio of drainage area to pond storage should not greatly exceed that necessary for dependable water supply.

Spillway design is important; many pond failures are due to inadequate spillway capacity or protection.

Ponds with small drainage areas (3 to 30 acres) can be designed with vegetated spillways around one or both ends. If this type of spillway is used, it is important to provide the proper width and depth to keep velocities within safe limits as set out in Chap. 40.

The present tendency in pond design is to go to larger drainage areas and larger ponds so as to be assured of a more adequate water supply. On these ponds a pipe through the fill is generally used for spillways as shown in Fig. 42-1. One or both

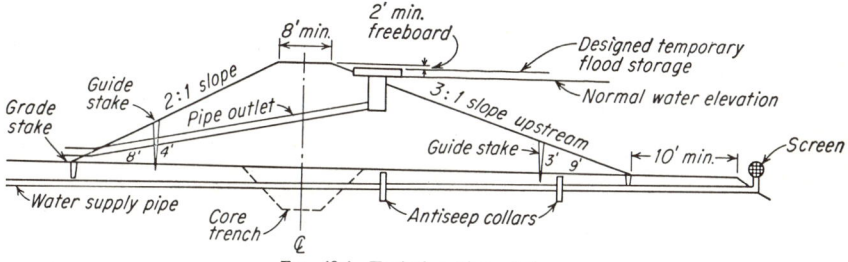

Fig. 42-1. Typical section of dam.

ends of the fill are left low for emergency flow when runoff exceeds the designed capacity. Generally the minimum pipe capacity is determined as follows:

Drainage area, acres	Minimum designed recurrence interval, years
Up to 30	5
30 to 80	10
80 and over	25

Pipe or concrete chute outlets should be provided on all ponds which are fed by spring flow. This is necessary since waterways would be extremely wet if used without mechanical outlets. Under this condition they would be damaged during periods of surface flow. The pipe capacity, freeboard, side slopes, top width, and other requirements for ponds with pipe or flume outlets are given in Chap. 41.

Site selection involves consideration of (1) the adequacy of foundation conditions, (2) availability of suitable fill material, (3) depth of water obtainable, (4) spillway possibilities, (5) drainage area to storage volume ratio (which usually can be modified up or down by diversion), (6) site economy (fill yardage per acre-foot of storage), (7) the condition of the watershed as affecting the expected sedimentation rate, (8) the stability of the drainageway below the site, (9) the convenience of the location for the grazing area to be served, and (10) water-holding qualities of the soils at the site being considered. Foundation investigations as explained in Chap. 41 should be made before site is finally selected.

Construction. Site preparation is the first step. Trees and brush should be removed from the pool area, and the area to be covered by the fill should be cleared of trees

[1] C. L. Hamilton and Hans G. Jepson, Stock-water Developments: Wells, Springs and Ponds, *USDA Farmers' Bull.* 1859, 1940.

or brush, roots, stones, and debris and stripped of sod and topsoil. Steep gully or channel banks should be sloped. The sod and topsoil should be stockpiled out of the way for later use in vegetating the fill and spillway. The entire foundation area should then be plowed or disked parallel to the axis of the dam.

Before beginning placement of the fill material the *center line should be staked and construction stakes set* at the toe of the slope, both upstream and downstream, on 50-ft centers. Guide stakes, as shown in Fig. 42-1, will be helpful. The top width of dam and side slopes must be known in setting the construction stakes, using the fill height at each point as determined from the surveyed cross section along the center line.

A 2-in. steel or 1-in. copper or plastic *pipe,* or larger, should be laid across the foundation in solid ground with enough grade so that it will be below frost line at the downstream toe of the fill as shown in Fig. 42-1. The pipe can be used to supply a stock water tank and to drain the pond if needed. At least two concrete seepage collars, 6 in. thick and 30 in. square and spaced 15 to 25 ft apart, should be *poured* around the pipe in the upstream one-half of the dam.

Fill material should be placed in uniform layers 5 to 10 in. deep, and the layers should be kept as nearly level as practical. In constructing the fill with a bulldozer, the fill material should be placed in uniform inclined lifts. Each layer should be placed over the full width and length and adequately compacted before the next layer is started. To obtain optimum compaction, it is necessary that the fill material be moist (Chap. 41). Sheep's-foot rollers produce the best compaction, but for ordinary farm ponds adequate results can be obtained by the systematic travel of teams or tractors over each successive layer of fill. When earth-moving equipment is used for compaction, all points of the fill should be carried at least 10 per cent higher than the designed height to allow for settlement. The top of the fill should have several inches of crown to prevent ponding or channeling.

After completion of the dam and spillway the stockpiled topsoil and sod should be spread over and compacted into the exposed surfaces to help establish a good vegetative cover for protection against erosion. Fertilization, seeding and mulching are also required to obtain a good cover quickly.

In *spillway layout and construction* care should be taken to avoid unnecessary destruction of existing vegetation or exposure of subsoil. Where excavation is necessary, the sod and topsoil should be stockpiled nearby for use in revegetation. Other excavated material may be used to form compacted dikes along the spillway to increase the channel capacity and reduce the required depth of excavation.

Fencing of the pond, dam, and spillway is recommended to prevent trampling by livestock. Livestock trampling around the margins of a pond keep the water muddy and kill vegetation, permitting soil to be carried into the pond by runoff from the surrounding area.

Fish-production possibilities of a small pond can be increased materially at little cost by eliminating shallow water areas during construction. Part of the material for the dam may be obtained from the sides of the pond just below spillway elevation to increase water depth at the edges. Minimum depth for a pond to be used for fish production in the South will be 4 ft if the water level will hold constant, although greater depth is preferred. A pond in the central states should be[1] 8 to 10 ft in depth and in the North 10 to 12 ft to protect fish from winter freezing.

Dug ponds are the only practicable type in relatively flat areas where the soil is impervious, giving low seepage losses. To avoid high construction cost and excessive evaporation losses, dug ponds are usually small in area and relatively deep. Usually diversion ditches are necessary to provide adequate inflow. The principles of hydrologic design are the same as for the impounding type of farm ponds.

Under ordinary soil conditions the side slopes of dug ponds should not be steeper than 2:1 to prevent sloughing; some soils are stable on 1:1 slopes. One side should be 4:1 slope to provide access for livestock; the remainder of the pond should be fenced.

---

[1] Verne E. Davison, Managing Farm Fish Ponds for Bass and Bluegills, *USDA Farmers' Bull.* 2094, 1955.

The excavated material should be placed in uniform dumps and sloped or spread to facilitate revegetation.

WATER SPREADING

The most widely used type of water spreading is the diversion of stream flow during flash floods and its spreading on pasture or range lands for irrigation. Water spreading is most efficient on deep, permeable soils on relatively uniform slopes of ½ to 1 per cent and usually is not considered practicable on slopes exceeding 2 to 3 per cent. Irrigation of range land by water spreading hastens plant recovery on range land in poor condition, increases the production of vegetation, and assists in controlling gully erosion.

Water Spreading for Irrigation. Ordinarily an earth dam is used to divert the water from the stream channel or waterway and it is carried to the spreading area by

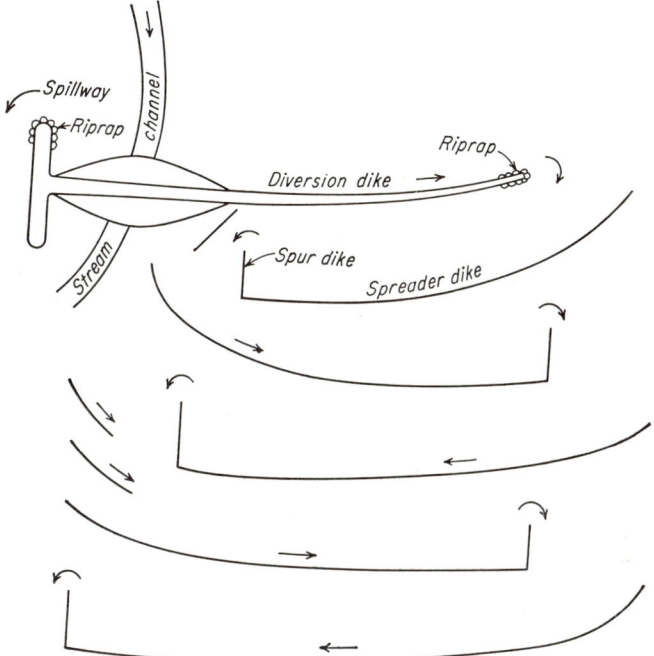

FIG. 42-2. "Syrup-pan" water spreading system.

a diversion ditch or dike. One of the common methods of water spreading, known as the *syrup-pan system*, is illustrated in Fig. 42-2. The water flowing around the end of the diversion dike is caught by a spreader dike or terrace, the lower end of which is partially closed by a spur dike. When the basin formed by the spreader dike becomes filled, the water flows around the spur dike and is picked up by the next spreader dike. The number of spreader dikes used depends upon their length and horizontal spacing and the size of spreading area desired.

Usually the *spreader dikes* are built about 18 in. high above natural ground and are level except for the graded upper ends and most of the dirt is obtained from the lower side. The horizontal spacing is such that water will be spread over as much of the area as possible. The spur dikes are built so that water will flow around them when the water level is 6 in. below the top of the spreader dike.

The desirable ratio of the size of the spreading area to the watershed area of the stream or waterway from which water is to be diverted varies widely, depending upon the precipitation intensities to be expected and the runoff characteristics of the watershed. For the range of conditions commonly encountered, an acre of spreading area should be provided for each 10 to 30 acres of watershed area. For good results in vegetative growth, an area on which water is being spread should receive at least one good irrigation annually. Maximum spreading and use of small or annual floods are desirable. The system also needs to be designed to function without major damage during large floods with up to 50-year recurrence intervals.

A diversion dike may be constructed to deliver water to more than one spreading area. A *rock weep* built in the dike, as illustrated in Fig. 42-3, rations water to each spreading system. This is desirable where the watershed area is large and no single spreading area of suitable size is available.

Water Spreading for Flood Control. Water spreading, in combination with floodwater-retarding structures, has been used effectively in arroyo flood prevention. A good example of such a system was constructed in the Camp Rice arroyo watershed, east of El Paso, Tex., to prevent flood damage to irrigated lands in the Rio Grande Valley.

Water Spreading for Ground-water Recharge. Water spreading for the replenishment of ground-water supplies has been practiced in California for 25 years and has

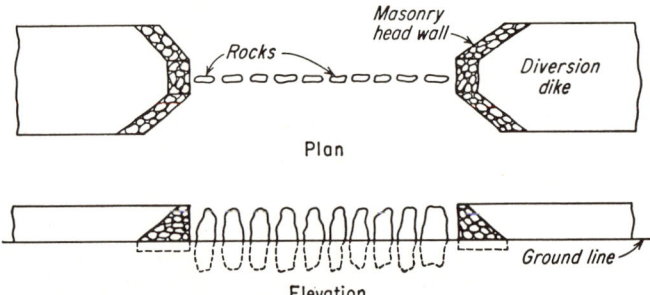

Fig. 42-3. Rock weep in diversion dike.

also been used successfully in Arizona and New Mexico. This type of spreading involves the application of large volumes of water over large areas for as long a time as water is available or a satisfactory rate of percolation can be maintained. Most of the spreading has been done on the debris cones of streams in southern California.

The three general methods of water spreading commonly used for ground-water recharge are the basin method, the furrow, or ditch, method, and the flooding method.

The *basin method* of spreading is commonly used where the surface of the spreading area is irregular. The layout and construction of the basins are such that the entire spreading area is submerged. The basins are formed by earth dikes, with pipes to carry the outflow, or by loose rock walls bound with heavy hog fencing and designed for overflow. The dikes or walls usually are built on the contour. The basin method of spreading is not suitable where the water contains silt or other fine materials which will cause sealing of the surface of the spreading area.

The *furrow, or ditch, method* is particularly suited for use in stream beds where other structures are not practicable because of the danger of damage by floods. Furrows are usually 8 to 10 in. deep and several feet wide, closely spaced to afford large percolation area. Water is fed to the furrows through pipes from a canal or ditch. The grade of the furrows should be sufficient to prevent deposition of silt or other fine material, which would reduce the percolation rate, but they should not be so steep that erosion results.

Flooding is the most efficient method of water spreading on gently sloping areas which are relatively free from gullies or depressions. After the water is released at the upper end of the spreading area, it may be directed by small ditches or dikes so that it will flow slowly in a shallow layer over as much of the area as possible. Since this type of spreading system can be installed with little disturbance of the native vegetation, it is easier to maintain relatively high spreading rates. Usually the best location and shape, or direction, of the directing ditches or dikes can be most easily determined as water is being applied. Thorough control of the water is essential for uniform spreading and percolation.

In the case of streams with wide shallow channels it is possible to spread water efficiently by constructing a series of level low weirs or dams entirely across the channel. This is considered a variation of the flooding method.

The most suitable areas available for spreading in some cases may consist of sandy loam or fine sandy loam soils. The major problem in these areas is to develop and maintain a satisfactory percolation rate, particularly where water is continuously available for spreading over long periods of time. The effect of various applications of trash and vegetation in increasing and maintaining percolaton rates is given in Chap. 34.

WATER-SUPPLY FORECASTING

Water-supply forecasting from snow surveys is concerned with the relationship between the water content of the snow cover at selected points on a watershed and the volume of runoff which may be expected to occur when the snow melts. Since elevation, aspect, vegetative cover, soil-moisture content, ground-water conditions, temperature and precipitation during the snow-melting period, and other factors are involved, this relationship must be determined for each watershed.

In 1923, George D. Clyde of the Utah Agricultural Experiment Station established a network of snow courses in Utah and carried out intensive studies relating to snow cover and runoff. These studies gave advance warning of the severe droughts of 1926, 1931, and 1934 and brought out the improved tubular scale for weighing the snow core, the aluminum snow-sampling tube, and mechanized oversnow vehicles. The program has grown rapidly, and now as many as 1,000 men make 2,000 to 2,500 snow surveys each winter on more than 1,000 snow courses in the 12 western states and British Columbia.

Forecasts. A preliminary seasonal forecast is issued Feb. 1, and forecasts are issued monthly thereafter until May 1. Seasonal forecasts are made on the basis that average conditions of temperature and precipitation may be expected during the runoff period. As the season progresses, variations from average conditions are considered in preparing subsequent monthly forecasts.

Snow-survey data and irrigation water-supply forecasts are of great value to irrigators throughout the western states. By knowing the water-supply situation 2 to 3 months in advance, irrigation districts can schedule and regulate their deliveries to make the most efficient use of the available water. With early information, farmers can adapt the kind and acreage of crops planted and their management in accordance with the expected water supply. Severe losses have thus been avoided even during drought years such as 1934 and 1946.

Water-supply forecasts are equally valuable to those concerned with municipal water supply, hydroelectric power development, mining, lumbering, agricultural credit, and crop estimates. Also, a knowledge of the water content of the snow cover is important in determining the flood potential on many western rivers.

Procedures. Each snow-survey course is mapped and permanently marked and usually consists of 10 to 12 stations at measured intervals of 50 to 100 ft. The course is located so that the average of the samples will be representative of the snow cover and water content for a particular elevation and exposure.

A sufficient number of *snow courses* must be selected on each watershed to indicate the variability of storms, as well as to reflect differences due to elevation and exposure. Snow courses should be at elevations where melting is not expected before Apr. 1, yet must be dependably accessible to permit regularity of measurement. Skis, snowshoes, and a variety of types of mechanized oversnow vehicles are used to reach the courses. To minimize the hazards of snow surveys, the men work in pairs or larger parties and shelter cabins are provided near the courses.

The *standard snow sampler* is duraluminum tubing, in 30-in. lengths, having an inside diameter of 1.485 in. One length is equipped with a cutting bit. The tube is pushed down through the snow to the ground surface, and graduations on the tube indicate the snow depth. The sampler is then withdrawn, containing the snow core, and weighed on an accurate scale of special design. The known weight of the sampler, ascertained before use, is then deducted, and for the tube of 1.485-in. diam, the weight of the snow in ounces is equal to the water content of the snow in inches.

Under certain conditions of snow–air-temperature divergence (particularly in the middle of the day) or snow water content, sampling difficulties occur because of the tube sticking in the snow or the snow core freezing in the tube. It has been found that by varnishing both the inside and outside of the tube, applying a thin coating of high-grade ski wax, and polishing the wax coat thoroughly to produce a hard glossy finish, samples can be taken successfully under most conditions.

It is possible to make a reasonably accurate water-supply forecast for a large watershed from surveys of a small number of snow courses because of the fact that most of the snowfall at high elevations occurs during a few major storms which are relatively uniform in intensity over large areas. Usually greater forecast accuracy can be obtained by using the maximum snow-water depth found for each snow course, regardless of the month this maximum depth occurs.

LEGAL PRINCIPLES OF WATER RIGHTS

A *water right* is a well-founded and acknowledged claim, recognized by law, to take possession of and put water to use. The two fundamental systems of water rights are the doctrine of *riparian rights* and the doctrine of *prior appropriation*. These two doctrines are diametrically opposite in principle, the former being based on the ownership of land contiguous to a stream, without regard to the time of use or to any actual use at all, and the latter on the time and amount of use and on actual beneficial use without regard to the ownership of land contiguous to the stream.

Riparian Doctrine. In the West the riparian doctrine has been recognized, although in varying degrees, by the courts in North Dakota, South Dakota, Nebraska, Kansas, Texas, California, Oregon, and Washington. In the eastern part of the United States the riparian doctrine generally applies, although at the present time eastern states are modifying it by water laws. It has been assumed to be in effect in Oklahoma, but its status has not been clearly defined in the courts. The riparian doctrine was recognized in Oregon but has since been virtually abrogated by statutes and court decisions, except for limited domestic purposes. It has been specifically abrogated in its entirety in Arizona, Colorado, Idaho, Montana, Nevada, New Mexico, Utah, and Wyoming.

According to the riparian doctrine, each owner of land contiguous to a stream has the right to make such use of the water as he needs for domestic purposes and the watering of farm livestock. The doctrine also has been interpreted to permit limited use of the water for irrigation or other purposes as is reasonable in consideration of like reasonable uses by all other owners of land riparian to the same stream. Thus it is a right in common.

Appropriation Doctrine. The doctrine of prior appropriation has been adopted in all 17 of the western states.

Under the appropriation doctrine the first user of water from a stream has the right to a given quantity at a given time and place and the right to continue such use so long as it is beneficial. He may use the entire supply if all of it is needed to satisfy the right which he has established by beneficial use. Thus this is an exclusive right of use.

Law of Diffused Surface Waters. Although all 17 of the western states recognize the right to appropriate water from a stream, there is wide variation in the laws relating to diffused surface waters arising from precipitation and snow melt. Some states have no specific legislation pertaining to diffused surface waters, but in most of these states it has been determined by court decision that the owner of the land on which the precipitation falls has first claim upon these waters.

Law of Ground Waters. Ground waters flowing in definite streams generally are considered subject to the same rules of law as surface streams. Percolating waters are ground waters which are not flowing in a definite underground stream. The statutes and court decisions of the several states recognize three diverse doctrines of ownership and use of percolating waters:

1. The English, or strict common-law, rule of ownership by the owner of overlying land without restriction as to amount, time, or place of use.

2. The American rule of reasonable use which recognizes ownership by the owner of overlying land but restricts his right to such use of the water as is reasonable with regard to the similar rights of all other owners of land overlying the same source of water supply. The American rule with respect to percolating waters is analogous to the riparian doctrine as applied to surface streams.

3. The doctrine of appropriation, which applies in the same manner as to surface streams, including the fact that the place of use need not be on lands overlying the ground-water source of supply.

Elements of Appropriative Rights

1. *Beneficial use.* Actual use of the water for beneficial purposes is essential to acquire and maintain an appropriative right. In 13 states failure to use water during a specified period, ranging from 2 to 5 years, constitutes abandonment or statutory forfeiture of rights. In Colorado, Kansas, Montana, and Washington, statutory provision is made for determination of abandonment.

2. *Quantity of water appropriated.* In most of the western states the quantity of water which may be appropriated for irrigation is limited by statute. Such statutes usually specify the maximum permissible rate of diversion in relation to the area irrigated, as well as the maximum volume of annual water use per acre irrigated. However, these standards are quite loose compared with sound irrigation practices.

3. *Period or season of use.* In North Dakota, South Dakota, and Oklahoma irrigation water may be delivered on the land only for a specified time each year.

4. *Point of diversion.* The point of diversion is specified in the appropriation, and it is there that the appropriative right attaches to the flow of the stream. Since the place of diversion may affect other appropriators of the flow of the same stream, it may be changed only if the rights of others are not affected adversely, and then only by prescribed procedures.

5. *Place of water use.* In most states, place of use is specified in the appropriation, is recorded, and cannot be changed except by following prescribed procedures.

6. *Purpose of use.* Laws of all western states specify the purposes for which water may be appropriated. In most of these states, domestic use is given preference over all other uses but the order of preference of other uses varies. This may be an important factor, since in periods of water scarcity a use of higher preference may take precedence over another right which is prior in time, although not without adequate compensation.

7. *Priority in time.* The essence of the doctrine of prior appropriation is that among users from the same source the first in time is the first in right and beneficial use is the measure of the right. When the available water supply is not adequate to meet the needs of all those holding rights to its use, the rights are supplied in the order of the dates on which they were acquired. The priority of a right dates from the time the first step was taken to acquire possession of the water rather than from the time the water was put to use, provided reasonable diligence was pursued in applying the water to beneficial use. The right may not extend beyond the original diversion capacity and may be less than that capacity if all the water was not put to beneficial use.

Advantages of Appropriative Rights. Concerning the right to use either surface streams or ground waters, the appropriation doctrine has several definite advantages, particularly in areas where the water supply is inadequate to meet the needs and opportunities for beneficial use. Some of these advantages are:

1. It puts greatest emphasis on the beneficial use and conservation of water. Appropriation provides greater latitude for maximum beneficial development, since it permits the transportation of water for use on distant lands which may be more productive than lands contiguous to a stream or overlying a ground-water area.

2. Appropriation affords great protection to investments dependent upon water

and thus encourages such investments. The "first in time, first in right" principle, although it imposes certain practical difficulties, nevertheless does afford adequate protection to vested rights.

3. Appropriation affords a means of orderly development through guidance provided by public agencies, state or local or both. Where water shortages exist, the limited supply must be managed through public control to assure the highest beneficial uses, and such control can be assured only under the appropriation system. The appropriation procedure facilitates administration and eliminates many of the uncertainties as to the nature and extent of each water user's right, since the right can be defined adequately as to quantity, purpose, and time.

The law of water rights is a highly specialized branch of the law, and in a number of states it has been changing rapidly during recent years. Before undertaking any development which involves the establishment or transfer of water rights, the applicable state laws should be consulted or advice obtained from an attorney familiar with water law.

FLOOD PREVENTION

Flood prevention is the maximum practicable management of storm precipitation on the land where it falls and of storm runoff in the small branches and creeks of the upper watersheds. It is called flood prevention, as distinguished from downstream flood control, because the measures and structures used are designed primarily to prevent the formation of floods and usually are located above alluvial-flood-plain areas. Obviously it is not possible to prevent all headwater floods by such means, just as it is not possible to completely control floods in downstream areas. But where physical conditions are favorable and it is economically feasible, a combination of land-treatment measures, watershed-stabilization structures, and systems of floodwater-retarding structures and associated works can prevent most of the present floodwater and sediment damages in tributary watersheds. The planning and installation of such work requires a thorough understanding of hydrology and hydraulics, a knowledge of the principles of structural design, and a familiarity with practical construction methods. An acquaintance with geology and plant-soil ecology is helpful. Because of its direct association with agricultural problems, the job is a logical field for agricultural engineers who have the necessary specialized training and experience.

Some of the *steps* in developing a comprehensive flood-control plan for a watershed are:

1. Land-treatment measures consisting of adequate rotations and proper fertilization should be applied wherever possible.

2. Supporting conservation practices such as terracing, waterway development, and farm ponds reduce runoff and should be planned where needed.

3. Needed structural gully-control measures should be provided to reduce soil loss, stop land destruction, and prevent silting of downstream works.

4. Areas of bottomland which would be benefited by floodwater detention and/or channel work should be delineated on a map. Possible detention-site locations as well as needed channel improvement with respect to these areas should be determined, and floodwater routed through these reservoirs by methods given in Chap. 41. A reservoir of this type is shown in Fig. 42-4.

5. Other downstream floodwater damages such as bridge and fence damage, deposition of infertile overwash, and silting of channels should be determined.

In making the study outlined it will be necessary to route runoff from floods of various recurrence intervals through downstream flood plains. Usually floods of 1-, 2-, 5-, 10-, 25-, and 50-year recurrence intervals are studied. Discharges can be computed using material given in Chap. 34. Reduction of the more frequent floods generally produces the greatest net benefit for agricultural areas. To study flood losses and reduction of flooding by land treatment and/or structural or channel measures requires cross sections of the channel and flood plain at such intervals as are necessary to compute water-surface profiles.

Water-surface profiles can be computed by Manning's formula using valley cross sections and average slopes. If more accuracy is desired, backwater curves can be

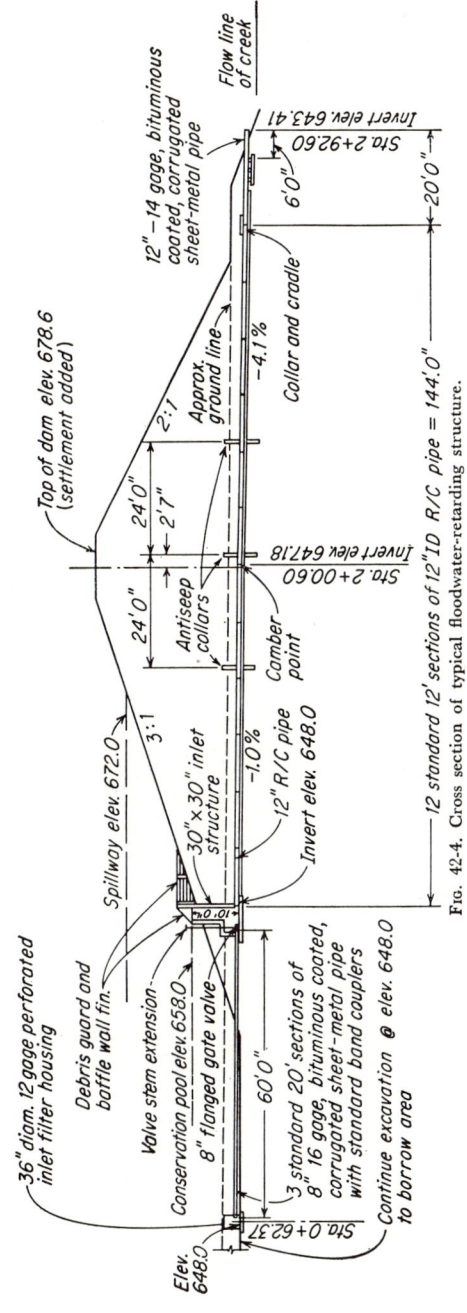

FIG. 42-4. Cross section of typical floodwater-retarding structure.

determined by use of Bernouilli's theorem. This is adequately covered in a number of handbooks.[1]

Objectives and Benefits of Flood Prevention. Tributary-flood prevention increases the protection afforded by major reservoirs and levees to flood plains along rivers and large tributaries. Although the large reservoirs and levees may provide adequate protection against main-stem floods, in many cases the areas behind levees are subject to frequent—and sometimes severe—floodwater and sediment damages caused by tributary overflows or runoff from adjacent hill land. Also, the flood protection afforded by major reservoirs decreases with distance downstream. Retarding storage in tributary watersheds may reduce the main-stem flood hazard materially.

Several types of *incidental benefits* may be realized from floodwater-retarding structures planned and built primarily for flood prevention. At favorable sites where large reservoir capacity may be developed economically, it may be feasible to include conservation storage for livestock water supply, irrigation, or for public water supply for towns. Even if no permanent conservation storage is provided, the storage volume reserved for eventual sedimentation may be used for livestock water supply, irrigation, fish production, and recreation for a considerable number of years.

BIBLIOGRAPHY

Andrews, Roy G.: Aids for Determining Runoff Probability, *Agr. Eng.*, 38:164, 1957.
Blaisdell, Fred W., and C. A. Donnelly: Hydraulic Design of the Drop Inlet Box Spillway, *USDA* SCS-TP-106, July, 1951.
———: The SAF Stilling Basin, *USDA* SCS-TP-79, 1949.
———: Report on Hydraulic Model Tests of Outlet Structure to Be Built at Lower Caney Lake, Minden, La., *Univ. of Minn. St. Anthony Falls Hydraulic Lab. Rept.* MN-R-3-31, September, 1946.
Busby, C. E.: American Water Rights Law, *S. Carolina Law Quart.*, vol. 5, 1952.
Chow, Ven Te: A General Formula for Hydrologic Frequency Analysis, *Trans. Am. Geophys. Union,* 32:231, 1951.
———:A Practical Procedure of Flood Routing, *Univ. of Illinois Civil Eng. Studies, Hydraulic Eng. Ser.* 1, 1951.
Clyde, George D.: Snow Melting Characteristics, *Utah State Univ. Agr. Expt. Sta. Tech. Bull.* 231, 1931.
———: Establishing Snow Courses and Making Snow Surveys, *Utah State Univ. Agr. Expt. Sta. Bull.* 91, 1930.
——— and R. A. Work: Precipitation-Runoff Relationships as a Basis for Water Supply Forecasting, *Trans. Am. Geophys. Union,* pt. III, vol. 43, 1943.
Criddle, Wayne, D.: Value of Midwinter Snow Surveys, *Trans. Am. Geophys. Union,* 28:888, 1947.
Croft, A. R.: Some Factors That Influence the Accuracy of Water-supply Forecasting in the Intermountain Region, *Trans. Am. Geophys. Union,* 27:375, 1946.
Davis, Calvin V.: "Handbook of Applied Hydraulics," McGraw-Hill Book Company, Inc., New York, 1952.
Frevert, R. K., G. O. Schwab, T. W. Edminster, and K. K. Barnes: "Soil and Water Conservation Engineering," John Wiley & Sons, Inc., New York, 1955.
Frost, W. T., and R. A. Work: The Use of Snow Survey Data for Agricultural Planning, *Agr. Eng.*, 29:490, 1948.
Garstka, Walter U.: Interpretation of Snow Surveys, *Trans. Am. Geophys. Union,* 30:412, 1949.
Hall, Warren A.: Theoretical Aspects of Water Spreading, *Agr. Eng.*, 36:394, 1955.
Hutchins, Wells A.: Selected Problems in the Law of Water Rights in the West, *USDA Misc. Publ.* 418, 1942.
Hydrology Handbook, *ASCE Manual Eng. Practice* 28, January, 1949.
King, Horace W., and Ernest F. Brater: "Handbook of Hydraulics," 4th ed., McGraw-Hill Book Company, Inc., New York, 1954.

---

[1] W. L. Cowan, Victor Mockus et al., "Engineering Handbook," Hydrology Supplement A, USDA SCS, 1957. Horace W. King and Ernest F. Brater, "Handbook of Hydraulics," 4th ed., McGraw-Hill Book Company, Inc., New York, 1954.

Marr, James C.: Snow Surveying, *USDA, Misc. Publ.* 380, 1940.
Miles, Wayne H.: Water Spreading, *Soil Conserv.*, 10(4):73, 1944.
Mitchelson, A. T.: Spreading Water for Recharge, *Soil Conserv.*, 15(3):66, 1949.
——— and Dean C. Muckel: Spreading Water for Storage Underground, *USDA Tech. Bull.* 578, 1937.
Monson, O. W., and J. R. Quesenberry: Range Improvement through Conservation of Flood Waters, *Montana State College Agr. Expt. Sta. Bull.* 380, 1940.
Muckel, Dean C.: Water Spreading for Ground Water Replenishment, *Agr. Eng.*, 29:74, 1948.
Semple, A. T., and B. W. Allred: Range Improvement by Water Spreading, *Soil Conserv.* 2(12):269, 1937.
Smith, Dwight D.: Storage Pond Design, *Agr. Eng.*, 36:743, 1955.
Strauss, Fred A.: The Use of a Temperature-Runoff Relationship in Water Supply Forecasting, *Trans. Am. Geophys. Union*, 31:879, 1950.
Teele, R. P.: The Western Farmer's Water Right, *USDA Bull.* 913, 1920.
Vanoni, Vito A., and James T. Rostron: Baffle Type Energy Dissipator for Pipe Outlets, *Agr. Eng.*, 25:301, 1944.
"Water Resources Law," Report of the President's Water Resources Policy Committee, vol. 3, 1950.
Work, R. A.: Snow-layer Density Changes, *Trans. Am. Geophys. Union*, 29:525, 1948.

Chapter 43

WIND EROSION AND ITS CONTROL[1]

Austin W. Zingg

Erosion of soil material by wind and its redistribution over the earth's surface is one of the important geological processes of soil formation operating over long time periods. Accelerated wind erosion is a problem associated with land use when the equilibrium between vegetation, soils, and climatic environment has been disturbed by the growing of cultivated crops or the overgrazing of lands. In general, it is associated with the former grasslands of arid to subhumid regions. This treatise deals with accelerated erosion on agricultural lands.

Processes of Erosion. Basically, two processes[1] are involved in erosion of soil material by wind. These are (1) *initiation* of movement of material and (2) its *transport* by wind force irrespective of the forms of erosion or the type of particle movement involved. Deposition of eroded material is not a process of erosion, but a product of the erosion phenomenon.

For heterogeneous soil materials there is no meaningful threshold velocity or shear at which bed movement is initiated and sustained. From a practical viewpoint, movement on bare and erodible field surfaces occurs normally when the atmospheric wind velocity 2 ft above the surface is 10 to 14 mph.

In wind-tunnel experiments where approximately uniform cohesionless beds of sand, ranging from 0.2 to 0.7 mm in diameter, have been studied, a saltation (see below) threshold for movement has been found to be[2]

$$\tau_s = 0.007 d$$

where τ_s = apparent shear of wind, psf of bed area
d = diameter of sand particles, mm

A minimum shearing force is apparently required to move beds of grains 0.10 to 0.15 mm in diameter. The average bed force required to move grains less than 0.10 mm in diameter tends to increase because of viscous characteristics of air flow over very small particles and possible cohesive effects.

The shear of the wind τ_o on a surface is proportional to the square of the velocity u at a given height above the surface. The rate of movement q of a given soil or sand material with respect to these factors is

$$q \propto \tau_o^{3/2} \propto u^3$$

Forms of Erosion. The forms of accelerated wind erosion are (1) leveling of the surface, (2) abrasion of soil material, and (3) removal of soil material. Consider a field that has been listed. After listing, angular clod masses and furrows are present. At first the surface has sufficient roughness to limit movement of soil material from

---

[1] *Contrib.* 492, *Kansas Agr. Expt. Sta.*, and USDA ARS (Cooperative Research in the Mechanics of Wind Erosion).

[2] A. W. Zingg, Wind Tunnel Studies of the Movement of Sedimentary Materials. *Proc. Fifth Hydraulic Conf. Iowa Inst. Hydraulic Research*, June, 1952.

surface projections to a lower elevation. Clods tend to disintegrate because of weathering actions, and the surface tends to be leveled by the action of wind, precipitation, and gravity on the exposed clods and ridges. With leveling of the surface, the force of the wind is gradually transferred from relatively stable surface projections to the smaller dislodged material. Finally a state is reached where erodible material constitutes a large proportion of the surface and the stage is set for particles to move more or less freely over or on the surface. The nature of particle movement is then such as to cause marked abrasion or breakdown of surface materials and the lowering of the surface through removal.

Types of Soil Movement by Wind. Grain motion takes place by suspension, saltation, and surface creep.[1] The materials capable of transport by *suspension* are those having a terminal velocity of fall less than the upward eddy currents of the average surface wind. In general, they are less than 0.1 mm in diameter, depending, of course, on their shape and apparent density.

Saltation is a term long used in geologic literature to describe the leaping and bounding action of sand grains above the surface of the ground in an air stream. *Surface creep* comprises the material driven along the immediate bed surface. While grains moving in saltation receive most of their forward momentum directly from the pressure of the wind, those in surface creep receive nearly all their momentum by impact from the saltation. Saltating grains have components of motion in the x, y, and z directions, and the characteristics of movement of an individual grain are governed by complex impact forces. In addition, high-speed photography has demonstrated that they have high but variable rates of rotation gained from elastic impact with other grains either upon the bed or by collisions in air. The axes of rotation are variable, and rotation may be in either direction about such axes.

Types of Wind Erosion. Mass removal of the soil materials from a land surface is a phenomenon associated with the finer-textured soils composed primarily of silt and clay. In these soils, all the primary particles may be suspended and removed by wind. They are usually loessal in origin. While the basic soil particles may be aggregated, little sorting of basic materials occurs from wind action. In general, the saltation movement of aggregates causes their breakdown and the soil materials may be transported far from the source area. In the Great Plains of the United States these soils are usually dark in color and are known as the "hardlands."

Selective erosion is the sorting of relatively fine materials from an eroding surface. The phenomenon is common to sandy lands containing some basic soil materials too large to be suspended by the wind and removed from the source area. Selective erosion results in the loss of the finer and more fertile portions of the soil from the eroding area. Local products are the accumulation of sand into hummocks or dunes. In some cases extremely coarse material remains in place to form a desert pavement. The soils are usually residual or formed from mountain outwash.

Climatic Factors. The climatic factors governing wind erosion are wind movement, precipitation, and temperature. Low precipitation, high winds, and high temperature are conducive to wind erosion. The seasonal occurrence of precipitation and wind at a location in the Great Plains of the United States is shown in Fig. 43-1. In this instance, the period of highest wind movement occurs near the end of the dry winter period and before the season of rainfall and active vegetal growth.

Damaging winds tend to come in the form of storms of a few hours to days duration. They vary diurnally and seasonally as well as from year to year. Directional influences and types of storms are also of consequence. Large differences in the scale of atmospheric turbulence are present for different winds, and storms of equal velocity vary greatly in their ability to erode soil.

Unfavorable combinations of drought, winds, and high temperatures tend to occur in irregular cycles. The severe dust storms of the 1930s in the Great Plains were the result of an extremely adverse combination of these factors. The probability of recurrence of the climatic condition present at that time is about once in a hundred years.

---

[1] R. A. Bagnold, "The Physics of Blown Sands and Desert Dunes," William Morrow & Company, Inc., New York, 1943.

Soil Factors. Soils vary in many properties; however, texture and structure appear to be pertinent factors from a wind-erosion viewpoint. When exposed to the wind, the physical state of the soil at the immediate surface is of primary importance. The clod and aggregate structure of a given soil vary with season, cultivation, freezing, thawing, wetting, and drying. The surface may further vary in roughness, degree of

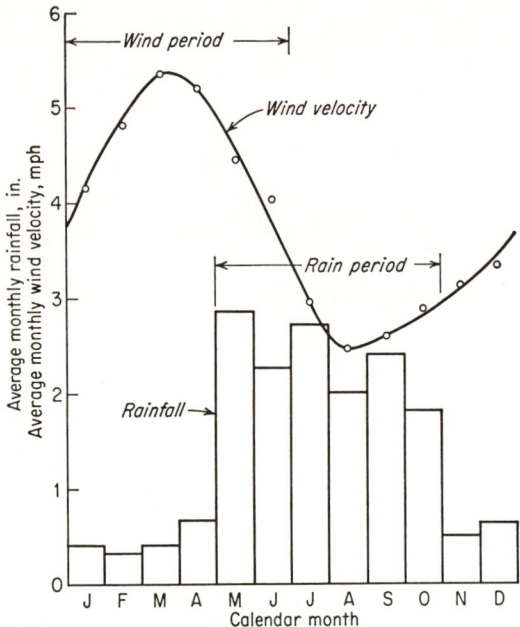

FIG. 43-1. Seasonal pattern of precipitation and wind movement at Portales, New Mexico.

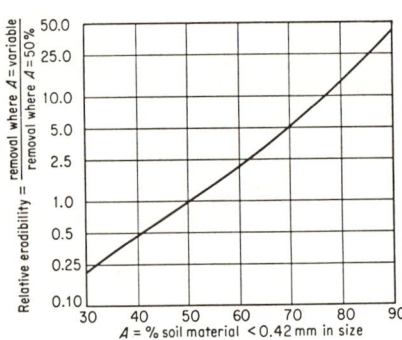

FIG. 43-2. Relative erodibility of soil based on the percentage by weight of dry soil material less than 0.42 mm in size.

compaction, and surface crusting or cementation. In wind-tunnel tests of the erodibility of soils it has been found that dry sieving of the surface inch of soil yields an index of the soil's erodibility at a given time.[1] A plot of the relative erodibility of soils with respect to the percentage of dry soil material less than 0.42 mm in size is shown in Fig. 43-2.

[1] W. S. Chepil, Factors That Influence Clod Structure and Erosion of Soil by Wind: I. Soil Texture, *Soil Sci.*, June, 1953.

Principles of Control. The control of wind erosion is based on two principles: (1) decreasing the direct wind force on erodibile soil material and (2) modifying the soil material to resist wind action or limit particle movement.

Plant covers and residues are capable of taking a portion of the direct force of the wind from the soil particles on the immediate surface. While the force of the wind is still transmitted to the soil mass, it is modified to bear on the plant or residue anchorage in the soil rather than on erodible grains at the surface.

Surface barriers, when normal to the wind direction, have a zone of influence within which the force of the wind on the soil surface is lessened. This protective device brings about changes in the pattern of air flow over the land surface.

Any cultural treatment, soil amendment, or device which tends to maintain non-erodible soil aggregates or clods on the soil surface utilizes the second principle of control. In addition, the roughness of ridged or cloddy soil surface serves as a reservoir to trap or limit the movement of erodible particles. The roughness of vegetal or residue cover has similar trapping characteristics.

Control Methods. The climax vegetation in which an approximate equilibrium between climate, soils, and vegetation has developed usually affords the greatest stability of the land surface to wind action. Generally, the more closely this condition can be approached under cultivated use of the lands, the better will be the control of wind erosion.

In dry-land areas moisture is often the limiting factor in the production of vegetation. The practice of *fallowing* to conserve moisture for subsequent crops is employed. Stubble-mulch systems of farming in which sweep-type plows are used have been devised to maintain a protective covering of crop residues over the soil during periods of fallowing for moisture storage and for crop establishment. The delayed fallow in which the land is not worked after harvest until the following spring is a further advancement in maintaining soil protection. This practice may be employed advantageously in dry years.

Field windbreaks, shelter belts of permanent shrubs and trees, afford mechanical protection of the barrier type and are adapted to deep soils having an ample supply of moisture. Spacing of fifteen to twenty times the barrier height are effective where the belts are oriented properly. A grid of belts in two directions appears necessary where damaging winds do not come from a prevailing direction.

The *planting of crops in a row direction perpendicular* to prevailing high-intensity winds is a device designed to increase the efficiency of vegetal protection to the soil. The employment of alternate strips provides further mechanical protection through their barrier effect and the elimination of contiguous erodible areas. Unfortunately, in some areas of the Great Plains of the United States, almost equal numbers of damaging winds come from east-west and north-south direction. In portions of the northern plains benefits are derived from plantings and establishing strips in a north-south direction. In portions of the southern plains the east-west-directional planting yields benefits.

Management techniques which promote soil stability and conserve and utilize moisture efficiently for plant growth are essential for good wind-erosion control in dry-land areas. Marked developments of more drought-resistant varieties of crops have been and are being made. The average water requirement of different crops also varies. For example, the sorghum plant is more efficient in the use of moisture than the wheat plant. In addition, it is a crop which can be planted following failure of winter wheat to regain a cover on the land quickly. Fertilizers also offer possibilities for more efficient use of moisture for crop production and increased quantities of plant growth for soil protection from wind.

Deep chiseling and subsoiling are employed to loosen relatively impermeable or consolidated soil materials below the depth normally cultivated to provide more favorable conditions for moisture penetration and root growth. At the same time these operations tend to bring consolidated cloddy soil material to the surface to provide relatively wind-stable soil conditions for some period of time. Again, deep plowing to depths of 12 to 18 in. has become a common practice to improve the wind resistance of sandy soils. Following the sorting action of wind, many field surfaces have become extremely sandy. Where they are underlain with finer-textured material, immediate benefits are associated with burying the erodible surface material and bringing silt and clay to the surface. Unless the practice is used as an aid to the establishment of relatively stable land-use systems, it is exploitative in nature.

For control of wind erosion, a tillage implement should leave the surface as cloddy as possible and at the same time avoid burying the crop residue.[1]

Sprinkler irrigation before and during heavy windstorms is used in some localities to control wind erosion on sandy or peat soils. The success of this method is dependent upon the complete coverage of the area during critical periods.

Emergency tillage to control erodible land surfaces is more or less a last-resort measure to keep land from going out of control or to minimize the abrading action of saltating soil grains on young plants. Under certain conditions it is capable of saving a crop. The object is to limit soil movement by roughening portions of a field to the point where it is nonerodible or will trap soil materials moving from intervening areas. Success of the operation is dependent upon its timeliness and the presence of relatively stable material below the immediate soil surface. Either consolidated soil material or moist soil capable of forming resistant clods must be present. Excessive working of the soil under certain conditions may do more harm than good.

The three R's of wind-erosion control appear to be residue management, roughness of the surface, and resistant soil material. Their nurture in areas where wind erosion is a hazard is capable of reducing the problem greatly. Where wind erosion and soil deterioration occur frequently with the best-known management practices, the land is not adapted to other than range use.

[1] N. P. Woodruff and W. S. Chepil, Implements for Erosion Control, *Agr. Eng.*, 37:751–758, November, 1956.

Chapter 44

IRRIGATION

WAYNE D. CRIDDLE

In many agricultural areas of the world, the amount and timing of natural precipitation are not adequate to meet the moisture requirements of crops. Rainfall varies widely even on lands presently irrigated. In many areas of the world, including southwestern United States, Egypt, West Pakistan, and India, there are years of no appreciable summer rainfall and nearly all the water needs of crops must be met by irrigation. Under more humid conditions, such as exist in eastern United States, only a few inches of irrigation water applied during the critical summer period may be needed. This water, however, may mean the difference between abundant production or crop failure. Thus irrigation is the practice of artificially applying the water necessary to meet a soil-moisture deficiency, in the production of crops.

Soils, within the root zone of the crop, are limited as to the amount of water they can store at any one time. Different soils have different water-holding abilities, and different crops have different root zones. There are many problems of when to irrigate, the quantity needed, and how to apply it. Therefore the science of irrigation must draw on many fields of endeavor. The engineer must team up with the soil scientist and the agronomist to work out the best irrigation principles and practices.

Successful irrigation requires good soils, water, and favorable climate, and any one of these may be a limiting factor. Thus, before any irrigation project is planned, an inventory must be taken of these factors. Standard investigational procedures have been developed and are followed by various state and Federal agencies within the United States and in other countries. Unless the factors measure up to certain minimum standards, irrigation should not be undertaken, since it is a costly practice.

WATER SUPPLY

The *source* of water for irrigation is usually precipitation or snow melt from areas above the area to be irrigated (Chap 42). In other cases, it may be from the ground-water reservoir immediately under the cropped land, recharged by rainfall that infiltrates into the ground. If the source is from natural stream flow, storage works must be developed to control the flow and make the water available as needed.

Water for irrigation must be of usable quality and be available when it is needed. Prior rights of others to the use of water must be considered before a source is considered adequate. Just because there is water flowing past a man's farm does not necessarily mean that he can divert such waters for irrigation. Owners of land down the stream may have already established rights to the use of such waters; and if the flow is insufficient to meet all needs, the first appropriators may expect to receive their water first. Later appropriators will get water only if and when surplus water is available. (See Chap. 42 for legal principles of water rights.)

Foreign material impairs *quality* of water whether dissolved or in solid form. If excess silt, sand, brush, or other debris is carried in the water, special treatment may be required to remove such material before the water can be used for irrigation. Such treatment may consist of simple screening or may require the use of expensive sand traps,[1] depending on the kind and amount of material the water carries. Under

[1] Ralph L. Parshall, Model and Prototype Studies of Sand Traps, *Proc. ASCE*, vol. 77, sep. no. 67, May, 1951.

510   SOIL AND WATER CONSERVATION

sprinkler irrigation, even an occasional occurrence of moss or small bugs may cause clogging of the nozzles.

Dissolved salts in the water must also be considered. Their total quantity and kinds, together with the characteristics of the crops, soils, climate, and irrigation practices, must all be considered in appraising water quality. As a general guide for appraising irrigation waters, Figs. 44-1 and 44-2 were prepared by the U.S. Department of Agriculture Salinity Laboratory. In general, the higher the salt content and the

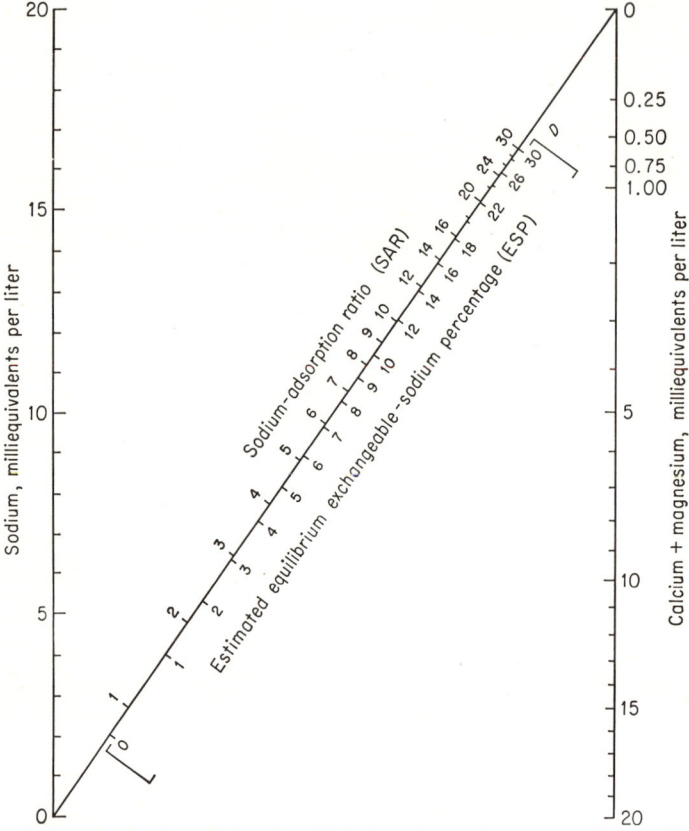

FIG. 44-1. Nomogram for determining the SAR value of irrigation water and for estimating the corresponding ESP value of a soil that is at equilibrium with the water. ("*Diagnosis and Improvement of Saline and Alkali Soils,*" USDA Salinity Lab. Handbook 60.)

higher the ratio of the sodium ion to the sum of the sodium, magnesium, and calcium ions, the poorer the quality of the water for irrigation.

As an *example* of the use of these figures, the chemical composition of typical waters used for irrigation in western United States is the following:

EC (electrical conductivity) $\times 10^6$ at 25°C, micromhos/
  cm (from Fig. 44-2) 1,210
Dissolved solids, ppm 981

Calcium, milliequivalents per liter 7.18
Magnesium, milliequivalents per liter 3.49
                                                        10.67

Sodium, milliequivalents per liter 3.47

Entering Fig. 44-1, it is found that the sodium absorption ratio (SAR) is 1.5. Using this SAR value and the conductance value on Fig. 44-2, it is observed that the water would be classed as C3-S1 quality. From Table 44-1 this would indicate that sodium would probably not cause much difficulty but that salinity trouble would undoubtedly develop on tight soils and special management practices would be necessary on the more permeable soils.

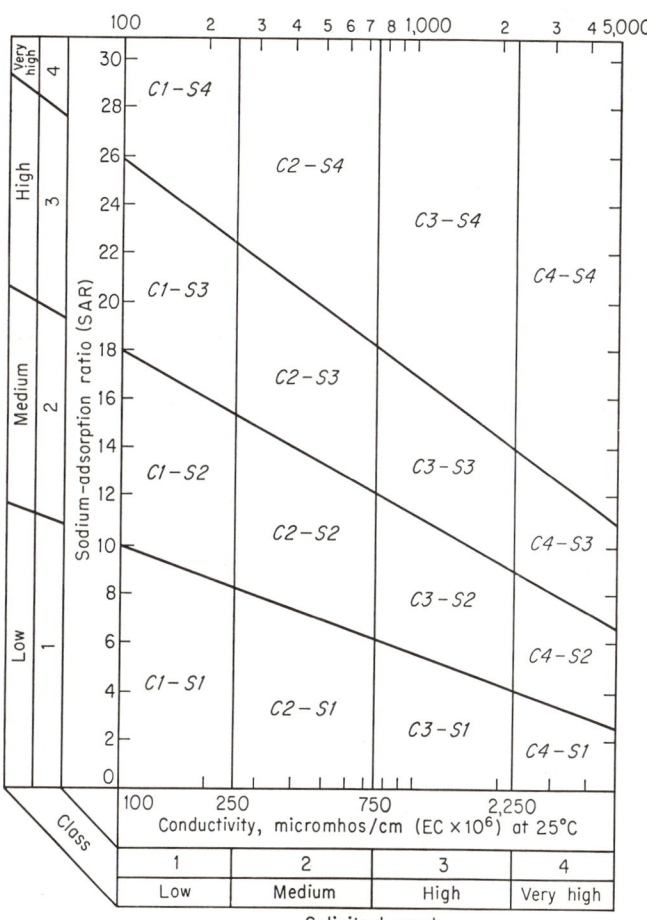

FIG. 44-2. Diagram for the classification of irrigation waters. ("*Diagnosis and Improvement of Saline and Alkali Soils,*" USDA Salinity Lab. Handbook 60.)

WATER DELIVERY AND CONTROL

Once a water supply is found, whether from surface or underground sources, various structures and physical controls will be needed. They include the development and control of the source, the canals and laterals to carry water to the farms, the farm distribution system, and all the structures necessary for the development and/or conveyance and control of the water supply. Canal linings, pumping plants, control structures, measuring devices, pipelines, siphons, check dams, head gates, and spiles are all part of the structures needed. This segment of irrigation is largely engineering and is not specifically related to the crops grown. It is related to the soils only in so far as the soils affect the stability of structures, seepage losses, and need for lining of conveyance channels.

TABLE 44-1. INTERPRETATION OF QUALITY-CLASS RATINGS OF WATER FOR IRRIGATION PURPOSES

Conductivity

Low-salinity water (C1) can be used for most crops and soils with little likelihood that soil salinity will develop. Some leaching is required, but this occurs under normal irrigation on all except tightest soils.

Medium-salinity water (C2) can be used if a moderate amount of leaching occurs. Plants with moderate salt tolerance can be grown in most cases without special practices for salinity control.

High-salinity water (C3) cannot be used on soils with restricted drainage. With adequate drainage, special management for salinity control may be required and plants with good salt tolerance should be selected.

Very-high-salinity water (C4) is not suitable for irrigation under ordinary conditions. If used, the soils must be permeable, drainage must be adequate, considerable excess irrigation water must be applied, and very tolerant crops should be selected.

Sodium

Low-sodium water (S1) can be used on almost all soils with little danger. Sodium-sensitive crops such as stone-fruit trees and avocados may accumulate injurious concentrations of sodium.

Medium-sodium water (S2) is hazardous for use on fine-textured soils having high cation-exchange capacity. This water may be used on coarse-textured or organic soils with good permeability.

High-sodium water (S3) may be harmful to most soils and will require special soil management: good drainage, high leaching, and addition of organic matter. Chemical amendments may be necessary except for gypsiferous soils.

Very-high-sodium water (S4) is generally unsatisfactory for irrigation purposes except at low salinity and where calcium from the soil or the use of gypsum or other amendments may make the use of these waters feasible.

SOURCE: Diagnosis and Improvement of Saline and Alkali Soils, *USDA Agr. Handbook* 60, 1954.

Storage reservoirs, both surface and underground, must be developed to regulate the flow of water to the user since the natural flow of the streams seldom corresponds directly with the irrigation needs of the crops.

Surface reservoirs may be developed for the purpose of long-time carryover, i.e., storage of runoff during years of heavy precipitation for use in years of light precipitation; they may be needed to store water from the wintertime till needed later in the summer; or they may be merely for the purpose of overnight storage so that all irrigation can be carried on during the daytime even though the supply is continuous.

Ground-water storage has also been found desirable and practical for irrigation water supplies. Artificial recharge and development of wells are discussed in Chap. 34 on Principles of Agricultural Hydrology.

Overdraft of ground-water supplies used for irrigation is creating serious problems. Water levels are dropping rapidly in such areas as the High Plains of Texas. In this section of the country, only limited surface streams and other sources of water are available for recharging the ground-water reservoir. Under such conditions, pumping of ground water becomes a mining process. Wells must be lowered deeper and deeper, and presumably the lift will some day reach the economic maximum, forcing many irrigators out of business.

In other areas where water for spreading is available, *artificial refilling of underground reservoirs* has been practiced for many years. Present spreading areas in southern California, exclusive of natural stream beds, indicates a development of over 12,000 acres with an over-all infiltration capacity exceeding 5,100 cfs. These spreading areas cover primarily the alluvial fans at the mouths of the canyons. See Chap. 42 for a discussion of water spreading for irrigation and ground-water recharge.

Water-delivery Systems. If surface streams are being used, the distribution system begins with a diversion structure from the natural stream to a canal. The canal delivers water to the laterals, which in turn deliver it to the farm ditches. Each of the components of the system must be capable of delivering the required quantity of water to meet the consumptive needs of the plants plus the necessary losses in conveyance and application. If the canals, laterals, and ditches are unlined and run through permeable soils, the conveyance losses may be heavy. In tight soils or lined channels, these losses may be minor.

Application losses occur by any known practical method of applying irrigation water to lands. Such losses may vary from 10 per cent to upward of 90 per cent, depending upon the adaptability of the method used in applying the water and the degree of land preparation, care taken of the water, the time water is allowed to run on each tract of land, and stream size–soil–area relationship. Under good conservation irrigation, it should be possible to reduce losses to 20 to 40 per cent.

In western United States, a modern system may be expected to lose no more than 10 to 15 per cent of the water in conveyance and administrative losses and may have *application efficiencies* of 65 to 75 per cent. Thus the *over-all irrigation efficiency* for the system may be from 55 to 70 per cent. If the water is conveyed in pipelines with adequate regulating reservoirs, the over-all efficiency could approach or equal the application efficiency.

On U.S. Bureau of Reclamation projects, canals and laterals are built to carry 1 cfs for each 40 to 70 acres of land served. This capacity increases as the area served by the canal or lateral becomes smaller since the smallest laterals must have capacity to serve all the farms under them simultaneously.

Delivery systems are usually considered in two categories. (1) *project delivery systems,* which deliver water to individual farms from a common source, and (2) *farm delivery systems,* which distribute the water over the farm to unit areas of the farm water-application system. However, either category must follow the basic principles involved:

1. It must deliver the required quantity of water to each segment of the system.
2. It must be accessible.
3. It should be flexible in operation.
4. It should not obstruct other farming operations.
5. It should not allow excessive losses of water in transit.

Canals or Ditches. Irrigation canals and laterals are often constructed between stable retaining *levees* in order that the water surface in the canal may be above the land surface. It is recommended that, at delivery points, water levels should be at least 1 ft higher than the land to be served. The levees should have ample crown

TABLE 44-2. RECOMMENDED MINIMUM CROWN WIDTHS FOR LEVEES OF DIFFERENT HEIGHTS
(Minimum side slopes = $1\frac{1}{2}:1$)

Fill height, ft	Crown width, ft	Freeboard, ft
0–3	2	$\frac{1}{2}$
3–6	3	$\frac{3}{4}$
6–10	4	1
10+	$2 + \frac{1}{4}$ fill height	1

widths and freeboard. Table 44-2 contains recommended widths and freeboards for different heights of levees.

The *gradient* of a canal must be such that there will not be excessive erosion. Where the farm ditch is merely used to convey water to a head gate or single turnout, the minimum slope should be 0.05 ft per 100 ft. A gradient of 0.1 ft per 100 ft, as used in Table 44-3, is common. Where the canal must deliver water to multiple turnouts, siphons, or spiles, the gradient must be flat enough to provide a uniform head at each of these structures. Canal gradients of 0.0005 to 0.001 ft per ft are recommended throughout the positions from which turnouts are to be operated simultaneously.

The canal should be constructed from *material* which can be readily compacted to prevent leakage and which does not slough or crack excessively. Where excess seepage losses occur, canal lining or pipelines should be recommended. Canals may be *lined* to reduce leakage, prevent erosion, or reduce friction loss. Friction losses are usually of little consequence, but erosion control or prevention of leakage may be very important.[1]

Concrete is one of the most satisfactory canal-lining materials if properly installed. It may be installed by plastering, spraying, or forming. Treated fabric linings or

[1] C. W. Lauritzen, O. W. Israelson, and W. W. Rasmussen, Lining Canals and Reservoirs to Reduce Conveyance Losses, *Utah State Univ. Agr. Expt. Sta. Circ.* 129, 1952.

plastics are effective where chance for rupture is slight and are easily installed and maintained. They should not be recommended where livestock have access to the canal. Soil cement, asphalt, and oil stabilizers may be used where practical. Although cheap and easily applied, these materials usually deteriorate rapidly. Clay material is often available for canal-lining purposes and will reduce leakage if properly applied.

Check, or *detention, structures* are needed to check the flow down the canal and permit diversion through turnouts or into other canals. Portable temporary checks may be used, but permanent gates or valves are generally more satisfactory.

Drop structures are needed wherever excessive grades are encountered. These structures should be permanent and designed to prevent damage to the canal above and below the structure.

Turnouts should be provided to conduct irrigation streams from the canal to each unit area. Portable or permanent turnouts such as siphons, spiles, gated tubes, or

TABLE 44-3. SIZE OF FARM DITCHES NEEDED TO CARRY DIFFERENT-SIZE STREAMS ON GRADE OF A 0.1 FT/100 FT (COMPUTED BY MANNING'S FORMULA)

Stream size, cfs	Unlined ditch*		Lined ditch†	
	Base width, in.	Water depth,‡ in.	Base width, in.	Water depth,‡ in.
0.5	12	6		
1.0	15	7	8	7
2.0	18	10	10	8
4.0	24	13	12	12
8.0	36	15	12	17
12.0	36	19	18	17
16.0	48	19	18	21

* Side slopes of 1½:1 and $n = 0.030$ used.
† Side slopes of 1:1 and $n = 0.015$ used.
‡ Does not include freeboard.

gated turnouts may be used. Cutting the canal levee to form an unprotected turnout is not recommended.

Head gates are used to divert water from a project canal into the farm delivery system or from a main canal into lateral canals. Head gates should be positive in action and either calibrated for flow or used in conjunction with a measuring device. Adequate protection to prevent scour in canals above and below the structure should be provided.

Division boxes are used to divide the flow of a main canal between two or more lateral canals. Where variations in flow are required, the division box should be fitted with head gates.

Pump basins are used to dissipate the energy of high-velocity streams from pumps discharging directly into the canal. Pump basins may sometimes be used as division boxes.

Relifts are pumping plants used in the canal system to raise the water to a higher elevation. Relifts should not be used unless sufficient saving can be made in cost of canal construction to offset the initial cost and operation of the relift.

Measuring devices should be an integral part of every farm delivery system. They should enable the operator of the irrigation system to ascertain rapidly the amount of water being delivered to any point of use.

Wherever the canal must cross roads or gulleys, structures such as bridges, culverts, flumes, or inverted siphons must be provided. These structures should have sufficient capacity to carry the required flow without damage to the canal by adverse intake or outlet conditions.

Traps are needed if the irrigation stream is carrying debris, sand, or silt in sufficient quantity to affect adversely the operation of the irrigation system.

Pipelines. Gravity and low-pressure pipelines generally are used where operating heads of less than 20 ft are required.[1] Concrete is the most common material used in the construction of permanent low-pressure pipelines.[2] *Portable* low-pressure pipelines are usually constructed of lightweight metal, treated canvas, rubber, or flexible plastic. They are ordinarily used on small systems or in place of temporary canals.

High-pressure pipelines are used where operating heads of more than 20 ft are required. Permanent high-pressure pipelines are usually constructed of steel, asbestos-concrete, or other reinforced-special-mix concrete. Portable high-pressure pipelines are usually constructed of lightweight metal with pressure-seal couplings. High-pressure hose may be used where maximum flexibility is required.

Head losses in the pipelines should not be so great as to interfere with the service requirements of the system. Standard pipe-flow formulas or charts may be used in

TABLE 44-4. LOSS OF HEAD DUE TO FRICTION, IN FEET PER THOUSAND FEET OF CONCRETE PIPE (COMPUTED FROM SCOBEY'S FORMULA)
$(Q = 0.00169 d^{2.625} H^{0.5})$

Stream, cfs	Loss of head in pipe with inside diameter of:					
	6 in.	8 in.	10 in.	12 in.	18 in.	24 in.
0.5	7.2	1.6	0.49			
1.0	28.0	6.3	2.0	0.75		
2.0	110.0	25.0	7.9	3.0	0.36	
4.0	.....	100.0	32.0	12.0	1.4	0.32
8.0	.....	.....	130.0	48.0	5.8	1.3
12.0	.....	.....	.....	110.0	13.0	2.8
16.0	.....	.....	.....	.....	23.0	5.1

calculating head loss for flow through pipelines. For concrete pipe, the head loss per 1,000 ft is shown in Table 44-4.

Control valves, vents, surge chambers, and other appurtenant structures should be located and designed to meet the service requirements of the irrigation system.

IRRIGATION-WATER REQUIREMENTS

Consumptive use, or evapo-transpiration, is one of the important elements in the cycle of water movement from the time precipitation falls on the land as rain or snow until it reaches the ocean. It is the best index of how much water will need to be supplied by irrigation for good crop production.

Consumptive use is defined as the sum of the volumes of water used by the vegetative growth of a given area for transpiration and building of plant tissue plus that evaporated from the adjacent soil, snow, or intercepted precipitation on the area in any specified time divided by the given area. It is usually expressed as inches of depth per a unit of time.

Actual measurements of consumptive use under each of the physical and climatic conditions of irrigated agriculture are expensive and time-consuming. Therefore some rapid method of transferring the results of measurements made in several areas to other areas is needed. Such a method has been developed by the Soil Conservation Service.[3] Details of this procedure are shown in Chap. 34, Principles of Agricultural Hydrology.

---

[1] Standards for Design and Installation of Non-reinforced Concrete Irrigation Pipe, *Agr. Eng. Yearbook*, 1959, pp. 149–152.
[2] "Irrigation with Concrete Pipe," Portland Cement Assoc., Chicago, 1948.
[3] Harry F. Blaney and Wayne D. Criddle, "A Method of Estimating Water Requirements in Irrigated Areas from Climatological Data," USDA, SCS, Division of Irrigation and Water Conservation, Logan, Utah, 1945 (later revised and published as *USDA SCS-TP-96*, 1950).

Tables giving *peak daily use of water* by various crops and frequency of irrigation for various soils have been prepared in many states by the Soil Conservation Service and state experiment stations and are available for use in irrigation design.

Effective Precipitation. There have been many estimates made of the portion of precipitation falling on cropped land during the growing season that is effective in reducing the needs of the plant for soil moisture. Some investigators have more or less arbitrarily assumed that showers of less than ½ in. have had little or no effect on satisfying the needs of the crops. Other values have been used for other conditions. Recent studies indicate the probable fallacy of this assumption. A shower, even though the rainfall is minor, does increase the humidity and cools the air around the plant, thus decreasing transpiration. Under most conditions, all the precipitation falling on an area, except that lost by surface runoff or deep percolation, can be considered effective.

This same phenomenon is undoubtedly true in areas of fog and heavy dew. Measurements of seasonal consumptive use of water by crops in the coastal areas of California show as much as 25 per cent less water consumption than might be expected from the temperature–daytime-hour vs. crop-coefficient relationships developed for interior valleys by Blaney and Criddle. If the amount of dew that condenses on the foliage and land surface were measured and added to the amount of water withdrawn from the soil, it seems quite possible that the total might well fit the same relationship as for interior valleys.

METHODS OF WATER APPLICATION

Many factors, including topography, source, amount, and quality of water supply, permeability of the soil, slope of the land, method of distribution of the water to the irrigator, precipitation during the irrigation season, and the crop grown will influence the choice of methods.

The only objective of applying water to the field is to supply sufficient moisture to meet the needs of the growing crop. The soil in which the crop grows is a storage reservoir. Application of water fills the storage space in the soil and makes water available to the plant through its root system. Therefore the objective of irrigation is to apply water uniformly throughout the root zone, in sufficient quantities to meet the plant needs, without erosion, with a minimum amount of deep percolation, and with a minimum amount of surface runoff. Probably the greatest need, both on old and new irrigated lands, is for more precise information as to the proper method of application and techniques to be used under different site conditions.

One problem of the irrigator is getting the water into the soil storage reservoir. The rate at which water will enter the soil (called the *infiltration*, or *intake*, *rate*) may vary from less than 0.1 in. per hr on some of the finer-textured irrigable soils to well over 5 in. per hr on some sandy soils. Engineers generally agree that efficient irrigation by surface methods is extremely difficult—if not impossible—when intake rates are much less than 0.1 in. per hr or greater than 3.0 in. per hr.

Movement of water through the soil, commonly called *percolation,* must also be considered. The permeability of the soil at each successive horizon beneath the surface will exert an influence upon the method of water application chosen. If soil characteristics controlling water intake are known, basic relationships between the size of the stream, the slope, the length of run, and the permeability can be established for each individual farm or for a group of farms having similar characteristics.

The method of water application finally adopted, whether it be surface, sprinkler, or subsurface, must satisfy the following *criteria:*

1. Uniform distribution of water
2. Minimum erosion or other damage to the land
3. Maximum efficiency in the use of water
4. Practical and economical performance from the standpoint of crop, labor requirements, and cost of land preparation and maintenance

In some climates, special consideration must also be given to protecting the land where heavy rains may occur during the irrigation season.

Prevention of Erosion from Rainfall on Irrigated Land. In different areas different sets of specifications have been prepared as guides for laying out irrigation systems. In the Southwest of the United States, the Soil Conservation Service uses a maximum slope limitation of 0.25 per cent for irrigation of row crops. In the Northwest, irrigation of row crops is allowed on slopes up to about 3.0 per cent. With cover crops, slopes as steep as 6.0 per cent are considered irrigable.

There has been considerable research work done to correlate some of the factors influencing the length of run, soil intake rates, and rainfall intensities with different cropping and site conditions. Yarnell has studied the *rainfall intensities* throughout the United States.[1] For example, he shows that in Kansas a 60-min rainfall of 1.50 in. might be expected once in 2 years, while in eastern Oregon and Washington, a maximum of 0.25 in. might be expected in 1 hr during a 2-year period. In general, the soils on the steeper lands in the Northwest are relatively pervious and have high *intake rates*. In the Midwest and Southwest many of the soils are relatively fine textured and have low intake rates.

In the Columbia Basin of Washington, with only 8 in. of rainfall well distributed throughout the year, it is quite likely that when it rains the soil is dry and will absorb the moisture readily. In the Midwest where rainfall is 25 in. or more per year and a major part comes during the summer months, it is quite probable that several high-intensity summer rains will occur while the soil is still wet from previous rainfall or irrigation and has a minimum absorption rate.

When these factors and their influences are combined, i.e., amount and intensity of rainfall and the inherent intake rates of the soils, it becomes quite obvious why the irrigation farmer in the Northwest is not concerned with erosion from rainfall; whereas under midwest conditions erosion from rainfall is of utmost concern.

From numerous measurements and observations, it appears that the most important erosion factors are the *size of stream* flowing in the furrows and the *slope* of the furrow. This holds for irrigation where the largest stream is at the upper end of the field or for rainfall runoff which is a maximum at the lower end. The empirical relationship which has been developed for determining the maximum allowable furrow streams for various slopes is

$$Q = \frac{10}{S} \tag{44-1}$$

where Q = maximum nonerosive furrow stream, gpm
S = slope of land, per cent

With an overflow rate (rainfall minus absorption) of 1.0 in. per hr, the overflow from 100 sq ft of land surface would equal 8.33 cu ft per hr, or 62.5 gal per hr, which is approximately 1 gal per min. Using the formula for maximum allowable furrow stream, the *safe length of run* can be found as follows:

$$L = \frac{1,000}{(I - A)WS} \tag{44-2}$$

where I = rainfall intensity, in./hr
A = absorption or infiltration rate of soil, in./hr
W = furrow spacing, ft
L = safe length of furrow, ft
S = slope of furrow, per cent

If $I - A = R$, then

$$L = \frac{1,000}{RWS} \tag{44-3}$$

where R = runoff, acre-in./(acre)(hr)

It is obvious from Eq. (44-2) that in the eastern Washington area, on soils that will absorb water faster than the maximum rainfall intensity of 0.25 in. per hr, the lengths of run can be infinitely long regardless of slope, since the intensity of the rainfall

---

[1] David L. Yarnell, Rainfall Intensity-Frequency Data, *USDA Misc. Publ.* 204, 1935.

would not exceed the rate at which the water would be absorbed and no erosion could occur.

Under Kansas conditions where rainfall intensity is frequently 1.5 in. per hr and the soil will take the water only at a rate of 0.25 in. per hr, the length of run in a cornfield with rows spaced 42 in. apart on a slope of 1 per cent should not exceed 228 ft. If the slope were only 0.25 per cent, the runs could be about 900 ft, and if it were only 0.1 per cent, the runs could be over 2,000 ft.

Thus can be seen the importance of maintaining flat grades in areas of heavy rainfall and tight soils, while steeper grades are entirely permissible where rainfall intensities are low, especially if intake rates of the soils are high.

Maintenance of Land Surface. One of the most difficult jobs in gravity irrigation is maintenance of the land surface. After a careful and expensive job of land leveling has been done, little will be gained if the surface is disturbed by indiscriminate plowing or by other operations. Land leveling is covered in Chap. 35.

The *two-way plow,* with a set of RH bottoms and a set of LH bottoms, is extensively used on irrigated lands since it is partially successful in maintaining a smooth surface. It is not necessary to plow in lands since the operator can turn, lower the alternate set of bottoms, and go back filling the furrow which has just been plowed. More recently many western irrigators favor tillage machinery using sweeps which simply cut through the soil, raise it slightly, and drop it back in the original location.

Various types of *land-smoothing machines* are available. Most of these are long and are provided with a central blade. Their operation is much like a carpenter's plane. They cut the high spots and fill in the low ones. Practically all these machines are now hydraulically operated and do not require a great deal of experience on the part of the tractor driver. Where gravity irrigation is to be practiced, some type of land-smoothing equipment should be operated over the land surface at least once each season. These machines range from 6 to 8 ft in width, and the smoothing operation is not particularly expensive. These same machines can also be used for maintaining wide drains in the more humid areas.

APPLICATION OF IRRIGATION WATER

High application efficiency and uniform distribution of irrigation water on the fields are prerequisites of good irrigation. *Surface irrigation* requirements are as follows:

1. Length of run must be based on the slope and soil-intake rate.

2. The allowable stream size from Eq. (44-1) is run down the row and then reduced to soil-intake rate after the water reaches the end of the furrow.

3. Water should be applied only as necessary to fill the soil to the depth of the root zone.

In addition to the last item above, a *sprinkler irrigation system* should be designed to apply water at or less than the soil-intake rate and in such a manner that crops are not damaged.

Water wasted by *excessive applications* leaches water-soluble plant food through the root zone of the soil. Excess water on higher land frequently waterlogs the rich lower-lying lands. Correction of this waterlogging usually requires an expensive system of drains, and frequently some of these lands are never truly productive again. On the field itself, poor distribution and inefficient use of irrigation water sometimes show up in poor yields at the top and bottom of the field and fairly good yields at the center, because the upper end was kept too wet and lost plant food by erosion and leaching while the lower end was too dry.

Determining When and How Much to Irrigate. The wide variations that occur in soils and in rooting habits of plants under different site conditions complicate this problem. For instance, in the Salt River Valley of Arizona, alfalfa will extract moisture from at least 10 ft of soil profile. Under eastern conditions, alfalfa roots even in the deeper soils may not go below 3 ft. Also, when annual crops are young, the root zone is limited to a few inches depth. Later in the season the roots may go much deeper.

Shockley[1] concluded that any root-zone depth could be divided into quarters. About

[1] Dale R. Shockley, Capacity of Soil to Hold Moisture, *Agr. Eng.*, 36:109–112, February, 1955.

40 per cent of the plant requirement is extracted from the upper 25 per cent of the root zone, 30 per cent from the second, 20 per cent from the third, and 10 per cent from the bottom quarter. He felt that this relationship holds regardless of crop, provided that adequate moisture is available to the crop at all times and that frequent showers do not tend to keep the topsoil moist and that the water table is well below the root zone.

Shockley further assumed that plant growth would be rapid until any one quarter of the root zone had lost its available moisture (field capacity to wilting point) and would then be retarded.

Thus for a soil with uniform profile, about 62.5 per cent of the total available moisture that could be held in the entire root zone could be withdrawn before exhausting the upper quarter.

Other investigators have suggested various methods of determining when and how much to irrigate. The appearance of the plant, cohesiveness and pliability of the soil in relation to texture, soil-moisture measurements by direct sampling, and indicators such as tensiometers and electrical-resistance methods have all been used successfully for specific purposes and conditions. All have certain faults and limitations for use in practical field irrigation, but will probably be used more as time goes on and the science of irrigation improves.

Furrow Method of Irrigation. Furrow irrigation is used for nearly all cultivated crops. Close-growing crops (small grains, hay, and pasture) on sloping soil that bakes or crusts badly after being wet are frequently irrigated by means of small furrows, often called *corrugations,* or rills. In irrigating by the furrow method, water is delivered from the farm lateral to a head ditch or pipeline along the upper edge of the field and is then diverted into furrows running down or across the slope of the land.

The *flow of water to individual furrows* may be controlled with siphons, furrow tubes, or spiles, gated pipe, or other devices. The chart in Fig. 44-3 can be utilized for estimating the size and number of siphons needed for a particular field. It can also be utilized for estimating flow through spiles providing they are round and have square ends. The maximum allowable furrow stream has been previously discussed, and the relationship to degree of slope is given in Eq. (44-1).

The *maximum allowable length of run* on a particular soil and slope is the distance that the maximum allowable furrow stream will run and still give reasonably uniform distribution along the furrow. It is recommended that the furrow streams reach the lower end of the field within one-fourth of the total time required for the irrigation. If this criterion is used, deep percolation at the upper end of the field will be held within reasonable limits. However, surface runoff will usually be heavy unless the flow of water into the furrow is reduced after the stream reaches the lower end of the field. In general, it is to the farmer's advantage to have the runs as long as allowable in order to reduce the turning of farm machinery and to eliminate extra setting of the furrow streams.

Furrow spacing in row crops is usually fixed by the type of crop or by equipment requirements, but can be varied in orchards or where corrugations are used in close-growing crops.

In medium-textured soils that are homogeneous throughout the root-zone depth of the crop, the wetted bulb of soil beneath the furrow has a width about the same as its depth. Under such soil conditions, maximum furrow spacing should not exceed the depth of root zone of the mature crop. If this maximum spacing is used, it is quite likely that while the plants are small with undeveloped root systems, irrigation for adequate lateral penetration will result in a considerable part of the water percolating below reach of the plants. For more mature plants, higher application efficiencies should be possible.

Generally, the soil profile is not homogeneous. A layer of less permeable soil within or near the root zone may cause a temporary perched water table, with a resulting wide horizontal movement of the water. An extremely permeable underlayer may have the opposite effect, limiting the amount of horizontal movement. Therefore, before a furrow spacing is selected, the site should be examined both by sampling the soil profile and by finding the pattern made by the percolating waters.

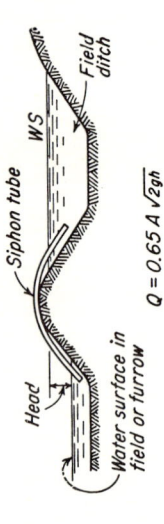
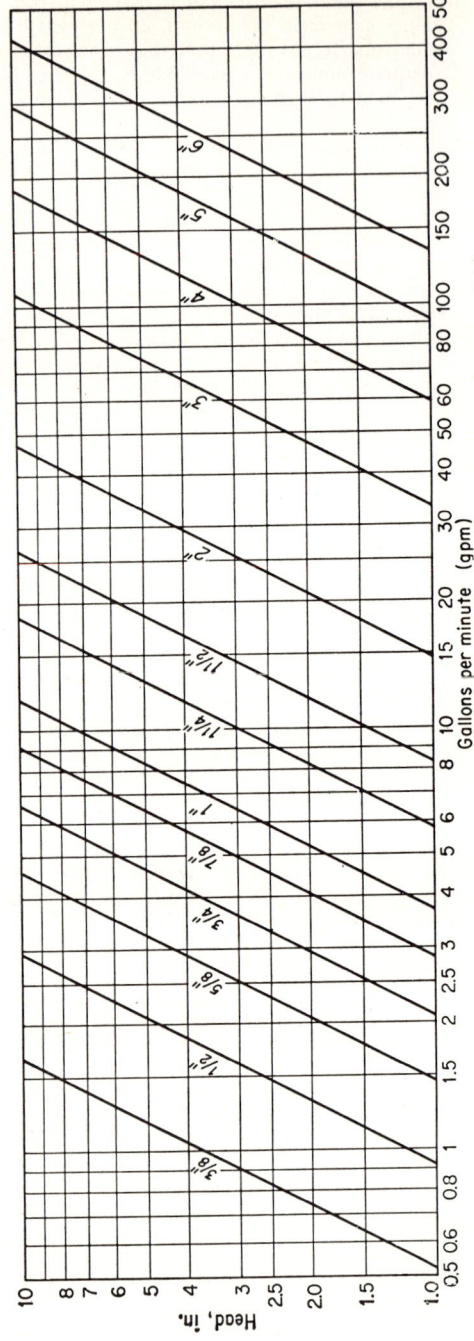

Fig. 44-3. Chart for estimating siphon discharge. (*USDA SCS, West Coast Area, January*, 1954.)

The *amount* of water needed is the difference between the amount already in the root-zone depth prior to irrigation and field capacity.

The *rate at which the soil absorbs water* during an irrigation decreases rather rapidly with time at first, and then after several hours it becomes nearly constant. When intake rate is plotted against time during a normal irrigation, the resultant curve has a general shape indicated by the formula

$$I = \frac{K}{T^n} \qquad (44\text{-}4)$$

where I = intake rate of soil, in./hr
T = time that water is on surface of soil, min
K, n = empirical values dependent upon soil characteristics and moisture conditions at time of irrigation

When intake rate is plotted on log paper as the vertical ordinate and time as the horizontal ordinate, K equals the intake-rate intercept at unit time and n equals the slope of the line (vertical scaled distance divided by horizontal scaled distance). On

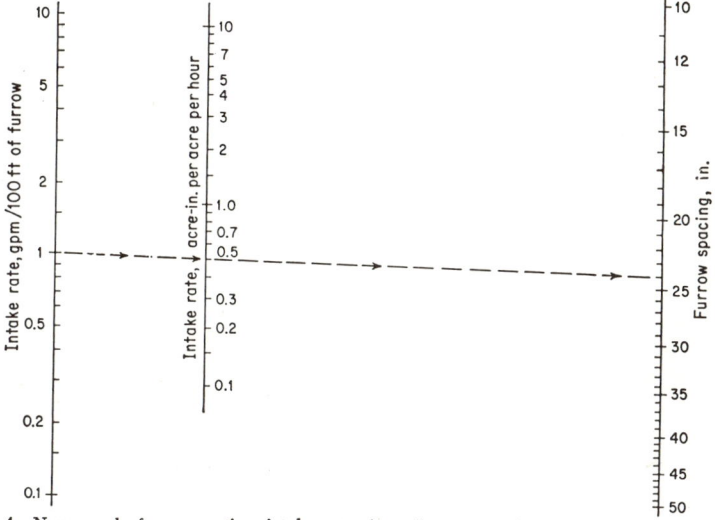

Fig. 44-4. Nomograph for converting intake rate in gallons per minute per 100 ft of furrow to acre-inches per acre per hour for various furrow spacings.

irrigated lands in western United States, K has been found to vary from 1 to 10 and n from about 0.2 to 0.8.

With furrow irrigation, only part of the surface of the land is in contact with the water and the equivalent field intake rate (the rate at which water is absorbed in acre-inches per hour) will vary, not only with the rate at which water is absorbed from the furrow, but also with the furrow spacing. Since, with furrow irrigation, the amount of water entering and leaving the furrow, or a section of the furrow, is often measured in gallons per minute, the furrow intake rate is commonly computed in gallons per minute per 100 ft of furrow. A rather simple relationship can then be used for converting this intake value to an approximate field intake rate. This relationship is that *the intake rate in gallons per minute per 100 ft of furrow divided by the furrow spacing in feet gives the approximate field intake rate in inches per hour*. Figure 44-4 is included for use in making the conversion.

Time required to refill the soil-moisture reservoir by applying D in. of water can be computed from Eq. (44-4) for intake rate. The area under a curve of intake rate vs. time constructed from this equation is proportional to the depth of water D absorbed by the soil. By integrating for the area under the curve, the following equation can be derived:

$$T = \left[\frac{60D(n+1)}{K}\right]\frac{1}{n+1} \qquad (44\text{-}5)$$

Because of both the decrease in the intake rate with time and the amount of ponding in the furrow, it is usually advisable to "cut back" the furrow stream at least once after it has reached or approached the end of the furrow. Unless the size of the furrow stream is reduced, loss of water by surface runoff may be heavy. With flat furrows of adequate storage capacity, cutting back may not be necessary.

Contrary to the belief of many, *increasing the size of the stream* in a bare V-type furrow on the steeper grades does not materially increase the rate at which the water is absorbed into the soil. A stream of 1 gal per min on slopes above 0.5 per cent will put about as much water into the soil per foot of furrow as will a furrow stream ten times as large. This relationship does not generally hold on the flatter slopes, nor where the furrows are broad or grass-covered. However, by using smaller streams, the irrigator usually can save both water and soil with a relatively small increase in the time of irrigation, provided the lengths of run are proper.

Furrow-irrigation relationships between slope, maximum allowable furrow stream, depth of water to be applied, and soil texture are given in Fig. 44-5. These relationships are based on general soil conditions encountered in western United States, and individual sites may vary widely if soils and farming operations are different from those tested.

Border Irrigation. For border irrigation, a field is divided into parallel strips, usually 50 to 100 ft wide, by throwing up borders (ridges of soil) to confine the water. Border irrigation can be used for all close-growing crops, some types of row crops, and orchards where topography and soils are suitable.

Border-irrigation layouts require relatively flat or level land of uniform grade with good land preparation. The strips must be level transversely. Border irrigation requires relatively large streams of water and permits efficient, rapid, and relatively easy irrigation, provided the borders are properly constructed. Labor requirements for this method of irrigation are low.

Border irrigation is suitable for most soil types. However, with finer-textured soils subject to surface baking, it may not be desirable unless natural precipitation or moisture applied by some other irrigation method can be used to get the crops established. Corrugations between the border ridges or sprinklers are sometimes used for this purpose.

As with furrow irrigation, *determination of the maximum allowable stream* for a particular site condition is still empirical. Careful field observations have indicated that on slopes above about 0.3 per cent, erosion may be significant. For fields lacking vegetative protection, streams larger than indicated in Table 44-5 should not be used.

Where a dense sod cover has been established on stable soils, border streams up to twice the size indicated in Table 44-5 have been safely used.

TABLE 44-5. SUGGESTED MAXIMUM SAFE IRRIGATION STREAMS, IN CUBIC FEET PER SECOND PER FOOT WIDTH OF BORDER STRIP*

Slope S, %	Maximum stream† Q, cfs	Slope S, %	Maximum stream Q, cfs
0.3	0.150	1.0	0.060
0.4	0.120	1.5	0.043
0.5	0.100	2.0	0.035
0.6	0.086	2.5	0.030
0.7	0.077	3.0	0.026
0.8	0.070	4.0	0.021
0.9	0.064	5.0	0.018

* For border strips without sod protection. Larger streams may be used with sod cover.
† $Q = 0.065 S^{0.75}$.

On slopes flatter than about 0.3 per cent, the maximum border stream will usually be governed by the height of the border ridges. With cover crops, streams of 0.15 cfs per foot width of border strip may be expected to have flow depths of 6 or 8 in. on

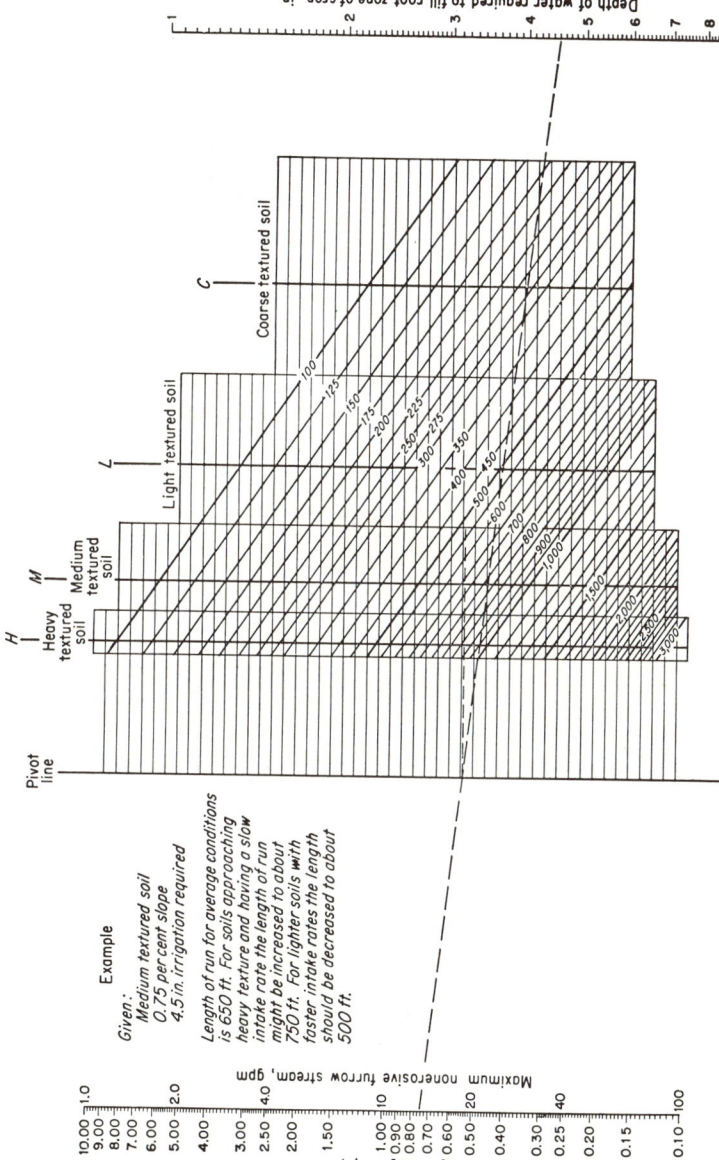

Fig. 44-5. Maximum lengths of irrigation runs for furrow or corrugations, based on average soil conditions resulting from conservation farming. (USDA SCS, West Coast Area, January, 1952.)

523

slopes under 0.3 per cent. Streams of 0.2 cfs per foot width of border strip may flow at depths exceeding 8 in. Since border ridges with settled heights in excess of 8 in. are usually difficult to construct and maintain, it is not generally advisable to design for the use of streams larger than about 0.12 to 0.15 cfs per foot width of border strip.

Under border or flood irrigation, it is necessary to consider the rate at which the sheet of water recedes as well as the rate at which it advances down the border strip or over the flooded land. With any type of flood irrigation, water should cover all portions of the field for equal time periods if uniform application is to be obtained.

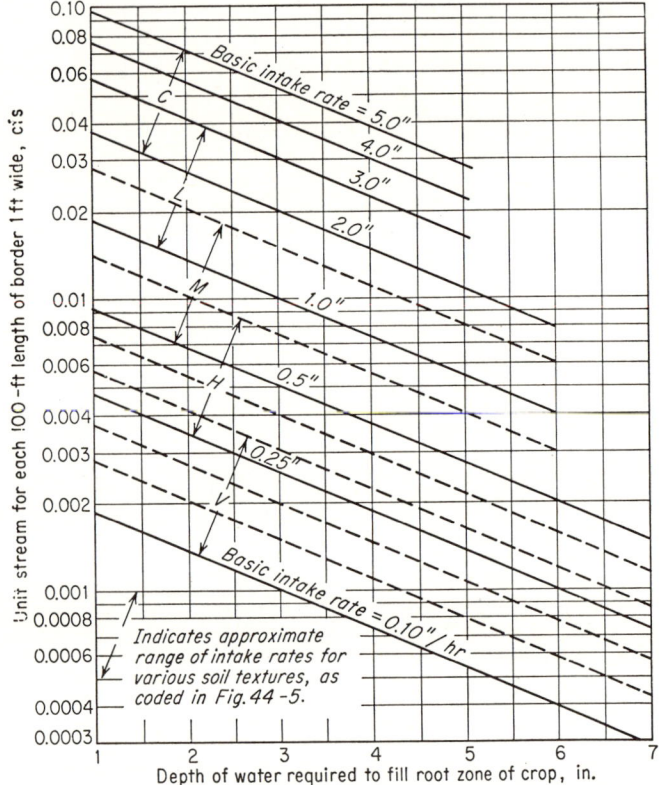

FIG. 44-6. Curves for estimating unit border streams on slopes of 0.5 per cent.

Although this is generally not entirely practical, it is possible to get fairly uniform application by properly balancing the intake rate of the soil with the stream size, area to be irrigated, depth of water to be applied, and slope of the land. Results of many tests over the western United States were used to develop a family of curves giving these relationships for average conditions. The tests were conducted in accordance with methods suggested by Criddle and Davis,[1] and the results are presented in Fig. 44-6.

Stream sizes are shown as *unit streams* (the stream required for each 100 ft of border strip 1 ft wide). Irrigating stream sizes for a slope of 0.5 per cent can be determined by multiplying the unit stream by the product of the border-strip width in feet times its length in hundreds of feet (border-strip area in hundreds of square

[1] Wayne D. Criddle, Sterling Davis, Claude H. Pair, and Dell Shockley, Methods for Evaluating Irrigation Systems, *USDA Agr. Handbook* 82, 1956.

feet). For other slopes, these unit streams should be multiplied by the slope factors shown in Fig. 44-7.

Intake rates used in Fig. 44-6 are *basic rates*, i.e., the final or nearly constant rate at which water will enter the soil after a period of several hours. *They are not the average rates* for the period of irrigation, nor are they necessarily the rate at the time an irrigation is completed. Under many conditions the required amount of water may be absorbed into the soil before the basic intake rate is reached. Also, in determining intake rates, it must be remembered that although texture may give an indication of the rate at which the soil will absorb water, the effect of other factors may drastically alter an intake rate estimated on the basis of texture alone. Thus it is extremely important that the design of any border irrigation system be based on *measured intake rates* under the existing field conditions.

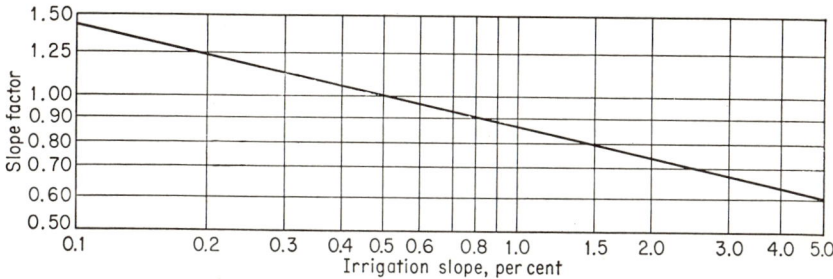

Fig. 44-7. Factors for use in adjusting unit streams from Fig. 44-6 for slopes other than 0.5 per cent.

Required Pump Capacity for Irrigation. The required pump capacity for an irrigation system can be computed by the formula

$$Q = 453 \frac{Ad}{FH} \qquad (44\text{-}6)$$

where Q = discharge, gpm
A = design area, acres
d = gross depth of irrigation, in.
F = number of days permitted for operation
H = average number of hours of operation per day

In a rotation system the capacity must be estimated for the peak season use. The following *example* is for a design area of 90 acres with crop acreages as follows:

10 acres of potatoes, last irrigation May 31:
  2.2-in. application lasts 12 days in May
30 acres of corn, last irrigation Aug. 20:
  2.2-in. application lasts 12 days in May
  2.6-in. application lasts 12 days in July
50 acres of alfalfa, irrigated all season:
  2.6-in. application lasts 12 days in May
  3.1-in. application lasts 12 days in July
Irrigation period 10 days in 12-day irrigation intervals, 16 hr per day

    Capacity requirement for May:

Potatoes: $\qquad Q = \dfrac{(453)(10)(2.2)}{(10)(16)} = 62 \text{ gpm}$

Corn: $\qquad Q = \dfrac{(453)(30)(2.2)}{(10)(16)} = 187 \text{ gpm}$

Alfalfa: $\qquad Q = \dfrac{(453)(50)(2.6)}{(10)(16)} = \underline{368 \text{ gpm}}$

$\qquad\qquad\qquad\qquad$ May total = $\overline{617 \text{ gpm}}$

Capacity requirement for July:

Corn: $$Q = \frac{(453)(30)(2.6)}{(10)(16)} = 221 \text{ gpm}$$

Alfalfa: $$Q = \frac{(453)(50)(3.1)}{(10)(16)} = 439 \text{ gpm}$$

July total = 660 gpm
Capacity required for this cropping system = 660 gpm

Sprinkler Irrigation. Sprinkler irrigation has spread rapidly in the United States since the early 1940s. Increasing recognition is being given to the effects of moisture stress in reducing crop yields and the necessity for having the moisture supply under complete control. Even in the more humid areas, rainfall distribution and water-holding capacities of the soils seldom meet crop requirements for maximum production. Sprinklers seem particularly adapted to areas where only small seasonal quantities of irrigation water are needed to produce high yields. Improved sprinkler equipment and power plants, as well as the benefits from more efficient use of water resources, have also helped to increase the use of sprinkler systems.

Sprinkler irrigation can be used where surface irrigation is inefficient or impossible because of excessive slopes, irregular topography, erosive soil, unfavorable intake rates and soil profiles, or combinations of these factors.

Regardless of the apparent advantages of sprinkler irrigation over surface methods for many conditions, the irrigation engineer should carefully analyze the annual costs and benefits from each system before advising the farmer to make large financial outlays for either type of irrigation.

A field evaluation method for sprinkler systems[1] measures the spray from the nozzles in the field by a number of pan gages set systematically over the "throw" area. The analysis takes into consideration the "overlapping" of spray patterns from adjacent lines and moves. The equipment needs and technical-help requirements are not great since two technicians can make about two complete evaluations per day.

In considering any sprinkler system there are seven main factors that should be checked to determine *adequacy of design and operation* and as a basis for possible adjustments in layout or use.

1. *Application rate.* Water should not be applied at a rate faster than the soil will absorb it, but fast enough to prevent excessive evaporation losses.

2. *Depth of application.* At the point of lightest application, the amount of water applied should not exceed the field capacity of the soil within the root zone of the crop. Greater amounts should be applied only when leaching to remove harmful salts is necessary.

3. *System capacity.* There should be enough equipment, and of sufficient size, to replenish the soil moisture at a rate at least equal to the peak rate of use by the crop, as determined above for pump capacity.

4. *Uniformity of application.* The point of lightest application usually should have a depth of application of at least 80 per cent of the average depth applied over the field. The uniformity of application is affected by variations between discharges of the individual sprinklers along a lateral and on different laterals. It is also affected by the uniformity of spray distribution within the effective area of individual sprinklers as a result of wind direction and velocity, and these should be taken into account in the design.

5. *Water losses.* These are a result of wind drift and evaporation between the sprinkler nozzle and the ground and are affected by water-drop sizes and rate of application. They should not exceed 10 to 15 per cent.

6. *Economical pipe sizes.* The distribution pipe size should be such that there is an economic balance between pipe cost and power cost.

7. *Crop damage.* Water must be applied in such a manner that it will not physically damage the crop.

[1] *Ibid.*

Designing the Sprinkler Layout. The following layout rules should be observed:

1. The system must have capacity to supply enough laterals, based on the time to set each lateral, to irrigate the area before soil moisture is depleted to an undesirable level.

2. Lateral lengths should be short enough so that practical pipe sizes will result and pressure along the lateral will not vary more than 20 per cent.[1]

3. Laterals should be placed across the slope as nearly level as possible to minimize pressure variation. For uphill layouts, the allowable length of lateral is materially reduced. Assuming that only 10 per cent of the pressure is lost to friction, the maxi-

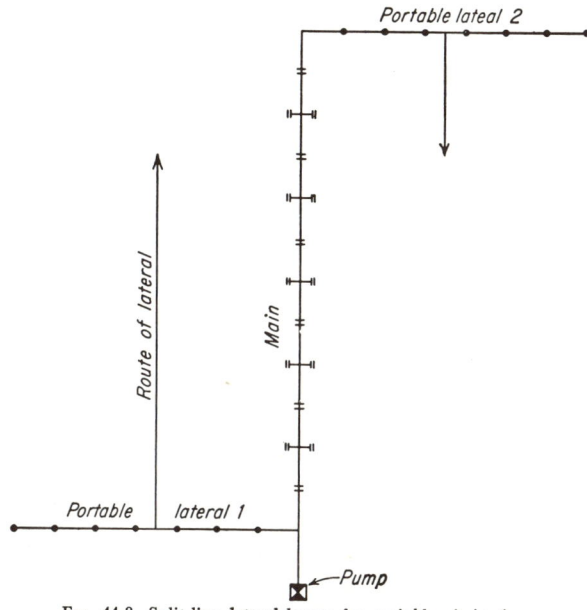

Fig. 44-8. Split-line lateral layout for sprinkler irrigation.

mum allowable uphill lift to stay within a 20 per cent over-all pressure variation is as follows:

Operating pressure	Max. elevation difference, ft
10	4.6
15	6.9
20	9.2
30	13.9
40	18.5
50	23.1
60	27.7
70	32.4
80	37.0

For downhill layouts it can easily be seen that lateral lengths can be increased. If the downslope of the lateral is approximately equal to the friction loss, the pressure in the lateral will remain almost constant.

4. Changes in pipe size necessary for pressure control should be made on the main line wherever possible.

5. Lateral pipes should be of only one size if possible, and never more than two.

6. Wherever possible, the source of water should be located in the field center for most economical use of pipe.

7. Split-line operation as shown in Fig. 44-8 should be used wherever possible. This

---

[1] ASAE Recommendation: Minimum Requirements for the Design, Installation and Performance of Sprinkler Irrigation Equipment, *Agr. Eng. Yearbook,* 1959, pp. 152–153.

528 SOIL AND WATER CONSERVATION

will help maintain equal pressure and will prevent long hauls of laterals back to the starting point.

8. Booster pumps should be considered for small areas in the design requiring high pressure in order to avoid carrying high pressures for these areas in the main system.

Types and sizes of sprinklers for practically all conditions are available. Intermediate pressure sprinklers (30 to 60 psi) are used for most field-crop irrigation. Sprinklers are available, however, that operate under pressures as low as 5 to 15 psi, and others that operate at 100 psi. For orchards, there are undertree sprinklers with a low water trajectory so as not to wet the fruit.

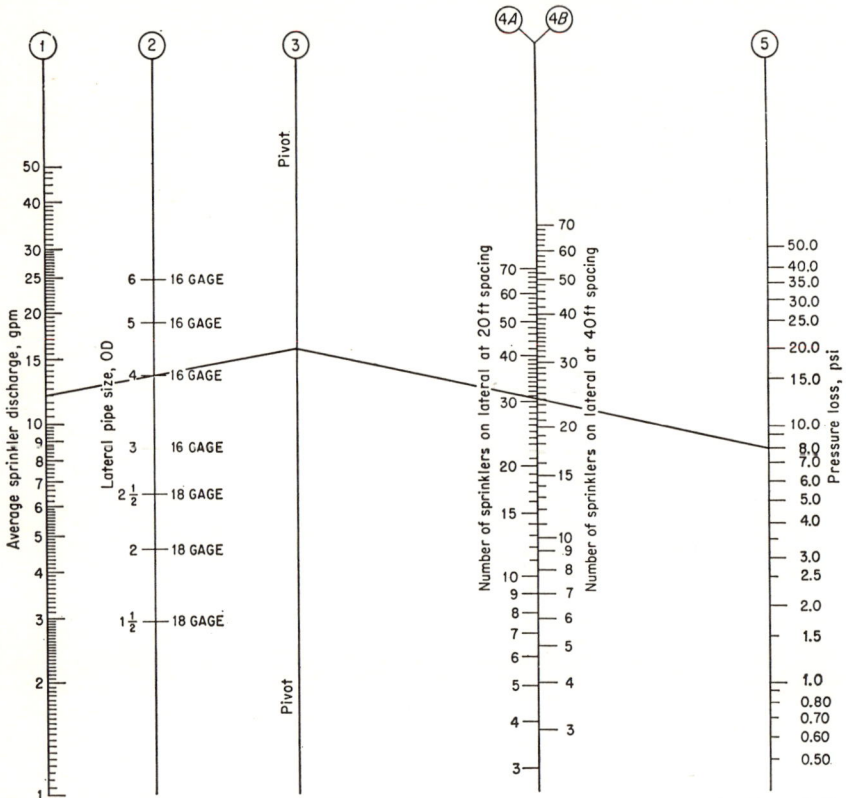

FIG. 44-9. Pressure loss in portable sprinkler laterals, based on Scobey's formula with values of K_s and multiple-outlet factors as recommended by J. E. Christianson. (*USDA SCS, Southeast Area, January,* 1951.)

Rates of discharge from the nozzles will depend on the nozzle sizes and pressures. Field application rate is also dependent on spacing of sprinklers along the lateral and distance between lateral spacing. Table 44-6 will assist in choosing the nozzle size, operating pressures, and spacings.

In addition to calculating capacity by Eq. (44-6), adjustments will need to be made for the number of sprinklers, their individual discharge, and system layout. Layout of laterals must be such that coverage can be made within allowable irrigation time.

Lateral Design. Determination of lateral pipe size, based on a maximum permissible head loss due to friction, can be made from Fig. 44-9. The example shown on this chart is based on a lateral 960-ft long operating at 40 psi with 24 sprinklers spaced 40 ft apart and discharging an average of 12 gpm. With 20 per cent allowable head loss, the pressure loss in Table 44-6, column 5, is 8 psi and 4-in.-OD 16-gage pipe is required.

Standards for hand move aluminum irrigation tubing are shown in Tentative ASAE Standard: Minimum Standards for Irrigation Equipment.[1]

Main-line Design. In the design of main lines, the amount of friction loss is ordinarily a matter of economics. Pipe sizes are selected by comparing the difference in fixed annual costs of the pipe sizes considered with the extra annual cost of pumping against the additional head of the smaller-diameter pipe. Local power costs and the average number of hours of pumping during the season greatly influence this. Figure 44-10 can be used for determining friction loss for main lines.

Example: Given 8-in. pipe, Q of 450 gpm, find friction loss for new aluminum pipe.

Draw straight line between given values on scale 1 and 2 to where it intersects scale 3. Project horizontal line from base line to where it intersects $K_s = 0.32$. Read 4-ft head loss per 1,000 ft of pipe. If pipe were portable aluminum with couplers, $K_s = 0.40$ and head loss would be 4.9 ft/1,000 ft of pipe.

TABLE 44-6. RECOMMENDED SPRINKLER SPACINGS AND PRESSURES*

(1) Sprinkler nozzle size, in.	(2) Recommended average pressure,* psi	(3) Discharge at recommended pressure, gpm	(4) Min. permissible pressure on last sprinkler, psi	(5) Max. spacing at recommended pressure,† ft	(6) Offsets recommended‡	(7) Theoretical rate of application under these conditions, in./hr
1/8	25	2.21	24	30 × 40	Double	0.18
				or 20 × 50	Single	0.21
5/32	30	3.78	28	30 × 50	Single	0.24
				or 20 × 50	Single	0.36
3/16	35	5.84	33	30 × 50	Single	0.38
				or 20 × 50	Single	0.57
5/32	30	3.89	28	30 × 50	Single	0.24
11/64	32	4.82	30	30 × 60	Single	0.26
3/16	35	5.93	33	30 × 60	Single	0.32
5/32 × 3/32	30	5.19	28	40 × 50	Double	0.25
11/64 × 3/32	35	6.50	33	40 × 60	Double	0.26
3/16 × 3/32	35	7.57	33	40 × 60	Single	0.31
3/16 × 1/8	35	8.52	33	40 × 60	Single	0.34
13/64 × 5/32	40	11.8	38	40 × 60	......	0.48
7/32 × 3/16	45	15.9	42	40 × 60	......	0.64
7/32 × 3/16	45	16.2	42	60 × 60	Double	0.44
				or 40 × 80	Single	0.49
1/4 × 3/14	45	18.8	42	60 × 80	Double	0.38
				or 40 × 80	Single	0.57
1/4 × 7/32	45	21.1	42	60 × 80	Double	0.43
				or 40 × 80	Single	0.64
9/32 × 7/32	55	26.8	52	60 × 80	Double	0.54

* Permissible variation is from 5 lb below to 10 lb above.
† For wind velocities above 4 mph, spacings should be reduced. This is especially true for the large sprinklers.
‡ "Single offset" refers to staggering of lateral line positions. "Double offset" includes staggering of both lateral line positions and sprinkler-head positions along the laterals. The purpose is to get a diamond-shaped pattern and less unnecessary overlap between settings.

Data and Design Summary. As an aid to systematizing sprinkler irrigation-system data and design, forms for *farm information, design and specification information,* and an *economic analysis* of irrigation are given in ASAE Tentative Standard: Sprinkler Irrigation Technical Data Sheet.[2]

Subirrigation. Subirrigation practices have developed in many areas of the United States, largely through trial and error. Most irrigation engineers have shied away from this method of irrigating, even where it is adapted, largely because they have not understood the principles of subirrigation or the limitations involved. Nevertheless,

[1] *Agr. Eng. Yearbook*, 1959, pp. 158–159.
[2] *Ibid.*, pp. 154–157.

Class of pipe	Scobey's K_s	Manning's n
New smooth-spiral welded steel	0.32	.009
New aluminum	0.32	.009
Transite	0.32*	.009
15-yr spiral welded steel	0.40*	.012
10-yr full reveted steel	0.44*	.014
Reclaimed invasion	0.52*	.017
New wood stave	0.40	.012
15-yr wood stave	0.44*	.014
Portable aluminum and couplers	0.40*	.012
Portable galvanized steel and couplers	0.42*	.013

* Recommended value to use for design.

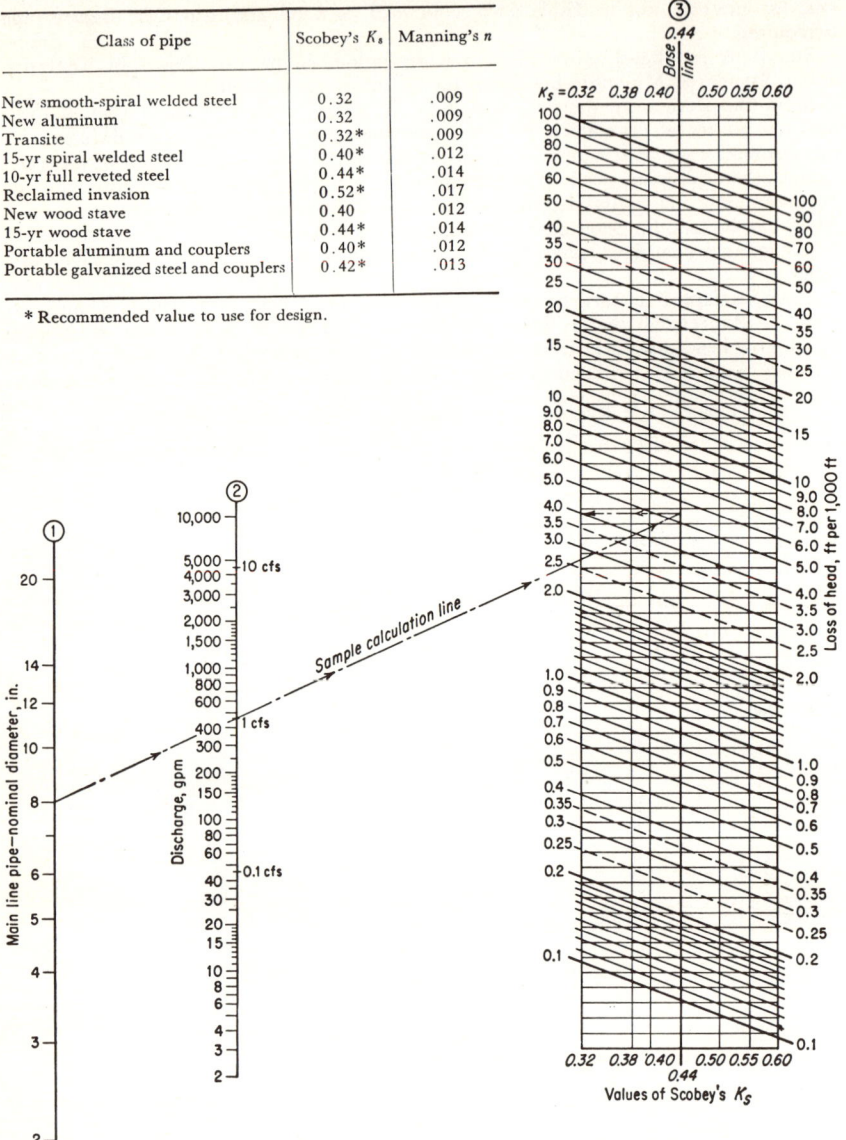

FIG. 44-10. Friction loss for sprinkler system main lines. (*USDA SCS, Southeast Area, January, 1951.*)

a number of tracts, generally small in area, have evolved where subirrigation systems do work satisfactorily. A good many other areas are believed to exist where subirrigation, if properly designed and operated, might be the best method available.

With subirrigation the water supply for the crop comes from underneath the surface of the land. Subirrigation depends on creating an artificial water table and maintaining it at some predetermined depth below the ground surface. Moisture then reaches the plant roots through capillary movement upward. This method requires that the water-table height be subject to rigid control; otherwise it can become too low or too high and either retard growth or stop it completely. One form of subirrigation is merely controlled drainage to maintain the water table at the desired height.

In order for subirrigation to be practical and successful, certain *natural conditions must exist,* as follows:

1. The surface soil must be of uniform texture, deep, and highly permeable.
2. There must be a natural high-water table or a "tight" or restricting subsoil layer, reasonably parallel with the surface, at some depth below the normal root zone of the crops upon which a perched or temporary water table can be developed.
3. The area to be irrigated must be large enough so that percolation losses in drainage channels will not be excessive in proportion to the water required for the entire area.
4. Both the soil and the water used for subirrigation must be relatively free of salts, particularly if the lateral movement of the water is limited and if excess water is not available for occasional leaching purposes.

Since subirrigation is dependent upon controlling the position of the water table, there must be some provision for getting the water into the soil as needed. Experience has shown that, under arid conditions, the use of parallel feeder ditches, run on the contour and spaced sufficiently close to assure proper control of the water table, is probably the best method. Water is maintained at a constant level in these feeder ditches and allowed to seep out and feed the water table. These feeder ditches are usually laid out on the contour, without any slope, and are closed at one end. If more water is desired in the soil, the depth of water in the ditches is held higher; if less is desired, the water is held lower or cut out entirely for a period. If this does not give the desired control, a change in feeder-ditch spacing may be necessary. Spacing of feeder ditches will depend on the depth of the water table and the permeability of soil around the ditch.[1]

[1] Jan Van Schilfgaard et al., Physical and Mathematical Theories of Tile and Ditch Drainage and Their Usefulness in Design, *Iowa Agr. Expt. Sta. Research Bull.* 436, 1956. Roy L. Fox et al., Design of Sub-irrigation Systems, *Agr. Eng.,* 37 : 103–107, February, 1956.

Chapter 45

FROST CONTROL

F. A. Brooks

Growing plants vary in their resistance to damage by subfreezing temperatures, as indicated by Tables 45-1 and 45-2.[1] Subtropical fruit such as citrus and avocados not only suffer fruit damage but may be killed by severe low temperatures. Deciduous orchards usually are located where the time of blossoming comes later than the average date of the last spring frost. Thus for the industry as a whole frost protection is usually uneconomical. Nevertheless, owners depending on orchard earnings every year may need protection from late frosts. Moderate frost-control practices greatly extend the areas adaptable to economic fruit and vegetable production, since temperature rises of 2 to 4°F can often be obtained in the case of radiation frosts. The frequency of freezing wind is limiting because this severe hazard cannot be controlled without expensive covers and heat.

RADIATION FROSTS

Radiation frosts occur on still clear nights when vegetation is cooled below freezing by radiation to the cold sky. The rate of heat loss [usually eq 20 to 25 Btu/(hr)(sq ft)] depends largely on the coldness of the sky, often equivalent to 0°F. The resulting cold air tends to drain into low areas, causing still lower temperatures. During the day, air temperatures are normally highest at ground level, because of heating from surfaces warmed by the sun's radiation. Under radiation-frost conditions, however, the air temperature 5 to 7 ft above the ground may be 10°F lower than 40 to 100 ft above the ground. This is called *temperature inversion,* and the greater the inversion, for a given minimum temperature, the better the chance of frost control because there is a shallower depth of cold air.

Natural factors which reduce temperature drop are:

1. Clouds which act as a radiation shield, particularly when thick and low.
2. Cold-air drainage away from the orchard which prevents the accumulation of a thick layer of cold air, which might, for instance, largely originate outside an orchard in an adjacent meadow of dry grass.
3. Large trees with dense foliage having high heat capacity for the exposed radiation surface. Such trees help prevent ground-level frosts because of the high chilling level. Citrus trees with their dense foliage are less easily damaged than bare deciduous trees which have their blossoms exposed to the sky and where the level of chilled air builds up from the ground.
4. Ground surface with good heat-conduction characteristics, so as to absorb more solar energy and at night give up stored heat with minimum temperature drop. Bare, firm ground with field capacity moisture content is much better, from this standpoint, than loose dry ground insulated by a dry mulch or cover crop. Orchards using oil spray for weed control seem to have a 1 to 2°F advantage over cultivated orchards.

---

[1] Floyd D. Young, Frost and the Prevention of Frost Damage, *U.S. Weather Bur. Publ.* 1484, December, 1940.

Temperature drop may be *reduced* by the following *frost-control devices:*
1. Flood irrigation and water sprinkling under trees
2. Covers for small plants
3. Numerous small heaters
4. Wind machines which mix warmer air from above with cold air near the ground
5. Fans to pump cold air out of low pockets or closed valleys

TABLE 45-1. AIR TEMPERATURE (IN THERMOMETER SHELTER) TOLERATED BY CITRUS FRUITS WITHOUT APPRECIABLE FROST DAMAGE FOR 30 MIN ON A TYPICAL RADIATION-FROST NIGHT

	Open blossoms, ¼″ green fruit	Green fruits ¾″ or larger	Ripe and nearly ripe	Min. safe orchard temp.
Lemons	29–30°F	27°F	28–29°F	30°F
Oranges:				
Navels		27°F*	26°F*	28°F
Valencias		27°F		
Grapefruit			26°F (24° dry)	28°F
Lemon trees:				
Defoliated			22°F	
Bark split			19°F	
Orange trees, defoliated			16°F	

* On nights of slow temperature fall, fruit damage often occurs 1°F higher.
SOURCE: Floyd D. Young, Frost and the Prevention of Frost Damage, *U.S. Weather Bur. Publ* 1484, 1940.

TABLE 45-2. TEMPERATURES TOLERATED FOR 30 MIN OR LESS BY DECIDUOUS FRUITS, SHELTERED THERMOMETERS

Fruit	Stage of development		
	Buds closed but showing color	Full bloom	Small green fruits
Apples	25	28	29
Peaches	25	27	30
Cherries	28	28	30
Pears	25	28	30
Plums	25	28	30
Apricots	25	28	31
Prunes, Italian	23	27	30
Almonds	24	26	30
Grapes	30	31	31
Walnuts, English	30	30	30

NOTE: The temperatures listed are of air at 4½ or 5 ft aboveground as measured by a thermometer shielded from the sky in a simple wooden shelter painted white. This temperature is not the same as the temperature of the fruit or blossom, which, when exposed to clear sky, is usually colder. The air temperature is, however, the most easily determined and the most familiar and is the primary ambient condition.
SOURCE: Floyd D. Young, Frost and the Prevention of Frost Damage, *U.S. Weather Bur. Publ.* 1484, 1940.

USE OF WATER FOR FROST PROTECTION

Moist ground is favorable for heat storage, but it should not be so wet that evaporation prevents surface warming during the day. Surface irrigation may thus minimize frost damage during a dry spell.[1]

A continuous application of water during a frost acts as a source of heat. The water gives up sensible heat as it cools to freezing, and it gives up latent heat as it freezes. A continuous application of $\frac{1}{10}$ to $\frac{1}{8}$ in. per hr is recommended for low-growing plants

[1] H. J. Franklin, Cranberry Ice, *Mass. Agr. Expt. Sta. Bull.* 402, 1943.

such as strawberries.[1] Sprinklers achieved a 3°F air temperature rise to 31°F, successfully protecting blueberries.[2]

Water sprayed on the trees may cause some ice breakage if the trees are leafed out. A fine spray promotes evaporation, absorbing heat from the air rather than supplying heat. In a dry wind above freezing temperature wet foliage can suffer frost damage by evaporative cooling. Open water, however, will add heat to the air even in a wind as long as the surface water temperature remains warmer than the air. Experience in Italy and to a limited extent in England and Michigan shows that with conventional irrigation equipment and large droplets, apple buds can be protected. Also, less icing occurs with large drops of water. The belief that a static ice coating will prevent the fruit from going below 32°F, however, is not correct because ice does not remain at 32° if the loss of heat continues. A layer of ice has some thermal-insulation value, but there is evidence that the presence of ice on the outside of the fruit affects the physiological condition of the fruit so that there might actually be more damage than if the surface were dry.

Tests with sprinklers *under* almond trees in a light frost near Denair, Calif., Mar. 7, 1956, indicated an advantage of 1 to 2°F, with damage to Drakes in the petal-fall stage decreasing linearly from 87 per cent on the border row (where cold air entered) to 60 per cent in the center and 40 per cent at the downdrift edge. The outside temperature minimum was 26°F. The damage to Nonpareils showed a similar pattern with much less injury.

In this orchard the sprinkler heads—$3/64$ No. 20—were spaced on 60- by 80-ft centers and wet about 35 per cent of the ground area at a rate of about 0.1 in. per hr. Although a good crop was obtained in this case, the same degree of protection would have been inadequate if a frost of the same severity had come later.

PLANT COVERS AND CLOUD SCREENS FOR FROST PROTECTION

Paper covers are extensively used to protect small plants.[3] The paper caps and shields used in the daytime cause a rise in soil temperature, and this contributes to the frost protection. To enclose a plant without including a heat source, such as the ground, can aggravate the frost hazard by depriving the plant of convective transfer of heat from the air while the paper radiates to the cold sky from an even greater area than that of the plant. Individual branches sacked for pollination control have frozen completely, while exposed blossoms suffered no frost damage. Covered nurseries and rows of low crops can be well protected by small heater blowers.

One of the oldest proposals for frost protection, which is yet to be developed, is artificial cloud screening to reduce the rate of heat loss to the cold sky. Military smoke-screen generators have been useless for long-wave (heat-radiation) screening.[4] This is because the natural 0.7-micron smoke-screen droplet is at least twenty times too small in diameter. Possibly plastic foam fogs[5] can have the necessary large particle size with low settling velocity (now being developed at Dow Chemical Co., Midland, Michigan, by Dr. W. Von Valkenburg). Deep fogs or natural water-droplet clouds at low levels are perfect frost protectors when a whole valley is covered and deep-soil temperatures are well above the freezing point. Occasional drifting of cloud screens across a highway is very hazardous to traffic.

FROST PROTECTION BY MANY SMALL HEATERS

In 1938 four million orchard heaters were in service.[6] Each one must be individually lighted and regulated, and all except pipeline heaters must be individually

---

[1] E. H. Kidder and John R. Davis, Frost Protection with Sprinklers, *Mich. State Univ. Ext. Bull.* 327, Rev. 1956.

[2] H. B. Schultz and R. R. Parks, Test of Frost Protection by Sprinklers, *Calif. Agr. Expt. Rept.*, Project 400-U, pp. 8-9, June, 1957.

[3] Young, *op. cit.* O. W. Kessler and W. Kaempfert, Die frostschaden Verhutung, Deutsches Reich Reichsamt fur Wetterdienst, *Wiss. Abhandl.* 6(2):57-64, 81-91, and pl. 3, 1940.

[4] J. O. Collins, Evaluation of Artificial Oil Fog as a Means of Frost Protection, *Standard Oil Develop. Co. Esso Lab. Rept.* PD-22T-46, Project, 22150, 1946.

[5] Betty L. Raskin, Smoke: Now It Comes from Plastics, *Chem. and Engr. News*, 37:66, Apr. 20, 1959.

[6] Floyd D. Young, California Heater Statistics, *Calif. Citrograph*, 23:371-388, July, 1938.

filled. During the severe freeze of 1937, about 2 million barrels of fuel oil and 17,000 tons of solid fuel (coal, coke, wood, and rubber tires) were burned in California during approximately 15 nights between Jan. 5 to 27. Present high costs of fuel and the shortage of labor have led to widespread use of wind machines, described later, but these occasionally need heater support especially around the borders.

Orchard heating is used to produce a rather small modification of outdoor temperature, namely, counteracting local radiation cooling to maintain a safe minimum temperature and raising the temperature of inflowing air near the ground a few degrees. Whether orchard heating should be used more extensively than at present is primarily an economic problem. In general, heating should be avoided if there is any doubt as to its being worthwhile.[1]

The *nocturnal atmospheric condition* is very different from that of the daytime in that a cold ground surface stabilizes the air whereas a warm ground surface promotes circulation. At night when the ground is colder than the air overhead, the ground air uniformly warmed a few degrees will still be heavier than the even warmer air overhead, and hence will rise only a short distance before it reaches a balance. The

TABLE 45-3. HEATING NECESSARY TO PRODUCE A TEMPERATURE RISE OF 1°F IN CITRUS ORCHARDS OF VARIOUS SIZES, USING LAZY-FLAME HEATERS

Orchard area, acres	Av. wind velocity at 20 ft, mph	Heat required to raise central temperature 1°F, million Btu/(hr)(acre)		
		4°F inversion	10°F inversion	15°F inversion
1	1.8	2.4	2.0	1.7
7	2.0	...	0.8	0.7
15*	1.6	...	0.6	0.5
28	1.2	0.9		
7–28†	1.4	0.9	0.6	0.5
600	...	...	0.6	

* Using Exchange Stack heaters.
† Omitting border heaters.

ideal method of orchard heating for a uniform temperature rise is the use of a very large number of small heating units distributed over the area to be protected. Experience in 1937 showed that without forced distribution of convective heat, a minimum of 35 heaters per acre burning about ½ gal per hr were necessary to provide reasonable protection; and of course more units would have been more effective. Larger combustion units may be more efficient as heat producers, but often waste more heat above the trees. Heating solely by radiation from two red-hot stoves per acre, mounted high above the ground, each burning 11 gal per hr, was found to be adequate for tomatoes in a 6°F frost but their operation was troublesome.

The temperature response to distributed heating in a 1-acre tract of oranges at Riverside, Calif., was studied by Kepner, Lorenzen, and Brooks during 1937 to 1941.[2] The effectiveness of various commercial heaters was also determined. With the common Lazy-flame heaters, the heat required to raise the temperature 1°F varied from 1.7 million Btu/(hr)(acre), with a strong inversion of 15°F (between 5 and 60 ft), to 2.5 million Btu/(hr)(acre), for a weak 3°F inversion. Naturally, much greater response was found in larger acreages. Table 45-3 shows the heating effectiveness for orchards of various sizes. Probably the ultimate in mass-heating practice is in a 600-acre orchard, where, by lighting heaters in every sixth row (130 ft), 1°F response is obtained for about 600,000 Btu/(hr)(acre) on nights of moderate inversion.[3] This

---
[1] W. R. Schoonover, F. A. Brooks, and H. B. Walker, Protection of Orchards against Frost, *Univ. Calif. Agr. Ext. Circ.* 111, 1939.
[2] R. A. Kepner, Effectivenes of Orchard Heaters, *Calif. Agr. Expt. Sta. Bull.* 723, 1951.
[3] D. A. Newcomb, Operating Orchard Heaters under Present Cost Conditions, *Calif. Citrograph*, 34:100 137, January, 1949.

example disregards border heating because of the large size of the orchard and the favorable foothill location. The direction of natural air drift is across the heater rows.

Figure 45-1 shows the typical temperature profiles in a citrus orchard with and without heating when there was a strong inversion. The tests were made with Return-Stack heaters when temperatures were slightly above the danger point so that off-on-off runs could be made without danger of damage to the crop. Colder air in the

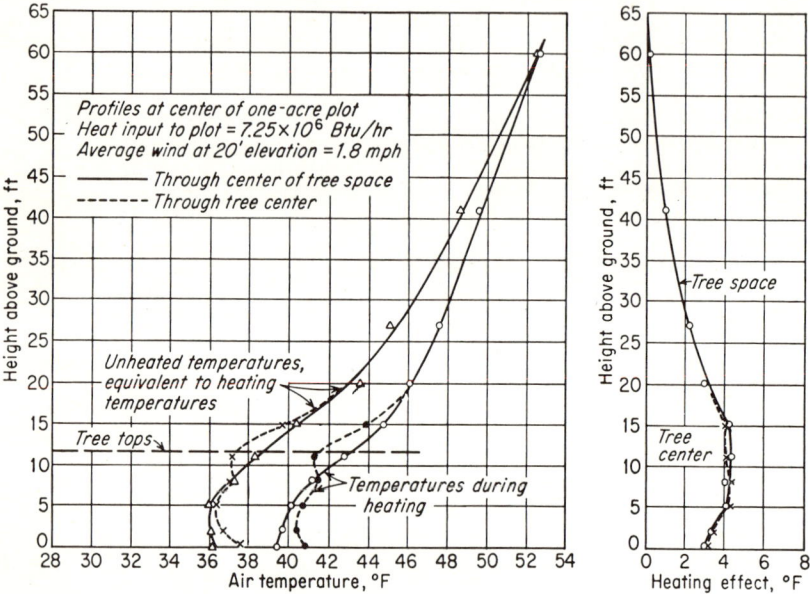

FIG. 45-1. Vertical profiles of temperature and heating effects at center of a 1-acre plot, for a night with a large inversion. Return-Stack heaters were used in the plot during this test.

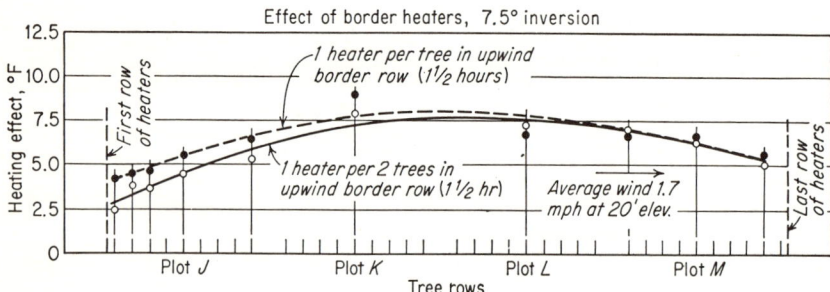

FIG. 45-2. Heating effects in center row across 15-acre heated area, measured at 5 ft above ground in tree centers. Average burning rate, 0.75 gal/hr/heater. (Lighting period is not included in averages.)

treetops than under the tree is typical, as is the change in shape of the tree-space profile when heating, the minimum temperature level dropping from 5 ft to ground level. As shown in Fig. 45-2, there is much more temperature response in the middle of a tract than at either end. The gain in temperature at the upwind edge by a moderate increase in border heating is shown also in Fig. 45-2. Ordinarily, extra heaters are also placed in the second tree row and occasionally in the third row. Since the heating response varies with distance in from the border, the heating results found for the 1-acre plot should be directly applicable to a strip around the outside of a larger orchard.

Response to Border Heating. When there is an inflow of cold air on an exposed frontage, extra heaters should be used on the upwind side to protect the first rows of trees. If the heaters are spaced one per tree on the exposed side, the radiant energy from the heaters plus the products of combustion will warm the near trees directly and thus afford some protection. The heat output of this first bank of heaters, however, is not adequate to warm the incoming air stream to the height where the air temperature due to inversion is naturally the same as that of the dispersed products of combustion.

FROST PROTECTION BY WIND MACHINES

Each propeller of a wind machine produces a jet of air which is directed a few degrees below horizontal.[1] The whole machine turns slowly on a vertical axis, acting as a turning jet. The diagram of Fig. 45-3 shows in nondimensional form the general shape and relative size and velocities of the various volumes of air in the turning jet.[2] It should be noted that although the machine is turning and the jet is described as a turning jet, the actual direction of flow is radially outward at all points. Also, the air velocities in the jet decrease much more rapidly in the outer parts than if the machine were not turning. The shearing in the turning air jet loses the advantage at continuous conical flow along an axis, and each part almost

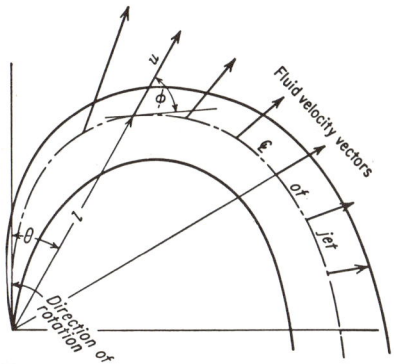

Fig. 45-3. Diagram of velocity vectors of a turning air jet, taken 10° or approximately 14 sec apart.

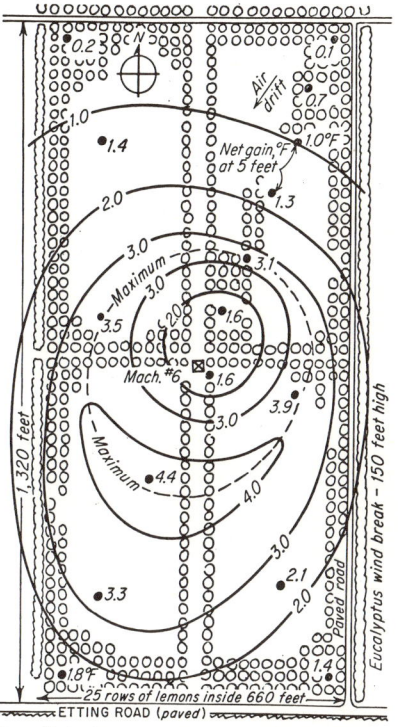

Fig. 45-4. Temperature response pattern produced by a large dual wind machine operating in a lemon orchard near Oxnard, California.

makes headway by its own momentum alone. Thus propeller thrust is the best single indicator of machine capability.

The most specific effect of the wind machine is its modification of the temperature inversion by mixing overhead warm air with the colder air in the trees. An example of this is shown in Fig. 45-4, which shows the results of a wind-machine test in a favorable location near Oxnard, Calif.[3] The heavy solid lines connect points of equal

---

[1] F. A. Brooks, C. F. Kelly, D. G. Rhoades, and H. B. Schultz, Heat Transfers in Citrus Orchards Using Wind Machines for Frost Protection, *Agr. Eng.*, 33:74–78, February, 1952; 33:143–154, March, 1952.

[2] A. S. Leonard, A. Preliminary Analysis of a Turning Jet, *ASME Paper* 56-7-19, September, 1956.

[3] F. A. Brooks, Climatic Environment: A Thermal System, *Univ. Calif. Assoc. Students Store Syllabus for Agr. Eng.* 106, Davis, Calif., 1951.

temperature rise for the 5-ft level within the tree foliage. The ring of maximum response, shown as a broken line, is found to be much warmer on the downdrift side. If 2°F temperature rise had been needed for "protection," Fig. 45-4 shows that this double-engine machine (68 bhp each) would have served about 14 acres; but if 1°F rise plus other effects are sufficient, a 20-acre rating would be warranted. The area response for tests of a 93-bhp machine of 1,050-lb thrust in a 14°F inversion showed protection of 2°F for 9½ acres, requiring about 10 hp per acre. There is quite a reduction in acreage with protection of 3°F because of the cold area around the base of the machine, but downdrift there is a strong protection (4°F) for 3.3 acres.

Two or more wind machines, if not located too far apart, will protect more acres per machine than one alone.[1] The principal reason is that there are substantial areas around a single machine which have some temperature rise but not enough for protection. With two or more machines, these areas overlap and the combined effects

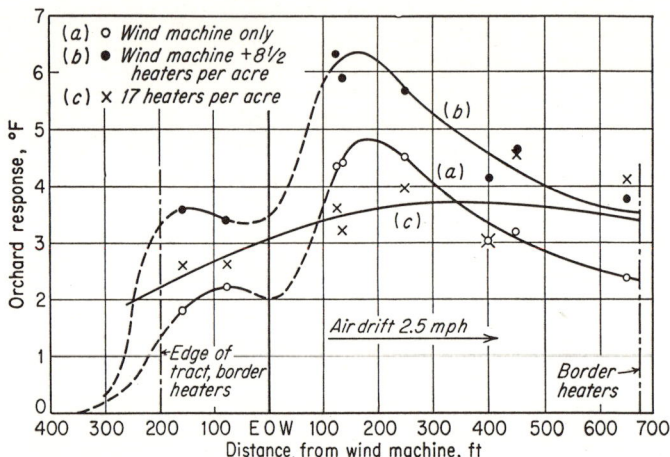

Fig. 45-5. Air-temperature increase in trees, produced by (a) one 90-bhp wind machine, (b) same machine supported by 8 heaters per acre, and (c) 17 heaters alone per acre. Within 350 ft, the machine seemed to be as effective as 17 heaters per acre. The inversion was 10°F.

result in adequate protection. Even though credit for these additional areas must be divided between several machines, the result is a sizable increase in the total area protected by each machine. The maximum allowable spacing for 75-bhp machines in the Riverside citrus area seems to be about 700 ft transverse to the direction of natural cold-air flow and 900 ft in the line of drift. With a multiple installation of four small machines of 300-lb thrust each (19 bhp on 600-ft spacing), the observed protection area of 2°F rise is about 5½ acres per machine and indicates 3½ bhp per acre, but this is not considered adequate for frost protection under typical conditions near Riverside, Calif. Probably the maximum practical spacing for 15-bhp machines is 400 ft transverse to the cold air drift and 550 ft in the line of drift. Heater support is essential for small machines and is only omitted with large machines in very favorable areas. Wind machines have not proved as satisfactory in deciduous orchards as in citrus, probably because of high losses in jet energy in the network of interlacing branches near treetop level. Wind machines of about 500-lb thrust operating in the rather open trunk space under the twig canopy can give about 2°F temperature gain, and nearly 3°F over about 10 acres when supported by heaters around the border and in an inner square about 200 ft from the blower.

WIND MACHINES AND HEATERS COMBINED

Wind machines depend more directly on temperature inversion for adequate temperature rise than do heaters. Tests at the Citrus Experiment Station, Riverside, Calif.,

[1] F. A. Brooks, D. G. Rhoades, and A. S. Leonard, Wind Machine Tests in Citrus, *Calif. Agr.*, pp. 8–10, August, 1954.

in 1951 to 1954 were made in a field with moderate-to-fast air drift. In a comparison between heaters and wind machines, it was found that the wind machines produced little change in air temperature at the 40-ft level, while the heaters raised it. The three response curves, shown in Fig. 45-5, for a single 90-bhp wind machine, the same machine supported by 8½ heaters per acre (with moderate border heaters), and 17 heaters per acre alone show that the single wind machine did about as well within 350 ft as 17 heaters per acre with an average inversion (10°F). The combined response of the machine plus 8½ heaters per acre was equivalent to about 22 heaters per acre. For zero inversion, there would be zero machine response but some heater response, although the wind machine should aid distribution of heat from the heaters. The distribution of response varies between the two systems, but both provide frost protection in degrees that can be evaluated on an economic basis.[1]

In general, frost-protection practices must make an area warmer than its surroundings and maintain this area against the tendency of the surrounding colder air to displace the warmer air in the protected area. This may be accomplished by continuous heating throughout the area, or by continuous turbulent air mixing in depth (forced modification of the temperature inversion) combined with outwardly directed jet momentum, or by a combination of the two. Heated air jets have more buoyancy and rise out of the orchard sooner than unheated jets. Conversely, heaters distributed through an orchard decrease the inversion and lower the equilibrium level of the jet.

[1] F. A. Brooks, A. S. Leonard, T. V. Crawford, and H. B. Schultz, Frost Protection by Wind Machines and Heaters, *ASAE Paper* 59-114, June, 1959. Todd V. Crawford and F. A. Brooks, Frost Protection in Peaches, *Calif. Agr.,* pp. 3–6, August, 1959.

SECTION III
FARMSTEAD STRUCTURES AND EQUIPMENT
CARL W. HALL, *Subeditor*

Chapter 46
STRUCTURAL REQUIREMENTS OF FARM BUILDINGS
MERLE L. ESMAY

FACTORS INVOLVED

Once the functional requirements of a farm building are known, the two basic considerations for determining the structural requirements are (1) the magnitude and direction of *loads* and *stresses* to which the building will be subjected and (2) the design specifications of *construction materials.* The structural requirements of farm buildings are unique in many ways and must be determined independently of the requirements for most other types of commercial buildings and public service buildings and for urban construction. Farm buildings are often subjected to *higher wind stresses* due to exposed locations. Farm buildings must be designed for varying loads of storage materials.

Farm buildings may be designed with smaller *factors of safety* than generally used for public buildings or other commercial types of structures. In the conventional type of framing for farm buildings practically all members, including studs, rafters, joists, and ties, are spaced 12, 16, or 24 in. on center. One, two, or a few of these members might fail in any one building without critically affecting the over-all structural strength of the building. Another consideration is that most farm service buildings shelter very few people at one time and therefore a complete failure would not incur disastrous loss of life as compared with public buildings.

The designer of farm service buildings should strive toward that point of *structural adequacy* at which the owner is not penalized by either excessive initial *investment* or excessive costs of *maintenance* and *repair.*

The use of *improved fastenings* in farm building framing is pointing the way toward more economical use of materials and stronger structural frames. Years ago the manufacture of wood-laminated arch ribs with predictable structural properties eliminated the sagging ridge lines on buildings. Performance tests and field service are showing that light wood *trusses* constructed with *glue* and *nail* joints or *bolt* and *ring* connector fastenings provide a means for economical buildings. Proper design saves material and can also provide clear-span roof frames for the building. Prefabricated steel buildings designed to meet farm conditions rather than urban codes are now available, with the design loads used specified on the erection drawings.

ESTIMATING LIVE LOADS

Snow Loads. Figure 46-1 shows the basic snow loads for the United States.

Flat roofs are subjected to maximum accumulations of snow and ice. Flat roofs are defined as those having a rise of 3 in. or less per foot. As noted in Table 46-1, the minimum uniformly distributed design flat-roof load for buildings anywhere in the United States is 15 psf, the maximum, 28 psf. Built-up roof coverings used on flat roofs require a more rigid base than shingle coverings used on sloped roofs if leaks are to be avoided.

The snow load on sloping roofs will be less than on flat roofs because less snow will be retained. The minimum uniformly distributed design vertical live roof loads can

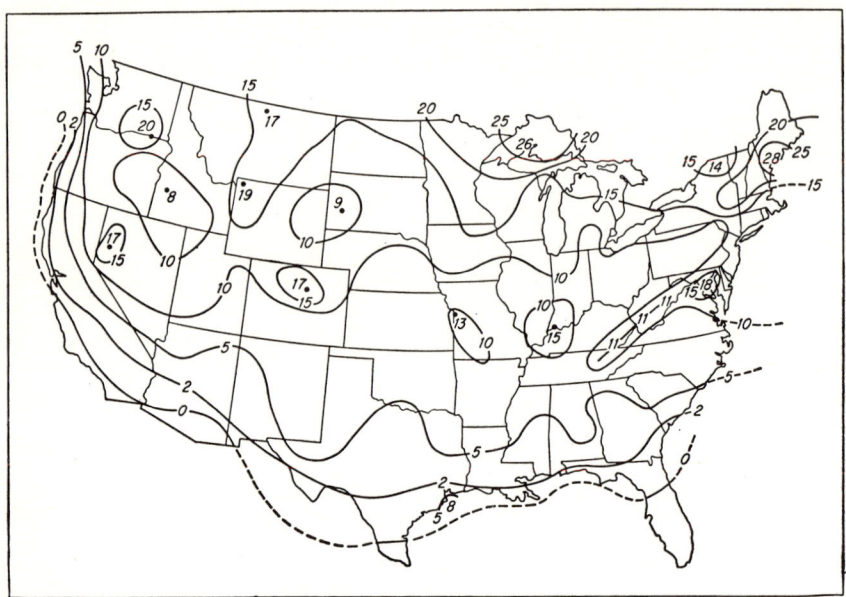

Fig. 46-1. Snow map for the United States. Isograms are basic snow load, pounds per square foot. (*Snow Loads on Farm Structures, USDA ARS-42-5, October, 1956.*)

TABLE 46-1. FLAT ROOFS
(Minimum Uniformly Distributed Design Vertical Live Roof Loads)

Pounds per square foot of horizontal projection

Flat roofs and roofs having a rise of 3 in./ft or less:	
Southern states	15
Central states	20
Northern states	25

TABLE 46-2. SLOPING ROOFS
(Minimum Uniformly Distributed Design Vertical Live Roof Loads)

	Pounds per square foot of horizontal projection, with slope:			
	3 in 12 or less	6 in 12	9 in 12	12 in 12 or more
Southern states	15	12	10	10
Central states	20	15	12	10
Northern states	25	20	15	10

be progressively reduced to a minimum operational sleet, ice, or snow load of 10 psf for roofs having a rise of 12 in. or more per foot. Information presented in Table 46-2 specifies minimum design loads for roofs of different slopes in the various areas in the United States.

For the purpose of farm building design, the basic design stress of the roof structural material can be increased by 33 per cent since loadings generally do not extend beyond 1 month. In some northern areas where there is winter-long continuous buildup of snow and ice, this increase should be reduced to approximately 25 per cent.

Wind Loads. The actual wind pressure to which a building is subjected depends upon its proportions and orientation with respect to the wind direction. As shown in Fig. 46-2, both positive and negative pressures must be considered.[1]

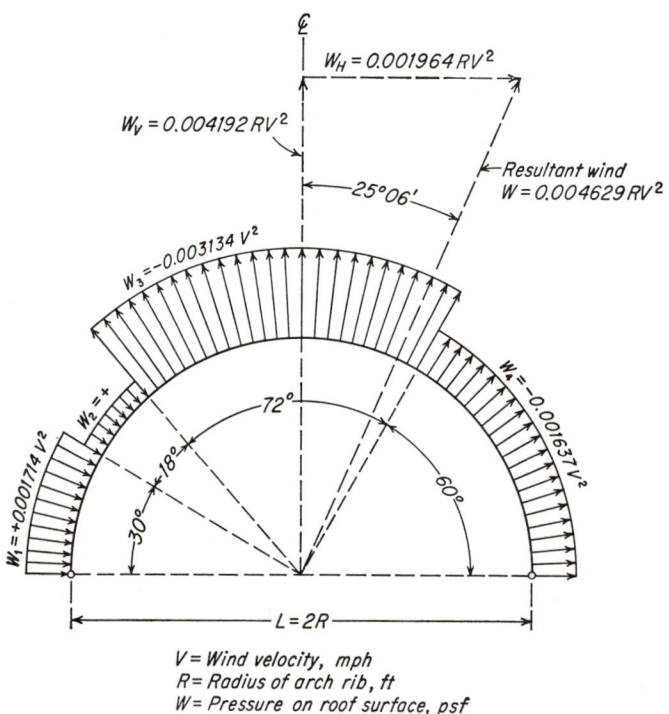

FIG. 46-2. Circular-arch-roof wind loading.

A system of force coefficients to be applied to the design wind-velocity pressure on a vertical surface has been used quite extensively in showing the theoretical pressures to which such building shapes are subjected.

$$P = CpA$$

where P = total pressure normal to surface, lb
C = force coefficient
p = psf/$0.00256W^2$, where p = velocity pressure on a surface perpendicular to air stream and W is air speed, mph
A = area of surface on which Cp acts, sq ft

Figure 46-3 shows graphically the direction and magnitude of the force coefficients on two similarly shaped gable-roof buildings. One has an open front, and the other is an entirely closed building. Here the additional magnitude of the upward stresses on the open-front building is significant.

---

[1] Preventing Storm Wind Damage to Farm Buildings, *USDA Agr. Infor. Bull.* 144, 1956.

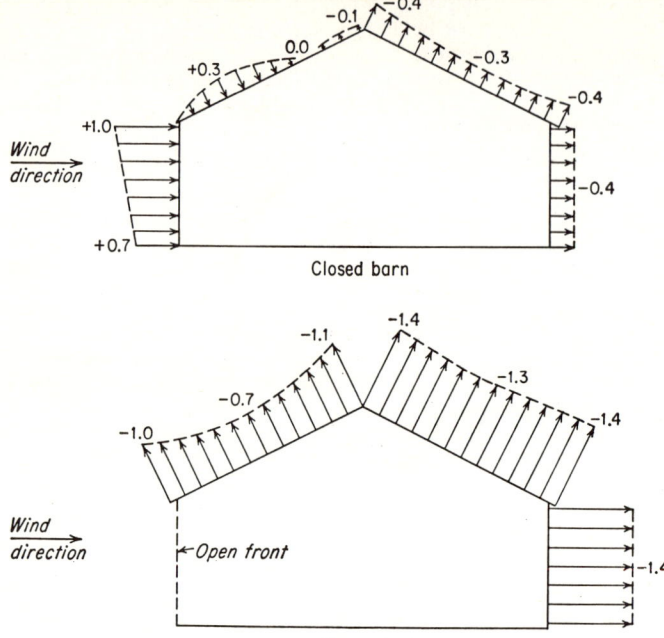

FIG. 46-3. Force coefficients for wind stresses on gable buildings. (*Fenton and Otis, Kansas State Univ. Eng. Expt. Sta. Bull.* 42, 1941.)

Recommended force coefficients for design wind loads on farm building roofs of different shapes are given in Table 46-3.

TABLE 46-3. RECOMMENDED FORCE COEFFICIENTS FOR DESIGN WIND LOAD FOR GAMBREL-, GOTHIC-, AND GABLE-ROOF FARM BUILDINGS

Surface	Gambrel roof		Gothic roof		Gable roof	
	Windward	Leeward	Windward	Leeward	Windward	Leeward
Wall............	+0.8	−0.3	+0.8	−0.5	+0.8	−0.4
Lower roof......	+0.7	−0.4	+0.6	−0.6	+0.3	−0.5
Upper roof......	−0.7	−0.7	−0.8	−0.8	−0.5	−0.5
End.............	+0.8	−0.6	+0.8	−0.6	+0.8	−0.6

SOURCE: F. L. Fenton and C. K. Otis, The Design of Barns to Withstand Wind Loads, *Kansas State Univ. Eng. Expt. Sta. Bull.* 42, 1941.

Figure 46-4 gives the wind-velocity pressures, including a gust factor of 1½, for the various areas of the United States (see Fig. 57-15 for probable recurrence periods of destructive winds). In designing for wind stresses, the following should be kept in mind:

1. Velocity pressure never is taken as less than 20 psf.
2. Wind may come from any direction; hence any surface member must be designed for the maximum force coefficients, both positive and negative.
3. Although a building may normally be closed, windows or doors may be left open, resulting in an increase of negative pressure on the leeward side.

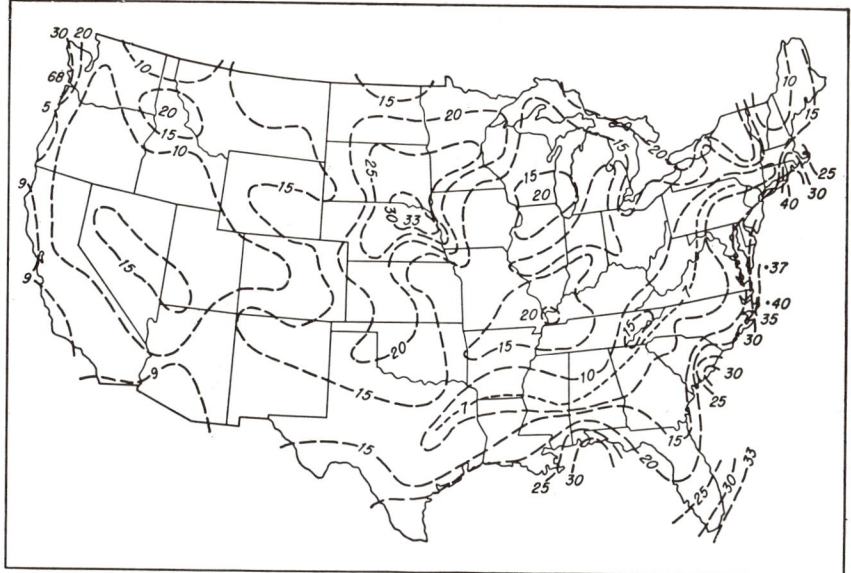

Fig. 46-4. Wind-velocity pressures in the United States (pounds per square foot of vertical surface for height aboveground of 30 ft; gust factor of 1.5 is included). [*American Standard Building Code Requirements for Minimum Design Loads in Buildings and Other Structures, Natl. Bur. Standards (U.S.)* M179, 1945.]

WOOD

Wood continues to be the primary structural material for farm buildings in the United States, although it is a nonhomogeneous natural product requiring judgment and critical appraisal. Full information on the physical properties and strength values of wood is necessary for designing purposes. For greater detail than is provided here, refer to the *Wood Handbook*.[1]

By classification *hardwoods* come from broadleaf deciduous trees and *softwoods* from coniferous evergreen trees having needlelike leaves.

Engineering Properties of Wood. The most important design factors are the strength and uniformity of the clear wood, the moisture content, the duration of load, and the size, number, and location of strength-reducing features such as knots, cross grain, checks, and splits.

The strength of *clear wood* is closely related to the weight or the density of the wood as shown by Fig. 46-5.

The moisture content of wood also affects its strength. When wood seasons, the loss of moisture is accompanied by a stiffening and strengthening of the wood fibers, but checking between fibers lessens resistance to shear or to tension across the grain. Some net increase in bending strength with drying is recognized in structural lumber less than 4 in. in thickness and in posts or columns of any thickness. Figure 46-6 shows graphically the equilibrium moisture content of wood in relation to temperature and relative humidity.

Structural lumber may be weakened by *knots, checks, shakes, wanes, pitch pockets, holes, slope of grain,* and various combinations of these.

Basic stresses are generalized safe ultimate-strength values for the clear wood of a species and are used as a basis for design stresses. The allowable design stresses may be smaller or larger than the basic stresses by some strength ratio, depending upon the type of wood and the loading. Basic stresses are obtained from laboratory tests and include a factor of safety for below-average variations of clear specimens but not for weakening by knots, etc. The basic stress of a species provides a basis for evaluat-

[1] *USDA Forest Serv. Bull.* 72, 1956.

ing the allowable working stress for any grade of that species. An example of laboratory-test results is shown graphically in Fig. 46-7. For these 791 tests, it is seen that the average strength of the test specimens was 7,450 psi for clear Douglas fir. The normal distribution of failure is also evident in this graph. The basic stress of Douglas fir, in comparison, is 2,200 psi. In obtaining these basic stresses, certain general principles were laid down as a guide:[1] (1) every individual timber must be capable of safely carrying its full design load; (2) the working stresses must be applicable to all conditions of use; (3) timbers must be capable of carrying the full design load for the life of the structure; and (4) it is assumed that workmanship, fabrication, and design are reasonably good.

The *factor of safety* for the most probable timber under the most probable service conditions was shown by a mathematical analysis made at the Forest Products Labora-

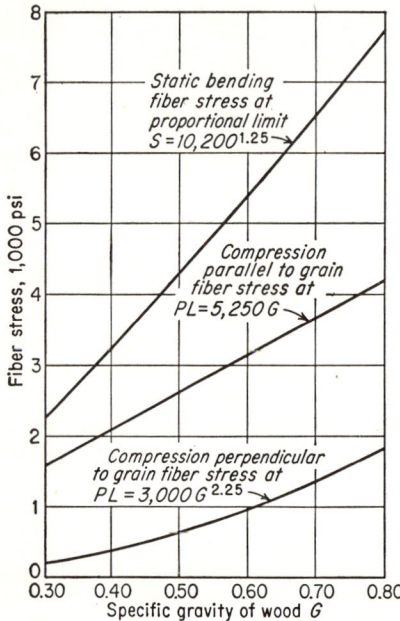

Fig. 46-5. Specific gravity–strength relationship of clear green wood. (*Wood Handbook*, USDA Forest Serv. Bull. 72, 1956.)

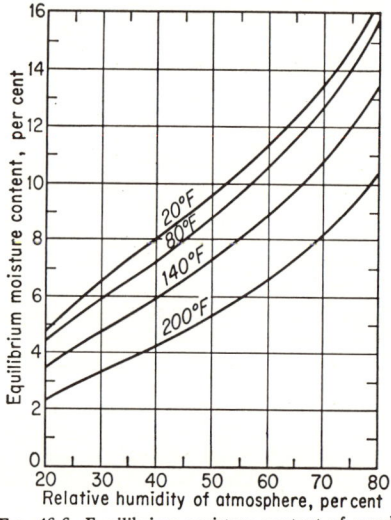

Fig. 46-6. Equilibrium moisture content of wood. (*USDA Forest Prods. Lab. Circ.* R-1651.)

tory to be $2\frac{1}{2}$ to 3 in bending and shear and $2\frac{1}{4}$ in compression. One timber in one hundred has a factor of safety as low as $1\frac{1}{4}$.

Table 46-4 provides the *basic stresses* for clear lumber under long-time service at full design load. Inasmuch as little of the lumber used for structural purposes is clear, some reduction factor or *strength ratio* must be applied to the basic stresses to obtain allowable design stresses. Thus a strength ratio of 75 per cent applies to a piece or a grade in which the maximum reduction in strength, due to knots, cross grain, etc., from that of clear wood is 25 per cent. To aid in classifying woods by strength, the structural woods have been *graded* in accordance with strength ratios. *Lumber grades* that have been set up on the basis of numbers and maximum sizes of knots can be appraised for strength ratios.

Duration of Load. Where maximum loads are of short duration such as wind and snow loads on roofs, the working stresses can be increased in accordance with the curve shown in Fig. 46-8.

[1] R. P. A. Johnson, "Working Stresses for Structural Lumber," Proceedings of Wood Symposium, 100 Years of Engineering Progress with Wood, Sept. 3–13, 1952, Chicago, Timber Engineering Co., Washington, D.C.

Inasmuch as wind loads are short in duration, the wind-load allowable design stresses may be increased 50 per cent above the level of the basic stresses provided the resulting structural members are not smaller than required for dead and live loads alone. For snow loads it is also apparent from this curve that the allowable design stresses can be increased from ¼ to ⅓ depending upon the location and duration of snow loads.

Softwood lumber is divided into three main classes: *yard* lumber, *structural* lumber (often referred to under the general term timber), and *factory*, or shop, lumber. For

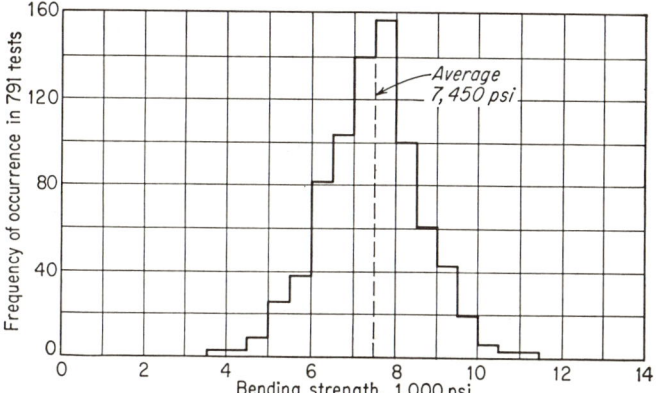

FIG. 46-7. Frequency diagram showing the range in bending-strength values in small specimens of clear green Douglas fir. (Wood, *"Basic Principles of Structural Grading,"* USDA Forest Products Laboratory, 1952.)

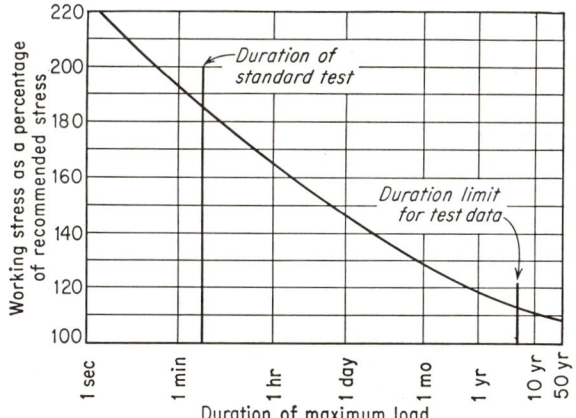

FIG. 46-8. Relation of working stress to duration of load. (*Wood Handbook,* USDA Forest Serv. Bull. 72, 1956.)

farm building construction, yard lumber is generally used. Yard lumber is divided into three subclasses entitled *finish, common boards,* and *common dimension.* The common-dimension material is most popular for farm building structural purposes. For years the common lumber has been graded into number classes, such as No. 1, No. 2, etc. There is in process, however, a changeover to a *name* grading system in place of the *number* system. The grade names that correspond to the past grade numbers of 1, 2, 3, 4, and 5 are, respectively, *select structural, construction, standard, utility,* and *economy.*

Some lumber associations provide a grade stamp for all lumber supplied by its member mills. The grade stamp gives the grade name, such as select structural, the

TABLE 46-4. BASIC STRESSES FOR CLEAR LUMBER UNDER LONG-TIME SERVICE AT FULL DESIGN LOAD,* FOR USE IN DETERMINING WORKING STRESSES ACCORDING TO GRADE OF LUMBER AND OTHER APPLICABLE FACTORS

Species†	Extreme fiber in bending or tension parallel to grain, psi	Maximum horizontal shear, psi	Compression perpendicular to grain,‡ psi	Compression parallel to grain, $L/d = 11$ or less, psi	Modulus of elasticity in bending, 1,000 psi
Softwood lumber					
Baldcypress (cypress)	1,900	150	220	1,450	1,200
Cedar:					
Alaska	1,600	130	185	1,050	1,200
Atlantic white (southern white cedar and northern white)	1,100	100	130	750	800
Port Orford	1,600	130	185	1,200	1,500
Western red cedar	1,300	120	145	950	1,000
Douglas fir:					
Coast type	2,200	130	235	1,450	1,600
Coast type, close-grained	2,350	130	250	1,550	1,600
Rocky Mountain type	1,600	120	205	1,050	1,200
All types, dense	2,550	130	275	1,700	1,600
Fir:					
Balsam	1,300	100	110	950	1,000
California red, grand, noble, and white	1,600	100	220	950	1,100
Hemlock:					
Eastern	1,600	100	220	950	1,100
Western (West Coast hemlock)	1,900	110	220	1,200	1,400
Larch, western	2,200	130	235	1,450	1,500
Pine:					
Eastern white (northern white), ponderosa, sugar, and western white (Idaho white)	1,300	120	185	1,000	1,000
Jack	1,600	120	160	1,050	1,100
Lodgepole	1,300	90	160	950	1,000
Red (Norway pine)	1,600	120	160	1,050	1,200
Southern yellow	2,200	160	235	1,450	1,600
Dense	2,550	160	275	1,700	1,600
Redwood	1,750	100	185	1,350	1,200
Close-grained	1,900	100	195	1,450	1,200
Spruce:					
Englemann	1,100	100	130	800	800
Red, white, and Sitka	1,600	120	185	1,050	1,200
Tamarack	1,750	140	220	1,350	1,300
Hardwood lumber					
Ash:					
Black	1,450	130	220	850	1,100
Commercial white	2,050	185	365	1,450	1,500
Aspen, bigtooth and quaking	1,300	100	110	800	800
Beech, American	2,200	185	365	1,600	1,600
Birch, sweet and yellow	2,200	185	365	1,600	1,600
Cottonwood, eastern	1,100	90	110	800	1,000
Elm:					
American and slippery (soft elm)	1,600	150	185	1,050	1,200
Rock	2,200	185	365	1,600	1,300
Hickory, true and pecan	2,800	205	440	2,000	1,800
Maple, black and sugar (hard maple)	2,200	185	365	1,600	1,600
Oak, commercial red and white	2,050	185	365	1,350	1,500
Sweetgum (gum, red gum, sap gum)	1,600	150	220	1,050	1,200
Tupelo, black (black gum) and water	1,600	150	220	1,050	1,200
Yellow poplar (poplar)	1,450	130	160	1,050	1,200

* These stresses are based on the strength of green lumber and are applicable, with certain adjustments, to lumber of any degree of seasoning or lumber used under any conditions of duration of load.

† Species names approved by USDA Forest Service. Commercial designations are shown in parentheses.

‡ Values given in previous editions of the Wood Handbook presumed some drying and were therefore at a higher level than these for green lumber.

SOURCE: Wood Handbook, *USDA Forest Serv. Bull.* 72, 1956.

species of wood, and the allowable design stress for the lumber of that grade. Select structural Douglas fir carries a design stress of 1,900 psi, construction grade of Douglas fir of 1,500 psi, standard grade of Douglas fir of 1,200 psi, and the utility and economy grades are unspecified as to design stresses.

The standard dimensions of the various yard-lumber products are given in Table 46-5. Yard lumber is usually carried in even foot lengths, although multiples of 1 ft are specified in the manufacturers' association grading rules. Common-dimension-grade lumber is usually available in lengths up to 18 ft, and some lengths beyond that can be obtained at a premium price.

TABLE 46-5. STANDARD DIMENSIONS OF YARD LUMBER

Light framing		Joists and planks	
2 × 2	1⅝ × 1⅝	2 × 6	1⅝ × 5⅝
2 × 3	1⅝ × 2⅝	2 × 8	1⅝ × 7½
2 × 4	1⅝ × 3⅝	2 × 10	1⅝ × 9½
3 × 3	2⅝ × 2⅝	2 × 12	1⅝ × 11½
3 × 4	2⅝ × 3⅝	3 × 6	2⅝ × 5½
4 × 4	3⅝ × 3⅝	3 × 8	2⅝ × 7½
		3 × 10	2⅝ × 9½
Center match		3 × 12	2⅝ × 11½
1 × 6	25/32 × 5¼ *	4 × 6	3⅝ × 5½
1 × 8	25/32 × 7¼ †	4 × 8	3⅝ × 7½
2 × 6	1⅝ × 5	4 × 10	3⅝ × 9½
3 × 6	2⅝ × 5	4 × 12	3⅝ × 11½
3 × 8	2⅝ × 7		
		Boards	
Shiplap		1 × 2	25/32 × 1⅝
		1 × 3	25/32 × 2⅝
1 × 6	25/32 × 5⅓	1 × 4	25/32 × 3⅝
1 × 8	25/32 × 7⅛	1 × 6	25/32 × 5⅝
1 × 10	25/32 × 9⅛	1 × 8	25/32 × 7½
		1 × 10	25/32 × 9½
		1 × 12	25/32 × 11½

* 5 seasoned.
† 7 seasoned.

Beam Design. Wood members should be designed so that they do not exceed the allowable design unit values for (1) the *extreme fiber stress* caused by flexural loads, (2) the *maximum horizontal-shear stress*, and (3) the stress in *compression* across the grain at the end bearings. *Deflection* should also be considered, particularly in buildings where this might cause trouble, as in plastered ceilings. Where the actual and nominal sizes of lumber differ, the allowable unit stress values should be applied to the actual size of members.

For a rectangularly shaped beam supported at the ends and uniformly loaded, the stress may be determined as follows:

$$S = \frac{3Wl}{4bd^2}$$

where S = stress in outermost fiber, psi
W = total uniformly distributed load, lb
l = length, in.
b = width, in.
d = depth, in.

For farm buildings, Fig. 46-9 provides graphical solutions for the common sizes and spans of dimension lumber under normal farm-type loadings. It will be noted that this graph is designed for an allowable fiber stress of 2,000 psi in bending. This is 100 lb above the basic design stress of select structural Douglas fir and 33 per cent above the design stress for construction-grade Douglas fir. This means that, for the construction grade, a strength ratio of 1.33 has been applied to the design stress of 1,500 psi. This is justified where the loadings are noncontinuous and temporary in duration, as are most live loads. Also, it is not economically practical to apply the minimum basic stress where the structural frames are made up of many interacting members.

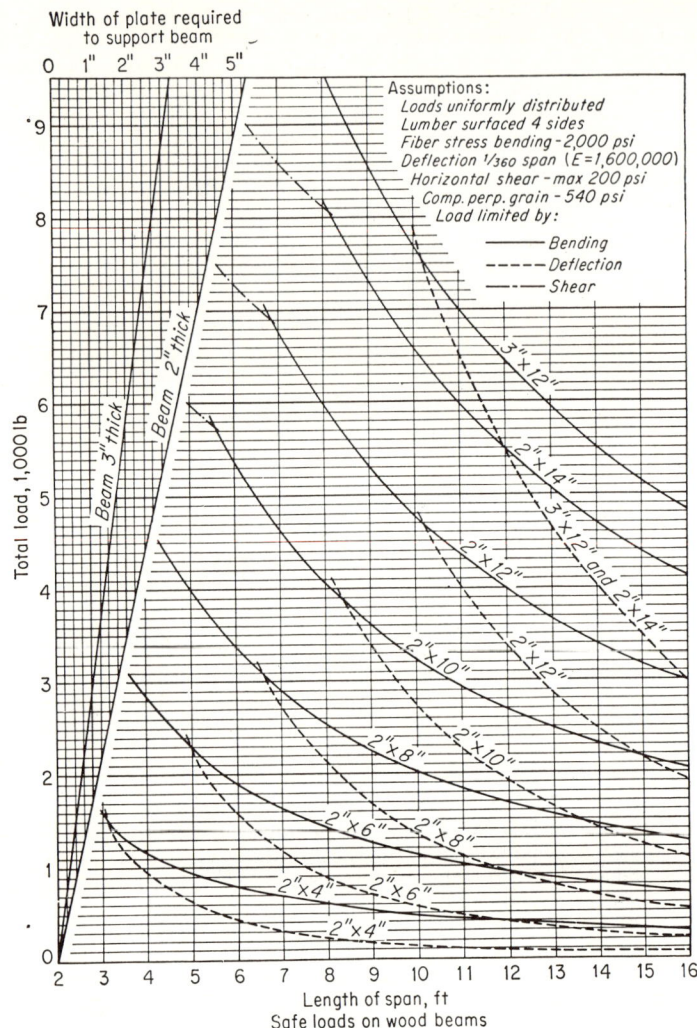

Fig. 46-9. Allowable spans for various beam sizes and loads. (*Giese, Iowa State University.*)

In *combined loading effects*, where bending loads are combined with end compression or tension, the combined stress is determined by algebraic addition.

Horizontal shear in beams is determined from the general formula

$$S_H = \frac{VQ}{Ib}$$

where S_H = maximum horizontal-shear stress, psi
Q = (area above or below neutral axis, sq in.) × (distance from neutral axis to center of gravity of area, in.)
V = vertical shear, lb
I = moment of inertia about neutral axis, in.^{4}
b = beam width at neutral axis, in.

For rectangular beams, this becomes

$$S_H = \frac{3V}{2bd}$$

where d = depth, in.

The area required for *end bearing* on a wood beam is governed by the allowable unit stress in compression perpendicular to the grain or by the allowable bearing stress on the support, whichever is smaller.

Wood Column Design. For design purposes there are three classifications of columns: (1) *short columns* with length not exceeding 11 times the least dimension, (2) *intermediate columns* with length between 11 and K times the least dimension, (3) *long columns* with length exceeding K times the least dimension.

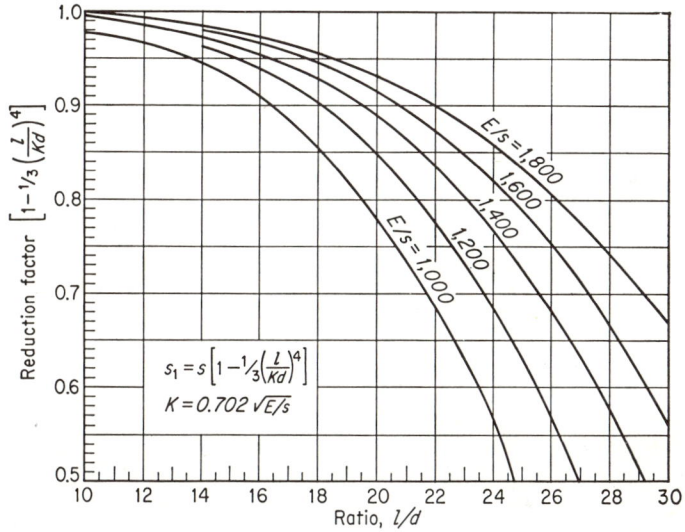

Fig. 46-10. Stress-reduction factor for wood columns of intermediate length (l/d between 11 and K). (Barre and Sammet, "Farm Structures," John Wiley & Sons, Inc., New York, 1950.)

Normally the ratio of l/d (unsupported length over dimension of narrow face) may not exceed 50. For design purposes, ends are assumed to be equivalent to pinned ends. Laminated members that are properly glued may be designed as fixed columns.

In the design of *short columns* where l/d does not exceed 11, the loading acts only in direct compression and the basic compressive stress parallel to the grain may be used to determine the required cross-section area.

Intermediate columns where l/d ratios are between 11 and K and not greater than 50 may be expected to fail through the combined effects of crushing and buckling. Allowance for buckling is made by reducing the allowable design stress in accordance with Fig. 46-10. The numerical value of K is obtained as follows:

$$K = 0.702 \times \frac{E}{s}$$

where E = modulus of elasticity

s = allowable compressive stress, psi, applying a reduction factor from Fig. 46-10.

Long columns where l/d ratios are equal to or greater than K but less than 50 are treated in a manner similar to intermediate columns except that the reduced allowable design is obtained from Fig. 46-11.

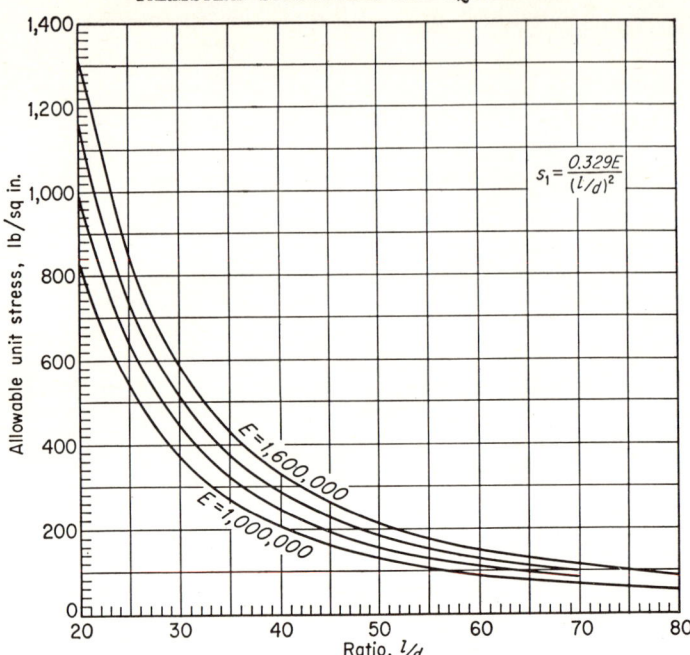

Fig. 46-11. Allowable stress for long wood columns (l/d greater than K). (Barre and Sammet, "Farm Structures," John Wiley & Sons, Inc., New York, 1950.)

PLYWOOD

Interior-type plywood has the veneers bonded with adhesive having a fair degree of moisture resistance and so retains its original form and practically all its strength when occasionally subjected to a thorough wetting and normal drying. *Exterior-type* plywood represents the ultimate in moisture resistance and will retain its original form and strength when repeatedly wetted and dried. It is suitable for permanent exterior use.

Within each type of plywood there are several *grades,* which are based on the quality of the exposed veneers. *Grade A* (sound) is free from knots, splits, pitch pockets, and other open defects. *Grade B* (solid) is free from open defects, but will admit neatly made circular plugs; knots up to 1 in. if sound and tight; tight splits; and slightly rough and other minor sanding and patching defects. The grade is paintable. *Grade C* (exterior back) may contain knotholes not larger than 1 in. in least dimension, open pitch pockets not wider than 1 in., splits not wider than 3/16 in., and worm or borer holes not more than 5/8 in. wide. *Grade D* (utility back, used only in interior-type panels) contains knotholes up to 2½ in. in maximum dimension, pitch pockets up to 2 in. wide by 4 in. long, splits up to ½ in. wide, and other characteristics that will not seriously impair the strength or serviceability of the panel. See Table 46-6 for recommended working stresses.

STRUCTURAL STEEL

Steel structural shapes may often be used economically in farm building construction for columns, girders, joists, beams, and other framing members. Allowable unit stresses for steel structural members are recommended by the American Institute of Steel Construction as shown in Table 46-7.

In the *design of columns* Fig. 46-12 shows graphically the working stresses for various l/r ratios, where l is the unsupported length in inches and r the radius of gyration

STRUCTURAL REQUIREMENTS OF FARM BUILDINGS   553

TABLE 46-6. RECOMMENDED WORKING STRESSES FOR PLYWOOD (DOUGLAS FIR)
(For Grades and Thicknesses Listed in US CS545-48)

In bending, tension, and compression (except bearing and 45° stresses) consider only those plies with their grain direction parallel to the principal stress.

Type of stress	EXT-DFPA·A-B	EXT-DFPA·A-B EXT-DFPA-Ply-shield·(A-C)	EXT-DFPA·Utility·(B-C) EXT-DFPA·Sheathing·(C-C) EXT-DFPA·Concrete Form·(B-B) Plyform·(B-B) Plyscord·(C-D)	Interior·A-A Interior·A-B Plypanel·(A-D) Plybase·(B-D)*
Dry location				
Extreme fiber in bending:				
Face grain ∥ to span, psi	2,188	2,000	1,875	100%
Face grain ⊥ to span, psi	1,875	1,875	1,875	80%
Tension:				
∥ to face grain (3-ply only†), psi	1,605	1,460	1,375	100%§
⊥ to face grain, psi	1,375	1,375	1,375	70%
45° to face grain, psi	496	472	460	80%
Bearing (on face), psi	405	405	405	100%
Shear, rolling, in plane of plies:‡				
∥ or ⊥ to face grain, psi	79	72	68	75%
45° to face grain, psi	105	96	90	75%
Shear, in plane ⊥ to plies:‡				
∥ or ⊥ to face grain, psi	210	192	180	45%
45° to face grain, psi	420	384	360	85%
Modulus of elasticity in bending:				
Face grain ∥ to span	1,600,000	1,600,000	1,600,000	100%
Face grain ⊥ to span	1,600,000	1,600,000	1,600,000	70%
Wet or damp location				

Where moisture content will exceed 16%, decrease by 20% values shown for dry location for following properties: extreme fiber in bending, tension, both ∥ and ⊥ to the grain and at 45°, and bearing. (No change in values for shear or modulus of elasticity.)

Example: The working stress in compression ∥ for Plypanel 5-ply (1,238 psi) is found by multiplying the value for EXT-DFPA·Plyshield 5-ply, 1,375 psi, by 90 per cent, the reduction factor shown in the last column and footnotes † and §.

* Apply these percentages to the stresses for the corresponding Exterior grade. See Example.
† For tension or compression, ∥ to grain, in 5-ply or thicker, use values for 3-ply, but in next lower grade.
‡ For certain conditions where stress concentrations exist, these working stresses for rolling shear should be reduced by 50 per cent. See *USDA Forest Prods. Lab. Bull.*, table 1, Approximate Methods of Calculating the Strength of Plywood.
§ For five or more plies, use 90 per cent.
SOURCE: "Technical Data on Plywood," Douglas Fir Plywood Association, Tacoma, Wash., 1948.

Table 46-8 gives the physical properties needed for designing columns of standard rolled H-beam sections. Table 46-9 gives the allowable column loads for various standard pipe sections. Two-angle members, used as columns, should be joined by units spaced 2 ft on centers (l/r for individual angles between rivets should not exceed ¾ l/r for the whole member).

In the design of steel *beams* for farm buildings, most of the steel sections may be selected from Fig. 46-13. If other than a uniformly distributed load is encountered, the bending moment must be determined and the beam critical section selected from more extensive tables of design properties of steel shapes in the AISC Steel Construction Manual. Normally, the beams are first selected for safe bending and then checked for shear, buckling, and end bearing, taking account of weakening by holes. Beam

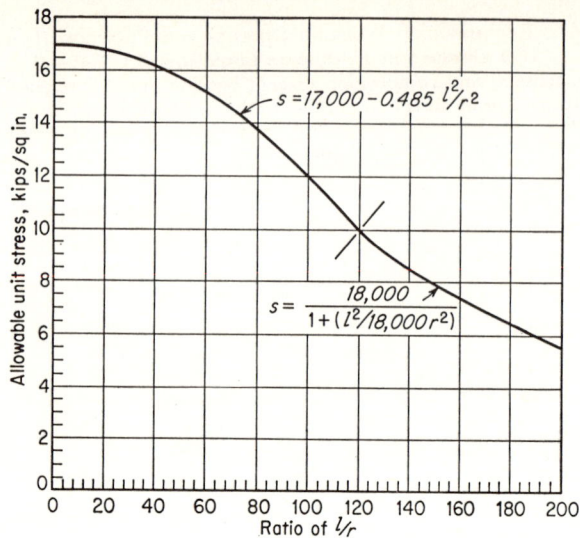

FIG. 46-12. Allowable unit stresses for steel columns. ("*AISC Steel Construction,*" American Institute of Steel Construction, New York, 1959.)

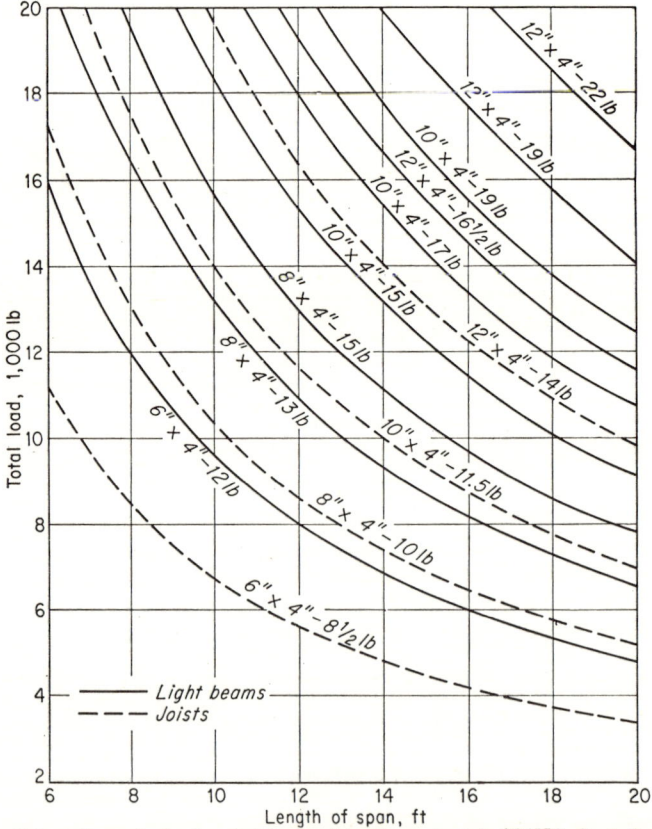

FIG. 46-13. Safe uniform loads for steel beams laterally supported. ("*AISC Steel Construction,*" American Institute of Steel Construction, New York, 1959.)

TABLE 46-7. ALLOWABLE UNIT STRESS FOR STEEL CONSTRUCTION

Structural steel, rivets, bolts, and welded metal

	Allowable stress, psi
(1) Tension	
Structural steel, net section..	20,000
Butt welds, section through throat...	20,000
Rivets, on area based on nominal diameter.................................	20,000
Bolts and other threaded parts, on nominal area at root of thread........	20,000
(2) Compression	
Columns, gross section	
For axially loaded columns with values of l/r not greater than 120......	$17{,}000 - 0.485 \dfrac{l^2}{r^2}$
For axially loaded columns (bracing and other secondary members) with values of l/r greater than 120..	$\dfrac{18{,}000}{1 + l^2/18{,}000 r^2}$
in which l is the unbraced length of the column and r is the corresponding radius of gyration	
Plate-girder stiffness, gross section.......................................	20,000
Webs of rolled sections at toe of fillet....................................	24,000
Butt welds, section through throat...	20,000
(3) Bending	
Tension on extreme fibers..	20,000
Compression on extreme fibers with ld/bt not in excess of 600............	20,000
In excess of 600..	$\dfrac{12{,}000{,}000}{ld/bt}$
in which l is unsupported length and d the depth of member, b the width, and t the thickness of its compression flange, all in inches	
Stress on extreme fibers of pins..	30,000
(4) Shearing	
Rivets...	15,000
Pins and turned bolts in reamed or drilled holes............................	15,000
Unfinished bolts...	10,000
Webs of beams and plate girders, gross section..............................	13,000
Weld metal	
Through throat of fillet weld...	13,600
Through throat of butt weld...	13,600
(5) Bearing (compression on projected area)	

	Double shear	Single shear
Rivets	40,000	32,000
Turned bolts in drilled holes	40,000	32,000
Unfinished bolts	25,000	20,000
Pins	32,000	32,000

SOURCE: "AISC Steel Construction," American Institute of Steel Construction, New York, 1959.

TABLE 46-8. PHYSICAL PROPERTIES OF ROLLED-STEEL H-BEAM SECTIONS FOR COLUMN DESIGN

Nominal size, in.	Wt. per foot, lb	Area of section, in.²	Depth of section, in.	Flange Width, in.	Flange Thickness, in.	Web thickness, in.	Axis $X-X$ I, in.⁴	Axis $X-X$ S, in.³	Axis $X-X$ r, in.	Axis $Y-Y$ I, in.⁴	Axis $Y-Y$ S, in.³	Axis $Y-Y$ r, in.
6WF 6 × 6	25.0	7.37	6.37	6.080	.456	.320	53.5	16.8	2.69	17.1	5.6	1.52
	20.0	5.90	6.20	6.018	.367	.258	41.7	13.4	2.66	13.3	4.4	1.50
	15.5	4.62	6.00	6.000	.269	.240	30.3	10.1	2.56	9.69	3.2	1.45
5WF 5 × 5	18.5	5.45	5.12	5.025	.420	.265	25.4	9.94	2.16	8.89	3.54	1.28
	16.0	4.70	5.00	5.000	.360	.240	21.3	8.53	2.13	7.51	3.00	1.26
4WF	13.0	3.82	4.16	4.060	.345	.280	11.3	5.45	1.72	3.76	1.85	.99
8 × 8	34.3	10.09	8.00	8.000	.438	.375	115.5	28.90	3.40	35.1	8.80	1.87
6 × 6	25.0	7.35	6.00	5.938	.500	.313	47.0	15.70	2.53	14.9	5.00	1.43
	20.0	5.88	6.00	5.938	.375	.250	38.8	12.90	2.57	11.4	3.80	1.39
5 × 5	18.9	5.47	5.00	5.000	.438	.313	23.8	9.5	2.08	7.8	3.1	1.20
4 × 4	13.0	3.82	4.00	3.937	.375	.250	10.4	5.2	1.65	3.4	1.7	.94

SOURCE: "AISC Steel Construction," American Institute of Steel Construction, New York, 1959.

TABLE 46-9. ALLOWABLE CONCENTRIC LOADS,* IN KIPS (1,000 LB) FOR STANDARD STEEL PIPE COLUMNS

Unbraced length, ft	Nominal diameter, in., and weight per foot, lb									
	12		10		8		6	5	4	3½	3
	49.56	43.77	40.48	34.24	31.20	28.55	24.70	18.97	14.62	10.79	9.11	7.58
6	246	217	200	169	154	140	121	92	70	50	42	33
8	244	216	199	168	153	138	210	90	68	47	38	30
10	243	214	196	166	151	136	118	86	64	44	35	26
12	240	212	194	164	149	133	115	82	61	40	30	21
14	237	210	190	161	147	129	112	79	56	34	25	18
16	234	207	187	158	144	125	109	74	51	30	22	16
18	231	204	182	154	141	121	105	69	45	26	19	13
20	227	200	178	151	137	115	100	63	41	23	17	
22	222	196	172	146	133	109	95	56	37	21	15	

* Loads below heavy lines are for l/r ratios between 120 and 200.
SOURCE: "AISC Steel Construction," American Institute of Steel Construction, New York, 1959.

deflection for simply supported members with uniform loads may be determined by the equation

$$D = \frac{5wL^4}{384EI}$$

where D = deflection at center, in.
w = load, lb/in.
L = length of beam, in.
E = modulus of elasticity
I = moment of inertia about neutral axis, in.^{4}

Members subject to stresses produced by wind may be proportioned for unit stresses 33⅓ per cent greater than those specified for continuous dead- and live-load stresses.

Sections made of any combination of flat elements formed from sheet or strip steel less than 3/16 in. thick require special consideration.[1]

CONCRETE

To give maximum service, concrete must be of *good quality*. The making of quality concrete involves the following steps:
1. Selecting ingredients
2. Proper proportioning
3. Careful mixing
4. Making, shaping, and adequate bracing of forms
5. Placing concrete in forms
6. Adequate finishing
7. Thorough curing

Cement is usually packed in cloth or paper sacks containing 94 lb. This is 1 cu ft of loose-volume cement. Cement quantities are often referred to in terms of barrels containing 376 lb, or the equivalent of 4 sacks.

The commonly used *aggregates* are sand, gravel, crushed stone, and blast-furnace slag. Cinders, burnt clay, and expanded blast-furnace slag are also used for lightweight concrete mixes. By definition, sand or *fine aggregates* include those particles which will pass a No. 4 screen. *Coarse aggregates* include those particles larger than a

---

[1] "Light Gage Cold-formed Steel Design Manual," American Iron and Steel Institute, New York, 1956.

No. 4 screen and smaller than the maximum allowable size of aggregate for the particular concrete job.

The grading of the coarse aggregate within a given size range may be varied over a wide range with little effect on the cement requirement if the proportion of fine aggregate is such as to give good workability. On the other hand, if the proportion of fine

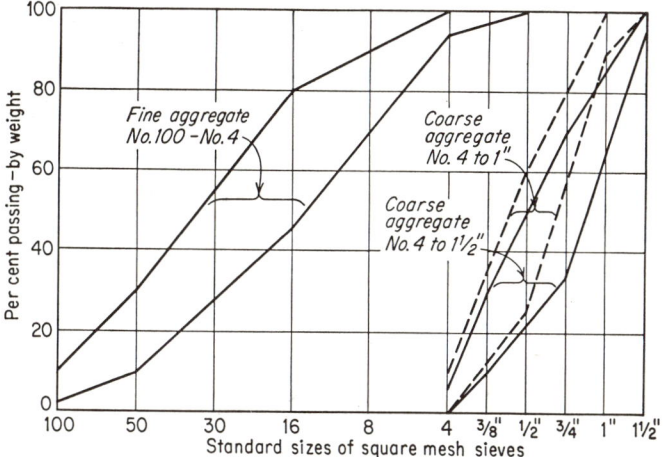

Fig. 46-14. Standard specifications for concrete aggregate (ASTM C33) for fine aggregate and for two sizes of coarse aggregate.

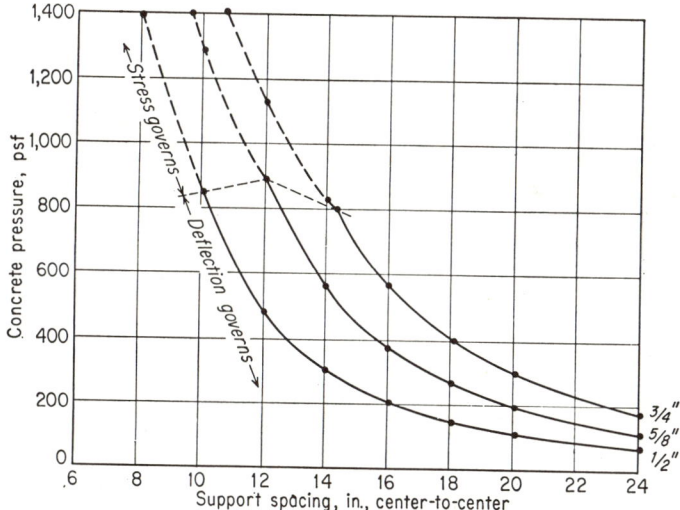

Fig. 46-15. Allowable support spacing for Plyform. Deflection limited to $\frac{1}{270}$ of span. Stress limited to 2,000 psi. Conditions: face grain perpendicular to supports; panels continuous across two or more spans. (*Douglas Fir Plywood Assn.*)

aggregate is held constant, variations in coarse-aggregate gradings will result in changes in the cement requirement, becoming uneconomic in some cases. For best results, an aggregate should be selected with a gradation falling within the limits indicated by the standard specification curves of Fig. 46-14.

Concrete mixtures should consist of the most economical and practical *proportions* which will produce the necessary workability in the fresh concrete and the required qualities in the hardened concrete. *Strength, durability,* and *watertightness* of concrete are controlled by the amount of water used per sack of cement. In general, the less water used the better the quality of the concrete, so long as the mixture is plastic and workable. The first step in proportioning a concrete mix is to determine the relative amounts of water and cement to be used. This must be based on the conditions to which the concrete is to be exposed and on the strength desired. Recommended quantities of water for different classes of work and suggested proportions of cement to fine and coarse aggregate to use in trial batches are given in Table 46-10.

TABLE 46-10. How to Select the Proper Mix
(Recommended Proportions of Water to Cement and Suggested Trial Mixes)

Kinds of work	Gallons of water to add to each sack batch if sand is:			Suggested mixture for trial batch			Materials per cubic yard of concrete*		
					Aggregates			Aggregates	
	Very wet	Wet (average sand)	Damp	Cement sacks	Fine, cu ft	Coarse, cu ft	Cement sacks	Fine, cu ft	Coarse, cu ft
5-gal paste for concrete subjected to severe wear, weather, or weak acid and alkali solutions (Maximum-size aggregate = ¾ in.)									
One-course industrial, creamery, and dairy plant floors, etc.	3¾	4	4½	1	1¾	2	8	14	16
6-gal paste for concrete to be watertight or subjected to moderate water and weather (Maximum-size aggregate = 1½ in.)									
Watertight floors, such as industrial plant, basement, dairy barn; watertight foundations; driveways, walks, tennis courts, swimming and wading pools, septic tanks, storage tanks, structural beams, columns, slabs, residence floors, etc.	4¼	5	5½	1	2¼	3	6¼	14	19
7-gal paste for concrete not subjected to wear, weather, or water (Maximum size aggregate = 1½ in.)									
Foundation walls, footings, mass concrete, etc.	4¾	5½	6¼	1	2¾	4	5	14	20

* Quantities are estimated on wet aggregates using suggested trial mixes and medium consistencies and will vary according to the grading of aggregate and the workability desired.

NOTE: It may be necessary to use a richer paste than is shown in the table because the concrete may be subjected to more severe conditions than are usual for a structure of the type being constructed. For example, a swimming pool ordinarily is made with a 6-gal paste. However, the pool may be built in a place where soil water is strongly alkaline, in which case a 5-gal paste is required.

The concrete mixture should be *placed* in the forms just as soon as possible after mixing (within 45 min at the most). In transporting concrete to the forms, care should be taken to prevent segregation or separation of coarse from fine particles. Concrete should be placed in level layers of not more than 12 in. deep at one time. Then tamp, spade, or vibrate just enough to make it settle thoroughly and produce a dense mass. Working the concrete next to the forms ensures an even, dense surface when forms are removed. Freshly mixed concrete exerts great pressure; therefore concrete *forms* must be substantial enough to retain their correct shape when filled. It is not sufficient that forms be strong; they must also be rigid. Estimates of lateral pressures to be expected from fresh concrete are given in Table 46-11. Figure 46-15 shows graphically the requirements for support spacing on forms.

The type of *finish* put on concrete floors depends on the expected usage. For livestock floors, paved yards, driveways, sidewalks, and other work where a nonskid surface is required, a wood float or a broomed surface is recommended.

Proper *curing* of concrete influences its strength, durability, and watertightness. Concrete hardens because of a chemical reaction between portland cement and water. Hardening continues as long as temperatures are favorable and moisture is present to hydrate the cement. Concrete damp-cured for 1 month will be approximately 100 per

TABLE 46-11. SUGGESTED CONCRETE DESIGN PRESSURES, PSF*

Rate of pour, ft/hr	Vibrated		Unvibrated	
	At 50°F	At 70°F	At 50°F	At 70°F
2	650	550	450	350
3	900	650	600	450
4	1,050	750	750	550
5	1,200	900	900	650
6	1,350	1,050	1,050	750
7		1,200	1,200	850
8		1,350	1,350	950
9				1,050

Depth, D, at maximum pressure, P:

$$D = \frac{P}{150} \text{ (vibrated)} \qquad D = 2 + \frac{P}{150} \text{ (unvibrated)}$$

* Test data indicate much variation in concrete pressures. The above values are suggested where better data are lacking. They are taken from M. L. Elkins, "Design Assumptions for the New Keely PlyForm Calculator," and include data furnished by the Universal Form Clamp Co. for vibrated concrete.

NOTE: In case form height is less than D (depth at maximum pressure), design pressure may be reduced by proportion of form height to D. For example, with a 5-ft-high wall poured with unvibrated concrete at 70° at 6 ft/hr, maximum pressure is 750 psf, and

$$D = 2 + \frac{750}{150} = 7 \text{ ft}$$

Then design pressure would be $\frac{5}{7} \times 750 = 536$ psf.

cent stronger than similar concrete in dry air. Damp curing may be done by covering with wet burlap, canvas, sand, or straw coverings as soon as it can be done without marring the surface. The covering material must be kept wet continuously.

Having determined the water-cement ratio and the fine aggregate–coarse aggregate–cement ratio, the *quantity* of materials may be accurately calculated by the absolute-volume method. Absolute volume of a material is defined as the actual volume of the solid particles only, with no voids.

$$\text{Absolute volume} = \frac{\text{weight of loose material}}{(\text{specific gravity})(\text{unit weight of water})}$$

The physical properties of the materials are as follows:

Material	Wt. in loose form, lb/cu ft	Abs. specific gravity
Cement	94	3.10
Fine aggregate	110	2.65
Coarse aggregate	100	2.65
Water	62.5	1.00

Example: Determine materials required for 100 cu yd of concrete, using a 5-gal water-cement ratio and a 1:2:3 mix.

First determine volume of concrete produced by 1 sack of cement.

Abs. vol. of cement: $\dfrac{1 \times 94}{3.1 \times 62.5}$ = 0.49 cu ft

Abs. vol. of fine aggregate: $\dfrac{2 \times 110}{2.65 \times 62.5}$ = 1.33 cu ft

Abs. vol. of coarse aggregate: $\dfrac{3 \times 100}{2.65 \times 62.5}$ = 1.82 cu ft

Abs. vol. of water: 5/7.5 = 0.67 cu ft

Vol. of concrete produced by 1 sack = 4.31 cu ft

For 100 cu yd, multiply above figures by (27/4.31)(100), or 625.

CONCRETE BLOCKS

Concrete-masonry building units are designated as:
1. Hollow load-bearing concrete block, 8 by 8 by 16 in., 8 by 12 by 16 in.
2. Solid load-bearing concrete block, 4 by 8 by 16 in.
3. Hollow non-load-bearing concrete block, 4 by 8 by 16 in., 6 by 8 by 16 in.
4. Concrete building tile, 5 by 8 by 12 in.
5. Concrete brick, 2 by 4 by 8 in.

The most common unit is the hollow load-bearing block of 8 by 8 by 16 in. nominal size weighing approximately 40 to 50 lb when made with heavyweight concrete and from 25 to 35 lb when made of lightweight aggregate. Heavyweight units are made with such aggregates as sand, gravel, crushed stone, and air-cooled slag. Lightweight units are made with coal cinders, expanded shale, clay or slag, and natural lightweight materials such as volcanic cinders, pumice, and scoria.

The *insulation* value of sand-and-gravel concrete block walls is low so that lightweight concrete blocks are used extensively. Eight-inch concrete blocks made from expanded slag, clay, or shale have a coefficient of heat transmission (U) 38 per cent below a regular sand-and-gravel block (Fig. 48-3). Additional insulation may be obtained by filling the cores with fill insulation, by building a cavity wall with two 4-in. tiers of blocks, by applying insulation board on one surface, and by furring out and plastering or using rigid insulation board.

In designing walls for concrete masonry, all horizontal dimensions should be multiples of nominal half-length units and all vertical dimensions multiples of nominal full-height units. Thus, with the nominal 8 by 8 by 16 in. block both horizontal and vertical dimensions should be designed in multiples of 8 in.[1] The standard 8- and 12-in.-width blocks will suffice for most farm building construction.

Lateral support for walls is one of the critical limitations. Lateral supports may be obtained by cross walls, pilasters, or buttresses. The limiting horizontal distances are as follows:
1. Walls of solid block or brick units, 20 times the nominal wall thickness
2. Walls of hollow concrete masonry units, 18 times the nominal wall thickness
3. Cavity walls, 14 times the nominal wall thickness

Quantities of materials required for concrete-masonry walls may be estimated from Table 46-12.

The proportioning of *mortar* for masonry walls is very important. Recommended mortar mixes by volume, *for ordinary service,* are as follows:

Cement (portions)	Hydrated lime (portions)	Mortar sand (portions)
1-masonry cement............	.....	2–3
1-portland cement............	1–1¼	4–6

[1] For detailed commercial design specifications see the American Standard Building Code Requirements for Masonry, A41.1-1944, U.S. Government Printing Office.

STRUCTURAL REQUIREMENTS OF FARM BUILDINGS 561

A mortar must be workable for the job at hand. Workability is obtained through the proper grading of the sand and by thorough mixing rather than through the use of excessive amounts of cementitious material.

For *watertight* wall construction, concave and V-shaped mortar joints are recommended in preference to struck or raked joints that form small ledges which may hold

TABLE 46-12. WEIGHTS AND QUANTITIES OF MATERIALS FOR CONCRETE-MASONRY WALLS

Actual unit sizes, width × height × length, in.	Nominal wall thickness, in.	For 100 sq ft of wall				For 100 concrete units, mortar, cu ft
		Number of units	Average weight of finished wall		Mortar, cu ft	
			Heavyweight aggregate, lb	Lightweight aggregate, lb		
3⅝ × 3⅝ × 15⅝	4	225	3,050	2,150	4.3	1.9
5⅝ × 3⅝ × 15⅝	6	225	4,550	3,050	4.3	1.9
7⅝ × 3⅝ × 15⅝	8	225	5,700	3,700	4.3	1.9
3¾ × 5 × 11¾	4	221	3,000	2,150	3.7	1.7
5¾ × 5 × 11¾	6	221	4,500	3,050	3.7	1.7
7¾ × 5 × 11¾	8	221	5,650	3,700	3.7	1.7
3⅝ × 7⅝ × 15⅝	4	112.5	2,850	2,050	2.6	2.3
5⅝ × 7⅝ × 15⅝	6	112.5	4,350	2,950	2.6	2.3
7⅝ × 7⅝ × 15⅝	8	112.5	5,500	3,600	2.6	2.3
11⅝ × 7⅝ × 15⅝	12	112.5	7,950	4,900	2.6	2.3

water. With modular-size masonry units, mortar joints should be approximately ⅜ in. thick. Concrete-masonry walls can be made weathertight by the application of two coats of portland-cement base paint. For basement construction, the earth side of the concrete masonry should be given two complete coatings of bituminous material for additional protection.

All door and window *lintels* should be designed to support the bearing load above.

FASTENERS FOR WOOD CONSTRUCTION

The strength and stability of any structure is no greater than that of the fasteners used to hold the parts together. Wood has been used successfully with many different types of fasteners, including *nails, spikes, screws, bolts, spike screws, drift pins, metal connectors* of various shapes, and *glue.*

For balanced design the fasteners should develop the potential strength of the wood members. Unfortunately, this is almost impossible with the conventional nailed construction used in most farm buildings.

The allowable design strengths given in the following section for various fasteners are applicable under conditions of long-continued or permanent loading. For loads of short duration the allowable loads can be increased as discussed under the topic of wood. The allowable loads for a particular loading condition are given for one fastener. Loads for more than one fastener of the same size or of miscellaneous sizes are the sum of the loads permitted for each, provided that the spacings and distances and edge margins are sufficient to develop the full strength of each unit.

Nails. There are many types, sizes, and forms of standard nails as well as many special-purpose nails, as shown in Fig. 46-16.

In general, nails provide stronger joints when driven into the side grain (perpendicular to the wood fibers) than into the end grain of wood. Nails should preferably be used so that the lateral resistance rather than the direct withdrawal resistance is utilized.

Withdrawal resistance of nails is greatly affected by such factors as type of nail point, type of shank, time interval since nail was driven, surface coating, and moisture-

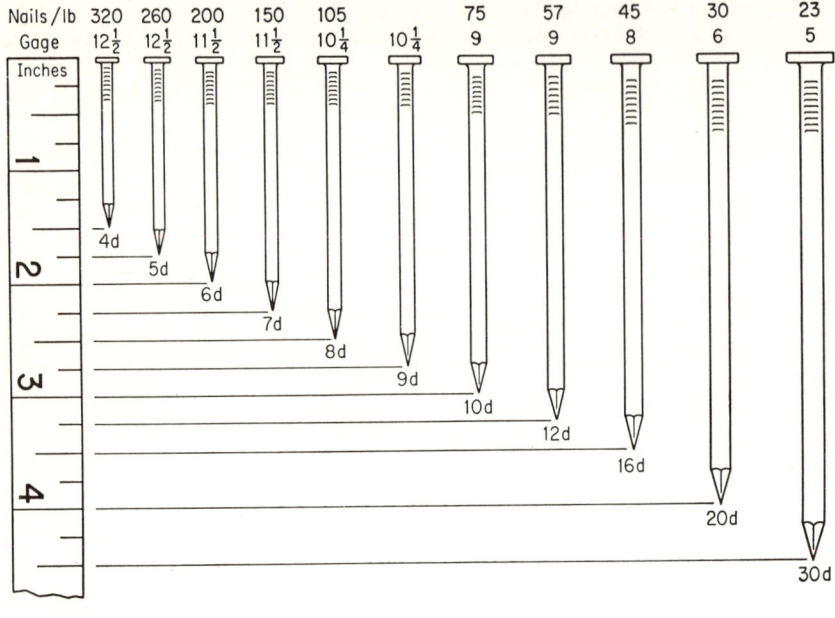

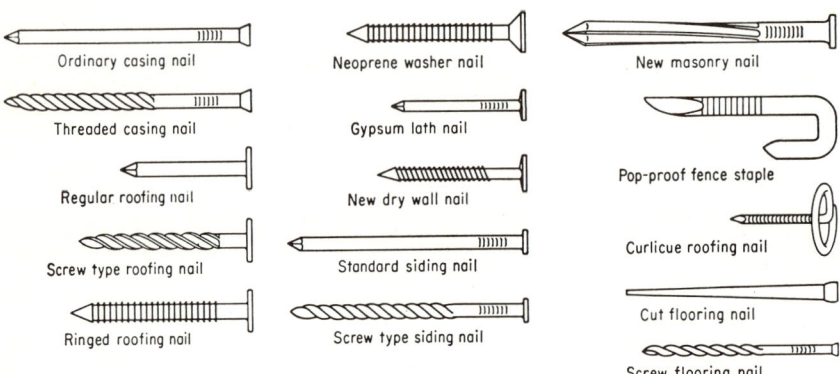

Fig. 46-16. Types and sizes of common nails.

content changes in the wood. The resistance of a nail to direct withdrawal is also intimately related to the density or specific gravity of the wood, diameter of the nail, and depth of penetration.

For bright common wire nails driven into the side grain of seasoned wood that will remain dry, or unseasoned wood that will remain wet, the *allowable withdrawal load* is given by the following formula:

$$p = 1150 G^{5/2} D$$

where p = allowable load (1/6 of ultimate load) per lineal inch of penetration of point into holding member, lb
G = specific gravity of wood based on oven-dry weight and volume
D = nail diameter, in.

The average allowable withdrawal *loads* under these conditions are shown in Fig. 46-17.

The common *surface treatments* for nails, such as properly applied cement coating, may double the resistance of the nails to withdrawal immediately after they are driven into the softer woods. The increase in withdrawal resistance of such nails is not permanent but drops off about one-half after a month or so in the softer woods. Special coatings such as zinc are primarily for corrosion and stain resistance.

Nail shanks have been modified to give an increase in surface area without an increase in nail weight. Special nails with barbs, spirally grooved, annular-grooved, and other irregular shanks are offered commercially. The form and magnitude of the deformations along the shank influence the performance of the nail in the various species. In general, annular-grooved nails sustain larger withdrawal resistance, and spirally grooved nails sustain greater impact-withdrawal values than the common smooth-shanked nails.

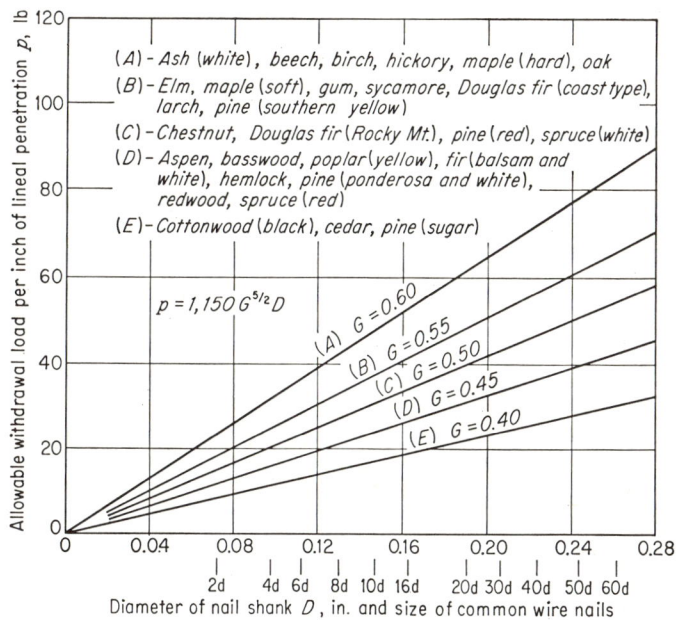

FIG. 46-17. Allowable withdrawal load per inch of penetration for common wire nails driven in the side grain of seasoned wood that remains dry or unseasoned wood that remains wet. Safety factor = 6. (*Wood Handbook, USDA Forest Serv. Bull.* 72, 1956.)

In dry or green wood a *clinched nail* provides from 45 to 170 per cent more withdrawal resistance than an unclinched nail when withdrawn soon after driving. In green wood that seasons after a nail is driven, the clinched nail gives from 250 to 460 per cent greater withdrawal resistance than an unclinched nail. Nails clinched across the grain have approximately 20 per cent more resistance to withdrawal than nails clinched along the grain.

Lateral Resistance. The allowable lateral load for a bright common wire nail when driven into the side grain (perpendicular to the wood fiber) of seasoned wood is expressed by the following general formula:

$$p = KD^{3/2}$$

where p = allowable lateral load per nail, lb
K = constant for wood species
D = nail diameter, in.

Figure 46-18 shows the allowable lateral resistance for conventionally sized nails driven in various species of wood, assuming point penetration of $10D$ in dense wood and $14D$ in lightweight wood.

Nails driven into green wood hold less after the wood is seasoned than those driven into seasoned wood. This erratic behavior of nailed joints with moisture change makes frequent observation desirable, with reinforcement when necessary.

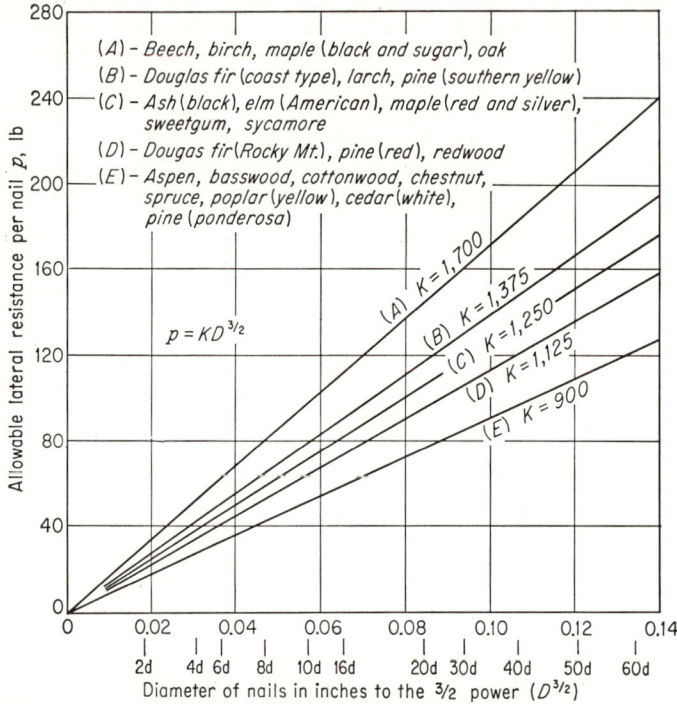

Fig. 46-18. Allowable lateral load of bright, common wire nails when driven into the side grain of seasoned wood. Safety factor = 6. (*Wood Handbook, USDA Forest Serv. Bull.* 72, 1956.)

Wood Screws. The allowable load and lateral resistance for wood screws in the side grain of seasoned wood is given by the formula

$$p = KD^2$$

where p = allowable lateral load, lb
K = constant for wood species
D = diameter of screw shank, in.

Screws should always be turned in. They should never be started or driven with a hammer because the practice tears the wood fibers and injures the screw threads, thereby seriously reducing the load-carrying capacity of the fastener.

Lag Screws. The *allowable direct withdrawal* load of lag screws from seasoned wood may be computed from the equation

$$p = 1500G^{3/2}D^{3/4}$$

where p = allowable withdrawal load (one-fifth of ultimate load) per inch of screw penetration, lb
D = shank diameter, in.
G = specific gravity of wood based on oven-dry weight and volume

STRUCTURAL REQUIREMENTS OF FARM BUILDINGS

Lag screws, like wood screws, require prebored holes of the proper size. Penetration of the threaded part to a distance of $7D$ in hardwoods and $10D$ to $12D$ in softwoods will develop the approximate ultimate tensile strength of the lag screw.

The allowable *lateral loads* for lag screws inserted in the side grain and loaded *parallel to the grain* of a piece of seasoned wood are shown in Fig. 46-19.

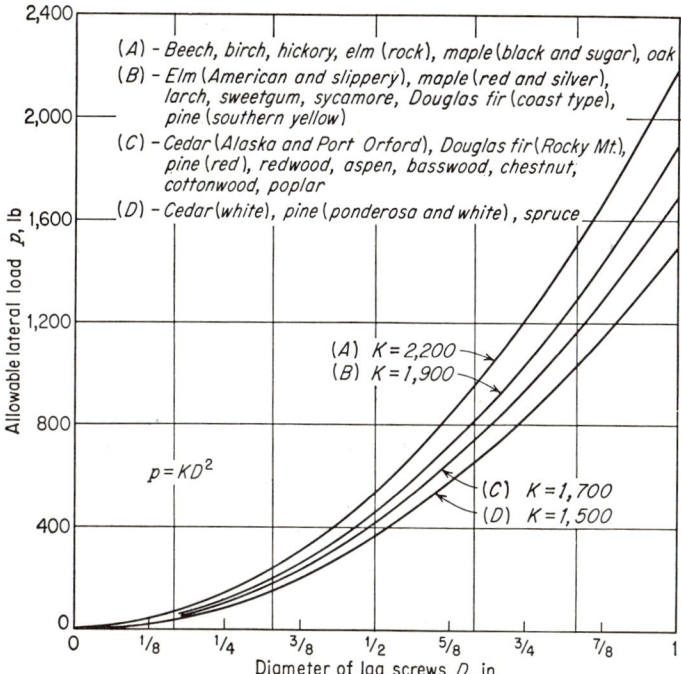

Fig. 46-19. Allowable lateral resistance of lag screws in the side grain of seasoned wood. Safety factor = 6. (*Wood Handbook, USDA Forest Serv. Bull. 72, 1956.*)

When the lag screw is inserted into the side grain of wood and the load is applied *perpendicular to the grain,* the allowable load given by the lateral-resistance formula should be reduced as follows:

Shank diameter, in	3/16	1/4	5/16	3/8	7/16	1/2	5/8	3/4	7/8	1
Strength factor	1.00	.97	.85	.76	.70	.65	.60	.55	.52	.50

The spacings in the edge distances and neck section for lag-screw joints should be the same as those for joints with bolts of a diameter equal to the shank diameter of the lag screw.

Bolts. The allowable basic loads for correctly spaced and aligned bolts, bearing parallel to the grain, are shown in Fig. 46-20. This graph is based on the allowable wood-bearing stresses parallel to the grain as determined and published by the Forest Products Laboratory, Madison, Wis. The various species of woods have been grouped for simplification purposes into four classifications.

For loading *perpendicular* to the grain, Fig. 46-21 has been developed to provide the allowable loads for bolts up to 2 in. in diameter. These loads are based on a diameter factor as well as on the allowable bearing stress perpendicular to the grain.

As the ratio of bolt length to diameter, L/D, increases, the effective unit strength of the wood decreases because of the load concentration as the bolt tips.

For designs with bolts *loaded parallel to the grain,* the following rules should be observed:

1. The edge distance should be $1\frac{1}{2}D$.
2. The end distance in a tension joint for softwoods should be at least $7D$ and for hardwoods at least $5D$. In a compression joint, the end distance should be at least $4D$ for all woods.
3. Center-to-center spacing between bolts should be at least $4D$.
4. Spacing of bolts across the grain is controlled by the reduction in area at the critical section. The net tension area remaining at the critical section should be 80 per

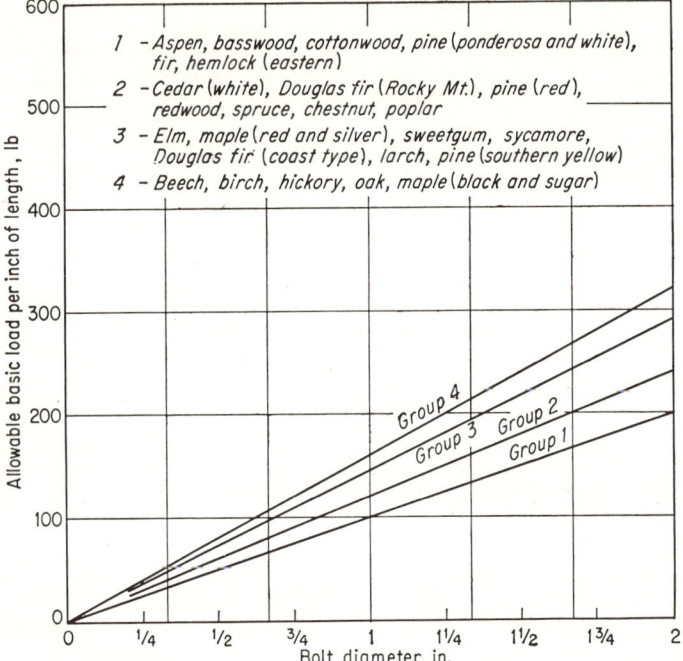

FIG. 46-20. Allowable basic loads for bolts per inch of length in the main member with loading parallel to grain. (*Wood Handbook, USDA Forest Serv. Bull.* 72, 1956.)

cent of the total area in bearing under all the bolts for softwoods and 100 per cent for hardwoods.

For loads perpendicular to the grain, the minimum distance from the edge toward which the load acts should not be less than $4D$. At the opposite edge the margin need only be sufficient to prevent splitting. The minimum center-to-center spacing across the grain for loads applied through metal side plates is governed by design requirements of the side piece and space for nut tightening. If wood splice plates are used, spacing is determined by design requirements for the splice plate.

Timber Connectors. There are three types of timber connectors that are used to some extent. They are (1) the *split ring,* (2) *toothed ring,* and (3) *shear plate.* The split-ring connector is, however, the most commonly used of the three and is the most adaptable to farm building construction. Split-ring connectors work particularly well in truss construction where the strength of the truss depends particularly upon the strength of the joints, but is not necessarily dependent upon whether the fastener pivots.

The safe load per connector is affected by (1) the type and size of connector, (2) the species of wood, (3) the thickness of pieces joined, (4) the spacing of connectors and the end and edge distances, (5) the direction of pull with respect to the grain, and (6) the number of faces of a piece which contains connectors.

The basic allowable loads per ring connector in high-strength woods have been determined primarily from performance tests and are given in Fig. 46-22. These loads must be modified by load factors from Table 46-13 for other woods.

Additional load factors must be introduced for joint dimensions based upon the spacing of rings and the edge and end distances. These load factors for joint design

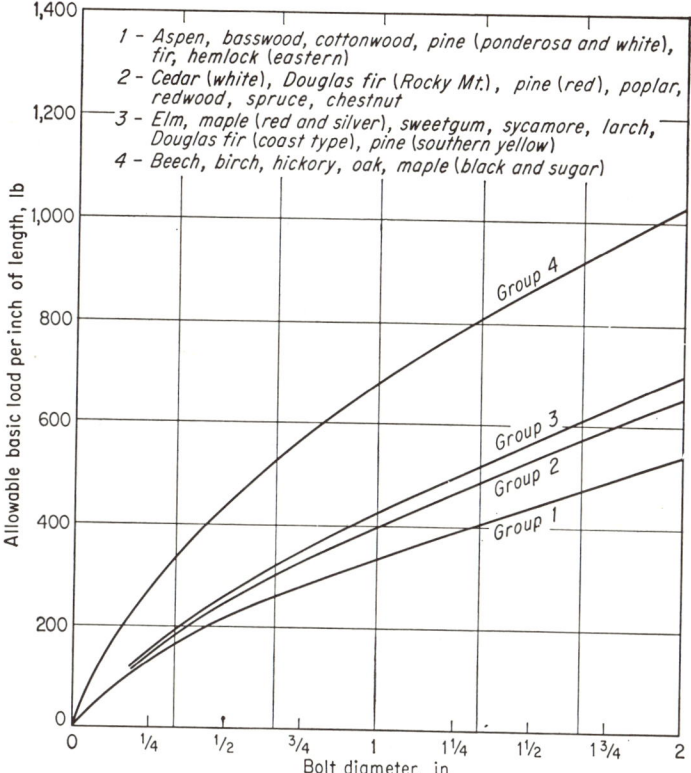

FIG. 46-21. Allowable basic loads for bolts per inch of length in the main member with loading perpendicular to the grain. (*Wood Handbook, USDA Forest Serv. Bull.* 72, 1956.)

are shown in Fig. 46-23. Load factors based on joint dimensions are not cumulative in their effect on the basic allowable load. Thus the smallest factor becomes the controlling load factor for that particular joint.

For most farm building construction, the 2½-in.-diam split-ring connector will be satisfactory. The 2½-in. ring connector can be used in 2- by 4-in.-dimension lumber. It will be noted from the edge-distance chart (Fig. 46-23a) that for a 2½-in. ring in 2- by 4-in.-dimension material, the load factor will be approximately 0.95 where the angle of grain is 15° or less. On the end-distance chart (Fig. 46-23b) it will be noted that 4½ in. from the end of the member is required for full strength where the angle of grain is 0°. Spacing between 2½-in. ring connectors should be at least 6¾ in. for full strength where the grain is parallel.

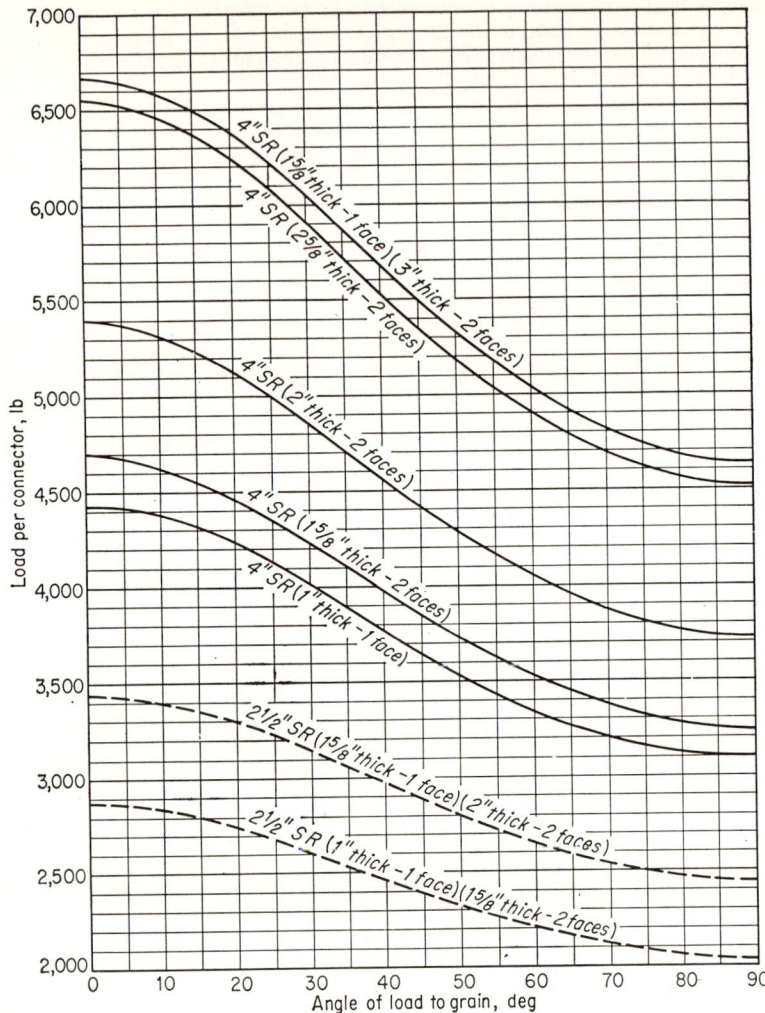

Fig. 46-22. Basic allowable loads for 2½-in.- and 4-in.-diam split-ring connectors (based on safe-load charts). (*TECO Design Manual for TECO Timber Connector Construction*, Timber Engr. Co., Washington, D.C., 1943.)

Glue. Glue was first used as a fastener or bonding agent in farm building construction in the fabrication of *laminated bent rafters*. In 1932, tests by Giese and Anderson[1] at Iowa State University showed that laminated rafters made of five 1 by 4's nailed together were about 3½ times as stiff when also glued. Bolts increased the stiffness of glued specimens only about 4 per cent, but increased unglued specimens about 40 per cent.

These studies were continued,[2] and it was concluded that design values for shear loading on glued joints parallel and perpendicular to the grain of the wood are 430 and 215 psi, respectively.

[1] Henry Giese and E. D. Anderson, Tests of Laminated Bent Rafters, *Agr. Eng.*, 13:11–13, January, 1932.

[2] Henry Giese and S. Milton Henderson, The Structural Application of Glue in Framing Farm Buildings, Part II, *Agr. Eng.*, 26:507–510, December, 1945.

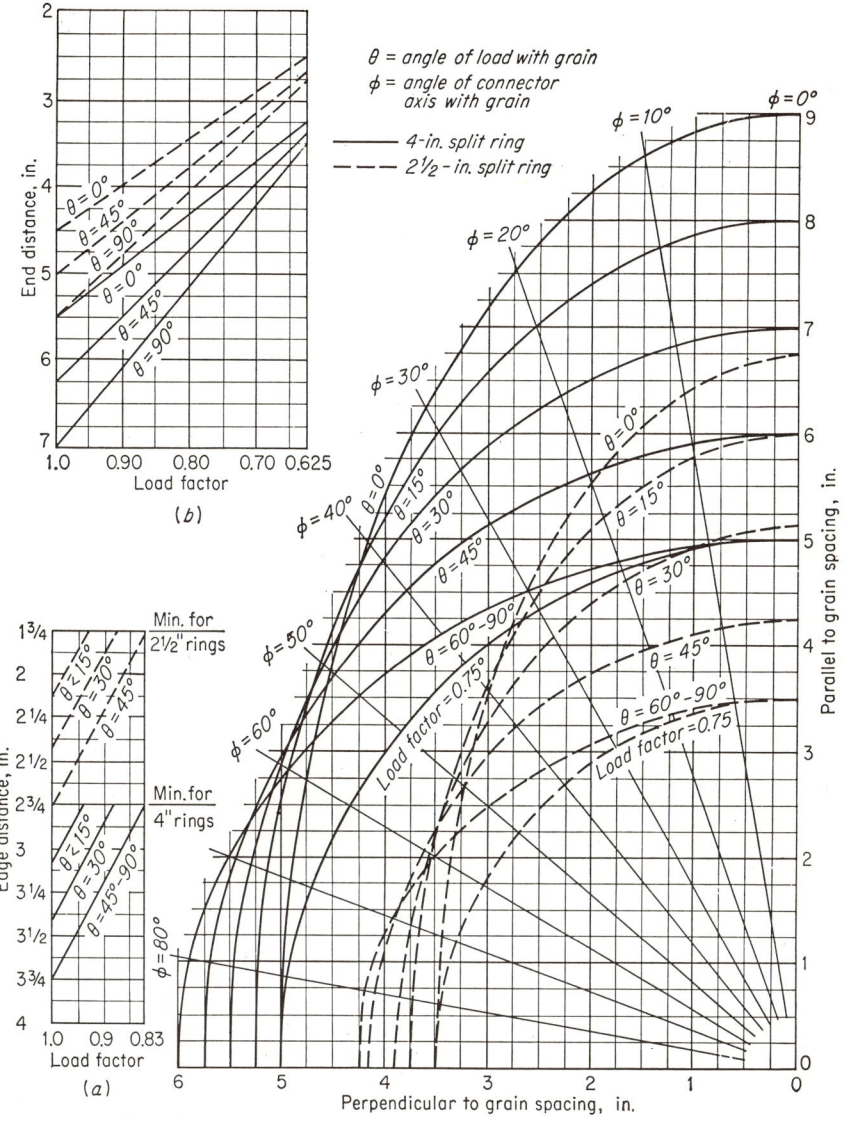

Fig. 46-23. Connector-joint details (based on spacing charts). (a) Edge-distance charts. (b) End-distance chart (compression). (For tension use curve $\theta = 90°$.) (c) Spacing chart. (*TECO Design Manual for TECO Timber Connector Construction, Timber Engr. Co., Washington, D.C., 1943.*)

Tests on the effects of moisture-induced dimensional changes in wood on the strength of glued joints[1] showed that, in lumber ranging up to 8 in. in width and 2 in. in thickness, strength was not significantly impaired.

Tensile stresses on the glue line caused by warping of the members are often great

---

[1] Henry Giese and Elwin D. Palmer, The Structural Application of Glue in Framing Farm Buildings: Part III, Effect of Dimensional Changes in Wood on Strength of Glued Joints, *Agr. Eng.*, 131: 455–464, September, 1950.

TABLE 46-13. SPLIT-RING CONNECTOR LOAD FACTORS FOR VARIOUS SPECIES

Connector group	Species			Load factor (split ring)
A	Ash, white Beech Birch Douglas fir (dense)	Elm, rock Hickory Pecan Maple, hard	Oak, red and white Pine, southern (dense)	1.00
B	Douglas fir (coast) Elm, soft	Larch, western Maple, soft	Pine, southern Gum, black or red	0.85
C	Cypress, southern and tidewater red	Hemlock, West Coast Pine, Norway Redwood	Poplar, yellow Spruce, eastern Spruce, Sitka	0.72
D	Cedar, western red Fir, commercial white Hemlock, eastern	Pine, ponderosa Pine, sugar Pine, eastern white	Pine, western white Spruce, Engelmann	0.62

SOURCE: "National Design Specifications for Stress Grade Lumber and Its Fastenings," National Lumber Mfrs. Assoc., Washington, D.C., 1944.

enough to cause partial failure of the wood fibers near the edges of glued joints made from 2 by 6 and 2 by 8 members.

Figures 46-24 and 46-25 provide some performance information on the use of casein glue as a fastener. Figure 46-24 indicates a linear relation between the average load

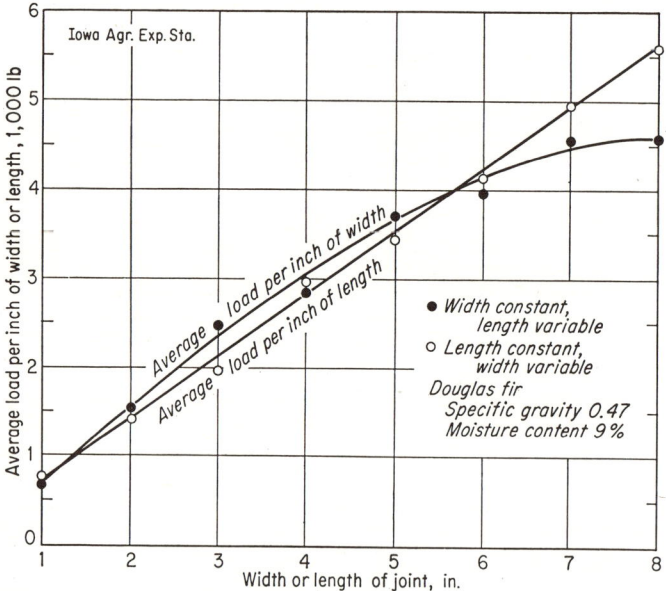

FIG. 46-24. Effect of width and length on the strength of glued joints. (*Giese and Hamlin, Agr. Eng., July,* 1952.)

per inch of length when the *length* of the spliced members are constant and the width variable. The other curve on this graph indicates that the average load per inch of width drops off significantly after the joint becomes 5 or 6 in. in length. Figure 46-25 shows the relationship between slippage and load for various types of fasteners. Here the significant characteristic of glued joints is emphasized by the fact that no slippage occurs in a glued joint until failure, which occurs at a considerably higher stress load than for any other type of joint.

Where extreme moisture conditions are anticipated in a building, waterproof glue may be justified. The resorcinal resin glues are more durable under adverse conditions, although higher-priced. The Grade A, or so-called aircraft casein, glues are recommended for water-resistant application.

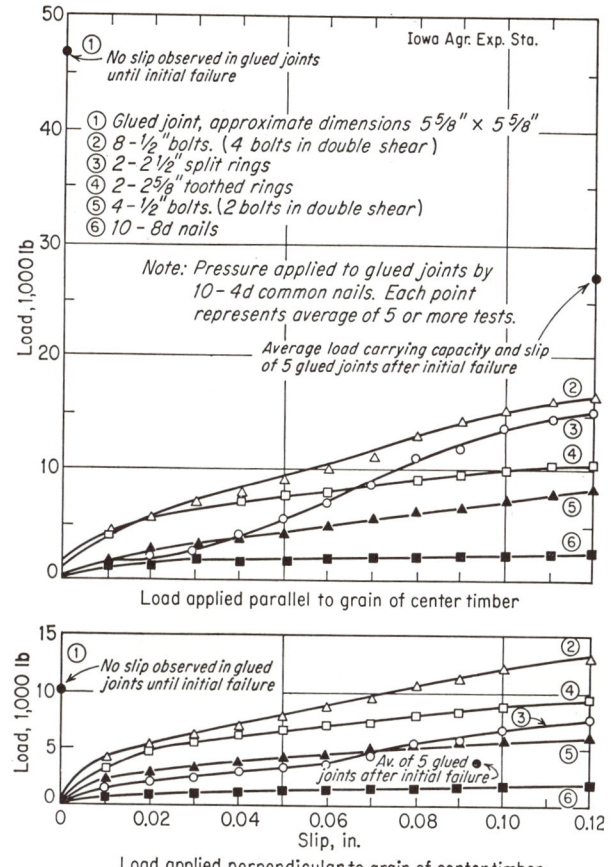

FIG. 46-25. Load-slip curves for timber joints. (*Giese and Hamlin, Agr. Eng., July*, 1952.)

FRAMING

Farm buildings must be framed to provide a shelter for storage and/or housing purposes. The trend is toward single-story buildings with clear-span roof construction. This provides a building that is entirely flexible in so far as utilization is concerned. The basic building shell should be adaptable for open-type housing or for insulated controlled-temperature housing with a ventilation system. The clear-span feature of the building shell allows maximum flexibility in the arrangement of mechanical equipment whether it be for dairy, poultry, hogs, or for storage purposes.

The building shell must be designed for rigidity and durability. In the past, many building frames were built with greatly oversize square wood beams and columns, but they were fastened together very poorly with mortise and tenon joints and dowel pins. Fastenings of this type were mainly compression-type joints which would develop only a fraction of the potential strength of the wood used in the various members. Engineered farm buildings provide balanced designs which are more economical in the use of materials and more rigid in resisting loads from all directions.

Prefabricated steel buildings have comparatively light frames covered with a single thickness of formed sheet metal. In some designs, the covering contributes greatly to the over-all strength of the structure. One type of steel building with an arched roof and spans of 40 ft or more in width is designed so that formed sheet-steel arch segments act as both frame and covering when bolted together. Small storage buildings with capacities up to 2,000 or 3,000 bu are constructed in the cylindrical form with practically no framing.

Walls. The old wood-post-and-beam construction provided a skeleton for the building, but additional framing was required for the application of siding and sheathing

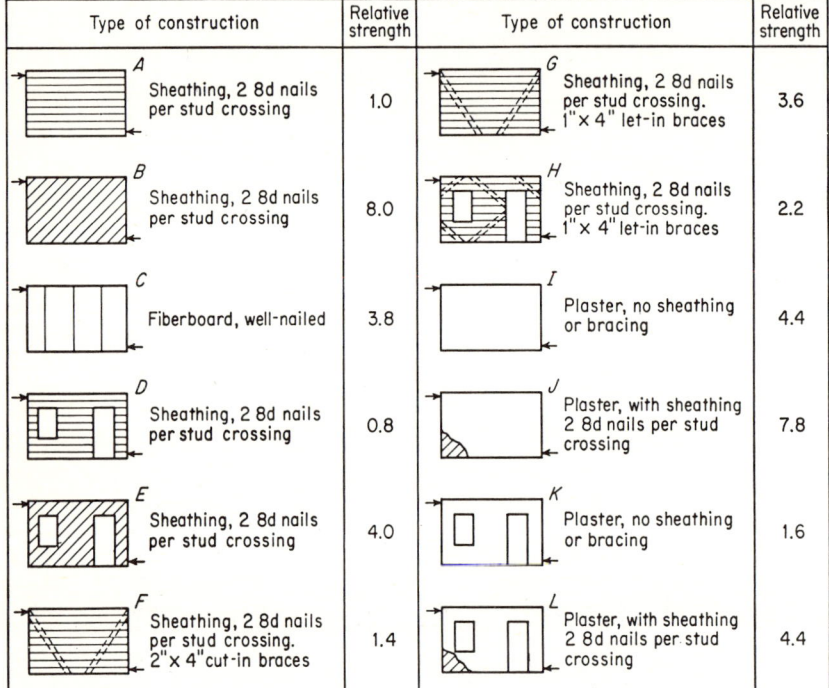

FIG. 46-26. Effect of various types of construction on relative strength of walls. *A, B, D-H,* 1- by 8-in. lumber; *C,* $29/_{32}$-in. fiberboard; *I* and *K,* plaster on wood lath only; *J* and *L,* plaster, wood lath, and 1- by 8-in. sheathing on opposite sides of frame. (*Abstracted by permission from A. D. Freas, Guides to Improved Frame Walls for Houses, Eng. News-Record, October,* 1946.)

materials. Versions of post-and-beam construction are still used in farm buildings, but for the most part, wall construction has changed to the conventional frame or stud type.

A conventional wood-frame wall is constructed of vertical members placed on a wood sill at the bottom, and normally topped with a double plate. The frame wall is conventionally anchored to a continuous concrete foundation wall. The entire wall frame and roof frame for two-story barns are constructed in this way with nothing other than 2-in.-dimensional lumber. With rectangularly spaced framing, rigidity parallel to the wall must be provided by diagonal braces, diagonal siding, or by sheet-type sheathing. Figure 46-26 shows the comparative relative strength of various types of frame-wall construction.

In frame construction, the anchorage of the wall-frame members to the sill and of the rafters to the wall frame is critical. Conventional toenailing is fairly good if done properly, but does not provide adequate resistance to the wind uplift and thrust stress common to light frame buildings. Figure 46-27 shows the comparative strengths of five other techniques of fastening wall studs to sills as compared with conventional toe-nailing. In frame-type grain-storage buildings, particularly, resistance to outward thrust on the studs at the sill line is critical and some additional type of fastening at these joints is required.

Figure 46-28 shows various ways of increasing the anchorage of the rafters to the frame wall plates.

With the trend toward simple one-story farm buildings, which must be depended upon to pay their own way economically, there is a renewed interest in *pole construction*. Pole construction is somewhat similar to post construction, but the poles extend down into the ground at least 4 ft instead of setting on individual concrete-pier-type

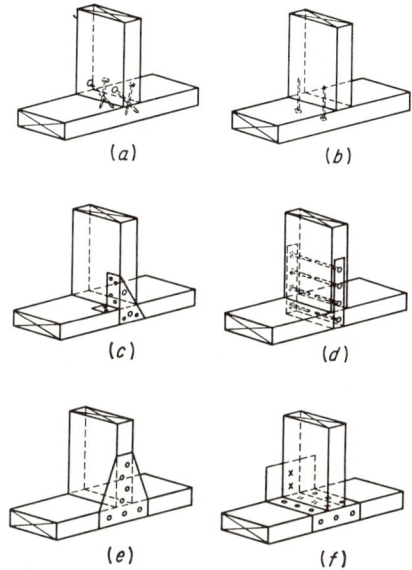

Fig. 46-27. Stud-to-sill fasteners. (a) Toenailing. Uplift = 100 per cent, thrust = 100 per cent. (b) End nailing. Uplift = 70 per cent, thrust = 75 per cent. (c) Trip-L-grip. Uplift = 200 per cent, thrust = 150 per cent. (d) U-strap iron. Uplift = 510 per cent, thrust = 170 per cent. (e) Gusset plate. Uplift = 220 per cent, thrust = 120 per cent. (f) Thrust plate. Uplift = 150 per cent, thrust = 200 per cent. (*Tests made by USDA Forest Products Laboratory. Scholten and Molander, Agr. Eng., November, 1950.*)

footings. The poles thus provide much rigidity for the building frame. Pole-frame buildings are of particular interest because they eliminate much of the concrete footing and foundation required for other types of construction and in general simplify construction of the building throughout. The poles must be pressure-impregnated with a preservative in order to secure a length of life near the expected life of the other materials placed in the building. Figure 46-29 illustrates a typical pole building which is open on one side and has a row of poles down through the center. This type of building might be used for machinery storage or for loose housing of cattle. The anchorage of the roof to the poles is of particular importance in this type of construction.

A pole building similar to the one shown in Fig. 46-29 can be made more flexible and functional by eliminating the center row of poles by using *trussed-roof construction*, as shown in Fig. 46-30. This type of construction provides an economical way of eliminating all center supports for widths up to 40 ft. In livestock loafing barns, it is important that mechanical cleaning equipment such as tractors and hydraulic scoops be used effectively. For machinery storage, the absence of any center obstructions allows maximum utilization of the floor area. Self-feeding hay-storage buildings with drying systems are more efficient and convenient if constructed without interior supports.

Concrete pads are necessary under the poles (Fig. 46-30) if straight eave and ridge lines are desired. Uneven settling of poles has resulted in many unsightly, poorly constructed pole buildings which reflect adversely on all pole-building construction. Pole

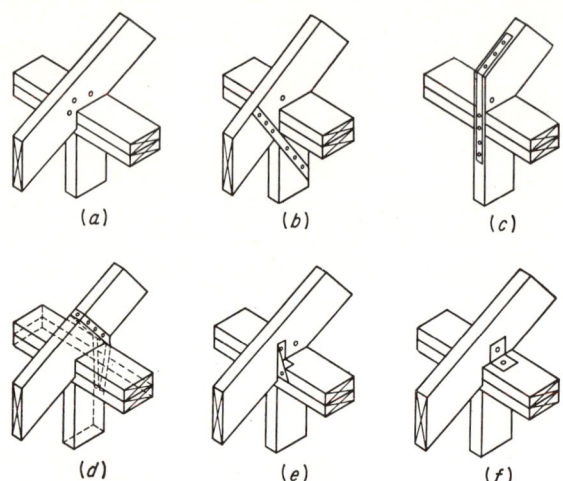

Fig. 46-28. Rafter-to-plate fasteners. (a) Conventional toenailing. Uplift = 100 per cent, thrust = 100 per cent. (b) Diagonal strap. Uplift = 140 per cent, thrust = 80 per cent. (c) End strap. Uplift = 110 per cent, thrust = 70 per cent. (d) U strap over top inside. Uplift = 280 per cent, thrust = 260 per cent. (e) Trip-L-grips. Uplift = 160 per cent, thrust = 80 per cent. (f) Clip angles and lag screws. Uplift = 300 per cent, thrust = 300 per cent. (*USDA Forest Products Laboratory.*)

Fig. 46-29. Typical pole machine shed. (*Michigan State University.*)

buildings need not be cheap, unsightly buildings, but may, with proper engineering and construction methods, be attractive, functional, economical, and long-lived structures.

Roof Framing. Wood rafters of nominal 2-in. thickness placed 2 ft on centers have been the conventional means for framing simple gable-type farm buildings. The two-story *gambrel-roof barn* found its way onto nearly every farmstead in the midwestern United States. This building served as a general-purpose barn, with horses and cows kept in stalls on the first floor and feed and bedding stored overhead in the loft. It was commonly 30 to 36 ft in width for the accommodation of two rows of stalls, with the double-rafter gambrel roof shown in the framing-detail illustration of Fig. 46-31. The dimensions of various members of the roof construction are given in the table of Fig. 46-31. This type of structure provided a large overhead storage space for loose hay and bedding, permitting feeding of the livestock housed on the first floor without going outside of the building.

Another version of the two-story barn is one constructed with a *gothic arched roof*. These buildings are constructed with arched laminated rafters. The introduction of glue for fastening the laminations was responsible for making this type of construction rigid, durable, and practical. Many of the earlier arched-roof barns constructed without glued rafters developed sagging ridge lines because of the lack of rafter rigidity.

The *clear-span wood-trussed roof* provides an economical one-story gable-roofed farm building. This type of construction can be competitive in cost, with convention-

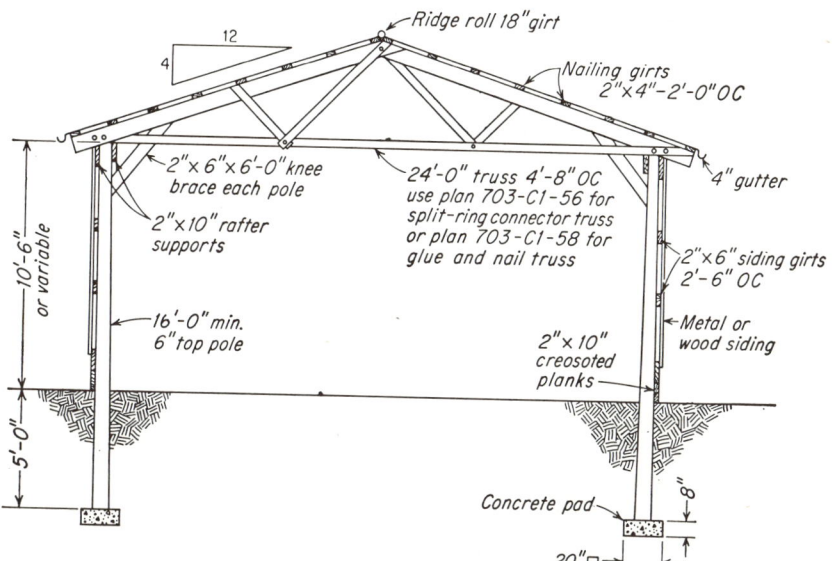

FIG. 46-30. Cross section of pole construction with clear-span roof trusses. (*Michigan State University*.)

ally framed buildings having posts or poles down through the center. Economy is made possible by balanced design, using fasteners which will develop the potential strength of the wood members.

Figure 46-32a shows a poultry house with a 36-ft roof truss constructed with bolts and ring connectors. Figure 46-32b shows a truss of the same dimensions, but with the joints held by plywood gusset plates which are glued and nailed to the members. This particular truss is illustrated on a masonry wall.

For farm buildings of from 24 to 40 ft in width, *roof trusses* designed for approximate 4-ft spacing are economical and practical. Four span widths, 24, 30, 36, and 40 ft with $\frac{1}{6}$ *pitch*, have been found adequate to fulfill the requirements of most types of farm buildings. A typical detailed truss plan based on 2,000 psi for the wood members is shown in Fig. 46-33. The trusses are readily adaptable to pole construction as well as to frame and masonry. Trusses designed for 4-ft spacing, for the most part, can be fabricated on the site in a horizontal position and raised into position on the building by two to four men or with a tractor loader. The approximate 4-ft spacing of trusses allows a purlin-deck construction of 2 by 4's laid flatwise 2 ft on center for the application of corrugated-sheet-metal roofing.

In *steel-fabricated farm buildings,* the clear-span feature is gained through the use of light steel trusses, rigid steel frames, arched steel ribs, or frameless arched-roof construction which gains its strength and rigidity from the shape of the roof sheets alone.

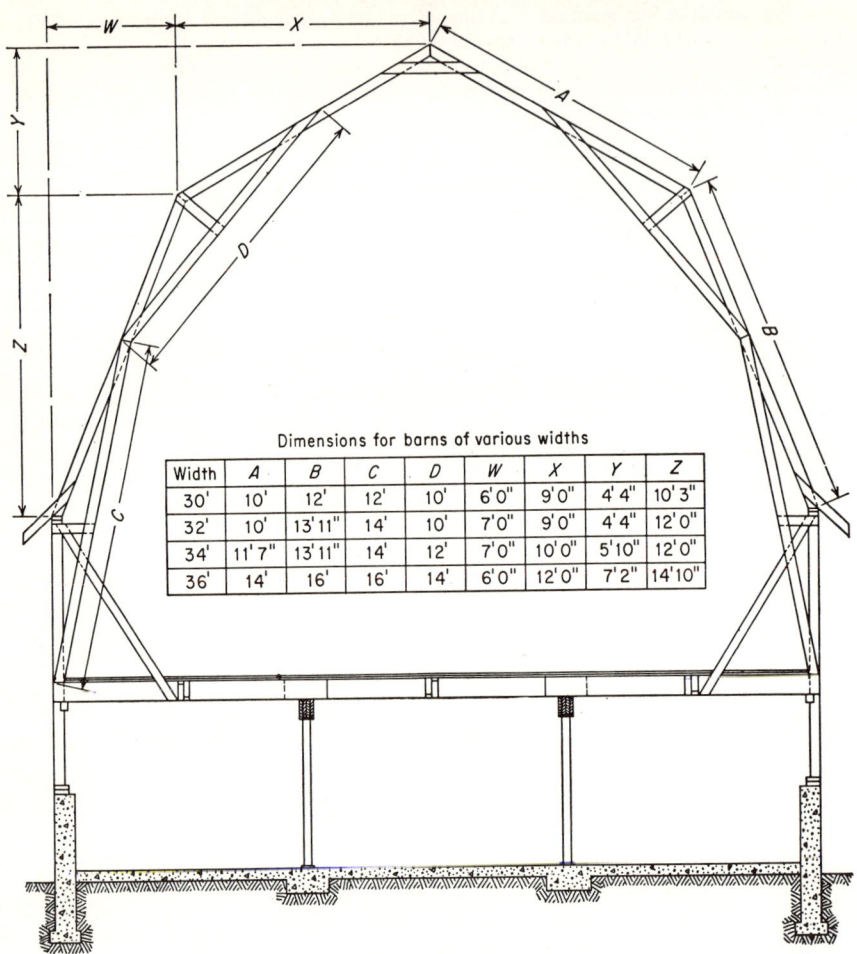

Fig. 46-31. Dimensions for gambrel-roof barns of various widths. (*Michigan State University.*)

Flat roofs, clear-span and of fireproof construction, may be obtained with *precast concrete joist systems.*

FOOTINGS

Knowing the structural load of a building, the required footing area can readily be determined by referring to Table 46-14.

TABLE 46-14. SAFE BEARING CAPACITIES OF SOIL FOR FARM SERVICE BUILDINGS

Class of soil	Carrying capacity, tons/sq ft
Soft clay, sandy loam, or silt	1
Ordinary clay or sand	2
Moderately dry clay or fine and dry sand	3
Hard dry clay or coarse firm sand	4
Gravel	6
Soft rock	8
Hard rock	15

For most frame-and-masonry-type farm buildings, a *continuous wall footing* of concrete or reinforced concrete will suffice. In general, it is satisfactory to make this continuous footing 8 in. wider than the foundation wall, projecting out 4 in. on each side, with a depth of twice the projection. In poorly drained locations or in soils with high coefficients of expansion and contraction with changing moisture content or otherwise unstable, this continuous footing should be reinforced with at least two ½-in. round reinforcing rods for all types of farm service buildings. Footings should extend below the normal frost line.

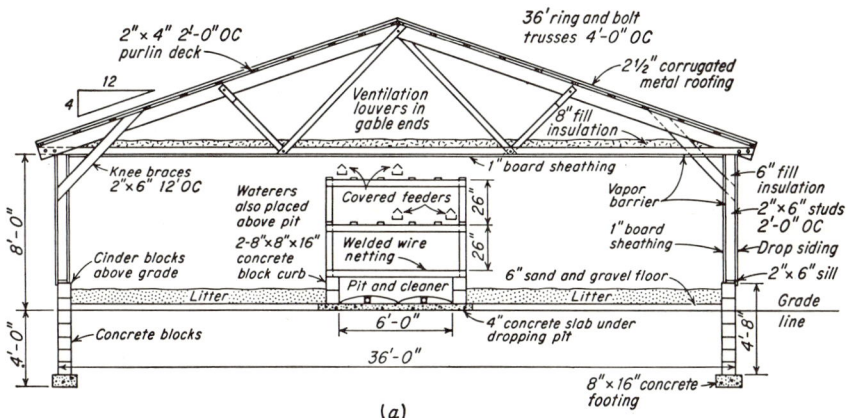

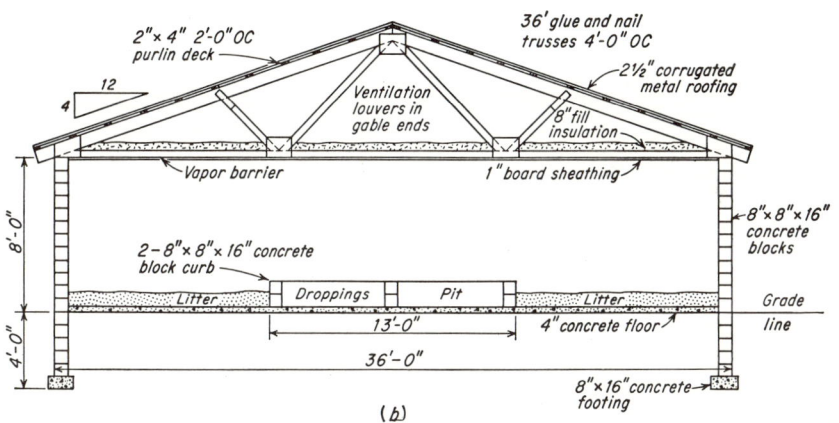

Fig. 46-32. Poultry-house cross sections: (a) frame wall with ring connector trusses; (b) masonry wall with glued and nailed trusses. (*Michigan State University.*)

In post-and-pier-type construction, separate footings are required. In the design of spread-type footings with plain concrete, the depth, as with the continuous footing, must be twice the horizontal outward projection from the foundation pier. For greater footing areas, the concrete should be reinforced with steel.

Footings are essential in pole construction in order to prevent uneven settlement of the poles, causing excessive stresses on many framing members and structural failure in some cases. Studies and experience indicate that concrete footing pads placed at the bottom of the holes underneath the poles will eliminate uneven settlement under most soil conditions. It is necessary to dig an 18-in.-diam hole for the poles in order to receive a pad of concrete 18 in. in diameter and 8 in. deep.

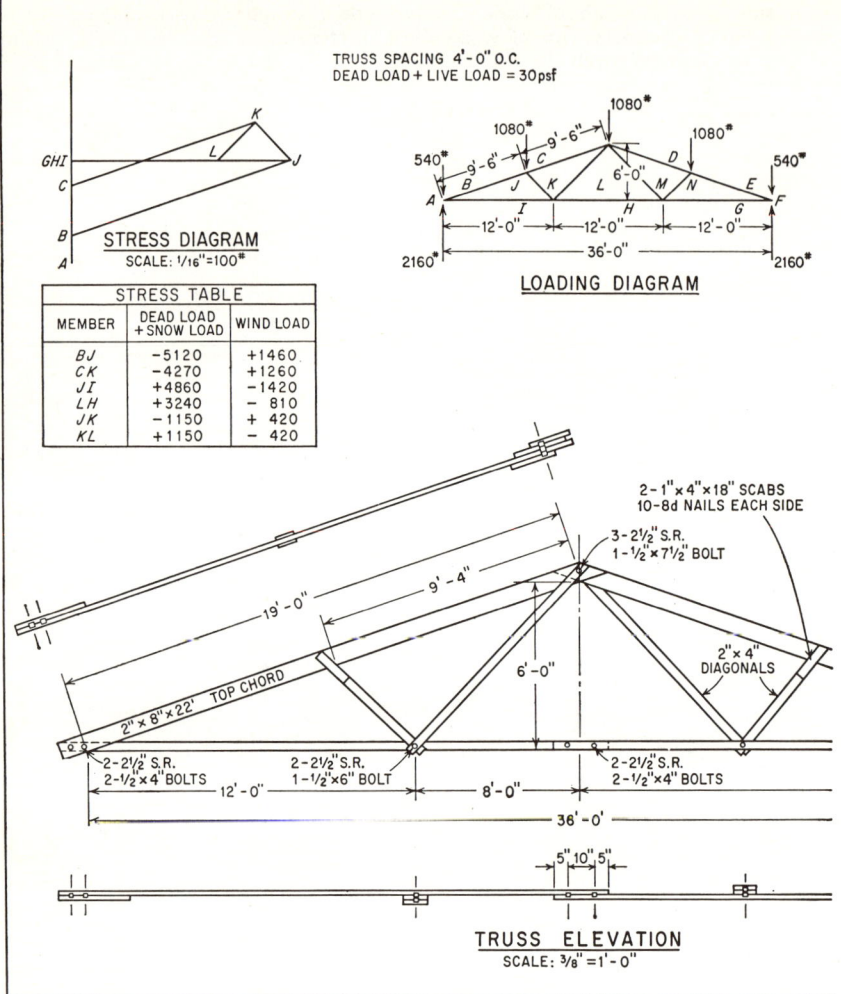

Fig. 46-33. Typical drawing

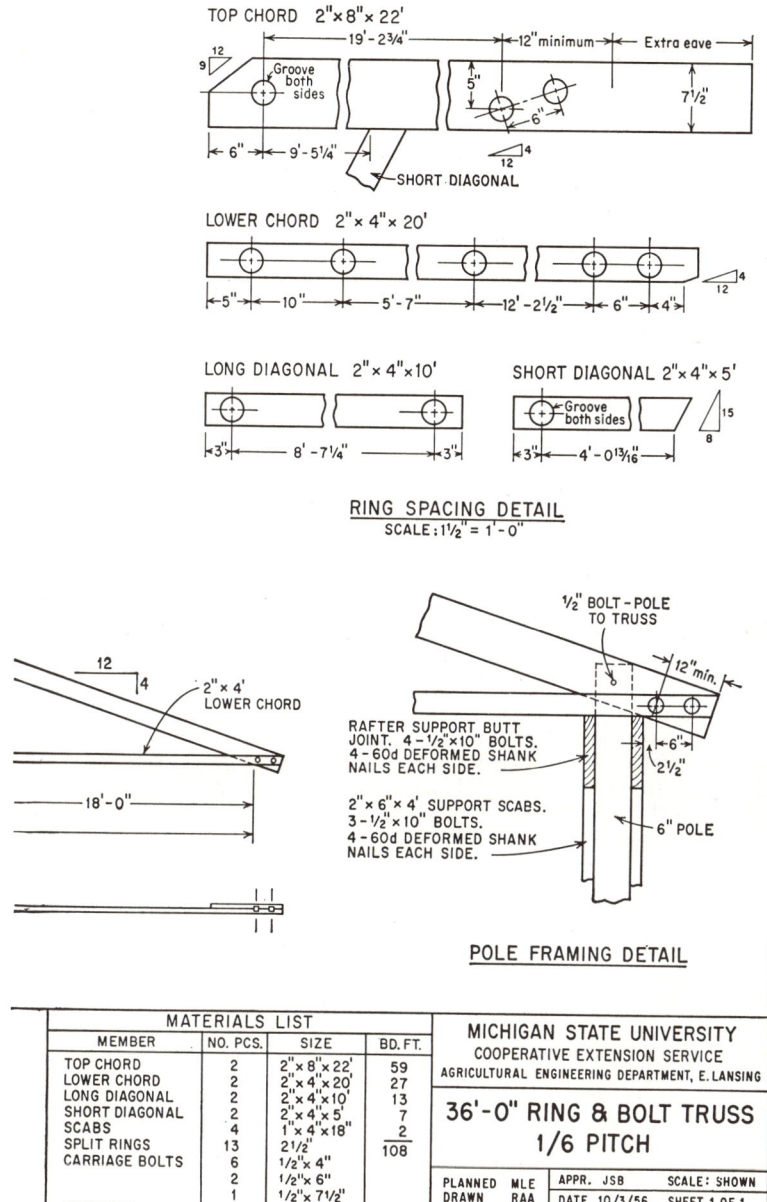

of roof-truss construction.

ROOFING MATERIAL

Any roofing material put on a permanent farm building should last at least 20 years with little maintenance. Lack of *wind resistance* is a serious weakness of some otherwise satisfactory roofing materials. Many farm buildings are located in unprotected places on open farmsteads, making wind resistance a critical factor. A study of wind damage to farm buildings in Iowa over a 20-year period showed that roofing losses comprised approximately one-third of the losses claimed and one-sixth of the total amount of wind damage of all types.[1]

The danger of roof fires caused by sparks from chimneys makes *fire resistance* an essential requirement for farm dwelling roofs. The annual cost is a much better criterion than first cost. Some materials require a solid deck, some a slatted deck, and some only a spaced purlin-type deck, affecting over-all cost accordingly. With the prevalent high cost of labor, ease of application is a significant cost factor.

Attractiveness should not be overlooked, since prestige and pride of ownership are important to all people.

Farm buildings are usually roofed with wood shingles, asphalt shingles, or sheet metal. There are various kinds of materials in each of the three types, along with a few other types used occasionally such as *cement asbestos, built-up roofing, roll roofing, tile,* and *slate shingles.*

Wood shingles are usually of western red cedar, redwood, or cypress. For best quality the shingles should be edge-grain to reduce the tendency to curl. They are cut in lengths of 16, 18, and 24 in., tapered, with a butt thickness of about ½ in. and of random widths from 2½ to 16 in. On roof slopes of 6 in 12, wood shingles 16 to 18 in. long should be laid not more than 4 in. to the weather; on slopes of 12 in 12 the exposure may be extended to 5 in. The corresponding exposure for 24-in. shingles is 6 and 7½ in., respectively. Wood shingles are generally applied on a slat-type deck spaced so as to provide a base for the row of nails in each course of shingles. High material and application costs, as well as lack of fire resistance, have reduced the use of wood shingles.

The most common type of *asphalt roofing* on farm buildings is the 12- by 36-in. *strip shingles* of the square-tab and hex-tab design. Asphalt shingles have become a popular farm roofing material, particularly on farm dwellings, because of their *fire resistance, attractiveness,* ease of application, and low first cost. They have, however, presented a problem of high maintenance due to lack of wind resistance. The study in Iowa[2] showed that, on a weighted basis, area for area, the probability of *wind damage* to asphalt-shingle roofing was 18.1 times greater than for damage to wood-shingle roofing. Further studies[3] showed that, through proper selection and application, much of the wind damage to asphalt shingles could be eliminated. The most popular type of asphalt shingle, the thick-butt, thin-top, is not designed for the wind resistance needed for farm building roofing. In a series of physical bending tests representing wind stresses, the uniform-thickness shingles performed better and made more efficient use of the materials from which they were manufactured than did the thick-butt types.

For wind-resistant roofing, thick-butt asphalt shingles should have the following minimum specifications:

1. Butt weight, 115 lb per 108 sq ft
2. Top weight, 75 per cent of butt weight
3. Thick portion, 7 in. long
4. Felt weight, 12 lb per 108 sq ft
5. Filler not over 35 per cent of coating by weight
6. Total mineral not over 50 per cent of total weight

They should be applied over smooth solid deck with six nails per strip shingle placed not more than 5⅝ in. from the butt edge.

---

[1] Merle L. Esmay and Henry Giese, Wind Damage to Farm Buildings, *Agr. Eng.*, 32:275–277, May, 1951.

[2] Roger M. Cleveland, "Wind and Hail Damage to Asphalt Shingles in Iowa," unpublished M.S thesis, Iowa State Univ., Ames, Iowa, 1948.

[3] Merle L. Esmay and Henry Giese, Factors Affecting the Wind Resistance of Asphalt Shingles, *Agr. Eng.*, 33:549–552, September, 1952.

Conventional *sheet-metal roofing* is of the *corrugated* type, with some having various spacing of *V crimps*. The "antisiphon" side-lap design is also common. Metal roofing is made of *steel* or *aluminum*. Both materials are well adapted for farm roofing. They are *economical* because they are easy to apply and can be used with *low-cost decks*. Both the 28-gage steel and the 0.024-in.-thickness aluminum of the 2½-in. corrugation variety can be applied on a 2- by 4-in. purlin deck spaced 2 ft on center. The 2- by 4-in. purlin deck requires rafter supports only every 4 ft on center if applied flatwise and 6 to 8 ft on center if on edge. This gives a low-cost but sound and durable roof if properly applied. Metal roofing materials are good conductors of heat, and insulation must be provided for enclosed warm-type buildings.

The gage, thickness, and weight of conventional *steel* roofing are given in Table 46-15.

TABLE 46-15. CORRUGATED-SHEET-STEEL ROOFING

Gage	Thickness	Weight per 100 sq ft, lb		
		Std. 1¼" corrugation 26" wide	Std. 2½" corrugation 26" wide	Std. 2½" corrugation 27½" wide
22	0.030	...	151	153
24	0.024	125	125	126
26	0.018	98	98	99
28	0.015	84	84	85
29	0.014	77	66	68

SOURCE: "AISC Steel Construction," American Institute of Steel Construction, New York, 1959.

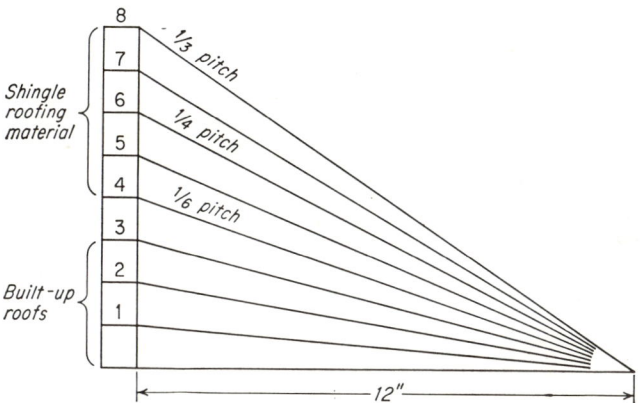

FIG. 46-34. Roofing requirements for various roof pitches.

The quality and durability of steel roofing depend largely on the effectiveness of its zinc coating in preventing rust. Many manufacturers produce steel roofing which carries the American Zinc Institute Seal of Quality. This signifies that the roofing is coated with at least 2 oz of zinc per square foot (1 oz per square foot on each side of the sheet). Sheets with standard commercial coatings will require periodic maintenance, depending upon the geographic area. A good coat of rust-inhibitive paint should be kept over the sheets.

Aluminum roofing is light to handle but is weaker than steel. Therefore not less than a 0.021-in. thickness, preferably in the 2½-in. corrugation type, is recommended for farm buildings. Only aluminum nails should be used with aluminum roofing to prevent electrolytic corrosion.

Sheet-metal roofing is manufactured in 26-in.-wide *sheets* that cover 24 in. of roof and in lengths from 6 to 12 ft. It should be applied with side laps of 1½- to 2½-in.

corrugations and at least 6 in. end lap. Deformed shank nails such as the screw or ring type have considerably more resistance to withdrawal than plain shanked nails.

Slope recommendations for roofing are summed up by Fig. 46-34. Wood shingles should not, however, be used on slopes of less than 6 in 12. Metal roofing may be used satisfactorily on slopes as low as 4 in 12. Built-up roofs are made of successive

TABLE 46-16. TYPE, SIZE, AND QUANTITY OF NAILS REQUIRED FOR DIFFERENT ROOFING MATERIALS

Roofing material	Type of nail recommended	Length, in.		Approx. pounds of nails per 100 sq ft	
		New roofing	Over old roofing	New roofing	Over old roofing
Wood shingles............	Hot-dipped galvanized	3–4d	5–6d	3½–4½	4½–6½
Asphalt:					
Roll roofing...........		1	1¾	¾	1⅛
3-tab square butt.......		1¼	1¾	1⅞	2½
Hex strip..............	Large-headed sharp-pointed hot galvanized wire nails	1¼	1¾	1¼	1⅜
Giant American........		1¼	....	1¾	
Dutch Lap............		1¾	....	1	
Indiana Hex...........		1¾	....	¾	
Steel:					
Corrugated or V-crimp..	Lead-headed or galvanized with lead washers	1¾	2–2½	1½	
Standing seam.........		1	1¾	1⅜	
Aluminum:					
Sheet.................	Aluminum with nonmetallic washers	1¾	2–2½	100 nails	
Roll..................	Aluminum	1	1¾		
Cement-asbestos shingles...	Galvanized needlepoint	1¼	2	1¾	

TABLE 46-17. FACTORS FOR CONVERTING HORIZONTAL PROJECTED AREA TO ROOF AREA

Rise, in./ft of run.........	4	5	6	7	8	9	10	11	12	14	16	18
Pitch, degree..	18°26′	22°37′	26°34′	30°16′	33°41′	36°52′	39°48′	42°31′	45°	49°24′	53°8′	56°19′
Pitch fraction...	⅙	5⁄24	¼	7⁄24	⅓	⅜	5⁄12	11⁄24	½	7⁄12	⅔	¾
Conversion factor........	1.054	1.083	1.118	1.157	1.202	1.250	1.302	1.356	1.414	1.537	1.667	1.803

TABLE 46-18. LABOR FOR APPLYING ROOFING

Type of roof	Man-hours/100 sq ft
Asphalt shingles:	
Strip...	2.0
Single..	2.8
Wood shingles:	
5 in. to weather, 720 shingles per square................	2.6
7 in. to weather, 515 shingles per square................	2.0
Roll roofing...	1.3
Built-up roofing:	
2-ply asphalt..	1.3
4-ply asphalt..	1.0
Sheet metal (galvanized iron)............................	1.0

SOURCE: H. J. Barre and L. L. Sammet, "Farm Structures," John Wiley & Sons, Inc., New York, 1950.

layers of roll roofing cemented together with tar and with gravel on top. They are seldom used for farm buildings but are popular for industrial low pitches and flat roofs.

Recommendations and estimating information for *roofing nails* are given in Table 46-16.

In estimating roofing requirements, the roof area must first be determined. Table 46-17 shows factors for converting the horizontal projected area to roof area when the roof slope is known.

Table 46-18 provides some information for determining labor cost.

LOSS PREVENTION

Fire. Figure 46-35 shows graphically some of the trends of Iowa fire and wind losses for a 20-year period beginning in 1930. A high percentage of farm buildings that catch fire burn to the ground. Figure 46-36 shows graphically the per cent of

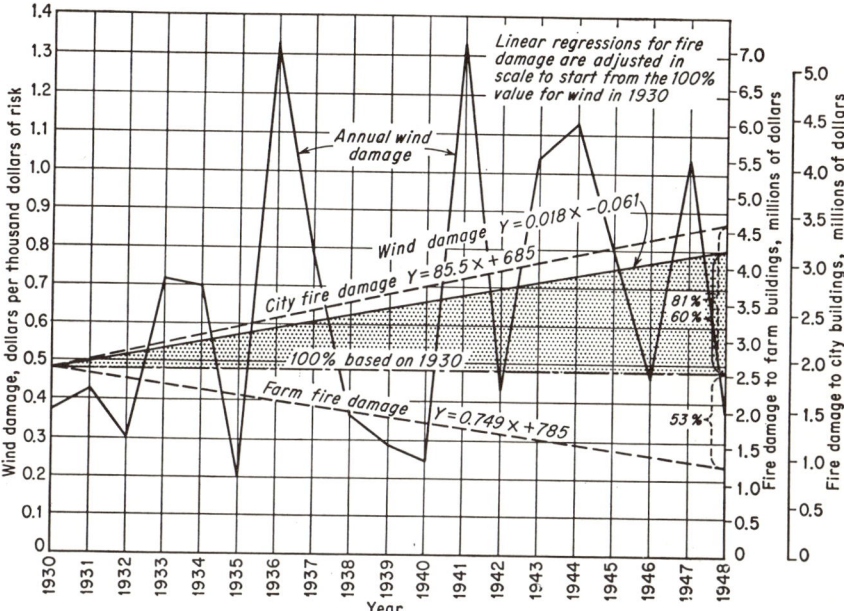

FIG. 46-35. Trend of annual wind and fire damage to Iowa farm and city buildings. (*Esmay and Giese, Agr. Eng., May, 1951.*)

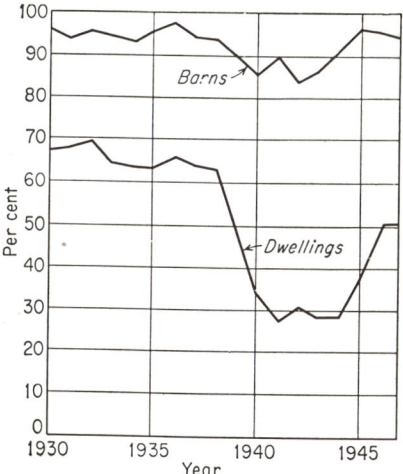

FIG. 46-36. Per cent of building burning to demolition. (*Iowa State University.*)

barns burning to demolition and also the per cent of farm dwellings. In spite of improvements in fire-fighting organization and equipment, the comparative isolation of farms still allows 90 per cent of the barns and 50 per cent of the houses to burn to the ground, emphasizing the importance of fire prevention.

The five major causes of farm-dwelling fires are flues, sparks on roof, heating systems, wiring, and oil and gas stoves. Figure 46-37 shows the trends in farm fire losses by cause. Fortunately, the two major causes of dwelling fires, defective flues and heating system and shingles vulnerable to sparks, can be recognized by a competent in-

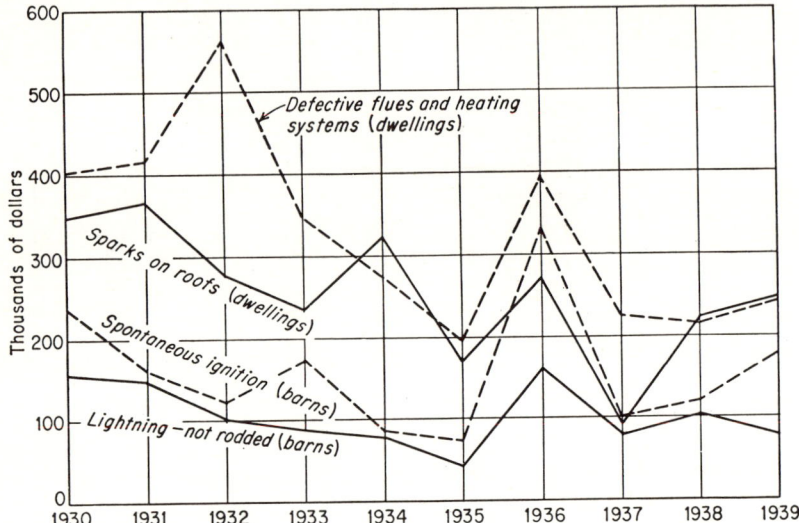

Fig. 46-37. Iowa farm-fire losses from the major causes. (*Iowa State University*.)

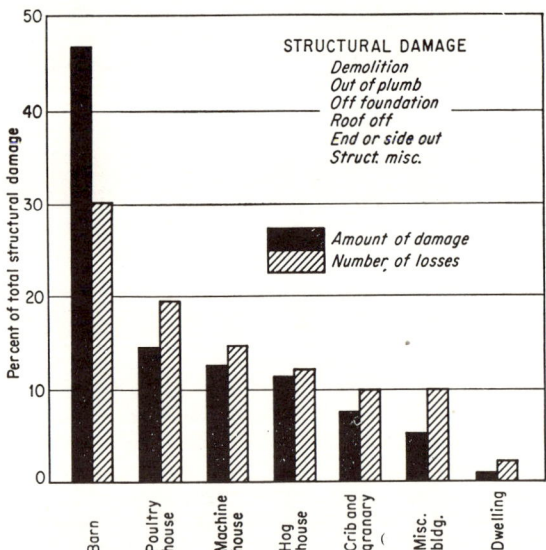

Fig. 46-38. Structural damage of buildings in Oct. 10, 1949, windstorm. (*Esmay and Giese, Agr. Eng., May,* 1951.)

spector. Such inspections made by insurance companies and other interested organizations have been successful to a remarkable degree, both in removing the hazard and in preventing recurrence. Wood shingles lying flat or prepared roofings well protected by mineral surfacing are not easily ignited by embers discharged from the house chimney. As to barn-loss prevention, properly installed lightning rods are credited with being almost perfect protection against lightning, and spontaneous ignition can be prevented through proper storage practices.

Specifications for installation of *lightning protection* on farm buildings as given by the U.S. Department of Agriculture and conforming with the rules given in the National Electric Code are as follows:

Conductors. Cable of commercially pure copper weighing not less than 3 oz per ft shall be used. If galvanized-steel star-section rod with copper-bronze coupling is used, the rod shall be ¾ in. in diameter. The cable or rod shall be coursed as directly as

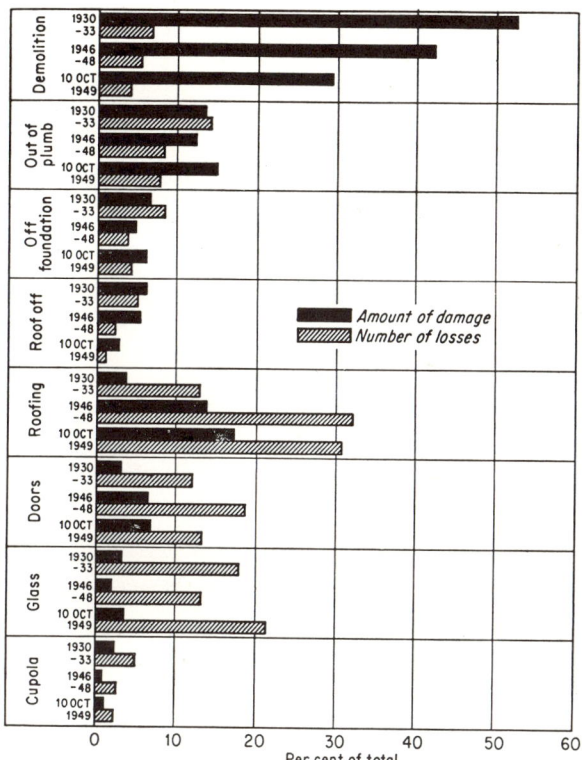

FIG. 46-39. Iowa farm-building wind damage, except hail. (*Esmay and Giese, Agr. Eng., May, 1951.*)

possible over the building with no sharp bends or loops and in such a manner as to connect each air terminal to all others.

Air Terminals. Tube elevation rods or star-section rods shall be used with an inverted Y connection to the main conductor. The walls of the tubing shall not be less than No. 20 gage in thickness, and the star-section rod not less than the diameter of the conductor. Solid-copper bayonet air-terminal points screwed to the upper end of the elevation rod shall be used. Air terminals shall be placed above ridges, gables, and flat roofs, above the tops of chimneys, peaks, and pointed parts of the building.

Ground Connections. Grounds shall be made by driving a ground rod to permanent moisture 8 to 10 ft deep, by drilling and then inserting the ground rod into the hole, or by means of buried copper plate or stranded cable. The down conductor shall be protected against injury by livestock or by other agents by surrounding it with a substantial guard at places designated.

Inner Connecting and Grounding of Metal Work. The gutters and downspouts should be connected with suitable clamps to the lightning conductors, using No. 6 gage copper wire or its equivalent. Pipes and other metal work within the building that extend for a considerable distance parallel to the lightning conductors shall be connected to them at one end and grounded within the building at the other end. No. 6 gage copper wire or its equivalent shall be used, and joints clamped. Small masses of metal in the barns shall be grounded to prevent sparking from induction.

Wind Losses. The trend of wind damage to farm buildings in Iowa has been upward, as shown in Fig. 46-35. Many observations and studies of wind losses to farm buildings established that poor fasteners are the critical weakness. Nails do not adequately tie the members together.

The conventional two-story barn with clear-span mow space for hay storage has been particularly vulnerable to wind damage. The barn is generally the tallest building on the farmstead and usually unprotected from the wind. The typical gambrel-roof construction has at the sill line, plate line, rafter joints, and ridge line nailed joints that may fail in tension on the leeward side of buildings subjected to high-wind stresses.

Figure 46-38 illustrates graphically the proportion of wind damage sustained by the various buildings on the farmstead in a 1949 Iowa windstorm. Barns suffered nearly 50 per cent of the dollar losses, whereas the dwelling loss was only 2 to 5 per cent of the total. This indicates that it is possible to design and construct buildings that are little damaged by winds other than tornadoes. Complete demolition of many barns has been observed where only superficial damage occurred to the house at the same time. Figure 46-39 shows graphically the major items of Iowa farm building damage from wind action.

Chapter 47

LIVESTOCK PRODUCTION FACILITIES

CARL W. HALL

The function of any system of handling livestock is to provide the animals with an environment favorable for efficient production with the least expenditure of human time and effort for animal care. Structures and equipment should be selected to contribute to the highest possible *net* profits. Because of the rapid change in agricultural technology and farm prices, *flexibility* is necessary to avoid losses resulting from obsolescence. The annual fixed cost of a farm livestock building is usually considered as 8 per cent of the original cost, which includes depreciation, interest, repairs, and taxes (D-I-R-T).

ENVIRONMENTAL FACTORS

Feed is the primary factor in animal production. Basic feed requirements for various animals have been well established by feeding trials, with much new information becoming available. Access to feed may be limited, as in batch feeding; unlimited, as in self-feeding; or largely dependent on the animal's exertion, as in pasturing. The objective in feeding is to secure the greatest quantity and quality of production for the cost of the feed consumed.

Water is essential, and free access at all times is normally desirable.

Temperature of the environment can be modified by the type of shelter used. There is considerable evidence that most animals do not need as much protection from cold weather as formerly considered necessary. The hardier strains of both dairy cows and poultry have shown a surprising ability to maintain production in cold weather with open-shed shelters, indicating that some of our buildings were designed more for the farmer's comfort than for the animal's. Experiments indicate that most animals are more adversely affected by 100°F weather than by freezing weather and that shade shelters are of considerable importance in hot weather. Feed efficiency is affected by extremes in temperatures.

Ventilation is no problem where open-shed shelters are used, but tight buildings without adequate ventilation to remove stale, humid, and warm air are definitely undesirable. Ventilation is needed at the ridge during winter to avoid condensation, and cross ventilation is required in summer if the structure is used for shade.

Sanitation requirements for animals or for their products are in some cases rigorous enough to dictate the basic type of handling, as with hogs or turkeys, which may periodically require fresh ground to avoid infection of young animals with virulent diseases, or dairy production where stringent health requirements must be met.

Exercise is an environmental factor controlled to a large extent by the handling and feeding system. Exercise may be undesirable for rapid fattening but essential for breeding stock.

Artificial lighting can be used to extend the length of day to the optimum for the particular animal and type of production. Egg production can definitely be stimulated by this means during the winter.

LABOR OPERATIONS

Livestock production requires more man-hours than crop production on most livestock farms. Eighty per cent of the labor for the production of eggs and milk is centered in and around the buildings, and 40 per cent for hogs. The major labor operations are as follows:[1]
1. Feeding
2. Watering
3. Bedding
4. Manure removal
5. Product collecting and handling, as with milk and eggs
6. Specialized operations on animals such as sorting, castrating, dehorning, dipping, vaccinating, etc.
7. Shifting, sorting, and transporting the animals

Materials handling requires the greatest amount of time and effort, as can be seen by an analysis of the above operations. Batch-type materials-handling processes, the kind most used on the farm, usually consist of the following steps:
1. Loading material into transporting container
2. Transporting
3. Unloading and distributing material

Methods of *loading* in common use are:
1. Manual. Using pitchfork, scoop shovel, bucket, or basket.
2. Elevator or blower. Reduces lifting even though manually fed.
3. Conveyer and elevator. Collects and lifts, as with barn cleaner or corn drag in crib.
4. Gravity. Requires overhead storage bin but little manual effort.
5. Tractor loader. Requires operator but reduces effort and is much faster.

Methods of *transporting* in common use are:
1. Manual. Pitchfork, bushel basket, or bucket.
2. Feed cart. On hard floor or tracks.
3. Wagon—pulled by animals or tractor.
4. Manure spreader.
5. Truck.
6. Tractor. With loader or rear platform.
7. Conveyer or blower.

Methods of *unloading* and distributing are:
1. Manual. Scoop shovel or pitchfork.
2. Gravity-feed cart on tracks above bunks. Dump body wagon or truck or tractor loader, dumping directly into bunks.
3. Power-unloading conveyor bottom with cross conveyer at end to unload into feed bunks or self-feeders. Manure spreader has scattering beaters.

The major *laborsaving principles* applicable to feed handling can be classified as follows:
1. Process material to reduce volume and facilitate handling such as baling or chopping hay, field-shelling corn.
2. Store as near to point of use as possible to eliminate further transport.
3. Self-feed directly from storage if possible. Corn cribs in hog pasture, ground-floor hay storage, horizontal silos.
4. Use conveyers from storage to feeding troughs, where justified.
5. Store feed in overhead bins for gravity removal and loading.
6. Combine operations such as grinding, mixing and loading feed.
7. Use overhead or bunk-track carriers for small or medium herds. Locate bunks between silo and grain storage.

[1] Wallace Ashby, What's Happening in Farm Buildings? *Agr. Eng.*, 26:101–103, March, 1945.

8. Use power-unloading wagon to deliver into fence-line feed bunks or self-feeders for large operations.

9. Use tractor loader to load manure, to take silage out of a trench silo, or for other heavy lifting jobs. Arrange buildings for efficient manure removal by tractor loader. Deliver silage directly to bunks by tractor loader.

10. Arrange feed storage, feed processing, and feeding areas for efficient circle routes.

11. Handle large batches to reduce travel time.

12. Self-feed or water directly from container used for transport.

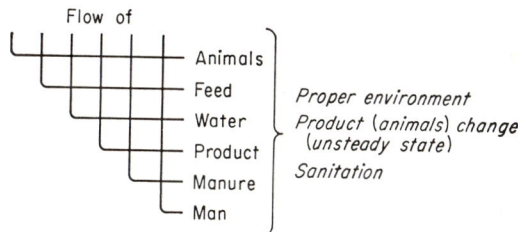

FIG. 47-1. Flow study of materials. (*Kelly, Bond, and Ittner, Engineering Design of a Livestock Physical Plant, Agr. Eng., 34:601-607, September, 1953.*)

A method of analyzing the requirements of a livestock feeding plant, making material flow charts of items shown in Fig. 47-1 and arriving at an engineered system for the most efficient operation, has been proposed.[1] The following steps are suggested:

1. Determine the problem.
2. Construct flow charts for materials.
3. Make preliminary arrangement.
4. Select equipment and machinery.
5. Evaluate primary design.
6. Make final selection of equipment and machinery.
7. Design facilities to provide correct environment.

SWINE PRODUCTION STRUCTURES

The hog is a short-haired animal which does not sweat. A minimum temperature of 40°F is desirable, but a mature hog is more tolerant of temperatures below 40°F than above 80°F. The most critical period of life is from birth to 40 lb weight. A temperature between 50 and 60°F is desirable for farrowing, the most critical period. (See Chap. 55 for additional information on environmental factors and on heat and moisture losses.) Because fattening hogs require large quantities of grain and supplement, the hog-raising area is concentrated in the Corn Belt. Hogs are excellent converters of feed to meat, requiring about 4 to 5 lb of grain and supplement for each pound of gain. A 200-lb hog requires about 2 gal of water per day. Only moderate quantities of roughages are consumed by hogs. The recent trend has been toward the meat-type hog. Hogs are susceptible to respiratory diseases and stomach worms. Thus a strict sanitation and medical program must be followed. The number of farrowings vary from one to three per year, one to two being most common. The average gestation period is 114 days. Space requirements are summarized in Table 47-12.

A *portable six-sow house* which combines portability and flexibility with the advantages of a central farrowing house as to cold-weather protection and laborsaving is shown in Fig. 47-2. It is built in two sections, each of which can be pulled through a 14-ft gate. For winter farrowing the sections are moved close to the farmstead but still left on rotation pasture. For northern climates, a large concrete slab may be laid near the farmstead and the sections placed on it for winter farrowing.

[1] C. F. Kelly, T. E. Bond, and N. R. Ittner, Engineering Design of a Livestock Physical Plant, *Agr. Eng.*, 34:601-607, September, 1953.

Central hog houses (Fig. 47-3) stay warmer than individual houses because there is less building surface area per sow. An electric heat lamp can be used for each litter of pigs, or a stove may be placed in the alley. Space should be provided for feed, water, and bedding, and the operator can work protected from the weather.

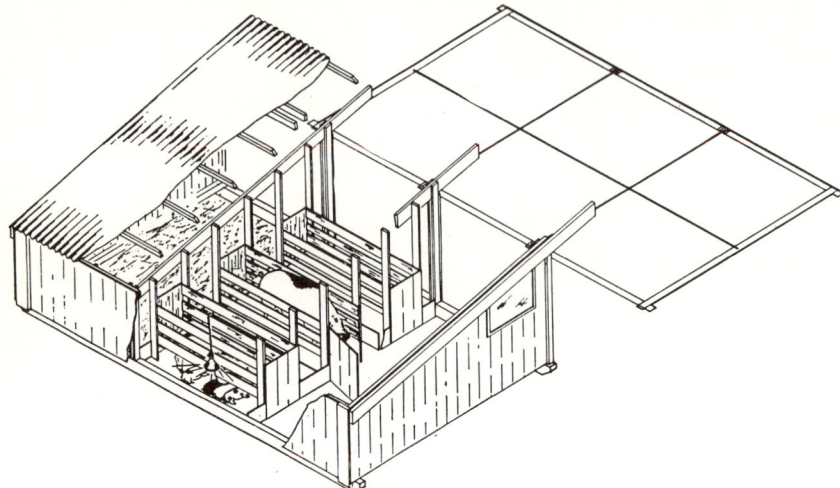

Fig. 47-2. Portable hog house accommodating six sows when two sections are faced together. (*By permission of Doane Agricultural Service, Inc.*)

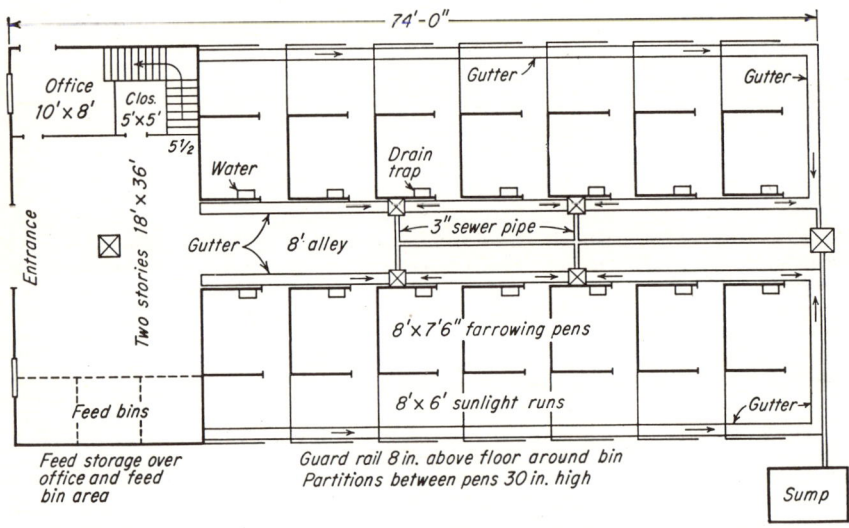

Fig. 47-3. Central farrowing house for 14 sows. (*Calif. Agr. Expt. Sta. Manual* 17, 1954.)

Southern areas should provide a larger area for housing of hogs than northern regions (Table 47-1).

The use of *infrared heat lamps* for brooding pigs is economical. About 40 million baby pigs die each year, and 50 per cent of these losses take place in the first 10 days, particularly in the first 24 hr.[1] The newborn pig has difficulty becoming acclimated

[1] T. A. H. Miller, Wallace Ashby, and J. H. Zeller, Hog Housing Requirements, *USDA Circ.* 701, 1944.

TABLE 47-1 SPACE REQUIREMENTS FOR HOUSING HOGS

Weight, lb	Northern areas, sq ft	Southern areas, sq ft
100	5–6	6–9
200	8–10	10–15
300	11–14	15–22
400	16–20	20–30

Farrowing sows, 8 by 8 ft.

SOURCE: T. A. H. Miller, Wallace Ashby, and J. H. Zeller, Hog Housing Requirements, *USDA Circ.* 701, 1944.

to its surroundings during its first 72 hr of life. The infrared lamp should shine directly on the pigs during their first 5 hr of life. After that the lamp can be placed in a protected cover. A 250-watt lamp will heat an area 24 in. in diameter when 36 in. off the floor (Fig. 47-4). After 1 week of age the auxiliary heat can be removed. It is recommended[1] that the infrared heating unit be 30 in. or more above the bedding, or 6 in. or more above the sow, whichever is the greater.

Partitions should be set up to form farrowing stalls to prevent sows from lying on their pigs. The partitions are later relocated to make a pig-feeding creep in the center.

A special narrow *farrowing pen* is often used, designed so that the sow cannot turn around. These stalls are available commercially or can be constructed on the farm (Fig. 47-2). The sides of the pen are open at the bottom so that the pigs may move to an area outside of the pen. The loss of small pigs is reduced, and they do not need to compete with the larger pigs for feed.

As the weather becomes warmer, the two halves of the portable house (Fig. 47-2) can be spread for more ventilation. In spring the sections are divided and faced to the south. All partitions and the floor sections are removable. In hot summer weather, the houses are faced to the northeast and all doors opened to provide cool shade. For cleanup the houses can be pulled aside, leaving the manure where it can be handled by a tractor loader. Hogs produce about 0.6 ton of manure each per year. Separate wood floor panels can be tipped for cleaning and sterilization. The inside walls of the open house are readily accessible for washing and disinfecting.

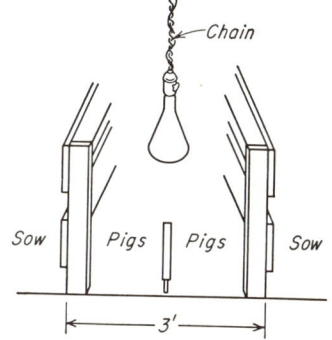

FIG. 47-4. Arrangement of heat lamp for pigs: 150- or 250-watt infrared lamp 30 in. above floor or 6 in. above sow.

The labor requirements for fattening hogs can be reduced greatly by proper selection and arrangement of equipment and proper management. Ten per cent of the cost of producing hogs is often for labor. Some farmers may work about 5 hr to produce a market hog, while a good operator with a large herd uses only 2 hr.[2]

The different methods of feeding hogs and comparative labor requirements are shown in Table 47-2.

The disease *vesicular exanthema* (VE) can be transmitted by feeding uncooked garbage. Most states require that garbage (not including domestic garbage) fed to swine be cooked by heating, usually at 212°F for 30 min. VE disease does not usually cause death of animals but causes slow growth. A typical cooking system consists of ½- to 1-in.-diam pipes placed in the bottom of a garbage truck and connected to a steam supply. The pipes have holes drilled in the sides for steam passage. For deep

---

[1] ASAE Recommendation. Installation of Electric Infrared Brooding Equipment, Approved 1955, *Agr. Eng. Yearbook,* 1960, p. 133.
[2] J. W. Oberholtzer and L. S. Hardin, Simplifying the Work and Management of Hog Production, *Indiana (Purdue Univ.) Agr. Expt. Sta. Bull.* 506, 1945.

TABLE 47-2. MAN-MINUTES OF LABOR TO FEED 1 TON OF CORN (FOR 100 HOGS)

Method	Time, man-minutes
Ear corn, self-fed from field crib	0.5
Ear corn, hand-fed from field crib	20.0
Ear corn, from permanent crib at barn	38.7
Shelled at farm, self-fed	71.5
Shelled at elevator, self-fed	93.5
Shelled and ground at farm, self-fed	97.5
Shelled and ground at elevator, self-fed	101.5

SOURCE: J. W. Oberholtzer and L. S. Hardin, Simplifying the Work and Management of Hog Production, Indiana (Purdue Univ.) Agr. Expt. Sta. Bull. 506, 1945.

truck bodies, downpipes with holes should be provided to give adequate cooking. A boiler horsepower of 7.5 to 10 for each ton of garbage to be cooked is desirable to heat the garbage to boiling in about 1 hr. One ton of garbage will produce 50 lb of pork. One boiler horsepower is required to evaporate 34.5 lb per hr of water at 212°F

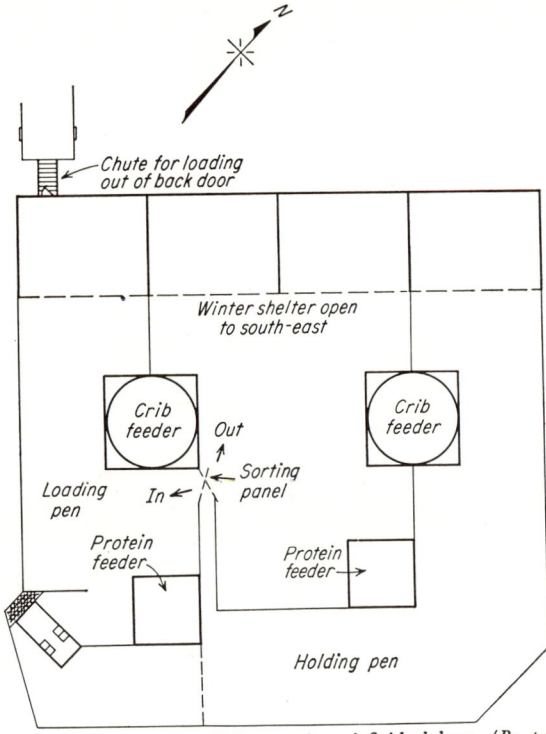

FIG. 47-5. Feed-lot layout arranged for efficient sorting of finished hogs. (*By permission of Doane Agricultural Service, Inc.*)

(33,480 Btu per hr). A heating surface (HS) of 5 to 10 sq ft is provided for each boiler horsepower.

Self-feeding of hogs saves much labor. Self-feeding equipment for grain, shelled corn, and ground feed has been used for many years and allows the feed to drop by gravity into the feeding trough as it is emptied. Design of the trough and feed inlet is such that the internal friction of the feed halts its movement before the trough overflows.

A self-feeding portable ear-corn crib can be utilized as a means of reducing feeding labor, or conventional cribs can be adapted for self-feeding. In the fall, a crib in the rotation pasture field which will be used for hogs the next year is filled with ear corn.

LIVESTOCK PRODUCTION FACILITIES 593

Sorting to select the hogs for market is a recurring operation which should be handled efficiently. The layout shown in Fig. 47-5 illustrates the principles involved in arranging a temporary holding pen, a sorting alley, loading pen, and "out" pen. The wood panels used for partitions in the houses can be wired together for temporary fencing.

The following *principles* and *practices* have been found to be most efficient for hog production:

1. Pigs are usually weaned at 8 weeks of age. With special management programs (pig hatcheries), the pigs can be taken from the sow at 1 to 2 weeks of age.
2. Pigs should be vaccinated for cholera at about the time of weaning.
3. Male feeder pigs should be castrated at 3 to 5 weeks of age.
4. A planned pasture program will reduce feed costs and decrease disease potential.
5. Self-feeding provides faster gains, with a saving of labor.
6. Cereal grains, except corn, should be ground. Use of slop requires much additional labor and usually is not justified.
7. With reasonably good performance 900 lb of grain will be required for marketing at 225 lb in 5½ months (including portion for breeding stock).
8. Gilts should be at least 8 months old when bred.
9. Provide plenty of shade or shelter for summer fattening.
10. Have water piped to hogs and available at all times.
11. Handle feed in bulk.
12. Individual farrowing houses for more than 20 sows become difficult to manage. A central house with a farrowing lot is more economical, if properly managed to prevent disease.
13. A good management program will result in an average of 1.8 to 2.0 litters per sow per year.
14. Consumer purchases of pork favor lean, lightweight hogs.
15. Silage can be used to reduce cost of grain for winter rations for sows.
16. On many farms, 80 per cent of the cost of swine production is for feed.

DAIRY-CATTLE PRODUCTION STRUCTURES

Production of milk in the United States is principally from European breeds, from large to small in body size, being Brown Swiss, Holstein, Ayrshire, Guernsey, and Jersey. These breeds do not sweat, even though some moisture is lost by transpiration through the skin. The large breeds are adversely affected by temperature above 80°F, and the small breeds by temperatures above 85°F. The respiration rate of the cow is increased as the temperature of the surroundings increases, thus reducing her efficiency of feed usage. A high relative humidity does not seem undesirable except above the critical temperatures. Cows are much more tolerant of cold weather than was formerly thought. Large breeds can tolerate cold temperature down to 0°F without decreasing production, but the production of the small breeds will decrease. (See Chap. 55 for additional information on environmental factors and on heat and moisture losses.) The average life span is 5 to 6 years, although many dairy cattle are still in good production at 10 years. Large breeds are first bred at 17 to 20 months of age; small breeds at 15 to 18 months. Artificial insemination is now used extensively. The gestation period averages 282 days. Basic space requirements for dairy cattle are summarized in Table 47-13.

The dairy cow is an excellent converter of roughage into milk. A minimum of about 4 lb of forage daily is required to make the most efficient use of grain.

Two principal *diseases* which can be communicated to man are tuberculosis and contagious abortion (Bangs), both of which are being eradicated. Mastitis, an udder disease, is a serious problem to dairymen and can be minimized by proper management. Animals are sold for meat after the milk-producing period.

There are two major systems of management and housing: the stall barn, where the cows are confined, and the loose-housing barn, where the cows are free to move from one area to another for feeding, milking, and resting.

Stall Barn. Cows are tied in stalls during most of the day and turned out to exercise periodically. The stall barn is particularly advantageous for a breeder selling cattle because the animals are normally tied and can be inspected easily. Carefully controlled tests comparing dairy production in a warm stall barn with that in a cold loose-housing barn were made over a 10-year period in Wisconsin. Production and general health were equal with both systems, although "the injuries observed in the stall housing (stepped-on teats, swollen hocks, stiffness and lameness) were almost completely absent in the loose housing barn."[1]

With *stall barns*[2] (Fig. 47-6) dairy cows are usually confined during the winter for all but a few hours during the day. Feeding and milking are usually done in the stall, although feeding of concentrates and milking may take place in a separate milking room (milking parlor).

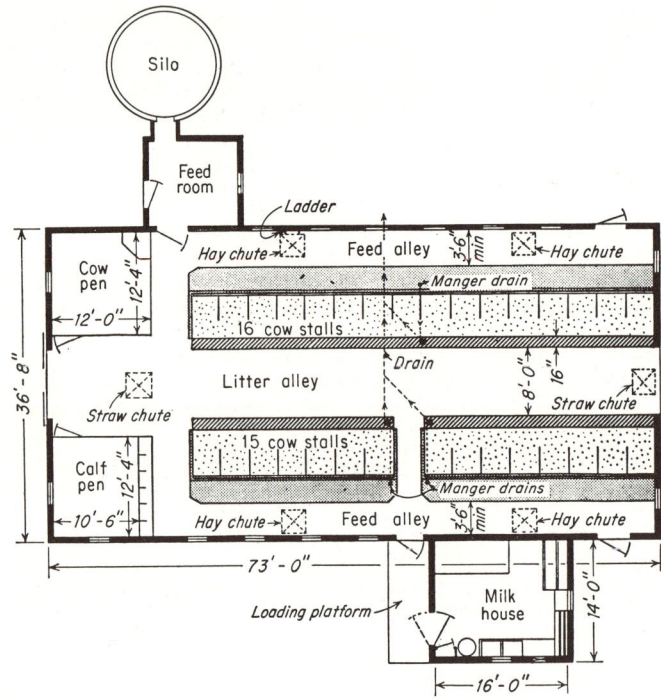

Fig. 47-6. Well-arranged plan for a two-story 31-stall barn. The litter alley shown is designed for drive-through cleaning. If a gutter cleaner is installed, the width of the alley may be reduced to 6 ft and the barn width to 34 ft. (*Cleaver, Thompson, and Yeck, Stall Barns for Dairy Cattle, USDA Infor. Bull. 123, 1954.*)

The most recently accepted stall design to reduce cow injuries and labor is shown in Fig. 47-7. This design provides (1) sweep-in rectangular manger, (2) stall column halfway between gutter and manger, and (3) gutter with same depth on both sides.

When planning a stall dairy barn, the following *principles* should be incorporated where possible:
1. Maximum labor efficiency is obtained when cows face out.
2. Sweep-in mangers make feeding and cleaning easier.
3. Provision should be made for expansion of herd and for increased feed storage.

[1] S. A. Witzell and D. W. Derber, Engineering Phases of Dairy Barn Research, 1941–1951, *Agr. Eng.*, 33:635–642, October, 1952.
[2] ASAE terminology, formerly often referred to as stanchion barns.

4. The milkhouse should be conveniently located.
5. Use mechanical equipment for handling materials.
6. Store material in a convenient location with chutes properly placed.
7. A minimum width of 32 ft is required for two rows of cows.
8. Provide adequate cross alleys to save steps.
9. Larger stalls aid in reducing injuries. By using cow trainers the cows can be kept cleaner in a longer stall.
10. Provide suitable pens for sick, injured, and maternity cows.

Barn Cleaners. Mechanical barn cleaners were first introduced in the early 1940s as a method of reducing labor requirements on the farm, particularly for dairy animals. A dairy cow produces 8.5 tons of manure per year; beef, 6.0 tons per year. With plenty of bedding, manure weighs about 25 lb per cu ft. The work required to move manure depends on the method of handling. With stall barns it is desirable to clean twice a day in the winter. In a controlled test in a barn housing 35 dairy cows, a barn cleaner moved and loaded 212 tons of manure with 82 kwhr of electricity.[1] The cost of operation of a barn cleaner is about 5 cents per year per cow for electricity. The principal requirement is that the unit must be designed to handle liquid without collection sumps. Gutters should be at least 10 in. deep and 16 in. wide to

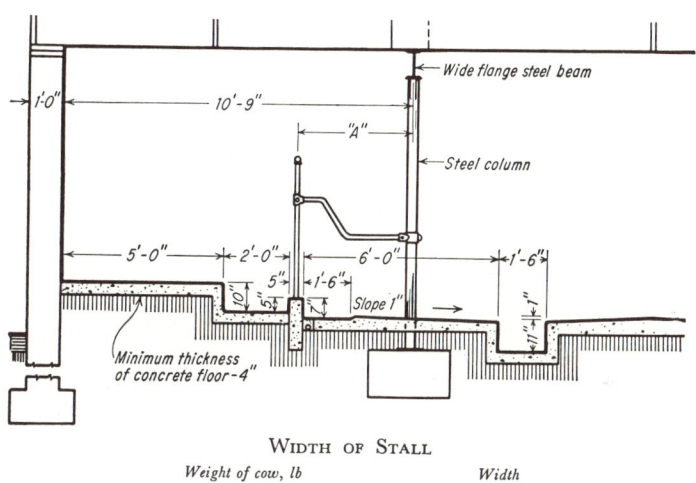

Width of Stall

Weight of cow, lb	Width
800	3'6"
1,200	4'0"
1,600	4'8"

Fig. 47-7. Designs for cow stalls.

satisfactorily accommodate a barn cleaner, and provisions must be made for moving the manure across alleys by using a metal or concrete cover to form a tunnel. These tunnels often create problems of manure accumulation if they are too shallow. The manure is moved at a speed of 10 to 30 ft per min. The various types are as follows:

Portable self-propelling. One type, consisting of a power-driven scoop which fits the gutter, is pulled by cable as the operator guides the shovel in the gutter to move a slug of manure. Another type consists of a conveyer-elevator unit which moves along the gutter, picking up the manure and placing it into a manure cart.

Pull-out. An apron with cleats or chains with flights moves the manure. The apron is wound on a main drum to move the manure out and then up an elevator (not over a 30° angle) into the spreader. A spider wheel holds the chain down at the incline

---

[1] R. L. Maddex and A. W. Farrall, Mechanical Barn Cleaners, *Mich. State Univ. Ext. Bull.* 265, 1951.

(Fig. 47-8). A 1½- to 3-hp motor is used for pulling out the manure. The apron or chain is then pulled back by hand or by means of a ¼- to ½-hp motor. A clutch is required to permit the motor to reach rated speed before engaging the conveyer. Units of this type use 2¼ to 4 kwhr per month.

Endless chain. Flights are fastened at one end to a single chain of special design, allowing flexing in a vertical direction as well as horizontal (Fig. 47-9). With this arrangement the chain and flight can turn corners and also go up an incline. The manure is moved to an elevating section or directly into a spreader for a sidehill

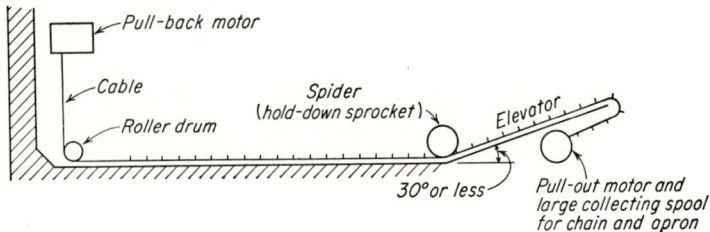

Fig. 47-8. Pull-out type of barn cleaner.

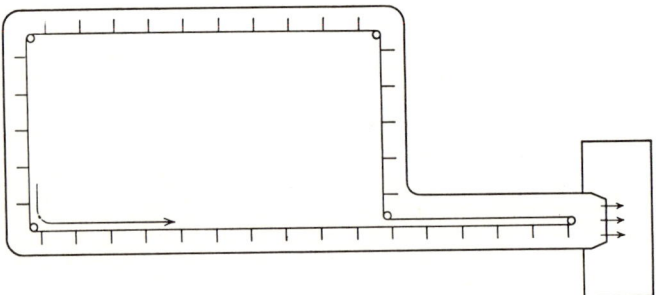

Fig. 47-9. Endless-chain type of barn cleaner in two-gutter barn.

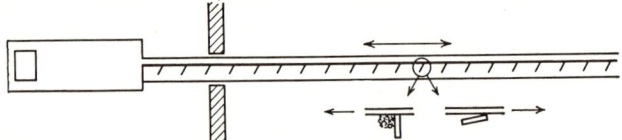

Fig. 47-10. Oscillating type of barn cleaner.

barn. The most satisfactory arrangements are those in which the elevating section and conveying section are continuous, rather than two separate units. Sprocket idlers are required for tightening the chain. A cleaning arm is provided for cleaning the flights. A pull of about 50 lb per cow is required to remove the manure in a straight section. A motor of 1½ to 5 hp operates the system, usually on a two-gutter arrangement, and uses 4 to 6 kwhr per month.

Oscillating. A rigid drive-bar member, held in the bottom on one side of the gutter, moves back and forth lengthwise in the gutter with about 8-ft strokes. At about 4-ft intervals hinged paddles are attached to the oscillating member. With one stroke the paddle moves the manure about 8 ft, and on the return stroke the paddle folds along the drive bar until the end of its stroke, and then engages the pile of manure previously collected by the adjacent paddle (Fig. 47-10).

For calf pens, the gutter cleaner should pass outside, or if the cleaner passes through the pen, it should have 16-in. clearance and be provided with cover plates over the gutter.

Loose Housing. In the loose-housing system the cows are brought consecutively to a central building or area for milking and usually for feeding of grain and concentrates. Hay and silage are fed in racks and bunks or self-fed by means of movable gates located in the storage structures. Except at milking time the cows are free to move between a paved feed lot and a sheltered resting area. The operator can do most of the work in a comfortable efficient milking room. Normally two or three milking units are handled per operator.

The labor required in caring for herds in the loose-housing barns is less than in the stall barn, and savings up to 35 per cent are possible with an elevated stall milking room and pipeline milking machine. The saving in labor is due primarily to:

1. Less operator travel
2. No handling of milk from milking area to milk-storage area
3. More bulk feeding
4. Reduced manure handling
5. Less operator fatigue, particularly with elevated milking stalls

In Kentucky, data for an efficiently arranged, equipped, and managed system for 15 cows showed that about 52 man-hr with 10.4 miles of travel was required per year per animal for loose housing, and 77 man-hr with 20.3 miles of travel for a stall barn.[1] Furthermore, the construction cost for a loose-housing system in this study was about one-third less.

Labor, of which 80 per cent is for milking and care of milking equipment, accounts for one-third of milk-production costs.[2] The proper planning of the milking area is of utmost importance for labor efficiency.

In addition to reducing labor requirements, a loose-housing arrangement with proper management can provide:

1. Better-quality manure
2. Milk of higher quality
3. Fewer udder and leg injuries
4. Higher production per animal through animal comfort
5. Reduced initial investment
6. Flexibility for adapting to increased size of herd
7. Less possibility of loss due to fire

Investigations at the University of Illinois indicated that the new cost of dairy buildings and equipment per cow was almost cut in half with loose housing compared with stall barns.[3]

The major disadvantages of loose housing are that more bedding may be required and that "boss" cows may try to keep others away from the feeding mangers. A stall barn, however, usually requires 6 to 8 lb per day of chopped straw, and this amount is adequate for loose housing if properly designed and managed. It is possible to operate a stall barn with practically no bedding, which is not possible with loose housing. The completely different management scheme for loose housing requires a complete break from the stall-barn system, and it is difficult for many farmers to make the change, both equipmentwise and habitwise. Practically all dissatisfaction with the loose-housing system can be traced to improper arrangement and management. A well-arranged loose-housing setup is shown in Fig. 47-11.

Several arrangements can be used for *milking rooms,* which are combinations of the following: level or elevated; abreast or tandem; side-opening, walk-through, or back-out. The speed of milking is about the same with the different arrangements, but the travel distance, stooping, and bending vary considerably. The most common arrangements are shown in Fig. 47-12.

---

[1] G. B. Byers, Effect of Work Methods and Building Designs on Building Costs and Labor Efficiency for Doing Chores, *Univ. Kentucky Agr. Expt. Sta. Bull.* 589, 1952.

[2] B. F. J. Cargill, "Methods Engineering of Loose Housing Dairy Barns," thesis for M.S. degree, Michigan State University, *Dept. of Agricultural Engineering,* 1952.

[3] Harry L. Garver and Wallace Ashby, Farm Buildings Animals Can Pay For, *Agr. Eng.,* 36:28–30, January, 1955.

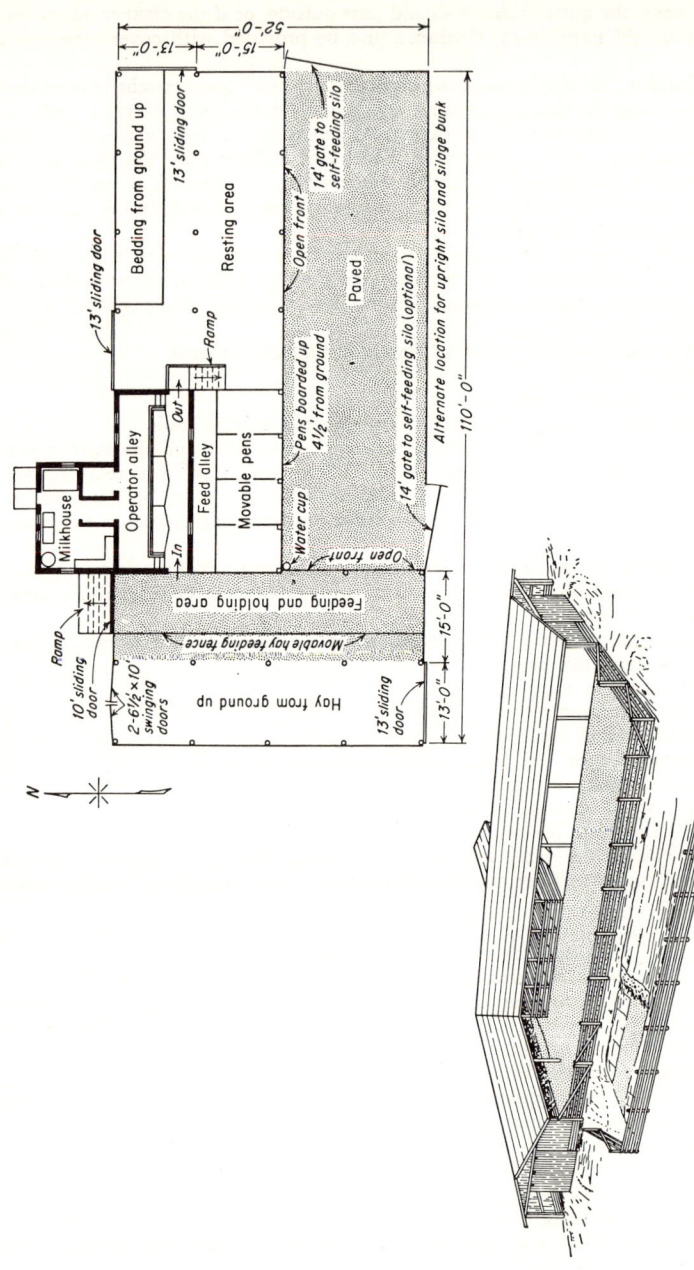

Fig. 47-11. An L-shaped loose-housing layout designed to give protection from winter winds. The feeding area slopes upward to the milking room. (*Cleaver and Yeck, Loose Housing for Dairy Cattle, USDA Infor. Bull. 98, 1953.*)

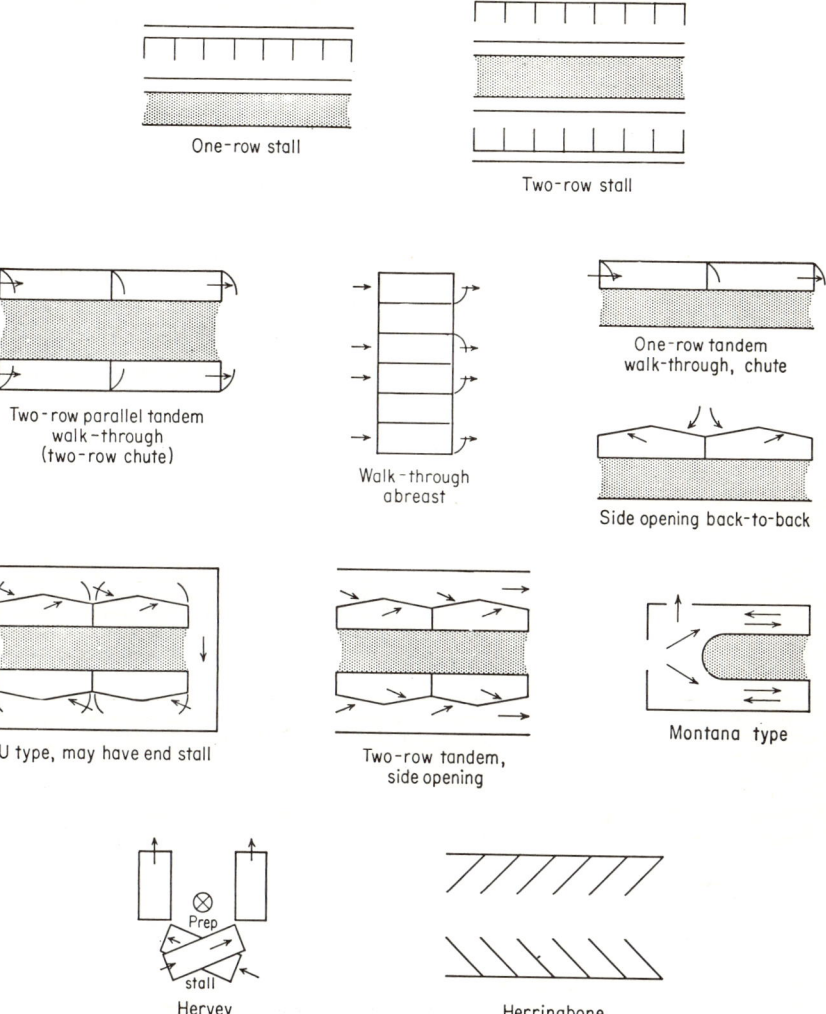

Fig. 47-12. Milking-room arrangements.

The following arrangements are recommended to obtain the best use of labor for the milking room:

Cows	No. of stalls in milking room	No. of operators	No. of milking machines	Position of floor
10	2	1	1	Level
10–15	2	1	2	Elevated (30–36 in.)
15–30	3–4*	1	2	Elevated (30–36 in.)
30–45	4	2	3	Elevated (30–36 in.)
Over 45	5	2	4	Elevated (30–36 in.)

* Three stalls in a line is maximum recommended for one operator.
SOURCE: Thayer Cleaver, Time-travel Studies on Dairy Farms, *Agr. Eng.*, 33:137–142, March, 1952

TABLE 47-3. AREA REQUIREMENTS FOR LOOSE HOUSING OF DAIRY ANIMALS

Major areas	Floor	Area, sq ft per animal	Principles and specifications
Barnyard.........	Paved, 4 in.	60 minimum, 100 recommended	Outside area located on south or east side of barn; slope ¼ in. per 1 ft away from barn and feed areas; place water cup in paved area to prevent unnecessary livestock travel; use two or three all-weather floodlights to illuminate important points.
Feeding..........	Paved	Self-feeding 18-in. manger per cow for each kind of feed; batch-feeding 30-in. manger per cow	For feeding silage and hay; located outside or inside barn, outside feeding preferred; self-feed as much as possible since this will increase corn-silage consumption and reduce hay and dry-grain requirements.
Resting (bedded)....	Dirt or paved	40 to 60	Lay out so cows do not walk across resting area to get to feed, water, or milking room; need 9½ ft ceiling for manure build-up; avoid long and narrow areas; straw storage should be close; need 6 to 9 lb straw per cow per day; can use 10 to 15 sq ft of resting area for straw storage at beginning of season; provide 1½ tons straw per animal in northern area; one 100-watt lamp for each 400 sq ft.
Milking..........	Paved	500 cu ft per animal	Entrance from barnyard or resting areas and exit to feed areas; will feed grain to animals in milking room from overhead distributor storage; steps for entrance with 6-in. rise and 18-in. run; ramp for exit with 2-ft rise in 8-ft run; overhead feed storage for 1-2-weeks supply; 30-36 in. height for platform of elevated stalls; 10% of floor area in windows or 12% in glass block; use a trapped-water-seal drain in milking room and milkhouse; provide lighting through floor of stall toward cow's udder.

NOTE: Codes pertaining to dairy farm structures may specify different space requirements in some areas. Consult local sanitarian or dairy fieldman for approval of plan.

Table 47-3 provides a summary of area requirements for loose housing of dairy animals. Feed-space requirements for stall barn or loose housing can be calculated from Tables 47-4 to 47-6.

The following *principles* and *practices* have been found to give maximum efficiency in loose-housing systems:

1. Locate bedded resting area away from traffic lanes to reduce bedding requirements and to keep cows clean.
2. Avoid concentrated traffic lanes.
3. Arrange buildings and board fences to act as windbreaks for the lot.
4. Have resting area open on the south or east side for good ventilation, sunlight, and easy access.
5. Provide no more than 60 sq ft of bedded area per cow for efficient use of bedding and a warm, firm manure pack.
6. Maintain manure pack until the end of the season because it provides warmth for the resting cows. A minimum ceiling height of 9½ ft is required for loose housing.
7. For minimum bedding requirements, turn under fresh droppings daily or place them along the unused edges of the manure pack.
8. Apply bedding evenly and only when needed.
9. Pave the feeding area and remove droppings daily, preferably by a tractor scraper to keep cows clean. Do not bed feeding area. Slope pavement away from feeding area, 1 in. in 4 ft. Approximately 1 cu yd of concrete is required per animal unit to pave the barnyard and feeding area.
10. Feed hay outside or from a self-feeding ground-level hay-storage building.
11. Feed silage outside, self-feed from a paved horizontal bunker silo, or feed in manger equipped with mechanical silo unloader for minimum labor. Dairy animals

given free access to corn silage under loose housing will eat much more than previously—up to 90 lb per day per animal.

12. Keep young stock and dry cows separate from the lactating herd.
13. Protect calves under 10 days old from drafts by individual box pens.
14. Calf pens should be covered by winter sunlight and have removable walls to allow cleaning by tractor loader.
15. Locate milking rooms so that cows enter from the lot or covered holding area but not from the bedded area.
16. Use elevated milking stalls to reduce operator fatigue.
17. If manure cannot be removed before fly season, permit it to crust over by keeping cows off.
18. The resting barn should be operated as a "cold" barn, well ventilated, without a draft. In the winter, the resting area may be only 5 to 9°F above the outside-air temperature.
19. Removal of horns from all animals under 2 years of age is desirable.
20. To avoid feed and manure odors in milk, keep cows off bedded area and off feed about 2 hr before milking.
21. Use of a preparation stall for each two milking stalls to increase efficiency.
22. Pipeline milkers and bulk coolers can improve efficiency.
23. Provide hand-washing facilities for milker and bulk-tanker operator.
24. Use a tractor loader to handle manure. Provide at least a 10-ft door height for loader.
25. Provide one all-weather automatic watering cup for each 25 cows.

TABLE 47-4. APPROXIMATE HAY REQUIREMENTS PER 1,000-LB COW FOR VARIOUS FEEDING PERIODS

Type of feeding	Feeding period, days	Amount of hay fed	
		Per day, lb	Total, lb
Hay alone...............	210	25	5,250
	175	25	4,375
	120	25	3,000
Hay and silage*..........	210	15	3,150
	175	15	2,625
	120	15	1,800

* Hay requirements will be lower if cows get all they want of top-quality silage.
SOURCE: Thayer Cleaver, Harold J. Thompson, and Robert Yeck, Stall Barns for Dairy Cattle, *USDA Infor. Bull.* 123, 1954.

TABLE 47-5. APPROXIMATE SILAGE REQUIREMENTS PER 1,000-LB COW FOR VARIOUS FEEDING PERIODS

Type of feeding	Feeding period, days	Amount of silage fed	
		Per day, lb	Total, lb
Hay and silage..........	210	30	6,300
	175	30	5,250
	120	30	3,600
Silage*.................	210	60	12,600
	175	60	10,500
	120	60	7,200

* Assuming that some good-quality hay is fed.
SOURCE: Thayer Cleaver, Harold J. Thompson, and Robert Yeck, Stall Barns for Dairy Cattle, *USDA Infor. Bull.* 123, 1954.

TABLE 47-6. APPROXIMATE CONCENTRATE REQUIREMENTS FOR COWS IN LACTATION*

Feeding rate	Milk production		Total concentrate requirement, lb
	Per day, lb	Per year,† lb	
1 lb concentrate to 3 lb milk.....	20	6,000	2,000
	25	7,500	2,500
	30	9,000	3,000
	40	12,000	4,050
1 lb concentrate to 4 lb milk.....	20	6,000	1,500
	25	7,500	1,875
	30	9,000	2,250
	40	12,000	3,000

* If high-quality roughages are self-fed, concentrate requirements will be reduced considerably.
† Lactation period of 300 days.
SOURCE: Thayer Cleaver, Harold J. Thompson, and Robert Yeck, Stall Barns for Dairy Cattle, *USDA Infor. Bull.* 123, 1954.

The distribution of annual dairy production costs per farm in a 1947 survey of 350 Illinois farms averaging around 30 cows per herd is shown in the tabulation in Table 47-7.

TABLE 47-7. DISTRIBUTION OF ANNUAL DAIRY COSTS PER FARM (30 COWS)

Cost item	Cash cost		Total cost	
	Dollars	Per cent	Dollars	Per cent
Buildings.............	147	8	720	10
Equipment...........	75	4	151	2
Stock................	158	8	310	4
Feed.................	1,252	65	5,139	70
Labor................	288	15	1,009	14
Total..............	1,920	100	7,329	100

SOURCE: Harry L. Garver and Wallace Ashby, Farm Buildings Animals Can Pay For, *Agr. Eng.*, 36:28–30, January, 1955.

"Labor costs were 40 per cent higher than building costs. With modern design and equipment, labor could be cut about in half and thus make a substantial addition to net income, or at least give the farm family more time for other things."[1]

Dairy Calf Housing. There is a tendency for dairymen to provide warm quarters with poor ventilation for dairy calves. Diseases are promoted and the growth of calves retarded in poorly ventilated buildings. In Missouri, open-type structures provided significantly lower losses of calves than closed structures.[2] Housing can be provided in the main dairy structure or in separate buildings. The calves are normally with their dams for 3 days, then fed milk for about 2 months, and then placed on dry feed entirely. Calves can be housed in individual or group pens. The pens can be cleaned regularly, or the manure permitted to accumulate. For individual pens, 36 sq ft is needed wtih solid partitions to prevent drafts and sucking of calves. Calves of the same age group can be placed in group pens but should be confined to

---

[1] Garver and Ashby, *op. cit.*
[2] M. L. Esmay, A. F. Williams, and B. F. Guyer, Young Dairy Calf Housing in Missouri, *Univ. Mo. Agr. Expt. Sta. Research Bull.* 527, 1953.

stanchions during and after feeding to prevent sucking. Calves need a dry environment, free from draft, with plenty of fresh air and sunshine.

The use of antibiotics such as aureomycin and terramycin gives increased rate of growth and improved health in the first 3 months of age.

Milkhouse. A milkhouse is required by law in most states on those farms which sell milk for bottling. The requirements vary considerably from one area to another. The

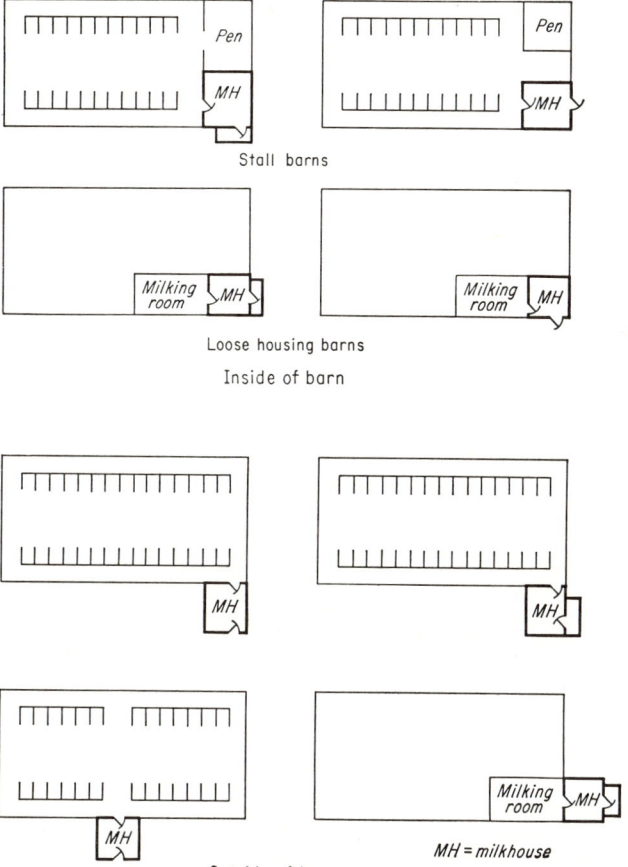

FIG. 47-13 Suggested locations for milkhouses. (*Boyd, Hall, Maddex, and Murray, Milkhouses: Planning and Construction, Mich. State Univ. Ext. Bull. 325, June, 1954.*)

material presented here is believed to satisfy most of the regulations. It is recommended that the local health inspectors and the dairy-plant fieldman be consulted before constructing a milkhouse.

Until recently, it was required that the milkhouse be located some distance from the dairy barn. However, this idea no longer prevails, and in most areas it is permissible to attach the milkhouse to the barn or place it inside the barn. The odors which milk absorbs come principally from the cow and not the manure or feed directly. The off-odors are a result of the dairy cattle breathing the fumes from the manure and other undesirable odor-causing products. The milkhouse should be located as close to the cows as the regulations will permit. It is undesirable to have the milkhouse next to cattle yards or manure piles, because of the possibility of contamination from flies. The milkhouse should be separated from the milking area by a swinging door that closes tightly. It is usually better to locate the milkhouse on the side of the barn rather than on the end, so that extension of the barn is not blocked. Suggested locations for the milkhouse, inside or outside the barn, are shown in Fig. 47-13.

The three major functions of the milkhouse are to provide (1) a place for storing and handling of milk, (2) a convenient place for cleaning of equipment, and (3) a place for storing clean utensils. To meet these three major functions the milkhouse must be of the proper size and in the proper location. There is considerable variation in the minimum size requirements of the milkhouse and the desirable or recommended size. The recommended sizes of milkhouses are given in Table 47-8 for can handling and for bulk milk handling with every-other-day (EOD) pickup. Note that the size of milkhouse *recommended* for cans would be *satisfactory* for bulk, but that if a new milkhouse is being planned, it should be made larger to meet the bulk-handling *recommendations*.

The milkhouse equipment consists of a utensil rack; a can rack, if cans are used; a wash vat; a water heater; a milk cooler (can or bulk); and a cabinet for storage of miscellaneous items. Space must be provided for keeping records. The milk cooler

TABLE 47-8. MILKHOUSE DIMENSIONS

Daily output, gal	Can requirements, absolute minimum for bulk, ft × ft	Satisfactory for bulk, recommended for cans, ft × ft	Future milkhouse for bulk, EOD,* ft × ft
Under 20	10 × 8	12 × 10	
20 to 50	10 × 10	12 × 12	12 × 14
50 to 100	10 × 12	12 × 14	12 × 16
100 to 160	10 × 14	12 × 16	14 × 16†

* D. L. Murray et al., Handling Milk on the Farm, *Mich. State Univ. Ext. Bull.* 342, 1957.
† Add 0.2 sq ft for each gallon over 160 gal capacity to get proper area.

should be located as close as possible to the entrance from the cow stable. The hot-water heater should be located to allow short pipelines from the heater to the wash tank. For a bulk-milk installation, hot and cold water must be readily available with an adequate supply for washing the tank. There is considerable variation in the space requirements for the same size of bulk-milk tanks sold by different manufacturers. The over-all dimensions of a 300-gal bulk-milk tank are from 7 to 9 ft long and from 3 to 4 ft wide. To provide means of installing and removing the tank, a 3-ft door should be set in a frame with a 1-ft removable panel to make possible a 4-ft opening. A few of the bulk-milk tanks are wider than 4 ft, but most of these can be turned on their side and moved through the door because their height is usually less than 4 ft. A satisfactory arrangement of equipment is shown in Fig. 47-14.

When changing from can to bulk handling of milk, considerable difficulty may be encountered in using the original concrete floors. The bulk-tank cooler is heavier since it holds the milk from 2 days production. A concrete floor should be laid over an 8- to 10-in. gravel bed. It should not be less than 5 in. thick for bulk-milk-tank support and should be reinforced with 6- by 6- in. No. 6 reinforcing wire. A good floor is especially important for a bulk-tank installation because the tank must remain level to utilize the graduated stick for determining the quantity of milk. There should be a cove between the sidewall and floor, and the floor should slope about $\frac{1}{4}$ in. per ft to the drain. The drain should be readily accessible for cleaning and maintenance and be provided with a 4-in.-diam. deep water-seal trap. In some states the drain must be at least 6 in. from the tank. A bulk-milk tank may also be supported by pillars beneath the legs of the tank. These pillars are in the shape of a cone or pyramid 8 in. wide at the top and 16 in. wide at the bottom and extending 30 to 40 in. into the ground, poured even with the floor. (See Chap. 51 for discussion of electrical milk handling and cooling equipment.)

Milkhouse ventilation can be provided by a gravity stack or by electric fan. A gravity ventilator should have a cross-section area of from 100 to 200 sq in., extend from the ceiling to the roof, and be provided with a damper to adjust the amount of air flow. Adjustable sidewall louvers or ventilators should be provided, particularly for summer use where bulk-milk installations are made. The cost of operation of refrigeration equipment, especially when an air-cooler condenser is used, is reduced considerably by keeping the milkhouse temperature low. The sidewall ventilator should be located near the air-cooler refrigeration condenser and be screened to prevent flies from entering. A more positive means of control is to use an electric fan. It is usually recommended that the system operate so that the milkhouse is pressurized,

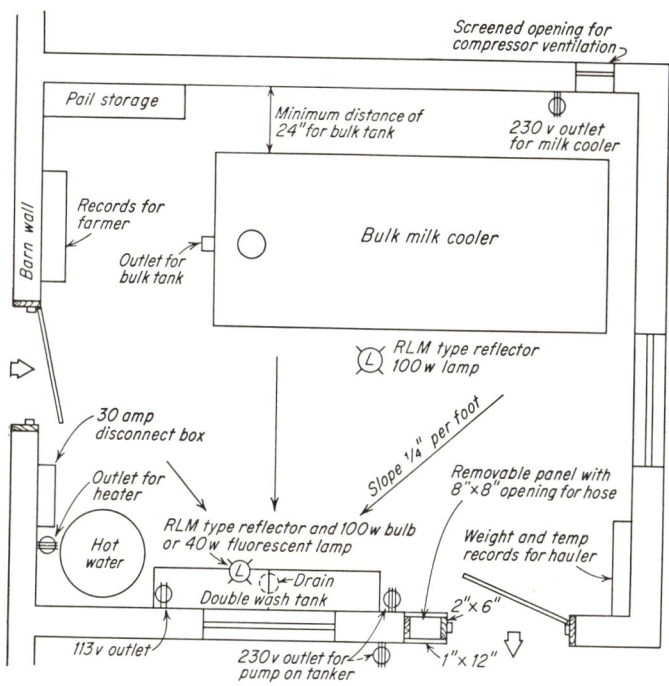

FIG. 47-14. A suggested layout for milkhouses with doors on adjoining sides. (*Boyd et al., Milkhouses: Planning and Construction, Mich. State Univ. Ext. Bull.* 325, *June,* 1954.)

by using air from a clean outside source, thus eliminating the possibility of undesirable odors being drawn into the milkhouse from the barn. A small fan of 100- to 500-cfm capacity is usually controlled by a thermostat to operate when the temperature gets above a certain level or to operate for about 15 min every 3 hr. The exact timing will depend on the size of the milkhouse, size of fan, proximity to the milking barn, temperature differences, etc. A milkhouse is adequately ventilated when the interior surfaces are dry, when the room is free of odors, and when the summer temperature is below or equal to the outside air temperature. Heat can be conserved by using insulation on the walls and ceiling (a minimum R value of 10 for northern climates), weather-stripping doors and windows, and using storm windows. Coal or wood heaters are not recommended in the milkhouse because of the dirt and the products of combustion. The most convenient method of heating a milkhouse is by a thermostat-controlled electric space heater requiring 1,500 to 3,000 watts. Various methods of using electricity for heating the milkhouse are summarized in Table 47-9.

Fuel-oil heaters probably provide the least expensive method of space heating but present more of a hazard and require a flue for the products of combustion. Combination hot water and space heaters which burn fuel or gas are available.

It is not desirable to maintain the milkhouse temperature over 40°F. The importance of utilizing the heat extracted from the water and given off by the milk cooler as a means of maintaining the milkhouse temperature above freezing is often overlooked. With an ice-bank refrigeration system the tank can be controlled by a time clock to build the ice bank and thus produce heat for 3 to 5 hr before the operator works in the milkhouse, providing the highest temperature during milking. Another unique arrangement consists of running well water into the ice-bank tank, causing the refrigeration system to cool the water and give off heat to the room. The water coming into the system can be controlled to provide the desirable room temperature. The milk-cooler principle of heating (heat pump) is economical if adequate water is available to supply the required heat. The refrigeration system has a coefficient of performance of approximately 3 so that approximately three times as much heat is given off to the milkhouse by using the heat pump rather than by using the electricity directly in a resistance heater. A well-insulated milkhouse of 1,000 to 1,200 cu ft can be expected to require about ¾ kwhr of direct heat per degree-day, while a milk cooler used as a heat pump will require ¼ kwhr per degree-day.[1]

TABLE 47-9. ELECTRICAL METHODS OF HEATING MILKHOUSES

Type	Size	Location
Space heater*	1,500 watts for 1,500 cu ft 3,000 watts for 1,500–3,000 cu ft	Directed on work area
Radiant panel*	3' × 4', 4' × 4' for 115 watts 4' × 8' for 230 watts	One-third of ceiling area
Heat lamp*	4–10, 250-watt infrared heat lamps (use heating cable on water pipes to prevent freezing)	Directed on work area
Milk-cooler heat pump	A 6-can size needs 60 to 90 gal/hr of water ½-hp unit (Michigan) ¾-hp unit (Minnesota)	Part of cooler; cooler will operate 4 to 6 hr more per day

* Control by thermostat at least 4 in. from outside wall and 3 to 5 ft above floor.

Electrical equipment predominates in the milkhouse. It is necessary to provide light outlets: 230-volt outlets for milk cooler, water heater, space heater, and pump on bulk-milk tanker, and 115-volt outlets for miscellaneous equipment (Fig. 47-14).

The principal features applying to milkhouse and room construction and equipment from the Milk Ordinance and Code, 1953, of the U.S. Public Health Service, Department of Health, Education and Welfare, are as follows:

"None of the milkhouse operations is conducted elsewhere. An exception may be made in the case of pipeline milkers which are cleaned and given bactericidal treatment in place. The floor may consist of concrete, brick, tile, asphalt-macadam, or other composition material. Floors must be impervious and drain properly. A grade of one-fourth to one-half inch per foot gives ample floor drainage. The finish of the floor should be as smooth as possible and the junction of the floors and walls should be curved and the joints rounded. For a framed construction milkhouse, all walls should be made of impervious material up to a height of 12 in. Floors with depressions in which liquids do not drain are unsatisfactory. Walls and ceilings shall be composed of smooth lumber, sheet metal or plaster board, well-painted with a light colored washable paint on tile cement block, brick concrete or cement plaster provided that the surfaces and joints are smooth. The window space must be not less than 10 per cent of the floor area and light must be reasonably evenly distributed. A minimum of 10 foot-candles of light shall be provided at all working surfaces from natural and/or artificial light sources. The milkhouse must be adequately ventilated to minimize odors and condensation on the floors, walls, ceiling and cleaned utensils."

BEEF CATTLE PRODUCTION STRUCTURES

Beef animals respond to environment similarly to dairy animals. The critical temperature is 80°F at 60 per cent relative humidity. (See Chap. 55 for additional in-

[1] C. P. Wagner and Marvin Nebben, Calculations for Milkhouse Heating Purposes, *Agr. Eng.*, 30:294–296, June, 1949.

formation on environmental factors and on heat and moisture losses.) It is quite common to breed and raise beef animals in the southern and western areas, then ship the steers to the feed-producing areas for fattening. The selection of beef-breeding stock is based more on body conformation than on ability to produce milk. A beef-breeding animal is normally productive to about 10 years of age. Female beef calves for breeding should weigh 450 lb when weaned and 650 lb when bred. Calves should be dehorned when young and at a time when they will not be subject to flies. Castration should be done before animals are 4 months old. Immunization for blackleg is common. Feeding steers and heifers are fattened to a weight of 900 to 1,000 lb. Approxi-

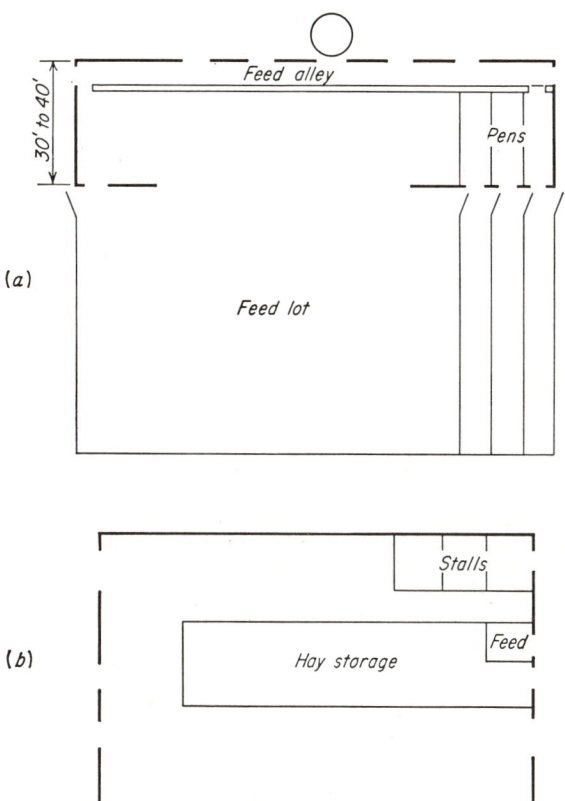

Fig. 47-15. Beef cattle housing: (a) open front; (b) combined storage and shelter.

mately 8 to 12 lb of feed, depending on the type of feed, is required for each pound of gain of steers. Feeders plan to put on a gain of 1½ to 2¼ lb per day per animal, depending on market conditions. Fattening animals will eat about 3 lb of feed per 100 lb of live weight early in the feeding period and decrease as feeding progresses. A corral should be provided for sorting. A beef animal will dress-out about 60 per cent of the live weight.[1] Basic space requirements for beef cattle are summarized in Table 47-14.

Open sheds provide adequate shelter, except possibly for new calves (Fig. 47-15). See Chap. 48 for a discussion of summer livestock shelters. Feeding is primarily a materials-handling problem (Fig. 47-16).

[1] H. R. Guilbert and G. H. Hart, California Beef Production, *Univ. Calif. Agr. Expt. Sta. Manual* 2, 1951.

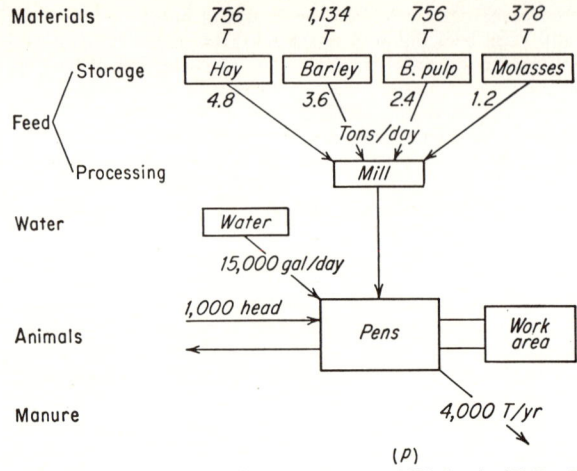

FIG. 47-16. Flow diagram of materials in beef feed lot for 1,000 head. (*Kelly, Bond, and Ittner, Engineering Design of a Livestock Physical Plant, Agr. Eng.*, 34:601-607, September, 1953.)

The following *practices*, in addition to those previously listed, have been found effective in reducing labor requirements for feeding beef cattle:

1. Confine the feed lot to approximately 100 sq ft per steer, including shelter area, for efficient conservation of manure.
2. Arrange shed and lot to be easily cleaned with a tractor manure loader.
3. Build a substantial loading and sorting corral next to the feed lot.
4. Store feed adjacent to feed lot if possible.
5. Mix grain and supplement at time of grinding.
6. Use large bin-type self-feeders.
7. Have large hay bunks when hay must be hauled.
8. Use portable feed house for field feeding to avoid hauling feed every day.
9. Feed hay in the field when roughing cattle through the winter.
10. Heavy cattle weighing 1,200 lb and over suffer a penalty on the market.

The feeding of cattle in Illinois required 10 hr of labor for each 100 lb gain. In Kansas 20 hr of labor is required annually per head for the cow herd[1] (Table 47-10).

TABLE 47-10. TIME REQUIRED AND DISTANCE TRAVELED PER TON OF FEED FOR BEEF CATTLE IN INDIANA

Task	Time, min		Distance, ft	
	High	Low	High	Low
Feed preparation............	47	15	6,346	109
Feed distribution............	37	10	11,300	273

SOURCE: Dean G. Carter, Progress in Farm Building Mechanization, *Agr. Eng.*, 33:707, November, 1952.

A *breeding beef* cow[2] should have 50 sq ft of inside floor area if she has access to outside yards. A maternity pen should provide from 100 to 120 sq ft. A breeding cow will consume 5 tons of silage or equivalent in a 200-day period. One ton of silage is equivalent to 800 lb dry forage.

[1] Beef Cattle Housing in the North Central Region of the United States, *North Central Regional Publ. 6*, June, 1946, as *S. Dakota State Coll. Agr. Expt. Sta. Bull.* 382.
[2] *Ibid.*

Paving Barnyards. A concrete paved area along barns and next to watering and feed areas provides an excellent means of improving sanitation and improving appearance and enables the farmer to move easily over areas that would otherwise be muddy. Less area is needed for the feed lot if paved.

The area to be covered with concrete should have the topsoil removed. The area should then be filled with 6 to 8 in. of coarse gravel and tamped to provide a base for the concrete. Concrete should be 4 in. thick, unless it is to be subjected to heavy tractor or truck loads, where 6 in. should be provided. The slabs should be poured in 10- by 10-ft sections. After sufficient drying, the concrete should be finished with a float, which will provide a rough surface. A slope of about 1 in. per 10 ft should be provided for drainage. An expansion joint made of a 2- by 4-in. plank oiled with crankcase oil should be placed next to buildings and watering tanks. For temporary steel posts a 4-in.-diam pipe can be inserted at the approximate location, but at least 6 in. from the pavement edge. The space between the pipe and post should be filled with asphalt or cement mix. Joints between each 10- by 10-ft slab, to prevent cracking, can be made with an ax or trowel and finished with a groover.

According to Portland Cement Association recommendations, concrete can be poured in winter by following these precautions:

1. Concrete should be between 60 and 80°F when placed in the forms. Water should not be hotter than 175°F.
2. Mix so that concrete is as stiff as possible and can still be handled.
3. Remove frost and ice from forms and thaw frozen ground.
4. Cover flat concrete with heavy building paper and 12 in. of hay or straw. Enclose other concrete members with canvas and keep warm.
5. Provide 5 days of curing at 50°F or higher. A shorter time is required if high-early-strength concrete is used.
6. By using 2 lb of calcium chloride per sack of cement, curing and hardening can be accelerated to 3 days instead of 5 days. Chemicals are not added to prevent freezing. The strength is not decreased by using calcium chloride.

POULTRY PRODUCTION STRUCTURES

Egg production does not vary greatly at temperatures between 45 and 65°F. Feed consumption increases at lower temperatures, and egg production falls off rapidly at temperatures above 80°F. Feed consumption per pound of eggs was least at 55°F.[1] (See Chap. 55 for additional information on environmental factors and on heat and moisture losses.) Hens should lay 160 to 180 eggs per year. Hens are commonly confined in laying houses on litter during the winter months. Wet litter is a serious problem because, without auxiliary heat in cold climates, it is apparently "not possible to move enough air through a poultry house to remove moisture and yet keep the temperature above freezing,"[2] for 2 to 3 sq ft per hen density. Daily removal of droppings from pits by adaptations of automatic gutter cleaners greatly reduces the moisture evaporated into the air and, in turn, aids considerably in keeping litter dry. Recent methods of housing using a higher density of population, such as one bird per square foot, permits maintaining temperatures above freezing. Wet litter is most serious with a vapor-sealed concrete floor and less serious with unsealed concrete, wood, or dirt floors. Hens are particularly sensitive to rapid changes in temperature. Poultry are particularly susceptible to respiratory diseases and pullorum. The incubation period of a fertilized egg is about 21 days. Cannibalism is a natural tendency in chickens. Basic space requirements for poultry are summarized in Table 47-15.

Laying Houses. Poultry laying houses may be either one- or multiple-story, the latter being used only with large flocks. In many cases, existing barns have been remodeled into multiple-story laying houses. One-story houses of pole construction having dirt floors built up with sand or gravel are being successfully used. The important factor with either type of house is to fit it to an efficient standardized method of production.

---

[1] Garver and Ashby, *op. cit.*
[2] W. J. Promersberger and Reece L. Bryant, Forced vs. Natural Draft Ventilation in Poultry Houses, *Agr. Eng.*, 33:782–783, December, 1952.

Conventional poultry houses for layers have 3 to 4 sq ft of floor area per bird. Considerable reduction of housing cost can be made by confining layers to 1 sq ft each or less. The heat production, coupled with mechanical air movement, helps provide a temperate atmosphere during cold weather. Feeders, waterers, and roosts can be stacked over a litter pit, approximately 6 ft wide, and cleaned with a mechanical cleaner (Fig. 46-32). Multidecked feeder roosts can be three high, although some report that the chickens cannot be kept clean. By cleaning daily, the moisture given off to the atmosphere is reduced. Nests can be placed next to an alley designed so that the eggs may be removed from the backside of the nests or the alley side or from a slanting bottom which permits the eggs to roll to a trough on the alley side. An egg conveyer could move the eggs to a grading center. In conventional units with over 200 hens, the annual labor is from 1.5 to 1.8 hr per hen.

Cage Operations. The use of wire cages for confining from 1 to 25 hens in a cage has developed rapidly in the southwest part of the United States and has been spreading to other areas. The operator of this system can keep the hens producing at a maximum and can eliminate those hens not producing. Culling of hens to maintain a high production is based on (1) egg-production records, (2) feed consumption and droppings, and (3) physical appearance of hens.

The wire cages are usually one high, provide water and feed, and deliver the eggs to the front of the cage. One operator can care for about 3,000 layers in this system.

Types of Houses. There is considerable variation in poultry-house design, depending principally on the climate, from totally enclosed buildings to relatively open buildings. One classification, based on closure, is as follows:[1]

Wall-less, or wire-walled, house, composed principally of a roof, used in southern climates of the United States.

Uninsulated, often with large front opening, which may or may not be closed with doors and windows, depending on the climate, used in southern climates of the United States.

Straw loft, an enclosed house, may have insulated walls in the far north, used in northern areas of the United States.

Warmhouse, insulated walls and ceilings, controlled ventilation, may have large insulated windows, used in northern areas of the United States.

Broiler Houses. Broiler production has grown from 0.6 billion lb in 1940 to over 4½ billion lb in 1957. From 0.75 to 0.80 sq ft per bird is provided for broilers. Dirt floors with built-up litter are generally used. Mechanical handling equipment is used throughout. Three to four batches of birds are raised per year. Buildings are 40 to 80 ft wide and in lengths to accommodate 10,000, 20,000, 30,000 birds, etc. Heating systems are used during the winter months to start the broilers.

Chicken Brooders. Infrared brooders may be used by large and small growers. Because of their simplicity, cleanliness, and economy, they are especially adapted for flocks of 500 chicks or less. A lamp of 250-watt size is used for approximately each 100 chicks in northern climates. Units for one to six lamps are commonly available. Bulbs are usually mounted at a 45° angle. Wafer-type thermostats are placed above the lamps and control one or two lamps in combination with one or two lamps operating continuously. Operating costs are reduced by using a thermostat. For brooding in February, the infrared brooder will use 1 to 1½ kwhr per chick. The lamps are placed 15 in. above the litter at the start of the brooding period and raised 2 in. per week up to 24 in. above the litter.[2]

The following *practices* apply to use of infrared brooders:
1. Suspend brooder from a chain or wire, not electric cord.
2. Build a cardboard or metal guard about 16 in. high on the floor around the brooder to confine the chicks for 2 weeks to prevent floor drafts.
3. Watch chicks to determine height and number of lamps. If cold, they will crowd under the lamps.

---

[1] Hajima Ota, Houses and Equipment for Laying Hens, *USDA Misc. Publ.* 728, 1956.
[2] Vernon H. Baker and James H. Bywaters, Brooding Poultry with Infrared, *Agr. Eng.*, pt. I, 32:316, June, 1951; pt. II, 33:15–20, January, 1952.

4. The maximum load should not exceed 80 per cent of the rated capacity of the electrical circuit.[1]

When comparing radiant infrared lamp, electric hover, underfloor electric heating (3.75 watts per sq ft), underfloor hot water heated with electricity, and bottled gas in an identical environment, there was no significant difference in the methods of broding as to grain, feed consumption, or feed efficiency. For 5-week tests, the electric hover (1,000-watt element) used 1.9 kwhr per chick; infrared heat lamp, 2.61; and underfloor electric, 5.71.[2] Heating equipment for broilers is compared in Table 47-11.

TABLE 47-11. COMPARISON OF METHODS OF HEATING BROILER HOUSES

Classification	Type: warm or cool room	Installation cost, cents per chick based on		Fuel cost per chick, cents	Characteristics
		500 chicks	10,000 chicks		
Individual units:					
Coal..........................	Warm	8	.....	1.9	More labor, fire hazard
Oil...........................	Warm	7	.....	2.3	Fire hazard
Gas..........................	Cool	9–11	.....	1.6	Litter damp, fire hazard
Electric resistance	Cool	10	.....	1.5	Litter damp
Infrared......................	Cool	6–10	18–20	2.7	High operation cost
Central heat:					
Indirect, steam, warm air (coal)....	Warm	....	12–18	1.0	Desirable for long distance
Direct hot air....................	Warm	....	16–20	1.0	Air-heated in furnace and distributed by ducts
Hot water* (coal)................	Warm	....	20–25	0.6	Hot water at 170–190°F distributed through 1-in. pipes, 6" apart, 14" above floor, under a hover
Hot water* (oil)..................	Warm	....	20–25	1.0	
Radiant floor....................	Warm	....	.....	...	High initial cost, high operating cost

* Known as the Shenandoah system.
SOURCE: Silas McHenry, Poultry Heating Systems, *Univ. Delaware Agr. Expt. Sta. Bull.* 50, 1950.

The following *principles* and *practices* have been found effective in achieving efficient poultry production:

1. Keep all hens in one building, if possible.
2. Keep hens in large pens to reduce work. Control disease by immediate removal of sick birds.
3. Use portable brooders and range shelters to raise healthy chicks.
4. Use self-feeders in hen house and on range. Hen-house feeders should hold several days' supply and have a chute through the wall for filling from the outside. For the range, feeders should hold a week's supply.
5. Store feed in the house if self-feeders are not used.
6. If home-grown feeds are used, balance with rotation pasture in summer and bright legume hay in racks in winter. Good ensilage is high in vitamins and may be valuable.
7. Pipe water to poultry house, and heat water in cold weather to prevent freezing.
8. Use deep built-up litter, and clean once or twice a year.
9. Use low roosts over boarded-off dropping pits which require cleaning only three or four times a year. Sprinkle phosphate over droppings to combine with ammonia and reduce odors. Use borax to control flies in summer.
10. Provide doors and layout which will permit driving a manure spreader through the house for cleaning or provide a mechanical cleaner.
11. Have trap doors and chutes for manure removal from multiple-story poultry houses.

[1] ASAE Recommendation. Installation of Electric Infrared Brooding Equipment, Approved 1955, *Agr. Eng. Yearbook*, 1960, p. 133.
[2] J. Roberts, J. S. Carver, and W. E. Matson, Comparison of Various Methods of Brooding, *Agr. Eng.*, 33:557–558, 560, September, 1952.

12. Permit hens to get plenty of fresh air.
13. Keep disease-carrying sparrows out by screening all openings.
14. Operate equipment the year around by designing range shelters with panels to cover the sides so that they may be used for brooder houses and pullet laying houses.
15. Put pullets in thoroughly cleaned pen separate from the old hens.
16. To obtain clean eggs use deep nests, filled with fine nesting material, and nests hinged to pull away from the wall for easy cleaning. Locate nests close to building entrances. Perches on nests, feeders, and waterers should not be over 2 in. wide to prevent the accumulation of droppings.
17. Use artificial light if necessary to give hens a 13- to 14-hr day for highest production. Use one 40-watt electric light per 200 sq ft of floor area (about 1 watt per hen). Lights which lengthen the day by turning on in the morning are preferred. Evening lights require dimmers.
18. Provide well-arranged storage and packing rooms close to the center of the plant.
19. Eggs stored at low temperatures at a relative humidity of 70 to 85 per cent will retain their grade longer than in dry air because moisture is not lost rapidly. Temperatures of 60°F are generally used for short summer storage to reduce surface condensation on eggs when removed.

SHEEP PRODUCTION STRUCTURES

Sheep subsist principally on roughages and can be used for weed and grass control. Loss of sheep killed by dogs and other predatory animals is often great. Sheep are particularly susceptible to parasites and foot diseases and require considerable attention, particularly at lambing. Sheep breeds are classified according to wool and mutton type—fine wool, long wool, medium wool dual-purpose, and medium wool mutton breeds. With the exception of the mutton types, sheep have a natural herding instinct. Equipment and shelter are rather inexpensive. In many areas shelter for protection from sun and driving rain and snow is sufficient for breeding stock. The daily water consumption is ½ to 1 gal per sheep. Basic requirements for sheep are given in Table 47-16.

Sheep Housing. The sheep flock is often placed in "space" not occupied by other animals at the time. However, the arrangement of buildings and equipment for labor efficiency and a suitable environment at an economical cost is as necessary for sheep as for other enterprises. The minimum size of flock for a profitable enterprise is 40 ewes.[1]

Sheep do best in a dry environment free from drafts. The buildings and equipment should be designed to provide a flexible arrangement and proper environment during:[1]

1. Inclement winter weather
2. Lambing season
3. Period after shearing
4. Hot summer weather

A building with an open front away from prevailing winds is desired. The open side can be partially closed with bales stacked about 4 ft high. Provisions should be made for small individual pens during lambing. One pen for each five ewes is sufficient, with tight walls to prevent drafts and an electric outlet for a heat lamp. The heat lamp should be at least 6 in. above the animals. A simple system for providing sunshine and protecting the sheep during the winter, and for keeping out the sunlight during summer with adequate ventilation is as follows: two sides of a shelter can be provided with woven-wire fencing about 3 ft high; above the fencing are windows: sliding or hinged doors are arranged to cover the bottom fence in winter.[2]

Housing is needed to protect wooled feeder lambs from rain, wind, and snow. Shelter is not needed for warmth.

The following *practices* and *principles* should be considered when designing a system for *feeder* sheep:

[1] J. C. Wooley, M. L. Esmay, and A. J. Dyer, Sheep Housing and Equipment, *Univ. Missouri Agr. Expt. Sta. Bull.* 655, 1955.
[2] *Ibid.*

1. Provide means of sorting lambs in different-size groups.
2. Before shearing feeder lambs, compare value of wool, cost of shearing, higher feed requirement, shelter requirements, and lower market price for lambs.
3. Wool should be kept clean. Provide adequate bedding, conveniently stored. Do not carry hay over sheep to feed racks.
4. Feeder lambs consume shelled corn efficiently without grinding. Grinding will help mix grains and hay.
5. The feeding of grain to sheep should be controlled to prevent overeating until they are placed on fall feed. A common feeding schedule starts with ¼ lb per lamb per day the first week, ½ lb the second week, ¾ lb the third, and so on up to the desired level.
6. Under full grain feeding lambs will gain ⅓ lb per day per lamb.
7. A 60-lb feeder lamb finished to the desirable weight of 100 lb will require approximately 120 lb of hay and 120 lb of grain.
8. Treatment of lambs for diseases and parasites is an important part of successful management.

The following *practices* and *principles* should be considered in designing a system for breeding sheep:[1]

1. The gestation period for ewes is 147 days. The normal breeding season is autumn, but in California the desirable breeding season is in the late spring or summer.
2. A yearling ram will serve 20 ewes; if full-grown, 30 to 50 ewes.
3. Ewes should not be fat during the breeding period. Ewes are often "flushed," i.e., fed so they are gaining weight at time of breeding.
4. Artificial insemination under farm conditions requires so much labor that it is not practical at present.
5. Size is an important criterion for determining whether a ewe lamb is ready to be bred. They should weigh at least 80 lb, which will normally be when they are 14 to 19 months of age.
6. Ewes are normally clipped of wool in the spring.
7. The wool on a ewe may mislead the observer regarding the condition of the ewe.
8. Docking and castrating should be done as soon as lambs are strong enough, preferably early in their life.

HORSE HOUSING

The number of draft horses has decreased rapidly over the past ten years, but the number of light horses has increased. The space requirements for draft horses in Table 47-17 should be modified according to the size of the horse. A box stall is preferred over a tie stall, since the horse is free to move. A box stall of 12 by 10 ft is recommended for a light or pleasure horse, and a 12- by 16-ft box stall for foaling mare or stallion.

Stall partitions are usually solid up to 4 to 4½ ft in height and slatted to the ceiling. The ceiling should be at least 8 ft high. The manger consists of a large box (3 by 2 by 3 ft) for hay with a feed box about 6 in. deep in one corner for grain. The door should be 4 ft wide, built in two sections so that the top half can be opened separately. A good floor consists of a packed layer of clay over tile covered with cinders.[2]

The following feeding *principles* should be considered when designing horse facilities.[2]

1. Water before feeding. Automatic waterers are often used. A light horse will consume 10 to 20 gal of water per day. Do not water heavily if horse is hot or before heavy work.
2. Feed hay before grain or feed chopped hay with grain. Musty or dusty hay should not be fed to horses. About 1 lb of hay per day will be consumed by a light-working horse for each 100 lb of live weight.

[1] William C. Wair and Reuben Albaugh, California Sheep Production, *Univ. Calif. Agr. Expt. Sta. Manual* 16, 1954.

[2] John M. Kays, The Care of Light Horses, *Univ. Missouri Agr. Expt. Sta. Circ.* 353, December, 1950.

3. Feed grain which is not moldy. Oats are the standard grain. Feed about 1 lb of grain per day for each 100 lb live weight for light work.

4. Do not work a horse after full feed or when exhausted.

REQUIREMENT SUMMARIES

Space requirements for livestock are summarized in Tables 47-12 to 47-17.[1]

TABLE 47-12. BASIC SPACE REQUIREMENTS FOR HOGS

Hogs require close conformity to proper conditions if the enterprise is to be successful. The requirements listed have been based on profitable hog units. Variations should be allowed only for unusual conditions.

Items	Sows	Pigs under 100 lb	Fattening hogs 100 to 200 lb	Boars
Pasture and lot area:				
Pasture............	1 to 3 sows per acre, average 2 (including their pigs for the season)	15 to 25 head per acre	10 to 20 head per acre	$1/8$ to $1/10$ acre per boar
Dry lot, surfaced.....	20 to 30 sq ft per sow	8 to 12 sq ft per head	12 to 20 sq ft per head	25 to 35 sq ft per boar
Buildings:				
Floor area...........	Before farrowing—20 to 35 sq ft. During and after farrowing—48 to 80, average 64 sq ft	3 to 6 sq ft	6 to 10 sq ft	20 to 65 sq ft
Guardrails..........	8 to 12" from wall 8 to 10" clearance above floor			
Pen partition, height..	36"	32"	36"	42"
Windows............	1 sq ft per 30 to 40 sq ft floor space	1 sq ft per 30 to 40 sq ft floor space	1 sq ft per 30 to 40 sq ft floor space	1 sq ft per 30 to 40 sq ft floor space
Doors..............	24" wide 36" high	24" wide 36" high	24" wide 36" high	24" wide 36" high
Shade..............	20 to 30 sq ft 4' high	5 to 8 sq ft 4' high	8 to 12 sq ft 4' high	20 to 30 sq ft 4' high
Sanitation..........	Portable shelter, feeding and watering equipment			
Feeding and watering equipment:				
Self feeder—ear corn, grain or carbohydrate feed	1 lin ft feeder space per 2 to 3 sows	1 lin ft feeder space per 6 to 10 head	1 lin ft feeder space per 4 to 6 head	1 lin ft feeder space per boar
Supplement (high protein).........	1 lin ft per 3 sows	1 lin ft per 10 head	1 lin ft per 10 head	1 lin ft per 2 boars
Feeding floor, with self-feeders (feeder space included)	3 to 4 sq ft per sow	1 to $1\frac{1}{2}$ sq ft per head	$1\frac{1}{2}$ to $2\frac{1}{2}$ sq ft per head	4 to 6 sq ft per boar
Hand feeding.......	15 to 20 sq ft per sow	4 to 6 sq ft per head	8 to 10 sq ft per head	15 to 25 sq ft per boar
Feed trough.........	18 to 20"	10"	14"	24"
Water per day.......	5 to 8 gal	$\frac{1}{2}$ to $1\frac{1}{2}$ gal	$1\frac{1}{2}$ to 3 gal	2 to 4 gal
Water-trough space...	1' per 10 to 15 sows	1' per 40 to 50 head	1' per 25 to 30 head	1' per 3 boars
Feed storage, 1-year supply:				
Grain..............	9 cu ft (7 bu) per 100 lb gained weight including feed space for sows, pigs, and boars, or 20 cu ft (16 bu) grain space per 225 lb hog raised; 40 cu ft if grain is ear corn			
Protein supplement...	1 cu ft (50 lb) per 100 lb gained, including feed space for sows, pigs, or boars			
Alfalfa meal.........	5 cu ft	1 cu ft	2 cu ft	3 cu ft
Bedding storage:				
Straw, loose........	250 to 400 cu ft (500 to 800 lb)	25 cu ft (50 lb)	50 cu ft (100 lb)	100 cu ft (200 lb)
Baled or chopped...	50 to 80 cu ft	5 cu ft	10 cu ft	20 cu ft

[1] Appreciation is expressed to the Doane Agricultural Service, Inc., St. Louis, Mo., for permission to reproduce these tables. The tables were originally published in December, 1947, and have been changed according to the suggestions of B. G. Perkins of the Doane Agricultural Service.

Table 47-13. Dairy Cows—Basic Space Requirements

Dairy cows are housed under two systems, stall and loose-housing barns. The loose-housing system is becoming more and more popular. Check local health and dairy codes. The following table covers both.

Items	Requirements
Lot area:	
Soil	300 to 1,200 sq ft per cow, heifers 300 to 1,000 sq ft, bull lot size optional
Surfaced	35 to 100 sq ft per cow, heifers 35 to 80 sq ft, bull lot size optional
Loose housing:	
Floor of resting area	40 to 100 sq ft per cow. Use smallest area when barn is supplemented with outside lot and all feeding and watering are outside away from loafing area. Heifers, 20 to 40 sq ft; bulls, 100 to 144 sq ft, or approximately 200 sq ft when using open shed
Width of resting area	Minimum for cows 16'; heifers 12'; bulls 10'; and preferably not over 30', except in extreme northern area
Width of open shed	24' to 32'
Height of ceiling	8' to 10'—clearance for power manure loader
Foundation or splashboard	3' above floor to allow for accumulation of manure
Floor	Well-drained dry soil is most satisfactory
Exposure	Open to south or east
Bedding	
Straw, loose	15 to 25 lb per cow per day, allow 1,500 to 2,000 cu ft (2 tons) per cow, 700 to 1,000 cu ft (1 ton) per heifer, and 1,500 to 2,000 cu ft per bull
Baled or chopped	400 cu ft (2 tons) per cow, 200 cu ft (1 ton) per heifer, and 400 cu ft per bull
Hay manger:	
Open type, length	3' to 5' per cow, 3' per heifer, 3' per bull
Partitioned, length	32" to 36" per cow, 24" to 32" per heifer
Horizontal silo, length	Depends on quantity of silage needed
Width	6 in. per cow for self-feeding
Water	Winter, under pressure, 1 sq ft open surface per 25 cows; water bowls in maternity, calf, and bull pens. Summer, tank 15 to 25 gal per head. Locate in lot away from loafing barn
Stall barn:	
Width of barn	34' to 36'
Floor area	60 to 80 sq ft per cow including stalls, mangers, feedway, and litter alley; heifers, varies according to how handled; bulls, 100 to 144 sq ft
Stall size:	
Holstein, length	4'10" to 5'8"
Width	3'6" to 5'0"
Ayrshire, length	4'6" to 5'6"
Width	3'6" to 4'6"
Guernsey, length	4'6" to 5'4"
Width	3'4" to 4'0"
Jersey, length	4'4" to 5'0"
Width	3'4" to 4'0"
Height, ceiling	7'6" to 8' in cold areas, 8' to 9' in moderate areas
Foundation or splashboard	Extend at least 12" above floor
Floor	Solid and impervious
Gutters, width	16" to 18"
Depth	8" at stalls, 6" at litter alley
Litter alley	8' wide—permit driving spreader through
Feedway	3'6" to 4'6" wide
Mangers	24" to 30" wide
Water	Under pressure, water bowl for every 2 cows, bowls in maternity, calf, and bull pens
Manure pit	Not recommended unless no way possible to haul manure direct to field. Capacity 1½ cu ft per cow per day. Separate tanks for liquid manure desirable but not always practical
Bedding:	
Straw, loose	8 to 12 lb per cow per day. Allow 500 to 750 cu ft (¾ ton) per cow, 250 to 400 cu ft (⅜ ton) per heifer, and 500 to 750 cu ft per bull
Baled or chopped	150 cu ft (¾ ton) per cow, 75 cu ft (⅜ ton) per heifer, and 150 cu ft (¾ ton) per bull
General:	
Maternity pens	12' by 12'
Bull pens	10' by 10' to 12' by 12'
Calf pens	24 sq ft per calf
Doors	Single, 3'6" to 4' wide; double, 8'
Light	Windows, 3 to 3½ sq ft per cow; electric, one 60-watt bulb per 4 cows
Feed storage	
Hay, loose, without silage	1,000 to 1,500 cu ft (2 to 3 tons) per cow or bull, 500 to 750 cu ft (1 to 1½ tons) per heifer
Baled or chopped, without silage	400 to 600 cu ft (2 to 3 tons) per cow or bull, 200 to 300 cu ft (1 to 1½ tons) per heifer
Loose, with silage	500 to 1,000 cu ft (1 to 2 tons) per cow or bull, 250 to 500 cu ft (½ to 1 ton) per heifer
Baled or chopped, with silage	200 to 400 cu ft (1 to 2 tons) per cow or bull, 100 to 200 cu ft (½ to 1 ton) per heifer
Grain, 3 to 4 weeks' supply	Allow 8 to 10 cu ft per cow, bull, and heifer
Supplement, 3 to 5 months' supply	Allow 6 to 8 cu ft per cow, bull, and heifer

Table 47-14. Beef Cattle—Basic Space Requirements

Beef cattle are handled under many different conditions. Requirements on individual farms and ranches will therefore vary accordingly. The table below lists the most generally accepted requirements under practical management.

Items	Cows or 2-year-old steers	Yearlings	Calves (400 to 500 lb)	Bulls
Feed lot:				
Soil	300 to 1,200 sq ft per head	300 to 1,000 sq ft per head	150 to 500 sq ft per head	Optional
Surfaced	35 to 100 sq ft per head	35 to 80 sq ft per head	30 to 50 sq ft per head	Optional
Open shed:				
Floor area	20 to 40 sq ft per head	15 to 25 sq ft per head	12 to 20 sq ft per head	200 sq ft per head. **Enclosed** building, 100 to 144 sq ft
Floor	Dirt preferred	Dirt preferred	Dirt preferred	Dirt preferred
Width (shed depth)	Minimum 20′, optimum 24′ to 32′	Minimum 20′, optimum 24′ to 32′	Minimum 20′, optimum 24′ to 32′	20′
Height, ceiling	8′ to 10′ Clearance for power manure loader	8′ to 10′	8′ to 10′	8′
Supports, open shed posts	Spans between posts lengthwise of building 12′—not less than 10′ nor more than 14′. Spans for width of buildings, no center pole needed if not over 24′ wide, gable roof, and on poles 4′ to 5′ in ground. Use 20′ rafter ties and 1 × 6's crisscrossed between the 2 × 6 rafters and ties forming a simple nailed truss. Width spans of 26′, 28′, 30′, and 32′ should have center post with rafters supported by side braces on posts to make unsupported portion of 2″ × 6″ rafters not over 14′; or truss may be used to eliminate center post			
Exposure	Open to south or east—except enclose 10% for calving in extreme northern areas			
Foundation or splashboard	2′ above floor to allow for accumulation of manure			
Feed storage	Varies greatly according to method of handling cattle			
Hay, loose without silage	500 to 1,000 cu ft per head (1 to 2 tons)	250 to 750 cu ft per head (½ to 1½ tons, depending on field roughage and grain fed)	125 to 500 cu ft per head (¼ to 1 ton, depending on grain fed)	500 to 1,000 cu ft per head
Baled or chopped without silage	250 to 500 cu ft per head (1 to 2 tons)	125 to 375 cu ft per head (½ to 1½ tons)	65 to 250 cu ft per head (¼ to 1 ton)	250 to 500 cu ft per head
Loose with silage	250 to 500 cu ft per head (½ to 1 ton)	125 to 375 cu ft per head (¼ to ¾ ton)	65 to 250 cu ft per head (⅛ to ½ ton)	250 to 500 cu ft per head
Baled or chopped with silage	125 to 250 cu ft per head (½ to 1 ton)	65 to 185 cu ft per head (¼ to ¾ ton)	30 to 125 cu ft per head (⅛ to ½ ton)	125 to 250 cu ft per head
Grain	Optional—none to 25 bu per head	30 to 45 bu per head	25 to 40 bu per head	None
Protein supplement	Optional—50 to 150 lb	100 to 300 lb per head	100 to 300 lb per head	None to 200 lb per head
Bedding storage:				
Straw, loose	300 to 1,000 cu ft per head (⅓ to 1 ton)	300 to 1,000 cu ft per head (⅓ to 1 ton)	250 to 750 cu ft per head (¼ to ¾ ton)	500 to 1,000 cu ft per head (½ to 1 ton)
Baled or chopped	70 to 200 cu ft per head	70 to 200 cu ft per head	50 to 175 cu ft per head	100 to 200 cu ft per head
Hay manger:				
Length, per head	24″ (partitioned)	20″	18″	30″
Height, at throat	26″	24″	20″	26″
Feed bunk:				
Length, per head	24″	20″	18″	30″
Height, at throat	30″ (Hogs quickly learn to jump in bunks less than 30″ high)	30″	24″	30″
Horizontal silo, length	Depends on feed needed for feeding program			
Width per head for self-feeding	4″	3″	2½″	Hand feed
Water:				
Lot feeding	1 sq ft open surface on pressure system per 25 head			
Pasture	350- to 600-gal tank supplies 25 head			

TABLE 47-15. POULTRY—BASIC SPACE REQUIREMENTS

Poultry must be closely controlled. Requirements vary with minor climatic conditions. The table lists the customary requirements for most conditions, but poultry raisers are advised to consult local authorities regarding their particular conditions. Most state colleges have devoted extensive study to poultry production, and county agricultural agents can supply results of their findings.

Items	Hens	Chicks
Lot and outside pen:		
Lot area..................	175 to 200 sq ft per hen	Open range—100 to 125 sq ft per chick
Outside pen, supplement floor area...	4 to 8 sq ft desirable where building floor area is less than 3 sq ft per bird	Where open range not available—1 to 3 sq ft to supplement brooder floor area as chicks become larger
Buildings:		
Floor area..................	2 to 3½ sq ft per hen	Baby chicks—⅓ to ½ sq ft Growing chicks—½ to 1½ sq ft Broilers—½ to ¾ sq ft
Roost space	7" to 10" per hen 13" to 15" apart	5" to 7" 8" to 10" apart
Roost height..............	12" to 15" above litter	10" to 15" above floor
Nests.....................	1 per 5 to 7 hens, 12" wide, 14" long, 14" high	
Windows..................	1 sq ft window per 25 to 50 sq ft floor	Approx. 1 sq ft window per 30 sq ft floor
Electric lights..............	50-watt bulb per 200 sq ft floor, give 12- to 14-hr day	
Doors for chickens..........	One 15" × 16" per 100 hens	One 10" × 12" per 100 chicks
Feed storage:		
Mash.....................	4 to 6 weeks' supply—10 to 15 cu ft (500 to 750 lb) per 100 birds	2 to 3 weeks' supply—2 to 3 cu ft (100 to 150 lb) per 100 birds
Grain (scratch).............	10 to 20 cu ft (500 to 1,000 lb) per 100 chicks	2 to 3 weeks' supply—2 to 3 cu ft (100 to 150 lb) per 100 birds
Feeding and watering equipment:		
Feed troughs...............	3" to 5" per hen	Baby chicks—1" per bird Growing chicks—3" per bird
Self-feeder.................	1" feeder space per hen	¾" to 1" feeder space per bird
Waterers...................	6 gal per 100 hens per day Farm flock—two 3-gal buckets on stands per 100 hens Commercial—one automatic waterer per 600 hens	Baby chicks—1 qt fountain per 50 chicks Growing chicks—1 to 2 gal fountain per 50 birds Broilers—1 automatic waterer per 600 birds

Table 47-16. Sheep—Basic Space Requirements

Sheep can be raised successfully with minimum conformity to requirements, but feeding for market should be done under favorable conditions. The requirements listed are more adaptable to the farm flock from which the lambs are fed out and to normal feeder-lamb operations than to large commercial units.

Items	Ewes (with lambs) or rams	Feeder lambs
Feed lot:		
Soil	25 to 50 sq ft	8 to 15 sq ft
Surfaced	15 to 20 sq ft	6 to 8 sq ft
Building:		
Floor area	15 to 22 sq ft	5 to 8 sq ft
Pens	Provide 1 pen to 5 ewes. Use two 4' gates hinged together set as pen along wall	
Partition fence	38" high—12" board at ground, 6" board on top. Posts 12' apart	space, 6" board, 6" space, and 6"
Width (shed depth)	20' to 24'	20' to 24'
Height	8' to 10'—clearance for power manure loader	8' to 10'—clearance for power manure loader
Supports	Spans between posts 12'—not less than 10' nor over 14'	
Exposure	Facing south or east. Open except 10% closed for lambing	Open to south or east
Feed storage:		
Hay, loose, without silage	150 cu ft (600 lb)	25 cu ft (100 lb)
Baled or chopped	60 cu ft (600 lb)	10 cu ft (100 lb)
Loose, with silage	75 cu ft (300 lb)	13 cu ft (50 lb)
Baled or chopped	30 cu ft (300 lb)	5 cu ft (50 lb)
Grain	1 to 2 cu ft (50 to 100 lb)	2 to 3 cu ft (100 to 150 lb)
Bedding storage:		
Straw, loose	50 to 80 cu ft (100 to 160 lb)	17 to 32 cu ft (35 to 65 lb)
Baled or chopped	10 to 16 cu ft (100 to 160 lb)	3½ to 6½ cu ft (35 to 65 lb)
Hay feeding rack:		
Length, per head	18" to 24"	12" to 14"
Width, feed both sides	20" to 24"	19" to 21"
Feed one side	14" to 16"	12" to 14"
	1" by 6" at 45° angle across lower inside corner	
Height, at throat	12" to 15"	10" to 12"
Hay feeding fence	12" board at ground, 8" open space, 8" board, 8" open space, and 6" board on top. Linear space per feeder lamb, 9" to 10"	
Grain trough:		
Length, per head	14" to 18"	10" to 12"
Width, feed both sides	20" to 24"	19" to 21"
Feed one side	14" to 16"	10" to 12"
Height, at throat	12" to 15"	10" to 12"
Self-feeder:		
Length, per head	Not used	3" to 4"
Width of feed area		8" to 10"
Height, at throat		10" to 12"
Depth, where lambs eat		5" to 6"
Creep, for lambs with ewes	Openings 8" to 12" wide	
Water	1 gal per day	½ gal per day

TABLE 47-17. HORSES (DRAFT)—BASIC SPACE REQUIREMENTS

Items	Mare	Colt, yearling, two-year-old
Stall size:		
Floor area	60 to 80 sq ft	
Width	5'	
Length, including manger	14' to 16'	
Box stall	10' × 10' to 12' × 14'	8' × 10' or 10' × 10'
Height	8'	8'
Hay manger:		
Width	2' to 3'	2'
Height at throat	38" to 42"	32" to 36"
Grain box:		
Width	12"	10" to 12"
Length	24"	12" to 24"
Height, sides	8" to 10"	6" to 8"
Above floor	38" to 42"	32" to 36"
Feed storage:		
Hay, loose	1,000 cu ft (2 tons)	750 cu ft (1½ tons)
Grain	50 to 100 cu ft (40 to 80 bu)	25 to 50 cu ft (20 to 40 bu)
Bedding:		
Straw, loose	500 cu ft (½ ton)	250 cu ft (¼ ton)
Baled or chopped	100 cu ft (½ ton)	50 cu ft (¼ ton)

Functional Requirements of Farm Buildings. The requirements of various farm buildings can be analyzed on the basis of function as it is affected by production. Thirteen functional requirements of buildings, such as appearance, cubage, clear span, exposure to manure, etc., have been evaluated for relative importance, as indicated by relative bar-graph length (Fig. 47-17). Thus a sidewall for a stall barn need be designed with slight consideration for exposure to manure, while for a loose-housing barn, exposure to manure is very important. Other functional requirements, such as sanitation, are important for certain buildings.

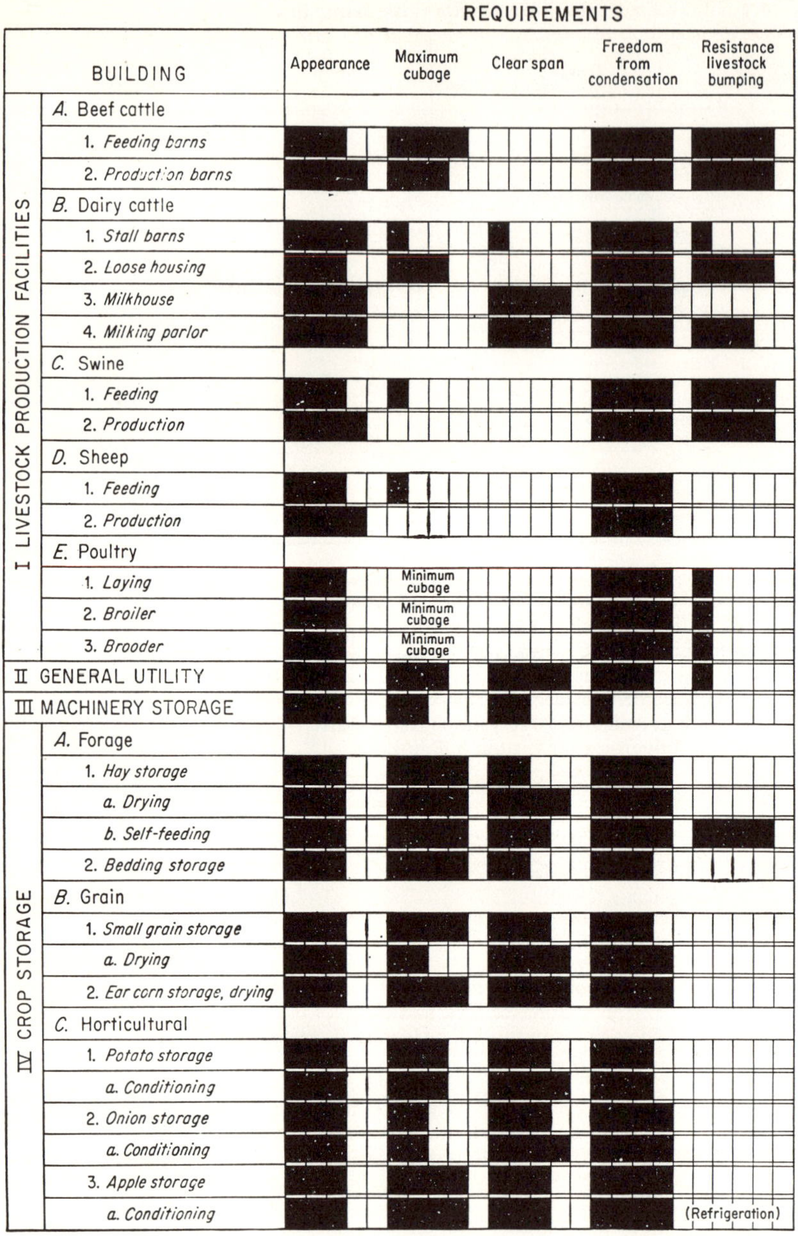

FIG. 47-17. Functional requirements of farm buildings.

LIVESTOCK PRODUCTION FACILITIES

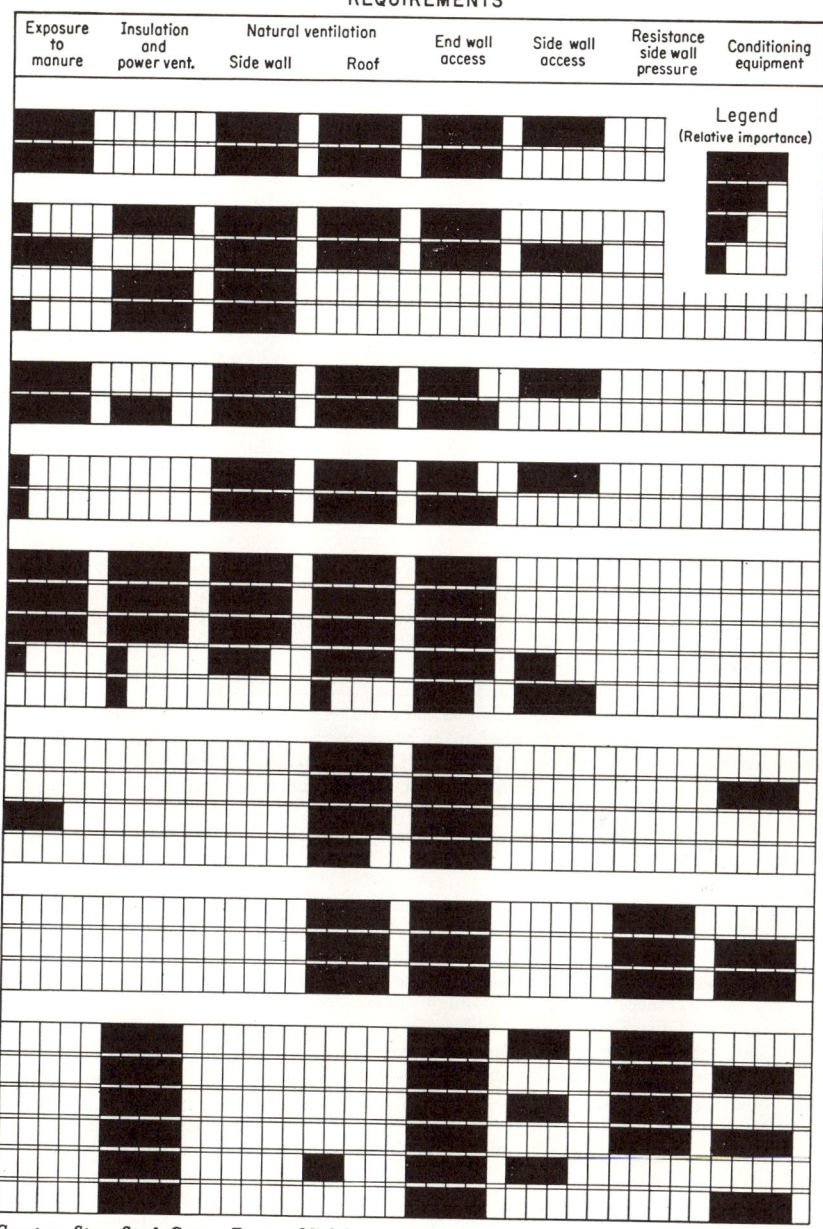

(*Courtesy Stran-Steel Corp., Ecorse, Mich.*)

Chapter 48

HEAT, AIR, AND MOISTURE

CARL W. HALL

The physical properties of air, water vapor, and heat are closely related to each other and are important in environmental control for livestock housing and drying and storing of crops. The relationships discussed are those that can be readily applied to field problems by engineers and farm operators. Some of the data are given in approximate values which are adequate for field use where most solutions must be approximated because the exact weather and crop conditions must be estimated in advance.

Psychrometric Chart. The various properties of air and air-water vapor mixtures are included in Fig. 48-1, known as the psychrometric, or humidity, chart. The psychrometric chart is based on standard barometric pressure. For most applications it is not necessary to correct for slight changes in atmospheric pressure.

If the initial condition of the air is known, it is possible to predict the final condition or calculate the change which will take place in order to obtain a desired final condition. For heating or cooling of air above the dew point, there is no change in vapor pressure because the change takes place along a horizontal line in Fig. 48-1, which represents constant vapor pressure. On some psychrometric charts the change in total heat can be read directly. Most psychrometric values are based on dry weight of air without moisture in order to provide a constant base.

The psychrometric chart is of particular value in applications of (1) ventilation of buildings, (2) crop drying, (3) condensation on building surfaces, and (4) air conditioning.

Vapor Pressure. Vapor pressure is the force that diffuses water vapor in all directions and is caused by the mass and velocity of vaporized water molecules. Vapor pressure is the sum of the forces exerted by individual molecules and consequently at any fixed temperature is proportional to the weight of water vapor in a cubic foot of air. At temperatures above the boiling point when the volume is confined under pressure, vapor pressures are termed steam pressures. Cold air has very little water vapor and consequently very little vapor pressure. Free water held in a closed vapor-tight container is said to have a vapor pressure for its temperature equal to the vapor pressure of saturated air space above the water surface.

Figure 48-1 shows the vapor pressure of water vapor at various humidities and the common temperatures. Thus, at 60°F and 50 per cent relative humidity, the pressure of the water vapor is 0.126 psi, represented by 38 grains of water per pound of dry air. As the weight of water remains constant and as heat is added, the humidity remains constant, the relative humidity decreases, and the vapor pressure remains constant.

Humidity. *Humidity* is the term which denotes the water vapor in the air, pounds water vapor per pound of dry air. *Relative humidity* is the percentage that the partial pressure of water vapor in the air is of water vapor in saturated air at the same temperature. Relative humidity is also defined as the percentage that the weight of water vapor present is of the weight of water vapor at saturation at the same temperature.

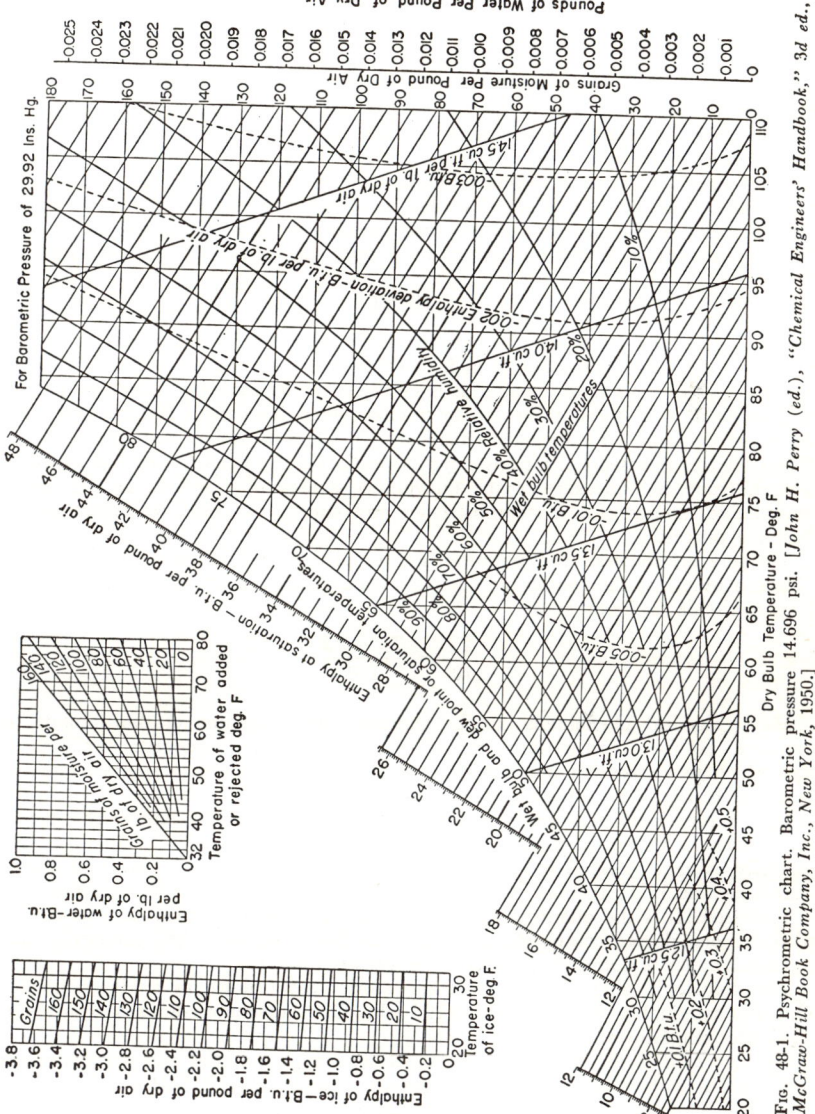

FIG. 48-1. Psychrometric chart. Barometric pressure 14.696 psi. [*John H. Perry (ed.), "Chemical Engineers' Handbook," 3d ed., McGraw-Hill Book Company, Inc., New York, 1950.*]

The *dew-point temperature* is reached when the temperature of the water vapor in the air cools to 100 per cent relative humidity. A further drop in air temperature or a drop in the temperature of a material exposed to water vapor at 100 per cent relative humidity will cause part of the water vapor in the air to be condensed, and it may appear as a liquid, as in sweating and rain, frost, or an increase in the moisture content of surrounding materials. Upon condensation, latent heat will be released. The relative humidity of the air is easily determined from its dew point. The dew point is most easily obtained by measuring the minimum temperature of a wet wick exposed to moving air. When the air is saturated at 100 per cent relative humidity, the temperature of the wet wick will be the same as the temperature of the air. If an ordinary thermometer is enclosed in the wet wick and exposed to moving air, the thermometer will indicate the dew-point temperature of the wick and the air, which is called a wet-bulb temperature. Figure 48-1 can be used to obtain the relative humidity from simultaneous wet- and dry-bulb thermometer readings. Wet- and dry-bulb thermometers mounted together for whirling in the air are called *sling psychrometers*. If mounted in an air stream, they are called hygrometers.

HEAT

Heat is the form of energy that provides differences in the temperature of molecular materials. Heat itself has no form and, like time, can be measured only by differences. The total heat content of a material is described as its *enthalpy*. Enthalpy is gained or lost when a material is heated or cooled, when it changes state between solid, liquid, or gas, when it is changed chemically, or when its atomic structure is changed. When heat is released, it is transferred by radiation, by conduction, and by convection. Entropy is a measure of the unavailable energy of a system. In practice, the unavailable energy of a system increases and energy available for work decreases during a process.

Temperature Scales. Temperatures are usually based on either the Fahrenheit (English units) or the centigrade (metric) scale. A convenient relationship to change from one scale to the other is

$$\frac{T_C}{100} = \frac{T_F - 32}{180}$$

Many physical phenomena can be expressed as a function of the absolute temperature. In the English units, the absolute temperature is degree Rankine (°R) equals degree Fahrenheit plus 459.69; and for the metric units, absolute temperature is degree Kelvin (°K) equals degree centigrade plus 273.16.

Heat Units. Engineers using the English language generally express the measurement of heat in terms of British thermal units, abbreviated Btu. *One Btu* is the quantity of heat required to raise the temperature of one pound of water one degree Fahrenheit at 58°F. In many branches of science the metric unit of heat, the calorie, is preferred. It is the quantity of heat required to raise the temperature of one gram of water one degree centigrade at 14°C. One Btu is equal to 252 calories. A larger heat unit called a *Calorie* is used which represents 1,000 cal.

The calorie is often designated gram-calorie; the Calorie as a kilogram-calorie. One Btu is equal to 778 ft-lb and 0.293 whr. One horsepower-hour is equivalent to 2545 Btu; one kilowatthour to 3413 Btu.

Sensible heat is the heat released or received by a material and accompanied by a change in temperature, but without change of state. *Specific heat* is the quantity of sensible heat required to raise the temperature of one pound of a material one degree Fahrenheit. The term *heat capacity* is often used instead of specific heat, with specific heat referring to the ratio of the heat capacities of the material to that of water. Table 48-1 provides useful average specific-heat values, some of which will show variations between individual samples. Average values are usually accurate enough for practical problems in the normal temperature range.

Latent heat is the heat gained or lost without a change in temperature when a substance passes from one state to another as a solid, liquid, or gas. When water freezes

TABLE 48-1. SPECIFIC HEAT (HEAT CAPACITY) FOR VARIOUS MATERIALS, BTU/(LB)(°F) OR CAL/(GRAM)(°C)

Material	Specific heat
Air at constant pressure	0.24
At constant volume	0.17
Aluminum	0.22
Asbestos	0.20
Bricks, stone	0.21
Butter	0.60
Cement	0.20
Coal	0.30
Concrete	0.16
Copper	0.09
Cotton	0.36
Flour	0.40
Fresh, high-moisture fruits, vegetables, and forages	0.90
Fresh meat, eggs	0.70
Glass	0.19
Hardwoods	0.57
Ice, 22°F	0.45
30°F	0.5
Iron, steel	0.12
Lead	0.03
Milk	0.93
Nickel	0.11
Nuts	0.25
Oils, petroleum, gasoline	0.50
Sand	0.20
Softwoods	0.65
Soil	0.17
Steam at 212°F	0.47
Water, 32°F	1.001
212°F	1.021
Wheat, ½ per cent moisture	0.30
15 per cent moisture	0.40
Wool	0.40

or when ice melts, 144 Btu per lb is given off or taken up without a change in temperature of the water or ice, and this is known as the *latent heat of fusion*. When free water evaporates at 212°F at atmospheric pressure, the heat taken up is 970 Btu per lb; this is known as the *latent heat of vaporization*. Condensation of water vapor liberates the latent heat.

HEAT SOURCES

Heat of Combustion. During combustion oxygen combines with the fuel to release heat. The quantity of heat released per pound of fuel is called the heating value, heat of combustion, or calorific value. The high, or gross, heat value refers to the amount of heat produced assuming that all the water vapor in the products of combustion is condensed. The low, or net, heat value refers to the amount of heat produced assuming that no water vapor is condensed.

Complete combustion occurs when all the carbon is oxidized to carbon dioxide. The presence of carbon monoxide, which is a poisonous gas, indicates incomplete combustion. There is from ½ to 1.5 lb of water given off for each pound of fuel completely burned. About one gallon of water is produced for each gallon of fuel oil burned. The high heat values of several agricultural products are listed in Table 48-2. See Table 9-1 for heat values of petroleum fuels.

Heat of Respiration. The heat of respiration varies considerably with internal chemical differences between two lots of the same product because of the stage of maturity, ripeness, and decay. All organic plant material, fruits, and seeds appear to produce some respiration heat at temperatures above freezing unless they are first treated with heat, chemicals, or radiation to inhibit biochemical activity due to internal enzymes or external bacteria, and molds. Optimum rates of production of respiration heat generally occur at temperatures above 80°F and up to 130°F, at high humidities and high internal moisture contents.

TABLE 48-2. APPROXIMATE HIGH HEAT VALUE OF SOME AGRICULTURAL PRODUCTS

Product	Btu per lb	Remarks
Bagasse (pressed sugar cane stalks)	8,000–8,700	40–55% moisture
Buckwheat hulls	7,550	14% moisture*
Butter	16,500	*
Cottonseed hulls	7,077	Weight of 15 lb per cu ft
Douglas fir	5,800	Wood waste
Garbage	7,000–10,500	63–85% moisture
Lard	16,700	From swine*
Oak	7,700	12% moisture
Oils of cottonseed, linseed, olive, grape	16,700–17,100	*
Sawdust briquets	8,100	Weight of 75–80 lb per cu ft
Shelled corn	7,800–8,500	For dry corn
Straw	5,000–6,500	Depends on moisture
Sugar maple	8,000	12% moisture content
Tanbark	2,600	Fibrous oak or hemlock residue
Yellow pine, white pine	8,100	Resinous wood

1 cord of wood is 8 by 4 by 4 ft = 128 cu ft, or 90 cu ft of solid material.
Conversion factor: 1 kg-cal per grain = 1,800 Btu per lb.
* "Handbook of Chemistry and Physics," Chemical Rubber Publishing Co., Cleveland, Ohio, 1952.
SOURCE: John H. Perry, "Chemical Engineers' Handbook," McGraw-Hill Book Company, Inc., New York, 1950.

RADIATION

Radiation is a general phenomenon, existing throughout the universe. Visible light makes up one section of the radiation spectrum. The other sections are not visible and contain the wavelengths of radio waves, infrared rays, ultraviolet rays, X rays, and other ultrashort wavelengths. Differences in radiation are due to the source, frequency, and wavelength. Frequency is the number of times per second a source emits a wave or pulse of radiation. Wavelength is the distance between waves or pulses. The product of the frequency and wavelength is the velocity of radiation. Hence, high-frequency waves are short and low-frequency waves are far apart.

The distance between various types of waves is measured in several units. Radio waves are usually measured in meters or fractions of meters. The wavelengths for infrared visible light and ultraviolet are usually expressed in angstroms. The angstrom (A) is a unit of distance equal to one ten-millionth of a millimeter. The wavelength might be expressed in millimicrons, which is one millionth of a millimeter.

The intensity of radiation is expressed by several different terms. For radio waves the intensity is expressed as amplitude and refers to the size of the waves. Intensity is a measure of the power in radiation and usually refers to the total energy striking on a given area. Intensity may be measured in watts, or in the case of visible light, in lumens, or candlepower. Regardless of the method of expression, the intensity is a measure of the total radiant energy on a specified area with a time interval such as a second, minute, or hour. The intensity of radiation varies inversely with the square of the distance from the source.

The longer wavelengths can be reflected by a suitable material and surface. Radio waves are so large that they require a large surface such as the surface of the earth, large structures, or mountains to reflect them.

Radiation travels in a straight line until it is absorbed, reflected, or refracted. Refraction takes place when radiation passes through the surface of a transparent medium of a different density. The familiar experience in which a straight pole partly submerged in water appears to be bent is an example of refraction.

Solar Radiation. Radiation from the sun is a combination of several wavelengths—infrared, light, and ultraviolet. About half of the energy in the sun's radiation is absorbed or reflected in the outer atmosphere on a clear day. The energy falling upon the earth when the sun is directly overhead is about 3 Btu/(sq in.)(hr) under the most favorable conditions. In Iowa during August,[1] an average solar-energy factor

[1] Alvin C. Dale and Henry Giese, Effect of Roofing Materials on Temperatures in Farm Buildings under Summer Conditions, *Agr. Eng.*, 34:168–177, March, 1953.

of 1810 Btu/(sq ft)(day) was found over a 10-day period on a roof facing south and sloped 45°.

Solar radiation on east and west walls is about the same through the seasons and reaches a maximum rate of about 1.3 Btu/(sq in.)(hr). In the winter, south windows receive most heat and the solar input may be slightly in excess of 1.0 Btu/(sq in.)(hr). In summer with the sun overhead, south windows receive little direct solar radiation, but may receive considerable sky and ground radiation from the warm temperatures that exist. However, ground radiation gives long wavelengths that do not penetrate window glass. In California's Imperial Valley, hot bare ground close to a livestock shelter radiated as high as 2 Btu/(sq in.)(hr) during the middle of the day in August and September.[1] Solar radiation is treated in more detail in Chap. 57.

Infrared Waves. Infrared rays have shorter wavelengths than radio waves and longer wavelengths than visible light. Slow-moving, low-temperature molecules emit the longer infrared wavelengths, and high-temperature, fast-moving molecules emit the short infrared wavelengths. Infrared radiation is a function of all molecular activity. All materials are constantly emitting and receiving infrared radiation. The emission of infrared energy reduces the temperature of a material, and the absorption of infrared increases the temperature. When the output and input of infrared are in balance, on an isolated body, its temperature remains constant. The rate at which materials absorb and emit infrared varies with the kind and the color of the surface. Only a few materials have a low absorption rate and high reflection rate. The best reflector is polished gold or gold plate which reflects about 95 per cent of infrared. Aluminum is a common material which reflects about 90 per cent of the longer infrared and about 75 per cent of the shorter infrared. Chrome plate may be used where aluminum is unsatisfactory because of temperature or corrosion and will reflect about 75 per cent as much infrared as aluminum. White paints are usually good infrared reflectors, white lead being nearly as good as aluminum.

Much of the infrared energy in sunshine is of a short wavelength which passes through window glass. In greenhouses, solar heating systems, and ordinary window systems, the short infrared from the sun passes through the glass and is converted to heat inside the building. The floors, walls, and furnishings within a building produce long infrared wavelengths which do not pass outward through window glass and are trapped within the building. Even though the infrared escape from the building may be small, heat may be rapidly lost by convection, conduction, or evaporation.

Materials do not usually absorb infrared energy and radiate or emit infrared at the same rate. Almost all bodies except a perfect black body reflect some of the infrared which falls on them. The rate of absorption can be expressed as a percentage of the radiant energy falling on a body. The remainder is reflected. Emission, however, is somewhat different. The rate of emission is usually expressed as a percentage of the rate of emission of a perfect black body or surface. The rate of infrared emission varies with the fourth power of the absolute temperature of the material.

Although conduction and convection methods are commonly used to move heat through a building in heating systems, much of the comfort of the heat is by infrared radiation from the walls and furnishings of the building. If the building is insulated so that the interior room surfaces are warm, the walls and ceiling radiate infrared energy back to human and animal bodies to make up most of that which is lost from the body. If the interior surfaces of a room are of polished clean aluminum and no furnishings are in the way, a man or animal will be kept comfortable by the infrared energy reflected back, even though the air in the room is cold. If, in addition, some infrared energy is supplied by heat lamps or radiation to make up for losses due to absorption, conduction, and convection, comfort can be achieved even though the air in the room is at 0°F or lower.

Solar Heating. The temperature to which surfaces are heated by the sun depends principally on the absorption coefficient and the ability of the surface to radiate heat. A black surface will absorb most of the radiant heat falling on it and hence reach higher temperatures than most other surfaces. A surface which will emit radiant

---

[1] C. F. Kelly, T. E. Bond, and N. R. Ittner, Thermal Design of Livestock Shades, *Agr. Eng.*, 31: 601–606, 1950.

heat (long-wave) will stay cooler than a similar surface which emits less heat. Work published by the National Bureau of Standards shows the following relationships among materials exposed to radiation from the sun:[1]

1. Aluminum foil is an excellent reflector, but because it has a poor emissivity at long wavelengths, it will become warmer than paints.
2. The temperature increase above atmospheric of aluminum paint with a spar-varnish vehicle was more than that of aluminum foil.
3. Of several colors of paints tested, the coolest to warmest were, with the *increase* in temperature *over* atmospheric: glossy white, 31°F; flat white, 35°F; ivory, 40°F; aluminum foil, 41°F; canary yellow, 45°F; pearl gray, silver gray, 54°F; aluminum paint, 60°F; light lead paint, 61°F; galvanized iron, 78°F; aluminum roofing shingles, 86°F; green roofing shingles, 90°F; red roofing shingles, 95°F; medium green, trim color, 96°F; black, lampblack, 100°F.

Design for Solar Heating. Energy from the sun can be utilized for heating or supplemental heating if the structure is properly designed and oriented. By facing a glassed side of the building toward the south, maximum solar heating can be obtained.

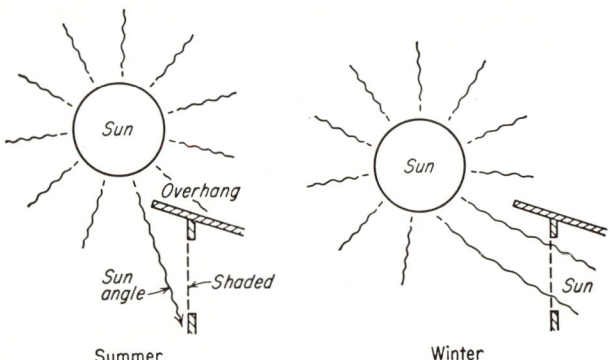

Fig. 48-2. Overhang and sun position.

By providing an overhanging roof, the solar heat is kept out in summer but can pass through the glass into the building in the winter (Fig. 48-2). Solar heating has been used for calf barns, milkhouses, drying barns, poultry houses, and hog houses. The amount of overhang needed depends upon the latitude, height of window, and time of year solar heating is desired (Table 48-3). The 30° latitude passes through

TABLE 48-3. NOON SUN ANGLES FOR DETERMINING ROOF OVERHANG TO SHADE WINDOWS FACING SOUTH

Lat.	From	Until	Sun angle from perpendicular
48°	May 11	Aug. 1	33°
44°	Apr. 26	Aug. 15	31°
40°	Apr. 14	Aug. 26	31°
36°	Apr. 3	Sept. 9	29°
32°	Mar. 23	Sept. 16	20°

For determining value of sun angle for other situations obtain "Sun Angle Calculator," prepared by Aeronautical Services, Inc., Washington, D.C., and published by Libbey-Owens-Ford, Toledo, Ohio, 1950.

southern Louisiana, and the 48° between the United States and Canada. See Chap. 57 for additional information.

Insulated Glass. Insulated glass, two panes with an air space between, is being used for structures, principally for solar heating on southern exposures. Insulated glass

[1] Herman V. Cottony and Richard S. Dill, Solar Heating of Various Surfaces, *Natl. Bur. Standards (U.S.) Rept.* BMS 64, 1941.

has less heat loss in winter than regular single-glass panes. Upon striking the walls and bedding in the structure, the short-wave (light) rays are converted to long-wave rays (heat). Less temperature variation occurs in structures with insulated glass, and less condensation takes place on the inside of the glass. Insulated glass designed for farm buildings is made of sheet glass and is less expensive than plate-glass windows used for homes. Standard sizes of insulated glasses are given in Table 48-4.

TABLE 48-4. STANDARD SIZES OF INSULATED GLASS FOR FARM SERVICE BUILDINGS

Type of frame and glass	Sash opening	Glass size	
These 7/8"-thick units for farm buildings are made with 3/16" sheet glass and 1/2" air space only. They are stenciled for use in farm buildings	....	36" × 44" 36" × 60" 44" × 72"	
These 1/2"-thick units for farm buildings are made with double-strength, A-quality window glass and 1/4" air space	Studs 24" on center for fixed windows with stops nailed to studs	22" × 28" 22" × 36"	
Steel sash, units 1/2" thick, made only with double-strength, A-quality window glass and 1/4" air space	32" × 36" 32" × 36"	31 1/2" × 35 1/2" 30 1/4" × 22" (top) tilting 31 1/2" × 11 3/8" (bottom) fixed	
Wood frame with louvers 1/2" thick, made with double-strength, A-quality window glass and 1/4" air space	49 3/8" × 104"	45 1/2" × 25 1/2"	
Wood ventilator sash for above	45 3/4" × 25 3/4"	42 1/2" × 22 1/2"	

SOURCE: Libbey-Owens-Ford Glass Co., Toledo, Ohio, 1950.

Radiation at Night. At night, certain surfaces in the presence of still air will radiate to clear skies, causing the temperature of the surface to go below the air temperature. This is called negative radiation. The radiation or cooling of surfaces at night is significant because the lower temperature might cause condensation and accumulation of frost on the inner side of uninsulated walls and the roofs and increase the heat loss from the structure at night. Materials which are good emitters of long-wavelength energy lose the most heat during nights with clear skies. The amount of cooling *below* atmospheric temperature was:[1]

Aluminum foil, 7°F
Aluminum paint, galvanized iron (new), 11°F
All-black surface and all surfaces covered with house paints, 13°F

These data represent new materials. Soiled, oxidized, or corroded surfaces will absorb more heat, making them warmer during the day, and emit more heat, making them cooler at night. Emissivity is a function of the character of the surface and not of the base material. See Chap. 45 for a discussion of radiation frosts.

Summer Livestock Shelters. The temperature of an animal under a shade will depend on the amount of heat radiation from the sun, sky, and ground, as well as upon air velocity and temperature. Low temperatures are desired in summer. The outside surface of the shelter will greatly influence the quantity of heat absorbed by the shade and its temperature. White paint is an excellent reflector of heat, even better than aluminum.[2] The amount of heat transferred to the animals from the bottom of

[1] *Ibid.*
[2] T. E. Bond, C. F. Kelly, and N. R. Ittner, Radiation Studies of Painted Shade Materials, *Agr. Eng.*, 35:389–392, June, 1954.

630 FARMSTEAD STRUCTURES AND EQUIPMENT

the shade is influenced by the undersurface of the shade. Black covering over the inside of either a steel or aluminum shade will decrease animal surface temperatures in open shelters. Animals under a shade on grass will collect 14 Btu/(hr)(sq ft) less than if on bare soil, equivalent to 12°F less than the environment.

To maintain low heat transfer to animals in enclosed structures it is desirable to have a material with high emissivity on the outside and low emissivity on the inside. This can best be accomplished by painting the exterior white and the interior with aluminum paint.[1] Spraying water over the roof area will reduce the temperature of the roof considerably, about 25°F in hot summer weather.

In southern and western areas the air temperature will reach 110°F and produce a high heat load on unprotected animals. Shade should be provided by trees or artificial means. An artificial shade is simply a hay-covered metal roof with aluminum or galvanized sheeting. For beef and dairy cattle the shade should be 10 to 12 ft high,[2] and for hogs 4 to 5 ft high. If the shade is too low, the animal cannot radiate much heat to the sky, and if the shade is too high, the animals must move to hot ground as the shade moves.

Tests in the Imperial Valley of California have shown that increased air movement over cattle in hot weather, either natural or from fans, promotes better weight gains and utilization of feed.[3]

See Chap. 55 for additional information on the effect of thermal environment on production.

HEAT TRANSFER

Thermal Insulation. Dry, still air is an excellent heat insulator. The insulating value of the air space between building walls is reduced by heat transfer because of convection air currents. Conduction insulation materials are designed to incorporate many small air pockets.

The rate of heat conduction through a building material is called the conductivity, k, in Btu/(hr)(sq ft)(°F) per foot of thickness (Table 48-5). Some tables give the conductivity per inch of thickness. The thermal conductivity of insulating materials, water, and gases increases with an increase in temperature. The thermal conductivity of most metals decreases with an increase in temperature, with the notable exception of aluminum and chromium alloys. The average value of k over the temperature range involved is generally used for calculations. The conductance C, in Btu/(sq ft)(°F), represents the heat transmission for a stated thickness of a material. The surface film conductance f, in Btu/(hr)(sq ft)(°F), is expressed in the same units as conductance and represents a certain thickness. The over-all coefficient of heat transfer is used to denote the rate of heat movement through more than one conduction medium and is represented by U, in Btu/(hr)(sq ft)(°F). Note that the thickness is not included in the units. The steady-state heat flow Q, in Btu per hr for a single material, is

$$Q = \frac{k}{x} A \Delta t$$

where x = thickness, ft (or in., depending on units of k)
 A = area, sq ft
 Δt = temperature difference between x_1 and x_2, °F

For several heat-conduction layers,

$$Q = UA \Delta t$$

where

$$U = \frac{1}{\dfrac{1}{f_i} + \dfrac{x_1}{k_1} + \dfrac{x_2}{k_2} + \cdots + \dfrac{1}{f_o}} = \frac{1}{R}$$

f_i, f_o = inside and outside surface conductance, respectively

---

[1] Dale and Giese, *op. cit.*

[2] N. R. Ittner, H. R. Gilbert, and F. D. Carroll, Adaptation of Beef and Dairy Cattle to the Irrigated Desert, *Univ. Calif. Agr. Expt. Sta. Bull.* 745, September, 1954.

[3] T. E. Bond, C. F. Kelly, and N. R. Ittner, Cooling Beef Cattle with Fans, *Agr. Eng.*, 38:308–321, May, 1957.

HEAT, AIR, AND MOISTURE

TABLE 48-5. THERMAL CONDUCTIVITY k OF VARIOUS MATERIALS, IN BTU/(HR)(SQ FT)(°F) PER FOOT OF THICKNESS

Material	k	Temperature and density
Air	0.014–0.018	32–212°F
Aluminum	117	Room
Asbestos	0.09–0.11	32–212°F, 36 lb/cu ft
Asphalt	0.43	Room, 132 lb/cu ft
Brass	60	Room
Brick, building	0.40	Room
Cast iron	27	Room
Cattle hair	0.022	Room, 12 lb/cu ft
Copper	224	Room
Corkboard	0.025	Room, 10 lb/cu ft
Corn	0.12	Room, 13% moisture
Corncobs, ground	0.03–0.04	
Cotton	0.035	100°F, 5 lb/cu ft
Fiber insulating board	0.028	Room, 14.8 lb/cu ft
Flax fiber	0.026	Room, 13 lb/cu ft
Gasoline	0.08	Room
Glass, ordinary	0.5	Room
Glass wool	0.024	Room, 4 lb/cu ft
Ice	1.3	32°F, 57.5 lb/cu ft
Kerosene	0.09	Room
Marble	1.2–1.7	Room
Mineral wool	0.0225	85°F, 9.4–19.7 lb/cu ft
Oats	0.075	Room, 13%
Paper	0.075	Room
Rubber	0.075–0.092	70°F, soft
Sand	0.19	70°F, 94.6 lb/cu ft
Sawdust	0.03	Room, 12 lb/cu ft
Shavings, wood	0.04	Room, 1.4 lb/cu ft
Silver	242	Room
Stainless steel, 18-8	15.0	Room
Snow	0.027	32°F, 34 lb/cu ft
Steel	26	Room
Straw	0.06	Chopped dry
Sugar cane fiber	0.028	Room, 15 lb/cu ft
Tin	36	Room
Wallboard	0.028	Room
Water	0.0343	32°F
	0.0363	100°F
	0.0393	200°F
Wheat	0.08	Room
Wood, across grain	0.087	34 lb/cu ft

Conversion: Multiply by 12 to obtain Btu/(hr)(sq ft)(°F/in.). Multiply by 0.00413 to obtain Cal/(sec)(sq cm)(°C/cm).

Because it is time-consuming and easy to make mistakes in calculations when determining U values, another designation, thermal resistance, R, is often used for heat-transfer calculations. The thermal resistance R is the reciprocal of the over-all coefficient of heat transfer U and conductance f. The conductance in Btu/(hr)(sq ft)(°F) refers to the heat transfer through nonhomogeneous materials or air spaces. For surface areas, surface conductance is used, generally as 1.65 Btu/(hr)(sq ft)(°F) for still air and 6.0 for a 15-mph wind (1,320 fpm). Thus, the over-all resistance can be obtained by adding the resistances to heat flow of the individual materials. Where a high degree of accuracy is required, it is important not to overlook air layers and paint films when calculating the quantity of heat transferred. Many designs for farm structures specify the insulation, which varies with the use of the building and its location (Fig. 48-3).

Reflective Insulation. At ordinary temperatures two-thirds of the total heat transferred across an air space is by radiation. The heat normally transferred by radiation can be reflected by silver, copper, and aluminum or their alloys. Aluminum sheets separated by an air space of ½ to ¾ in. are used commercially, with the number of sheets depending on the temperature difference. About 95 per cent of the heat is reflected.

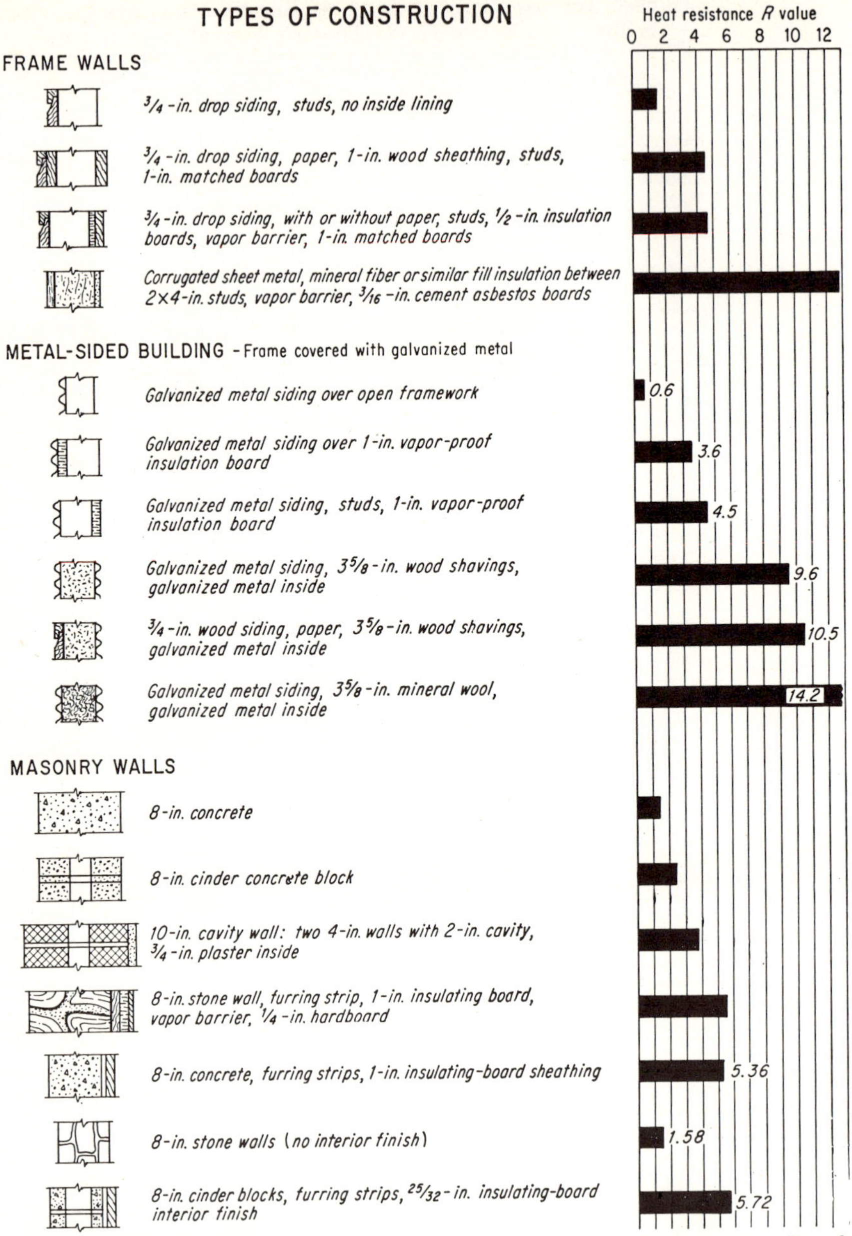

TYPES OF CONSTRUCTION

FRAME WALLS

Construction	R value
3/4-in. drop siding, studs, no inside lining	~1
3/4-in. drop siding, paper, 1-in. wood sheathing, studs, 1-in. matched boards	~3
3/4-in. drop siding, with or without paper, studs, 1/2-in. insulation boards, vapor barrier, 1-in. matched boards	~4
Corrugated sheet metal, mineral fiber or similar fill insulation between 2×4-in. studs, vapor barrier, 3/16-in. cement asbestos boards	~12

METAL-SIDED BUILDING — Frame covered with galvanized metal

Construction	R value
Galvanized metal siding over open framework	0.6
Galvanized metal siding over 1-in. vapor-proof insulation board	3.6
Galvanized metal siding, studs, 1-in. vapor-proof insulation board	4.5
Galvanized metal siding, 3 5/8-in. wood shavings, galvanized metal inside	9.6
3/4-in. wood siding, paper, 3 5/8-in. wood shavings, galvanized metal inside	10.5
Galvanized metal siding, 3 5/8-in. mineral wool, galvanized metal inside	14.2

MASONRY WALLS

Construction	R value
8-in. concrete	~1
8-in. cinder concrete block	~2
10-in. cavity wall: two 4-in. walls with 2-in. cavity, 3/4-in. plaster inside	~3
8-in. stone wall, furring strip, 1-in. insulating board, vapor barrier, 1/4-in. hardboard	~4
8-in. concrete, furring strips, 1-in. insulating-board sheathing	5.36
8-in. stone walls (no interior finish)	1.58
8-in. cinder blocks, furring strips, 25/32-in. insulating-board interior finish	5.72

Fig. 48-3. Comparative insulating value R of various types of construction. Heat transfer U is the reciprocal of R.

TYPES OF CONSTRUCTION

ROOFS

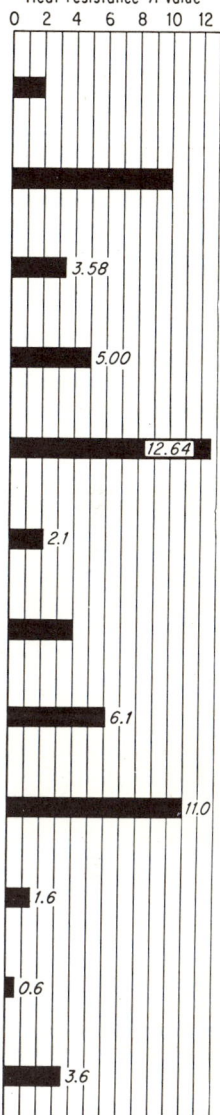

Construction	R value
Prepared roofing, 1-in. matched lumber, rafters	
Corrugated metal roofing, 3-in. fill or blanket insulation, vapor barrier, ¼-in. cement asbestos or other hardboard	
Roofing, boards, ½-in. insulation, and rafter	3.58
Roofing, boards, 1-in. insulation, and rafter	5.00
Roofing, boards, fill insulation, and sheathing	12.64
Wood shingles on 1 × 4-in. lath spaced 2-in. apart	2.1
Same as above with under side of rafters lined with paper and ceiled with ¾-in. matched boards	
Same as above except 1-in. insulation board in place of paper and ¾-in. matched boards	6.1
Same as above except rafters space filled with 3½-in. of loose insulating material and ceiled underneath with galvanized metal	11.0
Slate, asbestos, or asphalt shingle roofing laid over paper on ¾-in. wood sheathing	1.6
Galvanized corrugated metal roofing over open framework	0.6
Galvanized corrugated metal roofing, rafters ceiled underneath with 1-in. vapor-proof insulation board	3.6

Fig. 48-3. (*Continued.*)

TYPES OF CONSTRUCTION

CEILINGS

 1-in. matched lumber, top of floor joists (If covered with 18-in. of hay)

 1-in. matched lumber, top of floor, 2-in. fill insulation, vapor barrier, ¼-in. cement asbestos or other hardboard below

 1-in. boards, joists, ¾-in. insulating board

1-in. boards, joists, 1-in. shavings, paper, and 1-in. boards

BUILDING MATERIAL

Window glass, single thickness

Smooth glass blocks, 7¾ × 7¾ × 3⅞-in.

Factory-sealed, double-glazed window glass

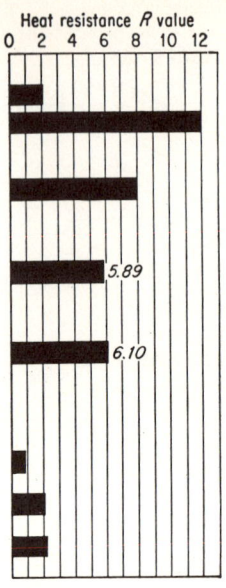

FIG. 48-3. (*Continued.*)

Over-all coefficients of heat transfer U, with the interior wall surface at 85°F and the outside wall surface as shown, are:

Outside temperature range*	Number of air spaces, ¾ in. wide†	U, Btu/(hr)(sq ft)(°F)
+30 to +40°F	4	0.075
+20°F	5	0.059
0 to +10°F	6	0.049
−10°F	7	0.04
−20°F	8	0.032

* Vapor sealing is important at the joints. Joints must permit expansion and contraction of metal sheets.
† For the ceiling, one-half the number of air spaces is used with wider sheet spacing of 1 to 1½ in.

A *rule of thumb* is to consider each reflecting surface as equivalent to ½-in. thickness of fill insulation.

Surface Condensation. It is desirable to prevent condensation on interior surfaces of buildings, particularly during the winter season in buildings where heat and moisture are produced by livestock. The interior wall-surface temperature must be kept above the dew point of the air next to the wall to prevent condensation.

Vapor Barriers. To keep insulation dry, free from condensed vapor, and freezing where the temperature is below 32°F, a vapor barrier is applied on or near the warm side of the insulation. The vapor pressure is usually highest on the warm side of the insulation, and the vapor attempts to move to the cold side, which is at a lower vapor pressure. To prevent the passage of moisture, a material which has a resistance to vapor is used. An asphalt material, aluminum foil, certain plastic film, and water-resistant paints perform as vapor barriers. The need for properly placed vapor barriers is evidenced by (1) peeling of paint from exterior walls, (2) damage to insulation where temperatures below 32°F are involved, (3) saturated insulation, thus destroying insulating effect, (4) stains on inside of heated buildings in winter, and (5) damage to structures, wood or metal. In practice, it is difficult to get a perfect vapor barrier because of installation problems at the joints and seams. Therefore it is desirable to permit the cold side of the wall to "breathe" or be vapor-permeable (not impermeable).

The permeability P to water vapor is expressed in grains/(sq ft)(hr)(psi). The vapor resistance is the reciprocal, $1/P$. Relative values of the permeability of various materials are given in Table 48-6. There is some variation among various investigations, and the data should be used to determine relative rather than absolute values.

Table 48-6. Permeability of Building Materials to Water Vapor*

Material	Thickness, in.	Permeability P, grains† / (hr)(sq ft)(psi)	Vapor resistance, $1/P$
Fiberboard	0.492	60.6	0.0168
1 surface asphalt, rolled	0.492	8.0	0.125
1 surface asphalt, dipped	0.63	17.3	0.0578
2 sheets, asphalt between	0.985	2.74	0.365
Wood:			
Spruce	0.563	3.48	0.287
	0.232	7.24	0.138
Pine	0.80	1.88	0.532
	0.315	5.55	0.180
	0.508	6.47	0.155
1 coat of aluminum paint	0.508	3.42	0.292
2 coats of aluminum paint	0.508	0.92	1.09
3 coats of aluminum paint	0.508	0.71	1.41
Kraft paper:			
1 sheet	0.00394	168	0.00595
3 sheets	0.00394	80	0.0125
5 sheets	0.00394	65.3	0.0153
Asphalt:			
Between 2 sheets of kraft	0.0071	0.946	1.06
On 1 surface of kraft	0.0071	8.6	0.116
Black building paper	0.0173	0.376	2.66
Asphalt felt, 15 lb	0.0319	13.5	0.0741
Corkboard, pressed	0.905	4.75	0.211
Plaster	1.34	27.1	0.0369
Plasterboard between heavy paper	0.37	70.2	0.0142
Masonite Presdwood	0.13	21.7	0.046
Tempered	0.13	9.76	0.102
Plywood, Douglas fir, ¼ in	......	8.72–13.1	0.115–0.0764
2 coats of asphalt	......	0.87	1.15
2 coats of aluminum paint	......	2.63	0.38
Sheathing paper	......	123.13	0.008
Asphalt-impregnated felt:			
15 lb for 100 sq ft	......	2.72	0.368
30 lb for 100 sq ft	......	1.80	0.556

* For further references see, for fiberboard, J. D. Babbitt, The Diffusion of Water Vapor through Building Material, *Can. J. Research*, 17:15–32, February, 1939; for plywood, L. V. Teesdale, How to Overcome Condensation in Building Walls and Attics, *Heating and Ventilating*, 36:36–40, April, 1939; for sheathing paper and asphalt-impregnated felt, H. J. Barre, The Relation of Wall Construction to Moisture Accumulation in Fill-type Insulation, *Iowa State Univ. Bull.* 271, 1940.

† 7,000 grains = 1 lb.

Source: Harold W. Wooley, Building Materials and Structures, *Natl. Bur. Standards (U.S.) Rept.* BMS 63, 1940.

VENTILATION OF BUILDINGS

Although there is much technical theory involved in the design of ventilation systems for livestock buildings, in practice the solution of problems is based on factors that can be determined by simple calculations.

The ventilation of a closed livestock building in warm weather differs from winter ventilation. As with other ventilation systems, an effective system for livestock must permit close control of the air flow by means of tight building construction with controlled openings for the entrance and discharge of air. Insulation and ventilation are closely related.

Summer Ventilation. During warm weather, the metabolism of animals declines in self-protection against overheating. Declining metabolism results in reduced milk, lower egg production, and reduced rate of gain. For hens and cattle various data indicate that temperatures above 70 to 80°F cause a sharp decline in production. Where possible, animals seek open shade during the heat of the day, but in many cases, dairy animals, steers, and poultry should be kept confined under a roof. In closed buildings with open doors and windows, the temperature around the animals may become excessive. The major objective of summer ventilation is to carry the body heat away from the animals.

One method of figuring the *amount of ventilation* required is to compute the heat produced by the animals and then calculate the quantity of air, cubic feet per minute,

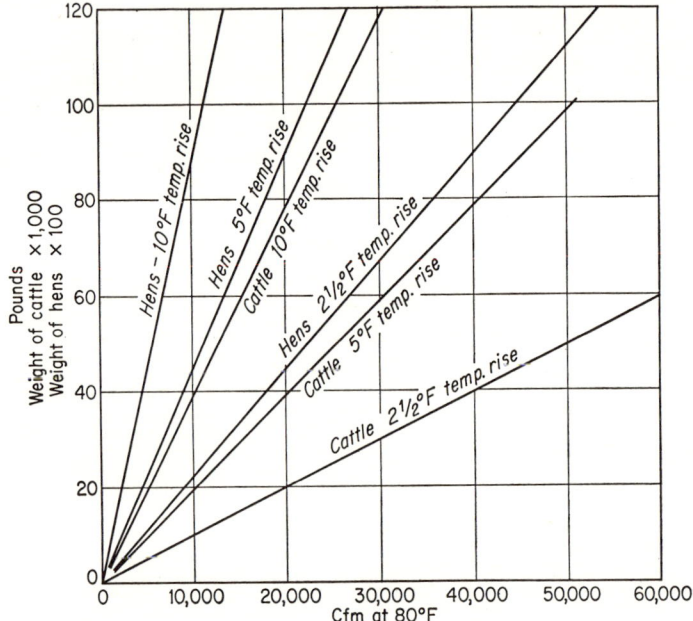

Fig. 48-4. Air flow for summer ventilation.

required to carry the heat away with a stated temperature rise. At a temperature of 80°F, 1 Btu will raise the temperature of 56 cu ft of air 1°F. Approximately 45 Btu per min is liberated by a 1,000-lb cow or feed lot steer. About 200 Btu per min is liberated by 1,000 lb of hens. These figures for heat production of animals are not exact, because the metabolism rate varies with the ration, age, and temperature. The heat to be dissipated may be increased by solar absorption and the production of heat in manure and litter. However, for general ventilation purposes, these heat-production values are suitable and have been used to compute the data in Fig. 48-4, from which calculations can be quickly made to determine the volume of air required. For example, 2,000 6-lb hens would have a total weight of 12,000 lb. From Fig. 48-4, 12,000 lb of hens will require about 27,000 cfm to limit the inside temperature to about 5°F above the outside air temperature. For 50,000 lb of cattle, 25,500 cfm would be required to limit the inside temperature to 5°F above the outside temperature. See Chap. 55 for additional information on heat and moisture production of various animals.

Open-sided sheds or shelters can be made cooler by blowing air through them.[1] The fan should be on the side of the prevailing winds, so that wind and fan will

[1] Bond, Kelly, and Ittner, *op. cit.*

combine to furnish ventilation. Placing a fan or fans in an open shed to stir the air is seldom of much real benefit. The air is mostly stirred rather than changed. It is of some benefit to place a fan or fans on the roof of an open shed to exhaust air up through an opening in the roof. Insulation under the roof of an open shed is of much benefit in excluding solar heat and may be of more benefit than a fan in an uninsulated shed.

Ventilation from the wind is often secured with open-front structures. The amount of air movement through an open-front building is closely related to the opening of the opposite side of the building. Considerably more air flow will be secured if the opposite partial wall opening is raised above the floor[1] (Fig. 48-5). An open side on the south will provide little air movement even with a south wind unless ridge or wall openings on the opposite side are provided. Continuous ridge and round ventilators

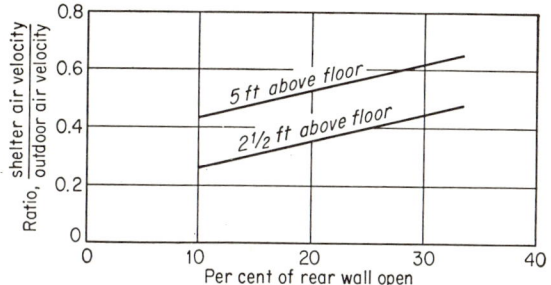

FIG. 48-5. Effect of rear wall openings on air motion through the open-front south-facing dairy-cattle shelter. (*Nelson and Berousek, Dairy Cattle Loafing Barns Designed for Summer Temperature Control, Agr. Eng.,* 33:290–292, May, 1952.)

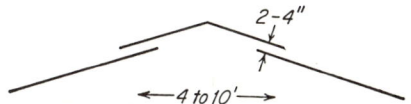

FIG. 48-6. Ridge ventilator commonly used to secure air movement.

aid ventilation of buildings (Fig. 48-6) and are commonly provided on buildings housing livestock.

Winter Comfort in Northern Climate for Warm-operated Buildings. The *functional requirements* of a ventilating system are:

1. Prevent excess condensation of moisture on walls and ceilings; usually maintain a relative humidity below 80 per cent.
2. Maintain dry bedding for litter, usually below 40 per cent, wet basis.
3. Maintain reasonable indoor temperature, usually above 32°F.
4. Maintain air purity and remove objectionable odors, usually accomplished if the air flow is sufficient to remove moisture.
5. Avoid draft and excessive air change.
6. Provide more comfortable environment during summer.

Closed buildings are ventilated by natural or mechanical draft. Natural ventilation requires more time and effort for supervision than mechanical ventilation and is not as positive and dependable in many cases for warm-operated buildings. Because of its low cost, positive action, and simplicity, electrically powered mechanical ventilation is most desirable. Natural draft ventilation tends to provide more ventilation when needed less, and less ventilation when needed more. One kilowatthour of electricity will move about 1 million cu ft of air for ventilation.

---

[1] G. L. Nelson and E. R. Berousek, Dairy Cattle Loafing Barns Designed for Summer Temperature Control, *Agr. Eng.,* 33:290–292, May, 1952.

The procedure for determining mathematically the *ventilation requirements* for heat and moisture removal is somewhat complex. A thorough ventilation analysis should include consideration of the following factors, in addition to the heat and moisture produced by the animals (see Chap. 55):

Secondary heat sources	*Secondary sources of moisture*
Solar heat through windows	Water spilled, leaked, or evaporated from waterers
Solar heat gained through walls and roofs	Water seepage through floors
Heat generated in litter	Water from entering rain or snow
Heat transferred from earth through floor	Moisture diffused through walls
Heat stored in building materials and equipment	Water resulting from decay of organic matter
Latent heat from condensing moisture	Water on eggs when laid
Heat from light bulbs	Water on equipment entering house
Heat from personnel	Drippage of moisture from walls, windows, or ceiling where moisture is condensed
Heat from water, feed, and equipment entering the house	

Numerous efforts have been made to devise structures, especially for dairy cattle, hens, and swine, which will provide enough comfort to prevent loss of production during cold weather. Animals usually provide enough heat from their bodies to maintain temperatures above freezing if they are housed in a tight building, adequately insulated to conserve the animal heat.

Although it is not difficult to construct a building that will reduce heat losses through ceilings, walls, and cracks, the tight construction traps large amounts of moisture and foul odors which must be eliminated.

Before electric power was available, ventilation was provided by *gravity systems* of intake ducts and one large exhaust flue, operation of which was dependent upon wind action and the difference in air temperature between the stable and the outside. Properly constructed, the gravity system of ventilation was reasonably satisfactory although it required adjustment to wind and temperature conditions.

The *air volumes* desirable for several kinds of livestock as recommended in different areas of the United States are listed in Table 48-7. Variable air flows are achieved by selecting a fan to move the larger amount of air and dampering it down in cold weather or with a two-speed fan, or by selecting two fans, one to operate at all times, the other to be controlled by a thermostat which will turn the fan on at a specified temperature.

Ventilation of Cold-operated Buildings. There is a tendency of many farm operators to close all the openings in a building designed to be operated as a cold structure. The heat and moisture given off by the animals are held in the building and condense on walls and ceilings, and undesirable odors are held in the building. The following principles should be followed for a cold-operated building:

1. Provide louvers or openings at the ridge of building for exhausting warm, moisture-laden air.

2. Provide air inlets in the sides of the building. The air inlets should be above the animals or should direct the air above the animals.

Insulation and Ventilation. A ventilation system is not capable of performing properly if the building has numerous air leaks and considerable heat loss. For stall dairy barns with one animal per 500 cu ft, a wall constructed of two layers of $\frac{3}{4}$-in. wood with heavy building paper between is usually adequate insulation, but fill insulation in hollow walls is more effective. Masonry walls in the northern states are seldom warm enough unless lightweight cinder concrete is used and the cores filled with insulation material. Solid masonry and concrete walls should be furred and plastered or paneled to obtain more insulation. One-story poultry and swine buildings generally should be constructed with fill insulation in the walls and ceilings except in areas of mild winter weather. Thick, insulated doors and storm windows are very important factors in preventing heat losses. The use of insulating glass in window areas is increasing. The R factor, $1/U$, should be above a certain value to prevent excessive wall condensation (Fig. 48-7).

The winter ventilation system is designed to operate when the temperature outside the building is colder than that desired inside, but in most regions, winter thaws and

TABLE 48-7. AIR-FLOW RECOMMENDATIONS FOR VENTILATION THROUGHOUT THE UNITED STATES

Animal	Cfm per animal	Conditions	Location	Source
Cow (1,000 lb).......	70–100	....	General	USDA Circ. 722, 1945
	60	....	Ohio	OSU Bull. 208, 1943
	60	500–700 cu ft/cow $R = 5$ or more	Mich.	MSU Ext. Bull. 310, 1952
	60	....	Ill.	Carter, 1954†
	100	Winter	Wis.	Univ. Wis. Spec. Bull. 4, 1954
	200	Summer (warm weather)*	Wis.	Univ. Wis. Spec. Bull. 4, 1954
	60	Below 20°F	Pa.	PSU Spec. Circ. 2, 1950
	200	Above 20°F	Pa.	PSU Spec. Circ. 2, 1950
	20–25	Set at 35–40°F	S. Dak.	SD Circ. 429, 1955
	100	Set at 45°F	S. Dak.	SD Circ. 429, 1955
Swine (300 lb).......	18	....	Northern U.S.	USDA Circ. 701, 1944
Fattening pig........	12	....	Ohio	OSU Bull. 208, 1943
Sow................	15	....	Ohio	OSU Bull. 208, 1943
Beef animal.........	50	....	S. Dak.	SD Bull. 382, 1946
Chickens:				
Layer............	3	Normal density, 2½–3 sq ft per bird	General	USDA Misc. Publ. 728, 1956
	1	....	Ohio	OSU Bull. 208, 1943
5-lb.............	5	3½–4 sq ft per layer	Pa.	PSU Spec. Circ. 13, 1953
	3–4	$R \geq 6$, southern part $R \geq 10$, northern part 3–4 times as much	New York	Cornell Ext. Bull. 947, 1955
	1	Winter	Minn.	Univ. Minn. Ext. Bull. 253, 1951
	2.5–3	In winter or ¾–1 cfm/sq ft, $R \geq 10$	Mich.	MSU Ext. Bull. 317, 1953
	4	....	Quebec	Choinière, 1953‡
	2–3	In winter, $R = 10$	Conn.	Univ. Conn. Ext. 55-51, 1955
	4–5	In summer, blow into pen		
Sheep..............	15	....	General	USDA Bull. 1393, 1924
Horse..............	70	....	General	USDA Bull. 1393, 1924

* Warm weather occurs when the average of the high and low readings for the 24-hr period exceeds the inside building temperature.
† D. G. Carter, "Farm Buildings," John Wiley & Sons, Inc., New York, 1954.
‡ J. A. Choinière, "Design and Performance of Fan Ventilation in Poultry Laying Houses," Michigan State University Dept. of Agricultural Engineering, M.S. thesis, 1953.

warm cycles occur which make it practically impossible for the ventilation system to prevent heat from accumulating in the stable in amounts that cause discomfort. Doors and windows can be opened, and often the animals are let outside where it is cooler, but this is not possible for many animals, especially hens. The result is a temporary or permanent decline in production.

Inlets and Outlets. There are several methods used for providing air passage into and out of buildings. Fans are usually installed to exhaust the air, and several inlets are provided, the number and location of inlets depending on the system used. For cattle housed in a stall barn an inlet of 60 sq in. is often used for every three cows and placed about 15 ft apart around a stable with an 8-ft ceiling height. Several types of inlets are shown in Fig. 48-8.

Some designers use a continuous-slot or bored-hole type of inlet around the edges of the mow floor to provide the required inlet area. This air inlet is not affected by the wind and draws tempered air from the hay mow and provides a uniform flow of air over the wall surface. The slot is easily constructed in new buildings, and holes can be placed in old construction (Fig. 48-8e). With this type of construction, the same air movement is required at all times.[1]

[1] C. N. Turner, Ventilate Your Dairy Stable with Electric Fans, *Cornell Univ. Ext. Bull.* 845, March, 1952.

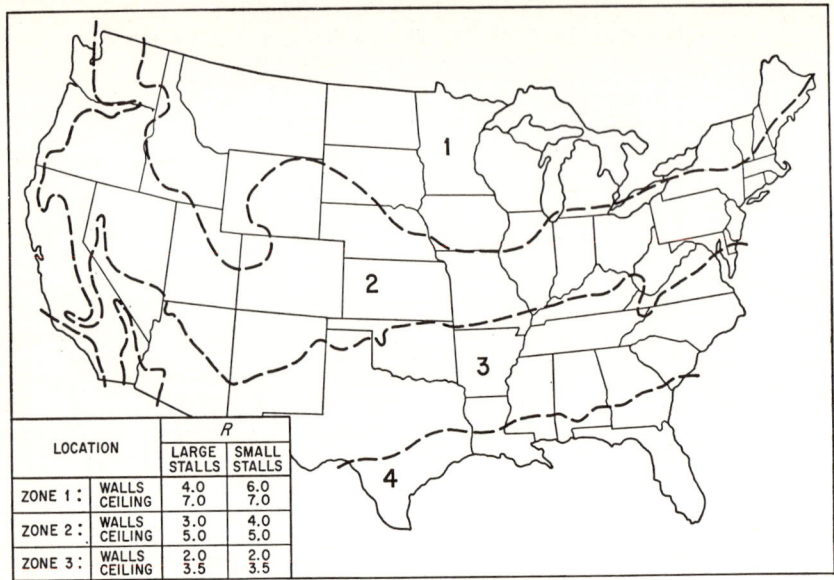

Fig. 48-7. Suggested minimum insulating values R for dairy stables. Map shows zones of similar winter temperature and relative humidity in the United States. (*Cleaver, Thompson, and Yeck, Stall Barns for Dairy Cattle, USDA Infor. Bull.* 123, 1954.)

If two fans are used, they are placed side by side. The fan must be equipped with automatic shutters which will keep out wind when the fans are not operating. Another arrangement consists of placing the fan in a box so that it will move the maximum amount of air and draw the air from the top of the stable in mild weather and by thermostat operation automatically close a damper and draw air from the floor during periods when the stable temperature falls below 50°F (Fig. 48-9). Because the stable temperature near the ceiling is often 4 or 5°F warmer than that near the floor, heat can be conserved during cold weather by drawing the air from near the floor.

Practices and Principles of Ventilation

1. Mechanical ventilation is more effective over a wide range of weather conditions than a gravity system for warm-operated buildings.

2. Continuous fan operation is generally favored over intermittent operation. For two- or three-fan systems, one fan can be kept in operation to maintain some air movement when others are stopped. Two-speed fans are used to obtain two different air flows for winter and summer operation. Infiltration through windows, doors, and cracks will often provide enough air movement for ventilation.

3. Automatic fan control is preferred over manual control.

4. Thermostats are preferred over humidistats because of their reliability, low cost, and ruggedness. The thermostat setting will vary with the amount of insulation, number of animals, temperature outside, etc., but is usually in the range between 40 to 50°F, with a 2°F differential.

5. The thermostat should be located away from walls, animals, stairways, or chutes and be about 5 ft above the floor.

6. Exhaust ducts should be placed near dropping pits for the poultry. The droppings of poultry contain 80 per cent of the water given off by the birds.

7. Locate the fresh-air inlets uniformly around the perimeter of conventional stall barns. Provide at least 20 sq in. of inlet area per 1,000 lb cow.

8. It is desirable to have enough heat available so that the air can be changed to remove the excess moisture without lowering the temperature greatly. The volume of

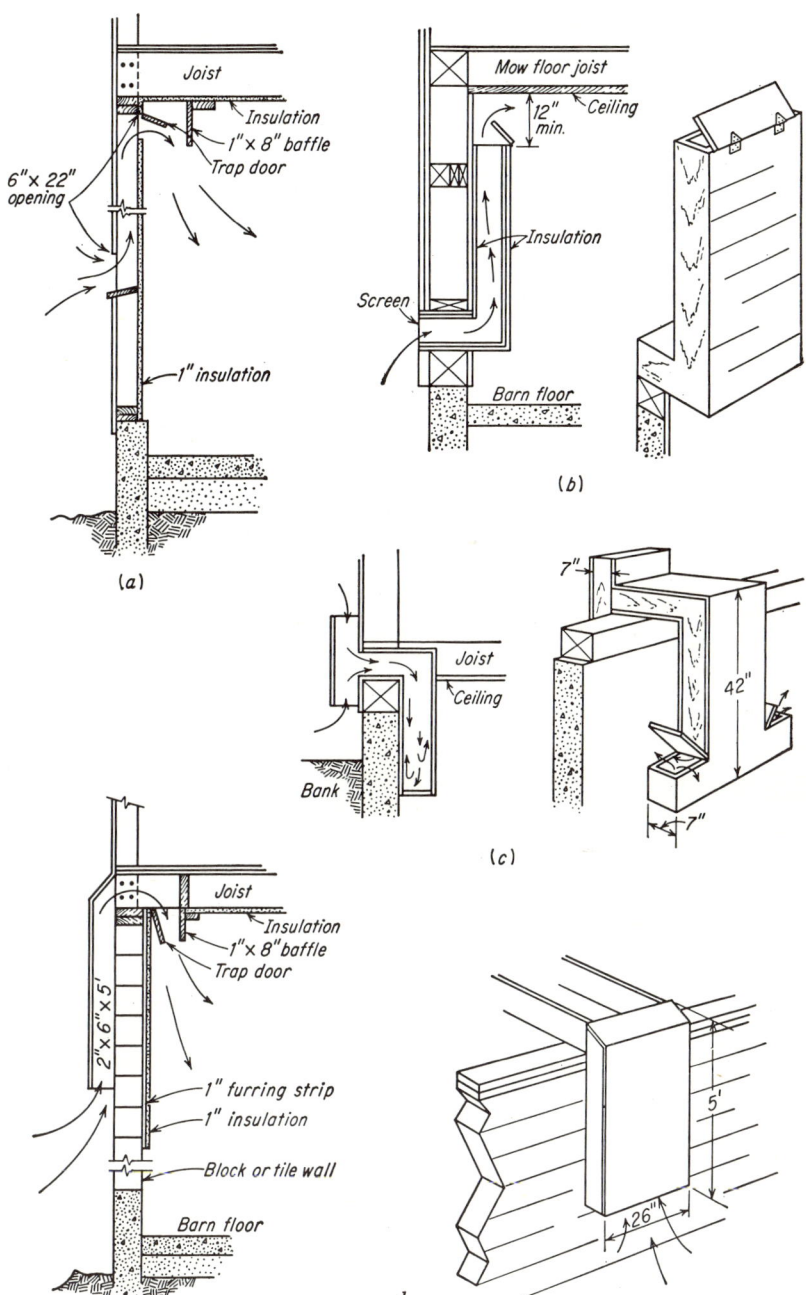

FIG. 48-8. Different arrangements for fresh-air ventilation inlets: (a) conventional frame wall; (b) old timber-frame barn; (c) bank barn with masonry walls; (d) barn with masonry walls above ground; (e) bored-hole intake through mow floor; (f) continuous slot intake through mow floor.

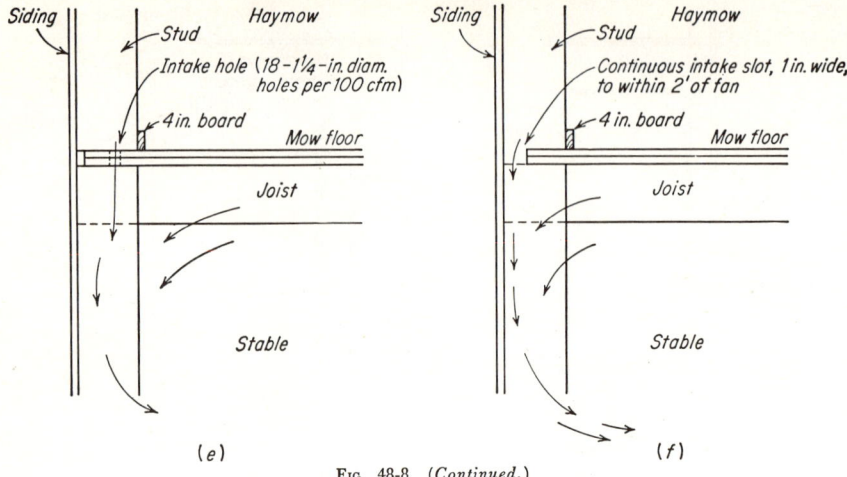

Fig. 48-8. (Continued.)

air required to remove the heat should be equal to, or greater than, the volume of air required to remove the moisture.

9. To maintain a desirable temperature, use insulation to reduce heat loss and keep building well stocked.

10. There are often some factors not accounted for by ventilation equations, and in some cases it is necessary to use a trial-and-error method to obtain a satisfactory ventilation solution. If condensation occurs at a certain area after a system is installed, place an inlet in the troublesome area.

Natural Air Movement in Structures. Natural air movement originates from (1) thermal forces and (2) velocity pressure of wind.

The thermal force is determined by the variation in the density of air. When two bodies of air exist side by side at equal atmospheric pressure but at different temperatures, the difference in density will cause one body to displace the other through available inlets and outlets. The pressure difference at a common elevation between two air masses of unequal temperature is proportional to the heights of the air masses and to the differences between the reciprocals of their absolute temperatures. The theoretical pressure difference at the base of the two air masses can be calculated as inches of water by using the following equation:

$$P_d = 2.96 F_h P_b \left(\frac{0.08636}{T_c + 460} - \frac{0.08636}{T_w + 460} \right)$$

where P_d = pressure difference between two air masses when no air is moving, in. of water
F_h = height of enclosed air mass, ft
P_b = local barometric pressure, in. of mercury
T_w = temperature of enclosed air mass, °F
T_c = temperature of unenclosed air mass, °F

If the enclosed air mass is colder than the atmosphere outside, then the pressure will be negative, indicating that the thermal force will be downward. If the enclosed mass is a gas other than air, the formula must be corrected by substituting the gas density at 0°F in place of 0.08636 in the equation. The pressure difference obtained by this formula is the maximum theoretical pressure and is often referred to as the draft pressure. The actual pressure difference between the two air masses will be affected by friction and wind forces when movement results. Thermal forces move air in chimneys, flues, fruit and vegetable storages, gravity heating systems, tobacco barns, and in livestock buildings and are often minimized by wind pressures. The equation can be used to calculate the thermal forces for theoretical design data. Thermal

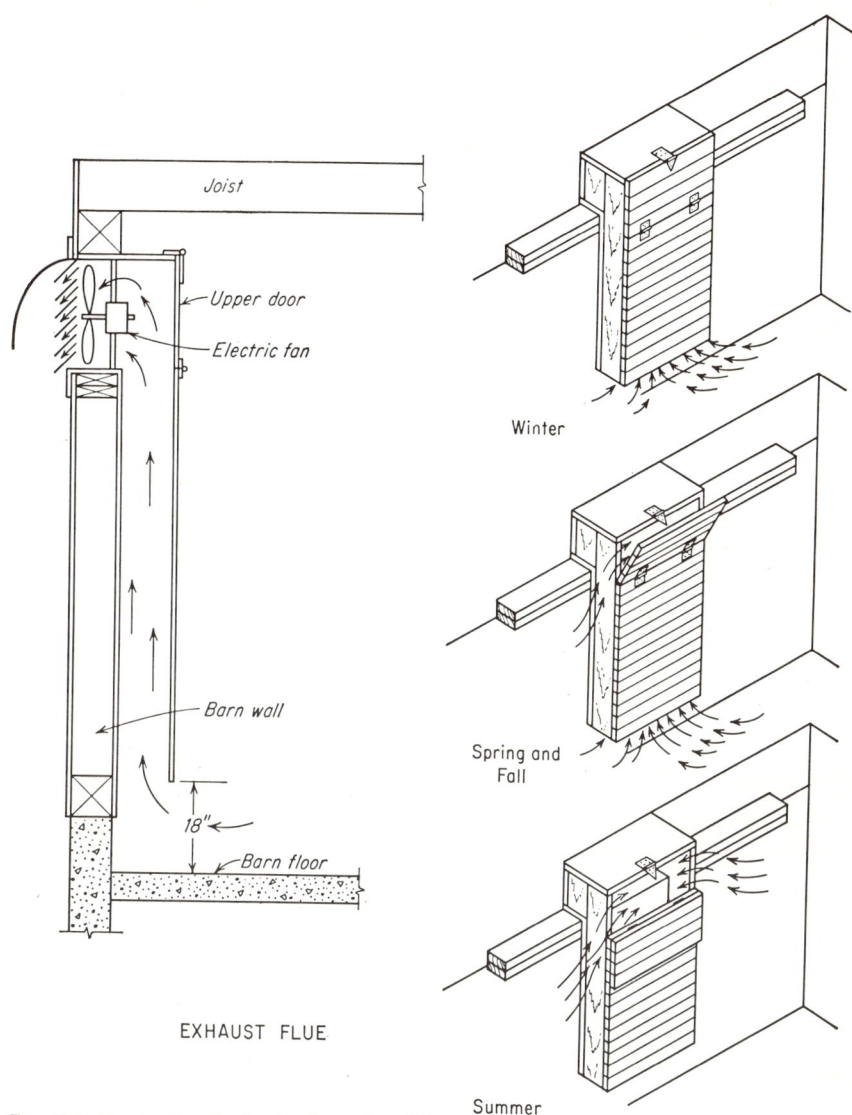

Fig. 48-9. Construction details of exhaust flue. Diagrams at the right illustrate how the door at the top of the fan duct is opened during mild weather, so that the fan draws the air off the ceiling, and is closed when outside temperatures are low, so that the air is drawn from the barn floor. (*Boyd and Maddex, Ventilation for the Modern Dairy Barn, Mich. Ext. Bull.* 310, 1952.)

flues must be insulated to prevent temperature changes between the bottom and the top of the flue. The dimensions of a thermal flue and of the air inlets and outlets must be balanced. Straight flues with smooth inner surfaces are most efficient. A vertical height of 20 ft or more should be provided. Figure 48-10 provides approximate performance data for straight flues as used in dairy barns and applicable to other situations where a temperature difference of 20°F or more can be maintained between the air inside and outside a building.

Thermal outlet flues must be protected from downdrafts at the top with roof vents

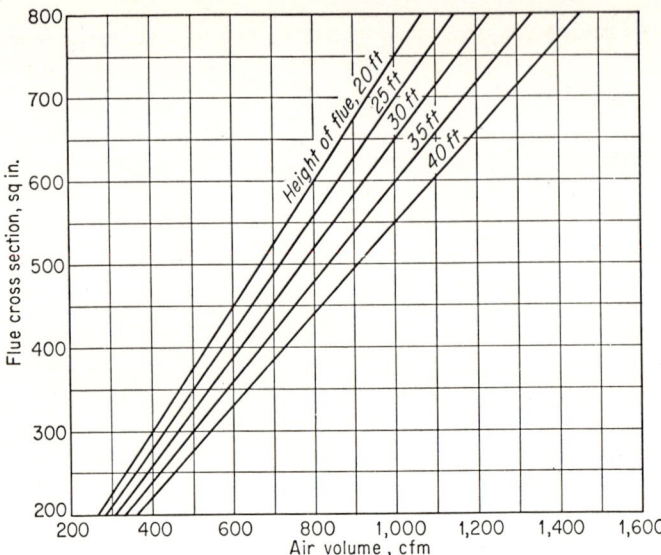

Fig. 48-10. Air volume through gravity flue. (*Based on Goodman, Dairy Stable Ventilation, Agr. Eng.,* 21:301–302, *August,* 1940.)

containing openings with an area equal to at least double the cross section of the flue. The top of the flue should be above nearby roofs and trees.

The inlets to a space ventilated by a thermal flue are equally important. The space must be tightly enclosed and insulated to prevent excessive infiltration and heat losses. The inlets are usually short duct sections with about 60 sq in. of cross section and spaced to introduce fresh air into all portions of the space. The total inlet area is usually equal to the cross section of the outtake flue.

Continuous ridge vents as shown in Fig. 48-11 also operate on the thermal principal in the absence of wind. The entire structure serves as the flue. Inlets are provided at or above the floor line. If wind is striking the structure, it may provide energy equal to, or greater than, the thermal energy. The continuous ridge vent with its large cross section makes it possible to move large volumes of air with slight pressures. Adjustment of intakes and ridge-vent outlet areas may be required to maintain interior temperatures and air flow.

Natural air movement in a poultry house is controlled by adjusting windows, ceiling flue ventilators, and drop doors on the side opposite the windows. Intake air is directed to the ceiling and interferes with outgoing air if the window is tipped in at the top. Intake air is directed to the floor and causes drafts if the window is tipped in at the bottom. Fresh air should be admitted through a raised window or specially designed air inlets. Air outlets are at the peak of the roof through ventilating flues or openings at the top of a shed roof. A 6-in.-wide door is usually provided on the opposite rear side of the building for summer ventilation.

Air Conditioning of Animal Shelters. Air conditioning refers to treatment of air to make it a more desirable environment. Ventilation refers to air movement, but when humidity and temperature are also controlled, air conditioning results. Other elements often controlled are dust, bacteria, and odors (addition of desirable or removal of undesirable). The adult human body at rest gives off 400 Btu per hr. The desired air condition, or comfort zone, varies under different conditions of air movement and seasons of the year. Generally speaking, the comfort zone is at 70 to 75°F and 30 to 70 per cent humidity. Where individuals must pass from the warm outside to an air-conditioned room, the temperature difference is kept at a minimum to prevent discomfort.

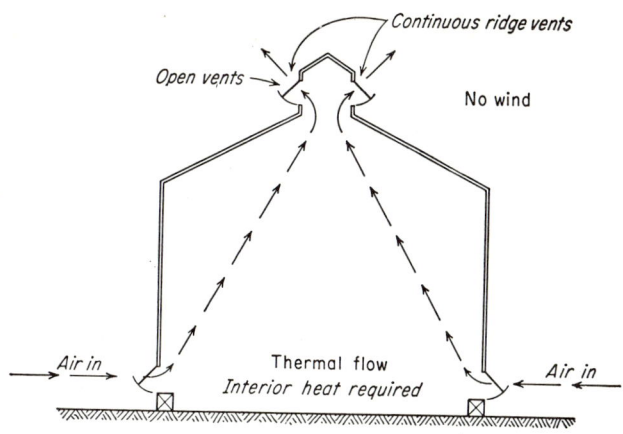
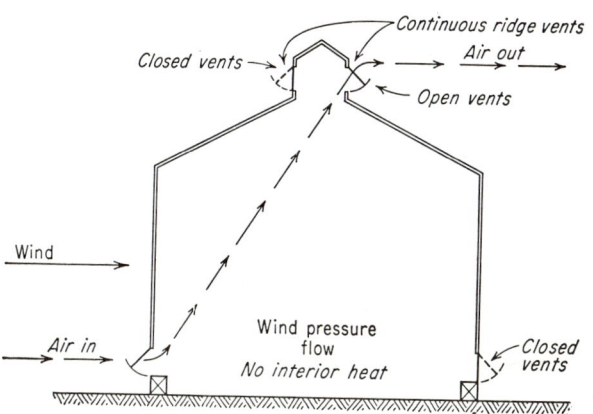

Fig. 48-11. Air flow through ridge ventilator.

The use of air conditioning has not become a general practice for livestock housing, but is receiving attention for swine, poultry, and cattle. To date the systems have not been generally accepted as economical. The desirable condition of air for meat animals is based on maximum gain in pounds per pound of feed. A 1-ton air conditioner would absorb 12,000 Btu per hr and require a 1-hp electric motor. Recent research on environmental conditions as they affect livestock production suggests that air conditioning of livestock shelters may become more common in the future.

Chapter 49

DRYING FARM CROPS

CARL W. HALL

An evaluation of drying must include harvesting, the operation which precedes drying, and storing, the operation which follows drying. Artificial drying of some harvested crops, such as rice, is essential because rice, to obtain best quality, must be harvested at a moisture content which is too high for safe storage. Early harvest and subsequent artificial drying are desirable for other crops to reduce field losses from weather, insects, or birds or to permit earlier use of harvested land for other purposes such as grazing or replanting. The average annual loss of grain crops is 5 per cent during harvest and another 5 per cent during storage. About 28 per cent of the hay crop in the United States is lost each year.

DRYING FACTORS

Moisture loss or gain can be induced by providing differences between the vapor pressure in the surrounding atmosphere and that of the water within a material or product. Reduction of the vapor pressure in the air is accomplished by condensing part of the water out of the air on chilled coils and plates, or by reducing the atmospheric pressure by exhausting air from a sealed plenum, or a combination of both. Occasionally, for small-scale operations, water vapor is removed from the air by passing it through a material such as concentrated sulfuric acid, which has a very low water-vapor pressure, or a material that can hold water in chemical bond. The alternative method is to increase the velocity and pressure of liquid-water molecules in the material by heating them until they can escape faster than they return from highly humid surrounding air. The most rapid results are obtained by heating the product to be dried and surrounding the product with low-humidity air to obtain rapid drying of porous or thin products, as in the drying of leaves and forages or the cooling of leaf vegetables. Heat for natural drying is supplied by solar radiation, by respiration of organic products, by the metabolism of bacteria, insects, and molds, or by sensible heat changes transferred by the convection of natural air flow. These natural sources of heat often result in a reduction or loss of quality in feeds and foods, except for products similar to cheese, where curing takes place.

When a material has a water content and temperature resulting in a vapor pressure equal to that of the water vapor in the atmosphere, it is said to be at its equilibrium moisture content (Table 49-1). If the temperature is held constant and the relative humidity increased, the product gains moisture; if the relative humidity is held constant and the temperature is increased, the product usually loses moisture.

Drying Rate. The rate of moisture removal at any time can be calculated quickly from the observed temperature drop within a drying bin, by using the following simple formula, which is useful for normal temperature ranges.

$$R = \frac{CT_d}{1,000} \tag{49-1}$$

where R = water evaporated by sensible heat removed from air, lb/hr
 C = air flow through drying system, cfm
 T_d = temperature drop through the crop, °F

TABLE 49-1. EQUILIBRIUM MOISTURE CONTENT, PER CENT, WET BASIS, 77°F

Product	Per cent relative humidity						Ref.
	15	30	45	60	75	90	
Barley...................	6.0	8.5	10.0	12.1	14.4	19.5	a
Corn, shelled yellow.........	6.4	8.4	10.5	12.9	14.8	19.1	a
Oats.....................	5.7	8.0	9.6	11.3	13.8	18.5	a
Sorghums, grain............	6.4	8.6	10.5	12.0	15.2	18.8	b
Soybeans.................	5.0	6.2	7.4	9.7	13.2	18.5	c
Wheat...................	6.4	8.5	10.5	12.5	14.6	20.0	a
Rice.....................	5.6	7.9	9.8	11.8	14.0	17.6	d
Flaxseed..................	4.4	5.6	6.3	7.9	10.0	15.2	a
Sugar beet seed............	...	8.1	9.8	11.8	14.0	18.2	e
Fresh-cut alfalfa...........	7	8	10	13	19	40	f
Dry alfalfa in mow.........	6	7	9	11	14.5	24	f

a D. A. Coleman and H. C. Fellows, Hygroscopic Moisture of Cereal Grains and Flaxseed Exposed to Atmospheres of Different Relative Humidity, *Cereal Chem.*, 2:275–287, 1925.

b D. A. Coleman, B. E. Rothgeb, and H. C. Fellows, Respiration of Sorghum Grains, *USDA Tech Bull.* 100, 1928.

c P. E. Ramstad and W. F. Geddes, The Respiration and Storage Behavior of Soybeans, *Univ. Minn. Agr. Expt. Sta. Tech. Bull.* 156, 1942.

d M. L. Karon and M. E. Adams, Hygroscopic Equilibrium of Rice and Rice Fractions, *Cereal Chem.*, 26:1–12, 1949.

e C. W. Hall, Drying Temperatures and Storage Problems of Sugar Beet Seeds, *J. Am Soc. Sugar Beet Technologists*, 1956.

f S. T. Dexter, W. H. Sheldon, and C. E. Huffman, Better Quality Hay, *Agr. Eng.*, 28:291–293, 1947.

For values of other products see *Agricultural Engineers' Yearbook*, published annually, and Carl W. Hall, "Drying Farm Crops," Edwards Bros., Inc., Ann Arbor, Mich., 1957.

Equation (49-1) applies only to drying with sensible heat lost from the drying air. If the drying temperature is in the summer temperature ranges at which green crop material respires rapidly, the heat of respiration may be adequate to double the drying rate observed from the temperature drop through the crop. At drying temperatures below 60°F, most of the heat comes from the air and the calculation from Eq. (49-1) will be reasonably accurate. A third source of drying heat of lesser importance is from sensible-heat losses when warm crop material is cooled to the dew point during the drying process or cooled with low-temperature air during autumn and winter. Drying by sensible-heat loss from crop material is sometimes used in grain-drying systems in which the grain is first heated, then cooled with unheated air.

STATIC PRESSURE AND AIR FLOW

Static Pressure. Static pressure is the difference between atmospheric pressure and the internal pressure within a plenum or air system if the air is not moving. It is this difference in pressure which causes the air to move. Static air pressure is expressed in inches of water or water column for most purposes, but may also be expressed in inches of mercury, or in pounds per square inch for pressure tanks. For common ventilation work, the static pressure is usually expressed in inches of water and is expressed that way in this handbook unless otherwise noted.

Velocity Pressure. Velocity pressure refers to the force exerted by moving air. Wind has velocity pressure but, because it is part of the atmosphere, no static pressure. However, when wind encounters a fence, tree, or building, part of the velocity pressure is converted to static pressure and a portion of the remainder is converted to heat. To obtain accurate velocity or velocity-pressure readings, it is desirable to average a number of spaced readings with anemometer, velometer, or pitot tube sampling equal areas across the width and height of a duct. The velocity pressure increases as the square of velocity ratio. Total pressure is the sum of the static and velocity pressures.

Estimating Static-pressure Drop for Grain. The static pressure against which the

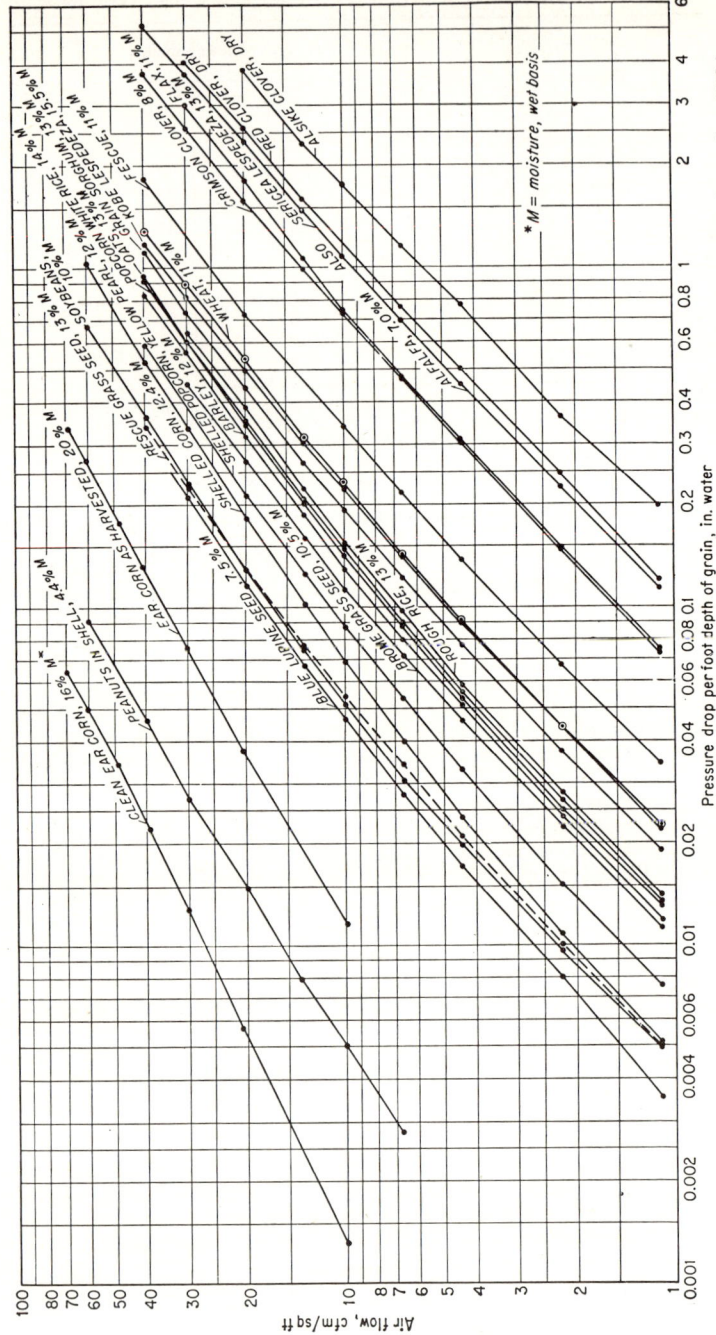

Fig. 49-1. Resistance of grains and seeds to air flow. (*Shedd, Resistance of Grains and Seeds to Air Flow, Agr. Eng.*, 34:616-619, *September*, 1953.)

fan must work is affected by the design of the duct and floor, depth of the grain, type of grain, and quantity of air flow through the grain.

The resistance of various grains and seeds to air flow has been summarized in Fig. 49-1.

To obtain the static-pressure drop through a drying system using the data in Fig. 49-1, the following procedure is used:

1. Determine the total air flow required based on recommended air flow for the quantity of product involved, cubic feet per minute.
2. Determine the area of storage, square feet.
3. Divide air flow by area to obtain the air flow of cross-section area, cubic feet per minute per square foot.

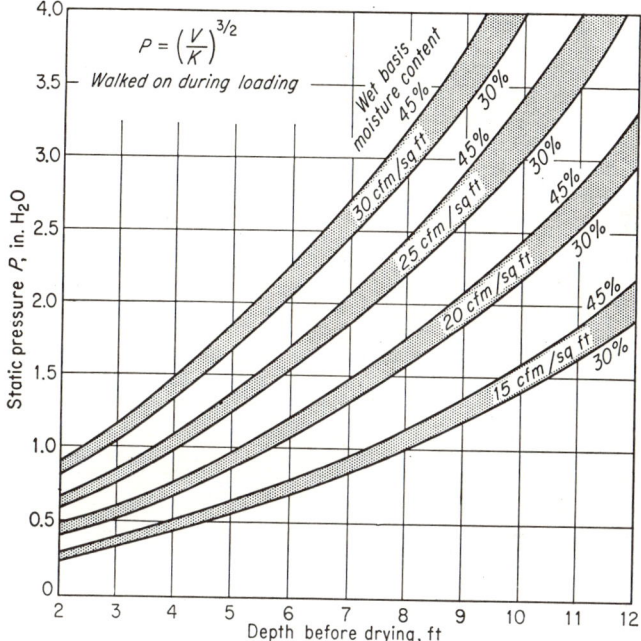

Fig. 49-2. Relationship of static pressure, depth of hay, air flow, and per cent moisture for long and chopped hay. (*Davis and Baker, The Resistance of Long and Chopped Hay to Air Flow, Agr. Eng., 32:92–94, February, 1951.*)

4. Read from Fig. 49-1 the static pressure per foot of depth.
5. Multiply the static pressure for 1-ft depth, as obtained from the chart, by the depth through which the air passes, to get the total static-pressure drop due to the grain or seeds.
6. Add the static-pressure drop from the duct and floor, usually about ¼ in. water if the air velocity is kept at 1,000 fpm or less.

Estimating Static-pressure Drop and Air Flow for Hay. The *resistance of hay* to air flow has been difficult to determine and is difficult to estimate because of variations in density of the hay mass. Many failures in mechanical ventilation have been caused by too little air or uneven air distribution through the hay. Formulas for *approximating* pressure drop for air flow through hay are available[1] (Fig. 49-2).

---

[1] Rene Guillou, Forced Air Flow in Drying Hay, *Agr. Eng.*, 27:514–520, *November*, 1946. Roy B. Davis and Vernon H. Baker, The Resistance of Long and Chopped Hay to Air Flow, *Agr. Eng.*, 32:92–94, February, 1951.

FANS

Mechanical ventilation systems derive their energy from fans. The velocity pressures delivered by fans are so much greater and more reliable than thermal and wind pressures that it is customary to ignore thermal effects and wind effects except for outlet and inlet locations. Most of the numerous fans that are available can be classed either as radial (centrifugal) or as axial (propeller) flow types. Some are intended for light-duty free-delivery work, such as exhausting air from buildings or stirring air within rooms. Other fans are constructed for heavy-duty industrial conditions.

Radial Flow Fans. Centrifugal fans draw air into the fan housing along the fan axis at one or both sides and accordingly are called single- or double-inlet fans. The fan wheel is confined by a housing so that air must pass into the center portion where the vanes of the wheel move the air outwardly and around the fan housing until it reaches the discharge opening.

Forward-curved-blade centrifugal fans are used primarily for heating and ventilating in commercial, public, and residential buildings. Forward-curved fans operate at low speed, and the outer edges of the vanes are curved forward, compelling the air to move faster than the velocity of the vane in order to escape from the fan. This results in a slow-speed quiet fan capable of delivering large volumes of air at little or no resistance. When resistance develops, the velocity of the air is dependent upon the centrifugal action developed. As an example, a forward-curved centrifugal fan with a 36-in.-diam wheel turning at 300 rpm will move about 18,000 cfm against $\frac{1}{4}$ in. of water resistance, with a power input of about 4.2 hp. If the resistance is increased to 1 in. of water at the same speed, the air delivery will fall to about 7,000 cfm, with a power input of about $1\frac{1}{2}$ hp. For duct work in the heating of buildings, this fan is quiet and ideal, because the air delivered and the power input automatically decline if outlets in various rooms are closed.

Straight-bladed centrifugal fans have as their principal use the delivery of materials, or a combination of materials and air, and are used for blowing grain, silage, and hay and in grain cleaners. In general, radial-bladed fans have not been used for crop drying, but have been used for aeration.

Backward-curved-blade centrifugal fans have a heavy wheel with large curved blades that lead away from the direction of rotation. This type of fan not only has centrifugal action but gives the air a push out of the housing as it leaves. It is quiet compared with the high-speed axial-flow fan. Its power and delivery characteristics are known as nonoverloading; i.e., the power load on the motor is about the same regardless of the resistance against which the fan works. Because of its nonoverloading characteristic, the backward-blade fan is suitable for use on rural lines for almost all agricultural purposes. A 36-in. backward-curved-blade fan of the same size as the forward-curved-blade fan above shows a performance of 17,000 cfm at 675 rpm with about 4.7 hp input, for a resistance of $\frac{1}{4}$ in. of water. At 1-in. resistance, the same fan at 675 rpm will deliver about 13,700 cfm with an input of 4.8 hp.

Axial-flow Fans. Axial-flow fans are called by various trade and subtype names and include blades similar to windmill vanes at one extreme and blades similar to those used for airplane propellers and wing sections at the other extreme. Propeller, vane-axial, and tube-axial fans belong to this category. All axial-flow fans receive the air along the fan axis and continue moving the air in the same direction for delivery. High-speed axial-flow fans are generally suitable for most agricultural purposes. Most have heavy-duty sealed ball bearings. The propellers have four or more blades shaped along aerodynamic principles and are usually one-piece aluminum castings.

A high-speed axial-flow fan with a 36-in. wheel operating at 1,740 rpm against $\frac{1}{4}$-in. resistance will deliver about 22,000 cfm with a power input of about 4.7 hp. If the resistance is increased to 1 in. at 1,740 rpm, the fan will deliver about 16,800 cfm and require 5.2 hp.

High-speed axial fans occupy little space, are light in weight, and are easy to install. A further advantage is that the speed and pitch of the blades can be selected for a particular job. Flat pitches at high speeds operate especially well against the higher resistances. For light-duty low-pressure jobs, the fan sheave can be changed to provide a lower speed. Current tapered sheave bushings are designed so that it is easy to change sheaves without pounding on the sheave or shaft. High-speed axial-flow fans have a comparatively high noise level because of the high tip speed, and they must be properly screened to prevent entrance of foreign matter into the fan. Axial-flow fans are usually direct-connected to the electric motor when operated at top speeds of 3,450 to 3,500 rpm.

Vane-axial fans are a variation of the axial-flow propeller fans and are intended for heavy-duty performance above the pressure range of propeller fans mounted in an ordinary orifice. The principal difference between a vane-axial fan, other than its

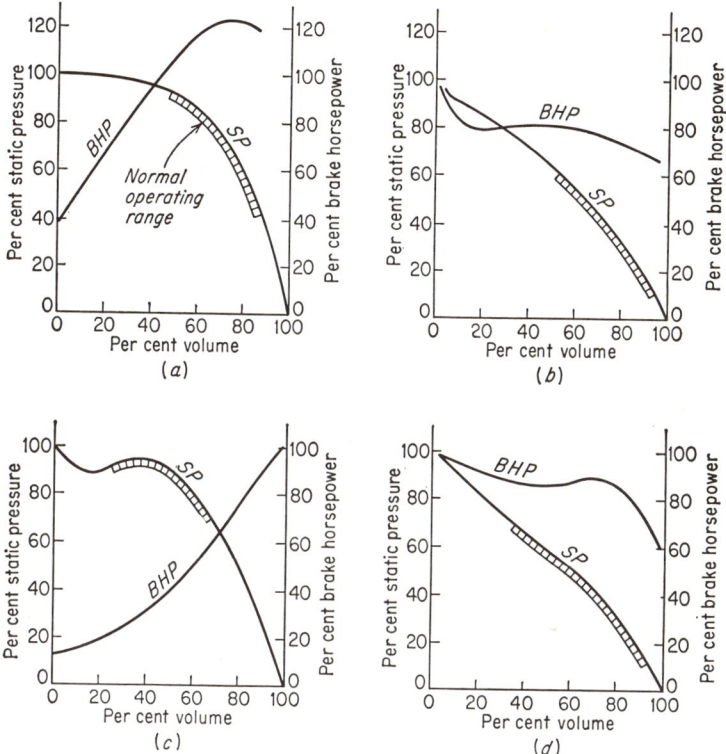

Fig. 49-3. Typical performance curves for fans used for drying: (a) backward-curved centrifugal; (b) high-speed propeller; (c) forward-curved centrifugal; (d) axial-flow.

rugged construction and heavy hub, is in the stationary vanes mounted in a duct section close to the fan blades. These vanes prevent swirling and reduce back pressure on the fan wheel and are a principal factor in the efficiency at higher resistances. Vane-axial fans are well suited for many agricultural jobs, but their higher cost and power are a limiting factor.

Fan Selection. Manufacturers usually list several fans that will produce a desired air volume. At high elevations, such as exist in the mountain states, barometric readings are lower. For high-altitude drying, it is necessary to add enough air capacity to make up for the reduced weight of air handled. This is done by increasing the speed to load the fan motor up to sea-level ratings or by selecting a fan with a larger power requirement than the motor would handle at sea level.

Typical performance curves for the various kinds of fans are shown in Fig. 49-3.

AIR-DISTRIBUTION SYSTEMS

Stack and mow dryers for hay and bin dryers for grain are of recent origin in the United States and have been facilitated by the concurrent development of rural electrification and suitable fan and system designs. A system consists of a fan, a duct system to distribute the air flow, and a structure to hold the crop. If heat is employed, a burner is included with suitable controls. The success of the drying system is determined by selection of proper fan capacity, suitable design of the duct system, and proper handling and loading of the crop. The duct system should have a low resistance, much less than the resistance due to the crop material. However, the duct system should enforce uniform air distribution, requiring that the ducts be designed with enough resistance to prevent unrestricted escape from the duct work, especially in forages and ear corn where the crop itself offers a variable resistance.

Main Ducts. Main ducts operating at relatively low static pressures less than 1 in. water, such as ear corn and hay, are designed for maximum air velocities of 1,000

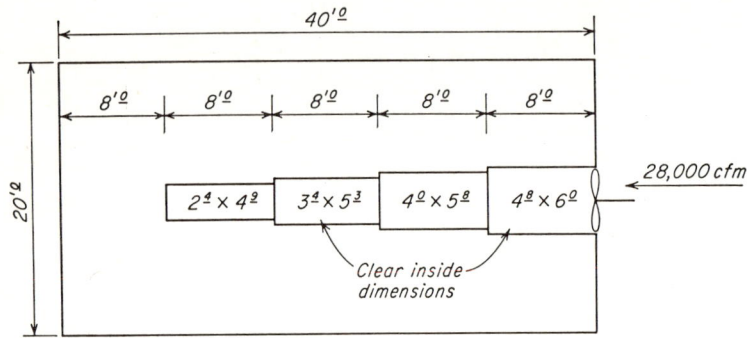

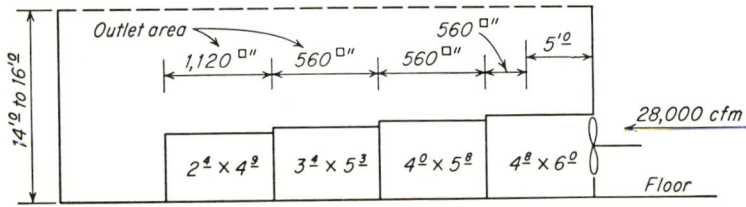

Fig. 49-4. Typical layout for central duct in mow, stack, or crib. For an open crib or stack, part of the air should be discharged at top and part at bottom of the duct. For a tight mow or bin, discharge all air at the bottom. Duct dimensions and outlet areas are in proportion to the amount of air moved to the crop in each section.

fpm requiring 1 sq ft of cross section for each 1,000 cfm flowing in the duct. Grain-drying systems operating at static pressures up to 3 in. of water column are often designed for higher velocities to avoid bulky duct sizes. Main ducts may be tapered or stepped down in cross section as portions of the air flow are diverted into lateral or submain ducts or may be constructed at a uniform cross section for the full length. The essential features of a main duct, in addition to maximum velocity limitations, are tight construction to avoid air leakage and the elimination of studs and bracing on the inside, especially near the fan outlet.

Risers are short sections of submain ducts connecting two floor levels. Because risers usually start and end with right-angle diversions of air flow, they are best constructed with 2 sq ft of cross section per 1,000 cfm flowing through the riser.

A main-duct design for a stack, bin, or mow is illustrated in Fig. 49-4, with laterals omitted. Outlets open directly from the main duct into the crop. Although the main duct in Fig. 49-4 is shown in sections of reduced cross-section area, it can be equally satisfactory if tapered or constructed without reduction in cross section. The crop should be loaded evenly around the duct so that air travel is about the same through all portions of the crop. Outlet areas in this type of duct are usually at the bottom of the duct, but a portion of the outlet area may be provided at the top of the duct if provision is made to keep the top outlets closed until covered adequately by the crop. Figures 49-5 and 49-6 illustrate the common positions of main ducts in mow and bin dryers. Plywood and building board make tight interior sheathing for main ducts. If boards are used, the interior of the main duct can be lined with strong paper or roll roofing to make the duct walls tight. If the duct system is laid on a wood

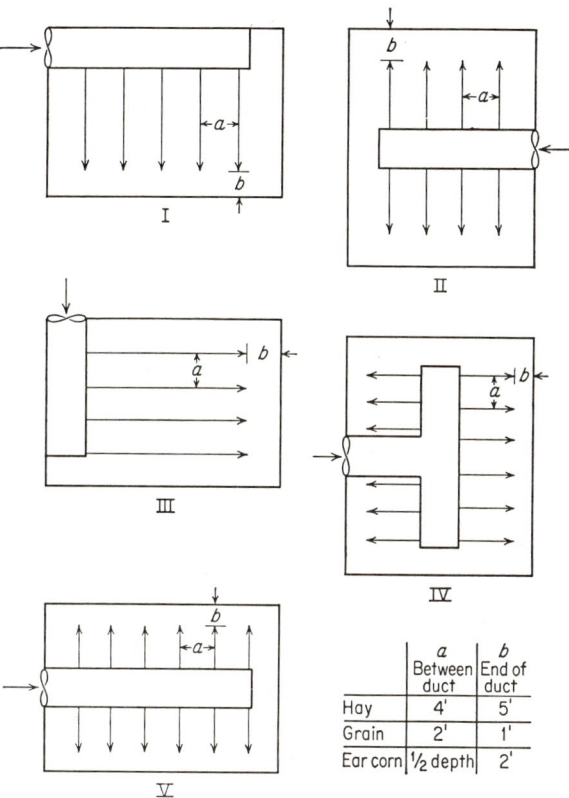

FIG. 49-5. Five common floor layouts for the main duct and laterals in mows and bins.

floor, it is usually best to cover the floor with sheets of building board or roll roofing to prevent the loss of air downward through the floor.

Lateral Ducts. Lateral ducts are usually designed to rest on the floor and dispense air from the underside (Fig. 49-7). They are constructed in lengths easy to handle and are usually made of uniform cross section or stepped down, although they may be tapered in sectional lengths. Laterals are usually butted or fitted together without permanent connections, because it is desirable to move them to clear the floor when the bin or mow is empty. In grain bins, hooks or latches are often provided to ensure that the laterals do not become parted and filled with grain. The laterals shown are made of wood, although metal of adequate strength may be used. Economics may dictate that prefabricated metal ducts be of uniform cross section, many of which are on the market today. Metal mesh is sometimes used and, when adequately strengthened, is satisfactory.

For crops which offer irregular resistance to air flow, laterals with limited outlet

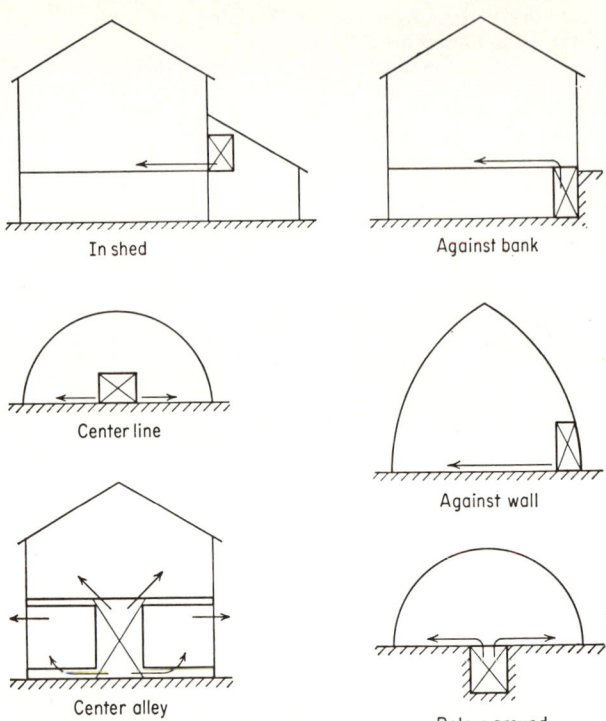

Fig. 49-6. Six common positions for the main air duct in bins and mows from end elevation.

areas are desirable to prevent most of the air from escaping through the least dense material without drying the heavier portions. In this case, the inlet end of a lateral duct is usually best designed with a cross section of 1 sq in. per 10 cfm, which provides an average lateral inlet velocity of 1,440 fpm. This velocity is chosen because of its self-balancing characteristic. At velocities of 1,500 fpm or less, the amount of pressure required to move air is small, but above that velocity the pressure required increases rapidly.

The *slatted-floor system* is a variation of lateral layout and is designed with 1,440-fpm velocity into the joist spaces under the slatted area. Figure 49-8 illustrates a slatted-floor system. The lateral joists are usually made of standard-dimension lumber of a width suitable to provide the required 1,440-fpm velocity out of the main duct. The surface under the slatted-floor joists must be airtight. Wood floors should be covered with heavy building paper to prevent air loss downward before the slatted floor is installed. Slatted floors are often made in sections that can be moved when the mow or bin is empty. The sections may all be of the same size, but it is more usual to make them in short stepped-down sections or to use tapered joists. If the joists are tapered, the minimum height at the outer end of the joist is 1½ in. Slatted floors are not extended to the walls for most crops. For ear corn, the openings should extend within 4 to 6 ft from the wall; chopped and loose hay, 5 to 8 ft; and baled hay, 2 ft.

Factors in selecting the type of air-distribution system—false floors or ducts—are cost, ease of cleaning, and multiuse features of the buildings, in which case nestability of the ducts is desired.

Duct Outlets. Outlet areas, through which the air flow escapes from the ducts or slatted floors into the crop material, are usually constructed with a cross section two to four times the cross section of the lateral. Coarse crops such as the forages, nuts,

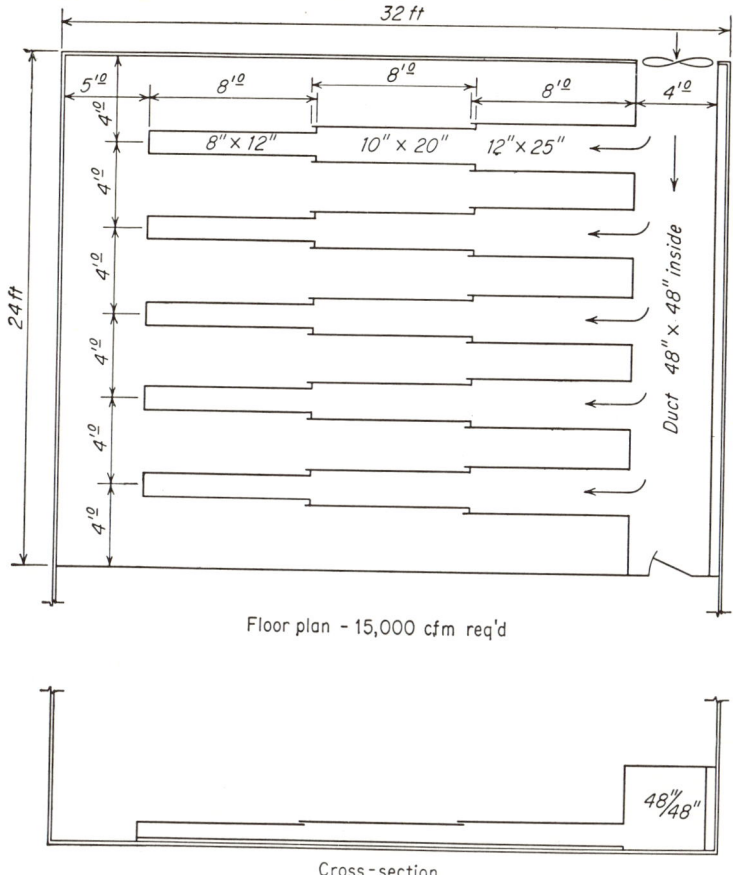

Fig. 49-7. Layout for typical side duct plus lateral system for hay requiring 15,000 cfm.

and ear corn have little internal resistance. Hence the outlets from the laterals are restricted to about 2 sq in. for each 10 cfm flowing in the lateral. Clean grains have comparatively uniform resistance and do not require lateral-outlet resistance; therefore the lateral outlets into grain are enlarged to at least four times the lateral cross-section area to prevent undue resistance in the outlets.

In general, if 7 per cent or more of the floor is in perforated openings in a metal sheet, the resistance of the metal openings is not excessive.[1]

Air-flow patterns within the crop material vary with the depth of the crop material and also with variations in the density and structure of the crop material. Except for clean grain, it is difficult to obtain uniform resistance in the crop material and an accompanying uniform rate of air flow through the crop material. The best air-flow patterns are obtained in bins in which the air is introduced evenly at the bottoms and travels vertically in a parallel flow pattern upward through the crop material, or in the reverse direction if the system is designed for exhaust from the bottom. In a parallel-flow bin, drying proceeds in a uniform plane across the bin, with the plane moving slowly in the direction of the air flow until the entire bin is dry. Although drying bins with parallel air flow give the best air-flow patterns, it is often conven-

---

[1] S. M. Henderson, Resistance of Soybeans and Oats to Air Flow, *Agr. Eng.*, 25:127–128, April, 1944.

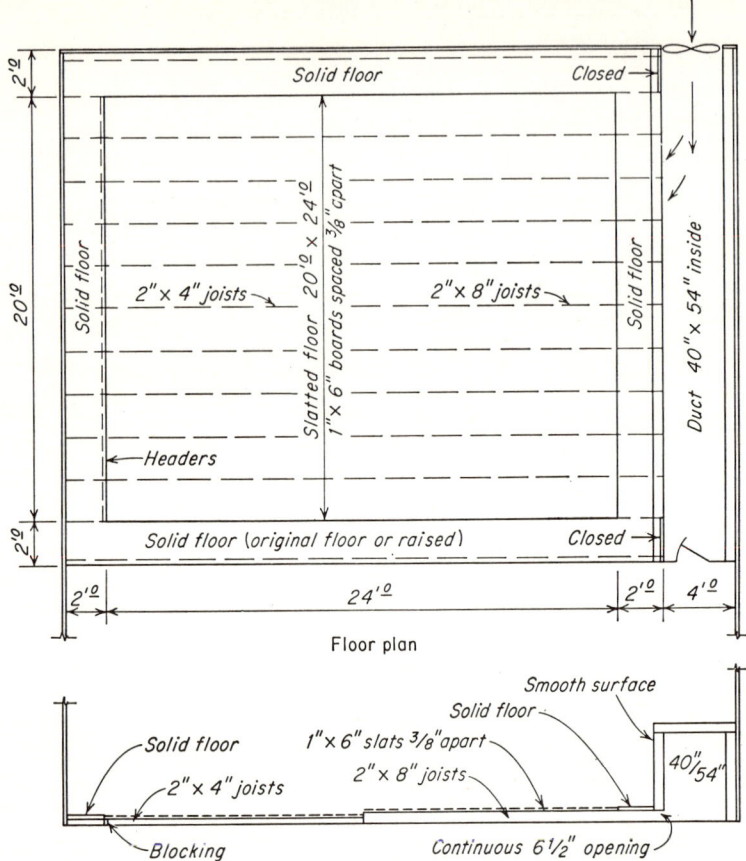

Fig. 49-8. Layout for typical slatted-floor system requiring 16,000 cfm for baled hay. Chopped or loose hay and ear corn are similar except for a wider (5-ft) solid border around slatted area.

ient to dry in stacks and mows without airtight confining walls. With careful design the air can be forced along routes of equal travel.

Reversible Air Flow. Reversible bin dryers permit control of humidity and final moisture content and eliminate the overdrying that often occurs with one-way heated air flow. Reversible designs are arranged so that the air flow may be directed either upward or downward (to right or left) by adjustments of dampers and doors. Humidity is controlled by the amount of air exhausted and recirculated. Most designs provide a furnace for control of temperature. With controlled temperature and humidity and frequent reversals of air flow, the crop material in the bins can be brought to a uniform selected equilibrium moisture content without the overdrying which impairs quality of some seeds and crops.

Vertical Ducts. Vertical ducts are often used in tall circular storage bins. In some designs, the vertical duct also serves as a central chute through which the crop is removed for feeding. In most designs, only one vertical duct is required. The vertical duct must be designed so that air can be forced into the bottom of the structure when the upper portion is empty. Figure 49-9 illustrates one type of vertical duct, using a pressure system in which a movable plug is raised or lowered as necessary. The plug consists of a circular frame fitted with a canvas bag which the air forces

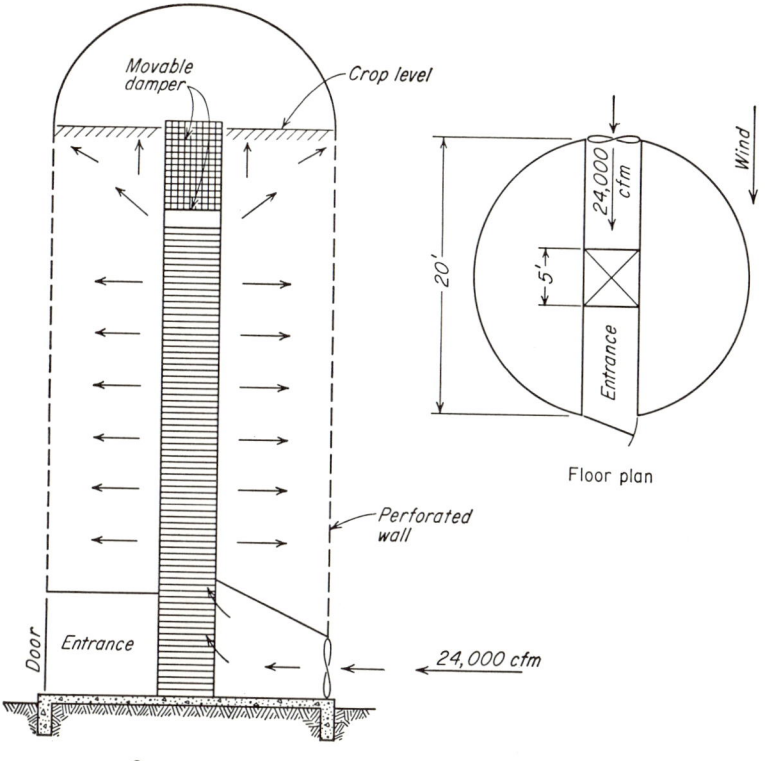

Fig. 49-9. Sections showing general layout for vertical storage 20 by 40 ft, requiring 24,000 cfm. Slatted vertical duct should have approximately 200 sq in. opening per foot of height. Perforations in the outer walls should also provide 200 sq in. opening per foot of height.

against the duct wall. If the storage is filled in two or more cuttings or harvestings, the lower portion of the duct in the first cutting should be shut off from drying with a paper lining which the air will force against the duct wall. An exhaust system can be used to advantage. An exhaust system prevents condensation on cold exterior metal walls at night, makes it easy to examine the last material to dry next to the vertical duct, and because it eliminates the sharp turning of air under pressure at the bottom next to the fan, less turbulence and more uniform static pressure are obtained.

Efforts to find a less expensive vertical self-feeding structure for chopped hay have led to the construction of vertical stacks in which the vertical duct is formed with the hay. Field stacks, 16 by 20 ft in diameter, have been built by forming the exterior with slatted-wire cribbing and forming the interior vertical duct with a cylindrical metal form lifted by a block and tackle from a center pole as the height of the stack increased.[1] Air passed from a portable fan to the vertical duct through a wooden horizontal duct laid on the ground before the stack was started. The vertical duct was covered after the form was removed by laying boards across the opening and covering the boards with several feet of chopped hay. Good-quality hay resulted.

Later, the search for a successful self-feeding storage for chopped hay resulted in the development of a tapered vertical chopped-hay storage.[2] An octagonal framework was constructed of slatted-wire cribbing supported on eight heavy poles rising 30 ft

---

[1] Ralph Lipper, Recommendations for Drying Chopped Hay in the Stack, *Kansas State Univ. CREA and Dept. Agr. Eng. Bull.*, 1947.

[2] Kenneth K. Barnes and Hobart Beresford, Self-feeding of Chopped Hay, *Agr. Eng.*, 35:515-553, August, 1954.

above the ground and with a 23-ft diameter at the base tapering to a 22-ft diameter at the top. The central vertical duct was formed by pulling a metal cylinder from the bottom to the top as filling progressed. Approximately 500 cfm of air per ton was introduced at the base of the vertical duct through a horizontal connection on the ground. The base of the structure was equipped with self-feeding mangers and gates. Good-quality hay resulted, and it self-fed satisfactorily. The hollow vertical center formed by the vertical duct eliminated one of the major self-feeding problems.

HEATERS AND HEATED AIR DRYERS

To increase the rate of moisture removal and make the drying operation less dependent on the weather, heated air dryers have come into use, particularly for commercial and large-farm operations. The use of heat must be limited to relatively shallow layers of the crop to prevent uneven drying.

Heaters for Crop Drying. Heaters for crop drying are classified as (1) direct and (2) indirect. With the direct heater the products of combustion pass through the product with the drying air. With the indirect system the heat-transfer surface separates the products of combustion and transfers heat to the air moved through the system, with the products of combustion being rejected to the outside atmosphere.

Several methods have been used for rating the capacity of heated air units. Rating of units on the basis of the bushels which can be dried per hour is meaningless unless the product and the range in moisture content are stated. Another method is to rate the unit on the basis of the number of pounds of water it will evaporate per hour, but this varies with the temperature of evaporation, level of moisture content, and the product. Unit capacities are also given on the basis of the number of gallons of fuel per hour consumed. This is directly related to the Btu output per hour, which is probably a more satisfactory way of rating heated-air units. No. 2 fuel oil contains about 140,000 Btu per gal, and portable heated-air crop dryers for farm use will burn from 3 to 10 gal per hr, depending upon the size. The fuel-burning capacity of a particular unit can be varied by changing the pressure of the burner unit and the number of nozzles. Coal-burning units consume up to about 50 lb per hr, with coal producing about 13,500 Btu per lb. Stationary heated-air crop dryers are generally available in sizes with outputs up to 3,500,000 Btu per hr. The portable heated-air units are powered by 3-, 5-, or 7½-hp motors or driven by tractor power take-off. The thermal efficiency of an indirect heater in transferring heat to the drying air is usually 75 to 85 per cent.

Batch dryers are bins with heaters designed for rapid drying of a charge of crop material. Loading and unloading should be done with mechanical conveyers. Heated air is employed to increase the rate of drying. The capacity of a batch dryer is usually designed to match the expected rate of harvest. In batch grain dryers, the material usually surrounds a perforated horizontal duct and the hot air passes out through it and escapes through a perforated outer wall. Hay bales are dried in a layer on a wagon with a perforated floor over a plenum chamber.

Portable batch bins for heated-air drying usually have a capacity of 250 to 450 bu. With air at 140°F moving at 100 cfm per bu, shelled corn can be dried from 28 to 13 per cent in 2 to 3 hr with an additional ½ hr for cooling.

Continuous Drying. Continuous drying occurs in a unit in which the crop material flows at a constant rate through a heated-air drying bin capable of operation 24 hr a day. Numerous continuous dryers have been built, of which five types are in considerable use. Three types, column, conveyer, and drum dryers, are adapted to general farm use. Two others, spray and roller systems, are used commercially, particularly for liquids.

A *column-type* continuous dryer is sketched in Fig. 49-10. It consists of vertical sections in rectangular or circular shape, enclosing a plenum space, with thin walls of grain carried in perforated, meshed, or baffled enclosures. Heated air introduced into the plenum is forced outward through the columns of grain as they move slowly downward to the base where the grain is conveyed away mechanically. The plenum space is divided into two parts. The upper, larger portion of the plenum space receives heated air, which dries by loss of sensible heat from the air. The lower, smaller

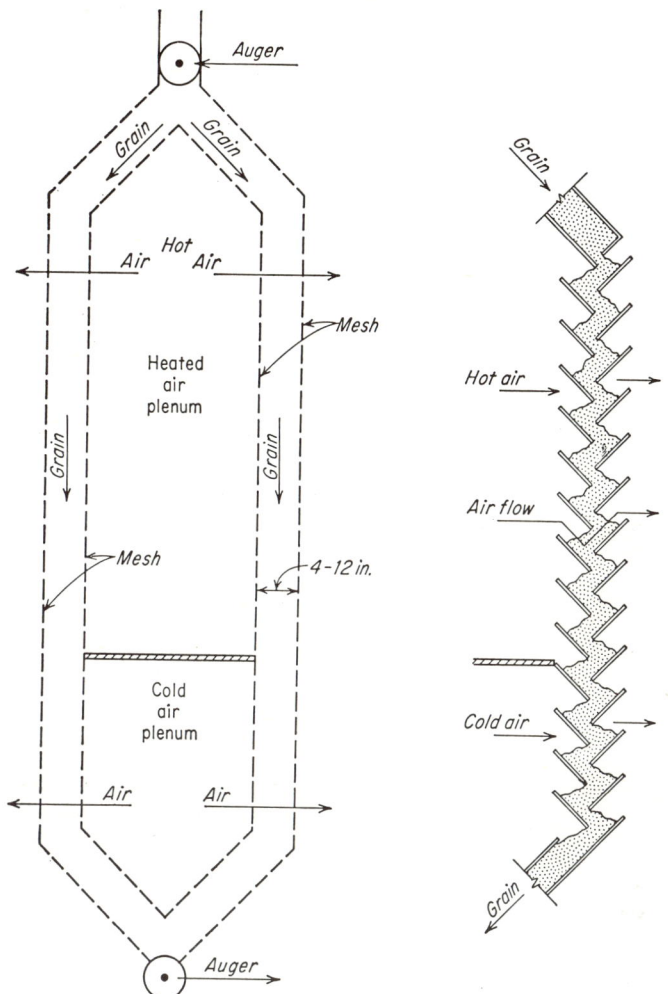

Fig. 49-10. Continuous-vertical-column dryer types. Left, vertical screened or perforated walls. If air is heated much above atmosphere, additional drying is obtained by forcing cold air through lower portion. Right, similar, except baffles are used instead of screen. Baffles mix grain and provide uniform drying.

plenum space receives unheated air for cooling and removes some moisture while removing sensible heat from the warm grain. Continuous column dryers are adapted to either large or small operations and can be used as batch dryers if both plenum spaces are first operated with heated air and then with cold air in the lower plenum space as the grain is removed. The depth of the grain columns can be varied. The thickness usually gives 4 to 18 in. of air travel through the grain, the thinner column sections being best for high-moisture grain sensitive to heat or rapid drying.

GRAIN DRYING AND PRESERVATION

Natural Ventilation of Grain. The quality of stored grain can be maintained, if only slightly above the moisture content considered safe for storage, by taking advantage of the natural forces of the wind and temperature changes to move the air. The method is limited by excessive static pressure which must be overcome when moving sufficient air through thick layers of grain and by the wind velocity. If the grain is high in moisture content, the moisture must be removed or the product cooled before mold growth or spoilage takes place.

A round metal crib, 8 ft in diameter by 12 ft high, with perforated walls and a perforated vertical center duct and rotating pressure cowl, reduced ear corn from 20.1 to 16.3 per cent, from fall to spring, in Columbus, Ohio.[1] By June, the corn was down to 13.5 per cent with no *visible damage. A large vane* is required on a pressure cowl to keep it headed into the wind. The same crib equipped with a rotating suction cowl was unsatisfactory. With a suction cowl, the windward side of the crib was dried but the leeward side was inadequately ventilated.

Rotating suction and turbine cowls develop about 35 per cent of the wind energy as negative static pressure compared with 89 per cent efficiency for the pressure cowls.[2]

In recent years, farmers have turned to the use of power-operated fans for ventilation of wheat and other small grains.

Horizontal ducts have been used occasionally to take advantage of *wind-pressure effects* for ventilation of ear corn, grains, and forages. In a test crib, corn averaging 22.0 per cent moisture at the start dried to 17.2 per cent by Apr. 1 in Ohio and continued to dry and keep in good condition until it was shelled in the fall after a year in storage.[3]

Horizontal flues in a circular metal bin filled with field-shelled corn at 18.3 per cent moisture indicated that the corn remained sweet and cool until late April, when bin temperatures began to rise, and it was necessary to use forced ventilation to avoid spoilage.[4]

Aeration or mechanical cooling systems are used to circulate air through crops in bins, stacks, and piles after the crop is properly dried or cooled. For dry crops such as feed grains, forages, wheat, and beans, unheated air is circulated in small amounts sufficient to dissipate any accumulation of respiration heat and moisture and to prevent uneven-temperature conditions caused by latent-heat exchanges and radiation

TABLE 49-2. APPROXIMATE MAXIMUM MOISTURE CONTENT FOR SAFE STORAGE FOR 1 YEAR AND RESPIRATION RATE

Grain	Moisture for safe storage, %, wet basis	Heat produced at 100°F, Btu/day, for each 100 lb dry matter*	Water produced at 100°F, lb/day, for each 100 lb dry matter
Corn, shelled.........	13.5	11.08	9.78×10^{-4}
Wheat...............	14.0	3.46	3.06×10^{-4}
Barley...............	13.5	2.77	2.45×10^{-4}
Oats.................	14.0	1.85	1.63×10^{-4}
Grain sorghum.......	12.0	3.00	2.65×10^{-4}
Rice, rough..........	14.0	4.62	4.08×10^{-4}
Soybeans............	11.0	3.69†	3.26×10^{-4}†

* C. H. Bailey, Respiration of Cereal Grains and Flaxseed, *Plant Physiol.*, 15:257–274, 1940.
† P. E. Ramstad and W. F. Geddes, The Respiration and Storage Behavior of Soybeans, *Univ. Minn. Agr. Expt. Sta. Tech. Bull.* 156, 1942.

NOTE: Moisture content for safe storage depends on temperature, humidity, microorganism infestation, and length of storage. For seed or long-time storage, the maximum safe moisture content is about 2 per cent below that listed. The respiration rate is approximately doubled for each 1 to 1.5 per cent increase in moisture content.

Moisture content should be lower by about 2 per cent in the southern states, where insects otherwise cause severe damage.

to or from the outer portions of the crop mass (Table 49-2). Aeration may be used to:

[1] G. R. Shier and R. C. Miller, Ventilation of Ear Corn in Metal Cribs. *Ohio State Univ. Agr. Expt. Sta. Bimonthly Bull.* 219, *pp.* 171–177, November–December, 1942.
[2] C. F. Kelly, M. G. Cropsey, and W. R. Swanson, Performance of Cowls for Ventilated Grain Bins, *Agr. Eng.*, 23:149–151, May, 1942.
[3] Shier and Miller, *op. cit.*
[4] G. R. Shier, R. C. Miller, and W. A. Junnila, Forced Ventilation of High Moisture Grains, *Agr. Eng.*, 24:381–383, November, 1943.

1. Equalize temperature.
2. Remove heat.
3. Reduce mold growth in storage due to reduced temperature.
4. Apply fumigants.
5. Remove storage odors.
6. Equalize moisture.
7. Add moisture.

For grain crops, air volumes of $\frac{1}{10}$ cfm per bu are desirable and much lower air-flow rates have been found satisfactory (Table 49-3).

TABLE 49-3. AIR-FLOW RATES FOR AERATION, CFM/BU

Type of storage	Northern states	Southern states
Flat	$\frac{1}{20}-\frac{1}{10}$	$\frac{1}{20}-\frac{1}{4}$
Upright	$\frac{1}{20}-\frac{1}{4}$	$\frac{1}{20}-\frac{1}{10}$
Farm	$\frac{1}{20}-\frac{1}{10}$*	$\frac{1}{20}-\frac{1}{4}$

* For continuous operation, $\frac{1}{60}$ to $\frac{1}{80}$ cfm per bu.

SOURCE: Leo E. Holman, Aeration of Grain in Commercial Storages, *USDA AMS Marketing Research Rept.* 178, 1957.

It is desirable to place air ducts on the floor of grain-bin structures before the grain is placed in the bin, allowing 1 sq ft cross section per 1,000 to 1,500 cfm in the ducts, or the bin can be provided with a slatted or perforated floor. Except for smaller air requirements, aeration systems are similar to drying systems. Perforated pipes may be forced downward in a vertical position into the centers from above to provide an aeration system. The ducts are perforated or covered with screen on the lower half, the remainder being solid. The horizontal duct, without laterals, is the most common system employed for flat storage. If insect infestation is present, insertion of a perforated pipe into the infected portion will cool and dry the grain enough to reduce further damage, provided that mold has not blocked the air spaces between kernels. If properly arranged, aeration systems are ideal for obtaining thorough fumigation. Provision is often made for recirculating the air and fumigant in an airtight system. Follow closely the instructions of the fumigant manufacturer.

Aeration systems are usually designed to operate as exhaust systems. Commercial systems are usually operated intermittently, first until the grain is cooled down to 35 to 40°F, and then periodically to maintain a low temperature. With low air flows, $\frac{1}{50}$ cfm per bu, the fan is often operated continuously. The major objective of aeration is to cool the grain, and not to remove moisture, although up to 1 per cent moisture might be removed during cooling.

Aeration systems for vegetables and fruits stored in bins are similar to those for grains except that care must be exercised to avoid undesirable drying and shrinkage.

Grain Drying with Unheated Air. The fan and air-distribution systems are designed and selected on the basis of the quantity of air, in cubic feet per minute, and the static pressure, in inches of water. The quantity of air is given in cubic feet per minute of air per unit volume of the product or cubic feet per minute of air per square foot of floor area.

From 1 to 5 cfm per bu is usually required for drying small grains with unheated air, depending upon the moisture content, type of grain (Table 49-4), and the geographical location.

Most *drying of ear corn* has been with forced unheated air or heated air in conventional cribs adapted to drying. Methods of adapting conventional cribs are shown in Fig. 49-11. The minimum air-flow rate recommended is 5 cfm per bu (1 bu = 2.5 cu ft), with 30 per cent moisture corn. If ear corn is placed on a forced-air system after the middle of October, it will be difficult to dry below 20 per cent moisture in humid areas. However, the corn will be preserved by cooling and moisture removed the following spring. Fans used for hay drying are suitable for ear-corn drying because the static pressure is usually less than 1 in. of water. Duct arrangements for

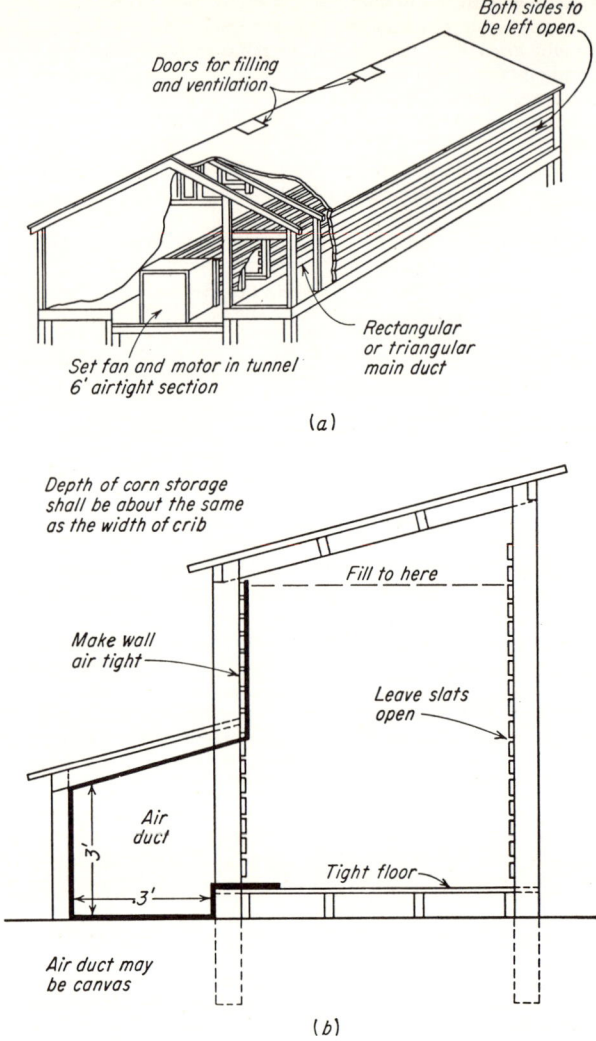

Fig. 49-11. Methods of adapting conventional crib for ear corn drying: (a) originally a crib with machinery storage in a center driveway; (b) narrow shed-type crib. (*Mich. Ext. Bull.* 316, 1954.)

hay are often used for ear corn. The concentration of husks, silk, shelled corn, and stalks under the elevator discharge will cause spoilage if not prevented.

Heated Air for Grain Drying. Heating the air before moving it through the grain increases the rate of drying and makes the rate of drying less dependent upon the atmospheric weather conditions. The quantity of air flow for heated-air installations varies from 10 to 200 cfm per bu. When heated air is used, the grain is dried in batches or continuously in layers not exceeding 2 ft in depth and more commonly less than 18 in. Plans for a batch bin are available in *USDA Leaflet* 314 (1951). Thin layers are required for heated air to prevent overdrying next to the air inlet of the bin. Some units recirculate the grain. An overdried product represents a loss to the farmer or elevator that sells the grain.

The operating cost for fuel and electricity for a heated-air drying system is only

TABLE 49-4. AIR AND FAN REQUIREMENTS FOR DRYING OATS, SHELLED CORN, AND WHEAT WITH UNHEATED AIR FROM DIFFERENT PERCENTAGES OF MOISTURE CONTENT AND AT VARIOUS PRACTICAL DEPTHS

Grain moisture content, %	Recommended minimum air-flow rate per bushel, cfm	Practical grain depths, ft	Static pressure,* in. water	Maximum quantity that can be dried per fan horsepower,† bu
Wheat				
20	3	4 6	1.2 2.3	830 440
18	2	4 8	0.8 2.5	1,880 600
16	1	8 10	1.3 2.0	2,300 1,500
Oats				
25	3	4 6	0.8 1.7	1,250 590
20	2	6 8	1.1 1.8	1,360 830
18	1½	8 10	1.4 2.0	1,430 1,000
16	1	8 12	0.9 1.9	3,330 1,580
Shelled corn				
25	5	4 6	0.7 1.6	860 380
20	3	6 8	0.9 1.5	1,120 670
18	2	6 8 12	0.6 0.9 2.2	2,500 1,670 680
16	1	8 12 16	0.5 1.0 1.6	6,000 3,000 1,880
Ear corn‡				
30	5	20	1.0	800
25	3	20	0.75	1,300

* Static pressure includes 0.25 in. allowance for loss from duct friction.
† Air flow, cubic feet per minute per horsepower, based on 3,000 cfm of air at 1-in. static pressure. *USDA Leaflet* 332, 1952.
‡ Calculated on the basis of a static pressure of ¾ in. water gage and a fan air-delivery rate of 4,000 cfm per hp. *USDA Leaflet* 334, 1952.

slightly greater than the cost of electricity for operating an unheated-air system. However, the initial investment for a farm-size heated-air unit is about twice as great as for an unheated-air unit, often being from $3,000 to $4,000 for a complete installation, including motor, fan, heater, and drying bin, with elevators for loading and unloading. The cost of fuel for removing each 1 per cent moisture from a bushel of corn or wheat is usually from ¼ to ⅓ cent. The units are usually designed to remove about 5 per cent of moisture by hot-air flow in 1 hr or less with an additional ¼ to

¾ hr for cooling the grain after drying. The equipment must be designed for (1) safe operation and (2) no damage to the product. Controls must be provided on the heater which will prevent overheating, explosion, or flames from spreading to adjacent buildings or equipment. Prefabricated bins are often used, and it is recommended that a commercially available heater with appropriate controls be used. When the heated-air unit is used within a building it is connected with a heat-resistant canvas duct at least 8 ft long. The maximum temperature that can be recommended for drying depends upon the use of the product. Temperatures up to and above 180°F are often used if the grain is to be fed to livestock. If the product is to be used for seed, temperatures are kept below 115°F. If the product is to be milled, it is normally recommended that the temperatures be kept below 140°F.

Some products are damaged by drying too rapidly. Damage consists of checking, cracking, and splitting of lumber, rice, and pea beans. Rate of drying may be controlled by:

1. Using a very limited amount of heat (supplemental heat)
2. Drying with heated air, placing in equalizing bins, and repeating until dry
3. Recirculating drying air
4. Using high-humidity air

Heated air may be used for *drying ear corn* to overcome the problem of poor drying weather during late fall harvest, although it is becoming common to shell at time of harvest (up to 30 per cent moisture content) and dry the shelled corn, because cob moisture is eliminated and less storage space is needed. Overdrying of corn next to a warm-air inlet is a problem. In a 15-ft crib, the driest ear corn may be 6 per cent when the wettest is 20 per cent. Dry to 13.5 per cent for storage; if for market, dry to moisture content on which price is based, usually higher than 13.5 per cent. Air heated to 110 to 115°F or less has been used many years for seed-corn drying. The air flow should be at least doubled (providing 10 cfm per bu or more) compared with unheated air to prevent excessive surface condensation. When first dried, the shelled corn shrinks on the cob. The cost of fuel and electricity for drying from 30 to 15 per cent moisture is about 5 cents per bushel.

Drying data for various grains are summarized in Tables 49-5 and 49-6.

Rice Drying. Rice is perhaps more difficult to handle without loss of quality than any other major grain. Most of the United States production of rice is in Texas, Arkansas, Louisiana, and California. The age-old method of handling rice was to harvest and store in shocks, stacks, or under cover until the rice moisture content declined to 14 per cent or less, after which the rice was threshed and stored in small lots. When mechanical equipment came into rice production, the same practices were continued until combining became general. Although the combine greatly reduced the labor and cost of rice harvesting, it created an acute storage problem. Rice is combined at an average moisture content of about 20 per cent, but some combined rice may have a moisture content as high as 25 to 28 per cent and drying should start within 6 hr after harvest. In addition to the usual field losses caused by storms, shattering, and weeds, excessive field drying reduces the yield of head rice because of cracking and checking of the grain. Head rice is that portion of the rice that mills out with a full-size grain. Under most favorable conditions, the yield of head rice is about 85 per cent, but the usual yield is less than 70 per cent.

The *drying of rough rice with forced air* in farm bins is similar to other grains. Continuous ventilation with unheated air produces satisfactory yields of head rice when the relative humidity is low enough for drying to the safe moisture content of 14 per cent and less.[1] Heating the air used for ventilation of deep bins may result in reduction in moisture content to less than 10 per cent for the rice in the bottom of deep bins, and if the air is heated 10°F or more above atmospheric, it may lower the milling quality of rice in the bottom of the bin. It is recommended that heated air be used only when necessary to lower the relative humidity of the air to provide a safe equilibrium moisture content of the rice.

Storages up to 20 ft in depth may be used for unheated-forced-air drying of rice.

---

[1] H. T. Barr and L. G. Coonrod, Bulk Drying and Storage of Rice on the Farm, *Agr. Eng.*, 23:158, 160, 162, March, 1952.

A minimum air flow of 2.0 to 2.5 cfm per bu is recommended for the Louisiana-Arkansas area. To dry rice to 13.5 per cent from 24 per cent in California, 1.0 to 2.0 cfm per bu is recommended.[1]

The internal structure of rice is such that, when *dried by forced ventilation with heated air*, the surface of the rice grain dries more rapidly than internal moisture can be transferred to the surface. This results in surface checking and cracking if the drying air is at a low relative humidity with a temperature above 100°F for continuous drying.[2]

Optimum heated air temperatures for drying rough rice vary somewhat with the quality of the rice as harvested and with the variety. The temperature of the drying

TABLE 49-5. TIME AND FUEL OIL FOR DRYING GRAIN WITH HEATED AIR

Grain	Moisture range, %	Temperature before heating, °F	Bin depth, ft	Capacity, bu	Air flow, cfm/bu	Direct-fired burner consumption of			
						5 gal/hr		10 gal/hr	
						Drying temp., °F	Drying time, hr	Drying temp., °F	Drying time, hr
Corn, shelled....	25–13	50	1	200	70	.....	.....	131	5.0
			2	400	28	102.5	13.7	155	5.0
			4	800	11	117	19	184	8.1
	20–13	50	1	200	70	.....	.....	131	3.2
			2	400	28	102.5	8.3	155	3.0
			4	800	11	117	10.8	184	4.8
Wheat*	20–13	70	½	100	137	113	4.1	156	1.5
			1	200	50	129	3.9	187	1.4
			2	400	20	148	5.1		
			4	800	7	172	9.2		
Oats*	20–13	70	½	100	150	109	4.1	148	1.4
			1	200	60	120	3.5	170	1.2
			2	400	22	137	3.4		
			4	800	10	157	5.4		
						Per 1,000 bu		Per 1,000 bu	
Corn, ear†	30–13	30	Over 6	...	5	70	100.0		
		30	Over 6	...	5	110	90.0	110	44
		60	Over 6	...	5	.....	.....	140	33
	25–13	30	Over 6	...	4	70	68.0		
		30	Over 6	...	4	110.0	62.0	110	30
		60	Over 6	...	4	.....	.....	140	23

* Adapted from *USDA Leaflet* 331, September, 1952.
† Adapted from *USDA Leaflet* 333, September, 1952.

air should not exceed a dry-bulb air temperature of 130°F for milling rice and 110°F for seed rice.[3] The temperature of the rice itself should not exceed 110°F as it leaves the dryer, and some evidence indicates that the maximum temperature of the rice should not exceed 100°F for the highest yield of head rice. These temperatures are for column dryers where the rough rice is exposed for 20 to 30 min to the drying air at rates up to 100 cfm per cu ft with a temperature drop of 20 to 30°F across a 6-in. rice column. The moisture loss under these drying conditions should not exceed 2 percentage points.[4] The rice should then be removed from the dryer to a tempering bin for a period not to exceed 12 hr before it is again passed through the dryer. The rice should go through the drying and tempering stage enough times

[1] S. M. Henderson, Deep Bed Rice Dryer Performance, *Agr. Eng.*, 36:817–820, December, 1955.
[2] Harold A. Kramer, Engineering Aspects of Rice Drying, *Agr. Eng.*, 32:44–45, 50, January, 1951.
[3] Barr and Coonrod, *op. cit.*
[4] Kramer, *op. cit.*

TABLE 49-6. SUMMARY OF RECOMMENDATIONS FOR DRYING GRAIN WITH NATURAL AIR AND HEATED AIR

	Ear corn	Shelled corn	Wheat	Oats	Barley	Grain sorghum	Soybeans	Rice	Peanuts
1. *Maximum* moisture content of crop at harvesting for satisfactory drying:									
With natural air, per cent†	30	25	20	20	20	20	20	25	45–50
With heated air, per cent†	35	35	25	25	25	25	25	25	45–50
2. *Maximum* moisture content* of crop for safe storage in a tight structure, per cent†	13	13	13 (12% for seed wheat)	13 (12% for seed oats)	13	12	11	12	13
3. Pounds of water per bushel which must be removed for safe storage when grain is harvested at moisture content of:									
30%	22.0	13.1							
25%	14.7	8.7	.....	5.0	.....	9.2	11.0	6.5	
20%	8.1	4.7	5.2	2.7	4.2	5.3	6.6	3.2	
18%	5.5	3.3	3.7	1.9	2.9	3.9	5.1	2.0	
16%	3.0	1.9	2.1	1.1	1.8	2.5	3.5	1.0	
4. *Maximum* relative humidity of air entering crop which will dry crop down to safe storage level when natural air is used for drying, per cent	60	60	60	60	60	60	65	60	75
5. *Maximum* safe temperature of heated air entering crop for drying when crop is to be used for:									
Seed, °F	110	110	110	110	105	110	110	110	90
Sold for commercial use, °Fa	130	130	140	140	105	140	120	110	90
Animal feed, °Fb	180	180	180	180	180	180			
6. Preferred depth‡ of crop for batch drying with heated air, ft	5–20 (not critical)	1½–2	1½–2	1½–2	1½–2	1½–2	1½–2	¾–1½	4–6
7. *Maximum* depth‡ of crop at different moisture levels for drying in tight structure with fans capable of delivering the required cubic feet per minute as listed in (8) below									
Moisture, %†	30, 25, 20, 15, 20, 20	30, 25, 20, 18, 4, 5, 6, 8	20, 18, 16, 4, 6, 8	25, 20, 16, 4, 6, 8	20, 18, 16, 4, 6, 8	25, 20, 18, 16, 4, 4, 6, 8	25, 20, 18, 16, 4, 6, 8, 10	25, 20, 18, 16, 4, 6, 8, 10	40–50
Depth, ft									6
8. *Minimum* air flow to dry crop at moisture level and depth as listed above in (7), cfm/bu									
Natural air	5, 5, 3	§5, 3, 2	3, 2, 1	4, 2, 1.5	3, 2, 1	§4, 3, 2	§4, 3, 2	4, 3, 2, 1	3
Heated air with not over 15°F rise in temperature	5, 5, 3	6, 5, 3, 2	3, 2, 1	4, 2, 1.5	3, 2, 1	5, 4, 3, 2	5, 4, 3, 2	4, 3, 2, 1	3

* If the products are to be stored for long periods, the moisture content should be 1 to 2 per cent lower than shown in this tabulation.
† Moisture contents on wet basis: *a.* Higher temperatures than those listed may be used when the corn is dried under carefully controlled conditions so that the maximum temperature of the kernels does not exceed 130°F at any time. *b.* If there is any possibility that the crop may be sold, use the lower temperature as listed for commercial use.
‡ Depths can be increased somewhat, especially at the lower moisture levels, provided fan capacity will meet the air-flow requirements in (8) at relatively high static pressures.
§ Not recommended because of high moisture.

SOURCE: Crop Dryer Manufacturers Assn, rev. Feb. 9, 1956.

to reduce it to 14 per cent moisture or less. After the last drying, unheated air should be passed through the rice to cool it to storage temperatures.

Tempering periods between periods of rapid drying are necessary with rice because the surface of the rice dries faster than the internal portion of the grain. During the tempering period, moisture moves out to the surface again and at the same time, respiration heat accumulates in the rice. Storage in tempering bins should be limited to a maximum of 12 hr for high-moisture rice unless the bin is equipped with a ventilation system for aerating the rice during the tempering period. Aerating systems are intended primarily to remove heat of respiration and to maintain uniform temperatures.

PRINCIPLES AND PRACTICES FOR DRYING OF GRAIN

Unheated Air

1. Drying of ear corn by natural ventilation is practical where the relative humidity is below 75 per cent and daytime temperatures exceed 60°F. Under less favorable conditions, supplemental heat is required. By ventilating for cooling, wet ear corn can be kept through the winter in northern climates and dried the following spring.

2. For aeration, an air flow of $\frac{1}{30}$ cfm per bu is recommended for continuous fan operation and $\frac{1}{10}$ cfm per bu for intermittent operation. Aeration is not a substitute for drying, but will help maintain the quality of grain up to 2 per cent over the generally accepted safe-storage standard. Air-exhaust systems are used for aeration.

3. For drying use an air flow of 1 to 5 cfm per bu, depending upon the moisture content and the kind of grain. A maximum depth of 6 to 8 ft is usually recommended for small grains to prevent more than 3 in. of water static-pressure drop. Deeper layers of grain and rice can be dried if the additional static pressure is considered in the fan selection. For ear corn, an air flow of 5 to 10 cfm per bu, for 20 and 30 per cent moisture content, respectively, is recommended, with 1 bu occupying 2.5 cu ft. Either the exhaust or pressure system may be used for grain and ear-corn drying. With the pressure system the last grain to dry is on the outside edge or top and is therefore more desirable from the farmer's standpoint. A 20-ft depth of ear corn has a static pressure of $\frac{1}{2}$ to $\frac{3}{4}$ in. of water with air flows of 5 to 10 cfm per bu.

4. Grain should be dried rapidly enough to prevent damage from mold formation. In summer, wheat should be dried within 2 weeks of harvest, and in fall, shelled corn within 4 weeks in midwestern areas.

5. From 5 days to 2 or 3 weeks may be required for drying, depending upon the moisture content of the product and weather conditions.

6. Electricity for operating the drying system will vary from 5 to 10 kwhr per 100 lb of water removed. To calculate the fixed cost, use an equipment life of 13 years.

7. Provide uniform lengths of air travel from the ducts to the outer surfaces.

8. Grain should be reasonably free of foreign material and uniformly distributed throughout the bin. Use a screen in the elevator to remove shelled kernels from ear corn, and move the elevator frequently to prevent concentration of husks, silks, etc.

9. Provide adequate inlets and outlets.

10. Operate the fan continuously if the wettest grain contains 15 per cent or more moisture and thereafter during the day only or when the relative humidity is below 75 per cent.

11. In the design provide 1 sq ft of main-duct area for each 1,000 cfm of air. (The shelling trench of an ear corn storage is not large enough for an air duct.) The exit from the main duct to the lateral should be designed for 1,440 fpm. Economics may dictate a compromise between the duct size and the power requirement.

12. Wet grain may be kept through the winter by ventilation for cooling, and drying completed in the spring.

13. Uniform air distribution is difficult to obtain until lateral ducts are covered by a depth of at least twice their center-to-center distance. Even then, spoilage of very wet grain may occur in the top layer between the ducts. Spacings of 2 to 4 ft are recommended, with the closer spacing more desirable.

14. Grain must be aerated (cooled) after drying to prevent moisture migration during the storage period.

Heated Air

1. Use an air flow of from 10 to 200 cfm per bu for drying batches 1½ ft deep or less. These air flows are necessary to prevent condensation in the grain or on the surface.
2. Static pressures of 1 to 5 in. of water are encountered in heated-air drying systems with high air flows and layers up to 24 in. thick.
3. Grain should be dried in layers less than 24 in. thick and preferably less than 18 in. Deep layers are satisfactory if the grain is turned during drying.
4. Drying in thin layers is accomplished in 2 to 3 hr, and the drying period must be followed by a cooling period of ¼ to ¾ hr before storing.
5. Fuel for a heated-air drying system ranges from 2 to 3 gal of fuel oil per 100 lb of water removed. To calculate the fixed cost, use an 8-year life.
6. Major consideration should be given to the handling system.
7. To estimate the fuel requirements:
    a. Determine moisture to be removed in pounds.
    b. Use 1050 Btu per lb of water to estimate the heat requirement.
    c. Estimate the gallons of fuel required based on the heat value of the type used.
    d. For a direct heater, use a thermal efficiency of 60 per cent for summer, 50 per cent for fall, and 35 per cent for winter. For an indirect heater use 45 per cent for summer, 36 per cent for fall, and 28 per cent for winter.
8. Use temperatures which will not harm the product for its intended use:
    a. Germination, under 115°F.
    b. Milling, under 140°F.
    c. Livestock feed, under 190°F, according to present practices.
    d. Cracking; some products as pea beans and rice are sensitive to heated air and crack, check, or split.
9. Grain must be aerated during storage period after drying to prevent moisture migration.

HAY DRYING

Hay Drying with Unheated Air. The usual air-flow recommendations for forced-air drying of hay are from 15 to 20 cfm per sq ft of mow space or 300 to 500 cfm per ton of wet hay. This air flow is needed for drying hay in the mow after it has been partially dried in the field, usually down to 35 to 40 per cent moisture content. Hay in the field will dry from 75 to 40 per cent in 1 day of good drying weather. For most installations on farms where both grain and hay are to be dried with forced air, there is much more water to be evaporated from a mow of hay than from a bin of grain. If properly selected, the same fan can be used for grain and hay drying. It is a common practice to use the same fan for hay and ear corn drying because of a similar range of air flow and static pressure. The fans used for hay drying deliver large volumes of air, often from 18,000 to 40,000 cfm, with static pressures of approximately 1 in. or less. For drying grain in a bin, the usual air flow in a farm installation might be 6,000 cfm against 3 in. of static pressure. Thus, for dual purpose, a fan would be required to deliver 20,000 cfm at ¾ in. static pressure, or 6,000 cfm at 3 in. static pressure.

The most popular duct arrangement for *long and chopped hay* is a center A frame or rectangular framework covered with slats or wire fencing for long hay and chopped hay. For mows wider than 36 ft, lateral ducts extend to within 6 ft of the walls and 6 ft of the ends of the hay mow and are placed from 4 to 5 ft apart. With the center duct system there should be about 1 sq ft of area opening into the hay for each 50 cfm of air which passes through the center duct. For deep narrow mows two levels of ducts may be used, with a second duct about 15 ft above the floor duct. Vertical flues extending from the center duct can be used. Also, center ducts may be provided with doors placed in the top which are closed for shallow depths and opened for deep mows.

When placing chopped hay on a dryer with a blower the leaves and stems tend to separate, making uniform drying practically impossible. The separation is overcome by using a conveyer to drop the hay over the dryer. The conveyer also requires less power, but the initial cost is higher than that of the blower.

The usual recommended maximum depth of storage for long hay over a dryer is 15 ft for long hay and 13 ft for chopped hay. With forced-air drying under good conditions, the hay will dry in about 1 week. A rule of thumb for placing hay on a dryer is that an average of 1 ft of wet hay may be placed on the dryer per day with no more than 3 ft of wet hay on any one day and no more than 6 ft of wet hay on the dryer at any one time. Otherwise, molding may occur before drying is complete.

It is difficult to move air through *bales* of hay. In order to successfully dry baled hay with forced air, without undesirable mold growth occurring, it is necessary (1) to make the bale as loose as possible, usually about 8 lb per cu ft and (2) to field-dry the hay to 35 per cent moisture content before baling (instead of the usual 40 per cent) for long or chopped hay. Bales may be placed in the drying system (1) tight-packed, (2) loose-stacked, and (3) helter-skelter. Tight packing with the bales on edge and staggered and with alternate layers turned 90° is the best system of stacking for obtaining good air flow with forced ventilation. Because of the problem of handling and placing bales on edge, some authorities recommend drying the bales flat. The bales may be used for forming the center duct of a center-duct system. A platform, or slatted lateral duct system, is normally used for a heated-air installation where considerable handling is involved. Bales may be piled up to seven layers deep.

Recommendations for unheated-air drying of hay are summarized in Table 49-7.

TABLE 49-7. SUMMARY OF RECOMMENDATIONS FOR DRYING HAY WITH UNHEATED FORCED AIR

	Long hay	Chopped hay	Baled hay
Quantity of air, cfm/sq ft of mow area	15–20	15–20	15–20
Cfm/ton	300–500	300–500	300–500
Maximum hay depth, as placed in mow	15 ft	13 ft	8 bales
Static pressure, in. of water	¾–1¼	¾–1¼	1–1½
Maximum initial moisture content for drying, per cent	40	40	35
Type of air-distribution systems used	Center main, main plus laterals, slatted floor	Center main, main plus laterals, slatted floor	Slatted floor
Distance between laterals, ft	4	4	Usually not used
Distance between laterals and ends and walls, ft	6	6	2–4
Density of hay, cu ft/ton	400–500	250–400	Twine tied, 200–250 cu ft/ton; wire tied, 125–200 cu ft/ton

The *installation cost* of a barn hay dryer using unheated air is about $15 per ton for 100 tons capacity (1959). The *operating cost* will vary, depending upon the amount of the product stored and the type of duct system selected. The electrical requirement is about 35 to 40 kwhr per ton when the hay is placed in the mow at 35 to 40 per cent moisture. For hay placed on the dryer at 60 per cent moisture, from 100 to 120 kwhr per ton is required for drying.

Hay Drying with Heated Air. It is not a safe practice to use heated air for drying hay in the mow because of the danger of fire. Heated-air drying of hay should be done in a separate structure, with the hay placed in the mow after drying. This is seldom done because hay is not as easily handled as grain. Considerable heat is produced by the respiration of the hay, and it is estimated that at 40 per cent moisture content, wet basis, 100 Btu is produced per pound of dry matter during the evaporation of a pound of water, supplying about 10 per cent heat of vaporization.[1] With the respiration there is a decrease in dry matter, but this is not considered too important if air is being moved through the hay to dry it as rapidly as possible. Because of the ease of handling, heated air has been used principally for drying of baled hay in buildings removed from the main storage structure. For heated-air drying, the air flow should be from 25 to 35 cfm per sq ft to prevent or reduce condensation on the

---
[1] R. B. Davis, Jr., G. E. Barlow, Jr., Supplemental Heat in Mow Drying of Hay, II, *Agr. Eng.*, 29:251–254, June, 1948.

surface layers. Bales are normally placed three deep on the dryer. The waste heat from an engine will increase the rate of drying by increasing the drying air temperature about 2 to 5°F. Wagon dryers are available and provide a separate structure for batch drying and reduce handling problems.

Alfalfa Dehydration. Dehydration of hay refers to drying below the equilibrium moisture content, usually from 5 to 8 per cent, wet basis. The carotene content of alfalfa can be preserved by removing the moisture within 2 hr after cutting. Single- or double-stage rotating drum dryers are usually used. Apron-type dryers are used to a limited extent. Oil and gas are commonly used fuels. From 60 to 70 gal of fuel oil is used for drying a ton of dehydrated meal or 8,000 to 10,000 cu ft of gas per ton.[1] The temperature of the drying air is maintained as high as 1500°F in some dehydrators. Handling of dehydrated hay is improved by pelleting, which costs from $2 to $3 per ton.[2]

Alfalfa dehydrators are located where 500 to 1,000 acres of alfalfa hay can be contracted for dehydration. Based on a 1949 study of 44 dehydration plants in Colorado, Kansas, and Nebraska, the average investment for one rotating drum dryer was $95,000 and the *total* cost of production was $37 per ton of dehydrated alfalfa.[3] Most kilns will evaporate about 6,000 lb per hr of water, with some large enough to evaporate 12,000 lb per hr. With present equipment it is not economical to dehydrate forage on the family-size farm.

PRINCIPLES AND PRACTICES FOLLOWED IN HAY DRYING

1. Hay should not be placed in the mow with an average moisture content exceeding 40 per cent or less than 35 per cent. Little shattering of the leaves takes place at this moisture content.
2. An air flow of from 300 to 500 cfm per ton of hay is recommended. The procedure for determining the air-flow requirements is as follows:
    a. Calculate the volume of the largest cutting which will be placed in the mow. The maximum allowable depths are 16, 13, and 12 ft for long, chopped, and baled hay, respectively.
    b. Calculate the quantity of hay in tons, using 400 cu ft per ton for long hay, 300 cu ft per ton for chopped hay, and 175 cu ft per ton for baled hay.
    c. Calculate the quantity of air required by multiplying the number of tons by from 300 to 500 cfm per ton.
    d. Select the fan on the basis of total air in cubic feet per minute required, at estimated static pressure, and be sure that the fan performance data show good delivery at the highest probable pressures.
    e. Estimate fan horsepower requirements on the basis of 1 hp for 7 to 10 tons of hay.
3. With unheated air the hay must be dried down to a safe level within 7 days to avoid mold.
4. The fan should be operated continuously for the first few days, even during showers, to dissipate the heat of respiration of the moist hay.
5. Hay should be placed uniformly over the system, and precautions taken to prevent leakage of air around posts, floor, etc., and excess packing in other spots. The floor should be airtight.
6. Provide air outlets and ventilation above the hay to allow the air to escape. At least 1 sq ft of air outlet from the building should be provided for each square foot of cross section of main tunnel.
7. The main tunnel is usually at least 5 ft high and contains 1 sq ft of cross section for each 1,000 cfm.
8. With the center-duct system, provide 1 sq ft *outlet area* from the duct for each 50 cfm of air entering the duct to keep the resistance low.
9. Chopped hay should be cut as long as possible. A 3-in. theoretical cut will give an actual length of from 3 to 10 in.

[1] Lacey F. Richey, Cooperative Alfalfa Dehydrators, *USDA FCS Circ.* 12, June, 1956.
[2] *Ibid.*
[3] Leonard G. Schoenleber, Operation and Performance of Alfalfa Dehydrators in Central United States, *Agr. Eng.*, 31:234–236, May, 1950.

10. To determine whether the hay is dry, turn the fan off for 12 hr and note whether warm air comes through when the fan is started again. If warm air is being discharged from the mow, continue to operate the fan for 12 hr and then repeat the check.

Costs of Using Heat for Drying Hay. The operating cost and time for drying 25 tons of alfalfa hay, dry weight, brought from the field at 40 per cent moisture, using 15,000 cfm of air flow and different amounts of heat, are given in Table 49-8.

TABLE 49-8. OPERATING COST OF DRYING ALFALFA HAY

Temperature, °F	RH, per cent	Heat added, °F	Time required, hr	Gallons of oil	Electricity, kwhr	Operating cost per ton
75.4	66	0	288	0	1,440	$1.15
84.5	49	10	110	147	550	1.20
94.5	35	20	72	193	360	1.29
104.5	26	30	55	220	275	1.36
114.5	19	40	45	240	225	1.43
144.5	8.5	70	30	280	150	1.58

Oil, 13¢ per gallon; electricity, 2¢ per kilowatthour.
SOURCE: Roy B. Davis, Jr., G. E. Barlow, Jr., and D. P. Brown, Supplemental Heat in Mow Drying of Hay, III, *Agr. Eng.*, 31: 223–226, May, 1950.

The fixed costs must also be considered for drying equipment and would be about twice as much for heated-air-drying as for forced-natural-air-drying equipment.

Chapter 50

STORAGE OF FARM CROPS

Carl W. Hall

SILAGE

Silage is the fermented state of forage crops as used for livestock food. Waste portions of various fruit and vegetable crops can also be made into silage. The first reaction in the silo is due to the presence of oxygen, enzymes, and microorganisms. The normal respiration processes continue, and molds and bacteria grow. If the silo is properly sealed against oxygen infiltration, the first stage comes to an end within a few hours when the oxygen mixed with the silage material has been consumed. At this stage, the silage will be warm and filled with carbon dioxide and perhaps a little sweeter because of the conversion of some starch to sugar. As the oxygen is consumed, anaerobic bacteria take over and produce acids and alcohol. When the acidity builds up to a pH of 3.5 to 4.5, the biochemical activity slows down and the silage remains in a pickled state. If properly sealed from oxygen and rain, silage will keep in good condition for many months. Making the first cutting of forage into silage provides an excellent means of preservation where poor weather conditions prevail for field drying of hay.

Most authorities agree that a *moisture content* between 65 and 70 per cent wet basis is the ideal range for making silage. Higher percentages of moisture result in leakage of juice. Lower moisture contents result in prolonged oxidation, dry-matter losses, higher temperatures, and excessive molding. Palatable brown silage can be made from alfalfa hay dried in the field to a 31 per cent moisture, but it was found necessary to store the silage in a silo in which the walls had been sealed with asphalt.[1]

Finely chopped material has greater density and entraps less oxygen. Fine chopping at ⅜ in. theoretical cut is recommended for material with less than 65 per cent moisture content and also for corn and other coarse forages. High-moisture material, especially the fine-stemmed legumes and grasses, can be cut up to 2-in. theoretical length without entrapment of excessive oxygen. Long cuts tend to reduce moisture leakage. The top layer of material should be cut fine to increase its density and sealing effectiveness. Fine-stemmed material will make good silage without chopping if it is thoroughly compacted with tractor wheels in a trench silo.

High-moisture legumes and grasses can be *wilted* in the swath or windrow for 1 or 2 hr to bring the moisture content below 70 per cent. Wilting reduces juice leakage, corrosion of the silo, and the lateral pressures on the silo structure. Because wilting proceeds rapidly, care must be exercised to prevent moisture reductions below 65 per cent. Material used to top a silo should not be wilted, and wilting is not necessary for horizontal silos or for small upright silos where low-density silage is made.

Dry materials such as hay, grain, hulls, ground cobs, and dried pulp may be added as an alternative to wilting. Dry material must be carefully mixed or it will result in moldy pockets. It should not be added in the top 10 ft. About 30 lb of dry material should be added per ton for each percentage point of moisture above 70 per cent.

[1] T. E. Woodward, Relation of Agronomic and Nutritional Factors to Engineering Problems and Farm Practices in Making Grass Silage, *Agr. Eng.*, 22:54–56, February, 1941.

Seals are helpful in reducing spoilage where surfaces are exposed on the top of upright and horizontal silos. The usual seal is a vaporproof paper containing asphalt between two layers of treated paper or plastic materials. The paper seal should be weighted with 2 or 3 in. of soil or ground limestone or 6 in. of sawdust. Seals must be arranged so that they settle with the silage without tearing. Excessive spoilage, over 1 ft deep, may occur at the top of upright silos because of shrinkage of the silage away from the silo wall. This can be avoided by thoroughly tramping the upper few feet while filling, before the seal is placed in position, and by the use of high-moisture finely chopped material in the top 10 ft. The blower should be operated to remove harmful gases, before a man enters a newly filled silo.

Molasses is added in amounts of 50 to 100 lb per ton or more as a *preservative* for high-protein low-carbohydrate materials. The molasses furnishes sugars for the use of microorganisms in the fermentation processes. Some authorities contend that if the silage is properly made and the material is suitable for silage making, molasses is not needed.

Acids are sometimes added to alfalfa and grass silages to quickly produce an acidity of a pH of 3 to 4. This acidity pickles the silage without much lactic acid fermentation and with very little heating and dry-matter loss. Phosphoric acid is often used as a preservative because it has some nutritional value. Adding a mixture of sulfuric and hydrochloric acids is known as the AIV process after its originator, A. I. Virtanen of Finland. The acid must be sprayed into the silage uniformly and is not recommended for concrete silos.

Compounds of sulfur are also used for preservation. *Sulfur dioxide* does not add to the feed value but provides good protection. *Sodium metabisulfate* has proved very successful and has the added advantage of improving the odor of the silage. Even distribution of preservatives is important for successful use.

Off-flavor is caused by undesirable fermentation in which butyric organisms are involved. The addition of molasses or ground grains helps the lactic acid organisms gain the upper hand in the fermentation cycle. Off-flavor is most common in materials with a low carbohydrate content. Silos without drains often make off-flavored waterlogged silage.

Dry-matter losses in silage are not easily determined. Factors increasing the dry-matter losses are excess oxygen content due to poor compaction, coarse chopping, low moisture content, and air infiltration through joints and cracks. Including top spoilage and juice leakage, the total dry-matter loss from a silo filled with alfalfa at 72.3 per cent moisture was 7 per cent, with about 23 per cent of the dry-loss matter in 1,807 gal of juice that drained. The largest loss was in the spoilage at the top of the silo. The dry-matter losses in wilted alfalfa ensiled at moistures of 60 to 65 per cent and chopped ¼ and ⅜ in. have been reported as 10 per cent when stored in smooth, tight-walled concrete-stave silos.[1]

SILOS

Silage occupies an important part in the feeding of roughages to dairy and beef livestock. There has been a considerable increase in the use of hay and grass for silage during the last ten years because of the saving of feed nutrients as compared with field drying of hay, the restrictions imposed upon farmers by corn-acreage allotments, and the advent of laborsaving equipment for handling forage. The amount of grass silage made in 1955 was close to 10 million tons.

Silos are classified as follows:

Horizontal silos:
    Below ground—trench
    Aboveground—bunker or surface
    Combination of the above two
Vertical silos:
    Below ground—pit
    Aboveground—upright or **tower**

[1] J. B. Shepherd et al., Experiments in Harvesting and Preserving Alfalfa for Dairy Cattle Feed, *USDA Tech. Bull.* 1079, 1954.

Horizontal Silos. The loss due to spoilage from horizontal silos is usually 20 to 35 per cent as compared with 10 to 20 per cent in upright silos. These losses are due to spoilage on top, weather damage, and air in the silage because of insufficient packing and sometimes because of the minimum depth of storage. Generally speaking, the horizontal silo is less expensive to construct than the vertical silo, but if built with a roof so as to prevent excessive loss to the silage, the cost per ton for construction would be approximately the same. Farm labor can be readily used for constructing a horizontal silo, reducing the cash cost. The horizontal silo can be filled without using a blower but must be packed with a tractor or some other piece of heavy equipment as it is filled. The surface horizontal silo can be used where there is a high water table and is suitable for flat land. Self-feeding from a horizontal silo is easily done and saves much labor at the expense of considerable waste.

The cross section of the horizontal silo should correspond in width and height to the quantity of silage which can be eaten by the animals per day, and the length should give the quantity of the silage needed for the feeding period. The height should be at least 6 ft and is normally from 6 to 8 ft. Animals are usually fed from 30 to 50 lb of silage (1 cu ft) per day, but will self-feed from 75 to 100 lb (2 cu ft) per day. The density of well-packed silage is 40 to 50 lb per cubic foot. To prevent spoilage, a 4-in. thickness should be fed from the cross-section area per day in warm weather and a 2-in. thickness in winter (Table 50-1). The bottom width of the hori-

TABLE 50-1. SILAGE AND FEEDING FROM A HORIZONTAL SILO

Bottom width, ft	Top width, ft	Depth after spoilage removed, ft	Weight of silage in 4-in. slice, lb	Animals which can be fed from 4-in. slice with daily allowance of:			
				30 lb	50 lb	70 lb	90 lb
12	13.5	6	1020	34	20	14	11
14	15.5	6	1180	39	23	17	13
16	17.5	6	1340	44	27	19	15
18	19.5	6	1500	50	30	21	17
20	21.5	6	1660	55	33	24	18
24	25.5	6	1980	66	35	28	22
14	16	8	1600	53	32	23	18
16	18	8	1813	60	36	26	20
18	20	8	2027	67	40	29	22
20	22	8	2240	74	45	32	25
24	26	8	2667	89	53	38	30
28	30	8	3093	103	62	44	34

SOURCE: J. R. McCalmont, Bunker Silos, *USDA Agr Infor. Bull.* 149, February, 1956.

zontal silo should be at least 10 ft so that tractors can be used for packing. The walls should be constructed on a slope, usually about 1½ in. per ft so that silage will not settle away from the wall. The cross section of the horizontal silo usually has the shape of a trapezoid.

The walls and floors for horizontal silos are designed the same for below- and aboveground installations with the exception that the walls must be braced for aboveground installations. The wall may be made of earth, wood, brick, stone, or concrete. Silos with earth walls can be cut into the ground for below-ground construction or soil moved to the silo for aboveground. Construction of aboveground horizontal silos with earth walls is not generally satisfactory without lining. Wood walls may be used for below-ground, aboveground, or partly aboveground horizontal silos. The wood should be treated with creosote to prevent deterioration, and dirt placed behind the wall to prevent air leakage. Substantial bracing is required for aboveground construction (Figs. 50-1 to 50-3). An earth wall usually has a slope of about 2 ft away from the vertical on an 8-ft height. For concrete walls, reinforced tilt-up concrete slabs 4 in. thick with dimensions 8 by 10 ft and weighing about 4,000 lb (Fig. 50-4) are formed on the level and then pulled into place and anchored for the silo walls. In below-ground construction the reinforcement rods are placed near the silage surface of the slab. In the aboveground tilt-up slab, the reinforcement should be placed near the surface of the slab opposite the silage surface (Fig. 50-5).

The following data will aid in designing horizontal silos:[1]

1. Lateral pressures of 125 to 150 psf are obtained at a depth of 1 to 2 ft below the surface during filling and packing.

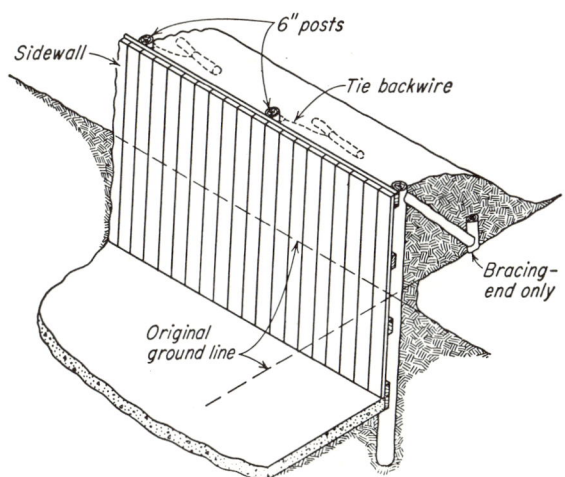

FIG. 50-1. Typical wood wall for below-ground or partly aboveground horizontal silos. (*Brevik, Friday, and Maddex, Horizontal Silos, Mich. State Univ. Circ. 723, 1955.*)

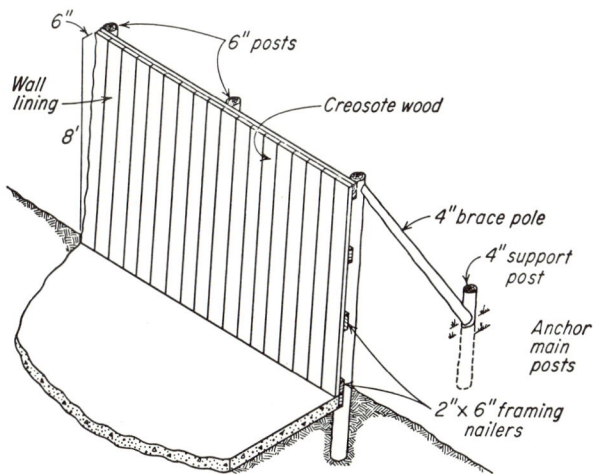

FIG. 50-2. Aboveground wood construction. The posts should be spaced 6 ft on centers and set to a depth of 3 ft. (*Brevik, Friday, and Maddex, Horizontal Silos, Mich. State Univ. Circ. 723, 1955.*)

2. The lateral pressure decreases after filling and relaxes to approximately 100 psf 2 ft below the surface.

3. $$\text{Overturning moment on wall, ft-lb} = 61.019 x^{1.774} \qquad (50\text{-}1)$$

where x is silage depth, ft.

[1] M. L. Esmay and D. B. Brooker, Lateral Pressures in Horizontal Silos, *Agr. Eng.*, 36:651–653, October, 1955. M. L. Esmay, D. B. Brooker, and J. S. McKibben, Design of Above-ground Horizontal Silos, *Agr. Eng.*, 37:325–327, 333, May, 1956.

4. A lateral concentrated load of 200 lb at the surface is caused by the packing tractor.

Floors for horizontal silos can be constructed of earth, concrete, asphalt, or wood. The use of earth floors should be discouraged for self-feeding silos or where the earth will not support tractor equipment used for removing the silage. A wet mud hole will be unsatisfactory. The floor should slope to one end with a grade of about ¼ in. per ft. A 6- to 10-in. depth of crushed rock or gravel should be used for an unfinished floor or as a base under either concrete or asphalt. The floor should also have a crown and slope of ¼ in. per ft to each side or from one side to the other side. Asphalt or concrete floors should be a minimum of 4 in. thick.

After filling and continuous packing the silo should be covered. An uncovered silo can have at least 10 in. of top-surface spoilage. The spoilage can be partially prevented by covering with 8 to 10 in. of sawdust, wet straw, poor-grade forage, soil, or

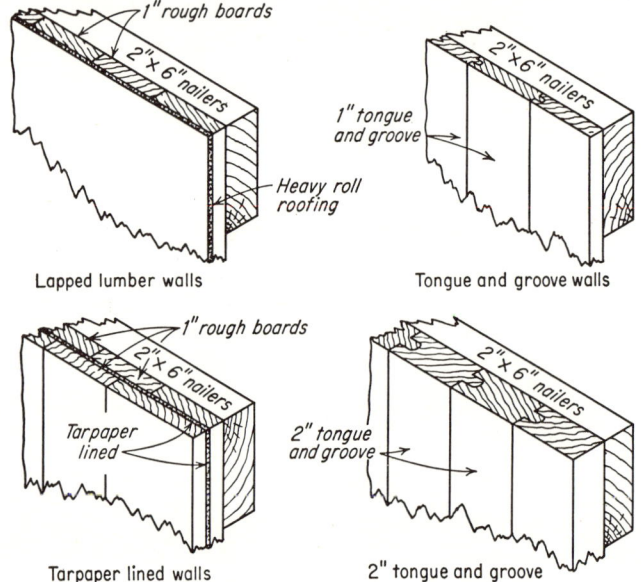

Fig. 50-3. Wall-construction details. (*Brevik, Friday, and Maddex, Horizontal Silos, Mich. State Univ. Circ.* 723, 1955.)

crushed limestone or with roll roofing, 15-lb felt building paper, or polyethylene or vinyl plastic held down with poles or earth.

A bunker silo can be erected from *portable wall sections* which permit changing location and size according to feeding schedule. For a more permanent location, the lumber silo walls are erected on a concrete slab. The portable wall sections are kept from sliding apart with ⅝-in.-diam rods which extend across the silo and bolts which hold the sections together (Fig. 50-6). The downward pressure of the silage on an 8-in.-wide ledge at the bottom of each section helps prevent overturning of portable walls.

Vertical Silos. The pit silo can be used only where there is a very low water table. For this reason, it is not in common use except in the semiarid regions, and in addition there is a greater possibility of poisoning from the fumes of the silage.

The structural requirements of the aboveground vertical silo are as follows: (1) The walls and door frames should be airtight. (2) Walls should be vertical to permit free settling. (3) Walls should be strong enough to support lateral pressure. (4) Both interior and exterior wall surfaces should be protected, metal from corrosion and

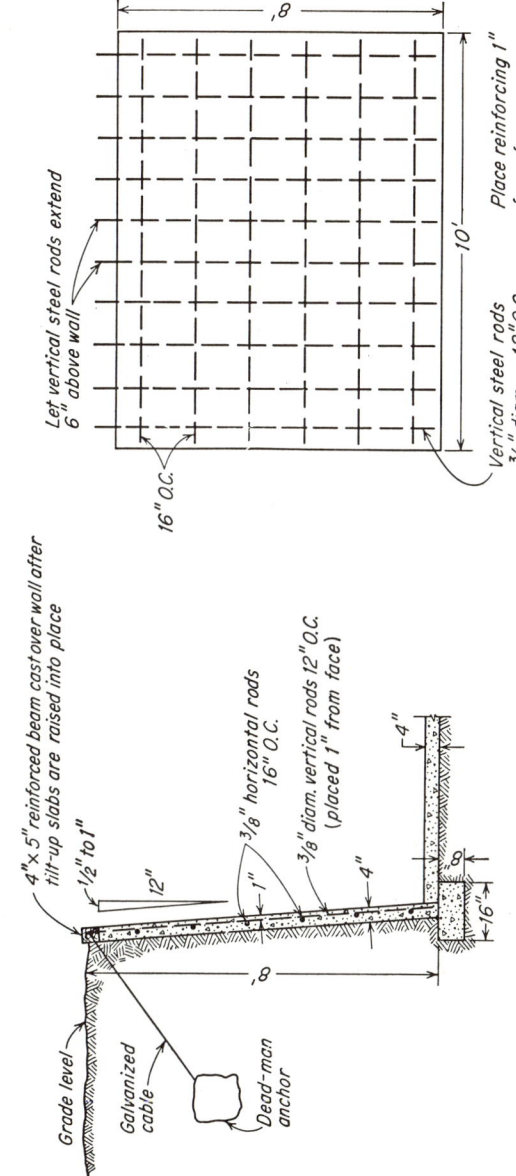

Fig. 50-4. Below-ground concrete wall construction. (Brevik, Friday, and Maddex, Horizontal Silos, Mich. State Univ. Circ. 723, 1955.)

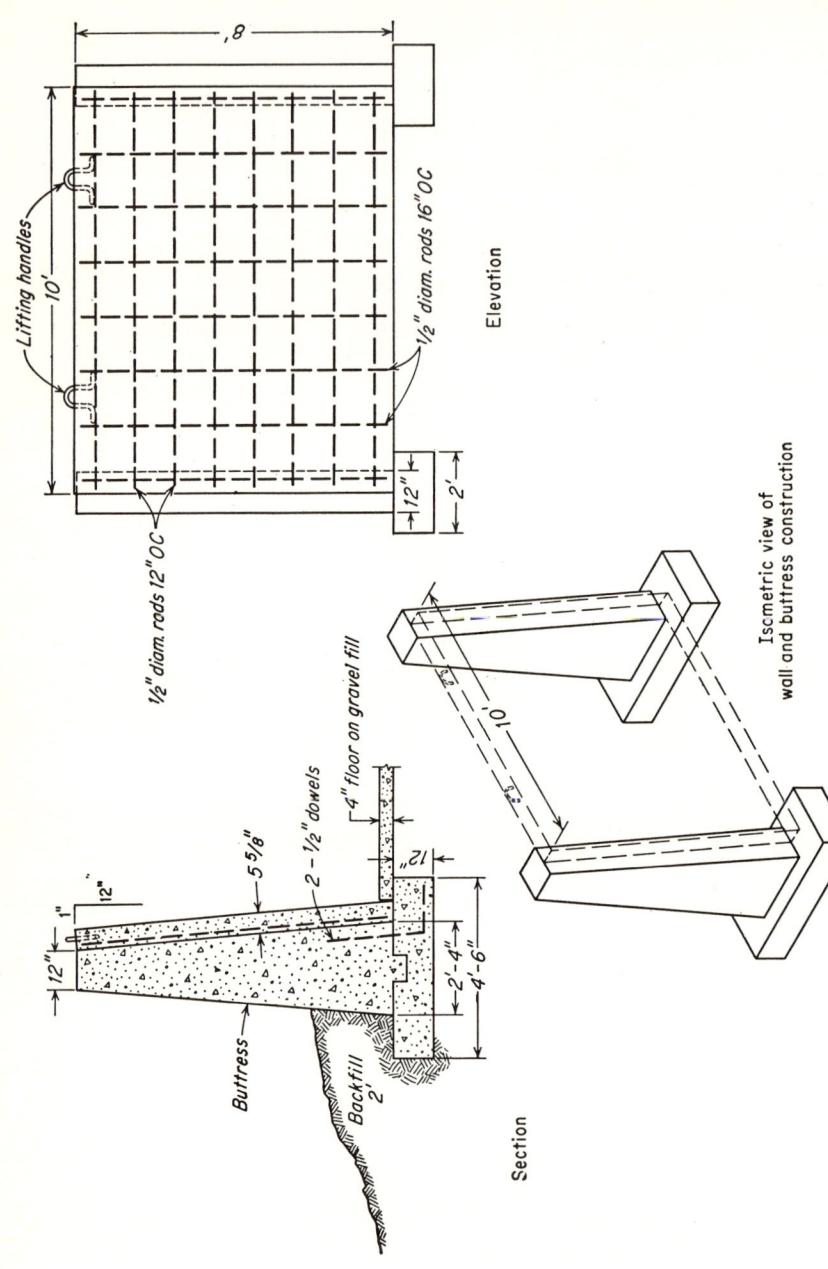

Fig. 50-5. Aboveground concrete wall construction. (*Brevik, Friday, and Maddex, Horizontal Silos, Mich. State Univ. Circ. 723, 1955.*)

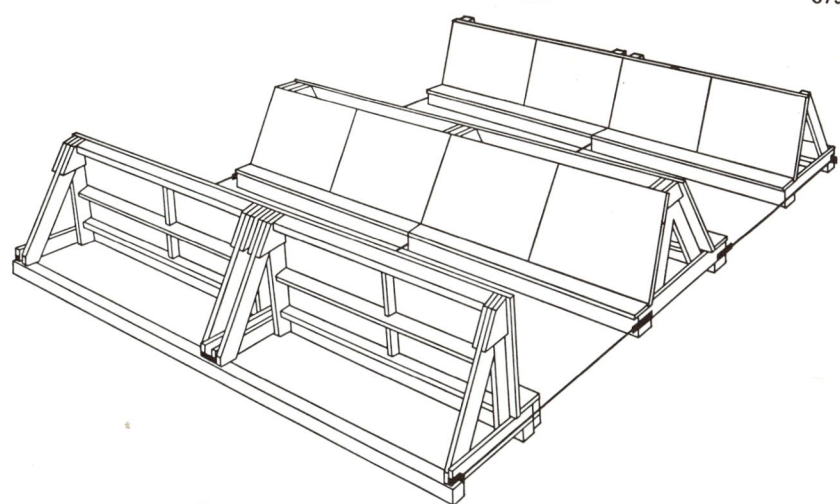

Fig. 50-6. Portable bunker silo. (*Midwest Plan* 77301.)

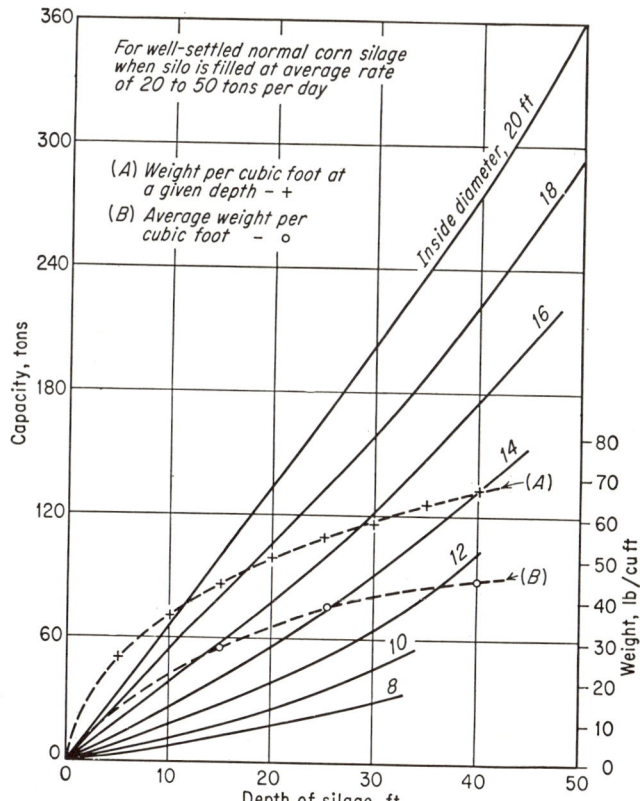

Fig. 50-7. Silage capacity and density. (*Based on data in USDA Farmers' Bull.* 1820, 1948.)

680   FARMSTEAD STRUCTURES AND EQUIPMENT

wood from excessive drying. (5) Access must be provided for filling and emptying the silo.

The silo should be sized according to the rate of feeding in order to prevent spoilage. It is necessary to feed about 4 in. per day in the summer or 2 in. during the winter to prevent spoilage. Most silos are cylindrical in shape with a circular cross section, and the use of other cross sections is discouraged. The density of silage varies with depth, length of cut, moisture content, amount of packing, and material. The approximate capacity of the silo in tons can be determined from Fig. 50-7 for selecting a silo size according to the feeding rate and for paying custom forage harvesters and fillers for silo filling.

The foundation wall for the vertical aboveground silo is extremely important. It must support as much as 60 per cent of the total weight of the silo, depending upon the silage and wall characteristics. See Table 50-2 for foundation load and size data. The foundation footing should extend not less than 3 ft below the ground surface where the ground freezes to a depth of 2 ft or more and 2 ft in southern climates.

TABLE 50-2. LOADS ON FOOTING PER FOOT OF CIRCUMFERENCE OF SILOS OF DIFFERENT HEIGHTS AND THE WIDTH OF FOOTINGS NEEDED ON SOIL OF 2½ TONS PER SQ FT BEARING CAPACITY

Height of silo wall, ft	Load and size of footing for silo walls of:								
	6-in. monolithic concrete			2½-in. concrete stave or 5-in. tile			Wood (2-in.) or metal		
	Load per foot of circumference, tons	Footing		Load per foot of circumference, tons	Footing		Load per foot of circumference, tons	Footing	
		Width, in.	Depth, in.		Width, in.	Depth, in.		Width, in.	Depth, in.
20	1.5	8	8	1.0	8	8	0.8	8	8
30	2.5	12	8	2.0	9	8	1.5	8	8
40	4.0	19	11	3.0	15	10	2.7	13	9
50	5.7	27	16	4.4	22	14			
60	7.8	37	23	6.5	31	19			

SOURCE: J. R. McCalmont, Silos: Types and Construction, *Farmers' Bull.* 1820, 1948.

The foundation should extend at least 6 in. above the ground. The footing should be placed on material with uniform bearing characteristics, to provide uniform settling. A pit of not more than 3 ft in depth is often included within the foundation to give additional capacity.

The floor is usually constructed of earth, although concrete is sometimes used. If a concrete floor is laid, it should include a drain to remove the liquid. If the earth floor is not porous, it is recommended that a concrete floor with a drain be provided.

The vertical load F, in pounds per foot of circumference at a distance h, in feet, below the top of the silo is given by[1]

$$F, \text{lb} = \int_0^h 5.5h^{1.08} = 2.64h^{2.08} \tag{50-2a}$$

The vertical component of the wall load f, in pounds per square foot, called the friction load, occurs because of the settlement of the silage.[2]

$$f, \text{psf} = 5.5h^{1.08} \tag{50-2b}$$

[1] William W. Gurney, Recommended Practice for the Construction of Concrete Silos (ACI 714–46), *Proc. Am. Concrete Inst.*, 43:149–164, 1947.
[2] *Ibid.*

The lateral pressure of silage is affected by the depth, moisture content, diameter, fineness of cut, speed of filling, type of silage, and preservative used. The lateral pressure increases as the moisture increases and as the diameter increases. A silo will ordinarily hold about the same tonnage of grass silage or corn silage.

The pressure build-up during settling reaches a maximum and then decreases slightly.

Corn is usually ensiled in the early dent stage when the moisture content is from 68 to 72 per cent. If less than 60 per cent, water should be added while filling. If the corn is too dry, high temperatures will be reached during storage and the product will be charred and damaged. Silage fermentation normally takes place at 80 to 100°F. Corn silage around 70 per cent moisture content will exert a lateral pressure of 8 psf and a vertical pressure of 5 psf for each foot of depth in a 14-ft-diam silo and 14 and 8 psf, respectively, per foot of depth in an 18-ft diam silo.[1] Grass silage with molasses exerted lateral pressure of 19 psf and vertical friction of 6 to 9 psf per ft of depth[2] (moisture content of 70 to 75 per cent). For grass silage the lateral pressure is about 40 per cent greater, requiring that a silo designed for corn silage only be reinforced with additional hoops, particularly at the bottom of the silo, to prevent bursting. The design pressures for 74 per cent moisture silage, safe for all types of silage materials, are given by the following equations.[3]

14-ft diameter or less: Lateral pressure, psf $= \dfrac{dh^{1.2}}{2.65}$ (50-3a)

16-ft diameter or more:[4] Lateral pressure, psf $= \dfrac{dh^{1.45}}{5}$ (50-3b)

where d = diameter of silo, ft
h = height of silo, ft

Laboratory studies show that a pressure of 576 psf is needed to obtain a density of 32 lb per cu ft and 1,000 psf for 40 lb per cu ft density.[5]

The hoops are usually made of $\frac{9}{16}$-in.-diam steel, with an allowable stress of 20,000 psi, and are normally placed about 30 in. apart up to a depth of 30 ft for a 14-ft diam silo or up to a depth of 20 ft for a 16-ft-diam silo, and for greater depths the spacing is 10 or 15 in. between hoops for the bottom half of the silo.

Silos are constructed with tile, monolithic (one-piece) concrete, concrete staves, concrete blocks, metal, brick, stone, wood staves, and plywood. Many of these silos require hoops in the construction material or around the outside. Additional strength is provided by adding reinforcing or using heavier-gage steel in prefabricated silos. For tile and concrete blocks, hoops are placed in the joints or between the walls of tile specially constructed for silos. Hoops with rolled threads are stronger than those with cut threads. Monolithic concrete is poured with reinforcing rods inserted. If the walls allow silage juices to leak through to the hoops, they may corrode until failure ensues and the silo collapses, since they cannot be readily inspected. Concrete-stave and wood-stave silos have the hoops on the outside. A wood-stave silo requires that the hoops be adjusted. They must be loose enough to allow the expansion of the wood as it soaks the silage juices and be tightened when the silo dries after emptying.

The metal silos not covered with an enamel or gloss lining must be painted with asphalt-base material or linseed oil to prevent corrosion from the action of the silage juices and air. The wood silo, concrete-stave silo, and metal silo must be properly anchored so that wind forces will not blow it over when it is empty. Wood staves should be creosoted on the interior to prevent deterioration and to reduce friction and should be painted regularly on the outside to maintain their condition. To prevent deterioration and to keep the silo airtight, concrete silos and wood silos must be treated

---

[1] Ibid.
[2] J. R. McCalmont and H. E. Besley, Silo Pressure and Temperatures with Corn and Grass Silage, *Agr. Eng.*, 20:227–230, June, 1939.
[3] J. R. McCalmont, Silos, *USDA Farmers' Bull.* 1820, 1948.
[4] It is believed that these relationships should be limited to silos 20 ft or less in diameter until more data are available.
[5] Nathan Rich, "Silage Removal from the Horizontal Silo," M.S. Thesis, Michigan State University, Dept. of Agricultural Engineering, 1953.

periodically. The inside of a concrete-stave silo can be finished with a cement plaster or cement wash. To maintain a smooth surface a linseed-oil, plastic, or asphalt base material can be placed over a cement finish. Several commercial finishes especially designed for silos are available. Some concrete staves are treated with a plastic sealer at the plant.

Doors are provided along one side of the silo for silage removal. With silos 16 ft in diameter or greater for large feeding operations, the possibility of placing doors on both sides of the silo should be considered to reduce the labor of manual removal.

Mechanical silo unloaders remove a uniform depth of silage, save labor, and eliminate the necessity for daily climbing the silo. Unloaders are equipped with 3- to 5-hp electric motors and may be used to unload from the top or bottom. Removal of chopped-grass silage requires more electrical energy and is slower than removal of corn silage. For top removal of 1 ton per hr of grass silage, 4.3 kwhr is required, and for bottom removal of 1.1 tons per hr, 4.0 kwhr; for top removal of corn silage at 1.6 tons per hr, 2.5 kwhr, and for bottom removal of 3.5 tons per hr, 1.2 kwhr.[1] The estimated daily cost of providing mechanical silage removal is 1 to 2 cents per animal.[1]

The problems encountered in *self-feeding* from a vertical silo consist mainly of (1) too much silage exposed, causing loss, (2) difficulty in vertical self-feeding of ensilage during freezing weather, and (3) feeding silage uniformly around the silo. Bridging problems have been alleviated by interior supports around the bottom wall which can be alternatively withdrawn to allow silage to feed down.[2]

A *gastight silo* which keeps air out and carbon dioxide in, with smooth inner surfaces and a bottom unloader, is finding considerable acceptance. A plastic breather bag with pressure-relief valve is placed in the top of the silo to compensate for pressure differences between inside and outside. The gastight silo has been used for successfully ensiling high moisture (above 25 per cent) shelled corn. After such storage the corn cannot be marketed, like dry corn, but must be fed quickly to avoid spoilage.

Temporary silos can be constructed from wooden snow fences or welded-steel fence lined with asphalt paper or a baled hay or straw outside wall held together with No. 9 wire or wire fence. With temporary silos the spoilage is high, usually 30 per cent, but decreases as the diameter is increased, provided that the silage can be fed fast enough for the diameter selected.

Safety. Many deaths and much body discomfort are attributed each year to "silo-gas" poisoning. Silo gas is principally nitrogen dioxide which is produced during the first few days of filling. It has a pungent odor, is yellow-brown in color, and is heavier than air. The concentration of nitrogen dioxide in a silo chute is often 115 ppm as compared with the industry safety-code specification of 15 to 25 ppm as hazardous.[3] The blower should always be operated before entering the silo. A victim of suffocation should be moved into fresh air, given artificial respiration, and treated by a respirator squad immediately.

Practices and Principles Followed in Silo Use:

1. Select silo size on the basis of quantity to be fed to prevent spoilage.

2. Animals that are self-feeding will eat considerably more silage, 75 to 100 lb per day per animal, as compared with normal consumption in a feeding ration, usually 30 to 50 lb per day per animal. Self-feeding from a vertical silo has been difficult. Top and bottom vertical silo unloaders are available.

3. It is necessary to pack a horizontal silo with a tractor when filling and continue packing daily for 1 week after filled. A force of about 40 psi is needed on top of the silage to properly remove and exclude the air.

4. Cover a horizontal silo after filling to exclude air and reduce spoilage.

5. A vertical wood silo should be covered to prevent loss of silage and to maintain the condition of the silo. Most insurance companies will not insure a wood silo without a roof against wind damage.

---

[1] Roger Asmus, Silo Unloaders on Ohio Farms, *Ohio State Univ. Ext. Bull.* 360, April, 1957.
[2] Mark E. Singley, Self-feeder Silo Controls Silage Flow, *Agr. Eng.*, 38: 84–93, February, 1957.
[3] United Press release published in *State Journal*, Lansing, Mich., March, 1957.

6. Adjust the metal hoops on a wooden silo seasonally to take care of the expansion and contraction of the wood staves.

7. Long or chopped forage can be placed in the horizontal silo.

8. Use an electric fence gate across the self-feeding horizontal silo to control feeding and prevent wastage. A self-feeding pushback feeder should be kept 18 in. from the silage face.

9. Ensile corn in the early dent stage. If the corn is too dry, wet it by pumping in water so that fermentation will take place and thus avoid excessive temperatures, which will cause charring and damage to the silage.

10. Preservatives or conditioners are not normally used for corn silage. Molasses is sometimes added to grass silage (about 50 lb per ton) to improve palatability and to supplement the ration. Other grass silage preservatives are ground corn, oats, beet pulp, barley, etc., sulfur dioxide (not over 5 lb per ton), and sodium metabisulfite. Felt paper should be placed on the inside of the silo over poor-fitting doors to keep air out and reduce spoilage.

HAY STORAGE

The recent emphasis on grassland farming has caused increased interest in and need for storage of roughage. Hay storage type and size depend on form (long, baled, or chopped), moisture content, method of feeding, condition, disposition, and farm equipment available for handling. The quantity of hay consumed by cattle per day varies from ½ to 3 lb per 100 lb of live weight. The storage space needed for different structures can be estimated from Tables 50-3 to 50-5.

TABLE 50-3. APPROXIMATE STORAGE-SPACE REQUIREMENTS FOR FEED, BEDDING, AND MATERIALS

Material	Lb/cu ft	Cu ft/ton
Loose hay:		
In shallow mows	3½–4	600–500
In deep mows	4½–5	450–400
Baled hay:		
Field-baled	8–14	200–250
Ordinary bales	10–15	200–135
Tight bales	12–20	100–175
Chopped hay:		
Long (2½ in. or more)	5½–8	360–250
Short (less than 2½ in.)	8–10	250–200
Loose straw	3⅔–4	600–500
Baled straw:		
Field bales	6⅔–8	300–250
Tight bales	10–13⅛	200–150
Chopped straw	4–5	500–400
Small grain	50–36	55–40
Oats	25–29	80–70
Mixed-feed concentrates	45	44
Wood shavings, baled	20	100
Lime	65	31
Ground phosphate	75	27

SOURCE: Based primarily on *USDA Infor. Bull.* 123, 1954.

Hay should be about 20 per cent moisture or below for safe storage without excessive mold development and browning. Above 25 per cent moisture spontaneous ignition may occur. Salt is often added to wet hay to prevent damage, but has not proved helpful under controlled laboratory conditions.

Large stacks of bales are often placed in the field and later sold or fed. Covering with canvas or bales of straw is often practiced to reduce leaching and bleaching of the exposed surfaces. Baled hay is particularly advantageous for commercial handling. Elevated storages that are adapted and used for baled hay should be checked for strength to support the weight.

Field stacks of long or chopped hay are easily adapted to self-feeding if placed close to the animals. To avoid excessive losses from self-feeding, the animals must be kept from tramping on the hay by movable feeding gates. Baled hay is often stacked within a self-feeding enclosure, and the bales are moved to the animals for a controlled-feeding program.

On-the-ground barn storage of hay is well adapted to self-feeding. It is essentially a stack protected by the building. Chopped hay is most popular for self-feeding. Drying equipment can be used for field stacks as well as for hay stored in buildings.

The lateral pressure exerted by hay is very slight while filling a mow. Siding on the outside of the studs will retain the hay. Upon settling, because of compaction and water removal, hay settles away from the barn wall. The side of a mow of chopped hay next to a driveway can be formed with a woven-wire fence.

TABLE 50-4. APPROXIMATE HAY CAPACITY FOR SEMICIRCULAR CROSS-SECTION BUILDING, TONS PER FOOT OF LENGTH

	Building width	
	32 ft	40 ft
Long hay	¾	1
Chopped hay	1	1½
Field-baled (twine)	1⅜	2
Field-baled (wire, medium-tight)	2	3
Loose straw	⅔	1
Chopped straw	⅞	1⅛

TABLE 50-5. APPROXIMATE HAY CAPACITY FOR STRAIGHT SIDEWALL STORAGES, TONS PER FOOT OF LENGTH

	Width 32 ft, height 10 ft	Width 36 ft, height 12 ft	Width 40 ft, height 14 ft
Long hay	0.8	1.1	1.4
Chopped:			
Long	1.0	1.3	1.6
Short	1.5	2.0	2.5
Field-baled:			
Twine	1.6	2.2	2.8
Wire	2.3	3.2	4.2

Spontaneous Heating and Ignition. The heat of respiration in partly dried grain and forages may lead to spontaneous heating and to ignition if adequate oxygen is available. The heat of respiration is furnished by enzymes, bacteria, and molds, in addition to the product. In grains, the major part of heat of respiration is furnished by the associated microorganisms (Fig. 50-8).

Little emphasis has been placed on heating of grain because spontaneous ignition does not usually occur, and for grain fed on the farm, a direct cash loss is not evident. Evidences of spontaneous heating of grain are (1) a low-percentage germination and (2) brown or discolored layers in the bin.

The initial heating of damp grain and hay reaches 120 to 130°F. If the heat can be dissipated as rapidly as it is produced, the temperature will eventually subside to normal storage temperatures. The secondary heating period results from chemical changes, and temperatures may reach 180 to 220°F, the critical temperature, particularly important for forages. The secondary heating period is usually avoided if the hay is below 25 to 30 per cent moisture when placed in the mow (without mow drying). The temperature rise will be only 1 or 2°F per day until the critical temperature is reached. The temperature may then decrease to a safe temperature or increase rapidly (20° per hour) approaching ignition. The critical temperature has not been established but undoubtedly varies with moisture content, density, chopped or long, size of mass, and infestation with microorganisms. The ignition temperature of hay varies with availability of oxygen and is generally considered to be near 400°F. Most fires from spontaneous ignition take place in 2 to 6 weeks after storage.[1]

A *probe*, made of a hollow pipe through which a thermometer can be inserted, or with a thermocouple on the probe point, can be used for determining storage temperatures. The insurance-company representative should be informed of heating. If the temperature starts to increase rapidly beyond the critical point, measures should

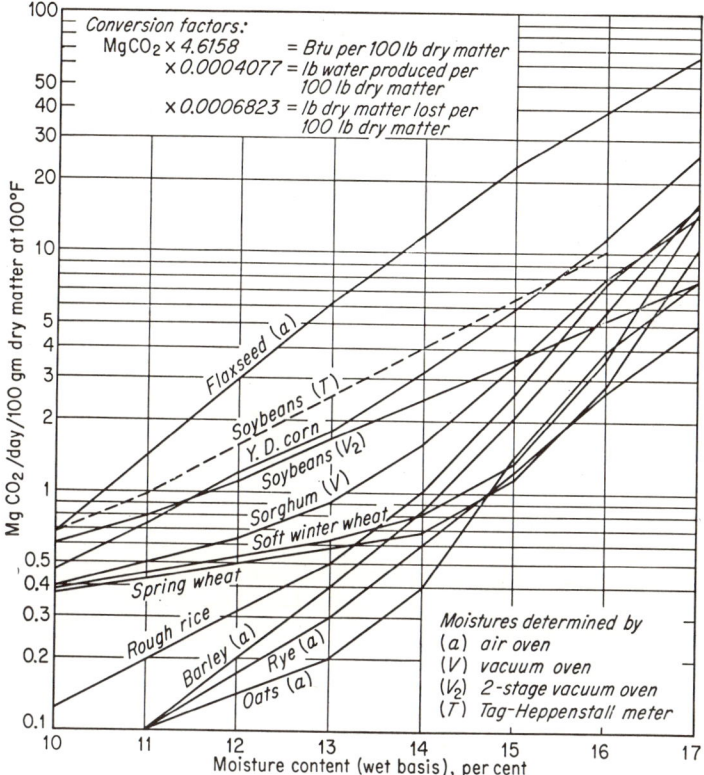

FIG. 50-8. Production of carbon dioxide by grain at 100°F. (*Reproduced from Agricultural Engineers' Yearbook, 1956. Bailey, Respiration of Seeds, Plant Phys., 15:257–274, 1940. Ramstad and Geddes, The Respiration and Storage of Soybeans, Univ. Minn. Agr. Expt. Sta. Tech. Bull. 156, 1942.*)

be taken to remove the hay to prevent loss of buildings and equipment. Admitting air (oxygen) to the "hot spot" will cause rapid burning. The mow should be opened in the presence of a fire truck, with water being sprayed over the heated area, and the heated portion removed. The heated hay will burst into flames when adequate oxygen supply is provided. Another procedure for preventing ignition is to "freeze" the hot spot by injecting liquid carbon dioxide into the area.

GRAIN STORAGE

Farm grain-storage buildings are usually constructed of wood or galvanized steel. The principal functions of a storage are to contain the grain, keep out moisture, and keep out rodents and birds.

With wood construction, wall studs properly sized and double-sheathed with waterproof paper between and fastened to a sole plate which is securely anchored to the

[1] Harry Doethe, Spontaneous Heating and Ignition of Hay, *Agr. Eng.*, 14:102–104, April, 1933.

floor will contain the grain. For a bin 8 ft deep, 2 by 4's 12 in. apart or 2 by 6's 24 in. apart are used for wall studs. Concrete floors should be 4 in. thick reinforced with 6- by 6-in. No. 10 wire and placed over a gravel base. To prevent moisture movement, an asphalt-base paper or plastic sheet is placed under the concrete floor. The sizes of joists for wood floors at different loadings are given in Table 50-6. These values apply to inside bins or to separate storage buildings. For plywood construction, use ¾-in.-thick walls and floors. Concrete blocks of standard construction will not withstand the lateral pressures of small grains without reinforcement.

To keep moisture out of stored grain:
1. All walls and roofs and openings should be tight.
2. Floors should be provided with a vapor barrier. Vertical joints should be calked and horizontal joints lapped 2 in.

TABLE 50-6. GRAIN DEPTHS CARRIED BY DIFFERENT SIZES OF JOISTS—24-IN. SPACING
(Grain can be twice as deep if joists are spaced only 12 in. apart.)

Size of joist, in.	Depth of grain for:			
	6-ft joist	8-ft joist	10-ft joist	12-ft joist
Joists supported at ends only				
2 × 6	3½ ft			
2 × 8	5 ft	3½ ft		
2 × 10	6 ft	4½ ft	3½ ft	
2 × 12	7½ ft	5½ ft	4½ ft	3½ ft
Joists supported at each end and at center				
2 × 4	3½ ft			
2 × 6	6 ft	4½ ft	3½ ft	3 ft
2 × 8	8 ft	6 ft	4½ ft	4 ft
2 × 10	10 ft	7½ ft	6 ft	5 ft
2 × 12	12 ft	9 ft	7 ft	6 ft

SOURCE: R. L. Maddex and J. S. Boyd, Construction of Farm Grain Storages, *Mich. State Univ. Farm Bldg. Circ.* 719, 1958.

TABLE 50-7. APPROXIMATE GRAIN CAPACITIES OF STANDARD SIZES OF CYLINDRICAL UPRIGHT STORAGES

Size (diam, ft × height, ft)	Capacity, bu
10 × 8	500
14 × 8	1,000
14 × 11	1,330
18 × 8	1,650
18 × 11	2,200
18 × 16	3,276

3. Junction between wall and foundation should be provided with an asphalt seal.
4. Foundation wall should not extend beyond the building. With wooden construction, the walls should extend beyond the foundation.
5. Overhang should carry water from building.

To keep rodents out of nonmetal construction:
1. Place woven-wire screen of ½ mesh or smaller over studs.
2. Cover top with woven wire or galvanized sheet metal.
3. Place an 8-in. strip of galvanized sheet metal around bin over 2 ft from ground to prevent rats from climbing wall.
4. Place metal strips around doors to prevent gnawing by rats.

Steel grain tanks are available in standard sizes up to 54 ft in diameter, beginning at 9, 12, 15, 18 ft, etc., in many heights. Metal circular bins available for farm storage are usually made of galvanized steel. Capacities are given in Tables 50-7 and 50-8 and Fig. 50-9.

TABLE 50-8. APPROXIMATE GRAIN CAPACITIES OF SEMICIRCULAR BUILDING PER FOOT OF LENGTH

Width, ft	Capacity, bu
16	85
20	100
32	250
40	420

Type of Floor. The type of floor construction for grain bins affects the grain losses. In tests in Kansas[1] during 1- to 2-year storages of wheat initially at 12 per cent moisture in 1,000-bu steel bins, the following floors gave safe storage without spoilage: steel on soil, gravel, or joists; concrete with aluminum-foil overlay or roll-roofing overlay; single and double wood flooring; and roll roofing on soil.

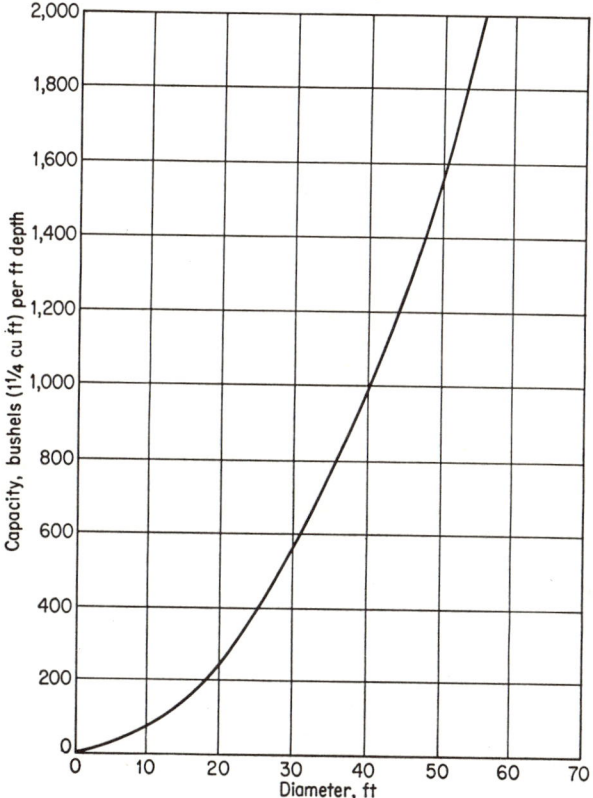

FIG. 50-9. Capacities of cylindrical storages, per foot of depth.

The amount of carbon dioxide (CO_2) produced by the grain and associated microorganisms is used to determine the heat and water produced and loss of dry matter (Fig. 50-8). The amount of carbon dioxide produced in milligrams per 24 hours for 100 grams of dry matter is a measure of the *rate of respiration*. The rate of respiration increases as the moisture content increases and as the temperature increases up to about 125 to 130°F, after which the respiration rate decreases.

DESIGN OF VERTICAL GRAIN-STORAGE BINS

The first step in designing a grain-storage bin after selecting the cross section and dimensions is to determine the lateral pressure exerted by the grain. The weight supported on the walls, foundation, and floor is then determined. The strength re-

[1] E. R. Gross, Performance of Grain Bin Floors, *Agr. Eng.*, 26:417, October, 1945.

quirement of the sidewall and hoops is based on the lateral (or horizontal) pressure exerted and the load supported by the wall. The foundation must support the load taken by the sidewall.

Storage structures are classified as:

Shallow: A line representing the plane of rupture drawn from a lower corner of the bin will cut the top surface of the grain.

Deep: A line representing the plane of rupture drawn from a lower corner of the bin will pass through the opposite side of the bin.

Practically, the plane formed by the angle of repose can be used instead of the plane of rupture. The plane of rupture x incorporates the effect of bin-wall friction and is represented by Eq. (50-4):[1]

$$\tan x = \mu + \sqrt{\mu \frac{1+\mu^2}{\mu+\mu'}} \qquad (50\text{-}4)$$

where x = angle with horizontal
μ = internal coefficient of friction of grain
μ' = coefficient of friction of grain against wall

For design purposes, it is desirable to know various physical properties of the grain and structural material. There is a dearth of information on physical properties of agricultural products, but by estimation, making simple tests, and using available equations, sufficiently accurate results for design can be obtained. Because of the danger to human life if failure occurs, designs must be adequate, often using a factor of safety as large as 10. The relationships discussed are based on the bin being filled level.

Vertical Grain Storages. Commercial storages are constructed of metal, wood, concrete, or tile. Metal and concrete have been the most popular. Galvanized steel is the most popular metal, and a few aluminum bins are in use. Wood is used for special applications such as storage of acids and other liquids. Concrete staves normally have a compressive strength of 5,000 psi. As the staves are placed, their edges are coated with a waterproof mastic. After erection, the insides and outsides are covered with a cement wash. A water-sealing compound is then placed over the outside. Hoops are normally of ½- or 9/16-in.-diam rods. An overhead reserve storage is often a part of the structure. Concrete staves are available which incorporate an air pocket for improved insulation. Hollow tiles with reinforcement at the joints are used. A split tile, composed of concrete center, with horizontal and vertical reinforcing rods, and tile face, is also used. Monolithic-concrete storages, poured in one piece, reinforced with vertical and horizontal rods, are used.

Design of Shallow Bins. If a bin has a smooth wall and the plane of rupture cuts the surface of the grain (a shallow bin), Rankine's equation[2] can be used for determining the pressure on a bin wall. If the wall is rough, Janssen's equation should be used. Rankine's equations (50-5) and (50-6) are commonly used for design of retaining walls because of their simplicity. Constants can be determined by simple tests.

$$L = wy \frac{1-\sin\phi}{1+\sin\phi} \qquad (50\text{-}5)$$

where L = lateral pressure at any point, psf
y = distance from top to point, ft
w = density of material, lb/cu ft
ϕ = angle of repose

$$P = \frac{1}{2} wh^2 \frac{1-\sin\phi}{1+\sin\phi} \qquad (50\text{-}6)$$

where P = total horizontal force per ft of wall length, lb
h = height, ft

Example. Given: Steel bin 10 ft deep by 12 ft in diameter filled with wheat.

[1] M. S. Ketchum, "The Design of Walls, Bins, and Grain Elevators," McGraw-Hill Book Company, Inc., New York, 1919, p. 318.
[2] *Ibid.*, p. 18.

Find: Lateral pressure at bottom and total force.
Solution: Shallow bin, smooth sidewalls; use Eq. (50-5).

(a)
$$w = 50 \text{ lb/cu ft} \quad \phi = 28° \quad \sin 28° = 0.48$$
$$L = 50(10)\left(\frac{1 - 0.48}{1 + 0.48}\right) = 176 \text{ psf}$$

(b) Total force on 1-ft width of wall is

$$P = \tfrac{1}{2}(50)(10)^2 \left(\frac{1 - 0.48}{1 + 0.48}\right) = 880 \text{ lb/ft}$$

Rankine's equation does not apply to deep bins.

Design of Deep Bins. Janssen's equation[1] is used widely for determining the lateral pressure of deep bins.

$$L = \frac{wR}{\mu'}(1 - e^{-k\mu'h/R}) \tag{50-7}$$

where L = lateral pressure of grain, psf
w = weight of grain, lb/cu ft
R = hydraulic radius of bin, cross-section area divided by circumference or perimeter, ft
h = depth of grain at any point, ft
V = vertical pressure of grain, psf
$k = L/V$, ratio of lateral to vertical pressure
μ' = coefficient of friction of grain against wall

The μ' and k values are difficult to obtain and must be determined experimentally or estimated. Available values are in Table 50-9. Note that the angle of repose ϕ is not included in the equation. However, the angle of repose is useful in estimating the k value by

$$k \cong \frac{1 - \sin \phi}{1 + \sin \phi} \tag{50-8}$$

The k value usually varies between 0.3 and 0.6. Bins designed for fluids may fail when used for grain because of the tremendous vertical load which a grain storage must support. The walls carry a load equal to $P\mu'$ lb per lineal foot of wall, where P is the total lateral pressure on the bin wall. Equation (50-9) is applicable where the depth is more than twice the diameter.

$$P\mu' \cong wR\left(h - \frac{R}{k\mu'}\right) \tag{50-9}$$

The load supported by the bin wall is obtained by multiplying the circumference in feet by $P\mu'$. The load supported by the floor is the difference between the total weight of grain and load supported by the vertical wall. Solutions of Janssen's equation for selected values of w, k, and μ' are given in Figs. 50-10 and 50-11.

The following relationships exist in deep-bin storage:

1. The k value is slightly greater for shallow depths. It is assumed as a constant in the equation.
2. There is an increase in lateral pressure of about 10 per cent for moving grain.
3. The lateral pressure is greater for very clean grain.
4. The lateral pressure is greatest just after filling and decreases for about a week. If the bin is filled rapidly, the pressure is slightly greater.
5. The maximum vertical pressure is obtained in the center at the bottom with minimum at the edges. Janssen's equation assumes V is uniformly distributed.
6. The vertical pressure exceeds the horizontal pressure.

[1] *Ibid.*, p. 314.

690 FARMSTEAD STRUCTURES AND EQUIPMENT

7. Janssen's equation can be used for determining lateral pressures in rectangular bins on the basis of hydraulic radius.

8. By increasing the coefficient of friction of grain on the wall, μ', the lateral pressure decreases.

9. If k is increased, the lateral pressure increases, or the vertical pressure decreases, or both.

10. The selection of k is not critical for usual situations. For a 100-ft silo, 20 ft in diameter, the lateral pressure increases less than 5 per cent as the k value is increased from 0.5 to 0.6.

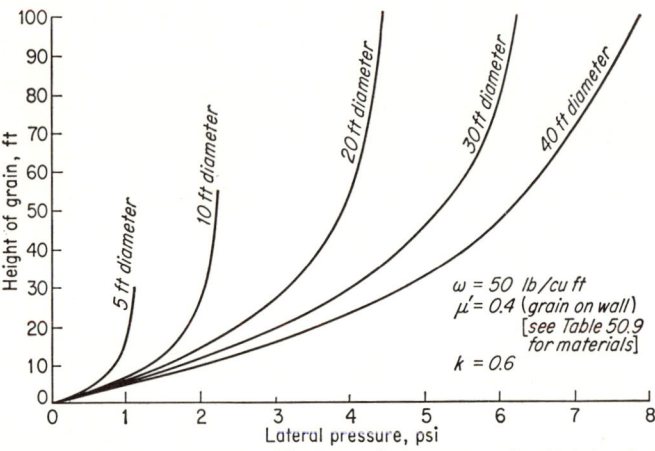

FIG. 50-10. Lateral pressure in cylindrical bin (or equivalent rectangular bin) based on Janssen's equation.

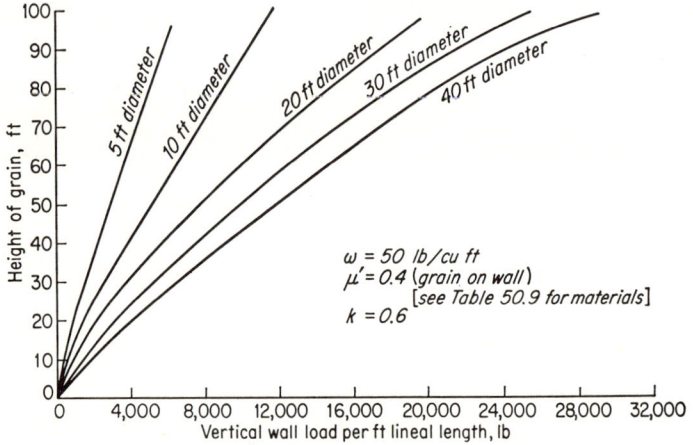

FIG. 50-11. Vertical wall loads on cylindrical bin (or equivalent rectangular bin).

Structural Strength. The lateral and vertical pressures are used for designing the structure. The structure is usually designed for wheat or corn even though it is planned to use the bin for a lighter grain. The μ, μ' and k values of these grains are readily available. For special cases where it is known that the bin will *always* be used for a lighter grain, such as oats, the appropriate constants for these grains are used. Figures 50-10 and 50-11 are based on values which suffice for corn, wheat, and other grains, with various bin-construction materials.

TABLE 50-9. STORAGE DATA ON SELECTED PRODUCTS (USE FOR JANSSEN'S EQUATION)

Product	Unit weight, lb/cu ft ω	Angle of repose, deg ϕ	Coefficient of friction on self, tan ϕ	Coefficient of friction μ'			Ratio of L/V, k	Ref.
				Steel	Smooth concrete	Smooth wood		
Barley.........................	38.4	28	0.53	0.376	0.452	0.325	....	f
Coal:								
Anthracite...................	52	27	0.51	0.287	0.509	0.466	0.3–0.4	d
Bituminous...................	50	35	0.70	0.325	0.70	0.70	0.3–0.4	d
Ear corn........................	28	...	....	.....	.....	0.62	....	
Flaxseed.......................	44.8	25	0.47	0.34	0.41	0.31	0.5	f
Grain sorghums...............	44.8	33	0.65	0.372	0.33	0.3	....	a
Hay (alfalfa-clover-grass mixture)								
Chopped 1.7", 10% moisture.....	.....	...	....	0.400	.....	0.454	....	b
Chopped 3.9", 11.2% moisture....	.....	...	....	0.396	.....	0.380	....	b
Oats...........................	25.6	32	0.62	0.412	0.466	0.369	....	f
Peas...........................	48	...	....	0.263	0.296	0.268	....	f
Rice, rough, 14% moisture........	41.6	36	0.73	0.41	0.52	0.44	0.48	c
Rye............................	44.8	26	0.49	0.406	0.35	0.33	0.4	d
Sand, dry......................	100	30	0.58	0.325	0.577	0.577	....	d
Shelled corn....................	44.8	27	0.51	0.374	0.423	0.308	0.65	d
Silage:								
Corn........................	30	...	....	0.70 (lower at high pressure)				g
Grass.......................	30	...	....	0.82 (lower at high pressure)				
Soybeans......................	48.0	29	0.55	0.366	0.442	0.322	....	f
Straw, chopped................	.....	...	....	0.22 (polished galvanized steel)				g
Sugar beet seeds, whole.........	.....	...	0.80	0.52	0.82	0.70	....	e
Vetch..........................	48.0	25	0.47	0.327	0.242	0.255	....	a
Wheat.........................	48.0	28	0.53	0.40	0.42	0.46	0.60	d
Tares..........................	49.0	29	0.55	0.364	0.36	0.394	....	d

a. Agricultural Engineers' Yearbook, ASAE, St. Joseph, Mich., 1957.
b. Max Christensen, "Investigation of Vertical Hay Feeder for Cattle and Sheep," unpublished M.S. thesis, Michigan State University, East Lansing, Mich., 1952.
c. Carl W. Hall, Drying Temperatures and Storage Problems of Sugar Beet Seeds, *J. Am. Soc. Sugar Beet Technologists*, 9:161–166, 1956.
d. M. S. Ketchum, "The Design of Walls, Bins, and Grain Elevators," McGraw-Hill Book Company, Inc., New York, 1919.
e. H. A. Kramer, Factors Influencing the Design of Bulk Storage Bins for Rough Rice, *Agr. Eng.*, 25: 463–466, December, 1944.
f. J. D. Long, Design of Grain Storage Structures, *Agr. Eng.*, 12:274–275, July, 1931.
g. Donald W. Richter, Friction Coefficients of Some Agricultural Products, *Agr. Eng.*, 35:411–413, June, 1954.

TABLE 50-10. HOOP SPACING FOR CONCRETE-STAVE SILOS USED FOR GRAIN STORAGE

Distance from top of silo, ft	Silo diameter, ft						
	10	12	14	16	18	20	24
	Hoop spacing for $9/16$-in.-diam reinforcement rolled-thread rods, in.						
0–5	30	30	30	30	30	30	30
5–10	30	30	30	30	15	15	15
10–15	30	30	15	15	15	15	10
15–20	30	30	15	15	15	10	10
20–25	30	15	15	15	10	10	7½
25–30	30	15	15	15	10	10	7½
30–35	30	15	15	15	10	10	7½
35–40	30	15	15	10	10	10	7½
40–50	30	15	15	10	10	7½	6

SOURCE: "Concrete Grain Storages," Portland Cement Assoc., Chicago, Ill.

Circular Steel Bins. The thickness of metal is determined from Eq. (50-10), using the lateral pressure L from Janssen's formula:

$$t = \frac{ld}{2Sf} \qquad (50\text{-}10)$$

where t = thickness, in.
  l = lateral pressure, psi (L from Janssen's formula, psf)
  d = bin diameter, in.
  S = allowable working stress of material; steel, S = 16,000 psi
    (The working allowable stress is usually used as one-half yield-point stress for ductile materials and would vary, depending on alloy.)
  f = efficiency of joint: 0.57 for single-riveted lap joint; 0.73 for double-riveted lap joint; 0.80 for double-riveted double-strap butt joints

Failure by collapsing has occurred in vertical circular steel bins next to a bin outlet located on the side of the bin. As the grain moves away from the side, the lateral pressure decreases, permitting the wall to collapse if it is not strong enough. Collapse of the bin wall can be avoided by running grain from the center of the bin.

Concrete. Concrete storages are always reinforced with steel. Horizontal reinforcing rods in either rings or spirals around the silo are designed to carry all the lateral pressure of the grain. The cross-section area of horizontal reinforcement can be calculated from Eq. (50-11).

$$A_s = \frac{LgD}{24f_s} \qquad (50\text{-}11)$$

where A_s = net area of steel rods for given depth of stored material, sq in.
  g = vertical spacing of rods
  L = grain pressure at a point midway between bars, psf
  D = bin diameter, ft
  f_s = allowable unit stress in rods, psi

The vertical reinforcement used for a monolithic-concrete silo must help carry the lateral load between the horizontal reinforcements, the weight of the wall, and grain carried by wall. The quantity of reinforcement can be calculated by using the theory of reinforced concrete and normally consists of ½-in.-diam rods spaced 9 to 18 in. or equivalent.

Steel reinforcement should be placed near the outside of the wall, but at least 2 in. from the outer edge so that the concrete will protect the steel from fire.

Openings for Grain Discharge. The quantity of wheat in bushels per minute which will flow through a horizontal opening can be estimated by the following equations.[1]

Circular: $\qquad Q = 0.1753 d^3 \qquad (50\text{-}12)$
Rectangular: $\qquad Q = 0.2232 W^2 L \qquad (50\text{-}13)$

where Q = bushels per min.
  d = diameter, in.
  W = width, in.
  L = length, in.

The flow from vertical openings is one-third the amount calculated by the above equations. The flow is independent of the head as long as the openings are flowing full. Sloped floors are usually designed with an angle of 10° more than the angle of repose to provide free flowing.

STORAGE OF EAR CORN

Ear corn storages are designed to permit air movement by natural ventilation to move through the crib. Ear corn is usually harvested and cribbed in the fall of the year when the kernel is at a moisture content above that for safe year-around storage. The quantity of air flow through a crib of ear corn depends on the velocity and

---

[1] Benton Stahl, Grain Storage Design, *USDA Circ.* 835, 1950 (based on data in Engineering Data on Grain Storage, *Agr. Eng. Data,* 1:1-11, 1948).

direction of the wind and the resistance to air flow of the ear corn. The resistance to air flow is dependent on the density of ear corn, foreign material, and width of the crib. The quantity of air flow which will move through ear corn in a crib can be found as follows:[1]

$$V = \frac{303(P)^{0.442}}{L^{0.540}}$$ (50-14)

where V = quantity of air flow, cfm/sq ft
P = wind static-pressure drop, in. of water = (pressure coefficient)$(0.00049w^2)$
(pressure coefficient = 1 for rectangular buildings, 0.8 for cylindrical buildings)
w = speed of wind, mph (usually 8 to 10 mph)
L = width of crib, ft

The recommended crib widths of rectangular bins to permit natural ventilation of ear corn is presented in Fig. 50-12. Round cribs without interior ventilators should not have a diameter greater than 1½ times the dimensions given in Fig. 50-12. One

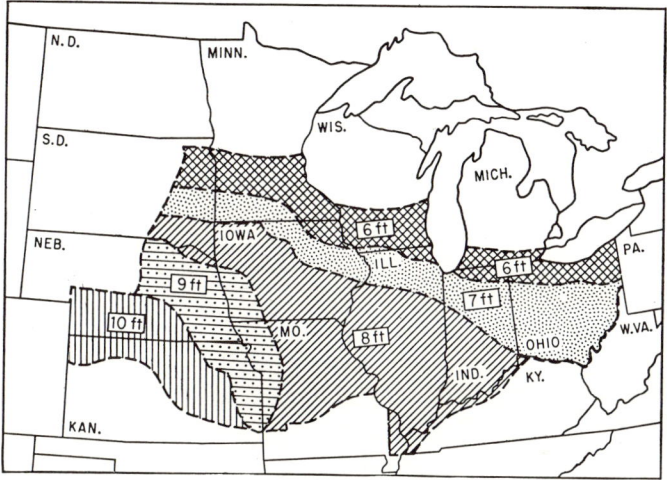

FIG. 50-12. Recommended crib widths for ear corn with natural ventilation. (*USDA Farmers' Bull.* 2010, 1949.)

bushel of ear corn is that amount which will yield one bushel of shelled corn. Two baskets (2-bu measure) are required to make one bushel. The volume occupied by one bushel of ear corn is 2.5 cu ft.

Government loans can be made for stored corn if it is under 20.5 per cent moisture, with a reduction in the loan if the corn is above 15.5 per cent moisture. When harvested, in north central United States, the kernels of ear corn will be from 18 to 20 per cent and the cob is about 35 per cent.

When corn is to be held over in storage more than 1 year, it is recommended that it be shelled and placed in a tight bin rather than in a crib.

Ear corn harvested by a mechanical picker contains damaged kernels, cobs, stalks, and trash. In order to permit air movement through the ear corn, debris should be removed and the elevator moved to evenly distribute trash and broken ears and kernels.

The use of ventilators aids air movement through the storage. Rectangular cribs utilize horizontal triangular-shaped ventilators, similar to those used for forced-air drying. The higher the duct extends into the corn, the greater the protection. The corn is protected only about 2 ft above the duct. A tall narrow duct which extends to within 2 ft of the corn surface is desirable.

[1] H. J. Barre and L. L. Sammet, "Farm Structures," John Wiley & Sons, Inc., New York, 1950.

Temporary bins may be constructed of wire fence or snow fence. A good foundation must be provided. Temporary corn cribs will tend to lean in the direction of prevailing winds because of settling during natural air drying. To avoid upsetting of temporary cribs, the height should not exceed 1 to 1½ times the width.

Circular cribs of metal, wood, and concrete are available commercially. Center vertical or square ventilators of 6- to 24-in. diam are desirable. The size of duct required for forced-air drying can be used as a guide to determine ventilator size for natural ventilation. One end of the ventilator should be open, with the other closed or covered with corn to direct the air through the corn in the crib.

In southern areas, the open crib does not protect ear corn from insect infestation.

Chapter 51

ELECTRICAL EQUIPMENT

EDITED BY DONALD P. BROWN
AND W. J. RIDOUT

Electricity was originally used on farms to supply light and to power household appliances, and in doing so it raised rural living standards to the same level as in the neighborhood towns. Electricity has, in addition, become the major source of energy for operations performed on the farmstead in connection with crop processing, crop storage, and livestock care. The proportion of farms receiving high-line electricity has increased from 7 per cent in 1920 to 95 per cent in 1957.

WIRING

MORRIS H. LLOYD

The proper functioning of all electrical equipment and appliances on the farm depends upon the adequacy of the wiring.

Planning Farm Wiring. Two factors must be considered when selecting sizes of conductors: (1) the connected or computed load to be served, figured in accordance with the National Electrical Code, and (2) the permissible limit of voltage drop for the probable maximum demand on the circuit or feeder.

The National Electrical Code is concerned primarily with the safety of wiring installations rather than with their adequacy (see Introduction to NE Code).[1]

It is important, in addition, to select sizes of conductors large enough to carry the probable maximum demand over the required distance without exceeding a voltage drop which will impair the operation of equipment and appliances.

This probable maximum demand may be, and usually is, less than the computed load on the circuit or feeder. However, considering maximum demand in relation to length of circuit and allowable voltage drop, it may indicate the need for larger conductors than are required by the Code for the computed load. For short runs such calculations may indicate a size of conductor smaller than that required by the Code.

The "Farmstead Wiring Handbook" contains detailed recommendations for wiring of farm structures and a summary of recommendations for residential wiring. The "Residential Wiring Handbook"[1] contains detailed recommendations for residential wiring. These two publications and the National Electrical Code will greatly assist in planning a farm wiring system.

The planning of a new wiring installation or the modernizing of an existing installation involves several steps.[2]

Types of Wires and Cables for Farm Wiring. Nonmetallic enclosures for cable assemblies are recommended by most authorities. The dampness prevailing in many farm buildings, and particularly the corrosive atmosphere often present in buildings housing livestock, makes a metal-clad system short-lived. Metal enclosures such as rigid conduit, electrical metallic tubing ("thin-wall"), flexible conduit ("Green-

---

[1] "National Electrical Code," National Fire Protection Association, Boston, Mass., 1959.
[2] "Farmstead Wiring Handbook," Sections 14 to 20 and Tables A to M, Industry Committee on Interior Wiring Design, New York, 1955.

field"), or armored cable should not be used except where local regulations or special conditions require such wiring methods for protection against physical damage.

Insulation on individual conductors used in cable assemblies and raceways for farm wiring are type R, code rubber; type RH, heat-resistant rubber; type RW, moisture-resistant rubber; also types RHW and RH-RW, combinations of heat- and moisture-resistant rubber. Types RU, RUH, and RUW are similar but made of latex rubber. Type T, thermoplastic, and type TW, thermoplastic, moisture-resistant, are used under similar conditions to types R and RW, respectively. Allowable current-carrying capacity for conductors with types RH, RUH, RHW, also RH-RW in dry locations, is based on 75°C (167°F) maximum operating temperature. Type RH-RW in damp locations and all other types listed are based on 60°C (140°F) maximum operating temperature. Other types of insulation such as asbestos and varnished cambric are rarely used in farm wiring. For overhead conductors between buildings, type WP weatherproof insulation is most common but type TW may be used.

Service Entrance Wiring. Service entrance cables are available in capacities up to and including 200 amp. For services above 200 amp, conduit and individual conductors must be used. Local regulations may require use of conduit for all services.

TABLE 51-1. TYPES OF SERVICE ENTRANCE CABLES

Type	Sheath	Use
SE, style U	Flame-retardant, moisture-resistant nonmetallic sheath	Aboveground, attached directly to surface wired over
SE, style A	Same as SE, style U, but with flat metal tape applied concentrically between neutral and sheath	Same as preceding
ASE	Metal interlocking armor under outer sheath	Same as preceding
USE	Rubber or neoprene outer sheath; individual conductors have type RHW or RH-RW insulation	Direct burial in earth; aboveground must be in raceway

In conduit-type services, where a bare neutral conductor is used, its carrying capacity is based on the rating of insulation of the ungrounded conductors. For example, if a service conduit contains two No. 0 type RH conductors and a No. 2 neutral, the neutral conductor carries the same rating whether it has type RH insulation or is bare.

Underground Wiring, Underground Services. Type USE cable, buried directly in earth. Individual type USE conductor, buried directly in earth; all conductors in same trench; made in sizes 14 to 1,000 MCM. Type RHW, RH-RW, DUW, or TW conductors in raceway. Neutral may be uninsulated.

Underground Feeders. Any of materials shown for underground services except neutral must be insulated.

Type UF underground feeder cable (approved in sizes 14 to 4, presently made in sizes 14, 12, and 10, two-wire and three-wire) buried directly in earth. Individual type UF conductors, buried directly in earth; all conductors in same trench; made in sizes 14 to 4.

Type UF cables have outer covering of flame-retardant, moisture-resistant, fungus- and corrosion-resistant thermoplastic material. Conductors have type TW, RW, RUW, or RHW insulation.

Underground Branch Circuits. Use any of materials mentioned for underground services or feeders. Insulated neutral required.

Overhead, Outdoor Feeders and Branch Circuits. Use type WP or TW conductors. Minimum permissible size is No. 10 for spans up to 50 ft, and No. 8 for longer spans.

Interior Wiring. Nonmetallic sheathed cables generally are preferred for most farm structures and are required by many inspection authorities.

Type UF or type NMC cable (nonmetallic, corrosion-resistant) is generally required in buildings housing livestock or where dampness is present. Presently available in sizes 14, 12, and 10, two- and three-wire, with or without separate, bare grounding conductors. Type NM cable with flame-retardant, moisture-resistant braided outer sheath is approved for normally dry locations.

TABLE 51-2. RATINGS OF SERVICE ENTRANCE CABLES
(75°C insulation on individual conductors; bare neutrals; all cables 3-wire except as noted.)

Copper conductors			Aluminum conductors		
Conductor size		Ampere capacity	Conductor size		Ampere capacity
Ungrounded	Neutral		Ungrounded	Neutral	
12*	12	20	10*	10	25
10*	10	30	8*	8	40
8*	8	45	6	8	50
6	8	65	6	6	50
6	6	65	4	6	65
4	6	85	4	4	65
4	4	85	3	5	75
3	5	100	3	3	75
3	3	100	2	4	100
2	4	115	2	2	100
2	2	115	1	3	110
1	3	130	1	1	110
1	1	130	0	2	125
0	2	150	0	0	125
0	0	150	00	1	150
000	0	200	00	00	150
000	000	200	000	0	170
			000	000	170
			4/0	00	200
			4/0	4/0	200

* Available in two-wire or three-wire cable.

TABLE 51-3a. PLUGS AND RECEPTACLES FOR GROUNDING PORTABLE APPLIANCES

Capacity	Type
15 amp or less, 125 volts, 2-wire	American standard parallel blade and U blade
15 amp or less, 250 volts, 2-wire	American standard tandem blade and U blade
20 amp, 125 or 250 volts, 2-wire	3-wire "crowfoot," or locking type
30 amp, 125 or 250 volts, 2-wire	3-wire L blade for grounding (clothes-dryer type)
50 amp, 125 or 250 volts, 2-wire	3-wire crowfoot
20 amp, 125 or 250 volts, 3-wire, or 250 volts, 3-phase	4-wire flat blade, or locking type
60 amp, 125 or 250 volts, 3-wire, or 250 volts, 3-phase	4-wire flat blade

TABLE 51-3b. FUSES, FUSIBLE SWITCHES, AND CIRCUIT BREAKERS

Fuses and fusible panels

Advantages	Disadvantages
Lower first cost	Can be bridged
Available in great number of ratings	Can be replaced by higher-rated fuse
Available in time-delay type for circuits serving motors	Must be replaced after operating

Circuit breakers

Advantages	Disadvantages
Not easily bridged or changed	Higher first cost
Nothing to replace after operation	Limited number of ratings

Service entrance cables may be used for branch circuits. If all conductors are insulated, use is the same as type NM or NMC. If neutral conductor is uninsulated, use is restricted to 115/230 volt range and clothes-dryer circuits or to 230-volt, two-wire circuits where neutral conductor is used only as grounding conductor.

Knob-and-tube work is still a good wiring method but is seldom used because of greater labor cost compared with use of cable described above.

Grounding of Portable Equipment and Appliances. The National Electrical Code permits the grounding of ranges and clothes dryers through the neutral conductor of the circuit. This may not be permitted by some local regulations. The Code also requires that a grounding-type receptacle be installed in the laundry or for laundry purposes and that this receptacle have slots to receive a parallel blade plug with U-shape grounding blade. Receptacles installed in or on open porches, breezeways, and residential garages and which may be used by persons standing on the ground shall be of this type also. Other than this, the Code has no specific requirements for grounding in residential occupancies. Whether farm buildings are part of a "residential occupancy" may depend upon local interpretation of the Code. Where grounding of portable equipment is necessary or desirable, it must be accomplished by means of a cord containing a grounding conductor and terminating in a plug cap having a special blade for connection to the grounding slot in a receptacle.

Where fuses are used for protection of circuits of 30 amp or less, it is recommended that they be the plug type rather than the cartridge type. Cartridge fuses are not as easy to change, are not always as readily available, and are somewhat higher in price. Cartridge fuses in sizes of 30 amp or less are not required on two-wire, 230-volt circuits or three-wire, 115/230 volt circuits derived from a three-wire, 115/230 volt system.

Where fuses are used to protect branch circuits, feeders, or services serving motor loads, it is recommended that they be the time-delay type to prevent needless blowing on motor-starting currents. Such fuses, however, will not provide running overload protection to a motor unless the circuit serves only one motor and the fuses are rated according to the motor full-load current. The matter of motor-overload protection will be more fully discussed in the section on electric motors.

ELECTRIC MOTORS

Robert H. Brown

A-C Motor Classifications. According to their principles and characteristics of operation a-c motors can be classed as synchronous, induction, or series-type motors. The induction motor is by far the most popular motor of this group. Classed according to the power source from which they operate, the divisions are single-phase motors and polyphase motors. The single-phase motors are the most important for the agricultural engineer since most farms are served with single-phase power. Horsepower output is the basis for a third method of classification, namely, fractional-horsepower electric motors and integral-horsepower motors. The fractional-horsepower group includes all motors built in a frame smaller than that having a continuous rating of 1 hp. Many farm electrical motors are in this group. The integral-horsepower classification includes motors 1 hp and larger. The various types of a-c- motors, grouped according to operating characteristics, are listed in Table 51-4.

TABLE 51-4. NAMES AND CLASSIFICATIONS OF A-C MOTORS

Induction		Series		Synchronous	
Single-phase	Polyphase	Single-phase	Polyphase	Single-phase	Polyphase
Split-phase Capacitor Capacitor-start induction Two-value capacitor Repulsion-start induction Repulsion induction Repulsion-start capacitor-run Shaded-pole	Three-phase wound-rotor induction motor Three-phase squirrel-cage induction motor Two-phase motor	A-c series Universal Repulsion Repulsion induction	Series	Single-phase synchronous motor Special-purpose synchronous motor	Three-phase synchronous motor

CHARACTERISTICS OF INDUCTION, SERIES, AND SYNCHRONOUS MOTORS

Induction Motors. This is the most important group of commercial electric motors, as indicated in Table 51-4. Their operating characteristics are constant speed, variable torque, variable horsepower, average power factors, and average efficiencies. These motors are so named because the input energy to the rotating member is accomplished through the process of induction.

Induction motors have two main parts, the stationary member, or stator, and the rotating member, or rotor. There are two types of rotors, the squirrel-cage and the wound-rotor. The squirrel-cage rotor appears outwardly as a solid mass but actually consists of a series of bars placed in slots around the periphery of the rotor. The conductors are short-circuited at each end by end rings. The wound-rotor type of construction appears quite different and is not used for single-phase motors. The windings are visible and are not short-circuited but are connected to slip rings located on one end of the shaft. The slip rings provide a means for connecting external resistance into the rotor circuit. The resistance changes the electrical characteristics of the rotor, namely, power factor and impedance, and therefore varies input current, torque, and speed of the motor. The wound-rotor induction motor is used only on polyphase systems and has a greater initial cost than the squirrel-cage.

Every induction motor rotates because the stator winding sets up a rotating magnetic field. Torque is developed when this magnetic field cuts across the rotor conductors. The rotor turns in the same direction as the rotating field and therefore must rotate more slowly. At full load the rotor speed is usually about 97 per cent of that of the rotating field. The speed of the rotating field, the synchronous speed, does not depend on voltage but is a function of frequency and the number of poles of the stator winding, as shown below.

Induction-motor Formulas. Some useful relationships for a-c motors running with induction characteristics are:

$$\text{rpm of rotating field} = \frac{120 \times \text{frequency}}{\text{number of poles}}$$

$$\text{Per cent slip} = 100 \, \frac{\text{rpm of rotating field} - \text{rpm of rotor}}{\text{rpm of rotating field}}$$

$$\text{Torque, ft-lb} = \frac{\text{hp} \times 5{,}750}{\text{rpm}}$$

Volt-amperes input (single-phase) $= EI$
Volt-amperes input (three-phase) $= \sqrt{3} \, EI$
Watts input (single-phase) $= EI \cos \theta$ (cos θ = power factor)
Watts input (three-phase) $= \sqrt{3} \, EI \cos \theta$

$$\text{Watts input} = \frac{\text{hp output} \times 736}{\% \text{ efficiency}}$$

Approximations:

  Rotor speed: ranges from 95 to 99% of rotating-field speed
  Efficiency: 70% fractional hp, 80% 2 to 5 hp; 85% above 5 hp
  Power input: 1,200 watts/hp up to ½ hp, 1,000 watts/hp for motor sizes ½ hp and larger
  Torque: varies as the square of the ratio of applied voltage; thus, reducing voltage from 120 to 80 volts reduces the torque to $(80/120)^2 = 44.4\%$ of its value with 120 volts applied

Series Motors. The a-c series motors operate with series characteristics and always have brushes, a commutator, and a d-c-type wound rotor. The rotor winding either is connected in series with the stator winding or is arranged electrically so as to yield the same result. The a-c series motor and the Universal motor have the stator field and the rotor connected in series, while the repulsion-type motors have no external electrical connections to the rotor. In the latter case the brushes are connected together to provide a closed rotor circuit and all rotor currents are induced currents.

Motors operating with series characteristics develop torque because of the interaction between rotor current and field flux. They do not depend upon a rotating magnetic field. These motors provide high starting torque, good running torque, variable speed, and approximately constant horsepower output. The no-load speed is very high, but the speed decreases rapidly as load is increased. They are best adapted for jobs such as food mixers, portable power tools, sewing machines, lawn mowers, and similar applications where an increase in load should be accompanied by a decrease in speed.

Synchronous Motors. The synchronous motor runs at an exact speed for all loads up to the pull-out point. If load is increased past this value, the motor will stop. In industry it is used because of this speed characteristic and because it can be made to take a leading current from the source, thereby improving the over-all power factor of combined electrical loads.

The synchronous motor is usually of the revolving-field type, requires a d-c source for exciting the field poles, and is provided with an amortisseur winding (in the pole faces) for self-starting. Other than the amortisseur winding the synchronous motor is identical with the a-c generator, and it can therefore be used as a stand-by generator for emergency use. For this reason it is adapted for use on certain types of farms.

CHARACTERISTICS OF THE CONSTRUCTION AND OPERATION OF A-C MOTORS

The main characteristics of motors which influence their initial selection and predict their performance under load are given in Table 51-5. The motor names are significant because they designate a type of motor designed to operate from a specific power source and to deliver a certain amount of starting torque, breakdown torque, etc.

TABLE 51-5. SUMMARY OF DATA FOR A-C MOTORS

Name of motor	Power supply	Usual range of hp sizes	Hp size for characteristic data	Starting torque, % full load	Breakdown torque, % full load	Starting current, % full load	Usual maximum overload, %	Operating characteristic
Three-phase squirrel-cage (design A and B)	3-phase	1/8–5,000	5	185	225	300	15	Induction
Three-phase wound rotor	3-phase	1/8–5,000	5	Varies with resistance (100 min)	200	Varies with external resistance (100 min)	15	Induction
Split-phase	Single-phase	1/20–1/3	1/3	175	250	600	35	Induction
Capacitor	Single-phase	1/100–3/4	1/3	60	150	300	35	Induction
Capacitor-start	Single-phase	1/6–5	1/3	375	250	400	35	Induction
Two-value capacitor	Single-phase	1/2–20	5	300	200	400	15	Induction
Repulsion	Single-phase	1/6–5	1/3	450	...	450	35	Series
Kepulsion induction	Single-phase	1/6–5	1/3	300	225	400	35	Series induction
Repulsion-start induction	Single-phase	1/6–10	1/3	500	200	375	35	Starts series, runs induction
Shaded-pole	Single-phase	Up to 1/6	1/20	60	180	175	15	Induction
A-c series	Single-phase	1/20–2,000	1/6	450	...	...	35	Series
Universal	D-c or single-phase	1/20–300	1/25	450	...	...	35	Series
Three-phase synchronous	D-c and three-phase	5–600	100	110	150	500	10	Synchronous
Special-purpose synchronous	Single-phase	Up to 1/20	...	...	...	...	10	Synchronous

ELECTRICAL EQUIPMENT 701

The performance of a motor cannot be predicted or evaluated until identification is made as to the type of motor involved. Motors can be identified, and a better understanding of the reason for their differences in performance is possible, by noting the major parts of the motor. The major parts, typical uses, and the methods for reversing the direction of rotation are included in Table 51-6.

TABLE 51-6. DATA ON CONSTRUCTION AND USE OF A-C MOTORS

Name of motor	Major parts	Typical use	Reverse direction of rotation by:	Remarks
Split-phase.........	Squirrel-cage rotor, CS,* starting and running windings	Seed cleaner, tool grinder	Interchanging either the leads to the starting winding or the leads to the main winding	Low initial cost, average starting torque
Capacitor	Squirrel-cage rotor, capacitor, two windings	Fans		Not very common on farms
Capacitor-start.....	Squirrel-cage rotor, capacitor, CS, starting and running windings	Water pump, automatic feeder, refrigerator compressor		Good for general-purpose use; high starting torque
Two-value capacitor	Squirrel-cage rotor, two capacitors, CS, two windings	Vacuum pump, feed grinder, crop dryer		Usually found in the larger hp sizes, high starting torque
Repulsion.........	Brushes, commutator main winding	Lathe	Shifting the brushes	Not very common, variable speed
Repulsion-start.....	Brushes, commutator main winding, CS	Air compressor		Very high starting torque/ampere
Repulsion induction	Brushes, commutator main winding, dual winding on rotor	Ensilage cutter		Popularity decreasing
Repulsion-start, capacitor-run	Brushes, commutator main winding, capacitor	Compressor, feed grinder		Very few in use to date
Universal.........	Brushes, commutator main poles	Portable tools, lawn mower	Interchanging the connectors to the brushes	Varying speed
Shaded-pole.......	Squirrel-cage rotor, main poles, shading coils	Desk fan, clock, small blowers	As a general rule no provision is made	Very low starting torque
Three-phase squirrel-cage	Squirrel-cage rotor, main winding	Feed grinder, mixer	Interchanging the connections of any two input lines	For use when three-phase supply is available

* Centrifugally operated switch.

The *schematic diagram* for each major type of a-c motor is given in Fig. 51-1. The input voltage of the source should be those values shown on the motor name plate. Whenever possible single-phase motors ½ hp and larger should be operated from 240 volts. Three-phase motor sizes from 1 to 25 hp should be served from a 240-volt source, and when possible 25-hp and larger sizes of motors should be connected to 480 volts.

GENERAL CONSIDERATIONS FOR SELECTION AND INSTALLATION

The *service factor* on the motor name plate is multiplied by the horsepower rating to give the maximum load that can be successfully carried by the motor if it is to operate continuously and remain within a safe temperature range. The percentage overload values for class A insulation, open frame, and 40°C rise are given in Table 51-5. These are approximate values since horsepower size and enclosure bring about small variations. Most integral-horsepower factors are 1.15, while fractional-horsepower motors have service factors ranging from 1.4 for 1/20 hp to 1.25 for ¾ hp.

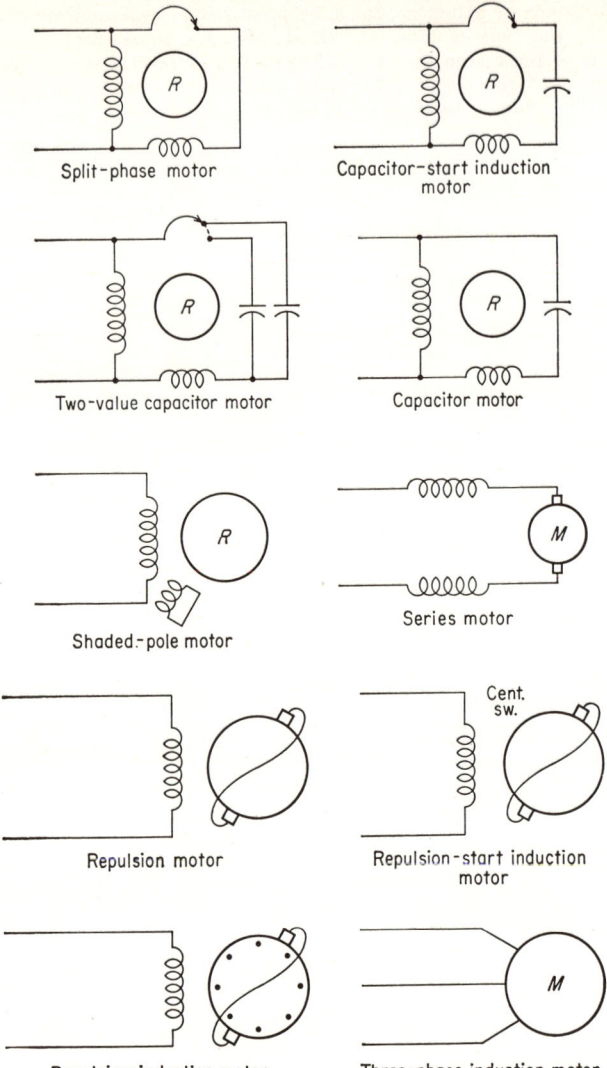

Fig. 51-1. Schematic diagrams for a-c motors.

Standard Horsepower Sizes. The standard horsepower ratings for single-phase induction motors are:

1/20	1/6	1/2	1 1/2	5	15
1/12	1/4	3/4	2	7 1/2	20
1/8	1/3	1	3	10	25

The standard horsepower ratings for polyphase induction motors are:

1/8	3/4	5	25	75
1/6	1	7 1/2	30	100
1/4	1 1/2	10	40	125
1/3	2	15	50	150
1/2	3	20	60	200

The horsepower ratings given above are for either continuous or intermittent operation and are based on a temperature rise above an ambient temperature of 40°C. The temperature rise is shown on the name plate. Motors designed for *intermittent-duty* periods of 5, 15, 30, or 60 min will operate within their temperature rating (40° + rise) for that period but will overheat if operated continuously. Motors rated on a continuous-duty basis should be selected for general-purpose farm motors.

Motor Enclosures. The National Electrical Manufacturers Association (NEMA) has established 18 standard classifications for enclosures. These classes can be broadly grouped as:

Open-drip-proof	Totally enclosed
Splashproof	Totally enclosed—special cooling
Open-guarded	Explosion-proof

The detailed specifications for each classification are given in the *NEMA Motor Standards Bulletin.* For farm applications the usual choice is made from one of these four types:

1. *Open-drip-proof.* For general-purpose use, the motor being located in dry locations which are relatively free of splashing liquids and dust particles. This is the most common type of enclosure.

2. *Splashproof.* For use out of doors if covered when not in use and especially for use on dairy farms and in processing rooms where washing down of equipment causes water to be directed toward the motor.

3. *Totally Enclosed.* For applications in those areas having above-average dust, lint, and foreign particles in the atmosphere. The enclosure furnishes protection for the motor, and its selection results in longer life and less maintenance than is possible with the open-drip-proof type.

4. *Explosion-proof.* This enclosure protects the internal parts of the motor from harmful atmosphere and in addition is designed so as not to cause ignition of explosive gases which may surround the motor.

Grounding. For added safety the frame of the motor and the controller case must be grounded. The grounding prevents a potential aboveground in case an ungrounded conductor accidentally comes in contact with a metal part. The grounding conductor should either be uninsulated or have green insulation. The grounding is accomplished by the proper selection of cables, plug caps, and receptacles which have provisions for bonding or by installing a separate conductor (usually bare) which ties all non-current-carrying metal parts to the earth.

Selection of Circuit Components. For proper installation of electric motors consideration must be given to the selection of a control for the motor, protection for the motor windings, the disconnecting means, the branch-circuit-wire size and its protection, and finally the feeder-wire size and its protection. Guides to the selection of these components, on a minimum basis, are included in the National Electrical Code. Alternative guides to the selection of these components which provide better-than-minimum sizes are given below. These recommendations are for all single-phase and for three-phase squirrel-cage motors.

Motor Controller. (Any switch or device for starting and stopping the motor.)

1. For motors $\frac{1}{6}$ hp and larger provide a controller having a rating in horsepower not less than the horsepower rating of the motor.

2. For motors smaller than $\frac{1}{6}$ hp the plug cap and receptacle or a switch with adequate ampere rating or a controller rated in horsepower may be used.

3. For motors up to 1 hp, except portable motors larger than $\frac{1}{6}$ hp, the disconnecting means may be used as the controller, provided it is protected by the branch-circuit overcurrent device.

Motor Overcurrent Protection. (May be a component part of the controller.)

1. For protecting the motor windings against excessive current, select an overcurrent device not greater than 1.25 times the motor name-plate current for 40°C rise motors and not greater than 1.15 times the current for 50° rise motors.

The overcurrent device may be a time-delay fuse, a circuit breaker, or a heater coil or strip which mounts in the controller, or in the case of smaller motors it may be built into the frame of the motor.

Disconnecting Means. (To disconnect from the source all ungrounded conductors serving the motor and the controller.)

1. For nonportable motors larger than ⅛ hp but smaller than 1 hp either a motor circuit switch with appropriate horsepower rating or a general-use switch having an ampere rating not less than fifteen or not less than three times the motor current shown in Table 51-7, whichever is greater, is to serve as the disconnecting means.

2. For nonportable motors 1 hp and larger a motor circuit switch or a circuit breaker having a rating in horsepower not less than the horsepower rating of the motor is to be used as the disconnecting means.

3. For any motor smaller than ⅙ hp and for portable motors use the plug cap and receptacle as the disconnecting means.

Motor Branch-circuit Wire Size. Select conductors on the basis of 2 times current of largest motor plus currents of other motors (from Table 51-7), or No. 12 conductors, or 2,000 watts per hp and 2 per cent voltage drop, whichever results in the largest wire size. (For three-phase motors ranging from 1 to 25 hp use 1,000 watts per hp instead of 2,000.) Individual branch circuits are suggested for each motor larger than ⅙ hp.

TABLE 51-7. FULL-LOAD CURRENTS FOR A-C MOTORS, AMP

Single-phase motors			Three-phase motors			
Hp	115 volts	230 volts	Hp	110 volts	220 volts	440 volts
⅙	4.4	2.2	½	4.0	2.0	1.0
¼	5.8	2.9	¾	5.6	2.8	1.4
⅓	7.2	3.6	1	7.0	3.5	1.8
½	9.8	4.9	1½	10.0	5.0	2.5
¾	13.8	6.9	2	13.0	6.5	3.3
1	16.0	8.0	3	....	9.0	4.5
1½	20.0	10.0	5	....	15.0	7.5
2	24.0	12.0	7½	....	22.0	11.0
3	34.0	17.0	10	....	27.0	14.0
5	56.0	28.0	15	....	40.0	20.0
7½	80.0	40.0	20	....	52.0	26.0
10	100.0	50.0	25	....	64.0	32.0

SOURCE: National Electrical Code, Tables 22 and 24, 1956.

Motor Branch-circuit Overcurrent Protection. (To protect the conductors, controls, and motor against short circuits and grounds.)

1. Select a protective device having a time-lag characteristic and a rating not greater than 1.5 times the rating of the branch-circuit conductors which were chosen by the method described above.

2. In the case of a branch circuit serving only one motor, the overcurrent protection for the branch circuit may also serve as protection for the motor provided it is a fuse or circuit breaker selected as described above for motor overcurrent protection.

Feeder Conductors

1. Select feeder-conductor sizes in the same manner described for branch-circuit conductors.

2. The overcurrent protection for feeder conductors should have a time-lag characteristic and an ampere rating equal to 1.5 times the ampere rating of the largest branch-circuit conductors served by the feeder plus the ampere rating of each additional motor served by the feeder. (Use Table 51-7 values for motor currents.)

EQUIPMENT FOR HANDLING MILK

CARL W. HALL

The milking machine, milk cooler, water heater, and pasteurizer are the principal pieces of milk-handling equipment operated primarily by electricity. The heating of milkhouses is discussed under milkhouse construction, and barn cleaners under livestock production facilities.

Milking Machine. A *piston*, or *vane*, type of pump is used for obtaining a vacuum of 10 to 14 in. of mercury. Vacuum pumps are rated for capacity as to number of milking units as 2, 3, 4, 5, 6, etc. Ordinarily a two-unit system will suffice in a stall barn, but if a weigh jar, milk pipeline from stall to milkhouse, and similar equipment are added, each will require an additional unit on the vacuum pump. A 1-hp motor with a five- or six-unit pump is recommended for a milking room. It is usually necessary to increase the size of a vacuum pump when changing from the conventional milking system to a piping system. A three-unit machine will usually have a $\frac{1}{3}$- to $\frac{1}{2}$-hp motor; a four-unit, $\frac{3}{4}$ hp; and a five- or six-unit, 1 hp.

A 1-in. galvanized pipe is used for a vacuum line. The motor to the vacuum pump is started under no load. A valve with weights on it is used for regulating and controlling the vacuum. There should be a dial vacuum indicator gage on the pipeline so that the operator can check the vacuum.

Milk is normally conveyed from the milker teat cups to a 5-gal pail or through a milk line to the milkroom. The milk line is made of $1\frac{1}{2}$-in.-ID glass piping (or stainless-steel tubing) and should be positioned so that the system will drain. Milkers can be of the short-tube type, which hang on the cow and hold about 38 lb of milk, or of the long-tube type, where the milk is collected in a pail.

Milk piping can be used in a stall barn or milking room. The installation cost of a sanitary milking line will often be $3,000 for a 30-cow herd. Portable milk lines from stall barn to milkroom are being used in which either a vacuum or milk pump moves the milk. The major problem in handling of milk in piping is rancidity, which can be prevented by following a few simple rules. Although much research is needed on this subject, the main consideration is that air should not be mixed with warm milk as it comes from the cow. The following *practices* tend to reduce air incorporation and prevent rancidity:

1. Milk pipeline should not have a slope of over 1 in. per 100 ft.
2. Keep the number of risers to a minimum.
3. Do not lift milk vertically with vacuum any more than necessary.
4. Prevent air leaks at valves in line.
5. Collect milk in pail and then move through line instead of milking directly into the milk pipeline.

A *pulsator* provides the formation and breaking of the vacuum to the teat cups. There are two vacuum lines to each teat cup, one of which is under constant vacuum. The other line changes from atmospheric to vacuum at the rate of about 45 to 55 pulsations per minute. The alternating vacuum takes place between the teat inflation and the teat-cup shell as controlled by the pulsator. Most units have two teat cups working together, and some all four.

Contamination of the milk can occur from spoilage organisms in the vacuum line which find their way into the milker pail. Some milkers have a device to prevent the sucking of air from the vacuum pipeline. The vacuum line should be kept clean and should be installed at an angle so that water can drain from it.

A *strainer* can be placed in the line between the teat cups and the milker pail. This strainer is usually single-service paper. A cloth strainer in the shape of a cylinder can also be placed in the milk line. Use of in-the-line strainers is not acceptable to many health inspectors. In these cases the milk can go from the pipeline to a strainer and then into the tank or can. The piping can be cleaned by flushing or recirculation with water and a germicidal agent at 160°F or above. It is important to recognize that the chemical solution, not the hot water, does the sanitizing.

The *vacuum releaser* permits the milk under vacuum in the pipeline to be rejected to the atmosphere without losing the vacuum in the line, as shown in Fig. 51-2. A vacuum releaser is not needed on vacuum milk tanks but is useful for washing the lines using the flushing method.

The following *milking practices* are recommended:

1. Prepare cows about 1 min before milking by washing udder with a warm chlorine solution strength of 100 ppm.

2. Use strip cup to remove first milk of high bacteria count and check for mastitis or other infection.
3. Leave milker on cow no longer than 4 min.
4. Machine strip to ensure complete milking.
5. Rinse milker in lukewarm water; then rinse in hot water with washing solution and chemical sanitizing solution.

Milk Cooler. The immersion-type and spray-type *can coolers* are the most prominent mechanical types of can milk-cooling units. With the immersion type the cans are placed in water cooled by an ice bank and the water is moved around the can with a circulating pump or agitator. The milk will cool from 90 to 55°F in the top of the can and to 45°F in the bottom of the can in 1 hr when immersed in water cooled to 35°F by an ice bank and agitator. The rate of cooling varies considerably from these figures, depending on the amount of water, ice bank, and agitation. With an air-cooled condenser about 1.5 to 1.8 kwhr of electricity is required to cool a 10-gal can

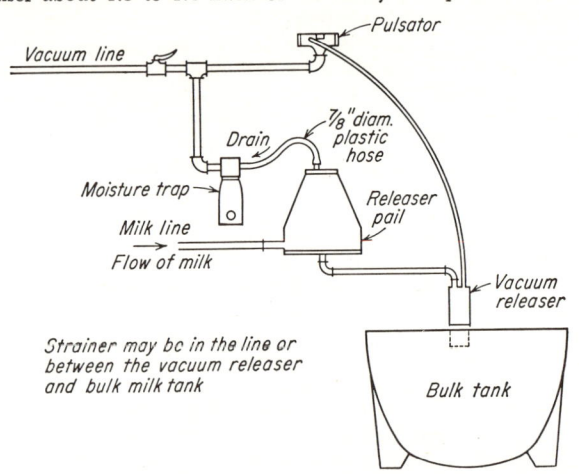

Fig. 51-2. Arrangement of milk and vacuum lines for atmospheric bulk tank.

to 40°F (86 lb of milk). The rate of cooling the milk is faster in a spray-type cooler, and this has side-opening doors so that it is easy to place the 10-gal cans of milk into the cooler. Hoists are available, either hand- or mechanically operated, to lift cans into an upright top-opening cooler.

The farm *bulk-tank cooler* has come into use quite rapidly in recent years. There were over 140,000 bulk-tank coolers in use on farms in 1960. These coolers are principally of two types or a combination of the two:

1. The ice-bank (**IB**) (sweet-water) system builds up refrigeration capacity in the form of ice and uses a pump for moving the cold water from the ice over the stainless-steel shell which holds the milk.

2. The direct-expansion (**DX**) cools the milk directly by the use of evaporator coils adjacent to the bottom and lower sides of the tank.

Bulk milk coolers are available in sizes from 80 to 1,000 gal. The tanks can be obtained as package units or with the tank and refrigeration units separately. The tank can be placed in the milkhouse, and the compressor in a remote location when required to adapt a small milkhouse to bulk handling. Many recommendations suggest that the size of the bulk milk tank be five times the average volume of one milking for every-other-day (EOD) pickup. The operating cost of the tank is almost as much when filled to 75 per cent capacity as when filled to 100 per cent.

The direct-expansion system requires about 1 hp of compressor motor capacity for each 50 gal of milk to be cooled at a milking. The ice-bank system requires about ½-hp compressor motor capacity for each 50 gal to be cooled. The original cost of an ice-bank system is generally less than the direct-expansion system; however, it requires more electricity than the direct-expansion system. Approximately 1 kwhr of electricity is required for each 100 lb of milk cooled with a direct-expansion system. The *total cost* of operation for the two types of tanks is approximately the same over a 10- to 12-year period. It is generally economical to use water for cooling the condenser with units having over a 3-hp compressor motor, provided that clean water is available at a reasonable cost.

Many different types of *agitators* are used in bulk tanks. The agitator operates when the compressor operates and can be operated separately to mix the fat and milk for test when the milk is obtained at the farm. The volume of milk in the tank is usually determined with a graduated stick and then converted to weight. The tank must be level in the milkhouse. Scales under individual farm tanks and flow meters may be used to determine the quantity of milk.

The 3-A Sanitary Standards for the *rate of cooling* in a bulk tank state that the milk must be down to 50°F within 1 hr after milking and 40°F within 2 hr after milking, during which time the milk is introduced in not less than five equal increments in not more than 1½ hr. When testing, air-cooled condensers must operate in not less than 90°F ambient air, and water-cooled units at not less than 120 psi head pressure. For everyday pickup the tank should meet the above cooling rate for one-half the capacity and for every-other-day pickup should maintain the required rate of cooling for one-fourth the capacity of the tank.

From 18 to 24 in. clearance on three or four sides of the bulk milk tank is required in most milk sheds.

Milk can be removed from bulk tanks and placed in cans by gravity feed if the tank is placed on legs. It can also be removed with the vacuum from the milker or with a separate milk pump.

Water Heater. A source of hot water in the milkhouse is required in practically all dairy milk sheds for washing and sanitizing of milking equipment and of the employees. Warm water may be used for drinking by calves and other livestock.

Nonpressure water heaters commonly have a capacity of 10 gal and require from 250 to 1,500 watts on a 115-volt service. They are inexpensive and do not require special plumbing, wiring, or water under pressure. However, because of the slow flow, slow reheat, and the possibility of contamination when opening the lid on top of the unit, it is not recommended for dairy use where over 15 gal of hot water is needed per day.

The *pressure type* is acceptable because it is automatic and relatively inexpensive to operate. A separate disconnect switch with fuses should be provided. The heater should be located as close as possible to the place where the hot water will be used, and all hot-water piping should be insulated. Practically all electrical farm water heaters are of the storage type, as distinguished from the continuous type.

The recovery of a water heater is the number of gallons it will raise from 40 to 140°F in 1 hr. The recovery of electric water heaters is about 4 gal per hr for each 1,000 watts of heater element. One kilowatt of electricity will heat 4.1 gal of water 100°F. About 75 per cent of the total capacity of an electric hot-water tank can be removed at nearly constant temperature. Quick-recovery electricity or gas units are available.

The cost of operation of an electric water heater can be reduced by using an off-peak control on the lower element. These rates specify a minimum size, usually 30 gal.

The size of water heater can be determined on the basis of the number of animals in the herd, often 1 to 1½ gal per cow plus additional miscellaneous needs, or the number of units of water required for various purposes. A single milker head requires one unit; strainer, pails, calf watering, ¼ unit; udder wash and machine rinse per cow, ¹⁄₁₀ unit. By determining the total number of units, the size of water heater can be selected on the basis of a capacity of 3 gal at 155°F for each unit.

Other types of water heaters use oil, natural gas, and bottled gas. These types require a flue connection for exhaust gases and should be installed according to the regulations of the insurance inspectors. Gas units should meet the standards of and be installed according to the recommendations of the American Gas Association. For a comparison of cost of heating water with different fuels, see Table 51-8.

Pasteurization of Milk. Milk *pasteurization* consists in killing most of the bacteria (about 95 per cent), including all those which are pathogenic (disease-producing). *Sterilization* is the process of killing all the microorganisms, but cannot yet be done without producing off-flavors. Milk pasteurization can be done by either a batch or a continuous process. In the batch or holding process, milk is heated to at least 143°F and held for at least 30 min. In the continuous process, it is brought to at least 160°F for at least 15 sec. For home pasteurization of milk by the holding method, a small 1- or 2-gal container is electrically heated with from 300 to 1,250 watts. The milk is placed in a container surrounded with water and heated by a thermostatically controlled immersion heater with a timer. About ½ kwhr per gal of cold milk (40°F) is required for home pasteurization. After pasteurization, the milk should be cooled quickly.

The holding method is generally employed by small processing centers. Hot water or steam is most often used. An electrical-resistance batch pasteurizer with 230 volts and 31 amp for a 12-gal synthetic or rubber-lined tank and using graphitized carbon electrodes has found limited use. It uses 0.3 kwhr per gal for pasteurizing from a starting temperature of 38°F. The continuous method (high-temperature–short-time pasteurization) should be considered for over 10,000 lb per day.

TABLE 51-8. COST OF HEATING 100 GAL OF WATER THROUGH 100°F

Fuel	Unit	Efficiency, per cent	Range in unit cost, cents	Btu per unit	Estimated range of cost per 100 gal, cents
Electricity	Kilowatthour	99	1–2½	3,415	25–62
Coal	Pound	56	0.4–0.7	12,500	5–9
Oil	Pound	60	1–3	18,000	8–24
Liquefied petroleum	Pound	88	6–10	21,650	27–45
Natural gas	Cubic foot	85	0.04–0.12	1,060	4–11

Sanitizing Dairy Equipment. Keeping dairy equipment clean and relatively free from bacteria is a major requirement in dairy operation. The cleaning and sanitizing of dairy equipment consists of the following steps:

1. *Prerinse.* After removing milk, flush equipment with lukewarm water to remove all milk residue.
2. *Wash.* Prepare washing solution, at 120 to 130°F, according to manufacturer's instructions, usually a chlorinated detergent, and brush all parts.
3. *Rinse.* Rinse all parts thoroughly with warm water to remove washing solution.
4. *Sanitize.* The equipment is sanitized just *before use* with a chlorinated solution (about 200 ppm), according to manufacturer's recommendations.

LAMPS AND LIGHTING

Karl A. Staley and J. P. Ditchman

Good lighting prevents accidents by enabling the farmer to see clearly and quickly. It also helps him to spot disease in his stock. It speeds up the chores around the farm and assists in maintenance, such as the repair of farm machinery. The better the light, the less time it takes to do a given piece of work.

A great number of farm buildings and work areas are poorly lit because good farm lighting has not been generally understood and because the eye is actually a poor judge of light.

The cost of light includes the cost of electric power, the cost of lamps and lighting equipment, and the labor to install and maintain the system. The principal factors—power cost and lamp cost—have been reduced steadily during the past two decades.

Modern lighting equipment gives the farmer a wide choice of bulbs, methods, and fixtures to solve lighting problems. Light bulbs are available with an inside white coating, which cuts glare. The lamps are white in appearance and are called *de luxe white*. The bulbs are coated on the inside with silica powder. Other types include *reflector* bulbs, which have parabolic shape and are silvered on the inside. They can be used to spotlight or floodlight any area. Another useful type is the *projector* lamp, made of thick glass. This lamp in most sizes will withstand weather without further protection. The larger sizes (200, 300, 500 watts) are available in narrow-, medium-, and wide-beam designs. Data for reflector and projector lamps are shown in Table 51-8.

Infrared lamps are filament lamps designed primarily for producing heat for pig and lamb brooders, dressing poultry, and similar uses. They are very long lived. With reasonable care, they will last several times as long as conventional bulbs, although after several thousand hours' use, the tungsten filaments become brittle and are subject to mechanical breakage.

TABLE 51-9. REFLECTOR AND PROJECTOR LAMP DATA
(120-volt)

Manufacturer's abbreviation	Description	Bulb	Base	Max. over-all length, in.	Approx. initial lumens (candle-power)	Approx. beam spread
75R30/SP	Reflector spot, light, i.f.†	R-30	Medium	5⅜	770	50°
75R30/FL	Reflector flood, i.f.	R-30	Medium	5⅜	770	130°
150PAR/SP	Projector spot*	PAR38	Medium skirted	5⁵⁄₁₆	1,730	30° × 30°
150PAR/FL	Projector flood*	PAR38	Medium skirted	5⁵⁄₁₆	1,730	60° × 60°
150R/SP	Reflector spot, light i.f.	R-40	Medium	6½	1,780	40°
150R/FL	Reflector flood, i.f.	R-40	Medium	6½	1,780	110°
200PAR46/3NSP	Narrow spot*	PAR46	Medium side prong	4	2,350	17° × 23°
200PAR46/3MFL	Medium flood lens cover*	PAR46	Medium side	4	2,350	20° × 40°
300PAR56/NSP	Narrow spot*	PAR56	Mogul end prong	5	3,650	15° × 20°
300PAR56/MFL	Medium flood*	PAR56	Mogul end prong	5	3,650	20° × 35°
300PAR56/WFL	Wide flood*	PAR56	Mogul end prong	5	3,650	35° × 65°
300R/SP	Reflector spot, light i.f.	R-40	Medium	6½	3,700	35°
300R/FL	Reflector flood, i.f.	R-40	Medium	6½	3,700	115°
300R/SP/1	Reflector spot, light i.f.*	R-40	Medium	6⅞	3,700	35°
300R/FL/1	Reflector flood*	R-40	Medium	6⅞	3,700	115°
300R/3SP	Reflector spot, light i.f.*	R-40	Mogul	7¼	3,700	35°
300R/3FL	Reflector flood, i.f.*	R-40	Mogul	7¼	3,700	115°
500PAR64/NSP	Narrow spot*	PAR64	Extended mogul end prong	6	.....	13° × 20°
500PAR64/MFL	Medium flood*	PAR64	Extended mogul end prong	6	.....	20° × 35°
500PAR64/WFL	Wide flood*	PAR64	Extended mogul end prong	6	.....	35° × 65°
500R/3SP	Reflector spot, light i.f.*	R-40	Mogul	7¼	6,400	35°
500R/3FL	Reflector flood, light i.f.*	R-40	Mogul	7¼	6,400	115°

All lamps listed have an approximate life of 2,000 hr.
* Special glass bulb—heat-resistant.
† i.f. = inside frosted.

Fluorescent lighting is growing in use in farm buildings. The advantages of this light are that it provides softer light, which is easier on the eyes. The lamp is about four times as efficient as a filament bulb. Equipment is somewhat more complicated and heavier, but once it is installed, and particularly in places where long burning hours are the rule, fluorescent lighting is superior from the standpoint of cost.

For most purposes, the *rapid-start* lamp is the best choice in the smaller sizes (40 watts); for long runs where more light is required, *power-groove* lamps (4-, 6-, and 8-ft lengths) are used. These lamps produce about three times as much light per foot as the rapid-start. See Table 51-10 for data on fluorescent lamps.

For outdoor use in cold weather, fluorescent tubes need to be protected from drafts and should be used with equipment designed for outdoor service.

Uniform Lighting. When incandescent lamps are used alone, uniform lighting is accomplished by applying the "1:1½ rule." By this rule, lamps at the ceiling are spaced not over 1½ times their height above the floor. This would space light fixtures

10 to 12 ft apart in the ordinary dairy barn, for example, about one light for every third stall.

With fluorescent fixtures, which are 4, 6, or 8 ft long, it is often more economical to run continuous rows of lights, attaching each row to one outlet box. This is much better lighting, for one thing, and there is a saving in outlet cost over small fixtures spaced several feet apart. Continuous lighting avoids shadows. See Tables 51-10 and 51-11 for specific measurements.

Walls of farm buildings should be flat white, if possible. This keeps light in circulation and reflects it to all sides of the stock, avoids dark corners, etc.

Where there is a fire hazard, vaporproof lighting fixtures are recommended. They are generally also rated as dustproof, which further increases their value and application. Dustproof fixtures are the safest type to install in haylofts. They should be used

TABLE 51-10. FLUORESCENT-LAMP DATA

General-line								
Nominal watts...............	15	15	20	25	30	40	40	90	100
Length, in..................	18	18	24	33	36	48	60	60	60
Bulb.......................	T-8	T-12	T-12	T-12	T-8	T-12	T-17	T-17	T-17
Average lamp watts...........	15.0	14.1	19.7	26.0	30.0	39.0	41.0	90	99
Lamp current, ma.............	300	330	380	490	355	430	425	1,550	1,520
Lamp volts..................	55	45.5	56	60	98	100	101	62	68
Lumens, cool white...........	730	620	1,000	1,600	1,890	2,500	2,500	5,150	4,850
Lumens, warm white..........	760	650	1,030	1,660	1,930	2,600	2,600	5,300	5,150

Power-groove		
Length, in.................	48	72	96
Base.......................	Recessed double contact		
Av. lamp watts..............	107	155	200
Lamp current, ma............	1,500	1,500	1,500
Lamp volts.................	84	120	160
Lumens, cool white..........	6,000	9,300	13,000

	Rapid-start	High-output		
Length, in.................	....	48	72	96
Base.......................	Med. bipin	Recessed double contact		
Av. lamp watts..............	39.0	60.0	85.0	105
Lamp current, ma............	430	800	800	800
Lamp volts.................	100	80	115	148
Lumens, cool white..........	2,500	3,250	5,200	7,300
Lumens, warm white.........	2,600	3,350	5,350	7,500

in any location where grain or feed products are ground up or where dust is present in quantities in which a fire or explosion hazard exists. Lighting recommendations are summarized in Tables 51-13 and 51-14.

Portable Lighting for Farm Operations. Low-voltage lamps may be operated by tractor-mounted generators for night harvesting. Farmers in many states have used electric light for night plowing, cultivating, and spraying. Night vegetable harvesting and night dusting are practical with simple headlighting, but are better with tractor-mounted generators equipped with floodlights for large-area operations. Because the moisture content tends to be higher at night than during the heat of the day, night-harvested celery, lettuce, and sweet corn break more easily from the stalk. They also retain a fresher appearance when displayed on a market stand.

Night dusting also has a practical advantage since the film of moisture on plants provides a surface for the powder to cling to. Extra work on wood lots can be accomplished by a night shift with adequate lighting.

TABLE 51-11. ILLUMINATION BY LAMP CHANNEL SYSTEM—WALL TO WALL
(Bare Channel at Ceiling,[a] 30 Per Cent; R-F, Walls, 10 Per Cent;
Av. Foot-candles[b] 30 In. above Floor)

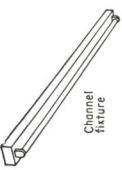
Channel fixture

Room A: 30 ft long, 15 ft wide; distance, floor to lamps, 8 ft; rows spaced 8 ft apart		Room B: 60 ft long, 30 ft wide; distance, floor to lamps, 10 ft; rows spaced 10 ft apart		Room C: 100 ft long, 50 ft wide; distance, floor to lamps, 12 ft; rows spaced 12 ft apart	
48″ Rapid Start lamp[c]—40 watts,[d] 2 rows,[e] 7 lamps/row	18	48″ Rapid Start lamp—40 watts, 3 rows,[f] 15 lamps/row	19	48″ Rapid Start lamp—40 watts, 4 rows,[g] 25 lamps/row	16
48″ High Output lamp—60 watts, 2 rows, 7 lamps/row	23	48″ High Output lamp—60 watts, 3 rows, 15 lamps/row	25	48″ High Output lamp—60 watts, 4 rows, 25 lamps/row	21
72″ High Output lamp—85 watts, 2 rows, 5 lamps/row	27	72″ High Output lamp—85 watts, 3 rows, 10 lamps/row	29	72″ High Output lamp—85 watts, 4 rows, 16 lamps/row	23
96″ High Output lamp, 2 rows, 3 lamps/row	22	96″ High Output lamp—105 watts, 3 rows, 7 lamps/row	26	96″ High Output lamp—105 watts, 4 rows, 12 lamps/row	23
48″ Power Groove lamp, 107 watts, 2 rows, 7 lamps/row	43	48″ Power Groove lamp—45 watts, 3 rows, 15 lamps/row	45	48″ Power Groove lamp—107 watts, 4 rows, 25 lamps/row	39
72″ Power Groove lamp—155 watts, 2 rows, 5 lamps/row	48	72″ Power Groove lamp—155 watts, 3 rows, 10 lamps/row	47	72″ Power Groove lamp—155 watts, 4 rows, 16 lamps/row	38
96″ Power Groove lamp—200 watts, 2 rows, 3 lamps/row[h]	40	96″ Power Groove lamp—200 watts, 3 rows, 7 lamps/row	46	96″ Power Groove lamp—200 watts, 4 rows, 12 lamps/row	40

[a] Average unpainted wood ceiling; if white, add 10 per cent.
[b] Values based on initial lumens—cool white lamps. Warm white, add 3 to 5 per cent. Maintenance factor, 70 per cent. For poor maintenance, reduce values 25 per cent.
[c] Same values for 48-in. general-service lamp.
[d] Lamp watts. For tulamp ballast watts add, for each 48-in. RS lamp, 6.5 watts; 48-in. HO, 5 watts; 72-in. HO, 15 watts; 96-in. HO, 27 watts; 48-in. PG, 8 watts; 72-in. PG, 30 watts; 96-in. PG, 40 watts.
[e] Spacing rows closer together raises fc values. For three rows, add 50 per cent; 4 rows, 100 per cent [same increase (100 per cent) for two-lamp channels, approximately].
[f] With four rows, add 33 per cent; 5 rows, 66 per cent; 6 rows, 100 per cent, approximately.
[g] With five rows, add 25 per cent; six rows, 50 per cent; eight rows, 100 per cent, approximately.
[h] For full-length row, one 72-in. PG lamp can be added. Average fc = 45.

TABLE 51-12. ILLUMINATION BY INDUSTRIAL REFLECTOR SYSTEMS—WALL TO WALL
(Semidirect Luminaire at Ceiling, 30 Per Cent; R-F, Walls, 10 Per Cent;
Av Foot-candles 30 In. above Floor)

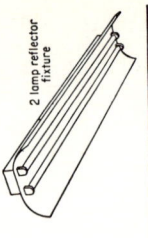

2 lamp reflector fixture

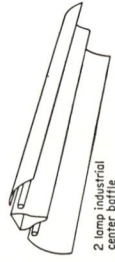

2 lamp industrial center baffle

Room: 30 ft long, 15 ft wide; distance, floor to lamps, 8 ft; rows 8 ft apart	
48″ Rapid Start lamp—40 watts, 2 rows, 7 lamps/row	26
48″ High Output lamp—60 watts, 2 rows, 7 lamps/row	34
72″ High Output lamp—85 watts, 2 rows, 5 lamps/row	39
96″ High Output lamp—105 watts, 2 rows, 3 lamps/row	33
48″ Power Groove—107 watts, 2 rows, 7 lamps/row	63
72″ Power Groove—155 watts, 2 rows, 5 lamps/row	60
96″ Power Groove lamp—200 watts, 2 rows, 3 lamps/row	50

Room: 60 ft long, 30 ft wide; distance, floor to lamps, 10 ft; rows 10 ft apart	
48″ Rapid Start lamp—40 watts, 3 rows, 15 lamps/row	25
48″ High Output lamp—60 watts, 3 rows, 15 lamps/row	33
72″ High Output lamp—85 watts, 3 rows, 10 lamps/row	35
96″ High Output lamp—105 watts, 3 rows, 15 lamps/row	74
48″ Power Groove—107 watts, 3 rows, 15 lamps/row	60
72″ Power Groove lamp—155 watts, 3 rows, 10 lamps/row	63
96″ Power Groove lamp—200 watts, 3 rows, 7 lamps/row	61

Room: 100 ft long, 50 ft wide; distance, floor to lamps, 12 ft; rows 12 ft apart	
48″ Rapid Start lamp—40 watts, 4 rows, 25 lamps/row	23
48″ High Output lamp—60 watts, 4 rows, 25 lamps/row	30
72″ High Output lamp—85 watts, 4 rows, 16 lamps/row	30
96″ High Output lamp—105 watts, 4 rows, 12 lamps/row	32
48″ Power Groove—107 watts, 4 rows, 25 lamps/row	55
72″ Power Groove lamp—155 watts, 4 rows, 16 lamps/row	54
96″ Power Groove lamp—200 watts, 4 rows, 12 lamps/row	57

TABLE 51-13. GENERAL LIGHTING RECOMMENDATIONS FOR FARM BUILDINGS

Location	Principal seeing task	Recommendations
Milkhouse*	Looking into containers	Locate fixtures at near edge of benches. Use large-area fixtures, corrosion-resistant
Barns	Care of stock—young, sick	Litter alley—continuous-row fluorescent or individual fixtures on close centers. Feed alley—space 1.5 times mounting height. Sheep barn—single row in narrow space; double in wide
General buildings for storing and processing feed; tool and machine sheds, grading, washing, packing	Grading	General-lighting fixtures over localized areas or points of work
Hay mow	Spotting moldy hay, wire, glass, nails	One or more dusttight fixtures. Extra light for chutes, stairs, ladders
Poultry houses, hen, brooder,* feed rooms	Morning or evening extension of daylight, inspection	Two 60-watt filament bulbs or equivalent per 20' × 20' henhouse; two 25-watt bulbs for all-night lighting; brooder, one 60-watt bulb
Silo	Seeing inside and on ladder	One PAR or reflector bulb at top of chute, tilted into silo
Shops—machine, forge, power tools, grinding	Power-tool work	Individual lights over each machine; general lighting according to best industrial lighting practice

* See Chap. 47, Livestock Production Facilities.

TABLE 51-14. FOOT-CANDLE RECOMMENDATIONS FOR INDUSTRIAL TYPES OF FARM WORK

Task	Foot-candles	Task	Foot-candles
Parking, protective	2–5	Fine woodworking, metal processing, power tools, repairing	50–100
Rough storage, stock storage, packing	5–10	Extra-fine inspecting, close detail, very critical seeing	200–500
Rough work, woodworking, foundry, forge work	30–50		

TABLE 51-15. INFRARED-LAMP DATA

Lamp watts	Manufacturer's abbreviation	Bulb	Base	Voltage range	Design volts	Over-all length, in.
125	125R40	R-40	Medium skirted	115–125	115	7¼
250	250R40/1	R-40	Medium screw	115–120	115	6½
250*	250R40/10	R-40	Medium screw	115–125	115	6⅞
250†	250PS30/33	PS-30	Medium screw	115–125	115	8 1/16

* Bulb of special heat-resistant glass.
† Brooder lamp with reflectorized pear-shaped bulb.

Insect Control. Electric lights have been successfully used to attract insects to electric traps. Corn borer moths were most attracted by a blue light in the near ultraviolet range with a peak wavelength of 3,250 angstrom units.[1]

Infrared Heating. Infrared heat lamps provide efficient and easily controlled heat for brooding of chicks and pigs.[2] Infrared lamp data are given in Table 51-15.

Radiation. The effects of various types of radiation on seeds, plants, insects, and animals have been investigated.[3]

[1] J. G. Taylor and H. O. Deay, Electric Lamps and Traps in Corn Borer Control, *Agr. Eng.*, 31:503–532, October, 1950. J. G. Taylor, II. O. Deay, and M. T. Orem, Some Engineering Aspects of Electric Traps for Insects, *Agr. Eng.*, 32:496–498, September, 1951. H. H. Beaty, J. H. Lilly, and D. L. Calderwood, Use of Radiant Energy for Corn Borer Control, *Agr. Eng.*, 32:421–426, August, 1951.

[2] V. H. Baker and J. H. Bywaters, Brooding Poultry with Infrared Energy, *Agr. Eng.*, 32:316–320, June, 1951. J. M. Stanley and V. H. Baker, Control Equipment for Infrared Poultry Brooding, *Agr. Eng.*, 34:751–753, November, 1953. J. G. Taylor et al., Fundamentals of Infrared Brooding of Pigs, *Agr. Eng.*, 33:213–215, April, 1952.

[3] R. S. Sheetz, The Application of Ultraviolet to Poultry and Livestock, *Agr. Eng.*, 32:208–210, April, 1951. V. H. Baker, Oscar Taboada, and D. E. Wiant, Lethal Effects of Electrons on Insects Infesting Flour and Wheat, pt. 1, *Agr. Eng.*, 34:755–758, November, 1953; pt. 2, *Agr. Eng.*, 35:407–410, June, 1954. D. T. Kinard and D. E. Wiant, Application of High-frequency Electricity to Young Chickens, *Agr. Eng.*, 35:865–869, December, 1954. V. H. Baker, D. E. Wiant, and Oscar Taboada,

HEAT PUMPS

Chester P. Davis and Ralph I. Lipper

Farm Applications. Major consumer acceptance of the heat pump has been in the field of commercial and residential space heating and cooling (Figs. 51-3 and 51-4). While the number of proven agricultural uses is not large, the heat pump has

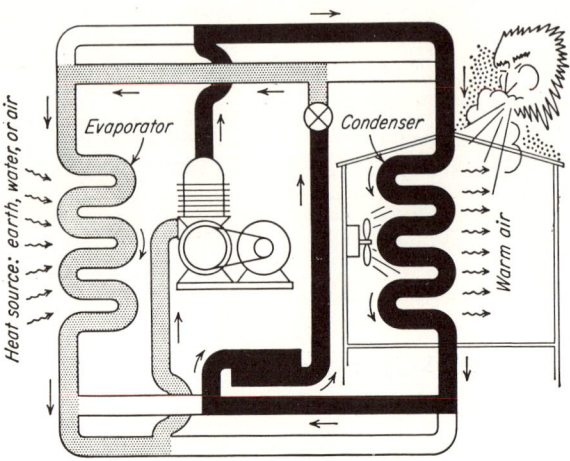

Fig. 51-3. Heat-pump flow diagram for heating.

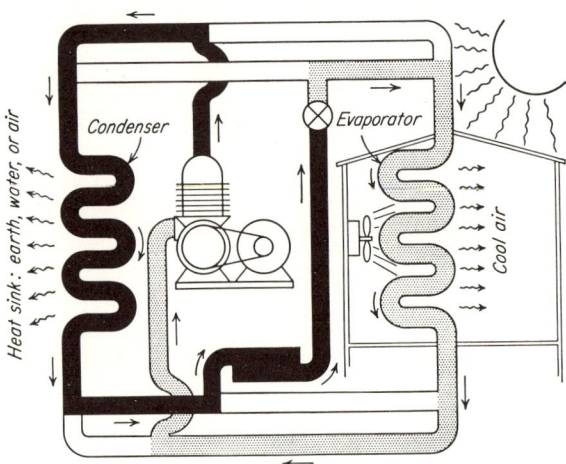

Fig. 51-4. Heat-pump flow diagram for cooling.

demonstrated functional feasibility for meeting specific requirements of air temperature and humidity.[1] Its most promising fields of application are where simultaneous utilization of the heating and cooling potentialities can be realized, where fully automatic operation over a range of temperatures involving both heating and cooling is desired, or where specialized control of temperature and humidity is required.

Effects of Electromagnetic Energy on Plants and Animals, *Agr. Eng.*, 36:808-812, December, 1955. O. A. Brown, R. B. Stone, Jr., and Henry Andrews, Low Energy Irradiation of Seeds, *Agr. Eng.*, 38:666-669, September, 1957. L. H. Soderholm, Effect of Dielectric Heating and Cathode Rays on Germination and Early Growth of Wheat, *Agr. Eng.*, 38:302-307, 1957.

[1] Chester P. Davis, Jr., Possible Farm Applications of the Heat Pump, *Agr. Eng.*, 34:323-325, 1953. Harold A. Cloud, A Study of the Possibilities of Heat Pump Applications in Agriculture, *Am. Gas Electric Serv. Corp. Rept.* 555, Nov. 15, 1954.

The heat pump has been used for environmental control in animal shelters. It has been applied as a dehumidifier in grain conditioning.[1] Supplemental cooling in conjunction with mechanical ventilation has extended storage life of vegetables, resulting in more orderly marketing. Initial cost and other limitations have restricted its use to certain seed grain and other special drying applications.

Development of water-heating applications has been under way for several years[2] (Fig. 51-5).

Milkhouse and dairy water heating in combination with milk cooling has been demonstrated to be economical on a research basis. Limited farm use has resulted.

Criterion of Effectiveness. The theory of heat-pump operation is simply that of the common vapor-compression refrigeration cycle.

Where heat rejection through the condenser for heating is the desired effect, the ratio of the heat rejected to the heat equivalent of the compressor work is called the

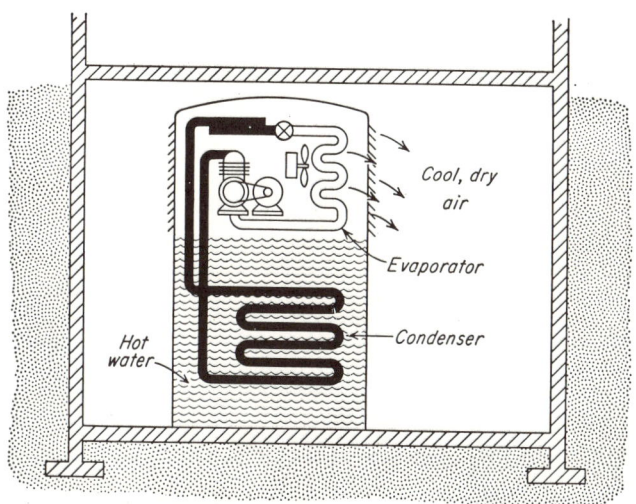

Fig. 51-5. Heat pump heating water and cooling air.

coefficient of performance (CP). The CP is greater for heating than for cooling since mechanical shaft work used to operate the compressor is rejected at the condenser as useful heat.

The Carnot cycle is a useful criterion for evaluating the inherent limitations of any heat-pump cycle. The Carnot CP for cooling is

$$CP = \frac{T_e}{T_c - T_e}$$

where T_e = evaporation temperature, °F abs.
T_c = condenser temperature, °F abs.

For heating,

$$CP = \frac{T_c}{T_c - T_e}$$

The Carnot CP for both heating and cooling decreases as the spread between evaporator and condenser temperatures increases. This is also true in practical cycles and is an important consideration in every heat-pump application.

[1] A. M. Flikke, H. A. Cloud, and A. Hustrulid, Grain Drying by Heat Pump, *Agr. Eng.*, 38:592-597, 1957. Chester P. Davis, Jr., "A Study of the Adaptability of the Heat Pump to Drying Shelled Corn," M.S. thesis, Purdue University, Lafayette, Ind., 1949.
[2] Chester P. Davis, Jr., "Heat Pump Water Heater Operational Tests," Twenty-ninth Annual Report of the Kansas Committee on the Relation of Electricity to Agriculture, Kansas State Univ., Department of Agriculture Engineering, 1953.

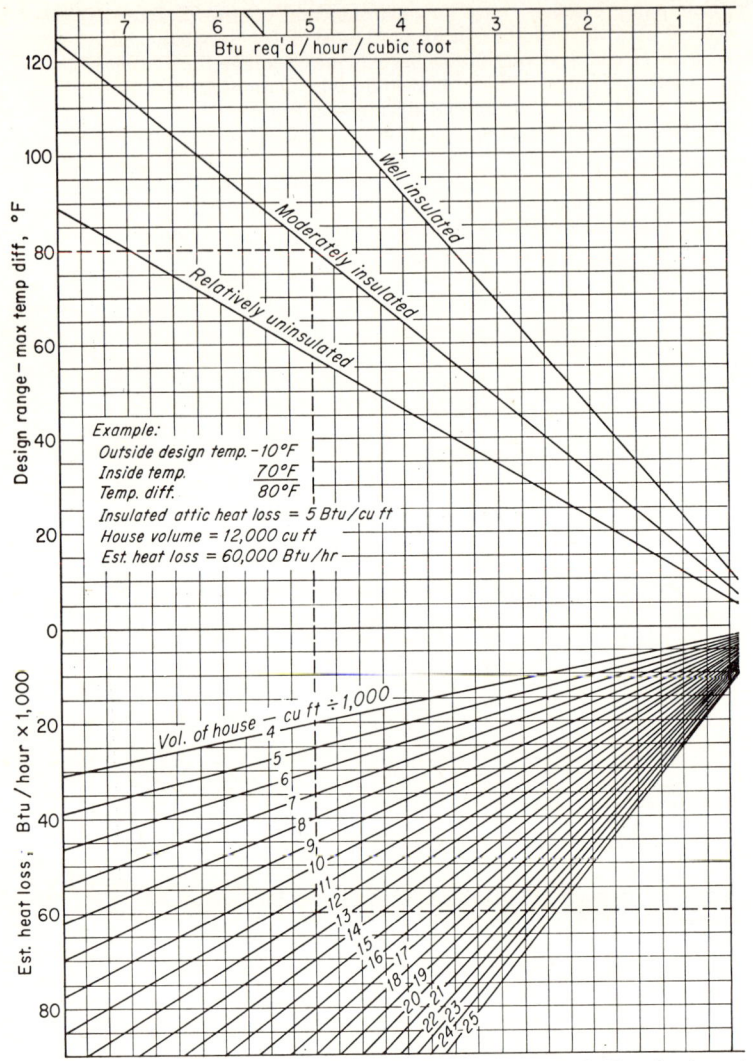

Fig. 51-6. Home-heating estimation

The actual CP based on the ratio of heat received or rejected to the heat equivalent of mechanical input may be computed for a given heat pump operating under stated conditions. In practice, another term is often required to rate over-all seasonal or period performance under existing variable conditions. Performance factor (PF) expresses the average performance for actual extended operation. For the water heater this may also be stated mathematically,[1] as follows:

$$PF = \frac{8.34(\text{gal})(T_o - T_i)}{3{,}413(\text{kwhr})}$$

where $T_o - T_i$ = weighted temperature differential between water entering and leaving the heater, °F

[1] Ibid.

ELECTRICAL EQUIPMENT

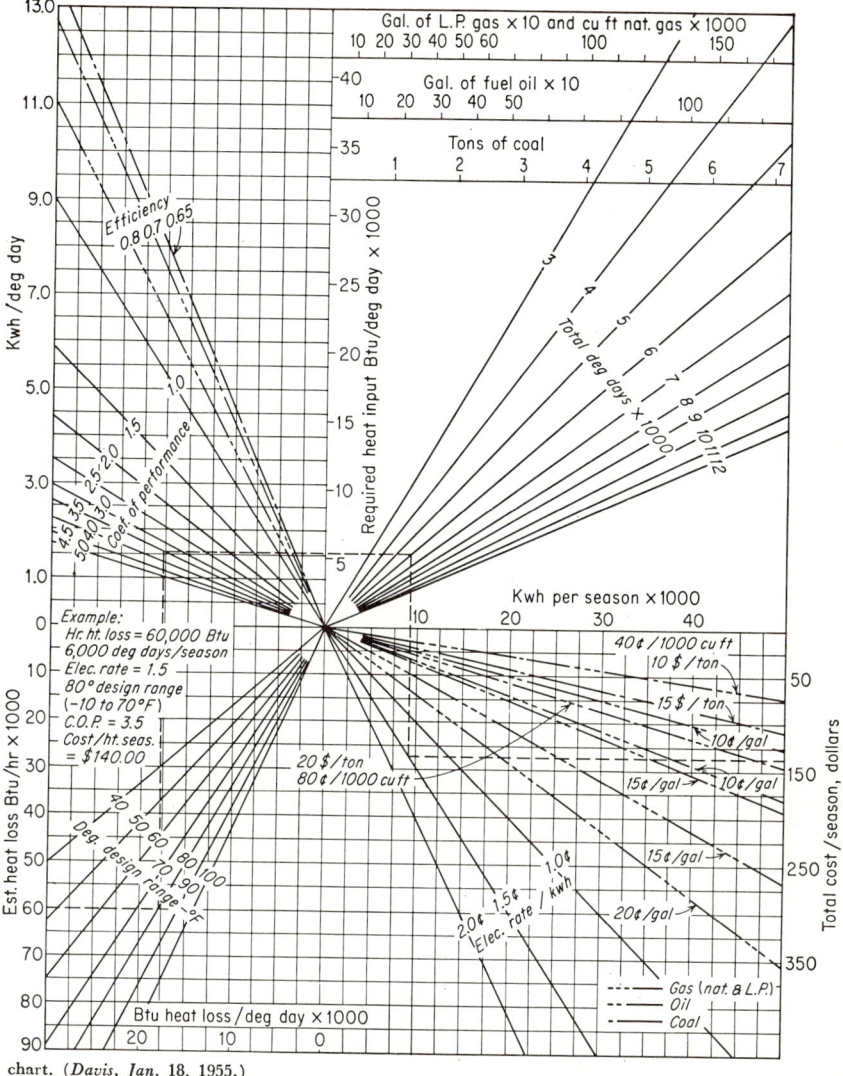

chart. (*Davis, Jan. 18, 1955.*)

Heat Sources and Heat Sinks.[1] Selection of a satisfactory heat source or heat sink, or both, is one of the major problems in any heat-pump design. Where water is available in sufficient quantities, its re-use for yard and garden irrigation or for livestock water may increase desirability.

Estimating Operating Cost. Figure 51-6 with the example calculation shows how to compare space heating cost with that of combustion-type equipment. Local information as to inside and outside design temperature in degree-days (Chap. 57), combustion efficiency and unit costs of fuels, and electric rates is required for cost

---

[1] Pertinent characteristics of various heat sources and sinks are listed in the "Heating, Ventilating, Air Conditioning Guide," chap. 37, published annually by The American Society of Heating & Air Conditioning Engineers, Inc., New York, 1955.

calculations. For electric-resistance heating, the coefficient of performance is 1.[1] Performance factors for heat pumps of current design should range from 2.5 to 3.25 for water sources and from 1.5 to 3.0 for air sources, depending on local conditions.

ELECTRIC HEATING

R. W. Kleis

There are several unique features of electric heating which are responsible for its extensive use in agriculture and related industries.

1. Ease and accuracy of control. Response to controls is practically instantaneous. Temperature may be maintained within narrow limits with rather simple control devices. Heat may be applied within a wide range of intensity.

2. Electric-heat generation does not involve combustion. The storage and handling of fuel and the disposal of products of combustion are eliminated.

3. Electric heating is clean and quiet.

4. The heating may be distributed, confined, localized, or subdivided as desired.

5. Equipment involved is compact and relatively inexpensive, may be portable, and may be installed in remote or concealed locations.

6. Efficiency is for all practical purposes 100 per cent.

7. Safety results from absence of combustion, combustible materials, and lethal products of combustion.

Methods of Electric Heating. Any one or combination of the three components of electrical impedance may be employed in the transformation of electrical energy into heat energy. Resistance, inductance, or capacitance may be utilized in the various methods of electric heating, which are:

1. Resistance heating (a-c or d-c)
2. Dielectric heating (high-frequency a-c)
3. Induction heating (high-frequency a-c)
4. Electric-arc heating (a-c or d-c)

The power transformed into heat by any of these methods is proportional to I^2R, where I is the current passing through a resistance R. The differences are in the nature of currents and resistances and in the physical arrangement involved.

Resistance Heating. This method of converting electrical energy into heat may be either direct or indirect. *Direct resistance heating* is accomplished by passing a current through the material to be heated. *Indirect resistance heating* converts electrical energy into heat in resistance units in a circuit apart from the material to be heated, and the heat is transferred by one or a combination of the three heat-transfer methods into the material.

The electrical energy converted into heat in resistance heating is

$$W = EI = \frac{E^2}{R_T} = I^2 R_T$$

and $\quad q = 3.413 \times W$
or $\quad Q = 3.413 \times whr = 3,413 \times kwhr$

where W = power, watts
I = current, amp, through resistance (effective current in a-c circuits)
R_T = resistance of heating unit at operating temperature
E = actual voltage applied across R_T
q = rate of heat generation, Btu per hour
Q = total heat generated, Btu
whr = watthours
$kwhr$ = kilowatthours

Metals and Alloys Used in Resistance-heating Units. It is desirable that a material for electric-resistance heaters have a high electric resistance, low temperature co-

---

[1] It is recommended on the basis of field experience and NEMA studies that estimated hourly heat-loss results obtained be multiplied by approximately 0.8 since individual room control of thermostats allows economy in operation.

efficients of linear expansion and electric resistance, high resistance to oxidation, high melting point, high mechanical strength, and reasonable cost.

Pure metals in general do not satisfy these requirements, and alloys of various types are used. Nickel-chromium alloys are the most generally used, while ferro-nickel, copper-nickel, and other alloys are used to a lesser extent.

The heat generated in a heating element is

$$q = 3.413I^2R_T = 3.413I^2R[1 + K(T - 20)]$$

where T = operating temperature of heating element, °C
R_T = resistance of element at operating temperature, ohms
R = resistance at 20°C, ohms
K = temperature coefficient of resistance per °C

The operating temperature for any given unit is dependent upon the rate of heat dissipation into the surrounding medium.

The *efficiency* of any electric process is

$$E = \frac{\text{net energy absorption (100)}}{\text{electric energy supplied}} = \frac{WC\,\Delta T}{3{,}413 \text{ whr}}$$

where W = weight of material, lb
C = specific heat, Btu/(°F)(lb)
ΔT = temperature rise of material, °F

Dielectric Heating. Poor conductors are heated internally by the application of a high-frequency dielectric field across the material. The material to be heated (charge)

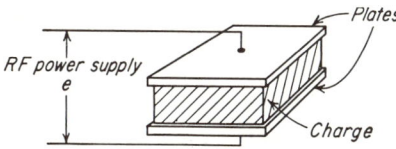

Fig. 51-7. Dielectric-heating arrangement.

forms the dielectric in a capacitor in a radio-frequency power circuit as shown in Fig. 51-7.

The *power* converted to heat in dielectric heating is

$$P = 1.41f\left(\frac{e}{t}\right)^2 K \cos\theta \times 10^{-12} \quad \text{watts/cu in.}$$

where f = frequency, cycles per sec
e = applied voltage
t = thickness of charge, in.
K = dielectric constant of material
$\cos\theta$ = power factor of material

The term e/t is often referred to as the *voltage gradient* and $K\cos\theta$ as the *loss factor*. Dielectric data for agricultural products are given in Table 51-16.

TABLE 51-16. Some Typical Values of Dielectric Constants and Power Factors for Agricultural Materials

	Dielectric constant K	Power factor $\cos\theta$
Air (dry)	1	0.00
Hay (dry)	1.8 (est.)	0.10
Water	81	
Wheat (13.6% moisture content)	5.51	0.10
Wood (air dry)	2.5–7.7	0.03

The voltage applied across the plates should as a rule not exceed 15,000, and the voltage gradient is normally between 2,000 and 5,000 volts per inch. These values are limited by the resistance of the materials involved to arcing and breakdown under high voltages.

Frequencies used in dielectric heating normally are in the range of 10 to 30 Mc. If the loss factor were independent of frequency, the power would increase directly as frequency. This is not the case, however, and optimum frequency is dependent on the electrical properties of the material to be heated.

The power-supply unit is usually a power r-f oscillator. A common unit is a grounded-grid Hartley oscillator.

Shielding of dielectric-heating apparatus is often necessary to prevent interference with communications systems. It is recommended that reference be made to rulings of the Federal Communications Commission.

Measurement of the *temperature* of the charge at any time is extremely difficult. Any unit placed in the charge for this purpose will be affected differently and independently by the electrostatic field and result in erroneous readings. Temperature

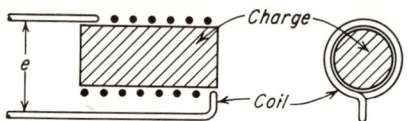

FIG. 51-8. Induction-heating arrangement.

must usually be calculated from time, the electrical parameters of the circuit and charge, and the thermal properties of the charge.

The *efficiency* may be calculated from the equation given above but is normally about 50 per cent.

Induction Heating. In induction heating the conductor to be heated is exposed to a high-frequency electromagnetic field. Heating results from the flow of eddy currents within the material and is sometimes called "eddy-current heating." The charge forms the core of a coil in the high-frequency power circuit, as shown in Fig. 51-8.

Because of the skin effect, the eddy currents are of greater magnitude near the surface of the charge. The degree of this concentration of current near the surface is a function of the frequency, and therefore the depth of heating can be controlled by appropriate selection of frequency.

The *power*, or heat-generating rate in induction heating, is

$$P = 0.126 \left(\frac{\pi N}{L}\right)^2 I^2 S(UfPa)^{1/2} \times 10^{-4} \quad \text{watts}$$

where N = number of turns in coil
L = length of charge, in.
I = current in coil
S = shape factor of charge
U = permeability of charge
f = frequency, cycles per sec
Pa = resistivity of charge, ohm-cm

The shape factor S is dependent upon the relationship between the physical dimensions of the charge, the electrical properties of the charge, and the frequency.[1]

Three types of *power supplies* are used for induction heating: high-frequency alternators (480, 960, 3,000 and 9,600 cycles per sec), spark-gap units (20 and 40 kc per sec), and electronic power oscillators (300 to 450 kc per sec). For economic reasons the one of the above methods and frequencies closest to the calculated optimum frequency is generally used.

As in dielectric heating, the efficiency of an induction-heating process can be calculated, but it is often about 50 per cent.

[1] Procedures for arriving at this factor are discussed in the "Standard Handbook for Electrical Engineers," 9th ed., McGraw-Hill Book Company, Inc., New York, 1957.

Arc Heating. Arc heating is used commonly in the production of alloys and in arc welding. Electric-arc furnaces for alloy production utilize either three-phase or single-phase alternating current, and the arc is maintained between either carbon or graphite electrodes and the charge being melted. The efficiency of arc furnaces is normally between 75 and 90 per cent, with a power factor of 70 to 90 per cent.

In arc welding an arc is maintained between a single electrode and the work. The size and composition of the electrode are dependent upon the size of the work, metals involved, position of joint, and purpose of weld. Filler metal from the electrode is

TABLE 51-17. SOME AGRICULTURAL APPLICATIONS OF ELECTRIC HEATING *
(Methods and Wattages)

Application	Heating method					Wattage demand
	Indirect Res.	Direct Res.	Dielectric	Induction	Arc	
Arc welder................................	...	...	...	...	E	2,000–3,000
Beehive heater............................	C	...	P	...	...	75–100
Calf dehorner.............................	C	P	...	...	P	200–300
Chick brooder.............................	E	...	...	...	...	2–4 per chick
Eave-trough de-icing....................	E	...	...	...	...	20 per ft
Engine heaters............................	C	P	P	...	...	1,000–1,500
Feed cooker...............................	C	P	P	...	...	
Greenhouse heater.......................	E	...	...	...	...	
Heating cable on water pipes..........	E	...	...	...	...	3–7 per ft
Home milk pasteurizer..................	C	P	P	...	...	300–1,250
Hotbed heater............................	C	P	...	...	...	5–13 per sq ft
Incubator..................................	E	...	...	...	...	½ per chick
Lamb brooder.............................	E	...	...	...	...	250 per lamb up to 10
Lamb docker..............................	C	P	...	...	P	200–500
Milkhouse heater.........................	E	...	...	...	...	1,000–3,000
Moisture tester (drying)................	C	P	P	...	...	500–5,000
Pail-type water heater..................	E	...	...	...	...	1,500
Paint remover.............................	E	...	...	...	...	
Pig brooder................................	E	...	...	...	...	125–250 per litter
Pipe thawing..............................	L	C	...	...	...	2,000–3,000
Plastic-container sealing...............	E	...	...	...	...	60–500
Poultry debeaker.........................	C	...	...	...	P	200–500
Poultry scalder...........................	E	...	...	...	...	5,000
Pumphouse frost protection............	C	P	...	...	...	100–500
Sidewalk heating.........................	E	...	...	...	...	50 per sq ft
Soil sterilizer.............................	C	P	P	...	...	3,000
Soldering gun.............................	E	...	...	...	...	
Soldering iron.............................	E	...	...	...	...	80–500
Stock-tank de-icer.......................	E	...	...	...	...	250–1,000
Storage heating..........................	E	...	...	...	...	
Utensil sterilizer.........................	C	P	P	P	...	1,000–6,000
Water heater (pressure)................	E	...	...	...	...	1,500 and up
Work-area heating.......................	E	...	...	...	...	250–3,000

* C = common usage; E = used entirely; L = limited usage; P = potential usage.

deposited on the molten base metal. The "penetration" of an arc weld is the depth to which the base metal is melted.

Either alternating current or direct current may be used in arc welding. Alternating-current units utilize a transformer and have an operating efficiency 85 to 90 per cent. Direct current is usually supplied by a motor-generator set with an over-all operating efficiency of 55 to 60 per cent. Welders are rated in ampere capacity, and a 180-amp unit is considered the minimum practical size for a farm shop.

A HOME FARM SHOP

T. J. WAKEMAN

It is almost impossible to operate a farm successfully without some type of farm shop. Many times its use has meant the difference in profit and loss in the farm business.

Unless the farmer has experience beyond the average, he should not attempt to perform specialized jobs such as rebuilding a tractor engine, checking the refrigerant in a cooling system, or adjusting the valves on his farm truck.

Providing a Farm Shop Building. Careful study of the following points is essential in either planning a new building or altering an existing building.

1. *Location.* The shop should be located near or attached to the farm machine-storage shed. It should be at least 150 ft from any other building (except the machinery shed) and so located that more machinery-storage space can be added when needed. The building should face the south on a well-drained location.

2. *Size.* The shop should be large enough to house the necessary shop tools and equipment and to accommodate the farm machinery, one piece at a time, and the construction and repair jobs usually done on the farm. Most farms need a shop 24 by 26 ft, 26 by 32 ft, or 32 by 40 ft; some need shops that are larger.

3. *Height.* The height to the bottom of the lintel above the equipment entrance should be at least 10 ft, more if trucks with cattle frames are to enter the shop. The machine shed should be the same height as the shop unless trucks loaded with baled hay, crates, etc., or machinery such as self-propelled combines are to be sheltered. In such a case there should be one section or more of sheds adjoining the shop constructed to meet these special needs.

4. *Depth.* The average 1½-ton farm truck with a 14-ft bed requires 24 to 25 ft of space. This does not allow any space for walking behind the truck or for a work bench in front. To work on the average farm truck with the shop door closed, the depth should be 28 ft.

5. *Width.* The shop should be wide enough to give a 2-ft 0-in. space on each side of the entrance door plus space for power equipment. The large entrance panel door should never be placed flush with sidewalls if work benches are to be placed along the walls.

6. *Windows.* Windows should be located in at least two walls to provide uniform lighting. This will also permit cross ventilation for fresh air and the removal of smoke and fumes.

7. *Doors.* There should be one large door for equipment entrance and one small door in each end or in alternative corners of the shop to prevent being trapped in case of a flash fire. One entrance door may be placed in the equipment entrance door. Never place a sliding door on the inside of the shop, as it takes up wall space. Do not have two equipment doors. Besides losing the wall space occupied by the doors, the space between the doors is usually void if both entrances are used.

8. *Floor.* The floor of the shop should be of 4-in.-thick concrete using a 1-2¼-3 mixture of cement, sand, and gravel. Three inches of gravel or cinders should be used under the concrete to permit drainage. Well-packed dirt will be satisfactory for a temporary floor in the machinery shed.

9. *Water.* Water is used in cooling hot metal, filling radiators, cleaning tools, fire protection, and for washing after working in the shop. It should be provided at a flat-rim sink, one faucet being a hose bib with garden hose connected to it at all times. The water pipe leading to the shop should have a "stop-and-waste" valve installed so that the pipe in the building can be drained in extremely cold weather.

10. *Heat.* In providing for heat, so that work can be done in cold weather, a chimney may be necessary since coal, wood, natural gas, or oil heaters are normally used. The chimney should be placed on the outside of the building or in the wall to give a straight inside wall surface.

11. *Electrical Equipment. Service equipment.* The shop should be provided with a 3-wire, 230-volt, single-phase service. The feeder wires to the shop should be large enough to serve the anticipated load with a voltage drop of not more than 2 per cent. Under no circumstances should the feeder wires or the entrance cable be smaller than No. 8 wire.

Entrance panel. The panel should be no less than 60-amp fused switch or circuit breaker to provide separate circuits for lighting, convenience outlets, power equipment, and range outlets for welders.

Light circuits. A maximum of six 300-watt bulbs should be placed on a circuit; 50 foot-candles is needed for bench work, 30 foot-candles for rough work. If reflectors are used, 150-watt bulbs take a 12-in. reflector, 200 watts take a 14-in. reflector, 300 watts take a 16-in. reflector. The reflectors should be placed high enough to clear the tallest piece of equipment. Spacing should be on 8- or 10-ft centers, starting 4 or 5 ft from the wall, making a total of 16 to 25 sq ft of floor space. If fluorescent lighting is used, compressed air should be available for cleaning.

Convenience outlets. Do not place more than seven outlets on a circuit. The 115-volt duplex convenience outlets should be located around the shop walls at a maximum distance of 12 ft. An outlet mounted on the side of the work bench is useful for plugging in portable tools and other smaller equipment. One outlet should be provided for each motor under ½-hp capacity. An outlet near each door is useful for plugging in extension lights and portable tools for outside work. An outlet should also be provided in each section of the machinery shed.

Power outlets. Each power outlet should be fused from individual wall switch boxes. Each motor should be equipped with a built-in automatic overload switch. A separately fused 230-volt outlet should be provided for each motor over ½-hp capacity. One three-wire polarized outlet should be provided near the large entrance door, and one near the small door leading into the machinery shed for the arc welder. This allows welding to be done in the shop, outside the shop, and in the machinery shed.

12. *Fireproofing.* There is the ever-present danger of fires in any farm shop. Open flames from gas welding, stoves, and the forge, along with sparks from arc welding and motors, contribute to the causes of fire. There may be shavings, rags, fuel, and grease around the shop that will burn rapidly. A structure made of masonry block and ceiled with asbestos board or metal eliminates much of the risk of great losses when fires do occur. These materials are inexpensive and can be used in building a shop by the average farmer.

Arranging Equipment. All shops should be planned on scaled paper, using blocks to represent machines to obtain the proper arrangement.

Generally the equipment is placed around the wall (except the woodworking equipment), leaving the center of the shop open. Leave space in front of the large entrance door for a service area. Place the 2- by 8-ft benches against the wall. The woodworking power equipment is placed in a 5- by 5- by 5-ft triangle if three pieces or more are used. If it is a large shop, the triangle may be 10 by 10 by 10 ft. Locate the equipment so that long boards can be planed and sawed as well as panels of plywood. A jointer, band saw, or planer may be located next to a post, building support, or wall.

Locate the metalworking equipment together. The arc welder is usually placed on one side of the combination welding table, and the acetylene on the other. Place the 10-in. grinder near the welding area. Place the tool-fitting grinder on a pedestal in the open space so that a 7-ft 0-in. mower knife can be sharpened.

ELECTRIC FARM EQUIPMENT DATA

J. P. Schaenzer

Introduction. The figures and operating data on electric farm equipment presented in Table 51-18 fall within the limits of average farming conditions. They may be larger or smaller, depending upon the size of the enterprise and the desires or wants of the farmer.

Explanation of Data. Kilowatthours cannot always be calculated for electric-motor-driven equipment even when the time of use and the horsepower of the motor are known, since it may be overloaded, loaded, or underloaded. In cases such as barn cleaning and wagon unloading, the power requirements decrease as the manure is removed from the barn or the hay or grain unloaded. Energy consumption also depends upon ambient temperature, for example, hotbed heating. Kilowatthours usually decrease per unit with the size of the enterprise. As an illustration, the larger incubators take fewer kilowatthours per 1,000 eggs set than the smaller farm units.

For a given piece of equipment the kilowatthours may be much higher or lower than the national average. This may be due primarily to the size of the farming enterprise and climate.

TABLE 51-18. APPLICATIONS OF ELECTRICITY FOR THE FARM

Item	Capacity	Motor, hp, or demand, kw	Kilowatthours	Remarks
		Crop-processing electric equipment		
Blower (see Chap. 33)	*Grain and ground feed:**	To maintain pipe air velocity:	1–3 per ton	Air velocity 4,000 fpm. Hp goes up slightly with whole and coarsely ground grain
	4″ pipe—3,500 lb/hr max.	1 hp/100 ft		
	5″ pipe—4,500 lb/hr max.	1¼ hp/100 ft		
	6″ pipe—6,500 lb/hr max.	1½ hp/100 ft Add ⅓ hp for each 1,000 lb/hr		
	Chopped forage	5–10 hp	1 per ton	
Corn sheller, hand model	Ear corn: 25 bu/hr	¼–2 hp	3–7 per 100 bu	
Crop dryer†	*Grain, small:* up to 5,000 bu	5–10 hp for fan	350–1,350 per 1,000 bu	Without heat
	With heat, 200- to 500-bu batch	Also motors for handling grain	200 per 1,000 bu	
	Ear corn: in crib, up to 6,000 bu	5–10 hp for fan	175 per 1,000 bu; up to 3 per bu with 40% moisture	Without heat
	Hay: up to 100 tons/season	3–10 hp	50 per ton; range 30–90 per ton, less with heat	Without heat
	Seed		2.5 per 1,000 lb	
Elevator (see Chap. 33)	*Cup:* 4″ and 5″ belts, 150–400 bu/hr, height of 10′–50′	¼–1 hp	10–15 per 1,000 bu	
	Portable: 16′ at 40° slope, 100 bu/hr	¼–1 hp	4 per 1,000 bu	
	140 bu ear corn/hr		2½ per 1,000 bu	
	180–360 bales hay/hr		1 per ton	
	Auger: 4″ pipe size, 10′–30′ length, 100–400 bu/hr	¼–⅓ hp		
	6″ pipe, 40′ length, height up to 30′	1–5 hp		
Ensilage cutter	6–10 tons/hr	5 hp and larger	1 per ton	Speed must be held at minimum
Feed mill (see Chap. 32)	*Hammer:* 1 hp, 400–1,300 lb/hr (Tex.)	1–7½ hp and larger	Sorghum, 1 per ton Oats, 8 per ton	Fan also used for elevating grain
	5 hp, 250 lb to more than 1 ton/hr (Wis.)		Corn, 4 per ton Oats, 36 per ton	
	Burr: 4″–10″ burrs—up to and more than 1 ton/hr	1–3 hp	Cracked corn, 3 per ton, oats 43 per ton	6″–8″ burrs best
	Automatic feed processing: combination grain grinding and mixing, 1,000 bu/hr (Wis.)	7½-hp motor plus 6 for crushing corn and conveying—total 13½ hp	10 per ton	
Feed mixer	500–1,000 lb/mix	1–3 hp	1 per ton	5–12 min./batch
Fruit and vegetable grader	Apples, 35 bu/hr	¼–½ hp	1 per 100 bu	
	Potatoes	½ hp and up	1 per 600–700 bu	
	Bulbs, 1,600 lb/hr	¼ hp		
Hoist, hay	7 min/truck or wagonload	1–5 hp	1 per 3 tons	Other jobs: lifting baled hay, grain, fertilizer, digging well, etc.
Stacking hay			1 per 4 tons	
Root cutter	500–1,000 lb/hr	¼–½ hp	1–2 per ton	
Seed cleaner	*Grain:* 20–125 bu/hr	¼–1 hp	1 per 100 bu	
	Seeds: 2–10 bu/hr		1 per 20–30 bu	
Sweet potato, curing and storage	Houses, 50–2,000 bu	Heaters 500–1,000 watts	1.5 per bu	
	Install 1 kw per 150–300 bu	115/230 volts; thermostatic control	Varies from 0.6 to 6 per bu	
Wagon unloader	7–8 min/3-ton load	¼–½ hp	⅓–½ per 20 tons	
		Dairy equipment for the farm		
Barn cleaner	Many types, 30-cow herd, 5–20 min/day	¾–5 hp	120 per yr or ½–1 per cow/mo	Saves 5–10 hr/cow/yr
Bottle washer, milk	1 or 3 brushes; 1 or 2 bottles	⅙–¼ hp	1 per 2,000 bottles	
Calf dehorner	10 or more sec/horn	200–300 watts		900–1000°F
Churn	2 qt to 10 gal Churning capacity about one-half of total	1⁄30–¼ hp	2 per 100 lb butter for larger sizes	

* *Univ. Ill. Rural Elec.* 6.

† Heated-air crop dryer can also be used to heat buildings.

TABLE 51-18. APPLICATIONS OF ELECTRICITY FOR THE FARM (Continued)

Item	Capacity	Motor, hp, or demand, kw	Kilowatthours	Remarks
\multicolumn{5}{c}{Dairy equipment for the farm (Continued)}				
Clipper, cattle		$1/10$ hp or less	1 per 10 hr operation	
Cream separator	Up to 1,400 lb milk/hr	$1/8$ hp and up	1 per 2,000 lb milk	
De-icer	Keeps stock tank from freezing over so that animals can drink	250–1,000 watts	150 per yr and up	5 kwhr/day possible
Feeder, mechanical	*Auger type:* up to 120'	$1/2$–$1 1/2$ hp		See Barn cleaner. Handles silage, hay, etc. Reduces labor to one-third and less
	Reciprocating: up to 150'			
	Endless chain or belt: up to 150', 30 head of cattle or more	$1/2$–3 hp		
Fence controller	Up to 20 miles of fence or 500 acres	10 watts	5 per mo	
Milk-can hoist	Used for removing milk cans from immersion coolers	$1/4$–$1/3$ hp	30 per yr	
Milk cooler	*Walk-in:* 50–400 gal milk/day	$1/2$–3 hp	1.3 per 10 gal	
	Wet storage: immersion or spray, 3–20 per 10-gal cans	$1/4$–1 hp	1 per 10 gal milk	Cooling capacity may be double can capacity
	Bulk storage: 100–1,000 gal and larger	$1/2$–$7 1/2$ hp plus fractional hp	0.9–1.3 per 100 lb milk; variable	
Milkhouse heating	40–45°F to prevent freezing. Heat with milk cooler-compressor, resistance heaters, water heaters, or infrared lamps	Compressor motor up to $7 1/2$ hp; resistance 1 to 3 kw; infrared lamps 250 watts each	500–2,000 per season; 800 av.	
Milking machine	Up to 60 cows/hr/man			
	Portable	$1/6$–$1/4$ hp/cow	1.5 per cow/mo	
	Pipeline	$1/8$–$3/4$ hp/cow	2.5 per cow/mo	
Pasteurizer Household	*Batch type:* 1–2 gal in up to $1 1/2$ hr	300–1,250 watts	0.3–0.75 per gal (both types) or 40 per mo	Thermostatic and/or time control. Temperature 143–162°F
	In-the-bottle type: 7 or 8 qt in about $1 1/2$ hr	1,500 watts		
Small bulk	25 gal, 50 min to 1 hr 50 gal, $1 1/4$ hr	12 kw and $1/4$-hp motor		143°F for 30 min
Pump jack	Water for stock tank; livestock drinking	$1/3$ hp and larger	1 per 1,000 gal	
Silage chipper	Loosens and pulverizes frozen silage 3 in. deep	$3/4$ hp	0.8 per hr	
Silo unloader	60–100 lb/min and more for 10'- to 20'-diam silos	2–5 hp	300 per yr	Time saved, up to 200 hr/yr. Top and bottom types
Stock-tank heater	*General-purpose:* 30 cows and 60 hogs. Also larger	200–1,000 watts, thermostatic control	200–1,200 per season	Water temperature held at 40–45°F
	Frost protection: heating cable wrapped around pipes			
Trainer, cow	One unit per animal	Equipped with fence controller		Reduces cow-cleaning time
Utensil sterilizer	2–10 10-gal cans plus all milk utensils. 10–70 cu ft	1–6 kw, thermostatic control	$2 1/2$ per 75–100 lb utensils/day	Operating temperature 180°F for 30 min
Ventilator	*Barn:* 10''–24'' exhaust or intake	$1/20$–$1/2$ hp, thermostatic control	240 per yr and up or $2 1/2$ per cow/mo	
	Milkhouse: 10'' and up exhaust or intake	$1/40$ hp and up		
Water heater	*Pour-in type:* 10 gal	250–1,500 watts, 115 volts	50–150 per mo	Both have thermostatic control
	Pressure type: NEMA wattages and sizes for household	1,500 watts and up, 115/230 volts	150–200 per mo	
Water-pail heater	10 qt hot water for dairy, etc.	1,500 watts	300 per yr	
Water system, pressure	*Shallow well*	$1/6$–1 hp and larger for both	1 to 7 per 1,000 gal, av. 2 per 1,000 gal	
	Deep well: 300 and more gal/hr			

TABLE 51-18. APPLICATIONS OF ELECTRICITY FOR THE FARM (*Continued*)

Item	Capacity	Motor, hp, or demand, kw	Kilowatthours	Remarks
Electric farm shop equipment				
Air compressor	Used on air hose, chuck, paint-sprayer gun, duster, grease gun, etc.	¼–½ hp	35 per yr	Rated at 80 psi and 1 cfm
Battery charger		600–750 watts	2 per charge for a dead battery	
Concrete mixer	3 cu ft per mixer load	¼ hp and up	1 per 2 cu yd	
Drill press	¼″ and ⅜″ drills	¼–1 hp	15 per yr	
Grinder, bench	Sharpening tools and sickles	¼–⅓ hp	⅓ per 3 hr	
Grindstone		¼ hp	1 per 3 hr	Surface speed 400 fpm
Hoist	250 lb; 7′ lift; 30 fpm			
Lathe		¼–1 hp	10 per yr	
Paint sprayer		⅒–½ hp	1½ per 1,000 sq ft	See Air compressor
Saw, shop	Band and circular types	¼–1 hp	15 per yr	
Soldering iron		200–300 watts; range 80–500 watts	¼ hr use	
Welder		150–200 amp	75 per yr	
Wood saw	20 cords cordwood/day, 4 or 5 men	2 hp and larger	0.5–1.2 per cord	
Horticulture (also see Crop-processing electric equipment)				
Hotbed cable	6′ × 6′ hotbeds or multiples thereof. Cable lengths 40′–120′ required per 6′ × 6′ bed	Supply 5–13 watts/sq ft, 115/230 volts, ½ ohm resistance/ft	1–6 per day per 6′ × 6′ hotbed	
Lights for plant growth				Follow instructions of research results
Soil sterilizer	1 cu yd	3,000 watts	30 per cu yd	
Livestock equipment				
Brooder	*Calf:* lamps hung over the pen and out of reach of calves	Infrared 250-watt lamps	6 per lamp/day	
	Lamb: 1 hover for several	Incandescent lamps 100 or 150 watts, infrared 250 watts	2.4–6 per lamp/day or 100 per 100 lambs	
	Pig: 1 hover/litter or hanging infrared lamp	Incandescent lamps for hover 100 or 150 watts, infrared 250 watts	24–60 per litter	
Sheep shears		⅒–¼ hp	2 per 100 sheep sheared	
Electric poultry equipment				
Beak-cutter cauterizer	200–300 hens/hr, 300–400 chicks/hr. Also used to clip wings	200–500 watts		
Brooder	*Battery:* 35–150 chicks/unit	100 watts	1 per chick	
	Hover: 25–500 chicks and larger	200–1,000 watts or 2 to 3 watts/chick	½–3 per chick	
	Infrared: 50–500 chicks or more	250-watt infrared lamps, 1–8	1–3 per chick	
	Underheat: heated concrete slab 5′ × 21′; hover 4′ × 20′, etc.	20 watts/sq ft	0.2–1 per chick	
Egg cleaner	*Dry type:* 125 doz/hr *Water:* 250 doz/hr	⅙–¼ hp 2,000–4,000 watts, ¼ hp and up	1 per 2,000 eggs	Both larger and smaller sizes

TABLE 51-18. APPLICATIONS OF ELECTRICITY FOR THE FARM (*Continued*)

Item	Capacity	Motor, hp, or demand, kw	Kilowatthours	Remarks
Electric poultry equipment (*Continued*)				
Egg cooler	*Evaporative*	$1/30-1/3$ hp	500 per 4-mo season	
	Mechanical: cool 10–40 cases daily	$1/4-1$ hp	300 per season	
Egg grader	4–10 cases/hr	$1/4$ and $1/3$ hp	45 per year	
Feeder, mechanical	Min. flock 500 birds; 40′–600′ and more; per ft of trough, 15 broilers, 8–10 hens, 10 poults, or 5–8 turkeys; 250 lb feed/hr; hopper 200–1,000 lb feed. Up to 12,000 broilers, 6,000 hens, or 3,000 turkeys	$1/6-1$ hp	250 per yr or 1 per 1,000 hens/day	Operated with time switch
Incubator	50–75,000 eggs and larger	$1/2$ watt/chick	180 per 200 eggs; 135 per 1,000 eggs; 40 per 1,000 eggs	
Lights for egg production	*Incandescent:* fall and winter up to 14 hr/day	60 watts/200 sq ft	5–8 per 100 hens/mo	
	Ultraviolet	300 watts/600 cu ft	7 per 100 hens/mo	Successful for brooding, growing for market, and eggs
Litter stirrer	To help dry out poultry litter	$1/4-1/2$ hp		
Picker	130 hens/hr	$1/2$ hp and larger	1 per 250 birds	
Scalder	100 gal, 100 turkeys/hr	5 kw	1 per 20–25 chickens	
Ventilator		$1/8$ hp and smaller	200 per season	
Water warmer	*Immersion, fountain*	150–700 watts	100–200 per yr or 60 per 100 birds/season	
	Base heater, fountain	150 watts and up		

There is a wide range in the capacities of electrically operated equipment, depending upon factors such as size, speed, temperature, and the degree of automatic or manual operation. The distance materials are moved may affect the volume handled per horsepower.

Chapter 52

FRUIT AND VEGETABLE HANDLING

JORDAN H. LEVIN AND DONALD H. DEWEY

Handling may be defined as the movement of a product or material in any direction whether it be horizontal, vertical, or any combination of the two. Whenever a material or commodity is picked up, moved, and set down, it is handled. Lifting, shifting, placing, stacking, and transporting are various phases of handling.

Fruits and vegetables are easily damaged and are perishable and therefore are handled mostly in containers of relatively small capacity, usually a bushel or less.

There is no standard-size container for fruits and vegetables. Table 52-1 indicates the containers commonly used. For most practical purposes the bushel is defined as a volume measurement of approximately 2,200 cu in. This is greater than the legal 2,150 cu in. per bu because fruit and vegetable containers are packed with a bulge at the cover.

EFFECTS OF HANDLING

Handling is important, not just because of the amount of work involved in moving the large quantities of fruits and vegetables, but because of its effect on costs, quality, and condition of the produce and on orchard management.

Costs. The cost of labor is the largest single item in producing and marketing fruits and vegetables. It is estimated that under present conditions 30 per cent of the labor employed in producing and marketing produce is utilized for handling purposes.

Efficient handling methods reduce the amount of labor needed and the costs involved. For example, detailed studies showed that in handling a 25,000-bu crop of apples, a fork lift truck and pallets saved approximately $1,200 over the cost of using hand methods and roller conveyers.[1]

Quality. Fruits and vegetables at the time of harvest are usually of good quality and appearance. During subsequent handling the quality and nutrient value should be maintained in order to preserve the monetary value of the crop. The prevention of mechanical injury is of prime importance since fruits and vegetables are easily bruised or scarred. The damaged areas are inedible and unsightly, are sources of excessive moisture loss, and serve as points of entry for decay organisms. The storage life of apples may be decreased 1 week for every day the fruit is held at temperatures of 60°F instead of at 40°F or lower.

PRINCIPLES

Produce handling, as is true for all material handling, involves the use of labor and equipment. If handling is to save time, money, and human energy, labor and equipment must be balanced so that the over-all operation is efficient. Time can be reduced by eliminating waiting and shortening travel distances and travel time and the number of times the material must be picked up and set down. Labor requirements

[1] J. H. Levin and H. P. Gaston, Fruit Handling with Fork Lift Trucks, *Mich. State Univ. Agr. Expt. Sta. Spec. Bull.* 379, 1952.

TABLE 52-1. CONTAINERS AND WEIGHTS COMMONLY USED FOR FRUITS AND VEGETABLES

	Field containers		Shipping containers		Weight per bushel, lb
	Type	Net weight, lb	Type	Weight, lb	
Fruits					
Oranges	Box and bulk	77–90	Fiberboard carton and box	45–90	60
Grapes	Lug	28	Western lug box	28	29
			Eastern basket	18	
Apples	Crate and box	42–48	Basket, box, fiber carton	42–48	42–48
Grapefruit	Box and bulk	65–80	Box and carton	65–80	
Peaches	½-bu basket and lug box	25	Box, lug, basket	48	48
			Western box	20	
Melons	Bulk and bushel basket	35–40	Crate and bulk	60–80	
Pears	Crate and box	50	Box, basket	50	50
Plums	½-bu basket	.....	Lug box	18	56
Lemons	Box	79	Carton	45	60
Apricots	Lug	35–40	Lug box	25	48
Cherries					
Sweet	Lug	25–30	Lug box	15–16	56
Sour	Lug	25–30			
Strawberries	Flats	12	Crate (16–24 qt), flat (12 pt)	36, 12	
Vegetables					
White potatoes	Crate, bag, bulk	.....	Bag	50–100	60
Tomatoes	Hamper, field box	25–30	Lug box	31	
			Crate	50	53
Sweet potatoes	Crate	50	Crate	50	55
Cabbage	Bag, bulk	.....	Bag	50	35
			Crate	83	
Sweet corn	Bag, bulk	.....	Bag	.....	68 (without husks)
Dry beans and peas	Bag, bulk	.....	Bag	2-5-7	60
Lettuce	Carton, crate	40	2-doz carton	40	
Onions	Crate, bulk	45–50	Bag	50	50–57
Celery	Crate	60–65	Crate	60–65	
Carrots	Crate, bulk	50	Crate	50	50
Snap beans	Hamper	30	Hamper	30	30
Green beans	Hamper	30	Hamper	30	30
Cucumbers	Crate	50	Basket	48	48
Cauliflower	Crate	.....	Crate	37	25
Spinach	Basket	.....	Crate	20–25	18
Beets	Bulk, crate	.....	Crate	.....	52
Asparagus	Bulk	.....	Crate	24–30	45

must be kept at a minimum by using proper layout, by handling large amounts of materials at once, and by using proper equipment.

The following *questions* bring out *principles* which can be used as aids to studying and improving a handling operation:

1. Can the number of times the material is handled be reduced?
2. Can the amount of material handled at one time be increased?
3. Can the distance the material is handled be reduced?
4. Can the speed of handling be increased?
5. Do employees have to bend or make unnecessary movements?
6. Do employees have to wait for materials?
7. Are there times when equipment is not used to full capacity?
8. Can gravity be employed to a fuller extent?
9. Can the operation be done mechanically at the same cost or for less cost?
10. Can layout be improved?
11. Can the handling system be made more flexible?
12. Can the handling system be made more continuous?
13. Does the handling system damage or waste some of the product?

730   FARMSTEAD STRUCTURES AND EQUIPMENT

14. Does the handling system reduce the quality of the product?
15. If labor is paid on an hourly rate, is it possible to pay on a piecework rate?

A "yes" answer to any of the questions means that the system can probably be made more efficient.

Handling Systems. There are three major systems of handling materials: (1) piece-by-piece method, (2) unit-load system, and (3) bulk system.

With the *piece-by-piece* method each container is handled individually. Examples of this method are apples handled in bushel field crates and onions handled in 50-lb bags. Handling by the piece-by-piece method can be done manually or with the use of mechanical aids. This system usually involves the least equipment expense and the largest labor costs.

The *unit-load system* handles a number of individual containers as a single, or unit, load. Equipment is available for handling unit loads that range in volume of fruits and vegetables from a few containers to 40 or 50. For example, a hand truck will move 4 or 5 boxes of fruit at a time, whereas a fork lift truck can handle 30 field crates of apples on a pallet at one time. This method has a greater equipment cost and a lower labor cost than the piece-by-piece method.

The *bulk system* or the semibulk system handles materials in large quantities, either without containers or in large containers. Usually the labor cost is lower for this method than for the others.

Potatoes, onions, cantaloupes, carrots, beets, citrus, and cherries are examples of crops handled in bulk.

EQUIPMENT

Equipment used for handling can be classified in several ways, such as by the type of service it performs, whether it is manual- or power-operated, by similarity of design or operation, and/or by products. Handling equipment is discussed here according to its general use (1) for piece-by-piece handling, (2) for unit-load handling, and (3) for bulk handling.

By this classification, the equipment most commonly used in handling fruits and vegetables is listed in Table 52-2. A brief description of handling equipment follows.

TABLE 52-2. HANDLING EQUIPMENT COMMONLY USED FOR FRUITS AND VEGETABLES

Piece-by-piece method	Unit-load method	Bulk method
Conveyers	Dollies	Belt conveyer
Belt	Wheelbarrows	Chute
Portable	Trucks	Bulk trailers
Stationary	Trailers	Bulk box pallets
Roller	Hand lift trucks	Bucket elevators
Wheel	Power lift trucks	Pumps
Gravity	Pallets and skids	Flumes
Power	Dumpers	Dumpers
Scraper	Hand trucks	Tank or bulk
Flight	Two-wheel	trucks
Overhead rail	Four-wheel	
Chutes	Stevedore type	
Trucks	Box type	
Trailers (farm and orchard)		

Gravity Conveyer. This equipment usually consists of a series of rollers or wheels. It is set level or on a slight slope. If rollers are used, it is often called a *roller* conveyer; if wheels are used, it is called a *wheel* conveyer. Such conveyers are available in joining lengths of 10 to 20 ft and in widths of 20 to 24 in. They are constructed either of steel or aluminum. The latter has come into widespread use because it is light in weight and easy to handle.

Belt Conveyer. The belt conveyer may be used to move individual containers of fruits and vegetables, or it may be used for handling loose fruits and vegetables in bulk. Belt conveyers are power-driven, with the drive at the discharge end. Many belt conveyers used in harvest and field handling are portable and sometimes driven by small gasoline engines. Belt conveyers for elevating fruits and vegetables in bulk usually have slat attachments on the belt and are often called *flight conveyers*.

Overhead Rail Conveyer. This equipment is sometimes called the trolley, or monorail conveyer. It consists of an overhead track with trolleys fastened together by chains. The chain and trolleys are usually power-driven. This stationary equipment is sometimes used in packing houses or processing plants for moving filled or empty field containers about the building to a number of loading and unloading points.

Dollies. A dolly is usually a rectangular platform supported on wheels or rollers, generally of the caster type. Several containers are placed on a dolly by hand and pushed to their destination. Dollies require a smooth surface wherever used.

Hand Trucks. This equipment is of variable design. It enables the movement of small numbers of individual containers as a unit load. The *four-wheel platform truck* has a deck on which the load is placed manually. It is of wood or metal construction and may have several arrangements of wheels and casters. The *two-wheel stevedore-type truck* carries the load at the lower end on a blade of metal. The *clamp-type two-wheel truck* has the same type frame, but instead of the stevedore lip, it has two arms which clamp in from the sides and grasp the boxes. The clamps are activated by a foot-operated lever. Hand trucks enable a worker to move six boxes or field crates as a unit load.

Skids. A skid is a platform elevated from the floor by legs, casters, or other special attachments. Because of the clearance from the floor, it can be lifted by hand or power lift equipment and moved.

Pallets. Pallets resemble skids but are designed to be placed on top of other loaded pallets for high stacking. Pallets are constructed of wood, metal, or other material. The nailed, wooden pallet is almost universally used for fruit and vegetable handling. Most pallets have an upper and lower deck, spaced and held together by wooden stringers.

The *pallet box* is essentially a combination pallet and box, holding 16 to 40 bu. The pallet forms the floor of the box and is an integral part of the unit. The sides and bottom are usually slatted. They are lifted, moved, and high-stacked with a fork lift truck.

Hand Lift Trucks. These are designed to raise and move skids or pallets containing the load.

Power Lift Trucks. These machines are designed to lift and move materials. The fork lift truck, most commonly used, is also designed for high stacking. It lifts heavy loads on pallets, moves them rapidly, stacks them high, and sets them down gently. Operation is simple and relatively inexpensive.

The *clamp lift truck,* which is used on a limited basis, raises the load by clamping several tiers of containers with two clamps, without the need for pallets. However, they are more expensive than fork lift trucks, the containers must be strong enough to withstand the pressure of the clamps, and space must be left between each stack of fruit for operation of the clamps.

Lift trucks are available for handling loads of 1,000 lb to many tons. Commonly, 2,000-lb-capacity trucks are used at the orchard or farm and 4,000-lb-capacity trucks are used by processing plants and large packing houses. Lift-truck *selection* should be based on capacity, tire type, power, exhaust, cost, and maintenance. Liquid-petroleum (LP) engines or electric engines are recommended where carbon monoxide gas may accumulate. A catalytic muffler is available to change carbon monoxide to carbon dioxide.

A serious limitation of regular fork lift trucks is that they are very heavy for their wheel size and cannot be operated on soft ground. There are lift fork attachments available for many makes of tractors for use in the orchard or field. There are four types available: (1) Forks which can be attached to tractors with three-point hydraulic hitches, at a cost of less than $50, can lift pallets or bulk boxes only 18 in. but are suitable for transporting. (2) Forks attached to a tractor-mounted loader or stacker (1,200- to 2,000-lb lift capacity), although not smooth-operating, can be used to raise pallets or bulk boxes 10 or 12 ft high. (3) Regular fork lift attachments are available for many tractors. They cost between $700 and $1,000 and are available for both the front and rear ends of the tractor. They are usually more satisfactory when attached to the rear. (4) Tractors with lift attachments on the rear are available with the steering and gear shift reversed.

Trailers. In moving fruit or vegetables out of the field or orchard, *trailers* pulled by farm tractors are generally used. Most trailers have low, flat decks for loading individual containers or containers on pallets. Bulk trailers have sides and provisions for quick unloading by conveyers or by tilting.

Motor Trucks. Several million motor *trucks* are in use on farms. Flat decks are used for handling produce on pallets, flat beds with side racks for piece-by-piece handling, and dump bodies for bulk handling. Trucks with tank bodies are sometimes used for handling cherries in water; others have detachable tanks mounted on flat decks. In either case, tanks may be open or closed.

Dumpers. Most fruits and vegetables are emptied from the field containers by tipping onto a grading or packing line. Although much of this work is done by hand, mechanical dumping equipment is often practical. Automatic equipment which picks up and dumps individual containers is also used.

Water Conveying. The bulk movement of produce by a rapid and large flow of water, sometimes referred to as hydroconveying, is sometimes used in processing plants. *Flumes* of various types are used, and usually they are on a slight slope, so that the movement of water and produce is by gravity.

Pumps are used for recirculating water and pumping produce from one level to another. The produce passes through the specially designed pumps without injury.

Other Handling Equipment. Other types of handling equipment which are sometimes used for fruits and vegetables and other agricultural products are block and tackle, hoists, elevators, tiering and stacking machines, pneumatic conveyers, screw conveyers, and ramps.

HARVEST HANDLING

The physical nature of fruits and vegetables and the method of harvest have a bearing on the handling procedures that are used following harvest.

Quality and Condition. Fruits and vegetables must be harvested at the proper stage of maturity so as to arrive at the market in the best possible condition. Some crops, like citrus or potatoes, may be harvested over a relatively long period of time; others, like strawberries or peas, must be harvested during a very short period for maximum quality. Frequent harvests are essential for some, especially berries, sweet corn, and asparagus. Many crops are susceptible to rapid deterioration at high temperatures and must be handled quickly.

Method of Harvest. The handling methods used in moving fruits and vegetables from the field are often dependent on the harvesting methods.

When harvest is performed manually the harvested produce must be disposed of by the worker within arm's length or a short distance. When bulk pallet boxes are used, they are placed at convenient points in the field or orchard, where they are filled by the workers.

Portable belt conveyers are sometimes used, whereby a number of workers harvest one row at a time. The fruit or vegetable is placed on a continuous belt which moves it to the end of the row and then into containers or a bulk truck.

Field to Dock. Once harvested, fruits and vegetables are moved out of the field to the farm packing house, storage, or loading dock, usually by truck or trailer. Trailers are preferable for fruit operations because their low clearance permits passage under the trees. Trailers usually hold 150 to 200 containers per load. Trucks are of variable size.

Tree fruits are usually harvested into picking bags or buckets. These hang at waist level on a harness which fits over the picker's shoulders. When a picker's bucket is full, its contents are transferred to the field containers. Vegetables are usually picked and placed directly into field containers when harvesting is done by hand.

When the piece-by-piece method of handling is used, the field containers are loaded manually onto the truck or trailer by a handling crew. Three workers and a trailer can pick up and haul out of the orchard about 1,000 field containers of fruit in a 10-hr day.

Pallets can be placed on the bed of the trailer or truck before it moves to the field. In the harvest area the containers are stacked on the pallets as a unit load. The loaded pallets are lifted off the truck or trailer with a fork lift truck and either stockpiled, moved to the packing line, placed in storage, or loaded onto outgoing trucks. Pallets sometimes are placed on the ground in the harvest area, loaded by the pickers, and then placed onto trucks or trailers for movement out of the field with tractor fork lifts.

Bulk boxes are coming into widespread use, especially for crops which are to be processed. If the grower is equipped with a field or orchard lift, the bulk boxes are placed in the harvest area for direct filling by the picker. The filled boxes are then moved by a tractor lift to the dock or loaded onto a truck or trailer and moved out of the field. One man and a tractor equipped with lifts on the front and rear of the tractor can handle two 20-bu bulk boxes at once and move 1,000 bu of produce out of the field in 8 hr. Bulk boxes of 40-bu capacity are usually used for vegetables. The capacity of the bulk boxes used for fruit is usually between 16 and 24 bu so as to minimize crushing and bruising.

Bulk boxes are also directly filled by moving them on a trailer or truck alongside a mechanical harvester equipped with a conveyer.

Bulk trailer bodies are commonly used for vegetables, such as onions and potatoes, which are mechanically harvested. They are loaded directly from the harvester, then unloaded with a conveyer belt built into the trailer floor. At the unloading area, the tail gate is lowered, a small electric motor is connected to the conveyer-belt drive, and the contents unloaded by this belt. Other bulk bodies are unloaded by tipping the whole vehicle with a hydraulic dumping mechanism or cable hoists.

STORAGE HANDLING

Storages Filled with Containers. Storages with capacities of 5,000 crates or less usually are limited to dollies or hand trucks and an inexpensive mobile rack on which the workmen stand for making high stacks. Storages of this size are often operated in conjunction with retail markets. The handling equipment in a storage holding 5,000 to 9,000 crates usually consists of a roller conveyer and a mobile stacking aid. The labor cost is approximately 4 cents per crate for moving apples in and out of a storage of this size. In larger storages, where the capacity is over 10,000 crates, the roller conveyer is often supplemented with a power elevator. With this piece of equipment, the labor cost may be reduced to 3½ cents per crate of apples handled.

Where more than 10,000 containers are handled, fork lift trucks have proved economical and have reduced the labor cost to ½ cent per crate.

Table 52-3 shows the total cost of handling stored apples. Costs for handling crates of stored vegetables should be similar.

Storages Filled with Loose Produce in Bulk. It is possible to store some fruits and vegetables in bulk. Onions and potatoes are often stored in this manner. If the produce arrives at the storage in field containers, the produce is poured out of the container onto a conveyer, then elevated and carried by a system of belts to the bin or desired location in the storage. The produce is then lowered by a chute onto the produce already in place. Care should be taken to limit the drop from the chute to less than several feet in order to avoid excessive bruising.

The most efficient handling in bulk storages is accomplished with produce that is harvested mechanically and loaded directly into bulk trailers or bulk trucks. The bulk vehicles are unloaded from the rear with conveyer belts on the bed of the body into bin loaders which move the produce to the top of the storage pile. The produce is fed from the truck bed into a large hopper and then elevated over a short wire-mesh belt with cleats that serves as a dirt eliminator. The produce then moves to an elevator consisting of a cleated rubber belt moving in a trough. The false floor needed for air circulation is laid down progressively as the storage is filled.

Two systems are widely used for *removing* produce from a bulk storage. In one system, permanent conveyers are exposed or portable conveyers laid down as sections of the false floor are removed. The produce falls by gravity, or is pushed onto the conveyer by a worker, and is moved to the grading and packing line. Some produce,

TABLE 52-3. FRUIT-HANDLING COSTS*

Capacity of storage, crates	Total annual cost, dollars				Annual per-crate cost, cents			
	Hand† trucks or dollies	Roller† conveyer	Roller‡ conveyer and power elevator	Fork§ lift truck	Hand trucks or dollies	Roller conveyer	Roller conveyer and power elevator	Fork lift truck
2,000	94.00	104.50	......	......	4.7	5.2	...	...
6,000	.....	271.50	......	......	...	4.5	...	...
10,000	.....	438.50	500.50	256.50	...	4.4	5.0	2.6
14,000	.....	615.50	647.50	291.20	...	4.4	4.6	2.1
18,000	.....	772.50	798.00	325.20	...	4.3	4.4	1.8
25,000	.....	1,063.00	957.00	366.80	...	4.2	3.8	1.5

* Includes labor, depreciation (at 10 per cent), interest (at 4 per cent), taxes and insurance (at 2 per cent each), and power.
† Labor calculated at 4 cents per crate.
‡ Labor calculated at 3½ cents per crate.
§ Labor calculated at ½ cent per crate.
SOURCE: J. H. Levin and H. P. Gaston, On-the-farm Refrigerated Fruit Storage, *Mich. State Univ. Agr. Expt. Sta. Bull.* 389, 1955.

such as potatoes, is flumed out of a bulk storage by water. For removal of the potatoes, the floor is progressively removed, exposing permanent channels. Water is then directed down through the produce and into the channel, carrying the produce through a series of flumes into a hopper, where it is elevated onto the grading-and-packing line. The water may be recirculated or wasted, depending on the supply.

Sorting and Sizing. Fruit and vegetable growers usually pack their products for fresh market according to standards for size and for grade. Raw produce for processing is also sorted for grade. Most fruits and vegetables, therefore, are sorted and sized before they reach market. *Sorting* is the act of separating the individual fruits and vegetables into lots to meet the prescribed standards of color, ripeness, blemishes, and other quality characteristics. *Sizing* is the separation of individual fruits or vegetables into lots according to size. *Grading* is a term that should be applied to the sorting operation for meeting a grade standard established by state and Federal agencies.

Lighting. Inasmuch as natural light is best for manual sorting, the walls surrounding the sorting area should, if possible, include an ample number of generous-sized windows. As a general rule, the window area should equal 10 to 20 per cent of the floor area and should be placed rather high in the walls. Areas with inadequate daylight should be supplemented by artificial illumination.

Packing houses are often operated during the late afternoon or evening, so that it is usually necessary to provide electric lights even though there is enough natural light during the day. Because of the shape of the tubes and the quality of the light emitted, many operators believe fluorescent lights are superior to other light sources for packing-house installations. The light tubes or lamps should be placed and shielded so that the light will fall onto the work and not into the eyes of the employee. A single fluorescent tube does not provide a satisfactory source of light since the so-called "stroboscopic effect" causes an unpleasant flickering. Fixtures containing two or more tubes should be used.

Sorting belts and facing stations require about 100 foot-candles of illumination. This light can be obtained by hanging fluorescent lighting fixtures directly above and within a few feet of the work areas.

Several studies have been made on the use of colored lights as an aid to illuminating fruits and to simplifying the sorting operation. Medium-purple light aided the sorting of red cherries and is used in some plants.[1] Present information, however,

[1] B. Parker and D. E. Wiant, Use of Chromatic Illumination and Methods of Produce Rotation for Sorting Red Tart Cherries, art. 36–50, *Mich. State Univ. Agr. Expt. Sta. Quart. Bull.*, May, 1954. D. C. White, "Maximizing Color Differences," Ph.D. thesis, Stanford University, Stanford, Calif., 1950.

indicates that daylight bulbs and black- or tan-background belts are satisfactory for general sorting operations.

Mechanical Aids. There are several types of conveyers for moving the produce past the sorting stations. Belt conveyers are often used, but they have the disadvantage that the entire surface of the individual fruits and vegetables cannot be seen unless the "sorter" rotates or picks up the produce. In some cases, rods are placed across the belts to rotate the produce.

Conveyers with moving rollers or spools that rotate the fruit are in common use for sorting. Optimum inspection efficiency occurs when the fruit or vegetable is rotated three revolutions per foot of translation.[1]

Sorting belts or conveyers should not be more than 4 ft wide. Conveyer belts or bins for disposal of the off-grade produce should be within arm's length of the sorters. Produce should move on the belts in a single layer and at a speed no faster than 30 fpm.

Automatic electronic devices have been developed for sorting fruits and vegetables on the basis of color differences. They are being used on a commercial basis for lemons.[2] An electronic device for determining the redness of tomatoes is also being tested.

Sizing Equipment. Produce can be sized by weight or by diameter. *Weight sizers* are used for apples, oranges, tomatoes, cantaloupes, and many other fruits and vegetables. The produce is placed into individual cups either by hand or by an automatic

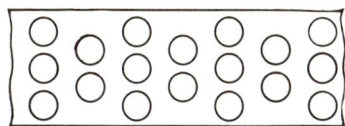

FIG. 52-1. Belt sizer, top view.

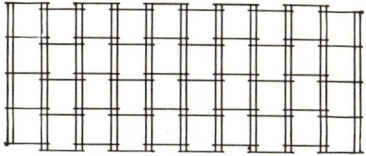

FIG. 52-2. Chain sizer, top view.

indexing feed. As the cups move through the unit they pass under spring-loaded trips that are adjusted by spring tension to variable weights. The tension becomes progressively weaker so that the heavier produce is sorted out first and the lightest last.

Weight sizers are more accurate and cause less bruising than diameter sizers and can be used for any shape produce. However, they are slower and cost more than diameter-sizing equipment.

Diameter-sizing equipment classifies the produce by measuring its diameter. The most common type is the screen conveyer utilizing a belt containing circular holes of the desired diameters. These belt sizers (Fig. 52-1) waste some space, and the size of the holes changes as the belt stretches or contracts.

A continuous screen of chain links, put together to provide square openings of the desired size, are used extensively for fruit (Fig. 52-2). Produce of a smaller diameter than the chain-links holes falls through, whereas the larger produce is carried over the screen.

The *diversion-belt sizer* consists of two narrow belts or ropes which diverge as they travel. The fruit or vegetable is carried between the belts until the opening is wide enough to allow it to fall through.

Other types of diameter sizers are designed with *rollers* and combinations of rollers and belts or wheels and belts. Roller sizers continually rotate the produce by the turning rollers, which measure the minimum dimensions. The rollers are set at a given distance apart so that fruits and vegetables of a smaller diameter fall through the rollers. The combination belt and roller or wheel utilizes the belt to move the produce. The rolls or wheels are set at fixed distances above the belt to knock or push off the

---

[1] Visual Inspection of Products for Surface Characteristics in Grading Operations, *USDA Misc. Publ.*, 1953.
[2] J. B. Powers, J. T. Gunn, and F. C. Jacobs, Electronic Color Sorting of Fruits and Vegetables, *Agr. Eng.*, 34:149–158, March, 1953.

belt the produce they contact. Mechanical injury during sizing and sorting can be reduced as much as 85 per cent with well-designed and adjusted grading equipment.[1]

Washing and Cleaning. Foreign material must be removed from fruits and vegetables before they are shipped to market. Light or loose material can be removed by a fanning mill or screen shaker cleaner, whereas the removal of dirt and spray residues may require brushing or washing. All produce for processing must be washed.

Almost all root vegetables are passed over screen shakers which remove some of the dirt from the produce. Such shakers are of the screen or chain type which carry the produce and allow the dirt to pass through. These devices are not used for fruit because of the possible damage by bruising or crushing as a result of the shaking action.

Most *washers* are continuous. Washing is accomplished by submerging the produce in water or by passing it under water sprays.

Soaker washers in which the produce is moved through a tank of water are not too effective in removing dirt, but are useful as a preliminary treatment before washing by sprays. Warm or hot water may be used. In washing asparagus, the hot water opens the spurs so that sand can be removed. Soaking softens dirt and sticky coatings and is especially useful on tomatoes.

When the water is *agitated* the efficiency of a soaker washer is increased. Compressed air may be used to agitate thre water. This works well for washing spinach. The water may be circulated by a pump or a propeller, and tree fruits are often washed in water agitated in this manner.

A simple *spray* washer consists of a series of nozzles under which the produce is passed on a belt or chain. Sprays may be low-pressure or high-pressure. When *rotary-drum* washers are used the produce is moved through a rotating drum by the motion and slope of the drum. As the produce tumbles in the drum, sprays of water hit all sides of the produce. Drum washers are used for beets, carrots, and spinach. *Shaker* washers employ a vigorous reciprocating motion which moves and tumbles the produce so that all surfaces are exposed to water sprays. Shaker washers are used for almost all types of berries.

Chemicals are frequently incorporated in the wash water to serve as disinfecting agents. Wetting agents are often used to increase the efficiency of the washing operation.

Some fresh-market fruits and vegetables are passed through a tank containing a *coloring* solution. Colored waxes may also be used. Certified dyes are used sometimes for coloring oranges and red potatoes. The dyes are harmless, improve the appearance, and result in a uniform product. However, products containing dyes must be labeled "color added."

The use of *wax* coatings on fruits and vegetables has increased in recent years. Over 75 per cent of all oranges for fresh market are treated by some waxing or polishing method. Waxing reduces the water losses and improves the appearance of fruits or vegetables. Apples, cucumbers, tomatoes, peppers, and eggplant are waxed commercially.

Polishing and *brushing* units consist of rotating brushes or rollers which press against the produce as it passes through the unit. The purpose of the brushers used without waxers, as used on apples, is to remove dirt and spray residues and to improve the appearance of the fruit.

Packing and Packaging. *Containers* used in shipping and marketing fresh fruits and vegetables can be divided into six principal classes: (1) baskets, (2) wooden crates or boxes, (3) corrugated cartons or boxes, (4) barrels, (5) bags or sacks, and (6) consumer-size prepackages such as film, mesh or paper bags, or cartons.

Wooden *boxes* and *crates* are of many sizes and may be of solid, paneled, or slatted material. Corrugated and solid fiber boxes are rapidly replacing wooden boxes for fruits and vegetables. Shippers of lemons and oranges have almost completely changed from wooden boxes to corrugated cartons, which are lighter in weight and less costly. They can be received "knocked down" and set up by machines at the packing plant.

---

[1] H. P. Gaston and J. H. Levin, How to Reduce Apple Bruising, *Mich. State Univ. Agr. Expt. Sta. Spec. Bull.* 374, September, 1951.

Bags and *sacks,* generally of burlap or treated paper, are used to handle onions, potatoes, green corn, and cabbage. Their size is usually designated by weight, such as the 50-lb bag or 100-lb bag.

Three main types of *pack* are in use: (1) the *jumble* pack, in which the produce is put into containers in no set order or arrangement; (2) the *face-and-fill* pack, in which the top layer is put in by hand so that it is uniform and has a symmetrical arrangement and the remaining space is jumble-filled; (3) the *set-arrangement* pack, in which each piece of produce is put into the container according to a set arrangement or compartments, as trays.

Overwrapping each piece of fruit or vegetable with a paper or other type of wrap, which is sometimes impregnated with a chemical to prevent decay, is a practice which is declining rapidly. It involves too high a labor cost, and some of the preventive chemicals can be incorporated in the master containers.

Almost all packing is done by hand. Workers remove the produce from a bin or belt and place it into the containers. The return-flow belt (Fig. 52-3) is coming into widespread use and is the best way to handle the produce for packing with a minimum amount of bruising. If workers miss a given piece of produce, they can pack it on the second time around. Several machines have been recently developed for automatically packing tray-pack containers.

Produce is packed according to diameter size or count (number of individual pieces in the container). Produce is usually packed by diameter size in areas east of

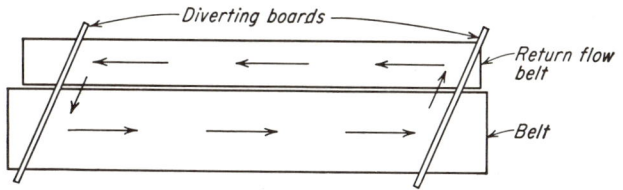

Fig. 52-3. Return-flow belt, top view.

the Mississippi. For example, apples are designated as 2¾- or 3-in. pack, etc. Citrus in both Florida and California and apples, pears, peaches, and plums in California and the Pacific Northwest are packed by count. For example, the standard four-basket plum crates contain between 176 and 340 plums and are designated as 176, 204, 340, etc., count; pears are packed in the standard pear box with counts varying from 76 to 245.

Prepackaging is performed either by the *hand* method, the *chute* method, or the *machine* method. In the hand method the worker fills and closes the consumer containers by hand. In the chute method the produce is diverted into a chute; the worker places a bag over the chute and removes it with the produce. In the machine method the bags are filled and weighed automatically; the worker merely holds the bag in position and removes it when full.

Packing-house Design and Layout. Produce is received, graded, sorted, sized, and packed at packing houses and then moved to storage or shipped to market.

The layout should be arranged so that produce can be moved directly to the packing line from the orchard or field, as well as from storage. The size and shape of the room will depend on packing methods, the daily volume handled, and the type of equipment used. Growers should allow a minimum of 3 to 4 sq ft of floor space for each bushel of produce packed in a 10-hr day.

Packing houses should be planned and equipped to be used for sorting and packaging more than one kind of fruit or vegetable. Flexibility is desirable since unforeseeable changes in volume, packaging, grading laws, and practices often make alterations necessary. Figure 52-4 shows a basic arrangement of a packing line which can be used for fruit packed jumble-fill, face-fill, by count, or prepackaged. Machine filling equipment can be placed against the return-flow belt if desirable.

The building should be ground-level and one-story high. A cold storage is desirable on the same floor level. Ceiling height should be 14 to 18 ft so that fork lift equipment can be used for high stacking. Good lighting should be provided. At least two sides of the packing area should be readily accessible by truck so that receiving and shipping do not interfere with each other.

Handling at Processing Plants. Recent figures show that more than 35 per cent of the vegetables and more than 55 per cent of the fruits marketed in the United States were processed. Although mechanization has increased in recent years, material handling still accounts for an estimated 30 per cent of the labor costs of food processing.

A great deal of emphasis is being laid on increased efficiency of handling in processing plants. The trend is toward bulk handling of the raw materials. At the plant, water conveying, or hydroconveying, is being used as a means of moving green peas, cut corn, cherries, Lima beans, string beans, and almost all produce up to 3 in. in diameter. Suitable pumps have been developed through which the produce may pass without injury.

Transporting. The important consideration in transporting fruits and vegetables by either rail or truck is to protect the produce against loss in quality during shipment. The temperature and humidity requirements for fruits and vegetables are given in

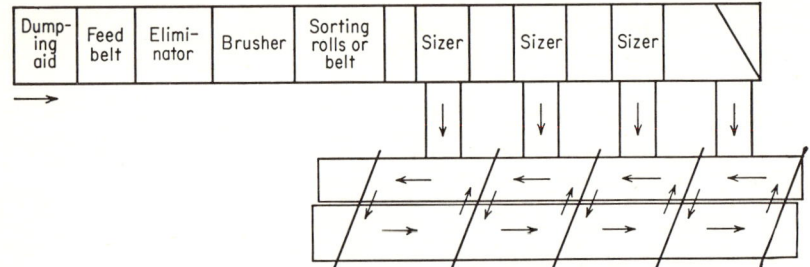

Fig. 52-4. Basic packing-line layout.

Chap. 53. These conditions are more difficult to obtain in transit than in storage rooms; however, they should be approached as near as practically possible, and especially if the transportation period is lengthy.

HANDLING PRACTICES

Citrus. Citrus makes up more than 60 per cent of the tree fruit grown in the United States. The percentage of citrus processed into juice, etc., has increased rapidly during the past six years, and in 1959 processing accounted for about 60 per cent of the total sales of citrus.

Citrus fruit is removed from the trees by workers paid on a piecework basis. Long straight ladders and picking bags which hold approximately 45 lb of fruit are used. The fruit is cut from the tree by a special type of clipper. There is a trend toward removal of the fruit by pulling, especially grapefruit. When a picker's bag is full, the fruit is transferred to field boxes. The boxes used in Florida hold approximately 100 lb, whereas California boxes are somewhat smaller. When filled, they usually are moved to a loading area or assembly yard by orchard trailers and transferred to large trucks for movements to the packing plant. At the plant the boxes are unloaded and either stacked in coloring rooms or dumped directly onto the packing line. The fruit is then sized, washed, sometimes dyed with color and treated with an antiseptic, dried, polished, graded, sized, and finally packed. Oranges may be stored for several months, and grapefruit a month before shipping.

Grapes. All grapes are picked by hand, although an experimental machine for harvesting wine and raisin grapes is under test.[1]

For efficient handling, table grapes are field-packed in lug boxes and placed on pallets holding 90 lugs. Fork lift equipment is then used for all subsequent handling

[1] L. H. Lamouria et al., Designing a Grape Harvester, *Agr. Eng.*, 39:218–236, April, 1958.

operations. The bulk principle is used for handling wine and raisin grapes. The grapes are harvested and moved directly into gondola trucks or bulk trailers.

Apples. Pallets and fork lift equipment are used for apples in many areas. The empty pallets are placed on the bed of the orchard trailer (or truck) and hand-loaded at the picking area with field crates of harvested apples. The containers are stacked so that a pallet and the packages piled on it can be handled as a unit load. The loaded trailer is then moved to a surfaced area in or near the orchard where fork equipment is used. The load of fruit is lifted off the trailer and either stock-piled, moved to the packing line, placed in storage, or loaded onto outgoing trucks.

Lift equipment is also used to place the empty crates on pallets which go back to the orchard.

Bulk pallet boxes are rapidly being adopted for handling all kinds of apples. The bulk boxes are placed in the orchard, filled directly from the picking bags, and handled thereafter with lift equipment. Mechanical tipping devices or flotation dumpers are used to empty the fruit onto the processing or packing line.

Red Tart Cherries. A new development in the handling of red cherries is transportation by bulk tanks containing water. During the 1959 season, over 50 million lb was handled in this way.

The cherries are picked by hand into 10-qt pails. The filled pails are checked, usually two at a time, at an assembly point in the picking area. They are loaded onto an orchard trailer and moved to a loading dock, where they are usually weighed and then poured onto a sorting table from which they move directly into the tank truck. While the tank is being loaded, cold water is usually circulated through the fruit. The water enters at the bottom of the tank, circulates through the fruit, and overflows at the top, cooling the cherries and carrying away a considerable amount of dirt, leaves, and stems.

Once the tanks are filled and the cherries cooled to a temperature of 60°F or less, the flow of water is discontinued. To save weight and prevent sloshing, the water level in the tank is then lowered to approximately 6 in. below the top level of the fruit. The tanks are either pallet tanks or tank trucks.

Upon arrival at the processing plant, the driver attaches a movable flume to the tank outlet, opens the valve, and directs auxiliary water into the tank. The cherries flow through the flume into a receiving boot and move to soak tanks or the processing line.

White Potatoes. Potatoes are harvested by tractor-drawn mechanical diggers which dig one or two rows at a time. (See Chap. 27 for information on potato diggers.) The digger is followed by workers who pick the potatoes off the ground and put them in sacks. Some areas use mechanical harvesters which not only dig the potatoes but move them over shaker cleaners and sorting rollers and convey them into bulk wagons or trucks. The potatoes are moved to storages and then to the packing shed, where they are washed, sorted, sized, and packed.

The trend is toward mechanical harvesting into bulk wagons or trucks or into 40-bu bulk pallet boxes. The potatoes are placed in bulk storages or stored in the bulk boxes. Upon preparation for market, many potatoes are being prepackaged in paper or mesh consumer bags holding 10 or 15 lb.

Tomatoes. For canning, tomatoes are harvested at the red-ripe state of maturity. Tomatoes for the fresh market are picked at the mature-green, pink, or red-ripe stages, depending on the distance to be shipped and their subsequent use. During the winter months, many tomatoes are picked in the mature-green stage for shipment to terminal markets, where they are ripened. Tomatoes are removed from the vines by hand and placed into lug boxes or hampers. These containers are moved to the canning plant or packing house. The tomatoes then are graded and packed in lugs and shipped to terminal markets, where a great percentage of them are prepacked in 1-lb cartons holding three or four tomatoes.

Sweet Potatoes. Upon reaching maturity, the first operation in harvesting sweet potatoes is to remove the excess vines either by hand, knife, or special vine-cutting cultivators. The potatoes are then lifted from the ground with a double moldboard plow. After drying for several hours they are usually gathered in piles and devined and graded simultaneously.

A minimum of handling is desirable since sweet potatoes are easily bruised. Curing at a temperature of 85°F and a relative humidity of 90 per cent for 10 days is essential for healing the abrasions and wounds incurred in harvesting.

Cabbage. Cabbage is hand-harvested by bending the plant to one side and cutting the head with a knife, just below the wrapper leaves. The heads are usually moved out of the field in bulk wagons and packed in crates holding from 25 to 35 heads or in 50-lb mesh sacks. They may be stored in bulk bins before packing.

Sweet Corn. Sweet corn must be handled quickly. For the fresh market, corn at the milk stage of maturity is picked by hand and promptly precooled. If the market is within several hours distance, it is often trucked in bulk with snow or chip ice. Otherwise it is packed in bags or wirebound crates and shipped under ice.

For processing, the trend is to harvest the crop with mechanical corn pickers, move it in bulk trucks to the plant, where it is dumped onto a floor. The corn is then pushed by bulldozer onto conveyers, which move it into the plant.

Lettuce. Head lettuce is cut just above the ground by hand, using a V-shaped blade which is attached to a long handle. The lettuce is trimmed of the nonwrapper leaves and field-packed into corrugated cartons which hold 2 dozen heads. The packed cartons of lettuce are precooled by vacuum (Chap. 53) and immediately loaded onto outgoing trucks or railroad cars which are refrigerated with ice or mechanically.

Onions. Onions are loosened in the soil by running some type of blade or cutter below the bulbs. Commonly, they are pulled by hand and then put into windrows for drying. After several days, the tops are removed by hand and the onions placed into field crates and allowed to cure for a few days. The crates are then moved to a storage and held 6 to 7 months before being packed and shipped to market. Packing consists of sorting for grade, sizing, and packing in sacks.

The trend is to harvest the crop with machines that dig, top the onions, and elevate them into bulk wagons or bulk pallet boxes (Chap. 27). They are moved to a bulk storage or stored in the bulk boxes. Curing is done in the storage by forcing air through them. Many onions are now prepackaged in 3- to 5-lb consumer bags.

Celery. In the last few years celery harvesters have been developed and are in extensive use in Florida. These machines cut off the roots, lift the stalks, and carry them over conveyers. Workers stationed on the machine inspect and trim the stalks and pack them in crates. Upon leaving the moving harvester, the celery is already crated for movement to railroad cars or trucks for shipment to market.

Chapter 53

STORAGE OF FRESH FRUITS AND VEGETABLES

Donald H. Dewey and Jordan H. Levin

Cold storage has an important role in the modern-day distribution and marketing of fruits and vegetables. Every commodity, whether fresh or processed, is subjected to one or more forms or types of storage. Storage may be extremely simple for products grown and marketed locally. For roadside marketing, storage may mean the mere holding of the crop in the shade for a few hours until it is purchased directly from the farmer. In other cases, storage may be complex, as illustrated by lemons harvested during the winter months, stored and treated in a modern storage plant, and then transported during the summer for several thousand miles to the market distributor, who will channel them to the consumer.

The *purpose* of storage is to preserve and possibly enhance the inherent edible-quality characteristics of harvested crops until they are processed or consumed in the fresh state.

Some crops can be stored for several months; others deteriorate rapidly and can be held for only a week or so. For the latter, storage is equally essential and desirable. The application of low-temperature storage to short-lived crops offers assurance of delivery to a processor or consumer.

The term *cold storage,* or *cooler storage,* as applied to fresh fruits and vegetables refers to storage in the temperature range of approximately 30 to 50°F. Storage at temperatures below 30°F employed for holding foodstuffs in the frozen state is known as *freezer storage.*

PRINCIPLES OF STORAGE

The methods of storage practiced today for fresh fruits and vegetables are aimed toward retarding deterioration caused by pathological infection and by metabolic changes of the tissue as a result of aging.

Decay. Decay caused by disease organisms is a primary problem in the storage of fresh commodities. Because the initial infection of produce frequently occurs in the field, much decay in storage can be avoided by a sound control program during the growing season. Other infections originate at harvest or shortly thereafter through skin breaks and other mechanical injuries associated with rough handling.

The most effective method of retarding decay development is by *temperature reduction.* Practically all decay organisms are slowed in their rate of growth and development when the temperature is reduced from field temperatures to near freezing. Some diseases are retarded adequately at temperatures near 45°F, i.e. Rhizopus soft rot, bacterial soft rot, and Fusarium rot on vegetables. Others are not retarded adequately to prevent spoilage at this temperature, and even lower temperatures of 30 to 32°F will not completely inhibit some organisms. Watery soft rot and gray mold rot will progress at these lower temperatures and seriously affect fruits and vegetables when stored for long periods. Decay organisms are not killed by holding at temperatures near freezing in many cases, but remain inactive until the temperature is raised.

A few commodities are susceptible to physiological injury because of storage at low temperatures above freezing. This so-called *chilling injury* results in a weakening of the tissues and an increased susceptibility to disease infection. Sweet potatoes stored at a temperature below 50° are affected in this manner, and the increased amount of decay that follows is frequently the most pronounced symptom of chilling injury.

Respiration. Fresh fruits and vegetables are living tissues that are not killed as a result of harvesting. They are useful for processing or fresh consumption only as long as they are maintained in a living condition. Upon harvest, the product is separated from its source of food materials and water, and continued life depends upon the supply of stored materials and the rate at which they are lost or utilized following harvest. As living plant tissues, fresh fruits and vegetables carry on respiration, the combining of the carbohydrates in the plant cells with the oxygen of the air to form carbon dioxide and water and to release energy, mostly as heat. As a general rule, the respiration rate is approximately doubled for each temperature increase of 18° from 30 to 90°F. The use of temperature reduction to reduce respiration is directly beneficial by slowing down the degradation of sugars and other carbohydrate material stored in the cells, and indirectly by reducing the refrigeration requirements.

The respiratory activity and, consequently, the heat production of the various kinds and varieties of produce are extremely variable. The *quantities of heat* evolved in units of Btu's per ton per 24 hr by many fruits and vegetables as a result of the respiration process have been tabulated by Wright et al.[1] From these figures, the crops have been classified in Table 53-1 according to their relative activity.

Table 53-1. The Relative Respiratory Activity of Fresh Fruits and Vegetables as Represented by Their Heat Evolution, in Btu/(Ton)(24 Hr) at 60°F

(1)	(2)	(3)	(4)	(5)
1000–4000 Btu	4000–7000 Btu	7,000–10,000 Btu	10,000–30,000 Btu	Above 30,000 Btu
Apples, fall	Bananas	Apples, summer	Beans, lima	Beans, snap
Cranberries	Cabbage	Beets, topped	Cherries, sour	Broccoli
Grapefruit	Oranges	Cantaloupes	Cucumbers	Corn, sweet
Grapes, Vinifera and American	Sweet potatoes	Carrots, topped	Pears, Bartlett	Lettuce
Lemons	Tomatoes	Celery	Raspberries	Mushrooms
Onions	Turnips	Peaches	Strawberries	Peas
Potatoes		Peppers		Spinach

The crops listed in columns 1 and 2 are mature plant parts and, for the most part, well suited to long-time storage. Those in column 5 are short-lived following harvest. Their perishability is related to the immaturity of the tissues, their leafy nature, or their high sugar content. Groups in columns 1 and 4 represent the extremes and merely illustrate the general relationship of respiratory activity to suitability to storage. There are many other factors determining ultimate suitability to storage; consequently, the crops in the intermediate groups cannot be readily classified by respiratorial behavior.

Chemical Changes. The changes associated with respiration are chemical in nature, and the factors affecting respiration have a similar effect on the many other chemical changes in harvested fruits and vegetables.

The depreciation of flavor because of *sugar losses* is often due to changes of sugar to starch as well as to the direct losses because of respiration. Although the general effect of low temperature is to decrease the rate of these changes, low temperature in some cases favors a conversion of starch to sugar. This conversion is desirable for parsnips, and consequently they are left in the field unharvested until after freezing weather in order to attain the sweetness required for high quality. White potatoes, on the other hand, are adversely affected in flavor by the conversion of starch to sugar. They become undesirably sweet when stored at temperatures below 40°F for several weeks or longer.

[1] R. C. Wright, Dean H. Rose, and T. M. Whiteman, The Commercial Storage of Fruits, Vegetables and Florist and Nursery Stocks, *USDA Agr. Handbook* 66, 1954.

The *softening* of fruits in storage is affected by the change of protopectin to soluble pectin. Such a change, characteristic of apples, pears, tomatoes, and numerous other crops, is slowed by temperature reductions.

Water Loss. The loss of water from most fresh fruits and vegetables is of serious concern to shippers and handlers since it occurs rapidly even under good handling and storage procedures and may markedly affect quality. Wilted products are unsightly and lacking in crispness. Water losses account for much of the weight shrinkage of sound products after harvest and during preparation for market. These weight losses alone may mean considerable financial loss.

The *water losses* of several vegetables under experimental conditions are illustrated in Fig. 53-1. The smallest losses occur for mature fruits, such as winter squash, and for root crops, such as topped carrots. Crops composed of immature and leafy tissues, like Brussels sprouts, have the greatest losses.

The *quantity of water loss seriously affecting* the appearance or condition of produce is also rather specific for different crops. Apples will develop noticeable wilt

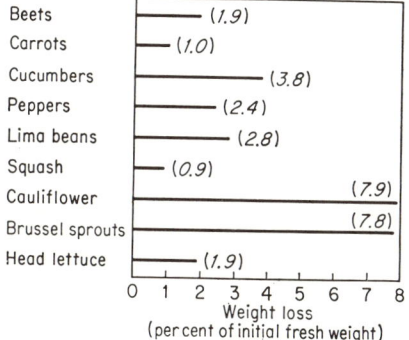

FIG. 53-1. Weight losses of vegetables held at 32°F and 98 per cent relative humidity for 3 weeks. (*Adapted from Platenius, Studies on Cold Storage of Vegetables, Cornell Univ. Bull. 602, 1934.*)

with a 5 per cent loss, and leafy vegetables are affected by a 2 per cent loss of their original fresh weight.

The important *environmental factors affecting water loss* are temperature and relative humidity. The reduction of transpiration by lowering the temperature or raising the relative humidity is simply explained on the basis of the movement of water vapor from areas of high vapor pressure to areas of low vapor pressure. Fresh fruits and vegetables contain a large percentage of water, normally in the range of 75 to 95 per cent (Table 53-2). Because of the high water content of the tissues, the air contained by the tiny spaces between the cells of the tissues is saturated with water vapor at any given temperature, or in other words, the relative humidity of the confined air is 100 per cent. This water vapor will move from the areas of high concentration within the tissues to areas of lower concentration outside of the tissues. The difference in concentration of water vapor is expressed as the saturation deficit, and the magnitude of the deficit directly affects the rate of water movement. The *saturation deficit* actually is the difference in vapor pressure of air at a given temperature at 100 per cent relative humidity and air at the same temperature at a lower relative humidity.[1]

Warm produce is extremely susceptible to water loss when it is placed in a cold room. The large saturation deficit due to the temperature difference between the produce and the storage-room air will cause excessive water losses even under conditions of 100 per cent relative humidity in the storage room. Water losses can be best minimized by rapidly cooling the produce. Air circulation, although it normally increases water losses, may actually reduce water losses under these circumstances because of its value in speeding the rate of cooling of the produce. Once cooling is

[1] A. D. Edgar, Studies of Potato Storage Houses in Maine, *USDA Tech. Bull.* 615, 1938.

TABLE 53-2. THE STORAGE REQUIREMENTS AND CHARACTERISTICS OF THE MORE IMPORTANT FRESH FRUITS AND VEGETABLES

Kind and variety or condition	Recommended storage conditions		Approximate storage life	Average freezing point, °F	Approximate water content, %	Approximate heat of respiration, Btu/(ton)(24 hr)	
	Temperature, °F	Relative humidity, %				At 32°	At 60°
Fruits							
Apples:							
Jonathan	34–36	85–90	2–4 mo	28.0		750	3000
McIntosh	36–38	85–90	2–5 mo	27.8			
Golden Delicious	30–32	85–90	3–6 mo	28.0			
Red Delicious	30–32	85–90	3–6 mo	27.7–29.7	84		
Northern Spy	30–32	85–90	4–6 mo				
Rome Beauty	30–32	85–90	4–7 mo	28.2			
Winesap	30–32	85–90	5–8 mo	27.0		300	2300
Apricots	31–32	85–90	1–2 wk	28.9–29.9	85		
Avocados	45 or 55[a]	85–90	4 wk	29.0–31.1	65–82		
Cherries:							
Sweet	31–32	85–90	10–14 days	26.2–28.2	80	1250	
Sour	32	85–90	2–3 days	28.9	84	1500	12,000
Cranberries	36–40	85–90	1–3 mo	29.7–30.3	87	650	
Figs, fresh	31–32	85–90	10 days	27.1	78		
Grapefruit	....[b]	85–90	1–3 mo	28.3–29.6	89	650	2600
Grapes:							
Vinifera	30–31	85–90	3–6 mo	26.5–27.7	82	400	2500
American	31–32	85–90	3–4 wk	29.4	82	600	3500
Lemons	50–58	85–90	1–4 mo	28.5–29.3	89	700	3650
Limes	48–50	85–90	6–8 wk	27.9–28.6	86		
Oranges:							
Florida	30–32	85–90	8–12 wk	26.7–29.7	87	750	4400
California	35–37	85–90	5–8 wk	29.2	87		
Peaches	31–32	85–90	2–4 wk	29.0–30.1	87	1100	8300
Pears:							
Bartlett	30–31	90–95	1½–3 mo	28.0	84	750	1050
Fall and winter	30–31	90–95	2–7 mo[e]	26.5–28.6	83		
Pineapples:							
Mature-green	50–60	85–90	2–3 wk	29.4–30.0	85		
Ripe	40–50	85–90	2–4 wk				
Plums and prunes	31–32	85–90	3–4 wk	26.5–29.8	86		
Strawberries	31–32	85–90	7–10 days	30.1–30.5	90	3250	17,950
Tangerines	31–38	90–95	2–4 wk	29.3–29.7	87		
Vegetables							
Artichokes, Globe	32	90–95	30 days	29.6	84		
Asparagus, green	32	85–90	3–4 wk	30.3–30.5	93		
Beans:							
Green or snap	45–50	85–90	8–10 days	30.0–30.2	89	5600	32,100
Lima	32–40	85–90	10–20 days	30.9	67[d]	2350	22,000
Beets:							
Topped	32	90–95	1–3 mo	28.6–30.2	88	2650	7250
Bunched	32	90–95	10–14 days	29.5–30.2			
Broccoli, Italian	32	90–95	7–10 days	29.9–30.9	90	7450	58,100
Brussels sprouts	32	90–95	3–4 wk	30.2	85		
Cabbage:							
Early	32	90–95	3–6 wk	30.0			
Late	32	90–95	3–4 mo		92	1200	4000
Carrots:							
Topped	32	90–95	4–5 mo	28.4–29.1	88	2150	8100
Bunched	32	90–95	10–14 days				
Cauliflower	32	85–90	2–3 wk	30.1–30.4	92		
Celery	31–32	90–95	2–4 mo	30.7–31.1	94	1600	8200
Corn, sweet	31–32	85–90	4–8 days	30.7–30.8	74	6550	38,400
Cucumbers	45–50	85–90	2–3 wk	29.9–31.0	96	1700	10,450
Eggplants	45–50	85–90	10 days	30.3–30.4	93		
Endive, or escarole	32	90–95	2–3 wk	30.7–31.7	93		
Garlic, dry	32	70–75	6–8 mo	26.3–30.3	74		
Lettuce	32	90–95	2–3 wk	31.0–31.5	95	11,300	46,000

TABLE 53-2. THE STORAGE REQUIREMENTS AND CHARACTERISTICS OF THE MORE IMPORTANT FRESH FRUITS AND VEGETABLES (Continued)

Kind and variety or condition	Recommended storage conditions		Approximate storage life	Average freezing point, °F	Approximate water content, %	Approximate heat of respiration, Btu/(ton)(24 hr)	
	Temperature, °F	Relative humidity, %				At 32°	At 60°
Vegetables (Continued)							
Melons:							
Watermelons	36–40	85–90	2–3 wk	30.3–31.2	92		
Cantaloupes	40–50	85–90	4–14 days	29.6	93	1300	8500
Honey dew	45–50	85–90	2–3 wk	29.8–30.0	91		
Onions, dry	32	70–75	6–8 mo	30.0–30.2	82	900	3000
Parsnips	32	90–95	2–4 mo	29.5–30.1	79		
Peas, green	32	85–90	1–2 wk	29.7–30.5	74	8200	42,000
Peppers, sweet	45–50	85–90	8–10 days	30.5	92	2700	8500
Potatoes:							
Early crop	50–70^e	85–90	^e	28.4–30.4			
Late crop	38–50^e	85–90	5–8 mo^e	28.7–30.4	78	700	2000
Pumpkins	50–55	70–75	2–6 mo	29.9	90		
Rutabagas	32	90–95	2–4 mo	29.7	89		
Spinach	32	90–95	10–14 days	31.3	93	4550	37,500
Squashes:							
Summer	32–40	85–95	10–14 days	29.8–30.9	95		
Winter	50–55	70–75	4–6 mo	29.8–30.0	88		
Sweet potatoes	55–60	85–90	4–6 mo	28.7–29.3	69		
Tomatoes:							
Ripe	50	85–90	8–12 days	30.2–30.8	94	1000	5650
Mature green	55–70	85–90	2–6 wk	30.6	95	600	6250
Turnips	32	90–95	4–5 mo	29.8–30.8	91	1950	5300

^a Most varieties at 45°, some West Indian varieties at 55°F.
^b Storage temperatures may vary from 32 to 55° by area; consult local authority.
^c Storage period variable for varieties; freezing points about 27°F.
^d Shelled Lima beans.
^e Variable temperatures and storage life according to areas and uses; consult local authority.

SOURCES: All information except water contents and freezing points from R. C. Wright, Dean H. Rose, and T. M. Whiteman, The Commercial Storage of Fruits, Vegetables and Florist and Nursery Stocks, *USDA Agr. Handbook* 66, 1954. Water contents from Charlotte Chatfield and Georgian Adams, Proximate Composition of American Food Materials, *USDA Circ.* 549, 1940. Freezing points from T. M. Whiteman, Freezing Points of Fruits, Vegetables, and Florist Stocks, *USDA Market Research Rept.* 196, 1957. (Range in average freezing points by varieties.)

accomplished, it is desirable to reduce air circulation to a level adequate to merely maintain uniform temperatures in the storage room. Although most fresh produce theoretically should be stored at 100 per cent relative humidity in order to prevent water losses and wilting, it is neither practical nor feasible.

The *recommended relative-humidity* levels for various crops are shown in Table 53-2. As a general rule, the leafy and root vegetables require a level of 90 to 95 per cent, and most of the other vegetables and practically all fruits require 85 to 90 per cent. Onions, as well as other vegetables which are stored dry, need a lower relative humidity.

Storage Disorders. Stored crops are susceptible to damage other than that caused by decay organisms. These storage disorders are commonly known as physiological disorders.

Freezing disorders are caused by exposure of the crop to low temperature. Freezing disrupts the cellular structure of the tissues and may actually kill the cells so that the tissues readily decompose. All crops have freezing points below 32°, the freezing point of water. Reduction of the temperature of the tissues below the freezing point does not necessarily mean the crop will be damaged, because some crops can be frozen a number of times without serious damage. Others may be damaged by a single freezing. The *USDA Agriculture Handbook* 66 has classified crops according to their susceptibility to freezing damage.

Many crops, in addition to being damaged by freezing, are injured by storage at temperatures above freezing. The crops listed in Table 53-2 with recommended storage temperature above 32°F, such as citrus, tomatoes, and sweet potatoes, are susceptible to *chilling injury,* a weakening of tissues making them very susceptible to decay after warming.

There are a number of factors other than freezing and chilling that are responsible for physiological disorders during the storage period. One of these is a lack of oxygen for normal respiratory needs of the product. *Suboxidation,* or *anaerobic respiration,* is responsible for black heart of potatoes, russeting of snap beans, smothering of cranberries, and albedo browning of lemons. The emanations, or gases, given off by fruits or vegetables sometimes cause injury, as for example apple scald or storage scald.

Insect and Rodent Damage. Insects and rodents are usually carried into storage rooms with the produce. Consequently considerable control, especially for insects, can be accomplished by their elimination in the orchard or field. Infestation of dried fruits and vegetables and nuts is a problem in common storage that is practically eliminated at lower temperatures in refrigerated storage.

HEAT REMOVAL

Sources of Heat. Stored commodities have two main sources of heat: (1) the field, or sensible, heat and (2) the respiration, or vital, heat.

Under ordinary conditions of storage whereby a crop is harvested and placed in cold storage or under refrigeration for transport purposes, the *field heat* is the greatest quantity of heat that must be dealt with in a short period of time. It is the heat that must be removed from a commodity in order to lower its temperature to the desired level.

The initial temperature will vary, depending upon environmental conditions of season and climate. Such practices as harvesting during the night or in the early part of the day and shading the harvested crops not only reduce the quantity of field heat to be removed in storage, but also reduce the rates of deterioration.

The *specific heat* of a commodity is dependent upon its water content. Fruits and vegetables have specific heats slightly below that of water, which means that their temperature can be changed 1°F by adding or removing less than 1 Btu of heat for each pound. The specific heat for commodities at temperatures above their freezing point is obtained with Siebel's formula, $S = 0.008a + 0.20$, in which S is specific heat, a is the percentage of water, and 0.20 is the specific heat of the solid constituents or dry matter.

The *field heat of the containers* holding the produce, trucks, or other handling equipment placed in the cooler must also be considered. Wood or paper containers have variable specific heats, in the range of 0.32 to 0.42, depending upon their water content.

The *respiration,* or *vital,* heat is the energy released from the tissues of the stored fruit and vegetables by the respiration process. The amounts of heat released by various crops at temperatures of 32 and 60° are listed in Table 53-2.

A constant source of heat that must be considered in refrigeration practice is *leakage.* This is the heat transmitted through the walls, ceiling, and floor of the structure and the heat carried by air exchanged by infiltration through cracks and crevices and by door openings.

Other heat sources are lights, electric motors, workmen and equipment, and the dead heat of the building. These, together with the heat leakage resulting from door openings, are sometimes estimated to be 10 per cent of the total of all other heat sources.

Precooling refers to the rapid removal of heat from commodities, generally prior to shipment, but also before storage, immediately after placement in storage, or prior to packing.

Air Cooling. Precooling by refrigerated air is accomplished in many ways. The simplest method is by rapid circulation of air cooled to about 32°F in specially designed rooms utilizing ice or ice with salt in bunkers. This is a relatively inexpensive method of warehouse precooling under conditions of short-season operation. Mechanical refrigeration is employed for precoolers operated in conjunction with cold-storage facilities. Precooling rooms are usually designed to provide two to four changes of the total volume of the air of a room per minute.

Tunnel coolers, utilizing high-velocity air blasts directed into the containers, are sometimes employed for rapid warehouse cooling of fruit such as grapes and cherries. Pentzer et al.[1] found that grapes in lug boxes could be cooled from a temperature of 75 to 80°F to 40 to 45° in 1 to 1½ hr by refrigerated air directed into the top of the unlidded containers at a rate of 400 to 500 lin ft per min. Commercial tunnel coolers are used in California for this purpose.

Precooling in railroad refrigerator cars is a common practice in many shipping areas. Refrigeration is often supplied from bunker ice with forced-air circulation provided by the built-in car fans driven by auxiliary motors or by special fan units temporarily installed for the precooling operation. Temperature reductions of approximately 40° for carloads of fruit in wooden containers in cars with air movement of approximately 8,000 cfm are attained in 14 to 18 hr of precooling.

Mechanical refrigeration from either trackside warehouses or individual portable or stationary trackside units is sometimes utilized for railroad-car and truck precooling.

Truckloads are more difficult to precool than railroad carloads because they are refrigerated from one end of the body, and spacings at the walls, floor, and between the containers are usually too limited for good air circulation.

Contact Icing. Crushed ice is applied usually as a means of transit refrigeration, but its values for precooling should not be overlooked. Much of the value of *package icing,* whereby crushed ice is placed in contact with the produce in the packing operation, lies in its immediate and rapid cooling effect. Package icing is used for broccoli, carrots, cauliflower, celery, sweet corn, and several other vegetables. Crushed ice in quantities up to 30 lb per package is applied between layers of the vegetables as they are packed.

Body icing, or *top icing,* refers to the application of crushed or snow ice over the whole load and is used for both precooling and in-transit refrigeration. It is used as a precooling method for cantaloupes.[2]

Hydrocooling. Water is used extensively for the removal of field heat from perishables in the United States. It is a rapid means of cooling, since water is an excellent heat-transfer medium and intimate contact with the produce is attained. Ice is the most common means of refrigeration, but mechanical refrigeration is also practical. This method of cooling is employed for such vegetables as asparagus, beets, broccoli, Brussels sprouts, carrots, celery, green onions, peas, radishes, spinach, snap beans, and sweet corn. Fruits are also cooled this way, especially peaches and sweet cherries.

Most hydrocoolers are constructed to expose the packed produce to a heavy shower or deluge of cooler water rather than to submerge the produce in an ice bath. The produce is conveyed by belt through the shower at an appropriate rate of travel to provide the desired amount of cooling. The ice or evaporator coils are located in the overhead, shallow, flat-bottomed tank, which is perforated to form the continuous shower of water. A large volume of water flow is essential, and a cooler 25 ft long and 8 ft wide should circulate over 2,000 gal of water per minute. Since the water is recirculated, there is a constant accumulation of dirt and decay organisms in the cooler. Fresh water should be used daily, or more often, as necessary to prevent excessive contamination of the produce. The water temperature should be maintained as near to 32°F as possible for best results.

The time required for hydrocooling varies with the kind of commodity and the initial temperature. Peaches may be cooled from 85 to 45°F in about 16 min, whereas smaller commodities such as asparagus and sweet cherries are cooled from 80 to 33°F in 9 to 12 min.

[1] W. T. Pentzer, C. E. Asbury, and W. R. Barger, Precooling California Grapes and Their Refrigeration in Transit, *USDA Tech. Bull.* 899, 1945.

[2] W. T. Pentzer et al., Top Icing Cantaloupes, *USDA Handling, Transportation and Storage Office Rept.* 185, 1948.

Hydrocooling by submersion is limited in use because of the slower rate of cooling and smaller capacity compared with the shower type.

Vacuum Cooling. Cooling by evaporation is a new concept of precooling produce for transit and storage. Vacuum cooling depends entirely upon the evaporation of water for removal of field heat from produce. It is based on the principle that a reduced pressure results in a lowered temperature of vaporization, or boiling point, of water. In order to boil water at 32°F, the normal atmospheric pressure of 29.92 in. of mercury must be reduced to 0.18 in., or in other words, a vacuum of 29.74 in. must be drawn. Since evaporation is a cooling process, heat can be removed from produce by encouraging water to evaporate from it in a vacuum chamber.

Large-capacity steam ejectors or mechanical pumps are required to produce the necessary vacuum. With steam ejectors, water-cooled condensers are used to aid in the withdrawal of water vapor, whereas with pumps, the water vapor from the produce is condensed between the vacuum chamber and the pump by mechanical refrigeration or ice.

The vacuum chambers or tubes generally hold ½ carload (approximately 300 two-dozen cartons) of lettuce and are loaded and unloaded mechanically on dollies or pallets. One-half hour is the usual cooling period for reduction of temperatures from 70 or 80 to 32°F. A vacuum-cooling plant with a capacity of 3 carloads per hr costs over $200,000. Consequently, a long period of operation each season is essential to justify the original investment in equipment. It is estimated that 90 per cent of the lettuce shipped from Arizona and California is vacuum-cooled. Other lettuce plants are located in Texas, Florida, Wisconsin, New York, and in Ontario, Canada.

KINDS OF STORAGE

The reduction and maintenance of temperature in a storage is dependent upon the prevention of heat movement into the storage area and on the removal of heat that accumulates in the storage. All storages are more or less protected against outside heat by insulation, but vary greatly in means of heat removal.

Air-cooled, or Common, Storage. Heat removal in the common, or air-cooled, storage is dependent on ventilation of the storage area with outside air. The outside air, when of a lower temperature than the air or commodities within the storage, is allowed to flow through the storage or may be forced through the storage by power-driven fans. Ventilation is cut off when the temperature of the outside air is above that of the commodities in the storage. Once the commodities are cooled to the desired temperature, ventilation is continued only to maintain this temperature or possibly to prevent freezing in the storage. In the latter case, outside air is admitted only when it is warmer than the commodity. In most common storages, however, inward heat leakage through the floor and the walls in submerged storages, plus the heat of respiration of the commodity, is adequate to prevent damage by freezing.

The simplest type of common storage is *field storage* in *trenches* and *pits* in which the produce is covered by straw and earth as protection against freezing after it is cooled to 35 to 40°F. This is temporary storage, but such crops as cabbage, celery, turnips, beets, carrots, parsnips, and potatoes can be held 1 to 2 months. There is no control of temperature and humidity; the produce can be removed only with great difficulty when the ground is wet or frozen; the produce may be injured by low temperature or by water when it is removed from the storage; and the labor costs of handling are large. This kind of storage, however, is always available, and construction costs are very low.

An improvement in field storage over the pit or trench type is the *aboveground* type. Ground that is level and well drained should be selected for this purpose. The produce is piled around an upright flue connected to ventilating trenches and covered with straw or hay and soil for protection against rain and cold weather. When there is danger of freezing, the ends of the ventilating trenches are closed and additional soil mounded over the pile of produce. All the produce should be removed once the pile is opened; consequently, the size should be adjusted to meet this requirement.

Permanent *outdoor cellars* are frequently used for the storage of root crops and potatoes. These cellars have earth on all sides, serving as insulation, and frequently the roof is covered with earth. A permanent entrance for loading and unloading is essential for easy access to the cellar. Provision must be made for adequate ventilation. Natural convection is possible where the storage is built into the side of a bank; otherwise forced air circulation with fans is essential.

The most widely used common storage building is *aboveground, with insulation* incorporated in the walls and ceiling. This type of structure is used for numerous crops, and especially for potatoes, sweet potatoes, root crops, onions, cabbage, and apples. The buildings may be new or adapted from existing structures such as barns. Regardless, the walls and ceilings must be adequately insulated against heat movement, the floors may be soil or concrete, and good ventilation must be provided since the outside air is the only provision for cooling the stored commodities.

Recommendations for *apples* in Michigan are that a storage of 10,000-bu capacity should be ventilated with four electric exhaust fans with a total capacity of 6,000 cfm located near the ceiling.[1] Eight air intakes, each with openings of 2.25 sq ft,

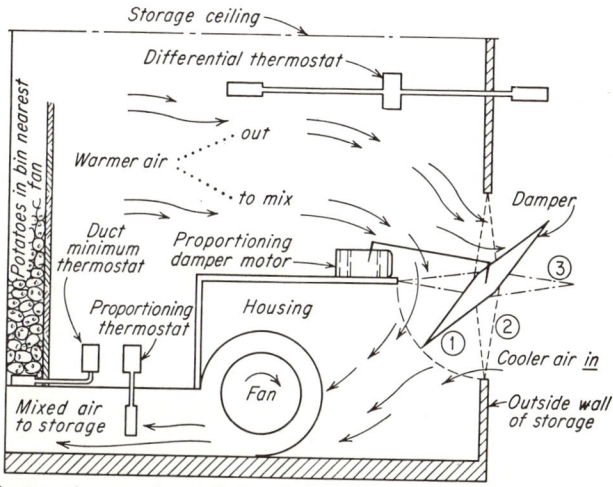

Fig. 53-2. Diagram of automatic ventilating device for a potato storage showing locations of fan housing, thermostats, air duct, and proportioning damper: (1) shows damper in partly closed position for ventilation with a mixture of outdoor and recirculated air; (2) in position for recirculation; and (3) in full-open position for ventilation. (*From Haynes, Jr., White Potato Storages for New Jersey, Long Island, and Southeastern Pennsylvania, USDA Market Research Rept.* 70, 1954.)

should be spaced around the base of the storage walls to distribute the air through the storage and to provide maximum cooling. Ventilation should be used to remove only the initial heat and to maintain the desired temperature. Further ventilation reduces the relative humidity and results in excessive wilting and shriveling of the fruit.

Air-cooled storages for *potatoes* are designed differently from those for apples because potatoes are held at higher temperatures (40 to 50°F) than apples (32 to 38°F), and potatoes are frequently stored loose in bulk or bins, whereas apples are stored in crates or boxes. The bulk potato-storage structure must be designed to withstand the pressure of the loose potatoes against the walls, and the ventilating system should not interfere with the loading and unloading operations. Often ducts are built into the floors so as to serve for both ventilation and unloading. Modern, large-capacity potato storages utilize forced-air systems which circulate air at the rate of approximately ⅓ cfm for each bushel of potatoes. They are often designed to automatically introduce outside air, mix outside air with inside air, or merely recirculate the inside air. Such a system is illustrated in Fig. 53-2.

[1] Roy E. Marshall, Construction and Management of Farm Storages with Special Reference to Apples, *Mich. State. Univ. Ext. Circ.* 143, 1951.

Common storage is used almost exclusively for the globe type of *onion* in the northern onion-growing states (Wisconsin, Indiana, Michigan, and New York). In the past, most onions were field-cured and stored in crates at temperatures down to 32°F and at humidities in the range of 70 to 75 per cent. Recently, bulk storage, either in boxes or loose, to a depth of 8 to 10 ft, has become popular. The onions are mechanically harvested and topped and immediately placed in storage. Curing is best accomplished by forcing naturally warm and dry outside air through the pile of onions. Artificially heated air is sometimes employed during rainy, cool weather, but temperatures over 80°F should be avoided. Once cured, the onions are cooled by forced circulation of outside air whenever the air temperature is lower than that of the onions. Recirculation of air in the storage room is often desirable after cooling in order to maintain uniform temperatures, to reduce moisture condensation, and to prevent possible freezing.

Home basements are unsatisfactory for the storage of fruits and vegetables because of their high temperature and low humidity. A small insulated room in the basement provides an ideal storage for the home. It should be air-cooled from an outside basement window through a ventilation flue extending from the outside of the house to the floor of the storage room. An outlet for warm air is provided by replacing one of the window panes with a vent. All sides of the room should be insulated. Sand on the floor, when moistened, aids in maintaining a high humidity.

Refrigerated Storage. An improvement over common or air-cooled storage is artificial refrigeration by bunker *ice*, usually in conjunction with air cooling. The ice is utilized during the initial cooling-down period to remove the large quantities of field heat and to hold the commodities at low temperatures until the outside air temperature is adequately low for storage purposes. The ice capacity of the storage is dependent upon the refrigeration needs and the frequency of ice replacement.

Practically all terminal or centralized cold storages are *mechanically refrigerated,* whereas most common or air-cooled storages are located at the farm. Many new farm storages are mechanically refrigerated, and in fact many air-cooled storages have been converted to artificial cooling in recent years. This trend has developed because of the need for better control of temperature and humidity, the relative decrease in the cost of mechanical refrigeration, and the greater access to farm storages by trucks through better highways.

According to recent data,[1] mechanically refrigerated farm storages cost about $1\frac{1}{3}$ times the cost of common storages of similar capacity and construction. They have been built and equipped with refrigeration for approximately $1 to $1.50 per bushel of capacity and operated at a yearly cost of 20 to 30 cents per bushel, depending on the size of the storage building.

Buildings with single floors are often preferable to multifloor structures because of the convenience in loading and unloading without the expense of elevators or conveyers. Single-floor structures are convenient for lift-truck operations in which the commodities are handled on pallets.

All new storages should be constructed for pallet handling, or at least for future conversion to pallets. This means that the ceilings should be at least 15 ft above the floor, the floors should be level and of concrete construction to handle a live load of at least 600 psf, the floor space should be free of pillars or posts that might interfere with the lift-truck operation, and the doors should be adequately large, at least $4\frac{1}{2}$ ft wide and 8 ft high, to accommodate the trucks and pallet loads.

The *space requirements* for palletized stock are slightly greater than for hand-stacked containers because of the spacing of the stacks and the larger aisles needed for maneuvering lift trucks. Normally, 2.5 to 2.8 cu ft of space is required for 1-bu boxes, without pallets, and 2.75 to 3 cu ft per bu with pallets. These allowances include the space needed for aisles, overhead clearance, and air circulation between the packages, as well as the space for containers. Hand-stacked 1-bu baskets require 3.5 to 3.8 cu ft of storage space for each container.

The *types of construction* employed for refrigerated storage buildings vary from

---

[1] H. P. Gaston and J. H. Levin, On-the-farm Refrigerated Fruit Storages, *Mich. State Univ. Agr. Expt. Sta. Spec. Bull.* 389, 1954.

single walls with plank types of insulation to double walls of masonry or timber with fill insulation. For cold storage at 32°, insulation equivalent to 4 in. of corkboard is adequate for walls in the northern part of the United States and equivalent to 5 in. in the southern part. Ceilings generally have an additional 1 in. of insulation. Fill types of insulation should be 1⅓ to 1½ greater in thickness than plank types. Reflective insulations have been used satisfactorily in the walls and in combination with other insulation materials in the ceiling.

Because of the marked difference in temperatures between the two sides of a refrigerated cold-storage wall, there is a strong tendency for water vapor to move through the wall. This water vapor is likely to condense and accumulate in the insulation and destroy much of its insulating value. Consequently, the insulation must be protected by a *vapor barrier*, or *seal*. The warm side of the insulation should be sealed, whereas the cold side should be unsealed. In a refrigerated storage the warm side of the insulation is considered to be the *outside*, even though the temperature inside the storage may be higher than the outside air during the colder days of the winter season. The vapor barrier, or seal, is usually applied to the inside of the outside wall and prior to application of the insulation. Such materials as asphalt-impregnated building paper and plastic-coated paper are satisfactory, provided the joints are thoroughly sealed. Sealing compounds that may be mopped into the surface of the wall are also satisfactory.

Many cold storages are constructed without floor insulation. Recent studies[1] show this to be feasible, provided there is good drainage and the normal ground-water level is 12 ft or more beneath the floor. Perimeter insulation extending down the wall to the footing is recommended.

REFRIGERATION SYSTEMS

Refrigerants. Freon-12 is widely used as the refrigerant for farm-operated storages which are not constantly attended by a mechanic or watchman. Ammonia is also used, especially in large storages, where someone is on constant duty to supervise and check the operation of the refrigeration equipment. Both serve satisfactorily for normal cold-storage needs. Freon-12 is selected primarily because it is nontoxic to human beings and will not damage produce if it escapes into the storage-room atmosphere. It is, however, more expensive than ammonia; it leaks easily from the system; and the leaks are difficult to detect because it is odorless. Ammonia is relatively inexpensive and comparatively efficient, but it will seriously affect many crops if it leaks into the storage room. It has a strong odor, however, and leaks in the system can be readily detected.

Cooling Systems. Either natural-air convection or forced-air convection may be used for transfer of heat from the produce or storage-room atmosphere to the refrigerant. For natural-air convection, direct-expansion bare pipe coils are mounted at the ceiling and sometimes also at the sidewalls of the storage. The warm air, upon contact with the coils, is cooled and moves downward. When warmed by the uptake of heat from the produce, it becomes less dense and rises to the coils. The slow air movement may result in uneven temperatures and slow rates of cooling. Drip pans should be installed to collect water from the coils as a result of condensation or defrosting.

Cold brine is sometimes circulated through bare-coil pipes instead of the refrigerant. This arrangement eliminates the danger of ammonia leaking into the storage room since the brine is cooled outside of the room. The brine system also serves to provide more uniform temperature regulation and acts as a reservoir of refrigeration in the event of a temporary malfunction of the refrigeration equipment.

Unit coolers with dry coils are popular in new storages. These units are designed with fins on the coils to provide a large surface area for contact with the air forced through the coils by built-in fans. The units are mounted on the floor at one end of the storage or overhead suspended from the ceiling. Defrosting is accomplished by periodic shutdown of the supply of refrigerant, by washing with warm water or

---

[1] G. F. Sainsbury, Heat Leakage through Floors, Walls, and Ceilings of Apple Storages, *USDA Mkt. Rept.* 315, 1959.

brine, or by hot gas, whereby the refrigerant is recirculated through the coils rather than to the condenser.

Air Circulation. Since the heat in the storage room is transferred to the cooling coils (evaporator) primarily by air, good circulation of air is essential for rapid cooling and for the maintenance of uniform temperatures of the stored produce. Forced-air systems provide better air distribution than the bare-coil systems, which depend on natural convection of the air. As a general rule, the capacity of the blower or fans should be about 1,000 cfm for each ton of refrigeration. For apple storages, the required amount of air in cubic feet per minute is roughly equivalent to one-third of the cubical content of the storage room or to 80 to 85 per cent of the room capacity in bushels of fruit.

Compressors. The selection of compressors is dependent upon the kind of refrigerant used, the operating temperature of the cold-storage rooms, and the total refrigeration load. The quantity of refrigeration required for fruit and vegetable storages varies greatly from season to season; consequently the compressors and entire system must be designed to handle the maximum load even though it occurs for only a short period of time. It may mean that much of the capacity is not used during the rest of the year. Two compressors, rather than one, are sometimes used to reduce operating costs and as a safety factor in the event of breakdown. Generally, one is rated about one-third the capacity of the other, so that one or both compressors can be used, depending on the cooling load. Compressor capacities are also varied by changing their speed of operation, but this should be done only on the advice of the manufacturer.

Condensers. Heat removed from the storage room must be dispelled from the refrigeration system outside of the storage with a condenser. Water, air, or a combination of air and water is used to remove heat at the condenser. Where water alone is used and is not recirculated, 75 to 120 gal per hr per ton of refrigeration, depending on the water temperature, is required. If water is limited, it can be recirculated over a cooling tower, where it is cooled by evaporation of some of the water. Such difficulties with water as a source of supply, pipe and tubing corrosion, and freezing in cold weather have encouraged many farm storage operators to install air-cooled condensers.

Temperature Control. The temperature in a mechanically refrigerated storage is usually regulated by a thermostat in the storage room. The location of the thermostat is critical to proper storage temperatures in that it must operate without other areas of the storage room becoming excessively warm or cold. The most desirable place is near the center of the room in an aisle where it can be readily checked and adjusted. Thermostats should not be mounted directly on outside walls. Accurate, easy-to-read thermometers should be permanently placed at several locations in the storage room, and in addition, the operator should periodically check temperatures at the floor and ceiling.

Humidity Control. Many of the factors necessary for good temperature control are also essential for good control of the relative humidity. Most important, however, is the temperature difference between the storage-room air and the refrigerant. High humidities are possible only with a large coil surface at the evaporator, so that the heat is removed from the air with a small drop in temperature and minimum condensation of moisture. Humidities may be lowered by decreasing the coil area and lowering the refrigerant temperature.

Determining Refrigeration Requirements. A number of factors must be considered in determining the amount of refrigeration required for the proper operation of a cold storage. The usual unit of measurement of refrigeration is tons per day. A ton of refrigeration is equivalent to 288,000 Btu of heat, which is the amount of heat absorbed by one ton of water ice as it changes from the solid state to the liquid state at 32°F. A refrigeration system rated at 10 tons has a theoretical capacity for freezing 10 tons of ice in 24 hr. On an hourly basis, a one-ton unit will remove 12,000 Btu per hr.

The sources of heat in a produce storage may be classified as heat leakage, field heat, respiration heat, and incidental heat.

Heat leakage refers to the heat transmitted through the ceilings, walls, and floor of the building. It is determined from the temperature difference between the inside and outside of the building, the total surface area, and the rate of heat transmission by the building materials and insulations (Chap. 48). The temperature difference is generally determined for the average conditions existing at the time when the maximum amount of refrigeration is needed, which in the case of an apple storage is at loading when the fruit must be cooled quickly to the storage temperature.

The *field heat* is the heat carried into the storage by the produce and containers. It is calculated by multiplying the total weight of the commodity and container by their specific heats and multiplying the resultant product by the degrees of temperature decrease desired per day. The specific heat varies with the kind of fruit or vegetable and kind of container and the water content of the container material. When a storage is loaded daily for a period of time, the heat load is equivalent to the amount of field heat moved into the storage each day. It is desirable to cool fresh produce as quickly as possible. For apples, all the field heat should be removed in 3 to 10 days.

The *heat of respiration* is the heat produced by the commodities as a result of their normal respiration process. The respirational heat varies with the kind and variety of fruit or vegetable and with its temperature (Table 53-2). Warm produce respires at a higher rate than cooled produce; therefore the best means of reducing the heat load resulting from respiration is by cooling.

Incidental heat includes all other sources of heat, particularly heat produced in the storage by fan motors, lift trucks, electric lights, and workmen and heat from air infiltration as a result of door openings. Frequently, the incidental heat is estimated as 10 per cent of the total of all the other sources of heat.

Although refrigeration requirements vary with individual storages, and from one time to another for a given storage, there are some *general guides* for estimating the heat load which can be used for checking the detailed calculations.

The general requirements for apple storages insulated according to minimum recommended standards for average loading conditions in New York State are shown in Table 53-3.

TABLE 53-3. APPROXIMATE TONS OF REFRIGERATION PER 1,000 BU REQUIRED FOR APPLE STORAGES LOADED AT VARYING RATES

Sources of heat	Rate of loading per day in percentage of total storage capacity		
	10%	7%	5%
Heat leakage.............	0.13 ton	0.13 ton	0.13 ton
Field heat...............	0.6	0.4	0.3
Heat of respiration.........	0.1	0.07	0.05
Incidental heat............	0.08	0.06	0.05
Approximate total........	1	¾	⅝

SOURCE: Harold E. Gray and R. M. Smock, Farm Refrigerated Apple Storages, *Cornell Univ. Ext. Bull.* 786, 1955.

These figures are for the period of maximum refrigeration requirement, which is at loading for an apple storage. Once the storage is loaded and the fruit cooled, the refrigeration requirement decreases greatly. Marshall[1] estimates that one ton of refrigeration will handle the heat load of 10,000 bu of apples at 32° during the late fall and early winter after the field heat has been removed.

STORAGE SUPPLEMENTS

Special treatments or methods of storage in addition to regulation of temperature and humidity are often employed to lengthen the normal storage life and to preserve

[1] Marshall, *op. cit.*

the condition of fresh fruits and vegetables. Many of the treatments involve the use of chemicals that are applied either directly or in proximity to products that are to be used for food, and caution must be exercised to comply with Federal, state, or local restrictions on their use. Chemicals used with foods must be approved by the Pure Food and Drug Administration of the United States government.

Controlled Atmospheres. Modification of the storage-room atmosphere may be used in conjunction with low temperature to prolong storage life. An increase in carbon dioxide and a decrease in oxygen content of the atmosphere slow respiration, delay ripening, and tend to retard development of decay.

In northeastern United States, many apples are being stored in controlled atmospheres. The leading variety, McIntosh, is stored at 38°F in an atmosphere maintained at 5 per cent carbon dioxide and 3 per cent oxygen and the remainder nitrogen. The other varieties require 3 per cent oxygen, but the carbon dioxide and temperature needs are variable, so that each variety must be stored separately. The recommended temperatures and carbon dioxide levels, respectively, are Jonathan, 32°F and 5 per cent; Red and Golden Delicious, 32°F and 2 per cent; Rome Beauty, 32°F and 2.5 per cent; and Northern Spy, 38°F and 8 per cent.

Accurate *measurement and control* of the gas concentrations in controlled-atmosphere storage rooms are essential to avoid injury to the fruit by either high carbon dioxide or low oxygen. The composition of the atmosphere must be determined at least once daily. Oxygen is added as necessary by ventilation with outside air, whereas the excess carbon dioxide is removed by passing the air through caustic soda (sodium hydroxide) or water absorbers.

The controlled-atmosphere room must be airtight, which means that all cracks and crevices must be sealed and tight doors must be installed and sealed. *Sealing* is best accomplished by lining the inside surface of walls and ceiling with sheet metal and caulking all joints of the sheet metal.

The costs of constructing and operating a controlled-atmosphere storage for apples are approximately double those of a regular refrigerated apple storage.

Air Purification. The values of air purification by filtering the storage-room air through activated coconut-shell carbon are controversial in respect to lengthening the storage life of fruit, but positive in respect to odor removal. Some researchers have found that apple scald is reduced and that fruit ripening is delayed by the use of filters. Others have found filters to be of no practical value in these respects.

The reduction of off-odors in a storage room is adequate to justify the use of air purifiers in many cases. Mustiness due to mold growths, pine odors from wood, and the cross transfer of odors from one crop to another, as for example from potatoes to apples, can be prevented.

Air-purification units consist of either canisters or flat metal-screen trays to hold the charcoal and fans to provide air circulation. As a general rule, 6 lb of activated carbon is used for each 1,000 bu of stored apples. It is estimated that air purification costs approximately 2 to 3 cents per bushel per storage season.

Plastic Covers and Liners and Waxes. Water losses can often be reduced by special handling methods and treatments so that wilt and shrivel are considerably delayed in storage and during the marketing period. Waxes have proved useful for this purpose and especially beneficial to citrus fruits, cucumbers, and rutabagas.

Recently, plastic films have been used for protecting commodities against water losses during storage and transit. Golden Delicious apples are stored in box liners of pliofilm, cellophane, or polyethylene, and bananas are shipped with a covering of polyethylene. The liners or covers are loosely closed, or if tightly closed, they are perforated to prevent injury to the fruit due to the accumulation of carbon dioxide or the depletion of oxygen.

Treatment with Chemicals. Various chemicals are employed on fruits and vegetables after harvest to control decay organisms, to retard the development of certain physiological disorders, and to kill insects and rodents. Unfortunately, there is no one chemical that serves as a general agent; rather, each specific crop and malady generally requires a special treatment. In all cases, the chemicals must be thoroughly tested prior to use as assurance against injury to the commodity, toxicity to the user of the product, and hazards to the applicator. The chemicals are applied to the products in the liquid state as dips and washes, in the dry form as dusts, and in the gaseous state as fumigants or slowly released volatiles from impregnated wrappers or pads of box liners.

MOISTURE CONDENSATION

The condensation of moisture on the surface of cooled products upon removal from cold storage frequently is troublesome, especially during periods of high humidity. Apples removed from storage in the fall and spring are likely to sweat so that normal cleaning by brushing is impossible until they are dried. In other cases, the film coverings of prepackaged items become fogged upon contact with warm and moist outside air so that visibility of the produce is impaired. Condensed moisture may also favor the development of decay on produce.

Condensation, or sweating, can be reduced or prevented upon removal from cold storage by gradually warming the produce, as for example by moving from 32 to 50° until warmed, and then to normal room temperatures. Grading and packing rooms often are maintained at an intermediate temperature of 50° to prevent sweating during the sorting and packing operation. A handler can determine whether or not sweating will occur under the existing conditions of humidity and temperature and make the necessary corrections accordingly by reference to a psychrometric chart (Fig. 48-1).

STORAGE SANITATION

Good sanitation in the storage room and plant is desirable for all food products. The high humidities generally required for the storage of produce are conducive to mold development, and precautions against its growth on the produce, containers, and storage walls must be taken. In addition to treating the fruits and vegetables against decay development, it is often desirable to wash the facilities with a sterilizing agent such as sodium hypochlorite. Other materials are equally suitable for this purpose, but in all cases the possible corrosive effects of the materials used and the carryover of undesirable odors from washing and disinfecting must be considered.

Chapter 54

SERVICE BUILDINGS AND EQUIPMENT

CARL W. HALL

WATER SUPPLY

Water can be supplied from wells, springs, and cisterns. The walls of the cistern should be tight to prevent contamination from seepage and loss of cistern water. The cistern should be designed so that the first few minutes of rainfall which washes dirt from the roof can be diverted. Wells can be driven, drilled, bored, or dug. Water from any of these sources can be moved to the point of use by a pump, preferably under an automatic system of electrical operation, or by a hydraulic ram from a spring.

A water-supply system consists of a source of water, usually a well, a pump, usually driven by an electric motor, a pressure tank, a distribution system of piping or tubing, and service equipment such as waterers, water heaters, etc.

Pumps. The characteristics of the different types of pumps are summarized in Table 54-1. A jet centrifugal pump can be used for a shallow well where the friction head, plus the lift of the water, does not exceed 25 ft. Part of the water is circulated through the ejector nozzle and into the venturi throat to provide pumping action. The *turbine* pumps which are used for heads up to 25 ft operate at 1,760 rpm and can be used for shallow wells. The allowable lift for a pump is usually specified at sea level, and it is necessary to reduce this by 1 ft for each 1,000 ft elevation. Thus, if a particular pump has a 22-ft maximum lift, it would have a maximum lift of 17 ft at 5,000 ft elevation. The total head consists of the vertical lift, the friction loss, the pressure, and velocity head, although the velocity head is usually neglected if the pipe is of adequate size. The horsepower requirements for shallow-well pumps of 50 per cent efficiency can be determined as follows.

$$\text{Horsepower for shallow-well pumps} = \frac{(\text{gal per min})(\text{head in ft})}{2,000} \quad (54\text{-}1)$$

Equation (54-1) can be used for deep-well *reciprocating* pumps by adding 50 per cent for a *single-acting barrel* or 15 per cent for a *double-acting barrel*. Most systems are designed to operate between 20 and 40 psi. To determine the head equivalent, multiply the pressure in psi by 2.3. Thus the contribution of a 40-psi pressure to the total head is 92 ft.

For deep wells the cylinder is placed within 22 ft of the water level. A piston pump can go as deep as 800 ft. A *jet-centrifugal* pump has a practical lift limit of 80 to 100 ft and requires two pipes. The deep-well ejector is a combination of *ejector* and *centrifugal* pump with a series of ejectors at different levels, with one ejector for each 44-ft lift. A *deep-well ejector* is usually installed in a casing with not less than 2½ in. diam. The *deep-well reciprocating* pump requires a casing of not less than 2 in. dia. The *deep-well turbine* is used for large-capacity discharges and requires a 4-in.-diam. or larger casing. Table 54-2 shows the amount of power required by a deep-well pump, which depends upon the rate of pumping, the total head in feet, and the mechanical efficiency of the pump. The mechanical efficiency of deep-well pumps is 40 per cent or less. Table 54-2 is based on an efficiency of 30 per cent

SERVICE BUILDINGS AND EQUIPMENT

TABLE 54-1. FARM POWER PUMPS

Pump	Practical lift	Working pressure, psi	Lift details	Delivery details

Reciprocating: Shallow well
  Low pressure
  Medium pressure
  High pressure
	22 ft	40 to 43
		Up to 100
		Up to 350
			Pulsating, pumps water having sand and silt, pumps air for pressure tank	Pulsating (air chamber evens pulsations), positive delivery

Deep well	Up to 875 ft. Suction lift below cylinder 22 ft	Normal 40	Pulsating, pumps water having sand and silt. Separate air pump for supplying air to pressure tank	Pulsating (air chamber and differential cylinder even flow), positive delivery
Centrifugal: Shallow well Straight centrifugal	Manufacturer's recommendations vary from 15 to 25 ft	40 normal, 70 usual max.	Nonpulsating, pumps water having sand and silt. Loses prime easily. Capacity decreases as lift increases. Special equipment needed for pumping air	Nonpulsating, high capacity with low pressure, low capacity with high pressure
Turbine type	Manufacturers figure 28 ft max. at sea level	40 normal, up to 100 available	Nonpulsating, capacity decreases as lift increases, unsatisfactory with sand and silt in water. Pumps air for pressure tank	Nonpulsating, high capacity with low pressure, low capacity with high pressure. Capacity does not increase as rapidly as straight centrifugal
Jet, shallow well and limited-lift deep well	Max. around 200 ft. More practical for lifts of 80 ft or less. Creates effective suction 15 to 20 ft below ejector	40 normal, up to 70 available	Nonpulsating, capacity decreases as lift increases, unsatisfactory with sand or silt in water	Nonpulsating, high capacity with low pressure, low capacity with high pressure
Rotary, shallow well	22 ft	100 (approx.)	Slightly pulsating, increased lift has little effect on capacity, unsatisfactory with sand or silt in water, pumps air for pressure tank	Slightly pulsating, increased pressure has little effect on capacity

SOURCE: M. M. Johns, Electric Water Systems for the Farm, *Univ. Tenn. Ext. Publ.* 260, 1942.

TABLE 54-2. POWER REQUIRED AND OUTPUT BY DEEP-WELL PUMPS

| Output, gal/hr | Depth to water ||||||
| | 50 ft || 75 ft || 100 ft ||
	With open tank, hp	With 40 psi pressure, hp	With open tank, hp	With 40 psi pressure, hp	With open tank, hp	With 40 psi pressure, hp
100	1/6	1/4	1/6	1/4	1/4	1/3
200	1/6	1/2	1/4	1/2	1/3	1/2
300	1/4	3/4	1/3	3/4	1/2	3/4–1
400	1/3	3/4–1	1/2	1	1/2–3/4	1–1 1/2
500	1/3	1	1/2	1–1 1/2	3/4	1 1/2
600	1/2	1	3/4	1 1/2	1	1 1/2–2
700	1/2	1 1/2	3/4	2	1	2
800	3/4	2	1	2	1 1/2	3

SOURCE: Composite of data from several manufacturers. Personal communication from W. H. Sheldon, Dept. of Agricultural Engineering, Michigan State University, 1955.

TABLE 54-3. Loss of Head, in Feet, Due to Friction, per 100 Ft
of 17-year-old Steel Pipe with Water

(For new pipe multiply readings by 0.6. For 25-year-old pipe multiply readings by 1.2.)

U.S., gpm	Pipe size											
	½	¾	1	1¼	1½	2	2½	3	4	5	6	8	10
1	2.1												
2	7.4	1.9											
3	15.8	4.1	1.3										
4	27.0	7.0	2.1										
5	41.0	10.5	3.3										
6	57.0	14.7	4.6	1.2									
7	76.0	19.5	6.0	1.6									
8	98.0	25.0	7.8	2.0									
9		31.2	9.6	2.5	1.2								
10		38.0	11.7	3.1	1.4								
11		45.0	13.3	3.5	1.7								
12		53.0	16.4	4.3	2.0								
13		62.0	18.7	4.9	2.3								
14		71.0	22.0	5.7	2.7								
15		80.0	24.2	6.4	3.0	1.1							
16		91.0	28.0	7.3	3.4	1.2							
17			30.5	8.0	3.8	1.3							
18			35.0	9.1	4.2	1.5							
19			38.2	10.0	4.6	1.7							
20			42.0	11.1	5.2	1.8							
21			45.5	11.9	5.5	2.0							
22			50.0	12.9	6.2	2.1							
23			54.0	14.0	6.6	2.3							
24			59.0	15.2	7.3	2.5							
25			64.0	16.6	7.8	2.7							
26			68.0	17.8	8.4	2.9							
27			73.0	19.0	9.0	3.1							
28			78.0	20.2	9.7	3.3	1.1						
29			83.0	21.7	10.0	3.5	1.2						
30			89.0	23.5	11.0	3.8	1.3						
35				31.2	14.7	5.1	1.7						
40				40.0	18.8	6.6	2.2						
50				60.0	28.4	9.9	3.3	1.4					
60				85.0	39.6	13.9	4.7	1.9					
70					53.0	18.4	6.2	2.6					
80					68.0	23.7	7.9	3.3					
90					84.0	29.4	9.8	4.1	1.0				
100						35.8	12.0	5.0	1.2				
120						50.0	16.8	7.0	1.7				
140						67.0	22.3	9.2	2.3				
160						86.0	29.0	11.8	2.9				
180							35.7	14.8	3.6	1.2			
200							43.1	17.8	4.4	1.5			
220							52.0	21.3	5.2	1.8			
240							61.0	25.1	6.2	2.1			
260							70.0	29.1	7.2	2.4			
280							81.0	33.4	8.2	2.8	1.1		
300							92.0	38.0	9.3	3.1	1.3		
325								43.5	10.7	3.6	1.5		
350								50.0	12.2	4.2	1.7		
375								56.0	14.8	4.6	1.9		
400								65.0	16.0	5.4	2.1		
425								72.0	17.2	5.8	2.4		
450								79.0	19.8	6.7	2.6		
475								87.0	21.6	7.3	2.9		
500								98.0	24.0	8.1	3.2		

TABLE 54-3. Loss of Head, in Feet, Due to Friction, per 100 Ft of 17-year-old Steel Pipe with Water (*Continued*)

U.S., gpm	Pipe size												
	½	¾	1	1¼	1½	2	2½	3	4	5	6	8	10
550									28.7	9.6	3.8		
600									33.7	11.3	4.5	1.2	
650									39.0	13.2	5.3	1.3	
700									44.9	15.1	6.1	1.5	
750									51.0	17.2	6.8	1.7	
800									57.0	19.4	7.7	2.0	
850									64.0	21.7	8.7	2.2	
900									71.0	24.0	9.8	2.4	
1000									88.0	29.2	11.9	3.0	1.0
1100										33.5	13.7	3.6	1.2
1200										39.3	16.1	4.2	1.4
1300										45.6	18.6	4.9	1.6
1400										52.3	21.4	5.6	1.9
1500										59.4	24.3	6.4	2.1

Based on Williams and Hazen formula with constant $C = 100$.
SOURCE: *F. E. Myers & Bro. Co. Manual* WSM 56, 1956.

and gives the recommended sizes of electric motors to use for various pumping rates and from various depths. With electricity at 3 cents per kilowatthour, the cost for pumping 1,000 gal with a shallow-well pump is from 2 to 4 cents and with a deep-well pump, from 4 to 7 cents. Pumping systems are generally considered to have a life of from 15 to 20 years. The size of pump should be selected on the basis of the maximum demand. Some manufacturers recommend that the pump should be selected on the basis of 2 hr running time to pump the water requirements for one day. Thus, divide the total daily requirements by 2 to obtain the pump capacity in gallons per hour. Reciprocating deep-well pumps normally use a capacitor or repulsion induction motor because of the high starting torque.

Tank. A *gravity system* of water supply is practical where there is a low-yield well. The water is pumped to a tank which is placed at an elevation to give the desirable pressure. A gravity tank at 50 ft elevation gives a pressure of 21.7 psi.

With a *pressure system,* a pressure tank is provided at the pump so that the pump does not start each time water is drawn. The pressure tank should be located as close as possible to the pump and normally has a capacity of 42 or 80 gal. The pressure tank is filled two-thirds with water and one-third with air. The water level is controlled by a float valve to control the air volume and a pressure control which starts the pump when the pressure is reduced to 20 psi and stops the pump when 40 psi is reached. About 20 per cent of the tank capacity is removed from stopping to starting of the pump. The air in the tank absorbs the pulsation from the discharge stroke of the pump. If too little air is provided, the pump starts and stops too often. Since some of the air is absorbed by the water, make-up air must be added. With deep-well pumps, the air is fed into the tank with the water on each plunger stroke, with excess air vented to the outside. With some types a diaphragm arrangement is used which puts a small charge of air into the tank each time the pump starts and stops. The tank should be about ten times the capacity of the pump in gallons per minute with standard sizes of 12, 42, 82, 120, 220, 315, and 525 gal. A larger tank is recommended where there is a low-yield well.

Piping and Tubing. Galvanized-iron pipe, copper tubing, or plastic pipe can be used for installations. Plastic pipe is now being used extensively and may prove to have a longer life due to greater corrosive resistance than galvanized iron or copper. Both copper tubing and plastic tubing are easier to install than galvanized-iron pipe. The loss of head in feet due to friction of pipe and fittings can be determined from Tables 54-3 and 54-4. Gate valves should be used in preference to globe (throttle) valves wherever possible. For pumping 350 gal per hr a distance of 200 ft, the use of 1-in.-diam pipe is normally recommended. Selection of pipe size is basically a matter of economics, balancing operating cost against investment in equipment. For distribution of water to water cups at the barn use ½-in. steel pipe or copper tubing for distances up to 50 ft and ¾-in.-diam piping or tubing for longer lines. Provisions should be made to prevent freezing by (1) permitting waste to drain out of the pipe into a sand pit when the valve is closed, (2) placing the pipes below the frost zone, (3) insulating the pipes where practical, or (4) providing electric-resistance-wire wrapping for heating those places not protected from the weather.

TABLE 54-4. FRICTION LOSS IN VALVES AND FITTINGS IN TERMS OF EQUIVALENT LENGTH, IN FEET, OF STRAIGHT PIPE
(See pipe friction table 54-3 for exact loss.)

Nominal pipe size	Gate valve, full open	45° elbow	Standard elbow, 90°	Standard T through side outlet	Close-return bend	Swing check valve, full open	Angle valve, full open	Globe valve, full open
½	0.35	0.78	1.7	3.3	4.1	4.3	9.3	18.6
¾	0.44	0.97	2.1	4.2	5.1	5.3	11.5	23.1
1	0.56	1.23	2.6	5.3	6.5	6.8	14.7	29.4
1¼	0.74	1.60	3.5	7.0	8.5	8.9	19.3	38.6
1½	0.86	1.90	4.1	8.1	9.9	10.4	22.6	45.2
2	1.10	2.40	5.2	10.4	12.8	13.4	29.0	58.0

SOURCE: *F. E. Myers & Bro. Co. Manual* WSM 56, 1956.

TABLE 54-5. PHYSICAL PROPERTIES OF PLASTIC

	Flexible, polyethylene	Flexible, butyrate	Rigid, polyvinyl chloride
Specific gravity	0.95	1.2	1.43
Tensile strength, psi	1,200	2,500	8,700
Modulus of elasticity, psi	15,000	100,000	330,000
Linear-expansion coefficient, °F	9×10^{-5}	12×10^{-5}	3.5×10^{-5}
Heat conductivity, Btu-ft/(hr)(sq ft)(°F)	0.171	0.12	
Compression-molding temperature, °F	300	300	
Compression-molding pressure, psi	200	1,000	
Specific heat	0.55	0.35	
Dielectric constant, 10^6 cycles	2.3	5	2.9
Trade name	Polyethylene	Tenite	PVC
Per cent water absorption in 24 hr, 25°C	0.01	1.0	0.27% at 75°F, 96 hr

SOURCE: "Handbook of Technical Data on Plastics," Plastic Materials Mfg. Assoc., Washington, D.C.,

Pipe and tubing of the same nominal size can be used interchangeably. The actual inside diameter of pipe is slightly greater than the nominal diameter. Tubing is designated by the outside diameter and is therefore smaller inside than the nominal size by twice the wall thickness. It is smoother, however, and 1-in. tubing has about the same internal friction as 1-in. pipe.

Plastic Pipe. Plastic pipe is finding considerable use for farm water systems because of its ease of installation, light weight, corrosion resistance, and expansion upon freezing without breaking. The National Plumbing Code does not approve plastic for water systems, although some states permit its use. Polyethylene and cellulose acetate butyrate are the major plastics used for flexible pipe. Physical properties of some plastics are given in Table 54-5. Farm systems normally use pipe designed for 75 psi or less. Pipe is available in sizes from ½ to 6 in. in diameter.

A large variety of fittings and adapters are available. Plastic fittings may be assembled with a plastic solvent. Special types of pipe include double conduits for jet pumps, a design for submersible pumps, and sprinkler-hose plastic pipe. The working pressure of plastic pipe is designated for a temperature between 70 and 80°F. As the temperature is increased, the strength of the plastic is decreased, so it is normally used for cold-water lines. Polyethylene pipe with a safe working pressure of 75 psi at 74°F will have a safe working pressure of 29 psi at 140°F and 15 psi at 160°F. Plastic pipe is not a conductor of electricity. Plastic can be used for natural gas, LP gas, and water. Undesirable characteristics of plastic for farm applications are that (1) it will burn; (2) it will not endure exposure to long periods of sunlight; and (3) it will flatten or cut or can be damaged if driven over or if rocks are placed on it.

The comparative properties of cellulose acetate butyrate plastic piping are as follows:[1]

1. In tests of ¾-in.-diam plastic and standard steel pipe, the plastic had 16 per cent more capacity because of greater inside diameter and 2½ per cent increased capacity because of less friction head.

2. The coefficient of expansion of plastic is about 12.4 times steel at ordinary temperatures (0.88 in. per 100 ft per 10°F).

Plastic pipe should be installed loosely in the trench with plenty of slack.

Installation of Livestock Waterers. Watering troughs are being replaced by waterers for cattle, hogs, chickens, and other livestock. One unit should be provided for 30 cows or 60 hogs. An electrically heated unit must be provided to prevent the bowl and pipes from freezing. The temperature in an electrically heated waterer is normally kept at 40 to 45°F. The water inlet to the bowl should be above the water level in the bowl to prevent back-siphoning. If back-siphoning occurs, contaminated water from the bowl moves back into the line and mixes with drinking water for human beings or with wash water. Water bowls which meet the nonsiphoning test of the Barn Equipment Association are designated by the letters *NS*.

Hydraulic Ram. A hydraulic ram, which uses no external power source except the water, can be used to supply running water for the farm, usually from a spring. With the hydraulic ram, water *falls* to the ram through a pipe, providing energy to lift a small portion of the flow to a storage tank *above* the source.

The following *practices* and *principles* apply to hydraulic rams:[2]

1. Minimum fall of 3 ft should be provided.
2. Have a minimum supply of 3 gpm.
3. Elevation should be at least ten times fall used.
4. Storage in gravity tank should be at least one-half daily need.
5. A hydraulic ram can rarely be justified if electricity is available to operate a shallow-well pump.

WASTE DISPOSAL

Waste disposal from farm buildings, other than the house, has generally been neglected. Contamination of streams from livestock droppings and toilets in the barn should be avoided and is illegal in many states.

Theory of Operation. Sewage flows into the septic tank, where it is held at least 24 hr, and the solids decomposed by anaerobic bacteria. The heavy material settles to the bottom for further decomposition, and the remaining liquid escapes through the outlet. The gas released escapes through a vent or through the disposal-field tile. A tank of sufficient size must be provided to handle the solids. If solid material leaves the tank before decomposition, the tile in the disposal field will become clogged and the system will fail. Adequate heat may not be provided during the winter in northern climates to maintain a desirable temperature for decomposition if the tank is too large.

Tank. A settling tank should be provided for separation and decomposition of solids from the milking room or stalls. A tank capacity of 150 gal per milking-room stall or 20 gal per milk cow should be provided. An opening should be provided for removing fats and other solids which do not decompose easily.

The tank may be made of poured-in-place concrete, of precast concrete blocks, or of asphalt-coated steel. Lumber companies, and in some cases the state extension service, have forms available for pouring concrete septic tanks. Capacities and dimensions of

---
[1] John F. Fugazzi, Plastic Tubing Use Found Practical, *Gas*, 28:49-52, October, 1952.
[2] Walter H. Sheldon, Hydraulic Ram, *Mich. State Univ. Agr. Expt. Sta. Bull.* 171, 1943.

single-chamber septic tanks are given in Table 54-6. A single-chamber tank (Fig. 54-1) is used for most farm installations of less than 1,000 gal. The tank and disposal field should be at least 50 ft from the well or any buried suction pipe.

In a large installation with a capacity over 1,000 gal and over 750 ft of tile lines, a siphon and dosage chamber can be provided to give intermittent effluent discharge to the disposal field.[1] The tank should not require cleaning more than every

TABLE 54-6. CAPACITIES, DIMENSIONS, AND CONCRETE MATERIALS FOR SEPTIC TANKS SERVING INDIVIDUAL DWELLINGS

Max. no. of persons served	Liquid capacity of tank, gal	Recommended inside dimensions				Materials for concrete 1-2½-4 mix		
		Width	Length	Liquid depth	Total depth	Cement, sacks	Sand, cu yd	Gravel, cu yd
4 or less	500	3'0"	6'0"	4'0"	5'0"	16	1½	2½
6	600	3'0"	7'0"	4'0"	5'0"	17	1¾	2¾
8	750	3'6"	7'6"	4'0"	5'0"	19	2	3
10	900	3'6"	8'6"	4'0"	5'0"	21	2¼	3¼
12	1,100	4'0"	8'6"	4'6"	5'6"	24	2¼	3½
14	1,200	4'0"	9'0"	4'6"	5'6"	25	2½	3¾
16	1,500	4'6"	10'0"	4'6"	5'6"	28	2¾	4¼

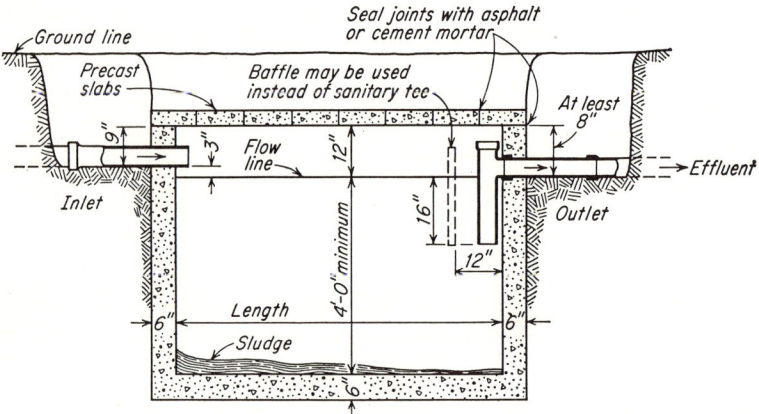

FIG. 54-1. Single-chamber septic tank.

3 years and in some cases may go 10 years. A tank should be cleaned when there is 2 ft of sludge in the bottom.

Disposal Field. The tile field for disposal of effluent from the septic tank is the most important part of the system. Unglazed drain tile 4 in. in diameter are placed in trenches at least 2 ft apart, 2 ft deep in northern areas, and 3 ft deep in southern climates.[2] For a 500-gal tank at least 150 ft of tile is required in sandy or gravelly soil. A simple method of determining the absorptive ability of the soil is to dig a hole about 1 ft in diameter and 18 in. deep. The hole is filled with water, and this water is permitted to seep away, after which 6 in. of water is added. Then if the water does not lower at least 1 in. in 30 min, the soil is not adequately porous for a disposal field. If 30 min is required for lowering the water, 325 ft of drain tile is

[1] W. F. Sheppard, G. Amundson, and W. H. Sheldon, Home Sewage Disposal, *Mich. Dept. Health Eng. Bull.* 2, 1958.
[2] John M. Hepler, George Amundson, Clare A. Gunn, and W. H. Sheldon, Septic Tanks, *Mich. State Univ. Ext. Bull.* 118, 1947.

needed instead of 150 ft. The layout of the disposal field depends on the terrain and shape of area available. The length of each tile line should not exceed 100 ft. The fall should be between 2 and 4 in. per 100 ft away from the tank. The trench in which the tile are placed should be at least 12 in. wide at the bottom, permitting the tile to be surrounded by 3 to 4 in. of gravel.

For low-porosity soils in which a disposal field will not function normally, the following methods may be followed:

1. Provide a dry well at the end of the effluent tile from the septic tank if there is no danger of the water table becoming contaminated.

2. Dig 6- to 8-in.-diam holes 4 ft apart and fill with gravel under the tile of the regular disposal field. The holes should extend down to a porous structure.

3. Lay the disposal tile in a layer of gravel with a filter trench and underdrain leading into the sewage pit (Fig. 54-2).

Practices for successful operation of septic tank:

1. Clean tank periodically every 3 to 5 years when 2 ft of sludge has accumulated in the tank.

2. Starters are not needed to promote growth of bacteria.

3. Inspect septic tank annually.

4. Dispose of sludge in a bury pit and cover with 2 ft of soil.

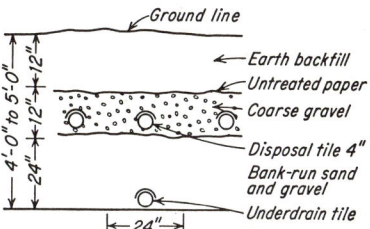

Fig. 54-2. Filter trench with three disposal tiles and one underdrain tile.

5. Do not light a match to inspect the tank because an explosion may occur.

6. Normal household soaps and detergents do not harm or hinder the operation of the tank.

7. Get a permit for installation and have installation inspected by public health authorities before covering with soil.

8. Do not add a large quantity of raw food products to the tank at one time. Two gallons of milk should be a maximum. Do not run water drainage from roof and yard into disposal system.

9. The usual family garbage can be added to the tank. Preferably, grind the garbage before placing it in the tank, and it is desirable to have the tank from $1\frac{1}{4}$ to $1\frac{1}{2}$ times the normal recommended capacity when it is used for garbage disposal.

10. See Table 54-7 for a summary of installation practices.

TABLE 54-7. SEWAGE-DISPOSAL-INSTALLATION REQUIREMENTS

Water to be handled	50 gal/person/day
From sink, lavatories	2-in. soil pipe, cast-iron
From stool	4-in. soil pipe, cast-iron
Sewage-disposal tile	4-in. unglazed drain tile
Distance of septic tank from well	50 ft or more and at a lower elevation
Joints for sewer vitrified tile	Mortar, 1 part cement, 2 parts sand
Joints for disposal drain tile	Upper half of joints covered with asphalt paper to keep out dirt
Near trees	4-in. soil pipe, leaded
Slope, drain tile to tank	$\frac{1}{8}$–$\frac{1}{4}$ in./ft
Slope, tank to distribution box	$\frac{1}{8}$–$\frac{1}{4}$ in./ft
Slope, disposal field	2 in./100 ft
Tank dimensions	5-ft depth of liquid, min. 4 ft
Inlet at tank	1–2 ft below ground surface
Outlet inside tank	10–16 in. below surface of liquid
Outlet at tank	2–3 in. below inlet
Absorption outlet	25–50 ft of tile per person; 12 to 24 in. deep in porous soils
Clean tank	Every 3 to 10 yr when sludge collected reduces liquid to 3 ft

Parts adapted from "Farm Book," Doane Agricultural Service, Inc., St. Louis, Mo., 1947.

Animal Disposal. Growers of small animals should provide a means of disposing of dead animals to prevent spreading of diseases and to improve sanitation. A pit for disposing of casualties from a 10,000-bird flock of chickens (or other small animals) can be constructed as follows:[1]

[1] Silas McHenry, Poultry Disposal Pits, *Univ. Delaware Agr. Ext. Folder* 22, 1950.

1. Dig a circular hole about 7½ ft in diameter and 6 ft deep.
2. Mark a circle 6 ft in diameter on the bottom of the hole and lay the first row of 8 by 8 by 16 concrete blocks outside of this line.
3. Nine courses of block (approximately 108 blocks) are needed to build the pit up to ground level. Each course should be set in 1½ in. toward the center (Fig. 54-3).

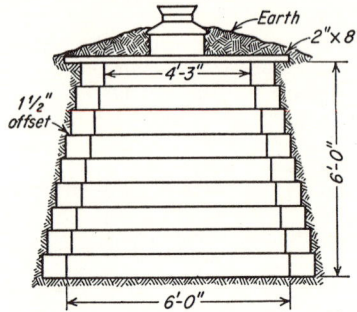

Fig. 54-3. Disposal pit. (*Univ. Delaware Ext. Folder* 221, 1950.)

4. Set an 8- or 12-in. tile with the bell end down on 2 by 8's laid across the top. An old milk can cut in half can also be used.
5. Cover top of pit with rough boards, or use a concrete slab 3 or 4 in. thick on top of the boards.
6. Cover the top with soil.
7. Build a cover for the tile or can.

An *incinerator* for burning animals is better than a pit if the soil drainage is poor. One of three types of incinerators is usually used:
1. Concrete-block fireplace.
2. Fifty-five-gallon oil drum, made up with doors, grate, and stack.
3. Brick incinerator with chimney, grate, and door.

FARM MACHINERY HOUSING

Machinery housing is an important part of the mechanized farm. A few years ago there seemed to be some doubt regarding the economics of constructing machinery housing. However, housing is necessary for the expensive machinery now used on the farms, with many intricate parts and controls. The machinery building should include the following areas:[1]
1. Storage for expensive machinery
2. Shelter for machinery which needs some protection
3. Space for servicing equipment
4. Shop for tools, repair, and maintenance

The *location* of the building site should have the following characteristics:
1. Good drainage
2. Space around the building: 30 ft from other buildings, 50 ft in front of building for turning, and space for enlargement
3. Conveniently located with respect to other buildings and fields
4. Openings protected from the weather
5. Utilities readily accessible

The size of the building is selected on the basis of the amount of machinery to be stored (Table 54-8). Usually it is best to select the most desirable width and then build the storage long enough to obtain the area needed. Common widths are from 26 to 40 ft in 2-ft intervals. To provide ease of movement of machinery, posts can be omitted from the inside of the structure by using a truss. An excellent method of determining the area requirements is to make templates of the various pieces of equipment to be stored and move the templates into place on a scale drawing.[2] The time

[1] Farm Machinery Housing, *North Central Regional Publ.* 31, *Univ. Illinois Circ.* 702, 1952.
[2] G. E. Henderson, "Planning a Machinery-storage Layout," Southern Association of Agricultural Engineering and Vocational Agriculture, Athens, Ga., 1956.

TABLE 54-8. MACHINERY STORAGE-SPACE REQUIREMENTS

Item	Requirements, ft		
	Width	Height	Length
Automotive:			
Car	7	6½	18
Tractor:			
One-plow	5	5	9
Two-plow	7½	5	12
Three-plow	7½	5	12
Truck:			
Pickup	7½	7	20
Stock rack	8	10	26
Grain bed	8	7	26
Semitrailer	8	9	26
Binder, grain:			
8-ft, reel off	10	5	16
10-ft tractor, reel on	12	8	19
Corn:			
One-row	7	7	12
Two-row	9	7	16
Baler	13	5–6	21
Bale sleds	6	.....	6
Combine:			
5–6 ft	9–12	8½	16–20
12 ft	13	10–12	22
16 ft	20	14	25
Cultivator, corn:			
One-row	5	4	6
Two-row, tractor	10	.....	7½
Four-row, tractor	15	.....	8
Two-row, 3-point hitch	10	4	4
Rotary hoe, two-row	10	3	6
Cutter:*			
Rotary field shredder	9½	.....	6½
Stalk cutter	6	.....	4
Digger, potato	5	.....	8
Drill, grain:			
8 ft, 14-7	11	5	6
10 ft, 16-7	13	5	6
14 ft, 24-7	18	5	7
Hammer mill	4	3	9½
Harrow:			
Spike-tooth	4	.....	6
Spring-tooth	3	.....	6
Disk, horse	8	.....	6
Disk, tractor, 10 ft	12	3	11
Disk, tractor, 7-ft lift	8	4	6
Harvester,* tobacco	9½	12–14	16
Loader:			
Hay	8	10	12–15
Manure	3–4	6–9	4–10
Cane loader	9	12	29
Mower:			
Horse, bar up	5	6	8
Tractor, trailer type 7 ft, bar up	5	8	4–6
Tractor, rear-mounted, bar down	12	2	6
Picker, corn:			
One-row, pull-type	8	6–7	12
Two-row, pull-type	16	6–7	17
Two-row, mounted	10	8	17
Cotton:*			
One-row	10	13	19
Two-row	11	13	20
Peanut*	6	.....	17
Planters, corn or cotton:			
Two-row (without hitch)	10	.....	6
Four-row (without hitch)	15	.....	6
Potato	6	.....	8

TABLE 54-8. MACHINERY STORAGE-SPACE REQUIREMENTS (*Continued*)

Item	Requirements, ft		
	Width	Height	Length
Plow:			
Walking	2	.....	8
Sulky	5	4	7
Two-bottom, horse	5	4	8
Two-bottom, tractor	5–6	4	9½–11
Three-bottom, tractor	6	4	11–13
Four-bottom, tractor	7	4	14
One-way disk	9	.....	10–14
Rack, hay	8	8	16
Rake:			
Dump, 10 and 12 ft	12–15	4½	6
Side-delivery	8–11	4½	12
Sweep, tractor	9–13	3–4	9–10
Tedder, 8-fork	9	5	6
Seeder, box-type, 11 ft	13	4	6
Silo filler	5	6	10
Sprayer, orchard and potato	8	.....	6
Spreader:			
Manure, horse	6	4½	15
Manure, tractor	6	4½	18
Lime,* 8 ft	10½	.....	4
Stripper,* cotton:			
One-row	2	10	12
Two-row	8	10	20
Tiller	12–16	.....	12
Thresher, grain separator	8–10	10	23–29
Wagon:			
Box and gear, high wheel	6	5½	14
Gear	6	3	9
Box and gear, rubber tire	6	4½	14

SOURCES: Doane Agricultural Service, Inc., St. Louis, Mo., and (for starred entries) G. E. Henderson and C. E. Turner, "Planning a Machinery-storage Layout," Southern Association of Agricultural Engineering and Vocational Agriculture, December, 1956.

of the year the equipment is used and the period of storage should also be considered. It is important that a doorway 10 ft high and 12 ft wide be provided for two-row equipment. There is a trend to larger-size equipment for four- and six-row operations and larger doors or entrances should be considered. For larger machines, one section of the machine shed can be built higher, with larger doors, or an end entrance with a gable roof can be utilized.

There are basically *two types* of machinery storage, according to movement through the building: (1) end entrance and (2) side entrance. The former is usually over 32 ft wide, with 40 ft preferred, and the latter under 32 ft wide. With the side-entrance arrangement, doors are on one side, with one or more doors on the opposite side, if the farmstead arrangement permits. Also, one or both sides of the building can be left open. The farm shop is usually located at one end of the side-entrance type or in one side of an end-entrance building.

Sliding or overhead doors can be used. Sliding doors are less expensive and should be hung on bird-proof track and roller bearings. The door rollers should be protected from weather so that rusting will not occur.

FUEL STORAGE

Practically all the fuel used for farm tractors is stored on the farm. It is important that fuel storage be designed to provide for safety and for maintenance of fuel quality. The major factor in reduction of quality is evaporation of the fuel. Fuel can be stored in underground or aboveground tanks. Aboveground storages should be at least 40 ft from farm buildings. An operator requiring over 300 gal of storage should strongly consider underground storage, in order to prevent excessive losses through evaporation. Storage should be provided for 2- to 3-week fuel requirements. Vaporization occurs continuously and reduces the quality of gasoline, particularly for starting. Condensation of water inside fuel storages is another problem which can be kept to a minimum with underground storage. The pump should be mounted in the tank in such a way that water which accumulates in the tank will not normally be removed through the fuel pump.

Liquefied-petroleum storage (LP gas) should be aboveground to permit detection of any leaks which might occur. The tank should not be filled over seven-eighths full to permit expansion of the liquid with an increase in temperature. These tanks are usually painted white or aluminum.

The following *practices* can be followed to reduce the losses in an aboveground gasoline tank:

1. Use a pressurized cap or tank outlet similar to pressurized radiator caps. The pressurized system should permit a pressure build-up of about 3 psi.
2. The pressurized cap should permit air to move back into the tank when the fuel cools.
3. Place storage in the shade.
4. Paint white or aluminum to reduce heat absorption.

FARM FENCES[1]

Fences are an important farm investment. Fencing is required to mark boundaries, confine livestock to fields, control livestock diseases, and to improve appearance. Field fencing can be constructed from rail, boards, stone, barbed wire, woven wire, or hedge. Normally lumber or concrete is used for barnyard fencing.

It is important that fields be laid out with the following objectives in mind:
1. Long rows for ease of cultivation
2. Land of similar soil type in the same field
3. Easy pasturing, with long lanes
4. Proper rotation by proper size of fields
5. As few fields as possible
6. Efficient control of rainfall runoff

Barbed wire has two- or four-point barbs spaced 4 to 5 in. apart on two smooth wires twisted together. Barbed wire can be used alone as a fence or placed above woven-wire fence. One, two, or three rows of barbed wire are used for livestock, with the number of wires and spacing adjusted for different livestock. It may be constructed on a permanent or temporary basis. Barbed wire and woven wire may be fastened to steel or wooden posts.

The labor requirements for installation are approximately the same for a good or poor fence.

Standardized style or code numbers are used for woven-wire fencing: No. 939-12-12½ has nine line (horizontal) wires, is 39 in. high, has stay (vertical) wires 12 in. apart, and is made of No. 12½ gage wire except for top and bottom strands.

Fencing recommendations for various purposes are given in Table 54-9.

Posts. The life of wooden posts varies with the type of wood (Table 54-10) and the acidity of the soil and the climate.

The life of wood posts can be greatly extended by impregnation with creosote. Pressure-creosoting is superior to dipping the posts. The heartwood, or center, of the tree is more durable than the sapwood, or new growth. Wooden corner posts should be at least 8 in. in diameter and set into the ground 3½ to 4 ft, anchored and braced (Fig. 54-4). The brace post, which is the one next to the corner, should be about 9 ft from the corner post. The spacing of the wood line posts should be from 12 to 14 ft. Line posts should have a minimum top diameter of 3 in. and should be set 2 ft into the ground.

Steel posts can be used for end, corner, or line posts or as braces. End and corner braces are from 7 to 9 ft long, and regular posts from 5 to 8 ft long in increments of 6 in. Corner posts and the corner braces should be set in concrete or provided with bracing (Fig. 54-5). Steel posts are normally 12 to 16½ ft apart along the fence and

[1] Henry Giese, "Farm Fence Handbook," Republic Steel Corporation, Chicago, 1942.

TABLE 54-9. FARM FENCES—WOVEN-WIRE AND BARBED-WIRE FOR LIVESTOCK

Heavy-weight to have all No. 9 wire; medium-weight to have No. 9 top and bottom wire and No. 11 filler wire. The spacing to be determined by application and chosen according to the following recommendations.

Type	Height of woven wire	Stay spacing	No. of barbed wires	Commercial designation	
				Heavy	Light
Cattle..............	47″	6 or 12″	1 (top)	1047-6-9 1047-12-9	1047-6-11 1047-12-11
Hogs..............	32″–39″	6″	1 (bottom)	....	832-6-11 939-6-11
Sheep and goats......	39″–47″	6 or 12″	1 (top)	939-12-9 1047-6-9 1047-12-9	939-6-11 939-12-11 1047-6-11 1047-12-11
Horses and mules......	47″	6–12″	1 (top)	1047-6-9 1047-12-9	1047-6-11 1047-12-11
General.............	47″	6″	1 (top)	1047-6-9	1047-6-11

SOURCE: Farm Fence Construction Standards Committee of ASAE, 1957.

TABLE 54-10. AVERAGE LIFE OF UNTREATED WOOD POSTS, IN YEARS

Osage orange............	30	White oak...............	11
Locust..................	24	Tamarack...............	10
Red cedar...............	20	Cherry..................	10
Mulberry................	17	Hemlock................	9
Catalpa.................	16	Sassafras................	9
Bur oak.................	15	Elm.....................	9
Chestnut................	15	Ash.....................	9
White cedar.............	14	Red oak.................	7
Walnut.................	12	Willow..................	6
Pine....................	11		

driven 2 ft into the ground. When there is a depression along the fence, steel line posts should be set in concrete to prevent them from pulling out. A corner post set in concrete should be used every 40 rods for long fences and have parallel braces to keep the fence straight and close to the ground.

Concrete posts must be reinforced with steel and properly cured before stretching the fence.

Staples should be long enough to adequately secure the fence to the wooden posts. Longer staples should be used for the softer wood, with a 1½-in. length for the hardwood and 2-in. length for the softwood.[1] Staples should not be driven into the post far enough to force the wire into the wood. Galvanized staples resist rusting and aid in preventing rusting of the attached fence.

Principles and practices for fence erection:
1. Don't drop a roll of fence on its end.
2. Tighten the fence until the bends in the horizontal wires are half-straightened. These tension bends prevent damage from contraction in cold weather and sagging in hot weather.
3. Stretch from corner posts, not around the corner.
4. Use 1½- to 2-in. staples.
5. Don't drive head of staple tightly against the wire.
6. Drive staples diagonally across the grain in wooden posts.
7. Bevel the tops of wooden posts to shed water.
8. Alternate wood and steel line posts on opposite sides of the fence if both are used.
9. Ground a wooden-post wire fence with steel posts every 150 ft for lightning protection.

[1] John R. Neetzel, Building Better Farm Fences, *Univ. Minn. Agr. Ext. Bull.* 272, March, 1953.

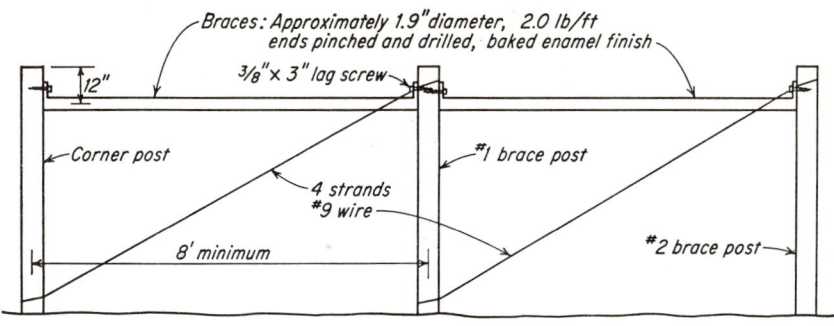

Detail 4
Wooden posts with steel braces

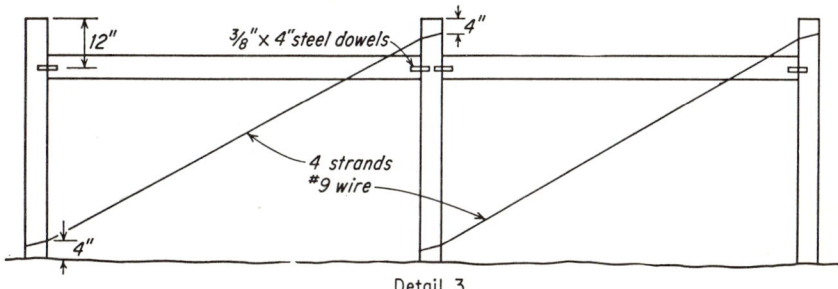

Detail 3
Wooden posts with wooden braces

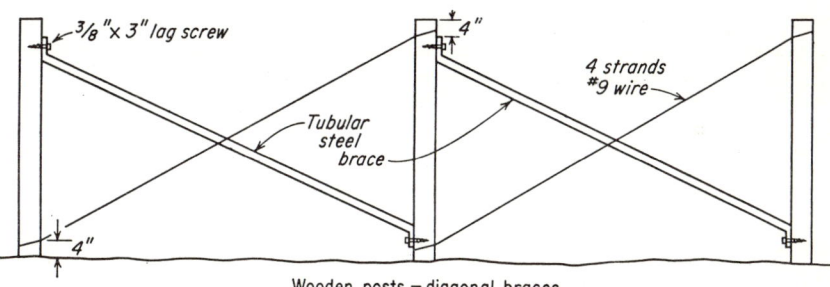

Wooden posts — diagonal braces

Alternate details

FIG. 54-4. Corner-post and corner-brace arrangement. (*Farm Fence Construction Standards Committee, ASAE,* 1957.)

10. Join wires with Western Union splices.
11. If the fence is over 32 in. high, stretch with double jack stretcher.
12. Barbed wire can be stretched with block and tackle.
13. In contour fencing the posts must be closer together to take the side thrust.
    a. The posts should be 10 to 14 ft apart on sharp corners and should lean outward so that they will be straight when tight.

770   FARMSTEAD STRUCTURES AND EQUIPMENT

  b. Stretch 10 to 15 rods at a time, but don't stretch two curves at the same time.
  c. The distance from the curvature of the exact radius and the straight line between line posts should not be more than 4 in. This rule can be used to determine how far apart the posts should be.
14. With one wooden post every rod, 5.6 lb of No. 9 gage 1-in. staples is needed for 80 rods of fence, or 8.1 lb of 1½-in. staples.
15. There are 20 rods to a regular roll of fence.

Electric Fence. Because of its low cost and simplicity, the use of electric fence on the farm has become almost universal. The electric fence consists, basically, of smooth or barbed wire mounted on wood or steel posts, separated from the posts by electrical insulators and connected to a controller. Electric fences are used for field fencing, dividing pastures, fencing marginal land where regular fencing is too costly, and inside and around structures for dividing pens for different usages, such as in loose housing.

Various methods of construction are used, depending on the terrain and the animals to be enclosed. One wire is used for horses, mules, and cattle. Two wires are

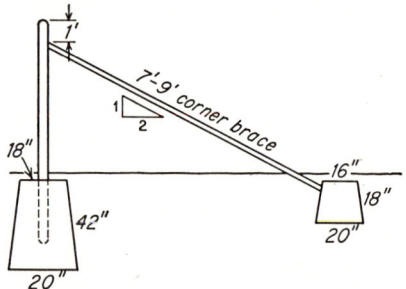

FIG. 54-5. Steel corner brace post.

recommended for swine. Two wires are required for sheep and goats. Energized gate wires require an insulated handle for hooking and unhooking the wire.

The controller provides the shock sensation for keeping the livestock within the enclosure. There are two types of shock: (1) the single-impulse shock as provided by batteries and (2) multiple-impulse shock, which is the more effective, provided from alternating current.

The controller includes a current interrupter which provides a shock of 0.1 sec duration with a period between shocks of 0.9 sec. The maximum permissible safe current is 10 ma. Electric fence controls should be approved by the Underwriters' Laboratory and the Industrial Commission of Wisconsin. Never, under any circumstances, should a homemade fence controller be connected to an ordinary lighting circuit. Many human deaths and livestock deaths are reported each year from shock by electric fence controllers. During shock the muscles which control breathing are paralyzed. Give artificial respiration to one who has been shocked, the same as in the case of asphyxiation by gas or drowning, and call a doctor. The controller must be grounded at the installation in order to provide a complete circuit through the soil and fence. The electric ground should be made of ¾-in.-diam pipe extending down to the level of permanent moisture, usually 8 ft or more. Spark-gap lightning arresters should be provided. Most controllers have a setting for wet and dry soil. A signal light is provided which indicates a short in the system, usually caused by weeds or brush moving against the fence.

After animals have been trained, there is little problem keeping them in the enclosure. An inexperienced person will usually place the wires too high. The electric-fence wire should be one-half to two-thirds the height of the animal above the ground.

SECTION IV
BASIC AGRICULTURAL DATA

Chapter 55

LIVESTOCK

C. F. KELLY

THERMAL ENVIRONMENT AND PRODUCTION

Farm animals and poultry are provided with a complex system of chemical and physical controls to maintain a constant internal temperature over a wide range of external thermal environments. As with man, a variation in body temperature of only a few degrees is serious.

Of the total food-energy intake by an animal or bird in production, approximately one-third is converted into heat and must be lost to the environment. Four avenues of heat dissipation are available to livestock for cooling: conduction, convection, radiation, and evaporation. The first three (or the nonevaporative methods) depend upon the difference between the surface temperature of the animal and the temperature of the air and the surroundings to control the heat-loss rate. As the environmental temperature increases, the animal's surface temperature also increases, but at a slower rate. The difference between the surface temperature and the environmental temperature then becomes less and less, and it becomes more difficult for the animal to lose, through the skin, the amount of heat necessary to retain a normal body temperature. In an effort to keep a normal temperature, the animal cuts down on its feed intake and increases its rate of respiration.

The thermal environment cannot be measured directly, as yet, by any one instrument. It is made up of the combined effect of air temperature, humidity, and velocity and the radiant temperature of the surroundings. A change in any one of these factors will change the response of the animal in so far as the proportion of total heat lost through any particular avenue is concerned.

With man, it is usual to think of thermal environment in connection with comfort. With farm animals the effect of the environment on comfort is measured by rate of production and by the amount of feed required for unit production. Measuring these effects is difficult because of the great variation in the efficiency of individual animals and the changes in production or rate of growth with stage of lactation, age, etc. While the experimental data that have been accumulated by various investigators are based on only a very small part of the animal "population," it does give some opportunity to evaluate the effect of environment on production.

Dairy Cows. The consensus is that 50°F is close to the optimum temperature for maximum production from dairy cows, which are better able to stand temperatures lower than 50°F than they are to stand temperatures higher. Kelley and Rupel [23] in tests in Wisconsin with Holstein cows concluded that under their conditions the optimum temperature appeared to be around 50° and that cows running loose in a pen barn withstood low temperatures better than when confined in stanchions. In Maryland, Buckley [10] concluded that low temperatures do not reduce the flow of milk. In California, Regan and Richardson [30] held cows in a psychrometric chamber at various air temperatures between 40 and 100°F and demonstrated that heavy milking animals withstand low temperatures (40°F) better than warm temperatures. Production started to decrease at temperatures above 80°F. In North Dakota, Dice [13] compared Holstein cows housed in an open shed during the wintertime with animals held in a barn, in tests covering four winters. The exposed animals were protected from the wind, snow, and rain, but the temperatures sometimes dropped below 0°F. His data indicate that dairy cows receiving an adequate ration under such environmental conditions will produce as well as cows kept in a conventional barn at 50°F, without requiring any more feed, and will tend to gain a little more in body weight. He calls attention to the fact that the exposed cows were always free to move around while the barn cows were confined in stanchions.

The most recent tests of the effect of thermal environment on dairy-cow production are by Ragsdale et al. [29] and Brody [8,9] at Missouri. These carefully conducted studies in the Missouri psychoenergetic laboratory covered a range of 0 to 105°F by 5 and 10° steps. Two air-conditioned chambers were used, each of six-cow capacity. The control animals were held in a chamber throughout a test at a constant temperature of 50°F. The experimental animals in the other chamber were subjected to various temperatures, but in most cases the temperature was not changed oftener than once in 2 weeks. The milk production of the Holsteins declined rapidly above 75°F but remained normal down to 10°F. The Jersey milk-production decline was slower at the high temperatures but declined at temperatures below 40°F. Results

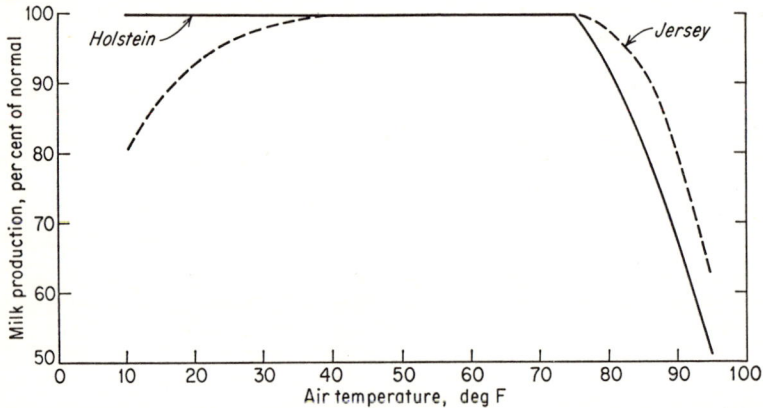

FIG. 55-1. Milk production of Holstein and Jersey dairy cows and air temperature. (*Yeck and Stewart, ASAE Trans.* 2, 1959.)

of studies between 10 and 95°F are shown in Fig. 55-1, as reported by Yeck and Stewart [52]. Feed consumption followed the same trend as milk production, decreasing above 70°F and increasing below 40°F. The Jerseys and Holsteins lost body weight above 75°F and gained at the lower temperatures.

Air velocity was found by Brody et al. [8] to not greatly affect the milk yield, feed and water consumption, or body weight of Brown Swiss, Holstein, or Jersey dairy cows in the range of 0.5 to 9 mph at air temperatures of 18, 50, 65, and 80°F. At 95°F the milk production and feed consumption of the Holstein and Brown Swiss cows were lower at the low air velocity than at the high velocities. In general, the studies indicated that the higher the milk production level and the larger the cow, the greater the effect of change in air velocity at 95°F.

Ragsdale et al. [29] found, in the psychoenergetic laboratory at the University of Missouri, that air *relative humidity* over a range of from about 40 to 80 per cent had no significant effect upon milk production and composition, feed and water consumption, and body weight in Holstein, Brown Swiss, Jersey, and Brahman cows when the atmospheric temperature was below 75°F. However, the depressing effect of humidity increased above about 75°. At 95°F Holsteins averaged only 11 lb of milk per day at low (40 per cent) humidity compared with 16 lb at high (80 per cent) humidity. At 75°F both groups of cows were producing about 27 lb of milk per day.

Beef Cattle. Fattening beef cattle do well in a cold environment, although protection from cold winds and rains is desirable. Feeding trials in various cold sections of the country indicate that enough heat is produced in the body as a by-product of digestion and assimilation of food to keep the animals warm in open sheds without increasing the feed intake per pound of gain. Cattle on a maintenance ration in cold climates may pay for a warmer shelter by savings in feed, but the savings will usually not justify more than the minimum expense. In tests in Oregon [28] and Idaho [18], fattening cattle, including calves, gained no faster, in the winter, when provided with shed shelter than when they had only windbreaks for shelter, and there was not sufficient saving of feed to justify the expense.

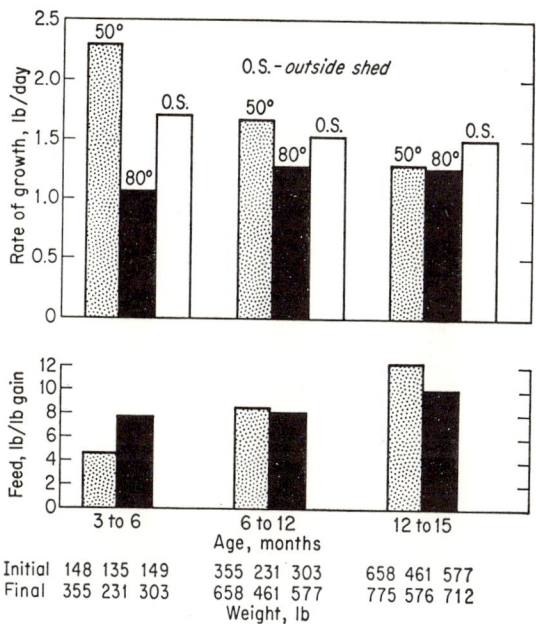

FIG. 55-1A. Upper section—Rate of growth of Shorthorn calves at constant air temperatures of 50 and 80°F and when housed in an open shelter in Missouri. (*Ragsdale et al., Univ. Missouri Agr. Exp. Sta. Bull.* 642, 1957.) Lower section—Feed used per pound of gain, under same conditions. Age and weights of animals also shown. (*Johnson et al., Univ. Missouri Agr. Exp. Sta. Bull.* 638, 1958.)

As with dairy cattle, high temperatures are more harmful to beef cattle than are low temperatures. Guilbert and Hart [15] state that "an average monthly temperature of 75°F may be considered about the upper limit for any period that does not result in depressive effects." The effects of environmental temperatures of 50 and 80°F upon growth rate, as reported by Ragsdale et al. [48], and upon use of feed, as reported by Johnson et al. [44], are given in Fig. 55-1A. In the warmer climates good shade is necessary for protection from solar and sky radiation. Tests in the Imperial Valley in California and in other hot areas [11,19–21,41,43,46,47] have shown the value of shade, showers, and cool drinking water where summer temperatures go well over 100°F. On the other hand, steers appear to graze better in certain climates when shade is not provided [4].

Bond, Kelly, and Ittner [6] report that Hereford beef cattle increased in average daily gains by as much as 80 per cent when the *air velocity* was increased to 3.7 mph from the normal average of 0.6 mph, during hot weather. These studies, in the Imperial Valley of California, were conducted under the average summer temperature of about 88°F, the average maximum temperature being over 102°F. The animals drank slightly more and made more efficient use of their feed at the higher air velocity.

Swine. The ability of swine to adapt themselves to a thermal environment depends to a certain extent on their size and age. The optimum temperature for growth is somewhat higher for the small pigs and lightweight hogs than for the heavier weights. Heitman and Hughes [17], and Heitman, Kelly, and Bond [42], in tests in a controlled-temperature chamber, found the rate of gain greatest and the amount of feed required to produce 100 lb gain least at an average temperature of about 75°F for hogs weighing 70 to 144 lb and approximately 60°F for hogs weighing 166 to 266 lb. As the air temperature was increased or decreased beyond these averages, rate of gain declined and utilization of food was lowered. The results of these tests are shown in Fig. 55-2. Small pigs appear to require a higher thermal environment. Bond, Kelly, and Heitman [5] observed that at 80°F small pigs seemed comfortable, at 70°F they

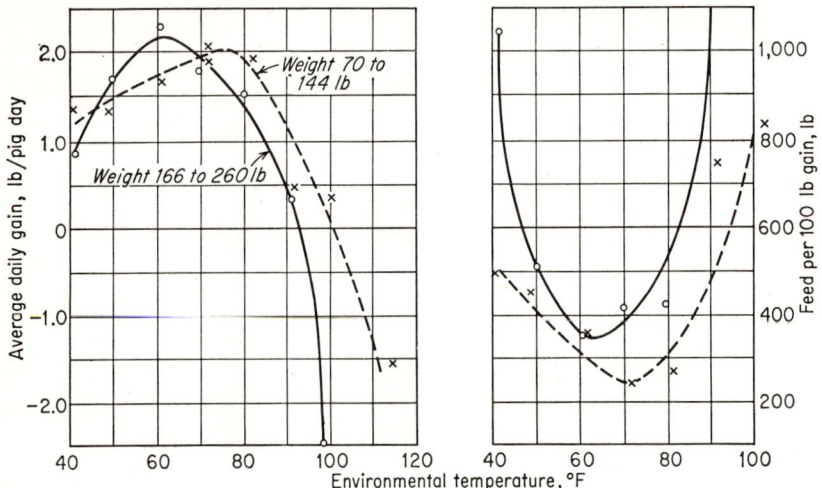

Fig. 55-2. Effect of environmental temperature on average daily gain and efficiency of gain by swine. (*Heitman and Hughes, J. Animal Sci., May, 1949.*)

shivered when standing by themselves, and at 60°F they never seemed to be warm enough. The sows were more comfortable at the lower temperatures.

Heitman and Hughes [17] found that at 96°F air temperature and 30 per cent air *relative humidity* swine weighing over 200 lb lost weight but survived for long periods. When the relative humidity was increased to 94 per cent over a period of 8 hr, the animals' body temperature increased 2.5° and the respiratory rate more than doubled. At an environment temperature of 90°F, the effect of these two humidity levels—30 and 94 per cent—on body temperature was not different.

Poultry. Laying hens are also sensitive to their thermal environment. Ota, Ashby, and Garver [26] found, in carefully controlled studies, that the optimum temperature for Rhode Island Red hens at low air velocities was about 55°F, production slowing down at temperatures higher and lower than this. Egg size and weight followed the same trend. These birds were on the floor (not in cages) and were managed under farm conditions as far as practicable. Wilson [36] indicates a rapid fall in production above 75°F, the lethal range beginning at about 100°F. These data are summarized in Fig. 55-3. Later studies by Ota, Wilson, and McNally [27] with White Leghorn hens housed in cages indicated about the same relation between environmental temperature and egg production (relative humidity 60 to 65 per cent), although the White Leghorns started to decrease in production at a higher temperature (about 90°F).

Light, aside from its thermal effect, is important to the egg production, feathering, and growth of poultry. Dobie et al. [14] state that 13 hr of Mazda incandescent light per day is necessary for maximum egg production. There was no significant difference in the effect of intensity of Mazda light varying from 1.0 to 31.3 footcandles on egg production, provided that the hens received 13 hr of light per day.

Chicks require a much higher temperature for brooding. Commercial *brooding practice,* based upon the work of Barott and Pringle [3], is to start day-old chicks at 95°F and drop the temperature 5° each week until 70°F is reached. These investigators use ambient temperature to define the environment. Baker and Bywaters [1] give amounts of infrared energy necessary to raise the mean radiant temperature of a brooder to offset lower-than-optimum ambient temperatures. Where the brooderhouse ambient temperature is not expected to go below 55°F, the estimated brooder capacity for a 375-watt lamp is 200 to 225 chicks; for a 250-watt lamp, 125 to 150 chicks; and for a 125-watt lamp, 50 to 75 chicks.

The effect of *humidity* and its relation to temperature were studied by Wilson and Edwards [37], using White Leghorn hens. Relative-humidity percentages of 28, 45,

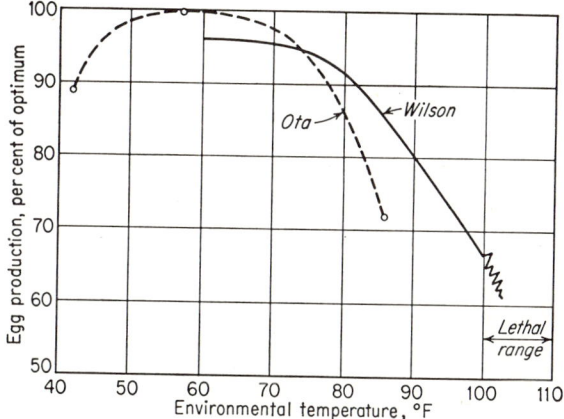

FIG. 55-3. Effect of environmental temperature on egg production. (*Ota et al., Agr. Eng., March,* 1953; *Wilson, Univ. Calif. Mimeo. Chart, February,* 1949.)

and 72 and temperatures of 70, 85, and 100°F were covered. The effect of high humidity (72 per cent) on respiration rate and body temperature was most pronounced at high temperature, 100°F. Skin and feather temperatures were observed to increase as relative humidity increased from 45 to 72 per cent, even at the lower temperatures. Heart rate was depressed more by high temperatures than by high humidity. No studies of the effect of humidity upon egg production were made.

HEAT AND MOISTURE LOSS

The well-being of livestock, the comfort of the human workers, and the useful life of the livestock shelter depend to a large extent on the temperature and moisture content of the air within the building. In order to maintain a healthful environment for animal, man, and building, some control of the ventilation is necessary. As the heat given off by the animals is usually expected both to warm the building and furnish motive power for moving fresh air through the structure, the amount of heat and moisture given off by the livestock must be known if a workable ventilation system is to be designed. All the sensible heat from the animal cannot be used to supply building and ventilation heat losses. Much of it will be used for evaporating liquids, at an expense of over 1000 Btu of sensible heat per pound of water removed. Where cooling is the problem, the heat and moisture loss from the animals must be added to the heat intake of the building when calculating the cooling load.

The heat loss from the various classes of livestock has been measured, calculated, or estimated by many investigators in the past. The bulk of the data collected has been from animals at rest, in a thermoneutral environment, and after fasting several hours. The heat loss under such conditions is considerably less than from active animals, on full feed, as would be the case under farm conditions. In so far as possible the data presented herein are from investigations where farm practices were followed in the housing and management of the livestock.

Dairy Cows. A series of tests has been made at the University of Missouri, cooperating with the U.S. Department of Agriculture, to determine the effect of thermal environment on heat and moisture loss from dairy cows [33,49,51,52]. Six cows were held in stanchions in an air-conditioned room, and the heat loss from the room measured. The air velocity ranged from 25 to 60 fpm, relative humidity was approximately 50 per cent, and the wall surfaces were a few degrees higher in temperature than the air. The animals were in full production, and feed and management were

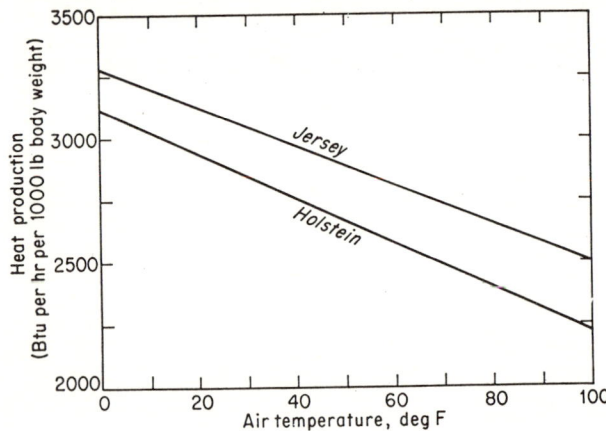

Fig. 55-4. Effect of air temperature upon total heat production of Holstein and Jersey dairy cows. (*Yeck and Stewart, ASAE Trans.* 2, 1959.)

similar to those of regular dairies. The average heat loss from Holstein and Jersey cows per 1,000 lb of body weight, as summarized by Yeck and Stewart [52], is given in Fig. 55-4. The curves include fermentation energies from animal rumen and manure, as well as animal metabolism. The data are presented as representative of cows in average production weighing approximately 1,000 lb each.

Beef. Nelson [47] and Kelly [46], in summarizing research of others, found total heat loss of Shorthorn beef cattle to be slightly below that of Holstein dairy cows of similar weight.

Swine. Studies of heat loss from swine have been made at the University of California, cooperating with the U.S. Department of Agriculture [5,22,39]. A group of four growing hogs, or a sow and her litter, were held in an air-conditioned chamber, and the heat loss from the chamber measured. The wall temperature was the same as the air temperature, air relative humidity about 50 per cent, and air velocity between 20 and 30 fpm. The total heat loss from growing, fattening, and mature hogs of different weights is given in Fig. 55-5. In Fig. 55-6 the average total heat loss from three sows and their litters is shown for an 8-week period. Each litter was held at a different temperature (60, 70, and 80°F), but since there was no significant difference due to temperature, only one curve is shown. Initial weight of the sows was 382, 335, and 390 lb. Litter sizes varied, averaging six pigs per litter.

Poultry. Studies are under way at the U.S. Department of Agriculture Research Center at Beltsville, Md., to determine the heat and moisture loss from poultry under usual management practices [26, 27]. Preliminary results indicate that the total heat loss from Rhode Island Red hens weighing 5 lb is 53 Btu per hr at an environmental temperature of 42°, 48 Btu at 56°, and 34 Btu at 86°F. In these tests air relative humidity averaged about 88, 83, and 57 per cent respectively. The hens

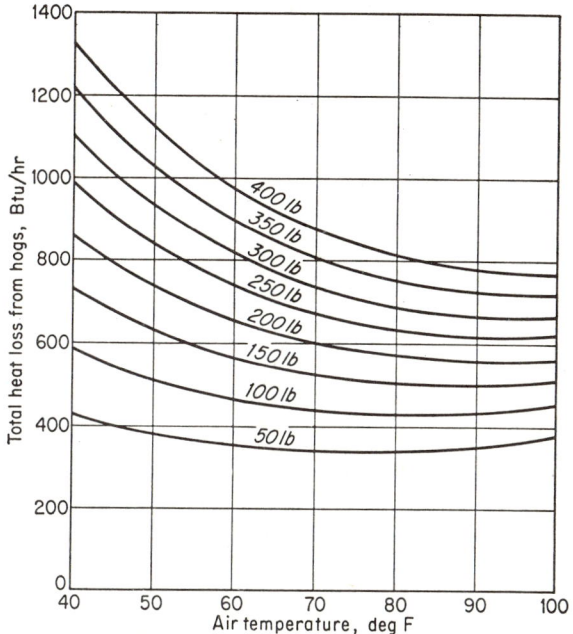

Fig. 55-5. Effect of air temperature upon total heat production of growing, fattening, and mature swine. (*Bond et al., ASAE Trans.* 2, 1959.)

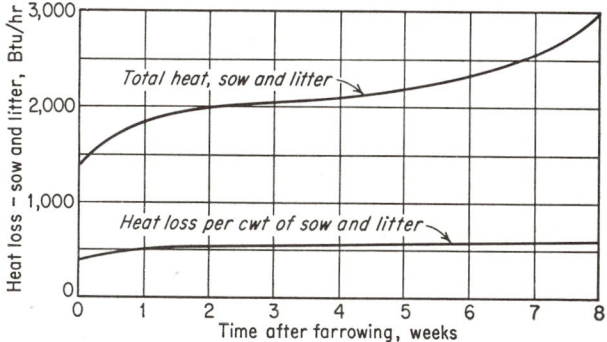

Fig. 55-6. Total heat loss from sow and litter during 8-week period following farrowing. (*Bond et al., Agr. Eng.*, March, 1952.)

were in production. Previous to this work Barott and Pringle [2] measured the total heat loss from Rhode Island Red females from hatch to maturity by means of a respiration calorimeter. The birds were kept in darkness during the one-day tests, without feed and water. Air relative humidity averaged between 50 and 60 per cent. None of these hens was laying. The results of both the Beltsville and Barott and Pringle measurements are shown in Fig. 55-7. The latter are included to indicate the proportionate effect of bird weight on heat loss and to point out differences that may be expected from experiments of the two types.

Mitchell and Kelley [24] estimated the heat loss from a 15-lb bronze turkey to be 119 Btu per hr; for a 5-lb duck, 57 Btu per hr; and for a 15-lb goose, 93 Btu per hr.

Evaporative Heat Losses. The percentage of the heat lost through *evaporation from the skin, lungs, and other internal surfaces* of livestock varies with the thermal environment, the proportion being greater at the higher temperatures. When the external surface temperature of the animal is the same as the environmental temperature, all the body heat is lost through the evaporation of moisture and none by radiation, convection, or conduction. The percentage of heat lost by evaporation, with respect to the total, is given in Fig. 55-8 for dairy cows, swine, and poultry.

In addition to water lost as vapor from the lungs and skin, a certain amount of *water from wet surfaces* in the stable will be removed in the ventilating air, the heat

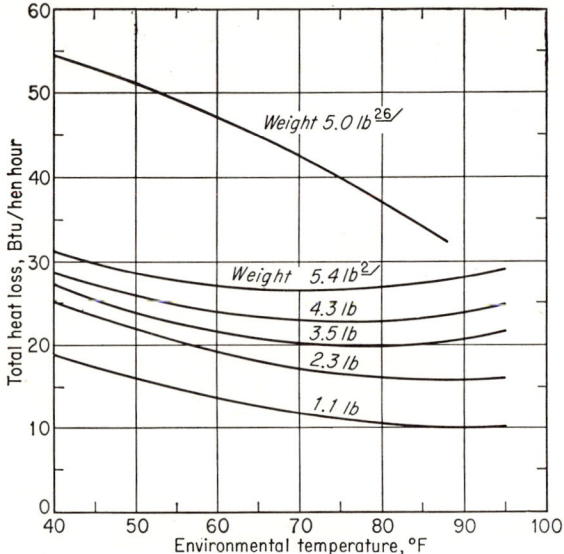

Fig. 55-7. Total heat loss from Rhode Island Red hens of various weights. (*Barott and Pringle, J. Nutrition, January, 1946; Ota et al., Agr. Eng., March, 1953.*)

from its evaporation being obtained from the animal sensible heat or from other sources. The amount will depend upon the area of wetted surfaces, such as floors, the velocity of the air over these surfaces, and the condition of the air. Thompson and Stewart [33] found that at air temperatures from 40 to 50°F the evaporation from *dairy-cow* stall surfaces ranged from 38 to 63 per cent of the total moisture load on the ventilating system and that from 9 to 13 lb of water was evaporated from the surfaces for each cow per day, the larger amounts being from the larger cows. The air relative humidity in these tests varied from 48 to 83 per cent, and the air velocity between 25 and 60 fpm.

Somewhat similar measurements by Bond et al. [5] with *swine* indicated that about twice as much water was removed in the ventilating air as was produced by the animals as vapor, up to an environmental temperature of 80°F, when the air relative humidity was about 50 per cent. The difference became less above this temperature. The air velocity ranged from 20 to 30 fpm. No bedding was used in the test chamber, and the concrete floor sloped to a drain. The effect of environmental temperature on the relationship between the water drunk by the pigs, the water removed in the ventilating air, and the water lost as vapor by the animals through the lungs and skin is shown for two weight classes of hogs in Fig. 55-9.

The use of water by 5-lb *laying hens* is shown in Fig. 55-10. The curve for vapor from the hen is based upon Fig. 59-8 and the upper curve in Fig. 55-7. The curves

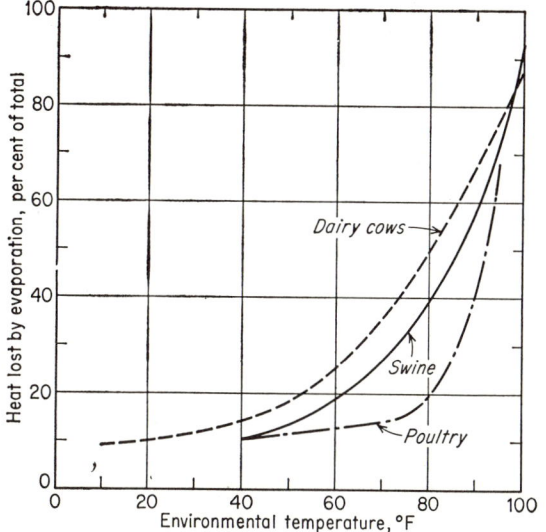

FIG. 55-8. Per cent of total heat lost by evaporation at various environmental temperatures by dairy cows, swine, and hens. Curves eye-fitted to data. (*Barott and Pringle, J. Nutrition, January, 1946; Bond et al., Agr. Eng., March, 1952; Thompson and Stewart, Agr. Eng., April, 1952.*)

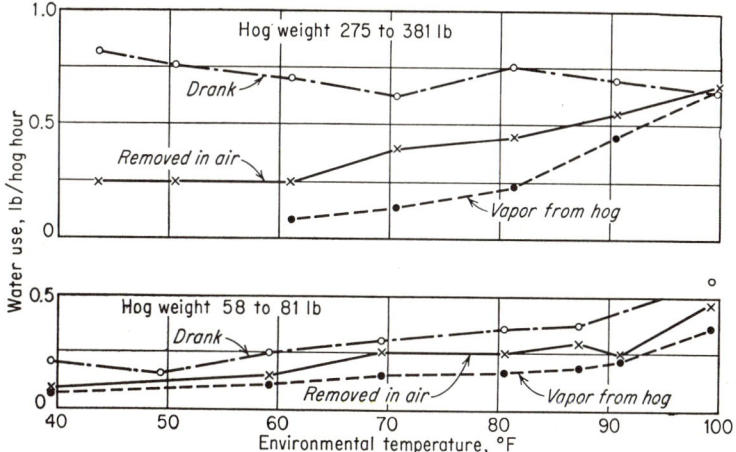

FIG. 55-9. Effect of environmental temperature on water use by swine. (*Bond et al., Agr. Eng., March, 1952.*)

for the amount of water removed and drunk by the birds are also from Ota et al. [26]. The reason that more water was removed from the chamber than was drunk by the birds is that the free water in the feed and the metabolic water created by the oxidation of the nutrients are included in the former. The litter, pine-oak sawdust and shavings, had a moisture content of about 50 per cent at the end of a 4-week test at 40°F, accumulating water at the rate of about 0.20 lb per hen day. At 55°F the rate of water accumulation by the litter was 0.174 lb per hen day, and the moisture content reached 36 per cent at the end of a 25-day test. At 80°F and over there was practically no moisture accumulation in the litter, and at the end of the test its moisture content was about 17 per cent. In all these tests ventilating air

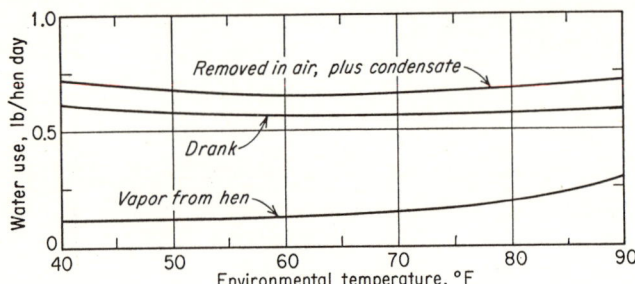

Fig. 55-10. Effect of environmental temperature upon water use by laying hens. (*Barott and Pringle, J. Nutrition, January, 1946; Ota et al., Agr. Eng., March, 1953.*)

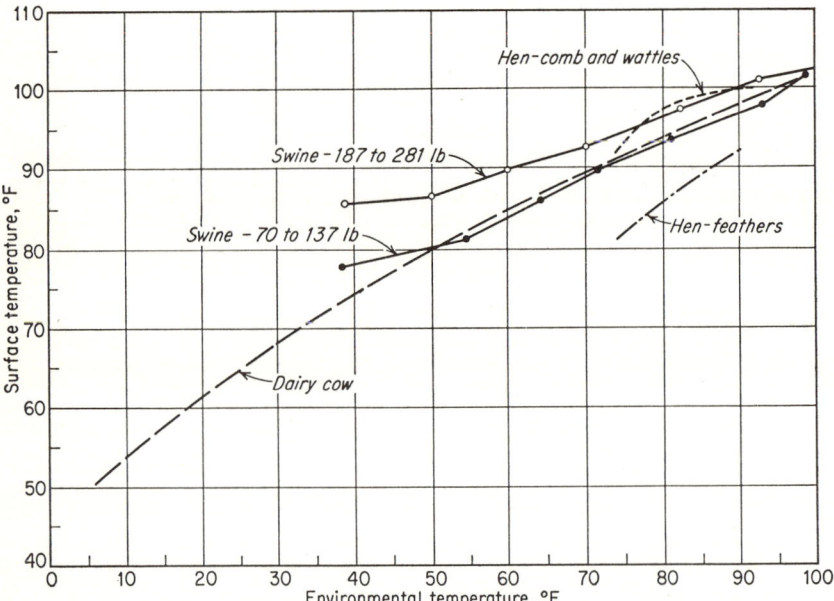

Fig. 55-11. Surface temperature of dairy cows, swine, and poultry, as related to environmental temperature. (*Kelly et al., Agr. Eng., December, 1948; Thompson et al., University of Missouri, 1951; Wilson and Plaister, Am. J. Physiol., September, 1951.*)

was supplied at the rate of between 4 and 5 lb of air per hen per hour. The rate of air movement within the chamber was about 30 fpm.

Surface Temperatures. Some measurements have been made of the surface temperatures of dairy cattle, swine, and poultry under various environmental conditions. These are plotted in Fig. 55-11. All measurements were made with the Hardy dermal radiometer [16]. The data for dairy cattle (Holstein and Jersey breeds) are from Thompson et al. [32]. The data for swine are from Kelly et al. [22], observations from seven different locations on the animals being averaged. The wall temperature is stated to be the same as the air temperature, the air relative humidity close to 50 per cent, and the air velocity between 20 and 30 fpm. The curves for feather and comb-wattle temperatures are from observations of Wilson and Plaister [35]. There are little data available on the surface temperature of other farm livestock at various environmental temperatures. For beef cattle, Kelly and Ittner [21] have measured the surface temperatures of animals in the shade and in the sun, at high air temperatures. The air velocity was approximately 2 mph. These data are given in Table 55-1.

TABLE 55-1. SURFACE TEMPERATURE OF CURLY-HAIRED YEARLING HEREFORDS
(Air Temperature = 100°F)

Location on animal	Average surface temperature, °F	
	Animal in sun	Animal in shade
Back.	125	103
Side (away from sun).	114	105
Side (toward sun).	131	
Belly.	109	

SOURCE: N. R. Ittner and C. F. Kelly, Cattle Shades, *J. Animal Sci.*, February, 1951.

Surface Area. The total heat loss from the surface of livestock is of course dependent upon the area involved, as well as upon its temperature and the thermal-environment conditions. Surface-area measurements have been undertaken by various investigators, and empirical formulas developed. The surface area of a given animal will change with every movement of the body and with environment temperature (due to shrinkage and expansion of the skin) and cannot be constant. The work of Brody et al. [7], who made careful studies of the surface areas of 482 dairy cattle, 431 beef cattle, 12 horses, and 12 swine, by means of a mechanical surface integrator, is shown in Fig. 55-12. The authors state that the use of a prediction formula

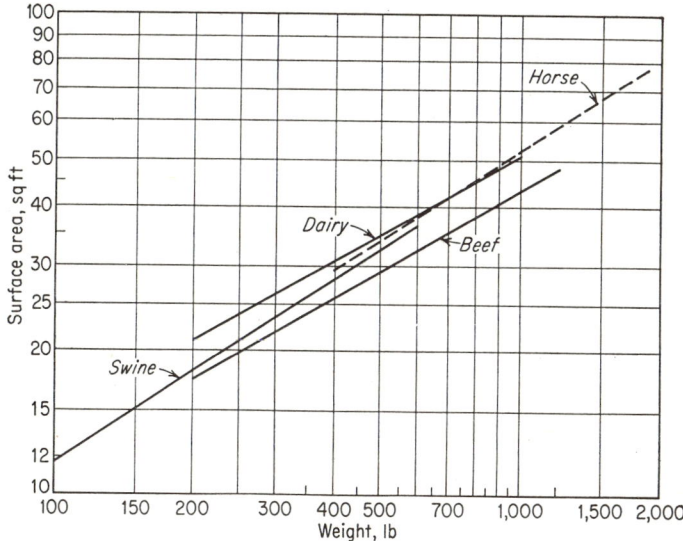

FIG. 55-12. Surface area of dairy and beef cattle, horses, and swine. (*Brody et al., Missouri Agr. Expt. Sta. Research Bull.* 115, 1928.)

(based upon the measurements of a population) for estimating the surface area of a particular animal may involve an error as high as 10 per cent of the true value. Deighton [12], after careful measurements of the surface areas of swine and cattle by several methods, concludes that the mechanical integrator gives the most accurate and consistent results.

Kelly et al. [22] have estimated that a single hog, lying down, has 20 per cent of its surface contacting the floor, 80 per cent losing heat by convection, and 75 per

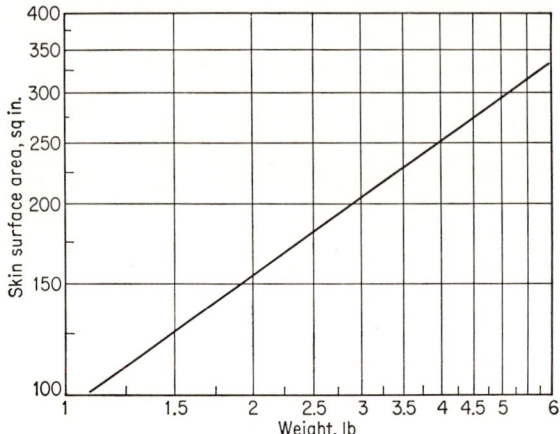

FIG. 55-13. Skin area of White Leghorn hens. (*Mitchell, J. Nutrition, March,* 1930.)

cent losing heat by radiation (because the inside of the ears, areas between the legs, etc., are facing areas of equal temperature).

The surface area of the skin of Single Comb White Leghorn chickens as determined by Mitchell [25] is shown in Fig. 55-13. The surface area of the bird losing heat by radiation and convection will be considerably greater, because of the feathers.

FEED AND WATER REQUIREMENTS

The general feed requirements for dairy and beef cattle, horses, sheep, swine, and poultry are given in Tables 55-2 to 55-8. The rate of feeding depends upon the

TABLE 55-2. FEED REQUIREMENTS OF DAIRY COWS

Milk produced daily, lb	Total pounds of concentrates to feed per day			
	Cows fed good hay*		Cows on good pasture†	
	4.0% fat in milk	5.5% fat in milk	4.0% fat in milk	5.5% fat in milk
20	5.0	8.4	2.0	4.7
25	7.2	10.9	4.2	7.4
30	9.3	13.7	6.3	10.0
35	11.5	16.3	8.5	12.7
40	13.7	19.0	10.6	15.3
45	15.8	21.6	12.8	18.0
50	18.0	....	14.9	19.5

* Cows fed 2 lb of good hay per 100 lb of live weight daily. If roughage is of very high quality, more will be consumed and amount of concentrate decreased. Good silage may be substituted for part of the hay at the rate of 3 lb of silage to 1 lb of hay.

† "Good" pasture is defined as being young, tender, abundant, and grown upon soils that are not seriously lacking in any of the essential mineral elements.

SOURCE: F. B. Morrison, "Feeds and Feeding," Morrison Publishing Co., Ithaca, N.Y., 1956.

TABLE 55-3. FEED REQUIREMENTS OF BEEF CATTLE, LB/DAY

Description	Body weight, lb	Expected daily gain, lb	Total feed, lb	Amount of hay (approx.), lb	Amount of grain (approx.), lb
Growing steers and heifers	400	1.6	12	9	3
	600	1.4	16	16	
	800	1.2	19	19	
	1,000	1.0	21	21	
Wintering weanling calves	400	1.0	11	11	*
	500	1.0	13	13	*
	600	1.0	15	15	*
Wintering yearling cattle	600	1.0	16	16	
	800	0.7	18	18	
Wintering pregnant heifers	800	1.3	20	20	†
	1,000	0.5	18	18	†
Wintering pregnant cows	800	1.5	22	22	†
	1,000	0.4	18	18	†
	1,200	0.0	18	18	†
Cow nursing calves, 1st to 4th months	900–1,100	0.0	28		
Fattening calves, short yearlings	400	2.0	12	4	8
	600	2.0	16	6	10
	800	2.0	20	7	13
Fattening yearlings	600	2.2	18	5–9	13–9
	800	2.2	22	7–11	15–11
	1,000	2.2	26	8–13	18–13

* With excellent roughage, no concentrates fed; otherwise 1 to 2 lb daily.
† Usually fed 1 to 2 lb of protein supplement.
SOURCE: "Recommended Nutrient Allowances for Beef Cattle," National Research Council, Washington, D.C., 1950.

TABLE 55-4. FEED REQUIREMENTS OF SWINE

Description	Body weight, lb	Expected daily gain, lb	Total air-dried feed,* lb
Market stock	25	0.8	2.0
	50	1.2	3.2
	100	1.6	5.3
	150	1.8	6.8
	200	1.8	7.5
	250	1.8	8.3
Breeding stock:			
Pregnant females and breeding boars	300	0.75	6.0
	500	0.50	7.5
Lactating females	350	....	11.0†
	450	....	12.5†

* Five types of diets made up from typical swine feeds contained 78 to 90 per cent grain, the remainder being supplements such as tankage, alfalfa meal, bone meal, and other materials ordinarily stored in sacks.
† Includes feed requirements of litter.
SOURCE: "Nutrient Requirements for Swine," *Natl. Research Council Publ.* 295, 1953.

work level, sex, age, weight, and stage of lactation or gestation, and these data should be used only as a guide in estimating the amount of storage space needed for particular cases, the design of mechanical feeders, etc. They are for the usual environmental conditions, where extra feed is not needed to provide heat in lieu of warmer housing. When silage is fed, most authorities suggest it may be substituted for hay at the rate of 3 lb of silage to 1 lb of hay.

Water Consumption. The amount of drinking water required by various farm livestock and poultry depends upon the environmental temperature, age, weight, production rate, etc. The average consumption of water by various classes of livestock and poultry at normal environmental temperatures is given in Table 55-9. Dairy cows in general require more drinking water than other animals, because about 87 per cent

TABLE 55-5. DRY-LOT FEED ALLOWANCES FOR SHEEP, LB/DAY

	Live weight, lb	Total* feed, lb	Approx. amount of hay, lb	Approx. amount of grain, lb	Approx. amount of silage, lb
Bred ewes, first 100 days gestation............	100	3.5	3.5		
Bred ewes, last 6 weeks before lambing........	110	4.0	3.5	0.5	
Ewes in lactation.........................	110	4.6	2.0	1.0	4.0
	120	4.7	3.3	0.7	2.0
	140	4.9	3.0	1.5	2.5
Fattening lambs, daily gain 0.35 lb............	70	2.7	1.5	1.2	

* Air-dry basis.
SOURCE: "Recommended Nutrient Allowances for Sheep," National Research Council, Washington, D.C., 1949.

TABLE 55-6. FEED REQUIREMENTS OF HORSES

Work level	Feed requirements per day
Idle..............	Principally roughage, with 8 to 12 lb alfalfa, clover, soybean hay, or timothy hay per day
Light work........	⅓ to ¾ lb grain and 1¼ to 1½ lb hay for each 100 lb live weight
Medium work......	¾ to 1 lb grain and 1 to 1½ lb hay for each 100 lb live weight
Heavy work.......	1 to 1¼ lb grain and 1 lb hay for each 100 lb live weight

SOURCE: A. B. Caine, "Care and Feeding of Horses," *Iowa State Univ. Ext. Serv. Circ.* 250, 1938.

TABLE 55-7. FEED REQUIRED BY CHICKENS OF DIFFERENT LIVE WEIGHTS FOR MAINTENANCE AND FOR PRODUCTION OF 0, 100, 200, AND 300 EGGS, RESPECTIVELY, PER YEAR

Average live weight, lb	Average total feed required per hen per year			
	0 eggs per year, lb	100 eggs per year, lb	200 eggs per year, lb	300 eggs per year, lb
4.0	57	71	85	99
4.5	61	75	89	104
5.0	65	80	94	108
5.5	70	84	98	112
6.0	74	88	102	116
6.5	78	92	106	120
7.0	81	96	110	124
8.0	89	103	118	132

SOURCE: L. E. Card, "Practical Poultry Feeding," *Univ. Illinois Ext. Serv. Circ.* 606, 1946.

of milk is normally water. Kelley and Rupel [23] found that the ratio of water consumed by the cow to milk produced varies from 3⅓ to 4 lb of water per pound of milk. Most of the water was drunk between 7 and 9 A.M. and between 6 and 8 P.M., after the feeding of hay.

The water consumption of beef cattle varies with thermal environment. The data given in Table 55-9 are for winter conditions in the Midwest. Ittner, Kelly, and Guilbert [20] found, in the hot Imperial Valley of California, that 900-lb Herefords drank approximately 15 gal per head daily and 700-lb Brahmans about 10 gal. The temperature of this water averaged 88°F. When the water was cooled to 65°F the consumption rate remained about the same (for the Herefords) but the rate of gain was increased.

The practice of self-feeding cattle with salt-cottonseed-meal mixtures affects the amount of water drunk. Savage and McIlvain [31] found that cattle in Oklahoma in the winter, self-fed the salt-meal mixture, usually consumed slightly more than twice as much water as comparable cattle fed cottonseed cake alone.

TABLE 55-8. ESTIMATED FEED REQUIREMENTS OF GROWING BIRDS
(Total Feed, Lb per Bird, to Age Shown)

Age, weeks	Growing pullets[a]	Broilers[b]	Broad-breasted bronze turkeys[c]	Pekins, ducklings[d]	Pheasants[e]
2	0.30	0.47	.....	1.7	0.33
4	1.00	1.65	1.8	6.27	0.65
8	3.60	5.71	8.3	19.42	3.17
12	7.65	11.76	18.9	26.07[f]	4.63
16	12.55	21.46	32.4	.....	7.66
20	18.25	30.76	47.7	.....	11.14
24	24.75	.....	64.9		
28	30.00[g]	.....	86.3		
32	.....	.....	110.7		
Mature males, lb/day	.....	.....	0.65	0.50	
Mature females, lb/day	.....	.....	0.50	0.50	

[a] L. E. Card, "Practical Poultry Feeding," *Univ. Illinois Ext. Serv. Circ.* 606, 1946. Data for White Leghorns.
[b] L. M. Potter and R. C. Ringrose, Growth and Feed Standards of New Hampshires, *Univ. New Hampshire Agr. Expt. Sta. Bull.* 401, 1953. Data for New Hampshire birds, mixed sexes.
[c] V. S. Asmundson and F. H. Kratzer, Turkey Production in California, *Univ. Calif. Agr. Ext. Serv. Circ.* 110, Davis, Calif., 1951. Data for mixed sexes.
[d] J. M. Hunter and J. C. Scholes, "Profitable Duck Management," 8th ed., The Beacon Milling Co., Inc., Cayuga, N.Y., 1950.
[e] Thomas Rae, "Profitable Game Bird Management," 6th ed., The Beacon Milling Co., Inc., Cayuga, N.Y., 1947.
[f] To the end of the tenth week.
[g] To the end of the twenty-seventh week.

TABLE 55-9. WATER REQUIREMENTS OF FARM LIVESTOCK AND POULTRY

Livestock	Water consumption, gal/day/head	Remarks
Dairy cows*	10–15	Heavy producers need more
Beef cattle:		
Average†	12	See text
800 lb weight‡	7	
Horses*	10–12	More in hot weather and at heavy work
Swine:*		
Fattening hogs	4 lb/cwt	
Weanling pigs	12 lb/cwt	
Brood sows, before farrowing	7.5 lb/head	
Sheep:		
Ewes, before lambing	1	On dry feed
Ewes, nursing	1.5	On dry feed
Fattening lambs	0.5	
Laying hens §	8–10	Per 100 birds
Turkeys: ‖		
10 weeks old	10	Per 100 birds
25 weeks old	14.6	Per 100 birds

* F. B. Morrison, "Feeds and Feeding," The Morrison Publishing Company, Ithaca, N.Y., 1936.
† Beef Cattle Housing in the North Central Region of the United States, *S. Dakota State Coll. Agr. Expt. Sta. Bull.* 382, 1946.
‡ Paul Gerlaugh, *Univ. Ohio Agr. Expt. Sta. Bimonthly Bull.* 159, 1932.
§ W. O. Wilson, High Environmental Temperatures as Affecting the Reaction of Laying Hens to Iodized Casein, *Poultry Sci.*, July, 1949.
‖ Neil F. Morehouse, Water Consumption in Growing Turkeys, *Poultry Sci.*, January, 1949.

Water intake of cattle as related to feed consumption and environmental temperature has been discussed in detail by Winchester and Morris [38].

High environmental temperatures affect the water consumption of swine, as shown by Fig. 55-9, and of laying hens. Wilson [34] states that the amount of water consumed by 100 White Leghorn and Rhode Island Red hens varied from 7.6 up to

10.3 gal per day, as ambient temperature increased from 70 to 100°F. After some of the hens were out of production, water consumption dropped to 5.8 gal for 100 hens in 70°F air temperature.

REFERENCES

1. Baker, W. H., and James H. Bywaters: Brooding Poultry with Infrared Energy, pt. I, *Agr. Eng.,* 32:316–320, June, 1951; pt. II, *Agr. Eng.,* 33:15–20, January, 1952.
2. Barott, H. G., and Emma M. Pringle: Energy and Gaseous Metabolism of the Chicken from Hatch to Maturity as Affected by Temperature, *J. Nutrition,* January, 1946.
3. Barott, H. G., and Emma M. Pringle: Effect of Environment on Growth and Feed and Water Consumption of Chickens, *J. Nutrition,* July, 1947; January, 1949; May, 1950.
4. Beef Cattle Housing in the North Central Region of the United States, *S. Dakota State Coll. Agr. Expt. Sta. Bull.* 382, 1946.
5. Bond, T. E., C. F. Kelly, and Hubert Heitman, Jr.: Heat and Moisture Loss from Swine, *Agr. Eng.,* 33:148–154, March, 1952.
6. Bond, T. E., C. F. Kelly, and N. R. Ittner: Cooling Beef Cattle with Fans, *Agr. Eng.,* 38:308–309, May, 1957.
7. Brody, Samuel, James E. Comfort, and J. S. Matthews: Further Investigations on Surface Area with Special Reference to Its Significance in Energy Metabolism, *Univ. Missouri Agr. Expt. Sta. Research Bull.* 115, 1928.
8. Brody, S. M., A. C. Ragsdale, H. J. Thompson, and D. M. Worstell: Environmental Physiology with Special Reference to Domestic Animals, *Univ. Missouri Agr. Expt. Sta. Research Bull.* 545, 1954.
9. Brody, Samuel: Climatic Physiology of Cattle, *J. Animal Sci.,* June, 1956.
10. Buckley, S. S.: Open Stables versus Closed Stables for Dairy Animals, *Univ. Maryland Agr. Expt. Sta. Bull.* 177, 1913.
11. Chiles, A. C., and O. F. Pahnish: The Effects of Shade Location on Summer Gains of Fattening Cattle, *Univ. Ariz. Rept.,* 1952.
12. Deighton, T.: The Determination of the Surface Area of Swine and Other Animals, *J. Agr. Sci.,* April, 1932.
13. Dice, J. R.: The Influence of Stable Temperature on the Production and Feed Requirements of Dairy Cows, *J. Dairy Sci.,* January, 1940.
14. Dobie, John B., J. S. Carver, and June Roberts: Poultry Lighting for Egg Production, *Wash. State Coll. Agr. Expt. Sta. Bull.* 471, 1946.
15. Guilbert, H. R., and G. H. Hart: California Beef Production, *Univ. Calif. Manual 2,* sec. II, Davis, Calif., 1951.
16. Hardy, J. D.: The Radiation of Heat from the Human Body, *J. Clin. Invest.,* July, 1934.
17. Heitman, H., Jr., and E. H. Hughes: The Effects of Air Temperature and Relative Humidity on the Physiological Well Being of Swine, *J. Animal Sci.,* May, 1949.
18. Hickman, C. W., E. F. Rinehart, and R. F. Johnson: Fattening Idaho Range Cattle, *Univ. Idaho Agr. Expt. Sta. Bull.* 209, 1934.
19. Ittner, N. R., and C. F. Kelly: Cattle Shades, *J. Animal Sci.,* February, 1951.
20. Ittner, N. R., C. F. Kelly, and H. R. Guilbert: Water Consumption of Hereford and Brahman Cattle, and the Effect of Cooled Drinking Water in a Hot Climate, *J. Animal Sci.,* August, 1951.
21. Kelly, C. F., and N. R. Ittner: Artificial Shades for Livestock in Hot Climates, *Agr. Eng.,* 29:239–250, June, 1948.
22. Kelly, C. F., Hubet Heitman, Jr., and Jack R. Morris, Effect of Envionment on Heat Loss from Swine, *Agr. Eng.,* 29:525–529, December, 1948.
23. Kelley, M. A. R., and J. W. Rupel: Relation of Stable Environment to Milk Production, *USDA Tech. Bull.* 591, 1937.
24. Mitchell, H. H., and M. A. R. Kelley: Estimated Data on the Energy, Gaseous, and Water Metabolism of Poultry for Use in Planning the Ventilation of Poultry Houses, *J. Agr. Research,* Nov. 15, 1933.

25. Mitchell, H. H.: The Surface Area of Single Comb White Leghorn Chickens, *J. Nutrition,* March, 1930

26. Ota, Hajima, Wallace Ashby, and H. L. Garver: Heat and Moisture Production of Laying Hens under Simulated Farm Poultry House Conditions, *Agr. Eng.,* 34:163–167, March, 1953.

27. Ota, Hajima, W. O. Wilson, and E. H. McNally: Calorimeter Studies on Caged White Leghorn Hens, *USDA Mimeo. Rept.,* Beltsville, Md., 1957.

28. Potter, E. L., and Robert Withycombe: Shelter and Warm Water for Fattening Steers, *Oregon State Coll. Agr. Expt. Sta. Bull.* 183, 1921.

29. Ragsdale, A. C., H. J. Thompson, D. M. Worstell, and S. M. Brody: Environmental Physiology with Special Reference to Domestic Animals, *Univ. Missouri Agr. Expt. Sta. Research Bulls.* 425, 1948; 449, 1949; 460, 1950; 521, 1953.

30. Regan, W., and G. A. Richardson: Reactions of the Dairy Cow to Changes in Environmental Temperature, *J. Dairy Sci.,* February, 1938.

31. Savage, D. A., and E. H. McIlvain: Self Feeding Salt Mixtures to Range Cattle, *USDA Southern Great Plains Field Sta. Rept.,* 1951.

32. Thompson, H. J., D. M. Worstell, and Samuel Brody: Environmental Physiology, pt. XV, *Univ. Missouri Agr. Expt. Sta. Research Bull.* 481, 1951.

33. Thompson, H. J., and R. E. Stewart: Heat and Moisture Exchanges in Dairy Barns, *Agr. Eng.,* 33:201–206, April, 1952.

34. Wilson, W. O.: High Environmental Temperatures as Affecting the Reaction of Laying Hens to Iodized Casein, *Poultry Sci.,* July, 1949.

35. Wilson, W. O., and T. H. Plaister: Skin and Feather Temperatures, *Am. J. Physiol.,* September, 1951.

36. Wilson, W. O.: The Effect of Temperature on the Hen, *Univ. Calif. Mimeo. Chart,* Davis, Calif., February, 1949.

37. Wilson, W. O., and W. H. Edwards: Interaction of Humidity and Temperature as Affecting Comfort of White Leghorn Hens, *Poultry Sci.,* September, 1953.

38. Winchester, C. F., and M. J. Morris: Water Intake Rates of Cattle, *J. Animal Sci.,* August, 1956.

39. Bond, T. E., C. F. Kelly, and Hubert Heitman, Jr.: Hog House Air Conditioning and Ventilation Data, *ASAE Trans.,* 2:1–4, 1959.

40. Bond, T. E.: Environmental Studies with Swine, *Agr. Eng.,* 40:544–549, 1959.

41. Garrett, W. N., T. E. Bond, and C. F. Kelly: Effect of Air Velocity on Gains and Physiological Adjustments of Hereford Steers in a High Temperature Environment, *J. Animal Sci.,* February, 1960.

42. Heitman, H., Jr., C. F. Kelly, and T. E. Bond: Ambient Air Temperature and Weight Gain in Swine, *J. Animal Sci.,* February, 1958.

43. Ittner, N. R., T. E. Bond, and C. F. Kelly: Methods of Increasing Beef Production in Hot Climates, *Univ. Calif. Agr. Expt. Sta. Bull.* 761, 1958.

44. Johnson, H. D., A. C. Ragsdale, and R. G. Yeck: Environmental Physiology and Shelter Engineering, *Univ. Missouri Agr. Expt. Sta. Bull.* 638, 1958.

45. Kelly, C. F.: Environmental Studies with Sheep, *Agr. Eng.,* 40:549–551, 1959.

46. Kelly, C. F.: Basic Requirements of Beef Housing, *Agr. Eng.,* September, 1960.

47. Nelson, G. L.: Effects of Climate and Environment on Beef Cattle, *Agr. Eng.,* 50:540–544, 1959.

48. Ragsdale, A. C., Chu Shan Cheng, and H. D. Johnson: Environmental Physiology and Shelter Engineering, *Univ. Missouri Agr. Expt. Sta. Bull.* 642, 1957.

49. Stewart, R. E., and M. D. Shanklin: Environmental Physiology and Shelter Engineering, *Univ. Missouri Agr. Expt. Sta. Bull.* 656, 1958.

50. Stewart, R. E., and C. N. Hinkle: Environmental Requirements for Poultry Shelter Design, *Agr. Eng.,* 40:532–535, 1959.

51. Yeck, R. G.: Environmental Research with Dairy Cattle, *Agr. Eng.,* 40:536–540, 1959.

52. Yeck, R. G., and R. E. Stewart: A Ten-year Summary of the Psychroenergetic Laboratory Dairy Cattle Research at the University of Missouri, *ASAE Trans.,* 2:71–77, 1959.

Chapter 56

SOIL

H. H. Krusekopf

SOIL FORMATION AND CLASSIFICATION

Soil science is that branch of natural science which deals with the soil.

The major divisions of soil science, as adopted by the American and International Soil Science Societies, are:
1. Soil physics
2. Soil chemistry
3. Soil biology
4. Soil fertility and plant nutrition
5. Soil morphology and classification (pedology)
6. Soil conservation and management (technology)

Soil is a complex material, which has different *definitions,* depending on the concept. In the pedologic concept, soil is defined as a natural body:

"The soil consists of the outer layer of the unconsolidated crust of the earth, ranging in thickness from a film to several feet, which differs from the material beneath in color, morphology, texture, structure, chemical composition and in physical and biological characteristics."

In the genetic concept, soil is defined as dynamic and changing:

"Soil is loose mineralogic material that has acquired definite characteristics due to soil-forming forces."

From an agronomic point of view:

"Soil is a natural body, composed of mineral and organic materials, on the surface of the earth, in which plants grow."

Soil is not defined by such terms as earth, ground, dirt, land, clay, mineral dust, etc.

Soil has a significance according to the interest of the individual or group:

Biologist: micro-animal life, organic matter
Chemist: colloids, pH, composition
Engineer: mechanical composition, subgrade, foundation properties
Geologist: mineral composition, origin
Farmer: crop production, fertility, tillage
Physicist: moisture capacity, porosity
Pedologist: a natural object which has its own science, laws, and principles that determine its character

Soil Formation. The earth's crust is composed largely (95 per cent) of igneous rocks, but about three-fourths of the earth's surface is covered with sedimentary rocks, of which 80 per cent are shale, 15 per cent sandstone, and 5 per cent limestone. It is estimated that 40 per cent of the world's continental land surface is covered with *unconsolidated deposits* that form soils. In the United States an even larger per cent of the land surface consists of soft unconsolidated material. The various types of such deposits and their geographic location are as follows:

Glacial drift—Northeastern United States approximately south to Ohio and Missouri Rivers
Marine deposits—Southeastern and Eastern United States Coastal Plain
Outwash deposits—Great Plains region east of the Rocky Mountains
Loess—North-central states, 5 per cent of North America
Volcanic ash—Rocky Mountain region
Alluvium—All parts of the United States

The process by which rocks and minerals are converted to soil material, by external forces, is known as *weathering*. It is a destructive process in which complex compounds are changed to simpler composition by the following processes:
1. Physical disintegration resulting from (a) changes in temperature, (b) fracture by freezing, (c) action of flowing water, (d) action of moving ice, and (e) action of winds
2. Chemical decomposition due to (a) hydrolysis, (b) solution, (c) oxidation, (d) carbonization, and (e) hydration
3. Biologic action due to effects of (a) roots, (b) animals, and (c) CO_2 from organic material

The great diversity of soils is indicated by the more than 5,000 soil series or units that have been recognized in the United States system of *soil classification*. Obviously, many of the series differ only in one or two minor features such as texture or depth, while others differ in basic morphological features of the profile or in mineralogic composition of the material. The identification of soil units and their classification must be done in the field; it cannot be done in the laboratory. It must be based on the features of the whole soil, and not the surface only. It must be recognized that the different horizons or layers that occur above or below each other belong together and that the entire profile satisfactorily characterizes the soil. For this reason, modern soil descriptions contain sketches and statements describing the profile. The question as to *how* and *why* soils differ is the problem of soil genesis and morphology, which are included in the broader subject of soil formation.

Soils differ in many features that include color, texture, structure, number and character of horizons, depth, mineralogic composition, etc. Soil examination must include the whole profile to depth of weathering—usually 3 to 6 ft. Most of the features used for identification are below the surface and are not seen. It is for this reason that soil differences are often not recognized and the basis for classification not understood. The *how* of soil difference is therefore based on many features of the entire soil and the detail with which these are considered for unit identification.

The question *why* soils differ involves the whole realm of soil science. Soil is a natural body. It is a dynamic system—not static. Soils have acquired characteristics imposed on them, generally over a long period of time, by the environment or soil-forming factors. These are climate (rainfall, temperature), vegetation, parent material, relief (topography), drainage, and time. Although in practice soils are defined on the basis of their own characteristics, yet for an understanding, the factors that produced them must be considered. A soil is no more independent of the environment than, for example, a living organism is independent of oxygen or water. The processes of soil formation extend from the decomposition of the rock to the development of the soil profile. A distinction is sometimes made between weathering and formation. The former refers to the chemical and physical changes from indurated rock to soil material. The latter applies to the development of the profile. The two processes overlap, and there is no clear line of separation.

Climate as a soil-forming factor is most apparent in regional differences. Thus the soils in regions of low rainfall are less leached and have a different structure and profile than soils in regions of high rainfall. Soils in cold climates have a shallow solum compared with the deep weathered zones in warm climates. Many other soil differences in color, clay content, reaction, etc., are determined by climate.

Related to climate is the *biotic* factor, mainly vegetation. Soils formed under forest are characteristically lighter-colored than soils of the grasslands. Subdivisions in the major vegetation types have corresponding differences in soils. Differences in natural vegetation are generally indicators of differences in soils.

Parent, or *soil-forming, material* in its endless variety is probably the most significant factor in determining soil differences both locally and regionally. Texture, thickness of soil mass, mineralogic composition, and relief are the most prominent conditions related to parent material. Loess and glacial till are the dominant soil-forming materials in the north-central states. The residuum of cherty limestones, sandstone, silt, and clay shales, etc., give rise to strongly contrasting soils. Differences in the rate of weathering of minerals affect the rate of soil formation. Parent material, therefore, is an important factor in determining fertility, moisture properties, and engineering qualities. Although not essential, knowledge of the geologic origin of a soil is highly desirable.

Relief, or topography, is a significant factor in determining soil differences. Erosion and the exposure of different soil-forming materials are related to the slope. Soils on level areas generally have more profile development than soils on slopes, even if derived from the same material. Differences in drainage as a result of topography always cause differences in soils.

Another soil-forming factor that is difficult to evaluate, yet highly significant, is *time.* Material that has been undergoing soil-forming processes for 10,000 years will be more leached and have a different morphology than the soils resulting from 1,000 years of weathering. Loess and till are relatively young geologic material compared with the residuum from indurated rocks. The time factor is generally expressed in the degree of soil development, or maturity. Three stages—young, mature, and old—are sometimes recognized, but have not been satisfactorily defined. It is assumed that maturity is evidenced by certain morphological characteristics that indicate the various stages of development. A soil in adjustment with the environment is *mature;* an *immature* soil retains some resemblance to the parent material. It should be noted that a distinct development of the profile is not the criterion of maturity. In general, lower fertility and a compact, brittle structure of the soil aggregates are associated with highly developed soils.

Soil formation is a complex process. It involves the decomposition of mineral and organic matter; the loss of some constituents by leaching; the movement and accumulation of some of the weathered material; and the synthesis of new material, mainly clay minerals. The relative importance of the soil-forming factors cannot be satisfactorily evaluated. An interpretation of the factors—the environment—is essential for an understanding of soils in their variety and form.[1]

The *types of process* by which a soil is formed depend primarily on climate and are classified as follows:

Podzolization is a type of weathering in a humid climate characterized by leaching. This process is most active under forest vegetation and in cold and temperate climate where incomplete decomposition of organic material gives rise to acids and removal of iron and aluminum. Podzolization tends to produce distinct horizons in profiles.

Laterization is a type of weathering in a humid, warm climate. Low acidity and rapid decomposition of organic matter favor solubility and removal of silica and accumulation of iron and aluminum. There is a tendency for red and yellow color and brittle consistency in soils.

Calcification is a process of soil formation in regions of low rainfall, where decomposition of minerals is slow and free carbonates tend to accumulate in the lower part of the profile.

Solonization is a soil-forming process generally applied to alkaline soils.

Gleization is a process confined to soils of very poor drainage or water-logged condition. Exclusion of air retards oxidation and reduction of iron compounds. Glei soils are characterized by steel-gray color, without horizons in the gleied zone.

Modern soil science is based on study of the *soil profile,* the vertical section from the surface down to the depth of active weathering, usually 3 to 4 ft in humid, tem-

[1] Ray W. Simonson, What Soils Are, "Soil," USDA Yearbook of Agriculture, 1957, pp. 17–30.

perate regions. Each layer, or section, in the profile is called a horizon. For convenience in description and to indicate genetic relation, horizons are designated by letters as:

A Surface soil, zone of eluviation (leaching)
B Subsoil, zone of illuviation (accumulation)
C Substratum, parent material, slightly weathered

Each principal horizon may have several parts. These are indicated by a numerical suffix to the horizon letter.

The differences in composition between horizons are shown by the analysis of Putnam silt loam in Table 56-1. This soil is representative of an extensive group of soils in Missouri, Illinois, Iowa, and Kansas. It is derived from loess and therefore has few particles larger than silt size—0.02 to 0.002 mm. It has a mature profile,

TABLE 56-1. COMPOSITION OF PUTNAM SILT LOAM

Horizon	Depth, in.	SiO_2, %	TiO_2, %	Fe_2O_3, %	Al_2O_3, %	MnO, %	CaO, %	MgO, %	K_2O, %	Na_2O, %	P_2O_5, %	SO_3, %
					Chemical analysis							
A_1	0–6	82.87	0.79	2.72	9.36	0.13	0.68	0.49	1.73	1.08	0.08	0.07
A_2	6–13	81.02	0.77	3.32	10.62	0.08	0.60	0.55	1.77	1.10	0.11	0.05
B	13–28	73.03	0.78	4.95	15.68	0.05	0.88	1.28	2.03	1.18	0.09	0.02
C	28–36	74.09	0.76	4.46	14.71	0.08	1.08	1.14	2.26	1.29	0.10	0.01

Horizon	Depth, in.	Gravel, 2–1 mm, %	Coarse sand, 1–0.5 mm, %	Fine sand, 0.5–0.1 mm, %	Very fine sand, 0.1–0.05 mm, %	Silt, 0.05–0.005 mm, %	Clay, 0.005 mm, %
				Mechanical analysis			
A_1	0–6	0.8	1.3	0.7	0.9	71.0	25.2
A_2	6–13	1.1	1.9	1.0	0.8	68.6	26.7
B	13–28	0.0	0.1	0.3	0.4	39.8	59.4
C	28–36	0.1	0.5	0.9	0.7	58.2	39.8

SOURCE: University of Missouri Agricultural Experiment Station, Columbia, Mo., 1952.

developed on level topography, under grass vegetation, and an annual rainfall of about 40 in. The high content of clay in the subsoil is correlated with a high aluminum content.

Soil Classification. Classification is the grouping of objects according to their similarities and differences. Soils, like all natural objects, can be classified. The first requirement in soil classification is the identification of soil units or types. This is difficult because the criteria used in classification are not readily apparent and therefore require examination of the soil to a depth of 3 ft or more.

Schemes of soil classification in the United States have undergone changes with increasing knowledge of soils. *Geologic* origin, once a basis for classification, is no longer used. It resulted in such groups as "limestone soils," "sandstone soils," "granite soils," etc., a petrographic viewpoint.

Classification according to *mode of formation* of the soil material, a geographic viewpoint, was made as follows:

1. Sedentary, or residual, soils (according to rock)
2. Soil material *transported* by water (alluvium, marine deposits), ice (glacial till), or wind (loess, sand)
3. Cumulose (organic accumulation)

Classification on the basis of original *vegetation* was made as follows:

1. Forest soils (elm-hackberry, post oak, pinewood)
2. Prairie soils (bluestem grass, buffalo grass)

This was not a classification of soil, but of vegetation on the soil. Vegetation has an influence on soil formation, but is considered an indicator only in soil classification.

Texture is one of the oldest and most precisely defined criteria in soil classification. It is the only feature recognized in soil-type name (for example, Marshall *silt loam*) and the only feature for which quantitative limits have been adopted.

Size limits of soil separates, as modified from a scheme adopted by the USDA, are of the following diameters, in millimeters: coarse sand (2.0 to 0.50), medium sand (0.50 to 0.25), fine sand (0.25 to 0.05), silt (0.05 to 0.002), and clay (0.002 or less). *Soil textural classes* and their relative composition are shown in Fig. 56-1.[1]

United States Scheme of Soil Classification. Almost every country has its own system of soil classification. The scheme adopted in the United States is a natural, or genetic, system. It is a classification based on the character of the soil itself, with

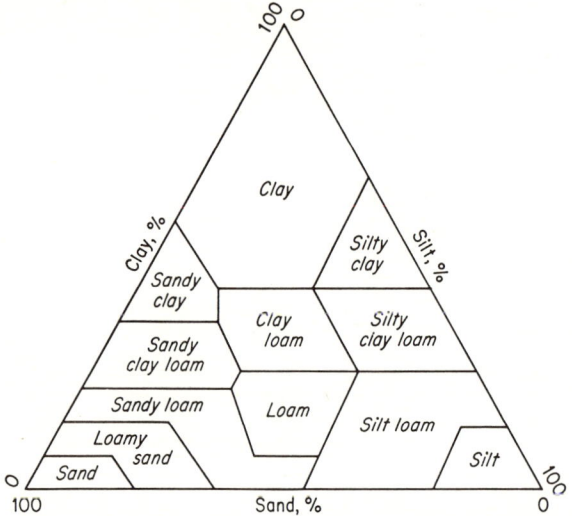

Fig. 56-1. Texture triangle showing relative composition of texture classes. (*Russell, USDA Yearbook of Agriculture*, 1957.)

consideration given to the factors that may have influenced soil development. Six categories are recognized, the three highest of which are shown in Table 56-2. The lower categories are soil *family*, soil *series*, and soil *type*. The unit used for *detailed* soil classification is the soil type. *Semidetailed* classification deals with minor soil groups and physiographic areas. *Reconnaissance* classification is general or regional and deals with Great Soil Groups or subdivisions.[2]

Soil surveys have been made of most of the cropland in the United States and county soil maps are available from the state agricultural extension services. Published soil maps vary in scale, depending on the detail of mapping. Most county maps are now on a scale of 1 or 2 in. to a mile.[3]

Land Classification (Land Capability Classification). Land may be classified according to its desirability for various uses, such as crops, forest, airfields, etc. Most land classifications are made from an agricultural standpoint, i.e., the desirability of the land for crop production, although this is not to be confused with soil rating. The number of classes usually varies from 5 to 10, ranging from the most desirable to the least desirable.

[1] M. B. Russell, Physical Properties of Soils, "Soil," USDA Yearbook of Agriculture, 1957, pp. 31–38.
[2] "Soils and Men," USDA Yearbook of Agriculture, 1938, p. 1019.
[3] Soil Survey Manual, *USDA Handbook* 18, 1951.

TABLE 56-2. SOIL CLASSIFICATION IN THE HIGHER CATEGORIES

Order	Suborder	Great Soil Groups
Zonal soils.........	1. Soils of the cold zone	Tundra soils
	2. Light-colored soils of arid regions	Desert soils
		Red desert soils
		Sierozem
		Brown soils
		Reddish-Brown soils
	3. Dark-colored soils of semiarid, subhumid, and humid grasslands	Chestnut soils
		Reddish Chestnut soils
		Chernozem soils
		Prairie soils
		Reddish Prairie soils
	4. Soils of the forest-grassland transition	Degraded Chernozem
		Noncalcic Brown or Shantung Brown soils
	5. Light-colored podzolized soils of the timbered regions	Podzol soils
		Gray wooded or Gray Podzolic soils
		Brown Podzolic soils
		Gray-Brown Podzolic soils
		Red-Yellow Podzolic soils
	6. Lateritic soils of forested warm-temperate and tropical regions	Reddish-Brown Lateritic soils
		Yellowish-Brown Lateritic soils
		Laterite soils
Intrazonal soils.....	1. Halomorphic (saline and alkali) soils of imperfectly drained arid regions and littoral deposits	Solonchak or Saline soils
		Solonetz soils
		Soloth soils
	2. Hydromorphic soils of marshes, swamps, seep areas, and flats	Humic-Glei soils (includes Wiesenboden)
		Alpine Meadow soils
		Bog soils
		Half-bog soils
		Low Humic Glei soils
		Planosols
		Ground-water Podzol soils
		Ground-water Laterite soils
	3. Calcimorphic soils	Brown Forest soils (Braunerde)
		Rendzina soils
Azonal soils........	....	Lithosols
		Regosols (includes Dry Sands)
		Alluvial soils

SOURCE: James Thorp and Guy D. Smith, Higher Categories of Soil Classification; Order, Suborder, and Great Soil Groups, *Soil Sci.*, vol. 67, February, 1940.

A land classification scheme of seven classes is as follows:

Class 1—Superior cropland
Class 2—Good cropland
Class 3—Medium cropland
Class 4—Marginal cropland
Class 5—Mainly pasture land
Class 7—Mainly forest land

SOIL PHYSICAL PROPERTIES

Don Kirkham

Basic agricultural data on soil physical properties will be presented under five headings, as to effects (1) on plant growth, (2) on drainage, (3) on runoff, (4) on erosion, and (5) on tillage. The subject matter to be considered thus takes a viewpoint opposite to that of most soil physics articles. They generally are concerned with the effects of plants, drainage, runoff, erosion, and tillage on the physical properties of the soil.

Effects on Plant Growth. Physical properties of soil which influence plant growth are those which affect soil moisture, soil air, soil solidity, and soil temperature.

Soil Moisture. Two important soil-moisture indices are the field capacity and the wilting point. The *field capacity* is the percentage of water held in the soil 2 or 3 days after a heavy rain. This index is not applicable to soils with a water table. The *wilting point* is the percentage of water in a soil when growing plants wilt to the extent that they will not recover when placed in a damp atmosphere. The field capacity and wilting point are both percentages referred to the oven-dry weight ($105°C = 221°F$) of the soil sample in question.

The presence of a layer of tight or sandy soil in a profile influences the field capacity throughout the profile. Antecedent moisture, ordinarily taken to be the wilting point, also influences field capacity. *Available water* is the field capacity minus the wilting point.

The *moisture equivalent,* which is the percentage of water held in a soil when centrifuged in a special type of centrifuge[1] at a force 1,000 times that of gravity, is very nearly equal to the field capacity for medium-textured soil [71]. For heavy-textured soils the moisture equivalent is somewhat higher (15 to 20 per cent) than the field capacity [41,78]; for sandy soils, lower [9].

Sandy soils have low field capacities and low wilting points; clay soils have relatively high field capacities and wilting points. The wilting point for all plants is about the same (Table 56-3) [70]. Laboratory determinations of the field capacity

TABLE 56-3. VALUES OF WILTING POINT FOR SOILS OF DIFFERENT TEXTURE AS DETERMINED WITH FOUR INDICATOR PLANTS

Indicator plant	Sand ME = 2.48	Sandy loam ME = 12.0	Clay loam ME = 27.4
Barley..............	....	6.3	14.2
Corn...............	3.1	6.5	15.5
Pea................	1.16	6.9	
Wheat.............	2.6	6.3	14.5

ME = moisture equivalent.
SOURCE: L. J. Briggs and H. L. Schantz, *USDA Bur. Plant Ind. Bull.* 230, pp. 27–32, 1912.

on small core samples of a soil profile will be unreliable if the soil profile contains, or is underlain by a layer of, slowly or highly permeable material [53].

Efforts have been made to mechanically determine the wilting point of soils. A *pressure membrane apparatus* [54] which expresses water from soil at 15 atmospheres (atm) pressure appears satisfactory. Letting F denote the percentage of moisture held by the soil at 15 atm pressure, the wilting point W is given by [56]

$$W = 0.85 + 0.96F$$

The expression is reliable for samples from both stratified and unstratified soils. A much-used expression [8],

$$W = \frac{M}{1.84}$$

where M is the moisture equivalent, is now known to be in considerable error for many soils [33,70,72,78].

Soil Air. Plant roots may be said to breathe. They absorb oxygen and give off carbon dioxide. The source of oxygen is the atmospheric air. The oxygen must pass through channels in the soil to reach the roots, and the carbon dioxide must return to the atmospheric air. In heavy, compacted, or undrained soil, movement of the oxygen and carbon dioxide may be seriously impeded.

There are a number of ways to measure soil aeration [60]. One useful device is the air pycnometer [32], which measures the percentage of air space in the soil. Adaptations of the pycnometer for field use have been made [59]. Poor soil management reduces aeration. Evans and Kirkham [15], 12 hr after a rain, measured soil air permeabilities for (1) continuous corn, (2) corn in a corn-oats-meadow rotation, and (3) meadow in the same rotation. Taking continuous corn as 1, the results were re-

[1] As obtainable from the International Equipment Co., Boston, Mass.

spectively 1, 18, 63. A simple apparatus for measuring air permeability in the field has been developed by Grover [20].

Soil Solidity. The weight per unit volume of oven-dry soil is commonly called the *volume weight, apparent density,* or merely density. A preferred term is *bulk density* [18]. Roots will not penetrate soils compacted to high bulk densities. Veihmeyer and Hendrickson [73] found no roots in soils where the bulk density exceeded 120 lb per cu ft. Roots failed to penetrate an Aiken clay loam at 91.2 lb per cu ft. These values may be compared with those of a fertile Webster loam and a much less fertile Edina silt loam: virgin Webster, 56.8; cultivated Webster, 71.2; virgin Edina, 65.0; cultivated Edina, 85.0. The values, all in pounds per cubic foot, are from Anderson and Browning [2] and are for the surface 6 in. of soil; at lower depths the bulk density increased.

Soil Temperature. Planting time for crops depends upon the temperature of the soil. Well-drained soils are warm soils. Wet soils are not warm, because the sun's heat is used largely to evaporate the excess water from the soil, rather than to increase the soil temperature. A simple physics calculation shows that the amount of heat required to evaporate one pound of water from soil could raise the temperature of 160 lb of drained soil (of specific heat 0.2) 10°F. More heat is required to raise a unit volume of wet soil 1°F than drained soil.

Mulches when left on the soil in winter help keep the soil warm [66]. Mulches also reduce the soil temperature in summer. In Arizona [68] a straw mulch reduced the temperature under young citrus from 7 to 10°F at 1-ft depth, compared with bare soil. In 3½ years the trees under mulch increased their diameters 320 per cent; those on the unprotected soil, 169 per cent. Soil microbiological activity is two to three times as rapid at 80°F as it is at 60°F [65]. Corn seedlings grow more than 3½ times faster at 80°F than they do at 60°F [64]. Temperatures under a darkened

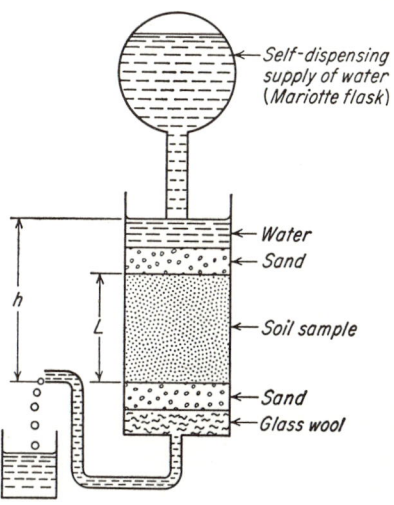

Fig. 56-2. Apparatus (permeameter) for measuring the hydraulic conductivity of soil.

straw mulch are higher than under undarkened straw [75]; reduced yields under mulch in Corn Belt soils have been attributed to reduced soil temperatures [75,61].

In considering soil moisture, soil air, soil solidity, and soil temperature, it should be recognized that these factors affect nutrient availability to plants [64]. Therefore soil physical properties governing moisture, etc., are also related, although indirectly, to nutrient availability.

Effects on Drainage. Permeability, or more specifically *hydraulic conductivity,* appears to be the most important property of soil influencing drainage [5,19,22,36,67]. Drainage here means removal of water by underground seepage flow. The term hydraulic conductivity, a term which is used in preference to *permeability coefficient, transmission constant,* etc. [55], is the coefficient k in *Darcy's law,*

$$v = ki$$

where v = quantity of water conducted per unit time across a unit area of soil surface
i = *hydraulic gradient*

Hydraulic gradient is the loss in head between two points a unit distance apart in the soil along the direction of seepage flow.

Figure 56-2 indicates *laboratory apparatus* for measuring hydraulic conductivity. The sample of soil should be an undisturbed core, and even then many samples will be needed if average field conditions are to be adjudged [14,77]. Let Q be the quantity of water which seeps through the sample in time t, and let A be the horizontal cross-section area of the sample. Let h be the difference in hydraulic head between the top and bottom of the soil sample; then (see the figure) the hydraulic gradient is h/L. Darcy's law now becomes, neglecting all losses of hydraulic head except those in the soil,

$$\frac{Q}{At} = \frac{kh}{L}$$

All quantities but k in the equation are measurable; hence k can be determined. In the laboratory it is convenient to take Q in cubic centimeters, A in square centimeters, h and L in centimeters, t in minutes; then k is in centimeters per minute. For practical applications in agriculture, better units are inches per hour, or feet per day, or gallons per square foot per day. In Fig. 56-2 the water outlet may be above the level indicated, but should not be below. If the outlet is below, air will then enter in the sand and glass wool beneath the soil and the formula will break down.

Table 56-4 gives the *average values* of hydraulic conductivity for several classes of soil material. The materials are disturbed samples and should therefore not be taken

TABLE 56-4. AVERAGE VALUES OF HYDRAULIC CONDUCTIVITY FOR SEVERAL CLASSES OF SOIL MATERIAL

Class of material		Hydraulic conductivity	
Name	Per cent by wt. of silt and clay	Ft/day	Gal/(sq ft)(day)
Silt and clay..............	25–71	0.071	0.53
Very fine and fine sand......	6–28	2.10	16.0
Fine and medium sand......	1–11	40.1	304
Coarse sand................	0–8	247	1,845
Gravel....................	0–6	1,800	13,400

SOURCE: C. H. Lee, *Trans. Am. Geophys. Union*, 1934, especially p. 543.

as indicative of soil conditions in the field. In the field the type of natural cleavage planes [49], cracks, root roles and wormholes, etc. [63], may govern seepage flow. Results for some undisturbed soils have been reported by Luthin and Kirkham [36]. They found hydraulic conductivities of 50 to 100 ft per day for a Webster silty clay loam and 10 ft per day for a Marion silty clay loam. For a Luton clay the values varied from about 0.15 to 5 ft per day. For a very fine sand, having less than about 20 per cent silt and clay, Kirkham and De Zeeuw [30] found hydraulic conductivities in the field of only about 0.3 ft per day. Here the silt and clay were apparently of just the size and amount to almost clog the pores between the sand.

The *field values* of hydraulic conductivity reported by Luthin, Kirkham, and De Zeeuw were made by rating the flow into cased and uncased auger holes bored into the soil below the water tables as they filled with ground water. The soil sample in these measurements consisted of a relatively large volume of soil about the test hole [52]. Simple procedures for applying this field method are described by Johnson, Frevert, and Evans [23]. An evaluation of a number of field methods for the determination of hydraulic conductivity has been given by Kirkham [31].

Often a tight subsoil layer may impede drainage. This problem has been analyzed theoretically for both an impermeable subsoil [27,28] and a subsoil of low but finite hydraulic conductivity [29]. The theory shows what should otherwise seem apparent, that tiles should not be placed in tight subsoils unless for a special reason, such as to obtain proper grade [26]. If it is necessary to place tiles in a tight subsoil, they should be covered with permeable backfill material. Well-aggregated surface soil, at least, should be used. The effect of backfill has been analyzed theoretically by Luthin and Gaskell [35].

Because of the great variability in hydraulic-conductivity measurements, efforts have been made to correlate other less variable, or easier to measure, soil properties with drainability. Such properties as particle-size distribution, pore-size distribution, and moisture equivalent may be mentioned. In the subsoil of the Imperial Valley, particle-size distribution has been useful [4]. Pore measurements should in general be more reliable. According to data of Kopecky presented by Baver [6], soils having 10 per cent or more of the total soil volume in large pores need not be artificially drained. The meaning of "large pores" is here not definite. Apparently they are the soil pores emptied when water-saturated soil attains the field capacity. Roe and Park [57], on more than 100 humid and dryland soils, found no certain relation, for disturbed soil samples, between moisture equivalent and hydraulic conductivity and concluded that moisture equivalent was not a reliable index for drainage. Neal [42] earlier, on four soils in a limited area, had observed that the moisture equivalent seemed to be related to depth and spacing of tiles. For some recent information on soil physical properties, as related to drainage, reference is made to a literature review by Childs and Collis-George [11]. An extensive monograph on agricultural drainage, with articles by agricultural engineers, soil scientists, and others, considers physical properties of soil [37].

Effects on Runoff. The two important properties of soil which influence runoff are ability to absorb water and ability to transmit water downward through the soil profile, once it is fully wet.

The ability of a soil to absorb water is called its *infiltration capacity*. A better term is probably *infiltration rate* [55], this rate ordinarily being expressed in surface inches per hour of absorbed water, when the water is supplied at a rate just short of that producing runoff. The term infiltration rate is quite general. It applies to the rate of intake of water when the soil is either dry or wet. It also includes effects of the whole wetted profile.

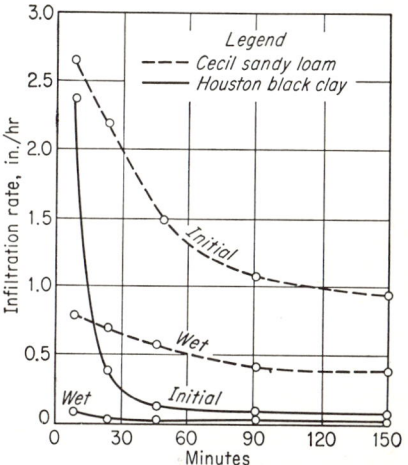

Fig. 56-3. Infiltration rates during initial and wet runs for Cecil sandy loam and Houston black clay. (*Free, Browning, and Musgrave, USDA, Tech. Bull. 729, 1940.*)

If the soil profile becomes water-saturated to a depth where there is free drainage (as in the artificial case of a gravel layer at the base of a lysimeter which is drained to the atmosphere), the infiltration rate becomes equal to the *percolation* rate for the soil profile, and this rate will stay essentially constant. If the soil profile is uniform down to a layer of free drainage, then the percolation rate is just equal to the hydraulic conductivity of the soil, a term which has been defined.

Soils often become tighter with depth. A tight soil layer may ultimately govern the rate of water movement through the profile and hence affect the rate of runoff. A tight layer at 7 to 14 in. depth has been observed to affect runoff during certain storms on silt loam watersheds at Coshocton, Ohio [62]. The hydraulic conductivity of these soils is less than one-tenth that of the surface soil.

Free, Browning, and Musgrave [16] report extensive infiltration studies. Infiltration rates were measured on 68 test sites over the United States. Thirty-nine Soil Series and six Great Soil Groups were represented. Galvanized-steel tubes 9 in. in diameter were jacked into the soil, 24 replicates at a site. Clear water was applied from a self-dispensing, calibrated burette, so as to maintain a ¼-in.-thick layer of water on the soil surface inside the tube. The amount of infiltration was measured a number of times after water was applied. Two runs were made for each tube, the first at the field-moisture content prevailing at the time, and the second, a wet run, 24 hr later. The wet run was thus made when the soil was approximately at field capacity. Figure 56-3 shows the relation between infiltration rate and time of water application. The rate is higher at the beginning of a run and approaches a nearly constant rate after 2½ hr. The rate for the initial run is always higher than for the wet run.

Tests were also made of the infiltration rate when artificial rain was applied at 13 sites. The artificial rain caused a turbid condition at the soil surface. On the average this turbidity reduced the infiltration rate to half the value of the clear-water rate.

A number of measured physical properties were found to be closely correlated with infiltration rate. Of the properties studied, *macroporosity* (Free, Browning, and Musgrave used the term noncapillary porosity) was the most closely correlated, and the amount of organic matter was next. Macroporosity here means the percentage of the soil volume which becomes air space when the soil, initially water-saturated, is drained to its field capacity.

The application of infiltration data to the computation of runoff from watersheds is complex. Nevertheless, progress in the problem is being made [3,12].

Effects on Erosion. It appears from work reported[1] that erosion is governed primarily by rainfall, slope of the land, and vegetative cover. After these come the physical properties of the soil itself. A large number of physical properties of soil have been measured to see just how these properties might be related to erosion.[2] So far no one property, or combination of properties into an index, has provided a reliable criterion for characterizing the erodibility of a soil. Physical properties which now seem most important are *water stability* and ability of a soil to absorb water and conduct it through the profile. Only water stability will be discussed here. Soil properties associated with water absorption and transmission have been discussed under runoff.

Water Stability. A soil is said to be water-stable if the aggregates or crumbs which normally comprise it are not easily broken down into fine material by water. Such fine material not only is easily carried away by water, but in addition seals the surface of the soil. A sealed soil surface prevents water from infiltrating into it, and hence increases runoff and consequent erosion. Lowdermilk [34] observed that use of muddy water reduced the rate of water movement into soil to $\frac{1}{10}$ the value obtained with clear water.

The *dispersion ratio* of Middleton [38] is one of the oldest and most widely used indices for measuring soil stability. Actually this index was designed to measure erodibility, and hence is an inverse measure of soil stability. The dispersion ratio is measured by shaking a sample of soil in water and then determining the amount of material smaller than a certain size. The ratio of the material which breaks down to the total amount of silt and clay in the soil is the dispersion ratio. Middleton specified 0.05 mm as the size of particles which were to separate the dispersed from the undispersed material, but other sizes have been used [50,51]. The sand fraction in the soil is not considered in this measurement.

In determining the dispersion ratio, care must be taken to perform all operations alike. The samples should all be at the same moisture content. But even with care it is not likely that the dispersion ratios will correlate more than qualitatively with measured field erosion values [39,58], especially from regions with different climates. Nevertheless, Anderson [1] has reported a high correlation between the dispersion ratio and soil erosion for a number of California watersheds. The amount of eroded material varied approximately as the 3.4 power of the dispersion ratio.

Another widely used method for evaluating water stability of soil is the *wet-sieving method* of Yoder [79], or one of its variations. In this method a sample of soil is placed on the top sieve of a nest of sieves, and the sieves are then slowly moved up and down in water. Yoder used six screens in a nest. Their openings were of sizes 5, 2, 1, 0.5, 0.25, and 0.10 mm. After the screens have completed the specified number of movements, the weight of soil on each screen is determined. Soils for which a large proportion of material stays on the larger screens are water stable. Wilson and Browning [76] found that the remaining material larger than 2 mm was most closely related to erosion losses on a Marshall silt loam. The greater the amount of material larger than 2 mm, the smaller were the erosion losses. Several indices have been proposed for combining the several sieve sizes and the soil remaining on them into a

---

[1] In *USDA Tech. Bulls.* 808, 837, 873, 888, 916, 959, 973, and 985.
[2] Results are reported in *USDA Tech. Bulls.* 178, 232, 316, 430, 461, 562, and 729.

single figure for specifying water stability. The *mean weight–diameter* of Van Bavel [69] has been used extensively. The *geometric mean diameter*, as described by Gardner [17], is superseding other indices. Bryant, Bendixen, and Slater [10] describe a two-screen wet-sieving procedure which yields a measure of water stability at less expenditure of labor than the multiple-sieve method.

Thus far the discussion has pertained to water erosion. For wind erosion, see Chap. 43 and a paper by Zingg and Chepil [80] and references there.

Effects on Tillage. The physical properties of soils which largely influence tillage are texture and moisture content. Organic matter, fertilizers, and certain other factors are also of interest. Reference in the following is to primary tillage. Much that is said will also apply to secondary tillage.

Texture. Clay soils offer higher resistance to the plow than sandy soils. Haines and Keen [21] observed in field tests an increase in drawbar pull from 1,025 to 1,590 lb, as clay content increased from 18.4 to 26.5 per cent. The increase in drawbar pull was linear with clay content for the range investigated.

The soil *plasticity constants* are largely influenced by clay content and are an index of the reaction of soil to tillage implements [44,45]. There are three plasticity constants. The *lower plastic limit* is that moisture content at which the soil can barely be rolled into a wire. The *upper plastic limit* is the moisture content at which the soil can barely be made to flow without rupture under an applied force. The *index of plasticity* is the difference between the two limits. These values are all expressed in per cent by weight of soil moisture referred to oven-dry weight of soil.

Moisture Content. Plow draft may increase or decrease with moisture content. Few numerical data are available on the point. Data by Ocock [48] indicate that the draft is least in soil in good plowing condition. Physical measurements for the limits of moisture for good plowing condition in soils have not been quantitatively specified.

TABLE 56-5. MOISTURE PERCENTAGE AT WET LIMIT* OF PLOWING RANGE FOR SEVERAL INDIANA SOILS

Soil type	Composition				Moisture,‡ %
	Sand	Silt	Clay	OM†	
Oaktown sand............	95	2.9	1.0	1.1	20.6
Warsaw loam.............	47.5	35.7	14.8	2.0	18.4
Mucky loam..............	24.1	46.4	18.4	11.1	37.6
Carrington silt loam.......	23.4	50.6	22.7	3.3	25.9
Brookston clay loam.......	21.8	47.4	27.4	3.4	23.9
Brookston silty clay loam....	15.0	53.5	27.1	4.4	27.0

* For criteria for the wet limit see text.
† OM = organic matter.
‡ Moisture percentage is referred to oven-dry soil.
SOURCE: M. S. Alleman and H. Kohnke, *Soil Sci. Soc. Am. Proc.* 1947, pp. 22–23.

Alleman and Kohnke list some upper-limit values (Table 56-5) and give four qualitative criteria for judging the upper limit.

1. A freshly cut surface should not glisten with moisture.
2. Kneading by hand should show no evidence of water.
3. The soil should scour freely from moldboard or spade.
4. The soil should be friable enough to break into aggregates rather than large chunks.

Weaver and Jamison [74] note an "optimum plowing moisture" and determine it by experience. For a Davidson loam and a Cecil clay the values were 9 and 27 per cent, respectively. The corresponding lower plastic limits were given as 12.3 and 27.4 per cent.

Organic Matter. Organic matter applied in the form of barnyard manure, except in large quantities, seems to have little effect on plow draft. Duley and Jones [13] noted that the average plow draft was 5.37 psi of furrow section for a plot manured at a rate of 6 tons per acre per year for 35 years. A comparison plot, with the manure omitted for the last 10 of the 35 years, gave a draft of 5.26 psi. Keen and Haines [24] made similar observations, noting in addition a "perceptible" increase in draft when 14 tons per acre per year of manure was added.

Organic matter in the form of colloidal humus can increase soil reaction to the plow. Noll's [47] data show that plots having the highest amounts of humus, approximately 3.2 per cent, gave an average draft of 455 lb; those with lowest humus, approximately 2.7 per cent, a draft of 440 lb. A *push soil* is an extreme example of a soil which does not plow easily because of excess organic matter in humic form. Such a soil is black and mucky and compacts considerably before it will shear [43,46]. As a consequence, the soil tends to push to the side rather than invert in the plowing operation.

Fertilizers and Lime. Duley and Jones [13] found no perceptible changes in draft for fertilized and limed soils as compared with a check plot. The fertilized plot received 500 lb of 3-10-4 fertilizer and 2 tons of lime every 4 years. The same treatment, except for omission of the lime, did not change the draft perceptibly. Five tons of lime per acre resulted in less than 1 per cent decrease in draft in an experiment by Haines and Keen [25]. They noted that 10 to 15 tons of chalk per acre applied in 1912 resulted in an "appreciable" decrease in plow draft 13 years later. At the Ohio Experiment Station it has been observed that addition of lime increased plow draft. The increase in this case was attributed to excess, undecayed root growth due to the lime [7].

BIBLIOGRAPHY

Baver, L. D.: "Soil Physics," 3d ed., John Wiley & Sons, Inc., New York, 1956.
Shaw, B. T. (ed.): "Soil Physical Conditions and Plant Growth," Academic Press, Inc., New York, 1952.
Various authors, *Soil Sci.*, vol. 68, no. 1, 1949.

REFERENCES CITED

1. Anderson, H. W.: Physical Characteristics of Soils Related to Erosion, *J. Soil and Water Conserv.*, 6:129–133, 1951.
2. Anderson, M. A., and G. M. Browning: Some Chemical and Physical Properties of Six Virgin and Six Cultivated Iowa Soils, *Soil Sci. Soc. Am. Proc.*, 1949, pp. 370-374.
3. Andrews, R. G.: Runoff Estimates Based on Infiltration Capacity, Antecedent Moisture Conditions, and Precipitation, *Agr. Eng.*, 31:26–30, January, 1950.
4. Aronovici, V. S.: The Mechanical Analysis as an Index of Subsoil Permeability, *Soil Sci. Soc. Am. Proc.*, 1946, pp. 137–141.
5. Aronovici, V. S., and W. W. Donnan: Soil-permeability as a Criterion for Drainage Design, *Trans. Am. Geophys. Union,* 1946, pp. 95–101.
6. Baver, L. D.: "Soil Physics," 2d ed., John Wiley & Sons, Inc., New York, 1948, pp. 272–273.
7. Baver, L. D.: "Soil Physics," 2d ed., John Wiley & Sons, Inc., New York, 1948, pp. 330–331.
8. Briggs, L. J., and H. L. Schantz: The Wilting Coefficient for Different Plants and Its Indirect Determination, *USDA Bur. Plant Ind. Bull.* 230, 1912.
9. Browning, G. M.: Relation of Field Capacity to Moisture Equivalent in Soils of West Virginia, *Soil Sci.*, 52:445–450, 1941.
10. Bryant, J. C., T. W. Bendixen, and C. S. Slater: Measurement of the Water-stability of Soils, *Soil Sci.*, 65:341–345, 1948.
11. Childs, E. C., and N. Collis-George: in A. G. Norman (ed.), *Advances in Agron.*, Academic Press, Inc., New York, 1950.
12. Cook, H. L.: The Infiltration Approach to the Calculation of Surface Runoff, *Trans. Am. Geophys. Union,* 1946, pp. 726–747.
13. Duley, F. L., and M. M. Jones: Effects of Soil Treatments upon the Draft of Plows, *Soil Sci.*, 21:277–288, 1926.

14. Edminster, T. W., et al.: Tests of Small Core Samplers for Permeability Determinations, *Soil Sci. Soc. Am. Proc.,* 1950, pp. 417–420.
15. Evans, D. D., and D. Kirkham: Measurement of the Air Permeability of Soil in Situ, *Soil Sci. Soc. Am. Proc.,* 1949, pp. 65–73.
16. Free, G. R., G. M. Browning, and G. W. Musgrave: Relative Infiltration and Related Physical Characteristics of Certain Soils, *USDA Tech. Bull.* 729, July, 1940.
17. Gardner, Wilford R.: Representation of Soil Aggregate-size Distribution by a Logarithmic-normal Distribution, *Soil Sci. Soc. Am. Proc.,* 20:151–154, 1956.
18. Gardner, Willard (chairman): Report of Committee on Terminology, *Soil Sci. Soc. Am. Proc.,* 1948, pp. 573–574.
19. Gardner, Willard, T. R. Collier, and Doris Farr: Groundwater: Fundamental Principles Governing Its Physical Control, *Utah State Agr. Coll. Agr. Expt. Sta. Tech. Bull.* 252, 1934.
20. Grover, B. L.: Simplified Air Permeameters for Soil in Place, *Soil Sci. Soc. Am. Proc.,* 19:414–418, 1955.
21. Haines, W. B., and B. A. Keen: Studies in Soil Cultivation, II and III, *J. Agr. Sci.,* 15:387–406, 1925 (especially pp. 392, 401).
22. Hooghoudt, S. B.: Tile Drainage and Sub-irrigation, *Soil Sci.,* 74:35–48, 1952.
23. Johnson, H. P., R. K. Frevert, and D. D. Evans: Simplified Procedure for the Measurement and Computation of Soil Permeability below the Water Table, *Agr. Eng.,* 33:283–286, May, 1952.
24. Keen, B. A.: "The Physical Properties of the Soil," Longmans, Green & Co., Ltd., London, 1931, p. 244.
25. Keen, B. A.: "The Physical Properties of the Soil," Longmans, Green & Co., Ltd., London, 1931, pp. 245–246.
26. Kirkham, D.: Reduction in Seepage to Soil Underdrains Resulting from Their Partial Embedment in, or Proximity to, an Impervious Substratum, *Soil Sci. Soc. Am. Proc.,* 1947, pp. 54–59. See also the preceding reference.
27. Kirkham, D.: Flow of Ponded Water into Drain Tubes in Soil Overlying an Impervious Layer, *Trans. Am. Geophys. Union,* 1949, pp. 369–382.
28. Kirkham, D.: Seepage into Ditches in the Case of a Plane Water Table and an Impervious Substratum, *Trans. Am. Geophys. Union,* 1950, pp. 425–430.
29. Kirkham, D.: Seepage into Drain Tubes in Stratified Soil, *Trans. Am. Geophys. Union,* 1951, pp. 422–442.
30. Kirkham, D., and J. W. De Zeeuw: Field Measurements for Tests of Soil Drainage Theory, *Soil Sci. Soc. Am. Proc.,* 16:286–293, 1952.
31. Kirkham, D.: Measurement of the Hydraulic Conductivity of Soil in Place, *ASTM Spec. Publ.* 163, pp. 80–97, 1955.
32. Kummer, F. A., and A. W. Cooper: Soil Porosity Determinations with the Air Pressure Pycnometer, *Agr. Eng.,* 26:21–23, January, 1945.
33. Lehane, J. J., and W. J. Staple: Desiccator Method for Determining Wilting Percentages of Soils, *Soil Sci.,* 72:429–435, 1951.
34. Lowdermilk, W. C.: Influence of Forest Litter on Run-off, Percolation, and Erosion, *J. Forestry,* 28 (pt. 1): 474–491, 1930.
35. Luthin, J. N., and R. E. Gaskell: Numerical Solutions for Tile Drainage of Layered Soils, *Trans. Am. Geophys. Union,* 1950, pp. 595–602.
36. Luthin, J. N., and D. Kirkham: A Piezometer Method for Measuring Permeability of Soil in Situ below a Water Table, *Soil Sci.,* 68:349–358, 1949.
37. Luthin, J. N. (ed.): Drainage of Agricultural Land, *Monograph Am. Soc. Agron.,* 1957.
38. Middleton, H. E.: Properties of Soil Which Influence Soil Erosion, *USDA Tech. Bull.* 178, March, 1930.
39. Middleton, H. E., and C. S. Slater: The Extent to Which the Erodibility of a Soil Can Be Anticipated by Laboratory Physical and Chemical Measurements, *Fifteenth Annual Meeting Am. Soil Survey Assoc. Bull.* 16, pp. 128–130, 1935.
40. Muskat, M.: "The Flow of Homogeneous Fluids through Porous Media," McGraw-Hill Book Company, Inc., New York, 1937, or J. W. Edwards, Publisher, Inc., Ann Arbor, Mich., 1946.

41. Neal, J. H.: Moisture Relationships of Soils in Situ, *Agr. Eng.*, 13:128–132, May, 1932.
42. Neal, J. H.: Spacing and Depth of Tile Drains, *Agr. Eng.*, 15:229–232, July, 1934.
43. Nichols, M. L.: Methods of Research in Soil Dynamics as Applied to Implement Design, *Ala. Polytechnic Inst. Expt. Sta. Bull.* 229, 1929.
44. Nichols, M. L.: Dynamic Properties of Soil Affecting Implement Design, *Agr. Eng.*, 11:201–204, June, 1930.
45. Nichols, M. L., and C. A. Reaves: Soil Structure and Consistency in Tillage Implement Design, *Agr. Eng.*, 36:517–522, August, 1955.
46. Nichols, M. L., and I. F. Reed: Physical Reactions of Soils to Moldboard Surfaces, *Agr. Eng.*, 15:187–190, June, 1934.
47. Noll, C. F.: Effect of Fertilizers on Soil Structure as Indicated by the Draft of a Plow, *Annual Rept. Penn. State College*, 1913–1914, pt. II, pp. 36–46, 1915.
48. Ocock, C. A.: The Draft of Plows, *Trans. ASAE*, 1912, pp. 13–30.
49. O'Neal, A. M.: Soil Characteristics Significant in Evaluating Permeability, *Soil Sci.*, 67:403–411, 1949.
50. Peele, T. C.: The Relation of Certain Physical Characteristics to the Erodibility of Soils, *Soil Sci. Soc. Am. Proc.*, 1937, pp. 97–100.
51. Peele, T. C., O. W. Beale, and F. E. Latham: The Effect of Lime and Organic Matter on the Erodibility of Cecil Clay, *Soil Sci. Soc. Am. Proc.*, 1938, pp. 289–292.
52. Reeve, R. C., and D. Kirkham: Soil Anisotropy and Some Field Methods for Measuring Permeability, *Trans. Am. Geophys. Union*, 32:582–590, 1951.
53. Richards, L. A.: Water Content Changes Following the Wetting of Bare Soil in the Field, *Soil Sci. Soc. Florida Proc.*, 15:142–148, 1955.
54. Richards, L. A.: Pressure-membrane Apparatus, Construction and Use, *Agr. Eng.*, 28:451–454, October, 1947. (This equipment is commercially available from Instrument Development & Mfg. Corp., Pasadena, Calif.)
55. Richards, L. A. (chairman): Report of the Subcommittee on Permeability and Infiltration, *Soil Sci. Soc. Am. Proc.*, 16:85–88, 1952.
56. Richards, L. A., and C. H. Wadleigh: in B. T. Shaw (ed.), "Soil Physical Conditions and Plant Growth," Academic Press, Inc., New York, 1951, pp. 156–157.
57. Roe, H. B., and J. K. Park: A Study of the Centrifuge Moisture Equivalent as an Index of the Hydraulic Permeability of Saturated Soils, *Agr. Eng.*, 25:381–385, October, 1944.
58. Rost, C. O., and C. A. Rowles: A Study of Factors Affecting the Stability of Soil Aggregates, *Soil Sci. Soc. Am. Proc.*, 1940, pp. 421–433.
59. Russell, M. B.: A Simplified Air-pycnometer for Field Use, *Soil Sci. Soc. Am. Proc.*, 1949, pp. 73–76.
60. Russell, M. B.: Methods of Measuring Soil Structure and Aeration, *Soil Sci.*, 68:25–35, 1949.
61. Schaller, F. W., and D. D. Evans: Some Effects of Mulch Tillage, *Agr. Eng.*, 16:731–734, October, 1954.
62. Schiff, L., and F. R. Dreibelbis: Movement of Water within the Soil and Surface Runoff with Reference to Land Use and Soil Properties, *Trans. Am. Geophys. Union*, 1949, pp. 401–411; see also *ibid.*, pp. 75–88.
63. Schiff, L., and F. R. Dreibelbis: Preliminary Studies on Soil Permeability and Its Application, *Trans. Am. Geophys. Union*, 1949, pp. 759–766.
64. Shaw, B. T. (ed.): "Soil Physical Conditions and Plant Growth," Academic Press, Inc., New York, 1952, p. 366.
65. Shaw, B. T. (ed.): "Soil Physical Conditions and Plant Growth," Academic Press, Inc., New York, 1952, p. 450.
66. Slater, C. S.: Winter Aspects of Soil Structure, *J. Soil and Water Conserv.*, 6:38–40, 42, 1951.
67. Slichter, C. S.: Theoretical Investigation of the Motion of Ground Waters, *U.S. Geol. Survey Ann. Rept.*, 1897–1898, 19(pt. 2):300–384, 1899.
68. Smith, G. E. P.: Control of High Soil Temperatures, *Agr. Eng.*, 17:383–385, September, 1936.

69. Van Bavel, C. H. M.: The Mean Weight-Diameter of Soil Aggregates as a Statistical Index of Aggregation, *Soil Sci. Soc. Am. Proc.,* 1949, 20–23.
70. Veihmeyer, F. J., and A. H. Hendrickson: Soil Moisture at Permanent Wilting of Plants, *Plant Physiol.,* 3:355–357, 1928.
71. Veihmeyer, F. J., and A. H. Hendrickson: Field Capacity of Soils, *Soil Sci.,* 32:181–193, 1931.
72. Veihmeyer, F. J., and A. H. Hendrickson: Some Plant and Soil Moisture Relations, *Am. Soil Survey Assoc. Bull.,* 15:76–80, 1934.
73. Veihmeyer, F. J., and A. H. Hendrickson: Soil Density and Root Penetration, *Soil Sci.,* 65:487–495, 1948.
74. Weaver, H. A., and V. C. Jamison: Effects of Moisture on Tractor Tire Compaction of Soil, *Soil Sci.,* 71:15–23, 1951.
75. Willis, W. O., W. E. Larson, and D. Kirkham: Corn Growth as Affected by Soil Temperature and Mulch, *Agron. J.,* 49:323–328, 1957.
76. Wilson, H. A., and G. M. Browning: Soil Aggregation, Yields, Runoff and Erosion as Affected by Cropping Systems, *Soil Sci. Soc. Am. Proc.,* 1945, pp. 51–57.
77. Wilson, H. A., F. F. Riecken, and G. M. Browning: Soil Profile Characteristics in Relation to Drainage and Level Terraces, *Soil Sci. Soc. Am. Proc.,* 1946, pp. 110–118. (The permeabilities in this paper are proportional to hydraulic conductivities.)
78. Work, R. A., and M. R. Lewis: Moisture Equivalent, Field Capacity and Permanent Wilting Percentage and Their Ratios in Heavy Soils, *Agr. Eng.,* 15:335–362, October, 1934.
79. Yoder, R. E.: A Direct Method of Aggregate Analysis of Soils and a Study of the Physical Nature of Erosion Losses, *J. Am. Soc. Agron.,* 28:337–351, 1936. (For a description of recent wet sieving apparatus, see a note by C. H. M. Van Bavel, *Argon. J.,* 1952, pp. 97–98.)
80. Zingg, A. W., and W. S. Chepil: Aerodynamics of Wind Erosion, *Agr. Eng.,* 31:279–284, June, 1950.

SOIL PLANT FOOD (NUTRIENTS)

O. T. COLEMAN

The approximate amounts of the more important plant foods removed by various quantities of the major field, vegetable, and fruit crops and livestock and livestock products are given in Table 56-6.

NITROGEN

Role in Nutrition. Nitrogen makes up a portion of all plant and animal tissues and is an important element in the composition of protein, a vital part of all plant and animal tissues. Its content is highest in those tissues that have to do with growth and reproduction. It has much to do with the rate of plant growth and the proper development of animals. A good supply of nitrogen is necessary for the abundant growth of stems and leaves. Large amounts in proportion to the other plant foods may make plants less resistant to certain diseases and cause excessive growth of stems and leaves at the expense of seed production, tending to delay maturity of seed-producing crops.

Origin in the Soil. Organic matter, or humus, in the soil is the chief source of nitrogen for plant nutrition. The original source is the air, which is about 79 per cent nitrogen. It is estimated that over every acre of land there is approximately 35,000 tons of nitrogen. Through the action of soil and legume bacteria, nitrogen from the air is fixed in the soil and made available to plants. The rotting down of these plants and the return of animal manures to the soil increase the supply of organic matter and nitrogen.

Lightning combines nitrogen with oxygen, uniting with moisture in the air to form either nitrous or nitric acid, which is carried down by rain or snow and enters the soil. Nitrogen which occurs as impurities in the air in the form of dust or gas may also be carried down by rain and snow. On the average, however, it is estimated that only 5 to 7 lb of nitrogen per acre per year is carried to the soil from the air, and this amount would by no means take care of the nitrogen requirements of crops.

TABLE 56-6. PLANT FOOD CONTENT OF VARIOUS CROPS AND AGRICULTURAL PRODUCTS

Crop or product	Yield or amount, air-dry	Nitrogen (N), lb	Phosphate (P_2O_5), lb	Potash (K_2O), lb	Lime (100% $CaCO_3$), lb
Corn, grain	60 bu	57	23	15	1
Fodder	2 tons	38	12	55	46
Total		95	35	70	47
Oats, grain	50 bu	35	15	10	4
Straw	1.25 tons	15	5	35	22
Total		50	20	45	26
Wheat, grain	30 bu	35	16	9	1
Straw	1.25 tons	15	4	21	14
Total		50	20	30	15
Barley, grain	40 bu	35	15	10	3
Straw	1 ton	15	5	30	16
Total		50	20	40	19
Rye, grain	20 bu	22	9	7	1
Straw	1½ tons	15	8	26	21
Total		37	17	33	22
Kafir, grain	50 bu	43	16	10	2
Fodder	3 tons	62	17	88	56
Total		105	33	98	58
Sweet sorghum, fodder	4 tons	81	25	132	98
Tobacco, 1,500 lb					
Leaves		55	10	80	50
Stalks		25	10	35	35
Total		80	20	115	85
Cotton, lint	500 lb				
Seed	1,000 lb	38	18	14	3
Stalks, etc.	1,500 lb	27	7	36	80
Total		65	25	50	83
Soybeans, grain	25 bu	110	35	40	8
Straw	1.25 tons	15	5	20	60
Total		125	40	60	68
Clover seed	100 lb	2.9	1.9	1.5	1.0
Alfalfa seed	100 lb	5.87	1.16	1.17	0.45
Cowpea seed	100 lb	3.78	1.10	1.76	0.25
Forages (air-dry):*					
Timothy	1 ton	26	10	30	13.5
Lespedeza	1 ton	43.5	10	23.5	49.5
Red clover	1 ton	40	10	35	60.5
Sweet clover	1 ton	37	9	33	73.5
Alfalfa	1 ton	46	12	45	71.5
Cowpeas	1 ton	62.5	12.5	45	56.5
Livestock and livestock products:					
Fat cattle	1,000 lb	25.0	16.1	2.4	32.0
Fat hogs	1,000 lb	18.0	6.6	13.0	11.3
Fat lambs	1,000 lb	20.0	11.2	1.7	23.5
Milk	1,000 lb	5.6	2.0	1.6	3.0
Butter	1,000 lb	1.6			
Eggs	100 doz (162.5 lb)	3.5	1.7	0.8	26.2
Chickens	100 lb	3.5	0.85		2.8
Fruits and vegetables:					
Apples, fruit	400 bu	20	7	30	5
Asparagus spears	5,000 lb	20	6	38	
Beans or peas:					
Seed	30 bu	73	23	24	9
Straw		22	7	31	
Total		95	30	55	
Beets:					
Roots	20,000 lb	36	14	34	
Tops		40	16	62	
Total		76	30	96	55
Blackberries, fruit	4,000 lb	6	4	8	
Carrots, whole crop	30,000 lb	120	50	240	
Cabbage, whole crop	15 tons	100	25	100	30
Cantaloupe, fruit	4,000 melons	57	16	100	

TABLE 56-6. PLANT FOOD CONTENT OF VARIOUS CROPS AND AGRICULTURAL PRODUCTS
(*Continued*)

Crop or product	Yield or amount, air-dry	Nitrogen (N), lb	Phosphate (P_2O_5), lb	Potash (K_2O), lb	Lime (100% $CaCO_3$), lb
Fruits and vegetables (*Continued*)					
Cherries, fruit	8,000 lb	18	6	22	
Grapes:					
Fruit	6,000 lb	8	4	15	
Leaves and canes	....	13	4	13	
Total	....	21	8	28	
Lettuce, whole crop	15,000 lb	41	17	71	
Onions, whole crop	600 bu	42	12	30	27
Peaches:					
Fruit	500 bu	30	15	55	7
Leaves and wood	....	55	10	45	
Total	....	85	25	100	
Pears, fruit	400 bu	18	6	17	
Plums, fruit	8,000 lb	15	2	20	
Potatoes (Irish):					
Tubers	300 bu	65	25	115	13
Tops	....	60	10	55	
Total	....	125	35	170	
Potatoes, sweet:					
Roots	300 bu	45	15	75	13
Vines	....	30	5	40	
Total	....	75	20	115	
Raspberries, fruit	3,000 lb	5	15	11	
Spinach, tops	12,000 lb	60	20	30	31
Strawberries, fruit	180 crates	9	6	18	
Tomatoes:					
Fruit	20,000 lb	60	20	80	25
Vines	....	40	15	95	
Total	....	100	35	175	
Turnips, roots	400 bu	51	31	69	29

* Approximately 70% of the N in inoculated legumes is fixed from the air.

SOURCE: Data principally from Missouri Balanced Farming Handbook, *Univ. Missouri Agr. Ext. Serv.* BF-5604. Analyses from various sources will vary considerably, but these probably best fit average Missouri conditions. Crops grown on better soils normally have higher composition than those grown on poorer soils.

Amounts in the Soil. The amount of nitrogen in the soil depends chiefly upon the amount of organic matter it contains. This amount may vary from less than 500 to 6,000 lb or more per acre in the surface layer to a 7-in. depth. Alluvial soils are usually high in nitrogen, because they are normally made up of topsoil, which is higher in organic matter than subsoil. Prairie soils of the Midwest average from 3,000 to 4,000 lb of nitrogen per acre in the surface 7 in., while timber soils average about 1,500 to 2,000 lb.

In a study made by Hans Jenny[1] it was found that for every 18°F fall in the mean annual average temperature, the average nitrogen content of the soil increased two to three times. This fact explains why the soils in the northern part of the United States are usually darker and higher in organic matter than those in the southern part, where higher temperatures favor the more rapid decomposition of organic matter.

Rate of Availability. When organic matter decomposes, it produces mobile nitrates that are approximately 100 per cent available to growing crops. Under average Midwest conditions, about 5 per cent of the soil organic matter is nitrogen. The amount of nitrates released from the organic matter varies with bacterial activity, which in turn is affected by moisture supply, temperature, aeration, and the amount and kind of organic matter present.

Table 56-7 gives estimates of the total pounds of nitrogen per acre in the surface 7 in. of soil and the approximate pounds of nitrogen released from different soils during an average growing season under semiarid conditions of the Midwest, based upon soil tests for the per cent of stable organic matter in the soil.

[1] Soil Fertility Losses under Missouri Conditions, *Univ. Missouri Agr. Expt. Sta. Bull.* 324, 1933.

Annual Losses Other Than to Plants. Cultivation of soil increases aeration and hastens the breakdown of organic matter, resulting in the more rapid release of nitrogen in a soluble, or mobile, form. Nitrogen in the ammonia form leaches less rapidly than the nitrate form. Additional losses of nitrogen take place through the burning of vegetation and by soil erosion. The annual loss of nitrogen from the soils of the United States by harvested crops, grazing, erosion, and leaching was estimated to be 22,899,046 tons.[1]

Average annual losses of nitrogen on a 3.68 per cent slope with different crops are shown in Table 56-8.

TABLE 56-7. ANNUAL RELEASE OF NITROGEN WITH VARIOUS PERCENTAGES OF ORGANIC MATTER IN DIFFERENT SOILS

Per cent organic matter	Stable organic matter, lb/acre	Total nitrogen, lb/acre	Annual release of nitrogen,* lb/acre		
			Silt loams	Clay and clay loam	Sands and sandy loam
0.5	10,000	500	7.5–15	6.25–12.50	20–30
1.0	20,000	1,000	15.0–30	12.50–25.00	40–60
1.5	30,000	1,500	22.5–45	18.50–37.00	60–90
2.0	40,000	2,000	30.0–60	25.00–50.00	80–120
2.5	50,000	2,500	37.5–75	31.25–62.50	†
3.0	60,000	3,000	45.0–90	37.50–75.00	
3.5	70,000	3,500	52.5–105	43.75–87.50	
4.0	80,000	4,000	60.0–120	50.00–100.00	
4.5	90,000	4,500	67.5–135	56.25–112.50	
5.0	100,000	5,000	75.0–150	62.50–125.00	

* It is estimated that on silt loams 1½ to 3 per cent of the total nitrogen may be released annually; on clays and clay loams, 1¼ to 2½ per cent; and on sands and sandy loams, 4 to 6 per cent.
† Sandy soils seldom have over 1½ per cent organic matter.
SOURCE: E. R. Graham, Testing Missouri Soils, *Univ. Missouri Agr. Expt. Sta. Bull.* 345, 1950.

TABLE 56-8. EROSION LOSSES OF PLANT FOOD ON A 3.68 PER CENT SLOPE 90 FT LONG, WITH VARIOUS CROPS

Cropping system	Annual loss, lb/acre				
	Nitrogen (N)	Phosphate (P_2O_5)	Potash (K_2O)	Calcium (Ca)	Magnesium (Mg)
Continuous corn	65.9	18.0	605	221	87
Continuous wheat	32.4	9.4	264	106	43
Corn-wheat-clover	26.4	6.2	214	86	29
Continuous bluegrass sod	0.6	0.2	2.7	1.0	0.25

SOURCE: M. F. Miller and H. H. Krusekopf, The Influence of Systems of Cropping and Methods of Culture on Surface Runoff and Soil Erosion, *Univ. Missouri Agr. Expt. Sta. Rev. Bull.* 177, 1932.

PHOSPHORUS

Role in Nutrition. Phosphorus is found in every living cell and is highly essential in both plant and animal nutrition. In plants the highest concentration is in the seeds, while in animals it combines with calcium to make up 75 to 80 per cent of the minerals found in the bones. An adequate supply of phosphorus in the soil favors the rapid growth and development of plants, hastens maturity, increases the proportion of grain to straw, and makes the grain more plump, thus adding to its quality and weight. Animals fed plants grown on soils low in phosphorus usually gain weight rather slowly and may develop diseases that can be corrected by adding phosphorus to the soil. On

[1] Jacob G. Lipman and H. B. Conybeare, Preliminary Note on the Inventory and Balance Sheet of Plant Nutrients in the U.S., *New Jersey Agr. Expt. Sta. Bull.* 607, 1936.

the soils of the Midwest, poor crop production is frequently due to lack of phosphorus.

Origin in the Soil. Most soils, especially those of the Midwest, were formed from rocks low in phosphorus. The continued removal of phosphorus through the growing of grain crops and high-phosphate legume hay crops further depleted the supply of this important plant food. The deficiency of phosphorus in the soil is indicated by the early use of bones and fish for improving the productivity of the soil.

In general, phosphorus is present in the soil in various combinations, primarily compounds of calcium and phosphorus, magnesium and phosphorus, or combinations of phosphorus with organic matter. Next in importance are compounds of iron and aluminum phosphate and the phosphorus present in the rocks from which the soils were originally formed.

Amounts in Soil. The phosphorus content of soils varies greatly. The Atlantic and Gulf Coastal Plain soils may contain less than 500 lb total phosphorus per acre in the surface 7 in. Most soils contain less than 1,200 lb total phosphorus in the surface 7 in., while those that contain 2,000 lb or more per acre are usually considered quite high. Our most fertile soils seldom contain more than 1,500 lb total phosphorus per acre in the surface 7 in. One pound of phosphorus (P) is furnished by 2.29 lb of phosphate (P_2O_5).

Rate of Availability. Only a very small portion of the total phosphorus in the soil is available to plants. It is relatively immobile, and the plant roots do not have a chance to contact it as readily as the more mobile nitrates. The percentage that is available to plants varies considerably with different kinds of soils, being highest in open sandy soils and lowest in heavy clay soils. Drainage and aeration also affect the release of phosphorus from the soil, especially that stored in the soil organic matter.

The most available forms are usually the compounds of calcium and phosphorus or magnesium and phosphorus, in the form of either monocalcium or dicalcium phosphate. Next in rate of availability are combinations of phosphorus with organic matter. Compounds of phosphorus with iron and aluminum are extremely low in availability, and where the phosphates exist only in this form, it can readily be assumed that phosphate fertilizer will be needed for the successful production of crops. Those phosphate compounds that are in the mineral or rock form are quite resistant to the action of water and are not available to plants until the rocks disintegrate. Studies of plant composition and growth have shown that liming acid soils will increase phosphorus availability. However, overliming may cause a reduction in its availability. These observations and studies indicate that its maximum availability in the soil is reached at a pH level of approximately 6.5. Below this it may be tied up by iron and aluminum, especially in clay soils. On the average, however, only about 1 per cent of the total phosphorus in the surface 7 in. of the soil may be used annually by plants.

Annual Losses Other Than to Plants. Since soil phosphorus is rather immobile in the soil, only a small part of it is lost by leaching. Much less is lost in this manner by phosphorus than by nitrogen. Phosphorus is found primarily in the finest particles in soils, and since these smaller particles are carried away most easily by water or wind erosion, its greatest loss from soil, other than through plants, is in this manner.

Erosion losses with different cropping systems are given in Table 56-8. The loss by erosion under continuous corn was the equivalent to the phosphorus removed in a 75-bu crop of corn, while where a rotation of corn, wheat, and clover was followed, the phosphorus loss was equal to that normally removed in a 25-bu per acre crop of corn. Lipman and Conybeare[1] have shown in their balance sheet of phosphorus in the soils of the United States that losses from erosion just about equal the amount removed by crops.

POTASSIUM

Role in Nutrition. Potassium aids in the formation of carbohydrates, sugar, starch, and cellulose and in the translocation of starch from one part of the plant to another. Unlike phosphorus, it may prolong the period of growth slightly, but seems to add to the plant's ability to resist certain diseases. Therefore, without sufficient available

---

[1] *Op. cit.*

potassium in the soil, plants may have less vigor; their growth processes may become impaired, and they may fail to develop normally. A balanced supply of available potassium improves the health and quality of the plants, adds to their efficiency in photosynthesis, and helps the plant utilize moisture to greater advantage. A good supply of potassium is necessary for well-filled kernels in grains and enough stiffness in straw to reduce lodging. It encourages the growth of the various legume crops and assists in the functioning of chlorophyll, being particularly helpful in the production of starch by sugar-forming plants. It may help overcome the slowing down of maturity on high-nitrogen soils, and may help prevent too rapid maturity in the case of too much available phosphorus. It seems to be especially needed for the successful production of potatoes and other tuber crops, as well as in the production of high yields of fibrous crops like cotton. Removal is heavy in the production of high yields of forage crops.

Origin in the Soil. Potassium is quite widely distributed in the earth's crust as a mineral. It usually occurs as orthoclase and microcline and in certain forms of mica, chiefly muscovite. Through temperature changes and the action of water, these and other potash-bearing minerals are broken down into clay, called kaolinite, and water-soluble potassium. In the earth's crust there is approximately 2.5 per cent potassium, while in ocean water there is about 0.04 per cent. Since soils are the result of breaking down of rocks through weathering, the potassium is in the form which occurred in these rocks. Potassium is naturally found in all animal tissues and secretions, and life is dependent upon it, as is the case with nitrogen and phosphorus.

Amounts in the Soil. The amount of potassium in the soil is relatively high when compared with nitrogen and phosphorus. There are, however, some open sandy soils with open subsoils and high organic soils, such as mucks and peats, that are quite deficient in total potassium. The range of potassium, usually designated as potash (K_2O), in the upper 7 in. of soil, is 1.5 per cent in the more sandy soils and up to 4 per cent or more in the clay soils. The average, however, is about 2 per cent, which means that around 40,000 lb of total potash is in the surface 7 in. as compared with approximately 3,000 lb of phosphate (P_2O_5) and 4,000 lb of nitrogen. Although it exists in relatively high amounts, it is mostly in an insoluble form. One pound of potassium (K) is furnished by 1.205 lb of potash (K_2O).

Rate of Availability. Usually the higher the clay, or colloid, content of a soil, the lower will be the rate of availability of the potash present in the soil. The greatest percentage is available to the crop in the light sands and sandy loam soils. Potash salts may readily dissolve in the soil, but are made less available when they unite with the colloids, the finest soil particles. Under these conditions, potash is rather immobile in the soil. In the sand and sandy loam soils, it may be readily carried down, while in the high-clay-and-silt soils this movement is largely restricted to the plow layer. It occurs in large amounts in plant tissues, especially the carbonaceous tissues, and in animal manures.

Annual Losses Other Than to Plants. Larger losses of potassium occur through leaching than of phosphorus, and there has been considerable wastage through improper use of barnyard manures, especially the liquid portions. There is also considerable loss of potash through burning of carbonaceous materials, since it is released in a form that can be carried away in the runoff water or leached down through the more open soils.

Erosion carries away large quantities of soil potassium, as shown in Table 56-8.

CALCIUM

Role in Nutrition. The growth processes of both plants and animals are dependent upon calcium. Calcium promotes early root formation, influences intake of other plant foods, and improves general plant vigor. People and animals living on plants grown in soil low in calcium suffer from calcium deficiency. Calcium is essential to normal body functioning, and where it is deficient in foods and feed, the bones and teeth may not develop normally.

Origin in the Soil. Most young soils, or soils that have not been subjected to excessive leaching, are relatively high in calcium, which comes from the original rock or basic material from which the soil was formed. As these soils become older, their calcium is carried down into the lower horizons by rain water, resulting in a lower content in the zone of the feeder roots. The amount of this leaching is determined by climate and the type, or texture, of the soil. In the more humid regions greater amounts of calcium are leached from the surface areas, but on the tighter high-exchange-capacity clay soils, there will be more resistance to this leaching, resulting in more of the calcium being retained in the upper layers.

Alluvial, or bottomland, soils vary considerably in calcium content, depending upon the material from which they were formed and the amount of leaching that has taken place since their deposition.

Amounts in the Soil. The amount of exchangeable calcium in the upper 7 in. of soil may vary from less than 2,000 lb per acre up to 10,000 lb or more. In the younger and tighter clay soils it is higher than in the older, more highly leached soils or the more sandy soils. Also, on eroded soil the content of calcium is usually higher in the remaining surface area because more of the former highly leached surface soil has been removed, and the calcium that has leached from the surface areas and caught in the lower depths of the soil is now nearer the surface.

Rate of Availability. The rate of availability of the calcium in limestone applied to the soil depends chiefly upon its fineness and hardness, how thoroughly it has been mixed into the soil, the texture, tilth, acidity, and organic-matter content of the soil, and climatic conditions. The softer the limestone, the more rapidly the calcium in it becomes available. In most cases, however, the limestone particles which are fine enough to pass through a 100-mesh screen (a screen having 100 meshes or openings to the linear inch) are usually considered available to crops within the first year. Under average conditions, that which passes the 40-mesh screen is usually considered available within 3 to 5 years after application. The more thoroughly limestone is mixed with the soil, the more will be its contact with the soil particles and the more rapidly its calcium becomes available. Limestone breaks down more rapidly in acid soils, but such soils are relatively lower in calcium and it may take longer to build up enough available calcium for maximum crop production.

Where the texture, tilth, and organic-matter content of the soil, as well as weather conditions, are such as to allow the aeration necessary for maximum bacterial activity, the availability of the calcium is proportionately increased.

Annual Losses Other Than to Plants. Rain water moving down through the soil removes lime, or calcium. In semiarid regions this loss may vary from 40 to over 100 lb of calcium per acre annually, being greater in the sandier, more open soils and less in the clay soils. Deep-rooted crops, like alfalfa and sweet clover, which are rather high in calcium, tend to bring calcium up from the lower regions and leave it nearer the surface when they decay.

Calcium, like other nutrients, is lost through erosion. This varies with the steepness and length of the slope, erosion-control measures used, and the cropping system followed, as shown in Table 56-8.

MAGNESIUM

Role in Nutrition. Magnesium promotes earlier and more uniform growth of crops by enabling them to make better use of other plant foods. It aids as a carrier of phosphorus to the growing and fruiting parts. It is necessary for seed development, promotes the formation of protein and oils in crops, and stimulates growth of soil bacteria. It increases the fixation of nitrogen by legumes and increases the resistance of plants to diseases.

Magnesium is an important part of chlorophyll (2.7 per cent), the green coloring matter of plants. A shortage of magnesium reduces this green coloring matter between the veins in the leaves. In the case of certain plants, such as blueberries or sweet potatoes, a reddening of the leaves may occur when there is a deficiency of magnesium. Yield, reduction can, however, be brought about by a deficiency of magnesium without any external symptoms being evident.

Origin in Soil. Magnesium in the soil comes mostly from the breaking down of dolomitic limestone, although most rock formations contain magnesium. It is usually most deficient in those areas where the residual material from which the soil was formed was almost pure calcium carbonate. The return of animal manures and other organic material to the soil helps keep the supply of magnesium in balance. Where soils are limed entirely with calcium limestone, the calcium-magnesium ratio may be thrown out of balance, thus increasing the need for magnesium.

Amounts in the Soil. The amounts of magnesium in the soil may range from less than 100 lb up to more than 1,000 lb per acre. Usually, however, magnesium deficiencies will show up where there is less than 200 lb per acre or where there is more than 20 lb of exchangeable calcium to each pound of exchangeable magnesium. Magnesium carbonate has approximately 19 per cent greater "sweetening" power from the standpoint of increasing the pH in the soil than does calcium carbonate.

Rate of Availability. Magnesium carbonate breaks down more slowly than does calcium carbonate, although it is influenced by the same factors. On soils of pH 6 or less, magnesium is relatively low in its availability to plants. This availability is increased by raising the pH to 6.5 or 7 but is again reduced if the pH is increased to 8.5 or more.

Annual Losses Other Than to Plants. In areas of high rainfall there is a heavy loss of magnesium through leaching. The more open the soil, the greater will be this loss and the more apparent are magnesium deficiencies in plants. It has been estimated that 5,180,000 tons of magnesium is carried away annually in solution in the Mississippi River alone. Erosion losses under various cropping systems are shown in Table 56-8.

USE OF FERTILIZER

Adapted by O. T. Coleman

Fertilizer Grades and Ratios (for Processed Materials and Mixed Goods). The first figure in a fertilizer grade or ratio is the pounds of nitrogen (N), the second figure the pounds of phosphate (P_2O_5), and the third figure the pounds of potash (K_2O) in 100 lb of fertilizer. A 1:3:1 ratio, such as 8-24-8, contains three times as much P_2O_5 as it

Table 56-9. Analysis of Basic Fertilizer Materials and Common Mixed Fertilizers

			Per cent by weight of N-P_2O_5-K_2O
Nitrogen (N) materials:			
Anhydrous ammonia, liquid			82-0-0
Aqua ammonia, liquid			23-0-0
Synthetic urea nitrate			45-0-0
Ammonia nitrate			33.5-0-0
Ammonium sulfate nitrate			26-0-0
Ammonium sulfate			20.5-0-0
Cyanamide			21-0-0
Sodium nitrate			16-0-0
Phosphate (P_2O_5) materials:			
Calcium metaphosphate			0-62-0
Triple superphosphate			0-45-0
Superphosphate			0-20-0
Potash (K_2O) materials:			
Muriate of potash			0-0-60
Manure salts			0-0-25
Sulfate of potash-magnesia			0-0-22
Common nitrogen-phosphate fertilizers			
16-20-0	11-48-0	13-39-0	17-7-0
Common phosphate-potash fertilizers			
0-10-20	0-15-30	0-20-10	0-30-20
0-20-20	0-25-25	0-30-10	0-24-15
Common complete fertilizers			
10-10-10	8-16-16	4-12-24	4-16-16	3-9-27
12-12-12	10-20-20	5-15-30	6-24-24	6-12-24
10-20-10	8-24-8	10-10-5	15-8-4	
12-24-12	10-30-10	15-10-10	4-24-12	

Each ton of barnyard manure contains the equivalent of 100 lb of 10-5-10 fertilizer but releases the approximate equivalent of 4-2-6 the first year and 2-2-3 the second year. Rock phosphate contains 30 to 34 per cent *total* phosphate, but the release rate varies with fineness of grinding, soil analysis, climate, etc.

does N or K₂O, or 8 lb N, 24 lbs P₂O₅, and 8 lb K₂O per 100 lb. A 1:3:1 ratio is also given by 6-18-6 or 4-12-4, but these are lower-analysis fertilizers and supply less plant food per 100 lb. An 82-0-0 fertilizer contains 82 lb of N per 100 lb, but no P₂O₅ or K₂O. An 0-62-0 fertilizer contains 62 lb of P₂O₅ per 100 lb, but no N or K₂O. An 0-0-60 fertilizer contains 60 lb of K₂O per 100 lb, but no N or P₂O₅. Table 56-9 gives the more common fertilizer materials and mixed goods, showing the usual percentages of plant foods they contain.

Indications are that the oxides presently included in analyses of phosphate and potash fertilizers may be dropped in some states and the grades and ratios given as the elements phosphorus (P) and potassium (K). When this is done, the trend will likely be to listing N, P, and K elements in similar ratios to those above, but possibly changed to whole numbers. One can, however, easily convert analyses to the element form by multiplying the pounds phosphate by 0.436 and the pounds potash by 0.83.

Soil Tests. The amounts of the different plant foods that will be available from a particular soil to various crops during their growing seasons can be determined by *soil tests* combined with information about the soil type (sand, silt, or clay), subsoil character, surface depth, drainage and slope, and recent soil treatment and cropping history. From the tests and other information, the added kinds and amounts of plant food needed to furnish an adequate, balanced supply for the most satisfactory production of quality crops can be calculated.

County or state agricultural extension offices can usually have soil tests made and should be consulted as to the recommended sampling technique for the area.

Soil-test measurements	*Materials applied to correct shortages*
Organic matter..............	Nitrogen (N in manure, legumes, mixed fertilizers, and straight nitrogen fertilizers)
Soluble phosphate...........	Phosphate (P₂O₅ in manure, rock phosphate, or in processed phosphate such as in mixed fertilizers and super and triple superphosphates)
Exchangeable potassium......	Potash (K₂O in mixed fertilizer, muriate of potash, manure salts, or manure)
Exchangeable calcium........	Limestone (calcium or dolomite)
Exchangeable magnesium.....	Dolomitic limestone (contains both magnesium and calcium) or soluble magnesium (sulfate of potash magnesia)
Exchangeable hydrogen......	Limestone (lowers hydrogen)
pH (acidity)................	Limestone (raises pH and reduces acidity)

Plant growth may also be limited by shortages of sulfur, iron, zinc, boron, copper, manganese, and "trace elements," that is, elements of which only a trace is required.[1]

A *corrective treatment* is designed to bring fertility up to the desired level in one series of applications. A *starter treatment* is used to promote early growth of seedlings. A *maintenance treatment* is a periodic replenishment of plant food removed during cropping after a corrective treatment has been made.

DETERMINING FERTILIZER REQUIREMENTS FOR CORRECTIVE TREATMENTS[2]

The elements calcium, magnesium, potassium, and hydrogen are held chiefly by the clay particles in the soil, and the total amount held, or exchange capacity, is chiefly dependent on the kind and amount of clay.

The *exchange capacity* of a given soil is measured in milliequivalents, or m.e., and can be found as follows:

$$\text{Exchange capacity, m.e./acre} = \frac{Ca}{400} + \frac{Mg}{240} + \frac{K}{780} + \frac{H}{1{,}000}$$

where Ca = exchangeable calcium, lb/acre
 Mg = exchangeable magnesium, lb/acre
 K = exchangeable potassium, lb/acre
 H = exchangeable hydrogen, lb/acre

The exchange capacity is considered *properly balanced* when it shows 75 per cent calcium, 10 per cent magnesium, and 2½ to 5 per cent potassium saturation. Table

---

[1] "Soil," USDA Yearbook of Agriculture, 1957, pp. 107-150.
[2] After *Missouri Balanced Farming Handbook* BF 5604, Soil Management section.

56-10 shows, for soils of various exchange capacities, the proper amounts of each of these elements in pounds per acre. The *deficiencies,* or corrective amounts to be applied, are determined from the differences between the measured amounts from the soil tests and the recommended amounts from Table 56-10 under the exchange capacity found for the particular soil.

Calcium. Liming is often the most important single thing in improving soil fertility because (1) it supplies calcium plant food; (2) it prevents the build-up of toxic aluminum and manganese which may occur in acid soils; (3) by the addition of calcium it tends to displace the phosphorus tied up by iron and aluminum, increasing phosphorus availability; (4) it prevents substitute consumption of potassium by plants growing where calcium and magnesium are low; (5) it neutralizes the acids produced by soil microbes which break down crop residues and fix nitrogen on legume roots, allowing them to function efficiently; and (6) it improves soil structure to admit more air and water through its mellowing effect.

TABLE 56-10. AMOUNTS OF CALCIUM, MAGNESIUM, AND POTASSIUM FOR PROPER BALANCE IN SOILS OF VARIOUS EXCHANGE CAPACITIES

Element	Total exchange capacity, m.e./acre										
	8	10	12	14	16	18	20	22	24	26	28
Calcium, lb/acre	2,400	3,000	3,600	4,200	4,800	5,400	6,000	6,600	7,200	7,800	8,400
Magnesium, lb/acre	192	240	288	336	384	432	480	526	576	624	672
Potassium, lb/acre	250	266	280	296	312	351	390	429	468	507	546

TABLE 56-11. EFFECT OF FINENESS OF GRINDING ON EFFECTIVENESS OF AGRICULTURAL LIMESTONE

Effectiveness, %	Fineness as indicated by percentages passing through various screens				
	8 mesh/in.	10 mesh/in.	12 mesh/in.	40 mesh/in.	100 mesh/in.
50	100	98	96	70	46
47–50	98	93	90	52	28
44–46	93	86	83	45	26
40–42	88	79	77	42	23
37–39	83	73	70	37	21
31–34	76	65	63	32	19

A *deficiency* of one pound of calcium requires 2½ lb of pure calcium carbonate ($CaCO_3$), 100 per cent effective, for correction. Agricultural limestone is often not pure $CaCO_3$, and the amount required must be corrected for the actual percentage of $CaCO_3$. In addition, an adjustment must be made for fineness of grinding, since this affects the rate at which the limestone will dissolve and become effective. As a result, the most finely ground agricultural limestone is only 50 per cent effective, as indicated in Table 56-11. The amount of agricultural limestone needed can be found as follows, using the deficiency indicated by the soil test and Table 56-10:

$$\text{Limestone needed, tons/acre} = \frac{12.5 \text{ (calcium deficiency, lb/acre)}}{(\text{per cent } CaCO_3)(\text{per cent effectiveness})}$$

The per cent effectiveness can be estimated from Table 56-11 by comparing the screen analysis with those shown in the table. Limestone which has been in the soil for less than a year may not show up in a soil test. In such cases, subtract the amount previously applied from any deficiency shown by the test.

Normally, *soil acidity* is automatically corrected by applying enough limestone to correct the calcium deficiency as calculated above. If the soil test shows acidity along with adequate calcium, add limestone as given in Table 56-12. A pH of 5.5 is seriously acid for crop production, since the ideal is a pH of 6.5 to pH 7.

A practical *minimum application* of limestone is 2 tons per acre, since this will not "overlime" any land that needs some lime, even if the need is less than 2-tons. If the calcium saturation is 65 per cent or above with a pH of 5.5 or above, about normal response can be expected from fertilizers on nonlegume crops without adding limestone.

Magnesium. A deficiency of 1 lb of magnesium requires the application of 2 lb of the magnesium found in dolomitic limestone. The magnesium contained per ton of dolomitic limestone varies with the percentage of magnesium carbonate ($MgCO_3$), amounting to 5.8 lb per ton for each per cent of $MgCO_3$. The amount of dolomitic limestone needed can be found by the following formula, using the deficiency from

TABLE 56-12. AMOUNTS OF LIMESTONE REQUIRED TO RAISE THE SURFACE 7-IN. LAYER OF SOIL TO A pH OF 6.5

Present pH	Limestone needed: 95% $CaCO_3$ and at least 50% through 40-mesh screen, tons/acre		
	Sandy loam	Silt loam	Clay loam
6.0	0.5	0.9	1.25
5.5	1.0	1.8	2.50
5.0	1.5	2.7	3.75
4.8	1.7	3.0	4.25

the soil test and Table 56-10 and the per cent effectiveness from Table 56-11, according to the fineness of grinding:

$$\text{Dolomitic limestone needed, tons/acre} = \frac{17.5 \text{ (magnesium deficiency, lb/acre)}}{(\text{per cent } MgCO_3)(\text{per cent effectiveness})}$$

As with calcium, applications made within the past year will not materially affect the soil test, and the deficiency should be corrected for such applications.

The calcium deficiency corrected by dolomitic limestone can be calculated as follows:

$$\text{Calcium, lb/acre} = 0.08 \text{ (dolomitic limestone applied, tons/acre)} (\text{per cent } CaCO_3)(\text{per cent effectiveness})$$

If dolomitic limestone is not obtainable, apply readily soluble magnesium (sulfate of potash-magnesium) annually at the rate of 30 to 35 lb MgO per acre on soils with a magnesium saturation of 5 per cent or less. In the case of extreme deficiency of magnesium in soils, some soluble magnesium should be applied the first year along with dolomitic limestone.

Potassium. Since potassium is held by clay particles, it does not move in the soil. It should be well distributed through the soil at a level of about 200 lb of available potash (K_2O) per acre in most soils to adequately supply crops with extensive root systems. Much potassium is unavailable, particularly in clay soils, and much is only slowly available. Muriate of potash (potassium chloride) may have a damaging effect on seedlings because of the chlorides, if placed in the row at high concentrations. Plowing under of broadcast applications is favored as a means of maintaining an adequate level.

The potassium deficiency indicated by the difference between the soil test and Table 56-10 should be corrected for the amount which will be released by crop residues during the coming season, if the soil tests were made during the latter half of the year. In such cases, subtract 40 to 80 lb per acre from the indicated deficiency, according to the amount of roots and crop residues being returned.

Potassium *requirements* vary with the type of crop as follows:

Alfalfa: Apply ⅔ lb of potash (K_2O) for each pound of potassium deficiency.

Pastures and corn with heavy nitrogen application: Apply ½ lb K_2O for each pound of potassium deficiency. Reduce this to ¼ lb for repeated crops if all residues are returned to the soil.

Small grains: Apply ⅓ lb K_2O for each pound of potassium deficiency for annual applications.

Phosphorus. Most of the phosphorus in soils is combined with the iron and aluminum in clays in unavailable form. Liming acid soil to a pH level of 6.5 helps increase phosphorus availability. Less fertilizer phosphorus is lost by fixation if it is concentrated in a band, preferably where seedling roots can feed on it, rather than being mixed through the soil. Phosphorus does not move in the soil, and the roots must find it. Crops with extensive root systems, such as the grasses, do a good job of gathering the available phosphorus, keeping it from being fixed, and their residues return much of it to the soil in available form.

Table 56-13 shows the amount of P_2O_5, in the form of processed phosphate (super and triple superphosphate, calcium metaphosphate, and the phosphates in mixed

TABLE 56-13. Amount of Phosphate (P_2O_5) Required for Different Phosphate Test Levels in the Soil

Soluble phosphate from soil test, lb P_2O_5/acre	Processed phosphate, lb P_2O_5/acre		Rock phosphate (applied every 6 to 10 years), lb P_2O_5/acre
	Applied annually	Applied every 3 to 5 years	
20	80	160	450
40	70	140	350
60	50	100	300
80	30	60	260
110	30	60	240
140	30	60	220
170	30	60	None

fertilizer) or rock phosphate needed for various levels of soluble phosphate in the soil as indicated by soil tests. If the crops for the next several years require a high phosphorus level, if little erosion is expected, and if the money is available, one of the longer-term applications will cost the least on an annual basis. An annual application of processed phosphate will give a maximum yield for 1 year, but will not carry over enough to raise the soil test until repeated several times. A corrective application of rock phosphate, according to soil test, will show in soil tests in succeeding years. The amount of rock phosphate required for a given amount of total P_2O_5 will depend on its analysis.

Nitrogen. There are four types of carriers for nitrogen fertilizer:

1. *Nitrates* (NO_3): ammonium, sodium, calcium, and potassium. Nitrates are water-soluble and immediately available to plants. They are also easily leached from the soil by water.

2. *Ammonia* (NH_3): ammonium sulfate, ammonium nitrate, and ammonium solutions. Ammonium ions are absorbed by soil colloids and not easily moved by soil moisture. Many plants can use nitrogen in this form, but others must wait until it is transformed into the nitrate form by soil microorganisms, which occurs rapidly in warm, moist, aerated soils. The fact that such nitrogen persists in the nonleaching ammonia form through the winter in the northern half of the United States makes fall applications practical in many cases. Nitrates do not start forming until the following growing season when they are needed.

3. *Synthetic organics:* urea and calcium cyanamide. These break down readily in the soil to form ammonia, discussed above.

4. *Natural organics:* manure, crop residues, and vegetable by-products. These must be decomposed in the soil before the nitrogen is available. Soil microorganisms transpose the protein nitrogen into ammonia nitrogen, but in the process of decomposing straw or stalks they need additional nitrogen food from the soil. Extra nitrogen fertilizer is often needed to supply the temporary deficiency when residues are plowed under.

Annual nitrogen (N) requirements vary with the percentage of organic matter in the soil (as shown by the soil test), the soil texture (sand, silt or clay), the crop to be grown, recent nitrogen applications, and the previous crop type, yield, and disposition. The organic matter releases nitrogen as it rots in the soil as shown in Table 56-7.

The additional nitrogen needed for 100 bu of *corn* per acre is shown in Table 56-14. *Grass* will require about the same amount. *Grass-legume* seedings, *small grains,* and new seedings of *alfalfa* will require about one-half this amount.

TABLE 56-14. ADDITIONAL NITROGEN NEEDED FOR 100 BU CORN PER ACRE, LB/ACRE

Soil texture	Percentage of organic matter by soil test									
	0.5	1.0	1.5	2.0	2.5	3.0	3.5	4.0	4.5	5.0
Clay................	150	140	130	125	120	110	100	90	80	70
Silt.................	130	125	120	110	100	90	80	75	60	50
Sand...............	125	110	95	80						

TABLE 56-15. NITROGEN FURNISHED BY PLOWING UNDER GREEN LEGUME CROPS

Crop	Nitrogen furnished, lb/acre	
	Entire crop plowed under	Pastured and plowed under
Alfalfa, sweet clover.......	40	30
Red clover, Ladino........	30	20
Lespedeza................	20	15

The amount of nitrogen furnished by plowing under good *legume* crops is shown in Table 56-15. *Plowing under* straw, cornstalks, weeds, and other nonlegume *plant residues* reduces the amount of nitrogen available for plant food, until decomposition is complete, because bacteria use the nitrogen in the rotting process. This nitrogen becomes available later, as well as additional nitrogen from the organic matter turned under, but there may be a temporary shortage. About 30 lb of nitrogen per ton of plant residue, dry weight, is usually needed. For example, if 3 tons of cornstalks is being turned under where 100 lb of nitrogen is being applied to meet soil-test requirements, no shortage should occur. On the other hand, if less than 90 lb was needed to bring the soil test up, additional nitrogen would be needed for rotting.

Manure furnishes approximately 4 lb of nitrogen per ton of manure the first year and 2 lb the second.

Methods of Applying Corrective Treatments. Heavy applications of *phosphate, potash,* and *limestone* are preferably plowed under or worked deeply into the soil. The material can be spread whenever the spreading equipment can be driven over the ground and worked into the ground as soon as convenient. Direct spreading from trucks saves much labor, but the ground should be dry or frozen to avoid excessive compaction.

Nitrogen may be furnished by the solid nitrogen carriers listed in Table 56-9, by anhydrous ammonia, aqua ammonia, or by liquid nitrogen solutions such as described in Chap. 16. Solid nitrogen may be plowed down, worked into the seedbed, and used as side- or top-dressing. Anhydrous ammonia should be placed in the soil at least 6 in. deep regardless of the time of year. Low-pressure nitrogen solutions or aqua ammonia should be plowed under or placed 3 in. deep in the soil to prevent loss. Nonpressure nitrogen solutions can be applied to the surface without loss. All these forms of nitrogen have been found to be equally effective as plant food when used according to recommendations.

For *corn*, it is preferable to plow down the nitrogen on silt loam or clay (gumbo), and preferably in the late fall or winter if carbonaceous material such as cornstalks or straw is turned under. On average sandy soils subject to nitrogen loss by leaching, plow down one-half to two-thirds of the total application and side-dress the remainder. On extremely sandy soil it may be desirable to side-dress all the nitrogen.

For *small grains*, work nitrogen into the seedbed or top-dress any time from seeding to 4 in. high. Sufficient nitrogen may also be provided in the starter fertilizer applied with the grain drill at the time of seeding. *Never apply a top-dressing to wet foliage.*

Pasture seedings may be treated like small grains. On established pastures, top-dress in early spring or late in August. *Do not apply to wet foliage.*

New alfalfa seedings should have the nitrogen worked into the seedbed or applied in the starter fertilizer.

Starter applications in general contain all three elements and are applied near the seed as it is planted to give the young plants a good start. Excessively heavy or improperly placed applications may damage seedings.[1] Row-crop applications made with planters should not ordinarily contain more than 20 lb of nitrogen or 40 lb of potash per acre.

Maintenance applications serve to replace the nutrients removed by cropping, erosion, and leaching; they maintain the desired fertility level after basic or corrective fertilizer applications have been made. When erosion is controlled, the amounts needed can be estimated from Table 56-6, which gives the amounts contained in various crops.

Plant Food Deficiencies. These can be determined by *plant-tissue tests* throughout the growing season. Deficiencies are also indicated by the following symptoms:

Nitrogen	Phosphorus	Magnesium	Potassium	Calcium	Boron
Slow growth Decrease in green color Yellowing Small stems Less branching	Small plants Lower yield of grain Delayed maturity Shrunken pods and grain	Streaked appearance of leaves—light green between plainly visible veins Yellowish green mottling Shedding older leaves Purplish red leaves with green veins	Whitish spots near leaf margins Scorching of leaf edges Ruffled or cupped leaves in legumes Lodging Short joints in corn, grains, grasses	Disappearance of clovers and alfalfa plants White spotting of leaves Stunted growth Drought or winter-killing easily Wilting of leaf stems Malformations	Yellowing or reddening of terminal leaves Shortened terminal internodes—rosette appearance Low seed production due to sterility Stunted root development

[1] R. L. Cook and W. C. Hulburt, Applying Fertilizers, "Soil," USDA Yearbook of Agriculture, 1957, pp. 216–229.

Chapter 57

FARM CLIMATES AND SOLAR ENERGY

F. A. Brooks

The comprehensive climatic information needed specifically for each area of interest must be developed from local climatic data impossible to cover in this short section. In a 5-year international bibliography, Wang and Barger [63] have classified 11,000 articles on agricultural meteorology. As a general guide, information is included here, usually in four concepts: (1) world-wide maps of seasonal and annual means, (2) tables and curves of monthly and hourly characteristic climatic factors, (3) formulas permitting specific determinations utilizing tabulated mean coefficients, and (4) United States maps of selected climatic factors. Naturally, all climatic mean magnitudes have to be interpreted against the natural daily weather, which is highly variable, so a very brief review is given of general atmospheric circulation, and specific information is included on solar energy as the primary source of life.

Long-term economic farming operations require frequency studies based on specific data (such as suggested in the list of annual graphs, Table 57-11) to determine:

1. The probabilities of temperatures, moisture, and light favorable for productivity
2. The probability of coincident soil and weather conditions suitable for timely operations
3. The improbabilities of destructive conditions

Then the current status of precipitation, soil temperature, etc., needs to be followed relative to long-term means to schedule field operations according to the present stage of the current season (Table 57-12).

CLIMATE CLASSIFICATION

Farm climate has two major aspects: (1) the climate of the geographical region, depending mostly on latitude, altitude, mountain barriers, and distance from the ocean, and (2) the local microclimate, depending mostly on exposure to sun, wind, and nocturnal air drift. For the first, measurements of air and sky conditions should be made 30 ft or more above the ground or treetops. For the second, measurements near the ground, including soil temperatures, are needed.

There are two well-recognized systems of classification of geographical climate, both drawn purposely to group general types of natural vegetation and soils.

Koeppen's broadest climatic groups are:

1. The tree climates A, C, and D (respectively, of tropical rain forest, warm-temperature rain woodland, and cold snow forest)

2. The drier climates BW, BS, and E (respectively, of the steppes, deserts, and polar regions)

Each of these major climatic groups has well-defined subdivisions devised primarily to delimit distinctive agricultural conditions. The most important factors in these subdivisions are the seasonal distribution of rainfall and the degrees of mean temperatures for the warmest and coldest months. The Koeppen classification of climates for the world is well described and illustrated in 15 colors by Haurwitz and Austin [24].

Thornthwaite's [46] 1948 classification of climates is based directly on the two most important factors in agriculture: first, on the annual relationships of rainfall to evaporation and transpiration [interpreted as "moisture index"; see Fig. 57-1, Eq. (57-17)] and secondly, on annual evaporation power, dependent on mean monthly temperatures and average length of day (interpreted as "temperature efficiency") shown in Fig. 57-2. These two factors are considered independent, although both use the concept of potential evapo-transpiration [33], namely, the water-loss rate when adequate moisture is always available. Thornthwaite's classification includes further subdivisions, mainly by summer conditions, in somewhat more detail than Koeppen's. For example, he rates Madison, Wis., $B_1rB_1'b_2'$, and Fresno, Calif., $EdB_3'b_3'$. Figure 57-1 shows that Thornthwaite's basic division line of zero index in the United States, separating the major regions of net moisture surplus from those of net moisture deficit, runs southward from western Minnesota. This is about 300 miles east of Koeppen's steppe climate border. Further discussion of Thornthwaite's moisture index is given later.

World-wide Distributions of Temperature and Precipitation. Differences in climates result naturally from regional variations in solar energy, rainfall, and flow of air. The resulting natural vegetation and soil erosion in regions of the same climate classification produce landscapes which look remarkably alike all over the world. But since climates are not easily *measured* by vegetation and specific information is needed to judge human comfort and agricultural limitations, the specification of each climate type is necessarily based on observed weather factors. The most universal measurements include only local daily rainfall and maximum and minimum temperatures, but taken year round, they reveal significant seasonal relationships which can be interpreted further by the known latitude and altitude of the place. The continual needs of water by plants in the growing season make very important differences in agriculture between areas where summers are dry and those where summers have frequent rain, even though mean monthly temperatures and annual precipitation are alike. Furthermore, within the narrowest conventional climate classification, differences in soil and in nighttime temperatures may lead to significant differences in vegetable and fruit production and in living comfort. The world-wide distributions of mean temperatures for January and July are given in Figs. 57-3 and 57-4. The damaging effects of extreme temperatures need to be studied from local data.

The magnitudes of all climatic factors vary markedly from year to year, the most important variation being in precipitation. The average annual world-wide precipitation is shown in Fig. 57-5.

Weather Belts; Air Masses. Explanation will not be attempted here of the complex meteorological factors leading to world-wide weather belts, and this modification by continents. These are vividly described by Trewartha [47]. The generalized world-wide distribution of weather in relation to latitude pictured by Petterssen [36] is reproduced in Fig. 57-6. The main departures from this generalized scheme are due to the irregular distribution of land and sea areas. The agricultural possibilities of a particular area are greatly affected by location relative to oceans and mountain ranges as well as by latitude.

It is the moving inland of maritime air masses that makes continental rainfall possible. Even the heat balance is strongly controlled by the properties of the overhead air mass, particularly its temperature, moisture, and velocity. Thus the condition of the existing air mass, usually described by weather type, is an essential part of the description of local environmental conditions.

The properties of an atmosphere determine the weather type and depend primarily on the surface temperature and moisture of the *source region* where the air mass has remained for a week or more. Thus there are major source classifications: *artic, polar,* and *tropical* and the qualifying classifications of *maritime* or *continental.* The air mass is further described, after it leaves its source region, by its thermodynamic temperature relative to the ground it is flowing over. This is designated by k for colder than the ground (a condition which promotes vertical circulations) and by w for warmer than the ground (which is a stable condition most noticeable in confining air pollutants). A fourth notation is used sometimes: s for stable and u for unstable temperature lapse rates in the air mass itself. The meteorological aspects of air flow are too complex for presentation here.

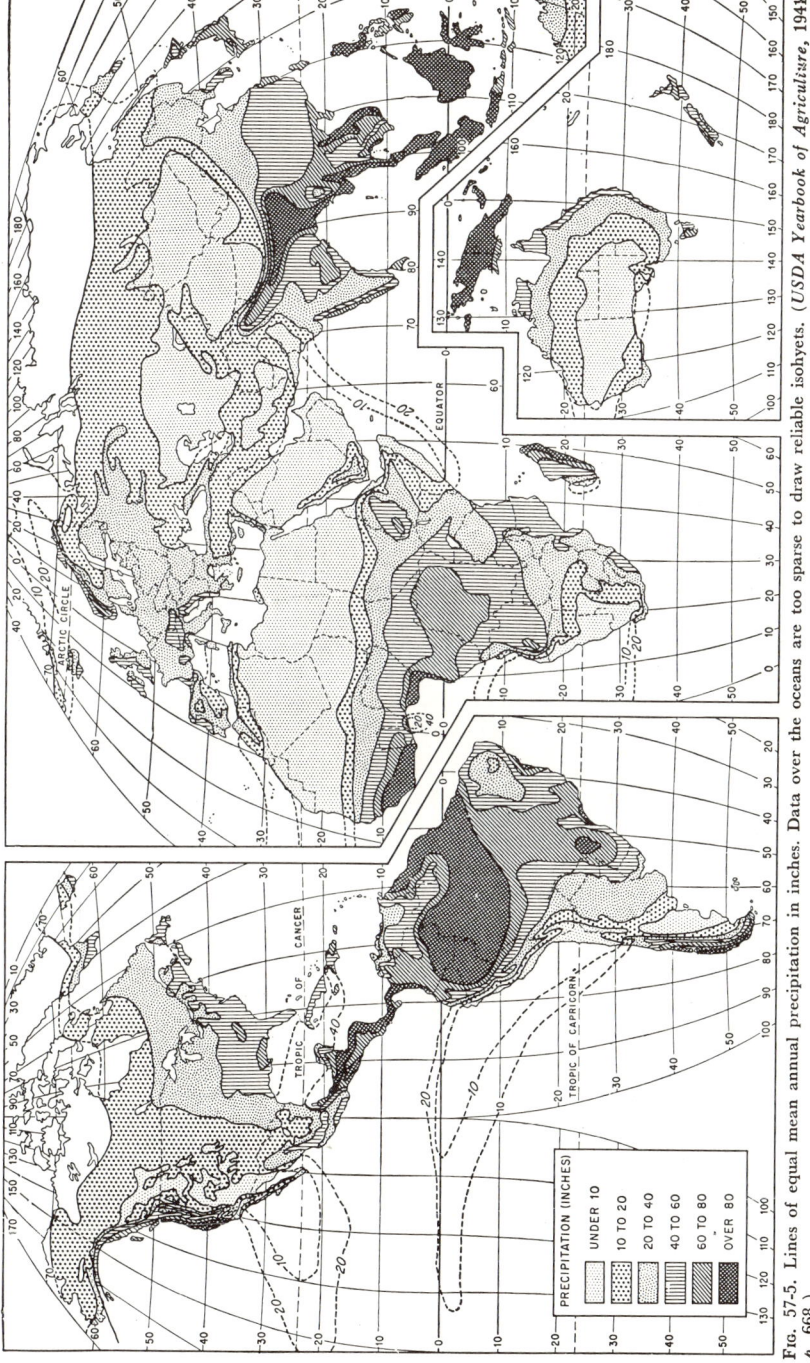

FIG. 57-5. Lines of equal mean annual precipitation in inches. Data over the oceans are too sparse to draw reliable isohyets. (*USDA Yearbook of Agriculture*, 1941, p. 668.)

Weather Types. Correlation of field observations with "mean" weather data, which is an average of all kinds of weather jumbled together, does not yield as significant results as would be possible if there were some classification by characteristic weather type. Even livestock behave differently in dry than in moist air. Meteorologists think of weather type in terms of cloudiness or of characteristic air mass and a recognizable pattern of pressure distribution. The cooperative observers who furnish the major information on agricultural climate classified weather (until July, 1948) as clear, partly cloudy, or cloudy. Neither of these systems is alone adequate for designating comparable days. Probably daily temperature range and dew point are the most specific indicators, but sky cover and air mass together should provide a better description. Such a combination, proposed by C. F. Brooks and F. A. Brooks [11] for noncoastal areas, is reproduced in Table 57-1. At any one location the many-year,

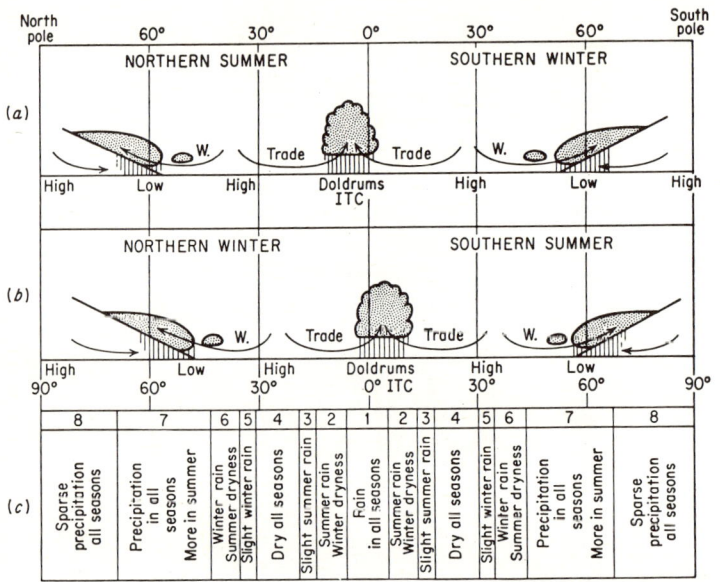

Fig. 57-6. Petterssen's schematic cross section through the atmosphere from pole to pole, showing the main weather zones due to general rising and settling air: (*a*) during the northern summer; (*b*) during the northern winter; (*c*) the belts of precipitation. The prevailing westerlies are roughly between 55 and 35° latitude, and the trade winds between 25°. and near the Equator, the whole pattern shifting almost 10° seasonally with sun. (*Trewartha, "An Introduction to Climate," 3d ed., McGraw-Hill Book Company, Inc., New York, 1954, p. 143.*)

monthly average of daily temperature range will depend mostly on the dominant weather type. In a climate of usually dry air the coming of moist air and the accompanying clouds will cause a large reduction in daily temperature range because of the blanketing effect. Conversely, in a location usually subject to moist air, the arrival of dry air will greatly expand the daily temperature range unless there is wind.

PRECIPITATION DUE TO NATURAL CONVECTION OR MOUNTAINS AND WEATHER FRONTS

Air near the ground takes up moisture from open water, from plant transpiration, and soil-moisture evaporation and acquires a dew point largely characteristic of the surface. Any action serving to cool the air below its dew point usually causes condensation of the water vapor to water droplets. When such cooling continues in free air, the droplets grow in size and lead to precipitation. The most effective means of cooling an air parcel is to lift it to elevations of less and less pressure, thereby causing

TABLE 57-1. APPROXIMATE CLASSIFICATION OF WEATHER TYPE FOR CONTINENTAL UNITED STATES SUITABLE FOR SELECTING SIMILAR DAYS IN SEASON

Designation			Visual indicators		Usual air-mass interpretation				Instrumental indication, Davis, Calif.					
Weather category (middle or low levels)	Weather type*	Sky cover (middle and low) average morning and noon, tenths	Diurnal cloud cycle or weather		Typical air mass or weather	Air-mass condition: k, colder than ground w, warmer than ground s, stable aloft u, unstable aloft			Diurnal range of dry-bulb temp., °F		Wet-bulb depression at 4 P.M., °F		Black globe av. daily max. temp., °F	
						Winter	Summer		Jan.	July	Jan.	July	Jan.	July
Clear or scattered clouds	0	0–2	Day: not windy, usually cloudless Night: cloudless, no fog		Continental Polar	Colder cPks	Colder cPku		24	42	7	26	72	110
	1	0–3	Day: windy Morning: usually Cu Noon: usually clear Night: no fog		Continental Polar	Colder cPku	Colder cPku		16	36	10	25	62	106
	2	0–4	Morning: usually Cu Noon: usually clear Night: tendency to fog		Continental Tropical	Warmer cTws	Warmer cTws		23	42	8	32	83	120
Partly cloudy	3	3–5	Day: Cu and/or Sc Afternoon: showers		Maritime Polar	Warmer mPws	Colder mPku		18	34	7	22	74	104
	4	5–8	Day: usually partly cloudy Afternoon showers Night: usually clear, no fog		Maritime Tropical	Warmer mTwu	Colder mTku		…	34	…	27	…	120
Largely cloudy or overcast	5	6–10	Mostly cloudy or overcast in middle or low levels, without rain or wind as in types 7 and 8		Mixed or indifferent Front approaching	Warmer Xws	Warmer Xwu		13	19	4	12	58	90
	6	8–10	Persistent rain without windshift or destruction as in type 8		Prewarm-front rain	………	………		9	18	3	10	47	91
	7	6–10	Windshift (cold front) usually with downpour		Cold-front rain	………	………		11	25	5	16	70	96
	8	6–10	Destructive wind		Gale, hurricane, violent squall, or tornado									

* Code number to be punched on IBM card.

cooling by expansion. The 5- to 10-mile depth of this vertical convection constitutes the troposphere, at the top of which air temperatures are about −65°F and the moisture content practically zero.

Moist air can be lifted from near the ground by thermal convection when the surface air is warmed and made less dense by sunshine heating the ground, or it can be lifted by forced flow up over a mountain or over a wedge of cold, dense air.

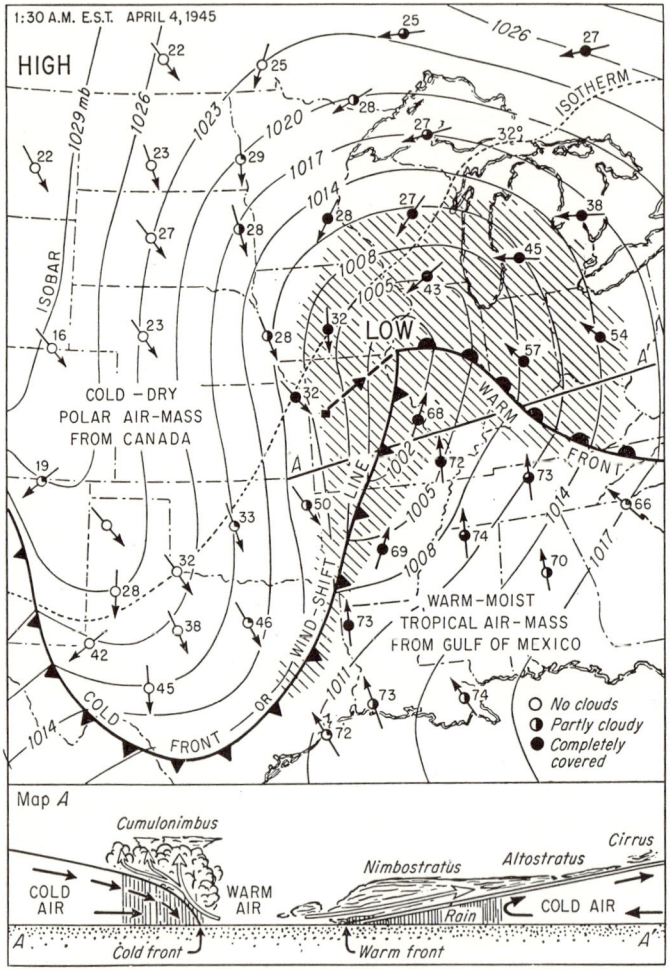

FIG. 57-7. Partial weather map and section of cyclonic storm idealized by Bjerknes and Solberg. (Strahler, "Physical Geography," John Wiley & Sons, Inc., New York, 1951.)

The climates of farms close to large areas of water or near mountain barriers are powerfully influenced by the direction of air flow. Besides a general wind, there may be definite 24-hr cycles of winds toward and away from the ocean and of upslope and downslope winds. In all cases, the flow leading to a cooling of the air either by advection over a colder surface or by expansion cooling in a lifting process will tend to produce cloud formation and, if cooling continues, precipitation. Air is warmed as it flows over a warmer, drier surface, and its relative humidity will decrease, resulting in clear, dry weather unless the temperature change in the lower levels develops instability in the air mass. Naturally, this depends on the moisture content of the air. The flow of air downhill causes warming by compression (this amounts to 5½°F per 1,000 ft of elevation if there is no heat loss), with no change in moisture content, so the "chinook," or foehn, wind is relatively warm and causes rapid drying of soil and plants. This dry air is the cause of the desert conditions in the Salt Lake intermountain basin.

Coastal weather is significantly different from continental weather mainly because of the moisture effect on air quality and the usual strong daily cycle of wind toward and away from the ocean. The Great Lakes also modify surrounding climate, which may be seen in the fruit and vegetable areas in Michigan and southern Ontario. Fog is an important seasonal factor, and the temperature of the ocean is a strong regulator, which, except for air moisture, decreases in effect directly with distance from the coast. These are regional influences and differ from local spot-climate variation due to orientation and exposure.

Weather Fronts; Maximum Rainfall. Farm climates in the zone of interaction between polar and tropical air masses are largely characterized by the recurring pressure

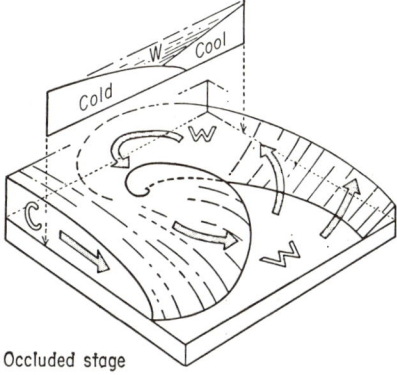

Fig. 57-8. Strahler's three-dimensional portrayal of the Bjerknes and Solberg idealized model of a moving cyclonic storm in the occluded state, greatly exaggerated in vertical scale. On the left the heavy cold air C is advancing under the lighter warm air W, while the latter at the right rides up over a retreating cool air mass. (*Strahler, "Physical Geography," John Wiley & Sons, Inc., New York, 1951.*)

patterns and rainfall resulting from the extensive, moving, cyclonic, and anticyclonic wind systems usually described as *lows* and *highs*. These are shown very clearly by the U.S. Weather Bureau [50] Daily Weather Maps. Figure 57-7 is Strahler's example map and section of a strong cyclonic storm. For the standard station model reporting all local observations, see the reverse side of a current U.S. Daily Weather Map. The weather caused by these extensive cyclonic storms varies principally with wind velocity and the slopes of the "fronts" separating contrasting air masses. Figure 57-8 is Strahler's [44] three-dimensional portrayal of the Bjerknes and Solberg idealized model of cyclonic storms given in section and plan in Fig. 57-7. This shows that in general a rather steady precipitation comes under the warm front from the gradual lifting of moist air as it overrides a wedge of colder, denser air having an irregular slope of about 1:50 to 1:100. Figure 57-8 also shows the line of more violent precipitation along the cold front where the advancing cold air abruptly lifts the warmer moist air.

The rate and duration of rainfall at a given locality thus depend on the weather front and on how long the station remains under that front. If the storm front stalls, there may be excessive precipitation such as occurs along fixed mountain barriers. The probabilities of maximum hourly rates naturally vary with both the area and the duration considered [42]; that is, a cloudburst is distinctly local and of short duration.

Fletcher's [15] analysis of extensive records of rainfall rate in relation to time and to area shows a combined envelope for maximum rates in the United States:

$$\frac{R}{\sqrt{D}} = 0.5 - \frac{266}{19.2 - A} \qquad \frac{R}{\sqrt{D}} > 1.0 \qquad (57\text{-}1)$$

where R = rainfall, in.
D = duration of downpour period, hr
A = area, sq miles

Various geographical locations differ significantly as to "cloudburst" rates which are due to slow travel of violent convection in deep moist atmosphere. The maximum recorded, 12 in. in 42 min at Holt, Mo., and nine other world's records are expressible by the envelope $R = 14.3 \sqrt{D}$. No such rates are recorded for the Pacific Coast, but a dozen maximums indicate a consistent regional envelope of $R = 2.9 \sqrt{D}$ at a given station.

Many studies of storm precipitation frequencies have been reported for various locations and can be applied in limited regions by hydrologists concerned with runoff

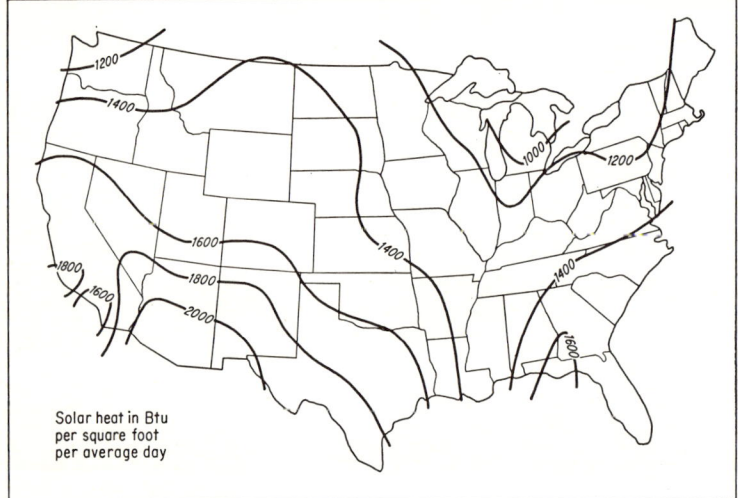

FIG. 57-9. Distribution of average daily solar energy received on a horizontal surface, Btu/(sq ft)(day). (*Hand, Heating and Ventilating, July,* 1953.)

and soil erosion. Not enough information is available yet for farmers to schedule field operations between storms on a monthly probability basis. Linsley's method of interpreting small basins is outlined later.

Estimates of probabilities of future weather events are becoming more numerous and useful. A good introduction is given by Thom [61] and by Van Bavel [62].

SUNSHINE ON HORIZONTAL, VERTICAL, AND SLOPED SURFACES

The primary factor in all climates is the daily receipt of solar energy. The amount varies importantly with season and cloudiness. Average annual curves of observed solar energy for 35 stations are reported by Hand [20]. Average magnitudes for June, December, and the whole year have been collected by Hand [22] for 49 stations in the United States and Canada and are given in Table 57-2, arranged in latitude order. To reveal the geographical distribution of average daily insolation, he constructed an approximate iso line map reproduced here as Fig. 57-9.

A cloud cover decreases the direct sunshine and increases the incoming diffuse short-wave radiation, and a continuous recording of insolation is rarely a smooth curve. Interpretations for average overcasts have been made by Haurwitz [23].

TABLE 57-2. AVERAGE DAILY SOLAR ENERGY RECEIVED AT 49 STATIONS IN THE UNITED STATES AND CANADA
(See Fig. 57-9 for map.)

City	North latitude		Average Btu/(sq ft)(day)		
	Deg	Min	June	December	Year
Miami, Fla.	25	49	1771	1085	1497
Brownsville, Tex.	25	55	2590	1192	1840
Gainesville, Tex.	29	39	1904	841	1471
Apalachicolas, Fla.	29	45	2192	959	1675
New Orleans, La.	30	02	1697	804	1250
Lake Charles, La.	30	13	2151	819	1568
El Paso, Tex.	31	48	2731	1207	2037
Fort Worth, Tex.	32	49	2332	867	1699
La Jolla, Calif.	32	52	2066	904	1526
Charleston, S.C.	32	54	2166	786	1575
Griffin, Ga.	33	14	2273	738	1578
Riverside, Calif.	33	32	2207	782	1558
Santa Maria, Calif.	34	56	2399	867	1817
Albuquerque, N.M.	35	03	2749	1085	1892
Hatteras, N.C.	35	15	2266	756	1594
Oak Ridge, Tenn.	35	55	2000	598	1307
Las Vegas, Nev.	36	05	2771	845	1822
Nashville, Tenn.	36	07	1934	465	1253
Stillwater, Okla.	36	08	2196	775	1467
Fresno, Calif.	36	46	2642	605	1670
Davis, Calif.	38	32	2642	576	1633
Washington, D.C.	38	56	1867	539	1234
Columbia, Mo.	38	57	2077	664	1388
Seabrook, N.J.	39	30	2007	535	1316
Grand Lake, Colo.	40	15	2362	613	1573
Salt Lake City, Utah	40	46	2192	443	1442
New York, N.Y.	40	46	1646	395	1054
Sayville, N.Y.	40	46	2066	498	1291
State College, Pa.	40	48	1845	424	1154
Lincoln, Nebr.	40	52	2052	613	1354
Upton, N.Y.	40	52	2081	487	1269
Cleveland, Ohio.	41	24	2214	406	1312
Newport, R.I.	41	30	1963	524	1258
Put-in-Bay, Ohio.	41	39	2092	399	1242
East Wareham, Mass.	41	46	1897	542	1204
Blue Hill, Mass.	42	13	1911	498	1243
Boston, Mass.	42	21	1823	406	1110
Medford, Ore.	42	22	2590	362	1575
Ithaca, N.Y.	42	27	1867	391	1119
Twin Falls, Idaho.	42	33	2303	450	1438
East Lansing, Mich.	42	42	1638	343	998
Madison, Wis.	43	05	1904	443	1218
Toronto, Ont., Canada	43	40	1926	347	1078
St. Cloud, Minn.	45	35	2066	561	1352
Caribou, Maine.	46	52	1963	391	1227
Spokane, Wash.	47	37	2269	280	1374
Seattle, Wash.	47	39	2280	229	1160
Glasgow, Mont.	48	11	2494	450	1455
Winnipeg, Man., Canada	49	53	2048	365	1216

SOURCE: I. F. Hand, Distribution of Solar Energy over the United States, *Heating and Ventilating*, 50:73–75, July, 1953.

For quick estimation of hourly solar radiation on clear days on horizontal and vertical surfaces, hourly charts are given in Fig. 57-10. These are from Hand [21], converted to English units. It is to be noted that the main curves for horizontal and vertical irradiation include diffuse radiation from the sky, but the diffuse is *not* included in the outer curve of beam radiation on an area normal to the sun's rays. This "full sunshine" curve is needed when calculating irradiation on sloped surfaces. The dashed line near the bottom gives the clear-sky diffuse radiation separately for a horizontal surface. Solar irradiation of a west wall is considered as the opposite to

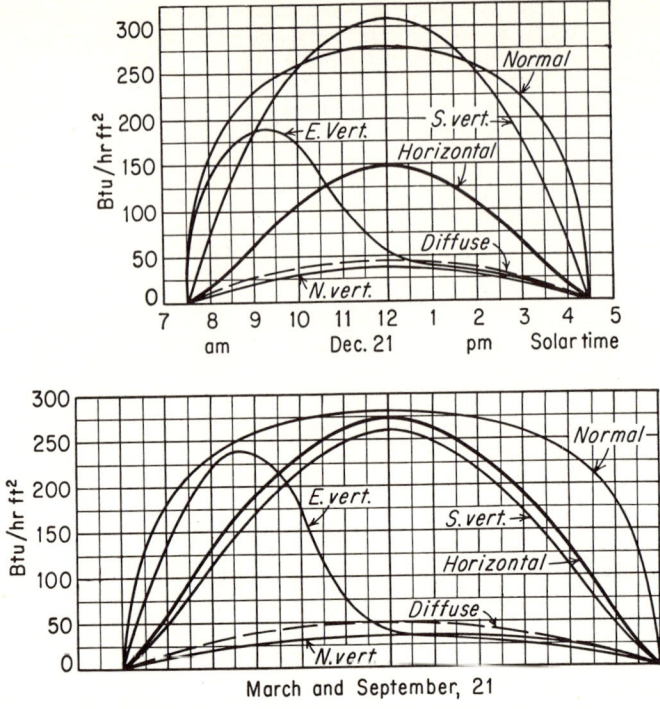

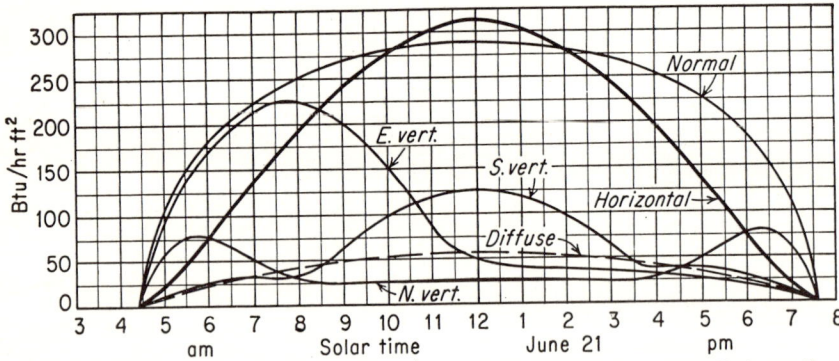

Fig. 57-10. Measured, clear-sky solar radiation on horizontal, vertical, and perpendicular surfaces. The "normal" curve does not include diffuse sky radiation. (*Converted to English units from Hand, Heating and Ventilating, January, 1950.*)

the curve given for an east wall. Irradiation on south walls in other latitudes is separately treated by Hand [19]. For vertical walls some reflected radiation, especially if the ground is covered with snow, should be added. For more specific treatment of beam sunshine by sun's altitude, see later paragraphs.

Standard Segments of the Solar Spectrum. Solar energy is only the short-wave portion of the total incoming radiation. The long-wave radiation from the earth's atmosphere adds 25 to 50 per cent, but this is more than counteracted by outgoing long-wave radiation from the ground, because its temperature is usually 20 to 50°F hotter than the equivalent temperature of the clear sky, excluding the sun. Therefore the whole-spectrum, net radiation exchange is *less* than the incoming solar radiation, and our general concept in cloudless weather should be of direct beam sunshine coming at a specific angle, plus diffuse sky radiation, *minus* diffuse long-wave radiation to the whole sky. This rather complex radiation exchange is measurable as a whole with an uncovered black radiometer. The three main portions can be determined separately because of the nearly complete separation of short- and long-wave spectral distributions [10] shown in Fig. 57-11. The diffuse short-wave sky radiation is measured separately by shading the glass-covered Eppley pyrheliometer. Then the total energies in the three spectral distributions can be used to estimate the spectral distri-

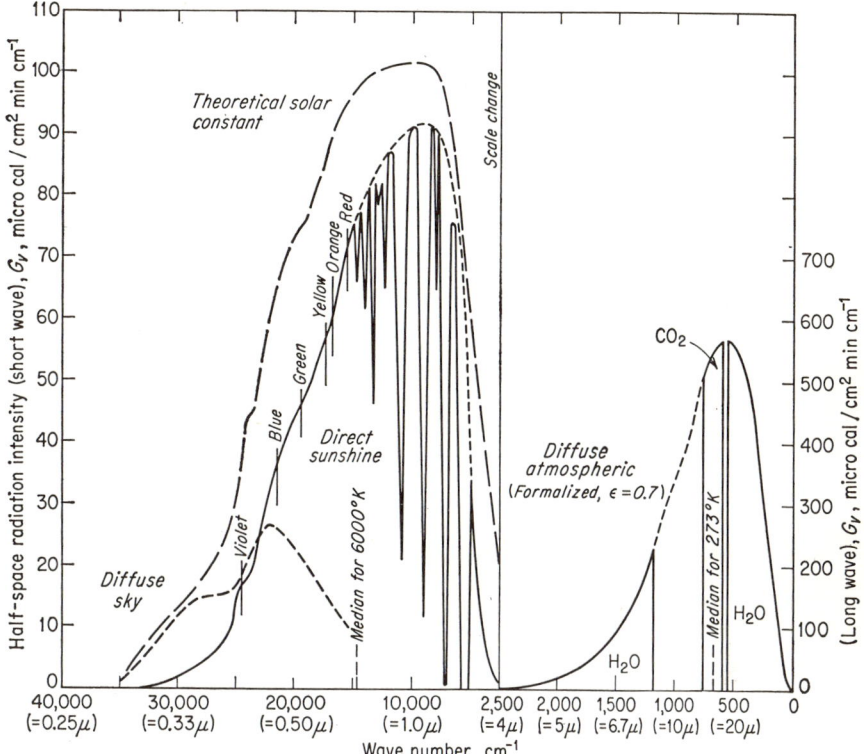

FIG. 57-11. Spectral distribution of all radiation downward. (*Brooks, "Solar Energy Research," 1954, courtesy of University of Wisconsin Press.*)

bution of irradiation. For an explanation of the spectral position of peak intensity and median line, see Hottel [27].

In the short-wave portion of Fig. 57-11 the upper curve shows the spectral distribution of energy per unit wave number in direct beam sunshine outside the earth's atmosphere as given by Johnson [28]. The lower curve is the spectrum after passing through the earth's atmosphere with the sun at an altitude of 30° as recommended by Moon [34] and adopted by the American Society of Heating and Air Conditioning Engineers. It is to be noted that although the spectrum is arranged with longer wavelengths to the right, the abscissa scale is in units of wave number to represent frequency instead of wavelength. This presentation suits the quantum definition $h\nu$ (which is basic for all photochemical reactions including eyesight), makes the shape of absorption lines symmetrical, and expands the spectrum in the short-wave region where more detail is of interest.

To meet the needs of plant scientists for interpreting photochemical responses which vary with wavelength, the Dutch Committee on Plant Irradiation [51] recommends the use of the following segments of the solar spectrum:

First band: Greater than 1.000 micron. "No specific effects of this radiation are known. It is acceptable that this radiation, as far as it is absorbed by the plant, is transformed into heat without interference of biochemical processes."

Second band: 1.000 to 0.700* micron. "This is the region of specific elongating effect on plants. Although the spectral region of elongating effect does not coincide precisely with the limits of this band, one may provisionally accept that the radiant flux in this band is an adequate measure of the elongating activity of the radiation."

TABLE 57-3. ENERGY DISTRIBUTION IN THE SOLAR SPECTRUM

Regions	Spectral segments			Energy in spectral segments			
	"Bands" and "windows"*	Interval boundaries			Johnson's solar constant outside atmosphere	Moon's recommended sea-level direct beam sunshine, sun's altitude 30° ($m = 2$)	
		Wave-length, microns	Wave number, cm^{-1}	Cumulative %	Cal/(sq cm) (min) in segment	Cal/(sq cm) (min) in segment	Foot-candles in segment
Ultraviolet, C	8th	0.0	∞	0.51	0.010	0.000	0
B	7th	0.280	35,713	1.97	0.029	0.000	0
A	6th	0.315	31,746	9.03	0.141	0.028	0
Visible, blue	5th	0.400	25,000	24.9	0.318	0.155	884
Green-yellow	4th	0.510	19,608	38.4	0.270	0.169	6,170
Red	3d	0.610	16,393	48.8	0.208	0.147	950
Far red	I	0.700†	14,286	66.5	0.354	0.249	4
Infrared	II	0.92	10,870	76.7	0.204	0.138	0
	III	1.12	8,929	85.5	0.176	0.092	0
	IV	1.40	7,143	93.02	0.150	0.070	0
	V–VIII	1.9	5,263	100.00	0.140	0.013	0
		∞	0.0				
Total	....	....	....	....	2.000	1.061	8,008‡

* Elder and Strong, "The Infrared Transmission of Atmospheric Windows," *J. Franklin Inst.*, 255(3):189–208' March, 1953.
† Being considered now by Dutch Committee on Plant Irradiation instead of 0.720.
‡ Average daily luminous efficiency in clear, natural atmospheres in Africa varies from 99 to 108 lumens per watt, namely, from 6,700 to 7,300 foot-candles per cal/(cm^2)(min) of whole-spectrum insolation (see Drummond [57]).

Third band: 0.700* to 0.610 micron. "This is almost the spectral region of the strongest absorption of chlorophyll and of the strongest photosynthetic activity in the red region. In many cases it also shows the strongest photoperiodic activity."

Fourth band: 0.610 to 0.510 micron. "This is a spectral region of low photosynthetic effectiveness in the green and of weak formative activity."

Fifth band: 0.510 to 0.400 micron. "This is virtually the region of strong chlorophyll absorption and absorption by yellow pigments. It is also a region of strong photosynthetic activity in the blue-violet and of strong formative effects."

Such energy segments should be used for determining photochemical efficiencies when comparing systems using artificial light with those using sunlight. The infrared

* Now being considered in place of 0.72.

windows I to VIII are bounded by strong water-vapor absorption band centers. Table 57-3 gives the energy fraction for each segment.

Sun's Hourly Position. In order to evaluate the solar energy received by sloped surfaces or by partially shaded surfaces, it is often necessary to determine the position of the sun. A very clear device for determining the sun's position and its angle of incidence and profile angles for windows facing any direction is obtainable from the Libbey-Owens-Ford Glass Co. [1] for every 4°N latitude. Similar charts of solar altitude and azimuth are given in the Smithsonian [43] Meteorological Table 170. The curves are for every 5° of latitude from 20 to 70° (and also for 0, 10, 80, and 90°) and for every 5° declination.

In many instances, shading by surrounding buildings, trees, or mountains is of importance. A valuable technique for determining the extent of shading by surroundings developed by Pleijel [37] consists of taking a vertical photograph directed downward on a convex mirror reflecting the entire sky hemisphere as seen from the given

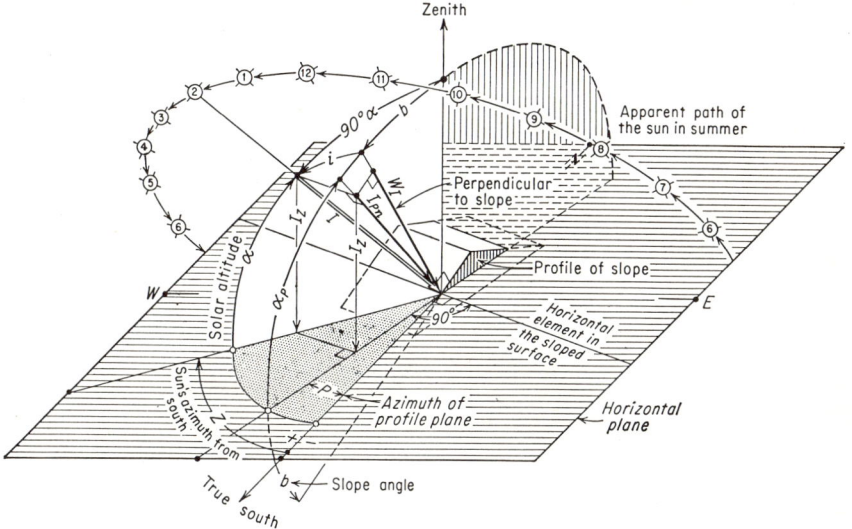

FIG. 57-12. Diagram of solar angles, azimuths, and slope.

spot. Azimuth angles can be shown around the edge of the hemisphere, and by superposing the polar-coordinate diagram of the sun's hourly position described above, the exact duration of shading by trees or building is seen vividly for any month.

The following formulas, based on Fig. 57-12, give the position of the sun's center when latitude, declination (seasonal), and time are specified:

$$\sin \alpha = \cos L \cos \delta \cos H - \sin L \sin \delta \qquad (57\text{-}2)$$

where α = sun's altitude above horizon, deg
 L = north latitude of place, deg
 δ = seasonal declination of sun, deg (Table 57-4)
 H = hour angle, deg (equals 15 times number of hours from solar noon, positive from 12 noon to 12 midnight)[1]

$$\alpha_{\text{noon}} = 90 - (L - \delta) \qquad (57\text{-}3)$$

$$\cot Z = \frac{\sin L \cos H - \cos L \tan \delta}{\sin H} \qquad (57\text{-}4a)$$

---

[1] Solar noon varies as much as ±15 min from mean solar noon. See Table 57-4.

where Z = azimuth angle of sun from due south, deg.

$$\sin Z = \frac{\cos \delta \sin H}{\cos \alpha} \qquad (57\text{-}4b)$$

$$\cos H_{\text{sunset}} = -\tan L \tan \delta \qquad (57\text{-}5)$$

$$\cos Z_{\text{sunset}} = \frac{-\sin \delta}{\cos L} \qquad (57\text{-}6)$$

Sun's Angle of Incidence on Horizontal, Sloped, and Vertical Surface. Knowing the position of the sun, or else the latitude, date, and time, it is possible to determine the angle of incidence of sunshine on any surface, and from that the instant rate of solar-energy input. Time integration is necessary to secure the input over a given time interval.

Basic formulas for angle of incidence i are as follows, again referring to Fig. 57-12.

When the sun's altitude and azimuth angles are known,[1] the angle of incidence for any sloped surface, as shown by Heywood [25], is given by

$$\cos i = \sin b \cos \alpha \cos (Z - P) + \cos b \sin \alpha \qquad (57\text{-}7)$$

where b = angle of slope, deg (positive if downslope is south, negative if north), also zenith angle of perpendicular to slope
P = azimuth angle of profile plane from due south, deg

For south-facing slopes ($P = 0$),

$$\cos i = \sin b \cos \alpha \cos Z + \cos b \sin \alpha \qquad (57\text{-}8a)$$

For east-facing slopes ($P = -90°$),

$$\cos i = \sin b \cos \alpha \sin Z + \cos b \sin \alpha \qquad (57\text{-}8b)$$

For vertical walls ($b = 90°$),

$$\cos i = \cos \alpha \cos (Z - P) \qquad (57\text{-}8c)$$

When solar radiation or shadows are to be calculated for any slope or wall, the sun's profile angle α_P, the altitude (with respect to a level surface) of the sunshine component directly facing the slope, is found from

$$\tan \alpha_P = \frac{\tan \alpha}{\cos (Z - P)} \qquad (57\text{-}9)$$

The hour angle H_{ob} for the beginning of irradiation on any sloped surface in the morning and ending in the evening can be determined from Eq. (57-10a), letting $i = 90°$.

[1] Knowing latitude, declination, and hour angle, the angle of incidence useful for time integration is

$$\cos i = \cos \delta \cos H (\cos P \sin L \sin b + \cos L \cos b) + \sin P \sin b \cos \delta \sin H$$
$$+ \sin \delta (\sin L \cos b - \cos P \cos L \sin b) \qquad (57\text{-}10a)$$

For surfaces sloped north or south ($P = 0$),

$$\cos i = \cos \delta \cos H \cos (L - b) + \sin \delta \sin (L - b) \qquad (57\text{-}10b)$$

For north-south slopes ($P = 0$),

$$\cos H_{ob} = -\tan \delta \tan (L - b) \qquad (57\text{-}11a)$$

where H_{ob} = hour angle in plane of slope, deg.

For vertical walls ($b = 90$),

$$\cos H_{ob} = \tan \delta \cot L - \frac{\sin H_{ob} \tan P}{\sin L} \qquad (57\text{-}11b)$$

From Eq. (57-4a), the end-point azimuth angle in plane of slope Z_{ob} is given by

$$\cot Z_{ob} = \frac{\sin L \cos H_{ob} - \cos L \tan \delta}{\sin H_{ob}} \qquad (57\text{-}12)$$

Direct-beam-sunshine Intensity. Direct-beam solar energy (on a surface perpendicular to the rays) can be estimated at all latitudes by considering the variation of its intensity with the altitude angle of the sun. The longer the path through the earth's atmosphere, the greater the depletion. The length of the optical air-mass path m is given in equivalent thicknesses of the earth's atmosphere; i.e., $m = 2$ for $\alpha = 30°$. The intensity values in Table 57-5 for standard and industrial atmospheres are published by the American Society of Heating and Air Conditioning Engineers [3], and the values for a horizontal surface under complete cloud cover are converted from Haurwitz [23].

The intensity of sunshine is depleted by dry air, water vapor, and dust. Outside the atmosphere, solar radiation is set by Johnson [28] as 2.00 cal/(sq cm)(min). At the earth's surface, Moon [34] recommends 420 Btu/(hr)(sq ft) [= 1.90 cal/(sq cm)(min)]. It seems advisable to continue using the 420 figure to allow for ultraviolet and other losses not dependent on path length m. On this basis, incident solar energy received on a surface perpendicular to the sun's rays can be found as follows:

$$I, \text{Btu/(hr)(sq ft)} = T_{pwd} \frac{420}{r_v^2} \qquad (57\text{-}13)$$

where r_v = sun's distance ratio (Table 57-4).

The over-all coefficient of transmittance T_{pwd} is integrated from Moon's monochromatic coefficients for depletion. The whole solar-spectrum transmittance can be expressed approximately by the product of three transmittances: (1) dry air, T_p, (2) water vapor, T_w, and (3) dust or aerosols, T_d:

$$T_{pwd} = T_p T_w T_d = [0.915^{\left(\frac{pm}{1,013}\right)^{0.75}}][0.84^{\left(\frac{wm}{20}\right)^{0.60}}][0.92^{(dm)^{0.9}}] \qquad (57\text{-}14a*)$$

where m = optical path length ($m = 2$ for $\alpha = 30°$)

p = local atmospheric pressure, millibars (1 mb = 1,000 dynes/sq cm and standard atmosphere, 29.92 in. Hg = 1,013.25 mb)

w = total precipitable water in atmosphere, mm (in continental U.S. midafternoon, $w \cong 1.88e$ in winter, and $w \cong 1.58e$ in summer, where e is vapor pressure, mb) [29]

d = dust factor

Figure 57-13 is a nomogram of Eq. (57-14a) for estimating atmospheric transmittance for direct beam sunshine. The main family of curves at the left is for dry air at sea level combined with various precipitable moisture contents. The small correction for other altitudes is given at the top. This increment is to be added to the moisture curves acting to increase transmittance for higher altitudes. Approximate average values for a nonindustrial sea-level area, 42°N, and cloudless sky (suitable for Hand's curves; see Fig. 57-10) are shown in Table 57-4.

Diffuse Solar Radiation. The incoming, diffuse short-wave sky radiation on a horizontal surface is also a function of solar altitude and air-mass quality. The American Society of Heating and Air-conditioning Engineers magnitudes for two types of clear atmosphere are included in Table 57-5. On moderate slopes, full horizontal intensity can be assumed, and on vertical walls, roughly two-thirds the horizontal rate can be assumed, except on walls directly facing the sun, because reflection from the ground can be high, the reflectivity ranging up to 0.40 for plain concrete pavements.

For a general concept Kimball [30] states: " . . . about half the radiation lost from incoming (direct) rays through scattering is finally received at the ground as diffuse radiation from the sky." This amounts to roughly 40 per cent of the entire depletion of insolation on a horizontal surface. The 1:6 ratio usually cited for diffuse to total received insolation is only for a nearly dust-free and rather moist atmosphere

---

*The equivalent exponential formula is

$$T_{pwd} = e^{\left[-0.089\left(\frac{pm}{1,013}\right)^{0.75} - 0.174\left(\frac{wm}{20}\right)^{0.60} - 0.083(dm)^{0.9}\right]} \qquad (57\text{-}14b)$$

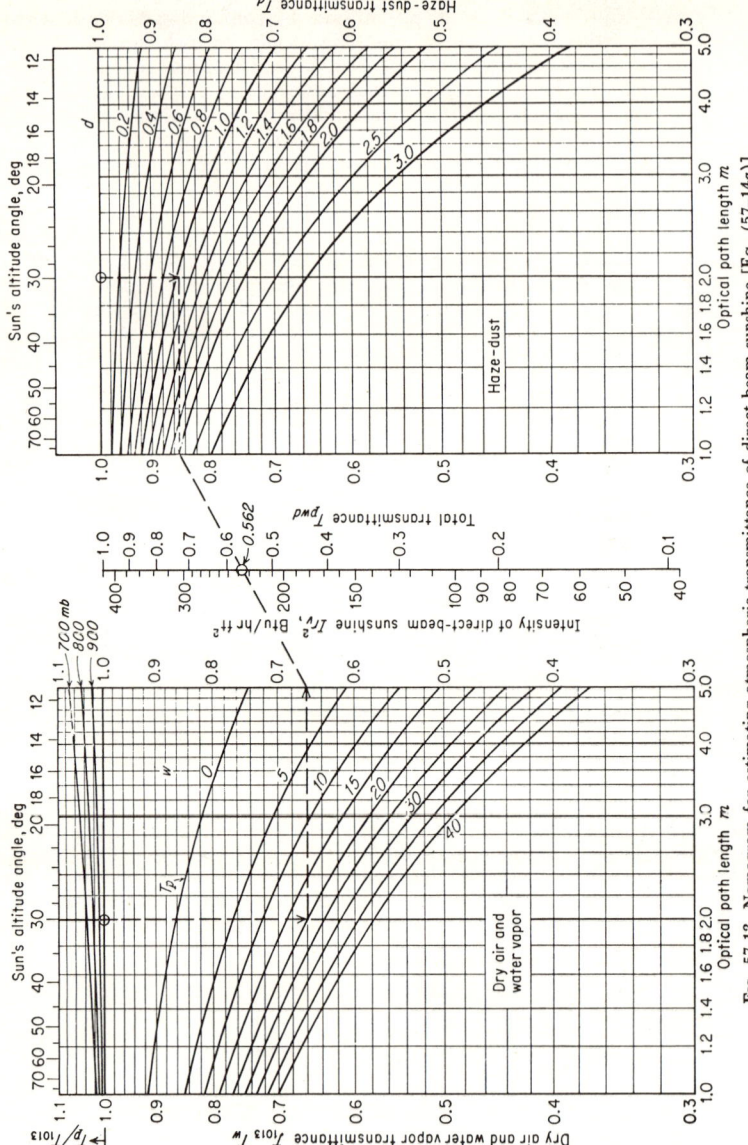

Fig. 57-13. Nomogram for estimating atmospheric transmittance of direct beam sunshine [Eq. (57–14a)].

TABLE 57-4. MONTHLY CONSTANTS FOR SUN'S POSITION, DISTANCE RATIO, TIME CORRECTION, AND APPROXIMATE TRANSPARENCY COEFFICIENTS (NONINDUSTRIAL, CLOUDLESS SKY, LAT. 42°N)

Date (1948)	Sun's distance ratio squared, r_v^2, nondim.	Declination, δ, deg	Time correction applied to mean solar time (std. time − 4 min per deg long. from std.) to obtain true solar time, min	Hour angle, noon to sunset, H_o, deg	Optical air path at median irradiation on surface sloped south 42°, m_m, nondim.	Approximate average values interpolated from Hand*		Approximate total, sea-level coefficient of transmittance for median m_m, T_{pwd}, nondim.
						w_m, mm	d_m, nondim.	
Jan. 21	0.9682	−20.2	−11.2	70.65	2.60	6	0.77	0.616
Feb. 20	0.9777	−11.2	−13.9†	79.73	1.98	9	0.93	0.635
Mar. 21	0.9925	0.0	− 7.5	90.00	1.55	14	1.04	0.648
Apr. 20	1.0094	+11.2	+ 1.1	100.27	1.31	16	1.08	0.669
May 21	1.0245	+20.2	+ 3.3	109.35	1.20	16.8	1.10	0.680
June 22	1.0330	+23.45	− 1.4	112.98	1.17	17	1.11	0.686
July 23	1.0321	+20.2	− 6.2	109.35	1.20	16.8	1.10	0.680
Aug. 24	1.0220	+11.2	− 2.4	100.27	1.31	16	1.08	0.669
Sept. 23	1.0065	0.0	+ 7.5	90.00	1.55	14	1.04	0.648
Oct. 23	0.9900	−11.2	+15.4‡	79.73	1.98	9	0.93	0.635
Nov. 23	0.9752	−20.2	+13.8	70.65	2.60	6	0.77	0.616
Dec. 22	0.9676	−23.45	+ 1.6	67.00	2.93	5	0.72	0.604
Spectrographic (from Moon [34])	......	....	....	......	2.00	20	1.05	0.560

* Varies hourly.
† Solar noon late.
‡ Solar noon early.

and does not apply at low solar altitudes. Klein [31] shows that reflection of insolation upward from the ground increases the downward diffuse radiation.

Combined Direct and Diffuse Solar Radiation on Slopes. Excellent diagrams, tables, and explanation of the effects of orientation and slopes on the amount of solar energy received daily are given by Orendorff [35]. For the New York–New Jersey area at latitude 40°N the average daily solar energy, both direct and diffuse, has been calculated for typical monthly cloud cover. Table 57-6 gives the results for south slopes of various degrees. This shows that the optimum slope for using solar energy in the winter should be nearly 15° steeper and for maximum summer utilization about 15° flatter than the latitude angle. The table also includes north slopes up to 30° to permit comparison of radiation on sloped ground, which leads to great differences in spot climate. Such tables should be developed for every major agricultural area, and in addition some indicator of probable departures from the average solar radiation is needed.

Radiant-energy Exchange. The amount of solar energy plus atmospheric radiation absorbed less the long-wave radiation emitted by a surface is the primary climatological factor. The partition of this net radiation at the ground or in foliage depends on rates of heat transfer by convection, evaporation, and conduction [12]. But since all these rates depend on surface temperature, the basic effect of net radiation is seen most clearly in the seasonal and diurnal cycles of soil temperature. The characteristic effects of daytime excess incoming radiation are well treated by Geiger [16] particularly in the temperature profiles in a stand of grain (shown in his Fig. 131). The characteristic effect of nocturnal heat balance on soil temperatures in a radiation frost is shown in Fig. 45-1.

The range in wavelength for solar radiation at the earth's surface is from 0.29 to about 4 microns, and except for red colors, the visual brightness of the surface is a fair indicator of total short-wave reflection. Incoming atmospheric radiation (long-wave) depends mostly on atmospheric moisture and on temperature, which is closely related to air temperature near the ground, as explained briefly in a later paragraph. All surfaces, however, lose energy by long-wave radiation, which cannot be judged by

TABLE 57-5. VALUES OF PERPENDICULAR AND DIFFUSE SOLAR RADIATION FOR CLEAR AND FOR INDUSTRIAL ATMOSPHERES* AND AVERAGE TOTAL INSOLATION THROUGH COMPLETE OVERCASTS OF VARIOUS CLOUD TYPES†

Solar altitude, α, deg	Optical air-mass path,‡ $m \sim \csc \alpha$	Standard, cloudless atmosphere			Industrial, cloudless atmosphere			Through complete overcasts, Blue Hill			
		Direct, perpendicular radiation, I, Btu/(hr)(sq ft)	Diffuse on horizontal, Btu/(hr)(sq ft)	Total on horizontal, Btu/(hr)(sq ft)	Direct, perpendicular radiation, I, Btu/(hr)(sq ft)	Diffuse on horizontal, Btu/(hr)(sq ft)	Total on horizontal, Btu/(hr)(sq ft)	Cirrostratus	Altocumulus	Stratocumulus	Fog
								Average total insolation on horizontal Btu/(hr)(sq ft)			
5	10.39	67	7	13	34	9	12	...	...	15	10
10	5.60	123	14	35	58	18	28	50	35	25	15
15	3.82	166	19	62	80	24	45	70	50	35	20
20	2.90	197	23	90	103	31	64	95	65	40	20
25	2.36	218	26	118	121	38	89	120	75	50	25
30	2.00	235	28	146	136	44	112	145	90	60	30
35	1.74	248	30	172	148	48	133	165	105	70	35
40	1.55	258	31	197	158	52	154	185	115	80	40
45	1.41	266	32	220	165	55	172	205	130	85	40
50	1.30	273	33	242	172	58	190	235	150	100	45
60	1.15	283	34	279	181	63	220	260	160	110	50
70	1.06	289	35	307	188	69	246				
80	1.02	292	(35)	(322)	195						
90	1.00	294	(36)	(328)	200 §						

* "Heating, Ventilating, and Air Conditioning Guide," American Society of Heating and Air Conditioning Engineers, New York, 1956, p. 286.
† B. Haurwitz, Insolation in Relation to Cloud Type, *J. Meteorol.*, 5:110–113, June, 1948.
‡ "Smithsonian Meteorological Tables," 6th ed., Smithsonian Institution Miscellaneous Collections, 1951, p. 422.
§ 192 would be more consistent with the curve from 70° down.

TABLE 57-6. TOTAL AVERAGE DAILY DIRECT AND DIFFUSE SOLAR RADIATION, BTU/SQ FT, INCIDENT ON VARIOUS SLOPES IN THE NEW YORK–NEW JERSEY AREA, LAT. 40°N

Slope	Radiation	Dec. 21	Jan. 21, Nov. 21	Feb. 21, Oct. 21	Mar. 21, Sept. 21	Apr. 21, Aug. 21	May 21, July 21	June 21	Annual average, Btu/(sq ft) (day)
30°N	Total	161	178	347, 358	633, 703	1094, 1144	1510, 1527	1619	773
	Diffuse	161	178	244	323	466	584	618	
20°N	Total	216	260, 265	496, 524	803, 901	1281, 1341	1663, 1682	1773	934
	Diffuse	169	190	254	362	515	616	656	
10°N	Total	342	394, 408	660, 702	965, 1094	1418, 1486	1760, 1781	1842	1071
	Diffuse	191	216	287	390	541	628	656	
Horizontal	Total	463	512, 636	809, 866	1105, 1258	1546, 1622	1851, 1873	1920	1195
	Diffuse	202	224	304	423	583	671	701	
10°S	Total	583	634, 666	952, 1022	1236, 1409	1618, 1698	1848, 1869	1912	1287
	Diffuse	220	242	328	465	599	685	712	
20°S	Total	695	741, 780	1079, 1160	1318, 1506	1680, 1762	1871, 1893	1898	1365
	Diffuse	239	260	362	481	632	693	710	
30°S	Total	790	831, 876	1177, 1266	1371, 1566	1668, 1749	1800, 1821	1806	1394
	Diffuse	258	279	385	499	634	671	693	
45°S	Total	914	947, 998	1205, 1296	1386, 1583	1622, 1702	1613, 1631	1586	1374
	Diffuse	298	310	398	507	607	631	628	
60°S	Total	985	990, 1045	1251, 1347	1317, 1505	1394, 1460	1351, 1365	1308	1195
	Diffuse	326	320	402	481	554	564	562	
75°S	Total	968	975, 1028	1177, 1267	1164, 1325	1133, 1184	1012, 1022	966	1102
	Diffuse	314	316	384	444	484	480	486	
Vertical S	Total	904	894, 944	1037, 1116	941, 1066	847, 881	674, 679	646	886
	Diffuse	291	289	341	383	424	421	430	

SOURCE: J. H. Orendorff, "Application of Climatic Data to House Design," U.S. Housing and Home Finance Agency, Washington, D.C., January, 1954.

eye. The essential gain or loss by net radiation depends largely, therefore, on the unseen magnitude of the "thermal" radiation. The emissive power of a surface depends on the 4th power of its absolute temperature and the emittance of the surface. Table 57-7 presents characteristic short-wave absorptances and long-wave emittances essential for determining the radiant-energy exchange. For surfaces outdoors the short-wave absorptance applies as a direct modifier of the intensity of incident solar energy as discussed in the foregoing paragraphs. Similarly, the long-wave emittance ε applies as a direct modifier for outgoing hemispherical radiation, usually expressed as

$$\frac{q}{A} = 0.173\epsilon \left(\frac{T}{100}\right)^4 \quad \text{Btu/(hr)(sq ft)} \tag{57-15}$$

where q = quantity of heat, Btu/hr
A = area, sq ft
T = absolute temperature, °R

Full treatment of thermal-radiation exchange between surfaces facing each other in various aspects and of heat transfer by conduction and convection is given in McAdams [27] and other texts.

Upward Reflection; Change of Albedo with Sun's Altitude. Upward reflection of direct and diffuse solar radiation and the upward thermal radiation from the ground on a hot day may be greater under a shade than the total downward radiation, even including the emissive power of the underside of the shade. The long-wave radiation

TABLE 57-7. SOLAR ABSORPTANCE (1.0 − ALBEDO) AND LONG-WAVE EMITTANCE OF VARIOUS SURFACES

Surface	Solar absorptance (0.3–2.5 microns)	Long-wave emittance ε (2.5 microns up)	References and page* (1)	(2)
"Hohlraum," theoretical perfectly black body	1.00	0.99+		h:56
Magnesium carbonate, MgCO₃ (white reference, solid)	0.04	0.79†	a:553	e:72
Water (1.0 − single-surface reflectance, i = 60°)	0.94	0.95–0.96	b:444	c:478
Sheet ice with sparse snow cover	0.31	0.96–0.97	b:443	e:103
Snow, ice granules (approx. 1/32 in. diam)	0.33 calc.	0.89	e:215	e:93
Freshest and brightest of fine particles like frost	0.13	0.82	b:443	e:93
Frozen soil		0.93–0.94		f:4
Dry playa sand, Monterey powdered	0.45†	0.84‡	k:585	e:69
Desert surface (see also Table 57-8)	0.75 } 0.82 }	approx. 0.90	b; see k b:442	
Sand, dry	0.91		b:443	
Wet	0.88–0.91 }	approx. 0.95	b:442	
Moist ground, 70–95% bare				
Ground, dry-plowed	0.75–0.80		b:443	
Grass, high and dry	0.67–0.69 }	approx. 0.9	b:443	
Common vegetable fields and shrubs	0.72–0.76 }		i:207	
Wilted	0.70		i:207	
Oak leaves (1.0 − reflectance, at 0.6 and 3 microns)	0.71–0.78	0.91–0.95	j:323	j:309
Alfalfa, dark green	0.97†	(0.95)	k:584	
Oak woodland	0.82 }	approx. 0.9	i:205	
Pine forest	0.86 }		i:205	
Paper, white	0.25–0.28	0.95	j:324	a:555
Plaster, white	0.07	0.91	j:288	c:478
Bricks, red	0.55	0.92	c:62	c:62
Concrete	0.60	0.88	c:62	c:62
Asbestos slate	0.81	0.96	c:62	c:62
Linoleum, red-brown	0.84	0.92	c:62	c:62
Wood, planed oak		0.90		c:477
Glass pane, absorptance and transmittance, i = 35°	0.90§	0.94	a:549	c:477
White paint (0.017 in. on aluminum)	0.20	0.91	g:73	e:94
Black paint (0.017 in. on aluminum)	0.94–0.98	0.88	m	e:94
Aluminum paint, bright and new	0.20	0.43	c:62	c:62
Aluminum foil	0.15	0.01–0.05	l:289	f:5; g:44
Aluminum commercial finish (at 0.6 micron), new	0.32	0.10‡	j:321	g:43
Galvanized iron, clean and new	0.65	0.13‡	a:553	g:46
Galvanized sheet iron, gray oxidized	(0.8)	0.28		c:476

* Column 1 applies to solar absorptance and column 2 to long-wave emittance.
† Calculated from spectral reflectance, assuming Moon's standard solar-energy spectrum for airpath = 2.0.
‡ Calculated from spectral reflectance, assuming that the 15-micron determination applies at all long wavelengths.
§ Absorption of solar energy in double-strength pane is approximately 4 per cent.

Emissivity References

a. Forsythe, W. E. (ed.): "Smithsonian Physical Tables," 9th rev. ed., Smithsonian Institution Miscellaneous Collections, vol. 120, 1954.
b. List, R. J. (ed.): "Smithsonian Meteorological Tables," 6th rev. ed., Smithsonian Institution Miscellaneous Collections, vol. 114, 1951.
c. Hottel, H. C.: in W. H. McAdams, "Heat Transmission," 3d ed., McGraw-Hill Book Company, Inc., New York, 1954.
d. Dunkle, R. V., and J. T. Bevans: An Approximate Analysis of the Solar Reflectance and Transmittance of a Snow Cover, *J. Meteorol.*, 13(2): 212-216, April, 1956.
e. Dunkle, R. V., J. T. Gier, and J. T. Bevans: "Final Progress Report, Snow Characteristics Project," University of California, Institute of Engineering Research, Berkeley, Calif., ser. 62, no. 5, Contract DA-11-190-ENG-3, Aug. 31, 1955.
f. Dunkle, R. V., J. T. Gier, et al.: "The Snow Emissivity Meter and Its Use," University of California, Institute of Engineering Research, Berkeley, Calif., ser. 62, no. 3, Contract DA-11-190-ENG-3, Dec. 15, 1953.
g. Dunkle, R. V., J. T. Gier, et al.: "Progress Report for the Year Ending June 27, 1953," University of California, Institute of Engineering Research, Berkeley, Calif., ser. 62, no. 1, Contract DA-11-190-ENG-3, June 27, 1953.
h. Dunkle, R. V., J. T. Gier, et al.: "Final Report, Thermal Radiation Project," University of California, Dept. of Engineering, Berkeley, Calif., Contract N7-onr-295, Task I, Sept. 1, 1950.
i. Thornthwaite, C. W.: in J. Mather (ed.), The Measurement of Evapotranspiration, *Johns Hopkins Univ. Publ. in Climatology*, vol. 7, no. 1, 1954.
j. Coblentz, W. W.: The Diffuse Reflecting Power of Various Substances, *Bur. Standards (U.S.) Bull.*, 9(2):283–325, April, 1913.
k. Ashburn, E. V., and R. G. Weldon: Spectral Diffuse Reflectance of Desert Surfaces, *Optical Soc. Am. J.*, 46(8):583–586, August, 1956.
l. Moon, Parry: "The Scientific Basis of Illuminating Engineering," McGraw-Hill Book Company, Inc., New York, 1936.
m. Drummond, A. J.: personal communication, Dec. 29, 1953, Eppley Laboratory, Inc., Radiation Laboratory, Newport, R.I.

can be calculated using emittances such as indicated in Table 57-7. The solar-radiation part, however, varies with wavelength and with the sun's altitude. Table 57-8 presents the measurements of Ashburn and Weldon [4] using a photosphere as receiver over wind-blown sand with sparse small brush. The range of wavelengths 0.4 to 0.6 micron does not cover the whole solar spectrum, but applying interval energies recommended by Moon, approximate magnitudes of albedos can be determined as added at the bottom of Table 57-8. This shows conclusively that the albedo (propor-

TABLE 57-8. VARIATION (WITH WAVELENGTH AND WITH SUN'S ALTITUDE) IN MICRONS OF DIFFUSE REFLECTANCE OF DESERT SAND* (SCALED) AND ESTIMATED ALBEDO (ASSUMING THAT THE 0.65-MICRON REFLECTANCE APPLIES TO ALL LONGER WAVELENGTHS)

Wavelength, microns	Sun's altitude									
	2°	6°	10°	15°	20°	25°	30°	33°	47°	60°
0.40	0.09	0.11	0.13	0.13	0.12	0.12	0.12	0.12	0.11	0.11
0.45	0.14	0.17	0.17	0.16	0.15	0.14	0.15	0.14	0.14	0.14
0.50	0.18	0.23	0.23	0.22	0.18	0.17	0.16	0.16	0.16	0.16
0.55	0.24	0.32	0.34	0.34	0.29	0.24	0.21	0.20	0.20	0.20
0.60	0.32	0.40	0.44	0.42	0.35	0.28	0.25	0.24	0.23	0.24
0.65	0.32	0.39	0.44	0.44	0.37	0.32	0.28	0.28	0.27	0.27
Estimated albedo.....	0.28	0.35	0.38	0.38	0.32	0.28	0.25	0.25	0.24	0.24

* According to E. V. Ashburn and R. G. Weldon, Spectral Diffuse Reflectance of Desert Surfaces, *Optical Soc. Am. J.*, 46(8): 583–586, August, 1956.

tion of incoming solar energy which is reflected) is much higher at low solar altitudes, and this correction should be considered in using the solar absorptance magnitudes listed in Table 57-7.

Atmospheric Radiation. The net difference between outgoing and incoming long-wave radiation can be determined from radiation charts [9,14] if the moisture and temperature distributions for the whole depth of the atmosphere are known from soundings. Since the outgoing long-wave radiation depends on surface temperature and the incoming atmospheric radiation is also strongly dependent on air temperatures near the ground, both vary together day and night. Therefore the simplest concept for net long-wave exchange at the ground is that the atmosphere has a transparency (mostly in the spectral "window" between 8 and 14 microns; see Fig. 57-11), depending on the total moisture in the whole depth of the atmosphere. The long-wave transmittance at sea level ranges from about 0.30 for rather dry air to below 0.10 in moist air and to zero with low overcast. For a rough estimate based on early-afternoon vapor pressure, Goss [18] has found the following empirical constants for cloudless days in the Sacramento Valley. Slightly different constants should be expected for other localities, such as greater transparency with increasing altitude.

For clear nights and $6.7 \leq e \leq 5.6$,

$$\frac{R}{\sigma T^4} = 0.660 + 0.040 \sqrt{e} \qquad (57\text{-}16a)$$

For cloudless days,

$$\frac{R}{\sigma T^4} = 0.642 + 0.052 \sqrt{e} \qquad (57\text{-}16b)$$

where R = downcoming atmospheric radiation from hemisphere, Btu/(hr)(sq ft)
T = absolute temperature, °F
e = vapor pressure, mb
σ = Stefan-Boltzmann radiation coefficient

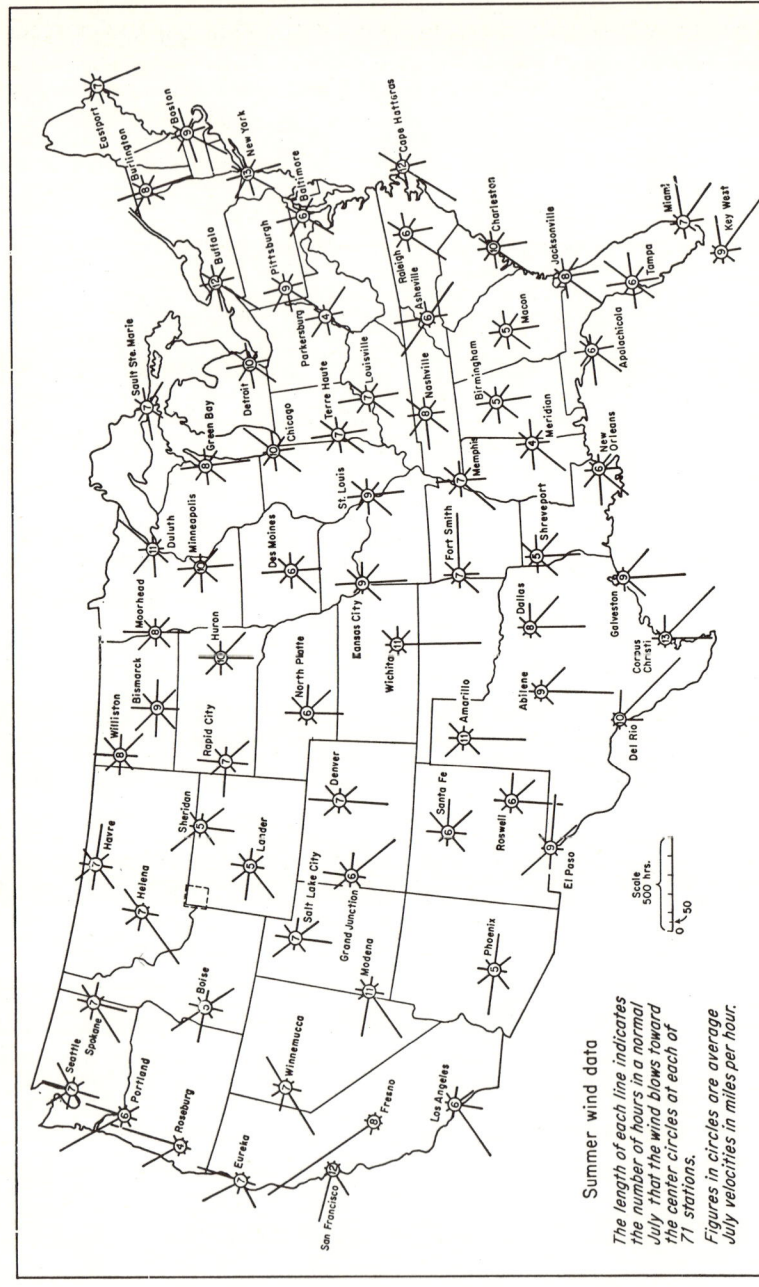

Fig. 57-14. Map of summer winds. (Albright, "Summer Weather Data," 1939, Courtesy of The Marley Co.)

SURFACE WIND AND PROBABILITIES OF MAXIMUM VELOCITIES

A considerable fraction of the energy in the thermodynamic "engine" of the atmosphere is in the kinetic energy of the wind. This natural power is mostly in the trade winds and in antarctic winds. In the middle latitudes cyclonic storms are the main sources of wind.

Golding [17] gives a thorough discussion of the characteristics of surface wind. In general, air flow is virtually frictionless at elevations above 2,000 or 3,000 ft, and

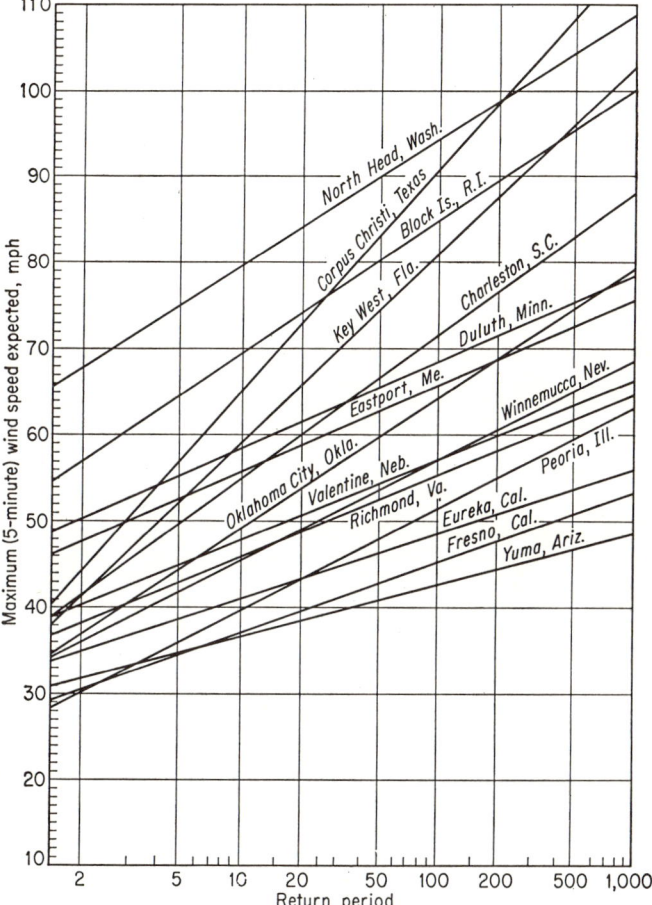

FIG. 57-15. Probable years for recurrence of destructive winds at various stations in the United States. (*Court, J. Franklin Inst., July,* 1953.)

because of the dynamics of the rotating earth, the direction of the gradient wind is not from high-pressure area toward low-pressure center but nearly at right angles, almost parallel to the isobars of the pressure pattern. The daytime wind near the earth's surface is primarily due to the drag of the overhead air on the surface air. With topographic roughness, however, the direction of surface wind tends to depart from the isobar direction 20 to 30° toward the high-to-low pressure-gradient direction. And in mountainous regions the surface-wind directions depend on reactions to the general flow over obstructions which can involve complete reversals, as in river eddies around rocks. The mean wind speeds for the United States in summer and frequency of direction are well presented by Albright [2] and are shown in Fig. 57-14.

At night the stability developed by cold, dense air, tending to stay close to the ground, acts to shield the ground air from a weak overhead wind. Therefore, with strong radiation chilling of foliage on clear nights, the usual air flow is due not so much to the overhead wind as to the gravitational tendency of cold air to flow downhill. On calm, sunny days the reverse often occurs, with warmed air flowing uphill. These gravity currents are a very important part of local climate.

Destructive winds are associated with closely spaced isobars, normally resulting in a straight blow of high velocity. Destruction can also come from convergence of two large air flows. The most frequent cause of high winds, however, is severe discontinuity along a cold front. The juxtaposition of two contrasting air masses, besides being a cause of strong pressure gradients, also creates a zone of great instability, along which the large energy available in a deep, moist air mass may occasionally develop tornadoes in which sharp pressure gradients, natural to the vortex whirl, produce winds of the highest known velocity. Destructive hurricanes (typhoons) are a much bigger fluid-dynamic vortex, originating over tropical oceans, but usually turning northward before dissipating their energy.

The standard wind-load design is based on a momentary velocity of 75 mph [45]. To design structures to withstand the maximum wind of any locality involves, however, not absolute magnitudes, but the probabilities of various high-velocity winds. This problem has been studied by Court [13] and Fig. 57-15 presents 15 of his 25 curves. These include hurricanes but exclude tornadoes. There is high correlation between maximum velocities and durations, which indicates that the 1-sec-gust velocities are 1.5 times the 5-min velocities treated in Fig. 57-15. Although wind force is proportional to the square of the velocity, its destructive power for flexible surfaces (flags, clothing) involves velocity cubed. To minimize structural damage in tornadoes, Reynolds [38] recommends vents of 1 sq ft per 1,000 cu ft and firm anchoring of buildings to foundations and of roofs to buildings.

SUMMER TEMPERATURES AND HUMIDITIES

Because summer temperature and humidity are of great importance to agriculture and for human comfort, the distribution of hourly dry-bulb temperatures excerpted from an important study by Albright [2] is given in Table 57-9. Many more tables are given by Albright, and a few examples of normal frequency curves for dry-bulb, wet-bulb, and dew-point temperatures. Designers needing more information than given in the above reference or in the Guide of the American Society of Heating and Air Conditioning Engineers should work with the hourly Weather Bureau data of the nearest major airport. Until such basic data are generalized in probability form, approximate estimates can be made by comparing average daily maximums with average daily means for July (United States maps, Figs. 57-20 and 57-21) and relative humidity (Fig. 57-23).

THORNTHWAITE'S EVAPO-TRANSPIRATION AND CLIMATIC MOISTURE INDEX

Thornthwaite's [46] potential evapo-transpiration is a standardized indicator for water need (see Chap. 34). Naturally, this varies from cold to warm climates, depending on how the net radiation energy is apportioned between conduction, convection, and evaporation. His examples, converted to English units for six areas, are shown in Fig. 57-16, relating mean monthly evapo-transpiration with mean monthly temperature based on 12 hr of possible sunshine per day. This figure can be used directly for making a rough estimate of potential evapo-transpiration from mean monthly temperatures at other localities by interpolation and correction for possible sunshine hours. More specific procedures are given in Mather [33]. Geographical rates of water use for various crops, such as those determined by Blaney and Criddle [7], are much more meaningful than the conventional figures of pounds of water per pound of dry crop because transpiration rates depend on wind and humidity, which are not directly related to plant growth.

TABLE 57-9. DISTRIBUTION OF TOTAL HOURS OF DRY-BULB TEMPERATURES WITHIN SPECIFIED CATEGORIES FOR 122 DAYS
(June to September)

State and city	Temperature categories, °F							
	60 and higher	60–69	70–79	80–89	90–94	95–99	100–104	105–109	110 and higher
Ala., Mobile	2925	98	1350	1269	205	3			
Ariz., Phoenix	2903	191	609	756	486	394	364	172	31
Ark., Little Rock	2875	249	1235	971	315	93	12		
Calif., Fresno	2712	672	686	610	260	254	178	48	4
Laguna Beach	2526	2124	390	12					
Oakland	1624	1202	326	82	14				
San Diego	2795	1705	974	116					
Williams	2468	635	630	551	257	202	130	58	5
Colo., Denver	2384	873	851	531	104	25			
Grand Junction	2684	784	848	698	248	92	14		
D.C., Washington	2809	695	1112	769	167	66			
Fla., Jacksonville	2926	113	1487	963	303	60			
Miami	2928	39	1303	1446	140				
Ga., Atlanta	2882	415	1348	834	222	56	7		
Idaho, Boise	2168	676	626	567	177	89	20	11	
Ill., Chicago	2570	1096	783	489	133	49	20		
Moline	2559	709	981	622	158	68	21		
Ind., Evansville	2798	570	1140	796	228	52	12		
Kans., Wichita	2701	447	835	735	300	186	121	72	5
La., New Orleans	2927	95	1343	1263	222	4			
Shreveport	2914	169	1187	986	402	146	24		
Mass., Boston	2574	1333	928	281	28	4			
Mich., Detroit	2520	862	1020	490	114	28	6		
Minn., St. Paul	2435	887	889	518	88	48	5		
Miss., Jackson	2880	261	1376	777	356	103	7		
Mo., Kansas City	2789	469	930	872	237	174	83	24	
Kirksville	2584	706	896	652	132	111	66	21	
St. Louis	2770	545	973	823	224	141	53	11	
Springfield	2862	494	1274	906	156	32			
Mont., Helena	1856	828	640	332	36	20			
Nebr., North Platte	2420	813	739	533	206	95	34		
Omaha	2667	610	961	714	197	119	54	12	
Nev., Elko	1871	551	580	510	155	66	9		
Reno	1870	592	568	450	180	70	10		
N.J., Camden	2726	912	1142	534	108	24	6		
Newark	2737	1006	1208	434	65	24			
N.Mex., Albuquerque	2695	864	869	700	222	40			
N.Y., Albany	2308	1093	861	307	39	8			
Buffalo	2188	972	880	304	32				
New York	2804	882	1325	511	61	22	3		
N.C., Raleigh	2899	469	1417	824	154	35			
Wilmington	2916	224	1652	962	76				
N.Dak., Fargo	2180	825	710	459	113	39	29	5	
Ohio, Cincinnati	2698	666	1120	689	145	56	19	3	
Cleveland	2570	859	1011	552	109	36	3		
Okla., Tulsa	2847	302	1014	913	281	209	112	16	
Ore., Portland	1949	1142	526	222	38	21			
Pa., Bellefonte	2820	469	1611	677	52	11			
Pittsburgh	2658	984	1047	532	72	23			
S.C., Charleston	2904	279	1718	718	165	24			
S.Dak., Huron	2471	826	816	542	155	121	11		
Rapid City	2327	846	730	510	148	93			
Tenn., Knoxville	2871	476	1314	855	200	22	4		
Memphis	2868	320	1259	875	283	109	22		
Tex., Amarillo	2817	710	910	745	329	113	10		
Brownsville	2928	35	1205	1483	201	4			
Dallas	2907	138	908	1041	436	305	76	3	
El Paso	2918	333	1056	904	415	199	11		
Houston	2923	131	1502	1010	226	50	4		
San Antonio	2927	76	1236	969	462	165	19		
Utah, Modena	2052	717	587	596	133	19			
Salt Lake City	2492	690	852	600	152	76	22		
Wash., Seattle	1620	1100	446	74					
Spokane	1927	810	608	354	109	35	11		
Wis., Green Bay	2340	1061	831	360	67	21			
La Crosse	2359	874	878	476	95	32	4		
Wyo., Rock Springs	1856	680	788	368	20				

NOTE: 24 hr per day for the four summer months totals 2,928 hr.
SOURCE: J. C. Albright, "Summer Weather Data," The Marley Co., Kansas City, Kans., 1939.

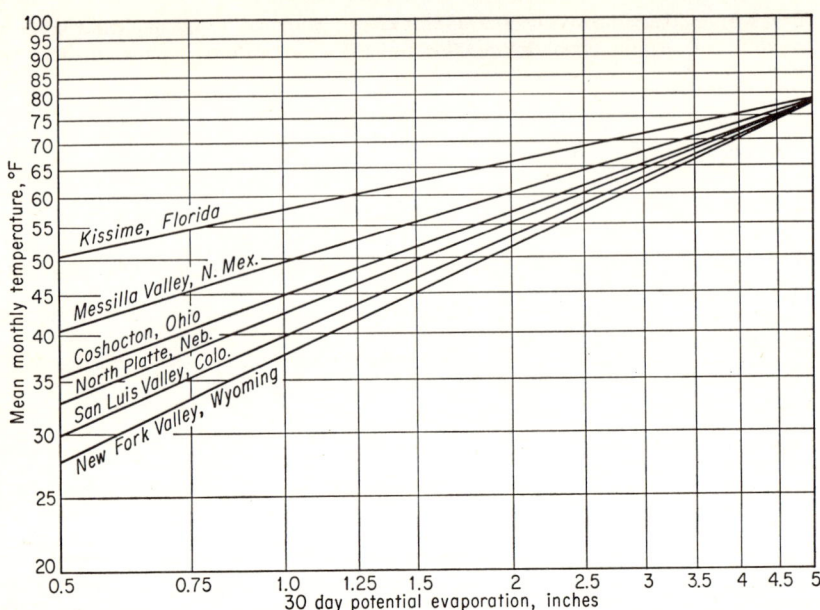

Fig. 57-16. Examples of monthly evapo-transpiration related to mean temperatures and for 12 hr of possible sunshine. (Converted to English units from Thornthwaite.) [*Courtesy of American Geographical Society, Geog. Rev.*, 38(1):55–94, 1948.]

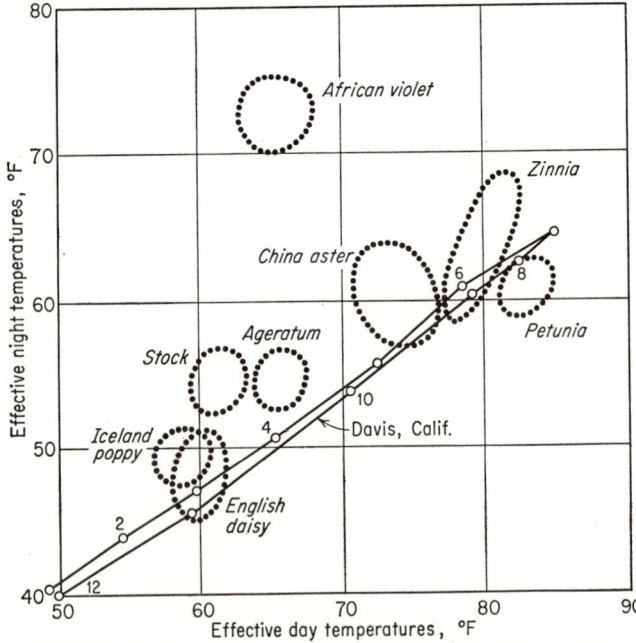

Fig. 57-17. Effective day-night temperature requirements for various flowers. (*Went, Nurserymen's Inst., January, 1957, unpublished.*)

Based on potential evapo-transpiration, Thornthwaite's climatic moisture index is a combination of 100 times the monthly precipitation surpluses, minus 60 times the monthly deficits, divided by the annual evapo-transpiration:

$$I_m = \frac{100 \sum_{surpl} [p - e - (4.0 - s)] - 60 \sum_{def} (e - p - s)}{\sum_{12} e} \qquad (57\text{-}17)$$

where I_m, nondim. = Thornthwaite's 1948 moisture index (see Fig. 57-1 for U.S. map)
  p = monthly precipitation, in.
  e = monthly evapo-transpiration (Fig. 57-16), in. (corrected for daylight hours)
  s = soil-moisture storage at beginning of month (not exceeding 4 in.; see below), in.
  $\sum_{surpl}$ = summation of surpluses, in.
  $\sum_{def}$ = summation of deficits, in. (usually some months have neither surpluses nor deficits)
  $\sum_{12}$ = total annual evapo-transpiration ("thermal efficiency"), in.

To use this formula it is necessary to determine the soil-moisture storage at the beginning of each month (assumed to range from 0 to 4 in.). This is done by starting at the close of a dry month when unirrigated plants would be wilted, $s = 0.0$, or else at the close of a month near the end of the rainy season when the soil is at field capacity, $s = 4.0$, and proceed by months adding p and subtracting e.

EFFECTS OF SURROUNDINGS, SLOPE, EXPOSURE, AND ELEVATION

The maps of temperatures, precipitation, etc., were determined by the U.S. Weather Bureau largely from the reports of local cooperators whose observation stations are necessarily near a house and usually are on level ground surrounded by some trees. Such an exposure, while not typical of open agricultural land, nevertheless indicates conditions for living areas even in mountainous terrain. A satisfactory method for estimating open-field weather from the reports of partially screened weather stations has not yet been devised. Schultz [40], comparing a lawn area and paved courtyard with standard exposure in the open, found that temperature departures varied with time of day and amounted to as much as 5°F where the wind was 2 to 5 mph. If the lawn station represents the ordinary garden exposure of Weather Bureau local cooperative stations, the true climatic condition in a large, level open field would normally be somewhat more severe as to maximum and minimum temperatures and wind. The daily maximum temperatures reported, however, are likely to be between bare-field and irrigated cropland maximums, the main variables being wind velocity and shading.

The *degree of slope*, and particularly the north or south orientation of sloped ground, has important effects on the amount of solar energy received, and consequently on the rate of soil-moisture loss and the degree of evaporation chill. As a first approximation, a farm on land with a general slope down to the south can be considered as if located in a latitude farther south by the same number of degrees as the average slope. This effect is spectacularly illustrated south of Yellowstone National Park, where in looking north one sees largely grass-covered hills but in looking south one has a feeling of being located in a complete forest area.

One of the clearest systematic interpretations of rough topography is that by Linsley [32] in estimating rainfall rates in small basins on the west slopes of the Sierra-Cascade mountain range. He uses six parameters defined as follows:

1. Index daily rainfall = $P_A/N_{0.25}$, in which P_A is the normal (average) precipitation September to April and $N_{0.25}$ the number of days on which the rainfall was ¼ in. or more

2. Land slope $= (E_{NE} - E_{SW})/50$, in which E is the average elevation at 25 miles radius in the NE (or SW) quadrant, the air flow usually being from the southwest during storms

3. Station elevation

4. Barrier height = average of maximum topographic elevations (within 45 miles of the station) along the seven rays at 15° intervals in the southwest quadrant

5. Orientation = azimuth bearing of the center line of the largest sector of a 20-mile-radius circle not containing a barrier more than 1,000 ft higher than the station location

6. Zone of environment: natural topographic boundaries (valley bottom, main west and east slopes of the Sierra range, coastal range, etc.)

This procedure yielded a correlation coefficient of 0.88 for 126 stations, so a similar approach might be very useful also for some other climatological factors.

The next most significant factor of local farm climate is *exposure to wind*. The daily solar energy received divides at the ground surface, part going into the ground by conduction and part into the atmosphere by convection and evaporation. The two latter modes of heat transfer increase with wind velocity, so daytime temperatures are lower in fields exposed to wind than in sheltered areas. The heat flows reverse at night, but often the surface winds fade out. Then the question of minimum temperature on a clear night depends largely on the velocity of the cold-air downhill drift.

Elevation of a local spot in the topography relative to that of the reporting weather station makes very little difference in its daytime temperatures unless the extra altitude is significant relative to the top or bottom of a characteristic fog layer. It should be borne in mind, however, that hillside air is warmer during the day and cooler at night relative to the temperature of the free atmosphere at that gravitational level. Free-air temperatures decreased about 3°F per thousand feet increase in altitude. Also, wind velocity increases with elevation, making exposure more important. The effect of elevation on nocturnal minimum temperatures, however, is of major agricultural importance. Farms located 200 or 300 ft above valley bottom are often described as being in a "thermal belt." (The air chilled by foilage under a clear, cold sky drains away on hillsides and keeps minimum temperatures closer to the free-air temperature at that level.) Hollows act as sinks for heavier-cold-air flow and are more susceptible to frost damage than hillside areas, but even in these "cold lakes" the slowness of air flow is often as much a cause of frost damage as the temperature of inflowing air. For this reason, large level plateau areas well above valley bottom may be subject to frost on clear calm nights even though on the map they may seem to be in the thermal belt. There is characteristic delay of phenological responses, such as dates of blooming and of harvesting, with increased elevation of valley-bottom areas above sea level. These have been analyzed by Hopkins [26], who has found 4 days average delay in rise of 400 ft or in a change of 1° latitude. His 5° longitude equivalence involves complex relationships between "spring weather" warm enough to cause blossoming and a winter which has been cold enough to complete the chilling requirement of the rest period. For some plants delayed foliation due to insufficient chilling is advantageous in delaying blossoming until after the average date of last spring frost. Several varieties of deciduous fruit are distinguished by differences in chilling requirement.

Heat Balance Indicated by Daily Soil Temperature. Concluding this brief discussion of expected variation of local farm climate from that reported from a nearby station, it needs to be emphasized that agricultural climate is the result of heat balance between (1) radiation from the sky, (2) conduction in the ground, (3) convection to the air, and (4) evaporation. The first two factors can be measured directly. The last is known approximately from the rate of soil-moisture loss. Eddy convection, both of heat and of moisture, is known to vary excessively with temperature gradient and wind velocity. The combined effect of all these factors, however, can be judged by soil temperature. This is because the soil has a large diurnal heat flow and its temperature must be that which provides a heat balance. Temperature at 4-in. depth shows the strong day-to-night cycle. The temperature at 20 in. responds strongly to the gain or loss from spells of weather and to the change of season. The soil temperature

TABLE 57-10. EFFECTIVE* DAY-NIGHT MONTHLY TEMPERATURES FOR AGRICULTURAL EXPERIMENT STATIONS IN THE UNITED STATES, °F

State and city	Jan.	Feb.	Mar.	Apr.	May	June	July	Aug.	Sept.	Oct.	Nov.	Dec.
Ala., Auburn:												
Day	53.4	56.3	62.2	69.9	77.6	83.4	84.5	84.3	81.2	71.3	61.0	53.7
Night	43.4	46.1	51.4	58.8	65.6	72.9	74.8	74.4	71.0	60.3	50.0	44.3
Alaska, Matanuska:												
Day	17.5	22.5	29.5	41.2	52.4	60.9	62.7	60.3	52.4	40.5	27.1	18.3
Night	8.6	13.7	20.4	31.8	41.4	49.5	52.4	50.7	43.1	32.7	19.6	10.2
Ariz., Tucson:												
Day	57.5	60.3	65.4	72.1	80.4	90.0	92.3	90.3	86.8	76.7	65.8	57.4
Night	42.3	40.5	49.3	54.9	62.7	72.4	78.9	77.1	71.8	59.7	46.9	42.5
Ark., Fayetteville:												
Day	43.1	45.5	55.4	64.2	71.9	79.7	83.9	84.1	78.2	67.0	55.1	44.8
Night	32.6	34.5	43.9	52.8	60.9	68.7	72.1	72.0	66.1	54.5	43.6	34.4
Calif., Davis:												
Day	49.1	54.4	59.6	65.3	72.4	78.4	85.0	83.5	79.2	70.5	59.3	49.8
Night	40.2	43.8	46.9	50.6	55.6	60.7	64.4	62.6	60.2	53.8	45.5	39.9
Calif., Riverside Fire Sta.:												
Day	58.9	60.6	63.4	67.9	71.8	78.2	84.7	84.7	81.5	73.8	66.9	57.8
Night	44.9	46.9	48.5	53.0	56.7	61.2	66.1	66.0	62.8	56.4	49.9	48.1
Colo., Fort Collins:												
Day	32.9	35.0	42.4	52.2	60.8	70.5	75.6	75.3	66.9	55.4	43.5	33.7
Night	18.9	21.0	28.7	38.2	47.0	55.6	60.8	59.7	51.1	40.0	28.8	19.8
Conn., Storrs:												
Day	29.7	29.3	39.2	50.5	62.0	69.8	74.5	72.3	66.2	56.5	44.2	32.7
Night	21.1	20.3	30.0	40.1	50.6	59.0	64.5	62.5	56.0	46.7	36.4	24.5
Del., Newark:												
Day	38.0	38.4	47.8	58.2	68.7	77.0	80.9	78.4	72.8	62.6	50.4	39.4
Night	29.7	29.1	36.8	46.0	56.6	65.2	69.6	66.9	60.6	50.5	39.9	30.5
Fla., Gainesville:												
Day	62.6	64.4	70.5	75.7	81.0	84.7	85.5	85.8	84.0	77.0	68.7	62.5
Night	51.8	53.2	58.8	63.5	69.2	74.2	75.7	76.1	74.2	66.5	57.3	51.7
Ga., Athens:												
Day	51.3	53.6	60.9	68.7	77.2	84.4	85.4	84.4	79.6	70.2	58.5	51.2
Night	40.2	41.5	47.9	54.9	63.5	72.2	74.1	73.4	68.2	57.0	45.7	39.7
Hawaii, Honolulu:												
Day	74.3	74.3	74.5	75.5	77.3	79.1	80.0	80.6	86.9	79.7	77.5	75.5
Night	69.6	69.6	69.9	71.0	72.8	74.6	75.7	76.3	76.0	75.2	72.9	71.1
Idaho, Moscow:												
Day	30.9	35.2	42.1	51.1	58.3	65.3	74.6	73.6	64.0	53.7	42.3	33.4
Night	24.6	28.0	39.9	40.9	46.9	52.7	59.6	58.7	51.3	43.1	34.2	27.6
Ill., Urbana:												
Day	29.7	32.8	44.9	55.5	66.5	75.2	80.3	78.1	71.6	59.6	46.1	33.2
Night	21.5	24.7	35.5	45.2	56.0	64.6	69.3	67.4	60.8	48.9	37.0	25.4
Ind., Lafayette:												
Day	34.1	35.7	45.3	56.8	68.2	78.5	83.1	80.7	73.2	62.7	46.6	35.5
Night	26.3	27.4	36.0	46.3	57.3	67.5	71.4	69.3	61.8	51.3	37.9	28.0
Iowa, Ames:												
Day	24.2	27.3	41.6	55.0	66.1	74.8	80.2	78.4	70.5	57.9	42.0	28.6
Night	14.1	17.2	30.9	43.2	54.1	63.4	67.6	65.9	58.2	45.6	31.6	19.3
Kans., Manhattan:												
Day	33.5	39.7	50.7	61.3	70.0	79.9	85.5	84.2	76.6	63.8	50.3	36.3
Night	22.8	28.3	37.7	48.7	58.4	68.1	72.7	71.1	63.4	51.0	38.0	25.8
Ky., Lexington:												
Day	36.5	38.6	47.1	57.8	67.9	77.4	81.0	79.8	74.5	62.9	48.2	38.9
Night	28.5	30.4	37.7	48.0	57.3	67.0	70.4	69.0	63.2	51.6	39.4	30.8
La., Baton Rouge:												
Day	56.8	60.0	65.0	71.6	78.2	84.2	85.6	85.2	82.3	75.0	64.2	58.1
Night	47.9	51.1	55.6	62.0	69.0	75.0	76.6	76.2	72.7	64.2	54.0	49.1
Maine, Orono:												
Day	22.3	24.3	33.9	52.7	59.0	68.0	73.5	71.6	63.9	52.4	39.4	26.6
Night	11.3	13.3	22.7	35.5	46.8	56.2	61.3	61.2	52.9	41.6	30.8	17.6
Md., College Park:												
Day	37.8	39.1	49.6	59.3	69.1	77.4	81.5	79.6	86.9	62.8	50.5	40.1
Night	27.8	28.8	38.1	46.6	56.2	65.2	69.7	67.7	61.8	49.6	38.8	30.2
Mass., Amherst:												
Day	28.6	28.4	38.9	51.3	62.9	71.2	76.3	73.9	67.4	55.4	43.4	31.1
Night	13.8	19.0	28.5	39.9	50.5	59.8	64.7	62.1	55.0	44.8	34.4	23.7
Mich., East Lansing:												
Day	27.3	28.0	37.4	51.7	62.0	72.2	76.7	74.6	67.0	55.4	41.4	30.2
Night	20.2	20.3	28.9	43.8	50.9	62.5	65.4	63.4	56.6	45.6	34.3	23.9

TABLE 57-10. Effective* Day-Night Monthly Temperatures for Agricultural Experiment Stations in the United States, °F (Continued)

State and city	Jan.	Feb.	Mar.	Apr.	May	June	July	Aug.	Sept.	Oct.	Nov.	Dec.
Minn., St. Paul:												
Day	17.0	20.9	33.6	50.6	63.0	72.3	77.8	75.2	66.1	53.6	36.0	23.2
Night	8.3	11.8	25.1	41.0	52.9	62.7	67.7	65.3	56.4	44.5	28.5	15.6
Miss., State College:												
Day	50.2	52.6	61.8	69.0	76.7	84.2	86.3	86.0	81.6	71.0	60.0	51.5
Night	40.1	42.4	50.0	57.9	64.6	73.2	75.7	81.6	70.1	58.8	48.6	41.6
Mo., Columbia:												
Day	52.6	54.4	61.1	72.7	77.8	85.3	86.6	85.2	81.2	71.0	59.9	52.6
Night	41.4	42.6	49.2	58.3	65.6	73.7	76.1	74.8	70.5	58.7	47.8	41.4
Mont., Bozeman:												
Day	25.0	27.8	34.3	46.3	55.1	63.6	71.4	70.4	59.3	48.8	36.5	27.2
Night	15.8	18.0	24.8	35.3	43.4	50.7	57.0	55.9	46.6	37.8	26.8	18.3
Nebr., Lincoln:												
Day	30.6	34.3	45.1	58.9	69.6	80.0	86.4	83.9	73.5	63.9	46.1	34.3
Night	20.3	23.8	33.6	46.4	57.2	67.8	73.2	71.3	62.3	50.9	35.1	24.9
Nev., Reno:												
Day	38.4	43.6	48.5	56.8	64.2	71.1	80.7	78.7	71.4	60.4	48.6	40.2
Night	23.9	28.9	32.6	39.1	46.3	51.8	58.4	56.1	49.6	41.0	31.8	26.1
N.H., Durham:												
Day	28.1	28.1	38.4	48.7	60.1	69.2	75.1	72.6	65.7	55.4	42.9	31.0
Night	18.3	18.5	28.6	33.6	48.7	57.4	63.7	61.2	54.3	44.6	33.9	22.0
N.J., New Brunswick:												
Day	34.5	34.4	45.4	55.6	69.3	74.7	79.5	77.2	71.8	60.5	48.3	37.1
Night	26.1	25.4	35.3	44.5	54.8	63.4	68.7	66.7	60.9	49.7	39.0	29.0
N.Mex., Experiment Farm:												
Day	40.4	45.6	52.6	62.5	70.9	79.3	83.8	81.7	75.1	64.0	50.2	41.9
Night	25.3	28.9	35.3	45.1	53.2	60.4	66.7	65.2	57.4	46.1	30.9	26.5
N.Y., Ithaca:												
Day	28.8	28.2	38.0	50.5	61.9	71.2	76.3	74.2	66.4	55.7	43.1	31.2
Night	20.6	19.5	28.2	39.8	50.2	59.8	64.2	62.5	55.1	45.1	35.2	23.4
N.C., Raleigh:												
Day	46.3	48.2	55.8	64.7	73.3	81.4	83.7	82.5	77.8	67.1	55.9	47.2
Night	36.5	37.9	44.3	52.3	61.3	69.9	73.3	71.9	67.3	55.0	44.8	37.5
N.Dak., Fargo:												
Day	12.0	15.8	29.9	47.6	61.2	70.7	77.7	75.5	65.0	51.8	32.0	17.4
Night	2.1	5.7	20.6	36.6	48.7	58.5	64.8	62.6	52.7	40.4	23.2	8.3
Ohio, Columbus (Univ.):												
Day	33.2	37.8	45.6	55.7	66.4	75.0	79.6	77.6	73.2	60.2	46.3	35.2
Night	25.3	25.6	36.1	45.2	55.1	64.1	62.7	66.5	59.0	48.7	37.5	27.8
Ohio, Wooster:												
Day	33.5	33.7	42.5	53.6	65.6	75.7	79.3	77.5	70.8	59.6	44.9	38.4
Night	25.5	24.8	32.6	42.3	53.5	63.8	67.2	65.4	58.4	48.0	36.4	26.9
Okla., Stillwater:												
Day	42.1	44.3	56.4	65.4	73.0	81.9	86.2	88.7	79.4	66.0	55.3	34.7
Night	30.5	34.4	43.5	47.2	62.2	71.4	74.9	75.4	66.9	54.2	43.1	32.3
Ore., Corvallis (St. Coll.):												
Day	42.0	45.7	50.1	55.6	61.0	66.5	73.0	73.5	66.9	58.5	49.1	43.3
Night	35.9	34.7	41.0	44.6	49.1	53.6	58.0	58.2	53.3	47.3	41.5	37.3
Pa., State College:												
Day	30.0	30.6	41.6	51.9	63.5	71.0	75.6	73.2	67.8	57.4	43.6	32.5
Night	22.3	22.4	32.4	42.1	52.9	60.7	65.3	62.9	57.3	45.7	35.7	25.5
R.I., Kingston:												
Day	31.6	31.1	39.6	49.3	59.8	68.6	73.9	72.5	66.9	56.7	45.2	34.5
Night	23.0	22.5	30.6	39.5	49.6	58.4	64.1	62.7	56.9	46.9	36.2	26.9
S.C., Clemson:												
Day	48.1	50.2	58.3	65.9	74.4	80.6	83.4	82.3	78.1	67.6	57.0	48.8
Night	37.9	39.2	47.1	54.0	62.9	70.0	73.2	72.6	67.5	55.9	45.1	38.5
S.Dak., Brookings:												
Day	18.2	21.7	35.7	51.5	62.5	71.2	76.9	75.4	66.7	54.0	37.3	23.0
Night	7.6	10.8	24.7	38.6	49.6	59.1	64.2	62.3	53.6	41.1	26.4	12.8
Tenn., Knoxville:												
Day	45.1	47.5	54.9	64.9	73.2	81.4	83.7	82.4	77.8	66.4	53.7	45.7
Night	35.8	37.4	43.9	53.1	61.6	70.1	73.0	71.5	66.3	54.2	43.0	36.3
Tex., College Station:												
Day	55.8	59.4	66.7	73.3	79.6	86.5	89.6	90.7	86.0	76.3	65.5	56.9
Night	45.6	48.6	56.1	62.9	69.5	75.6	78.1	78.5	74.2	64.2	54.2	47.1

TABLE 57-10. Effective* Day-Night Monthly Temperatures for Agricultural Experiment Stations in the United States, °F (Continued)

State and city	Jan.	Feb.	Mar.	Apr.	May	June	July	Aug.	Sept.	Oct.	Nov.	Dec.
Utah, Logan:												
Day	28.2	32.4	41.8	52.8	61.0	70.8	79.4	77.8	67.9	55.3	41.9	28.9
Night	20.0	24.2	32.1	41.4	49.1	57.4	65.5	64.4	54.9	43.8	32.9	21.2
Vt., Burlington:												
Day	22.7	23.2	34.1	47.6	61.3	71.5	76.4	73.9	65.6	53.4	40.3	27.0
Night	13.1	13.0	24.4	36.9	49.4	59.5	64.4	62.2	54.1	43.0	32.4	18.6
Va., Blacksburg:												
Day	38.3	38.9	48.8	57.1	66.4	73.4	76.7	75.6	70.9	59.9	48.6	39.5
Night	28.4	28.6	37.4	44.6	53.4	61.2	64.8	64.1	58.0	46.8	36.0	29.7
Wash., Pullman:												
Day	30.9	35.0	42.7	51.6	58.3	66.0	75.3	74.4	63.0	53.6	40.9	33.2
Night	25.2	28.8	34.9	41.6	47.6	54.1	60.9	59.9	51.9	43.8	34.9	28.1
W.Va., Morgantown:												
Day	36.9	37.0	47.8	57.6	68.0	75.0	79.3	77.8	73.6	59.8	48.6	38.6
Night	27.7	27.8	37.3	46.2	56.2	63.5	67.9	66.3	61.8	48.9	39.3	30.0
Wis., Madison:												
Day	23.5	26.4	37.1	51.1	63.4	73.3	79.3	76.9	67.8	55.9	39.7	26.9
Night	14.7	17.3	27.9	40.2	51.5	61.5	66.6	64.4	56.4	44.9	30.9	19.0
Wyo., Laramie:												
Day	27.6	28.4	35.1	43.7	52.9	63.7	70.2	68.9	60.7	48.8	37.7	27.5
Night	16.7	18.1	24.0	32.0	40.7	49.7	56.0	54.7	46.4	35.8	26.3	16.6

* Arbitrary mean photoperiod temperature and nyctotemperature defined by Went [55] in quartiles of mean daily range: effective day = mean maximum minus ¼ range; effective night = mean minimum plus ¼ daily range. This interpretation of natural diurnal temperature is comparable to the square-wave cycle used in phytotron experiments.

therefore follows local differences in the net effect of all impinging heat rates and can be used as a simple indicator of spot climate.

The effects of soil temperature on plant growth, however, have not been as well evaluated as the effects of air temperature. The growth rate of most plants seems to be almost proportional to temperature excess above some critical degree, so a record of "degree-hours" is very useful if correction is made for hours of excessive temperature. Conversely, most plants also require considerable winter chilling, usually below 40°F.

Within the regional climatic pattern very narrow zones are recognized by agriculturalists as defined rather clearly by fruit quality or production. Locally these seem to follow roughly the topographic contours and variations in soil quality. Such climatic subdivisions indicated by plants (adequately irrigated) depend mainly on susceptibilities to temperatures at specific stages of growth. Went [53] describes laboratory tests on the four-dimensional relationships for plant response to environment:

1. Photoperiod
2. Temperature during photosynthesis, called here "effective day temperature"
3. Temperature during dark period, called here "effective night temperature"[1]
4. Stage of growth

Intensity of illumination and several other factors are shown by Withrow [64] to also influence growth very strongly. These effects on plant growth in cloudy climates may be comparable to day/night temperature effects under clear skies.

Most significant for flowering and fruit production is the mean daily relation of responses 2 and 3, often with progressive lowering of effective temperatures toward harvest time, especially for tomatoes. In Fig. 57-17, Went [54] shows the optimal growing conditions for a few flowers. Such optimal day-night temperatures are found only in certain months in any single location and explain how for annuals the same plants (lettuce, carrots) grown in the winter near El Centro are grown during the spring near Salinas, and later in northern California.

For a general treatment on environmental effects see Specter [60]. It is to be noted that many growth phases sensitive to climatic environment are listed alike for both

[1] The word "effective" has been used to avoid confusion with the specific climatological meaning of "average" and to avoid the problem of the actual length of the "photoperiod."

TABLE 57-11. SUGGESTED LIST OF GRAPHS OF ANNUAL CYCLES OF CLIMATIC FACTORS
(Useful for Detail Comparison of Climates and to Agricultural Operators and Industrial Designers)

Graph No.	Title and main reasons for selection and discussion	Preferred observations		Comparable curves from published data	Elements most closely related
		Height and unit	Curves (plotted monthly)		
1	Air temperature, dry bulb. Primary thermal condition of the air in which agriculture exists. Observations higher aboveground are more representative per acre (less influenced by irregularities close to the thermometer), but published observations are mostly for 4½ to 6 ft.	80 in., °F	5-yr max. Mean monthly max. Mean daily max. Mean daily Mean daily min. Mean monthly min. 5-yr min.	10-yr high for month Av. highest for month Av. daily max. for month Av. daily min. for month Av. lowest for month 10-yr low for month	Air temperature at other level, or globe temp. corrected for radiation and convection
2	Black-globe temperature. Temperature of equilibrium between radiation and convection on a 6-in. sphere. This is the temperature black objects tend to reach except for evaporative cooling or internal heating. It is the best single measure for animal comfort. This is a combination response instrument by which radiation can be estimated if air temperature and wind speed are measured at the same height.	80 in., °F	Same as for no. 1	....	Air temperature, radiation, wind
3a	Wet-bulb temperature. The most general measure of air moisture. It is a direct measure of heat content of the air and the limiting temperature in direct evaporative cooling.	80 in., °F	2 P.M. mean monthly max. 2 P.M. mean daily 4 A.M. mean daily 4 A.M. mean monthly min.	Noon mean daily for month	Dew point or relative humidity
3b	Dew point. This temperature is an absolute measure of air moisture now reported on the daily weather map. It also has direct significance for the formation of dew and the level of cumulus clouds. The early afternoon moisture near the ground is strongly correlated with overhead moisture because of fully developed thermal convection.	°F	2 P.M. mean monthly max. 2 P.M. mean daily 4 A.M. mean daily 4 A.M. mean monthly min.	Noon mean daily for month	Wet- vs. dry-bulb temperature
3c	Relative humidity. This is the least useful measure of air moisture content but has direct significance for the equilibrium moisture content of cellular material important for drying and fire hazard. It is usually measured by hair hygrometer.	%	2 P.M. mean daily 4 A.M. mean daily	4 A.M., 10 A.M., 2 P.M., 8 P.M. mean daily	Wet- vs. dry-bulb temperature

#	Description	Units	Values	Related
4	Soil temperature. Primary thermal condition of seed and root environment. Maximum and minimum are needed in this case to describe the diurnal cycle and the mean temperatures at shallow depths. The seasonal lag is shown by temperature at 40″ depth or more. Shallow-soil temperatures are most significant for germination, soil-pest behavior, and minimum nocturnal air temperature and represent the combined effects of radiation, wind, and evaporation.	−4″ under bare soil, °F	Mean monthly max. Mean daily max. Mean daily av. Mean daily min. Mean monthly min.	6″ av. high for month 6″ daily mean for month 6″ av. low for month 72″ mean for month	Mean daily air temperature
5a	Wind speed. Primary condition of forced convection, which also indicates hazard of wind damage.	26′, mph	Fastest 5 min. 2 P.M. mean monthly max. 2 P.M. mean daily 4 A.M. mean daily	Fastest mph Av. speed	Black-globe vs. air temperature and radiation
5b	Wind direction. Basic factor for exposure, shielding, and natural ventilation. For these the time duration of direction is more important than velocity of air flow because gusts provide most air flow, but do not describe cold-air drainage in frost hazard or the difficult conditions for natural ventilation.	26′, % time 16 pts.	Mean for max. winds 2 P.M. prevailing 4 P.M. prevailing	Direction max. velocity Prevailing monthly 04, 10, 16, 22 hr	Humidity
6	Solar energy. Primary measure of daytime heat input. This is measured usually by glass-covered thermopiles, but the uncovered black plate is probably better since this measures total net radiation relative to natural air temperature at the instrument. An additional curve of hourly range (midday) would be useful to indicate temporary shading by low clouds.	80″, Btu/(day)(sq ft)	Mean daily 10-yr min. daily	10-yr max. daily Clear-day av. Daily average	Sunshine hours, black-globe temperature, air temperature, and wind velocity
7	Capillary evaporation. Measurement by physical instrument most nearly indicating natural evaporation and transpiration from soil and plants with plentiful moisture (Thornthwaite's potential evapo-transpiration). The Livingston atmometer is a calibrated, porous, ceramic sphere which has a combination response to radiation, vapor-pressure deficit, and wind. It is the best single instrument to measure the environmental conditions for plant life.	80″, cc/day	10-yr recurrent max. Mean daily 10-yr recurrent min	Av. pan evaporation	Pan evaporation, solar energy, wet-bulb depression
8	Days per month of frosts in mild climates. This chart indicates the frequency and severity of frosts. The average growing season is determined by median dates of last spring frost and first fall frost, and the variability is indicated by the 25 and 10% probability dates.	80″, days/mo	10-yr recurrent max. 0, 1° mean total 2, 3° mean total 4, 5° mean total 6, 7° mean total 8, 9° mean total	Av. days 32°F or lower Last and first dates 28, 24, 20, 16°F	Air temperature

tree crops and grain but that the time periods vary greatly. The phenology of corn is well described by Shaw and Thom [59].

Following the suggestion by Went [55] that the natural diurnal temperature cycle be interrupted by taking for the "effective" night temperature the daily minimum plus one-fourth the range from minimum to maximum and for the "effective" day (photoperiod) temperature the daily maximum minus one-fourth the daily range, Kimball [58] has found this to give rather consistent climatological description of the California plant climatic zones. Table 57-10 is included here to show this day-night temperature regime at the agricultural experiment stations across the United States, because for irrigated agriculture this annual ellipse of temperature is probably even more significant than the temperature-precipitation climographs given in Berry, Bollay, and Beers [6], and the monthly mean maximums and minimums are not given in the USDA 1941 Yearbook. Other daily and yearly cyclic data are also needed to explain differences in plant growth and for detailed comparison of local climates. Important annual curves are described in the next paragraph.[1]

ANNUAL GRAPHS OF CLIMATIC FACTORS

When the local climates of specific localities are to be compared, the climatological data of the U.S. Weather Bureau [49] must be consulted. The importance of each climatic factor will vary for different crops and livestock, so there is no fixed scale of economic value of farm climate. If monthly precipitation and temperature provide a sufficient climate comparison, climographs are very effective, as shown by Berry et al. [6]. As mentioned in the beginning, most climatic data for agriculture need to be organized first in mean annual curves, second in probabilities, and third in cumulative curves, comparing the current year with the long-record means.

Only a few of the variable characteristics have been expressed in terms of probabilities, but a series of mean magnitudes, including the daily maximum and minimum, and these per month and per 10 years for that month, give a fair idea of probable variation.

Table 57-11 is a descriptive list of annual curves which would define most of the aspects of average daily climate significant for agriculture. Such averages include, of course, the natural storm sequences which, in the United States, often repeat every 6 or 7 days. The diurnal cycles are also superimposed on the mean annual curves and vary in intensity, as previously discussed, with weather type (Table 57-1).

The whole make-up of local farm climate is thus a complex of many factors, usually integrated subconsciously by the experienced, observant farm operator, yet definable in an engineering sense by annual and diurnal graphs.

Cumulative curves of precipitation, degree-days, and solar energy (Table 57-12) would provide a basic background of long-term averages on which the plotting of the current year's record reveals at a glance how existing conditions are behind or ahead of the average. The only comparable cyclic curve is that for seasonal progression in soil temperature, as explained previously.

ESSENTIAL CLIMATIC MAPS FOR THE UNITED STATES

Specific estimates of local climates are best made by direct use of long-period records of temperature, moisture, and wind. It is not possible, however, to grasp a general concept of climate from complex numerical tables. To proceed, therefore, from the general to the specific, the first maps considered at the beginning of this chapter give the world-wide distributions of temperature, weather belts, and air circulation.

Climatic maps of the United States [5,8] can well be studied for sharper climatic detail, but so many excellent maps of specialized information are available that a reader seeking to compare two or three regions loses a general concept of the climate in each area. A limited selection of United States maps [48] is presented, therefore, in Figs. 57-18 to 57-34 of those few mean magnitudes having geographic distribution which can best be viewed as a spread to portray the typical climate at any locality.

---

[1] A discussion of diurnal curves and instrumentation is given in F. A. Brooks, "An Introduction to Physical Microclimatology," ASUCD Store, University of California, Davis, Calif., 1959. This includes Kimball's California map and tables.

TABLE 57-12. CUMULATIVE ANNUAL GRAPHS TO BE USED AS AVERAGE BACKGROUND FOR PLOTTING CURRENT OBSERVATIONS

Graph No.	Title and main reasons for selection and discussion	Preferred observations		Comparable curves from published data for some stations	Elements most closely related
		Height and unit	Curves (ploted weekly)		
A	Total precipitation. Cumulative, gross measure of water received.	40", in.	10-yr recurrent max. Mean total by weeks 10-yr recurrent min.	10-yr highest monthly Mean monthly 10-yr driest, monthly	Soil moisture
B	Capillary evaporation. Cumulative measure of water loss from continually moist surface.	80", cc	10-yr recurrent max. Mean total by weeks 10-yr recurrent min.	Av. pan evaporation monthly	Pan evaporation, solar radiation
C1	Mean degree-days above 50°F. This cumulative total affords a rather good base for correlating plant growth except in very hot weather. Temperatures of 104°F cause actual damage, so a degree-hour curve above 95°F is often needed. Daylight hours may also have to be considered.	(80") °F day	10-yr recurrent max. Mean total 10-yr recurrent min.	10-yr highest day—max. deg Av. total day—max. deg 10-yr total day—max. deg Av. days 90°F or above	Mean daily air temperature
C2	Mean degree-days below 50°F. This cumulative total indicates fairly well the extent of chill experienced which is an essential factor for fruiting. Temperatures below 40°F may not contribute extra to the chilling response.	(80") °F day	Same as no. C1	....	Mean daily air temperature
D	Cumulative solar energy.	(80")	10-yr recurrent max. Daily mean 10-yr recurrent min.		Evaporation

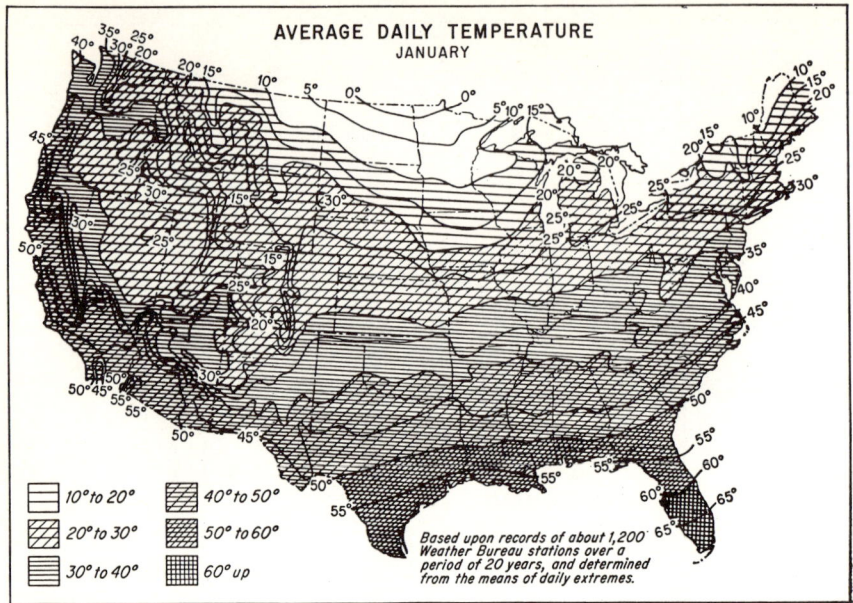

Fig. 57-18. United States average daily temperature, January. (*U.S. Weather Bureau, Daily Map* 17, *December,* 1951.)

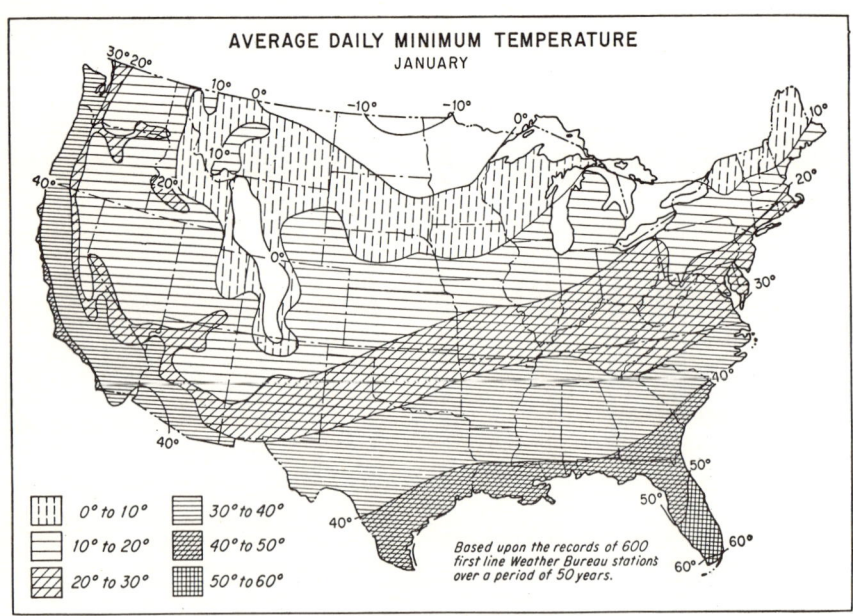

Fig. 57-19. United States average daily minimum temperature, January. (*U.S. Weather Bureau, Daily Map* 17, *December,* 1951.)

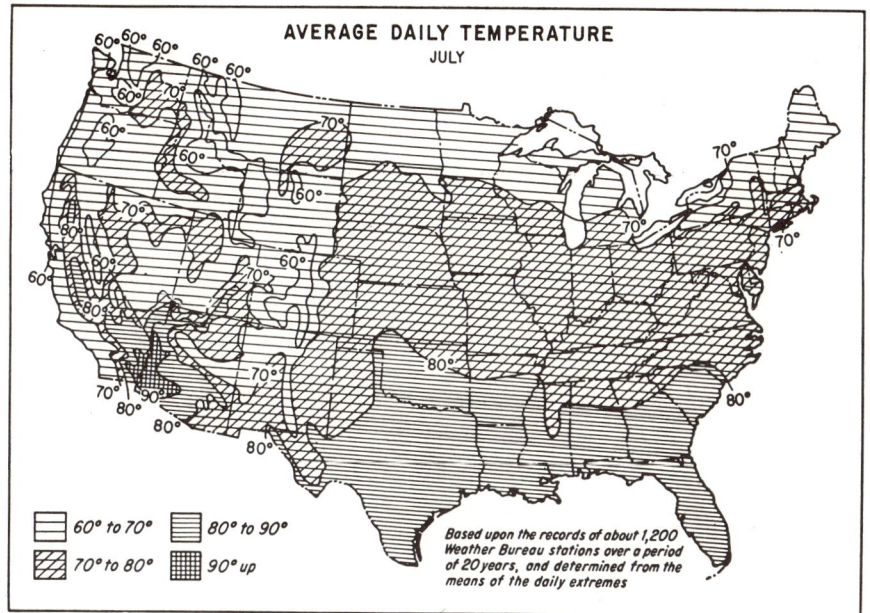

Fig. 57-20. United States average daily temperature, July. (*U.S. Weather Bureau, Daily Map* 18, *June*, 1951.)

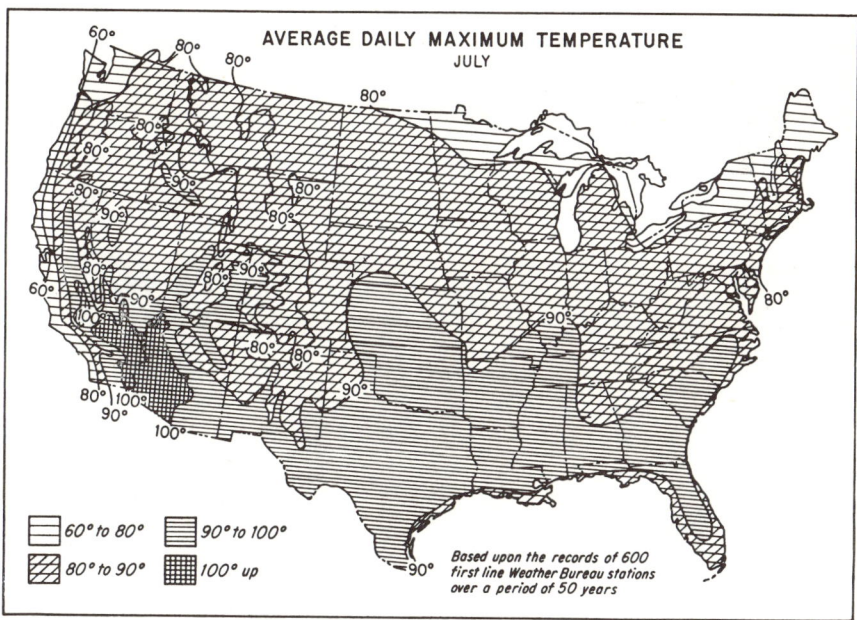

Fig. 57-21. United States average daily maximum temperature, July. (*U.S. Weather Bureau, Daily Map* 18, *June*, 1951.)

858 BASIC AGRICULTURAL DATA

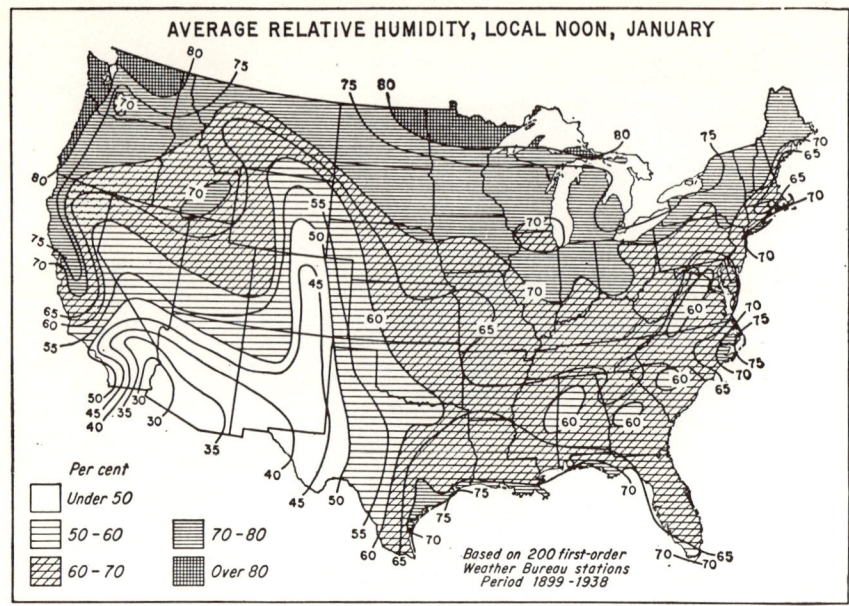

FIG. 57-22. United States average relative humidity, local noon, January. (*USDA Yearbook of Agriculture*, 1941, p. 733.)

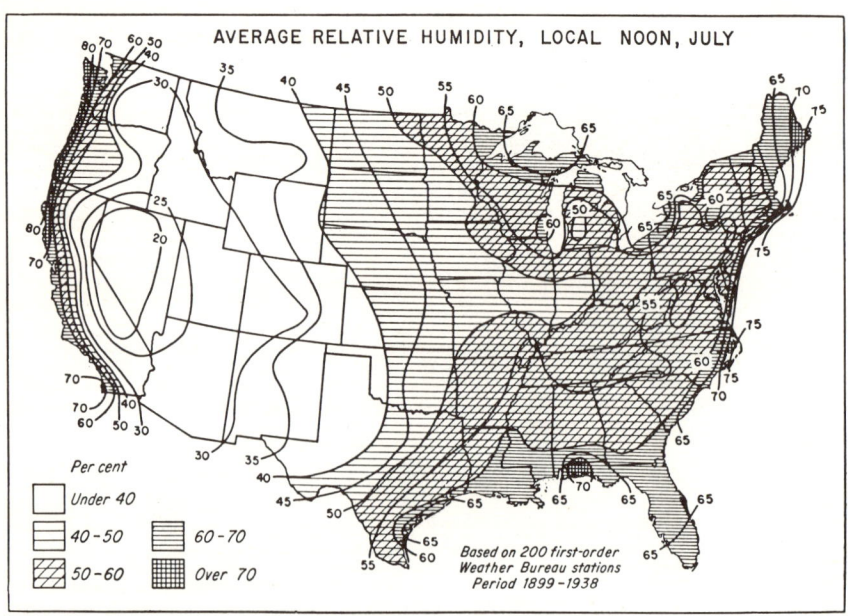

FIG. 57-23. United States average relative humidity, local noon, July. (*USDA Yearbook of Agriculture*, 1941, p. 734.)

FARM CLIMATES AND SOLAR ENERGY 859

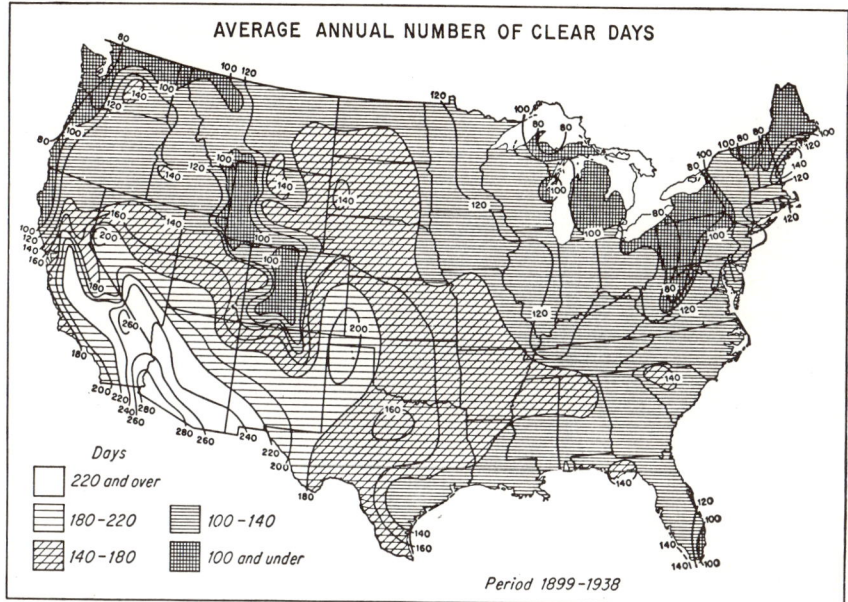

FIG. 57-24. United States average annual number of clear days. (*USDA Yearbook of Agriculture*, 1941, p. 742.)

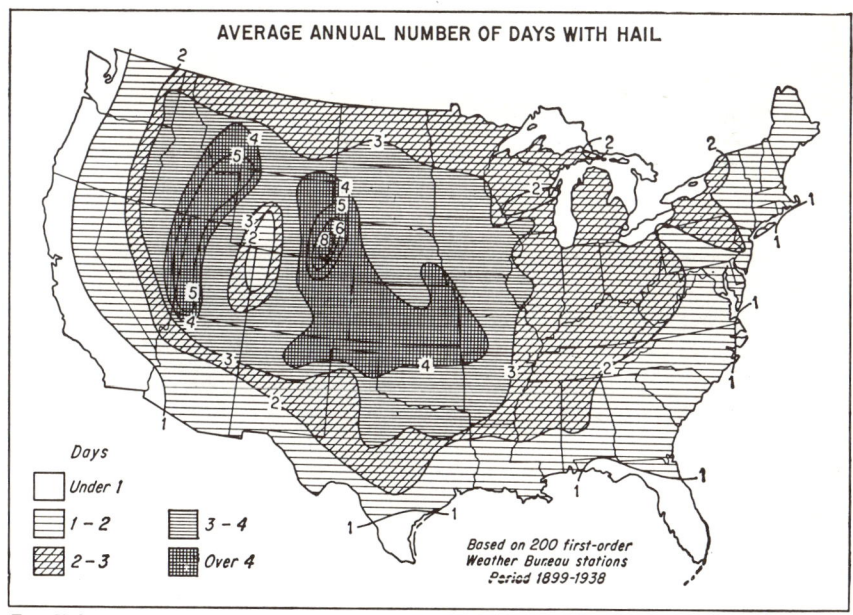

FIG. 57-25. United States average annual days with hail. (*USDA Yearbook of Agriculture*, 1941, p. 730.)

860 BASIC AGRICULTURAL DATA

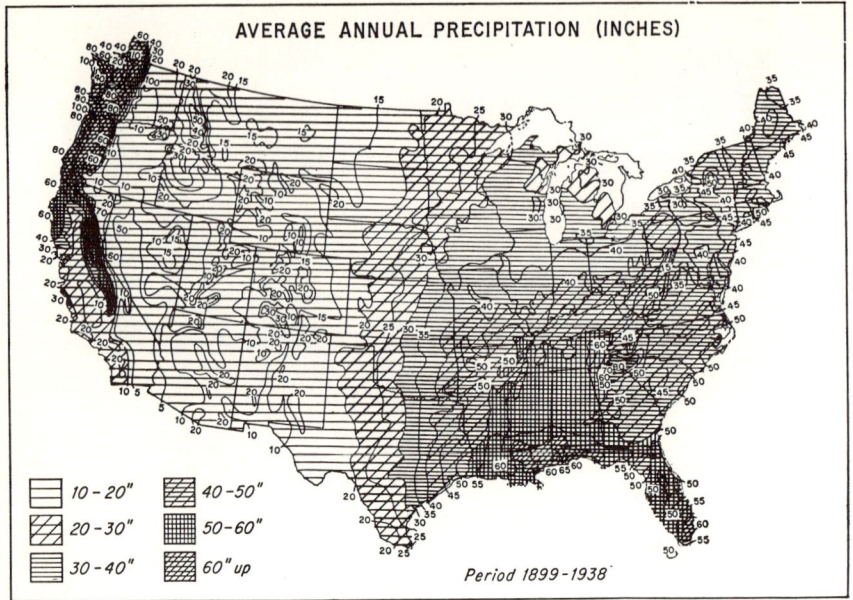

Fig. 57-26. United States average annual precipitation. (*USDA Yearbook of Agriculture*, 1941, p. 711.)

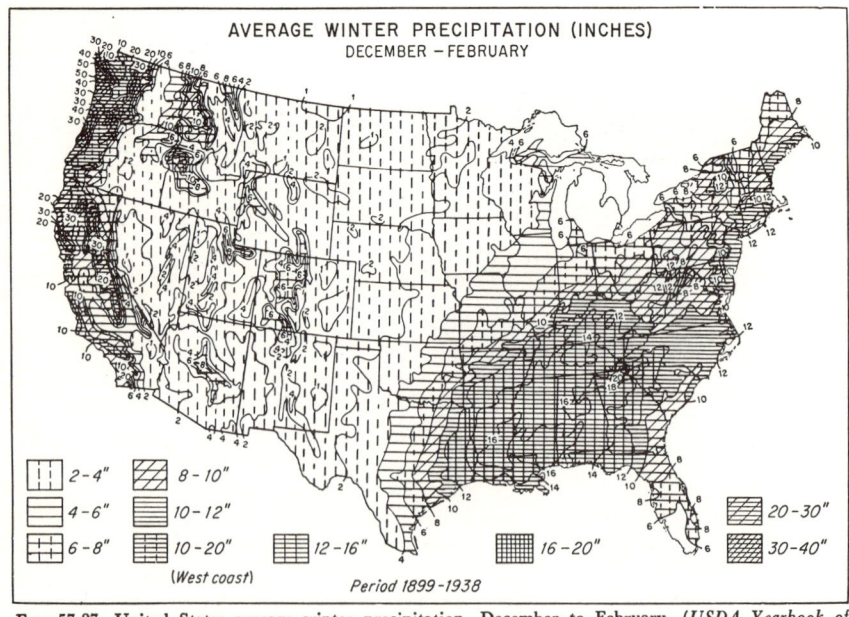

Fig. 57-27. United States average winter precipitation, December to February. (*USDA Yearbook of Agriculture*, 1941, p. 713.)

FARM CLIMATES AND SOLAR ENERGY 861

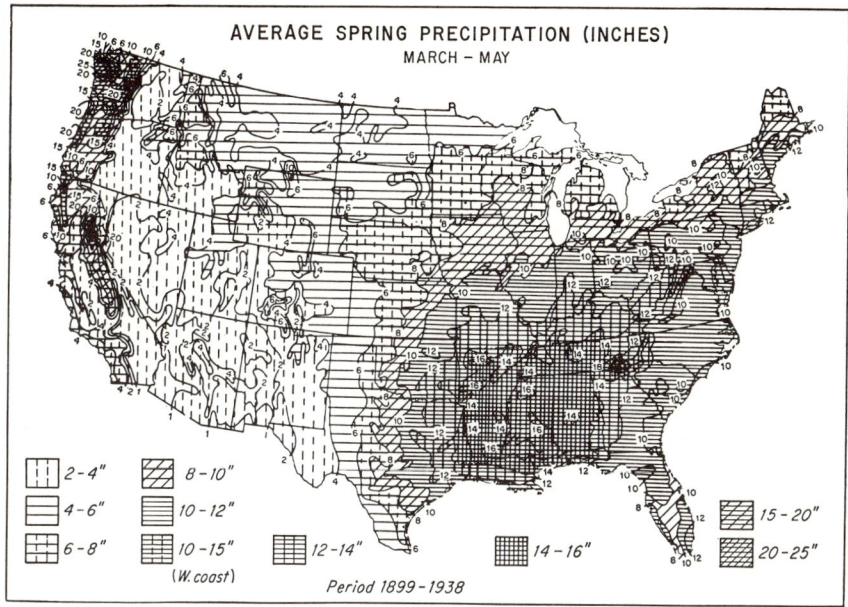

FIG. 57-28. United States average spring precipitation, March to May. (*USDA Yearbook of Agriculture*, 1941, p. 714.)

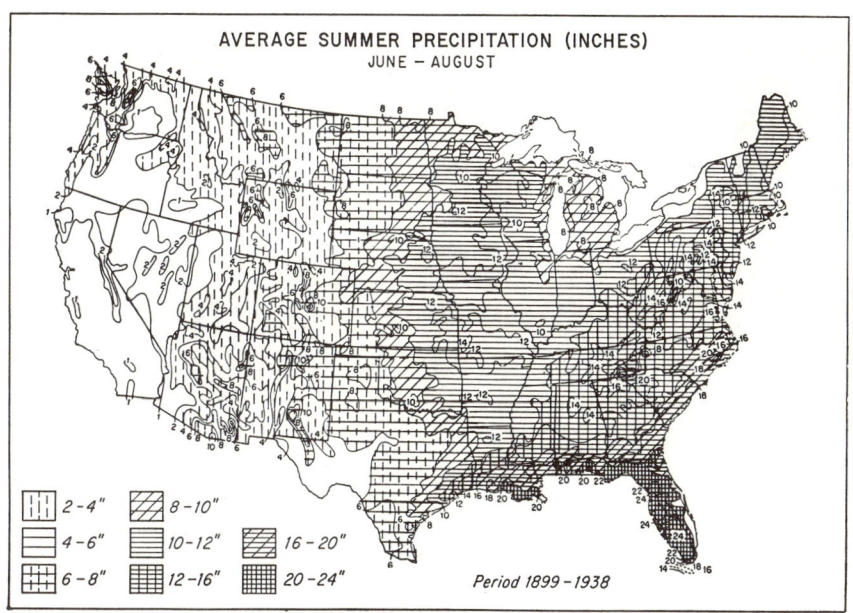

FIG. 57-29. United States average summer precipitation, June to August. (*USDA Yearbook of Agriculture*, 1941, p. 715.)

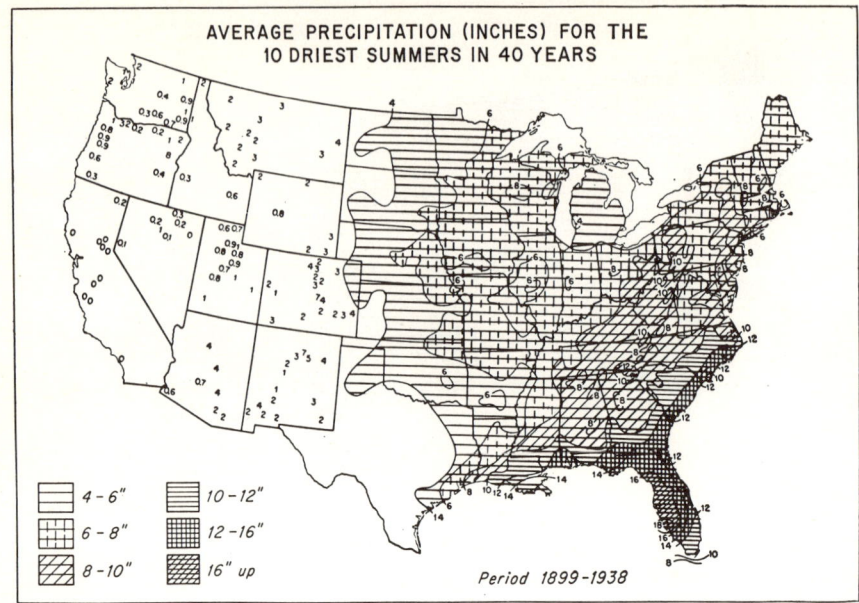

Fig. 57-30. Average precipitation for the 10 driest summers in 40 years. (*USDA Yearbook of Agriculture*, 1941, p. 720.)

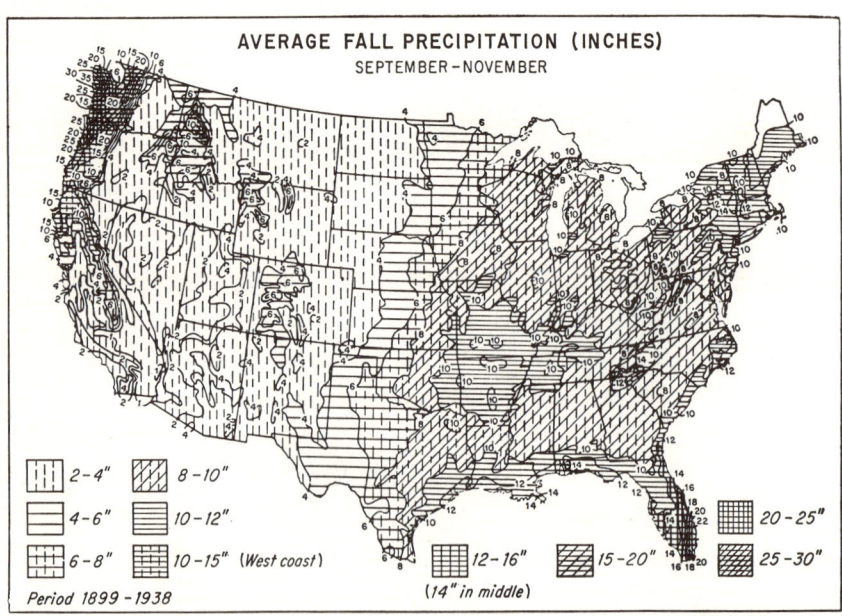

Fig. 57-31. United States average fall precipitation, September to November. (*USDA Yearbook of Agriculture*, 1941, p. 716.)

FARM CLIMATES AND SOLAR ENERGY 863

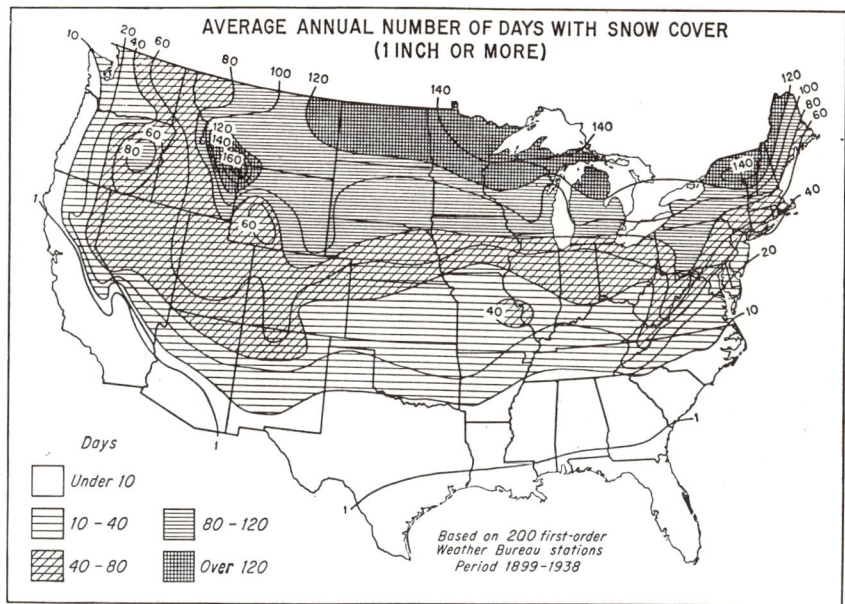

Fig. 57-32. United States average annual number of days with snow cover (1 in. or more). (*USDA Yearbook of Agriculture*, 1941, p. 728.)

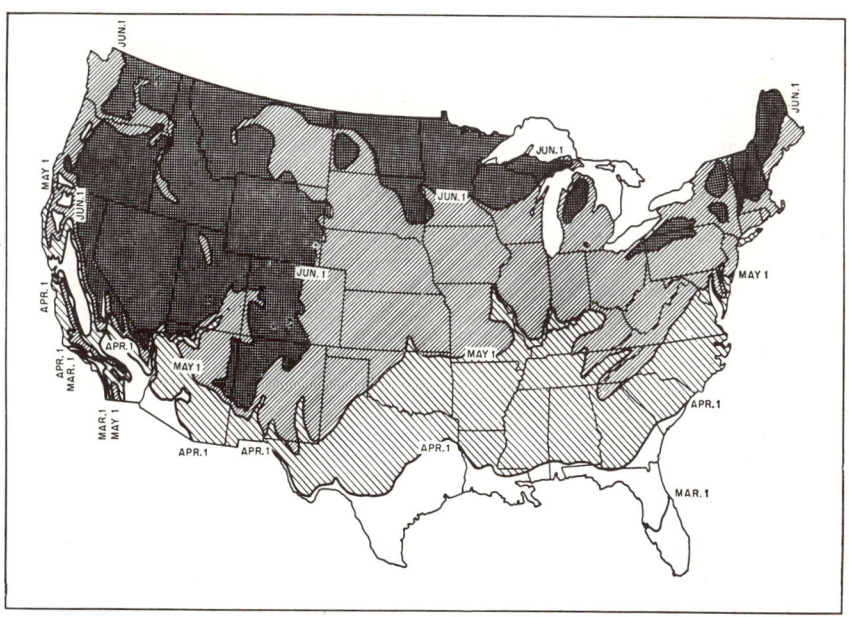

Fig. 57-33. United States latest killing frost in spring, 10-year recurrence interval (*Baker, USDA Atlas of American Agriculture*, 1936.)

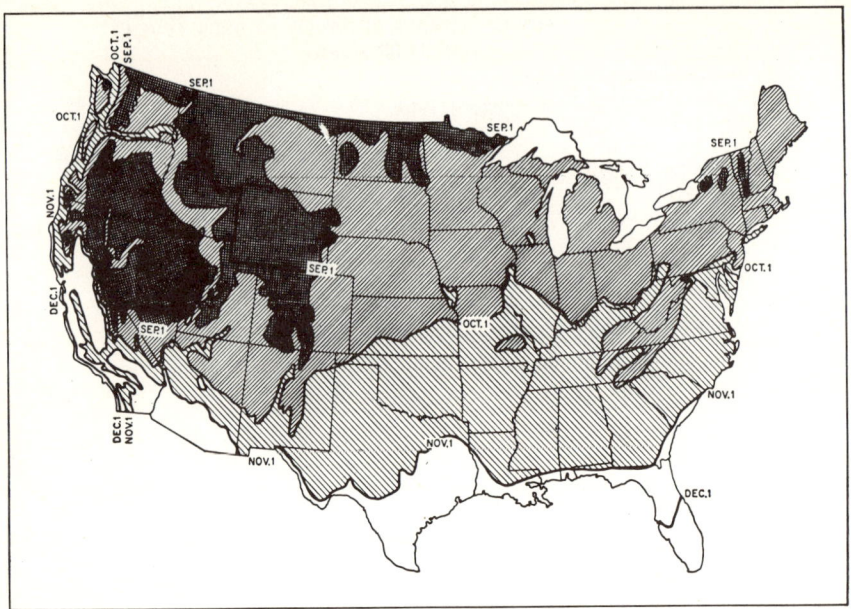

Fig. 57-34. United States earliest killing frost in fall, 10-year recurrence interval. (*Baker, USDA Atlas of American Agriculture*, 1936.)

REFERENCES

1. Aeronautical Services: "Sun Angle Calculator," Libbey-Owens-Ford Glass Co., Toledo, Ohio, 1951.
2. Albright, J. C.: "Summer Weather Data," The Marley Co., Kansas City, Kan., 1939.
3. "Heating, Ventilating, and Air Conditioning Guide," American Society of Heating and Air Conditioning Engineers, New York, 1956, p. 286.
4. Ashburn, E. V., and R. G. Weldon: Spectral Diffuse Reflectance of Desert Surfaces, *Optical Soc. Am. J.,* 46(8):583–586, August, 1956.
5. Baker, O. E.: "Atlas of American Agriculture," USDA, 1936.
6. Berry, F. A., E. Bollay, and N. R. Beers: "Handbook of Meteorology," McGraw-Hill Book Company, Inc., New York, 1945, pp. 978–986.
7. Blaney, H. F., and W. D. Criddle: Determining Water Requirements in Irrigated Areas, *USDA SCS Tech. Publ.* 96, 1950.
8. Brooks, C. F., A. J. Conner, et al.: "Climatic Maps of North America," Harvard Blue Hill Meteorological Observatory, Milton, Mass., 1936.
9. Brooks, F. A.: Atmospheric Radiation and Its Reflection Upward from the Ground, *J. Meteorol.,* 9(1):41–56, February, 1952.
10. Brooks, F. A.: Notes on Spectral Quality and Measurement of Solar Radiation, sec. II, pt. 2, in Daniels and Duffie, "Solar Energy Research," University of Wisconsin Press, Madison, Wis., 1954, pp. 19–29.
11. Brooks, F. A., and C. F. Kelly: Instrumentation for Recording Microclimatological Factors, *Trans. Am. Geophysical Union,* 32(6):833–848, December, 1951.
12. Brooks, F. A.: "An Introduction to Physical Microclimatology," ASUCD Store, University of California, Davis, Calif., 1959.
13. Court, Arnold: Wind Extremes as Design Factors, *J. Franklin Inst.,* 256(1):39–56, July, 1953.
14. Elsasser, W. M.: Heat Transfer by Infrared Radiation in the Atmosphere, *Harvard Meteorol. Studies,* no. 6, Milton, Mass., 1942.
15. Fletcher, Robert D.: A Relation between Maximum Observed Point and Areal Rainfall Values, *Trans. Am. Geophys. Union,* 31(3):344–348, June, 1950.

16. Geiger, R.: "The Climate near the Ground," Harvard University Press, Cambridge, Mass., 1950.
17. Golding, E. W.: "The Generation of Electricity by Wind Power," E. & F. N. Spon., Ltd., London, 1955.
18. Goss, J. R., and F. A. Brooks: Constants for Empirical Expressions for Downcoming Atmospheric Radiation under Cloudless Skies, *J. Meteorol.*, 13(5):482–488, October, 1956.
19. Hand, I. F.: Solar Energy for House Heating, *Heating and Ventilating*, 44:80–94, December, 1947.
20. Hand, I. F.: Weekly Mean Values of Daily Total Solar and Sky Radiation, *Weather Bur. Tech. Paper* 11, 1949.
21. Hand, I. F.: Insolation on Clear Days at Time of Solstices and Equinoxes for Latitude 42°N., *Heating and Ventilating*, 47:92–94, January, 1950; also, 51:97, February, 1954.
22. Hand, I. F.: Distribution of Solar Energy over the United States, *Heating and Ventilating*, 50:73–75, July, 1953.
23. Haurwitz, B.: Insolation in Relation to Cloud Type, *J. Meteorol.* 5:110–113, June, 1948.
24. Haurwitz, B., and J. M. Austin: "Climatology," McGraw-Hill Book Company, Inc., New York, 1944.
25. Heywood, H.: Solar Energy, Past, Present and Future Applications, *Engineering (London)*, 176:377–380, 409–411, Sept. 18–25, 1953.
26. Hopkins, A. D.: Bioclimatics: A Science of Life and Climate Relations, *USDA Misc. Publ.* 280, p. 9, and figs. 1, 55, 1938.
27. Hottel, H. C.: in W. H. McAdams, "Heat Transmission," McGraw-Hill Book Company, Inc., New York, 1954, p. 60.
28. Johnson, F. S.: The Solar Constant, *J. Meteorol.*, 11(6):431–439, December, 1954.
29. Kimball, H. H.: Variations in the Total and Luminous Solar Radiation, *U.S. Monthly Weather Rev.*, 47:769–793, 1919.
30. Kimball, H. H.: Intensity of Solar Radiation at the Surface of the Earth and Its Variation with Latitude, Altitude, Season, and Time of Day, *U.S. Monthly Weather Rev.*, 63:1–4, 1935.
31. Klein, W. H.: Calculations of Solar Radiation and the Solar Heat Load on Man, *J. Meteorol.*, 5(4):119–129, August, 1948.
32. Linsley, R. K.: Relation between Rainfall Intensity and Topography in Northern California, *U.S. Bur. Public Roads, Contract CPR*-113185, *Research Rept.* 1, June, 1956.
33. Mather, J. R. (ed.): The Measurement of Potential Evapo-transpiration, *Johns Hopkins Univ. Publ. in Climatology*, 7(1):1–225, 1954.
34. Moon, Parry: Proposed Standard Solar-radiation Curves for Engineering Use, *J. Franklin Inst.*, 230(5):583–617, November, 1940.
35. Orendorff, J. H.: "Application of Climatic Data to House Design," U.S. Housing and Home Finance Agency, Washington, D.C., January, 1954.
36. Petterssen, S.: "Introduction to Meteorology," 2d ed., McGraw-Hill Book Company, Inc., New York, pp. 208, 133 (discussion pp. 74–78, 121), 1958.
37. Pleijel, G.: "The Computation of Natural Radiation in Architecture and Town Planning" (Statens Nämnd för Byggnadsforskning), Victor Pettersons Bokindustri Aktiebolag, Stockholm, 1954.
38. Reynolds, G. W.: Venting and Other Building Practices as Practical Means of Reducing Damage from Tornado Low Pressures, *Am. Meteorol. Soc. Bull.*, 39(1): 14–20, January, 1958.
39. Russell, R. J.: Climate of California, *Univ. Calif. Publ. in Geography*, 2(4): 73–84 and map insert, 1926.
40. Schultz, H. B.: Exploratory Observations of Temperature Differences in Various Exposures, *Univ. Calif. Agr. Expt. Sta.*, Project 701-J, pp. 122–125, December, 1956.
41. Shands, A. L.: Mean Precipitable Water in the United States, *U.S. Weather Bur. Tech. Paper* 10, 1949.

42. Shands, A. L., and D. Ammerman: Maximum Recorded United States Point Rainfall, *U.S. Weather Bur. Tech. Paper* 2, April, 1947.
43. List, R. J. (ed): "Smithsonian Meteorological Tables," 6th ed., Smithsonian Institution Miscellaneous Collections, vol. 114, Washington, D.C., 1951.
44. Strahler, A. N.: "Physical Geography," John Wiley & Sons, Inc., New York, 1951.
45. Thompson, G. N. (chairman): American Standard Building Code Requirements for Minimum Design Loads, *Natl. Bur. Standards (U.S.) Misc. Publ.* M 179, 1945.
46. Thornthwaite, C. W.: An Approach toward a Rational Classification of Climates, *Geog. Rev.*, 38(1):55–94, 1948.
47. Trewartha, G. T.: "An Introduction to Climate," 3d ed., McGraw-Hill Book Company, Inc., New York, 1954.
48. "Climate and Man," USDA Yearbook of Agriculture, 1941.
49. Climatological Data, *U.S. Weather Bureau* (current).
50. Daily Weather Map, *U.S. Weather Bureau* (current).
51. Wassink, E. C. (chairman), Dutch Committee on Plant Irradiation: Specification of Radiant Flux and Radiant Flux Density in Irradiation of Plants with Artificial Light, *J. Hort. Sci.*, 28:177–184, July, 1953.
52. Went, F. W.: The Role of Environment in Plant Growth, *Am. Scientist*, 44(4):378–398, October, 1956.
53. Went, F. W.: "Experimental Control of Plant Growth," Chronica Botanica Co., Waltham, Mass., 1957, pp. 227–236.
54. Went, F. W.: Most Favorable Day and Night Temperatures for Some Garden Flowers, *Nurserymen's Inst. Agr. Ext. Serv.*, January, 1957 (unpublished).
55. Went, F. W.: personal communication to Mr. H. Kimball, June 4, 1958, restated at the National Conference on Agricultural Meteorology, Kansas City, May 19, 1960.
56. Barger, G. L., and H. C. S. Thom: Evaluation of Drought Hazard, *Agron. Jour.*, 41(11):519–526, November, 1949.
57. Drummond, A. J.: Notes on the Measurement of Natural Illumination, II, Daylight and Skylight at Pretoria, *Archiv. Meteor. Geophys. u. Bioklim.*, ser. B., vol. 9, no. 2, 1958.
58. Kimball, M. H., and F. A. Brooks: Plantclimates of California, *California Agriculture*, 13(5):7–12, May, 1959.
59. Shaw, R. H., and H. C. S. Thom: On the Phenology of Field Corn: The Vegetative Period, *Agron. Jour.*, 43(1):9–15, January, 1951.
60. Specter, W. C. (ed.): "Handbook of Biological Data," chap. VII, Environment and Survival, W. B. Saunders Co., Philadelphia and London, 1956.
61. Thom, H. C. S.: The Analytical Foundations of Climatology, Washington, D.C., U.S. Weather Bureau, Jan. 20, 1954, 25 pp., 4 figs., 8 refs. Presented at the 50th Anniversary Meeting of the Association of American Geographers.
62. van Bavel, C. H. M.: A Drought Criterion and Its Application in Evaluating Drought Incidence and Hazard, *Agron. Jour.*, 45(4):167–172, April, 1953.
63. Wang, J. Y., and G. L. Barger: "Bibliography of Agricultural Meteorology," University of Wisconsin Press, Madison, Wis., in press.
64. Withrow, R. B. (ed.): Photoperiodism and Related Phenomena in Plants and Animals, *Publ. Am. Assoc. Advance. Sci.*, no. 55, Washington, D.C., 1959, particularly pp. 440–471.

INDEX

Acceleration formula, 26
Ackerman steering, 288, 289
Acme harrow, 158
Adiabatic compression, 110
Air conditioning of animal shelters, 644
Air flow for drying farm crops, 647–658
   distribution systems, 651–658
      duct design, 652–658
      slatted floor, 654
   fans, 650, 651
   requirements, 663, 669
   resistance, of grain and seeds, 648, 649
      of hay, 649
   reversible, 656
Air standard efficiency, 111
Airplane dusting and spraying, 198, 199
Albedo, 839–841
Alfalfa dehydration, 670
Alfalfa pollinating machine, 279
Alter shield, 315
Ammonia, anhydrous, 814
Angle of repose of grains, forages, etc., 691
Asparagus harvester, 282

Back hoes, tractor, 297
Bale loaders, 226
Bale sizes, 216
Bale throwers, 226
Baler components, 216–224
   bale-length control, 219, 220
   chamber, 217
   density control, 218, 219
   feeding mechanism, 217
   pick-up, 216, 217
   plunger, 217, 219
   ram stop, 223
   shearing knife, 219
   tying mechanism, 220–223
      holder, 221
      knife, 221
      needles, 219
      twine-knotter bills, 221, 222
      wire-twister gears, 222
      wire-twister hook, 221

Balers, plunger force, 218
   power requirements, 224, 225
   round, 225
Baling twine, 220
Baling wire, 220
Bangs disease, 593
Barn (gutter) cleaners, 595, 596
Barns, beef-cattle housing, 607
   horse housing, 613, 619
   stall (*see* Stall barns)
Bearing seals, 34
Bearings, characteristics of various types, 33
   loads due to chains and belts, 34, 41–44
   lubrication, 34
Bedding, 376, 377
Beef cattle, effects of thermal environment on, 773
   air velocity, 773
   shades, 629, 773
   feed and water requirements, 783–785
   heat loss, 776
   labor for feeding, 608
   production structures, 606–609
   space requirements, 616
   surface area, 781
   surface temperature, 781
Beef-cattle housing, 607
Beef production and feeding, principles and practices, 608
   time and distance for feeding, 608
Beet harvester elements, lifters, 272
   separation from dirt, 272–274
   toppers, 270–272
Beet harvesting, losses, 274
   power requirements, 274
Bending stress formula, 26
Bin unloaders, 307
Bins (*see* Grain storage bins)
Birefringent coatings, 88
Blowers, impeller, forage (*see* Forage blowers)
   grain, 308
Blueberry harvesting, 282
Bluegrass strippers, 276

867

868 INDEX

Bmep (brake mean effective pressure), 108
Bolts, allowable loads in wood, 565, 566
 heads and nuts, dimensions, 28
Break-back protective mechanisms, 57–59
Bridges, farm, 370, 371
Brinell hardness, 23
Brittle-coating stress analysis, 75
Broadcast seeders, airplane, 179
 disk tiller, 180
 endgate or spinner, 179
 wide-hopper, 179
Brooders, chicken, 610, 611, 775
 pig, 590, 591
Buck rakes (sweep rakes), 214, 215
Buildings (*see* Farm buildings)
Bulk handling of fruit and vegetables, 733
Bunk feeders, auger, 311
Burr mills, 298, 299

Calcium plant food, 808–816
Calibration, fertilizer distributors and grain drills, 164
Capacity of field machine, formulas for estimating, 10–12
 optimum, 15–17
Carbon dioxide production by stored grain, 685
Castor bean harvesting, 280
Cetane number, 113
Chain, stripping of chain fingers, 229
Chain drives, dimensions and strength, 39
 formula for length, 40
 pitch modifications, 38
 velocity fluctuation, 38
Check rowing, 180, 182, 183
 adjustments for accuracy, 182, 183
 check-wire handling, 182, 183
 valve action, 182
Chinook wind, 827
Chisels, 149
Circuit breakers, 697
Climate, air masses, 818
 classification, 817–824
 by moisture index (Thornthwaite's), 817–819
 by vegetation (Koeppen's), 817
 elevation effects, 848
 local factors, microclimate, 852–855
 slope effects, 847, 848
 weather belts, 818
 weather types, 824, 825
 (*See also* Weather)
Combine efficiency, 249, 250

Combine elements, 238–247
 baling press attachment, low-density, 245
 cleaning shoe, action, 247
 chaffer, 246
 sieve, 246
 concaves, clearance settings, 243, 278
 closed-type, 243
 open-type, 243
 corn head, 257
 cylinder, 241–244
 rasp bar, 241, 242
 speed settings, 242, 243, 278
 spike-tooth, 241, 242
 spring-tooth, 242, 275
 grain tank, 247
 header, 238–241
 auger, 238
 cutter bar, 238
 divider, 238
 pick-up guards, 238
 reel, 237, 238
 undershot conveyor, 241
 windrow pick-up attachment, 240, 241
 recleaner or rotary screen attachment, 247
 straw-cutter attachment, 245
 straw rack, 244, 245
 straw-spreader attachment, 245
Combines, efficient operation, 250
 hillside, 247
 losses, cracking, 242, 272
 cutter bar, 249, 272
 cylinder, 249, 272
 rack, 249, 272
 shoe, 249, 272
 power requirements, 247, 248
 self-propelled, 247, 248
Compression ratio, 113, 114
Concrete, 556–560
 aggregate, 556, 557
 blocks, 560–561
 forms, 557–559
 materials, quantity calculations, 559, 560
 proportioning mixes, 558
 winter pouring, 609
Condensation, surface, 634
Consumptive use of water (*see* Irrigation water)
Contour farming, 414
Contour strip cropping, 414
Conveyers, auger, 306, 307
 capacity, 307
 power requirements, 307
 flight, 305, 306
 capacity, 306

INDEX

Conveyers, flight, chain pull, 306
    power requirement, 306
  portable, 305
  pneumatic (*see* Feed conveying)
  vertical bucket, 307
Corn binders, 259
Corn combining, 257
Corn cribs, 692–694
  natural ventilation, 693
  recommended widths, 693
Corn-picker elements, 251–256
  cleaning fan, 255, 256
  ear retarders, 254, 255
  elevators, 256
  gathering chains, 251
  husk conveyer, 255
  rolls, husking, 254
    snapping, 251–253
    stalk-ejector, 253
Corn pickers, losses, 256–258
  power requirements, 258, 259
  types, 256
Corn planters, 180–183
Corn shellers, 259, 260
Corn snappers, 253
Cotton, cultural practices for mechanical harvesting, 268
Cotton gin turnouts, 261
Cotton grades, 261, 262
Cotton pickers, spindle, 262–266
  losses, 264–266
  picking rates, 265
Cotton planters, 183, 184
  fertilizer attachments, 184
  plateau profile, 184
Cotton strippers, double-roll, 267
  finger, 266
  losses, 267, 268
  single-roll, 266, 267
Cover crops, 416
Critical depth, definition, 429
  emergency spillway, 465
  tables for various discharges, 444
Critical flow, 429
Critical shaft speed, 54
Crop production, costs, 2
  estimating procedure, 12–15
  man-hours, 2–3, 13
Crops, plant food content, 804, 805
Cucumber harvester, 282, 283
Cultivators, field, 149
  flame, 154
  lister, 153, 154
  row crop, 151–154
Culverts, capacity calculations, 364–369
  (*See also* Spillways for erosion control, closed-conduit)
  head loss coefficients, 447, 448
  monolithic, capacity, 446

Cutting energy for forage, 201, 232, 233
Cyclonic storms, 827

Dairy-calf housing, 602
Dairy cattle, effects of thermal environment on, 771, 772
  air velocity, 772
  relative humidity, 772
  temperature, 771
  feed and water requirements, 601, 602, 782, 784, 785
  heat and moisture loss, 776
  production structures, 593–606
  space requirements, 615
  surface area, 781
  surface temperature, 780
Dairy production costs, 602
Delayed lift, 122, 123
Densities of grains, forages, etc., 683, 691, 693
Depreciation, 4–6
Detonation, 113–115
Dew-point temperature, 624
Dielectric constants of agricultural materials, 719
Diesel combustion chambers, 115–116
Disk blades, action, 138
  draft, 20, 140
  forces on, 138–140
  sizes, 138–140
  steels for, 138, 139
Disk harrows, bearings for, 143
  offset, 144, 145
  lift-type, 143
  trailed-type, 141–143
  single, 141
  tandem, 141–143
Disk plows, 147
Disk tillers, 144–147
Ditches (*see* Diversion ditches; Drainage, open-ditch; Irrigation canals and ditches; Waterways)
Diversion ditches, capacity, 413, 422
  construction, 413
  design, 413
  maintenance, 413
  staking, 413
  system planning, 412
Dog clutch, 117
Dorroh curve, 320
Draft control, 122
Draft requirements, farm machines, 19–22
Drainage, effects of soil characteristics on, 795–797
  enterprises and districts, 399, 400
  investigation and surveys, 356
  irrigated lands, 398

Drainage, laws on, 399
  mole, 383
  open-ditch, 357–374
    bridges, 370, 371
    construction, 371–373
    construction tolerances, 372
    culvert capacity, 366–369
    design, 357–371
    ditch capacity, 363–364
    excavation, 372
    hydraulic grade line, 362
    investigation and surveys, 356
    leveling spoil banks, 373
    maintenance, 373
    Manning roughness coefficients, 362
    maximum velocity, 361, 362
    runoff coefficients, 358
  pumping plants, 393–397
    automatic operation, 393
    capacities, 393–394
    head, total, 394, 395
    head losses, 397
    power-type, 397
    pump selection, 395
    suction and discharge pipe, 396, 397
  surface, 376–382
    bedding, 376, 377
    ditch capacity, 382
    ditch systems, 378–381
      cross-slope, 378, 380
      parallel, 378, 381
      random, 378, 379
    field ditches, 380–382
  tile, 382–393
    auxiliary structures, 392, 393
    benefits, 382
    blinding, 392
    breathers, 393
    capacity, 385
    coefficients (rate of removal of water), 386
    depth and spacing, 386–389
    grade limits, 389, 390
    gridiron system, 383
    inlets, 392
    junction boxes, 393
    loads on, 387–390
    minimum tile size, 389
    permissible tolerance in laying, 392
    random system, 384
    seepage interception, 384
    specifications, 391
    staking, 391
    tree removal, 393
    trenching equipment, 383
Drifting of spray and dust, 187, 188
Dryers, heated-air, batch, 658
    column, 658, 659

Dryers, continuous, 658
  heaters, 658
    direct, 658
    indirect, 658
Drying rate, 646
Dusters, airplane, 198, 199
  fans, 198
  metering, 197
Dynamometers, 107

Ear corn drying, 661, 664, 666
Ear corn storage (see Corn cribs)
Earth dams (see Embankments)
Earthwork quantities, land leveling or grading, 353
Electric heating, 718–721
  agricultural applications, 721
  arc, 721
  dielectric, 719, 720
  induction, 720
  infrared (see Infrared heat lamps)
  resistance, 718, 719
Electric lamps, 708–713
  data, 709
  fluorescent, 709, 710
  illumination by, 711, 712
  infrared, 709, 713
  insect control, 713
  radiation effects, 713
  types, 708, 709
Electric motors, a-c, 698–704
  characteristics, 699–701
    induction, 699
    series, 699, 700
    synchronous, 700
  construction, 701, 702
  enclosures, 703
  overcurrent protection, 703, 704
  service factors, 701
  standard hp sizes, 702, 703
  types, 698
Electrical wiring, 695–698
  current capacity, 697
  grounding, 698
  planning, 695
  types of wires and cables, 695–697
  underground, 696
Electrically operated farm equipment, data, 723–727
Elevators (see Conveyers)
Embankments, core trench, 490
  design types, 482
  drains, 483
  foundation investigation, 483
  phreatic line, 488
  rock riprap, 490–491
  sealing against seepage, 489
  site condition, 485

Enthalpy, 624
Entropy, 624
Equilibrium moisture content, 647
Erosion effect on, of cropping, 403
    of soil characteristics, 798, 799
    estimating soil loss, 402
    gully, 406
    irrigation, 517
    soil erodibility factors, 405
    water, mechanics of, 401
    wind (*see* Wind erosion)
Evaporation, annual amount, 334
    formula for, 333
Evapo-transpiration, 326–332
    consumptive use coefficients, 330
    formula, 328
    maximum daily, 331
    potential, 332, 818, 844, 846, 847
Extensometers, 77

Factors of safety, 25
Fans, performance curves, 651
    types, 650, 651
Farm buildings, fire losses, 583, 584
    fixed costs, 587
    footings, 576, 577
    framing (*see* Framing)
    functional requirements, 619–621
    lightning protection, 585, 586
    pole construction, 573–575
    roofing, 580–582
    steel, 572, 575, 686, 692
    wind losses, 583–586
Farm-machinery housing, 764–766
    space requirements for machines, 765, 766
    types, 766
Farm machines, cost of use, 3–10
    design requirements, 18–19
    draft requirements, 19–22
    percentage of nonoperating time, 10
    power requirements, 19–22
Farm-mechanization statistics, 1
Farm ponds (*see* Ponds)
Farm shop, 721–723
Farrowing pen, 591
Fatigue testing, 89
Feathering action of tooth bars, 208, 209
Feed consumption of livestock, 782–786
Feed conveying, pneumatic, capacity, 310
    fluidized, 311
    grain damage, 309
    power requirements, 310
Feed grinders, burr mills, 298, 299
    combination or roughage mills, 303, 309
    hammer mills, 299–302

Feed grinders, hammer mills, cyclone feed collector, 301
    drives, 300
    fans, 300, 301
    hammers, 299, 300
    power requirements, 301
    screens, 300, 302
    roller mills, 304
Feed grinding, advantages, 298
    automatic, 304
    fineness modulus, 299
Feed lot layout, beef cattle, 608
    dairy cows, 598
    hogs, 592
Feeding, livestock, automatic, 311
Fences, 767–770
    construction, 768–770
    electric, 770
    posts, 767–769
    staples, 768
    types of wire, 767, 768
Fertilizer, application methods, 815, 816
    deficiency indications by plants, 816
    determining requirements, 811–815
    dry, 160–161
        hygroscopicity, 161
        kinetic angle of repose, 161
        metering mechanisms, 161–165
            bottom-delivery, 162–164
                agitator feed, 163
                auger feed, 164
                endless-belt feed, 164, 165
                revolving-plate feed, 163, 164
                star-wheel feed, 164
        side-delivery, 162
        top-delivery, 161, 162
    placement, 167
    specific gravity, 161
    elements, 803–815
        amounts in soil, 803–810
        availability, 805–810
        losses, 805–810
        nutritive role, 803–809
        origin in soil, 803–810
    grades, 810, 811
    liquid, 167–171
        anhydrous ammonia, 167–170
        applicator blades, 170
        aqua ammonia, 170
        distributor, 169–171
        mixed, 170, 171
        nitrogen solutions, 170
        storage tanks, 170
    materials, 810, 811
Fertilizer attachments, corn and cotton planters, 166, 167
    cultivators, 166, 167
    dual-level applicators, 167
    grain drills, 166

872

INDEX

Fertilizer distributors, central-hopper, 165–166
   liquid, 167–171
   wide-hopper, 165
Field capacity, soil moisture, 793
Field tillers, 149–150
Finger weeders, 158
Fire losses, farm buildings, 583, 584
Flexural center of various sections, 24
Flood control, 496
Flood flow, graphical solution, 473
   prevention, 500
   routing, 468
Floodwater, detention structures, 429
   retarding structures, 501
Foot-candle requirements, 713
Footings, 576–577
Forage blowers, 235–237
   delivery pipes, 237
   peripheral velocities, 235
   unloading action of blades, 236
Forage harvester, capacity, 232
   cylinder-cut, 231, 232
   flail-type, 234, 235
   flywheel-cut, 229–231
   power requirements, 232–234
Forage-harvester elements, 227–232
   cutter bar, 227
   cutter head, 229
      cylinder, 231, 232
      flywheel, 229–231
      knives, 230, 231
   feed rolls, 229
   impeller blades, 231
   reel, 227
   row-crop attachment, 227–229
   shear bar, 231
Forage-seed harvesting, 276–279
   baling for shatter prevention, 279
   combine adjustments, 276–278
   combining losses, 277
   direct combining, 277, 278
   stripper, 276
   vacuum, 278, 279
   windrow combining, 276
Force efficiency, traction wheel, 67
Fork lift trucks, 731, 733
Foundations, soil evaluation, 475, 483
Four-stroke cycle, 110
Four-wheel-drive tractors, 70
Fracture-pattern analysis, 92
Frame design, farm machines, fabrication, 22–23
   forces on trailed machines, 287
Framing, wood construction, 571–576
   fasteners, 573, 574
   roof trusses, 573–579
   wall, 572, 573
Freeboard, 428

Friction, coefficients of, axle bearing, 60
   grain, forages, etc., 691
Frost, killing, dates of, United States maps, 863, 864
   radiation, 532, 533
Frost control, cloud screens, 534
   heaters, 534–539
   plant covers, 534
   water spray, 533, 534
   wind machines, 537–539
Fruit harvesting, 281, 282
Fruit and vegetable cooling, 746–748
   air, 746, 747
   contact icing, 747
   hydrocooling, 747
   vacuum, 748
Fruit and vegetable handling, automatic electronic color sorting, 735
   containers and weights, 729
   conveyors, 730–732, 735
   equipment, 730–732
   handling costs, 734
   harvest, 281–283, 732, 733
   lighting for sorting, 734
   packing and packaging, 736–738
   practices, 738–740
      apples, 739
      cabbage, 740
      celery, 740
      cherries, 739
      citrus, 738
      grapes, 738
      lettuce, 740
      onions, 740
      potatoes, 739
      sweet corn, 740
      tomatoes, 739
   sizing equipment, 735
   storage, 733, 734
   washing and cleaning, 736
Fruit and vegetable storage, controlled atmosphere, 754
   deterioration processes, 741–746
      chilling or freezing injury, 742, 745, 746
      decay, 741, 742
      softening, 743
      sugar loss, 742
      water loss, 743
   moisture condensation on removal, 755
   recommended conditions, 744, 745
   storage characteristics, 744, 745
Fruit and vegetable storage structures, 748–751
   air-cooled or common, 748–750
      apples, 749
      onions, 750
      potatoes, 749
      ventilation, 750

Fruit and vegetable storage structures,
    air purification, 754
    refrigerated, 750, 751
Fuel storage, 766
Fuels, tractor, diesel, 111–113
    gasoline, 111–113
    LPG (liquified petroleum gases), 111
Furrow openers, grain drills, and planters, 177, 178
Fuses, 697

Gage shoes, 177
Gage wheels, cultivator, 152
    plow, 134
Gangs, cultivator, 152, 153
Garbage cooking for hogs, 591
Gears, 35–37
    cast-iron, strength of, 36
    causes of failure, 35–36
    life factors, 37
Glued joints, wood, strength of, 568–571
Grain blowers, peripheral speeds, 308
    power requirements, 308
Grain drill feeds (*see* Seed-metering mechanisms)
Grain drills, fertilizer, 180
    one-horse, 180
    pasture-renovating, 180
    plow-press, 180
    press, 180
Grain drying, heated air, 662–668
    principles and practices, 668
    time and fuel requirements, 665
    safe moisture content, 660
    summary of recommendations, 666
    unheated air, 661–663
        air and fan requirements, 663
        principles and practices, 667
Grain storage bins, 687–692
    capacities, 686, 687
    concrete, 692
        hoop spacing, 691
    design of walls, 687–692
        deep, 689, 690
        shallow, 688, 689
    discharge openings, 692
    floors, 687
    lateral pressure, 688–690
    steel, 686, 692
    vertical load, 689–690
Grain ventilation for preservation, 659–661
    mechanical aeration, 660, 661
    natural, 659, 660
Granular insecticides, 198
Grape harvester, 282
Grass-seed harvesting (*see* Forage-seed harvesting)

Green-manure crops, 416
Ground-water depletion, 337
    artificial, 336
    natural, 335
Ground-water movement, 335

Hail, United States map, 859
Hammer mills (*see* Feed grinders)
Hardness conversion scale, 23
Hay crimpers, 206
Hay crushing or conditioning, 204–206
    clogging, 205, 206
    field losses, 205, 206
    increase in drying rate, 204
Hay drying, duct systems, 651–658
    heated air, 669, 670
        cost of operation, 671
        principles and practices, 670
        slatted-floor system for bales, 654
    unheated air, 668–670
        air flow requirements, 669
Hay harvesting methods, changes in, 213
    labor and cost comparisons, 214
Hay loaders, 214
Hay rakes, dump, 212
    side-delivery, 207–212
        cylindrical reel, 207–208, 211
        side-stroke or roller-bar, 207, 211
        wheel, 207, 208
    strippers, 211
    suspension, 212
    teeth, 211
    tooth bars and bearings, 211
Hay raking action, analysis of, 208–211
Hay stackers, 215
Hay storage, capacities, 684
    self-feeding, 683, 684
    space requirements, 683
    spontaneous ignition, 684, 685
Hay wafers or pellets, 225
Heat, of combustion, 625, 626
    latent, 624, 625
    of respiration, 625
    sensible, 624
    specific, 625
Heat loss of livestock, 775–780
Heat pump, 714–718
    coefficient of performance, 715, 716
    operating cost, 716–718
Heat transfer, 630–635
Heaters for crop drying, 658
Hens (*see* Poultry)
Hill dropping, 174
Hog houses, central, 590
    farrowing, 589, 590
    portable, 589
    ventilation, 591

Hog production and feeding, principles and practices, 593
Hogs (see Swine)
Horse housing, 613
  space requirements, 619
Horsepower, belt, 108
  drawbar, 108
  formulas, 26
  hydraulic, 26, 118
Horsepower-hours per gallon, 108
Housing cost for machines, 7
Humidity, 622
  relative, United States maps, 858
Husker-shredders, 259
Hydraulic control systems, 118, 121–123
  desirable capabilities, 118
  lift linkage, 122, 134
Hydraulic couplings, 123, 124
Hydraulic equivalents, 338
Hydraulic pumps, 118, 119
Hydraulic ram, 761
Hydraulic remote cylinder, 123
Hydraulic valves, automatic flow-dividing, 120, 121
  check, 120, 121
  control, 121
    closed-center spool, 120, 121
    open-center spool, 121
  relief, 120
  unloading, 119
Hydraulics, definition of, 313
Hydrographs, 342–346, 468
Hydrologic cycle, 313
Hydrology, definition of, 313

Incinerator for burning animals, 764
Infiltration, 322
  soil, rate of, 797
Infrared heat lamps, 590, 610, 611, 709, 713
Infrared waves, 627
Insulated glass, 628
Insulation, reflective, 631
  thermal, 630, 631
  values of various walls, 632–634
Interest cost, 7
Irrigation, sprinkler, 526–530
  lateral design, 528
  layout design, 527, 528
  mainline friction loss, 530
  nozzles, 528, 529
  sub-, 529
  surface, application criteria, 516
    border, 522–525
      intake rate and time, 524, 525
      maximum stream size, 525

Irrigation, surface, furrow or corrugation, 519–523
  intake rate and time, 521, 522
  maximum length of run, 519, 523
  siphon discharge, 520
Irrigation canals and ditches, 513, 514
  prevention of erosion, 517
  pump capacity, 525
  structures, 514
Irrigation water, application losses, 512, 513, 526
  crop requirements, 327, 332, 515, 518, 519
  delivery systems, 512–515
  forecasting supply, 497
  quality, 509–512
  storage and supply, 511–515
Isobars, 843
Isothermal compression, 111

Janssen's equation for bin wall pressures, 688

Kinetic-energy formula, 26
Knife-tooth harrow, 158

Laborsaving principles, 588, 729, 730
Land capability classification, 792, 793
Land leveling, computing earthwork quantities, 353
  design methods, 349–353
    contour adjustment, 349, 350
    plan inspection, 350, 351
    plane, 352, 353
    profile, 351
  equipment and construction, 354
  for irrigation, 347, 518
  maintenance, 355
  for surface drainage, 348
  surveys and staking, 348
Levees and dikes, 374, 375
  (See also Embankments)
Lighting, portable, 710
Lighting recommendations, 709, 713
  fruit sorting, 734
Lightning protection, farm buildings, 585, 586
Lime (calcium plant food), 808–816
Lime spreaders (see Fertilizer distributors)
Lister planters, 177
Listers, 136, 137
Livestock, environmental factors, 587, 771–780
  feeding plant layout, 589, 592, 608
  labor operations, 588, 589

Livestock, summer shelters, 629, 630
Loaders, tractor, bucket controls, 295
  hydraulic components, 296, 297
  lifting force, 296
  rear-mounted, 297
  tractor stability, 295
Loose housing, dairy cows, 597
  area requirements, 600
  L-shaped layout, 598
  principles and practices, 600, 601

Magnesium plant food, 809–813
Manning formula, 363
  graphic solution by, drainage channels, 363–364
    surface drainage, 382
    waterway channels, 322–323
Manning roughness coefficients, drainage, 362
  waterways, 421
Manure spreaders, beaters, 292, 293
  box, 294
  capacity, 294
  conveyor, 293
Mastitis, 593
Materials of construction, properties of, 27
Materials handling, 588, 728–730
  loading, 588
  transporting, 588
  unloading, 588
Middlebusters, 136, 137
Milk coolers, 706, 707
Milk piping, 705
Milkhouse, 603–606
  dimensions and layout, 604, 605
  equipment, can and bulk, 604
  heating, 606
  location, 603
  Ordinance and Code for, 606
  ventilation, 605
Milking machines, 705, 706
Milking-room layouts, 597
Moisture content for safe storage, ear corn, 693
  grain, 660
  hay, 204, 683
Moisture equivalent of soil, 794
Moisture loss of livestock, 775–780
Moldboard plows, draft of, 20, 129, 130
  effect of speed, 130
  soil factors, 130
  sources of, 130
  forces, 130–132
  materials for, 136
  mounted, 134
  scouring, 129
  traction for, 134, 135

Moldboard plows, trailed, hitching of, 132–134
  trash coverage, 135, 136
  types of bottoms, 128
Moment of inertia, approximation for formed sections, 23
  for various sections, 29–32
Mowers, cutting action, 200–202
  double-knife, 202
  hydraulic-driven, 203, 204
  knife forces, 202, 203
  materials in cutter bar parts, 204
Mulch planter, 127
Mulch tillage, 417

Nails, 561–564
  allowable loads, 562–564
  roofing requirements, 582
  types and sizes, 562
National Electric Code, 695
Nebraska tractor tests, 109
Nitrogen plant food, 803–815
Nut harvesters, 281–282

Obsolescence, 5, 6
Octane number, 113
Onion harvesters, 275
Openers, furrow, 177, 178
Oscillogram, 84
Outlets, water, cantilevered, 456
  impact-type, 457
  SAF, 459

Pallet handling, 733, 739, 750
Pasteurization, milk, 708
Paved barnyards, 609
Pea-and-bean seed boxes, 184
Peanut combines, 275
Peanut digger-shakers, 274
Peanut threshers, 275
Permeability to water vapor, 635
Phosphorus plant food, 806–815
Photoelasticity, 85
Photoperiods, plants, 851
Pipe, concrete, head loss due to friction, 447, 448, 515
  irrigation, friction loss, 528, 530
  plastic, 760, 761
  steel, columns, allowable loads, 556
    loss in head due to friction, 759
    properties of, 24
Pipelines, irrigation, 515
Plant population for various spacings, chart, 181
Planter feeds (*see* Seed-metering mechanisms)
Planting, seed factors affecting, 172

Plastics for pipe, physical properties, 760
Plows, disk, 147
  moldboard (see Moldboard plows)
Plowshare impact strength, 57
Plowshare wear, 135, 136
Plywood, 552, 553
Pneumatic conveying (see Feed conveying)
Pole construction, buildings, 573–575
Polytropic compression, 111
Ponds, farm, 492–494
  construction, 493
  dug, 494
  fish production, 494
  hydraulic design, 430
  staking, 494
  water storage, 492
Position-responsive control, 121, 122
Potassium plant food, 807–815
Potato diggers, 269, 270
Potato harvesters, bruise prevention, 270
  separation from dirt, 270
Potato planters, 184, 185
Potato spinners, 270
Potential evapo-transpiration, 332, 818, 844, 846, 847
Poultry, brooding practices, 775
  disposal pits, 763, 764
  effects of thermal environment on, 774
    humidity, 775
    light, 775
  feed and water requirements, 784, 785
  heat and moisture loss, 776–780
  production structures, 609
  space requirements, 617
  surface area, 782
  surface temperature, 780
Poultry housing, broilers, 610, 611
  cage operations, 610
  laying, types, 609, 610
Poultry production, principles and practices, 611, 612
Power efficiency, traction wheels, 67
Power requirements, farm machines, 19–22
Power take-off, implement drives, 54, 55
  torsional loads, 21
Precipitation, 313, 824–828
  effective, 516
  gages, 314, 315
  maximum probable rate, 828
  measuring equipment, 314, 315
  United States maps, 860–862
  world map, 823
  (See also Rainfall)
Press wheels, corn planter, 181
  grain drill, 180
  seed, 178, 179
  transplanter, 186

Probe for hay temperature, **685**
Prony brake, 107
Psychrometric chart, 622, 623
Pulsator, milking machine, 705
Pulverator, 155
Pumping plants (see Drainage, pumping plants)
Pumps, hydraulic oil, 118, 119
  sprayer, 189, 192
  water, 756, 759
    output and hp, 756, 757
    types and characteristics, 756, 757

Radiant-energy exchange, 837, 839
Radiation, 626–630, 828–841
  atmospheric, 841
  effects of, 713
  frosts, 532, 533
  long-wave emission, 837–841
  night, 532, 533, 629
  sunshine (see Solar energy)
Radius of gyration, various sections, 29–32
Rainfall, accumulation graph, 316, 320
  antecedent, 321
  extension of design storm, 318
  intensity-frequency data, 317, 321
  point, 319
  run-off curve numbers, 323–324
  six-hour point, 316–319
  watershed average, determination, 315
  Yarnell data, 318
Rakes (see Hay rakes)
Rankine's equation for bin wall pressures, 688
Raspberry harvester, 282
Refrigeration, cold storage, 751–753
  components, 751, 752
  determining requirements, 752, 753
  refrigerants, 751
Relative humidity, United States **maps,** 858
Release hitches, 56, 57
Repair cost, 5, 6
Resistance to air flow, of grains and seeds, 648
  of hay, 649
Respiration, of fruits and vegetables, 742, 744, 745, 753
  of grain, rate, 685, 687
Rice drying, 664–667
Riprap, rock, 490–491
Rockwell hardness, 23
Rod weeders, 158
Roller mills, 304
Rollers, 159

Rolling resistance, coefficients of, 60–64
    rubber tires, 62–64
    steel wheels, 60–62
    tandem wheels, 64
Roof trusses, 573–579
Roofing, 580–582
    asphalt shingles, 580
    labor requirements, 582
    nail requirements, 582
    pitch recommendations, 581, 582
    sheet metal, 581
    wood shingles, 580
Rotary hoes, 158, 159
Rotary tillers, 154–157
Roughage or combination mills, 303, 309
Runoff, average annual, 337, 338
    effects of soil characteristics on, 797, 798
    estimating from rainfall, 326
    hydrographs, 342–346
    maximum rates, 339
    peak rates, 345
    pumping-plant capacity, 394
    rational formula, 340
    snow, 337, 338
    summation method, 341
    surface-drainage coefficients, 358
    time of concentration, 342–344

Safety, design precautions, 19
    tractor, 95–97, 102, 104
Saltation, 505
Sanitizing dairy equipment, 708
Scouring, 129
Screw threads, dimensions, 27
Screws, wood, allowable loads, 564, 565
Seals, bearings, 34
Section modulus, formula, 26
    various sections, 29–32
Seed covering devices, 178, 179
Seed delivery passage, 176, 177
Seed-metering mechanisms, 172–176
    agitator, 172
    belt feed, 176
    cup feed, 174
    double-run or internal force feed, 173, 174
    fluted or external force feed, 172, 173
    horizontal plate, 174–176
        cell fill factors, 174, 175
        cutoffs, 175
        knockouts, 175
        picker-wheel, 183, 184
    inclined plate, 174
    vertical plate, 174
Seed press wheel, 179
Seedbed requirements, 125
Seeders (see Broadcast seeders)

Selective lift, 123
Self-feeders, hay, 657, 658, 684
    hogs, 592
    silage, 682
Septic tanks, 762
    capacities, dimensions, and materials, 761, 762
    design, 762, 763
    disposal fields, 762, 763
    operation, 761, 763
Shearing stress formula, 26
Sheep, feed and water requirements, 784, 785
    production principles and practices, 612, 613
    production structures, 612
    space requirements, 618
Shields, cultivator, 153
Shore hardness, 23
Shovels, draft of, 20, 148
    steels for, 149
    types, 153
Shredders, stalk and brush, blades, 285
    cross-row, 285
    cylinder or hammer-mill, 284, 285
    horizontal-blade, 285
    power requirements, 286
Silage, 672, 673
    density, 674
    dry-matter losses, 673
    moisture content, 672
    preservatives, 673
Silo-gas poisoning, 682
Silo unloaders, 682
Silos, gas-tight, 682
    horizontal, 674–678
        bunker, 674, 676, 678
        feeding rate from, 679
        lateral pressures, 675, 676
        trench, 674, 675, 677
        walls, 674–679
    principles and practices, 682, 683
    temporary, 682
    vertical, 676, 680–682
        foundation design, 680
        lining, 681, 682
        wall design, 681
            lateral pressures, 681
            vertical loads, 680
Siphon discharge, irrigation, 520
Sizing equipment, 735
Sling psychrometers, 624
Slip clutches, 59
Slippage, driving wheels, 67
    transport wheels, 64
Snow, measurement of, 315
    moisture equivalent, 314
    roof loads, 541, 542
    runoff, 337, 338

Snow cover, United States map, 863
Sod chutes, 423
Soil, acidity, 813
  air, 794
  bulk density, 795
  classification, 483, 789–792
  compaction by traffic, 71–72
  definitions, 788
  fertility, 803–816
  footing loads, 576
  formation, 788–791
  foundation evaluation, 479, 483
  heat balance, 848
  hydraulic conductivity, 795–797
  infiltration rate, 797
  moisture, 793, 794
  optimum moisture content for compaction, 487, 488
  physical properties, 484, 793–803
  plastic limits, 799
  profile and horizons, 790, 791
  push, 800
  shearing strength, 66, 486
  temperature, 795, 848, 851
    effect of mulches, 795
  tension computation, 42–44
  water stability, 798, 799
  (*See also* Soil mechanics)
Soil mechanics, compaction, 487
  grain size distribution, 476
  laboratory tests, 487
  unified classification system, 478
Soil structure, 125, 126
Solar energy, atmospheric transmittance, 835, 836
  bands, spectral segments, 832
  diffuse radiation, 835–838
  direct-beam intensity, 835, 838
  distribution, hourly, 830
    spectral, 830, 831
    United States, 626, 627, 829
  sloped surfaces, 837
Solar heating, materials, 627, 628
  roof overhang, 628
  (*See also* Solar energy)
Spike-tooth harrows, chain-type, 157, 158
  draft of, 20, 157
  flexible, 157
Spillways for erosion control, antiseep collars, 433
  auxiliary, 428, 462
  box-inlet drop, 441
  camber, 435
  cantilevered outlet, 456
  chute, 454
  closed-conduit, 446
  critical depth flow, 429
  cutoff walls, 433
  emergency, 420, 462

Spillways for erosion control, erosion control, 426
  expansion joints, 435
  inlet protection, 429
  island-type, 432
  material for construction, 468
  outlets (*see* Outlets)
  plans for, 427
  principal, 437
  SAF stilling basin, 459
  straight drop, 437
  surveys for, 427
  toewalls, 425
  trapezoidal weir box, 442
  trash racks, 433
Spontaneous heating and ignition, 684
Spray droplet size, 187, 188
Sprayers, airplane, 198, 199
  field, 194–196
    booms, 195
    calibration, 196
    nozzle height and spacing, 195, 196
  hose, 194
  nozzles, 189, 190
    fan-type, 189
    hollow-cone, 189
    shear plate, 189
  orchard, 197
  pressure regulators, 192
  pumps, 189–192
    centrifugal, 190
    diaphragm, 191
    efficiency, 192
    gear, 191
    piston, 190, 191
    plunger, 191
    single-screw, 192
    vane-type, 192
  strainers, 194
  tanks, 193, 194
    tank agitation, 193, 194
Sprigging, 425
Spring teeth, 151
Spring-tooth harrows, 150, 151
Stall barns, dairy cows, 594
  planning principles, 594, 595
  stall design, 594, 595
Static-pressure drop of air, 647
Steel, bend allowance, 25
  farm buildings, 572, 575, 686, 692
  properties, 27
  structural, beam and column design, 552–556
  weights of sheets, 27
Stoichiometric air-fuel ratio, 113, 114
Stopping force, formula for, 56
Strain gages, 77–85
  electric circuits, 83

Strength of materials, 27
Stress concentration, 85
Stresscoat, 75
Strip cropping, 414
Subsoilers, 148–149
Subsurface tillage, 417
Sugar cane harvesters, 281
Sun, angle of incidence, 834
  hourly position, 833, 834
Sunshine, annual clear days, United States map, 859
  possible, by latitude and month, 332
  (See also Solar energy)
Surface area of livestock, 781, 782
Surface temperature of livestock, 780, 781
Swathers, 240
Sweep blades, 148, 150, 153
Sweep rakes, 214, 215
Sweet-potato harvesting, 275
Swine, effects of thermal environment on, 774
  relative humidity, 774
  feed-lot layout, 592
  feed and water requirement, 783
  heat and moisture loss, 776, 779
  labor requirements for production, 591, 592
  production structures, 589–593
  space requirements, 614
  surface area, 781, 782
  surface temperature, 780, 781

Taxes on machines, 7
Telemetering, 85
Temperature, absolute, 624
Temperatures, climatic, summer, United States, cities, 844, 845
  effects on plant flowering, 851, 854
  U.S. Agricultural Experiment Stations, 849–851
  United States maps, 856, 857
  world maps, 821, 822
Terminal velocity of spray droplets, 188
Terraces, channel velocities, 410
  construction, 412
  cross sections, 411
  diversions (see Diversion ditches)
  grades, 409
  level, 411
  maintenance, 412
  outlets, 407
  planning of system, 407
  spacing, 410
  staking, 411
  types, 408
Thermal conductivity, 631
Thinners, mechanical, 159
Threads, screw, 27

Three-point hitch linkage, 134
Tile drainage (see Drainage)
Tillage, deep, 126
  effects of soil characteristics on, 799, 800
  minimum, 126, 127
  mulch, 127, 149, 150
  ridge, 127
  rotary, 155–157
  stubble-mulch, 149, 150
  sweep, 127, 149, 150
Tillers, field, 149–150
Tilth, 125
Timber connectors, allowable loads, 566–570
Tires, pneumatic rubber traction, life, 72
  performance factors, 67–71, 73
  selection, 72–73
  sizes, 73
Tomato harvester, 283
Torque formula, 26
Traction, coefficient of, 67–71, 73
Tractive efficiency, 67–71, 73
Tractor, force reactions, 94–102
  horizontal plane, 101, 102
  vertical plane, 94–101
    drawbar load, 97
    loader, 100
    midmounted implements, 99
    semimounted implements, 98, 99
    three-point hitch, 97, 98
Tractors, degree of loading, 8, 9
  estimating fuel consumption, 8
  lateral instability, 102
  operating costs, 8, 9
  performance factors from Nebraska tests, 109
  requirements for various operations, 105
  types of configuration, 104–106
Trailers, two-wheel, dynamic forces on, 287
  trailing characteristics, 288
Transducers, 83
Transpiration, 326–332
Transplanters, 185
  mechanical setting devices, 186
  tree, 186
Trash racks, 433
Travel reduction, 67
Tree-crop harvesting, 281, 282
Truck-crop harvesting, 282, 283
Truck spreaders, 165, 166
Two-stroke cycle, 110

Universal joints, constant velocity, 56
  secondary couples, 53, 54
  velocity fluctuation, 52, 53

V-belt drives, adjustable speed, 46–51
  service factors, 41
  service-life computations, 44–46
  tension computations, 42–44
Valves and fittings, water, head loss due to friction, 760
  (*See also* Hydraulic valves)
Vapor barriers, 634, 635
Vapor pressure, 622
Vegetable harvesting, 282, 283
Vegetables (*see* entries under Fruit and vegetable)
Velocity pressure of air, 647
Ventilation of livestock housing, 635–645
  air-flow recommendations, 636, 639
  fans, 640
  gravity or natural air flow, 638, 642–644
  inlets and outlets, 639, 641
  insulation requirements, 638, 639
  open-front, 636, 637
  principles and practices, 640, 642
  ridge, 637, 645
  summer, 636, 637
  warm-operation requirements, 637, 638
Vesicular exanthema, 591
Vibratory fruit harvesting, 281

Wagon boxes, flare, 290, 291
  flat-deck platform, 291
  power-unloading, 291, 292
Wagon gears, four-wheel, brakes, 290
  draft, 290
  flexibility, 290
  steering arrangements, 288, 289
  telescoping tongue, 290
  turning radius, 289, 290
Water (*see* Irrigation water)
Water consumption of livestock, 785
Water heaters, milkhouse, 707
Water rights, laws on, appropriation doctrine, 498, 499
  riparian doctrine, 498
Water spreading, 495
Water systems, 756–761
  gravity, 759
  head losses due to friction, 759–760
    pipes, 759
    valves and fittings, 760

Water systems, piping and tubing, 759–761
  pressure, 759
Water units of measurement, 338
Waterers, livestock, 761
Waterways, capacities, 422
  classification of cover, 420
  construction, 424
  dimensions, 421
  permissible velocity, 421
  seeding, 425
  shapes, 419
  sodding, 425
Wave action, 428
Weather, belts, 818, 824
  coastal effects on, 827
  effect on machine capacity requirements, 15–17
  fronts, 827
  maps, 826
  mountain effects on, 827, 828
  types, 824, 825
Weight transfer, tractor, 95
Wheel standards, 73
Wheel weights, 70
Wilting point, soil moisture, 794
Wind, Chinook, 827
  destructive, probable recurrence, 843, 844
  loads, buildings, 543, 545
  losses, farm buildings, 583, 586
  maximum speeds, 843, 844
  summer direction and speed, 842
Wind erosion, control, 507, 508
  factors, soil and climate, 505, 506
  mechanics of, 504, 505
  types, 505
Winter chilling requirements of plants, 851
Wood, 545–553
  basic stresses, 545, 548
  beam design, 549–551
  column design, 551, 552
  connectors, 566–570
  dimensions, standard, 549
  factor of safety, 546
  framing construction (*see* Framing)
  glued joints, 568–571
  grades, 546–549
  plywood, 552

S 675 R5
Richey, C.B.
 Agricultural engineers'
  handbook